Multisensor Fusion

NATO Science Series

A Series presenting the results of scientific meetings supported under the NATO Science Programme.

The Series is published by IOS Press, Amsterdam, and Kluwer Academic Publishers in conjunction with the NATO Scientific Affairs Division

Sub-Series

I. Life and Behavioural Sciences	IOS Press
II. Mathematics, Physics and Chemistry	Kluwer Academic Publishers
III. Computer and Systems Science	IOS Press
IV. Earth and Environmental Sciences	Kluwer Academic Publishers
V. Science and Technology Policy	IOS Press

The NATO Science Series continues the series of books published formerly as the NATO ASI Series.

The NATO Science Programme offers support for collaboration in civil science between scientists of countries of the Euro-Atlantic Partnership Council. The types of scientific meeting generally supported are "Advanced Study Institutes" and "Advanced Research Workshops", although other types of meeting are supported from time to time. The NATO Science Series collects together the results of these meetings. The meetings are co-organized bij scientists from NATO countries and scientists from NATO's Partner countries – countries of the CIS and Central and Eastern Europe.

Advanced Study Institutes are high-level tutorial courses offering in-depth study of latest advances in a field.
Advanced Research Workshops are expert meetings aimed at critical assessment of a field, and identification of directions for future action.

As a consequence of the restructuring of the NATO Science Programme in 1999, the NATO Science Series has been re-organised and there are currently Five Sub-series as noted above. Please consult the following web sites for information on previous volumes published in the Series, as well as details of earlier Sub-series.

http://www.nato.int/science
http://www.wkap.nl
http://www.iospress.nl
http://www.wtv-books.de/nato-pco.htm

Series II: Mathematics, Physics and Chemistry – Vol. 70

Multisensor Fusion

edited by

A.K. Hyder
University of Notre Dame,
Notre Dame, Indiana, U.S.A.

E. Shahbazian
Lockheed-Martin Canada,
Montreal, Canada

and

E. Waltz
Veridian Systems, Ann Arbor,
Michigan, U.S.A.

Kluwer Academic Publishers

Dordrecht / Boston / London

Published in cooperation with NATO Scientific Affairs Division

Proceedings of the NATO Advanced Research Workshop on
Multisensor Data Fusion
Pitlochry, Perthshire, Scotland
June 25 – July 7, 2000

A C.I.P. Catalogue record for this book is available from the Library of Congress.

ISBN 1-4020-0722-1

Published by Kluwer Academic Publishers,
P.O. Box 17, 3300 AA Dordrecht, The Netherlands.

Sold and distributed in North, Central and South America
by Kluwer Academic Publishers,
101 Philip Drive, Norwell, MA 02061, U.S.A.

In all other countries, sold and distributed
by Kluwer Academic Publishers,
P.O. Box 322, 3300 AH Dordrecht, The Netherlands.

Printed on acid-free paper

Printed in the Netherlands.

CONTENTS

CONTENTS (Cont'd)

3. MULTISENSOR DESIGN AND IMPLEMENTATION

CONTENTS (Cont'd)

CONTENTS (Cont'd)

APPENDICES

PREFACE

A NATO Advanced Study Institute (ASI) on Multisensor Data Fusion was held at the Atholl Palace Hotel, Pitlochry, Perthshire, Scotland, from June 25 through July 7, 2000. This publication is the Proceedings of the Institute.

The NATO Advanced Study Institute Program of the NATO Science Committee is a unique and valuable forum, under whose auspices almost one thousand international tutorial meetings have been held since the inception of the program in 1959. The ASI is intended to be primarily a high-level teaching activity at which a carefully defined subject is presented in a systematic and coherently structured program. The subject is treated in considerable depth by lecturers eminent in their field and of international standing. The subject is presented to other scientists who will already have specialized in the field or possess an advanced general background. The ASI is aimed at an audience of approximately post-doctoral level. This does not necessarily exclude graduate students and may well include senior scientists of high qualifications and notable achievement in the subject of the ASI or in related fields.

This ASI emphasized the theory and applications of multisensor data fusion. For some time, all branches of the military have used a wide range of sensors to provide data for many purposes, including surveillance, reconnaissance, target detection and battle damage assessment. Many NATO and Partner nations have also attempted to apply these sensors to a variety of civilian applications such as crop monitoring, agricultural disease tracking, environmental diagnostics, cartography, ocean temperature profiling, urban planning, and characterization of the Antarctica ozone hole.

The organizers believed that the time was appropriate for a tutorial conference on this issue of data fusion. The recent convergence of several important technologies has made possible new, advanced, high-performance sensor-based applications relying on the near-simultaneous fusion of data from an ensemble of different types of sensors. These are:

- The emergence of advanced sensor technologies that offer high-resolution and three dimensional imagery across the electromagnetic spectrum, and for phenomena based on, for example, acoustic, ultrasonic and olfactory sensing that have not been exploited previously;
- The development of robust, competent, and flexible methods and algorithms for combining information from multiple sources;
- The continued rapid increase in the availability of high-performance computing, from processor speeds of 20 MHz only six years ago to affordable processors operating above 1 GHz today, and with faster speeds emerging almost daily;
- The reduction of the cost of digital memory by two orders of magnitude in a decade together with an increase in the density of memory devices by three orders of magnitude in the same time period;

- The combined development of innovative sensors that will enable new space and terrestrial applications particularly as data fusion techniques are developed to fully exploit the information acquired by ensembles of sensors and as the cost of space access continues to decrease.

The ASI examined the underlying principles of sensor operation and data fusion, the techniques and technologies enabling the process, including the operation of 'fusion engines.' It sought to present the fundamental theory and the enabling technologies of data fusion in a systematic and accessible manner. The ASI concluded with presentations on current interest, including applications in medicine, meteorology, BDA and targeting, transportation, cartography, the environment, and agriculture, as well as manufacturing and process control. Students and lecturers were encouraged to explore innovative areas of the application data fusion techniques and technologies. Wherever possible, the ASI included practical guidelines on the use of data fusion technology and emphasized how new applications are leveraging the research and development investments made over the past two decades by the military both in NATO and Partner nations.

The theme of the Institute was the enhancement of scientific communication and exchange among academic, industrial, and government laboratory groups having a common interest in the development of sensors, their modeling, and the synergistic benefits that flow from the fusion of their data.

The first session opened with an overview of the basics of data fusion and the fundamentals of reasoning. Following that the ASI focused on the mathematical basis of the fusion process, and fusion at levels zero through four. This included a series of tutorials on detection and association, still a critical deficiency in the fusion process. Session three focused on identification, including object recognition. The fourth session dealt with methodologies and higher-level fusion approaches. The fifth and final session covered applications from medicine to undersea detection to BDA.

The initial pace of the Institute allowed ample time for informal discussion sessions to be organized and scheduled by the participants and lecturers with the encouragement and assistance of the Directors. As the Institute progressed, the interest and demand for these additional sessions grew and they consumed much of the unscheduled time. A departure from the normal format of an ASI which greatly enhanced this Institute was the opportunity for participants to contribute poster papers in two evening sessions. The posters remained in place during virtually the entire ASI and served as a catalyst for technical interaction among the participants during all of the breaks. Eleven of these poster papers have been selected by the editors for inclusion in the proceedings and are presented in the appendix.

The Institute was attended by ninety participants and lecturers representing Canada, Russia, France, Italy, Czech Republic, Ukraine, Hungary, Bulgaria, Poland, Belarus, Denmark, Uzbekistan, Israel, Greece, Norway, Portugal, Belgium, Turkey, the United Kingdom, and the United States. A distinguished faculty of lecturers was assembled and the technical program organized with the generous and very capable assistance of an Advisory Committee composed of Dr. Vitaly Bystritiskii (Russia), Dr. Alain Appriou (France), Dr. Phillip Greenway (UK), Dr. Elisa Shahbazian (Canada), Professor Giorgio

Tacconi (Italy), Mr. Ed Waltz (USA), Dr. Peter Lombardo (USA) and Dr. Anthony K. Hyder (USA).

The value to be gained from any ASI depends on the faculty - the lecturers who devote so much of their time and talents to make an Institute successful. As the reader of these proceedings will see, this ASI was particularly honored with an exceptional group of lecturers to whom the organizers and participants offer their deep appreciation.

We are grateful to a number of organizations for providing the financial assistance that made the Institute possible. Foremost is the NATO Scientific Affairs Division which provided not only important financial support for the Institute, but equally important organizational and moral support. In addition, the following sources made significant contributions: the U. S. Strategic Defense Initiative Organization, the U. S. Air Force Office of Scientific Research, the U. S. Army Research Office, the U. S. Office of Naval Research, EOARD and ONR in London, Laval University (Canada), Lockheed Martin of Canada, the U. S. Army Communications-Electronics Command, Utah State University (USA), Defense Research Establishment Valcartier (Canada), Veridian Systems (USA), and the University of Notre Dame (USA).

We would also like to thank, Dougal Spaven, the general manager of the hotel hosting the ASI; his wife, Sally; and all of the staff at the Atholl Palace Hotel for a truly enjoyable and memorable two weeks in the Scottish Highlands. Their warm friendliness made all the attendees feel welcomed and the Atholl Palace accommodations, meals, service, and meeting facilities were superb.

A very special acknowledgement goes to Susie M. Anderson and her colleagues of Dahlgren, Virginia. Susie undertook the very challenging task of centrally retyping the lecturers' manuscripts and of producing a camera-ready document to Kluwer Academic Publishers. Thank you for your long hours and hard work.

And, finally, all of our thanks go to the people of Perthshire, Scotland, who certainly displayed, in every way, 'Ceud Mile Failte' (Gaelic for 'A Hundred Thousand Welcomes').

Elisa Shahbazian
Montreal, Canada

Ed Waltz
Ann Arbor, Michigan USA

Anthony K. Hyder
Notre Dame, Indiana USA

December, 2001

INFORMATION FUSION IN THE HUMAN BRAIN

The Complexity of De-Centralized Symphonic Collaboration

R. VON HANWEHR
CEO Solus, Inc.
700 New Hampshire Avenue, N.W. #1008
Washington, DC 20037

1. Introduction

There is growing interest in the defense sensing community with regard to higher neurocognitive function, including issues of information flow and potential information fusion processes in the brain. In particular, models of higher functional-organizational brain activity may be useful comparative paradigms in neural-biomimetic robotic systems development and in multi-sensor systems design [1-4]. Methods for understanding neural processing and possible fusion of multi-modal sensory imagery is highly relevant for colleagues in the defense sector concerned with innovations in visual neuromorphic robotics [5], 3D stereospatial neural network architectures, particularly for high-throughput, automated, analysis of satellite imagery. Finally, the area of biomimetics and neurobiomimetics has received considerable attention in recent years. Although most work is centered on emulation of insect based neural systems [6], there is gradually evolving interest in mimicry of mammalian neural processing paradigms [7], in particular for the purposes of exploring new approaches toward the design of workable quantum computing models. This direction is prompting renewed interest in the question of how the human brain: 1) regulates afferent information flow, 2) fuses data and imagery, and how it 3) processes and manages the complexity of multi-dimensional information transfer and utilization, particularly at higher infrastructural and organizational levels of neural function.

2. The Central Question: Neural Information Fusion As A Hypothesis

The riddle of whether and how the human brain fuses multi-modal sources of information is a complex and controversial topic, which may be phrased in the form of a central entry question which it is the objective of this paper to discuss:

How Does Our Brain Manage Information Flow, and Does This Evolved Biocomputational System Fuse Multimodal Information in a Conventional Context.

A. K. Hyder et al. (eds.), Multisensor Fusion, 1–36.

3. The Neural Landscape

At the outset, it is useful to briefly describe the neural landscape through which information flow is evidently managed in the brain. The adult intracranial vault --- which contains the cortex, subcortex, central axonal pathways, brainstem, and cranial nerves --- comprises significantly more than 10^{12} neurons [8]. Structurally, at least, these are the basic constructional elements of the more complex computational system we know as the brain. However, the cytological terminals that interconnect neurons --- the synapses -- are the functional and thus determinant building blocks of the brain viewed as a system. The 1×10^{12} or more neurons comprising the brain substance are further supported in a cellular foundation of at least $5\text{-}10 \times 10^{12}$ astroglial cells [8], a heterogeneous population with many specialized subcomponent cell groups such as the oligodendrocytes that manufacture myelin, the basic insulation substance which envelopes nerve sheaths and that ensures efficient electrical axonal conductivity. This maze of an organized cellular infrastructure is termed the neuronal tissue microenvironment [9].

The interconnected and multi-specialized subpopulations of neurons within this brain infrastructure are exposed to an immense spectrum of incoming information, sensory input, and biological signals not only from the nervous system, but from countless other organ system components throughout the human organism. Such neural input directed to the cerebral cortex from a host of sources is globally termed 'afferent multi-modal neuronal input'. This is to be distinguished from 'efferent neuronal output', such as the outgoing neural commands from our brain to peripheral target sites, output that activates our muscle bundles for achieving bipedal gait, for instance. In a wider context, the concept of bidirectional afferent and efferent input extends to the interchange of data and imagery between dendritic (afferent) and axonal (efferent) elements and their synaptic connections throughout the brain, i.e. the 3D neuronal network.

Viewed as a whole, the 'Afferent' (incoming input) component of this 'multi-modal neural interchange' emanates from diverse, if not countless, 'neural sensor' sources. Also, whereas there is true incoming *de novo* sensory data, there is also a voluminous loop of re-processed, exchanged and modulated data that flows, multi-directionally and multi-dimensionally, from neuron to neuron and between specialized neuronal subpopulations. The mystery of how this afferent multi-sensor data is converged and utilized constitutes the essence of the question as to how the human brain effectively fuses, de-assembles, and re-assembles multimodal information packet input.

A variety of afferent data input may be categorized. These include abstract linguistic-recognition-coupled Lexical Data or the more nebulously-defined and symbolically-based Semiotic Data, historically termed '*Gestalt*'. Today, this concept is recognized as an oversimplification. Although still taught in medical school, another oversimplification is that, because the majority of individuals are right-handed and left-brain dominant, abstraction must reside on the left hemisphere, and symbolic patterned processing on the right hemisphere. Hence, creative individuals are by definition left-handed (right-brain dominant) or ambidextrous. This is clearly incorrect. In reality, the interconnectivity of the hemispheres via the central axonal switching pathway known as the corpus callosum eloquently parallels the synaptic interconnectivity of neurons at the level of the tissue microenvironment.

Other sources of 'Afferent Multi-Modal Input' are from sensory organs, which may be considered multi-sensor arrays. These include vision, hearing, taste, smell, tactile touch, etc. The brain views and patterns these as imagery --- for instance, sound is processed as a neurally-projected image map. Hence, sensory input is increasingly defined as imagery, whereas lexical vs. semiotic content is generally defined as data. However, at the systems level of synaptic neurotransmission, afferent neuronal input is likely managed as quanta or packets of chemoelectrically transmitted information.

Other afferent information content that reaches the brain --- or is exchanged between neurons --- includes bioelectromagnetic and trophic cell to cell signals, as well as neurochemical and neurohumoral signals that allow the brain to interact at a biochemical level with distant organs such as the endocrine system, with cardiac rhythm centers, and with the immune system. Additionally, there is direct axonal and ganglionic peripheral interconnectivity with organ systems (somatic), with key autonomic functions, as well as with peripheral nerve, neuromuscular and neurocutaneous target receptors. In part, the sympathetic and parasympathetic chains of the ganglionic, spinal and peripheral nervous systems support these functions.

The neuronal landscape of the human brain thus represents a unique multi-sensor imagery input network and data packet fusion processing system.

4. The Primary Modular 'Information-Flow' Units

The search for an answer on whether and how the brain fuses information begins with consideration of brain structure and function. Analogous to a computational engineering systems approach, one helpful way of viewing the neural information fusion problem is to identify the primary modular structural and functional unit(s) that regulate information flow in the brain. Contemporary neuroscience is in general agreement that this structural building block is the neuronal cell (the neuron), while at a functional level the modular unit it is the connective interface between neurons, defined as the synapse or neurosynaptic terminal.

In essence then, the simplest overall primary cytostructural functional unit in the brain is composed of two neurons interconnected by a synapse. This satisfies the requirements of defining structure as well as function, in a convergent and unitary manner. Neuronal cells project onto each other via their synaptic axonal output fibers which ultimately arborize as dendritic input fibers that project onto a set of synapses, thus yielding an exponentially expansive structural network. the dendritic input receptor fibers of other neurons. At the risk of some simplification therefore, at a neuronal cell level, outgoing information flow ('efferent' output) from neuronal synaptic terminals exits via axons, while incoming information flow ('afferent' input) is routed via dendritic projections onto neuronal synapses which then represent specialized junctions that regulate information flow. The potential for synaptic inerconnectivity, however, is not as simple, and may in fact be nearly infinite. It is here that the riddle of information (data and imagery) fusion in the brain begins.

It may be somewhat astounding to the non-neuroscientist, that greater than 10^{20} Synapses are present in a highly developed human brain [8]. Together with at least two parent neurons, the synapse is the primary unit responsible for the immense capabilities of the brain as regards plasticity (remodeling) and interconnectivity. Neuronal networks have an apparently limitless ability to sprout dendritic connectivity and the neurochemoelectromagnetic switches we call synapses. Thus, the interconnectivity-dependent phenomenon of neural plasticity is analogous to a 'self-assembling computational architecture' that responds to environmental adaptive challenge, sensory input, trophic influences, threat, pleasure, learning, new experiences or déjà vu, to physiological challenges, and to the process of growth-maturation-aging. Even in later years, recent work suggests that the ability to sprout and re-allocate synaptic interconnectivity may be unbounded. The synaptic architecture thus constitutes the primary modular functional unit essential for adaptive neural plasticity and interconnectivity, and thus for regulation of information-flow and fusion in the brain.

5. Interconnective Synaptic Plasticity And Information Fusion

Our anatomical, as well as neurochemical, electrophysiological, biomagnetic and molecular understanding of neuron and synapse is extensive. Equally impressive is our appreciation of how individual sensory input pathways and specialized afferent sensory processing systems in the brain are structured and how they function. However, understanding interconnectivity between these systems at a higher organizational, and ultimately cognitive, level is proving an elusive challenge.

Traditional neuroanatomical and neurophysiological models of hierarchical structural organization are proving inconsistent with newly emerging data on complex neural function. As live biomagnetic anatomic-functional imaging data is becoming available on normal subjects as well as on individuals with neural deficits, it is now clear that contemporary models of neural processing and transmission are incredibly simplistic. This is because traditional and reasonably contemporary models have been structured on the concept that information is transmitted from an origin sensor (for instance the optic or auditory system) to target neurons in a specialized region of the brain, or from one set of neurons in the brain to a specific target region of other neurons elsewhere in the brain. As we will see, newly emerging information reveals that, at a higher level of neurocognitive and neurosensory perceptive function, this fixed directional view of organization hierarchy is thus not consistent with reality. Rather, axes of origin-to-target (sensor-to-target) communication are highly diversified and modulative in near real-time, depending on the afferent information content, its context, environmental or systemic physiological factors, and so on.

Since synapses and dendrites cannot sprout dynamically, this indicates a fact, now accepted, that the dormant synaptic and dendritic (neuron receptor) reserve is far greater than might be assumed, and that recruitment of neuronal-synaptic clusters can occur in real-time and with astounding plasticity and flexibility. This includes recruitment of dormant neuronal-synaptic clusters, under-utilized clusters, or in many emerging empiric models also includes re-allocation of clusters away from existing functions, to demands-of-the-moment type functionality. Thus, a higher ordering of priorities exists, and begins at the trans- and supra-synaptic level, but its organizational and infrastructural makeup remains largely unkown.

Although hierarchical in nature, this higher ordering of priorities does not appear to be coordinated by definable regulatory centers in the brain, but rather in a collaborative symphonic manner, as in a modern philharmonic orchestra without a conductor. Indeed, as an analogy, the Vienna, Berlin, and London Philharmonics can perform effectively without a conductor, but the superior creativity and perfection of unity does require a central regulator -

-- the maestro or impresario --- an element of the brain that remains to be defined. Furthermore, whether the brain improvises rather than follows a predetermined synaptic electrofunctional map for each scenario, whether it is analogous to a 'conductorless' jazz musician or to the complex organized template that constitutes the musical score of the apotheosis in Prokofiev's Cinderella ballet (which requires a conductor, stage manager, and choreographic director), is also unkown.

6. Brain Systems Analysis And Information Flow Modeling

With the need for revisions in conceptual modeling of brain systems analysis emerging, it is evident that structural and functional units may not be equivalent at any point in time during neural processing. Components of one larger structural unit composed of, say 5×10^2 neurons and their synaptic connections may participate in different simultaneous functional activities, i.e. in a setting of organized-disorganization. Although there is central structure to the orchestra, the key to its functionality lies in the versatility (plasticity) of its components at pursuing de-centralized tasks in tandem and in parallel, i.e. de-centralized symphonic collaboration. Again analogous to an orchestra, precisely the same note structure may be input into the system, but the arrangement and harmonics between the different instrumental units in any one set of 'sound functions' in the orchestra will yield a totally different total output of sound to the audience. In essence, it is the orchestration of functional elements in the brain, and of the synaptic pathways, that represents the likely key to understanding the process of information flow and information fusion.

In raising the issue of a conductor-less collective consciousness or collective sentience that executes impeccable symphonic collaboration of its structural-functional units, the brain does better than even the Vienna or London Philharmonic would. Namely, the human brain achieves real-time correlation of stored, incoming, and ongoing reprocessed information at a neurophysiological packet transmission rate which remains to be reliably quantified in the laboratory. If this fusionesque neural matrix system functions in a leaderless manner, which is far from proven, it is a system without comparison or parallel.

Thus, the problem at hand may be summarized in the following fashion. Our understanding of the structural and functional basis of neural plasticity and interconnectivity continues to be an elusive goal. Traditional models of hierarchical structural organization are of limited value in explaining complex neural function. Contemporary models of target-directed neural processing and transmission pathways are frequently inconsistent with emerging biomagnetic imaging observations of higher-level, real-time, brain function. The difficulty in modeling plasticity and interconnectivity at a structural-anatomical level is that the latter may represent a potentially misleading view of functional organization. A more realistic paradigm for explaining information flow and fusion in the brain may involve the 'decentralized, but symphonically-collaborative and correlative processing' of stored, revised/re-modulated, and new incoming (afferent) information. At the level of neurocomputational design, such a neural matrix model represents the inherent paradox of 'organized disorganization'. However, it is known from electroencephalographic and biomagnetic studies that at a suprastructural and suprafunctional level, there is a dome of exquisitely stable electromagnetic rhythmicity to brain function. Yet, wavelet analysis and related techniques have already revealed that the components of this rhythmicity indeed appear to contain elements of asymmetric disorganization and dysynchrony, but in a repertoire of patterns that likely have deliberative symmetry which contemporary science has yet to translate. The Rosetta stone needed for this translation of the neuronal information fusion mystery remains to be encountered.

7. Restructuring The Central Question

The question at hand can thus be reconsidered in the following manner:

A revisionist view of information fusion theory argues that the brain performs information fusion via processes that manipulate fused information packets which contain stored and/or dynamically-modulated 'multi-modal afferent information input', or

- A more novel view is that of '*relativistic correlation of information*', where neural flexibility prefers non-fused information used in a highly-coordinated, irregularly-populated data matrix with many possible solutions for any given multi-dimensional set imagery data.
- The question is thus whether: the neural matrix prefers relativistic correlation of information with an equilibrium between fused and non-fused information, and with a constant exchange of information packets within a highly coordinated, but variably-populated data matrix. Such an arrangement would constitute a system with many possible solutions for any given set of data. The latter seems the idealized approach, certainly from an adaptive evolutionary standpoint

8. Understanding Neural Information Fusion

In addressing the goal of synthesizing a unified conceptual framework for Neural Information Fusion, this paper discusses four interrelated bodies of knowledge, which include:

1. Core neuroanatomical and neurofunctional principles relevant to neurosynaptic organization, infrastructure, plasticity, and interconnectivity, inclusive correlates of the neurosynapse viewed as a modular structure-function unit in the brain
2. The nature, categories, and scope of afferent information input to the brain
3. Novel information derived from state-of-the-art brain imaging methods designed for tracking information flow and collaborative information fusion behavior in the 'dynamic' brain
4. Traditional vs. evolving paradigms that provide a template for originating neurocomputational models of neural information flow and packet information fusion. Ideally, such paradigms, and their resulting models will afford an enhanced understanding of how the human brain fuses information, data, and multi-modal afferent imagery.

9. Neuroanatomic, Neurofunctional And Neurosynaptic Correlates

Using a progressive building block approach, one may consider the major features of structure/function features that characterize multi-hierarchical organization in the brain, primarily from the standpoint of their context *vis ˉa. vis* information flow and putative information fusion.

Neurons are among the largest cells in the human organism and can exhibit the most significant longevity. Neurons demonstrate significant adaptive plasticity in changing functionality, priorities, and behavior, as well as in exhibiting truly remarkable capabilities for synaptic networking and for generating and projecting ever increasing synaptic interconnectivity. Despite this, they are among the least effective replicative cell elements known. In other words, although at times their metabolic rate may vary significantly, once matured they are not readily dividing cells in the conventional sense of the concept. Notwithstanding, this view may become increasingly dated. For instance, neuronal de-differentiation and replenishment of neuronal elements can indeed by stimulated within the neuroglial tissue microenvironment, and there is tantalizing work to this effect, particularly in the spinal area. However, for the purposes of this discussion regarding dynamic information flow and data/imagery fusion, cell division is not a directly implicated real-time mechanism.

Despite the limited native capacity to procreate, neuronal population numbers in the brain are far in excess of the number of mature neurons we require for even peak functionality, and neurons exhibit an extensive palette of trophic changes in cell behavior, in sprouting dendritic protrusions, and in establishing new synaptic terminal connections. The emerging implication of this is the likelihood that, not infrequently, the success of the brain's design is potentially dependent not on cellular replacement through division and new progeny, but on cellular repair, functional re-prioritization and task-specific or capability-specific re-adaptation in large part through the versatility of ongoing synaptic interconnectivity. Although subject to eventual revision by laboratory data to the contrary, the prevailing view is that, collectively, these non-replicative phenomena constitute the dominant contribution to neuronal plasticity.

Plasticity is thus made possible in large part through the mechanism of dendritic sprouting and generation of new or re-directed synaptic interconnectivity. Thus, if in an area of brain is damaged, a significant potential of neurons survive, even if these had tasks and designated functions other than those of the neurons lost. Plasticity can be harnessed to re-direct those neurons to take over some or all of the tasks of their former neuronal brethren. If no regional neurons are available, then recent work confirms strongly that distant neurons in distal centers with often convoluted pathways of connectivity to the area of damage can in fact assume the burden of re-establishing some aspects of lost function. It is axiomatic that these processes and the resulting capabilities are presently viewed as central to information flow and fusion of information content in the brain.

In order to better understand these capabilities, we need to consider three aspects of brain architecture, as illustrated in Table 1.

TABLE 1. Aspects of the Brain Architecture

Architecture	Anatomical Elements	Functional Elements
ANATOMY	Molecular / Subcellular	Genu Expression
	Ultrastructural Anatomy	Signal Transduction
	Neuronal CytoArchitecture	Synaptic Conductive Elements
	Tissue Microenvironment	Neuronal Interconnectivity
	CNS Nucleii / Axonal Pathways	Specialized Neurocommunication
	Macroanatomical Topography	Higher Cottical Organization
NEUROCHEMISTRY	Synaptic Architecture	Neurotransmitter Projection Pathways
	Neurotransmitter Packets	Information Transmission Quanta
BIO-MAGNETISM	Neurosynaptic Matrix	Neurochemical to EMF Conversion
	Cortical / SC Nucleii - Axons	Electrobiomagnetic Interconnectivity
	Grey-White Matter Organization	Chemo Biomagnetic Information Flow
SUPRAORGANIZED	Higher-Level Complexity	Neural Information Fusion

Table 1 divides these aspects of brain architecture into four key areas: anatomy, neurochemistry, biomagnetism and supraorganization. As depicted and presented in turn, from the most fundamental to the most complex supraorganizational level, brain structure begins with the genome and its regulation of critical subcellular organelles, and then with the neuronal cell itself. For neurons to base function and resiliency not on division and progeny, but on cellular longevity, on re-directed adaptation (plasticity), and on a limitless capacity for neurosynaptic expansion (given the appropriate input stimuli), they need to have highly developed and complex gene-regulated subcellular machinery, i.e. in order to manage the signal transduction capabilities between cell compartments and the cell membrane that faces the outside environment or communicates with other cells.

The left-hand scheme in Figure 1a [10] illustrates the subcellular ultrastructure of a single neuron, including the nucleus containing the chromosomal genomic components, as well as the cytoplasmic compartment that contains sub-cellular organelles such as a mitochondria. These are energy-regulating generators required by the cell to convert chemical and energy-management processes at a molecular level. Input refers to the dendritic receptor end of synaptic interconnectivity for a neuron. Output refers to the axonal output fiber for the neuron, allowing the neuron to, in turn, project on other neurons' dendrites via synaptic terminals. Given the structural and functional demands at a molecular level, it is not surprising that the coupling of gene regulation and the specialization of subcellular functionality in neurons is of the most complex form known. Thus, and as was shown in Table 1, the functional equivalent of genomic and subcellular structure is gene expression and gene-regulated signal transduction.

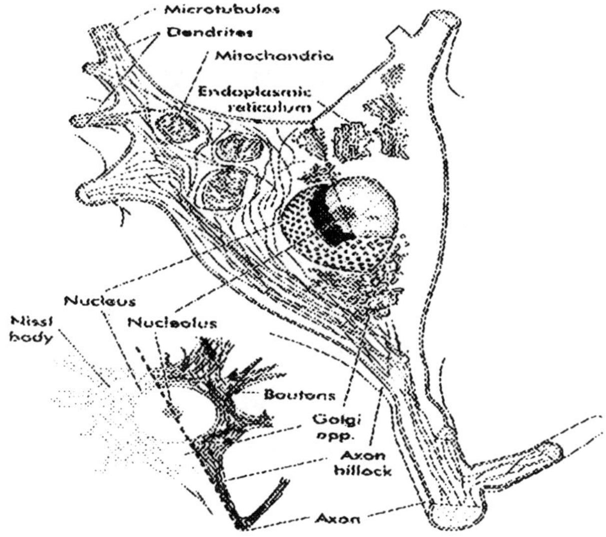

Figure 1a. Subcellular ultrastructure of a single neuron

The next level of structure is at the neuronal pair level. At the level of this, the most basic building block of the neuronal cytoarchitecture, two neurons comprise the functional equivalent of a synaptic grid. The right-hand Figure 1b [11] shows two neurons interconnecting via a synapse or terminal, as labeled on the illustration (*). The output axon(shown with its Nodes of Ranvier and myelin sheath) exits the upper neuron at the top of the schematic. The output axon fiber becomes myelinated and thus insulated for efficient long-tract electrical conductivity. At its output point close to the second neuron below, the axon sprouts and projects via a synapse onto two dendrites, one belonging to the smaller second neuron. At each dendrite an axon projects to, there are one or more synaptic terminals, which are transit pathways to the next neuron.

Therefore, Synapses are neuroelectrochemical communication terminals that connect output on the axonal side of one neuron with input on the dendritic side of another neuron. The functional purpose of a synapse is to convert (transduce) an input electric potential signal into a chemical event, and back into an output electrical signal. As shown in Figure 2 [12] the axonal output terminal can be visualized as a bulb that is bounded by a presynaptic membrane, and separated from the dendritic input by a gap or cleft. The start of the dendrite side of the synapse is then bounded by a post-synaptic membrane known as a density. As shown, the presynaptic axonal bulb of the synapse contains neurotransmitter-filled vesicles which in essence are neurochemical packets of information. These are discharged into the synaptic cleft, as a result of an incoming axonal electrical signal. At the post-synaptic dendritic side of the cleft, the action of the discharged neurotransmitter chemicals induce propagation of a new electrical impulse into the receiving neuron. The reason that Figure 2

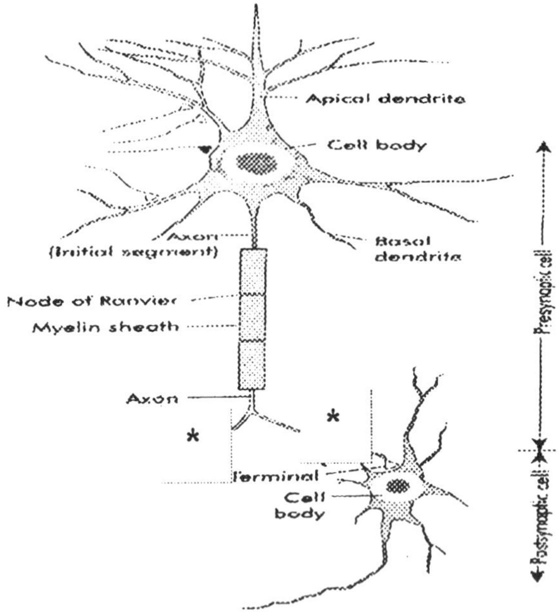

Figure 1b. Interconnection of neurons

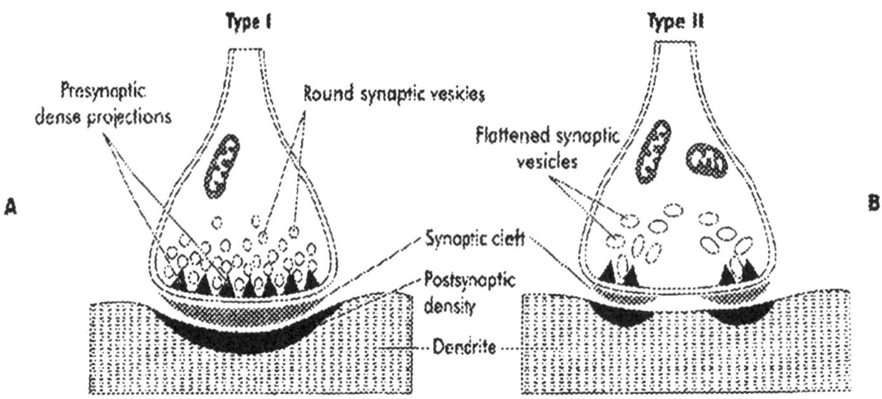

Figure 2. Axonal output terminal

depicts two types of synaptic bulbs is to demonstrate that a synaptic bulb can be specialized to one neurotransmitter with a symmetrical cleft (A), or can manage multiple packets of neurotransmitter vesicles within different cleft regions (B) specialized for different types of neurotransmitter signals, some excitatory, others inhibitory.

The left-hand Figure 3a [13]demonstrates the same concept by the direction of an action potential down the axon and across the synaptic cleft to the dendritic end. On the right, Figure 3b [13] provides an electronmicroscopic imaging-derived ultrastructural view of an actual neuron in cross- section, demonstrating an axonal protuberance in the upper right hand corner, and terminating in a synapse with associated neurotransmitter packet vesicles. Another synaptic terminal is found in the lower-right hand corner, where the vesicles may be more easily recognized.

Utilizing Table 1 summarizing brain architecture, it is useful to recall that structure at the single cell level is known as neuronal cytoarchitecture, while the next level of higher structural organization is the arrangement of neuronal cell populations into a synaptic network, which includes supporting neuroglial cells --- a structural arrangement termed the neuronal tissue microenvironment. The functional equivalents are the synapse as a conductive neurotransmission element, and ever increasing levels of neuronal interconnectivity.

Neuronal cell types that make up the brain are highly specialized. Different combinations of discreet types of neurons thus form different tissue microenvironments. This cytograph in Figure 4 [14] contrasts cellular morphology of just a few types of neurons. Shown is the comparatively linear nature of the pyramidal cortical neuron on the left hand side, as well as this cell's distance-extended dendritic tree, which enables effective communication with neurons across different layers of the cortex. Various such specialized neurons, with very linearly extended dendritic trees, assemble to make up interlocking layers of the cerebral cortex. This is depicted on the extreme right, where *S* stands for the cortical surface and *D* stands for depth. Other types of neurons pictured in Figure 4 have starburst dendritic trees suitable for clustered populations of cells. One is an associative neuron from the thalamus, that controls critical pain sensory input. The Purkinje cell to its right is a specialized cell of the hindbrain or the cerebellum which has a linearly projecting axon merged with a starburst dendritic tree, again meeting a set of specialized requirements for that neuron.

Different neuronal cell sub-populations in the brain form the discrete cellular layers of the brain cortex, while other functionally-specialized neuron clusters are found in subcortically-located subpopulation centers called nuclei. Furthermore, the axons of neuronal cell layers and clusters are grouped into a maze of discrete and complex fiber projections known as axonal pathways or axonal tracts. Figure 5 [15] demonstrates these features in the stained cross-section of brainstem microanatomy. This imagery points out an important feature. Clusters of neurons in nuclei or in cortical regions are known as grey matter because if its cell-rich matrix, appearing grossly as a gray scale. Areas where axonal fibers congregate and form tracts are known as white matter, because of their comparatively blanched appearance due to the myelin coating on axons. Figure 5 includes a medullary cross-section view that depicts concentrations of functionally-specialized neurons (nucleii), contrasted with areas of axonal fiber tracts such as the pyramids. As is typical for this type of anatomical imagery, the inverted grey scale image distortion of the histological stain used renders the grey matter paradoxically blanched in comparison to the white matter tracts which appear

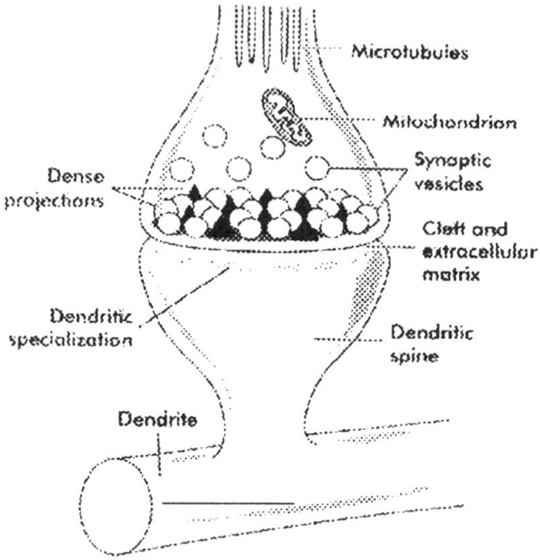

Figure 3a. Action across the synaptic cleft

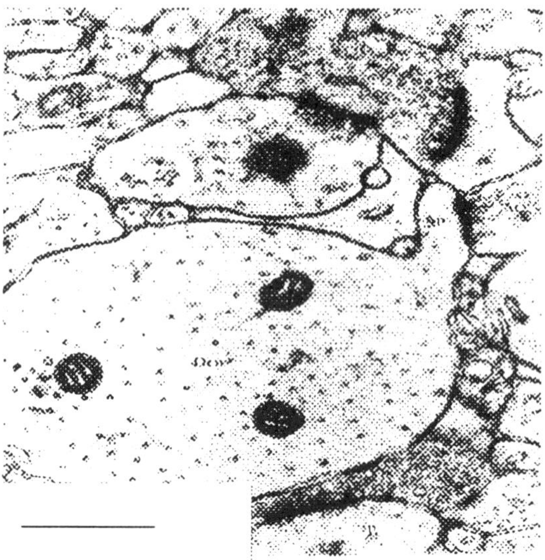

Figure 3b. Ultrastructural view of a neuron in cross-section

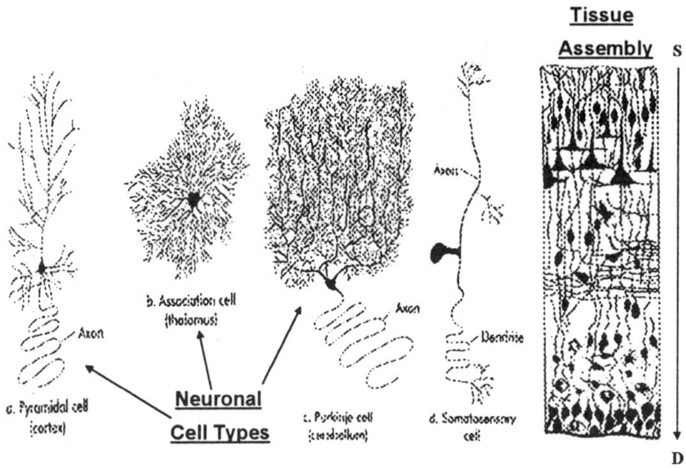

Figure 4. Neuronal cell types – cellular building block specialization

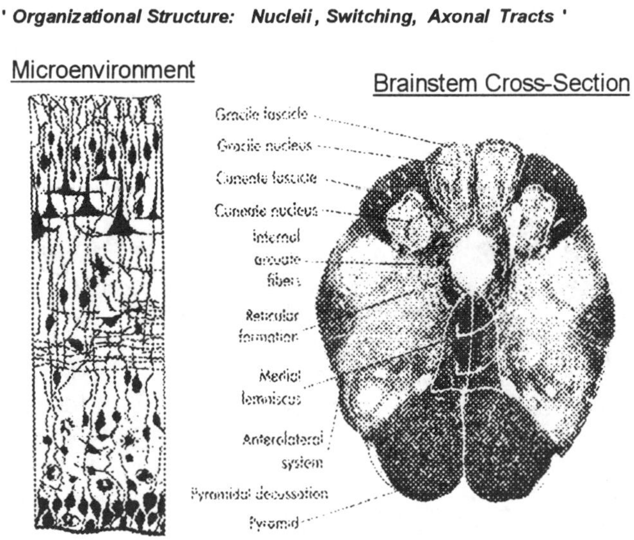

Figure 5. The tissue microenvironment

dark. In a real world photorealistic setting, the shades would be reversed. In contrast to axonal white matter tracts, dendrites are more intimately associated with neurons, are partially unmyelinated, and thus fragile. Light or fluorescence microscopy with special neural enhancing stains is required to highlight dendritic architecture, in part since dendritic arbors are more diffusely distributed and are not always easily distinguished as separate regions on cross-sectional macroanatomical imagery.

Having assembled the brain by components, starting from the single cell level, the discussion of the anatomical, functional, and synaptic components of neural information fusion reaches the a higher level of specialization, which is best introduced using macroanatomical topography. Figure 6 [16] depicts characteristic whole brain macroanatomy on the right, with the division into lobes of the cortex and central subcortical regions such as the diencephalon, leading down to cerebellum and medulla. The projection on the left of Figure 6 is that of a classic Brodmann's Functional Cytoarchitectonic Map, which has been phenomenologically constructed over many decades of experience, tracking comparative anatomy with neurological function in thousands of patients and other mammals. This map reveals the patterns for incoming afferent (multimodal information input) and outgoing efferent (command) neurological functions, which are localized to the cortical surface. At a structural level, the higher neurocognitive and neurosensory faculties relevant to information flow and fusion in the brain are reflected in our knowledge of macroanatomic topography, that maps major areas of function to different brain regions, as is shown in Figure 6.

' Functional Topography vs. Structural Anatomy '

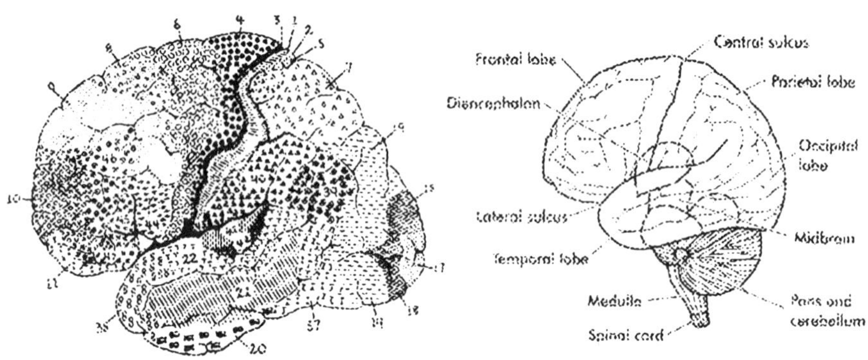

Functional Cytoarchitectonic Map

Pure Structural Macroanatomy

Figure 6. Macroanatomy and function

Correlated with dynamic neuroimaging observational data, such cytoarchitectural and topographical specialization maps are an entry approach the speculative realm of understanding afferent information flow and imagery data fusion.

Having examined all the levels of structural organization relevant to a consideration of information fusion in the brain, one may reconsider the many transitional options, originally depicted in Table 1, for extending an understanding of structure to one of function. In particular, the interface of neurochemistry and biomagnetism with the brain's basic functional unit, the synapse, is central to this consideration. As Table 1 shows, as one proceeds from the neurochemistry to the biomagnetism category, one proceeds structurally from the level of the synaptic building block to the whole brain. Equivalently, and in a functional context, one proceeds from individual neurotransmitter packets (that carry information quanta) to chemoelectrotransmission (conduction) and to the generation of bioelectromagnetic field matrices in the brain. Information fusion is thus in essence ultimately dependent on electromagnetic information flow. With the synapse functioning as the core processing unit for assembling and transmitting information packets, it is the key to the generation of the biomagnetic field patterns that constitute higher brain function and sentience. Thus, the synapse is also at the root of modulation in biomagnetic field patterns that subtends the process of dynamic information flow and afferent imagery data fusion in the brain. This conversion of electrical impulses to synaptic neurochemical events and then to biomagnetic field vectors is demonstrated in Figure 7 [17]:

1. As an axonal electrical action potential is inputed down an axon in the direction of the synapse, it initiates the release of neurotransmitter packets at the synaptic junction.
2. The neurotransmitter packet release is in turn associated with movement of large concentrations of anions or cations along voltage dependent channels on the presynaptic (axonal) side of the synapse. This results in the generation of an electrical gap or potential across the synapse, which, depending on the individual neurotransmitter and ion combination, is either stimulatory or inhibitory.
3. When the neurotransmitter packets reach the opposite (post-synaptic) side of the cleft, they neurochemically activate ion exchange that serves to propagate that electrical potential into an action potential that travels down the dendritic arborization to the next neuron's axonal fibers, and so on.
4. It is the generation of the electrical potential across a synapse that gives rise to biomagnetic fields at right angles to the vector of the electrical potential, and this is the basis for electromagnetic conduction in the brain.

These concepts are again summarized in Figure 8 [18]:

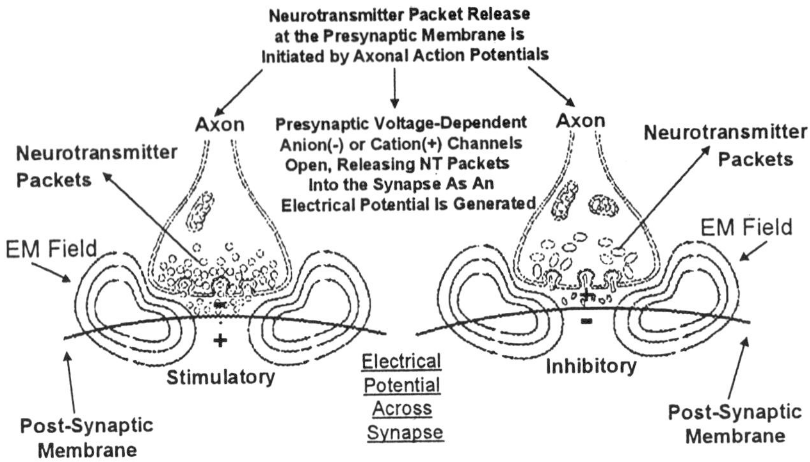

Figure 7. Back to the synapse
(The Core Structural and Functional Element)

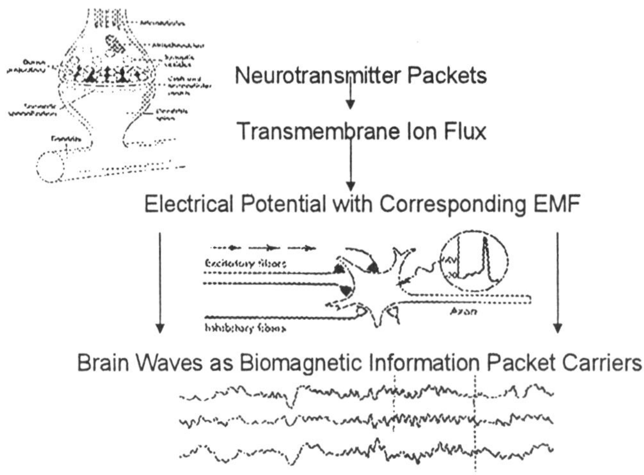

Figure 8. Converting neurochemistry to brian waves
(Synaptic Molecular Transduction)

9.1 AFFERENT NEURAL INFORMATION INPUT: IT'S NATURE, CATEGORIES, AND SCOPE

As noted, the mammalian brain is exposed to an enormously multimodal/multisensor information-rich environment that is saturated with diverse forms and sources of afferent multi-modal input. This afferent input consists of three broad categories, summarized in Table 2.

9.1.1 Lexical and Semiotic Data Multi-Sensor Type ImageryRegulatory Signal Input

In the neurocognitive field, the traditionally most important form of multimodal information and data content is lexical data. *Lexical Data* entails both linguistically-related and abstract information, such as some mathematical concepts. *Semiotic Data* and content is symbolic in nature and ranges from appreciation of shapes, forms, and patterned meaning to intuitive and emotively-modulated or imprinting-influenced information. It also covers more abstract and lexical content that is coupled to conceptual or symbolic elements. An example would be the visualization of a mathematical concept as a graphic representation. The spectrum of data input includes these *Lexical and Semiotic Data* information elements in various, called mixed assembly packets. Information can be extensively processed and modulated by numerous factors combinations (Table 2, left-hand column), including experience- or prior learning-based input, as well as neuroregulatory, humoral-regulatory, reflexive and somatic neural input originating from the endocrine or immune system, and from the peripheral nervous system.

TABLE 2. Categories of Afferent Inputs

Data Input	**Multi-Sensor Imagery**	**Regulatory Signal Input**
Lexical Data Content	Visual Imagery	Neurotransmitter Pathways
Semiotic Data Content	Auditory Imagery	Cell Signal Transduction
Mixed Data Assembly	Olfactory Imagery	EM Potential Transmission
	Gustatory Imagery	Pituitary-Hypothalamic Axis
	Tactile & Texture Mapping	Neurohumoral Regulation
	Form and Shape Mapping	Pineal Axis - Neurocircadian Axis
Data Modulation	Thermal / Cryo Mapping	Neuropeptidergic(endorphinergic) Axis
	Dermo-Photonic Input	Imidazolinergic Neurotransmitter Axis
Emotive	Proprioception / Vibration	Other Neurotransmitter Axes
Experience	Truncal-Rotational Mapping	Selected Trophic Neurochemistries
Imprinting	Diadodokinetic Mapping	Neuroparacrine & Neurocardiac Axes
Intuitive	Vestibular Mapping	Neuroimmunological Axis
Subliminal		CNS / CBF Autoregulation
Subconscious		Somatic & Secretory Regulation
Reflexive		Sympathetic / ParaSympathetic Chains
Regulatory		Reticular Activating System
Somatic		Autonomic Regulatory Centers
Autonomic		Biomagnetic Field Patterns

9.1.2 Multi-Sensor Type Imagery
Entails a broad range of afferent information incoming to the neural synaptic network of the brain. The middle column of Table 2 lists examples of such multi-sensor input that includes the major sensory modalities, all of which map onto cortical regions of the brain in a manner analogous to the projection of a pixel image or voxel image. A few definitions are useful. Gustatory input refers to taste, olfactory input to smell, and tactile/texture/form/shape mapping denotes the fact that our skin and visual sensors are patterned in a dermatomal manner that is projected onto the relevant area of the cortical surface topography. Other sensory modalities are thermal/cryo input, which refers to perception of temperature, and dermo-photonic input, which refers to the pineal axis sensitivity to light photons impacting on the skin. This only recently defined axis between skin receptors, the nervous system, and pineal hormones is one component of a biological circadian clock regulatory system in the human organism. Proprioceptive sensory input denotes the ability to tell position independent of visual input. Truncal input refers to central equilibrium, while diadodokinetic input refers to equilibrium and precision in the complex movements of our extremities.

Another type of information the brain receives is Regulatory Signal Input, as shown in the right hand column of Table 2. This category of afferent multisensor information includes a broad range of input from other elements in the human organism, from peripheral branches of the nervous system, to a diverse range of vascular system receptors, and to input from various hormonal and immunological regulatory axes. Even biomagnetic fields in organs outside of the brain may constitute a form of input sensed by the brain and the nervous system, in general.

Many of these individual sensory input modalities have been extensively studied, and for some modalities their natural systems design in the brain is fairly well understood - from the input sensor organ to the cortical destination of the imagery-based information. The schematic in Figure 9 [19]demonstrates one multi-modal sensory information input system, namely visual perception, orientation, and visual reflexes. This schematic is intended to demonstrate the point that, with many systems like with the visual neuronal apparatus, the functional anatomy and nuances of information packet distribution are extremely well characterized, but only for a sensory system in isolation. The schematic also demonstrates that one sensory system may have multiple input components. Here, in the visual system, we have two information transport pathways with different targets. Although the two pathways exist on each side of the brain, i.e. for each retinal projection, only one is depicted on each side of the schematic, this for the need to maintain clarity in the drawing.

1. The axonal fiber bundle projection for visual information denoted in black is the geniculostriate pathway that transmits direct complex visual imagery from the retina to the primary occipital visual cortex. This pathway additionally transmits information on the visual imagery elements of form, edges, patterns, depth of focus, movement, and color. From the visual cortex in turn (not pictured), there are secondary projections to the temporal lobe that contain color and form information, and to the parietal lobe related to extraction of information on spatially-visualized motion of objects.
2. The projection for visual information denoted in grey delivers information important to visual reflex function and visually-coupled orientation. This pathway

is important for orienting us to visual stimulii, and it projects to the region of the superior colliculus in the posterior midbrain known as the tectum. From there, other fibers (not pictured) project to the thalamic nucleus and to visual association areas in various distant cortical regions.

3. Visual information from the right field of vision projects to the left brain and vice versa. Furthermore, as demonstrated on the left side of the schematic, the retinal visual field (the circle at the top) is projected in segmental fashion to equivalent and Mapbase zones on the topography of the optical cortex, near the posterior midline known as the calcarine fissure. The more the visual information is at the center of the visual field, the closer its projection is to the midline of the calcarine cortex.

What we know about vision has largely been learned from the contribution of macaque monkeys, but we only know about one system in isolation and not about how vision information, for instance, fuses with information from balance sensor input or from the auditory system, etc. However, Figure 9 illustrates an important aspect of the visual sensory system viewed in isolation: the fact that different sub-components of visual information are not fused but rather separated into distinct pathways. This theme is repeated for many sensory input systems studied in isolation, and the fusion of afferent sensory information, must therefore occur at a higher cortical level, where attempts at characterization in this author's view have proven inconclusive to date.

In order to consider afferent sensory information flow from the vantage point of potential information fusion at higher cortical levels of systems organization, a more complex problem is illustrated utilizing the schematic in Figure 10. In the last few years, interest has been directed at studying the only recently recognized phenomenon of *'lexical thought'* [20,21].

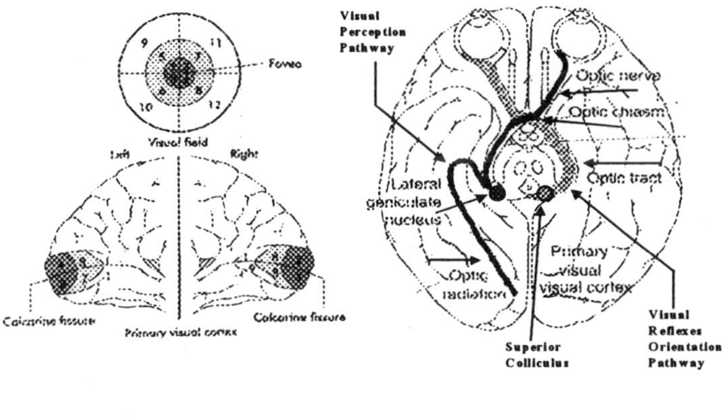

Figure 9. Visual imagery input model (Structure-Function Well Characterized)

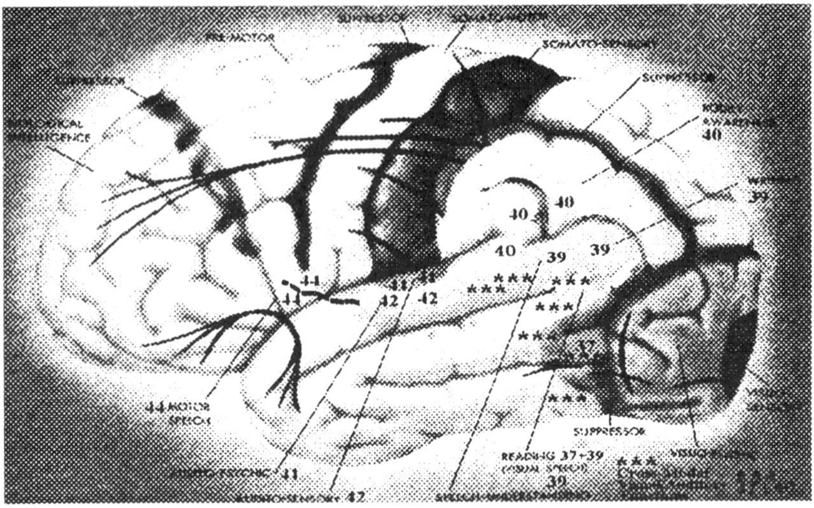

Figure 10. Modeling neuronal information fusion (The Mystery of Lexical Telepathy)

At the risk of entering the realm of extreme science, this phenomenon is also being 'termed lexical pseudotelepathy'. Potentially present in only discrete individuals, the facility for volitionally-controlling lexical thought processes can possibly be acquired through recruitment of symbolic learning coupled to ongoing lexical (linguistic) abstraction. The phenomenon may be described thus: when thinking silently of a short sentence, but only when silently mouthing the words at the lips, a unique set(s) of patterned EEG (electroencephalographic brain wave) signals are generated in an area of the temporal lobe yet to be clearly defined spatially. Obviously, the ability to detect bandwidth-specific EEG and biomagnetic signal patterns with a device worn on the ear lobe (and equipped with a pattern recognition system that assigns alphanumeric or diacritical letters, words and even sentences to specific bioelectromagnetic patterns) would permit telepathic dictation or commands to be realized. This would represent a biomedical and commercial revolution, but is expected to entail substantive technological challenge, in part because standard electrodermal sensors or even optrodes have insufficient signal-to-noise resolution and do not directly detect biomagnetic signals. However, with the emergence of magnetic source imaging and, more recently, normal temperature micro-SQUID technology, such recording techniques using portable equipment may become feasible within one decade.

The model of lexical telepathy is one which requires the concept of information fusion be invoked, in order to explain the observed phenomena. Clearly, the fusion process appears to involve: lexical and linguistically linked information content, speech and meaning recognition, awareness of tactile sensation (in this case peri-oral lip musculature) awareness, and silent auditory recognition processing facility (subject-variable temporal and parietal lobe regions). On the Cytoarchitectonic Map in Figure 10, the various topographic regions of the cortex involved in such a phenomenon are noted. In particular, these are motor speech required for 'mouthing of words' (Area 44) of the frontal operculum, the adjacent arcuate fasciculus axon pathway drawn as a black arch, the auditory-psychic (silent auditory recognition) and auditory sensory cortex regions (Areas 41-42), the areas responsible for reading-based, written and more general speech and linguistic understanding (Areas 37 and 39), and the region responsible for body awareness (Area 40). It is clear in viewing this map that the phenomenon of lexical thought is multi-functional, in terms of involving multimodal afferent inputs and efferent outputs. From the 2D electroencephalographic data available on this type of phenomenon, it is already clear that information flow for this functionality must require some type of fusion capability at a higher level of neural organization in the brain of selected individuals. Further studies of this phenomenon with MSI technology should yield at least a partial insight into patterns of information packet flow and fusion behavior. However, at present, MSI data on dynamic information flow is available only for less challenging and mostly phenomenological models of brain function.

9.2. NEUROIMAGERY TRACKS DYNAMIC FLOW AND COLLABORATIVE DATA FUSION

Having analyzed the fundamentals of the brain's structural and functional makeup that are central to consideration of how the brain may control information flow and achieve multimodal information fusion, the issue of practical correlation is then germane. State-of-the-art neuroimaging techniques facilitate the study of multimodal information fusion by assisting in the development of methods for tracking dynamic EM and biomagnetic field and activity flow --- these in turn are observational modalities that reflect processes likely to be involved in higher-level aspects of collaborative data and imagery fusion in the brain. Recently, new functional magnetic resonance imaging (*f*MRI) [22,23], positron emission tomography (PET) [24], and 3D magnetoencephalography (3D MEG) [25-27] technologies --- now collectively termed magnetic source imaging (MSI) --- have emerged [22-27], and are proving increasingly useful in pioneering the frontier of imaging fusion. MSI provides a literally phenomenal capability to track biomagnetic activity source points within all areas of the brain, both cortical, subcortical and localized to individual nuclei. This can be achieved in near live fashion, and the signal data superimposed with stereospatial accuracy onto volume brain anatomy concurrently imaged via magnetic resonance scanning. Figure 11 provides an example of 3D MSI/MRI imagery. Techniques like MSI will thus increasingly allow neuroscientists to track the biomagnetic filed fluctuations that reflect information flow and information fusion in a dynamically awake brain.

Figures 12-14 [28,29] demonstrate how the study of biomagnetic activity in the human brain can be utilized to study patterns of information packet flow. In a range of normal subjects, as well as those having specific neural deletional syndromes or deficits, patterns of brain activity can be studied with precise anatomical localization, and this can be performed while manipulating multi-modal sensory input to the brain. With such methods, serial patient responses and those of groups of patients can be contrasted and compared under different conditions.

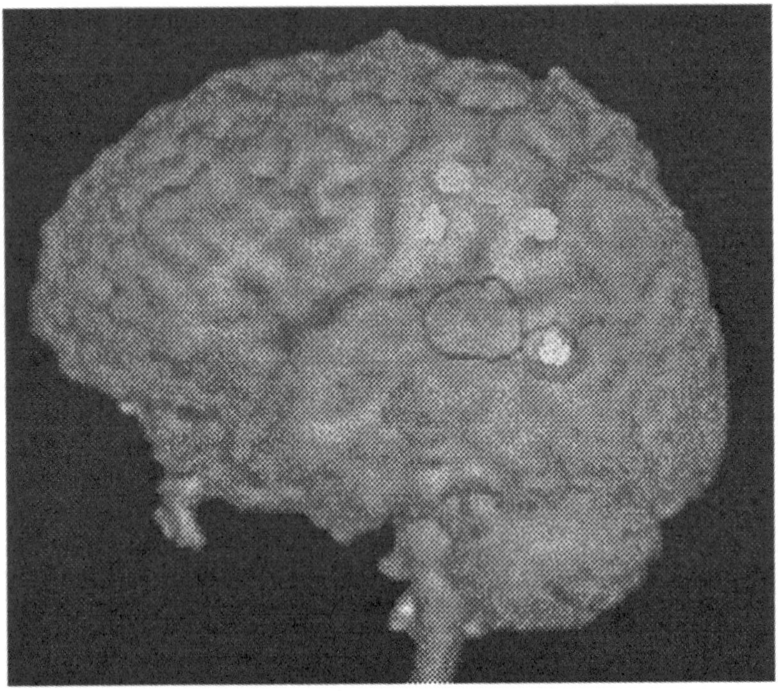

A Window into the Brain

Figure 11. Magnetic source imagine technology (Tracking Biomagnetic Activity)

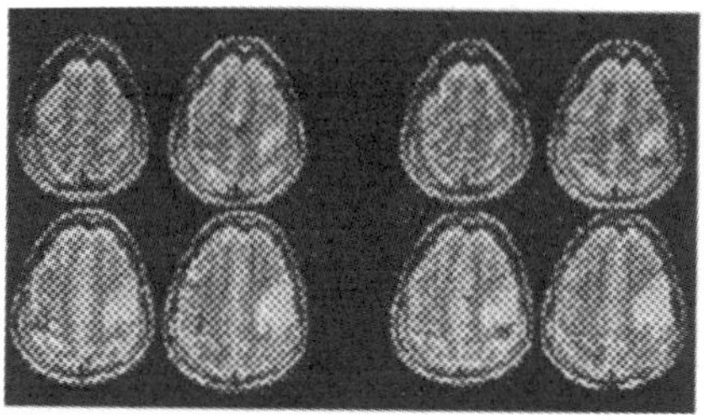

'Biomagnetic Packet Information Flow'

Figure 12. Magnetic source imaging technology (Tracking Biomagnetic Activity)

Figure 12 reflects a patient presented with slight variations in sensory input relating to unpleasant sensation and associative conceptual patterns that trigger exposure to those sensations. Please note that in this type of axial cross-sectional brain imagery, the right side is always projected as the left side of the image, i.e. one always sees the image from the feet up to the head. This patient has a lesion in the left posterior frontal motor strip and the anterior parietal lobe, as can be seen by the large white zone on the right posterior aspect of the image. This lesion has resulted in the patient losing some ability to concretely abstract or deduce aspects of pain perception, and related associative conceptual patterns which the patient sensory apparatus is exposed to in the test. With this type of patient, one can encounter intriguing findings during MSI, namely that only slight differences in sensory and conceptually-coupled pattern input result in entirely different patterns of biomagnetic signals. This is accentuated in patients with lateralizing sensory deficits (as opposed to normal subjects) and likely indicates a re-adapting brain rapidly trying out and sampling new and otherwise unusual paradigms for routing and prioritizing information packet flow. These studies are difficult to implement, because with repetitive testing, the issue of induced symbolic learning and patterning of behavior becomes a factor, the impact of which varies from subject to subject. However, the influence of patterned learning is less of an issue in patients with deficits such as the one just presented, because the random re-distribution of information packet signaling frequently continues over many MSI testing runs.

Figure 13 demonstrates another information flow phenomenon which is encountered in MSI research on patients with ictal disorders and post-traumatic deficits, namely the appearance of discrete vs. clustered biomagnetic signal packet distribution. Particularly when this is evident in the same patient under comparable test design conditions, and with only slight but definite variation in sensory input provided to the patient, this observation offers strong circumstantial evidence for information fusion as depicted in the images on the right, and information packet disassembly as depicted in the images on the left. Furthermore, in many patients with degenerative deficits or post-traumatic attention deficit disorder, or even with apparently normal structure on MRI scanning, diverse variations in test conditions and sensory input will yield: 1) redistribution of signal activity between different anatomical loci, and 2) evidence of disassembly vs. fusion of signal activity, reflecting neurosynaptically-transmitted information packets. This is shown in Figure 14.

In many cases, given knowledge of which brain centers and pathways should be involved in biomagnetic signal activation with a given set of sensory input during an MSI session, it is noted that MSI signal activity is found in regions of the brain which, given our putative anatomic-functional knowledge, should not be involved in a particular afferent sensory information input stimulation process or task scenario. The foregoing indicates not only that our knowledge is inadequate, but that 3) both plasticity and flexibility of information flow in the brain is incomprehensibly immense at a higher level of organization, and that 4) our knowledge of structure function relationships in the brain is an inadequate core of information for explaining the complexity of dynamic, real-world, neural function, even in a loose sense. Given what contemporary neuroscience is likely ignorant of, as regards higher cortical integration occurring at levels of superorganization required to implement complex information tracking and fusion behavior in the brain, present day models that describe mechanisms that regulate information flow (and that promote processes of information fusion) in the brain, may be entirely mis-constructed.

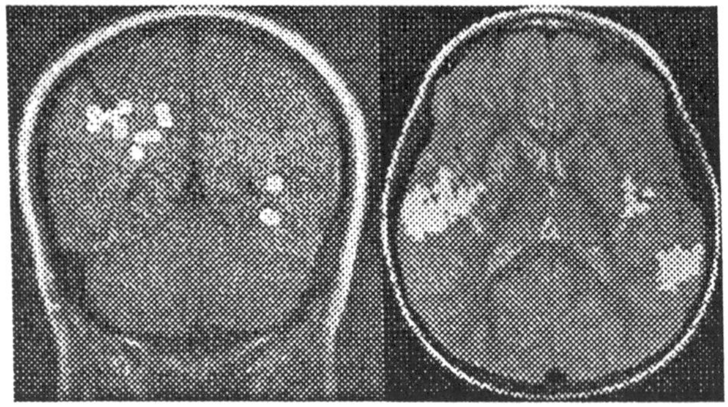

'Discrete vs. Clustered BM Packet Distribution'

Figure 13. Tracking dynamic information flow
(Is This Consistent with Fusion)

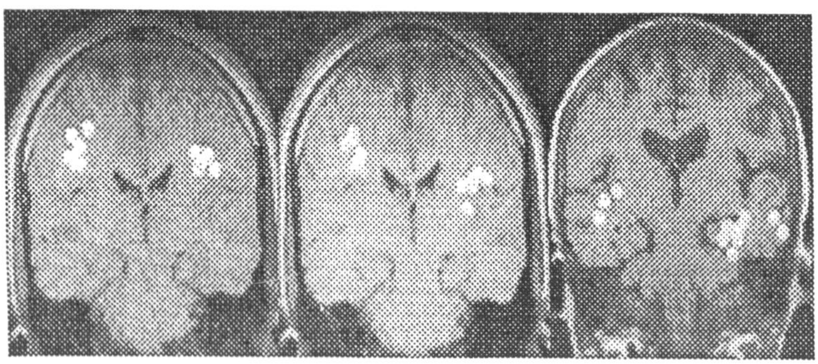

'Filtration and Re-Sampling of Biomagnetic Packets'

Figure 14. Tracking dynamic information flow
(Is This Consistent with Fusion?)

It may also be pointed out that MSI studies with a range of patients --- having deletional or anatomical isolation syndromes, where various components of the brain can be studied in partial isolation --- will prove highly illuminating as regards to observations on fusion-type biomagnetic signal patterns elicited under a variety of multi-modal sensory input. This includes patients with right-left hemisphere and interconnectivity-specific disconnection syndromes; those with either cortical and sub-cortical lesion and sub-cortical deletion syndromes, and patients with specific psychosurgical lesions or focal psychomotor-neurocognitive dysfunction. Another exciting innovation is transcranial magnetostimulation (TCMS) technology [30-32], whereby normal patients or lesion/deletional deficit patients can be subjected to transient induced suppression of cortical electromagnetic conduction in very precisely targeted regions of the brain. This dynamic TCMS probing technology results in a host of artificially-induced, transient, pseudo-deletional effects, occurring over short periods of time. Particularly in larger animals and when combined with MSI [25-27], *f*MRI [22-23], and positron emission tomography [24] (PET) neuroimaging, such models will provide greater insight into patterns of information flow and fusion in the brain.

The foregoing poses the question of whether the emerging neuroimaging observations in dynamic patients provide a rational basis for tracking definable electrobiomagnetic field patterns, which in turn would need to serve as specific signposts useful for unraveling the complexity of neural information flow and afferent data imagery information fusion in the brain. In order to facilitate such an understanding, and to begin translation of imaging and electrobiomagnetic pattern data into a meaningful set of models that explain information flow and data/imagery fusion in the context of known anatomical and functional neural systems in mammals and humans, a range of adaptive neurochemical and neurocomputational approaches are increasingly being explored. In this manner, the phenomenology of imaging observation and electrobiomagnetic field patterns --- combined with the more rigidly constrained knowledge on anatomy and neural function --- is being correlated with evolving neurochemical and neurocomputational models that seek to explain various findings in terms of likely operant processes of information fusion in the brain.

9.3 TRADITIONAL AND EVOLVING NEUROCOMPUTATIONAL PARADIGMS

Translating imaging observations and the dynamics of electrobiomagnetic field patterns into dataforms that provide relevance to potential mechanisms of information flow and afferent sensory data and imagery fusion in the brain necessitates sophisticated neurocomputational paradigms.

One such area of research involves the development of artificial neurocomputational algorithms and paradigms designed, or rather intended, to model real mammalian brain function with fidelity. To date, few such approaches even begin to explain the complex dynamics of symphonic decentralized collaborative patterns of synchronicity and dys-synchronicity that occur between diverse brain centers including those involving vision [26]. These dynamics can vary substantially between identical brain centers, depending on task relevance, sensory input, reward-linked attentional status [34], and de-centralized clustering patterns --- such as those putatively associated with information fusion. Because many prevailing neural network algorithmic designs are inadequate for characterizing the interaction of memory and heuristics (particularly in terms of visual recognition and response functionality), newer approaches are sought.

Not surprisingly, and particularly in the higher cognitive function modeling area, a range of algorithmic and adaptive systems modeling approaches are sprouting [35-42]. Some of these are targeted at better modeling aspects of complex neurocognition, such as relationally-relevant information recall via pattern completion, error-driven task learning and Hebbian model learning (for instance in conjunction with properties of inhibitory competition, distributed representation, bi-directional activation propagation, and biological realism taken into account). Presently, there are no resiliently-comprehensive models which concurrently address all of these important neuro-computationally relevant functions. In addition, more practical aspects of neurocognitive function also remain inadequately accounted for in many such models. Examples are non-linear conditioning, habituation, contextual learning, recognition-triggered recall, and retrograde amnesia. This includes amnestic phenomena that are selective and reversible, but 'irregularly irregular' in response to patterning with different combinations of stimuli, information input, environmental factors and even facial emotive responses --- that reflect non-linear impact on learning and memory-coupled image recognition processes. Certainly, in considering the inadequacy of present neurocomputational modeling efforts, one can note the paradox encountered when attempting to explain diverse sensory fusion phenomena --- namely, that fusion efficiency involving complex multimodal visual imagery, as well as comprehensional sensitivity of biological multipedal motion have actually both been demonstrated to increase with the amount of inflowing visual data, with no apparent threshold for data saturation identified [43]. No available neurocomputational model is known that explains such afferent sensory fusion phenomena in the mammalian brain, and in fact some models exist that eloquently explain why such an observed capability cannot in fact exist [46].

In this regard, a very small subset of models with real-world neuro-observational relevance nevertheless, and arguably, may offer a significant degree of correlation between real-world neural functional observations relevant to information fusion, vs. results obtained from evolving neurocomputational algorithms. A few approaches are demonstrating sufficient maturity for consideration of their adaptive transfer to neuroimaging and artificial multisensor heuristics technologies. In the defense community, such approaches may eventually have relevance to various applications. Within the area of defense intelligence, these could include: improved methodologies for enhancing analyst performance, novel neurovisual science-inspired designs for advancing human computer interface capabilities, and automated artificial heuristics technology for defense imaging and image analysis needs. With a view beyond the defense intelligence community, other areas of defense science would benefit from such neuroscience-inspired technology. These would include next generation human systems interfaces for real-time decision support and force protection, new designs for battle space and high altitude intercept information flow environments, enhanced predictive asymmetric threat modeling capabilities, and a range of adaptive defense-relevant biomimetic R&D. Breakthroughs in neuroscience-inspired neurocomputational advanced multisensor heuristics technologies would open extensive markets in the area of automated and operator-enhancing image analysis and procedural image guided functions throughout the biomedical field. Other applications could extend to planetary remote sensing, to meteorology and the commercial geospatial community, as well as across an imposing range of scientific and industrial technology sectors involving complex image understanding, as well as, biomimetically-modeled [35] complex adaptive systems, and neuromorphic sensor-driven robotic control systems [5,44].

9.4 EVOLVING MODELS RELEVANT FOR UNDERSTANDING CENTRAL NEURAL SENSORY FUSION PHENOMENA

Most of the models of interest for neurocomputational characterization of potential information fusion phenomena in the human and animal brain relate to multimodal visual afferent sensory input and the fusion of such input sources. Vision is generally not thought of as a fusion domain, but in fact visual information represents extremely rich imagery mapping content, including: shape and form definition, linguistic recognition, texture, depth, edge detection, color, hues, shading, multidimensionality, rotational and translational motion, to name a few components. Many of these components are in fact compartmentalized in different neural pathways, as was shown in Figure 9.

With visual data and imagery mapping type fusion phenomena representing among the most complex examples of afferent multisensor fusion activities that the brain engages in, some of the most promising experimental, emerging neurocomputational paradigms may be found in this sub-domain of neuroscience.

9.4.1 Synthetic Visual Scene Reconstruction via Lateral Geniculate Nucleus Ensemble Responses

In order to arrive at a capability for modeling visual fusion phenomena, it will be essential to understand the meaning of neural messagery via methods for translating sensory processing data. Such understanding of live neurovisual processing has progressed to the extent that one can utilize the live mammalian central nervous system, inclusive dynamic recording of cortical and subcortical region and cranial nerve pathway neuronal flow, to artificially extract information for reconstruction of 'live' visual imagery as it is likely perceived by the visual system, this from end organ receptors to the visual cortex.. This represents the ability to extract visual information on the outside world from complex sensory circuits. Recently, and for the first time, this capability was reported [45]. Recording the ensemble responses of 177 lateral geniculate nucleus neurons in the cat, linear decoding techniques were used to reconstruct ongoing spatiotemporal visual input from dynamic natural scenes as perceived by the feline central nervous system. This capability is still at its infancy, and "the quality of reconstruction depends on the number of cells. For each point in space, the quality of reconstruction begins to saturate at six to eight pairs of on and off cells, approaching the estimated coverage factor in the lateral geniculate nucleus. Thus, complex visual inputs can be reconstructed with a simple decoding algorithm, and these analyses provide a basis for understanding ensemble coding in the early visual pathway [7,45].

9.4.2 Human Cortical and Subcortical Specialization Relevant to Image Fusion Processing

Increasingly, structural and functional fields within the mammalian brain are being modeled at a level relevant to understanding neural information fusion and the unconventional parallel processing paradigms which may be applicable for explaining neurovisual fusion capabilities [46]. The emerging pattern of organization and function that characterizes neural synaptic interconnectivity, particularly as relates to visual processes, appears to indicate a decentralized symphonic collaborative pattern. For instance, both visual cortical areas and somatosensory cortex regions have separate functional streams for the detection of the micro- and macro-geometry of both 2D imagery, 3D stereospatial visual scene imagery, and 4D/5D multivariate 'superimposition mapping-perceived' informational imagery content. Interestingly, mapping of the mammalian auditory projection brain space reveals similar functional stream interconnectivity in the region of the superior colliculus that indicates visual auditory integration and overlap in projection of visual and auditory space maps.

In an attempt to explain these types of findings, some intriguing neurocomputational methods are being developed, including the 'Cortical Field Activation Hypothesis' --- which states that if two brain tasks make concurrent use of identical or overlapping neural fields, errors or increases in latency result, thus restricting the plasticity of parallel processing [46]. Proponents of this non-neural net based approach to neurocomputational characterization of the mammalian brain have recently described methods for transposing traditional Brodman's cytoarchitectural cortical maps to microstructural/functional area defined population maps that are extended to include 3D representation of subcortical nuclei. This parcellation has yielded a set of increasingly refined neuronal population maps of functional brain fields that require neurocomputational analysis. Due to the limitations of cognitive subtraction design paradigms, neuronal subpopulation 'image cluster' intersection analysis methods have been applied to better understand the functional interrelationships between overlapping function maps.

This approach generates a nearly convincing, albeit paradoxically-predictive models of why the brain cannot achieve what neurophysiologists are increasingly demonstrating it can, and to a phenomenal degree (i.e. as already noted, fusion efficiency involving complex multimodal sensory input can increase with the amount of inflowing sensory data, with no apparent threshold for data saturation being identified [43], thus probably indicating that neural overlap summation behavior is space-time non-linearly integrative --- and reorganizationally-adapts to the nature of the external visual stimulus). Moreover, there is functional MRI (fMRI) evidence that the regions within the V5 area of the medial superior temporal cortex respond dynamically and selectively to changes in afferent optic flow [23], with distinct subspecialization as regards translation of imagery and perception of circular/radial patterning. These pathways respond concurrently or separately, and with different overlap patterns simply depending on a stimulus set's visual content. Thus, neuronal field map-based cortical field activation models cannot explain these observed attributes of neuronal visual function. Hence, some do not prefer this approach for understanding the de-centralized collaborative behavior observed in multiple overlapping synchronous or desynchronous neural information stream datafusion in the brain --- phenomena that are increasingly recognized as characterizing mammalian neurovisual processing.

9.4.3 Biomagnetic Microstates and Transcortical Cell Assembly Models of Multi-Sensory Processing

Another approach to modeling the complexity of visual and other sensory information flow and information fusion in the brain, is to view neuronal function in terms of frequency cluster organization --- for instance, as applied to biomagnetic source data or standard electroencephalographically-recorded neuronal output. Such approaches greatly facilitate the inclusion of neuronal pathway communication data patterns, such as neuronal cluster and encoding synchrony or dys-synchrony phenomena --- and particularly those which reflect functional interconnectivity between frontal and parietal regions, and between hemispheres during convergent and divergent thinking tasks [47]. With respect to the modeling of visual image fusion, another approach is to describe those aspects of multimodal visual imagery processing that are believed to be most relevant to higher neurocognitive function, for instance in terms of 'spontaneous thought' building blocks referred to as 'momentary conscious mind states'. These, in turn, are represented as electromagnetic 'brain microstates' which are recognized as discrete elements that together summate to produce an electric field in any specific area of the brain. 'Brain microstates' are increasingly being referred to as the 'atoms of thought'. Although a totally phenomenological approach that utilizes EEG monitoring of human subjects, this work has already yielded promising results. These suggest that the continuous stream of consciousness consists of 'electromagnetic microstates' that seamlessly and rapidly interconnect into larger patterns, but which individually implement different, discreet modes, actions or functions. This might be analogous to the packet flow of Morse code in a message transmission. It is continuous, but actually consists of a stream of packets with breaks, where the packet content, size, duration, amplitude, frequency, as well as the packet cyclicity and geometry of intrapacket delays, all contribute to the overall description of an individual or group of cortical microstates.

Thus, field maps derived from this model are functional and transiently reconfigurable, and not structurally defined. This work has revealed that different modes of visual imagery input utilize different brain mechanisms, involving many cortical and subcortical regions both within and outside of the traditional occipital visual cortex. Input-driven imagery tasks initiate left-sided occipital event potential related maxima, as correlated with PET studies on attentive task imagery solutions that induced stronger left-sided activity. Memory-coupled recognition and recall tasks, especially if information-less, were preferentially right-sided. Non-linear, unpredicted and non-repetitive, mental rotation tasks were also predominantly right-sided, similar to imagery examination conditions.

Recognition of the importance of microstate modeling has led to the concept of transcortical cell assemblies [48], a form of cellular automata relevant to neurocomputational systems. The most interesting and image fusion-relevant work in this regard relates to the study of gamma-band spectral EEG and MEG (MSI) responses, considered as physiological signposts of higher neurocognitive and sensory data processing in the brain. In regard to visual fusion activity, the dynamics of γ-band spectral responses [49-50] are indicators of visual analysis tasks that are framed in a lexical or abstract perceptual and recognition context, i.e. 'thinking about imagery'. Different patterns of γ-band activity, reflecting many populations of transcortical cellular assemblies, may be found in terms of stimulus onset and locking, cortical topography, and various levels of centralized synchrony, or decentralized and variably-synchronous collaborative activity, between distant brain centers. The results of 3D MEG (MSI) studies, designed to relate psycholinguistic processing models of picture naming to the dynamics of cortical activation, generally confirm the pattern topography obtained from γ-band pattern modeling [27]. These processes are lexically-coupled and quite different from shape, pattern, and movement perception type image content (that can predominate independently of an abstract context). 3D MEG (MSI) or EEG frequency patterns, synchrony-encoded clusters or desynchronous (de-centralized) collaborative neuronal output cascades involved in perceiving object shape --- and the cortical-subcortical regions they subtend --- all reflect distinct set of regions and interrelationships. One such example of a cortical-subcortical region is the lateral occipital complex, which is specialized for recognition of objects within imagery defined by different visual cue identifiers (such as grayscale photographs vs. line patterns) [51].

9.4.4 Memory-Associative Heuristics, Relational Recognition/Recall, Emotion and Decisions

Contemporary neurocomputational modeling is rapidly addressing at least the imperative of considering a series of inter-related components to higher cognition and multi-sensory input processing and information utilization. Contrary to classical convention, the orbitofrontal cortex is no longer considered the absolute center of decision-making. Rather, and depending on the decision-related content and the resulting abstract to emotive balance induced by a given information set, structures such as the amygdala, the somatosensory/insular cortex, the hippocampus, and the visual associative cortex, are intimately interconnected across a range of decision processes, including those based on interaction with visual imagery [52]. The impact of temporal factors on memory-associative heuristics (which in turn impact on processes of discontiguous relational recognition and recall, including those involving visual analysis tasks) is increasingly recognized as an important aspect of hippocampal function[22,53]. Exquisitely sensitive to metaplastic effects from stress [54], the hippocampus serves as the specialized brain region responsible for, among other functions, associative/discriminative processing and retrieval of temporally and spatially discontiguous heuristic memory elements [53]. In an attempt to explain this multimodal complexity on a neurocomputational basis, different variants of Heebian 'long-lasting activity-dependent variation in synaptic efficacy' learning models, as well as Bienenstock-Cooper-Munro synaptic-plasticity theory have been applied to neural fusion phenomena, although with uneven success.

By contrast, and on a non-computational modeling basis, neurophenomenologists have advanced a number of alternative theoretical frameworks. Most notable of these are the somatic marker hypothesis [55] and time-locked multiregional retroactivation [56] models. Although not neurocomputational methods per se, these approaches offer valuable conceptual insight into visual and neurocognitive aspects of information fusion-type processes in the human brain.

The main premise of the somatic marker hypothesis is that 'bioregulatory marker' signals influence the process of neuronal response to stimuli at multiple levels of operations, both overt (conscious) and covert (non-minded). These signals arise in the brain's somatic representation of the body which, by definition, is somatic and therefore anatomically imprecise. With respect to vision, this process would apply more to reasoning-coupled visual analysis and higher cognitive decision-making, than to non-lexical pattern recognition. Although the ventromedial prefrontal cortex is the 'anatomical/functional brain center' paradigm for this model, its application is not restricted to the frontal cortex and has potentially far broader relevance.

The time-locked multiregional retroactivation model is intended to improve the conceptual modeling for understanding neural substrates of relational recall and recognition of either lexical or visual content and tasks. Basically, this model proposes a 'neuron ensemble architecture, located in multiple and anatomically unrelated sensory association cortex regions that house representations of dynamic feature fragments contained in memory. These feature fragments can be visual imagery-specific. Recall of entities (images) and events occurs when neuron ensembles, and thus feature fragment information flow, are activated in a synchronous time-locked fashion. This activation is modeled as taking place in convergence zones between neurons storing or reactivating demand for feature fragments, and between first order sensory association cortex- vs. perceptuomotor cortex-specialized neurons. These neuron ensembles are located in separate and even widely-distant and unrelated regions, but function in a de-centralized and highly collaborative (symphonic) manner, essentially as ensembles. Mediated by feedback neuronal projections --- and specific to the convergence zones between neuronal ensemble members (that are relevant to a given contextual recognition and recall substrate --- synchronous neuron firing elicits retrieval of information desired and relevant feature fragments. In the case of neurovisual processing, feature fragments are various imagery components. Many neuroscientists correlate this locked multiregional retroactive model [56] with transcortical cell assemblies, and with the concept of brain Microstates discussed previously [47-50].

Although no practically useful neurocomputational correlate of this work exists, this model is highly relevant to defense imaging and imaging analyst tasks and challenges. The model defines clearly, perhaps for the first time in neuroscience, that perceptual (for instance imagery) experiences depends on neural activity in multiple regions activated concurrently, rather than in a single region where experiential integration would occur. This implies that complex multimodal sensory perception is in fact fusion dependent. During free recall or recall generated by perception in a recognition task, the multiple region activity necessary for experience occurs via retroactivation --- namely, the recall of experiences depends on reactivation close to original neuronal input and dynamically-requested output sites (convergence sites). The model thus rejects a single anatomical site or storage region for integration of memory-heuristic coupled activity (inclusive imagery processing, relational recognition/recall, and retrieval). Rather, meaning and decision is achieved by collaborative symphonic (independent but synchronously correlated and coupled) retroactivation of widespread fragment records. These processes appear governed equally by lexical and abstract centers, as by perceptive and emotive centers --- the latter as the metaplastic sensitivity of hippocampal memory recall to stress confirms [54].

9.4.5 Visual Imprinting, Awareness/Attention, Task Modulation & Memory Recognition Performance

The understanding thus gained, largely via phenomenological models, points to a number of attractive areas for applying neurocomputational designs to the understanding of memory optimization and heuristically-coupled recognition processes relevant to information fusion in the brain. It is increasingly accepted that distinct cortical and subcortical anatomical/functional regions on the one hand [57], or distant and only transiently-related neuronal ensembles on the other hand [56], can impose attentional sets for different aspects of neurosensory (including visual imagery analysis-relevant), task-switching-intensive [58], and cognitive demand-intensive [59] functionality. In support of this view, correlative neurochemical evidence exists for the role of cholinergic [60] and dopaminergic systems [34] in governing such control of awareness and attention to, among others, visual imagery and visual analytical tasks. Lately, and as already discussed in regard to gamma frequency responses [48-50], both neuronal firing and frequency output patterns relevant to modeling of 'transcortical assembly brain microstates' demonstrate distinctive patterns related to different attentional sets and task switching demands.

Specifically, time-pressure effects on success vs. error in visuomotor task performance can be effectively modeled by comparing patterns of theta, mu, and gamma frequency in EEG studies. In such study data, time and performance pressure for visual analysis tasks specifically increased frontal midline theta activity, gamma midline, and gamma right frontal activity. Furthermore, when correlated with the theta frequency patterns, right parietal alpha activity patterns could accurately predict the success vs. failure of visual task performance [61]. Thus, as task load increases, the variance in frequency output reflects neuronal ensemble sampling of different task-relevant/task-irrelevant neurocognitive processing and decentralized collaboration strategies --- a phenomenon reminiscent of dynamic self-configuring functionality in computational models based on cellular automata. Thus, it is becoming evident that neural mechanisms of memory-coupled recognition, and processes of visual analysis-relevant imprinting, may be interrelated. Recent work has lent further weight to this postulate, by specifically localizing visual imprinting to an anatomically-defined polysensory afferent input-specialized region in avian species, known as the intermediate medial hyperstriatum ventrale [62]. It remains to be shown that similar systems exist in man. Yet, both neurochemical and neurogenetic assays designed to correlate learning, memory and visual imprinting responses with responses at the cellular and molecular level, indicate that at least the capability to utilize memory and heuristic-coupled recognition processes --- for the purpose of visual imprinting --- appears inherent to the design of many animal brains.

9.5 HOW DOES THE BRAIN FUSE INFORMATION?

This brings us to the central question of whether the brain fuses information. As has been discussed, recent studies by a number of groups have indicated that, if fusion processes are involved with information flow, these will be somewhat decentralized and collaborative in nature, as in the symphonic model. A number of key elements appear essential to higher brain fusion of information:

1. Intact distant and short term lexical and semiotic memory coupled with normally developed symbolic learning processes
2. Required memory elements include relational recognition and recall of lexical and semiotic data, as well as reorganization and incorporation of new and previously-experienced Multi-Modal sensory input.
3. Embedding of afferent information with these elements occurs in a fluid manner, and allows for management of information packets in disassemble form, partially linked form, or in a variety of fused formats.
4. Diffuse collaborative behavior between brain centers is subject to dynamic rearrangement, re-assembly, and disassembly of information fusion packets.

It is of interest to note that there is a precedent in nature for this type of arrangement at a molecular level, namely disassembly and re-assembly of genetic information packets in a prokaryotic biological system. It is known to occur at a genetic level in certain extreme environment-compatible bacteria who achieve survival against incredible environmental challenges, for instance to a pulse of 1 million rads of high energy ionizing radiation. These coccobacilli, *deinococcus radiodurans*, readily disassemble their entire DNA genome into constituent nucleotide bases upon encountering a challenge. Once the challenge is past, in a few hours the genetic code is reassembled with a remarkably minimal mutational error rate, and the bacillus resumes replication. In this case, the analogy between DNA computing and higher functional cortical organization is a reasonable one.

Thus, it may be cautiously concluded that interconnectivity patterns and dynamically-fluid neural information flow in the brain depend on real-time synaptic plasticity --- and are driven by memory, symbolic learning, and by context which can involve any combination of lexical, semiotic, afferent sensory, or emotional information content.

Contemporary evidence gained from MSI studies in a variety of subject settings is providing an increasing basis for arguing that information flow and fusion processes in the brain follow a pattern of ongoing symphonic re-arrangement. This in turn implicates a constant re-alignment of participating brain centers and rearrangement of synaptic neurochemical transmission and biomagnetic field activity patterns. This in turn reflects a likely constant re-allocated exchange of information packets in a de-centralized collaborative setting regulated by complex relational mechanisms.

The absence or presence of a symphonic conductor is unknown.

If this arrangement is indeed consistent with information fusion, then it appears to involve a unique and evolutionarily superior paradigm that is characterized by the presence of a dynamic equilibrium between disassemble and fused information packets flowing through the brain's complex stereospatial synaptic network.

10. References

1. Arkin, R., Collins, T., and Endo, Y., (1999) 'Tactical Mobile Robot Mission Specification and Execution,' on line publication: www.cc.gatech.edu/aimosaic/robot-lab/mrl-online-publications.html.
2. Parker, L., (1995) 'The Effect of Action Recognition and Robot Awareness in Cooperative Robotic Teams', Proceedings of the 1995 IEEE/RSJ International Conference on Intelligent Robots and Systems (IROS'95), Volume 1,pp. 212-219.
3. Balch, T., and Arkin, R., (1995) 'Motor Schema-based Formation Control for Multiagent Robot Teams', Proceedings of the First International Conference on Multiagent Systems, San Francisco, CA.
4. Kube, C., Zhang, H., (1996), 'The Use of Perceptual Cues in Multi-Robot Box-Pushing', IEEE International Conference on Robotics and Automation, pp. 2085-2090.
5. Indiveri, G; Modeling selective attention using a neuromorphic analog VLSI device. Neural Comp ut. 2000 Dec; 12 (12):2857-80.
6. Bonabeau, E., Dorigo, M., and Theraulaz. G., (1999) Swarm Intelligence: From Natural to Artificial Systems. Oxford University Press Inc, New York, NY. ISBM 0-19-513158-4.
7. Dan, Y; Alonso, (1998) JM; Usrey, WM; Reid, RC: Coding of visual information by precisely correlated spikes in the lateral geniculate nucleus. Nat Neurosci. 1998 Oct;1(6):501-7.
8. Lewine, Jeffrey D. '*Introduction to Functional Neuroimaging: Functional Neuroanatomy*', in: Orrison, William W; Lewine, Jeffrey D.; Sanders, John, A.; and Hartshorne, Michael F.; eds.: Functional Brain Imaging, Ch. 2, pp.13-95, Mosby Year-Book Publishers, St. Louis, Mo 1995.
9. Cserr, Helen E., (1986) ed.: The Neuronal Microenvironment, in: Annals of the New York Academy of Sciences, Volume 481, Dec. 30, 1986, The New York Academy of Sciences, NY, NY.

10. Adapted from: Bullock, TH; Orkand, R., and Grinnell, A., (1997) Introduction to Nervous Systems, Freeman, San Francisco.
11. Adapted from: Lewine, Jeffrey D., (1995) *'Introduction to Functional Neuroimaging: Functional Neuroanatomy'*, in: Orrison, William W; Lewine, Jeffrey D.; Sanders, John, A.; and Hartshorne, Michael F.; eds.: Functional Brain Imaging, Ch. 2, pg. 15, Mosby Year-Book Publishers, St. Louis, Mo.
12. Adapted from: Lewine, Jeffrey D., (1995) *'Introduction to Functional Neuroimaging: Functional Neuroanatomy'*, in: Orrison, William W; Lewine, Jeffrey D.; Sanders, John, A.; and Hartshorne, Michael F.; eds.: Functional Brain Imaging, Ch. 2, pg. 21, Mosby Year-Book Publishers, St. Louis, Mo.
13. Adapted from: Peters, A; Palay, Sl; Webster, HD, eds. (1976) The Fine Structure of the Nervous System. Saunders, Philadelphia.
14. Adapted from: Kolb, B.; Whishaw, IQ, eds. (1980) Fundamentals of Human Neuropsychology. Freeman, San Francisco, and Sidman, RL; Rakic, P. Brain Res 62:1-35, 1973
15. Adapted from: Sidman, RL; Rakic, P: Brain Res 62:1-25, 1973; and Martin, JH, Neuroanatomy: text and atlas, Appleton and Lange, Nowalk, Ct., 1989
16. Adapted from: Brodmann, K., (1995) Vergleichende Lokalisationslehre der Grosshimrinde in Ihren Prinzipien Dargestellt auf Grund des Zellenbaues, Barth, Leipzig, 1909; and Lewine, Jeffrey D. *'Introduction to Functional Neuroimaging: Functional Neuroanatomy'*, in: Orrison, William W; Lewine, Jeffrey D.; Sanders, John, A.; and Hartshorne, Michael F.; eds.: Functional Brain Imaging, Ch. 2, pg. 33, Mosby Year-Book Publishers, St. Louis, Mo.
17. Adapted from: Lewine, Jeffrey D. (1005) *'Introduction to Functional Neuroimaging: Functional Neuroanatomy'*, in: Orrison, William W; Lewine, Jeffrey D.; Sanders, John, A.; and Hartshorne, Michael F.; eds.: Functional Brain Imaging, Ch. 2, pg. 31, Mosby Year-Book Publishers, St. Louis, Mo.
18. Adapted from: Peters, A; Palay, Sl; Webster, HD, eds. (1976) The Fine Structure of the Nervous System. Saunders, Philadelphia, and Lewine, Jeffrey D. *'Introduction to Functional Neuroimaging: Functional Neuroanatomy'*, in: Orrison, William W; Lewine, Jeffrey D.; Sanders, John, A.; and Hartshorne, Michael F.; eds.: Functional Brain Imaging, Ch. 2, pg. 32, Mosby Year-Book Publishers, St. Louis, Mo 1995; and Hughes, JR: EEG in Clinical Practice, Butterworth, Boston, 1982,
19. Adapted from: Lewine, Jeffrey D. (1995) *'Introduction to Functional Neuroimaging: Functional Neuroanatomy'*, in: Orrison, William W; Lewine, Jeffrey D.; Sanders, John, A.; and Hartshorne, Michael F.; eds.: Functional Brain Imaging, Ch. 2, pp. 74-75, Mosby Year-Book Publishers, St. Louis, Mo.
20. Kubler, A., Kotchoubey, B., Hinterberger, T., Ghanayim, N., Perelmouter, J., Schauer, M., Fritsch, C., Taub, E., and Birbaumer, N., (1999) The thought translation device: a neurophysiological approach to communication in total motor paralysis. Exp Brain Res.124(2):223-32.
21. Birbaumer, N., Kubler, A., Ghanayim, N., Hinterberger, T., Perelmouter, J., Kaiser, J., Iversen, I., Kotchoubey, B., Neumann, N., and Flor, H., (2000) The thought translation device (TTD) for completely paralyzed patients. IEEE Trans Rehabil Eng. (2):190-3
22. Stark, C.E., and Squire, R.L., (2000) Functional Magnetic Resonance Imaging (fMRI) Activity in the Hippocampal Region During Recognition Memory. J Neuroscience, 20(20):7776-7781
23. Morrone, M.C., Tosetti, M., Montanaro, D., et al: (2000) A Cortical Area that Responds Specifically to Optic Flow, Revealed by fMRI. Nature Neuroscience, 3(12)1322-1328
24. Cabeza, R., and Nyberg, L., (2000) Imaging Cognition II: An Empirical Review of 275 PET and fMRI Studies. Journal of Cognitive Neuroscience 12:1, pp. 1-47.
25. Bundo, M., et al: (2000) Human visual motion areas determined individually by magnetoencephalography and 3D magnetic resonance imaging Hum Brain Mapp 11(1):33-45
26. Barinaga, M., (1998) Listening in On the Brain. Science, 280:376-378
27. Levelt, W.J., et al: (1988) An MEG Study of Picture Naming. J Cog Neuroscience 10(5):553-567
28. Adapted from Lewine, J.D., Orrison, WW: *'Magnetoencephalography and Magnetic Source Imaging'*, in: Orrison, William W; Lewine, Jeffrey D.; Sanders, John, A.; and Hartshorne, Michael F.; eds.: Functional Brain Imaging, Ch. 9, pg. 409, Mosby Year-Book Publishers, St. Louis, Mo 1995; and Lewine, J.D.; Orrison, WW; Halliday, A. et al: *'MEG Functional Mapping in Epilepsy Surgery'*, in: Cascino, G.D., Jack, C.R., eds.: Neuroimaging in Epilepsy: principles and practice, Butterworth and Heinemann, Boston, 1994.
29. Adapted from Lewine, J.D., Orrison, W.W., (1995) *'Magnetoencephalography and Magnetic Source Imaging'*, in: Orrison, William W; Lewine, Jeffrey D.; Sanders, John, A.; and Hartshorne, Michael F.; eds.: Functional Brain Imaging, Ch. 9, pp. 410-411, Mosby Year-Book Publishers, St. Louis, Mo.
30. Schurmann, M., Nikouline, V,V,. Soljanlahti, S., et al: (2001) EEG responses to combined somatosensory and transcranial magnetic stimulation. Clin Neurophysiol;112(1):19-24

31. Hong, K., Lee, S.K., Kim, J., Kim, K., and Nam, H., (2000) Visual working memory revealed by repetitive transcranial magnetic stimulation. J Neurol Sci;181(1-2):50-5
32. Hallett, M., (2000) Transcranial magnetic stimulation and the human brain. Nature; 406(6792):147-50
33. Gilchrist, I.D., (2000) Refixation frequency and memory mechanisms in visual search. Curr Biol;10(19):1209-12
34. Schultz, W., et al: (1997) A neural substrate of prediction and reward. Science, 275:1593-1599
35. Beer, R.D., (1990) Intelligence as Adaptive Behavior: An Experiment in Computational Neroethology, in: *Perspectives in Artificial Intelligence,* Volume 6, Academic Press, London.
36. MacGregor, R.J., (1987) Part II. Computer Simulation of Neuronal Systems, in: *Neural and Brain Modeling*, Academic Press, London.
37. O'Reilly, R.C., (1998) Six principles for biologically-based computational models of cortical cognition. Trends in Cognitive Sciences, 2:11:455-462
38. O'Reilly, R.C., (2000) Rudy JW: Computational principles of learning in the neocortex and hippocampus Hippocampus;10(4):389-97.
39. Cohen, J.D., Braver, T.S., and O'Reilly, R.C., (1996) A computational approach to prefrontal cortex, cognitive control and schizophrenia: recent developments and current challenges. Philos Trans R Soc Lond B Biol Sci;351(1346):1515-27.
40. McClelland, J.L., McNaughton, B.L., and O'Reilly, R.C., (1995) Why there are complementary learning systems in the hippocampus and neocortex: insights from the successes and failures of connectionist models of learning and memory. Psychol Rev;102(3):419-57.
41. O'Reilly, R.C., and McClelland, J.L., (1994) Hippocampal conjunctive encoding, storage, and recall: avoiding a trade-off. Hippocampus;4(6):661-82
42. Farah, M.J., O'Reilly, R.C., and Vecera, S.P., Dissociated overt and covert recognition as an emergent property of a lesioned neural network. Psychol Rev; 100(4):571-88.
43. Neri, P., Morrone, M.C., and Burr, D.C., (1998) Seeing Biological Motion. Nature, 395(29):894-896.
44. Indiveri, G., and Douglas, R., (2000) Neuromorphic Vision Sensors. Science, 288:1189-1190.
45. Garrett, G.B., Li, F.F., and Dan, Y., (1999) Reconstruction of Natural Scenes from Ensemble Responses in the Lateral Geniculate Nucleus. J Neuroscience, 19(18):8036-8042.
46. Roland, P.E., and Zilles, K., (1998) Structural divisions and functional fields in the human cerebral cortex. Brain Research Reviews, 26:87-105.
47. Razoumnikova, O.M., (2000) Functional organization of different brain areas during convergent and divergent thinking.. Cognitive Brain Research, 10:11-18.
48. Pulvermuller, F., and Mohr, B., (1996) The concept of transcortical cell assemblies. Neuroscience and Biobehavioral Reviews, 20(4):557-566.
49. Pulvermuller, F., Lutzenberger, W., Preissl; Birbaumer, N., (1995) Spectral responses in the gamma band. NeuroReport, 6:2059-2064.
50. Pulvermuller, F., Birbaumer, N., Lutzenberger; F., and Mohr, B., (1997) High-frequency brain activity. Progress in Neurobiology, 52:427-445
51. Kourtzi, Z., et al., (2000) Cortical regions involved in perceiving object shape. J Neurosci, 20(9):3310-3318.
52. Bechara, A., et al., (2000) Emotion, DecisionMaking and the Orbitofrontal Cortex. Cerebral Cortex, 10:295-307.,
53. Wallenstein, G.V., et al., (1998) The hippocampus as an associator of discontiguous events. TINS, 21(8):317-323.
54. Kim, J.J., et al., (1998) Stress, metaplastic effects in the hippocampus. TINS, 21(12):505-509.
55. Damasio, A.R., The somatic marker hypothesis and the possible functions of the prefrontal cortex. Phil Trans R Soc Lond B 1996, 351:1413-1422.
56. Damasio, A.R., (1989) Time-locked Multiregional Retroactivation. A systems level proposal for the neural substrates of recall and recognition. Cognition, 33:25-62.
57. Banuch, M.T., et al., (2000) Prefrontal regions play a predominant role in imposing an attentional 'set': evidence from fMRI. Cognitive Brain Res, 10:1-9
58. Sohn, M.H., Ursu, S., and Anderson, J.R., (2000) the role of prefrontal cortex and posterior parietal cortex in task switching. PNAS November 21, 97(24):13448-13453.
59. Duncan, J., and Owen, A.M., (2000) Common regions of the human frontal lobe recruited by diverse cognitive demands. TINS, 23(10):475-483
60. Gill, T.M., et al., (2000) Sustained Visual Attention Performance-Associated Prefrontal Neuronal Activity: Evidence for Cholinergic Modulation. J Neurosci, 20(12):4745-4754.

61. Slobounov, S.M., et al., (2000) Neurophysiological and behavioral indices of time pressure effects on visuomotor task performance. Cognitive Brain Res, 9:287-298
62. Horn, G., (1998) Visual imprinting and the neural mechanisms of recognition memory. TINS (21)7:300-305.

FUNDAMENTALS OF REASONING AND MULTISENSING

Introduction to the Foundations of Multisensor Data Fusion

E. WALTZ
Veridian Systems
1975 Green Road
Ann Arbor, MI 48105
US

1. The Fundamentals of Reasoning

The study of human reasoning processes and the automation of those processes by computing machines is a broad area of study that encompasses many disciplines and has contributors that span well over two millennia. From the abstract considerations of human thought (metaphysics) to the very practical and pragmatic development of human decision-making processes (game theory, decision theory, risk management and psychology), the study of reasoning is essential to our understanding of ourselves, the world about us, and our ability to create. The objects of the study of reasoning, defined in human, not machine terms include:

- Cognition refers to the human mental mechanisms that provide self-awareness and understanding about the environment, including: perception, attention, reasoning, judging, learning, memory, thought or reflection, concept formation, and problem solving.
- Reasoning is the use of the logical, rational and analytical faculty of the mind to form conclusions, inferences, or judgments by analyzing evidence and arguments.
- Inference is the reasoning process that specifically manages (creates, modifies and maintains) beliefs.

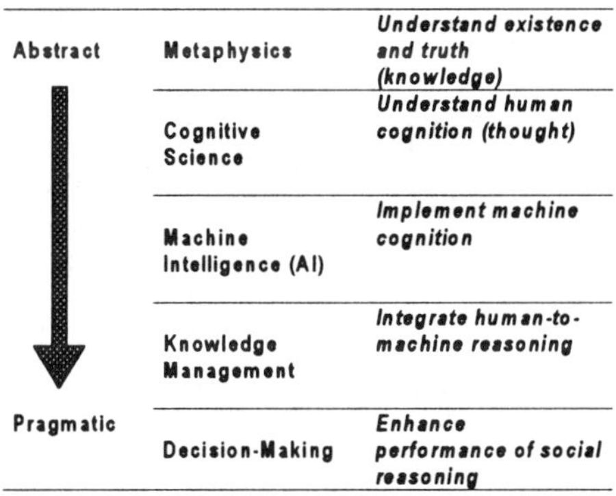

Figure 1. The Disciplines of Thought

A. K. Hyder et al. (eds.), Multisensor Fusion, 37–70.

1.1 COGNITION AND REASONING PROCESS

Aristotle, an empiricist philosopher, discussed the process of cognition in his *Metaphysics*, written in 350 B.C. He distinguished physical objects, and the abstractions (ideas) that the mind creates once the senses observe the object. He further distinguished the words that the mind creates to symbolize or represent the abstractions of the mind. He described three processes of the intellect that manipulate these domains:

- *Apprehension* is the process by which the mind perceives and understands the sensed physical object, and creates a mental abstraction. (Physical-to-cognitive object mappings are formed.)
- *Predication* is the process of making declarations or propositions about the object – characterizing the object and its behavior in a human language abstraction. (Cognitive-to-symbolic mappings are created.)
- *Reasoning* is the process, then, of applying logical principles to the propositions to create new conclusions, or syllogisms. Here, Aristotle recognized the methods of deduction and induction. (Symbolic logic draws new conclusions about cognitive and physical objects.)

This model can be compared to American pragmatist philosopher William James' (1842-1910) concept of cognition. James wrestled with the issue of determining agreement between reality and our mental idea of that reality. Because truth is the agreement between idea (knowledge or beliefs, expressed in statements, concepts and theories) and reality (facts about objects or events) James modeled the processes that intervene between these two. He recognized the importance of our predisposition and preconceptions that influence the process of apprehending sensed information.

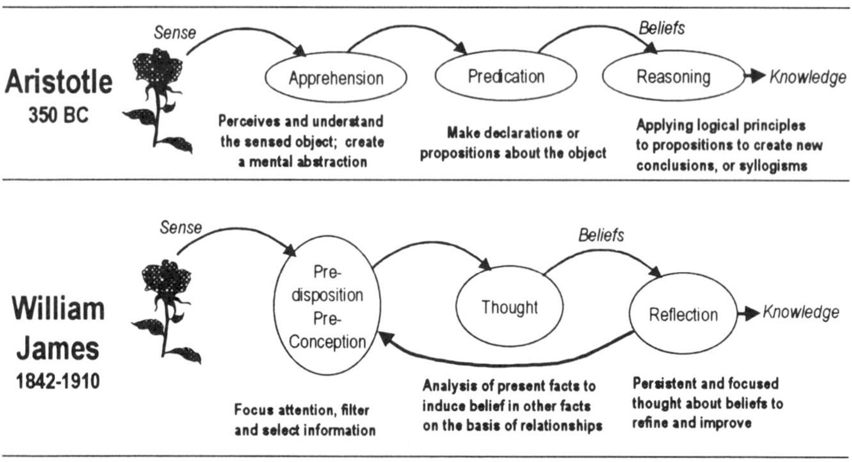

Figure 2. Two basic Models of Human Thought

While these two basic models are only representative of many philosophical models of the cognitive process, we must also recognize that the emerging field of cognitive science has contributed a significant refinement to our understanding of human reasoning. A very simplified cognitive science model (Figure 3) [1] illustrates four sub-processes of human cognition (the arrows provide the flow paths (from sensation to response) that perform judgment:

- Sensor data is accepted by a perceptual-declarative process that performs the apprehension process of Aristotle; this process is predisposed and preconceived, as in the James' model. It characterizes and interprets sensor data, creating concepts.
- A conceptual-declarative process accepts the perceived concepts and performs the objective reasoning processes, conceiving cause-effect, association, and logical inferences to derive beliefs. The first two constitute what is often referred to as the *declarative* side of human judgment [2]. The beliefs derived from this side are passed to the imperative (subjective, "gut-feel") side of the model.
- The Imperative-Cerebral processes apply more subjective (and less deliberate) assessment and valuation of the beliefs.
- Finally the visceral-imperative process responds with decision and actions that are not directly driven by pure beliefs – even as the perceptual-declarative process filters the sensed inputs, this process filters and conditions the output actions.

Notice that there are two paths to judgment in this model; the first short path of visceral response (e.g. an automatic flinch to evade a punch to the face) is shorter, faster and less deliberate that the longer path of cautioned reasoning.

A more process-oriented model of cognition (Figure 4) describes the cognitive process in structure terms more familiar to computer scientists [3]. In fact, models such as this form the basis for the development of machine (computer) counterparts to perform automated reasoning. This model is based on the recent concepts described by authors Newell and Anderson, who are attempting to unify the various models of cognition into a unified model.

1.2 CLASSIC PRINCIPLES OF REASONING

Reasoning processes that *infer*, or perform *inference* are those that create, manipulate, evaluate, modify and assert belief. Within the family of inference processes, we may distinguish several different categories by several characteristics, including:

- The direction of inference – that is by the direction in which they move to assert beliefs, from or toward more general or more specific beliefs.
- The fallibility of inference –inferences may be characterized as fallible (e.g. probably true, true with 0.99 probability , or more likely true that false), or infallible (e.g. the inference, if deduced by prepositional logic may be as true as all the premises.)

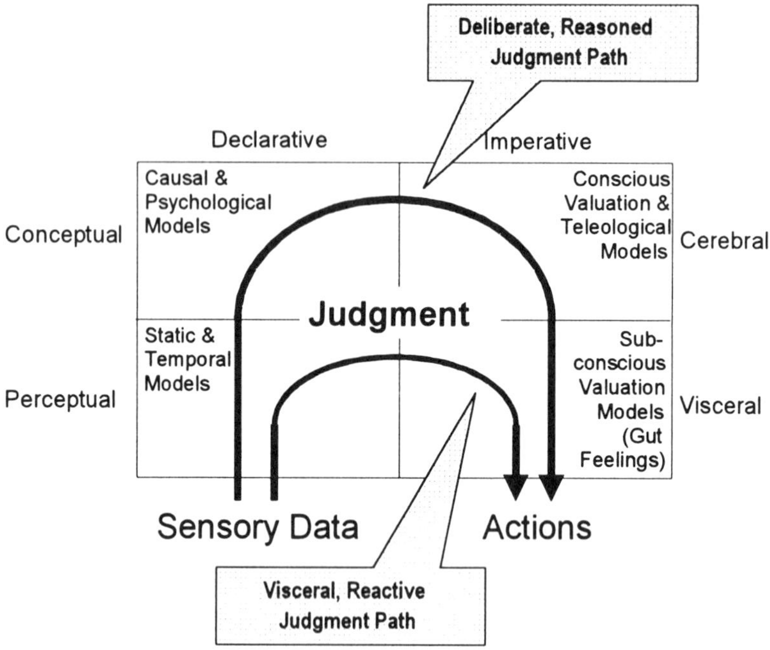

Figure 3. A Basic Cognitive Science Model of Cognition to Perform Judgement

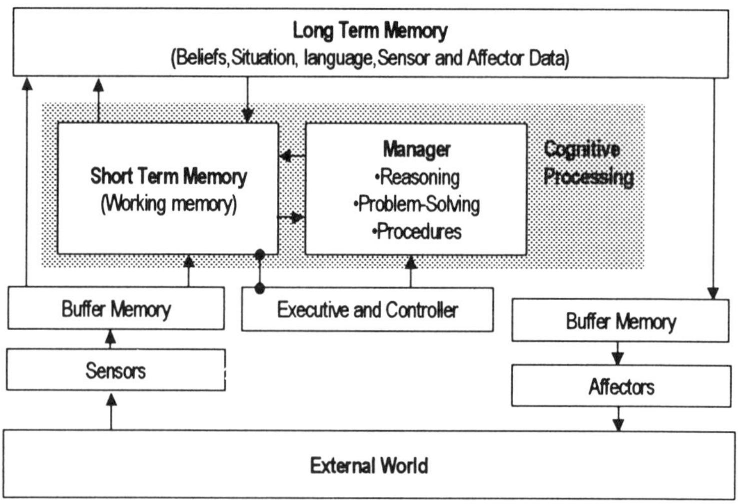

Figure 4. A Functional Process-Oriented Cognitive Model

A basic taxonomy of the inferential reasoning processes (Figure 5) distinguishing three basic classes: Induction, abduction, and deduction. The figure defines four fundamental types of inference, defining each in common terms and by syllogism form that shows the premises and conclusions. Induction and deduction are the methods accepted by classical philosophers, while abduction is the more pragmatic formalism introduced by mathematician and logician C.S. Peirce (1839-1914). The following sections introduce each of these methods.

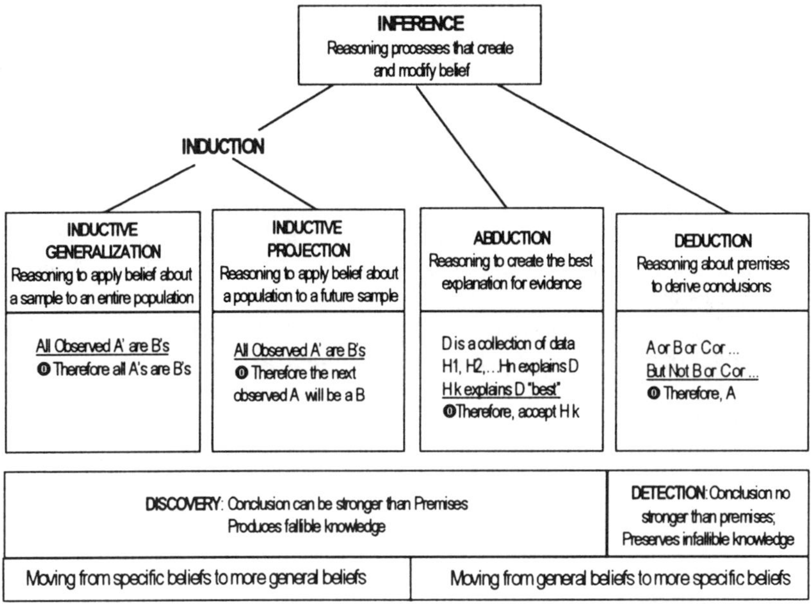

Figure 5. Taxonomy of the Fundamental Methods of Inference

1.2.1 *Deduction*

Deduction is the method of inference by which a conclusion is inferred by applying the rules of a logical system to manipulate statements of belief to form new logically consistent statements of belief. This form of inference is infallible, in that the conclusion (beliefs) must be as certain as the premise (beliefs). It is belief preserving in that conclusions reveal no more than that expressed in the original premises. Deduction can be expressed in a variety of syllogisms, including the more common forms of propositional logic (Table 1).

Texts on formal logic present the variety of logical systems that may be defined to provide foundations for deductive inference. [4] The propositional logic system (or calculus) described above is the basic deductive tool; predicate calculus is the logical system of mathematics that extends deductive principles to the quantitative realm.

TABLE 1. Several Basic Deductive Argument Forms

Argument Type	Logical Form	Example
Modus Ponens	$A \rightarrow B$ $\underline{A}$ $\therefore B$	If it rains, then it is wet It is raining Therefore it is wet
Modus Tollens	$A \rightarrow B$ $\underline{\text{Not } B}$ $\therefore A$	If it rains, then it is wet It is not wet Therefore it is not raining
Disjunction	$A \vee B$ Not B $\therefore A$	It is either raining or sunny It is not raining Therefore it is sunny
Logical Symbols:; $\rightarrow$, Condition (example If A then B) v , Disjunction (example, A or B) $\therefore$, therefore		

1.2.2 *Induction*

Induction is the method of inference by which a more general belief is developed by observing a limited set of observations or instances. Induction moves from specific beliefs about instances, to general beliefs about larger and future populations of instances. It is a fallible means of inference.

The form of induction most commonly applied to extend belief from a sample of instances to a larger population, is *inductive generalization*;

All Observed A's are B's
$\therefore$ Therefore all A's are B's

Inductive Prediction, extends belief from a population to a specific, future sample:

All Observed A's are B's
$\therefore$ Therefore the next observed A will be a B

In addition to these forms, induction can provide a means of analogical reasoning (induction on the basis of analogy or similarity), and inference to relate cause and effect. The basic scientific method applies the principles of induction to develop hypotheses and theories.

1.2.3 *Abduction*

Mathematician C.S. Peirce (1839-1914) introduced the concept of abduction as a pragmatic syllogism that can be expressed in the following form:

D is a collection of data
Hypotheses $H_1, H_2, \ldots H_n$ all can explain D
H_k explains D "best"
$\therefore$Therefore, accept Hypothesis H_k as the best explanation [5]

or, in a more practical example, consider the syllogism:

Data shows that Mr. J. was at the scene of the robbery, has gun powder stains on his hand, and has unexplained cash in his pocket. It can be explained by three hypotheses:

> Hypothesis (H_1) – Mr. J is a rich gun collector who happened on the scene,
> Hypothesis (H_2) – Mr. J. happened on the scene and was framed by the real robber who smeared gunpowder on his hand and secretly placed cash in his pocket
> Hypothesis (H_3) – Mr. J. is the armed robber who shot the gun
> <u>Hypothesis H_3 best explains the evidence</u>
> ∴Therefore, accept Hypothesis H_3 as the best explanation (arrest Mr. J.)

Unlike infallible deduction, abduction is fallible because is subject to errors (there may be other hypotheses not considered; or other hypotheses, however unlikely, may be correct). But unlike deduction, it has the ability to extend belief *beyond the original premises.* Peirce contended that this is the logic of discovery, and is a formal model of the process that scientists apply all the time. The stages of Peirce's abductive method of scientific investigation of a process include abduction, induction and deduction:

(a) Observe the process that is not explained; collect data
(b) Apply abduction to create feasible hypotheses that are able to explain the process
(c) Apply induction to test the hypotheses in experiments
(d) Apply deduction to confirm that the selected hypothesis is able to properly predict the process and new observed data

We will see that all of these methods are also used in the basic data fusion process.

1.3 REASONING PROCESSES IN THE BASIC DATA FUSION MODEL

We can now illustrate the implementation of these basic inference processes in both single and multiple sensor inference systems (Figure 6). Consider first a simple analog sensor system that directly measures one variable, q(t) in order to estimate the value of another variable, P(t). In the pre-operational phase (before deployment of the sensor system) field data are collected to measure the relationship between the dependent variable, p(t), and the independent measurable variable, q(t). By inductive generalization, the sample population is used to derive a general function, p=P(q) , that is believed to apply across the entire population of all conditions and measurements. This general function is then inductively projected to apply to future measurements. This allows the function to be applied in the sensor operationally to convert values of p to estimates of q, by deductive or abductive inferences. If the p-to-q transformation is based on a straightforward table lookup process (i.e. if q=5, then p=45) the process may be considered to be deductive. If, on the other hand, there can be multiple hypotheses about p (e.g. the value of P may also be dependent on other variables, such as signal-to-noise ratio) and other factors may modify the estimated value of p, the process may be considered to be abductive.

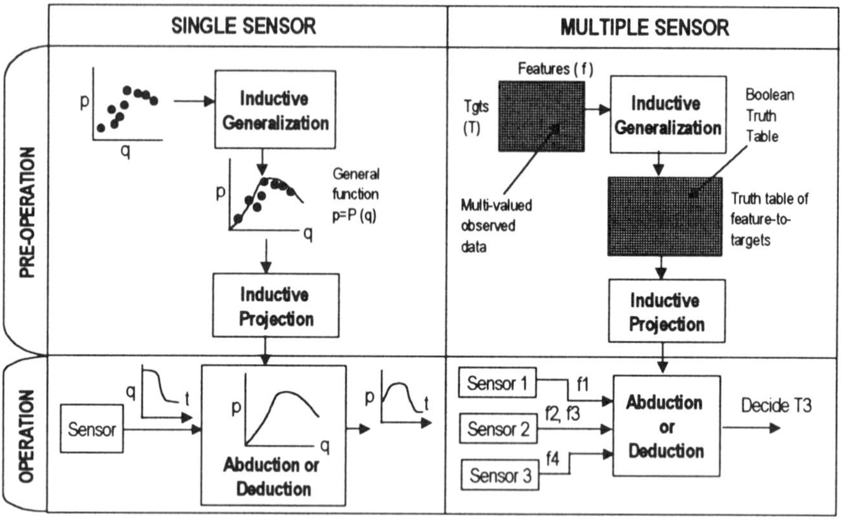

Figure 6. Implementation of the Inference Processes in Common Sensor Models

1.3.1 *Implementing Reasoning Processes*

The formal reasoning methods may be organized to enable application and implementation in automated machine processes, often referred to as "cognitive services." These processes include the functions that transform data to knowledge, where data refers to the raw observed measurements, information refers to sets data that are organized (associated and related) organized, and knowledge refers to information explained. These transformation processes emulate and complement the human cognitive processes, automating three fundamental reasoning or "knowledge creating" activities (Table 2):

- Discovery - These functions discover, or learn unique patterns or characteristics in data. The functions may be supervised (i.e. "training" – the search for common patterns in sets of data known to represent a specific target category) or unsupervised (i.e. the search for patterns in data not necessarily known to contain any targets). Commercial data mining tools perform many of these discovery functions and provide means for semi-automated induction, in which discovered patterns are presented to human analysts. The analyst determines the *meaning* of the pattern- validating it as a discovered signature of a target, for example, or recognizing it as an unusable artifact of the data.
- Detection – The basis of detection is deduction - correlation of a previously known pattern (also referred to as a matched filter template or a signature) to a similar pattern in data to deduce the presence of a target. Data fusion is primarily a

deductive process, applying known templates to multiple data sources at multiple levels to deduce signals, objects, situations and impacts on goals.

- Explanation – The abductive inference process focuses on "explanation" of observations by reasoning to create and select the best explanation. This process combines both synthesis (creation of plausible hypotheses) and analysis (assessment of the feasibility of hypotheses to choose the best explanation).

These goals of machine learning, and artificial intelligence are to automate these reasoning functions to "augment the human intellect" [6] of the humans to comprehend complex information situations, to plan and then respond.

We can now illustrate the role of the basic inference processes in a standard model of the data fusion process.

TABLE 2. The Fundamental Reasoning Processes

Reasoning Process		**Reasoning Mechanisms**	**Contributing Technology** ***and application emphasis***
Formal	*Discovery*	*Concept Learning* – Clustering and induction of decision trees	**Data Mining** – Automated induction processes, discovery of knowledge in databases. Discovery of patterns, signatures and correlations within large databases (information warehouses) across many dimensions of data categories.
		Logic Learning – Production of logical rules	
		Inductive Inference – *Develops new knowledge by analogous reasoning that establishes relationships*	
	Detection	*Apprehension* (Perception and Conception) – abstraction, extraction of features, and assignment of meaning	**Data Fusion** – Automated deduction processes, detection of known patterns or signatures in real-time or stored data. Perform alignment, filtering, correlation, categorization and abstraction of large volumes of raw (or preprocessed) data.
		Judgment – Construction and adjudication of propositions	
		Deductive Inference – Provides new knowledge by establishing relationships between propositions	
Common-Sense	*Explanation*	*Explanation* – Generalization to develop a fruitful family of explanatory hypotheses that explain observations	**Information Understanding**- Automated synthesis, ranking and selection of comprehensive sets of plausible explanations of evidence. Analogous and case-based reasoning, with the evaluation of multiple perspectives and collaborative reasoning across diverse views.
		Abductive Inference – Creates new relevant hypotheses to explain evidence with provision to select most plausible one(s)	

1.3.2 *The Joint Directors of DoD Laboratories (JDL) Model*

The U.S. DoD Joint Directors of Laboratories (JDL) have developed a standardized and general model of the data fusion process that applies the methods of inference to reason about data from multiple sensors and sources.[1] The model identifies five "levels" of activity that apply reasoning processes. [7] The first four of these levels form a forward reasoning chain from sense to knowledge, while the fourth level is a feedback chain that adapts the reasoning chain and controls the sensing process:

- Level 0 - Sub-Object Data Association and Estimation: At this level signals are association and characterized to estimate and detect, via abduction or deduction, individual objects or events of interest.
- Level 1 - Object Refinement: Once individual object are detected observation-to-track association, continuous state estimation (e.g. kinematics) and discrete state estimation (e.g. target type), is performed using deduction and abduction. Prediction (next-observation prediction) is performed using inductive prediction.
- Level 2 - Situation Refinement: Objects are clustered and abduction-deduction is applied to detect aggregate situations or object groups. Predictive induction may be applied to project future states or behaviors.
- Level 3 - Impact Assessment: Consequences of current situations are predicted using inductive prediction. Abduction and deduction perform susceptibility and vulnerability assessment.
- Level 4 - Process Refinement: The data acquisition and reasoning chain processes are adapted to optimize the knowledge creating process in accordance with some optimizing function. The process is adaptive, and therefore includes inductive (generalization and predictive) processes, as well as deductive processes that apply pervious knowledge (premises) to apply control laws.

It should be noted that although the Level 0 to 3 reasoning chain is essentially a deductive-abductive process, any such adaptive process may include inductive (learning) processes throughout all stages. The JDL model provides for database services (equivalent to the short and long term memory of cognitive models) and interfaces with sensors (input) and human users (output).

1.4 REASONING TOOLS IN KNOWLEDGE MANAGEMENT SYSTEMS

Knowledge Management (KM) includes the processes for acquiring, creating, maintaining and applying knowledge to achieve defined business objectives. Encompassing a broad collection of data processing and organizational capabilities, KM

[1] We withhold, here, a full description of data fusion but introduce the concept to illustrate the role inferential processes in the model. The JDL defines data fusion as "a process dealing with the association, correlation, and combination of data and information from single and multiple sources to achieve refined position and identity estimates, and complete and timely assessments of situation and threats, and their significance. The process is characterized by continuous refinements of its estimates and assessments, and by evaluation of the process itself, to achieve improved results." See section 2 of this chapter for more general definition.

includes an estimated commercial market in excess of $5B (year 2000) including consulting services, hardware, and software. The commercial KM discipline, and it's military counterparts are a major drivers of technology development and implementation of automated reasoning processes in business and government.

A taxonomy of the functions of KM and information enterprise systems is illustrated in Figure 7, where the distinction between functional capability (knowledge management, at the top) and processes/technologies (the enterprise, at the bottom) is compared. The major elements of KM include organizational, human-machine, and automated processing functions and related processes/technologies.

KM Organizational Elements – The purely human aspects of KM focus on enabling learning and knowledge creation within the organization. Organizational structures and activities that enhance the creation and exchange of tacit knowledge (instinct, skill, intuition, experience, etc.) among members of real and virtual teams enhance the effectiveness of the organization. Concepts of learning within the organization [8] include methods to solve problems, enable virtual team formation and operation, and capture and manage tacit knowledge. KM tools include training support, virtual workspace software, problem-solving support (e.g. "thinking tools" that aid hypothesis creation, evidence accumulation, and evaluation), and automated means to capture and maintain profiles of behavior that represent experience.

KM Human-Machine Elements – The interaction between humans (analysts, decision-makers) and the computing elements of the enterprise provides the transaction between tacit and explicit knowledge (explicitly represented knowledge within machines). This is provided by intelligent software functions (agents) that accept and perform high-level tasks for humans, tools that both model and represent knowledge to humans, and displays that allow the knowledge to be visualized.

KM Processes and Technologies – The purely machine aspects of KM focus on the acquisition of explicit data (sensor measurements, images, signals, text reports, video, voice, etc.) and its transformation to explicit knowledge. This process includes the capture and warehousing of data sources and the subsequent exploitation for presentation to human analysts. The exploitation process is automated by two categories of transforms: 1) Organizing processes that index, correlate and link data, placing the data in context to transform it into organized sets, or information, and 2).

Reasoning processes (abductive, deductive and inductive) that transform the information to knowledge (or foreknowledge) by modeling and explaining the underlying processes and behaviors that give rise to the observed data.

The KM discipline describes the process of transferring knowledge between tacit and explicit forms as the "knowledge spiral" [9] in which humans and machines interact to create new knowledge useable by the humans. The spiral process (Figure 8) includes four areas of exchange, each requiring a form of conversion and collaboration (each category numbered in the Figure) [10].

<table>
<tr><td rowspan="3">Focus</td><td colspan="9">Knowledge Management
Acquiring, creating, maintaining and applying knowledge to achieve defined business objectives.</td></tr>
<tr><td colspan="3">Explicit Knowledge</td><td colspan="3">Explicit <-> Tacit</td><td colspan="3">Tacit Knowledge</td></tr>
<tr><td colspan="3">Data Transformation to Knowledge</td><td colspan="3">Knowledge Transaction</td><td colspan="3">Learning Organization</td></tr>
<tr><td rowspan="3">Functions</td><td>Data Capture, Represent, Warehouse</td><td colspan="2">Cognitive (Reasoning) Support</td><td>Human-Machine Collaboration</td><td>Human-computer Interface (HCI)</td><td>Knowledge Presentation</td><td>Problem-Solving</td><td>Collaboration, Virtual Teams</td><td>Learning, Tacit Capture Transfer</td></tr>
<tr><td rowspan="2">Data Representation Index, correlate, and link</td><td>Deduction</td><td>Induc-tion Abduc-tion</td><td rowspan="2">High- level abstract interaction between human & machine</td><td colspan="2" rowspan="2">Data, Information and high dimensionality know-ledge presentation to humans</td><td rowspan="2">Identify, articulate and solve complex problems</td><td rowspan="2">Virtual team establish-ment and support of time and space</td><td rowspan="2">Auto Training Auto experience capture</td></tr>
<tr><td>Pattern Match Hypothesis Test</td><td>Hypothesis Create and Evaluate</td></tr>
<tr><td>Technology</td><td>Data Ware-housing</td><td>Data Fusion Process-ing Process Modeling & Simulation</td><td>Data Mining Processing ,On Line Analytic Processing</td><td>Tasked Agents (intelligent, with mobility autonomy)</td><td>Displays Nat'l Lang. Process (NLP)</td><td>Multi-Dim Visualization tools Model Tools</td><td>Reasoning Support Tools, Problem tracking</td><td>Collaboration Tools Auto contact finding, network creation</td><td>User Profiling, Push (broad, narrow casting)</td></tr>
<tr><td rowspan="2">View</td><td colspan="6">Technical and System Architecture</td><td colspan="3">Operational Architecture</td></tr>
<tr><td colspan="9">The operations, systems infrastructure and technical elements that enable people using networks and computing machines to manage knowledge.
Enterprise Information Architecture</td></tr>
</table>

Figure 7. The Components of Knowledge Management

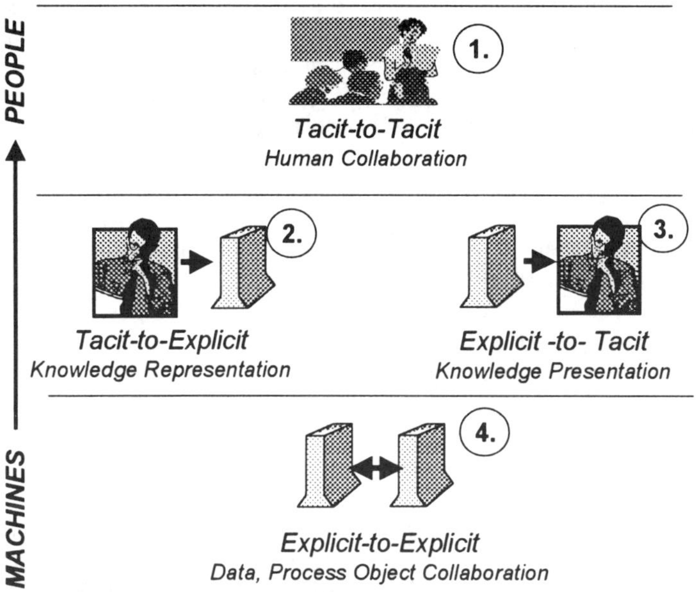

Adapted from: *The Knowledge-Creating Company : How Japanese Companies Create the Dynamics of Innovation* by Ikujiro Nonaka, Hirotaka Takeuchi, Hiro Takeuchi , Oxford University Press, 1995

Figure 8. Four modes of knowledge Transaction

(a) Tacit-to-Tacit activities apply KM tools to support human collaborative analysis and the exchange of tacit knowledge between humans. (In the process, some tacit knowledge may be captured in explicit form.)

(b) Tacit-to-Explicit activities transfer tacit knowledge to explicit representations. KM tools in this area can passively observe human interactions (e.g., monitoring electronic collaboration, e-mail activity, or computer interaction behavior) to infer tacit knowledge, or can actively query humans to create explicit knowledge representations. An example of such an activity is the creation of signature or profile templates for automated deduction by experienced analysts.

(c) Explicit-to-Tacit activities present explicit data, information, or knowledge to the human for examination and analysis. The primary mode of presentation is visualization (highest bandwidth) of models that present the highest level (of abstraction) representation of the object of interest, rather than presentation of raw data. These views of knowledge should allow for drill-down to the underlying information and data that drives the model.

(d) Finally, at the machine level, Explicit-to Explicit knowledge exchange provides for collaboration across services, components, and data objects. This is the normal exchange of fully represented data using custom protocols (e.g. JTIDS Link-16, or ASCII), formal languages (HTML), or more general exchange languages with content tagging (e.g. XML.)

1.5 REASONING TECHNOLOGY

The automation and refinement of inferential reasoning technologies will continue to move in the following sequence:

- Reasoning about limited, special cases, to
- Reasoning about a wider range of many cases, to
- Reasoning about general cases, and then,
- Reasoning based on general principles

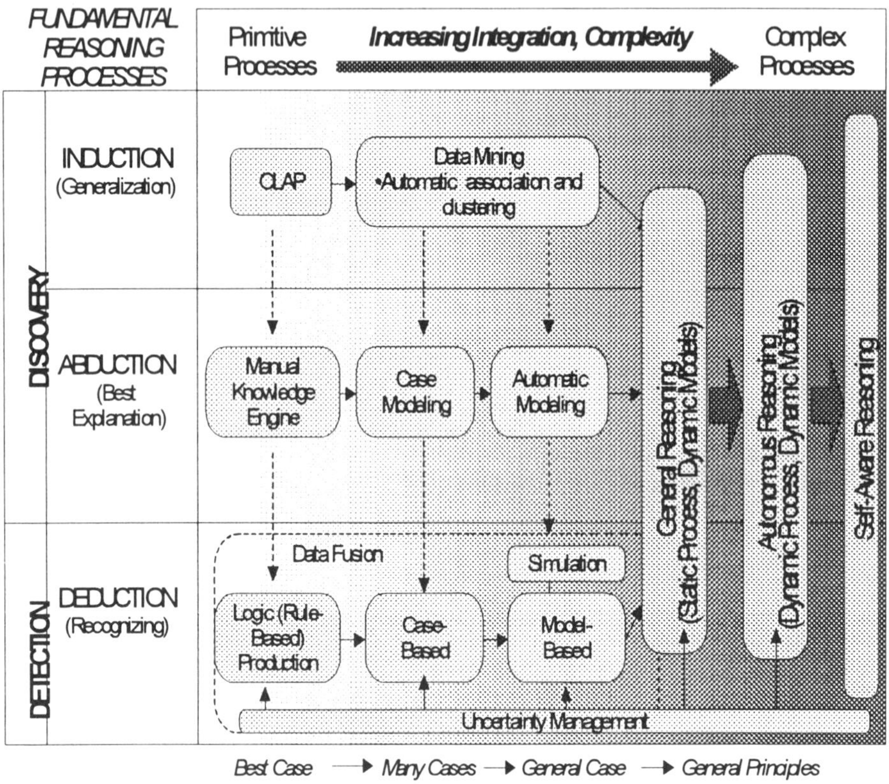

Figure 9. Technology Path toward Comprehensive Automated Reasoning

In addition, the inferential processes will merge (Figure 9) to provide robust discovery of new patterns, conception of ideas and explanations and detection of previously explained patterns. The transition will move from *specific* to more *general reasoning* as deductive, abductive and inductive processes increase in complexity and integration.

Current capabilities exhibit limitations due to the narrow range of applications over which they operate. Data mining and online analytic processing (OLAP) are based on correlation procedures that uses to generalize observations, but are limited in the ability to differentiate true cause and effect. Manual expert systems (knowledge engines) use case-based or default models as the basis for inferential reasoning but often fail to capture generality or true invariance to context. Cased-based systems and expert systems using production rules are sometimes used for pattern detection at the higher inferencing levels, but are limited in their ability to represent variability.

Future refinements will bring automatic learning with greater generality, and automated model synthesis and simulation for to support more general model-based detection using far more general classes of models. More sophisticated models that accommodate variability will enhance robustness.

In the long-term, the formal distinctions between discovery and detection may be blurred as models dynamically constructed from many atomic pieces become part of a general reasoning strategy.

2. Fundamentals of Multisensing

The coupled processes of *multisensing* and *data fusion* deliver knowledge derived from multiple sources of data, generally technical sensors and non-technical (e.g. human reports) sources. Specifically,

- *Multisensing* is the process of sensing phenomena related to an object or event using multiple sensors
- *Data Fusion* is an adaptive knowledge creation process in which diverse elements of similar or dissimilar observations *(data)* are aligned, correlated and combined into organized and indexed sets *(information)* which are further assessed to model, understand and explain *(knowledge)* the makeup and behavior of a domain under observation [11] .

The basic principles of automating multisensing and fusion with military applications have been developed and described in numerous texts beginning in the early 1990's. [12-18] The similar multisensing process that combines biological senses and reasons about the implications, efficiently designed by our Creator, can be observed across the entire natural realm. Introducing this process requires an brief description of sensing before applying the methods of inference to combine sensory data.

2.1 PRINCIPLES OF SENSING

The principal source of data in most multisensing systems is the technical sensor that observes and measures some physical phenomena that can be related to a target object or event. Radar, sonar, electro-optical cameras, multispectral scanners, radio frequency receivers, and nuclear detectors are but a few of the remotely sensing instruments that we refer to with the generic term "sensor". The representative sensor model (Figure 10) includes the following elements in the sensing chain:

- The aperture collects energy, radiated from the physical phenomena and focuses it on the receiver-detector element to convert it to a measurable phenomenon. The lens aperture of an electro-optical camera, for example, focuses light energy on a charge-coupled device array to convert it to an electrical current. The antenna of a radio receiver focuses radio frequency (RF) energy on a tuned circuit to convert a weak electromagnetic field to an electrical current.
- The receiver detector converts, amplifies and often limits the electrical current, to provide a quantifiable variable that can be proportionally related to the sensed phenomena.
- The signal processor then performs operations to enhance the signal and separate true signals from interfering noise. Noise cancellation, automatic control of gain, and filtering operations are performed at this stage, using prior (statistical) knowledge of signal and noise properties. The signal processor also extracts signal features (derived characteristics) that will be used for target detection.
- The decision rule is applied at the end of the signal chain to decide when target object signals are present. The decision is based on extracted features of the signal (e.g. signal amplitude, frequency, rate of change, pulse duration, modulation, etc.)

Notice that the most basic statistical decision rule is a binary decision between two hypotheses, H_0 and H_1. (The hypotheses may be, for example: signal present or no-signal present, or target type one or target type two.) The figure illustrates the binary decision between two possible signals, each characterized by a Gaussian distribution along a single measurement variable. The decision line between H_0 and H_1 lies midway between the mean values of the two distributions, separated by D. This distribution and separation of hypotheses is characterized by the receiver operating characteristic (ROC) curve that plots detection versus false alarm probabilities. The ROC curve on the figure illustrates a specific operating point for a chosen decision line. Notice two important aspects about the ROC:

- *The ROC curve* is a function of the value of D, the separation distance between the two signals; as D increases, the curve moves toward the upper left corner, providing improved detection performance.
- *The detection operating point* on the curve is determined by the location of the decision line between the two signal distributions. (One would expect the nominal location of his line mid-way between the distributions, favoring neither hypothesis and assuring balanced decisions.) As the decision line is moved toward H_0 (the no

signal hypothesis), the ROC operating point moves upward on the curve, favoring the signal present decision and rendering a higher detection rate and higher false alarm rate.

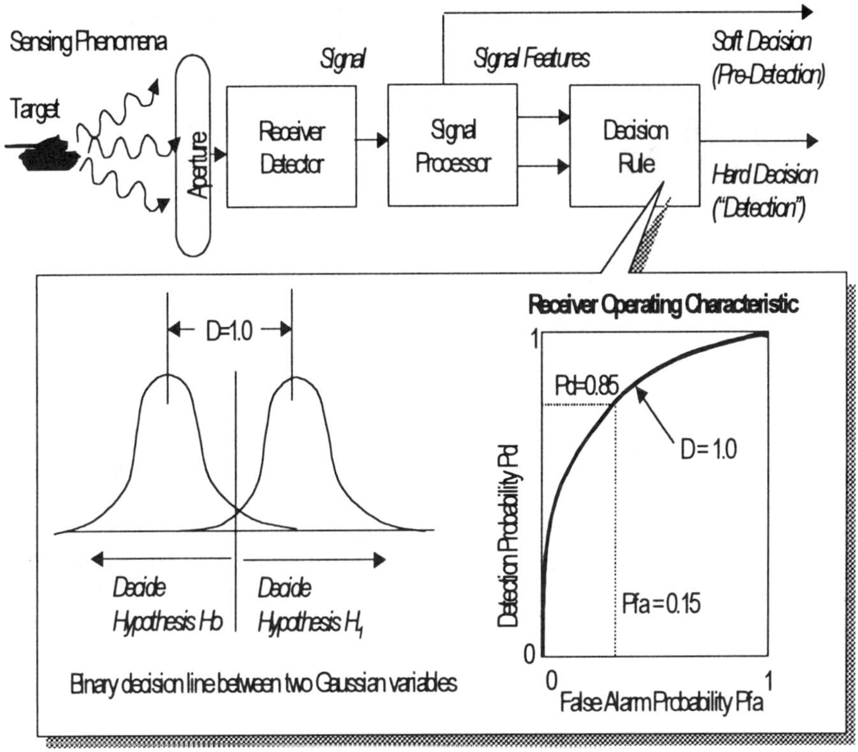

Figure 10. The General Sensing Model

2.2 MULTIPLE SENSOR COMBINATIONS

Multiple sensors can be orchestrated, in a network, to provide robust surveillance intelligence and reconnaissance (ISR), as in the military application depicted in Figure 11. The figure illustrates the flow of sensor data, through the forward JDL levels of fusion described earlier. The multisensor system transforms the flow of sensed data to information, and then to knowledge (intelligence) that explains the situation on the observed battlefield. The output of the entire system is a common operating picture (COP) of the battlespace that can be shared by all participants, and that utilizes data provided by all sources. The JDL level 4 process provides feedback to manage the collection of data by the sensors, and to manage the information chain to focus processing on critical areas of interest, and on critical targets.

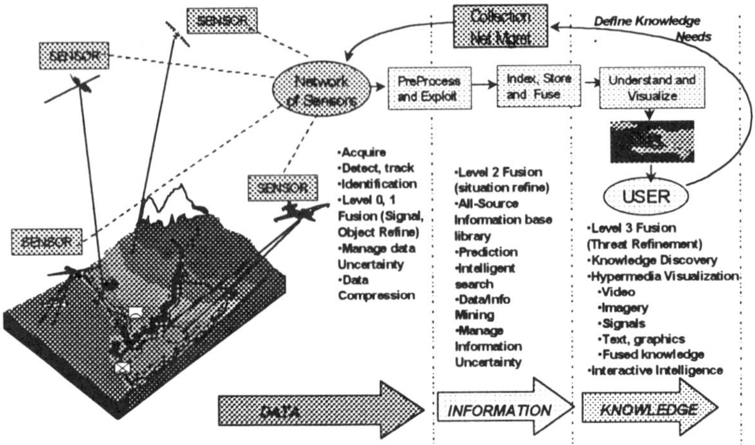

Figure 11. Multisensing with Military Intelligence, Surveillance and Reconnaissance

The feedback nature of the multi-level process (Figure 12) includes a comparison of the mission objectives (needs for knowledge for the COP) with the derived knowledge to provide management feedback to allocate sensing network resources and processing resources to complete the picture. The network-centric approach to warfare requires a "sensor grid" network that enables all sensors to be accessed by the "information grid" - to enable command and control processes to access any sensor fo- surveillance or fire control [19].

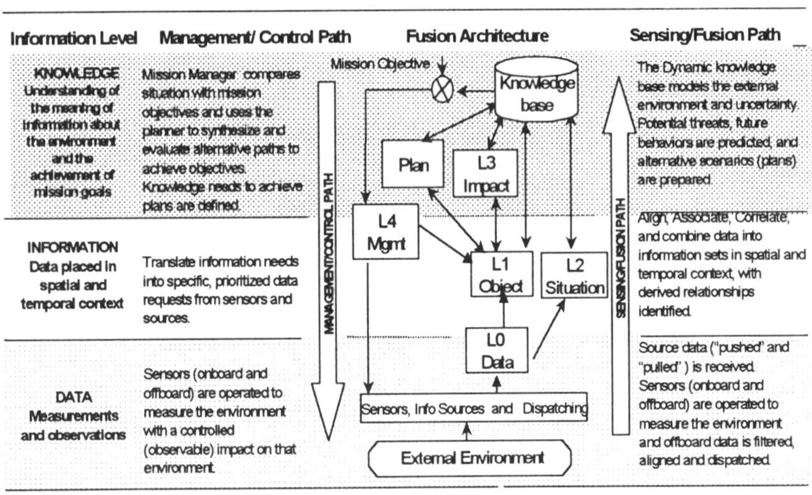

Figure 12. The Data Fusion Feedback Loop

2.3 MOTIVATIONS FOR SENSING BY MULTIPLE SENSORS

The strongest motivations for applying multiple sensor systems is to achieve improved performance over that achievable by single sensor alternatives. These performance gains can occur in any of the three basic functions of the multisensing system:

- *Detection* – the process of deciding that a target object (or event) exists.
- *Tracking* – the process of dynamic modeling and estimating the state of an existent target (or event). The static location or dynamic motion variables (derivatives of the static state) make up the common components of the state of a spatial target. This process maintains a closed loop dynamic model of the target state, usually referred to as the tracking filter. The model produces a state vector, whose components usually represent position and time derivatives, allowing prediction of the next sensor observations. These predictions are used for comparison (correlation) with future observations to associate observations-to-current tracks. Once a time series of observations from an individual target are associated, the model represents the target "track."
- *Classification* – the process of discriminating the class (or type) of target by measurable features (spectral, spatial, or temporal behavior).

In addition to these performance gains, multisensing may also offer system gains in the areas of overall coverage (spatial, temporal or spectral), reliability, or operational flexibility. The overall impact of these performance gains is the delivery of improved information that provides user benefits in the areas of information timeliness, accuracy or completeness.

The following subsections describe the design considerations for implementing multisensing systems, followed by the means by which performance gains are achieved in each of the three basic areas of sensing. The discussions are focused on JDL level 1 (individual targets) but the principles apply equally are Levels 0, 1,2 and 3 because all of these levels are essentially abductive-deductive inference processes that detect, estimate state and estimate identity of targets (whether signals, object-events, groups or situations).

2.4 MULTISENSING DESIGN CONSIDERATIONS - PERFORMANCE

From the preceding fundamentals, we see that the designer is faced with a number of data fusion system design alternatives, each offering a different level of performance, in several different areas of functional operation (e.g. in detection, tracking and classification). The most fundamental performance considerations are based on the degree to which multiple sensors are combined to detect, track or identify a target (or event). The four basic design alternatives can be expressed in design taxonomy (Figure 13) that moves from lower to higher performance in the following sequence:

(a) Sensor Cueing – does not combine the data from multiple sensors, but allows sensors to alert, handoff targets to, and even control other sensors. Cueing allows two (or more) similar sensors to manage a larger coverage volume, or target density. It also allows dissimilar sensors to observe a common target (though at different times) and subsequently combine the data using one of the methods that follows.

(b) Post-Detection Combination – associates sensor-detected objects, then estimates state and infers (classifies) identity by combining the results from each sensor. (The combining process is *after* sensor detection.) The detection process of the sensors limits the performance of this process; an object not detected by the sensors cannot be discriminated.

(c) Pre-Detection Combination – achieves the highest performance by combining sensor data that is prior to full detection (i.e. signal levels that are below the normal single-detector detection threshold). This can be performed at several lower levels of signal abstraction before object detection. The two most common levels are : 1) Feature level – signal features (characteristics) are quantified for combination across sensors, or 2) Pulse/Pixel Level – raw signals are combined (e.g. by addition, coherent comparison, etc.) and the detection decision is made on the combined signal.

The functional implementation of post-detection (JDL Level 1 decision, or Object) combination and the two types of Pre-detection combination are illustrated in Figure 14 and the characteristics of each approach are summarized. Notice that the critical design trade considerations include:

- *Performance vs. Computation* – The highest performance is achieved by combination at the lowest level of signal abstraction, but at the highest level of computation.

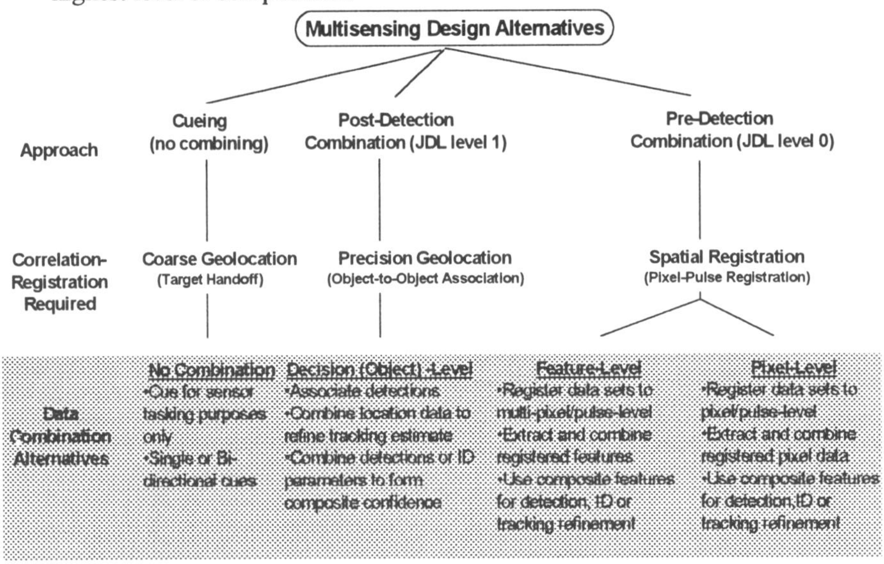

Figure 13. Performance Level Combination Alternatives

- *Spatial accuracy vs. Spectral Accuracy* – Spatial accuracy requirements for post-detection combination demand that detected objects must associate, therefore they are on the order of the object size. Pre-detection methods require a higher spatial accuracy because sub-object signals (e.g. pixels or pulses that are sampled below the spatial dimensions of the object) must be associated for combining[2]. (Note that for a fixed spatial accuracy, pre-detection performance may be less than post-detection due to a high level of association errors that allow measurements from different object to be combined.)
- *Distributed vs. Centralized Detection* – Post-detection combination (distributed detection) offers the minimum sensor-to-fusion communication capacity and the simplest combination process. Pre-detection centralizes the combination process at the expense higher capacity to send feature or raw signal data to the central fusion node.

<table>
<tr><th>Level of Fusion</th><th>Data Flow and Level of Combination</th><th>Characteristics of Approach</th></tr>
<tr><td>Level 1
DECISION
(OBJECT)</td><td>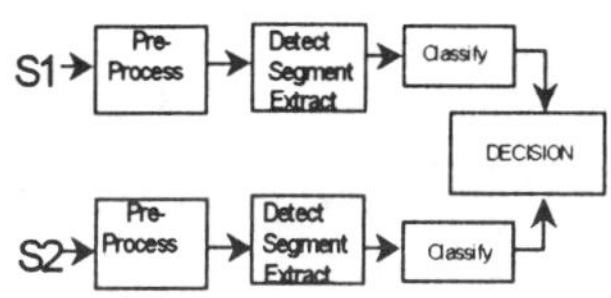
</td><td>•Presumes independent detection, geolocation, tracking and classification in each sensor domain
•Combines independent sensor hard or soft decisions on target objects using Boolean AND, OR or heuristic rules or soft combination methods
•Simplest computation
•Composite final decision space</td></tr>
<tr><td>Level 0
FEATURE</td><td>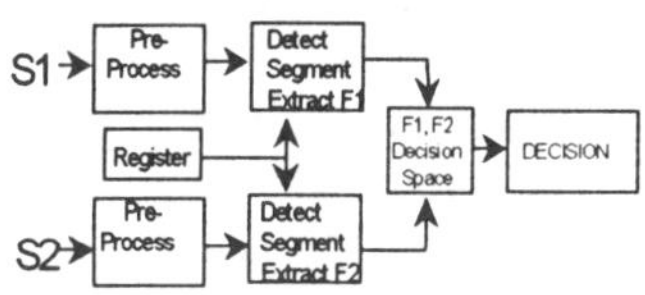
</td><td>• Presumes independent detection in each domaain
•For each potneial object, features are extracted in each domain
•Combines extracted features in composite decision space for detection or classification
•Composite feature decision space</td></tr>
<tr><td>Level 0
SIGNAL</td><td>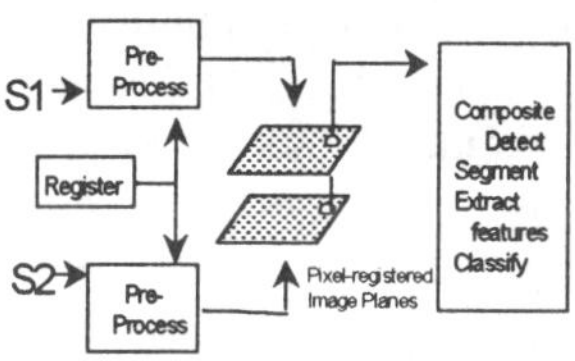
</td><td>•Highest potential detection and classification performance
•Demands accurate spatial registration of data to assure correct extraction of composite features; registration errors directly affect combination performance
•Composite data space</td></tr>
</table>

Figure 14. Implementation of Pre- and Post-Detection Combination

[2] Here, we presume a spatial sensor (e.g. an imaging sensor) that requires spatial registration of the sensor data before two detected objects (or features, or raw pixel samples) can be associated. This description also applies to time-based sensors (e.g. RF receivers), where the registration accuracy is defined in the time dimension rather than space.

2.5 MULTISENSING DESIGN CONSIDERATIONS - ARCHITECTURE

These performance considerations lead to system architecture implications as the choice of pre- or post- detection combining (centralized or distributed) can be applied (independently) to each of the detection, tracking or classification functions. The two combining alternatives and the three fusion functions generate six basic design architecture alternatives (Figure 15).

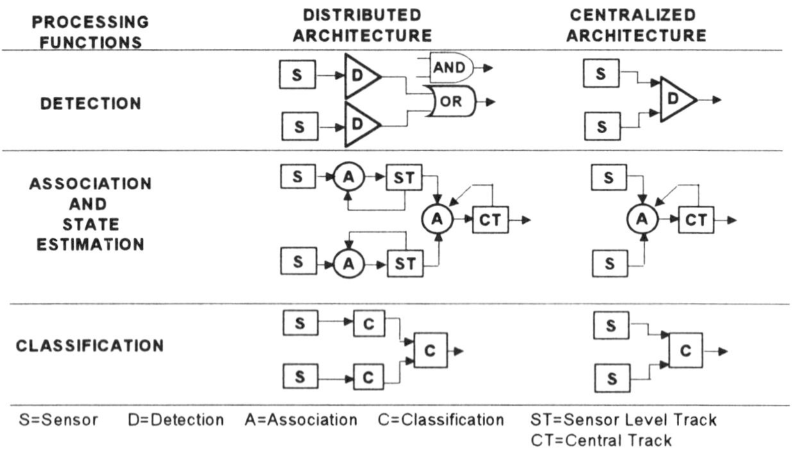

Figure 15. Six Basic Combining Architectures

The two detection structures either make independent detection decisions at the sensors, then combine those decisions at a central node (distributed detection), or they perform the detection decision on the basis of all sensor data a common node (centralized detection). When distributed detection makes a "hard" binary decision (H_0, target present; H_1, target not present) the combination is a deductive process, applying a propositional or predicate logic. The figure illustrates the simple Boolean alternatives: Logical AND (IF sensor A declares target AND sensor B declares target, THEN target exists) or logical OR. The alternative to hard decision methods include such "soft decision" methods such as Bayesian or Dempster-Shafer inference, or multi-valued logic (e.g. fuzzy logic). In these approaches, a soft decision is made at the sensor to derive a vector of variables representing degree of belief in the candidate target classes (e.g. a conditional probability for Bayesian inference, a evidential interval for Dempster-Shafer). The appropriate inference algorithm at the central node then combines these sensor report vectors.

The association and state estimation (tracking) structures are distinguished by the location of tracking functions. Distributed tracking requires a tracking function at each sensor, and these tracks (not raw observations) are associated, then combined to a central track at the central node. Centralized tracking associates and then combines raw observation data from all sensors at a common point, and maintains a central tracking model.

Classification, like detection, can be distributed at the sensors or centralized. Each sensor may maintain an independent classifier, and the target classifications may be combined by hard decision logic, or by soft-decision inference methods.

We now consider two, two-sensor design examples (Figure 16) to illustrate how the six alternatives can be combined in various ways to achieve performance requirements for a given pair of sensors. The "highly distributed" example maintains independent detection, tracking and classification processes at each node, combining the track files and target classifications at a central node. The 'centralized" example detects the presence of targets at each node and, for detected targets only, sends the raw data reports to the central node for central state vector estimation and target classification. Notice that in this example, the central detection node provides feed back to the sensors, adapting the detection thresholds to allow maximum target reporting (based on detection statistics and channel bandwidth).

HIGHLY DISTRIBUTED	CENTRALIZED

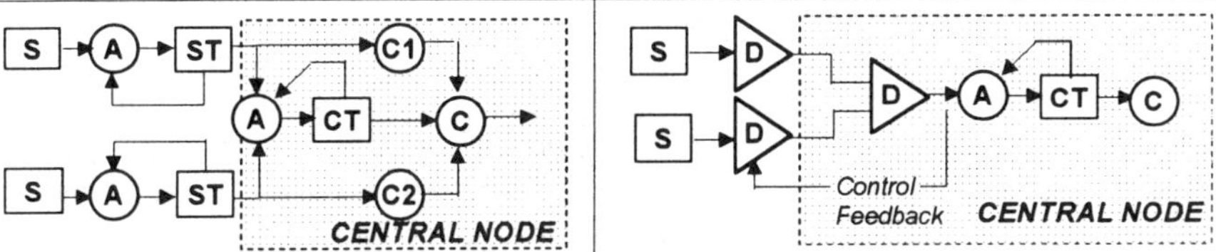

HIGHLY DISTRIBUTED
- Combination of hard decisions
- Uncertainty not passed from sensors to fusion node
- Sensor-sensor alignment is simplified
- Modular architecture- simplest interfaces
- Reduced global processing requirements
- Local sensor optimizations, but not global performance optimization

CENTRALIZED
- "Measurement fusion"
- Uncertainty managed at the central node
- Sensor-sensor alignment is critical
- Requires higher bandwidth sensor-to-node communication channel
- Most severe global processing requirements
- Minimizes loss of information to local decisions

Figure 16. Structures of Two Example Fusion Nodes

Hierarchical Data Fusion Architectures – Notice that the design decisions, above, enable the configuration of a single fusion node to combine any number of sensors to produce a common report of detected targets, their states and identities. A large scale, practical data fusion system, however, may be constructed of a large number of such fusion nodes organized in a hierarchical fashion based upon operational and system engineering considerations, such as:

- Overall system performance requirements
- Sensor viewing geometries and target access windows
- Communication link bandwidths
- Processor capabilities
- Location of sensors on platforms (Satellites, aircraft, ships, submarines, ground stations, etc.)

In a practical distributed example, the ELINT receivers on an aircraft and the onboard radar may be fused in a local node, then reports are forwarded to a second tier (theater-wide) node to be fused with sensor reports from other similar aircraft and ships. The reports of this second tier node may be forwarded to a third tier (national) node that combines theater-wide and space data to present a global common operating picture for distribution to all participants.

In such networked data fusion systems, it is important to assure that derived (fused) information is forwarded, but never fed back into predecessor tiers such that results are input as to be fused with themselves or their contributions. This system error violates the assumption that data to be fused are independent, and introduces bias errors in the process, with potential cascading effects throughout a networked system. The solution to this problem includes careful architectural design, and the use of pedigree[3] metadata with target reports. Pedigree checking processes can assure that feedback does not occur within a networked system.

2.5.1 *Performance Considerations*

It is important to recognize that the processes of detection, tracking and classification are coupled and overall performance includes critical interactions between these processes. Figure 17 illustrates the basic relationship, in which detection performance (P_D, P_{FA}) directly influences the performance of the association process. Poor detection in one sensor, degrades the ability of a correlator to distinguish between targets, potentially introducing false associations ($P_{miscorrelate}$, and P_{Loss}). False associations (due to $P_{miscorrelate}$) corrupt the state estimation and classification processes as different target data sets are mixed in estimates, producing erroneous state vectors and classification results. Failures to associate (due to P_{Loss}) degrade the state estimation and classification processes by withholding sensed data.

The following subsections describe the major performance factors that influence composite detection, tracking and classification.

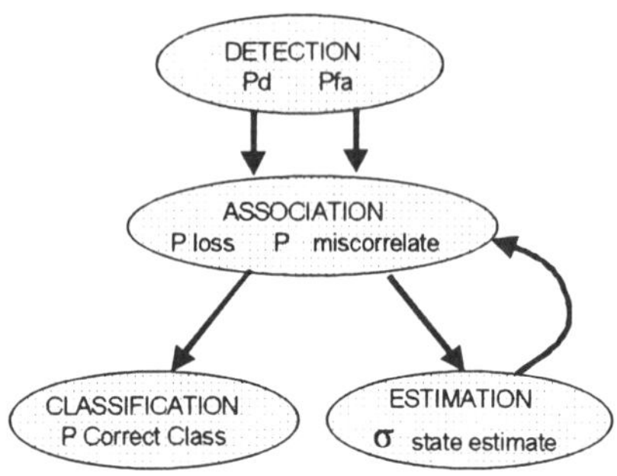

Figure 17. Performance Relationships

[3] Pedigree refers to data that describes the origin and intervening combination processes used to derive an information product.

2.5.2 *Detection*

Multisensing provides improved detection performance by combining data or decisions from more that one sensors, observing a common object. By signal integration the combined object signal is increased over that of uncorrected noise, raising the composite multisensor signal-to-noise ratio, increasing the detection probability (P_D) and reducing the false alarm probability (P_{FA}) for given decision threshold. The relative detection performance improvements of distributed and centralized combination of sensors are illustrated in Figure 18 on the standard receiver operating characteristic (ROC) plot of P_D and P_{FA}.

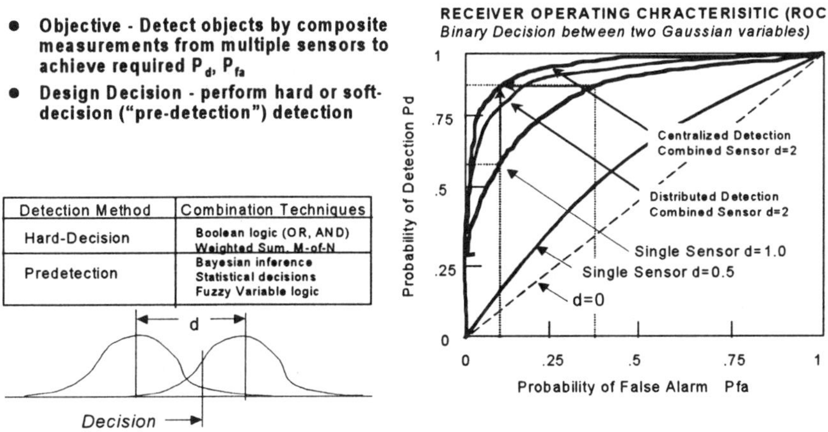

Figure 18. Detection Gains Attributable to Multisensing

2.5.3 *Tracking*

The designer is faced with two basic alternatives for combining data from two or more tracks: 1) distributed tracking in which independent track models (filters) are maintained at each sensor; in this case the association is performed from "track-to-track", or 2) centralized tracking in which all raw sensor reports are associated and modeled at a common node; the "report-to-track" association process compares each report to a single central tracking model.

The relative performance of four basic approaches to implement these alternatives is illustrated in Figure 19. The four approaches, in order of increasing relative performance, are:

(a) State Vector Fusion – compares the state vector estimates of two sensor tracking models, and once associating two reports that are derived from a common target, the state vectors are linearly combined with appropriate weighting to compute the combined state estimate.

(b) Track Selection Fusion - also compares the state vector estimates of two sensor tracking models, and once associating two reports that are derived from a common target, the "best" state vector is chosen as the combined state estimate.

This heuristic approach uses a metric (e.g. the state vector error covariance) to make the selection when the accuracy of the two estimates are dissimilar, and may switch to state vector fusion when the sensor measurements are comparable.

(c) Measurement Fusion – again makes the association based on state vectors, but combines the raw, associated sensor reports to derive a central track estimate.

(d) Centralized Fusion – maintains no independent track files, and raw sensor reports are all associated and combined to maintain a common, global track file. The approach is often implemented as a batch process, and has the potential to achieve the highest performance, because no sensor data is lost or filtered by intervening sensor-level tracking models.

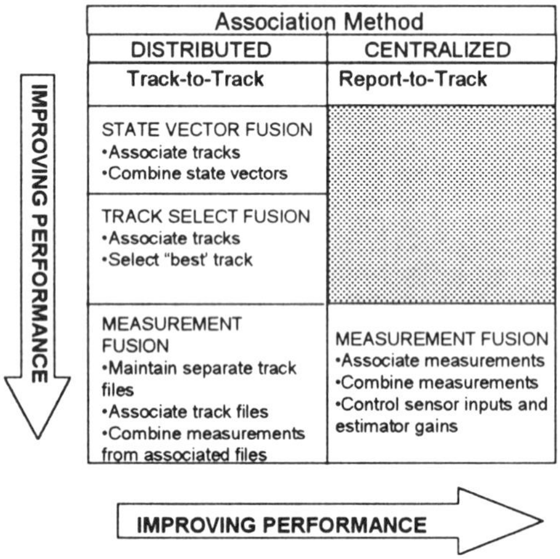

Figure 19. Top-Level Tracking Alternatives

2.5.4 *Classification*

Target classifiction (identification) uses target features to distinguish between target classes, generally using deduction or abductive (often statistical) automatic target recognition processes . The addition of multiple sensors can improve discrimination peformance in two ways:

(a) Discimination Dimensionality – Adding sensors can add features that increase the dimensionality of the decision spaece. Of course, the degree to which additional features aid discimination peformance is a function of the phenomena being sensed and its relation to the characteristics of the target classes being disciminated.

(b) Signal Integration – Classification peformance can also be marginally improved by adding similar sensors. Figure 20 illustrates the marginal gain in classification performance that can be attributed to adding identical sensors, using Bayesian inference to combine sensor conditional probabilities [20]. The relative gain is plotted for three cases, each being the gain attributable to adding two sensors to a baseline number of sensors (e.g. N=1 case plots the gain of three sensors over a baseline of one sensor). The curves quantify intuition: there is a greater marginal gain when combining sensors with lower performance (i.e. low Pn), that when combining sensors with higher performance.

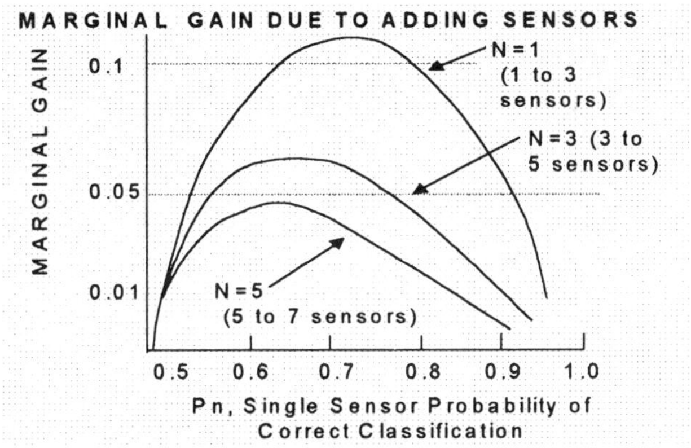

Figure 20. Marginal Gain attributable to the addition of sensors

2.6 QUANTIFYING THE VALUE OF MULTISENSING

Over the past decade, significant progress has been made in developing and deploying intelligence, surveillance and reconnaissance (ISR) support systems which utilize multiple sources of data and provide composite results to users. A 1996 conference sponsored by the Air Force TENCAP (Tactical Exploitation of National Capabilities) Program Office provided insight into the diversity of nearly fifty correlation systems that have been conceived (and many developed, fewer deployed) to exploit multiple intelligence sources [21]. In spite of this development, and the general recognition of the benefits of multiple source fusion and evaluation, there remains much less understanding about the relative performance of these systems and their overall effectiveness to military missions. The current emphasis on information operations and the value of information to the military mission has refocused attention on the need to quantify and assess the performance and military effectiveness of *information*, just as we assess weapon systems effectiveness.

In order to quantify the value of one byte of information, we must relate that information, or a marginal improvement in the information, to military effectiveness or utility. Consider several examples:

- Increased information on *target location* (increased 3D accuracy) can be directly related to its influence on weapon performance (target kill probabilities) and therefore on military effectiveness.

- Improved information on the *relationships* between forces can improve the understanding of their most likely behavior, thereby allowing improved prediction and greater warning time - reducing vulnerability or increasing offensive lethality, increasing military effectiveness.
- More timely information on *target behavior* (increased sensor revisit, more accurate tracking and identification) can be related to the speed with which targeting and attack can occur, and the resultant influence on military effectiveness.

In each case cited above, information must be quantified and mathematically related to the effect on weapon or warning systems effectiveness to measure the effects of improved information on military utility. Yet, there are no standard measures of relative information gain of stages in new emerging intelligence correlation systems - and no standard way of comparing alternative subsystems for relative performance or competing systems for overall effectiveness. The progress in implementing correlation (or fusion/evaluation) must be matched by corresponding progress in the quantitative assessment of information systems - and the margin improvements in information performance they provide.

As the value of information in warfare has grown, so has the need to measure and compare the processes that convert data to processed intelligence, enhancing the value of that information. In particular, program managers require cost/benefit analyses to evaluate alternative solutions to choose migration paths to implement ISR systems. In order to provide measures of relative benefits, system analysts must provide effectiveness measures to compare contributions of ISR sources and the elements that combine data to produce composite outputs. Three categories of issues must be addressed: element performance, system effectiveness and relative cost/benefit (Table 3).

Typical Military User Values – Five primary values of intelligence user values are compared to metrics (Table 4) that can quantify relative achievement of in each area. These metrics include basic performance measures (e.g. Detection and false alarm probabilities), practical measures (e.g. Pedigree of data presented to user; the sources and means of deriving fused solutions), and operational measures (e.g. compatibility of presentation to the human user).

Methods to Quantify Information Value - Performance and effectiveness measures, and methods of modeling processes for information-based systems have been defined for command and control [22] and data fusion [23] systems in classic texts. These methods distinguish hierarchical levels of metrics that quantify information and its impact on human judgment and weapon systems. Four levels are distinguished:

- Dimensional parameters describing data (e.g. pixels, pulses, Bit error rate, time delay, SNR) are related to measures of information performance (MOP's) that measure system information performance (e.g. accuracy, variance , detection/false alarm statistics, coverage, etc.) This relationship (transfer function) is influenced the processing stages that filter, align, associate, combine and level the information.

- *Measures of Performance* - MOP's are then related to the functional performance of the information system - measures of information effectiveness.
- *Measures of Effectiveness* - MOE's, which can be directly related to the systems they support - defensive indications and warning systems or offensive weapon systems.
- *Utility* - Finally, the overall military utility to the warfighter is determined as a function of many attributes of information - and it's influence on the military systems it supports.

The metrics can be related to the JDL data fusion model (Figure 21) to illustrate how source data is converted to information, and the measurable gains that can be attributed to the first level of fusion. Individual objects (and events) are detected and uncertainty is reduced as data is correlated among sources and as it accumulates over time. The next stage of fusion, Situation and Threat Assessment derives aggregate information and discovers knowledge about the relationships between individual objects providing and understanding of group behavior. The Evaluation stage *applies* the understanding of the current situation, predicted feasible courses of action of enemy forces, and possible actions (by friendly forces) to derive alternative plans for human decision-makers. It is at this last step that knowledge is coupled to military decisions - and ultimately to military effectiveness and campaign-level utility.

Table 4 summarizes typical MOP and MOE metrics for the three levels of data fusion in the model, illustrating the increasing levels of abstraction from understanding of individual objects, to understanding of group behavior and finally implications for impact on military missions.

TABLE 3. Three categories of critical operational analysis questions

ISSUE AREA	CRITICAL QUESTIONS
Processing Element Performance	• How can information value be quantified for INT data contributors? • What is the relative information contribution of each source? • What are the processing and delay factors influencing value? • Which processing elements are most critical in terms of information contribution? • What is the information gain (reduced uncertainty) provided by each stage of processing?
Intelligence Chain Effectiveness	• What is the overall effectiveness of the all-source fusion process? • What are the relative effectiveness contributions of each processing element? • What is the effect of cumulative delay on effectiveness? • What is the effect of correlation accuracy on effectiveness? • What is the trade space for delivering uncertain data (or multiple hypotheses) sooner or higher confidence data later?
Cost/Benefit	• Where in the information chain can we achieve the greatest gain for a fixed investment? • What are the relative cost benefits of alternative fusion, evaluation or dissemination technologies?

TABLE 4. Intelligence User Information Priorities, Values, and Metrics

User Priority Area	*User Values*	*Representative Metrics*
1 Accuracy of information	• **Accuracy of all information:** • **Geo-location,** • **State estimate,** • **identification,** • **Pedigree**	• Probability of Detection, Probability of False Alarm • Unit Mean Square Error (MSE) Location Accuracy, state vector • Detected target purity • Detected target fragmentation
2 Timeliness of information	• **Time from first message receipt-to-target declaration** • **Time-to-unit recognition**	• Time-to-Correct Detection Identification DDS Function Time • Time-to-Pull Data
3 Dependability & Explanation of Sources	• **Audit trail to pedigree of target, unit declaration** • **Drill-down to source data**	**The utility of this area is qualitatively assessed by user response to product finishers**
4 Accuracy and control of models	• **Accuracy of models to represent targets** • **Flexibility and user control**	• Model Quality • Model Performance • FE improvement with model (Detections, FA, MSE) • PF improvement
5 Presentation Compatibility	• **Compatibility with map, product standards** • **Consistent symbology**	**The utility of this area is qualitatively assessed by user response to product finishers**

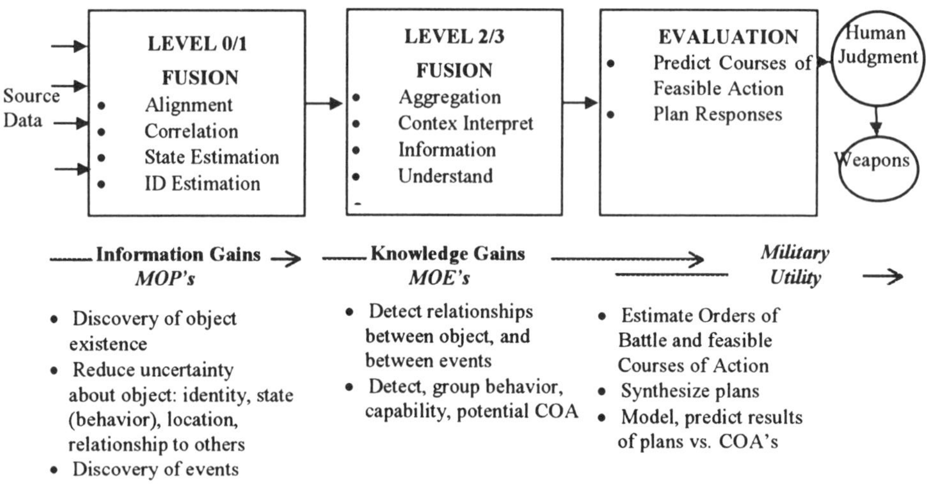

Figure 21. Information Gains provided by Fusion/Evaluation Processes

TABLE 5. Measures of Performance and Effectiveness at Three Levels

		METRICS	
JDL Fusion Level	Objectives	MOP's *Technical Performance*	MOE's *Effect on Operational tility*
Level 3 Threat Assessment	***FOCUS: Future Situations*** • Understand behaviors as: 1) threats (defense interest), 2) opportunities (offense interest) • Predict possible COA's and responses	Group behavior metrics: • Threat detection • Opportunity detection • COA analysis • Response analysis	Threat assessment influence on operational utility: • I&W-Response to threats • Attack Planning
Level 2 Situation Assessment	***FOCUS: Present Situation*** • Understand multi-object relationships and model aggregate behavior Understand constraints to behavior and alternative COA's	Group association metrics • Group detection • Group association • Relationship modeling • Constraint	Situation nderstanding influenceon operational utility: • ISR Tasking • Aggregate Targeting • Force Planning • I&W
Level 1 Object Refinement	***FOCUS: Individual Objects*** • ID and model object behaviors • Correlate, associate and combine obs. of common objects • Detect, discriminate objects	Object refinement metrics: • Detection • Correlation • Tracking • Identification	Object information influenc on operational utility: • Sensor Tasking • Object Targeting • BDA, I&W

Evaluating Multisensor Data Fusion Systems - The practical measurement of performance, effectiveness and utility in operational systems requires analysis of metrics along the information path, and assessment of the changes (marginal gains) in metrics over a wide range of system and environment variables. Quantifying the information contribution in a typical system (Figure 22) distinguishes at least four categories of metrics:

- *Context Metrics* measure the characteristics of the sensed data stream (Signal-to-noise ratio, target density, clutter factors, etc.) to assess the relative difficulty of data sets to be fused. These metrics may also be used in the data fusion process to adaptively set detection parameters, or to select fusion engines.
- *Absolute Performance Metrics (Absolute MOP's)* compare the results of the fusion process with "truth." Useful in simulations and wargames (where truth can be independently generated or observed, respectively), these metrics can be used to baseline and evaluate performance of systems under lab or operational test conditions.
- Relative Performance Metrics (Relative MOP's) – Relative metrics compare the sensor "message truth" with fusion results to evaluate the performance of the fusion process relative to optimal methods of combining the messages (or to absolute metrics).

- *Mission Utility Metrics (MOE's and Utility)* – At the highest level are operational effectiveness and Utility metrics that measure the impact of improved information on military effectiveness. These metric account for *the way the military systems use the information* and must account for the often complex transformation of information into action.

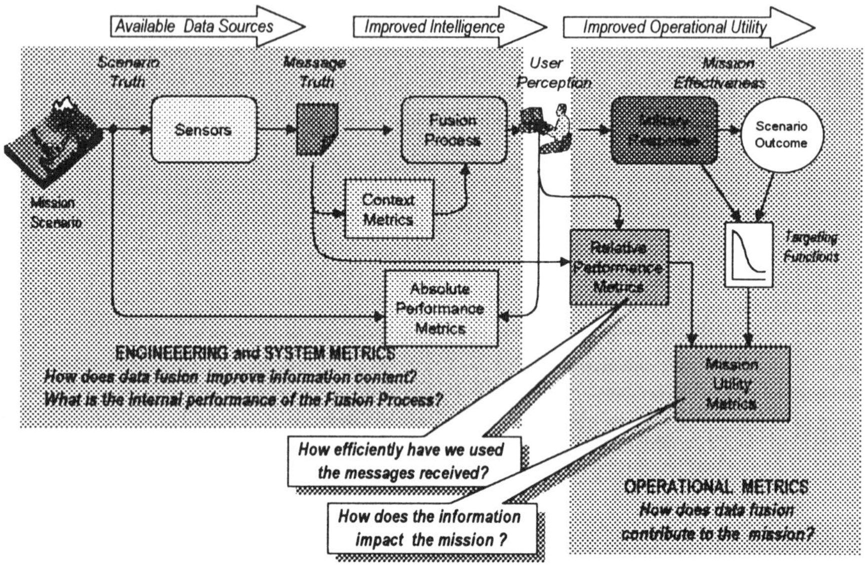

Figure 22. Measuring Information Gain in an Operational System

To illustrate the way utility is a function of operational effectiveness measures, which are in turn a function of dimensional engineering parameters, consider the influence of fused information on the Suppression of Enemy Air Defense (SEAD) mission example, Figure 23. This mission requires accurate and timely location of enemy air defense units, when they become active. The mission utility, U, is a function of detection performance (P_D), location accuracy (mean square error of location coordinates in X,Y,Z, or MSE) for targeting, and time between emitter ON and the precision location for targeting (T):

$$U = U(P_D, MSE, T) \qquad \{U: 0 \rightarrow 1]$$

The relationship between lower level metrics and Utility (U) form a three dimensional surface (for a fixed level of P_D) as a function of targeting accuracy (MSE) and timeliness (T). The surface has high utility within the region of sufficient accuracy and timeliness to rapidly target and kill the air defense unity before it becomes a threat to friendly forces. The utility drops off rapidly as the accuracy exceed the CEP of the selected weapons (requiring more weapons to kill the target) and as the time-to-detect enables the air defense unit to increase it's lethal period of operation, or escape.

The figure illustrates the hierarchy of engineering parameters that are used to derive performance metrics, and then the relation of those metrics to the utility function, U, that determines overall utility for the SEAD mission example.

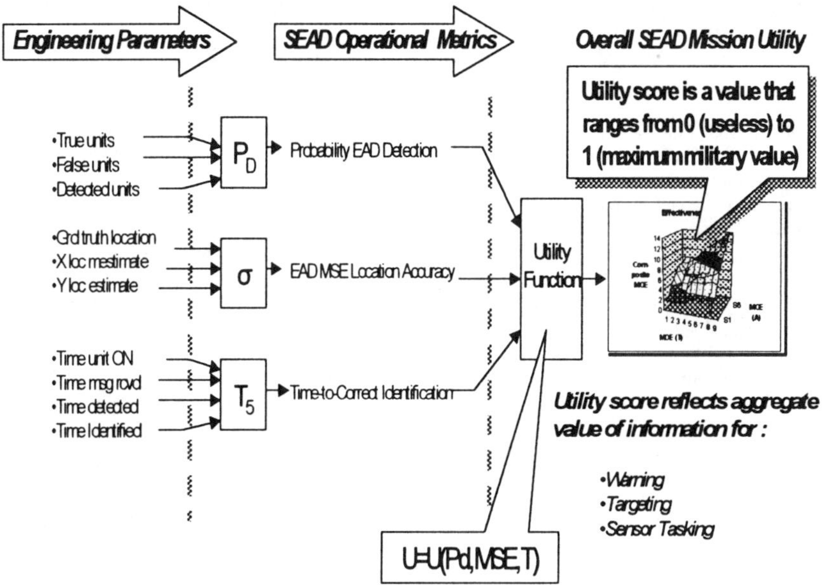

Figure 23. Example Utility Surface for Suppression of Air Defense (SEAD) Targeting

3. Summary

This chapter has introduced the fundamental concepts of reasoning processes and their application in multisensor data fusion systems that acquire and process data from multiple sources to create knowledge. Data fusion is a critical component process in networked knowledge management systems, with applications to military Intelligence, Surveillance and Reconnaissance (ISR) systems, as well as commercial business knowledge management (KM). The basic processes underlying multisensor data fusion include reasoning processes that organize multiple sets of data into organized information sets, then account for the information to create knowledge that explains the underlying processes that give rise to the sensed data.

The fundamentals introduced in this chapter provide a foundation for the design of hierarchical or networked data fusion architectures that produce knowledge from disparate sensing sources.

4. References

1. Waltz, Davenport, Johnson and Oxborrow, "The Critical Role of Cognitive Models in Next Generation Intelligence Architectures", in *Proc. of 8th Annual AIPA Symp.*, Washington DC, 23-24 March 1998. This model was contributed by Dr. Dan Davenport.
2. Anderson, J., *The Adaptive Character of Thought*. Hillsdale, NJ: Erlbaum. 1990.
3. Lambert, D..R., *A Cognitive Model For Exposition of Human Deception and Counterdeception*, Technical Report 1076, SPAWAR, Oct. 1987.
4. Barwise, Jon and Etchemendy, John (2000) Language, Proof, and Logic, (Seven Bridges Press, NY).
5. Josephson, J.R. and Josephson, S.G. (eds.), (1996) *Abductive Inference*, Cambridge:Cambridge Univ. Press.
6. Engelbart, D.C. (1962) *Augmenting the Human Intellect: A Conceptual Framework*, Summary report AFOSR-3223, Stanford Research Institute, Menlo Park, CA. Page 1: "By 'augmenting the human intellect we mean increasing the capability of a man to approach a complex problem situation, to gain comprehension to suit his particular needs, and to derive solutions to problems."
7. Steinberg, A. and White, F., (1999) "Community Status Report and Proposed Revisions to the JDL Data Fusion Model", Proc. of the National Symposium on Sensor and Data Fusion, Ottawa.
8. Senge, P.M. (1990) *The Fifth Discipline*, New York: Doubleday.
9. Nonaka, I. (1998) "The Knowledge-Creating Company" in Harvard Business Review on Knowledge Management, Boston MA, page 26.
10. Nonaka, I., Takeuchi, H., and Takeuchi, H., (1995) *How Japanese Companies Create the Dynamics of Innovation*, adapted from: The Knowledge-Creating Company, Oxford University Press.
11. Waltz, E., and Buede, D. (1998) *Data Fusion*, McGraw-Hill Encyclopedia of Science and Technology.
12. Waltz, E., and Llinas, J. (1990) *Multisensor Data Fusion*, Boston: Artech.
13. Waltz, E. (1998) *Information Warfare Principles and Operations*, Boston: Artech.
14. Hall, D.L., (1992) *Mathematical Techniques in Multisensor Data Fusion*, Artech House, Boston.
15. Antony, R. (1995) *Principles of Data Fusion Automation*, Boston, Artech House.
16. Steinberg, A.N., (1993) *Sensor and Data Fusion*, "The Infrared and Electro-Optical Systems Handbook", IRIA, Vol. 8 Ch. 3, pp. 239-341.
17. Bar-Shalom, Y. and Fortmann, T.E. (1988) *Tracking and Data Association*, Academic Press, San Diego, CA.
18. Blackman, S.S., (1986) *Multiple-Target Tracking with Radar Applications*, Artech House, Boston.
19. Alberts, D.S., Garstka, J.J., and Stein, F.P., (1999) *Network Centric Warfare*, Washington DC, CCRP, second edition.
20. Nahin and Pokoski, (1980) "The Marginal Gain of Data Fusion", IEEE Aerospace Electronic Systems, AES-16, May.
21. Proceedings of Combat Information/Intelligence Correlation Symposium, TENCAP Project Correlation, Peterson AFB, 16-18 January 1996.
22. Wohl, J.G., Gootkind, D., and D'Angelo, H., (1981) *Measures of Merit for Command and Control*, MTR-8217, MITRE Bedford, 15 January.
23. Waltz, E., and Llinas, J., (1990) *Multisensor Fusion*, Artech House. See Chapter 11 for a description of Multisensor Modeling and Analysis.

INTRODUCTION TO DF: MODELS AND PROCESSES, ARCHITECTURES, TECHNIQUES AND APPLICATIONS

E. SHAHBAZIAN
Director of R&D
Lockheed Martin Canada
6111 Royalmount Ave
Montréal, QCm H4P 1K6
Canada

1. Introduction

Data Fusion (DF) technology is young, although some of its elements were developed in the early 1960s [1]. In the early 1980s, the U.S. military services began extensive research activities in this domain to automate and enhance the information processing capabilities of various systems. To ensure synergy between various research activities within the Department of Defence (DoD), the Joint Directors of the U.S. DoD Laboratories (JDL) convened a DF subpanel which started to meet in 1984. The subpanel observed that it was necessary to provide a common frame of reference to the wide variety of researchers who did not have a uniform view/definition of the DF process and scope. The subpanel concluded that DF is a discipline in its own right and developed a DF model, as well as a corresponding taxonomy and lexicon. Figure 1 depicts the JDL DF Subpanel Definition of DF [2].

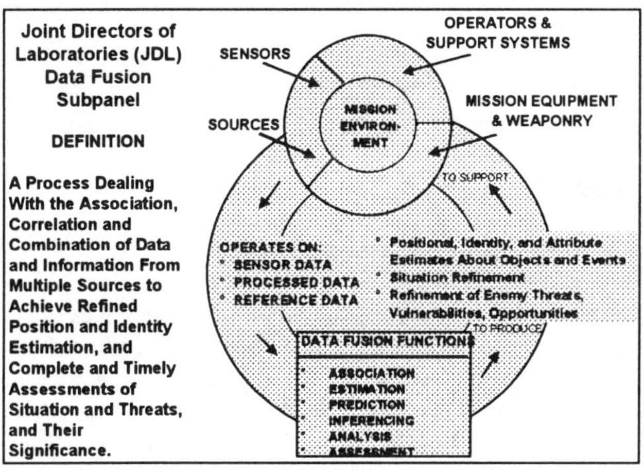

Figure 1. JDL DF Subpanel Definition of DF

A. K. Hyder et al. (eds.), Multisensor Fusion, 71–97.

Since the publication of the JDL DF subpanel's first taxonomy for DF, many books [1,3-15] were published, covering the whole discipline [1,7], or focusing on sub-domains, level of processing, algorithms and techniques, architectures, applications, etc., while the IEEE transactions, various IEEE proceedings, various SPIE proceedings, the JDL tri-service DF conference proceedings, and many other defence and scientific conferences and symposia proceedings have continuously published the results of ongoing advanced research in the DF discipline.

These lectures summarize the:

(a) **Models** used for structuring and organizing the DF discipline
(b) **Processes** (or functions in JDL lexicon) within the DF phases (processing levels) and **Techniques,** algorithms and methods used within the DF processes
(c) Alternatives for choosing the **Architecture** of a DF system for an application
(d) Example **Applications** of DF

The summary of theory on models, process techniques and architectures in these lectures are, in most cases, directly referenced from the preceding books and publications, while the analyses of applications are based on the experience gained through building DF systems at Lockheed Martin Canada (LM Canada) over the last decade.

2. Models

All systems that must make decisions by synergically using multiple observations and existing/accumulated knowledge in their reasoning perform DF. DF is a very diverse field which applies methods of mathematics, probability, operational research, industrial engineering, theory of computation, mathematics of computing, statistics, communication and decision theory, fuzzy logic, uncertainty management, estimation theory, image processing, digital signal processing, computer science and artificial intelligence, etc. in its functions. To categorize, structure and organize the DF discipline, the following models have been proposed by various authors:

(a) The JDL Model
(b) The Boyd Control Loop
(c) The Waterfall Model
(d) The Intelligence Cycle
(e) The Dasarathy Model
(f) The Omnibus Model.

These models identify the processes, functions, and techniques applicable to DF as information flows from the sources to the human operator. Different models group and map the processing and functions that combine the observations and knowledge, perform reasoning and make decisions differently. The detailed description and mapping of these models are provided by Bedworth [16] . The following sections concentrate on the two most broadly used models, the JDL and the Boyd Control Loop, and the most recent, the Omnibus model [16].

2.1 THE JDL MODEL

Initially most of the research in the domain of DF was done for defence applications, and in more recent years this technology was applied for:

(a) Medical diagnostics
(b) Remote sensing
(c) Control systems for large plants (factories, power stations, etc.)
(d) Air traffic control
(e) Communication networks
(f) Security systems
(g) Robotics, etc.

The initial JDL model, which was primarily geared to defence applications, has evolved [17,18] to accommodate other domains where DF can be applied.

The initial JDL DF process model was published in 1986 and is described in detail by Waltz and Llinas [1]. The JDL model 1992 version is shown in Figure 2 [1,7,19]. Hall and Garga [19] provide a good summary of the functionality in each level of processing and element in this model.

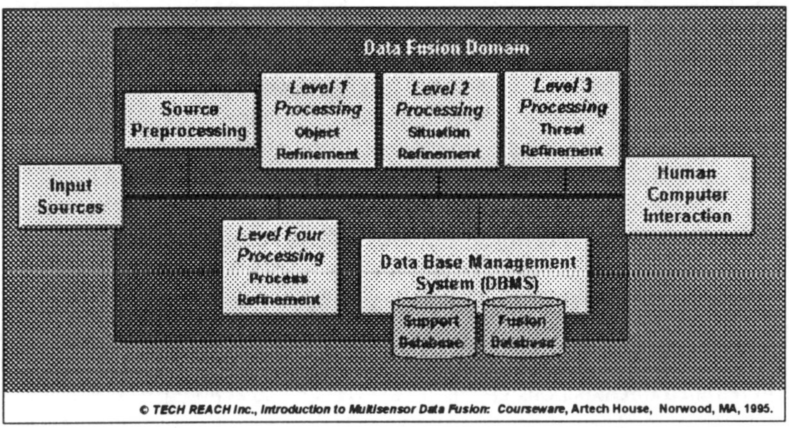

Figure 2. The JDL DF Process Model

Human Computer Interaction (HCI) – The right side of Figure 2 shows the HCI function for fusion systems.

Source Preprocessing – Source preprocessing pre-screens data and reduces the DF system load by allocating data to appropriate processes (e.g., location and attribute data to Level 1, alerts to Level 3 processing, etc.).

Level 1 Processing (Object Refinement) – This process combines location, parametric, and identity information to refine the representations of individual objects (e.g., emitters, platforms, weapons, or geographically constrained military units).

Level 2 Processing (Situation Refinement) – Level 2 processing develops a description of the current relationships among objects and events in the context of their environment.

Level 3 Processing (Threat Refinement) - Level 3 processing projects the current situation into the future to draw inferences about enemy threats, friendly and enemy vulnerabilities, and opportunities for operations.

Level 4 Processing (Process Refinement) – Level 4 processing may be considered a meta-process, i.e., a process concerned about other processes. Level 4 processing (1) monitors DF process real-time performance is monitored; (2) identifies what information is needed to improve the multi-level fusion product (inferences, positions, identities, etc.); (3) determines the source-specific requirements to collect relevant information (i.e., which sensor type, which specific sensor, which database, etc.); (4) allocates and directs the sources to achieve mission goals (may be within the Resource Management (RM) functionality).

Data Management – A major support function required for DF is data management [11]. This collection of functions provides access to, and management of, DF databases, including data retrieval, storage, archiving, compression, relational queries, and data protection.

The above descriptions are specific to defence application. The most recent revisions to the JDL DF model have further reduced the defence context from the definitions 17,18]:

Level 0: Sub-Object Data Assessment – Estimation and prediction of signal/object observable states on the basis of pixel/signal level data association and characterization.

Level 1: Object Assessment – Estimation and prediction of entity states on the basis of observation-to-track association, continuous state estimation (e.g., kinematics) and discrete state estimation (e.g., track ID).

Level 2: Situation Assessment – Estimation and prediction of relations among entities to include force structure and cross force relations, communications and perceptual influences, physical context, etc.

Level 3: Impact Assessment – Estimation and prediction of effects on situations of planned or estimated/predicted actions by participants; to include interactions between action plans of multiple players (e.g., assessing susceptibilities and vulnerabilities to estimated/predicted threat actions given one's own planned actions).

Level 4: Process Refinement – Adaptive data acquisition and processing to support mission objectives (an element of RM).

Figure 3 shows the data flow between the DF levels, as described above.

With the JDL model, the RM portion that deals with impacting the environment (rather than information sources and the DF processes) is considered to be outside of the DF domain.

Within the JDL model, the DF levels are presented as a sequential flow. In reality there is a lot of parallel/concurrent activation of the functions within a DF system in all levels, and more significantly for levels above 0 and 1. This is further discussed in the application sections.

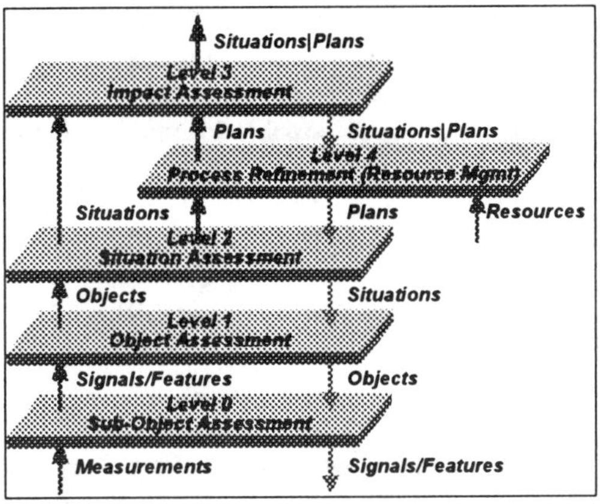

Figure 3. Logical Flow Across the DF Levels

2.2 THE BOYD CONTROL LOOP

The Boyd control loop [20]' also called the OODA loop, consists of four phases: Observe, Orient, Decide and Act. Mapping of the OODA loop to JDL DF levels is shown in Figure 4.

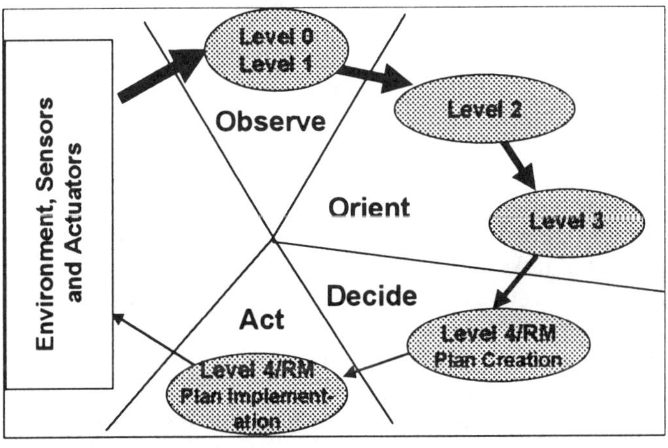

Figure 4. OODA Loop and Mapping to JDL DF Processing Levels

This mapping is currently used by many Canadian DF researchers, however not all authors agree with this mapping of the JDL with OODA[16]. The width of the arrows in this figure also indicates the amount of the data flow in the loop. As data gets further refined in the loop (becoming information), the amount of data passed to the next phase (or DF level) decreases.

The attractiveness of this model is the fact that it closes the loop, acting on its environment and sensors, manifesting the cyclic nature of the DF. It lacks the ability to show the impact of the Decide and Act phases (level 4) on the other phases of the loop. It also implies a sequential execution of the phases. It is easy to imagine an OODA loop (or a few OODA loops) within every phase of this loop performing reasoning to assess different specific information. It is also possible to imagine that these loops act on each other.

2.3 THE OMNIBUS MODEL

The Omnibus model was introduced by Bedworth and O'Brien [16]. The Omnibus model flow chart is shown in Figure 5.

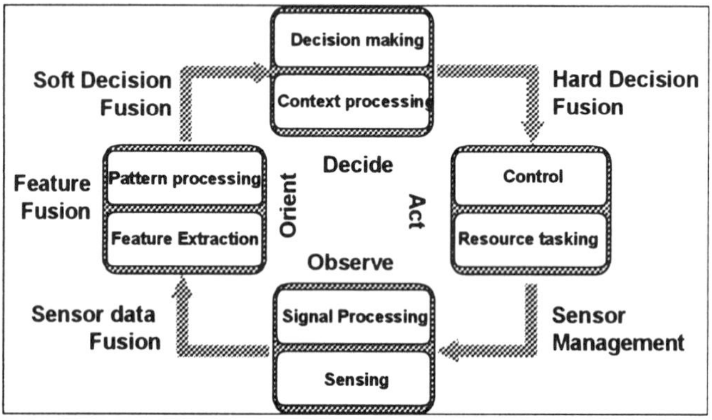

Figure 5. Omnibus Model Data Flow

This model is built around the OODA loop and tries to unify a number of earlier models. The authors propose to use the model in two ways:

(a) To characterize and subdivide the overall system with the aim to provide an ordered list of tasks
(b) To use the same structure to organize the functional objectives of each such task.

The main consequence of this model is that it makes the loop within the loop concept explicit and it also has no defence context.

The drawback of this model is that it combines too many concepts (e.g., DF model, level of processing, architecture), which may cause the misconception that a specific architecture (soft decision, hard decision) is appropriate for only a certain specific phase of the loop.

3. PROCESSES AND TECHNIQUES

This section describes the DF processes (functions) within each processing level as defined in the JDL model. It also lists the techniques used in these processes.

3.1 LEVEL 0 PROCESSING (SUB-OBJECT ASSESSMENT)

The processes within the Level 0 processing include:

- Advanced signal processing
- Image processing
- Synthesis of complex array data to create synthetic information
- State estimation.
- Moving Target Indicator (MIT) algorithms.

The output of this processing may be a **contact**, defined as (a) a set of detections/measurements/patterns/features believed to be associated with a single target, or (b) an object detected by one sensor, or a **track**, defined as measured course, speed and attributes of an object (in space) of tactical significance. The detections/measurements may be **soft,** i.e., sensors provide an m-ary decision, with a measure of uncertainty or confidence, or **hard**, i.e., the sensors make a binary decision, without a measure of uncertainty or confidence.

The type and level of processing in this level greatly determines the methods/approaches appropriate for Level 1 DF processing.

3.2 LEVEL 1 PROCESSING (OBJECT ASSESSMENT)

The Level 1 process is often called Multi-Sensor (or Multi-Source) Data Fusion (MSDF). It deals with object kinematics and identification estimation. Processes within Level 1 include:

- Alignment
- Association
- State Estimation.

Figure 6 shows the high level diagram of the MSDF process.

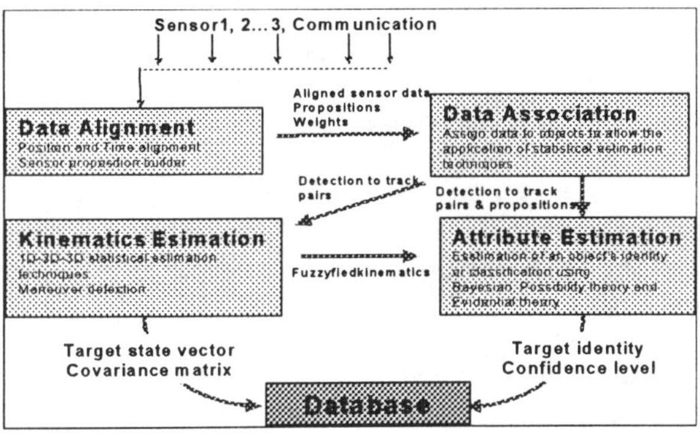

Figure 6. MSDF Process

3.2.1 *Data Alignment*

In **Data Alignment**, incoming information is taken and aligned to the language of fusion. The incoming information must be transformed into a common reference frame and coordinate system and in a format that is understood also by the Attribute Estimation function.

For positional data, alignment can:

- Refine and extend in time the estimates of an object's position - dead reckoning and correction (time alignment)
- Transform sensor data into a consistent set of units and coordinates - rotation and translation, triangulation, (spatial alignment)
- Extrapolate data using maximum likelihood estimation techniques applied on history of contact data
- Assess uncertainties for data that are received without this information.

For attribute data alignment can:

- Apply Fuzzy Logic on subjective information
- Translate attributes into ID propositions
- Combine various information from sensors that provide complex reports
- Assess the confidence level for data that are received without this information.

Note that the last item in both position and attribute alignment is the assessment of uncertainties/confidence levels, which would not be required in a theoretical (ideal) MSDF system, where Level 0 processes is specifically designed to feed a MSDF process. This is mainly done to compensate for missing information in the existing sensor interfaces.

3.2.2 *Association*

Association is the process of matching kinematic, attribute, or time history data to the appropriate objects to allow the application of statistical estimation techniques.

This process requires two algorithms:

(a) The first measures (provides a score) the degree of similitude (correlation) given two sets of information (one from the sensor and one from the object (track)
(b) The second decides which similitudes are the best (assignment).

The inputs of this process are a list of contacts with positional and covariance data, a list of contact propositions and confidence level ID data, a list of track positional and covariance data, and a list of track propositions and confidence level ID data. The output is a list of contact-track pairs, a list of proposed new tracks, and a list of unassociated tracks.

Various approaches for handling multiple target association problems are described in the literature. A number of books and review papers specifically address this subject [3,6,10.21]. The essence of these analyses is the conversion of the single sensor association methods and techniques into a multiple sensor configuration. The main techniques and methods of association are:

(a) Nearest neighbour
(b) Jonker-Volgenant-Castanon (JVC)
(c) Gating
(d) Splitting track
(e) Probabilistic Data Association (PDA)
(f) Maximum likelihood
(g) Joint PDA (JPDA)
(b) Multi-hypothesis method.

The nearest neighbour association method, which was developed in the 60s (also called nearest neighbour filter), is applied when more than one contact is obtained during a complete sensor scan. For very low target density, the method selects the contact that has the smallest statistical distance (ratio between squared physical distance and the covariance) to the predicted track measurement. In a multiple target environment, a contact assignment method has to be used. The resolution of the assignment matrix can be performed in several ways. The simplest methods are either optimal but computationally extensive (e.g., by calculating all possible pairings and taking the solution that minimizes the total cost, which is of order $m!/(m-n)!$, where m and n represent respectively the largest and smallest number between the number of observations N_o and the number of tracks N_t), or suboptimal. The most popular optimal method is the Munkres algorithm. In environments where false alarm density is moderately high, the tracking system that utilizes nearest neighbour methods can easily lose the target over time.

More sophisticated methods exist that allow an optimal resolution of a 2-D assignment matrix, at a lower computational cost (typically of order n^3). Several such methods are well-known and often quoted in the context of target tracking (e.g., Hungarian method, modified Munkres, Jonker-Volgenant).

In a study comparing the processing time of a number of assignment algorithms [22], Drummond et al. found that the fastest optimal method for unique 2-D assignment of an arbitrary n x m sparse matrix was, by a large factor, David Castanon's version of Jonker-Volgenant's algorithm (up to a factor of 100 for 400 scenarios with targets). This code, has become widespread and is now known as the JVC assignment algorithm.

The JVC algorithm is actually performed in two stages, combining the advantages of two optimal methods:

(a) Auction type algorithm (complex nearest neighbour), which is a fast starter but slow when finishing
(b) Modified Munkres algorithm, which is slow at the beginning but finishes faster.

JVC enjoys the best of both worlds by having the Auction algorithm perform an initial (non-optimal) assignment. The initial solution is then passed to a Munkres algorithm, which finds the optimal assignment.

Gating is a technique used to withdraw the unlikely associations between the measured contact and the predicted contact [23,24]. A gate is usually a hyper-ellipsoid (it has the dimension of the state vector) surrounding the predicted contact. The use of a gate, with certain rules of selection, increases the performance of the filter and avoids spurious associations. The rules are set in relation to the association method used, but all methods require at least the following rule: when only one measurement is within the gate of a single track, the contact is associated with the track.

The splitting track method is based on the fact that if one or more measurements are found in the gate region of an existing track, new tracks are formed by splitting that track [25]. For example if two contacts are inside the gate region of a specific track, three tracks will be formed from this track. Each new track represents a hypothesis, which respectively assumes that the first contact number belongs to the target of interest and the second to noise, the second contact belongs to the target of interest and the first to noise, or both contacts are noise. The likelihood function (summation of all squared statistical distances of the contacts that form the track) of each track is then evaluated to withdraw the unlikely tracks.

The PDA method evaluates the posterior probability that in the gate region of a particular track the candidate measurements are associated with the particular track [26]. These posterior probabilities are used to form a weighted sum of innovations (residuals) for processing in a filter. A simple version of a PDA algorithm assumes that there is only one target, which is observed in the presence of additional measurements that originate from other targets or clutter. The incorrect measurements are assumed to be independent and identically uniformly spatially distributed random variables. The resulting PDA filter yields state estimates and covariances that account for the measurement origin uncertainty.

The maximum likelihood method evaluates the most likely arrangement of contacts in order to establish the tracks corresponding to an unknown number of targets [27,28]. The selection of the possible trajectories is a huge combinatorial problem. To simplify this partitioning problem it can be assumed that each measurement belongs to some track (no false alarm) and each measurement belongs to only one track. The maximization of the appropriate likelihood function of the partition set becomes a typical binary programming problem. Once the most likely set of target trajectories have been chosen from this method, the state vector and covariance matrix of the corresponding tracks are computed from standard filter (Kalman, MLE, etc). This approach is ideally non-sequential and can be appropriate for track initiation. The partitioning process involves the use of parallel programming processing that could increase the processing performance in dense target environments.

The JPDA method is an approach in which a set of established targets is used to form gates in the measurement space and compute posterior probabilities [29,30]. This method is an extension of the PDA method for the multiple target tracking problem. This method associates probabilistically the various measurements to existing tracks and then updates the tracks in a suitable manner. The computation of the probabilities is done jointly for all targets and measurements that form a cluster. Then the weighted innovation is computed for all tracks and in turn is used to update the tracks. This method provides powerful multi-target tracking performance but is not appropriate for track initiation of new targets.

The multi-hypothesis method of association is a method where the decision that a new single contact belongs to an existing single track or is a new track can be deferred [29,3]. When deferred to the next data scan, the method becomes basically the same as JPDA. This method has the great advantage of successfully processing new contacts that originate initiation of new tracks. Theoretically this method is optimal and ultimately will become the method used by the multi-target tracking systems when the problem of real-time implementation of a high computational processing demand is resolved.

Since their initial development, these techniques have been continuously evolving and numerous publications can be found [31-33] (see also references to these examples) describing advanced association methods in various proceedings.

In the specific DF application at LM Canada, an algorithm where the correlation is given by the product of the statistical distance probability for positional part and the evidential agreement belief for the attribute part has been developed and is being used. For the assignment, LM Canada prefers the JVC algorithm applied on the logarithm of the correlation probability.

3.2.3 *State Estimation*

State Estimation refines the estimation of an object's position (velocity, acceleration, etc) and identity or classification. It consists of two processes, which can (should) execute concurrently:

- Kinematics Estimation (or Position Fusion)
- Attribute Estimation (or Attribute Fusion).

In Figure 6 there is input from Kinematics Estimation to Attribute Estimation. This is specific to the LM Canada application where fuzzyfied kinematic variables of the object are used as attributes for identification [34,35]. At the same time the results of Attribute Estimation are used in the association process.

Kinematics Estimation is a statistical process used to evaluate the kinematics of a static or dynamic target based on collected measurements. The kinematics refers to the position, velocity and acceleration of an object in space. The kinematics estimate of an object is normally implemented with linear statistical estimators [1]. Note that the term "Kinematics" is used in these lectures to avoid confusion in terminology. Some authors use the term "state" to mean only position, velocity and acceleration [1] some to mean identification and kinematics [7] and some kinematics and other parameters [4].

In tracking processes there are two general methods for processing the information given by sensors to estimate the kinematics of the target. These are the batch and recursive methods.

Given a set of scans collected by a sensor, the batch method processes all the measurements together to form target tracks and kinematics estimates. The batch processing techniques are usually used for static location problems. Batch processing demands high computational capabilities since it requires the storage and process of a lot of information at once, as more and more data are received. Batch processing is viewed as an optimal solution for estimation problems but simplification of this method using recursive technique can be made for real-time implementation [1,3,4,7]. Examples of batch estimators are the weighted least squares estimator and the maximum likelihood estimator.

The recursive methods are generally used for dynamic targets. They process the measured information received from the sensor at each scan using only data received on that scan to update the results of previous processing [1,3,4,7]. Examples of recursive processing filters are the recursive form of the least squares estimator and the Kalman filter.

The Kalman filter is a statistical estimator and is most often applied to dynamic tracking [1]. The inputs into the filter are the initial values of the kinematic vector and estimation error covariance matrix and the measurements. The outputs from the filter are the kinematic vector estimate and the current value of the estimation error covariance matrix. Figure 7 shows the Kalman filter diagram [1].

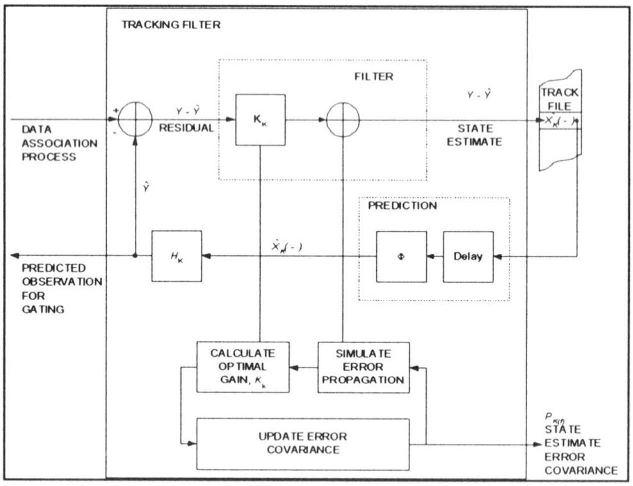

Figure 7. Kalman Filter Diagram

In most common cases, tracking algorithms use some practical models of target motion in order to estimate the present and future target kinematic quantities such as the position, velocity and, in certain cases, acceleration. If the target behaviour changes more rapidly than expected by the model, misassociation or loss of continuity of the measurement may occur.

There are different approaches in the literature for handling manoeuvre detection and manoeuvre adaptive filtering. The most general approaches use:

(a) Manoeuvre detection and parameter adjustment
(b) Manoeuvre detection and state augmentation
(c) Several parallel filters.

Other approaches also use a combination of two or three of these methods. Target manoeuvre algorithms mainly use the Kalman filter since it provides optimum estimates of target position and velocity [1,3].

The first approach is to modify the Kalman filter parameters after manoeuvre detection. This adaptive technique results in an "adaptive filter" capable of monitoring its own performance and making appropriate parameter adjustments when performances deteriorate. There are many variants of adaptive filters reported in the literature.

The second approach is the augmentation of the filter state model upon manoeuvre detection. When there is no manoeuvre, a constant velocity target model is assumed. After manoeuvre detection, tracking is performed with an augmented state model that uses an acceleration state. This method requires the storage of previous measurements, which are used for filter reinitialization after manoeuvre detection. The augmented state model is used until the magnitude of the acceleration estimate is determined to be insignificant, then use of the initial state modes is resumed [3].

In the third approach, multiple Kalman filters are used to operate in parallel and simultaneously on the measurement sequence, each with a differing manoeuvre model. Thus the filter with the best output can be chosen. The simplest example for this method is to have just two Kalman filters, in which one filter assumes a straight line motion while the other is matched to a worst case manoeuvre condition. The non-manoeuvre case is assumed until a detector is triggered. Then the exchange of filters is made to the one with manoeuvre ability. Thus upon manoeuvre detection, adaptation and correction are applied together by switching to a more responsive filter whose output is more representative of the true target state [3]. Several variations of this concept exist.

Another interesting method, which is a variant of parallel filters, is the Interacting Multiple Model (IMM) filter [36]. The IMM filter assumes that the target dynamics can be exactly described at a specific time by one of the "M" possible dynamic (or motion) models. The output of the IMM algorithm is a weighted sum (or linear combination) of the output of each model.

Attribute Estimation can be done using many different approaches including neural nets, pattern recognition, knowledge based and rule based, fuzzy logic and statistical based algorithms [1,7]. Typical attribute data are not variable on a continuum as are the positional data, but are subjective information to which a probability value or a level of confidence is assigned. The role of the target **attribute estimation and fusion** process is to provide the necessary algebra in order to manipulate the evidence and extract the identity of the target under tracking with an appropriate level of confidence.

Waltz and Llinas [1] and Hall [7] have provided a detailed summary of a large number of approaches for identifying objects using the large variety of attribute data that may be available within a system. Table 1 includes a summary of these approaches [1].

Figure 8 depicts the summary of the Bayesian Fusion and Figure 9 shows the summary of the Dempster-Shafer Fusion [1].

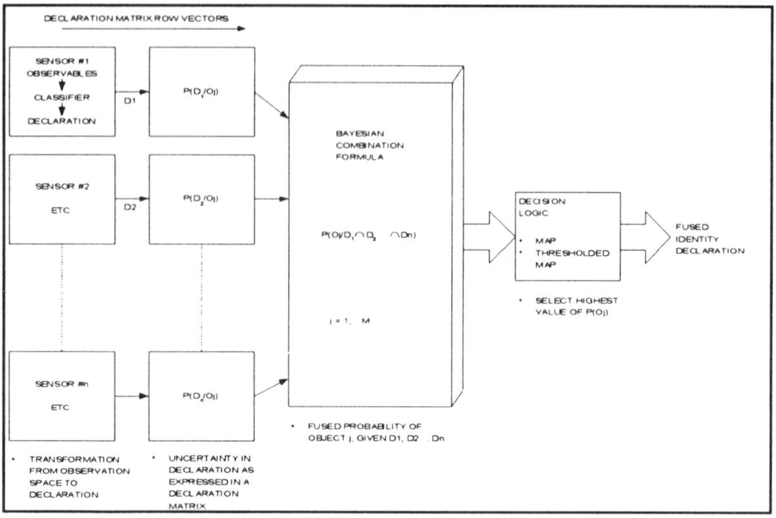

Figure 8. Summary of the Bayesian Fusion

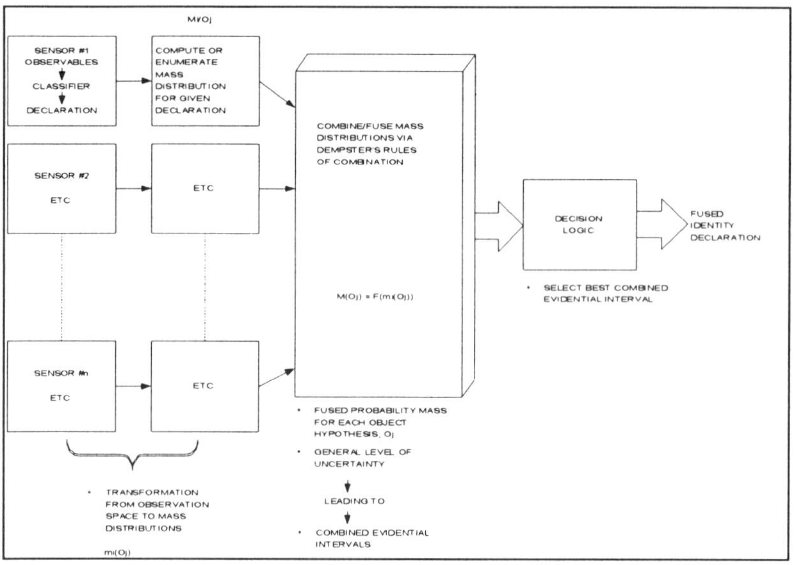

Figure 9. Summary of Dempster Shafer Fusion

Table 1. Overview of Analytical Methods for Identity Fusion.

Method	Kernel Process	Required Input Characteristics	Characteristics of Output
Classical Inference	Pr (Observation/H)	Empirical Probability Population Distribution for Statistic	Pr (Error/Dec. on H)
Bayesian Inference	"A Posteriori" Pr (H/Evidence) (Updated Belief on H/Given new data)	Empir/Subj. Probability Exhaustive Definition of Causes "A Priori" Pr (Causes) Mutually Exclusive Clauses	Updated Likelihood of the Occurrence of an Event
Dempster-Shafer Method	"Pr (H/Mult. Evidence) - Gen'l Uncertainty	Empir/Subj. "Probability" Exhaustive (Incl. Disj) Pr. (H/Evidence)	Updated Likelihood of the Occurrence of an event and level of Ignorance
Fuzzy Set Theory	Set Algebra where Set Elements have Membership Function	Subjective Membership Functions for all Set Elements	Profile of Goal Set Elements and Membership Function
Cluster Analysis	Sorting of Obsvns into "Natural G Groups" based on "Similarity Measure"	Parametric or Subjective Data	Cluster Elements and Similarity Measures
Estimation Theory	"Best State" (Identity) Estimate for given Obsvns. (Least Squares)	Quantitative Obsvns., State/ Obsvns. Model	State (Identity) Vector
Entropy	Computes measure of Information Content	Empirical or Subjective Probability	Identity Estimate which Maximizes Info. Content
Figure of Merit	Computes Degree of Similarity between Para-meters and Entities	Attribute Vectors from Sensor Obsvns.	Numerical Value of Similarity of Candidate Object to Postulates Class
Expert Systems	Computer Program to Mmic Human Inference Process	Observation Data to Support Inferences	Declaration of Inference (Object Identity) that best Satisfies Logic Conditions
Templates	Pattern matching Techniques for Complex Associations	Observed Data Records	Declaration that Data Supports (Matches) a Template-hence Object ID
Adaptive Neural Nets	Non-Linear Transformation from Observation to Recognition Space	Observed Parametric Data (Features)	Match of Input Data to Pre-Specified Categories

3.3 LEVEL 2 AND 3 PROCESSING (SITUATION AND IMPACT ASSESSMENT OR SITUATION AND THREAT ASSESSMENT (STA))

Level 2 and 3 processing is combined together because together they perform the analysis of the situation, as defined and estimated by Level 1 processing. While Level 0 and 1 processing can be, to a significant extent, considered sequential (of each other and internal functionality, except for kinematic and ID estimation), Level 2 and 3 processes can perform their analyses independently of each other. They may be activated concurrently and in parallel with Level 0 and 1 processes.

Figure 10 shows a diagram of an example of DF Level 2 and 3 processing for a Defence Application [37].

In Figure 10 the functions are listed in a legend box. This is just a subset example including functionality to cover Level 2 and 3 processing as follows:

(a) Perform estimation and prediction of relations among entities to include force structure and cross force relations, communications and perceptual influences, physical context, etc. (DR, BA, KE, HA, HV) for Level 2

(b) Perform estimation and prediction of effects on situations of planned or estimated/predicted actions by participants; to include interactions between action plans of multiple players (TE, TS, HG, HV) for Level 3. Note that the Situation Interpretation functions (HG and HV) analyze hypotheses on both levels of fusion.

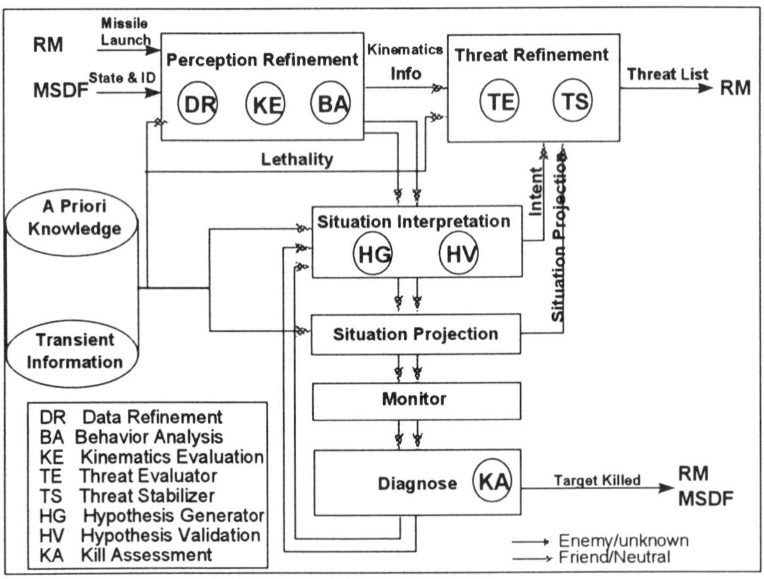

Figure 10. Example DF Level 2 and 3 Processing Data Flow for a Defence Application

The methods used within these functions can be statistical, rule-based, based on pattern recognition, fuzzy logic, etc. They can start their execution as soon as the DF Level 1 Processing has recorded some fused states and will continuously refine their evaluation as new information is added. It is possible that more than two functions providing the same functionality, using different approaches can concurrently refine the system knowledge about the situation and impacts.

3.4 LEVEL 4 PROCESSING (PROCESS REFINEMENT)

Level 4 Processing (Process Refinement) – is an element of RM. It deals with the refining the fusion process performance in terms of real-time control and long-term performance as well as the amount and quality of information derived/inferenced. It also evaluates source specific requirements (sensor data, database) and directs these sources to fulfil these requirements. RM performs similar monitoring/ planning/reasoning roles to refine and control the execution of the actuators (weapons). A combined DF Level 4 and RM diagram is shown in Figure 11.

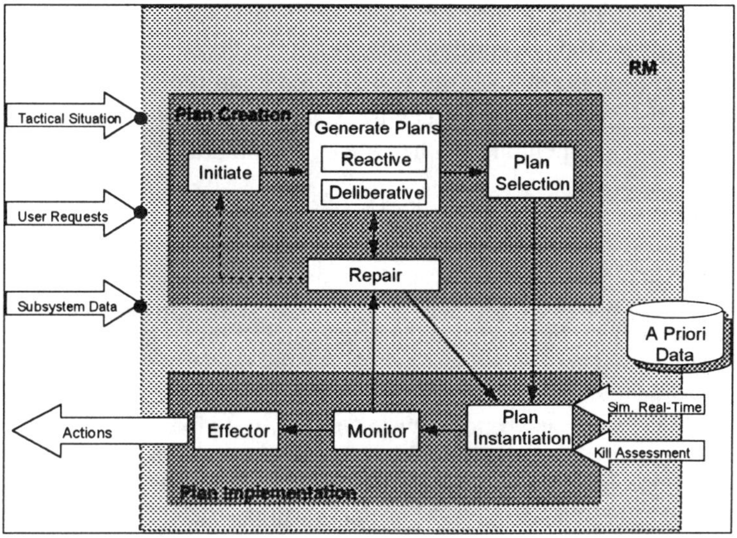

Figure 11. DF Level 4 Processing/RM Diagram

4. Architectures

The topic of DF architectures is widely discussed in the literature [1,3,7,10,12,38,39]; however, these discussions almost exclusively address the DF Level 1 processing architecture (or MSDF architecture). Since DF Level 0 processing provides the data that the MSDF must fuse, these discussions also address the characteristics of this data, i.e., the processing that takes place within the DF Level 0.

The term "MSDF architecture" is used in the literature to indicate the general method (philosophy) of combining the sensor data. Three types of fusion architectures have been identified, based on the level at which the sensor data is processed:

(a) Central-level, report-to-track or measurement fusion, centralized tracking, where sensor contact data (or raw sensor reports) are used in the MSDF processes to establish the state and identification of the objects.

(b) Sensor-level, or track-to-track fusion, distributed or autonomous tracking, where each sensor creates its own tracks from its sensor data and passes them as tracks or probabilities to the MSDF function to fuse and establish a master track file.

There are two fundamentally different approaches of integrating data from multiple sensors within a sensor-level fusion architecture: **hard fusion** and **soft fusion**. In the hard fusion architecture, sensors measure signals and return yes/no responses (declarations) based upon decision criteria within each sensor. In the soft fusion architecture, sensors return a measure of confidence (such as a probability) that quantifies the uncertainty in detection and/or identification [40].

(c) Hybrid, where the sensor systems maintain their track files, while the MSDF function uses either track data or contact data in its different processes, depending on the type of hybrid architecture that is selected.

Each one of these architectures has certain advantages and drawbacks:

<u>Central-level</u>:

The major advantage of the central-level architecture is that all sensor data is provided to the MSDF function without delay in the sensor system for tracking or identification, enabling the fusion function to make full use of the complementarity of the data from dissimilar sensors. This results in extended sensor detection range, faster identification of the target and better precision of the target kinematic variables.

The major drawbacks of the central-level architecture are increased communication bandwidth and processing requirements, and the fact that if one of the sensors is degraded, it may deteriorate the MSDF function performance.

<u>Sensor-level</u>:

The major advantages of the sensor-level fusion architecture are reduced data bus bandwidth and processing requirements, filters specifically suited for each sensor, and the fact that, if one of the sensors is degraded, its observations will not deteriorate the overall MSDF function performance.

The major drawbacks of the sensor-level fusion architecture are lower accuracy and reduced sensor detection range since the central fusion function receives filtered data from each sensor, and some sensor information is lost. Also more complex fusion algorithms are required, since the assumption of error independence between sensors from one update period to another is not valid.

<u>Hybrid</u>:

The major advantage of the hybrid architecture is that it can benefit from the advantages of both sensor-level fusion (e.g., lower data bus bandwidth requirements) and central-level fusion (e.g., availability of raw measurements for correlation ambiguity resolution) architectures [41].

The major drawback of the hybrid architecture is again, as in central-level architecture, increased processor requirements.

The above discussions assume that there is one MSDF fusion centre which, depending on the selected fusion architecture, fuses the contact data (or raw sensor report) or the sensor track data. The Distributed Sensor Network (DSN) fusion is another way of combining data from multiple sensors [42-44]. The DSN is composed of a set of nodes that can communicate to each other via a communication network. Each DSN node contains a fusion function (acting as a fusion centre) collecting data from one or more sensors. This fusion function performs tracking using the local sensor data, and communicates the processing results to other nodes according to some communication strategy. This fusion function is also responsible for maintaining the fused results from all fusion centres. Different types of fusion architectures can be selected for each node.

In each of these architectures, the fusion results can be fed back into the sensor subsystems to provide cueing, threshold control, etc., for the purpose of optimizing the overall response [45].

The aim of the system designer is to select an optimal MSDF architecture and a sensor cueing mechanism within the Command and Control System (CCS), minimizing the effects of the drawbacks and fully benefiting from the advantages of the selected fusion architecture. However, the selected MSDF architecture greatly depends on whether Level 0 processing is also open to choice of design alternatives. Most often an MSDF architecture has to be designed given a set of existing sensors and sources of information, and also given a variety of other requirements on computer resource, execution speed, environmental and mission constraints etc. Therefore the MSDF architecture selection is made based on a number of criteria, which include:

(a) Characteristics of input data
(b) Computer resource constraints
(c) Timing constraints
(d) Environmental and mission constraints
(e) Cost versus performance trade-offs, etc.

Figure 12, Figure 13 and Figure 14 show three alternative MSDF architectures that were designed for a future upgrade of the HALIFAX Class Frigate in 1992. These figures show different choice of algorithms depending on the level of pre-processing assumed in the sensors (DF Level 0).

After some years of theoretical trade-off analyses and prototyping, the current choice is a hybrid architecture that is still evolving.

The following section describes some current applications and the architecture trade-offs for their design.

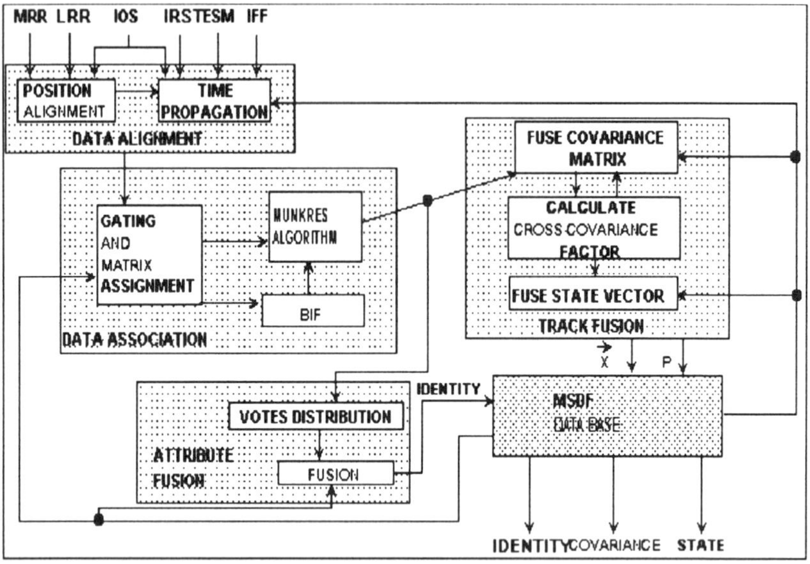

Figure 12. Sensor Level MSDF Architecture Design for a Future HALIFAX Class frigate

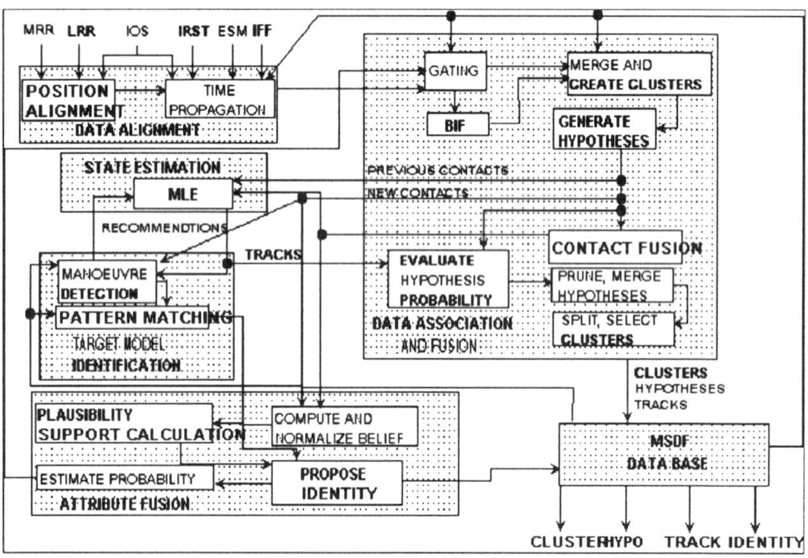

Figure 13. Central Level MSDF Architecture Design for a Future HALIFAX Class frigate

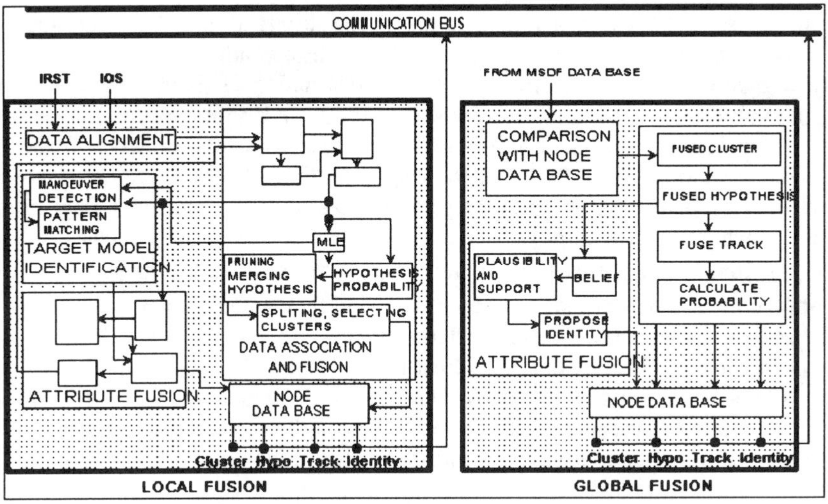

Figure 14. DSN MSDF Architecture Design for a Future HALIFAX Class

5. Applications

The very first DF application that was developed at LM Canada was the single platform local sensor near-term prototype MSDF for the HALIFAX Class frigate. The local sensors of the frigate that were fused are the SPS-49 Long Range Radar (LRR), SG-150 Medium Range Radar (MRR), an Electronic Support Measures (ESM) sensor and an Identification Friend or Foe (IFF) transponder. Considering that the sensors were developed long before the DF discipline became conceivable for integrating and automating the tactical situation establishment, the sensor interfaces were not designed to provide all the data the MSDF processes require. Therefore compromises were made in the MSDF design.

Since all sensors on HALIFAX Class provide already processed confirmed track data, the MSDF architecture is constrained to be sensor-level hard fusion, which is tailored to fit the type of data provided by the sensors. Although this is not the most optimal fusion architecture for an MSDF function, it is expected that it will still enhance the HALIFAX Class Combat System tracking and identification performance considerably.

The two radar sensors provide track data to the MSDF function; however they do not provide all track information (e.g., complete covariance matrices) required by the true track-to-track MSDF fusion processes. Therefore conventional sensor level fusion techniques are not applicable, and other fusion approaches must be used. The fusion selected is contact-to-track, with contacts made of the positional information of the sensor tracks (state vector) and the appropriate variances constructed from the sensor accuracy variance.

The time of arrival of the radar data is essential for track management and the association process of the MSDF function. The beam number of the SG-150 radar beam gives some indication of target elevation. Since the SG-150 radar is the most accurate sensor of the HALIFAX Class, and its position on the ship is used as the reference origin for the MSDF function. Therefore the range and the azimuth of the SPS-49 is being spatially aligned to the reference frame of the SG-150 radar. This is obviously a minor correction except when targets are at very short ranges. Also the radar target tracks are fused only if a new contact has been processed to update the sensor target track. The tracks which are estimated by the radar tracking algorithms due to a "miss" do not have statistical interest for the fusion since the state estimation process of the MSDF function performs its own track estimation.

The positional information such as range, azimuth, σ_R and σ_β (range and azimuth errors) are the main input into the fusion process, and they are considered as contact data for this process. Since the radars send tracks that are not statistically independent (already correlated), a phenomenological treatment is applied to decorrelate the range and azimuth without compromising the accuracy of the MSDF data association process. This treatment applies to the variance data. Instead of using the reported radar track variance, 2/3 of the sensor variance is used. For tracks recently initiated this is probably too low (by about 33%), for established tracks this is probably too high (by about 20%), but on average, the final result should be an acceptable approximation in this first implementation. The speed and its variance information are not being fused but are used in the data association process and in the state estimation process for track initialization.

ESM sensor on HALIFAX Class provides azimuth contacts and identity declarations. The azimuth information has a good accuracy but the sensor updates only when a large variation of azimuth has been detected. Therefore the passive tracks that are passed to the MSDF function are handled considering this restriction in the data association process. The azimuth information is used for association purposes only. The attribute information from the ESM sensor is by far the most important information for MSDF. The ESM sensor generates a list of potential target identities with associated confidence levels. The ESM sensor detects directly electromagnetic waves and matches the measured wave parameters with a list of emitter types (attribute data). From the emitter types another match is performed on a list of possible target identities. These target identities are high level processed information. For optimal fusion it is more advantageous to deal directly with the emitter types but the ESM sensor operator can provide either the platform or the emitter data. For the MSDF demonstration model it is assumed that the ESM sensor provides the emitter data only. The simulation system simulates emitter data from an emitter list prepared from Jane's, which is then used for attribute fusion processes.

The IFF interrogator sensor provides information about a target when a friendly target has responded to the interrogation. When a response is obtained, unambiguous positional information, range and azimuth are available for data association purposes. If no answer is obtained only azimuth information can be used. Due to enemy deceptive operation, friend responder status, or any other situation it is necessary to consider the IFF response as an attribute information, while IFF positional data is used for track association.

From the Command and Control System, the MSDF function needs information relative to the ship movement when establishing the target state vector. The MSDF function writes all its tracking results onto the MSDF database.

This first model was completed in 1994 and continued evolving until 2000. The current version fuses all surface and above water sensor data, using, however, the same contact-to-track fusion approach. It uses JVC for association, IMM for kinematics estimation and a modified version of the Dempster-Shafer (Truncated Dempster-Shafer) developed at LM Canada [34,35,52].

Having built and demonstrated an initial Level 1 DF application, fusing a set of detections using dissimilar sensors but similar data, a number of questions can be asked:

(a) What type of architecture should be used for higher level DF implementation?
(b) How to fuse imaging data with non-imaging data?
(c) How should the overall MSDF design evolve to fuse more dissimilar non-imaging data?
(d) How should the fusion of off-board stale data be processed?
(e) What type of architecture could support co-operation of multiple platforms with their own DF system?

The works in progress attempting to answer these questions have been presented elsewhere [37,47-53]. The current conclusions to these questions are listed below.

5.1 WHAT TYPE OF ARCHITECTURE SHOULD BE USED FOR HIGHER LEVEL DF IMPLEMENTATION

A blackboard based architecture was demonstrated to be best suited for higher level DF implementations [37,47]. The testbed architecture where higher level fusion has been demonstrated is shown in Figure 15.

Here the STA and RM functionality are implemented as agents around the blackboard. This architecture specifically is suited to the concurrent execution of multiple reasoning functions to achieve an incrementally better understanding of the situation and impact (or threat).

The blackboard infrastructure intrinsic communication layer is also very well suited for applying the multiple platform cooperation. This has not yet been demonstrated, but is currently planned as the next phase demonstration system.

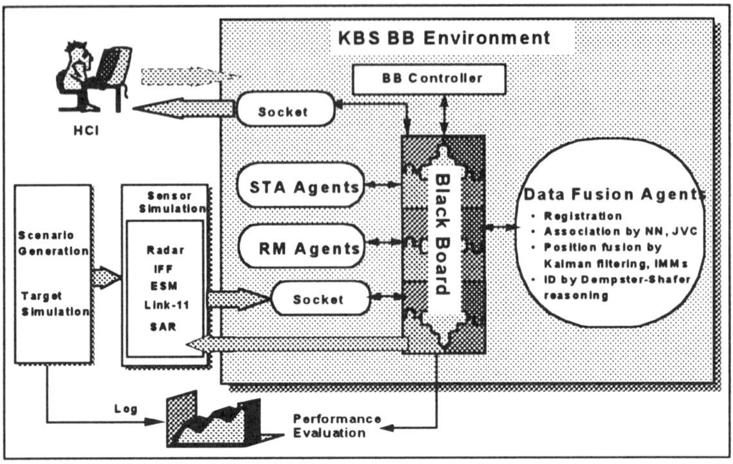

Figure 15. Level 2, 3 and RM Demonstration for the HALIFAX Class Frigate

5.2 HOW TO FUSE IMAGING DATA WITH NON-IMAGING DATA

The fusion of imaging data with non-imaging sensors has been evaluated for Canada's CP-140 airborne surveillance aircraft. This aircraft's next upgrade will have a Synthetic Aperture Radar (SAR) and a Forward Looking Infrared Radar (FLIR) in addition to the other usual sensors such as surveillance radar, ESM and IFF. Considering that all non-imaging sensors for this application again provide pre-processed tracks, it was decided that the most appropriate approach would be to pre-process images to extract features (attributes) (as shown in Figure 16) then to provide it to the Truncated Dempster-Shafer algorithm for fusion with the rest of the attribute data received from the non-imaging sensors [49,52].

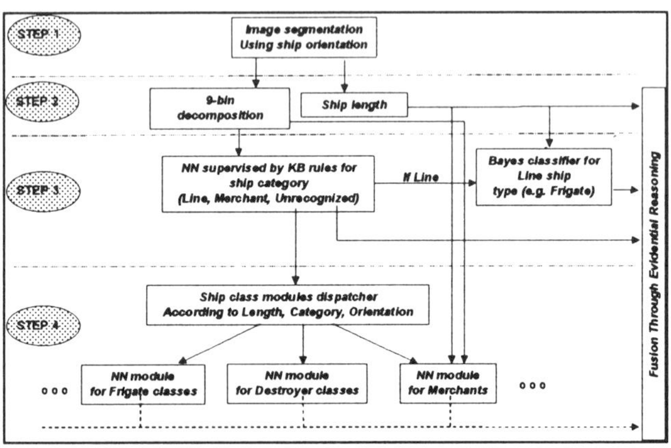

Figure 16. Hierarchical SAR Classifier

5.3 HOW TO FUSE VERY DISSIMILAR NON-IMAGING DATA

Assuming that the HALIFAX Class sensors are not going to be modified to provide all the data an ideal MSDF application would require, and knowing that dissimilar data from additional sources should be incrementally integrated with the existing prototype, the analyses conclude that [53]:

- A multiple fusion centre approach with specialized algorithms to handle fusion of each type of data will be more appropriate, even if not the most optimal, shown in Figure 17, and
- The data driven approach based on the blackboard infrastructure will support the incremental upgrading and will be most efficient, as shown in Figure 18.

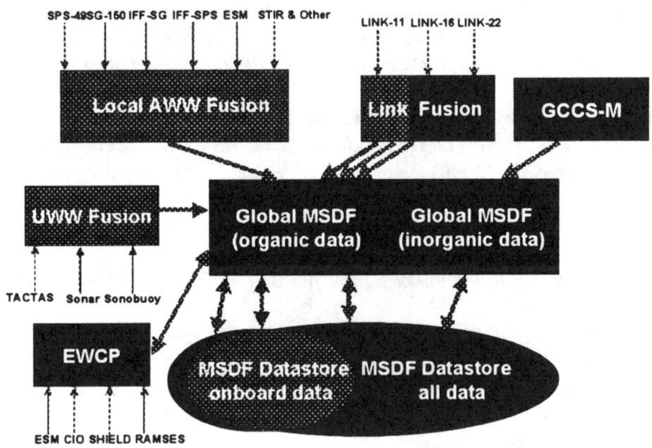

Figure 17. The Multiple Fusion Centre Architecture

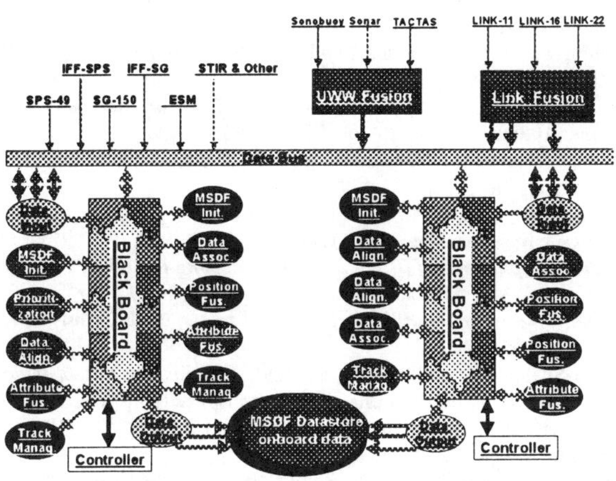

Figure 18. The Data Driven Muti-Processor DF Architecture

6. References

1. Waltz, E., and Llinas, J., (1990) *Multisensor DF*, Artech House.
2. White, F.E. Jr., (1987) *DF Lexicon*, Joint Directors of Laboratories, Technical Panel for C3, DF Subpanel, Naval Ocean Systems Center, San Diego.
3. Blackman, S. (1986) *Multiple-Target Tracking with Radar Application,* Artech House.
4. Bar-Shalom, Y., and Fortman, T.E., (1988) *Tracking and Data Association*, Academic Press.
5. Bar-Shalom, Y., (1990) *Multitarget Multisensor Tracking: Advanced Applications*, Artech House.
6. Bar-Shalom, Y., (1992) *Multitarget Multisensor Tracking Applications and Advances*, Artech House.
7. Hall, D.L., (1992) *Mathematical Techniques in Multisensor DF*, Artech House.
8. Bar-Shalom, Y., and Li, Xiao-Rong (1993) *Estimation and Tracking: Principles, Techniques, and Software*, Artech House.
9. Dasarathy, B.V. (1994) *Decision Fusion*, IEEE Computer Society Press.
10. Klein, L.A. (1993) *Sensor and DF Concepts and Applications*, Society of Photo-optical Instrumentation Engineers.
11. Antony, R.T., (1995) *Principles of DF Automation,* Artech House.
12. Varshney, P., (1996) *Distributed Detection and Data Fusiom*, Springer-Verlag.
13. Goodman, I.R., Mahler, R.P.S., and Nguyen, H.T., (1997) *Mathematics of DF*, Kluwer Academic Publishers.
14. Waltz, E., (1998) *Information Warfare Principles and Operations*, Artech House.
15. Blackman, S., and Popoli, R., (1999) *Design and Analysis of Modern Tracking Systems*, Artech House.
16. Bedworth, M., and O'Brien, J., (2000) *The Omnibus Model, A new Model of DF?*, IEEE AES Systems Magazine, pp 30-36.
17. Steinberg, A., Bowman, C., and White, F., (1999) *Revisions to the JDL DF Model*, SPIE vol. 3719, pp. 430-441.
18. Steinberg, A.N., (1999) *New Developments: FUSIAC and JDL DF Model Revision*, NASA Fusion and Data Mining Conference.
19. Hall, D.L., and Garga, A.K., (1999) *Pitfalls in DF (and How to Avoid Them)*, Applied Research Laboratory, Pennsylvania State University, EuroFusion99.
20. Boyd, J., (1987) *A Discourse on Winning and Losing*, Maxwell AFB lecture.
21. Blackman, S., and Broida, T., (1990) *Multiple Sensor Data Association and Fusion in Aerospace Applications*, Journal of Robotic Systems.
22. Drummond, O.E., Castanon, D.A., and Bellovin, M.S., (1990) *Comparison of 2-D assignment algorithms for sparse, rectangular, floating point, cost matrices*, Journal of the SDI panels on tracking, # 4, pp. 4-81 to 4-97.
23. Sittler, R.W., (1964) *An Optimal Data Association Problem in Surveillance Theory*, IEEE Trans. Military Elec., Vol. MIL-8, pp. 125-139.
24. Singer, R.A., (1970) *Estimating Optimal Tracking Filter Performance*, IEEE Trans. on Aerospace & Electronic Systems, Vol. AES-6, pp. 473-483.
25. Smith, P.L., and Buechler, G., (1975) *A Branching Algorithm for Discriminating and Tracking Multiple Objects*, IEEE Trans. Automatic Control, Vol. AC-20, pp 101-104.
26. Bar-Shalom, Y., and Tse, E., (1975) *Tracking in a Cluttered Environment with Probabilistic Data Association*, Automatica, Vol. 11, pp 451-460.
27. Stein, J.J., and Blackman, S.S., (1975) *Generalized Correlation of Multitarget Data*, IEEE Trans. on Aerospace & Electronic Systems, Vol. AES-11, pp 1207-1217.
28. Morefield C.L. (1975) *Application of 0-1 Integer Programming to Multi-Target Tracking Problems*, IEEE Trans. Automatic Control, Vol. AC-22, pp 302-312.
29. Reid, D.B., (1979) *An Algorithm for Multiple Tracking Targets*, IEEE Trans. Automatic Control, Vol. AC-24, pp. 843-854.
30. Fortmann, T.E., Bar-Shalom, Y., and Scheffe, M., (1983) *Sonar Tracking of Multiple Targets Using Joint Probabilistic Data Association*, IEEE Journal of Oceanic Engineering, Vol.OE-8, pp. 173-184.
31. Barker, T.N., Persichetti, J.A., Poore, Jr., Aubrey, B., and Rijavec, N., (1995) *Method and system for tracking multiple regional objects*, US patent #5,406,289.
32. Poore, A., and Yan, X., *Partial Branch and Bound Algorithms for Improved Data Association in Multi-frame Processing*, Fusion 99, Vol II, pp. 1037-1044.

33. Drummond, O.E., *On Features and Attributes in Multisensor, Multitarget Tracking*, Fusion 99, Vol II, pp. 1045-1053
34. Simard M.A., Valin P. and Shahbazian E. (1993) *Fusion of ESM, Radar, IFF and other Attribute Information for Target Identity Estimation and a Potential Application to the Canadian Patrol Frigate*, AGARD 66th Symposium on Challenge of Future EW System Design, Ankara (Turkey), AGARD-CP-546, pp. 14.1-14.18
35. Bégin, F., Boily, E., Mignacca, T., Shahbazian, E., and Valin P. (1994) *Architecture and Implementation of a Multi-Sensor DF Demonstration Model within the Real-Time Combat System of the Canadian Patrol Frigate*, AGARD symposium on Guidance and Control for Future Air-Defence Systems, Copenhagen, AGARD-CP-555, pp. 28.1-28.8.
36. Bar-Shalom, Y., Chang, K.C., and Bloom, H.A., (1989) *Tracking a Maneuvering Target Using Input Estimation Versus the Interacting Multiple Model Algorithm*, IEEE Transactions on Aerospace and Electronic Systems, Vol. AES-25, No.2, pp. 296-300.
37. Bergeron, P., Couture, J., Duquet, J.-R., Macieszczak, M., and Mayrand, M., A. (1998) *New Knowledge-Based System for the Study of Situation and Threat Assessment in the Context of Naval Warfare International Conference on Multisource-Multisensor Information Fusion*, FUSION 98, Las Vegas, Vol. I, pp. 388-395.
38. Drummond, O., and Blackman S., (1989) *Challenges of Developing Algorithms for Multiple Sensor*, Multiple Target Tracking, SPIE Vol. 1096, Signal and Data Processing of Small Targets.
39. Krzysztofowicz, R., and Long, D., (1990) *Fusion of Detection Probabilities and Comparison of Multisensor Systems*, IEEE Trans. On System Man & Cybernetics, Vol 20, No.
40. Buede, D., and Waltz, E., (1989) *Sensor Architecture*, SPIE Vol. 1096, Signal and Data Processing of Small Targets.
41. Llinas, J., Hall, D., and Waltz, E., (1989) *DF Technology Forecast for C3MIS*, 3rd Int. Conference on Command, Control Comm. & Management Information Sys. IEE London UK.
42. Chong, C., Chang, K., and Mori, S., (1986) *Distributed Tracking in Distributed Sensor Networks*, Advanced Decision Systems, Mountain View, CA.
43. Chong, C., Miltonberger, T., et al, (1984) *Fusion in System Architecture Analysis and Synthesis*, Rome Air Development Center: Air Force System Command.
44. Fahmy, A.M., and Titus, H.A., (1987) *Horizontal Estimation and Information Fusion in Distributed Sensors Networks*, IEEE International Symposium on Circuits and Systems, Vol 1, P.243-245.
45. Reiner, J., (1985) *Application of Expert Systems to Sensor Fusion*, IEEE.
46. Jouan, A., Jarry, B., and Michalska, H., (2000) *Tracking Closely Maneuvering Targets in Clutter with an IMM-JVC Algorithm*, accepted for presentation at Fusion 2000, Paris, France.
47. Shahbazian, E., Duquet, J.-R., Macieszczak, M., and Valin, P., (1998) *A Generic Expert System Infrastructure for Fusion and Imaging Decision Aids*, International Conference on Data Fusion, EuroFusion98, Great Malvern, Vol. I, pp. 167-174.
48. Chalmers, B.A., Carling, R., Bossé, E., and Bergeron, P., (1997) *Towards a Generic Model for Situation and Threat Assessment, in Digitization of the Batthefield II*, SPIE Aerosense 97, Orlando, Proc. Conf. 3080, pp. 171-182.
49. Jouan, A., Valin, P., and Bossé, E., (1999) *Testbed for Fusion of Imaging and Non-Imaging Sensor Attributes in Airborne Surveillance Missions*, FUSION 99, Sunnyvale, CA, Vol. 2, pp. 823-830.
50. Shahbazian, E., Bergeron, P., Duquet, J.-R., Jouan, A., and Valin, P., (1999) *Data Fusion Applications for Military and Civilian Purposes Developed on DND/LM Canada Decision Support Testbed*, in Proc. 33rd Asilomar Conference on Signal, Systems and Computer, Pacific Grove, CA.
51. Simard, M.-A., Helleur, C., and Lefebvre, E., (2000) *Multisource information fusion applied to ship identification for the recognized maritime picture*, in SPIE Aerosense Conf. on Sensor Fusion: Architectures, Algorithms and Applications IV, Orlando, Florida, pp. 67-78.
52. Simard, M.-A., Valin, P., and Lesage, F., (2000) *The ID box: a multisource attribute data fusion function for target identification*, in Proc. SPIE Aerosense Conf. on Sensor Fusion: Architectures, Algorithms and Applications IV, Orlando, Florida, vol. 4051, pp. 180-190
53. Shahbazian, E., (2000) *Practical Trade-offs in Fusion Architecture Design*, accepted for presentation at Fusion 2000, Paris, France.

THE FUSION OF DECISIONS FOR DISTRIBUTED RECOGNITION

Hard Decision Fusion and Soft Decision Fusion

M.D. BEDWORTH
University of Central England
Perry Barr
Birmingham
B42 2SU
UK

1. Introduction

Data fusion is now a widely accepted approach for increasing the performance of data analysis when the data sources are distributed (see [1] or [2] for example). The combination of the opinions from multiple experts (or sensors) is an attractive and conceptually simple problem in multi-sensor data fusion. The task arises in a diversity of situations, from non co-operative target recognition [3] to medical diagnosis [4] and machine condition monitoring [5]. In each case the underlying problem is similar – how should decisions from two or more separate classifiers be combined to provide a fused decision of higher quality? At this stage we intentionally leave the measure of quality undefined, since accuracy, timeliness or robustness (amongst others) will each play a part depending on the specific application.

We can, in principle, treat the combined data from all sensors as a single measurement and proceed with the use of a standard pattern classification algorithm. However, the problem becomes more interesting, and of more practical value in a distributed system, if such an approach is prohibited on the grounds of communication or processing constraints. Information such as this may be distributed because of geographical location or area of expertise.

We sub-divide the field of decision fusion into two broad categories: *hard decision fusion* in which each expert provides only their best guess as to the class and *soft decision fusion* where additional information is provided regarding the confidence that the expert places in their decision. In both cases uncertain information needs to be brought to bear on the problem. There are several approaches to representing and inferencing with uncertain information – probabilistic approaches generally place greatest emphasis on stochastic uncertainty, whereas possibilistic representations admit the concept of vagueness and evidential reasoning allows for ignorance. Here we concentrate on the probabilistic approach although in various circumstances the others have much to commend them and have been widely used by data fusion practitioners (see Abdulgahfour [6] for an overview).

A. K. Hyder et al. (eds.), Multisensor Fusion, 99–124.

2. Hard Decision Fusion

In its present form hard decision fusion was introduced by Nahin and Pokoski in [7] and was formulated as a distributed detection problem by Tenney and Sandell in [8]. The basic arrangement is shown in Figure 1 with each sensor having a separate decision-maker. The fusion centre fuses these hard decisions.

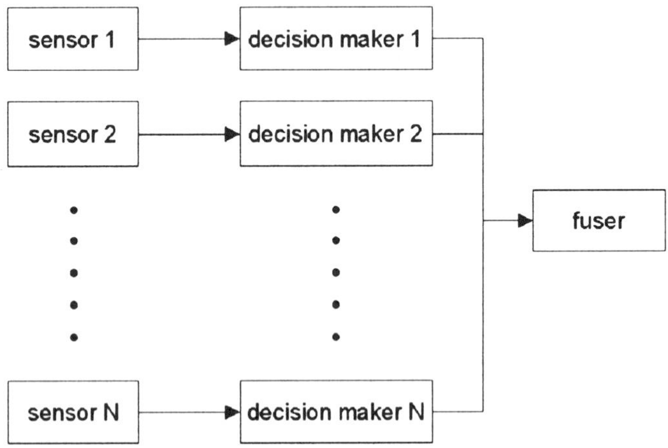

Figure 1. Hard decision fusion.

There has since been much research into hard decision fusion with significant developments in both the fusion rules and the local decision rules. Key papers in this respect are those by Chair and Varshney (including [9] and [10]). We shall constrain ourselves to simple hard decision problems in which the local decisions are conditionally independent (see Kam [11] for a method for fusing correlated decisions) and are made synchronously (see Chang [12] for a treatment of asynchronous decision fusion). Furthermore we assume that minimisation of the misclassification rate is the overriding criterion for both the local decision-maker and the fusion centre. These assumptions greatly simplify the treatment of hard decision fusion. Even so, we shall see that the exact decision rule depends on the information available to the system designer.

We begin by establishing some notation. Consider a recognition task with a set of C target classes, $\omega \in \Omega = \{\omega_1, \omega_2, \cdots \omega_C\}$. The classes are assumed to have a prior distribution, $P(\Omega) = \{P(\omega_1), P(\omega_2), \cdots P(\omega_C)\}$, which is either known or can be estimated. The fusion system comprises a set of N sensors, which collectively produce sensor data $X = \{x_1 \cdots x_N\}$. Each sensor is provided with a local decision-maker that produces a class label, $y_i = f(x_i) \in \{\omega_1 \cdots \omega_C\}$, based solely on the local sensor data.

2.1 MAJORITY RULE

If no further information about the characteristics of the sensors is available then a simple fusion rule must be used. The majority rule fits the bill admirably since it is both intuitively simple and straightforward to apply. The majority rule is also conservative since it only outputs a decision when a majority occurs.

The performance of the majority rule depends on the characteristics of the sensors and the local decision-makers. For the moment we shall characterise these only by their probability of error, $P(E) = P(y_i \neq \omega^*)$, and further assume that the errors are evenly distributed between classes. As an example let us consider a system with three sensors for which a majority will occur if two or more local decision-makers are in agreement. We may calculate the probability of such a majority corresponding to the correct class. First we determine the probability that exactly two sensors correctly classify the entity under scrutiny:

$$P(N^* = 2) = (1 - P(E))^2 \cdot \left(\frac{P(E)}{C-1}\right) \cdot {}^N C_2 \cdot (C-1) \qquad (1)$$

The first term is the probability that two sensors will be correct. The second term is the probability that the remaining sensor will be wrong. The third term is the number of situations with exactly two sensors correct and the final term is the number of ways that one sensor can be wrong. In the general case of n sensors being correct, this simplifies to:

$$P(N^* = n) = (1 - P(E))^n \cdot P(E)^{N-n} \cdot {}^N C_n \qquad (2)$$

Which, we may note, is independent of the number of classes. The probability that the majority rule will result in a correct classification may now be computed as:

$$P(N^* > \frac{N}{2}) = \sum_{n=\frac{N+1}{2}}^{N} (1 - P(E))^n \cdot P(E)^{N-n} \cdot {}^N C_n \qquad (3)$$

Examples of this computation are shown in Table 1 for various values of $P(E)$ and N.

It can be seen that the largest proportional benefits are obtained when the local probability of error is low. When the $P(E) = 0.5$, the majority rule offers no improvement in performance.

As mentioned earlier, the majority rule is a conservative fusion rule since it will defer its decision if an insufficient number of sensors agree on a classification. For instance, of the 10% of instances that are not correctly classified for $P(E) = 0.2$ and $N = 3$ (see

TABLE 1. The expected fused correct classification rate using the majority rule for various local error probability values and number of sensors.

	$P(E) = 0.1$	$P(E) = 0.2$	$P(E) = 0.3$	$P(E) = 0.4$
$N = 3$	0.97	0.90	0.78	0.65
$N = 5$	0.99	0.94	0.84	0.68
$N = 7$	0.100	0.97	0.87	0.71
$N = 9$	0.100	0.98	0.90	0.73

Table 1.) not all will be misclassified. To determine the misclassification rate we must calculate the probability that a majority is reached that is associated with the wrong class. In this case the number of classes is important since the majority must consist of the same, wrong class ω^{λ} and not just a set that does not include the correct class, ω^{*}. In similar fashion to equation (2) we may write:

$$P(N^{\lambda}=n)=\left(\frac{P(E)}{C-1}\right)^{n}\cdot\left(1-P(E)+\frac{C-2}{C-1}P(E)\right)^{N-n}\cdot{}^{N}C_{n}\cdot(C-1) \qquad (4)$$

Which simplifies to:

$$P(N^{\lambda}=n)=\left(\frac{P(E)}{C-1}\right)^{n}\cdot\left(\frac{C-1-P(E)}{C-1}\right)^{N-n}\cdot{}^{N}C_{n}\cdot(C-1) \qquad (5)$$

As before the first term is the probability that n local decisions are wrong (and all the same). The second term is the probability that the other local decisions are either correct or wrong in a different way. The third term is the number of such situations that can arise and the final term is the number of wrong classes available (since any one of them will suffice to cause a misclassification). The misclassification rate is merely the sum of equation (5) over appropriate values:

$$P(N^{\lambda}>\frac{N}{2})=\sum_{n=\frac{N+1}{2}}^{N}\left(\frac{P(E)}{C-1}\right)^{n}\cdot\left(\frac{C-1-P(E)}{C-1}\right)^{N-n}\cdot{}^{N}C_{n}\cdot(C-1) \qquad (6)$$

This is given in Table 2 for various values of C and N. The proportion of n-tuples that do not achieve a majority is simply the shortfall after adding the result of equation (3) to equation (6). For example for a three class problem with three sensors each having $P(E)=0.3$ the correct classification rate is 78%, the misclassification rate is 12% and the deferral rate is 10%.

TABLE 2. The expected fused misclassification rate using the majority rule for various numbers of classes and sensors, in all cases $P(E)=0.3$

	$C=2$	$C=3$	$C=4$	$C=5$
$N=3$	22%	12%	8%	6%
$N=5$	16%	5%	3%	2%
$N=7$	13%	2%	1%	0%
$N=9$	10%	1%	0%	0%

2.2 WEIGHTED VOTING

If the probability of error, $P_i(E)$, for each local decision-maker is available at design time then a slightly more sophisticated rule may be employed. In this case the votes for a particular class are cast as follows – all the local decision-makers, D^{+}, that decided on that class make a weighted vote for it and all the local decision-makers, D^{-}, who chose a different class make a weighted vote against it. The final vote is biased according to the prior probability of the class. Specifically we have:

$$V(\omega_j) = \sum_{i \in D^+} \log(1 - P_i(E)) \quad - \quad \sum_{i \in D^-} -\log P_i(E) \quad + \log P(\omega_j) \tag{7}$$

With the class having the largest vote being declared as the fused decision.

The reason for this weighting scheme is as follows. We would like to declare as the fused decision the class having the maximum posterior probability given the local decisions. First we assume that the decisions are conditionally independent:

$$P(Y \mid \omega_j) = \prod_{i=1}^{N} P(y_i \mid \omega_j) \tag{8}$$

Then applying Bayes' rule we have:

$$P(\omega_j \mid Y) = \frac{\prod_{i=1}^{N} P(y_i \mid \omega_j)\, P(\omega_j)}{P(Y)} \tag{9}$$

Now the likelihood of each output given the class depends only on the local probability of error since:

$$P(y_i \mid \omega_j) = \begin{cases} 1 - P_i(E) & \text{if } y_i = \omega_j \\ \\ P_i(E) & \text{if } y_i \neq \omega_j \end{cases} \tag{10}$$

Defining D^+ to be the set of $i : y_i = \omega_j$ and conversely for D^- we have:

$$P(\omega_j \mid Y) = \frac{\prod_{i \in D^+} (1 - P_i(E)) \cdot \prod_{i \in D^-} P_i(E) \cdot P(\omega_j)}{P(Y)} \tag{11}$$

Since we only need know for which ω_j this is maximised we may drop the normalising constant and take logarithms:

$$\log P(\omega_j \mid Y) \propto \sum_{i \in D^+} \log(1 - P_i(E)) \quad - \quad \sum_{i \in D^-} -\log P_i(E) \quad + \log P(\omega_j) \tag{12}$$

This justifies the weighted voting scheme of equation (7).

The fused probability of error for weighted voting can be computed by summing the likelihood of all *n*-tuples, Y, that give rise to a correct fused decision.

The technique may be illustrated using a simple example. Consider a system comprising three sensors, the first has probability of error $P_1(E) = 0.10$, the second $P_2(E) = 0.15$ and the third $P_3(E) = 0.35$. There are two equally likely classes ω_1 and ω_2. There are $2^3 = 8$ possible triples representing the local decisions. For each triple we find the likelihood of that triple and the vote for both classes. This computation is summarised in Table 3.

TABLE 3. The intermediate results in the computation of the weighted voting rule and the fused probability of error for the example described in the text.

y_1	y_2	y_3	$P(Y \mid \omega_1)$	$P(Y \mid \omega_2)$	$V(\omega_1)$	$V(\omega_2)$
1	1	1	0.497	**0.005**	-0.604	-2.581
1	1	2	0.268	**0.010**	-0.873	-2.312
1	2	1	0.088	**0.030**	-1.358	-1.828
1	2	2	**0.047**	0.055	-1.627	-1.559
2	1	1	0.055	**0.047**	-1.559	-1.627
2	1	2	**0.030**	0.088	-1.828	-1.358
2	2	1	**0.010**	0.268	-2.312	-0.873
2	2	2	**0.005**	0.497	-2.581	-0.604

Summing $P(Y \mid \omega_1)$ for those triplets for which $V(\omega_1) < V(\omega_2)$ and vice versa for ω_2 (marked in boldface in table 2) and weighting by the priors (equal in this case) gives the fused error rate as 9.2% which, in this example, is only slightly better than the best local decision-maker. As can be seen from Table 3, the fused decision is the same as y_1 unless both of the other two decision-makers disagree with it.

Note that weighted voting produces the same fused decision as majority rule in this case. In general, if $P_i(E) = P(E) < 0.5$ and the class priors are equal, then weighted voting is equivalent to majority rule.

2.3 MAXIMUM A POSTERIORI DECISION FUSION

The method described in section 2.2 may be extended to include more information about the characteristics of the local decision-makers. Let us now specify a transition matrix, T_i, for each local decision-maker. The matrix T_i is defined such that the element $t_i(j,k)$ gives the probability that the ith decision-maker will output $y_i = \omega_k$ when the entity is actually of class $\omega^* = \omega_j$.

With this extra information we proceed, as before, to calculate the posterior probability of each class given all possible local decisions. This is given by Bayes' rule as:

$$P(\omega_j \mid Y) = \frac{P(Y \mid \omega_j)\, P(\omega_j)}{P(Y)} \tag{13}$$

We must again assume conditional independence to decompose $P(Y \mid \omega_j)$ into the contributions from the separate decision-makers:

$$P(\omega_j \mid Y) = \frac{\prod_{i=1}^{N} P(y_i \mid \omega_j)\, P(\omega_j)}{P(Y)} \tag{14}$$

From the definition of $t_i(j,k)$ this is:

$$P(\omega_j \mid Y) = \frac{\prod_{i=1}^{N} t_i(j, y_i)\, P(\omega_j)}{P(Y)} \tag{15}$$

We may again drop the normalising term when computing the fused decision since it remains the same for all classes.

Again, a worked example will suffice to illustrate the method. Take a system with three sensors each of which aims to classify the same entity as belonging to one of the two equally likely classes ω_1 or ω_2. Assume that the transition matrices for the three decision-makers are:

$$T_1 = \begin{pmatrix} 0.95 & 0.05 \\ 0.15 & 0.85 \end{pmatrix} \quad T_2 = \begin{pmatrix} 0.90 & 0.10 \\ 0.20 & 0.80 \end{pmatrix} \quad T_3 = \begin{pmatrix} 0.75 & 0.25 \\ 0.45 & 0.55 \end{pmatrix}$$

Just as in the previous section there are $2^3 = 8$ possible triples representing the local decisions. For each triple we find the likelihood of that triple given each of the classes. These values are summarised in Table 4.

The transition matrix for the fused decision-maker is then found by summing appropriate values of $P(Y \mid \omega_1)$ and $P(Y \mid \omega_2)$:

$$t_F(1,1) = 0.641 + 0.214 + 0.071 = 0.926$$

$$t_F(1,2) = 0.024 + 0.034 + 0.011 + 0.004 + 0.001 = 0.074$$

$$t_F(2,1) = 0.014 + 0.016 + 0.054 = 0.084$$

$$t_F(2,2) = 0.066 + 0.077 + 0.094 + 0.306 + 0.374 = 0.916$$ [1]

Giving a fused error rate of 7.9%, the fused transition matrix for *maximum a posteriori* (MAP) fusion is:

$$T_F = \begin{pmatrix} 0.926 & 0.074 \\ 0.084 & 0.916 \end{pmatrix}$$

TABLE 4. The intermediate results in the computation of a decision using maximum *a posteriori* fusion and the fused probability of error for the example described in the text.

y_1	y_2	y_3	$P(Y \mid \omega_1)$	$P(Y \mid \omega_2)$	y
1	1	1	0.641	0.014	1
1	1	2	0.214	0.016	1
1	2	1	0.071	0.054	1
1	2	2	0.024	0.066	2
2	1	1	0.034	0.077	2
2	1	2	0.011	0.094	2
2	2	1	0.004	0.306	2
2	2	2	0.001	0.374	2

[1] The alert reader will have noticed that this sum appears incorrect. In fact the underlying arithmetic is correct but both the arguments and the result have been separately rounded to three significant figures.

2.4 DEFERRAL

We saw in section 2.1 that the majority rule deferred its decision when a majority was not achieved. This makes the majority rule a conservative fusion technique that does not declare a fused decision when the evidence does not clearly support a particular class.

The same deferral option may be built into the other decision fusion schemes in a quite straightforward and intuitive manner. Since weighted voting is really only a special case of MAP fusion, we shall only deal with the latter.

Previously we omitted the normalising term from equation (15) since we only needed to know the class having the maximum posterior probability. To implement deferral at the fusion centre we now need to know the actual value of this probability. We do this by summing over classes:

$$P(\omega_j \mid Y) = \frac{\prod_{i=1}^{N} t_i(j, y_i)\, P(\omega_j)}{\sum_{k=1}^{C} \prod_{i=1}^{N} t_i(k, y_i)\, P(\omega_k)} \tag{16}$$

The user then specifies a decision threshold, τ, such that a fused decision is only declared if the conditional probability of the most likely class exceeds τ, *i.e.* defer if $P(\omega_j \mid Y) < \tau\ \forall j$. Table 5 is similar to Table 4 but additionally includes intermediate results required to compute deferrals. In this case the decision threshold was set to $\tau = 0.7$.

TABLE 5. The intermediate results in the computation of a decision using maximum *a posteriori* fusion and the fused probability of error for the example described in the text.

y_1	y_2	y_3	$P(Y \mid \omega_1)$	$P(Y \mid \omega_2)$	$P(\omega_1 \mid Y)$	$P(\omega_2 \mid Y)$	y
1	1	1	0.641	0.014	0.98	0.02	1
1	1	2	0.214	0.016	0.93	0.07	1
1	2	1	0.071	0.054	0.57	0.43	?
1	2	2	0.024	0.066	0.27	0.73	2
2	1	1	0.034	0.077	0.31	0.69	?
2	1	2	0.011	0.094	0.11	0.89	2
2	2	1	0.004	0.306	0.01	0.99	2
2	2	2	0.001	0.374	0.00	1.00	2

It can be seen from table 5 that with $\tau = 0.7$ two triples result in a deferred decision. The fused transition matrix is then:

$$T_F = \begin{pmatrix} 0.855 & 0.040 \\ 0.030 & 0.839 \end{pmatrix}$$

In this case 10.5% of instances from ω_1 and 13.1% of instances from ω_2 lead to the fusion centre deferring its decision. However the misclassification rate has correspondingly dropped from 7.9% to just 3.5%. Depending on the circumstances of how the final decision is used this may be judged as more effective. In practice the decision threshold τ would be carefully selected by the system designer to minimise the cost of the decisions and deferrals.

As well as including a deferral option in the fusion rule it is possible to include local deferral in the computations for each of the techniques described. The strategy is to augment the set of possible values for y_i with an additional deferral option so that $y_i = f(x_i) \in \{\omega_1 \cdots \omega_C, \omega_?\}$, where $\omega_?$ represents deferral. The remaining calculations proceed as before treating $\omega_?$ as an additional class label.

2.5 BEHAVIOUR KNOWLEDGE SPACE

If a sufficient database of local decisions pertaining to the same set of entities is available then a look-up table mapping local to fused decisions may be estimated. Each entry in the table corresponds to an n-tuple of local decisions and the relative occurrence of classes in the design data that gave rise to that n-tuple are used to determine the best fused decision. Such a scheme (called *Behaviour Knowledge Space* fusion) requires a large training set to give good performance. It is conceptually very simple and will not be further examined here.

3. Soft Decision Fusion

In the previous section we allowed for the average characteristics of the local decision-makers by calculating the probability of the given (hard) decision under various assumptions. This oversimplifies the decision making behaviour since there will be many different inputs that elicit the same decision. In some cases great confidence may be placed in the decision while at other times the decision may be decidedly unsafe. For example consider the digitised numerals in Figure 2. One might imagine a hypothetical character recognition machine that can correctly classify more than 80% of such instances, yet alone this does not capture the obvious difficulty in classifying the middle row, or the relative ease of classifying the first and last rows. By assigning a confidence measure to each decision (rather than to the decision-maker) it is possible to use such information and to thereby improve fused performance.

Soft decision fusion has found application in two quite distinct areas: classifier combining (see Haykin [13] for example) and multi-sensor fusion [14]. While techniques developed for one area may be readily applied to the other we shall see that the underlying assumptions are really quite different.

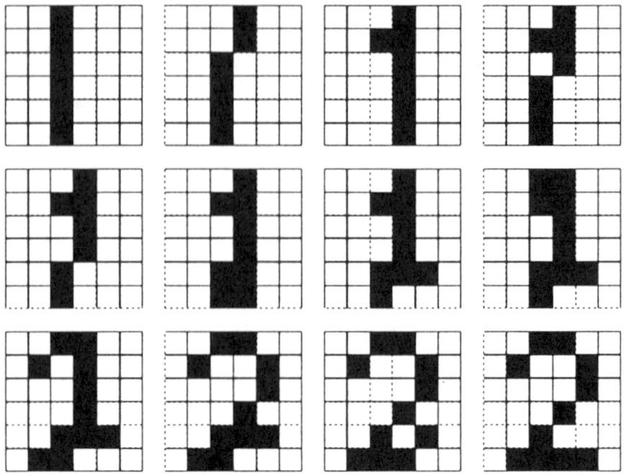

Figure 2. Examples of hand-written numerals from classes ω_1 and ω_2.

3.1 CLASSIFIER COMBINING

We first establish some notation before describing a method of combining the results of multiple soft-decision classifiers. Consider that we wish to determine the conditional probability, $y = P(\omega \mid x)$, that a given measurement, x, was made on an entity of a particular class, ω. Now in general we do not know the functional relationship between x and $P(\omega \mid x)$ but take the practical step of approximating this relationship using a classifier model, m, from the space of all possible classifier models, M, and settings its parameters using a design database, D. By the law of total probability we may write:

$$P(\omega \mid x, D) = \int_{m \in M} dm \quad P(\omega \mid x, D, m) P(m \mid x, D) \tag{17}$$

Realistically we are likely only to have access to a small number, say N, instances of classifier models that will not cover M completely. Nonetheless we may approximate the required conditional probability by summing over these instances:

$$P(\omega \mid x, D) \approx \sum_{i=1}^{N} P(\omega \mid x, D, m_i) P(m_i \mid x, D) \tag{18}$$

That is we sum the outputs of the individual classifier models weighted by the probability that each model is appropriate given the design data, D, and the current measurement, x. This basic formulation makes possible an increase in classification performance by combining multiple classifier models each operating on the same sensor data. It is widely known as *classifier combining* and can be regarded a form of model fusion. This type of fusion is shown diagrammatically in Figure 3.

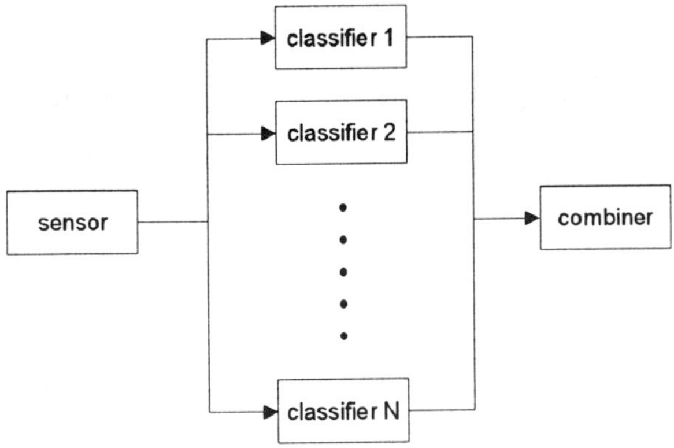

Figure 3. Classifier combination.

Given this basic summation rule it only remains to determine how individual outputs, $y_i = P(\omega \mid x, D, m_i)$, are used and how the weights, $a_i = P(m_i \mid x, D)$, are computed. There are a number of approaches for each. Weighting schemes fall into two broad categories depending on the conditioning used. If the conditioning on both x and D is retained then the method is termed *dynamic* since the weights will vary during execution as x changes. Of most relevance from a distributed fusion viewpoint is the analogous situation when only that part of x which is encapsulated in the output of the classifier model, y, is used. If the conditioning on x is dropped altogether then the weights depend only on the classifier models themselves (and the design data). In this case the weights remain fixed and the method is termed *static*. In either case the weights may be calculated directly or iteratively adapted. We shall only consider non-adaptive weighting schemes.

3.1.1. *Individual outputs*

There are three main options for using the outputs of the individual classifier models; winner-takes-all, rank score and belief score. The winner-takes-all approach quantizes the outputs to binary values with the greatest output being set to unity and all the others set to zero. In the rank score scheme the classes are ordered according to the classifier outputs. The most favoured class (*i.e.* the one corresponding to the greatest output) is given an initial score of C, equal to the number of classes. The second most favoured class is given a score of $C-1$, and so on until the least favoured class, which is given a score of one[2]. These initial scores are then normalised by the factor $\frac{1}{2}C(C+1)$. In belief scoring the outputs are used directly as if they were probabilities. This may require the application of an affine transformation (offset and re-scaling) to ensure that all of the outputs are

[2] Note that sometimes a scoring scheme from C-1 to zero is used. However, the author has found that this strategy may be less robust, particularly when more sophisticated combination rules are employed.

positive and that they sum to unity. In some cases this is an unnecessary step since the combination technique may not make explicit use of these constraints.

It can be seen that of these schemes the winner-takes-all approach retains least information about the relative probability of each class and the belief score retains the most (and potentially all) of this information. In general the belief score is the approach of preference *if the model design is of sufficient quality to warrant it.* If the models are poor (because of inappropriate functional form, sub-optimal parameter estimation or inadequate design data) then one of the other schemes should be considered. We shall use belief scoring in the remaining examples.

3.1.2. *Static Weights*

Static weights may be equal or inversely proportional to the probability of error. Equal weights are the simplest case, with the weights are fixed for all classifier models at:

$$a_i = 1/N \tag{19}$$

Inverse error weights give more credence to classifier models that have a low probability of error, $P_i(E)$. In this case the weights are given by:

$$a_i = \frac{1 - P_i(E)}{\sum_{j=1}^{N} \left(1 - P_j(E)\right)} \tag{20}$$

3.1.3. *Dynamic Weights*

A number of dynamic weighting schemes have been proposed including certainty weighting, entropy weighting and consistency weighting. In all cases the weights are normalised to sum to one:

$$a_i = \frac{\delta_i}{\sum_{j=1}^{N} \delta_j} \tag{21}$$

Certainty weighting sets the weights to be proportional to the difference between the highest and second highest output for each classifier:

$$\delta_i = y_i(1) - y_i(2) \tag{22}$$

(Assuming that y_i has been sorted into descending order).

Like certainty weighting, entropy weighting favours decisive classifier models but assigns weights according to the entropy of the whole output vector:

$$\delta_i = \frac{1}{\sum_{k=1}^{C} \frac{y_i(k)}{\log y_i(k)}} \tag{23}$$

or the entropy squared:

$$\delta_i = \left(\frac{1}{\sum_{k=1}^{C} \frac{y_i(k)}{\log y_i(k)}} \right)^2 \tag{24}$$

Consistency weighting, on the other hand, gives less weight to classifier models whose output is different from the others:

$$\delta_i = \frac{1}{\sum_{j=1}^{N} \sum_{k=1}^{C} \left(y_i(k) - y_j(k) \right)^2} \tag{25}$$

3.1.4 *Computer Experiments*

We now compare the classifier combination weighting schemes described above by testing them on a common pattern identification task. The synthetic task comprised four, equally likely classes with the data from each being drawn from a bivariate Gaussian distribution. The configuration of the classes and the spread of the data are shown in Figure 4, derived from the character recognition problem introduced earlier (Figure 2).

It can be seen from Figure 4 that given sufficient data a quadratic classifier using separate covariance matrices for each class should provide a good classification performance. For illustrative purposes, however, we shall make the task more difficult.

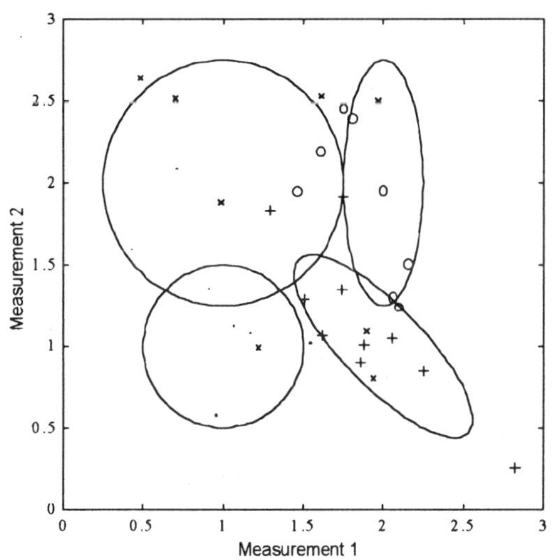

Figure 4. The configuration of the four classes in the classifier combination experiments.

Firstly we shall severely restrict the number of samples in the design database to just ten from each class (the actual samples are also shown in Figure 2). Secondly we shall train five quadratic classifiers with different covariance matrix properties but we exclude the case of separate, full covariance matrices already identified as the most suitable. The individual classifiers and the classifier combination algorithms were evaluated on a separate test database of 1,000 patterns from each class. The results are summarised in Table 6 and Table 7, showing comparable performance for all methods.

The tables show a statistically significant reduction in error rate over the individual classifiers. In this case all of the weighting schemes used gave similar results.

TABLE 6. The performance of the individual Quadratic classifiers on the test database.

Covariance matrix properties	Test Error rate
Shared, spherical	33.5%
Shared, diagonal	34.4%
Shared, full	35.2%
Spherical	35.2%
Diagonal	33.1%

TABLE 7. The performance of the classifier combination methods on the test database.

Weighting scheme	Test error rate
Equal	31.6%
Inverse error	31.7%
Certainty	31.5%
Entropy	31.6%
Entropy^2	31.6%
Consistency	31.6%

3.2 MULTI-SENSOR FUSION

While classifier combining concentrated on fusing different models of the same type of data (e.g. multiple, identical sensors), we now return to the idea of fusing models of different data (e.g. different sensor types). We take a similar approach to that outlined in section 2 as illustrated in Figure 5. Consider the same recognition task of C target classes, $\omega \in \Omega = \{\omega_1, \omega_2, \cdots \omega_C\}$ with prior distribution, $P(\Omega) = \{P(\omega_1), P(\omega_2), \cdots P(\omega_C)\}$. The fusion system comprises a set of N sensors, which collectively produce sensor data $X = \{x_1 \cdots x_N\}$. Each sensor is provided with a local classifier that produces a conditional probability, $y_i = P(\omega \mid x_i)$, conditioned only on the local sensor data.

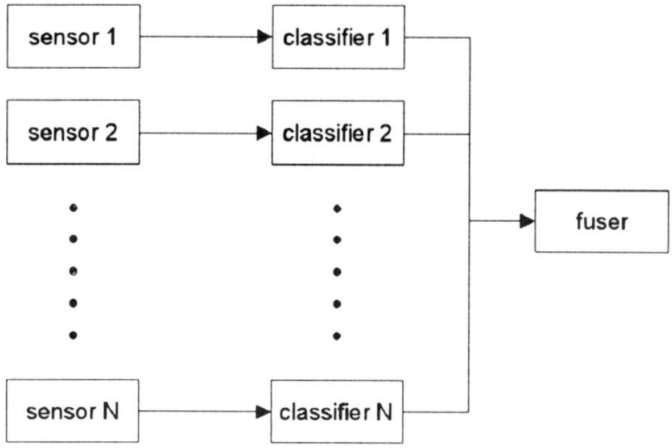

Figure 5. Multi-sensor fusion.

3.2.1. *Product Rule*

In the simplest case the x_i are conditionally independent, that is given the class information the joint probability of the data is equal to the product of the separate probabilities:

$$P(X \mid \omega) = \prod_{i=1}^{N} P(x_i \mid \omega) \tag{26}$$

This is not the same assumption to that made in section 2 when we assumed the decisions to be conditionally independent. Note also that this does not, in general, imply that the x_i themselves are independent since (for example) the means of the distributions for each class can be highly correlated across sensors (see Figure 6).

As long as the measurements are conditionally independent then the local estimates, y_i, may be combined in a straightforward manner using repeated application of Bayes' rule since:

$$P(\omega \mid X) = \frac{P(X \mid \omega)P(\omega)}{P(X)} \tag{27}$$

and from the conditional independence assumption contained in equation (26) we obtain:

$$P(\omega \mid X) = \frac{\prod_{i=1}^{N} P(x_i \mid \omega) \quad P(\omega)}{P(X)} \tag{28}$$

Then applying Bayes' rule again we have:

$$P(\omega \mid X) = \frac{\prod_{i=1}^{N} P(\omega \mid x_i)P(x_i)/P(\omega) \quad P(\omega)}{P(X)} \tag{29}$$

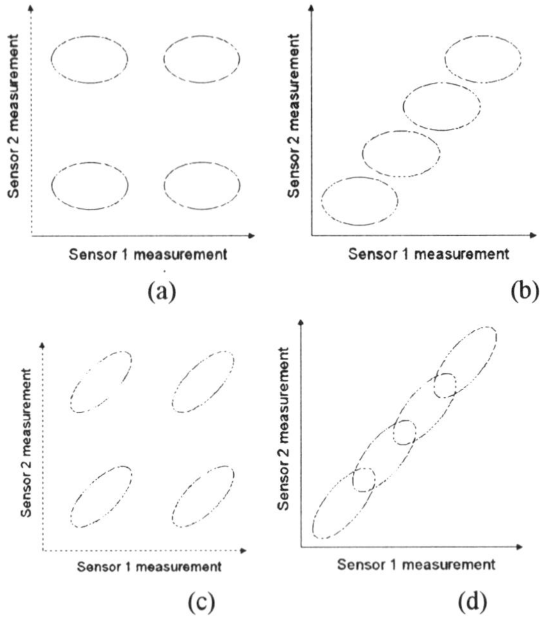

Figure 6. In (a) the measurements are both independent and conditionally independent whereas in (b) the measurements remain conditionally independent but are unconditionally dependent. In both (c) and (d) the measurements are conditionally dependent with (c) displaying almost no unconditional dependence and (d) strong unconditional dependence.

After cancelling the class priors and isolating the terms independent of the class we finally have:

$$P(\omega \mid X) = \frac{\prod_{i=1}^{N} P(\omega \mid x_i)}{P(\omega)^{N-1}} \cdot \frac{\prod_{i=1}^{N} P(x_i)}{P(X)} \tag{30}$$

The first term is just the product of the local conditional probabilities appropriately weighted by the class prior. The second term is independent of the class and acts merely as a normalising constant. It may therefore be omitted from calculations as long as the fused soft decisions are subsequently normalised.

By way of an example consider a three-sensor system that outputs the following probability vectors corresponding to the confidence that the entity under scrutiny is one of three classes:

$$P(\omega \mid x_1) = \{0.2, \quad 0.3, \quad 0.5\}$$
$$P(\omega \mid x_2) = \{0.5, \quad 0.3, \quad 0.2\}$$
$$P(\omega \mid x_3) = \{0.3, \quad 0.4, \quad 0.3\}$$

We may then proceed to calculate:

$$P(\omega \mid X) \propto \frac{\prod_{i=1}^{N} P(\omega \mid x_i)}{P(\omega)^{N-1}} = \frac{\{0.2\times 0.5\times 0.3,\quad 0.3\times 0.3\times 0.4,\quad 0.5\times 0.2\times 0.3\}}{\{\frac{1}{3},\ \frac{1}{3},\ \frac{1}{3}\}}$$

$$= \{3\times 0.030,\quad 3\times 0.036,\quad 3\times 0.030\} \qquad (31)$$

$$= \{0.090,\quad 0.108,\quad 0.090\}$$

Renormalising gives the fused output probabilities of $P(\omega_1 \mid X) \approx 0.31$, $P(\omega_2 \mid X) \approx 0.38$ and $P(\omega_1 \mid X) \approx 0.31$. Note that using this scheme class ω_2 would be selected as the fused decision despite the fact that only the third sensor considered it the most likely class (and then not very confidently).

This fusion rule is widely used and warrants a little further investigation. Consider the simplest situation – a two-class problem with two sensors. Figure 7 shows the fused probability of one of the classes, as a function of the separate probabilities for that class.

The amplification of confidence in the fused decision is more clearly demonstrated by assuming that the sensors produce equal probabilities. Figure 8 shows the fused probability versus the set of identical individual probabilities for 2, 3, 4 and 5 sensors.

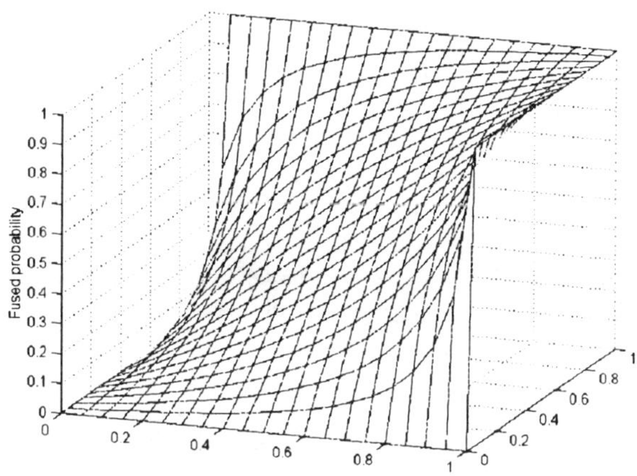

Figure 7. The amplification of confidence using the product rule. On the horizontal axes are the individual class probabilities. The fused probability is plotted on the vertical axis.

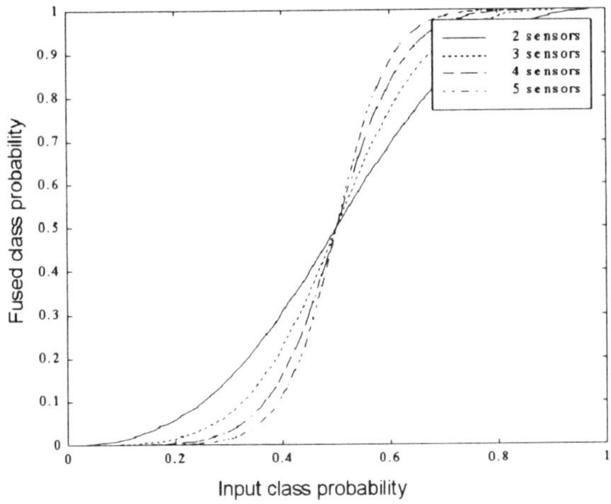

Figure 8. The amplification of confidence using the product rule for identical sensors. On the horizontal axis is the individual class probability which is the same for all sensors. The fused probability is plotted on the vertical axis.

Therefore a relatively small change in the individual probabilities can result in a large change in the fused probability. Particularly when one of the sensors outputs a class probability near zero. This is called the *veto effect* and we shall return to it later.

3.2.2 *Computer Experiment*

The benefits of fusing soft decisions rather than hard decisions has been widely reported (see Buede and Waltz [13] for instance). Here we illustrate the advantages of soft decision fusion using a simple synthetic problem. The problem comprises the classification of data from two classes. The data from each sensor for each class is distributed according to a normal distribution and the means are the same for each sensor. 10,000 instances from each class were used to determine the individual and fused performances. Table 8 shows the fused performance using MAP decision fusion and the product rule for soft decisions, as the separation between the distributions and the number of sensors are varied.

TABLE 8. The performance of hard and soft decision fusion rules for synthetic problems of varying difficulty and numbers of sensors.

		3 sensors		5 sensors		7 sensors	
Separation	Individual performance	MAP decision fusion	Product rule	MAP decision fusion	Product rule	MAP decision fusion	Product rule
0.5	60%	67%	69%	71%	75%	75%	79%
1.0	70%	81%	85%	87%	91%	90%	94%
1.5	77%	91%	93%	95%	97%	97%	99%
2.0	84%	96%	97%	98%	99%	99%	100%

Figure 9 shows a graph of the hard and soft decision fusion for three sensors as problem difficulty is decreased. It can be seen that soft decision fusion using the product rule always outperforms hard decision fusion (at the cost of having to transmit more information).

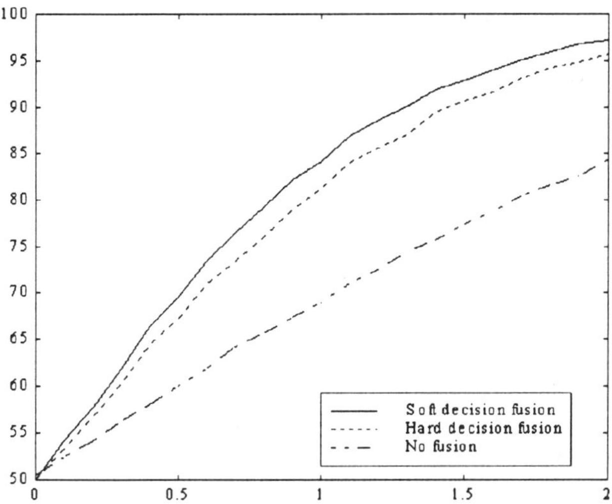

Figure 9. On the vertical axis the performance as percentage of test measurements correctly classified as the separation between the means of the two classes increases. Shown is the performance for a single sensor plus hard decision fusion and soft decision fusion using the product rule for three sensors.

3.2.3. *Relaxing the independence assumption*

The independence assumption represented by equation (26) is rather severe and may not be justified in practice. By taking the logarithm of equation (30) and dropping the normalising term, we may note the similarity to the ensemble-averaging rule for classifier combination:

$$\log P(\omega \mid X) \propto \sum_{i=1}^{N} \log P(\omega \mid x_i) \quad -(N-1)\log P(\omega) \tag{32}$$

We may then conjecture that a weighted sum may provide us with a means of allowing for different sets of sensors. For instance, if all the weights were unity then we would be back with our independent fusion rule and if the sensors were completely correlated then the weights could be set to $\frac{1}{N}$. Of course, the weights applied to the logarithm of the class probabilities manifest themselves as exponents in the original formulation, which we shall refer to as the generalised product rule [16]:

$$P(\omega \mid X) \propto \frac{\prod_{i=1}^{N} P(\omega \mid x_i)^{a_i}}{P(\omega)^{N-1}} \tag{33}$$

For the moment let us focus on the two-class problem as this is by far the easiest to derive a rule for. The log probabilities of the two classes are:

$$\log P(\omega_1 \mid X) = k + \sum_{i=1}^{N} a_i \log P(\omega_1 \mid x_i) \quad - \log P(\omega_1) \tag{34}$$

and correspondingly:

$$\log P(\omega_2 \mid X) = k + \sum_{i=1}^{N} a_i \log P(\omega_2 \mid x_i) \quad - \log P(\omega_2) \tag{35}$$

Where in both cases k is the same constant to ensure that the log probabilities sum to zero. Subtracting equation (35) from equation (34) we obtain:

$$\log \frac{P(\omega_1 \mid X)}{P(\omega_2 \mid X)} = \sum_{i=1}^{N} a_i \log \frac{P(\omega_1 \mid x_i)}{P(\omega_2 \mid x_i)} \quad - \log \frac{P(\omega_1)}{P(\omega_2)} \tag{36}$$

Now to obtain a correct fused decision we only require that this quantity be positive for examples of ω_1 and negative for ω_2. Since we either know, or can estimate the class priors we only need a labelled set of design data to determine a suitable set of weights. This design data should comprise the outputs of the local classifiers operating on the same instances of X (plus, of course, the identity of the correct class in each case).

The a_i may be determined using one of two methods. An iterative error-correcting learning algorithm along the following lines may be used. Start the weights, a_i, at some convenient value (unity being a reasonable initial value for all the weights since this is equivalent to assuming conditional independence). Repeatedly select instances from the design dataset and adjust the weights. If the sign of the RHS is in agreement with the class label (that is no error) then the weights remain unchanged. If the sign of the RHS is negative and the example is of class ω_1 then the weights are increased in proportion to the logarithm they are multiplying. An equivalent decrease is used to reduce the value of the RHS when required[3]. Alternatively, the weights may be found using least squares regression with suitable values selected for the LHS of equation (36).

If there are more than two classes then the same, simple formulation is not possible. In this case two approaches may be taken, both involve decomposing the C-class problem into a set of $\frac{1}{2}C(C-1)$ two-class problems. In the first approach the learning rule is applied for each labelled input pattern using the set of $(C-1)$ equations that involve that class. The resulting weights are a compromise for all of the classes. The second approach is to determine one set of weights for each of the two-class problems and to treat the resulting classifiers as a set of specialists. The output of these specialists may be combined at run time using one of the techniques from section 3.1.

[3] Some readers will recognise this simple rule as an example of the perceptron learning algorithm adopted by Minsky and Papert in the 1960's. It carries with it a proof that if the classes are linearly separable using this formulation then a set of appropriate weights will be found in finite time.

3.2.4 *Computer experiments.*

To illustrate the usefulness of the generalised product rule we apply it to some simple synthetic data. The data comprised examples from two equally likely classes. Data from each class was drawn from a bivariate Gaussian distribution. The standard deviation along the major and minor axes was the same for each class (being 2.0 and 0.1 respectively). The mean for class ω_1 was at (1,1) and the mean of class ω_2 was at (2, 1½). A number of different experiments were carried out with the orientation, θ, of the major axes varied from one experiment to the next. In each case θ was the same for both classes). The design dataset was composed of 1,000 instances of each class and the test dataset of 10,000 instances of each class. Figure 10 shows the configuration of the data for one of the experiments with $\theta = 50°$.

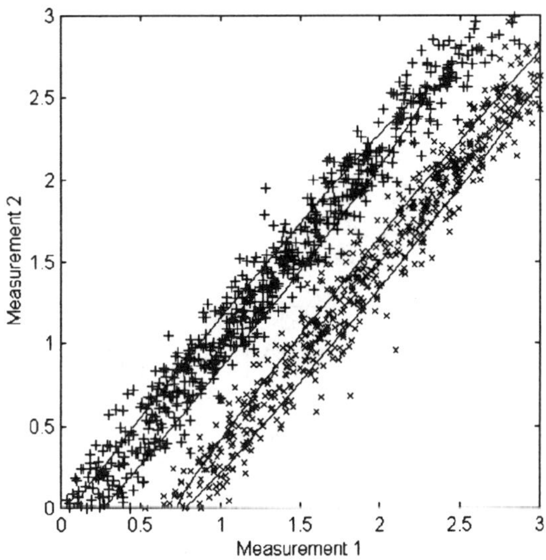

Figure 10. The design data for the experiments using the generalised product rule. Note that separate classifiers were designed for the x and y values – this joint measurement space was not directly available during the experiments.

Two separate classifiers were designed for the x and y values and the outputs of these classifiers were fused using both the product rule (PR) and the generalised product rule (GPR). Table 9 shows the classification performance of separate and fused classifiers as the angle θ was varied.

It can be clearly seen that if the data provided by the different sensors is not conditionally independent then the use of the product rule may be quite poor (in fact it can be worse than the best individual sensor). The use of the generalised product rule increases performance in some circumstances – particularly, as in the case here, where the measurement covariance is shared between classes.

TABLE 9. The performance of individual and fused classifiers on a synthetic problem of varying degree of correlation.

θ	x	y	PR	GPR
0°	62%	99%	99%	99%
10°	61%	76%	72%	92%
20°	59%	63%	62%	66%
30°	60%	60%	60%	61%
40°	62%	57%	61%	73%
50°	66%	57%	64%	93%
60°	70%	57%	67%	100%
70°	76%	56%	73%	100%
80°	92%	55%	91%	100%
90°	100%	54%	100%	100%
100°	91%	56%	93%	100%
110°	77%	54%	82%	100%
120°	69%	57%	79%	100%
130°	67%	57%	78%	100%
140°	62%	58%	83%	100%
150°	63%	61%	97%	100%
160°	60%	63%	96%	99%
170°	58%	76%	88%	100%
180°	59%	99%	99%	99%

3.2.5. *The Veto Effect*

At the end of section 3.2.1 we alluded to the sensitivity of the product rule to inaccuracies in the individual class probabilities. We shall illustrate the issue with a specific example. Consider a hypothetical fusion system with six sensors, which attempts to determine the class of an entity. The two classes are equally likely *a priori*. Assume that, for a particular instance of class ω_1, five of the sensors output a probability of 60% for this class, but that the sixth sensor outputs only 10%. From equation (30) we may calculate the fused probability of the two classes thus:

$$P(\omega_1 \mid X) \propto \frac{0.6 \times 0.6 \times 0.6 \times 0.6 \times 0.6 \times 0.1}{0.5^5} \approx 0.249$$

$$P(\omega_2 \mid X) \propto \frac{0.4 \times 0.4 \times 0.4 \times 0.4 \times 0.4 \times 0.9}{0.5^5} \approx 0.295$$

Normalising these quantities gives a fused decision of ω_2 with associated probability 54%. It can be seen that the high confidence (in the wrong class) shown by the sixth sensor has overridden the decisions of the remaining classifiers. The main reason why a classifier might output a highly confident wrong decision is that it has been presented with atypical data – either because of inherent randomness in the signal, sensor failure or because of intentional data quality corruption (for example by spoofing or jamming).

The propagation of overconfident wrong decisions through a data fusion system is known as the veto effect. In such situations the data fusion maxim "you can't fool all of the sensors all of the time" may not hold.

Several heuristics have been suggested for dealing with the veto effect. In the distributed approach the responsibility is placed on the local classifiers, each of which has to ensure that its output is not falsely overconfident. In the centralised approach the local classifiers are allowed to produce soft decisions as normal but the fusion centre applies a correction to the confidence estimates before fusion (see Bedworth [17]).

Distributed Veto Correction. If the individual classifiers use maximum likelihood parameters that have been determined using a small amount of data then the classifier will not usually generalise well and will often display overconfidence. In the case of Gaussian distributions a Bayesian approach results in a straightforward correction. Instead of using the sample mean and sample variance we integrate over the unknown population mean and population variance:

$$P(x_i \mid D) = \int_{-\infty}^{\infty} d\mu \int_{0}^{\infty} d\sigma^2 \; P(x_i \mid \mu, \sigma^2) P(\mu, \sigma^2 \mid D) \tag{37}$$

It can be shown that the sample mean, $\hat{\mu}$, the sample variance, $\hat{\sigma}^2$, and the size of the sample, N, are sufficient to encapsulate the information in the design data, D. Furthermore the integral is analytically tractable and results in the "Student" t-distribution:

$$P(x_i \mid \hat{\mu}, \hat{\sigma}^2, N) = \frac{\Gamma\left(\frac{N}{2}\right)}{\hat{\sigma}\sqrt{(N-1)\pi}\;\Gamma\left(\frac{N-1}{2}\right)} \cdot \left(\frac{(x_i - \hat{\mu})^2}{(N-1)\hat{\sigma}^2} + 1\right)^{\frac{-N}{2}} \tag{38}$$

Note that the t-distribution has much heavier tails than the Gaussian and the asymptotic behaviour of a ratio of t-distributions is quite different to that for Gaussians. Figure 11 shows the likelihood and the probability of class ω_1 for a simple case.

Note that the class probability for outlying data for the t-distribution nears 0.5 whereas the same quantity using maximum likelihood parameters is either zero or one.

Centralised Veto Correction. If the individual classifiers continue to output overconfident class probability estimates (as for example can be seen in figure 11c) then the fusion centre itself must make an appropriate correction. Of course, it cannot reproduce the correction that could have been made by the local decision-maker since it does not have access to the data, x_i, but only the soft decisions, y_i.

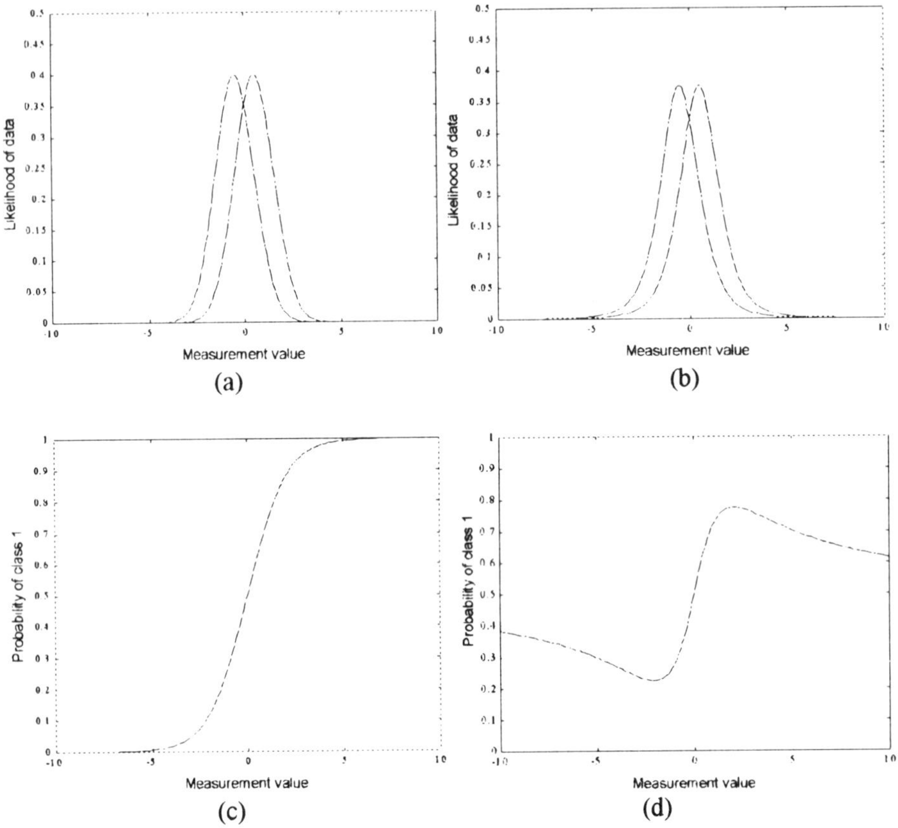

Figure 11. The use of Gaussian distributions and t-distributions for classification. In (a) is shown the data likelihood for the maximum likelihood Gaussians. The corresponding probability of one of the classes is shown in (c). Two t-distributions are shown in (b), which yield less extreme class probabilities for outlying data (d).

Figure 12 shows a plot of the overconfident soft decisions (from Figure 11c) versus the corrected soft decisions (from Figure 11d). It can be seen that over much of the range of values there is little difference.

A simple means of correction that can take place at the fusion centre is therefore to limit the values that the class probabilities may take:

$$y_i' = \begin{cases} \lambda_i & \text{if} \quad y_i < \lambda_i \\ 1-\lambda_i & \text{if} \quad y_i > 1-\lambda_i \\ y_i \quad \text{if} & \lambda_i \le y_i \le 1-\lambda_i \end{cases} \tag{39}$$

The value of the limit, λ_i, for each local decision-maker may be determined adaptively using a suitable error correcting learning algorithm. However, in practice it seems that setting $\lambda_i = \lambda$ using trial and error is almost as good a strategy. A typical value for λ would be 0.2. For illustration three local decision-makers were designed using just 10 samples from two equally likely classes. The distribution for class ω_1 had a mean of zero and for class ω_2 a mean of one. Both class distributions had unit variance. Performance was assessed using a test database of 10,000 samples with the local decisions were fused using the product rule. By limiting the confidence of the local decisions using $\lambda = 0.2$ the performance increased from 74% to 78%.

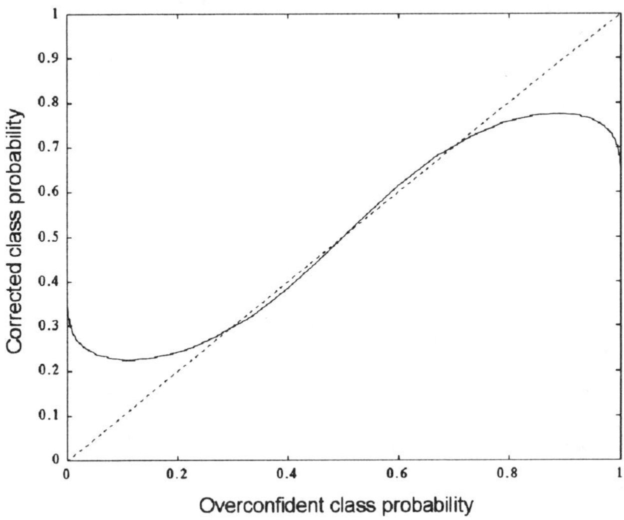

Figure 12. The corrected class probability from figure 11d as a function of the uncorrected class probability of figure 11c.

4. Summary

Decision fusion algorithms can be broadly categorised as making use of class labels only (hard decisions) or class labels together with associated confidence estimates (soft decisions). The fusion of hard decisions has the advantage that it is relatively straightforward to implement, requires minimal communication bandwidth and in many cases provides perfectly adequate classification performance. The fusion of soft decisions generally yields higher classification performance at the expense of increased communication bandwidth. This chapter has detailed three basic algorithms for fusing hard decisions including majority voting, weighted voting and maximum *a posteriori* fusion. The ability to defer a fused decision was also covered. Methods for soft decision fusion either for classifier combining using weighted averaging or multi-sensor fusion using the product rule were examined. The product rule was generalised to allow for sensors that are not conditionally independent. Finally the veto effect and a simple means of correcting for it when using the product rule was described.

5. Acknowledgements

The author thanks Prof. David Lowe from the University of Aston, UK for commenting on the draft typescript.

6. References

1. Bedworth, M.D. (in press) *Data Fusion Engineering*, Springer-Verlag, London.
2. Harris, C.J., Bailey, A. and Dodd, T.J. (1998) Multi-sensor Data Fusion in Defence and Aerospace, *The Aeronautical Journal*, May issue.
3. Waltz, E.L. and Buede, D.M. (1986) Data Fusion and Decision Support for Command and Control, *IEEE Transactions on systems, Man and Cybernetics, Volume* **16**, *Number 6*, pp. 865-879.
4. Debon, R., Solaiman, B. Cauvin, J.-M., Peyronny, L. and Roux, C. (1999) Aorta Detection in Ultrasound Medical Image Sequences using Hough Transform and Data Fusion, *Proceedings of FUSION'99, Sunnyvale, CA, USA, Omnipress*, pp. 59-66.
5. Hansen, R.J., Hall, D.L. and Kurtz, S.K. (1995) A New Approach to the Challenge of Machinery Prognostics, *Transactions of the ASME, Journal of Engineering for Gas Turbines and Power, Volume* **1117**, pp. 320-325.
6. Abdulgahfour, M., Goddard, J. and Abidi, M.A. (1990) Non-determinstic Approaches in Data Fusion – A Review, *Proceedings of AeroSense'90, SPIE Volume* **1383**, *Sensor Fusion III: 3-D Perception and Recognition*, pp. 596-610.
7. Nahin, P.J. and Pokoski, J.L. (1980) NCTR Plus Sensor Fusion Equals IFFN, *IEEE Transactions on Aerospace and Electronic Systems, Volume* **16**, *Number 3*, pp. 320-337.
8. Tenney, R.T. and Sandell, N.R. (1981) Detection with Distributed Sensors", *IEEE Transactions on Aerospace and Electronic Systems, Volume* **17**, *Number 4*, pp. 501-510.
9. Chair, Z., and Varshney, P.K. (1986) Optimal Data Fusion in Multiple Sensor Detection Systems, *IEEE Transactions on Aerospace and Electronic Systems, Volume* **22**, *Number 1*, pp. 98-101.
10. Chair, Z., and Varshney, P.K. (1988) Distributed Bayesian Hypothesis Testing with Distributed Data Fusion, *IEEE Transactions on Systems, Man and Cybernetics, Volume* **18**, *Number 5*, pp. 695-699.
11. Kam, M., Zhu Q. and Gray, W. (1992) Optimal Data Fusion of Correlated Local Decisions in Multiple Sensor Detection Systems, *IEEE Transactions on Aerospace and Electronic Systems, Volume* **28**, *Number 3*, pp. 916-919.
12. Chang, W. and Kam, M. (1994) Asynchronous Distributed Detection, *IEEE Transactions on Aerospace and Electronic Systems, Volume* **30**, *number 3*, pp. 818-826.
13. Haykin, S. (1999) Neural Networks: A Comprehensive Foundation 2nd edition, *Prentice-Hall.*
14. Demirbas, K. (1988) Maximum A Posteriori Approach to Object Recognition with Distributed Sensors, *IEEE Transactions on Aerospace and Electronic Systems, Volume* **24**, *Number 3*, pp. 309-313.
15. Buede, D.M. and Waltz, E.L. (1989) Benefits of Soft Sensors and Probabilistic Fusion, *Proc. AeroSense'89, SPIE* **1096** – *Signal and Data Processing of Small Targets*, pp. 309-320.
16. Heading, A.J.R. and Bedworth, M.D. (1991) Data Fusion for Object Classification, *Proc. IEEE Systems, Man and Cybernetics*, pp. 827-830.
17. Bedworth, M.D. (1998) Less Certain, More Infallible, *Proceedings of FUSION'98, Las Vegas, NV, USA, CSREA Press*, pp. 572-580.

REEXAMINING DATA FUSION PROCESSING AT LEVELS 2, 3, AND 4

J. LLINAS
Center for Multisource Information Fusion
Department of Industrial Engineering
State University of New York at Buffalo--Amherst Campus
342 Bell Hall
Buffalo, New York 14260
USA
llinas@eng.buffalo.edu

1. Introduction

1.1 DEFINITIONS, DEFINITIONS, DEFINITIONS.....

Historically, one of the impediments to the maturation of the field if data and information fusion (DF/IF) was its lack of a standard taxonomy (classification conventions) or nomenclature. (naming conventions). The community has made great strides in these areas for Level 1 (L1) processing but has not addressed the terminology associated with Levels 2 and 3 (L2, L3) in nearly so rigorous a fashion. So one of the first topics that should be addressed as we contemplate formalization of technical approaches is that of language. The following are simple extractions from [1] but the reader is referred to [2] for a collection of some 24 different definitions of the phrase "situational awareness" drawn strictly from military or aerospace-type literature. The taxonomy of the US Joint Directors of Laboratories Data Fusion Group ("JDL/DFG"), [3] the primary US data fusion technology oversight organization, is another useful reference for fusion-related terminology.

Situation: the way in which something is placed in relation to its surroundings; position with respect to conditions and circumstances; the sum total of internal and external stimuli that act upon an organism within a given time interval; relative position or combination of circumstances at a certain moment.....

Threat: an indication of something impending; an expression of intention to inflict evil, injury, or damage....

Impact: the force of impression of one thing on another; an impelling or compelling effect.

Words are important—they are the means by which we convey concepts to one another. These words, above, will motivate a large part of the discussion in this chapter, and so we offer these dictionary meanings of them in their abstract context, for readers to contemplate.

But our interpretations are neither abstract nor context-free; in fact, they are totally contextually-dependent. In the modern (post-Soviet) world, this contextual dependency

A. K. Hyder et al. (eds.), Multisensor Fusion, 125–161.

will greatly complicate the case-specific meanings of the above terms, largely because there are more cases, i.e. more problem variability.

This contextual dependency can in fact be appreciated when examining a Soviet-based model of dynamic decision-making as shown in Figure 1. This model shows that there must be specified dependencies and interrelationships among "situational elements", "decision elements", and mission goals and objectives. That is, what comprises important "elements" or data-items that together form an information vector that we call a "situation" or "situational abstraction" are those informational components that can be justified on the basis of their relationship to (i.e. supporting the formulation of) certain decision elements that in turn are justified as helpful toward the satisfaction of mission goals or objectives.

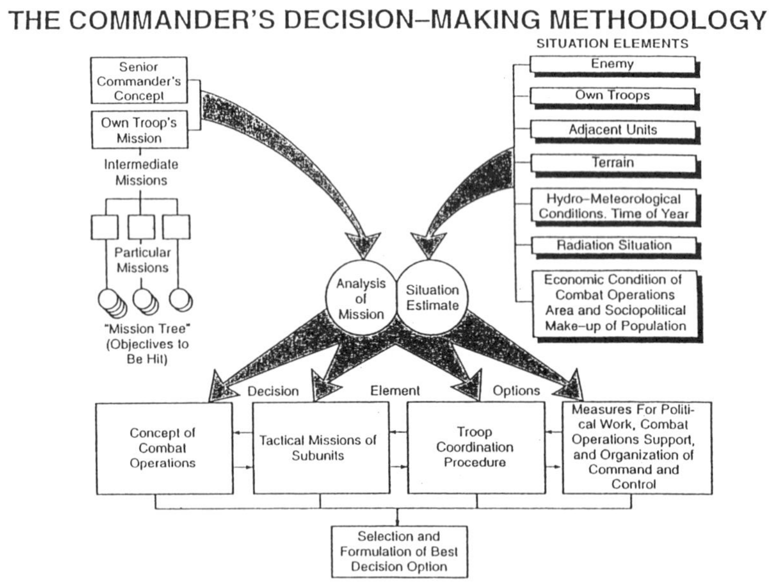

Figure 1. The Commander's Decision-Making Methodology: Relationships Among Situation Elements, Decision Elements, and Mission Goals and Objectives

This same interpretation can be extended, to a degree, toward the conceptual understanding of the terms "Threat" and "Impact". However, these terms have a particular and complicating distinction from the term "situation"—they attempt to be *predictive* or to forecast a *potential* future condition, and in that sense are notions much more difficult to estimate. It could be argued that "threat" is a special type of situation (a "threatening" situation) or that the current situation will have some "impact" on something of interest (presumably friendly forces). One interpretation of the difference between situation and threat notes that the notion of threat has special components, involving the assessment of "lethality" (an ability to do harm, as per the dictionary definition), "intent", and "opportunity"; all three must be present to determine that a threat condition exists. Each also has a particular meaning in the context of the 25 missions listed above; what a threat is, and what its lethality component is, are considerably different in Information Warfare and in, say, Chemical/Biological Warfare.

1.2 RELATION TO LEVEL 1 PROCESSING

The Level 1 Object Refinement Data Fusion processes of the JDL model can be thought of as producing what a mathematician might call a "labeled set" for which the "labels" reflect the improved, fused estimates of *kinematic parameters* and *identity-related parameters* (and an identity estimate or "declaration" as well) for a set of hypothesized objects considered to be present as a result of the multiple measurements and observations. This information product, however, is free of contextual interpretation in the sense of the meaning of these data in the framework of the general problem being analyzed. Typically, drawing contextual meaning from this labeled set requires further analysis steps which: (1) attempt to cluster the objects into functionally significant groups--i.e., to aggregate the data, (2) identify events and activities of interest--which typically requires a temporally-based interpretation, and (3) assess the impact of the "environment" of the problem on these interpretations, where environment can mean weather, terrain, or "rules of behavior" or other constraints that may be applicable to the particular domain (e.g., so-called "rules of engagement" for military problems). Collectively, these analytical and estimation-type processes result in a "situation assessment", an informational product of considerably higher abstraction than the very quantitative, algorithmically-derived position and identity estimates of Level 1 operations. Figure 2 attempts to convey these distinctions where it can be seen that the Level 1 processing provides the "labels" as shown, whereas the focus of Level 2 processing is to "interpret" those results in a contextual framework.

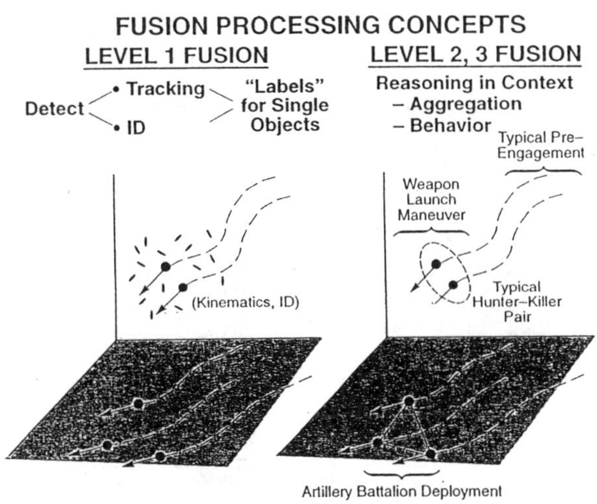

Figure 2. Processing and Product Notions for Data Fusion Levels 1, 2, and 3

Figure 3 shows another notional interpretation of the nature of the L1 to L2, L3 relationship. Here, we see the L1 product, the "labeled set" serving as a *pointer* into the multi-dimensional domain of space and kinematics, with these pointers allowing "location" of this point into various a priori or dynamic data bases such as order of battle, weather, and tactics data (or knowledge) bases. This then notionally achieves the purpose of "putting the L1 product in context", i.e., the context associated with these different data and knowledge sets. The figure indicates that processes are "spawned" i.e., brought forward or instantiated, from these kinematic and ID pointers, meaning that these parameters place the associated L1 object into the appropriate context in the software-based reasoning processes for L2 or L3. Now then ideally there would also be an interplay (feedback/feedforward) between L1 and L2, L3 processing operations but we are first addressing the basic concepts here in a feed-forward sense. Consistent and rigorously-defined definitions of a "situation" or a "situation assessment" necessary to support automated implementations of a situation-estimating process are difficult to find, as has been noted above. By and large, these definitions all relate to the idea of "knowing what's going on", i.e. an analysis or estimation goal to hypothesize and describe a collection of behaviors in the real world, and their significance to the observer. In this context, Level 3 fusion processes then are those associated with "special" situations of interest, such as threat conditions in the military case, or life-threatening medical conditions in the medical-application case, etc. Thus, the fusion processes applied at both Levels 2 and 3 are similar in nature, and produce similar results with special but subtle distinctions between the two. Figure 4 shows the primary functional operations for Level 2 processing within the JDL process model framework. It can be seen that the primary functions are to aggregate information both on objects and behaviors, execute a contextual interpretation, and to do this within a "multi-perspective" approach. The multi-perspective idea is that, ideally, these interpretations would be done (by the human and automated DF processes working together) not only from the traditional view of the "blue force" commander and his mission goals but also from the point of view of the "red force" commander (i.e., as if he had the same information available), as well as in the "white" context which judges the effects of environmental factors on interpretation.

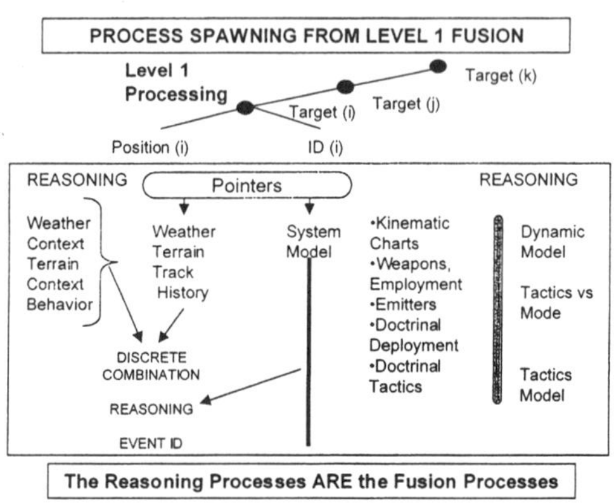

Figure 3. Process "Sprawning" from Level 1 fusion

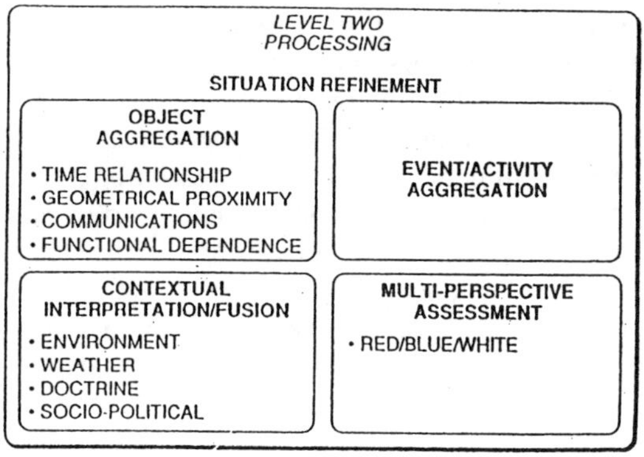

Figure 4. Level 2 Processing Functions

1.3 NATURE OF L2, L3 PROCESSING

These automated L2, L3 Data Fusion processes are those usually called "knowledge-based" or symbolic methods. All such methods attempt to encode applicable human knowledge and inferencing processes into "adequate" representations in software. This knowledge acquisition process has been characterized as "situational abstraction". (These models are not neuron-firing-by-neuron-firing models; using a variety of knowledge representation techniques and inference control processes drawn from the fields of artificial intelligence and automated reasoning, they capture the essence of an interpretative or diagnostic procedure.) It should be noted however that these processes are often accompanied by additional algorithmic i.e., numeric Data Fusion techniques to support the reasoning processes, and so these Level 2 and 3 operations are most generally of a hybrid nature in this symbolic-numeric sense. Broadly speaking, the data and information fusion processes occurring at these Levels happen as a result of defining the logical, temporal, functional, and various other *dependency-constituency relationships* in the (multiple) sets of data through the process of encoding the pertinent human reasoning into software. That is, the fusion processes here are *embodied in the data and information relationships* in the human knowledge encoded into software. Note that such knowledge represents "patterns of *expected* behavior", and that strategies to accommodate unexpected behaviors, sometimes employing so-called "meta-knowledge", knowledge about the reasoning process itself, or "slack" in the parametric data within the models, are typically required[1]. Real-time estimation of situations thus occurs, in essence, by fitting the multiple data streams to these pre-defined behavioral expectation patterns, in what is essentially a deductively-based approach. As might be imagined, efforts to implement such automated processes in a reliable, robust fashion for complex real-world problems is a very difficult process indeed.

Whether for defense or industrial or other applications, what drives the complexity of automated situation assessment process development are the collective quality-related attributes of the data, information, and models available. Just as Level 1 kinematic

[1] This chapter will not address the issue of integrating learning-based approaches, including exploitation of data mining methods, into the L2, L3 processing framework, but these methods should not be ignored.

estimation techniques require valid statistical models of observation noise processes to deal with those specific uncertainties, encoding of the much more complex and abstract reasoning processes at Levels 2 and 3 requires equal attention to the issue of uncertainty both in the broad mix of information to be fused, and in the applied knowledge itself. Additional complexities stem from the fact that these reasoning processes depend on the availability, accuracy, completeness, etc. of considerable a priori data base information (e.g. to define the "environmental" conditions necessary to support the reasoning operations). In fact, a distinguishing feature of the Level 2 and 3 fusion processes is their dependency on such extensive data base information. While military problems have the special concerns in reasoning about a hostile adversary (e.g. concerns about deception, concealment, psychological warfare, etc.), non-military problems can also have quite complex noise backgrounds, poor sensing and instrumentation, and weakly-understood behaviors, among other complicating factors.

2. A Framework for Situational Understanding

2.1 CURRENT-DAY MILITARY MISSIONS

Military organizations around the world are wrestling with the definition of new charters and responsibilities, the essence of which is embodied in new “missions”. In the United States, at least three broad categorizations of new mission types can be enumerated:

(a) Operations Other Than War (OOTW)
(b) Gray Area Phenomena (GAP)
(c) Asymmetric Threats (AT)

Lists of the types of missions or problems associated with each of these broad categories are shown in Tables 1-3.

TABLE 1. Operations Other Than War (OOTW)

Non-Combatant Evacuation Operations
Arms Control
Support to Domestic Civil Authorities
Humanitarian Assistance and Disaster Relief
Security Assistance
Nation Assistance
Support to Counter-Drug Operations
Combating Terrorism
Peacekeeping Operations
Peace Enforcement
Show of Force
Support for Insurgencies and Counter-Insurgencies
Attacks and Raids

TABLE 2. Representative Gray Area Phenomena (Gap)

Ethno-Religious-Nationalistic Conflict
Weapons Proliferation-Conventional and NBC
Conflict Over Scarce Resources
Infectious Disease Control Issues
Globalization of Organized Crime
Drug Trafficking
Economic Warfare-Conflicts Over Technology
Emigration Policies and Control
Famine-Related

TABLE 3. Asymmetric Warfare Mission Categories (AS)

Weapons of Mass Destruction (NBC)
Terrorism, Counter-Terrorism
Information Warfare

Note that OOTW comprises some 13 missions, GAP 9 missions, and AS 3 mission types. While there are some overlaps among these mission categories, collectively they amount to 25 missions. It can be appreciated that within each mission type that there would be some range of "situations", "threats", and "impacts" of interest. Appreciating that no matter what the approach to fusing information is in order to estimate such states, and that any approach will require some type of modeling activity, the scope of such efforts can be expected to be significant. Justification of the definitions of situational states etc in each case will require analyses of the three-way mapping hinted at in Figure 1, which provides another sense of the effort that will be involved. The problem of course is that the dimensionality of the information and its possible interrelationships for any given mission and situation have increased dramatically in relation to the dimensions at Levels 0 and 1. Yet further complications can arise when considering such issues in a distributed Data Fusion environment or architecture. In the US and in other countries, much has been said about the goal of achieving "Common Operating Pictures" in distributed-force cases (i.e. a common situational understanding) but the achievement of this capability involves even broader issues than directly associated with the mission requirements. When considering the heavy dependence of any modern mission on the quality of relevant information, the goal of "Information Dominance", defined as having superior situational awareness in relation to your adversary's awareness, becomes a significant challenge.

2.2 DECISION-MAKING SUPPORT

We have briefly discussed one factor, the mission factor, in the depiction of Figure 1. The decision factor also should be contemplated. To do so generally requires a model of decision-making; this is itself a complication in that there is a multiplicity of such models, each having slightly different components and reflecting slightly different viewpoints.

Such models periodically fall into or out of favor; at the time of this writing the fashionable model is the "OODA Loop" model, the acronym reflecting the main elements of "Observe-Orient-Decide-Act" in this dynamic model. In the same way that there are specific situations within each mission category, it can be expected that there is a range of decision-types as well. The full extent of the mission-situation-decision triad of course cannot be dealt with here; we will necessarily abstract and generalize these aspects. In the OODA Loop model, it is the Orient phase in which a mental construct we could call a situational model is formed, and it is in the Decide phase in which a determination of the need to react, and the type of reaction is made. It is in the Decide phase that a "mapping" occurs as hinted at in Fig 1, where important elements of the situation are each associated to elements of a "decision vector", in the context of some aspect of the mission at hand. How this really happens in the human mind is certainly beyond our discussion here, and it should be appreciated that this process is extremely complex. Current-day military goals involve getting "the right information to the right person at the right time"; this means that an adequately-detailed model of the process just described needs to be formed, and must involve notions of expected reaction time in the face of a given situation, as well as defining notions of relevance. Figure 5 gives one idea of the timing aspects among real-world phenomena and C2 system and decision-making activities (one can lump "Process + Compare" in this figure to "Orient" in the OODA sense). This figure only speaks to "events" which are discrete states that occur without consuming any meaningful time; situations however are the result of conditions and relationships that evolve over time, i.e. they consume time. Thus, there is an *inherent* time for the true situation to evolve, and there is the *estimation* time for the (imperfect) working of the estimation algorithm + human mind to conclude (i.e. estimate) that the situation exists. Thus, one can conceive of a notion of "situation detection", analogous to the idea of a threshold crossing in signal detection theory, and perhaps think carefully about the difference between *detection* of a situation and the *assessment* of a situation.

In this chapter we are focusing on *automated methods* for estimating these states. Given the complications and dimensionality issues mentioned so far, as well as the usual military requirement that a human be the ultimate decision-maker[2], it is very important that serious consideration and analysis be applied to the "functional partitioning"[3] of interpretive responsibilities between man and computer-algorithms. It is fair to say that this latter issue is rarely addressed in the best way, and is driven in any particular case by designers biased to one side or the other. However, it is a necessary part of the design approach and framework for developing automated aids for situation, threat, and impact estimation.

[2] At least for the foreseeable future; in light of evolving ideas about "uninhabited combat air vehicles" and the like, it may be that truly automated decision-making is at least on the planning horizon but this topic will not be included here.

[3] This is a formal term from the field of human engineering and human factors analysis.

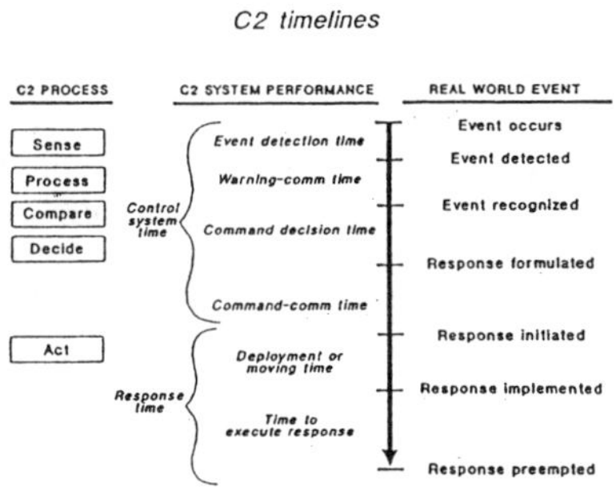

Figure 5. C2 Timelines for C2 Processing—C2 Performance—Real World Events

2.3 THE MODERN INFORMATION ENVIRONMENT

In today's world, there is another concern when thinking about the ideas being discussed here; this is the concern for information manipulation or for information warfare. While it is true that IW is designated as a individual mission area, the ubiquity of information distribution, processing, and management capability that has proliferated through major regions of the world results in an environment where, for sensibly any mission, it can be expected that an adversary has the means to somehow corrupt information that may be available to his opponent. This can be the result of "perception management" resulting largely from deception and related real-world activities (i.e. actions in the physical world), or by manipulation of an adversary's internal information processes which occur in a virtual world, and as enabled by hacking, communications corruption, and subversive software embedded in commercial or other integrated software components. That is, it is healthy to contemplate these possibilities when designing any Level 2 or 3 Data Fusion process and associated algorithms, etc., i.e. to think defensively. Such thinking will necessarily lead to what could be called "multiple-hypothesis" approaches to Level 2 and 3 techniques, where designers anticipate a yet-broader range of possible situations beyond the already complex set associated with "benign" information environments. This is tricky business, as the target of such attacks is both the DF algorithm and the human mind [4,5], and the manner in which they work together to generate the situation estimate is complex.

Another complexity of the modern environment is that consistent situational estimates will need to evolve within the framework of a highly-distributed but highly-connected information architecture. This is the vision of achieving the "Common Operational Picture" or COP. It is associated with the visions of the future battlefield information environment that depict this highly-distributed, highly-connected system. However, the data and information fusion communities have hardly addressed the issues associated with the inherent challenges of this framework. Further, there are issues associated with design responsibility in these distributed systems (e.g. responsibility for bandwidth utilization, etc) that must also be settled. It would seem that the prudent approach for the DF/IF research communities is to assume that they have control of such decisions, and formulate solutions which are sensitive to bandwidth utilization, etc but with the idea of imposing DF/IF requirements on the communications/data linking infrastructure.

2.4 FEASIBLE AND AFFORDABLE SOLUTIONS

This portrait of the needs and issues surrounding the formulation of situational, threat, and impact estimates based on fusion of various types of data and information is a daunting one, and the DF/IF research community will need to think carefully about strategies for developing solutions which are minimally feasible but also satisfying (or satisficing), and also affordable. The next section will examine a number of aspects related to this challenge:

--The nature of the required solution
  --models of the battlefield
  --satisfying or satisficing solutions
  --best-so-far and anytime solutions
  --notions of trustworthiness and evaluation of L2, L3 processes
--Communicating ambiguity and uncertainty
--The employment of formal methods from automated reasoning
--Issues and approaches to distributed situational awareness
--Aided adversarial decision-making; SA within the OODA Loop
--Reexamining notions of intent

3. Examining the Issues that Influence Solution Characteristics

3.1 THE NATURE OF THE REQUIRED SOLUTION

Situational analysis is (mostly) either associated with, i.e. supportive of, either analysis or interpretation or with decision-making. Thus, it can be argued that the specification of the characteristics of any situational estimate would follow from a top-down analysis of such functions or tasks. As was mentioned above, an analysis of this type would necessarily incorporate the important factor of required reaction time associated with various decisions, and also carefully partition subtasks between the human operator and automated DF/IF processes. Important also in this analysis would be the understanding of the timelines of data and information collection; what is important is the temporal evolution of a situation in the real world, not the temporal evolution of the (latent) collected data—i.e. the time lags in the collection system must be understood in relation to the models of the temporal dynamics of battlefield events.

3.1.1 *Models of the Battlefield*

Developing deductively-based approaches to automated situational estimation (the historical approach) is no different in concept than developing a Kalman Filter; the quality of the solution revolves about the appropriateness of the underlying models. Thus, the DF/IF process designer must take a stand on how to view the battlefield. Clearly this will (should) require close cooperation with military operations analysts, at least ideally, to form this view. The most important part of this dialog will be that which determines if there is a limit to what can be done in DF/IF-based situational estimation. In particular, such discussion should determine if it is considered feasible to model force-on-force combat interactions. Some analysts (e.g., [6]) consider that SA is feasible during force build up, commitment, and through initial contact of forces but question whether combat interactions—considered chaotic by some—can in fact be analyzed. Figure 6, from [6], depicts these relationships.

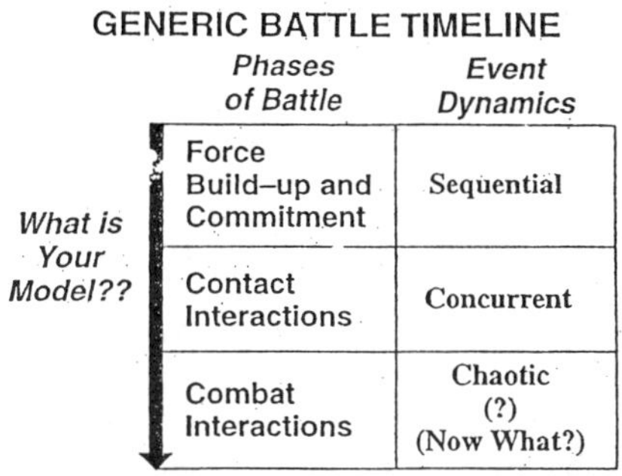

Figure 6. Notions of Event Dynamics as a Function of Battle Phases [6]

In this work, Egan [6] challenges us to think about how far into the battle timeline we can carry our model-based, deductive approach. If, as he suggests, combat interactions, i.e., the behaviors of fully-engaged forces, is chaotic, then we must build a data fusion and situational estimation process that realizes that such engagements are occurring, and to signify to the user that the system is at its operating limit. This would imply that the use of automated aids of this type is limited to surveillance, reconnaissance, up to the point of initial force contact operations. We are not suggesting Egan is correct; we are simply suggesting that a position on such points needs to be taken for whatever mission is of interest.

3.1.2 *Satisfying or Satisficing Solutions; Best-so-far and Anytime Solutions*

The issues discussed in this section largely revolve about considerations of time and complexity. It has to do, in part, with the temporal issue mentioned above in terms of the lag between real-world events and the fusion system's depiction of those events. Even if the DF/IF process incorporates high sampling rates, achieving highly accurate solutions may be constrained by levels of complexity, uncertainty, ambiguity, and incompleteness of information. Careful consideration must be given to the notion of what can be done in light of the time available and in light of probable complications in the information available. This analysis must also consider the distinctions between detection of the situation, the provision of an estimate adequate for decision-making (satisfying or satisficing), and the provision of an in-depth estimate of the estimated "truth" of the situation. These ideas are sketched in Figure 7.

This figure suggests that there are phases and degrees of situational understanding that will typically evolve in accordance with the incorporation of additional information collected over time, and in accordance with the evolution of the true, underlying situation. By analogy to signal processing, we typify a "detection" phase, an "awareness" phase (analogous to parameter extraction), and an "estimation" phase.

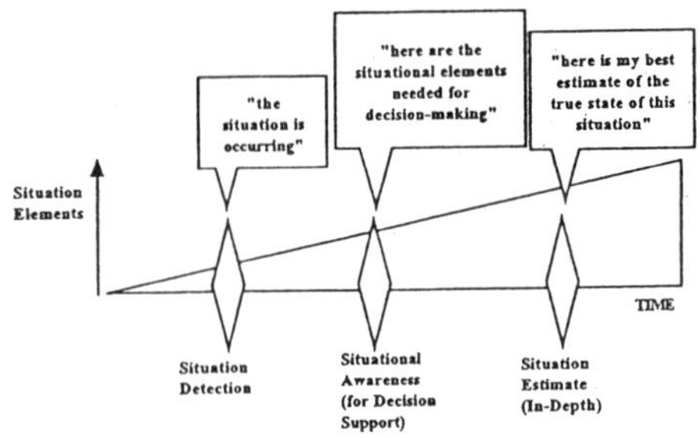

Figure 7. Notions of "Stages" of Situational Estimation

In considering general approaches to developing the best solution to these complex problems, one could assert that the way to go is to employ the most powerful computing approach possible, either the fastest machines or multiple machines. But this may not be the most affordable approach or the overall best approach, nor in some cases even feasible. Instead, one could explore approaches that trade solution quality for execution speed; given the military's high level of concern for timely solutions, this may be a viable if not very attractive approach. "Anytime Algorithms" are one family of algorithms that exchange execution time for quality of results. Since many computational tasks are too complicated to be completed at real-time speeds, anytime algorithms allow systems to intelligently allocate computational time resources in the most effective way, depending on the current environment and the system's goals.

Currently, anytime algorithms are being used in a variety of applications, such as: sensor interpretation and path planning [7] model-based diagnosis [8], and evaluation of belief networks [9,10]. Anytime algorithms iteratively work on problems to provide an interruptible way of producing results that improve in quality over time. Many current search-based algorithms can be converted into anytime algorithms with little trouble.

Anytime algorithms can be executed in two ways: either by being given a *contract time* (a set amount of time to execute), or by employing an interruptible method (the anytime algorithm may be halted at any time). Anytime algorithms can also be continued once they have halted (either from the point where the contract time is up, or an interruption) and continue to improve the solution. According to Grass [11]: "Anytime algorithms are currently being implemented in a variety of ways, but most anytime algorithms can be divided into three components that work together to produce a complete anytime algorithm. These components are:

An iterative improvement function.

This function improves the quality of the result during each execution. Many algorithms can easily be converted into an iterative improvement function (e.g. iterative deepening search, Monte Carlo algorithms, sorting algorithms) by removing looping structures so that the function represents one step in the process of completing the algorithm.

A result evaluator.

This function evaluates the result and the iterative improvement function to return information about the quality of the result.

A performance profile.

This function returns an estimate of the quality given the amount of time that the anytime algorithm will be executed and information about the input data. This information about the input data is termed the input quality, but could be any type of meta-information about the initial data set, from the size to distribution information. If the performance profile uses the quality of the input data it may also be called the conditional performance profile.

Grass [11] distinguishes anytime algorithms from other algorithms in that they return not only a functional result but also a quality measurement, cast in various possible ways, such as:

"*Accuracy.*

This measurement tells the system how close the result returned is guaranteed to be to the actual answer. Much like many numerical approaches to approximating transcendental functions with infinite series, a plus or minus error range is returned along with the result.

Coverage Area.

This measurement tells the system how much of the potential search space or database has been explored.

Certainty.

This measurement states the correctness of the results returned by the algorithm: it can be a set of possible results, a probability distribution, or other approaches.

Resolution.

This measure returns the Level of detail. For example, in a vision algorithm, this could represent the dimensions of the image that have been processed. The whole image is processed, but the dimensions may increase in size over time until the dimensions are equal to the image resolution."

The DF/IF community has not explored such methods in its efforts to identify possibly-attractive algorithms for L2 and L3, and such methods may hold promise for those environments in which there is considerable time-based stress.

3.1.3 *Notions of Trust, Trustworthiness and the Evaluation of L2, L3 Processes*

We argue that, in an analysis or decision-making task context employing automated aids of some kind, final human decision and action depend on the degree to which the human trusts the automated-aid output. No matter how developed, corrupted, or displayed this product may be, it is that degree of trust which may be the deciding factor in proceeding with an analytical assessment or a decision and consequent action. The notion of trust is complex and multi-dimensional, and rather little work has been done on the particular subject of Trust in Automation. The starting point turns out to be, as one might expect, the sociological literature, about trust among humans. The dimensions and factors affecting human trust do seem to carry over to the case of Trust in Automation, perhaps because of the human-like metaphor (e.g. "Hal") especially ascribed to computers.[4] Since so few experiments have been done, there is correspondingly not much work on the subject of measures and metrics. Our assertion that Trust in Automation is possibly a deciding issue in determining the final decisions and actions performed by humans implicitly depends on what has been learned or observed to date on how human cope with distrust in automation. While there has been little work to support or disprove this assertion, that work that has been done shows that there is a significant hysterisis loop that develops when humans suspect malfunctioning decision aids. In the tactical context of IW, this means that an adversary could conceivably take an opponent "off-line" thru IW actions that lead to distrust in the opponents decision-aiding system. If this was done at some opportune time, the payoff to tactical operations could be significant. So there is at least limited evidence that the effects of trust can indeed be quite influential in adversarial decision-making.

Table 4 shows some notions of the attributes of trust and related IW features, i.e. how the feature or quality of trust is developed (basis for trust development), what types of information the feature is dependent on, what the vulnerabilities of that information are, and how the feature may be measured. For instance, Reliability, as a feature or quality of trust, is based on repetitive functionality and is "conditioned" in the mind of the user; in an automated system this quality is achieved if the software "perceives" given cases as similar and therefore develops (computes) the same result. If an adversary creates false parameters of this type, he can influence the repeatability and thus the reliability. This can be measured using a Bayesian approach or one using notions of utility.

Bisantz [12] has been conducting research on notions of trust, metrics with which to gauge trust, and the conceptualization of an experimental framework within which meaningful assessments of a decision-aids (i.e. data-fusion-based decision-aid) trustworthiness could be evaluated. Summary remarks on this work follow.

3.1.3.1 *Measures Of Trust And Related Notions.* Previous research in the area of trusth as used two basic types of measures for assessing trust in different situations: personal rating measures, and process and performance measures. Rating techniques, such as questionnaires, may seem to be the most direct way to measure a subjective feeling or state of mind such as the amount of trust one has in another person, or in the information provided about a complex system. However, more objective measures can also provide information relevant to the assessment of trust. For instance, people's decision-making or control strategies may change based on the trust they have in the available information, or the quality of various systems components.

To identify potential similarities and differences among concepts of general trust, trust between people, and trust between humans and systems, and between trust and distrust, Bisantz carried out a three-phased, empirical study, modeled after that used by Zhang, Helander, and Drury [13]. The first phase collected various words related to concepts of trust and distrust. The second phase investigated how closely each of these

[4] Some argue that human-like behavior of computers clearly follows from the fact that humans program them, and so this relationship is more than a metaphor; it is real (but this is a separate subject).

Table 4. Notions Of Trust And Related Iw Features
Abbreviations: DA=decision aid, SW=software, T=trust or trustworthiness

Trust Attribute	Characteristics	Basis for Trust Development	Information Dependencies	Information Vulnerabilities	Measures and Metrics
Reliability	Repeated, consistent functioning	Conditioning	DA performance Depends on "characteristic" or "problem-encoding" parameters-i.e., if SW "reads" problem as "same", given the input data	Adversary creating false values of problem-characterizing parameters, causing DA to switch problem-modes	•Bayesian approach • Reliability can also be linked to Utility: U=Rel x Usefulness
Robustness	Similar to Competence; good performance under varying problem conditions	Conditioning; often achieved with a small sample of cases (and potentially unwarranted)	DA performance depends (more critically than for Reliability) on "characteristic" or "problem-encoding" parameters	Adversary creating false values of problem-characterizing parameters, causing DA to switch problem-modes	Degree of trust as a function of inter-problem performance (or output) variance is one idea; i.e. T = f (Perf Var)
Familiarity	Employment of familiar/friendly/natural procedures, terms, etc.	Inherited from the culture of domain (or life experience); can be unwarranted, irrational	Familiar notation and Symbology	Creation of unfamiliar notation/symbology	Degree of Intervention related to adjustments in notation and symbology
Understandability	Not the same as Familiarity or ease of use; supports formation of mental model and predictive framework	Some notion of "visibility" into inner processes. Distrust and Alienation formed when inner workings not understood.	Ability to Achieve the "visibility" mentioned—availability of Information/data related to inner Processes	Any actions that lead to bewilderment in viewing inner processes, or reductions in "communicativenes" of the system	Degree of intervention to access inner processes or to reformulate system output
Explica-tion of Intention	Not same as Understandability in that intention is *overtly stated*; Understandability allows *inferring* intent	Execution of Actions per Stated intent--in a sense, follow-up on intended/stated actions	Stated intent	False indications of intent, or failures in acting as (presumably) intended	Degree of inquiry into intended (but not visible) follow-up actions. Queries relatedto disparities in inferred vs stated intent
Usefulness	Classical notion of utility	From use of a "focused" or "directed" system that supports assigned work.	Information that Supports efficient rate of progress toward work-goal	Information that impedes work-flow	Degree of intervention to alter flow of operations
Dependence	Reliance on system to support human performance	Can be "Catch-22" type relationship--a type of *forced* trust on a system you (must) depend on	Information that Prevents alternative Ways of conducting Necessary work	Information that suggests that alternative ways of doing work are possible	Number of work-mode changes (e.g. automated to manual possibly)

words was related to trust or distrust in order to evaluate whether or not trust and distrust were opposites or represented completely different concepts, and whether or not concepts of trust and distrust were similar for general trust, trust between people, and trust between humans and systems. The third phase, yet to be done, will conduct a paired comparison study to identify multiple, independent factors of trust and distrust.

In addition to subjective measures such as the rating scales described above, one can also investigate trust by considering performance and process measures as measured on operators or decision-makers while carrying out their tasks. Intuitively, if people hold an inappropriate level of trust in a situation, combined human-system performance may suffer. Additionally, as people's trust in various system aspects changes, they may act in qualitatively different ways.

The above experiments provide results that are important to the development of an empirically developed measure of trust. First, high negative correlations of ratings of trust and distrust were observed and statistically qualified, which indicates that these concepts can be treated as opposites, lying along a single dimension of trust. In previous studies, this has been assumed, but not empirically tested. In practical terms, this implies that it is not necessary to develop questionnaires to measure high and low levels of distrust, as well as high and low levels of trust. Second, patterns of ratings were similar across the conditions of general trust, human-human trust, and human-machine trust, as indicated by a high degree of similarity in sets of words related to trust. This indicates that future work on the development of trust measures might not have to treat these types of trust differently, and that results from studies of human-human trust (e.g., those that examine stages in the development of trust; e.g. Rempel et. al, [14]) may indeed have applicability to situations of trust between humans and automated systems. Again, this transfer of trust concepts from the sociological to human-machine domain had not previously been empirically tested.

Future work (Phase Three) has been initiated to develop a multidimensional scale of trust based on the results of these two phases. Participants are being asked to rate the similarity between all pairs of words that were highly positively and highly negatively related to trust, from the results of Phase Two. The results from this paired comparison study will be analyzed using factor and cluster analysis techniques to identify a set of factors that comprise a scale of trust. A multi-dimensional scale of trust will then be developed based on these factors. It is planned to use this final scale, possibly coupled to performance and/or process measures as a measurement basis for future human-in-the-loop experiments. As part of this future work, quantification of the relationship between a "trust-word-metric" and the implied degree of acceptability of the result or estimate from the DF/IF process being evaluated must also be done. This latter point then gives guidance to DF/IF process designers as to how accurate or timely or whatever their process results need to be.

3.1.3.2 *Test and Evaluation Methods for L2, L3 Processes.* Most would agree that most of the inputs to a DF or IF process comprise random variables; this is especially true if we focus on sensor measurements, and often qualifies human input as well. Further, the phenomenon of interest being observed may itself be a stochastic occurrence, even if the observation processes were fully deterministic. It is well-known that any function that operates on a random variable produces a random variable, no matter the complexity or nature of that function. Following this argument, and considering the largely knowledge-based processes used to fuse L2, L3 input as "functions", it can be asserted that the products of L2, L3 fusion operations are themselves random variables, not unlike the L1 results such as Kalman Filter estimates which are well-accepted as random variables. Thus, a compelling argument can be made that testing procedures for L2, L3 fusion processes should be statistically grounded, employing, ideally, statistically-designed experiments, concepts from sampling theory, Monte Carlo techniques, analyses of variance, etc. Any survey however of fusion works at L2, L3 would rarely show the use of such methods. The lack of such test procedures can also be argued as contributing to the construction of fusion systems whose behavior is not well-understood and therefore susceptible to distrust in operation. Thus, no matter what the metrics of trust will be (although we need them to be qualified as well), proper test procedures are also needed with which a linkage can be made between qualified tests and qualified metrics; both are needed to properly interpret the results and to provide deep understanding of performance as a function of "problem-space" parameters, i.e. parameters which influence the nature of the problem.

The other point to be made here is that, given we have trust measures, the desirable trends of those measures as a function of problem-space parameters will need to be specified. Said otherwise, what characteristics of trust are "good", as a function of problem conditions? Such behavior would then qualify as "trustworthiness" in the evaluated DF-based decision-aid[5]. One possibly desirable behavior would be that trust improves with the severity, criticality, or some otherwise-specified important characteristic of the situation being estimated (this presumes that uniformly-high trust could not be achieved over all operating conditions). Thus, a "trustworthy" system would knowingly degrade in performance as the problem or situation being estimated became "less important" in some sense. This is only one "trust profile" that can be imagined. Another tough question is: how does one design a DF/IF process to achieve such characteristics? Answering this latter question will require more insight to design factors than exists today; nevertheless, one important point being put forth here is that there is room for significant improvement in both test procedures and in the definition and study of test metrics for L2, L3 processes and their products[6].

3.1.3.3 *Communicating Ambiguity and Uncertainty.* The last factor discussed here that influences trust in a DF/IF system is that of visualizing ambiguity or uncertainty ("visualizing" should perhaps be replaced by "perceiving", as what is intended here is any multi-modal approach to communicating). The idea here is that no matter the scale or metric of trust, no matter the trend depicting trustworthiness, the last step in conveying the quality of the situational estimate is how the communication of that estimate is presented to the user. The visualization aspects are notionally factored into the candidate design being evaluated but we break this aspect out separately for discussion. Historically, the degree of uncertainty was presented for L1 products in the fashion of a

[5] One possible definition of trustworthiness might be the characterization of a process and its results such that the likelihood of failure (or success) as a function of important problem-space parameters is well understood.

[6] The so-called "spiral" approach to software and system development could be considered consistent with engineering trustworthy systems as defined here, in that this process is risk-sensitive i.e. focusing higher levels of design and development energy on the most crucial and risky issues.

surface or contour of equi-probability (e.g. the well-known, "CEP" or covariance ellipse, etc). This of course conveyed a notion of uncertainty in the context of a thresholded probability.

Another way in which the graphical form of information presentation could be used to represent uncertainty is through the use of degraded or distorted images. Lind, et al [15] provide evidence that the form of displayed information may affect the use of uncertain data. In a study to investigate the extent to which the graphic depiction of weather systems could be degraded (due to technical limitations) and still be acceptable to general aviation pilots, Lind et al. found that pilots' estimates of weather hazards increased as the graphical distortion increased. In this case, the distortion took the form of larger polygon/ellipse shaped depictions of weather patterns, in contrast to the non-distorted continuous, fine-grained representation. This increase in perceived risk might indicate a decrease in subjects' confidence of their understanding of the current specific weather patterns.

Thus, there is some indication that iconic representations based on degraded or distorted images may be used to convey the uncertainty associated with a decision aid estimate. Finger and Bisantz [16] investigated properties of distorted and blended icon sets intended to convey uncertain information about an object's identity as either potentially hostile or friendly. A second experiment investigated the impact of a subset of these icons, selected based on the pilot study results, on a decision-making task. A sample of the distorted-icon set used in these experiments is shown in Figure 8.

Figure 8. Distorted-Icon Strategy to Communicate Uncertainty

Task performance results indicated that displays using degraded icons to convey uncertainty performed as well as degraded icons annotated with numeric probabilities, or in a situation where the only information regarding uncertainty was conveyed via numeric probabilities. That is, participants were able to perform equally well when their only information regarding object uncertainty was conveyed via the degraded images: the presence of numeric probabilities did not provide an advantage in the tasks studied. This result is significant, because it indicates both that people are able to understand uncertainty conveyed through such a manner, and thus that the use of distorted or degraded images may be a viable alternative to convey situational uncertainty. This may be particularly advantageous on displays, such as military tactical displays, where many objects may need to be displayed, and annotations of numeric probabilities may additionally clutter the display space.

3.2 THE EMPLOYMENT OF FORMAL METHODS FROM AUTOMATED REASONING

This section attempts to re-examine the symbolic-based reasoning approaches, as formal logics, and to assess their suitability and capability for situational estimation for defense type problems. Part of the rationale for this re-examination is the premise that if inferencing systems could be built on these formal foundations they would require less testing on the one hand, and would also exhibit more consistent inferencing behavior which would make them more acceptable to users; i.e. they would be more cost-effective. The aspect of consistent (and correct) inferencing behavior is also extremely important in shaping user acceptance and trust in the Data Fusion system. So another motivation is simply to help assure that the system is *used*; there are many examples where such inferencing/decision-aiding computer-based systems are ignored or shut off by the users, in part because of distrust or because of some type of erratic behavior.

There are many approaches to the study of imperfect-information processing. There are also many ways in which information can be *imperfect*. It can be uncertain, incomplete, imprecise, inconsistent, ambiguous, or some combination of these; Table 5 shows a list of some of the type of information qualities addressed herein.

Although a multiplicity of approaches to automated reasoning exists, no single approach is comprehensive in addressing all these factors, and no consensus of the requirements and principles for reasoning with incomplete and uncertain information has arisen. In fact, a reasonable argument can be made that there is vigorous disagreement over fundamental issues, such as whether approaches should be based on logic, measures, or some combination of the two, and whether the measures, if used, should pertain to probability, confidence, belief, or degree of membership in a set.

In building practical systems with imperfect/incomplete information, one strategy to overcome the requirement for universally quantified data is to *assume typicality* of the input. If we have an assertion that "birds fly" and know that not *all* birds fly, when we are told "Tweety is a bird," we can (with some risk) assume Tweety is a typical bird that flies i.e., in the absence of information about Tweety's atypicality, we assert a *default assumption* that Tweety is typical. On learning something new about Tweety

TABLE 5. Taxonomy of Imperfect Information Cases

Qualification Problem	Can't enumerate all qualifying criteria
"Frame" Problem	Really the problem of temporal reasoning — i.e., over what time frame are conditions constant?
Time Constraints	A combination of (1) and (2)—how to select time epochs/intervals for analysis
Diminishing Returns	Dealing with the satisficing result
Inaccessible Info	Good example is multi-level security
Random Behaviors	Problem of developing a stochastic model
Ambiguity	Complete data but multiple interpretations possible
Imprecision	Fuzzy quantification of parameters, problem to model effects of errors
Inconsistent/Conflicting Info	What to do — e.g., assume both true, resolve later as more data come in

(Tweety is an ostrich), we learn our default assumption is invalid as is our conclusion that Tweety can fly. The inferencing process here is *non-monotonic*. On learning a new fact we retracted our prior conclusion.

Default reasoning gives rise to four requirements (see Post and Sage, [17]) that exacerbate the problem of developing a practical system:

(a) A mechanism to make default conclusions based upon reasonable assumptions
(b) The reasoning process must be nonmonotonic
(c) When conclusions are withdrawn or changed it is necessary to maintain, or readjust, the truth of other related conclusions accordingly; and
(d) Default reasoning can give rise to multiple extensions or solutions in cases where different combinations of feasible defaults are in contention, so it is necessary to *order the extensions* to identify the preferred solution(s). (One could call this a need for an "assumption/inference tradeoff mechanism" to handle the case of multiple lines of reasoning and multiple possible conflicts with *sets* of assumptions.)

These requirements make default reasoning systems generally more difficult to implement than monotonic reasoning, which is itself a difficult problem.

In the face of these many and complex issues involved with the application of formal methods, the choice for a practical approach can be quite difficult. Post and Sage [17], in arguing their position about the need for hybrid formal logics, offer the summary replicated here as Table 6 which describes whether and to what degree any of a number of formal automated or approximate reasoning systems satisfy the requirements for default reasoning shown above.

TABLE 6. Assessment of the State-of-the-Art in Automated Reasoning [17]

STATE OF AUTOMATED/APPROXIMATE REASONING

SATISFACTION OF REQUIREMENTS BY SELECTED REASONING SYSTEMS

Systems or Model	Default	Nonmonotonic	Truth Maint.	Ordering of Extensions	Practicability
First-order predicate calculus	N	N	N	N	Semi-decidable
Production system	N	N	N	N	Exponential[a]
Prolog	N[a]	N[b]	N	N	Exponential[a]
OPS-5	N	N	N	N	Exponential[a]
Art, KEE	Y[c]	Y[c]	Y[c]	N	Exponential[a]
Statistical classifiers	Y	N	N	N[d]	Polynomial
Fuzzy set theory	Y	N	N	N[d]	Polynomial
Mycin	Y	N	N	N[d]	Exponential[a]
Bayesian hierarchical	Y	Y[c]	Y[f]	Y	Polynomial
Pearl's Bayesian net	Y	Y[c]	Y[f]	N[d]	Polynomial
Pearl's net with loops	Y	Y[c]	Y[f]	N[d]	Exponential
Pearl's net, composite	Y	Y[c]	Y[f]	Y	Exponential
Boltzmann machine	Y	Y	Y	Y	Exponential
Dempster Shafer	Y	Y	Y	N[d,g]	Exponential
Reiter's default logic	Y	Y	N	N	Non-decidable
Autoepistemic logic	Y	Y	N	N	Non-decidable
Circumscription	Y	Y	N	Y[g]	Non-decidable
ATMS	Y	Y	Y	N	Exponential

[a] Given that the rules do not contain endless loops.
[b] Only Y if one considers the closed-world assumption to satisfy.
[c] Through multiple contexts or viewpoints.
[d] Extensions are not ordered, but individual propositions are.
[e] *New* information cannot be linked in to the nets. However, priors can be *updated*.
[f] Adjusted evidence will percolate through the tree or net.
[g] Partial ordering.

We have studied the arguments about the various logic-types as presented in [17] and developed a sort of "decision-tree" or guideline that shows what the important tradeoffs might be in utility vs opportunity cost as a function of the nature of input information. This decision-tree is shown in Figure 9. It is simply representative of the type of engineering guidelines needed to deal with problems of this sort.

This figure attempts to relate, in a very preliminary way, how the nature of input information and a priori knowledge affect the nature of the tradeoff between expected utility of any automated reasoning approach and the opportunity cost or difficulty of implementation. With regard to nonmonotonic reasoning, from the beginning of the 1980s to the present there has been an explosion in research in this field, and, in particular, a great deal of work on logics that capture this type of reasoning. Researchers in the field of nonmonotonic reasoning have developed *formal methods* that enable an intelligent system to operate adequately when faced with incomplete information, providing *rigorous mechanisms* for withdrawing conclusions that turned out to be wrong and for deriving new, alternative conclusions instead. As stated previously, it is these rigorous mechanisms that we feel warrant further exploration for Data Fusion processing for Level 2 and 3 type applications, in spite of the fact that they are complex and have their own limitations; we consider that exploratory research is warranted to assess their applicability and feasibility. The argument here is not a passionate one; it is an argument calling for fair evaluation of the viability of such methods, motivated by the prospect of a number of possible benefits.

Complications in the default-based approaches typically arise for two reasons. First, because defaults often conflict, it is difficult to say what conclusions should hold in particular cases. The second is because known approaches to the problem generally induce yet another level of computational complexity to already-hard problems. Because nonmonotonic reasoning involves appeals to what is not provable, as well as to what is, it is inherently harder than standard logical deduction. This is in marked contrast to the original motivating arguments that launched research into nonmonotonic reasoning. Originally, it was thought that nonmonotonic reasoning should be more tractable than monotonic reasoning because exceptions would not have to be explicitly considered.

The extent to which nonmonotonic reasoning can be useful for practical purposes is yet an open question, as commented on by Post and Sage. However, since the seminal works of McCarthy in his theory of Circumscription [18], McDermott's nonmonotonic (modal) logics [19, 20], autoepistemic logic [21], and Reiter's defaults logic [22], many researchers have faced this question either from the point of view of studying the representational and computational limitations of the existing formalisms, or, from a more constructive approach, suggesting algorithms and architectures oriented to improve the applicability of nonmonotonic reasoning in real-world situations.

In the same spirit as [14], we too feel that hybrid solutions may be required but not necessarily new, formal hybrid logics having to be developed from scratch. Instead, it may be feasible to integrate a hybrid solution, at least to a degree that would hopefully require limited new development. The idea of this is shown in Figures 10 and 11, the latter a continuation of the former.

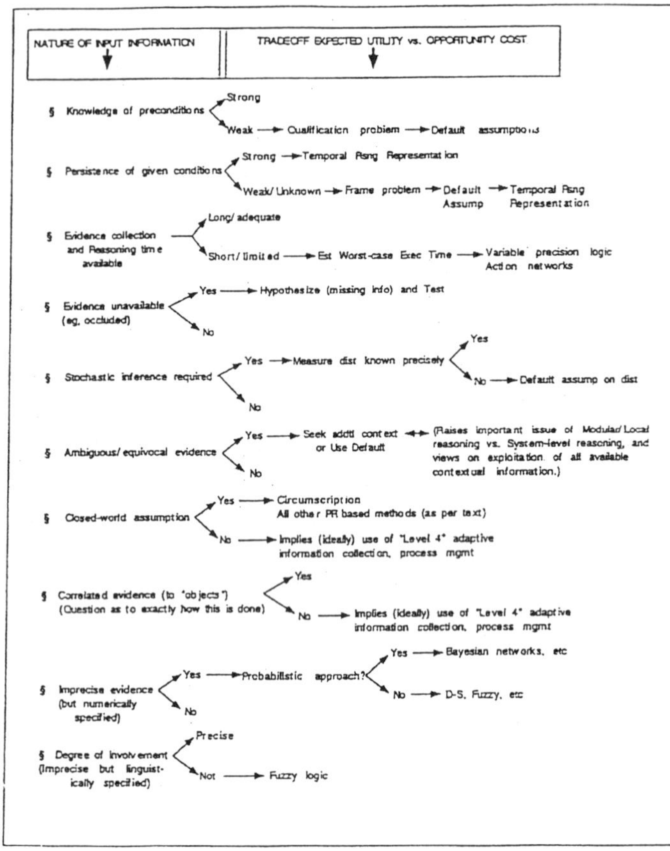

Figure 9. Preliminary Concept for a Decision-Tree Selection Strategy for Methods of Automated Reasoning

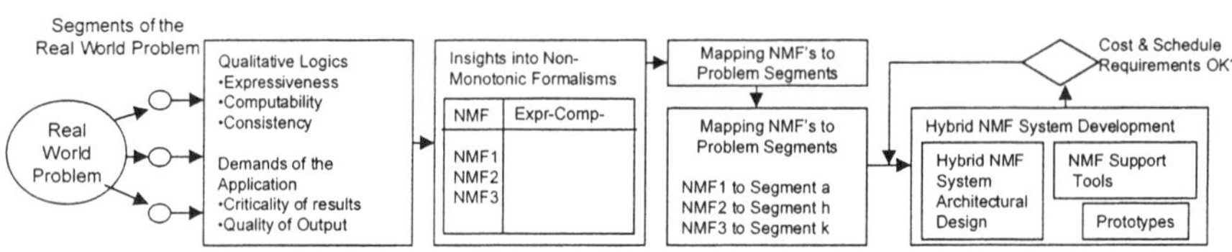

Figure 10. Generic Implementation Concept for a Non-Monotonic Logic

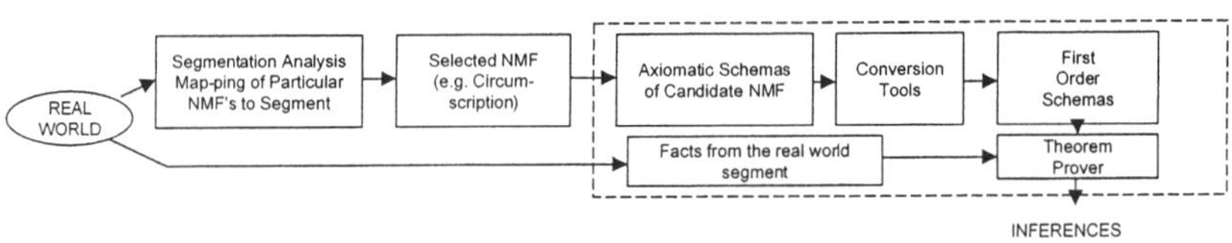

Figure 11. Conceptual Strategy for Implementation of a Circumscription-Based Logic

With good domain insight, it may be possible to partition the domain into what we have called "segments". What these are in turn depends on what type of Logic-Quality is desired (e.g. computability etc as shown in Figure 10), and on the demands of the application such as the criticality of results. These factors are seen as influencing the choice of nonmonotonic formalism (NMF), according to the insight that the designer has about the NMF's themselves, and how wide-ranging he will be in admitting the wide variety of NMF's into the feasible solution space. As can be seen, it is envisioned that a "mapping" of NMF's to various problem or domain segments is then carried out. Then, an integrated hybrid system must be put together, both architecturally (see "architectural designs" on the figure) and otherwise, supported by NMF-related tools. The initial choices have to be balanced as usual against cost and schedule concerns, an iterative process, and final decisions then made.

An instance of this design process can be realized in some detail for the case where Circumscription is selected as the NMF approach (for some segment of the problem or say for the entire problem)[7]. We show this in Figure 11, where the first "segmentation" process has resulted in the selection of a Circumscription-based NMF. In this case there are tools mentioned in the literature which convert Circumscription-axioms to a first-order system (i.e. DLS and SCAN), which can then be used with an available first-order theorem-prover to invoke a complete logic, which can be tested against real-world facts as shown.

Of course this is all notional but we believe this type of thinking should be fostered in the Data Fusion community for at least exploratory purposes and to seriously evaluate the potential of these formalized methods.

3.3 ISSUES AND APPROACHES TO DISTRIBUTED SITUATIONAL AWARENESS

When thinking about the definition and design of DF processes for distributed L2, L3 operations, the direct coupling between those processes and the communications/data-linking network in which they are embedded immediately comes to mind. That is, the design of such processes, located at specific, possibly all nodes in a network, cannot be carried out without incorporating: (1) the features and constraints of the network in the topological sense, (2) the features and constraints of specific data-links to which any node can communicate, and (3) the strategy employed for information-distribution, since the node can only fuse its local information with information received from external nodes. This coupling raises a question about "design authority", i.e. the question of authoritative control of the network design. It is not likely that the DF/IF process designer will be in charge of such designs, so the approach to designing node-specific DF/IF processes may

[7] Circumscription (meaning to delimit or confine) is one of the possible formalizations for defeasible reasoning is to reason only with a specific set of models, by using the criterion of minimality with respect to the set of true atoms (i.e. atomic, or basic expressions). The idea behind minimality is to address the problem of incomplete reasoning by assuming that a fact is false whenever possible. Such criterion allows one to deal with the qualification problem by representing only true statements of a theory, saving or eliminating the work associated with the explicit representation of all the false ones.

have to be done in frameworks in which (1) close collaboration will be carried out by network and DF/IF designers (the ideal case) or (2) with the network infrastructure and features as a "given", into which the distributed DF/IF processes must fit, or (3) as a result of a network design that has at least considered DF/IF requirements developed by DF/IF designers. Each mode can significantly affect the approach to the design of a distributed DF/IF process, as well as the nature of what that distributed DF/IF process will be. Since the DF/IF and network, communications system communities are disparate, it is felt here that the best approach to be taken by the DF/IF designer for distributed process design is (3) above, where the fusion engineer makes rational choices in network design factors, develops appropriate DF/IF processing designs, and communicates the desired features of the network to the network designer who will (hopefully) give them serious consideration. The point is that these issues, while non-technical and socio-political, can markedly affect what can be done in the distributed DF/IF process design.

Some important design issues for distributed L2, L3 process design can be defined by analogy to similar concerns at L1: for example, there is the important "double-counting" problem in which, due to possible cycles in the network, previously-sent messages return to a node as if new; we call this "overt" double-counting. However, there is also what we call "implicit" double-counting, which is the result of incorporating data or information (e.g. other estimates) into in node's fusion process that is not statistically independent from similar information available locally at the node. Such situations cause havoc and destabilization of Kalman Filters, for example, as well as Bayesian-based ID fusion processes. In generating fusion-based estimates at L2, L3 that are based on symbolic (reasoning) methods, these double-counting issues must be considered in the mechanisms used for "truth maintenance system (TMS)" or for "belief revision (BR)" operations at the receiving node. That is, in such cases there is no statistical formalism (at least usually) in the TMS or BR system, and some logical formalism will need to account (consistently) for the effects of redundant information. If so-called "uncertainty management" techniques are employed in the L2, L3 processing, then in fact there may exist local statistical formalisms at the receiving node, through which appropriate adjustments to the generation of fusion estimates will need to be made.

There are many possible pathological scenarios for distributed DF/IF processing along these lines. Some important results affecting the viability of good designs of distributed DF/IF processes are described in a dissertation produced at Oxford by Utete [23]. This is an important piece of work, which should be reviewed by DF/IF process designers. One line of development in this work first argues for the need for so-called "arbitrary" network topologies, meaning in essence unspecified or irregular, to include cycles or loops in the structure. This is based on the significant disadvantages of either Fully-Connected (FT) or Tree (T) topologies (egg lack of scalability for FC, brittleness (to fragmentation) of the T). However, the FC and T topologies do lend themselves to allowing communication policies that can prevent double-counting problems, so they are attractive in that sense. However, one key result (a graph-theoretic proof) in [23] is that for Arbitrary Network topologies, there is no communication policy that can guarantee either a consistent (globally-fused) fusion based estimate achieved in finite time, nor one that can definitively prevent the detrimental effects of double-counting on the fused estimate.

In the face of these and yet other complexities in designing and implementing well-functioning, reliable, and trustworthy distributed DF/IF processes, the goal of achieving a "common or consistent operational or tactical picture" will be difficult indeed. It may be required that achieving consistency in inter-nodal situational estimates will have to be the result of an "adjudication" process carried out largely by the humans involved at each node. This will itself consume a degree of bandwidth but is also a process that should be a candidate for automated support, given the surge in the capability of technologies associated with group problem-solving, cooperative work environments, etc.

3.4 AIDED ADVERSARIAL DECISION-MAKING; SA WITHIN THE OODA LOOP

This relatively brief section is inserted to discuss a concept for the development and evaluation of processing methods for L2 and L3. The idea stems from assertions that if friendly forces are able to process their OODA Loops faster than an adversary, reliably good consequences will evolve on the battlefield. While this assertion has intuitive appeal, there seems to be little quantification of such it that are necessary to make important design decisions about SA quality vs timeliness and the attendant cost/investment decisions for system development. What has apparently not been studied are the typical dynamics in the evolution of situations when there is an intelligent, reactive adversary with different technology (to include a DF/IF-based decision aid) and from a different culture with a different value system. While it is categorically true that relatively little research involving human-in-the-loop experiments has been done at all in the DF/IF community, it is especially true at Levels 2 and 3 where, ironically, the most significant involvement of the human lies. Here we propose that such experimentation is sorely needed, and that the evaluation of the nature of a situational or threat/impact estimate and its timeliness as produced by a prototype DF/IF process be evaluated in the context of "aided, adversarial decision-making (AADM)", involving two-sided, human-in-the-loop experiments. It is not argued that this is the only way to conduct L2, L3 evaluations but that such methods are part of an overall test and evaluation approach, leading to the production of well-tested, trustworthy L2, L3 techniques. This notion is depicted in Figure 12.

This figure shows a group of friendly and a group of adversarial decision-makers working in concert with their respective Data Fusion-based decision-aids against a common "world" or tactical environment. Each is dependent on the respective quality of their situation and threat-estimating capabilities produced by the automated DF aids. This section is recommending that experiments be conducted that are reflective of such environments (with real humans involved) as part of a comprehensive approach to the evaluation of L2 and L3 capabilities of DF-based aids.

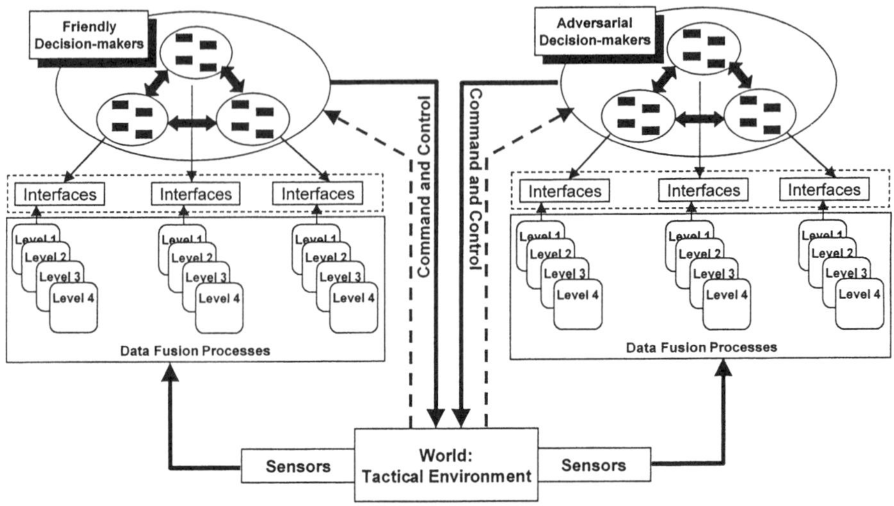

Figure 12. Notional Concept of Aided, Adversarial Decision-Making

3.5 REEXAMINING NOTIONS OF INTENT

Regretfully, one does not have to look very far today to appreciate the difficulty of understanding a person's intent, particularly when it involves the conduct of harmful acts. This difficulty extends well beyond that related to defense and military problems; in the US, incidents of totally unexpected terrorism among school-age children have occurred that have baffled experts from law enforcement to psychology. Such agencies, as well as defense intelligence agencies, have attempted to "profile" the behaviors associated with such acts but this methodology leaves much to be desired. The basic problem with these methods is that they result in too many false alarms, since there are many parameters or indicators involved. In a review of basic concepts and literature on this subject, Llinas and Deutsch [24] reexamined a large body of work spanning various disciplines, ranging from psychology to philosophy to law, and to intelligence analysis, about this subject.

Interestingly, the past three decades have seen an explosion of analysis and research on intentions and intentionality, although serious assessment of this work does not seem to have been carried out for DoD purposes. A number of factors have contributed to this rapid growth:

- The exhaustion of both behaviorist and Freudian movements in psychology.
- Empirical advances in biology and the neurosciences.
- The advent of digital and optical computing as a paradigm and tool for research.
- The "naturalization" of logic, language, and mathematics.

Moreover, interest in intentionality or "intentional systems" has transcended the traditional boundaries of philosophy, linguistics, psychology, sociology, biology, and ethology. Studies on intentionality draw upon an ever-expanding universe of theoretical and empirical research ranging from abstract logic systems to field studies on chimpanzees.

There are many paradigms about the notion of intent, and there is a need to distinguish the subtleties between intent, intentional acts, intentionality, etc—one needs to be precise in what is being discussed. Military intelligence analysis typically relies on extrapolations of recent behavior to estimate what likely future actions may be; sensibly all the literature argues against such methods, mostly on the basis of the combinatoric relationships between recent and imminent behavior, and also because intent is *conditional*; this means understanding it involves the so-called "qualification" problem that makes automated reasoning difficult—i.e. that it is difficult (impossible?) to enumerate all the conditional requirements to make an inference correct. Such extrapolation methods are also risky because decisions about imminent actions are conditional, possibly on many factors. Related to this is the "partial plans" approach [25] in which it is argued that humans are constantly adjusting their plans based on prior intentions, driven by "beliefs and desires", to decide next steps and plan-filling options. There are also arguments about whether intentions form a conscious state, since some intentions seem to lie dormant; these aspects argue against intention having *duration*, which is a characteristic of conscious activities. Cohen and Levesque [26] depict intent as comprising choice plus commitment; the choice of action is a goal-related choice and the commitment stems from the belief in the viability of the intended action, which in this model is also conditional.

Space does not permit the development of the various ideas that led to one possible model from which to operationalize automated aiding of intent estimation but we show this very preliminary model in Figure 13.

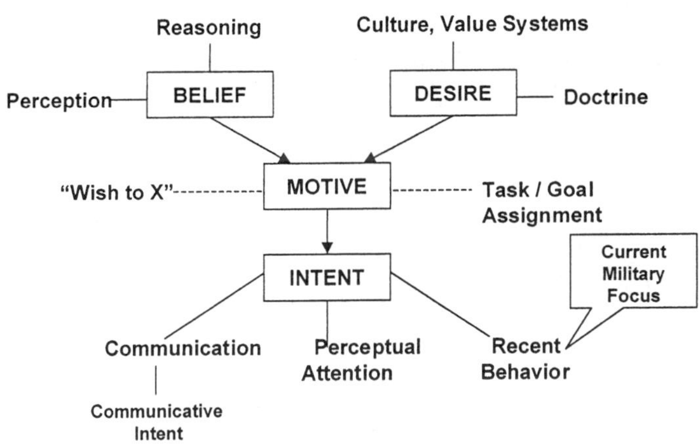

Figure 13. One Possible Conceptual Structure for Intent-Estimation

This figure shows that intent may be the result of contemplations on Belief and Desire, which are deep-seated notions, which may lead to a Motive, which underlies Intent. Belief is in turn influenced by Perception (observation in the broad sense), Desire by Doctrine or philosophy, and Motive by assigned Tasks or Goals. In considering what, from that which can possibly be observed, may be used to estimate Intent, it is not only Recent Behavior as shown on the figure, but perhaps also what is communicated and also where perceptual attention is focused. But this is too simplistic; all of the possibly influencing factors in this figure bear on the actual formation of intent. What has to be understood is the symbolic interaction between all behaviors and relations between behaviors.

In this chapter section we can only review a few salient points of the work in [24], and the interested reader is referred to that work (and its references) for a more thorough review of this very important topic.

3.5.1 *Fundamentals of Intentionality and Intentions*

According to John Searle [27, 28]:

> "Intentionality is that property of many mental states and events by which they are directed at or about or of objects and states in the world."

Intentionality should not be confused with "intention". Intentional states are directed; they are about something, not simply about intending to act. Intention -- in the sense of intending to act -- is merely one of many types of intentional states. Beliefs, desires, fears, hopes, doubts, suspicions, etc. are some of the many kinds of mental states that can be called intentional. What distinguishes intentional from other kinds of mental states is their content -- intentional states are "about something." In this view, only some mental states can be characterized as Intentional states which have content, an "aboutness".

> "While the boundaries between intentional and non-intentional states are obviously fuzzy and subject to fierce debate in the philosophical, psychological, and ethological literature, the directedness of intentional states can be interpreted or represented with "that" or "of". "

Intentional states have content and a psychological mode or direction -- e.g., believe, desire, hope, etc. Representations -- pictorial, linguistic, symbolic, etc. -- are clearly intentional in nature, but not all intentional states can be represented by means of images, language, or symbols. Intentionality should not be confused with consciousness. According to Searle [27],

> "Intentionality is not the same as consciousness. Many conscious states are not Intentional, e.g., a sudden sense of elation; and many Intentional states are not conscious, e.g., I have beliefs ("about" something) that I am not thinking about at present and I may never have thought of. "

Many beliefs and desires survive and thrive beneath the surface of conscious awareness and we consciously experience many sensations that cannot be called intentional. Finally, what are the scope and domain of intentionality? Table 7 summarizes Searle's views on five key "forms" of intentionality.

These five "forms" could be taken to define the core scope of intentional states. While most authors would agree on the centrality of beliefs and desires in the understanding of intentionality, the intentional basis of memory and perception is more controversial and the literature on the intentionality of action is large, growing, and fraught with disagreement over fundamentals.

The domain of intentionality refers to the class of objects – persons, animals, systems, etc. – that display or be treated as displaying – intentionality. Obviously, debates over the intentionality of non-human systems – higher-order animals on the one hand and complex computer programs on the other – are endemic and irreducible in many respects. Nevertheless, the domain of research and application is inexorably growing as researchers are developing and using increasingly sophisticated means to test for intentionality in animate systems and build intentional states into inanimate systems (an interesting point when considering future concepts of warfare).

TABLE 7. Summary of Searle's Forms of Intentionality

	Nature of Inten-tional Component	**Presentation or Representation**	**Direction of Fit**
Believing	belief	representation	mind-to-world
Desiring	desire	representation	world-to-mind
Remembering	memory	representation	mind-to-world
Seeing (Perception)	visual experience	presentation	mind-to-world
Intentional Action	intention in action	presentation	world-to-mind
Prior Intention	prior intention	representation	world-to-mind

3.5.2 *Intentional Stance and Intentional Systems*

While philosophical excursions on the nature of intentionality and the relationship between intentionality and mind date back to the nineteenth century, it was Daniel Dennett [29,30,31] who popularized the concept of intentionality and freed it from the narrow bounds of philosophical discussion. Dennett's writings are accessible to a broad audience and were primarily responsible for the rebirth of interest in intentionality and intentional systems.

Dennett's principal contribution was to move the focus of debate from the "essence" of intentionality to the benefits of using intentional terms to explain and predict the behavior of people, animals, and systems of any kind. Dennett's approach was instrumental -- intentional terms are useful in predicting the behavior of all kinds of systems regardless of the intrinsic intentionality of the system itself.

Dennett distinguishes three broad "stances" or strategies for explaining and predicting the behavior of a system: physical, design, and intentional.

(a) **Physical Stance**. The observer determines the physical constitution of a given system and its physical interactions with its environment and then uses the appropriate laws of physics to predict the outcome for any input. The physical stance is generally useful for predicting the behavior of fairly simple systems with limited numbers of elements and states. For complex systems, the utility of the physical stance declines rapidly. However, the physical stance is most useful in explaining breakdowns or malfunctions in both mechanical and human systems.

(b) **Design Stance**. The observer "ignores the actual (possibly messy) details of the physical constitution of an object, and, on the assumption that it has a certain design, predicts that it will behave as it is designed to behave under various circumstances." "The essential feature of the design stance is that we make predictions solely from knowledge or assumptions about the system's functional design, irrespective of the physical constitution or condition of the innards or the particular object." "Different varieties of design stance predictions can be discerned, but all of them are alike in relying on the notion of *function*, ...That is, the design of a system breaks it up into larger or smaller functional parts, and design-stance predictions are generated by assuming that each functional part will function properly." The design stance informs explanations of computer systems and human brains alike – one can identify the various modules and their respective functions and even track their activities in progress, but one does not need or seek to know the actual configuration of states for any given operation at any given time. Design-stance predictions are limited, however, to the designed behavior of a system in its normal or designed environment. Changes in the environment, attempts to perform undesigned functions, or unanticipated interactions of designed functions will typically produce system breakdowns -- failures often best explained by means of the physical stance.

(c) **Intentional Stance**. When the design of a system is exceedingly complex or inaccessible to the observer, one can adopt the intentional stance. "One predicts behavior in such a case by ascribing to the system the possession of certain information and supposing it to be directed by certain goals, and then by working out the most reasonable or appropriate action on the basis of these ascriptions and suppositions." "Whenever one can successfully adopt the intentional stance toward an object, I call that object an *intentional system*. The success of the stand is of course a matter settled pragmatically, without reference to whether the object *really* has beliefs , intentions, and so forth." Thus, complex computer systems like Deep Blue can be considered intentional systems in Dennett's sense, because their behavior can be successfully predicted by ascribing beliefs and desires. "When should we expect the tactic of adopting the intentional stance to pay off? Whenever we have reason to suppose the assumption of optimal design is warranted, and doubt the practicality of prediction from the design or physical stance."

By placing the burden on the observer, Dennett ducks the ontological issues of what really constitutes an intentional system. Intentionality is in the eye of the beholder.

3.5.3 *Orders of Intentionality*
Dennett further elaborates on his concept of intentional systems by distinguishing between different orders of intentional systems.

(a) ***Zero-order*** intentionality – simple stimulus-response behavior or reflex reaction without intentional content.
(b) A ***first-order*** intentional system has beliefs and desires (etc.) but no beliefs and desires *about* beliefs and desires.
(c) A ***second-order*** intentional system is more sophisticated; it has beliefs and desires (and no doubt other intentional states) about beliefs and desires (and other intentional states) – both those of others and its own.
(d) A ***third-order*** intentional system is one that is capable of such states as Tom wants Mary to believe that Tom believes he is successful.
(e) A ***fourth-order*** system might want you to think it understood you to be requesting that it leave.

This progression may go on forever, at least in principle. Indeed, in intelligence contexts, high orders of intentionality are involved when opponents are attempting to psych-out or manipulate each other. In most social contexts, however, five or six orders of intentionality are the most that human participants can effectively manage.
However, orders of intentionality become a critical issue in evaluating the complexity or intelligence of a given system.

> These orders ascend what is intuitively a scale of intelligence; higher-order attributions strike us as much more sophisticated, much more human, requiring more intelligence. Grice [32] and other philosophers (see especially Bennett, [33] have developed the view that genuine communication, speech acts in the strong human sense of the word, depend on at least three orders of intentionality in both speaker and audience. . .

Thus, orders of intentionality provide a framework for measuring the intelligence of an intentional system and attributing the presence or absence of truly human-like communication in animals. Notions of orders of intentionality are important to define, discern and understand in military problem settings, where adversaries are working toward psyching each other out.

3.5.4 *Dennett's Legacy*
However, this strategy liberates the analysis of intentionality from the prison of philosophy and challenges scientists in a wide variety of disciplines to come up with empirical standards and tests for intentionality. Intentionality has different standards, tests, and applications for primatologists, psychologists, neurobiologists, and jurists. Dennett's writings constituted an open challenge for specialists in all fields to operationalize intentionality to serve their own agendas.

4. Level 4: Process Refinement

The fourth Level of the DF process[8] is called Process Refinement and, as originally conceived, implied any and all possible functions and processes associated with enabling adaptive feedback into the overall DF process. If one draws an analogy (a proper analogy, as argued below) to a classical feedback control process, this is the place where the "Control law" exists which determines the feedback action on the basis of some "Performance Index (PI)" upon which the quality of the processing results are based. It is acknowledged that this operation is part of an overall Resource Management function but whether and to what degree resources other than information central to DF operations is managed within L4 is case-by-case dependent, involving both technical issues and also issues of authority over the control of those resources (e.g. weapons, countermeasures, etc).

In the case of military applications, runtime adaptation is at the heart of the ability to adjust the Data Fusion information processing to the temporally varying problem environment, and so is at the heart of the ability to achieve optimality in such processing. Therefore, requirements for adaptive Data Fusion processing are derived from mission-level requirements and must reflect the military-users perspectives on maintaining the delicate balance between mission accomplishment and survivability. Achieving adaptive Data Fusion requires one or more of the following: agile sensor systems, hospitable architectures, accessibility to "control points", i.e. places in the processing stream where adaptation is desired, defining a "cost-to-adapt" function so that cost vs effectiveness can be assessed, and the ability to sense the "problem state", i.e. to estimate the nature of the real-world problem at hand, because such estimate often forms the basis for or at least a partial basis for adaptation. In this last regard, there can be an "observability" problem in that one must have an ability, from sensed data alone, to compute such a "problem state" estimate as an input to the control law, i.e. as a problem-context estimate. Ideally, the strategy for adaptation, i.e. the "control law" *should be developed from or at least incorporate control-theoretic aspects*, such as concern for stability in adaptation, good transient performance and response-time characteristics balanced with good noise suppression, etc. To our knowledge, no studies and experiments in Data Fusion process adaptation have taken such a view; we give an example of one project where this was done. From a control point of view, it is believed that Data Fusion process adaptation requires an approach based on a hybrid (numeric + symbolic), hierarchical, learning controller strategy; such methods are on the edge of the state of the art in intelligent control techniques but should offer promise for future studies and experiments in adaptive fusion processing.

There are two basic questions to be answered in developing the "control law" or the logic contained in L4:

(a) What is the notion of "Quality" that will be defined for the overall fusion process (i.e. the "Performance Index")? and

[8] This could be the fifth Level, the result of adding a "Level 0" to the overall process model, which has been suggested in [34]

(b) Upon what controllable factors does this Quality depend?

For instance, if it is control of the sensor operations that will be the core of L4 (typically called "sensor management", then), we must have a quantitative or logical understanding of the way in which the process Quality is affected by sensing operations and the capture of particular observables or measurements. Furthermore, there is the important notion that time is another resource to be managed within L4. That is, we may, on the basis of understanding the Quality-Sensing relationship, have a desired strategy to conduct certain sensing operations but these operations may take different amounts of time and contribute in different ways to the overall Quality of the fused result. These issues are not very different from those faced by designers of phased array/electronically-steered radars, and much can be learned from that prior and ongoing research for L4 application. But L4 design is much broader in an informational sense than radar operations are, and the degree of carry-over of that knowledge will be helpful but not definitive. Figure 14 shows the type of issues being discussed here; it can be seen that the fusion process has other characteristics that will influence the L4 design:

--it is stochastic
--it involves both quantitative-type and qualitative-type information, whose Quality is to be optimized
--metrics for defining such Quality will need to be defined
--a balance must be struck in L4 design between desired Quality and desired

Timeliness in processing; in essence solving a dual-objective-function type problem.

The output of L4 for sensor management applications will be a sensor tasking schedule, and the complexity of developing that schedule will depend in part on the task scheduling constraints, such as whether the sensing task has a hard deadline (must be done by a certain time), or the degree of slack in the task schedule, etc. This can lead to so-called "hard scheduling tasks" in certain cases.

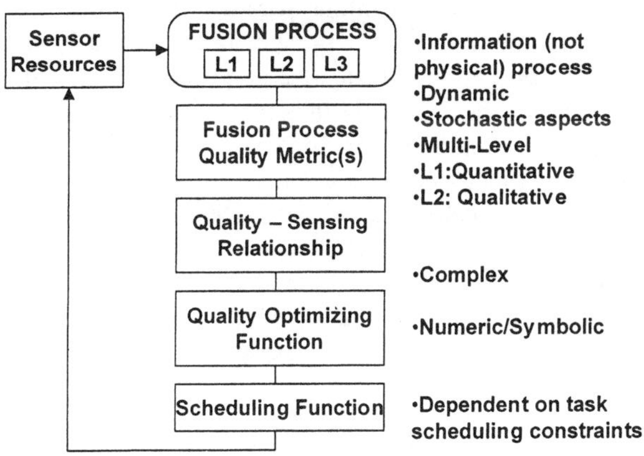

Figure 14. Conceptual Issues in Sensor Management Design

By analogy to control-theoretic methods, the features listed above that an L4 process must deal with would suggest that the type of control-theoretic approach required here would be a Hybrid (numeric/symbolic) Hierarchical (multi-fusion-level) Stochastic Controller. The degree to which formal methods from control theory of this category can be applied to Process Refinement operations is yet to be determined but it is asserted that the extent of such formal methods should be fully explored before resort is made to heuristic techniques.

At the Center for Multisource Information Fusion (CMIF) at the State University of New York in Buffalo, New York in the USA, there was an opportunity to study such issues in a submarine-tracking problem of interest to the Navy [35,36]. Further, because of the fundamental nature of the problem, this application involved limited but realistic complexity for the exploration of control-theoretic methods. In particular, the problem was: fundamentally mathematical (initially involving fixed-coefficient α-β tracking estimators), of limited dimensionality (two-dimensional, initially), and involving limited target dynamics (slowly moving/maneuvering submarines). The fixed-coefficient approach was driven by the fact that the sensor buoys on which such processing was occurring had limited computational power as well as limited electrical power. However, the α-β approach compromised performance when submarines maneuvered. Thus, a situation arose where an inquiry into defining an adaptable logic that would alter the coefficients during maneuvers was warranted. That adaptation (L4) logic could not, however, significantly increase the computational burden—hence a simple yet effective approach was sought.

It turns out that approaching the adaptive control of an α-β filter had been studied some considerable time ago. Sklansky [37] in his seminal paper analyzed the behavior of an α-β filter. His analysis of the range of values of the α-β smoothing parameters which resulted in a stable filter constrained the parameters to lie within a stability triangle. The depiction of a stability triangle for the α-β case is shown in Figure 15. This figure shows the viable limits and interrelationships between α and β to maintain a stable solution, and the curve (tangent to $\alpha = 1$) is the critically-damped α-β relationship. He also derived closed form equations to relate the smoothing parameters for critically damped transient response and the ability of the filter to smooth white noise using a figure of demerit which was referred to as the noise ratio. Finally he proposed, via a numerical example, a procedure to optimally select the α-β parameters to minimize a performance index, which is a function of the noise-ratio and the tracking error for a specific maneuver. Following his work, Benedict and Bordner [38] used calculus of variations to solve for an optimal filter that minimizes a cost function that is a weighted function of the noise smoothing and the transient (maneuver following) response.

Singh and Tenne's work at CMIF [36] extended this work to the α-β-γ filter. It can be shown that the transfer function relating the evolution of the predicted position of the target to the observed position is given by the equation:

$$G(z) = \frac{\alpha + (-2\alpha - \beta + \tfrac{1}{4}\gamma)z + (\alpha + \beta + \tfrac{1}{4}\gamma)z^2}{z^3 + (\alpha + \beta + \tfrac{1}{4}\gamma - 3)z^2 + (-2\alpha - \beta + \tfrac{1}{4}\gamma + 3)z + \alpha - 1}$$

This equation can now be used to determine the bounds for stability. For this complex system, the Jury's Stability Test [39] is used to determine the region of stability. It turns out that α-β-γ can be algebraically related to parameters labeled a, b, and c and that there is a depiction of the stable region for the α-β-γ filter analogous to the stability triangle which we call the "stability prism"; this is shown in Figure 16, which has a straightforward analogy to Figure 15 for the α-β case. From these relationships, derived using classical control-theoretic principles, it was possible to evolve a fuzzy-logic-based control law for the L4 processing on the sensor buoys, with little compromise in the processing load and with measurable improvement in tracking performance (see [35,36, 39] for the details).

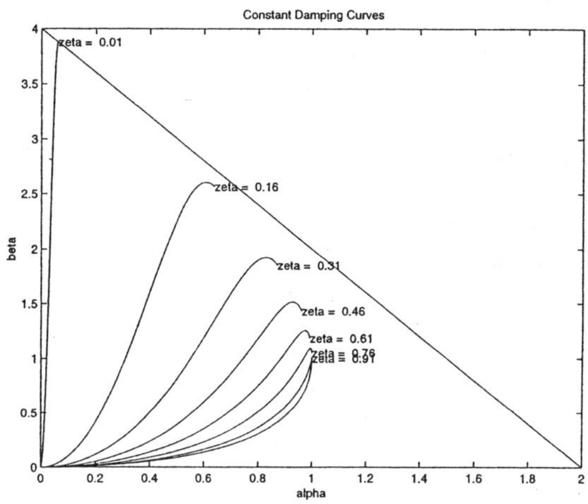

Figure 15. Stability Triangle for the Alpha-Beta Filter

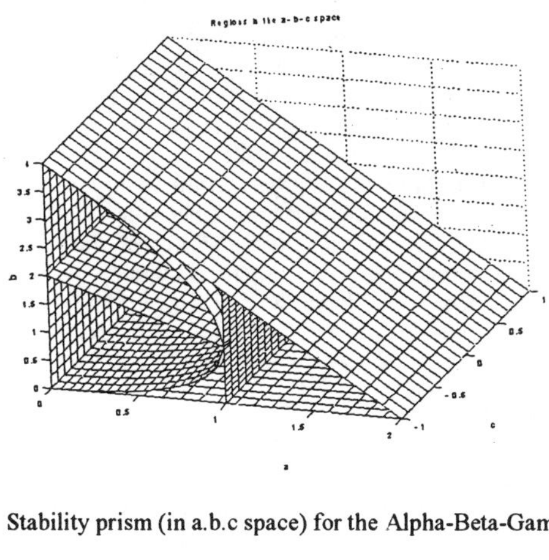

Figure 16. Stability prism (in a.b.c space) for the Alpha-Beta-Gamma Filter

As has been already remarked, the extent to which such formalisms can be applied to more complex L4 process designs is unclear, but the overall spirit of this chapter is an argument for the necessary exploratory research to in fact better understand these bounds.

5. Summary Remarks

This chapter has attempted to reexamine some of the issues and concerns when dealing with either understanding the nature of L2, L3, and L4 DF/IF processing, or when faced with the difficult challenge of designing efficient and effective processes to carry out the associated functions. This is an area that most in the DF/If community would acknowledge is generally under-researched, and it is especially under-researched if one considers the employment of formal, scientifically-grounded methods. This chapter has also thus had a hidden agenda, to pique the interest of readers in the possible issues involved in implementing such formal methods but also in realizing the benefits that may result, and the dangers that lurk if process designs are largely heuristic. While it is recognized that good systems can be developed on the basis of solid experience and empirically-based techniques, the overarching theme of this chapter is that the DF/IF community should at least explore the potential of formal methods, and understand the tradeoffs between the use of such methods and those which are largely informal.

6. References

1. Anon., (1981) *Webster's New Collegiate Dictionary*, G&C Merriam Company, Springfield, MA, USA
2. Llinas, J., (1992) *Situation Awareness*, Lecture Materials, Multisource, Buffalo, NY, USA
3. Doig, W., (Ed.) (19XX) *Data Fusion Taxonomy,* Joint Directors of Laboratories Data Fusion Group, San Diego, CA., USA
4. Szafranski, R., (1995) *A Theory of Information Warfare*, Airpower Journal, USA, Spring
5. Stein, G.J., (1995) *Information Warfare*, Airpower Journal, USA, Spring 1995
6. Egan, J.T., (1990) *Non-Sequential Considerations in Attaining Situation Awareness*, US DoD (JDL/DFG) Data Fusion Symposium, Johns Hopkins University, Laurel, MD USA
7. Zilberstein, S. and Russel, S.J. (1993) *Anytime sensing, planning and action: A Practical model for robot control.* In Proceedings of the Thirteenth International Joint Conference on Artificial Intelligence, pp. 1402-1407, Chambery, France
8. Pos, A. (1993) *Time-Constrained Model-Based Diagnosis.* Masters Thesis, Department of Computer Science, University of Twente, The Netherlands
9. Horvitz, E.J., Suermondt, H.J., and Cooper, G.F. (1989) *Bounded Conditioning: Flexible inference for decision under scarce resources.* In Proceedings of the 1989 Workshop on Uncertainty in Artificial Intelligence, pp. 182-193, Windsor, Ontario, Canada
10. Wellman, M.P. and Liu, C.L. (1994) *State-Space Abstraction for Anytime Evaluation of Probabilistic Networks.* In Proceedings of the 10th Conference on Uncertainty in AI, Seattle, WA.USA
11. Grass, J., (1996) *Reasoning about Computational Resource Allocation: An introduction to anytime algorithms,* http://www.acm.org/crossroads/xrds3-1/racra.html
12 Bisantz, A.M., Llinas, J., Seong, Y., Finger, R., and Jian, J., (2000) *Empirical Investigations of Trust-related System Vulnerabilities in Aided, Adversarial Decision-Making*, Center for Multisource Information Fusion (Rpt no to be assigned), State Univ of NY at Buffalo, USA
13. Zhang, L., Helander, M.G., and Drury, C.G., (1996) *Identifying Factors of Comfort and Discomfort in Sitting*, Human Factors, 38
14. Rempel, J.K., Holmes, J.G., and Zanna, M.P., (1985) Trust in Close Relationships, Jl. Of Personality and Social Psychology, 49, 1, USA
15. Lind, A. T., Dershowitz, A., Chandra, D., and Bussolari, S. R. (1995). *The effect of data link-provided graphical weather images on pilot decision making.* In IFAC proceedings

16. Finger, R. and Bisantz, A.M., (2000) *Utilizing Graphical Formats to Convey Uncertainty in a Decision Making Task,* Proc. Of the IEA 2000/HFES 2000 Triennial Congess, San Diego, CA, USA
17. Post, S. and Sage, A.P., (1990) *An Overview of Automated Reasoning*, IEEE Trans. SMC, 20, 1, Jan/Feb, USA
18. McCarthy, J., *Applications of Circumscription to Formalizing Common-Sense Knowledge,* Artificial Intelligence, 28, USA
19. McDermott, D. and Doyle, J., (1980) *Nonmonotonic Logic I,* Artificial Intelligence, 13, USA
20. McDermott, D., (1982) *Nonmonotonic Logic II: Nonmonotonic Modal Theories*, Jl. ACM, 29, 1, USA
21. Moore, R.C., (1988) *Autoepistemic Logic,* in Non-Standard Logics for Automated Reasoning, P. Smets, E. H. Mamdani, D. Dubois, and H. Prade (eds.) Academic Press, NY, USA
22. Reiter, R., (1980) *A Logic for Default Reasoning,* Artificial Intelligence, 13, USA
23. Utete, S. and Durrant-Whyte, H. F., (1994) *Reliability in decentralized data fusion networks,* IEEE International Conference on Multisensor Fusion and Integration for Intelligent Systems, Piscataway, NJ, USA.
24. Llinas, J. and Deutsch, R.D., (1999) *Intentionality and Intentions: Approaches and Applications; Part II: Perspectives of Intent from Various Bodies Of Literature-Moving Toward New Models of Intent-Estimation and the Development of Decision Aids,* Center for Multisource Information Fusion, State University of New York at Buffalo, Report CMIF-2B-99, Buffalo, NY, USA
25. Bratman, M.E., (1987) *Intentions, Plans, and Practical Reason*, Harvard University Press, Cambridge, MA, USA
26. Cohen, P.R. and Levesque, H.J., (1990) , *Intention is Choice with Commitment,* Artificial Intelligence, 42, 2, March, USA
27. Searle, J.R., (1979) *Expression and meaning : studies in the theory of speech acts,* Cambridge University Press, Cambridge, Eng.
28. Searle, J.R., (1983) *Intentionality, an essay in the philosophy of mind,* Cambridge University Press, Cambridge, Eng.
29. Dennett, D.C. (1969). *Content and Consciousness.* Humanities Press, New York, USA
30. Dennett, D.C. (1978). *Brainstorms : philosophic essays on mind & psychology.* Bradford Books, Montgomery, Vt., USA
31. Dennett, D.C. (1987) *The Intentional Stance.* MIT Press, Cambridge, Mass., USA
32. Grice, H. P. (1989) *Studies in the Way of Words.* Harvard University Press, Cambridge, Mass., USA
33. Bennett, J. (1976) Linguistic Behaviour. Cambridge University Press. Cambridge, Eng.. 34. bowman and steinberg on revised model
34. Steinberg, A.N., Bowman, C.L., and White, F.E. (1999) *Revisions to the JDL data fusion model ,* Proceedings of Spie, v 3719 p 430-441
35. Llinas, J., Singh, T., Chen, L., Tenne, D., and Wang, S. (1998) *Fuzzy Mathematics and Logic for Multisource Data Association and Target Tracking,* Center for Multisource Information Fusion, State Univ of New York at Buffalo, Report CMIF-3-98, Buffalo, NY, USA, Aug.
36. Llinas, J., Singh, T., Chen, L., and Tenne, D., (1998) *Fuzzy Mathematics and Logic for Multisource Data Association and Target Tracking (Continuation),* Center for Multisource Information Fusion, State Univ of New York at Buffalo, Report CMIF-4-98, Buffalo, NY, USA, Sept.
37. Sklansky, J., (1957) *Optimizing the Dynamic Parameter of a Track-While-Scan System,* RCA Laboratories Journal, Princeton, NJ, USA, June
38. Benedict, T.R. and Bordner, G.W., (1962) *Synthesis of an Optimal Set of Radar Track-While-Scan Smoothing Equations,* IRE Trans on Automatic Control, Vol. AC-1, USA, July
39. Yeung, K S. (1982) *On the Schur-Cohn-Jury Criterion,* International Journal of Control. v 35 n 3 Mar, p 573-574

AN INTRODUCTION TO DISTRIBUTED DETECTION THEORY

P.K. VARSHNEY
Syracuse University
Electrical Engineering and Computer Science Dept.
121 Link Hall
Syracuse, NY 13244 USA

The quality of decision is like the well-timed swoop of a falcon which enables it to strike and destroy its victim.

Sun Tzu (6th-5th century B.C.), Chinese general

1. Introduction

Everyone faces decision-making problems at each step of his or her life. A timely decision of high quality can spell the difference between success and failure. In a decision-making scenario, observations regarding a certain phenomenon are available and one needs to select a particular course of action from several possibilities. Such problems arise in many applications. For example, in a radar detection context, a decision is to be made regarding the presence or absence of a target based on the radar return. In a digital communication system, one of several possible waveforms is transmitted over a channel. Based on the received noisy observation, we need to determine the symbol that was transmitted. In a biomedical application, based on a smear of human tissue, one needs to determine if it is cancerous. In a pattern recognition problem, one needs to determine the type of aircraft being observed based on some aircraft features. In all of the above applications, the common underlying problem is to make a decision among several possible choices. This is carried out based on available noisy observations. The branch of statistics dealing with these types of problems is known as statistical decision theory or hypothesis testing, [1-2]. In the context of radar and communication theory, it is known as detection theory, [3-6].

Conventional decision theoretic problems involved a single decision-maker and centralized data processing. There are many real world situations where group decision making structures are found. Application areas include financial institutions, air-traffic control, oil exploration, medical diagnosis, military command and control, electric power networks, weather prediction, and industrial organizations. For example, a medical doctor may order multiple diagnostic tests and seek additional peer opinions before a major surgical procedure is carried out, or a military commander may use data from radar and IR sensors along with intelligence information while deciding whether or not to launch an

A. K. Hyder et al. (eds.), Multisensor Fusion, 163–182.

offensive. In many applications, multiple decision makers arise naturally, e.g., managers in an industrial organization. In many other applications, additional decision makers are employed to improve system performance. For example, deployment of multiple sensors for signal detection in a military surveillance application improves system survivability, results in improved detection performance or in a shorter decision time to attain a prespecified performance level, and may provide increased coverage in terms of surveillance region and number of targets.

A variety of organizational structures for group decision-making systems can be formulated. The specific structure to be used depends upon the application and factors such as the distribution of decision makers over the region of interest, computational ability of each decision maker, and communication or organizational constraints. In general, individual decision makers coordinate their actions to achieve the goal of the team decision-making system. In a commonly used organizational structure multiple team members observe a common phenomenon, process them, and send their processed results over limited bandwidth channels to a fusion center that yields the global inference. This topology is known as a parallel fusion network. Information conveyed to the fusion center can be in the form of hard decisions, soft decisions (multi-level decisions instead of binary decisions), likelihood ratios, or quantized observations. Since the fusion center of such a system has access to less information compared to a system in which raw observations are available at the fusion center, system performance with such an architecture degrades. But its many advantages that include survivability and reduced communication requirements make it an attractive alternative for distributed detection systems. Other distributed detection network topologies include a serial network, a hierarchical tree network, or a combination of these basic topologies. Feedback may also be introduced.

In order to design optimum (or at least near-optimum) distributed detection networks, an extension of conventional statistical decision/detection theory is required. This extension is non-trivial and the design is computationally intensive. This article introduces the design problem by solving two specific problems. Solutions of other more complex problems and a more detailed presentation of the field can be found in [7]. In Section 2, the conventional signal detection problem is briefly presented. It serves to introduce the notation and methodology employed in Section 3. Two relatively simple distributed detection problems are solved in Section 3. Some final comments are made in Section 4.

2. The Signal Detection Problem

Consider the simple binary hypothesis testing problem with the two hypotheses denoted by H_0 and H_1. The noisy observation is denoted as y so that the conditional densities under the two hypotheses are $p(y \mid H_i)$, $i = 0, 1$. These are assumed known. The a priori probabilities of the two hypotheses are denoted by P_0 and P_1 respectively. In the binary hypothesis testing problem, four possible courses of action can occur. Two of these correspond to correct decisions and two to errors. We assign costs to each of these possibilities. Specifically, C_{ij}, $i = 0, 1$, $j = 0, 1$, represents the cost of declaring H_i true when H_j is present. In the Bayesian formulation, the decision rule that minimizes the average cost is determined. The average cost or Bayes risk function, $\Re$, is given by

$$\Re = \sum_{i=0}^{1} \sum_{j=0}^{1} C_{ij} P_j \int_{Z_i} p(y \mid H_j) dy, \tag{2.1}$$

where Z_i is the decision region corresponding to hypothesis H_i, i.e., hypothesis H_i is declared true for any observation falling in the region Z_i. Let Z be the entire observation space so that $Z = Z_0 \cup Z_1$ and $Z_0 \cap Z_1 = \varphi$ the null set. We may expand (2.1) which after some manipulations can be written as

$$\begin{aligned} \Re = P_0 C_{10} + P_1 C_{11} + \int_{Z_0} \{[P_1 (C_{01} - C_{11})\, p(y \mid H_1)] \\ - P_0 (C_{10} - C_{00})\, p(y \mid H_0)]\}\, dy \end{aligned} \tag{2.2}$$

The first two terms on the right hand side are fixed. The risk $\Re$ can be minimized by assigning those points of Z to Z_0 that make the integrand of (2.2) negative. Assuming $C_{10} > C_{00}$ and $C_{01} > C_{11}$, the minimization results in the likelihood ratio test (LRT)

$$\frac{p(y \mid H_1)}{p(y \mid H_0)} \underset{H_0}{\overset{H_1}{\gtrless}} \frac{P_0 (C_{10} - C_{00})}{P_1 (C_{01} - C_{11})}. \tag{2.3}$$

The quantity on the left hand side is known as the likelihood ratio denoted by $\Lambda(y)$ and the quantity on the right hand side is the threshold η. A special case of the Bayes test occurs when we set $C_{00}=C_{11}=0$ and $C_{10}=C_{01}=1$, i.e., the cost of a correct decision is set to zero and the cost of an error is set to unity. In this case,

$$\Re = P_0 \int_{Z_1} p(y \mid H_0) dy + P_1 \int_{Z_0} p(y \mid H_1) dy, \tag{2.4}$$

which is just the average probability of error. In this case, the Bayes test simply minimizes the average probability of error. The threshold η, in this case, is given by P_0/P_1. The resulting receivers are called minimum probability of error receivers.

As indicated earlier, two types of errors can occur in decision making. When H_0 is true and H_1 is declared true, an error of the first kind is said to occur. In radar terminology, it is known as a false alarm, and the associated conditional probability of error is known as the probability of false alarm, P_F. An error of the second kind, a miss, is said to occur if H_0 is declared true when H_1 is present. The associated conditional probability of error is known as the probability of miss, P_M. Then,

$$P_F = P(\text{Decide } H_1 \mid H_0 \text{ present}) = \int_{Z_1} p(y \mid H_0)\, dy, \tag{2.5}$$

and

$$P_M = P(\text{Decide } H_0 \mid H_1 \text{ present}) = \int_{Z_0} p(y \mid H_1)\, dy, \tag{2.6}$$

We also define the probability of detection P_D as the probability of declaring H_1 true when H_1 is present, i.e.,

$$P_D = 1 - P_M = \int_{Z_1} p(y \mid H_1)\, dy. \tag{2.7}$$

The performance of LRTs can be conveniently described in terms of these probabilities and a graph known as the receiver operating characteristic (ROC). This is a plot of the probability of detection P_D against the probability of false alarm P_F. The ROCs of all continuous LRTs are concave downward and lie above the $P_D = P_F$ line.

3. Decentralized Signal Detection Problems

Next, we consider decision-making problem where multiple detectors solve a common signal detection problem in a cooperative manner. The detectors may be geographically distributed or may be confined to a smaller area. They are connected in a network whose topological and communication structures are pre-defined. Here we present solutions to two distributed detection problems. More details on these solutions and solutions of other problems are available in [7].

3.1. DISTRIBUTED DETECTION WITHOUT DECISION COMBINING

We consider a two-detector system consisting of two decision makers DM1 and DM2 as shown in Figure 1. Both detectors observe a common phenomenon. They make local decisions that are not combined to yield a global inference. Decisions at the two detectors are coupled; although the decisions are independent, the cost assignment is a function of

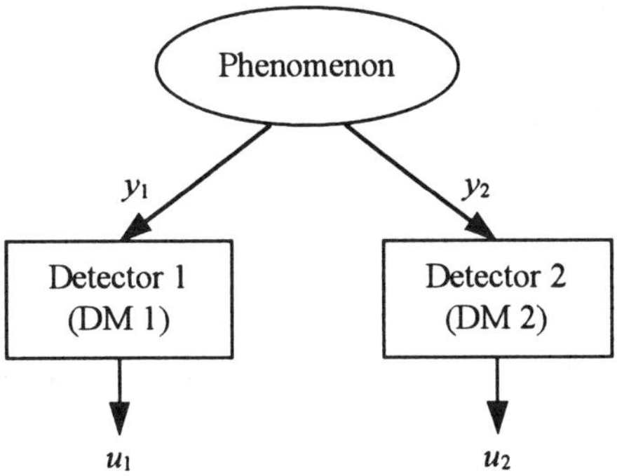

Figure 1. A two-detector parallel fusion network without fusion.

both decisions. Optimization is carried out in a system-wide manner. As before, let H_0 and H_1 be the two hypotheses with P_0 and P_1 denoting the associated a priori probabilities. The observations at the two detectors are denoted by y_1 and y_2, respectively. The joint conditional density functions under the two hypotheses are $p(y_1, y_2 | H_i)$, $i = 0, 1$. The decisions at the two detectors are denoted by

$$u_i = \begin{cases} 0, & H_0 \text{ is declared present.} \\ 1, & H_1 \text{ is declared present.} \end{cases} \tag{3.1}$$

The local decisions u_i are based only on their local observations y_i, i.e., there is no communication between the detectors. The costs of different courses of action are denoted by C_{ijk}, $i, j, k = 0, 1$, where C_{ijk} represents the cost of detector 1 deciding H_i, detector 2 deciding H_j when H_k is present. The objective is to obtain decision rules at the two detectors that minimize the average cost of decision making. The Bayes risk function is given by

$$\begin{aligned} \mathfrak{R} &= \sum_{i,j,k} \int_{y_1,y_2} p(u_1, u_2, y_1, y_2, H_k) C_{ijk} \\ &= \sum_{i,j,k} \int_{y_1,y_2} P_k C_{ijk} \, p(u_1, u_2 \mid y_1, y_2, H_k) p(y_1, y_2 \mid H_k) \; . \end{aligned} \tag{3.2}$$

Note that the local decisions u_1 and u_2 are independent and are made based only on their local observations y_1 and y_2, respectively. Furthermore, they do not depend upon the hypothesis present. Therefore,

$$\mathfrak{R} = \sum_{i,j,k} \int_{y_1,y_2} P_k \, C_{ijk} \; p(u_1 | y_1) \; p(u_2 | y_2) \; p(y_1, y_2 | H_k) \; . \tag{3.3}$$

We expand the right-hand side by explicitly summing over u_1 and get

$$\Re = \sum_{j,k} \int_{y_1,y_2} P_k p(u_2 \mid y_2) p(y_1,y_2 \mid H_k) \times \left[[C_{0jk} p(u_1 = 0 \mid y_1) + C_{1jk} p(u_1 = 1 \mid y_1) \right] . \tag{3.4}$$

Using $p(u_1 = 1 \mid y_1) = 1 - p(u_1 = 0 \mid y_1)$, and ignoring a constant term with respect to u_1 we express $\Re$ as

$$\Re = \int_{y_1} p(u_1 = 0 \mid y_1) \sum_{j,k} \int_{y_2} P_k p(u_2 \mid y_2) p(y_1, y_2 \mid H_k)[C_{0jk} - C_{1jk}] . \tag{3.5}$$

$\Re$ is minimized if we set

$$p(u_1 = 0 \mid y_1) = \begin{cases} 0, & if \ \sum_{j,k} \int_{y_2} P_k p(u_2 \mid y_2) p(y_1, y_2 \mid H_k)[C_{0jk} - C_{1jk}] \geq 0. \\ 1, & otherwise . \end{cases} \tag{3.6}$$

Note that the decision rule at detector 1 is a deterministic rule (not a randomized rule) regardless of the costs, decision rule at the other detector, and the joint conditional density of the observations. Therefore, we can express (3.6) as the decision rule

$$\sum_{j,k} \int y_2 P_k p(u_2 \mid y_2) p(y_1, y_2 \mid H_k) \left[C_{0jk} - C_{1jk} \right] \underset{u_1 = 0}{\overset{u_1 = 1}{\gtrless}} 0 . \tag{3.7}$$

We make the assumption that $C_{0j0} < C_{1j0}$, i.e., the cost of detector 1 making an error when H_0 is present is more than the cost of it being right regardless of the decision of detector 2. After some manipulations, the decision rule at DM1 can be expressed as

$$\Lambda(y_1) \underset{u_1 = 0}{\overset{u_1 = 1}{\gtrless}} \frac{P_0 \sum_j \int y_2 \ p(u_2 \mid y_2) p(y_2 \mid y_1, H_0)[C_{1j0} - C_{0j0}]}{P_1 \sum_j \int y_2 \ p(u_2 \mid y_2) p(y_2 \mid y_1, H_1)[C_{0j1} - C_{1j1}]} , \tag{3.8}$$

where $\Lambda(y_1)$ represents the likelihood ratio at detector 1 and is given by

$$\Lambda(y_1) = \frac{p(y_1 \mid H_1)}{p(y_1 \mid H_0)} \ .$$

Note that the right-hand side (RHS) of (3.8) is not a simple constant threshold. It is data dependent, i.e., it depends on y_1 due to the appearance of the term $p(y_2 \mid y_1, H_k)$, k = 0, 1. Also, the right-hand side is a function of the decision rule at the other detector due to the term $p(u_2 \mid y_2)$.

The situation simplifies considerably if y_1 and y_2 are assumed to be conditionally independent, given any hypothesis. In this case, the right-hand side of (3.8) is no longer data dependent. Under this condition, the right-hand side of (3.8) reduces to a threshold t_1 which still depends on the decision rule of the other detector, i.e., its threshold t_2. The threshold t_1 is given by

$$t_1 = \frac{P_0 \sum_j \int_{y_2} p(u_2|y_2)\, p(y_2|H_0)\, [C_{1j0} - C_{0j0}]}{P_1 \sum_j \int_{y_2} p(u_2|y_2)\, p(y_2|H_1)\, [C_{0j1} - C_{1j1}]} \ . \tag{3.9}$$

which can be expressed as

$$t_1 = \frac{P_0 \int_{y_2} p(y_2|H_0) \left\{ [C_{110} - C_{010}] + p(u_2 = 0|y_2)[C_{100} - C_{000} + C_{010} - C_{110}] \right\}}{P_1 \int_{y_2} p(y_2|H_1) \left\{ [C_{011} - C_{111}] + p(u_2 = 0|y_2)[C_{001} - C_{101} + C_{111} - C_{011}] \right\}} \ . \tag{3.10}$$

The threshold t_1 is a function of $p(u_2 = 0 \,|\, y_2)$ which specifies the decision rule at the second detector, and, thus, it is a function of t_2. Therefore,

$$t_1 = f_1(t_2), \tag{3.11}$$

where $f_1(\cdot)$ is the function defined in (3.10). Similarly,

$$t_2 = f_2(t_1), \tag{3.12}$$

where $f_2(\cdot)$ has a form similar to $f_1(\cdot)$. Equations (3.11) and (3.12) specify necessary conditions that the thresholds t_1 and t_2 must satisfy if they are to be optimal. These conditions are not sufficient. Their solution provides locally optimum solutions. When there are several local minima, each must be examined to determine the globally optimum solution. It is important to observe that the two thresholds are coupled and this is a result of our objective of system-wide optimization. In general, the resulting thresholds are not the same as would be obtained if each local detector was optimized independently.

Example 1: Consider the cost assignment

$$C_{000} = C_{111} = 0,$$

$$C_{010} = C_{100} = C_{011} = C_{101} = 1,$$

$$C_{110} = C_{001} = k. \tag{3.13}$$

This cost function is symmetric in u_1 and u_2. All cases of equal number of errors are assigned equal costs. Let us assume that the two local observations are conditionally independent, identical and are Gaussian distributed. The conditional densities are given by

$$p(y_1 \mid H_j) = \frac{1}{\sigma\sqrt{2\pi}} \exp\left(-\frac{(y_i - m_j)^2}{2\sigma^2} \right), \; i = 1,2 \; ; \; j = 0, 1; \; -\infty < y_i < \infty \tag{3.14}$$

Assume that the a priori probabilities P_0 and P_1 are equal. The likelihood ratio at the two detectors is given by

$$\begin{aligned} \Lambda(y_i) &= \frac{p(y_i | H_1)}{p(y_i | H_0)} \\ &= \exp\left(\frac{1}{2\sigma^2} [2y_i (m_1 - m_0) - (m_1^2 - m_0^2)] \right), \quad i = 1,2 \end{aligned} \tag{3.15}$$

The likelihood ratio test at each detector is given by

$$\Lambda(y_i) \mathop{\gtrless}\limits^{u_i = 1}_{u_i = 0} t_i \, , \tag{3.16}$$

which reduces to

$$y_i \underset{u_i=0}{\overset{u_i=1}{\gtrless}} \frac{2\sigma^2 \log t_i + (m_1^2 - m_0^2)}{2(m_1 - m_0)}, \quad i = 1, 2\ . \tag{3.17}$$

The thresholds t_1 and t_2 can be obtained from (3.10) and its companion equation. The threshold t_1 is given by

$$t_1 = \frac{P_0 \int_{y_2} p(y_2 \mid H_0)\{(k-1) + p(u_2 = 0 \mid y_2)[2-k]\}}{P_1 \int_{y_2} p(y_2 \mid H_1)\{1 + p(u_2 = 0 \mid y_2)[k-2]\}} \tag{3.18}$$

$$= \frac{(k-1)\int_{y_2} p(y_2 \mid H_0) + (2-k)\int_{y_2} p(u_2 = 0 \mid y_2) p(y_2 \mid H_0)}{\int_{y_2} p(y_2 \mid H_1) + (k-2)\int_{y_2} p(u_2 = 0 \mid y_2) p(y_2 \mid H_1)}\ .$$

It reduces to

$$t_1 = \frac{(k-1) + (2-k) p(u_2 = 0 \mid H_0)}{1 + (k-2)\, p(u_2 = 0 \mid H_1)}\ . \tag{3.19}$$

For Gaussian observation statistics, the probabilities $p(u_2 = 0 \mid H_j)$, $j = 0, 1$, can be easily obtained to be

$$p(u_2 = 0 \mid H_0) = \operatorname{erf}\left(\frac{\sigma \log t_2}{m_1 - m_0} + \frac{m_1 - m_0}{2\sigma}\right),$$

and

$$p(u_2 = 0 \mid H_1) = \operatorname{erf}\left(\frac{\sigma \log t_2}{m_1 - m_0} - \frac{m_1 - m_0}{2\sigma}\right),$$

where

$$\operatorname{erf}(x) = \int_{-\infty}^{x} \frac{1}{\sqrt{2\pi}} \exp\left(-\frac{y^2}{2}\right) dy\ .$$

Substituting these probabilities into (3.19) and its companion equation,

$$t_1 = \frac{(k-1) + (2-k)\,\mathrm{erf}\left(\frac{\sigma \log t_2}{m_1 - m_0} + \frac{m_1 - m_0}{2\sigma}\right)}{1 + (k-2)\,\mathrm{erf}\left(\frac{\sigma \log t_2}{m_1 - m_0} - \frac{m_1 - m_0}{2\sigma}\right)}, \tag{3.20}$$

and

$$t_2 = \frac{(k-1) + (2-k)\,\mathrm{erf}\left(\frac{\sigma \log t_1}{m_1 - m_0} + \frac{m_1 - m_0}{2\sigma}\right)}{1 + (k-2)\,\mathrm{erf}\left(\frac{\sigma \log t_1}{m_1 - m_0} - \frac{m_1 - m_0}{2\sigma}\right)}, \tag{3.21}$$

The solution of (3.20) and (3.21) yields the thresholds t_1 and t_2. As noted before, multiple solutions may result, and each one must be examined to determine the best solution. Consider the solution of (3.20) and (3.21) for different values of k. Assume that $m_0 = 0$, $m_1 = 1$, and $\sigma = 1$. For $1 \le k < 4.528$, there is only one solution $t_1 = t_2 = 1$ that yields the minimum value of $\Re$. For $k \ge 4.528$, there are three solutions. One of the solutions is $t_1 = t_2 = 1$ but it does not yield the minimum value of $\Re$. The other two solutions need to be used in a pair, i.e., one threshold is set equal to one solution and the other threshold is set equal to the other solution. This setting of unequal thresholds yields the minimum value of $\Re$. Threshold values as a function of k are shown in Figure 2. Note that an interesting situation has arisen in this example. Even though observations at each detector are identically distributed and the cost function is symmetric, unequal thresholds are optimal for $k \ge 4.528$.

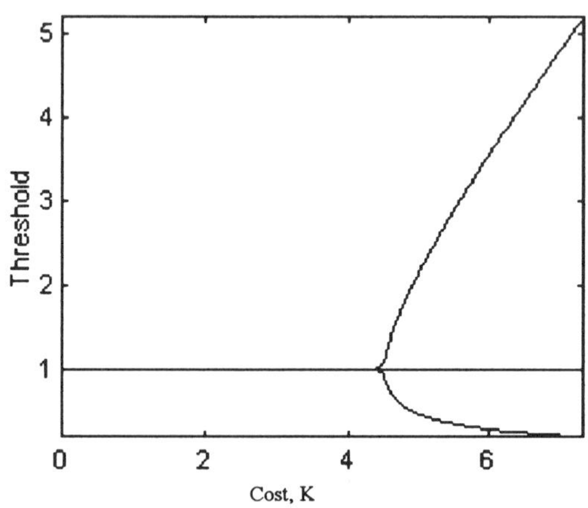

Figure 2. Threshold values as a function of k for Example 1

This is due to the large cost assigned to double errors, and the optimum action on the part of the detectors is to employ different thresholds.

The results obtained in this section can be extended to cases with multiple detectors and involving multiple hypotheses.

3.2 THE DECISION FUSION PROBLEM

In Section 3.1, the conventional hypothesis testing problem was extended to a decentralized scenario where multiple detectors solved a common decision-making problem based on their own observations. Decisions were made locally and no attempt was made to combine them. Here, we consider the problem of combining multiple decisions that have been made. We do not consider the problem of decision-making itself. We focus only on the problem of decision combining or decision fusion.

Consider the binary hypothesis testing problem with hypotheses H_0 and H_1 and their associated prior probabilities P_0 and P_1. There are N local decisions u_i, $i = 1, \ldots, N$, where

$$u_i = \begin{cases} 0, & \text{if detector } i \text{ decides } H_0, \\ 1, & \text{otherwise.} \end{cases} \tag{3.22}$$

As shown in Figure 3, local decisions form the input to the fusion center which combines them to yield the global decision u_0,

$$u_0 = \begin{cases} 0, & \text{if } H_0 \text{ is decided,} \\ 1, & \text{otherwise.} \end{cases} \tag{3.23}$$

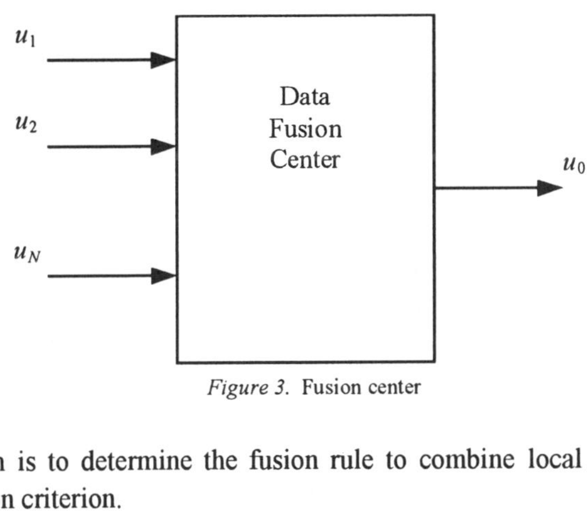

Figure 3. Fusion center

The problem is to determine the fusion rule to combine local decisions based on some optimization criterion.

The fusion rule is essentially a logical function with N binary inputs and one binary output. In general, there are 2^{2^N} fusion rules when there are N binary inputs to the fusion center. For example, there are 16 possible fusion rules for combining two binary local decisions as listed in Table 1. Note that commonly used logical functions AND and OR are 2 of the 16 logical functions. The fusion rule f_2 represents the AND rule in which $u_0 = 1$ only when both of the local decisions are 1, i.e.,

$$u_0 = \begin{cases} 1, & \text{if } u_1 = 1 \text{ and } u_2 = 1. \\ 0, & \text{otherwise.} \end{cases} \tag{3.24}$$

TABLE 1 Possible Fusion Rules for Two Binary Decisions

Input		Output u_0															
u_1	u_2	f_1	f_2	f_3	f_4	f_5	f_6	f_7	f_8	f_9	f_{10}	f_{11}	f_{12}	f_{13}	f_{14}	f_{15}	f_{16}
0	0	0	0	0	0	0	0	0	0	1	1	1	1	1	1	1	1
0	1	0	0	0	0	1	1	1	1	0	0	0	0	1	1	1	1
1	0	0	0	1	1	0	0	1	1	0	0	1	1	0	0	1	1
1	1	0	1	0	1	0	1	0	1	0	1	0	1	0	1	0	1

The OR rule is given by f_8 in which $u_0 = 1$ as long as at least one of the local decisions is 1, i.e.,

$$u_0 = \begin{cases} 0, & \text{if } u_1 = 0 \text{ and } u_2 = 0, \\ 1, & \text{otherwise.} \end{cases} \tag{3.25}$$

Several of the sixteen functions may not be suitable as fusion rules. For example, the all zero function f_1 and the all one function f_{16} totally disregard the two inputs and, therefore, may not be preferred in many situations. Similarly, some other functions that may be unsuitable totally disregard one of the inputs, e.g., the function f_4 totally disregards u_2. In most cases, only monotic fusion rules need to be considered. For a definition, See [7].

An ad hoc approach to the design of fusion rules would be to arbitrarily select a fusion rule from the set of commonly used logical functions, such as the AND, the MAJORITY, and the OR functions. An arbitrary choice of the fusion rule may provide satisfactory results in many instances, but optimality is not guaranteed. Next, we consider the design of optimum fusion rules in a Bayesian framework. Each input to the fusion center is a binary random variable characterized by the associated probabilities of false alarm and miss. Let P_{Fi}, P_{Di} and P_{Mi} denote the probabilities of false alarm, detection and miss of detector i respectively, i.e.,

$$P_{Fi} = P(u_i = 1 \mid H_0),$$

$$P_{Di} = P(u_i = 1 \mid H_1),$$

and

$$P_{Mi} = P(u_i = 0 \mid H_1).$$

The probabilities of false alarm, detection, and miss of the overall system, i.e., at the fusion center, denoted by P_F, P_D, and P_M, are given by the following, respectively:

$$P_F = P(u_0 = 1 \mid H_0),$$

$$P_D = P(u_0 = 1 \mid H_1),$$

and

$$P_M = P(u_0 = 0 \mid H_1).$$

The objective is to determine the fusion rule or decision rule at the fusion center that minimizes the average cost. The problem can be viewed as a two hypothesis detection problem with individual detector decisions being the observations. The optimum fusion rule is given by the following likelihood ratio test:

$$\frac{P(u_1, u_2, \ldots, u_N \mid H_1)}{P(u_1, u_2, \ldots, u_N \mid H_0)} \underset{u_0 = 0}{\overset{u_0 = 1}{\gtrless}} \frac{P_0 (C_{10} - C_{00})}{P_1 (C_{01} - C_{11})} \triangleq \eta, \tag{3.26}$$

where C_{ij} denotes the cost of global decision being H_i, when H_j is present. Due to the independence of local decisions, the left-hand side (LHS) of (3.26) can be written as

$$\frac{P(u_1,u_2,\ldots,u_N \mid H_1)}{P(u_1,u_2,\ldots,u_N \mid H_0)} = \prod_{i=1}^{N} \frac{P(u_i \mid H_1)}{P(u_i \mid H_0)}$$

$$= \prod_{S_1} \frac{P(u_i = 1 \mid H_1)}{P(u_i = 1 \mid H_0)} \prod_{S_0} \frac{P(u_i = 0 \mid H_1)}{P(u_i = 0 \mid H_0)} \tag{3.27}$$

$$= \prod_{S_1} \frac{1-P_{Mi}}{P_{Fi}} \prod_{S_0} \frac{P_{Mi}}{1-P_{Fi}} ,$$

where S_j is the set of all those local decisions that are equal to j, $j = 0, 1$. Substituting (3.27) in (3.26) and taking the logarithm of both sides,

$$\sum_{S_1} \log \frac{1-P_{Mi}}{P_{Fi}} + \sum_{S_0} \frac{P_{Mi}}{1-P_{Fi}} \underset{u_0=0}{\overset{u_0=1}{\gtrless}} \log \eta , \tag{3.28}$$

or,

$$\sum_{i=1}^{N} \left[u_i \log \frac{1-P_{Mi}}{P_{Fi}} + (1-u_i) \log \frac{P_{Mi}}{1-P_{Fi}} \right] \underset{u_0=0}{\overset{u_0=1}{\gtrless}} \log \eta . \tag{3.29}$$

Thus, the optimum fusion rule can be implemented by forming a weighted sum of the incoming local decisions and, then, comparing it with a threshold. The weights and the threshold are determined by the reliability of the decisions, i.e., by the probabilities of miss and false alarm of the local detectors. The threshold also depends on the prior probabilities and the costs.

Example 2:

Consider the fusion of two local decisions. Let the probabilities of false alarm and detection be given by $P_{F_i} = 0.25$ and $P_{M_i} = 0.3$, $i = 1, 2$. It is desired to minimize the system probability of error.

Out of the sixteen possible fusion rules shown in Table 1, six are monotonic and need to be considered further. Since the two sensors are statistically identical, u_0 should not change if u_1 and u_2 are exchanged. Therefore, only four fusion rules need to be considered: the ONE rule, the AND rule, the OR rule and the ZERO rule. These rules are listed in Table 2. The optimum fusion rule is determined for different values of prior probability P_0. The result is given in Table 3. The minimum probability of error Pe is plotted against P_0 in Figure 4. We notice that the ONE and the ZERO rules are optimal for certain values of P_0.

TABLE 2. Fusion rules for two binary decisions

Input	Output u_0			
u_1 u_2	ZERO	AND	OR	ONE
0 0	0	0	0	1
0 1	0	0	1	1
1 0	0	0	1	1
1 1	0	1	1	1

TABLE 3. Optimal fusion rules for Example 2

Range of values for P_0	Optimal fusion rules
0 – 0.137	ONE
0.137 – 0.528	OR
0.528 – 0.886	AND
0.886 – 1	ZERO

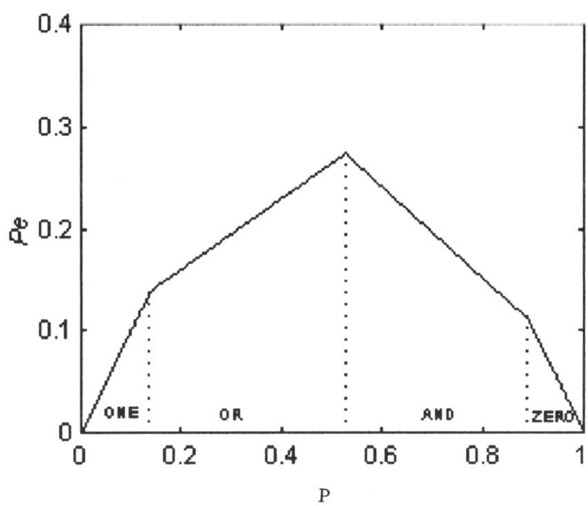

Figure 4. Probability of error vs. prior probability for Example 2

Example 3:

Consider the fusion of multiple local decisions. Let the probabilities of false alarm and detection be given by $P_{F_i} = P_{M_i} = 0.1$, $i = 1, \cdots, n$. The objective again is to minimize the system probability of error.

Since the sensors are statistically identical, the optimal fusion rules are K-out-of-N rules [7]. By a K-out-of-N rule, we mean that if K or more incoming decisions are H_1 then the global decision is H_1. For fixed $P_0 = 0.5$, the optimal fusion rules are given in Table 4 with N ranging from 1 to 5.

Because multiple sensors are used, the fusion system has some robustness against degraded sensor decisions. This degradation includes sensor defects, communication link failures and environment changes. We consider four levels of degradation. In the normal case, there is no degradation. In the sensor defect case, we let $P_{F_i} = P_{M_i} = 0.3$. In the neutral sensor case, we let $P_{F_i} = P_{M_i} = 0.5$. In the negative sensor case, we have $P_{F_i} = P_{M_i} = 1$. With the fusion rule given in Table 4, we compute the probability of error that is associated with one degraded sensor decision. In Figure 5, these probabilities of error are plotted against the number of sensors. One can see that the effect of degradation in a sensor decision diminishes when the number of sensors increases. In fact, the fusion system that employs five sensors always works at least as well as a single good sensor.

TABLE 4. Optimal fusion rules for Example 3

Number of sensor N	Optimal K-out-of-N rules
1	1-out-of-1
2	1-out-of-2
3	2-out-of-3
4	2-out-of-4
5	3-out-of-5

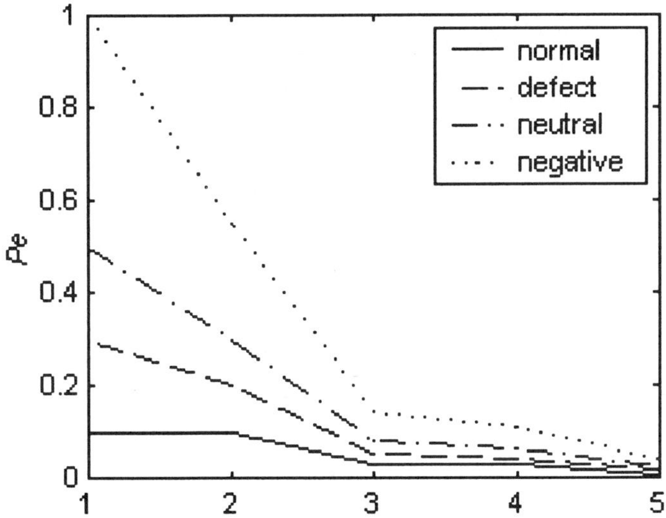

N, Number of Sensors

Figure 5. Probability of error vs. number of sensors

3.3 FUSION RULE FOR THE SOFT DECISION CASE

In this section thus far, we have considered the design of fusion rules for the case when the local detectors could make only hard decisions, i.e., u_i could take only two values 0 or 1 corresponding to the two hypotheses H_0 and H_1. Next, we generalize the design of fusion rules for the binary hypothesis testing problem to include the case in which the local detectors are allowed to make multilevel or soft decisions. Assume that the observation space at each local detector is partitioned into J mutually exclusive regions so that, if the observation at detector i lies in the partition j, we set $u_i = j$, $j = 0, ..., J - 1$. These soft decisions are transmitted to the fusion center which combines them to yield the global decision u_0 as defined earlier in (3.23). Let us define the following probabilities:

$$\alpha_{ji} = P(u_i = j \mid H_0) , \tag{3.30}$$

and

$$\beta_{ji} = P(u_i = j \mid H_1) , \tag{3.31}$$

The fusion rule that minimizes the average cost is still given by (3.26) but now the LHS can be expressed as

$$\frac{P(u_1,u_2,\ldots,u_N \mid H_1)}{P(u_1,u_2,\ldots,u_N \mid H_0)} = \prod_{i=1}^{N} \frac{P(u_i \mid H_1)}{P(u_i \mid H_0)}$$

$$= \prod_{j=0}^{J-1} \prod_{S_j} \frac{P(u_i = j \mid H_1)}{P(u_i = j \mid H_0)} \tag{3.32}$$

$$= \prod_{j=0}^{J-1} \prod_{S_j} \frac{\beta_{ji}}{\alpha_{ji}} ,$$

where S_j is the set of all those local decisions u_i that are equal to j, $j = 0, 1, \ldots, J\text{-}1$. Substituting (3.32) in (3.26) and taking the logarithm of both sides, we obtain the minimum average cost fusion rule as

$$\sum_{j=0}^{J-1} \sum_{S_1} \log\left(\frac{\beta_{ji}}{\alpha_{ji}}\right) \mathop{\gtrless}_{u_0=0}^{u_0=1} \log \eta . \tag{3.33}$$

This fusion rule is an obvious generalization of the fusion rule for the hard decision case. Once again, the fusion rule depends on the reliabilities of the local detectors, i.e., their probabilities of errors.

3.4 FUSION RULE WITH DIRECT OBSERVATIONS

The fusion center considered thus far accepts the set of hard or soft decisions from the local detectors. It can not observe the phenomenon directly. Next, we consider the case where the fusion center also receives direct observations regarding the phenomenon. Let y_0 denote the direct observations at the fusion center. The inputs to the fusion center are a continuous random variable y_0 and the discrete random vector u. The optimum fusion rule in this case is given by the following likelihood ratio test:

$$\frac{p(y_0 \mid H_1)}{p(y_0 \mid H_0)} \frac{P(u_1,u_2,\ldots,u_N \mid H_1)}{P(u_1,u_2,\ldots,u_N \mid H_0)} \mathop{\gtrless}_{u_0=0}^{u_0=1} \eta , \tag{3.34}$$

Using (3.32) and denoting the likelihood ratio test for the observation y_0 as $\Lambda(y_0)$, it can be written as

$$\Lambda(y_0) \prod_{j=0}^{J-1} \prod_{S_j} \frac{\beta_{ji}}{\alpha_{ji}} \underset{u_0=0}{\overset{u_0=1}{\gtrless}} \eta . \tag{3.35}$$

This decision rule can also be expressed as

$$\Lambda(y_0) \underset{u_0=0}{\overset{u_0=1}{\gtrless}} \eta \prod_{j=0}^{J-1} \prod_{S_j} \frac{\alpha_{ji}}{\beta_{ji}} , \tag{3.36}$$

where the threshold on the RHS is a function of the decision vector u and can take J^N possible values. Taking the logarithm of both sides, we may express the fusion rule in another form, a generalized version of (3.33),

$$\log \Lambda(y_0) + \sum_{j=0}^{J-1} \sum_{S_j} \log \left(\frac{\beta_{ji}}{\alpha_{ji}} \right) \underset{u_0=0}{\overset{u_0=1}{\gtrless}} \log \eta \quad , \tag{3.37}$$

This fusion rule minimizes the average cost in the case where the fusion center can observe the phenomenon directly, and it also receives soft decisions from the local detectors.

4. Concluding Remarks

Distributed detection theory represents an extension of conventional statistical decision theory that is applicable to the situation where a team of decision makers solve a hypothesis testing problem in a cooperative manner. The theory is particularly applicable to distributed multisensor data fusion problems with constrained data links. This topic was introduced here by discussing two problems in the area of decentralized decision-making. These problems represent two components of the overall distributed detection network using a parallel topology. The first problem focused on decision making by the individual team members while the second one considered the fusion of decisions made by the team members. Observations were assumed to be conditionally independent which simplified matters considerably. Optimization of the complete system in which both components are considered together is much more involved even under the conditional independence assumption. When this assumption is not valid, the problem has been shown to be NP-complete. Development of efficient computational algorithms for distributed detection problems is ongoing. Several algorithms such as the Gauss-Seidel cyclic coordinate descent algorithm have been employed successfully for system design. For a more complete understanding of these issues, the reader is referred to [7].

The focus here was on the parallel fusion network topology and a Bayesian formulation of the problem. Distributed detection problems for other network topologies and under other formulations such as the Neyman-Pearson criterion have been discussed and solved [7]. Open problems in this area include system design for the dependent observations case, and development of computational algorithms and paradigms to solve multiple hypothesis distributed decision problems for applications such as object recognition.

5. Acknowledgements

This work is supported by DARPA under Grant No. N66001-99-1-8922.

6. References

1. Ferguson, T.S. (1967) Mathematical Statistics: *A Decision Theoretic Approach,* Academic Press: New York.
2. Lehmann, E.L. (1986) *Testing Statistical Hypotheses*, John Wiley & Sons: New York.
3. VanTrees, H.L. (1968) *Detection, Estimation and Modulation Theory*, Part I, John Wiley & Sons: New York.
4. Poor, H.V. (1988) *An Introduction to Signal Detection and Estimation*, Springer-Verlag: New York.
5, Helstrom, C.W. (1975) *Elements of Signal Detection and Estimation*, Prentice-Hall: Englewood Cliffs, NJ.
6. Srinath, M.D., Rajasekaran, P.K. and Viswanathan, R. (1995*) An Introduction to Statistical Signal Processing with Applications*, John Wiley & Sons: New York.
7. Varshney, P.K. (1967) *Distributed Detection and Data Fusion*, Springer-Verlag: New York.

DISTRIBUTED SENSOR NETWORKS AND NEURAL TREES FOR MULTISENSOR DATA FUSION IN COMPUTER VISION

G.L. FORESTI
Department of Mathematics and Computer Science (DIMI)
University of Udine, Via Delle Scienze 208 – 33100 Udine (Italy)
E-mail: foresti@dimi.uniud.it

1. Introduction

This chapter describes a Distributed Sensor Network (DSN) that is applied to integrating data coming from multiple sensors in the context of important computer vision tasks such as object recognition and scene understanding.

The concept of multisensor data fusion is not new, and the advantage of using several sensors has been demonstrated in several application domains [1,2]. Different sensors are sensitive to different properties of the environment, each one of which can contribute significantly in interpreting the environment itself. Humans and animals have evolved the capability to use multiple senses to improve their ability to survive. The human brain is an excellent example of a data fusion system that performs extremely well. For example, sight, sound, smell, taste and touch data are combined to determine whether a food item is rotten. Sight and sound data are integrated to make decisions regarding safety in an unknown environment.

Multisensor data fusion can be defined formally as a complex process which refers to acquisition, processing and combination of information gathered by various knowledge sources and sensors to provide a better understanding of the phenomenon under consideration [1]. In recent years, thanks to the development of new sensors and processing hardware which make possible the real-time fusion of data, multisensor data fusion has been an area of intense research activity with applications in numerous fields. The applications can be classified into two groups, namely military and nonmilitary applications. Military applications includes: automated recognition of military entities such as ships, aircraft, weapons and missiles, guidance of autonomous military vehicles, battlefield surveillance, air-to-air and surface-to-air systems, and so on. Nonmilitary applications includes air traffic control, law enforcement, robotics, manufacturing, medical diagnosis and remote sensing.

One of the main problems of multisensor data fusion, however, is to integrate in a meaningful fashion, the data collected by the various sensors. In order to manage the large amount of data generated by such sensors, different integration and fusion approaches have been proposed in literature, such as neural networks [3,4], Bayesian networks [5], Kalman filter architectures [6], knowledge-based systems [7,8], weighted decision methods [9] and so on. An accurate overview of such methods can be found in [1,2,10].

A. K. Hyder et al. (eds.), Multisensor Fusion, 183–196.

An important issue of multisensor data fusion is the level at which fusion takes place. A three level hierarchy is widely accepted, i.e., data-level fusion, feature-level fusion and decision-level fusion [1]. To this end, the characterization of a data fusion system involves the transformation between observed data (e.g., sensor signals) and a decision or inference level that represents a more abstract concept [9]. These transformations are necessary if sensor data are not commensurate. For example, row sensor data can be fused by means of classic detection or estimation techniques if they are measuring the same physical phenomena such as two optical sensors or two acoustic sensors. Conversely, if sensor measures are not commensurate, data must be fused at higher levels (feature level or decision level). Feature-level fusion involves the extraction of representative features from each sensor observation and their integration into a single concatenated feature vector which can be used as input of neural networks, clustering algorithms or template methods. Finally, in decision level fusion, each sensor makes decisions based on its own data. These decisions are then combined to yield the final inference. Example of decision level fusion methods include voting methods [11,12], classical inference and Bayesian inference.

In order to unify the terminology related to data fusion, the Join Director of Laboratories (JDL) Data Fusion Working Group, established in 1986, proposed a process model for data fusion (Figure 1) which identifies the processes, functions, categories of techniques and specific techniques applicable to data fusion.

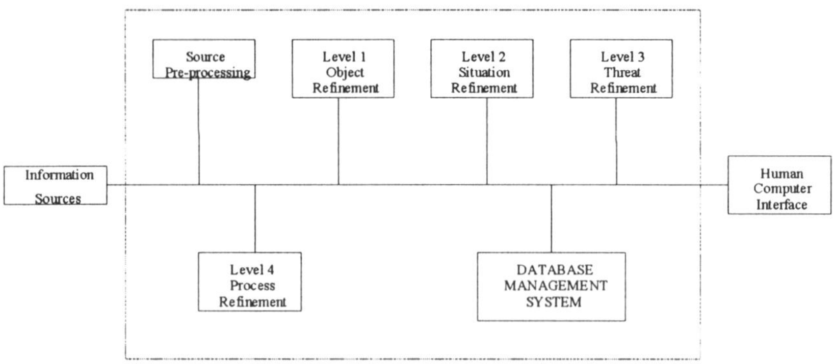

Figure 1. Data fusion model proposed by the Join Director of Laboratories (JDL) Data Fusion Working Group

The input/output of the system is represented by information sources which include sensors and/or other a priori information and by a human computer interface which provides a mean for human computer interaction (e.g., alerts and display of results). An initial process (*source pre-processing*) allocates data to appropriate processes and performs data pre-screening. The first level processing (*Object Refinement*) performs four key actions: (1) *data alignment* which transforms sensor data into a common reference system, (2) *tracking* which refines the position, velocity and other object attributes by means of different techniques such as Kalman filtering, batch estimation, maximum likelihood, (3) *data association* which correlates data with objects by means of gating techniques or nearest neighbour techniques, and (4) *identification* which refines the estimation of an object's identity by means generally of feature-based techniques such as neural networks, cluster algorithms and pattern recognition. This level represents the most mature area of data fusion processing. The second level processing (*Situation Refinement*) develops a description of current relationship among objects, events and a priori information by employing contextual information. The third level (*Threat Refinement*) projects the current situation into the future to draw inferences about the global situation. The second and third levels are currently dominated by knowledge-based methods. Several prototypes have been build, but few can be considered robust operational systems. Finally, the last level (*Process Refinement*) controls the fusion process by monitoring its performances and it identifies information needed to improve the system performances.

One important issue that needs to be addressed is that of topological structure of sensors. Parallel configuration is the most common topological structure: different sensors make observations concurrently. Another popular structure is the serial topology. In this case, sensor observations are temporally separated, e.g., target moving through the field of view of different sensors. An important additional gain to classical data fusion architectures can be obtained by considering systems based on a distributed organization of computational resources: multiple processors gather and process information provided by multiple information sources to analyze observed situations. This approach leads to the concept of Distributed Sensor Networks (DSN) [13], which are basically systems composed of connected, co-operating information sources organized into highly structured, message-exchanging networks. In a DSN, sensors may be either physical (i.e., to generate information by translating observed physical quantities into electrical signals) or virtual (i.e., to produce new information from existing one).

In this paper, a DSN is applied for integrating data coming from multiple sensors in the context of an important field of computer vision, i.e., object recognition and scene understanding. A new neural network model, called *neural tree*, is applied to fuse information at different levels of the DSN. A real example employing fusion of different kinds of sensors, e.g., ultrasonic sensors, infrared sensors, visual cameras, tactile sensors, will be presented in the context of autonomous mobile robots.

2. Distributed sensor network

The main goal of a DSN is to process data acquired by different sensors, to integrate them, thus reducing the degree of uncertainty naturally associated with the acquired information, and to produce an interpretation of the environment observed. A DSN is characterized by the following properties:

(a) Each local node of the network has a degree of intelligence and must know only a part of the whole problem to be solved. So the information representation on each node is structurally simpler than the corresponding information in a centralized system.

(b) Imposing local constraints at the lower levels reduces the number of possible solutions from the lower to the higher levels of the distributed system. At the same time, the network performance may improve if an appropriate fusion mechanism is used.
(c) The dimensionality of information is reduced as only significant information is propagated through the network.

Figure 2a shows the general architecture of the proposed DSN.

This network is organized into a hierarchical tree structure which consists in a root node (*decision level*) and as many leaf nodes (*sensor level*) as the number of physical employed sensors, e.g., optical cameras, infrared sensors, ultrasonic sensors, laser range finder sensors, etc. The intermediate levels (*feature extraction level*) are represented by specific modules, called virtual sensors, that are devoted to extract features directly from Sensor data or from data computed by lower virtual sensors. Data fusion can be performed at the root node or at intermediate nodes in six different modes [2].

(a) *Data in – data out fusion*. Raw observations u_i from each sensor are combined. Examples of this type of fusion are fusion of images acquired from different visual sensors, i.e., image registration.
(b) *Feature in – feature out fusion*. Feature vectors f_i are extracted from sensor observations and then are subsequent fused. For example, feature fusion can be used to integrate information coming from optical and infrared sensors.
(c) *Decision in – decision out fusion*. Decisions d_i taken from raw data or feature extracted from different sensors are integrated to obtain a final decision. This type of fusion is performed when the sensors available are non commensurate.
(d) *Data in – feature out fusion*. Raw data are combined and result in features associated with the environment under consideration. Example is the fusion of visual data from stereo cameras, resulting in depth information.
(e) *Feature in – decision out fusion*. Features from different sensors are considered as input and a decision is generated as output, e.g., object recognition.
(f) *Data in – decision out fusion*. Raw data from different sensors are considered as input and a decision is generated as output.

In the literature exist several methods for data in – data out fusion and feature in – feature out fusion. These methods have been applied to different kinds of sensors and have been demonstrated into different fields. Although it is true that more information is available at lower levels and it might make more sense to fuse at lowest possible level, one should also consider that data are more noisy at lower levels, especially in real application domains. This fact has oriented several researchers to perform fusion at a higher level.

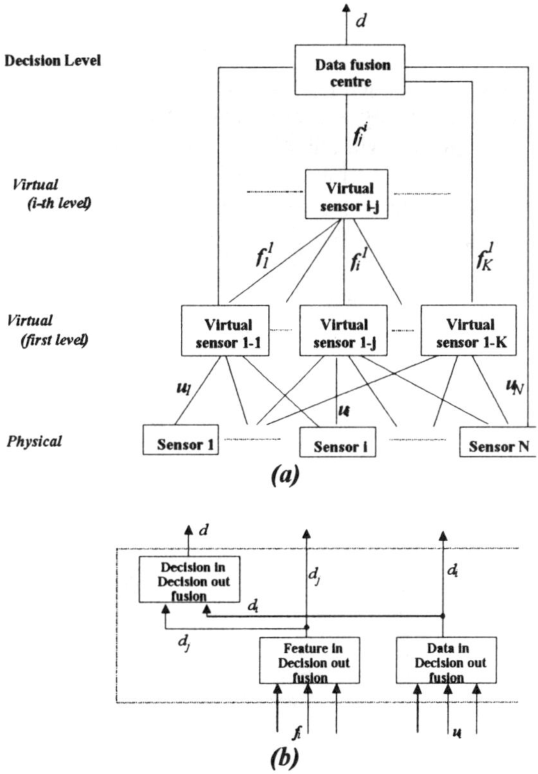

Figure 2a. General Architecture of the Proposed DSN.
Figure 2b. Detailed View of the Data Fusion Module

3. Data fusion strategies

The proposed DSN architecture has been designed for two well known computer vision tasks: object recognition and scene understanding. The main goal of the system is to produce a decision about certain situations, e.g., the presence of mobile objects, the classification of a given object, the detection of a normal/dangerous situations, etc. To this end, the fusion process is performed according to a limited subset of the six possible fusion modes. In particular, the modes (c), (e) and (f) from the previous discussion have been considered as they produce as output a final decision about the considered problem.

The data fusion module is organized into three different submodules (Figure 2b), each one characterized by appropriate fusion strategies. A neural tree-based approach is applied to fuse information at different levels of the DSN. Data fusion benefits from multiple attributes of neural networks. The first is adaptive fusion inference, e.g., neural networks can infer the relationship between the desired fusion output and the multisensor input. The second is generalization capability of incomplete information. Other attributes are the non-linear capability of filtering the noise and the parallel data computing.

3.1 NEURAL TREES

A new neural network model, called neural tree (NT) [14-18], is applied for sensor fusion. The proposed NT is characterized by a tree structure whose nodes are basic artificial neural networks, e.g., perceptrons without hidden layers and with activation function characterized by a sigmoidal behavior. Each perceptron takes $(N+1)$ input, i.e., a pattern vector $\boldsymbol{p}=[u_1,..,u_N]$, (the last input is the *bias* value with weight θ_i) and generates K outputs, i.e., an output vector $o=[o_1,..,o_K]$, called activation values (Figure 3a).

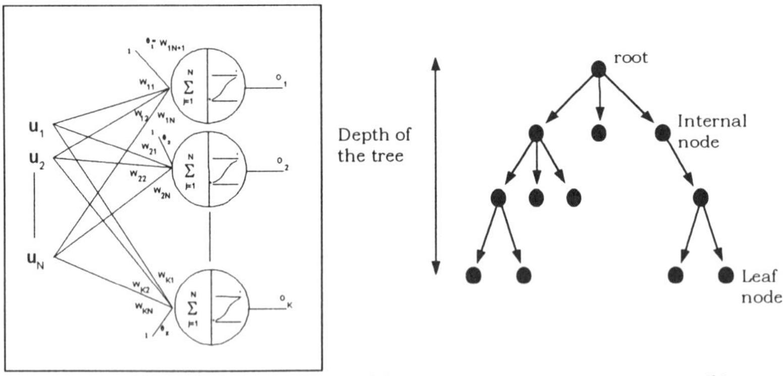

(a) ***(b)***

Figure 3a. Internal Representation of a Perceptron with N Input and K Output,
Figure 3b. General Structure of a Neural Tree with Internal and Leaf Nodes.

The weight matrix W will be as follows:

$$\mathbf{W}=\begin{bmatrix} w_{11} & w_{12} & .. & w_{1,N} & \theta_1 \\ w_{21} & .. & .. & .. & .. \\ .. & .. & .. & .. & .. \\ w_{K1} & .. & .. & w_{K,N} & \theta_N \end{bmatrix} \tag{1}$$

The activation function of the i-th neuron generated by the q-th pattern of the training set will be:

$$o_i^q=\frac{1}{1+exp\left(\sum_{j=1}^{N} w_{ij}p_j^q-\theta_i\right)} \tag{2}$$

The NT structure is characterized by L levels, $l\in[1,L]$ (L is the depth of the tree) (Figure 3b). The root node is on level l=1. A node h on level l may have links at maximum to $k\leq K$ nodes on level $(l+1)$ which are the children of h. The NT grows during the learning phase and it not requires neither any a-priori information about the number of neurons, hidden layers, hidden neurons, etc. nor an exhaustive search as required by classical decision trees.

The goal of the proposed NT architecture will be reached by performing a convenient training process (*learning phase*) using well defined surface patterns then by performing the recognition of each pixel of a testing surface (*classification phase*).

3.1.1 *Learning algorithm*

Let $\Omega = \{\omega_1,\dots,\omega_M\}$ be the set of the problem classes and let $TS_{root} = \left\{(p^q,\omega_{i_q}) \mid p^q \in \Re^{N+1},\ \ \omega_{i_q} \in \Omega,\ \ q = 1,\dots,Q\right\}$ be the training set, where Q is the number of patterns.

(a) Create a single perceptron and set TS=TS_{root}. The weight matrix W is inizialized in a random way.

(b) Start the training of the current perceptron by applying the *perceptron training rule*.

  (1) If the TS is correctly classified (TS is linearly separable), the current perceptron ends with M children leaf nodes, one for each class. Go to (4).

  (2) If the TS is split into M groups (M>1), TS_1, .., TS_M, a new level of M children perceptron nodes is created and each TS_i is assigned to the corresponding child node for the training process. Note that a pattern is assigned to the group depending by the highest output value and that one or more groups could be empty. Go to (2) with TS=TS_i.

(c) If the TS cannot be divided in any group (the perceptron repeats the same classification of the parent node), a *splitting rule* is applied [18]. The TS is divided into at least two groups, TS_r and TS_l, which are assigned to corresponding perceptron child nodes. Go to (2) with TS= TS_r.

(d) If all current nodes are leaves, the algorithm ends. Otherwise go to (2) to train the remaining perceptrons.

According to the *perceptron training rule*, the current perceptron is trained with the patterns of the training set until the average error $\bar{\delta}$ computed on the all output neurons and on the all patterns keeps in the range $\left[\bar{\delta} - toler, \bar{\delta} + toler\right]$ for more than a given number *wait* of epochs. The average error is defined as :

$$\bar{\delta} = \frac{1}{QM}\sum_{q=1}^{Q}\sum_{i=1}^{M}\delta_i^q \tag{3}$$

where $\delta_i^q = t_i^q - o_i^q$, $\forall$ i=1,..,M and $t_i^q = \begin{cases} 1 & if \quad i = i_q \\ 0 & otherwise \end{cases}$ is a generic element of the target vector. The threshold *toler* should be selected close to 0, while the threshold *wait* should be chosen according to the complexity of the problem (e.g., $wait \in [10,50]$).

3.1.2 *Classification algorithm*

The classification algorithm rule follows:

(a) Present an unknown pattern to the root of the NT.
(b) Move through the tree in a top-down way following the path suggested by the classification given by each considered node.
(c) When a leaf node is reached, label the pattern with the classification provided by that node.

3.1.3 *Complex TS distributions*

Sankar and Mammone [16] demonstrate that particularly complex distributions of the patterns of the TS does not allow the convergence of standard training algorithms. In these cases, the current perceptron is unable to divide the TS, so it assigns the whole TS to the same class by repeating the classification obtained by the parent node. This occurs, for example, when the TS is composed by few patterns of one class surrounded by an extremely large number of patterns from another class resulting in a "donut" shape. The solution proposed by Sankar and Mammone foresees to classify all patterns into the class with the highest number of training elements. The consequence of this choice is that the TS is not correctly classified by the NT and that only a suboptimum solution is obtained.

The proposed NT naturally overcomes this problem, but it requires a large use of splitting nodes which reduce the capability of generalization and increase the NT dimensions (depth and number of nodes). To maintain a satisfactory generalization capability of the NT, a modified learning strategy has been proposed. If the current perceptron is unable to divide the TS, its weights $w_{i,j}$ are initialized by forcing the splitting hyperplane to pass for the barycenter of the TS. After performing several experiments on a large set of TS (about 50) containing pattern distributions with donut shape, we have observed that the number of splitting nodes reduces considerably (about 60%). This can be obtained by generating the weights $w_{i,j}$ in a random way and computing the bias values (stored in the $(m+1)^{th}$ column of the W matrix) as follows:

$$w_{i,N+1} = -\sum_{j=1}^{N} w_{i,j} p_j \qquad i=1,..,M \tag{4}$$

Then, in order to avoid wrong effects (e.g., unbalance of the tree) generated by the different cardinality of the classes belonging to the TS, the equation for updating the weights is modified as follows:

$$\Delta\overline{w}_{i,j} = \frac{\hat{N}}{N_{i_q}} \eta\, \delta_i\, o_i^q\, (1-o_i^q)\, p_j^q \tag{5}$$

where N_{i_q} represents the global number of patterns belonging to the class ω_i and $\hat{N}$ the cardinality of the class with the minimum number of elements.

3.2 FUSION PROCESS

Let us consider a 3D scene containing both fixed and mobile objects in which a robot equipped with several sensors navigates by avoiding obstacles. Raw data are extracted from different sensors and stored into appropriate vectors, u_i. The size of each data vector depends on the kind of considered sensor, i.e., the number of measures produced by the sensor itself. Three different kind of sensors have been considered: visual cameras to monitor the whole scene, infrared sensors to detect obstacle presence within a short range and ultrasonic sensors to determine the distance to objects within a long range. In this case, the raw data vectors generated by the sensors are respectively: (a) $u_1=[x_1,..x_{NxN}]$, where x_i represents the grey-level value of the i-th pixel and N the image size, e.g. N=256 for visual cameras, (b) $u_2=[y]$, where y represents the estimated distance between the infrared sensor and the obstacle and (c) $u_3=[z]$, where z represents the estimated distance between the ultrasonic sensor and the obstacle.
Three different fusion strategies have been chosen: data in – decision out fusion, feature in - decision out fusion and decision in – decision out fusion. In all cases, a binary NT has been learned with different input data and with the same output classes, i.e., the decision about the presence or not in the scene of possible obstacles for the robot.

3.2.1 *Data in – decision out fusion*

The first fusion strategy we consider is the data in – decision out fusion. Visual sensors (optical or infrared cameras) which generate several measures (image pixels) cannot be used in the context of this fusion strategy. In fact, neural networks with a very high number of input (e.g., NxN) should be considered, but this became a problem computationally intractable [19].

On the contrary, data coming from multiple infrared or ultrasonic sensors can be fused in a similar way. For simplicity, we consider the case of an object moving in the scene far (close) from the sensors, so that only ultrasonic sensors (infrared sensors) generate data. In this case, the input pattern for the NT, $p=[y_1,..,y_P]$, is composed by a set of P distance measures, one for each employed sensor (e.g., P=4,8,16,..). The sensors are spatially separated in the scene, so the distance measures can be assumed independent.

A TS composed by 10^3 patterns taken from 10 different objects placed at different distances from the sensor has been build up. The TS has been completed also with data acquired by scenes without objects. A NT composed by 19 internal nodes, 4 split nodes and 12 leaf nodes disposed on L=5 levels has been obtained. The output of the NT is the decision whether an object is present or not in the observed scene.

3.2.2 *Feature in – decision out fusion*

The second fusion strategy we consider is the feature in – decision out fusion. This approach allows the use of sensors as visual cameras which generate a large amount of data. To this end, virtual sensors are used to extract feature vectors f_i from row sensor data and used to build up the input pattern of the NT. As regards visual sensors, a Kalman-based approach is applied: a sequence of K images is processed to estimate the distance of moving objects from the camera [20]. For simplicity, we consider only scenes containing one moving object. First, a change detection procedure based on the simple difference method [21] is used to identifies mobile objects in the scene by separating them from the static background. Then, characteristic points of detected blobs (e.g., barycenter, quartiles [22], etc.) are matched into successive frames and used as observations of an extended Kalman filter which estimates the 3-D depth of such points. Finally, a feature vector, $f_1=[\xi_1, \xi_2,.., \xi_K]$, composed by a set of estimated distances of the object from the camera, computed at different time instants, is generated. As regards infrared and ultrasonic sensors, they are processed by appropriate virtual sensors which divided them into more groups, sg_j, $j=1,..,M$, each one composed of three neighboring sensors. With this sensor arrangement, the distance, dy_j (dz_j), measured by the j-th sensor group from the sensor to the object is expressed as:

$$dy_j = \min\left(y_i \middle| i = 3j-2, 3j-1, 3j\right) \quad for\ j = 1,..,M \tag{6a}$$

$$dz_j = \min\left(z_i \middle| i = 3j-2, 3j-1, 3j\right) \quad for\ j = 1,..,M \tag{6b}$$

The estimated distances, dy_j (dz_j),, are computed by the virtual sensors at different time instants and are used to generate the feature vectors, $f_2=[dy_1, dy_2, \ldots, dy_K]$ and $f_3=[dz_1, dz_2, .., dz_K]$.

In this case, the input pattern of the NT, $p=[f_1,f_2,f_3]$, is composed by three sets of features, one for each type of employed sensor. A TS composed by 30 patterns taken from 20 different scenes containing moving objects has been build up. The TS has been completed also with feature data extracted by 10 different scenes without objects. A NT composed of 12 internal nodes, 2 split nodes and 10 leaf nodes disposed on L=5 levels has been obtained.

3.2.3 *Decision in – decision out fusion*

The last fusion strategy we consider is the decision in – decision out fusion. This approach allows the use of non-homogeneous sensors. Individual decisions are first computed from each sensor data sets, and then a NT is learned to fuse the different decisions into a final system decision. Decisions can be obtained both by processing directly raw sensor data or by first extracting appropriate features and then fuse them. In this case, the input pattern of the NT, $p=[d_1,d_2,..,d_K]$, is composed by the final decisions, d_i, taken by each sensor. A TS composed by 10 patterns taken from 8 different scenes containing moving objects and 2 without objects has been build up. A NT composed of 5 internal nodes, 0 split nodes and 6 leaf nodes disposed on L=4 levels has been obtained. The output of the NT represents the final decision of the system.

4. Experimental results

A PIONEER II mobile robot (Figure 4) is used for testing the proposed multisensor data fusion framework in the context of a real task, e.g., navigation. The robot is equipped with four sensing kinds of sensors: tactile, infrared, ultrasonic and visual. Its three wheel drive mechanical system enable it to move at a maximum translation speed of 0.5 m per second and a maximum rotational speed of 45 degrees per second. The tactile system is a bumper composed by pressure sensitive sensors detecting physical contact when colliding obstacles. The infrared system is a set of four sensors detecting obstacle presence within a short range. 16 ultrasonic sensors are useful for determine the distance to objects within a long range, from 1 meter to 7 meters. The visual system is composed by a CCD colour camera for image generation.

The goal of the proposed system is to move the robot through an unknown environment, e.g., an indoor area of the University of Udine composed by narrow corridors and offices, by detecting and avoiding obstacles. The illumination conditions of the test site are very different as some offices and/or corridors can be dark, artificially illuminated or natural illuminated.

A DSN is applied to fuse data collected from different physical and virtual sensors to detect the presence of obstacles in front of the robot. The general architecture of the proposed DSN is shown in Figure 5.

Physical sensors are placed at the first level of the system. The tactile sensor gives directly a binary output t, which assumes the value 1 whenever an obstacle is touched otherwise it assumes the value 0. Optical images $I(x,y)$ represent the input of three virtual sensors, i.e., optical flow [23], depth form motion [24] and depth from focus [25], which are placed at the second level of the DSN. The optical flow extracts from pairs of successive images the displacement vectors $dv=(\Delta x, \Delta y)$ (on the image plane) of some characteristic moving points (e.g., corner points, etc.). The depth from motion virtual sensor uses the displacement vectors computed from pair of successive images and the motion parameters of the robot (speed and direction) to estimate the distance dm of the detected objects from the sensor (Figure 6a).

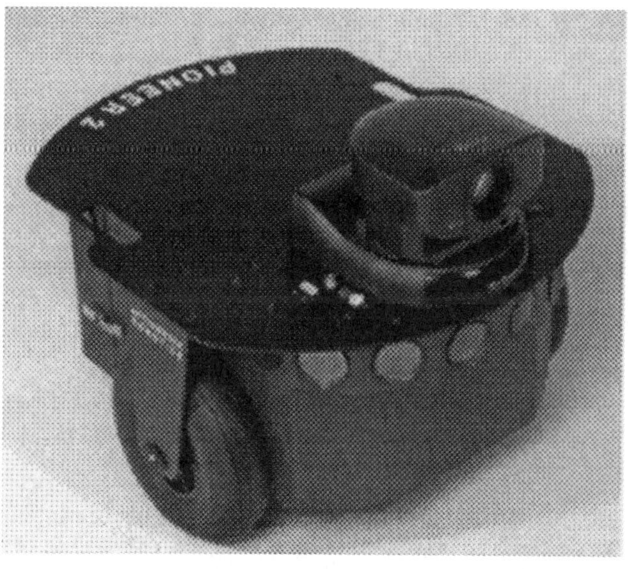

Figure 4. Pioneer II robot used in the experimental tests.

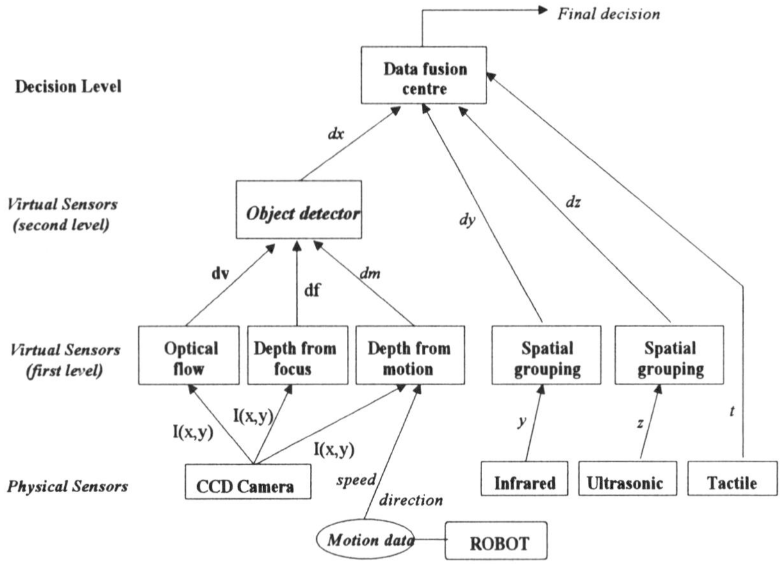

Figure 5. General architecture of the proposed DSN for multisensor data fusion.

Finally, the depth from focus search for the best focus image from a sequence of images acquired with different focus parameters to estimate the distance *df* of the object from the sensor (Figure 6b).

The second level of the system is completed with two virtual sensors which integrate (according to equations 6a and 6b) the distance measures generated by both infrared and ultrasonic sensors. Two confidence values, $cy, cz \in [0,1]$, are also computed. At the third level, an object detector virtual sensor, based on a voting strategy [12], has been used to fuse the distances computed by the optical flow, depth form motion and depth from focus virtual sensors into a unique distance value, *dx*. A confidence value $cx \in [0,1]$ is also computed on the basis of the congruence of the obtained distance measures.

At the last level of the DSN architecture, the final data fusion module, based on a NT network, integrates the four different distance measures, *dx*, *dy*, *dz* and *dt* in order to decide about the presence or not of an object in front to the robot. Measures with low confidence values (e.g., <0.5) are not considered. In this case, **the input pattern** of the NT, $p=[dx,dy,dz,dt]$, is composed by four distance measures. A TS composed of 10 patterns taken from 8 different scenes containing moving objects and two scenes containing any object has been build up. Situations in which sensors measures are not reliable have been also considered. A NT composed of 10 internal nodes, 1 split node and 12 leaf nodes disposed on L=4 levels has been obtained.

Table 1 shows the performances of the proposed approach expressed in terms of correct percentage of obstacle detection. A test set of 50 real scenes (40 containing an object in front of the robot and 10 without obstacles) has been considered.

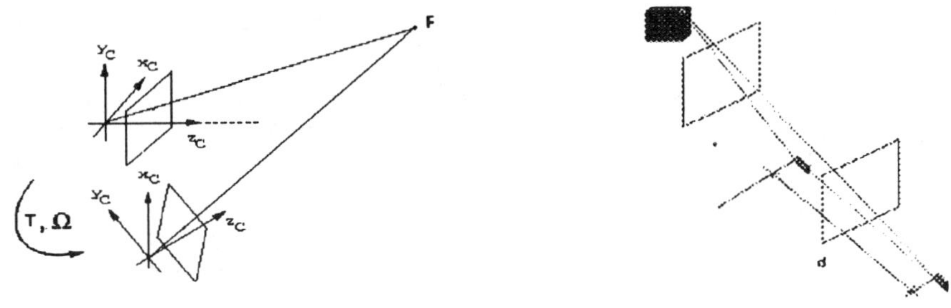

(a) (b)

Figure 6. Geometric representation of the (a) depth from motion and (b) depth from focus virtual sensors.

TABLE 1. Comparison of the correct percentage of object detection among the proposed data fusion system and single sensors versus environments with different illumination conditions.

	DSN Approach	*CCD camera*	*Infrared*	*Ultrasonic*	*Tactile*
Normal illumination	92	84	73	85	100
Artificial illumination	87	76	73	85	100
Dark	81	21	73	85	100

Different environment conditions (e.g., dark, natural illumination, artificial illumination, etc.) and objects with different surface characteristics (e.g., rough surfaces, smooth surfaces, etc.) have been employed to generate the test set.

In normal and artificial illumination, the DSN approach out performs any single remote non-tactile sensor allowing the robot to avoid obstacles before contact. In darkness, the single ultrasonic sensor out performs DSN because the fusion process allows "noise" from the CCD and IR sensors to reduce the combined detection performance; this can be eliminated in future DSN algorithms.

5. References

1. Hall, D. L. and Llinas, J. (1997) An introduction to multisensor data fusion, *Proceeding of the IEEE* **85**, 6-23.
2. Varshney, P.K. (1997) Multisensor data fusion, *Electronic and Communication Engineering Journal*, 245-253.
3. Pearson, J.C., Gelfand, J.J., Sullivan, W.E., Peterson, R.M., and Spence, C.D. (1988) Neural network approach to sensor fusion, in *Proceedings of SPIE on Sensor Fusion*, Vol. 931, C.W. Weaver, Ed., Orlando, FL, 103-108.
4. Jouseau, E. and Dorizzi, B. (1999) Neural networks and fuzzy data fusion – Application to an on-line and real time vehicle detection system, *Pattern Recognition Letters* **20**, 97-107.
5. Luo, R.C., Lin, M., and Scherp, R.S. (1988) Dynamic multisensor data fusion system for intelligent robots *IEEE Journal of Robotic Automation* **3**, 386-396.

6. Regazzoni C.S. and Tesei, A. (1996) Distributed data fusion for real-time crowding estimation *Signal Processing* **53**, 47-63.
7. Hanson, A.R., Riseman, E.M. and Williams, T.D. (1988) Sensor and information fusion from knowledge-based constraints, in *Proceedings of SPIE on Sensor Fusion*, Vol. 931, C.W. Weaver, Ed., Orlando, FL, 186-196.
8. Pau, L. (1989) Knowledge Representation Approaches in Sensor Fusion Automatica **25**, 207-214.
9. Hall, D. (1992) *Mathematical Techniques in Multisensor Data Fusion*, Artech House, Boston, MA.
10. Luo, R.C. (1989) Multisensor Integration and Fusion in Intelligent Systems, *IEEE Transactions on System, Man, and Cybernetics* **19**, 901-928.
11. Sakar, S. Boyer, K.L. (1994) A computational structure for preattentive perceptual organization: graphical enumeration and voting methods, *IEEE Trans. on System, Man, Cybernetics* **24**, 246-266.
12. Foresti, G.L. and Regazzoni, C.S. (2000) A hierarchical approach to feature extraction and grouping, *IEEE Transaction on Image Processing* **9**, (in press).
13. Iyengar, S.S., Kashyap, R.L and Madan, R.N. (1991) Distributed Sensor Networks - Introduction to the Special Section, *IEEE Transactions on System, Man, and Cybernetics* **21**, 1027-1031.
14. Utgoff, P.E., (1988) Perceptron tree: a case study in hybrid concept representation, in *Proceeding of the VII National Conference on Artificial Intelligence*, 601-605.
15. Sankar, A. and Mammone, R.J. (1993) Growing and pruning neural tree networks, *IEEE Transaction on Computers* **42**, 291-299.
16. Song, H.H. and Lee, S.W. (1998) A self-organizing neural tree for large set pattern classification, *IEEE Transaction on Neural Networks* **9**, 369-380.
17. Foresti, G.L. (1999) Outdoor Scene Classification by a Neural Tree Based Approach, *Pattern Analysis and Applications* **2**, 129-142.
18. Foresti, G.L. and Pieroni, G.G. (1997) Exploiting Neural Trees in Range Image Understanding, *Pattern Recognition Letters* **19**, 869-878.
19. Ripley B.D. (1996) *Pattern Recognition and Neural Networks*, Cambridge University Press, UK.
20. Foresti, G.L. (1998) A Line Segment Based Approach for 3D Motion Estimation and Tracking of Multiple Objects, *International Journal of Pattern Recognition and Artificial Intelligence* **12**, 881-900.
21. Skiestad, K. and Jain, R. (1989) Illumination independent change detection for real world image sequences, *Computer Vision Graphics and Image Processing* **46**, 387-399.
22. Foresti, G.L. (1999) Object Recognition and Tracking for Remote Video Surveillance, *IEEE Transaction on Circuits and Systems for Video Technology* **9**, 1045-1062.
23. Nagel, H.H. (1987) On the Estimation of Optical Flow: Relations between Different Approaches and Some New Results, *Artificial Intelligence* **33**, 299-324.
24. Dalmia, A.K. and Trivedi, M. (1996) Depth extraction using a single moving camera – an integration of depth from motion and depth from stereo, *Machine Vision and Applications* **9**, 43-55.
25. Rajagopalan, A.N. and Chaudhuri, S. (1994) A variational approach to recovering depth from defocused images, *IEEE Transactions on Pattern Analysis and Machine Intelligence* **19**, 1158-1164.

ARCHITECTURES FOR EFFICIENT DATA FUSION

P. VANGASSE AND D. NICHOLSON
BAE Systems
Advanced Technology Centres – Sowerby
Filton, Bristol BS34 7QW, UK
Email: Paul.Vangasse@baesystems.com,
David.Nicholson2@baesystems.com

1. Introduction

This chapter reviews architectures for efficient multi-sensor data fusion (MSDF). The following definitions are used:

- **Data Fusion:** This is the process of combining sensed and contextual data through increasing levels of abstraction in order to detect, classify and track objects, and to form a situation assessment and infer its probable impact. *The chapter focuses only on sensed data fusion and specifically on track fusion.*
- **Architecture:** This refers to the structural organisation of an MSDF system with respect to its sensing, communications and data processing sub-systems. There are two fundamentally different types of MSDF architecture: centralised and decentralised systems. *The chapter emphasizes decentralised systems, as these are generally the more efficient architecture.*
- **Efficiency:** This is a measure of the performance of an MSDF system as a function of its sensing, computational and communications resource requirements. *Information theory is a principled framework for decentralised data fusion and systems resource management.*

The motivations for multi-sensor data fusion systems are numerous. Relative to a single sensor system they include: an extension in spatial coverage, an extension in temporal coverage, increased confidence, reduced ambiguity and improved system reliability [1]. The generic MSDF system architecture comprises three main sub-system components: sensors, communications and data processors. The structural organisation of these sub-systems defines the MSDF architecture and gives rise to the following set of fundamental issues:

(a) **What data is communicated?**
This question refers to the type of information that is transmitted around the network. For example, is it raw sensor information (e.g., plots) or is it processed information (e.g., tracks)? Is it an absolute or a relative measure of information? If it is processed information, is its processing history also communicated?

A. K. Hyder et al. (eds.), Multisensor Fusion, 197–221.

(b) **Where is the data received?**
This question refers to the destination of the transmitted information. For example, is the information received at a central site only or is it received at several distributed sites? Can the same item of information be received at a particular site more than once?

(c) **Where is the data combined?**
This question refers to the MSDF system's processing chain. For example, is the data combined centrally or is the fusion of data distributed across the network? If the data fusion processing is distributed, is the processing chain hierarchical or non-hierarchical?

(d) **When is the data combined?**
This question refers to the frequency of the data combination process relative to the sensing and communications updates. For example, if the data fusion process is distributed, is it synchronised across the network? Are local fused estimates updated at the same rate as global fused estimates?

(e) **How is the data combined?**
This question refers to the mechanics of the data fusion algorithms. For example, how is the decentralised Kalman filter implemented? How do the algorithms deal with redundant data? How robust are the fusion algorithms?

There is no single overarching answer to all these questions. In practice, a number of different trade-off issues have to be considered. These typically revolve around the desired quality of the fused product with respect to various sensing and processing constraints and operational requirements on system resilience and flexibility. In this lecture it is argued that, in general, decentralised system architectures offer the most useful set of trade-offs.

Section 2 specifies the MSDF system architecture taxonomy, indicating why decentralised systems are more efficient than centralised systems and noting two current application areas. Section 3 reviews information based decentralised estimation, and outlines a number of architectural and algorithmic approaches to the rumour propagation problem. Section 4 provides a snapshot of BAE SYSTEMS current research in decentralised data fusion. Section 5 concludes the highlights of several promising areas for future research.

This review of MSDF architectures draws on a number of published resources, to which the authors strongly recommend the interested reader. Specifically, the information based decentralised data fusion paradigm is described in the textbooks by Manyika and Durrant-Whyte [2] and Mutambara [3]. The hierarchical and distributed estimation architectures are discussed in the review papers by Chong, Mori and Chang [4] and by Chong [5]. Tutorials and papers on Covariance Intersection can be accessed over the internet [6]. In addition, there are several articles on decentralised estimation in journal publications such as *IEEE Transactions in Aerospace and Electronic Systems*. The SPIE conference proceedings *in Signal and Data Processing of Small Targets* and *Signal Processing, Sensor Fusion and Target Recognition* also provide a useful indicator on current research topics.

2. Architectures

This section reviews the MSDF system architecture taxonomy. There are two basic types of MSDF architecture: centralised and decentralised systems. These types are distinguished by the extent to which they distribute the data fusion processing and communications load between different sites. Some of the options are specified here and their pros and cons are indicated.

2.1 CENTRALISED SYSTEMS

The centralised system architecture is displayed, using Chong's information graph representation [5], in Figure 1. This architecture is characterised by the presence of only one fusion site (denoted in Figure 1, at each fusion time step, by an open square). This site is responsible for fusing data provided by the multiple sensors (denoted by the solid squares). The sensors and the fusion centre may all be located on a single platform (e.g., a group carrier), or the sensors may be distributed amongst a number of platforms that communicate their local sensor data directly to a command and control centre. The sensor data can be combined at either the 'raw' signal level or at the feature level.

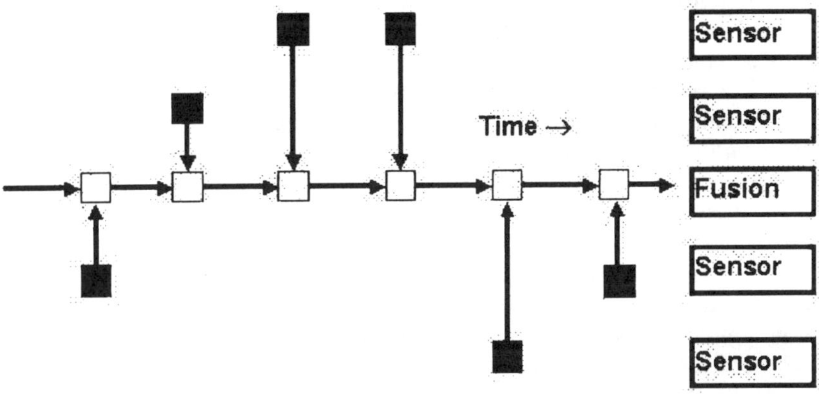

Figure 1. Information graph for a centralised system architecture

If a centralised system performs its data fusion at the signal level, it suffers minimum information loss and therefore provides an accurate fused product. However, such systems are generally not efficient. They require commensurate sensors (i.e., sensors which observe the same physical manifestation of an object) and access to significant system resources. While it is possible to achieve some trade-off between performance and efficiency by implementing fusing data at the feature level, the centralised architecture is still subject to a number of other key disadvantages. The main ones are:

(a) **Vulnerability:** If there is a failure in the central processor or the central communications facility, the entire system fails.

(b) **Non-modularity:** If more sensors are added to a centralised system, some re-programming is generally required and communications bandwidth and computer power would need to increase (assuming a fixed total processing time constraint) in proportion to the amount of extra data. In effect, the centralised system is difficult to expand.

(c) **Inflexibility:** If the application changes, the centralised system may not easily adapt. For example, the central processor might not be powerful enough to handle the new application.

In summary, centralised systems fuse data "at its source" and this yields the optimal result. However, in order to achieve this optimality, centralised systems lack fault-tolerance, modularity and flexibility. This may not be too important in some application areas, but in a dynamic battlespace environment it can impose some very severe operational constraints indeed.

2.2 DECENTRALISED SYSTEMS

Decentralised systems are characterised by the presence of a number of fusion sites, i.e., the fusion load is distributed across the network. Consider a decentralised tracking system. As a minimum, each sensor site would generate a local track based on its own sensor data. Some, or all, of the sites are then responsible for fusing these individual tracks into a global track file. In fact the detailed mechanics of decentralised tracking can be rather more complicated than this implies, as we shall discuss in the next section. Before embarking on that discussion, however, it is important to discriminate carefully between *partially* and *fully* decentralised data fusion architectures.

2.2.1 Partially decentralised architectures

The partially decentralised data fusion architecture is characterised by *unequally* devolving the data processing load between different sites. Specifically, the platforms are differentiated into low and high level fusion nodes and there may be more than two levels. The low-level nodes are responsible for forming local tracks, say, and the high-level nodes are responsible for combining them into a global picture. This architecture clearly alleviates two major problems that exist with the centralised architecture. First, it is more robust. This is because there is no now longer a single point of failure. Second, the architecture suffers less from communications and processing bottlenecks. This is because the information flow is reduced between levels, and the data processing load is distributed.

The simplest partially decentralised architectures are the hierarchical systems. The information graph for a two level hierarchical system architecture is displayed in Figure 2. The low and high level fusion nodes are marked "L" and "H" respectively. A prime is used to indicate the state of the fusion sites at an earlier time. In the architecture shown there is no propagation, or feedback, of data from the high-level to the low-level nodes. This is a good example of the type of trade-off issue that underpins the design of efficient data fusion architectures. No feedback clearly cuts down on communications, but it also misses an opportunity to enhance the lower level estimates.

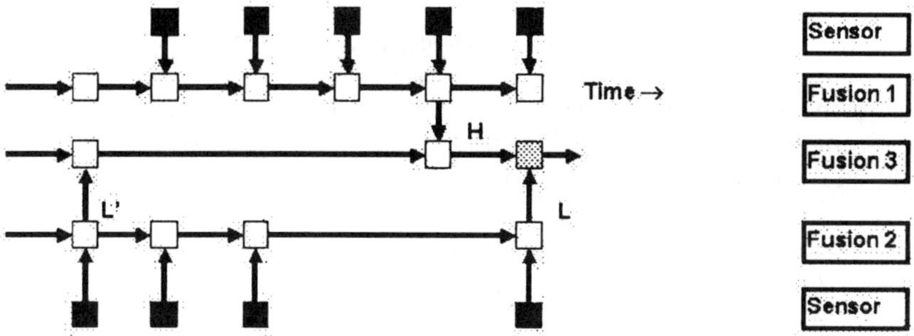

Figure 2. Information graph for a hierarchical system architecture without feedback

The information graph for a two level hierarchical system architecture *with feedback* between the levels is displayed in Figure 3. This system should result in an enhanced lower level estimate, but at the expense of increased communication traffic. In fact, due to the local trackers sharing a common process noise, this may not actually be the case. See section 3.3.5 for details.

A general partially decentralised architecture is likely to have a more complex communications structure than the simpler two level hierarchical systems. While this can raise the system's tolerance to local communications or processor failures, it also creates a problem. Specifically, there is now an increased potential for redundant data to proliferate in the system. If these redundancies are not recognised and removed, the fused estimates can become inconsistent with the underlying data from which they are formed. It is only possible for each fusion node to recognise the correlated component of its input if it has prior knowledge of the global network topology, or if it is communicated the entire processing history of its input estimates. Either of these solutions will limit the efficiency of the system. The first impacts flexibility and expandability, the second requires more communications bandwidth and more computer storage/processing than the simpler hierarchical system architectures.

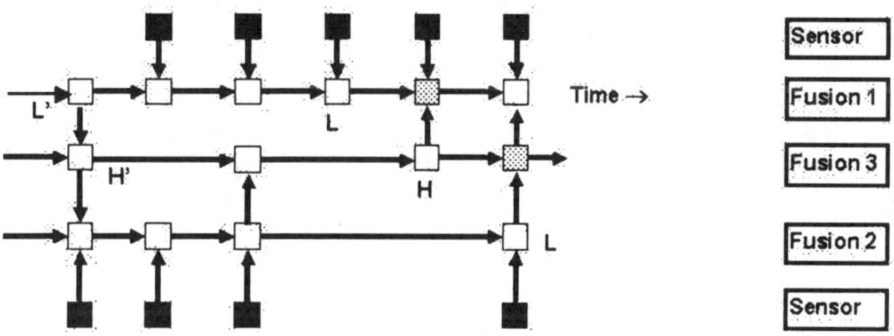

Figure 3. Information graph for a hierarchical system architecture with feedback

2.2.2 Fully decentralised architectures

The fully decentralised architecture is characterised by *equally* devolving the data processing load between different sites. In this case there is no differentiation between the fusion sites and each is responsible for an identical set of operations. These operations are: (1) fusing local sensor measurements, (2) formulating an information packet for communicating to other sites, and (3) assimilating information communicated from other sites.

A fully decentralised, partially connected, system architecture is displayed in Figure 4.

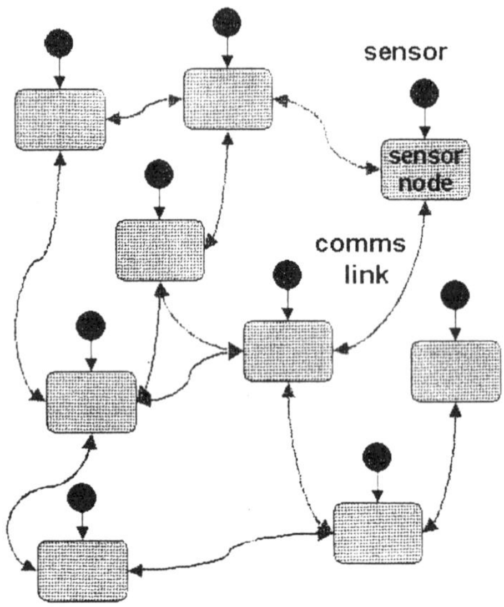

Figure 4. A fully decentralised (partially connected) MSDF system architecture

The fully decentralised system represents the most general architecture from which the others can be derived as special cases. In addition to the absence of special fusion sites, an additional requirement on fully decentralised systems is that they should have no common communications facility. Specifically, the nodes are not allowed to broadcast their data and communications must be kept on a strictly point-to-point basis. Finally, and in spite of the redundant data problem that is therefore a key issue in fully decentralised systems, the nodes are denied access to the global network topology. They are only permitted to know about their nearest-neighbours.

These constraints endow the fully decentralised system with the following useful attributes:

(a) **Fault tolerance:**
System performance degrades gracefully when individual nodes or links are damaged. This is in complete contrast to the centralised system, where a central failure is catastrophic. The performance degradation in partially decentralised systems can also be significant if one of the higher level nodes should happen to fail.

(b) **Scalability:**
New nodes and sensors may be added without the development of communication bottlenecks or computational processing overload. Again, this is in marked contrast to the centralised case. The partially decentralised architecture is also less scalable due to asymmetries in its communications and data processing structure that are inherent in its design.

(c) **Flexibility:**
The fully decentralised architecture supports on-line addition or deletion of nodes, i.e., it is a "plug-and-play" system. This flexibility assumes that there is a solution to the redundant data problem, as communicating the entire processing history of the data is very inflexible. Fortunately, as we shall discuss in the next section, there is a solution to this problem that allows the flexibility attribute of a decentralised architecture to be preserved.

(d) **Modularity:**
The nodes within a fully decentralised system can be constructed and programmed in a modular fashion. This is because of the parity in the data fusion processing operations that at performed at each node. The partially decentralised system is less modular due to the typically unequal functions of the low and high level fusion nodes.

2.2.3 Applications

Two potential applications of the decentralised data fusion architecture, one military and the other civil, are outlined below. These are in the early stages of development, but promise significant returns when they are implemented:

- **Network-Centric Warfare (NCW).** NCW is "an information superiority-enabled concept of operations that generates increased combat power by networking sensors, decision-makers, and shooters to achieve shared awareness, increased spccd of command, higher tempo of operations, greater lethality, increased survivability, and a degree of self-synchronisation [7]." The perceived power of NCW is obtained by empowering all the decision- makers in the battlespace, rather than just one or a few. This paradigm is clearly consistent with the definition of a fully decentralised decision-making architecture. The concept is illustrated, in a schematic way, in Figure 5.
- **Air Traffic Management Systems (ATMS).** Increasingly busy runways and crowded skies require new thinking about the function of ATMS. This has led to a concept known as "Free Flight [8]." This proposes a partially decentralised ATM decision support system that transfers much of the current functionality of air traffic control to each individual aircraft. Free flight would minimise fuel consumption, since each aircraft is able to optimally plan its own trajectory. This, in turn, could reduce delays in and around airports. Free Flight would also help aircraft to avoid turbulence, since the aircraft would not have to wait for clearance from air traffic control to avoid rough weather patches. Obviously the other advantages of the decentralised architecture, fault tolerance in particular, are equally important.

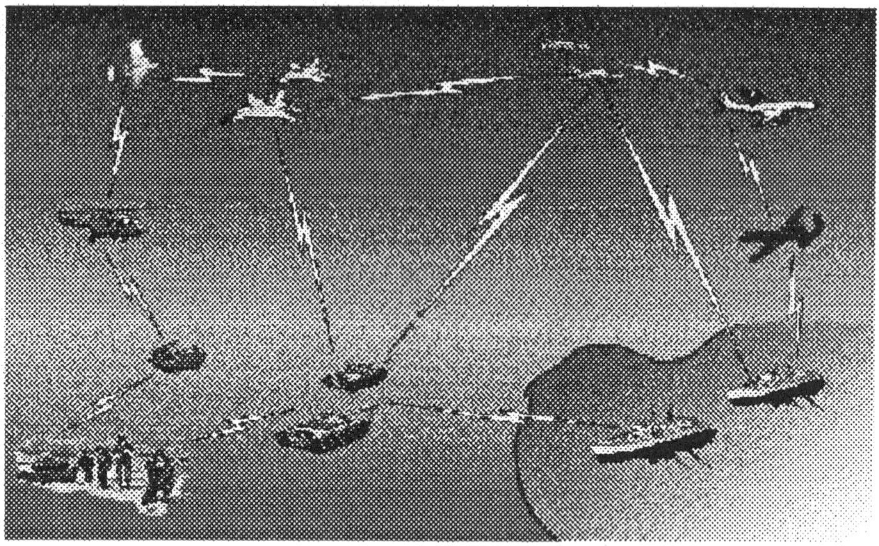

Figure 5. The Network-Centric (Decentralised) Warfare concept

3. Decentralised Estimation

This section considers decentralised estimation in the context of multi-platform track-to-track fusion. This has been an active focus of research for around twenty years. The early effort was mostly concentrated on decentralised filter theory, e.g., [9,10], but recent attention has turned to practical implementation issues, such as systems resource management, e.g., [2,11]. The purpose of this section is to outline first an information theoretic framework for the decentralised estimation problem. A number of architectural and algorithmic approaches to the problem are then considered.

3.1 INFORMATION SPACE

Information space provides a more natural and direct basis for decentralised estimation theory than does conventional state space. This is because the information form of the Kalman filter has some useful properties that can be exploited for developing efficient algorithms. The following table relates the familiar state space variables ($\mathbf{z},\mathbf{R},\mathbf{x},\mathbf{P}$ [12]) to their information space equivalents ($\mathbf{i},\mathbf{I},\mathbf{y},\mathbf{Y}$ [2]), where $\mathbf{H}$ is the observation model.

TABLE 1. State space variables and their information space equivalents

Description	State space symbol	Infm.Space symbol	Relationship
Sensor measurement	$\mathbf{z}$	$\mathbf{I}$	$\mathbf{i} = \mathbf{H}'\mathbf{R}^{-1}\mathbf{z}$
Sensor uncertainty	$\mathbf{R}$	$\mathbf{I}$	$\mathbf{I} = \mathbf{H}'\mathbf{R}^{-1}\mathbf{H}$
State estimate	$\mathbf{x}$	$\mathbf{Y}$	$\mathbf{y} = \mathbf{P}^{-1}\mathbf{x}$
State uncertainty	$\mathbf{P}$	$\mathbf{Y}$	$\mathbf{Y} = \mathbf{P}^{-1}$

The information form of the Kalman filter is well known [13]. What is not as widely known is its utility in decentralised data fusion. The decentralised information filter (DIF) is equivalent to the decentralised Kalman filter (DKF) and yields identical estimates [14]. However, the DIF is much easier to partition among multiple sensor nodes than the DKF. In particular, information states are easy to update and don't require the computation of a gain matrix. Suppose a node receives N synchronised partial estimates at time step k (denoted $\mathbf{m}(k \mid k)$ and $\mathbf{M}(k \mid k)$, which we define later) from its adjoining communications channels. In information space these are simply added to its current prediction state in order to form the new combined state:

$$\mathbf{y}(k \mid k) = \mathbf{y}(k \mid k-1) + \sum_{i=1}^{N} \mathbf{m}_i(k \mid k)$$

$$\mathbf{Y}(k \mid k) = \mathbf{Y}(k \mid k-1) + \sum_{i=1}^{N} \mathbf{M}_i(k \mid k)$$

Prediction is more complex in information space but, because it relies on a propagation coefficient that is independent of the observations, it is also simpler to partition and decentralise amongst a network of sensor nodes [23].

The computational advantages of the DIF are important, but perhaps more important at a fundamental theoretic level is the fact that this formalism can be developed directly from probabilistic information update [2]. The information update paradigm also leads to an intuitive method for addressing sensor management [2] and communications management [11]. Further details are provided in section 4.

3.2 INFORMATION FLOW

The communications structure of a decentralised network is typically more complicated than it is for a centralised system. It is therefore no surprise that the key issue in decentralised estimation revolves around information flow. Specifically, what information should be sent down the communication channels, to whom should it be sent, and how should it be fused upon reception? These questions must be resolved in such a way that the multiple counting of redundant information is either strictly avoided or compensated for in some way. If redundant data is allowed to proliferate in the network, and repeatedly fused under an erroneous independence assumption, the fused estimates can become quickly inconsistent. Specifically, their calculated covariance will be less than their true covariance. Overconfidence is obviously a very undesirable aspect of an estimator and it must be eliminated.

There are three main sources of redundant data counting in decentralised networks:

(a) Consider a node's information state estimate at time step k, i.e., $\mathbf{y}(k \mid k)$. This estimate is conditionally dependent on, or `backloaded' with, all the sensor measurements that have contributed to it up to this time. Therefore, if the estimate is passed to a node that has received it an earlier time, some of its integrated sensor information will be redundant.

(b) If two sensor nodes are tracking a target, their tracks will be correlated as a result of sharing the same underlying process noise. This non-trivial dependence was first noted by Bar-Shalom [15], who derived a corrective fusion equation. Chang and colleagues have recently explored this problem analytically by employing the information update equations [16,17]. Specifically, they study hierarchical systems with and without feedback. They find that the feedback system actually performs *poorer* than the non-feedback system as an adverse consequence of the common process. This important result is discussed in more detail in section 3.3.5.

(c) The source of the redundant data identified in (1) above is a consequence of a temporal loop in the system. This causes an item of information from time step $k-n$ to be "re-used" at time step k. Redundant data counting can also arise as a consequence of *spatial loops* in the network. This case has been examined by Utete [18]. It is also interesting to note the possible connections here with Turbo codes. These have recently been analysed in terms of "loopy" Bayesian Networks [19].

All three sources of redundant data counting are of prime concern in the design of decentralised data fusion architectures and their associated algorithms. It is essential to counter the threat of inconsistency that is posed by this problem, otherwise the decentralised data fusion paradigm may be undermined. The remainder of this section outlines some proposed measures.

3.3 PARTIALLY DECENTRALISED ESTIMATION

The simpler hierarchical systems are considered here. As specified earlier, these have two or more fusion levels with a single communications path between them. Let us, for the moment, assume that the trackers' process noise is small and the fusion intervals are short. The effect of relaxing this assumption is discussed in section 3.3.5. Also, for notational convenience and ease of explanation, we assume that a single target is being tracked in an uncluttered environment with ideal data association. Distributed data association in a cluttered environment is investigated by, e.g., [4,20].

3.3.1 Hierarchical estimation without feedback

In the zero feedback system (Figure 2) there is a one-way propagation of information to the high-level fusion nodes from the lower level nodes. This has the potential to create redundant data counts due to temporal loops in the system. However, it is a reasonably straightforward matter to identify this source and remove it prior to high-level fusion. Consider a low-level estimate at time k, represented by the information vector/matrix pair $\mathbf{y}_L(k \mid k)$ and $\mathbf{Y}_L(k \mid k)$. Suppose the last update at the high-level node was at time $k-n$, when it was transmitted the low-level estimate $\mathbf{y}_{L'}(k-n \mid k-n)$ and $\mathbf{Y}_{L'}(k-n \mid k-n)$. The updated high-level estimate, as a result of the new communication, is calculated as follows:

$$\mathbf{y}_H(k \mid k) = \mathbf{y}_H(k \mid k\text{-}n) + \mathbf{y}_L(k \mid k) - \mathbf{y}_L(k \mid k\text{-}n)$$

$$\mathbf{Y}_H(k \mid k) = \mathbf{Y}_H(k \mid k\text{-}n) + \mathbf{Y}_L(k \mid k) - \mathbf{Y}_L(k \mid k\text{-}n)$$

The low-level node forward predicts its previous communicated estimate to the current time step and subtracts it from its current estimate. The resulting information differential was referred to earlier as a partial estimate, and it was denoted in the fusion equations by $\mathbf{m}(k \mid k)$ and $\mathbf{M}(k \mid k)$. The high-level node forms a global estimate by forward predicting its past update to the current time, and then adding to it the sum of its received partial estimates.

The only implementation issue in this case is where to compute and store the partial estimates. This can be performed at either the low or high level nodes.

3.3.2 Hierarchical estimation with feedback

In the feedback system (Figure 3) there is a two-way flow of information between the low-level and the high-level fusion nodes. This now means that the de-correlation of redundant data has to be performed at each level. In common with the no feedback case, the fusion equations are very straightforward to implement in information space. Suppose the low-level (L) nodes perform their fusion operations after every m time steps and the high-level (H) nodes after every n time steps. The fusion equations for the information matrices at the current time step k read as follows,

$$\mathbf{Y}_L(k \mid k) = \mathbf{Y}_L(k \mid k\text{-}m) + \mathbf{M}_H(k \mid k)$$

$$\mathbf{Y}_H(k \mid k) = \mathbf{Y}_H(k \mid k\text{-}n) + \mathbf{M}_L(k \mid k)$$

The equations are of an identical form for the information vectors. Although the partial estimates, $\mathbf{M}_H(k \mid k)$ and $\mathbf{M}_L(k \mid k)$, are not written down explicitly here, they are calculated in a similar way to the no feedback case. Specifically, previous estimates from the last fusion update time are extrapolated forward to the time of the current estimate, and then removed.

There is an implementation issue here with regard to where the partial estimates are calculated and stored. The main question is when to communicate between levels with respect to where and when the partial estimates are calculated. This question must be resolved by considering the processing/bandwidth trade-off that each option implies. See Chong [5] for further details.

3.3.3 Tracklets

As noted previously, a current track estimate is `backloaded' with all the previous sensor measurements that have contributed to its formation. A tracklet is basically a `bit of track' maintained between communication intervals. So if a track has accumulated over k time steps, the tracklet corresponding to the next n time steps is created by initialising a new track and updating it with sensor measurements acquired in the time interval *k-n*. The track that had been calculated up to the initialisation time is removed from memory. Tracklets have been investigated in the decentralised track fusion context by Drummond [21]. The information graph for a hierarchical system employing tracklets is displayed in Figure 6.

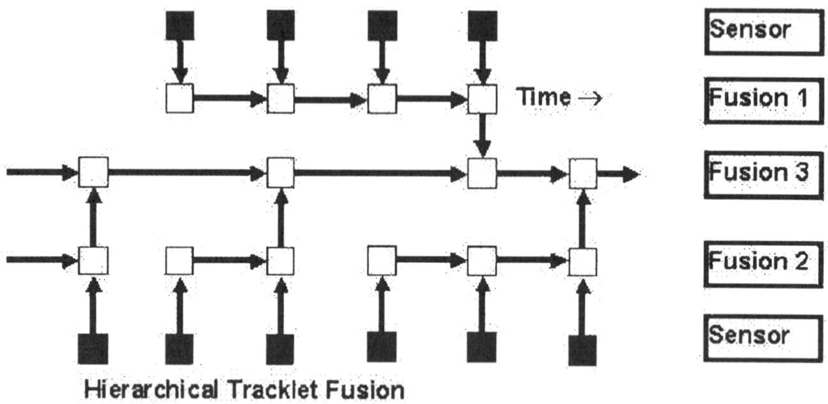

Figure 6. Information graph for a hierarchical system employing tracklets

Here the low-level estimates are initialised from scratch after every communications to the high-level fusion node. In this way, the high-level node is guaranteed to not receive the same sensor measurement more than once. The price to pay for this simple implementation strategy is a drop in performance, because the algorithm does not use all the available sensor data. The magnitude of this performance degradation will depend, e.g., on the trackers' process noise.

3.3.4 Federated filters

The so-called Federated (Zero-Reset) Filter for decentralised estimation is characterised by two sets of filters per sensing platform [22]. The principal features of the architecture are displayed in Figure 7.

Each sensing platform has a primary and a secondary track filter. The primary filter contains the total information in the track. The secondary filter contains only the information acquired since the last fusion update time. In fact the secondary filter maintains tracklets because the track information content is zeroed after every communication to the global track filter. The latter is responsible for fusing information across the network from other primary/secondary filters. If a fully connected network or broadcast communications is assumed, the global track filter can simply fuse this information by summing a set of partial estimates.

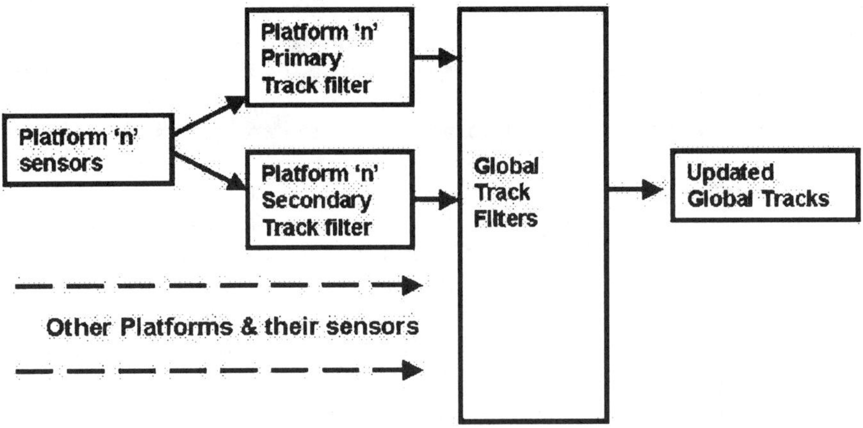

Figure 7. The Federated (Zero-Reset) Tracker Architecture

The mathematics of federated filtering is similar to the hierarchical estimators, but with some key implementation differences. In particular, the federated architecture is completely symmetric and does not differentiate between low-level and high-level fusion nodes, as do the hierarchical architectures. This is important because it allows the platforms to operate autonomously with no need for synchronised operations or master/slave designations. The main deficiency of the federated filter architecture is shared with the other partially decentralised architectures i.e., its broadcast communications facility means that the total system communications load is not invariant with respect to the number of platforms. The federated filter is described in more detail in [22] and references therein.

3.3.5 The effect of process noise

The previous fusion equations, in sections 3.3.1 and 3.3.2, and are only approximate. They assume small process noise or full-rate communications. If these conditions are relaxed, the effect of the common process noise that exists between the trackers is felt. The tracks are cross-correlated and therefore optimal track-to-track fusion is non-trivial. This problem has been investigated by Chang [17], remarkably in terms of an analytic closed-form analysis. The analysis considers information updates in a hierarchical fusion architecture. It is found that, due to the approximate nature of the fusion equations, the no-feedback architecture can produce tracks that are more accurate than the feedback architecture. This is because the estimates being fed back to the low-level fusion nodes are *wrong* when the process noise is large or the fusion update intervals are infrequent.

3.4 FULLY DECENTRALISED ESTIMATION

The fully decentralised estimation (tracking) architecture is displayed in Figure 8. It is fully modular i.e., each platform performs an identical set of estimation and communications functions. The function of the estimation process is to assimilate all its local platform's sensor information with the information it receives, via the communications process, from other platforms. The communications process is responsible for message management and maintaining a recursive estimate of the platform's incoming and outgoing information. For an overview of the decentralised architecture, see [25].

In fully decentralised estimation there are two new concerns to be addressed:

- **Information propagation**:
  As noted earlier, communications in fully decentralised systems is on a strict point-to-point basis. In a fully connected system, each node has access to all the information generated elsewhere in the network. In a partially connected system, employing the information update equations, this is not the case. An explicit mechanism is therefore required for propagating information, at no extra communications cost, to distant branches of the network. This function is performed by channel filters [23].

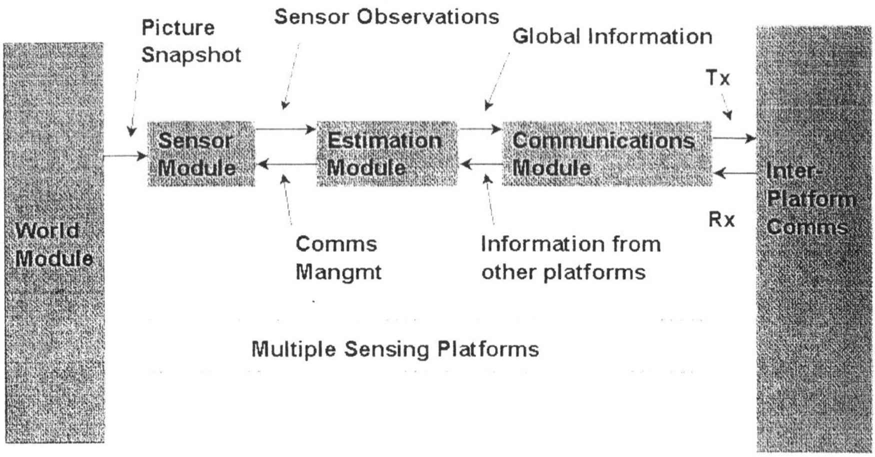

Figure 8. The fully decentralised tracking architecture

- **Spatial network loops**:
  If information propagates globally and the network is not restricted to a single communications path, there is a potential for redundant data to proliferate around spatial network loops. In fully decentralised systems, nodes do not have access to the global network topology. They are also unable to get this information by having the system propagate the entire processing history of each data item. This would remove the operational advantages of the fully decentralised architecture. Consequently, some way must be found of safeguarding the consistency of fused estimates in arbitrary decentralised networks. A promising new solution to this problem is Covariance Intersection [24], which is described further in subsection 3.4.2.

Each of these concerns is described briefly in the remainder of this section.

3.4.1 Channel filters

Channel filters are an explicit mechanism for maintaining a recursive estimate of the information that is propagated up and down a communications link [23]. They serve two key functions in fully decentralised data fusion architectures:

- The removal of redundant data prior to fusion in *singly connected* networks (e.g., tree networks).
- The propagation of information across a partially connected network at no extra communications cost.

Consider a communications channel connecting two nodes labelled i and j. Denote the channel estimate associated with this link at time k as $\mathbf{y}_{ij}(k \mid k)$ and $\mathbf{Y}_{ij}(k \mid k)$. In a non-dynamic network, $\mathbf{Y}_{ij}(k \mid k) = \mathbf{Y}_{ji}(k \mid k)$ so only one channel filter is required per link (this symmetry obviously also holds for the channel filter information vectors - for brevity, we only present the information matrix equations in this section). For full rate communications, the channel filter is updated after a synchronised two-way transmission between the nodes according to the following equation:

$$\mathbf{Y}_{ij}(k \mid k) = \mathbf{Y}_{ij}(k \mid k\text{-}1) + [\mathbf{Y}_i(k \mid k)\text{-}\mathbf{Y}_{ij}(k \mid k\text{-}1)] + [\mathbf{Y}_j(k \mid k)\text{-}\mathbf{Y}_{ij}(k \mid k\text{-}1)]$$

$$= \mathbf{Y}_i(k \mid k) + \mathbf{Y}_j(k \mid k) - \mathbf{Y}_{ij}(k \mid k\text{-}1)$$

The two terms in square brackets are partial estimates. The first term is used to update j's estimate and the second term is used to update i's estimate.

The channel filters create a computing overhead, but this is far outweighed by their stated advantages. In effect, channel filters allow decentralised estimates to be generated that are equivalent to those of a centralised system, with some time delays due to finite propagation times across the network [23]. The reader is reminded that the ability of channel filters to propagate information *and* remove redundancy is restricted to singly connected networks.

3.4.2 Covariance Intersection

It is impossible to either identify or prohibit redundant data, in arbitrary decentralised networks, without relaxing the operational constraints imposed on the system [18]. One technique involves the creation of a dynamic spanning tree for the network, by artificially breaking links. Such dynamic reconfiguration has been investigated in [26]. It requires prior knowledge of the network topology and therefore sets limitations on the flexibility of the architecture.

Given that redundancy in the data is generally impossible to identify and remove, a different approach to the problem is required. Specifically, what is needed is a fusion mechanism that is insensitive to cross-correlation terms in the following sense: given N items of data, which are individually consistent, the mechanism should provide a consistent fused product without requiring the cross-covariance of the data to be specified. It turns out that there is indeed such a mechanism, Covariance Intersection (CI), developed by Uhlmann [24].

CI is simple to explain in terms of geometry. Consider the two mean estimates, (**a**,**b**), and their associated 1σ error covariance matrices ($\mathbf{P}_{aa}$,$\mathbf{P}_{bb}$).
The covariance ellipses are displayed, for coincident means, in Figure 10. Now, suppose that these estimates are optimally combined to form a new estimate given by (**c**,$\mathbf{P}_{cc}$). The optimal combination requires the cross-covariance, $\mathbf{P}_{ab}$, of **a** and **b** to be specified. The left side of Figure 9 displays $\mathbf{P}_{cc}$ for a range of values of $\mathbf{P}_{ab}$. Note that the combined covariance ellipses are circumscribed by the intersect points of the individual ellipses. This implies that, if we can fit an ellipse through these intersect points, it will provide an upper bound on the optimal uncertainty. There are, of course, an infinity of such ellipses. A subset of them is shown on the right side of Figure 9. One of these is selected on the basis of minimising some cost of **P**cc, such as its trace or determinant.

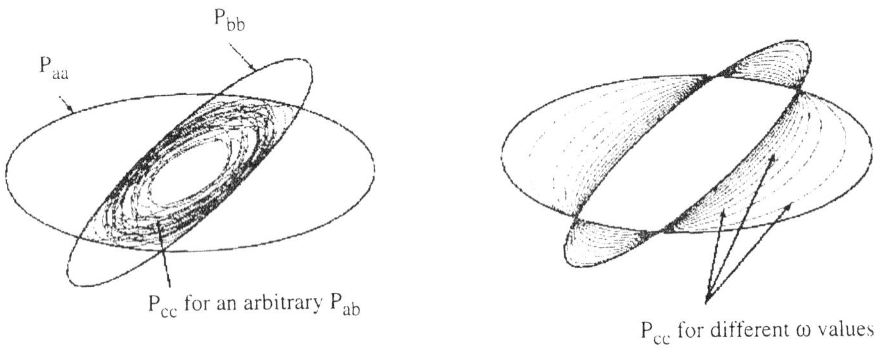

Figure 9. Covariance Intersection geometry

The CI geometry translates into the following simple fusion equations:

$$\mathbf{P}_{cc}^{-1} = \omega\mathbf{P}_{aa}^{-1} + (1-\omega)\mathbf{P}_{bb}^{-1}$$

$$\mathbf{P}_{cc}^{-1}\mathbf{c} = \omega\mathbf{P}_{aa}^{-1}\mathbf{a} + (1-\omega)\mathbf{P}_{bb}^{-1}\mathbf{b}$$

These equations introduce a scalar weighting parameter, $\omega \in (0,1)$. This is obtained by solving a convex optimisation problem with respect to a chosen norm of the combined covariance matrix. (It is worth noting that CI with ω=1/2 corresponds to a Kalman update with its covariance inflated by a factor of 2). If there are only two estimates to be combined, the optimisation can be performed using a standard technique such as bisection or the Newton-Raphson method. Sophisticated techniques, such as Boyd *et al.*'s semi-definite and convex programming methods [27], are required for batch optimisation over multiple estimates. The efficiency of such methods, and their accuracy with respect to sequential pair-wise CI, still needs some further investigation.

CI's proven consistency, with respect to unknown correlations, lend it an obvious appeal in the context of decentralised estimation. Julier and Uhlmann have reported a preliminary investigation [28]. This work was extended by one of the current authors (DN), who formulated a hybrid decentralised CI/Kalman algorithm [29]. This was applied to the problem of track-to-track fusion in dynamic communication networks. Further details of this work are presented in section 4.

4. BAE Systems Research Snapshot

The authors are with BAE SYSTEMS Advanced Technology Centres - Sowerby, based in the UK. Here the System Autonomy and Advanced Embedded Systems Group has been actively involved in data fusion research for over a decade. This work has concentrated on an assessment of efficient data fusion architectures. Specifically, in collaborations with the University of Sydney (Hugh Durrant-Whyte) and the Naval Research Laboratory (Jeffrey Uhlmann), the Group has undertaken a thorough investigation of the decentralised data fusion architecture and its associated algorithms. This section provides a selective snapshot of some recent and on-going research.

4.1 COMMUNICATIONS MANAGEMENT

In dense multi-target tracking environments there may be insufficient communications bandwidth to support the propagation of large numbers of track estimates (means and their covariances) over a decentralised network. Even if there is sufficient bandwidth, it may be operationally desirable to manage this resource in such a way that the volume of traffic is minimised. These situations require a strategy for communication resource management.

The application of information theoretic measures to the general sensor management problem in MSDF systems has been explored by a number of researchers [2,30,31]. In particular, Manyika and Durrant-Whyte [2] have formulated a decentralised information theoretic approach for data fusion and sensor management. BAE SYSTEMS has shown how a communications management strategy can also be formulated in terms of decentralised information theory [11]. This has been assessed, in simulation, for a simple decentralised identification, tracking and joint identification/tracking system.

Consider the type of N-on-M air engagement scenario shown in Figure 10. The friendly aircraft system (to the left of the picture) forms a fully connected decentralised network. The enemy aircraft (to the right of the picture) are in track and we would like to calculate some overall measure of the tracking performance of the system when its communications bandwidth is restricted.

Our performance measure is the total steady state uncertainty volume in the tracks. This is calculated for a range of inter-platform communications bandwidth values, expressed as a percentage of the total number of targets, i.e., $bw = (M/5)\times 100\%$. Performance is assessed with respect to two communication management strategies: information-based and round-robin.

The communications management problem, given that each network link supports the transmission of only m-out-of-M tracks ($m<M$), is to decide which subset of tracks a node should communicate to another node at each time step. The information-based approach attempts to maximise the total information (or equivalently to minimise total

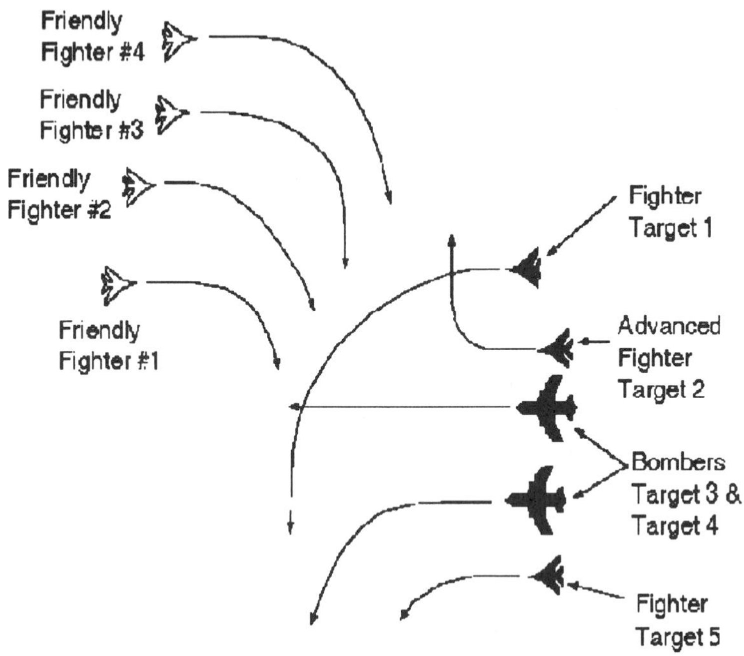

Figure 10. Battlespace scenario used for communications management study

uncertainty) at the receiving node over a single communications time step. The round-robin approach, which is not based on any optimality criterion, selects its target subset from a sequentially ordered list, e.g., $(2,3,4)_{c=1}$ $(5,1,2)_{c=2}$, $(3,4,5)_{c=3}$ *etc.* for bw=60%., where c indexes a node's communications update time step. A detailed description of these algorithms and their analysis is provided in [11,32].

Notice from the scenario diagram (Figure10) that the enemy aircraft are of mixed type with respect to their manoeuvrability. A constant velocity tracker was employed with a manoeuvre-dependent process noise. It was found that the information-based algorithm communicated on the most manoeuvrable targets more frequently than the less manoeuvrable targets. This is understandable in terms of the bigger reduction in information that is caused by fusing two of the manoeuvring target tracks relative to two of the non-manoeuvring target tracks.

The performance of the communications management algorithms, with respect to the uncertainty measure indicated earlier, is displayed in Figure 11. The information-based algorithm is equivalent to the round-robin algorithm at zero and full communications bandwidth. In the partial bandwidth situation, the information-based algorithm outperforms the round-robin algorithm, by almost a factor of two at bw = 40%. The statistical error bars on these curves were established by Monte-Carlo simulations, and are small.

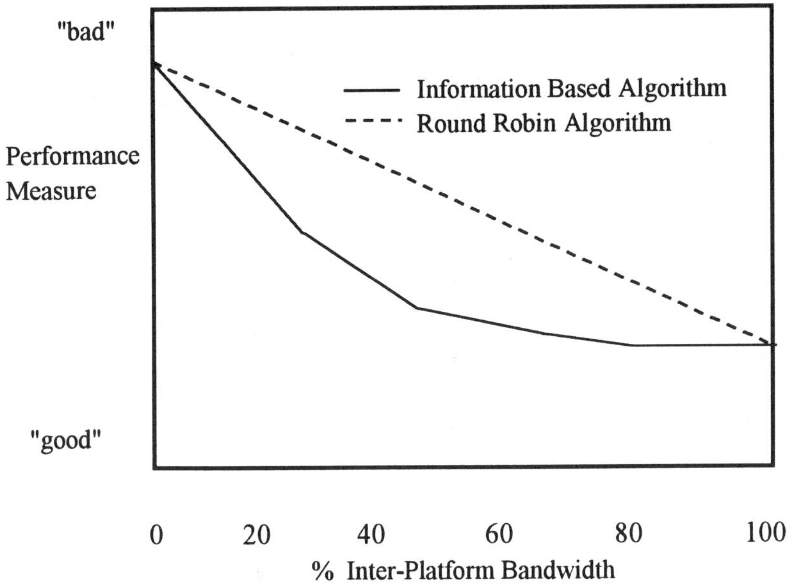

Figure 11. Example performance assessment results for communications management

Results such as these demonstrate the extent to which advanced information processing can be used to manage a limited communications bandwidth resource. They may have a decisive role to play in improving the efficiency of a decentralised architecture with minimal information loss. At a fundamental theoretic level, our analysis shows communications management can be formulated within the probabilistic information update framework that underpins decentralised estimation and sensor management.

4.2 DYNAMIC NETWORKS

BAE SYSTEMS is currently supporting the U.S. DoD Naval Research Laboratory's Decentralised Data Fusion (DDF) project. This is an ONR-funded effort to assess CI in the context of data fusion problems that are relevant to the U.S. Navy. Specifically, BAE SYSTEMS has been applying CI to the problem of robust decentralised estimation in arbitrary, dynamic, battlespace networks.

The current status of this work is described best in terms of an example. Consider the battlespace scenario displayed in figure 12. The friendly force (left side of picture) is tracking an enemy force (right side of picture) based on simulated radar sensor measurements. One of the enemy aircraft (T4) flies overhead the blue force and selectively jams its communication links. The network architecture is also shown in Figure 12. At four equal intervals of forty simulation time steps, the communication links into P5 are jammed. These phases are indicated (b) to (e) in the diagram. During phase (a) of the scenario, i.e., its first forty time steps, the network is fully connected. During phase (e) of the scenario, P5 is completely jammed out of the network.

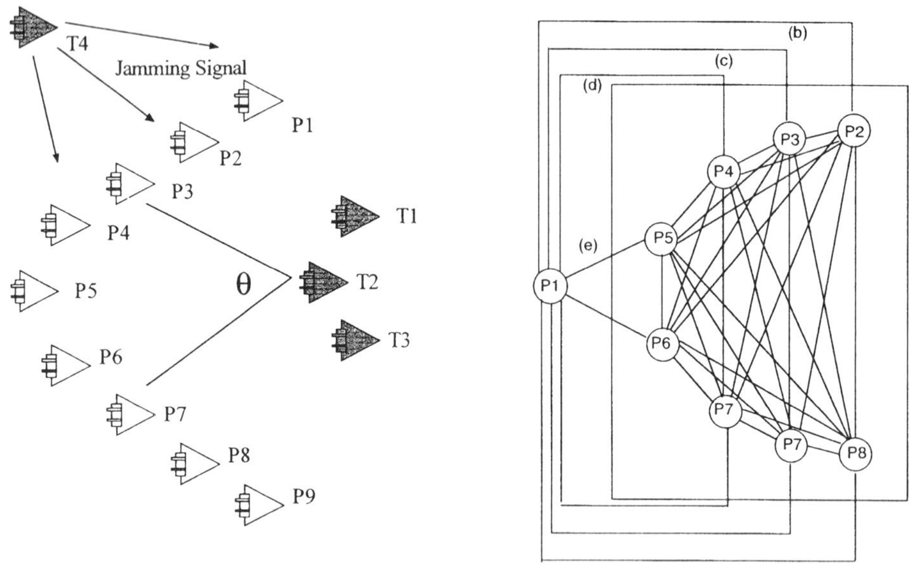

Figure 12. Battlespace scenario and network architecture used in the dynamic network study

The result of this scenario focuses on P5's fused track covariance matrix with respect to target T2. The decentralised information filter is linear with a constant velocity process noise. Ideal data association is assumed.

Three decentralised data fusion algorithms were considered in this particular study. The first was CI, optimising ω with respect to the fused track uncertainty volume. The second algorithm employed channel filters and a Kalman update rule. This set-up prevented P5 from obtaining information that is generated outside its own neighbourhood, i.e, P5 is only able to fuse its track with the tracks that are generated at P4 and P6. Finally, a third option was considered. This is a hybrid formulation of the other two.

Figure 13 (a) and (b) display the cross-range and range components of P5's fused track covariance respectively, across the full scenario time. In each case, the results for the local Kalman algorithm exhibit a noticeable degradation at each phase transition. This is because, for local Kalman, P5 has to be *directly* connected to a node for it to receive its information. The degradations are only slightly noticeable for the CI algorithm. This is because the CI algorithm allows information from distant nodes to propagate through to P5 indirectly. In doing so, this information accrues extra process noise due to its increased path length. The process noise was not set to a high value in this simulation, so CI does not suffer by much in this example as a result of the network's jammed communication links. In effect, the local Kalman algorithm is fusing impoverished information optimally and the CI algorithm is fusing a full information set conservatively. Consequently, the local Kalman algorithm outperforms CI up until the third phase of the scenario, after which CI performs best. The hybrid Kalman / CI algorithm is the best performer overall, because it combines the best of each algorithm.

These results indicate that CI has an important role to play in decentralised estimation. Specifically, CI enables a consistent and globally fused tactical picture to be maintained, even when the network is changing in an assumed unknown way. These changes could be imposed by a jamming signal, as in the example, or they may be a result of network reconfiguration due to platforms entering and exiting the battlespace. The results also indicate that the Kalman filter should be employed where possible, i.e. for fusing components of the information that can be safely assumed to be independent.

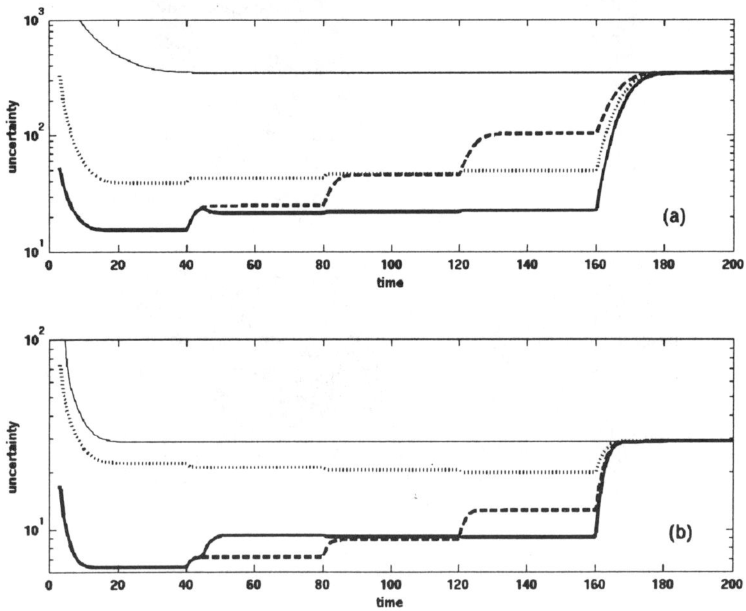

Figure 13. Dynamic network results. (a) Cross-range position error component of track; (b) Range position error component of track. Thin solid line = no track fusion; Dotted line = CI; Dashed line = local Kalman; Thick solid line = CI/Kalman hybrid

4.3 DEMONSTRATORS

The maturity of the decentralised data fusion paradigm can now be asserted with some degree of confidence. Based on a model of uniform information flow, several efficient algorithms have been derived. The next major step is to back up this confidence with a range of practical hardware demonstrations. The Signal and Knowledge Integration with Decisional Control for Multi-Sensory Systems (SKIDS) project, in collaboration with the University of Oxford, was a first step in this direction. This succeeded in demonstrating, under laboratory conditions, real-time decentralised tracking and surveillance [14]. This section outlines two current hardware projects.

4.3.1 Decentralised Data Fusion Demonstrator

The Decentralised Data Fusion Demonstrator (DDFD) is a BAE SYSTEMS laboratory facility for researching, under a fairly benign set of conditions, a range of practical issues in decentralised data fusion. The DDFD infrastructure consists of four wall-mounted cameras, four laptop computers connected via a radio LAN, and a number of mobile ground-based robots. A selection of the DDFD hardware is displayed in Figure 14.

Figure14. DDFD hardware components

The construction of the DDFD framework has in itself provided a valuable exposure to some of the real-world issues that are often side stepped by simulators. One example is how to handle asynchronous (out-of-sequence) data caused by communication delays in the network. Other examples relate to real-time operation, offset-induced correlation, multi-platform registration, and distributed data association. The DDFD is a very useful experimental test-bed for recognising and trying to understand these problems and many others that are likely to be uncovered.

4.3.2 ANSER

The Autonomy, Navigation, Sensing and Experimental Research (ANSER) project is a collaborative programme involving BAE SYSTEMS (in the UK and Australia) and the University of Sydney, Australia. A goal of ANSER is to field a multi-platform decentralised data fusion system in order to demonstrate its claimed benefits (i.e., fault-tolerance and flexibility), under considerably more challenging conditions than are offered by the laboratory.

ANSER involves four flying sensor platforms, of which one is displayed in Figure 15. The test site for these platforms is located in an isolated area near Sydney. The aircraft have a 3m wingspan and a payload in the range 5-8kg. They will each be equipped with a multi-sensor suite, including a camera, GPS and INS. Communications between the aircraft will be on a point-to-point basis.

The ANSER project should provide compelling demonstrations of single platform multi-sensor data fusion, multi-platform multi-sensor data fusion, and autonomous navigation. The project is in its early stages at the time of writing, but results are expected to be forthcoming within the next two years.

Figure15. The ANSER flight platform

5. Conclusions

This chapter has reviewed MSDF system architectures and algorithms. The centralised architecture, although it generates the optimal fused product, can pose such severe computational and communication bottlenecks in doing so that it is generally not practicable. Consequently, decentralised data fusion architectures are drawing considerable interest. A potential military application area is network-centric warfare. New ("Free Flight") air traffic management systems are an interesting area of potential civil application.

Decentralised architectures can be either partially or fully decentralised. The simplest partially decentralised architectures are the hierarchical systems. Fully decentralised systems are subject to a number of additional operating constraints, but these make the systems more resilient, flexible, and scalable.

Decentralised estimation requires more sophisticated algorithms than centralised estimation. This is to counter the problem of rumour propagation, which has spawned a range of architectural and algorithmic designs. Examples include the hierarchical estimators with and without feedback, tracklets, federated filters, channel filters, and Covariance Intersection. These algorithms are efficiently implemented in information space. Indeed probabilistic information update provides an underpinning theoretical framework for decentralised estimation and systems resource management.

Decentralised data fusion is now a reasonably mature field of research. While there is still some theoretical and algorithmic research to be done, it is our belief that the balance should be shifting to practical demonstrations. BAE SYSTEMS is currently involved in two demonstrator programmes.

This chapter has only considered architectural efficiency in the context of sensor data fusion, i.e., the picture compilation phase of data fusion. An important, but relatively under researched issue, is the impact of decentralised architectures on the situation and threat assessment problem.

6. Acknowledgements

The authors would like to thank their colleagues and collaborators at the BAE SYSTEMS Advanced Technology Centres (Sowerby), the University of Sydney (Australia) and the Naval Research Laboratory (Washington DC). Any errors, omissions, or oversights are, of course, entirely our own.

7. References

1. Hall, D., (1992) *Mathematical Techniques in Data Fusion*, Artech House.
2. Manyika, J., and Durrant-Whyte, H.F. (1997) *Data Fusion and Sensor Management: A Decentralized Information-Theoretic Approach*, Ellis Horwood.
3. Mutambara, A. (1998) *Decentralized Estimation and Control for Multisensor Systems*, CRC Press.
4. Chong, C.Y., Mori, S. and Chang, K.C. (1990) Distributed Multitarget Multisensor Tracking, *in Multitarget Multisensor Tracking: Advanced Applications*, Y. Bar-Shalom, ed., Artech House, pp. 247-295.
5. Chong, C.Y. (1998) Distributed Architectures for Data Fusion, in *Proceedings of the International Conference on Multisource-Multisensor Information Fusion (Fusion '98)*, H.R. Arabnia and D. Zhu, eds., CSREA Press, pp. 84-91.
6. See http://ait.nrl.navy.mil/people/uhlmann/CovInt.html
7. Alberts, D.S., Gartska, J.J. and Stein, F.P, (1999) *Network Centric Warfare, Developing and Leveraging Information Superiority*, CCRP, (Available in pdf format from www.dodccrp.org/ncw.htm)
8. Tomlin, C., Pappas, G.J., Lygeros, J., Godbole, D. and Sastry, S. (1996) A next generation architecture for air traffic management systems, in *Hybrid Systems IV, Proceedings of the 4th International Conference on Hybrid Systems, Lecture Notes in Computer Science*, Springer Verlag.
9. Speyer, J.L., (1979) Computation and transmission requirements for a decentralised linear-quadratic Gaussian control problem, *IEEE Trans. On Automatic Control*, AC-24:2.
10. Levy, B.C., Castanon, D.A., Verghese, G.C. and Willsky, A.S. (1983) A scattering framework for decentralized estimation problems, *Automatica*, 19:4.
11. Deaves, R.H., Nicholson, D., Greenway, P. and Vangasse, P. (1997) Communications management in battlespace data fusion, in *Proceedings of the Agard Symposium on ulti-Sensor Systems and Data Fusion for Telecommunications, Remote Sensing and Radar*, Lisbon.
12. Bar-Shalom Y. and Fortmann, T.E. (1988) *Tracking and Data Association*, Academic Press Inc.
13. Maybeck, P.S. (1979) *Stochastic Models, Estimation and Control*, Vol I, Academic Press.

14. Rao, B., Durrant-Whyte, H. and Sheen, A. (1991) A fully decentralised multi-sensor system for tracking and surveillance, *Int. J. Robotics Research*, 12(1):20-45.
15. Bar-Shalom, Y. (1981) On the track-to-track correlation problem, *IEEE Trans. Automatic Control*, AC-26:571.
16. Chang, K.C., Tian, Z. and Saha, R.K. (1998) Performance evaluation of track fusion with information filter, *Proc. First International Conference on Multisource-Multisensor Information Fusion*, pp.648, Las Vegas.
17. Chang, K.C. (2000) Evaluating hierarchical track fusion with information matrix filter, *Proc. Third International Conference on Multisource-Multisensor Information Fusion*, Paris.
18. Utete, S.W. (1995) *Network Management in Decentralised Sensing Systems*, PhD thesis, University of Oxford.
19. Weiss, Y. (2000) Correctness of local probability propagation in graphical models with loops, *Neural Computation*, 12:1-41.
20. Pao L.Y., and Kalandros, M.K. (1997) Algorithms for a Class of Distributed Architecture Tracking, in *Proc. American Control Conference*, Albuquerque, NM, pp. 1434-1438.
21. Drummond, O.E. (1997) A hybrid sensor fusion architecture and tracklets, in *Proc. SPIE Signal and Data Processing of Small Targets*, Vol. 3163.
22. Carlson, N.A. (1999) Federated filter for multiplatform track fusion, in *Proc. SPIE Signal and Data Processing of Small Targets*, Vol. 3809, pp. 320-331.
23. Grime, S. and Durrant-Whyte, H.F. (1994) Data fusion in decentralized sensor networks, *Control Eng. Practice*, Vol.2, pp. 849-863.
24. Uhlmann, J.K. (1996) General data fusion for estimates with unknown cross covariances, *Proc. SPIE Aerosense*, Vol. 2755, 4.
25. Durrant-Whyte, H.F., Deaves, R. and Greenway, P. (1998) Decentralised multi-platform data fusion, in *Proc. SPIE Aerosense*, Vol. 3393.
26. Utete, S.W. and Durrant-Whyte, H.F. (2000) Local communication constraints in decentralised sensing networks, *IEEE Trans. On Systems, Man and Cybernetics*, in press.
27. Vandenberghe, L. and Boyd, S. (1996) *Semidefinite programming*, SIAM Review.
28. S. Julier, S. and Uhlmann, J.K. (1997) A nondivergent estimation algorithm in the presence of unknown correlations, in *Proceedings of the 1997 American Control Conference*
29. Nicholson, D. and Deaves, R.H . (2000) Decentralised track fusion in dynamic networks, in *Proc. SPIE Signal and Data Processing of Small Targets*, Vol. 4048.
30. Hintz, K.J. and.McIntyre, G. (1998) Information instantiation in sensor management, in *Proc. SPIE Aerosense*, Vol. 3374, pp. 38-47.
31. Kastella, K. (1997) Discrimination gain to optimise detection and classification, *IEEE Trans. on Systems, Man, and Cybernetics*, Vol. 27/1.
32. Deaves, R.H. (1998) *The Management of Communications in Decentralised Bayesian Data Fusion Systems*, PhD thesis, University of Bristol.

REASONING FRAMEWORKS

P. VALIN
Advisory Member of R&D
Lockheed Martin Canada
6111 Royalmount ave
Montréal, QC, H4P 1K6
Canada

1. Introduction

Reasoning over attributes or situations plays a big role both in military and commercial domains. The focus of this lecture is reasoning frameworks for uncertain and incomplete information that can lead to a correct understanding of the military picture at all levels of fusion, leading to refinements in single object (Level 1), situation (Level 2) and implication (Level 3). An occasional venture into more plebian commercial domains is made through appropriate examples.

The attributes, over which one has to reason at Level 1, can originate from imaging or non-imaging sensors and be kinematical, geometrical or relate more directly to the ultimate obtention of a unique identification (if those come from intelligent sensors such as Identification Friend or Foe (IFF), Electronic Support Measures (ESM), or imagery classifiers). The concept of a unique identification (ID) itself contains an ambiguity related to the desired taxonomy, over which reasoning occurs, and this choice is discussed with respect to the operator's need for a decision aid tool. This taxonomy spans all levels of fusion and appropriate attributes must be found that correspond to the traits of target agglomerations for Levels 2 and 3. An example set of databases is discussed that contains platform characteristics, emitter-to-platform mapping, geopolitical allegiance information, contextual and doctrinal information for situation assessment, etc.

Physical attributes often require fuzzification for proper treatment by evidential reasoning, and fuzzy logic is just one of many reasoning frameworks. The following reasoning frameworks are introduced:

- Fuzzy logic
- Neural Networks (NNs)
- Bayesian approach (with a priori information)
- Dempster-Shafer (DS) approach with its variants.

A useful measure of performance is defined that can be applied to simulated real-world examples. The need for a real-time performance of the identification estimation (or attribute fusion) function of a Level 1 data fusion module are shown to lead to various levels of approximation, depending on which scheme is used, and declassified examples

A. K. Hyder et al. (eds.), Multisensor Fusion, 223–245.

are presented. A truncation scheme is presented that will cure the exponential increase of proposition in the DS approach. This lecture provides real-world examples from LM Canada projects [1-14]. Shafer-Logan theory is covered in another lecture [15] and is not addressed here.

2. Reasoning at Different Levels of Data Fusion

According to the updated 1999 JDL classification of Data Fusion (DF) levels, one can expect reasoning at correspondingly different levels:

- Level 0: *sub-object assessment* should require only pre-processing possibly selected from a priori knowledge of possible acquisition problem or difficult situations
- Level 1: *single object refinement* should involve evidential reasoning over single object kinematics and attributes, towards the goal of obtaining the best individual platform identity [ID] or at least some hierarchical level of the desired taxonomy tree
- Level 2: *situation refinement*, often referred to as Situation and Threat Assessment (STA) should involve reasoning over groups of objects and proceed by higher inference rules involving doctrinal and contextual information
- Level 3: *implication refinement*, should involve reasoning over plan alternatives, suggest plan decisions and act on the best plan selected
- Level 4: *process refinement* should involve reasoning over ownship and environmental conditions in order to perform better sensor management and thus close the loop for the next step in this Observe, Orient, Decide, Act (OODA) loop.

This lecture concentrates mostly on single object evidential reasoning (mostly algorithmic, fuzzy...), and touches on reasoning over groups of objects (mostly knowledge-based or involving Artificial Intelligence (AI) reasoning.

3. Attributes Characteristic Of Platforms At Levels 1 And 2 Of Fusion

As a preamble, one should be aware that the format in which the databases should be encoded is an active topic of research, with the experts still debating whether relational databases or Object-Oriented (OO) databases are preferable. Compromises often have to be made and the use of Commercial-Over-The-Shelf (COTS) software (SW) may in the future make as many inroads as COTS hardware (HW) is currently doing throughout the world's military establishment. This section only describes the content of these databases, not their optimal architecture nor how complex their interrelations are woven. The discussion is at a relatively high level, without any real-world examples, since these are presented in another lecture [15].

3.1 THE PLATFORM DATABASE AND RELATED NAMELISTS

For ID estimation to be properly achieved in Level 1 DF, all possible attributes that can be measured by all of the sensors must be listed in the Platform Database (PDB). The attributes to be catalogued in the PDB can be split into three groups [4] (a more detailed explanation can be found in [5]):

(a) *Kinematic attributes*: which can be estimated through tracking in the positional estimation function of DF, and through reports from IFF and datalink. The maximum acceleration, the maximum platform speed, the minimum platform speed and the cruising speed either serve as bounds to discriminate between possible air target IDs or suggest the most plausible IDs. The maximum altitude that a platform may reach can serve as a bound for altitude reported by the IFF or deduced from a tracker in 3-D from sensor reports in 2-D.

(b) *Geometrical attributes*: which can be estimated by algorithms which post-process imaging information from sensors such as Forward Looking Infra-Red (FLIR), or Electro-Optics (EO) and Synthetic Aperture Radar (SAR). Classifiers that perform such post-processing can be thought of as Image Support Modules (ISM) performing much the same functionality as the ESM does for the analysis of electromagnetic signals. These ISMs need the three geometrical dimensions of height, width and length (for FLIR and EO), and also Radar Cross Section (RCS) of the platform seen from the front, the side and the top (for the SAR and radiometric radar). In addition, the distribution of relevant features may be needed for classifiers, but may be considered part of the algorithms that generate plausible IDs. The RCS values are empirically much larger than the geometrical cross-section obtained by the product of the two relevant dimensions, since metallic objects offer strong radar backscatter when compared to the geometrical cross-section.

(c) *Identification attributes*: which can be directly given by the ESM, or as outputs of the FLIR and SAR ISM. The ESM requires an exhaustive list of all the emitters that are carried by the platform. However the ISMs are usually designed to not only to provide the best single ID possible but also to estimate confidence in higher levels of an appropriate taxonomy tree. This taxonomy tree is usually derived from some standard, either STANAG 4420 or MIL-STD 2525 (A or B), or can even be customized depending on the application (naval, airborne or helicopter platforms). This is indeed the case for the applications studied by Lockheed Martin Canada (LM Canada), namely the Halifax Frigate (HFX), the maritime surveillance aircraft CP-140 Aurora (a P3-C with class S-3B avionics), and the LAMPS helicopter, which will perform roughly the same tasks as the Maritime Helicopter Program (MHP) helicopter (that will replace the ageing Sea Kings on the CPF).

(d) A *taxonomy tree*: which will be described in the next section.

Some sensors measure attributes quite directly. For example the ESM will provide an emitter list with some confidence level about the accuracy of the list that reflects the confidence in its electromagnetic spectral fit. However an IFF response can lead to an identification of a friendly or commercial target but the lack of a response does not necessarily imply that the interrogated platform is hostile. One has to distribute the lack of a response between at least two declarations: the most probable foe declaration and a less probable friendly or neutral declaration that allows for an IFF equipment that is not working or absent.

Some complications arise when dealing with kinematic parameters reported occasionally by the tracker in *positional estimation.* First, each physical quantity has a different dimension (speed, acceleration) and an accurate determination is not necessarily needed for fusion. Indeed it is convenient to bin the attribute "speed" into fuzzy classes like "very fast", "fast", "average", "slow" and "very slow" (separately for air and surface targets). The same can be done for length in bins of 40-meter width, for example. Membership in each class is a measure of how well the measured value fits into the descriptor as described below. Thus fuzzyfication is necessary even though fuzzy logic may not be used as a reasoning framework. Similarly defuzzification may be used to present results to the operator through the Human Computer Interface (HCI).

Other complications arise in DF with respect to correlating tracker speed values with the speed attribute in the PDB. Indeed, speed reports must be fused only if they involve a significant change from past historical behaviour in that track. The reason is two-fold. First, no single sensor must attempt to repeatedly fuse identical ID declarations, otherwise the hypothesis that sensor reports are statistically independent is violated. Second the benefits of the fusion of multiple sensors is lost when one sensor dominates the reports. Furthermore, a measured value of speed only indicates that the target is capable of that speed, not that it corresponds to either the maximum or minimum speeds listed in the PDB. It is a reasonable working hypothesis to fuzzify the value reported by the tracker into adjacent "bins" to account for the target being at, say, only 80% of its optimal speed (a "very fast" target can occasionally travel "fast"), or travelling with a strong tailwind (a "fast" target can occasionally appear as "very fast"). Finally the concept of binning can be generalized to continuous membership functions of a fuzzy set. An alternative way of proceeding is to bin around the cruising speed, which can be very different from 80% of the maximum speed for fighters on afterburners.

Although the PDB is the most important of the databases and is immutable, there are many supporting databases which are necessary and flexible, such as the Geo-Political Namelist (GPL), which contains acronyms of the countries that own platforms in the PDB. In the PDB, the acronym is used by the attribute fusion function to link the PDB platform with the country allegiance indicated in the GPL: friendly, foe or neutral (depending on the mission). In case language communications can be intercepted, the languages spoken in different countries are also listed. The Emitter Namelist (ENL) offers a correspondence [7] between the index number of each emitter and its name (actual or given by NATO), various characteristics (Pulse Repetition Frequency (PRF), frequency, detection range, etc.) and can offer (optionally) a cross-listing of platforms on which each emitter can be found.

3.2 CHOICE OF A TAXONOMY TREE

The breadth and width of a taxonomy tree can easily grow out of proportion. This is where the decision on the database architecture plays a major role. As currently implemented at LM Canada, fields are added to the width of the PDB to account for a limited taxonomy. Figure 1 illustrates this fact for a taxonomy tree [1] of depth 2 (not counting the final ID level).

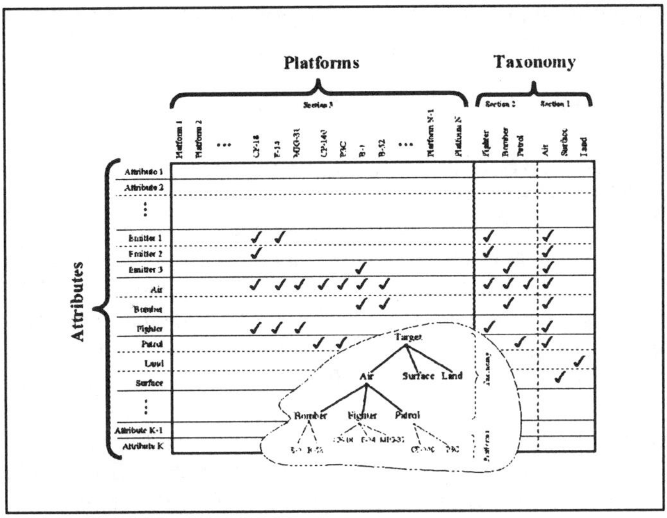

Figure 1. PDB with Platforms and the Taxonomy Tree

In practice, there are many more levels in a taxonomy tree. As a pedagogical example (not used in any of the applications involving LM Canada yet), the first level of detail should choose between AIR, SEA SURFACE, SUB-SURFACE and LAND (or GROUND). The next finer levels (e.g., class and sub-class) could offer the following branches, shown in Table 1 for ground and sea-surface nodes only.

TABLE 1. Taxonomy Example for Ground and Sea Surface Targets

For	GROUND	for	SEA SURFACE
Level 2	Level 3	Level 2	Level 3
Vehicle	Armoured	Combatant	Line
	Engineer vehicle		Amphibious warfare ship
	Train locomotive		Mine warfare vessel
	Civilian vehicle		Patrol
Weapons	Missile launchers		Hovercraft
	Single rocket launcher		Station
	Multiple rocket launcher		Navy group
	Antitank rocket launcher	Non-combatant	Underway replenishment
	Other		Fleet support
Sensor	Radar		Intelligence
	Emplaced		Service & support harbour
Special	Laser		Hospital ship
	NBC equipment		Hovercraft
	Flame thrower		Station
	Land mines	Non-military	Merchant
Installation	material facility		Fishing
	ship construction		Leisure craft
	government leadership		Law enforcement vessel
	airport		Hovercraft
	other		Emergency or Hazard

The branch terminations above serve as nodes for finer branches, e.g., specific classes, leading eventually to the single platform ID. Noteworthy examples are line ships separating into Frigate/Corvette, Destroyer, Cruiser, Battleship, Aircraft Carrier, and merchants separating into Cargo, Roll-on/Roll-off, Oiler/Tanker, Ferry, Passenger, Tug, etc.

3.3 DATABASE FOR HIGHER LEVELS OF FUSION

It seems natural that Level 2 fusion accesses a different database than the PDB relevant for Level 1. The STARM-DB should contain all the platform parameters relevant for STA as well as RM, i.e., since missiles (number and detailed properties) on enemy ships are relevant for STA, while the same information on possible own-platforms is relevant for RM. In a Cooperative Operating Picture (COP) context, the lethality of enemy platforms (the red force) is important for STA, and the lethality of cooperating Participating Units (PUs) is relevant for RM within the blue force. A non-exhaustive list contains at least four SEA SURFACE platforms, with elements pertaining to:

(a) Platform and Mission: Displacement (in tons under full load), number of hulls (sea-worthy examples of the ID), list of hull numbers, list of hull names, platform type, amplification, and role, crew (for full operation), operational or standby flags
(b) Armament: Number of Surface-to-Surface Missiles (SSMs), Surface-to-Air Missiles (SAMs), Close-In Weapon Systems (CIWSs), guns, torpedo tubes, troop complement (number of special force people for assault or landing), lethality, and flags for Electronic Countermeasures (ECM), acoustic CM, infra-red CM (e.g., chaff) equipment
(c) Sensors: Number and list of radars, sonars (e.g., Hull-Mounted Sonar (HMS), towed-array, sonobuoy, tethered sonar), and imaging sensors (e.g., EO, Forward Looking Infra-Red (FLIR), IRST, SAR)
(d) Air Platforms on Deck: Number of helicopters, aircraft, with PDB index list and similarly for AIR platforms with the same decomposition (item 4 cannot exist):
    - Platform type, amplification role, number of wings (examples in flying conditions), list of wing number, range, crew (number of people for full operation), operational or standby flags
    - Number of Air-to-Surface Missiles (ASMs), Air-to-Air Missiles (AAMs), CIWSs, conventional bombs, troop (number of special force for assault or parachuting), lethality, and flags for ECM and infra-red CM equipment
    - Number and list of radars, number and of imaging sensors (e.g., EO, FLIR, SAR), guidance

4. Fuzzy Logic Reasoning

Fuzzy logic deals with approximate modes of reasoning. In standard logic, a proposition is either true or false. In fuzzy logic a proposition has a parameter value, called a "membership value", ranging from 0 (completely false) to 1 (completely true). Zadeh's fuzzy logic is a well-defined formalism that describes the fuzzy propositions and combination rules to create syllogisms and inferencing for using fuzzy probability. Fuzzy logic application to the ID estimation problem is not as well documented in literature as are the Bayesian or DS approaches.

Information from the ESM sensor can be considered fuzzy. The ESM measures few parameters and the results of the measurement process are compared with a database of emitter characteristics. The measured parameters can be corrupted by countermeasures, making the comparison uncertain. Also, information in this database can be considered fuzzy, to a certain extent, due to inherent imprecision. Bayesian or DS approaches presuppose that input data follow strict logic rules of statistics or probability. If sensor declaration "noise" is a little different from the measurement noise or if the sensor declarations are not completely independent, the Bayesian or DS approaches produce results that are quantitatively inexact but qualitatively correct. In other words, they are neither false nor true as per Boolean logic. This is exactly the character of a fuzzy quantity.

In this section, despite the absence of literature about the attribute combination for ID estimation, a simple fuzzy logic algorithm that combines the ESM sensor evidence is developed as a pedagogical example. The algorithm below is an *ad hoc* algorithm that imitates the phenomenology coming out of a fuzzy logic estimator. The example does not cover all particularities of fuzzy logic rules since the information comes from a single ESM sensor and is always of the same form. Nevertheless, the example permits appreciation of the fuzzification of the input and the defuzzification of the output processes with their interpretation.

A typical fuzzy logic algorithm (used here as an estimator) is composed of three fundamental processes: the input fuzzification process, which combines the fuzzy variables, the output defuzzification process together with the knowledge base source that contains the inference rules, and the logic combination rules. One has previously seen how to fuzzify continuous physical variables like kinematical variables such as speed and acceleration, and geometrical quantities, such as length, width, height and the different RCS.

The input data from the sensor are first interpreted and put in a fuzzy variable form. The process has to perform normalization and interpretation of the input data. When the data are all numerical, all of the same type and from a single source, the interpretation task is relatively easy. The combination process performs the arithmetic or the algebra that is necessary to combine the input fuzzy variables, under the recommendation of the inference rules or logic combination rules. The defuzzification process performs the opposite task of the first process: it transforms the fuzzy statements into the appropriate information type (usually the same type as the initial input type). These three processes are linked to a database, which contains the "know-how" about the interpretation of the output and input data and the combination of the information.

4.1 EXAMPLE OF FUZZIFICATION RULES

To be able to apply this method, the ESM sensor is required to output a list of candidate emitter identities with a Confidence Level (CL) associated with each emitter. This CL should be a numerical representation of the quality of the fit between the emitter characteristics and the measured parameters. The number appearing in the second column for each emitter in Table 2 can be seen as an appropriate CL. A fuzzy logic estimator considers the CL as an indication and categorizes the CL to some fuzzy statement. For each pre-determined CL range a fuzzy statement is associated. As ranges descend, the fuzzy statements are less and less categoric about the possibility of the ESM declaration. The third column shows the numerical weight which is given to the fuzzy statement and used by the combination process.

TABLE 2. Fuzzification Rules for an ESM Sensor

ESM Computed CL	Associated Fuzzy Statement	Numerical Weight
1≥Conf>0.99	Extremely Possible	3
0.99≥Conf>0.95	Highly Possible	2
0.95≥Conf>0.70	Very Possible	1
0.70≥Conf>0.50	Possible	0
0.50≥Conf>0.25	Moderately Possible	-1
0.25≥Conf>0.02	Slightly Possible	-2
0.02≥Conf>0.005	Almost Impossible	-3
0.005≥Conf≥0	Impossible	-4

As already mentioned in Paragraph 2.1, the fuzzification process can be generalized in terms of membership functions, where the crude binning of the example above is replaced by overlapping functions which sum up to one, usually of triangular or trapezoidal shape. An example of a 5 term decomposition into triangular membership functions is shown in *Figure 2* below for the physical attribute of air speed (VS = Very Slow, S = Slow, M = Medium, F = Fast, VF = Very Fast). The overlapping regions measure the fact that experts need not agree on a precise definition of linguistic terms for speed.

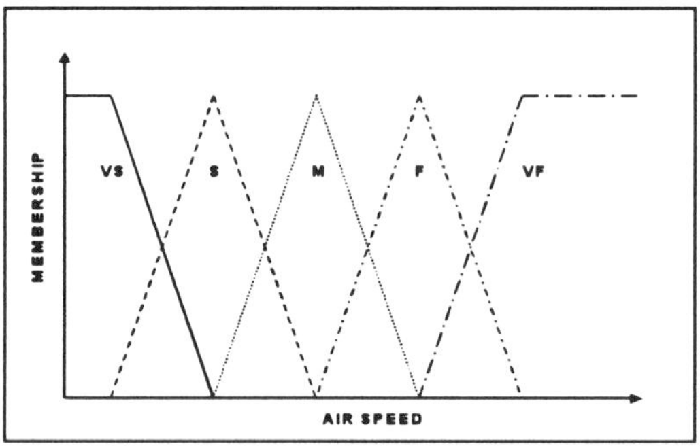

Figure 2. Fuzzy Membership Functions for Speed

4.2 EXAMPLE OF COMBINATION PROCESS

The combination process is performed in two steps. First, the input possibility is combined with the previous combined possibility and second, the last five combined possibilities are averaged to get the track level output possibility. For each emitter E_i, the track level possibility at any time t_n is computed by

$$Copos(E_i,t_l) = Copos(E_i,t_{l-1}) \otimes InPos(E_i,t_l)$$

$$Pos(E_i,t_n) = \frac{1}{h}\sum_{l=n-h+1}^{n} CoPos(E_i,t_l) \qquad h = Smaller(5,n)$$

where the last equation gives the initialization condition. *Pos, CoPos* and *InPos* are the output track level possibility, the combined possibility and the input possibility (contact level), respectively. The index h, which is the lesser of n and 5, restricts the average to the last five values of combined possibility (this number is of course chosen as an example and does not reflect true ESM design).

$$Pos(E_i,t_1)=CoPos(E_i,t_1)=InPos(E_i,t_1)$$

The circle multiplication symbolizes the application of the combination rule which is represented by the following inference matrix.

The matrix in Table 3 provides the result (using the numerical weight) of the combination of the new input information with the previous one. The rationale concerning the choice of the elements is essentially system performance dependent and based on the experience of the designers with the problem of ID estimation. As the symmetry among the numbers in Table 3 shows, the combination rules favour the new input data more than the previously combined data. However, this is counterbalanced by the averaging process over the last five samples.

TABLE 3. Inference Rules Combining the Last CoPos with the Input Possibility

	Last CoPos							
InPos	3	2	1	0	-1	-2	-3	-4
3	3	3	3	3	2	1	0	-1
2	3	3	3	2	1	0	-1	-2
1	3	3	2	1	0	-1	-2	-3
0	2	2	1	0	-1	-2	-3	-4
-1	1	1	0	-1	-2	-3	-3	-4
-2	0	0	-1	-2	-3	-3	-3	-4
-3	-1	-1	-2	-3	-3	-3	-3	-4
-4	-2	-3	-3	-4	-4	-4	-4	-4

4.3 EXAMPLE OF DEFUZZIFICATION RULES

The result of the application of the averaging over five samples is a number. It is from this number that the defuzzification is done as per the rules shown in Table 4 below. A pre-determined range of computed output possibility values receives an associated fuzzy statement, which does not necessarily have the same interpretation as the fuzzy statement defined during the fuzzification process. Again the rationale concerning the selection of the interval in the first column depends on the situation. The last column represents an equivalent translation into the same language used by the ESM sensor to output CL. For the CCS operator the fuzzy statements are usually sufficient.

TABLE 4. Defuzzification Rules Returning the Possibility Into the ESM Sensor Language

Computed output possibility value	Significance in terms of Fuzzy statements	ESM language Output Confidence
3≥Pos>2.75	Extremely Possible	99 %
2.75≥Pos> 1.75	Highly Possible	95 %
1.75≥Pos> 0.75	Very Possible	75 %
0.75≥Pos>-0.50	Possible	50 %
-0.50≥Pos>-1.75	Moderatly Possible	25 %
-1.75≥Pos>-3.75	Slightly Possible	5 %
-3.75≥Pos>-4	Almost Impossible	1 %
Pos=- 4	Impossible	%

5. Neural Networks (NNs)

NNs are particularly useful when one has a large volume of data to reason over, such as a large collection of image features that are to be used for the purposes of classification. This is the case of FLIR and SAR imagery. A typical NN application starts out by extracting features from typical imagery, e.g., invariant moments and auto-regressive model parameters for FLIR ship images. Depending on the size of the training set and the independence of the input attributes, a number of hidden layers containing a selected number of neurons with intricate interconnections is chosen and the NN is trained on a fraction of the available imagery. The remaining imagery is kept for the validation and test sets.

A practical example of such a FLIR classifier is shown in Figure 3 where 11 inputs are used to identify eight types of ships obeying a taxonomy tree similar to the one discussed previously, namely: destroyer, frigate, cruiser, destroyer with guided missiles, landing assault tanker, auxiliary oil replenishment, civilian freighter, and cargo/container. The taxonomy tree regroups the first four as "line" ships, the last two as "non-naval", while "line" and the landing assault tanker form a "combatant" group (borrowing from STANAG 4420 terminology). The auxiliary oil replenishment ship is the lone example (in the imagery dataset available) of the non-combatant type.

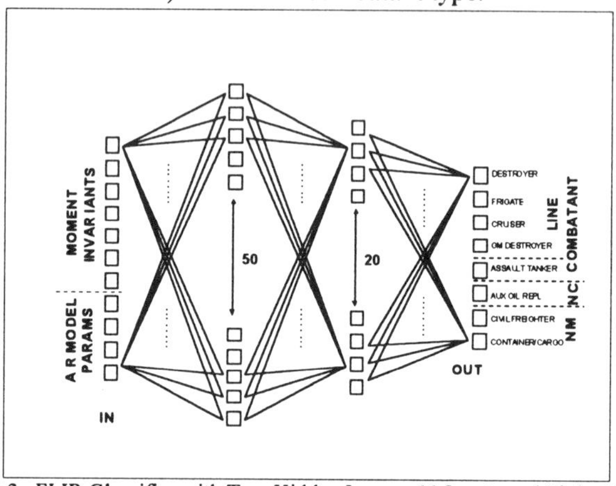

Figure 3. FLIR Classifier with Two Hidden Layers, 11 Inputs and Eight Outputs

An example of the training, validation and test steps of NN classifiers is shown below for the SAR category classifier for ships into line or merchant categories. One first decomposed the range profile of ships into 9 bins, then selected 32,211 profile vectors which, according to classified knowledge based rules, correspond to the following ship distribution:

(a) 6,259 line combatants
(b) 5,922 merchant ships
(c) 10,030 others.

The 9-bin profile vectors were normalized such that the sum was one and no given cell exceeded 0.4. Following the usual claims that a single hidden layer NN can classify any problem as long as enough neurons are taken, one has chosen to optimize a single layer NN by varying the number of hidden neurons. The next step was to vary the training set size, given a fixed (but randomly chosen) validation set of 1000 profile vectors. The values used for the training set size were thus varied through 100, 200, 500, 1000, 2000, 5000, 10000. Figure 4 shows a few training set curves. The top curve shows the generalization error and the lower curves the learning error. The upper left and right sub-figures, which correspond to 100 training examples, show a marked difference between the two curves, which is an indication that the system has memorized the examples. One can conclude that 100 examples for training is too little. However, the lower left and right sub-figures, which correspond to many more examples (5000), have very similarly shaped curves and the generalization curve (upper curve) flattens out rather than increases, which is the desired behaviour.

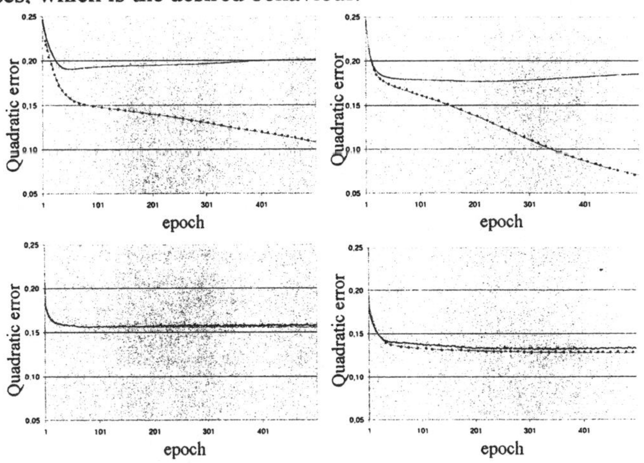

Figure 4. **Training Set Curves for 100 or 5000 Examples and 6 or 19 Neurons**

The best category NN had 29 hidden neurons on a single hidden layer and was trained on 100 examples. This seemed to provide the best compromise between approximating the knowledge based rules, whose performance is shown in the top part of Table 5 and good overall performance shown in the bottom part of Table 5.

TABLE 5. Performance of the NN for Category

KBS perf.	Line combatant	Merchant	Other
Line combatant	105	1	5
Merchant	2	3	5
Other	9	7	7
NN perf.	Line combatant	Merchant	Other
Line combatant	109	1	1
Merchant	4	3	3
Other	13	4	6

6. Bayesian Reasoning

Bayesian reasoning is at the foundation of many tracking and evidential reasoning approaches. The tracking aspect and its limitations will be discussed in my other lecture [16]. I will concentrate here on its application as an alternative classifier to the NN approach outlined in the previous section, for ship type (rather than category), extracted from SAR imagery.

If the confidence level on a "line" declaration is high enough (lets say > 50%) from the NN, then an estimate of the line ship type should be initiated. This is performed using a Bayes classifier based on the frigates, destroyers, cruisers, and aircraft carriers (porte-avion in French) length distributions (again borrowing from STANAG 4420 terminology). Ship lengths for this statistical analysis have been obtained by browsing Jane's Fighting Ships and their probability density distributions approximated by the curves shown in Figure 5 with the types listed above from left to right (not enough battleships were available to form a separate type).

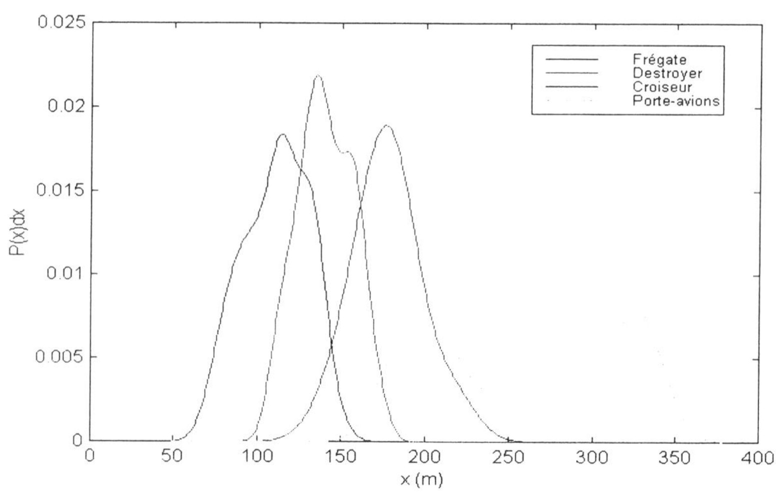

Figure 5. Probability Density Distribution Model for Line Ship Length

Given that a ship length range has been evaluated from the ship end points in the imagery, one calculates the mean *a posteriori* type probability $P_{avg}(t|s)$ that a ship s belongs to type t by averaging the standard Bayes rule over the entire ship length range,

$$P_{avg}(t|s) = \underset{length\ range}{Avg}\left(\frac{p(s|t)P(t)}{p(s)}\right) \qquad p(s) = \sum_i p(s|t_i)P(t_i)$$

where $p(s|t)$ and $P(t)$ are the probability density distribution and the *a priori* probability of type t, respectively. Mean *a posteriori* type probabilities are re-normalized, in order that their sum is unity, and multiplied afterwards by $O_{category}$. The resulting number O_{type} serves to set the CL on ship type declarations that have been empirically chosen as 0.98 O_{type}. Note that the merchant and unrecognized categories CLs are not modified. Naturally the *a priori* probability $P(t)$ depends on the context and could be set by the radar operator prior to the mission.

Given the four following a priori distributions of Table 6:

TABLE 6. A Priori Probabilities for Four Types of Line Combatants

Figure	Frigate	Destroyer	Cruiser	Aircraft Carrier
a	0.25	0.25	0.25	0.25
b	0.10	0.10	0.70	0.10
c	0.10	0.70	0.10	0.10
d	0.15	0.70	0.15	0.0

one obtains the following probabilities shown in Figure 6.

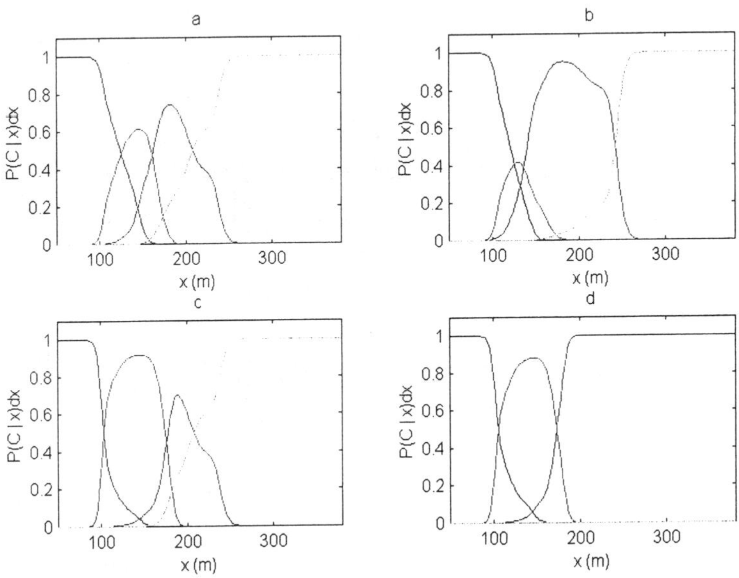

Figure 6. Probability Distributions of Line Combatants for Different A Priori Probabilities

The performance of this Bayesian classifier on the actual database of 145 ships used, given an equal 0.25 a priori probability of occurrence of any particular type, is shown in the confusion matrix of Table 7.

TABLE 7. Bayesian Line Type Performance for Equal A Priori Probabilities

	Frigate	Destroyer	Cruiser	Aircraft Carrier
Frigate	37	13	-	-
Destroyer	9	34	4	-
Cruiser	-	3	13	2
Aircraft Carrier	-	-	10	10

7. Complex Problems May Require Hybrid Methods

The above discussions about NN and Bayes techniques show that they can compete in the designer's mind, but it should be reminded that they can also be complementary. Indeed both have different required assumptions that must hold, preferred regions of applicability, and performance measures which vary both in classification accuracy and in computing time, stability and complexity. Hence a hybrid design can exploit the strengths of both approaches while avoiding any shortcomings. Such a hybrid design is the final design that was selected as the best compromise for a hierarchical SAR classifier at LM Canada [3, 9]. The design is presented in Figure 7.

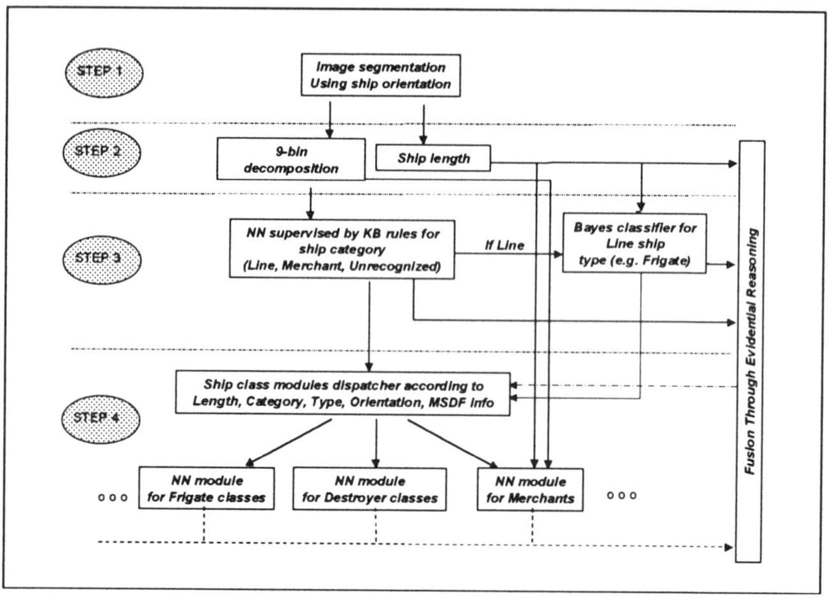

Figure 7. Hybrid SAR ISM Classifier

In Step 1, target segmentation from the ocean clutter is achieved by noise removal and merging of small regions. The resulting segmented image is made binary and sent to the ship centerline detection algorithm. Ship end-points can be obtained by estimating the ship centerline from the maximum peak of the Hough transform of the segmented image. This technique is more robust than a least-square-fit through the target or principal axes detection, because a large amount of cross-range scatterer spreading in some images (especially ISAR) tend to bias the centerline estimation. Ship length in Step 2 is obtained by identifying the target's end-points, using an empirical formula, and then sent to the Bayes length classifier. Simultaneously, the range profile is decomposed into 9 bins (by counting the number of pixels within that bin) and sent to the NN supervised by the classifed knowledge based rules. Figure 7 shows that the current design (not implemented due to the lack of fine detail in the imagery at the this time) has the fourth step of the ISM further refining the target ID up to the ship class (e.g., Mackenzie-class frigate, Belknap-class cruiser). This necessitates much more sophisticated classifiers. Rather than training a single huge classifier, which would most probably lead to convergence problems, a modular approach has been selected that uses small dedicated classifiers (NN sub-modules), each specialized in recognizing a subset of the ship database within a small viewing angle range. During the recognition process, only modules corresponding to the acquisition geometry (Step 1 of the classifier) and most probable ship types (Step 3 of the classifier) will be launched.

8. Dempster-Shafer (Ds) Reasoning

The mathematical formalism of the DS evidential theory requires that propositions pertain to the set of all possible propositions that can be output. More precisely, using the language of the set theory, a proposition is a set where each irreducible element (elementary proposition) pertains to a complete set named the "frame of discernment" usually noted Θ. If "n" is the cardinality of Θ (number of elements) there will be 2^n possible propositions that can be raised from Θ. The set of all possible propositions is usually noted as 2^Θ. The next section presents the mathematical formalism of the evidential theory, i.e., it describes how the propositions are combined to create other propositions.

8.1 MATHEMATICAL FORMALISM

The Basic Probability Assignment (BPA), also known as mass, is a value associated by a sensor with the proposition that indicates the level of confidence or certainty given by the sensor. There are three fundamental axioms about the mass that form the spine of the evidence theory. Let $m(A_i)$ be the mass in a proposition represented by subset A_i, Θ the frame of discernment and $\varnothing$ the empty set. The axioms are:

- $m(\varnothing)=0$
- For all the $2^{|\Theta|}$ subsets A_i included in Θ

$$\sum_{i=1}^{2^{|\Theta|}} m(A_i)=1$$

- For a collection of subset $\{A_1, A_2, \ldots, A_n\}$

$$m(A_1 \cup A_2 \cup \ldots \cup A_n) \geq \sum_{I \subset \{1,2,\ldots n\}} (-1)^{|I|+1} m(\bigcap_{i \in I} A_i)$$

An important consequence of the third axiom concerns the negation of a proposition

$$m(A) + m(\neg A) \leq 1$$

where the $\neg$ symbol stands for the logical negation. This results in a consequence of the fact that a part of the mass, called the ignorance, is directly associated with the set Θ. When a sensor assigns a non-zero belief to the proposition Θ (which represents the union of every proposition), this must be interpreted as a situation where the sensor is unable to state definitely about some specific propositions and prefers to state about everything globally. This ability to represent ignorance reduces the likelihood of erroneous sensor declarations. A sensor can declare only what it knows about some specific proposition without having to speculate about others.

The mass is a quantity output by a sensor which infers information only about the proposition "A" it has deduced and nothing at all about the propositions which are subsets of A. A mass close to one is a strong indication that this proposition is probably true. However, a mass close to zero does not necessarily mean the opposite, since this depends on the mass given to other alternative propositions, including the ignorance. In DS all the information about a proposition is obtained from a pair of quantities which are computed from the masses of various "related" propositions. The DS approach defines the concept of an evidential interval, denoted *[Bel, Pls]*. The lower bound, the belief, is computed by

$$Bel(A) = \sum_{i=1}^{2^{|A}} m(A_i) \qquad \text{where all } A_i \subset A$$

and the upper bound, the plausibility, is obtained from

$$Pls(A) = 1 - Bel(\neg A)$$

The belief in a proposition "A" represents the minimal commitment which can be extracted by the masses from various sensor-declared propositions B_j. All B_j that are subsets of A contribute to that minimal commitment. The plausibility of a proposition "A" represents the maximal commitment from the sensor declarations. All B_j that have at least an element of Θ in common with A contribute to the plausibility. Consequently the Bayesian Probability of the proposition A satisfies

$$Bel(A) \leq Prob(A) \leq Pls(A)$$

As an example, let Θ be the set $\{A_1, A_2, A_3, A_4, A_5\}$ and let the sensor declarations be $m(A_1)$, $m(A_2)$, $m(A_1 \cup A_2)$, $m(A_4 \cup A_5)$, $m(A_1 \cup A_2 \cup A_4)$ and $m(\Theta)$. All the other subsets of Θ to which the sensor does not report belief are considered as having no belief. From these data, the support and plausibility to the proposition $A = A_1 \cup A_2$, are:

$$Bel(A_1 \cup A_2) = m(A_1) + m(A_2) + m(A_1 \cup A_2)$$

$$Pls(A_1 \cup A_2) = m(A_1) + m(A_2) + m(A_1 \cup A_2) + m(A_1 \cup A_2 \cup A_4) + m(\theta)$$

Finally, there are other evidential intervals which can be defined in a way analogous to *[Bel, Pls],* such as the Expected Utility Interval (EUI) covered in another lecture [15].

8.2. COMBINATION RULES

Conventional literature allows for the traditional combination rules where conflict can exist between the latest sensor declaration about a target and the existing cumulative knowledge about that platform in the track database [2,6,8,12,14]. More recent literature [1] allows for not declaring a conflict at a finer level of the taxonomy tree but rather for throwing back the "conflict" to the first coarser level of taxonomy where conflict does not exist. These combination rules are referred to as the "orthogonal sum" and the "hierarchical orthogonal sum", respectively, and are discussed below.

8.2.1 Orthogonal Sum and Renormalization Due to Conflict

Suppose A_i and B_j are members of two lists of sensor propositions which are statistically independent, i.e., their mass values are not correlated or computed from the a priori knowledge of each other. The numbers of members in each list are α and β respectively. As with Bayes formula, which combines independent probability measurements, DS evidential theory has a way to combine independent mass measurements to get new mass data on the output propositions and also some propositions obtained from combined propositions. The frame of discernment Θ should be the same for both lists of declarations. If the masses associated with propositions A_i and B_j are independent, the new belief of the intercept of both propositions is obtained from DS's rule of combination, a.k.a. the orthogonal sum:

$$m(C_k) = \sum_{i,j \text{ where } A_i \cap B_j = C_k} \sum \frac{m(A_i) m(B_j)}{1 - K}$$

$$K = \sum_{i,j \text{ where } A_i \cap B_j = \varnothing} \sum m(A_i) m(B_j)$$

where K is called the conflict. The applications of these two formulas are not straightforward when the amounts α and β are large. When the number of reported propositions is high, the combination rule has a tendency to increase the number of propositions by creating new ones. The problem is an NP-hard one and a truncation scheme has to be devised. As an example, an ESM sensor reports at two different times the data indicated in the left column and first row of Table 8:

TABLE 8. Example of Propositions Combined with Dempster-Shafer Combination Rule

	A_1, m=0.5	A_1 or A_2, m=0.1	A_4 or A_5, m=0.1	Θ, m=0.3
A_1, m=0.4	A_1	A_1	K	A_1
A_2, m=0.2	K	A_2	K	A_2
A_1 or A_2, m=0.1	A_1	A_1 or A_2	K	A_1 or A_2
A_2 or A_4, m=0.1	K	A_2	A_4	A_2 or A_4
Θ, m=0.2	A_1	A_1 or A_2	A_4 or A_5	□

This table shows the 20 combinations that are possible under the rule. Some combinations lead to the same proposition as those given by the sensor, as in the example of proposition "A_1". A new proposition, A_4, from which the sensor did not report any belief, is born from the combination. Only one combination leads to the ignorance Θ and this fact is a general feature of the combination rule. In this example there are five elements in the frame of discernment, which leads to a total of 32 possible propositions. If a sensor does not report any belief on most of them in the approach they are considered as having a belief value of zero. There are the five combinations that contribute a term in the summation of the conflict. Using the data of the example, the conflict K is:

$$K = 0.4 * 0.1 + 0.2 * 0.5 + 0.2 * 0.1 + 0.1 * 0.1 + 0.1 * 0.5 = 0.22$$

The new combined belief for proposition A_1 is obtained from the sum of the five combinations.

$$m(A_1) = (0.4 * 0.5 + 0.4 * 0.1 + 0.4 * 0.3 + 0.1 * 0.5 + 0.2 * 0.5)/0.78 = 0.65$$

For all the propositions the calculation is similar, and the propositions which do not appear in the table have a mass of zero, a belief of zero and a plausibility at least larger than b(Θ). Note that if the total conflict becomes large, the remaining propositions that do not contribute to the conflict are renormalized from a small value to a much larger one. Successive conflicts of different origins, such as would come from a malfunctioning sensor or fusion algorithm, would cause an oscillation between different propositions, each of which has little direct evidence to support it. In other words, despite the fact that the orthogonal sum takes conflict into account in a mathematically correct way, this may not be sufficient for the stable operation of DF in a decision aid role.

8.2.2 Hierarchical Orthogonal Sum (HOS) and the Resolution of Conflict

In the case of the existence of a taxonomy tree, one can do better than the orthogonal sum's resolution of conflict. In order to apply the DS evidential combination method efficiently, information has to be mapped into proposition bitstreams. For example, if the report Air is received from a sensor, and the platforms or taxonomy elements #1, 2 and 3 are members of set "*Air*", then the bits #1, 2, and 3 of the proposition will have the value "1". All other elements that do not support the input information shall have their corresponding bit set to *off* (i.e., "0"). Figure 8 shows an example where an input attribute *Air* is supported by 21 elements.

0	1	1	0	1	1	1	•••	1	0	•••	0	1	0
bit 1	bit 2	bit 3	bit 4	bit 5	bit 6	bit 7		bit 22	bit 23		bit 1022	bit 1023	bit 1024

Figure 8. Proposition from an Input Attribute for 1024 Platforms and Taxonomy Elements

The peculiarity of the HOS is that one implements bitstreams based not only on the platforms but also on the whole taxonomy tree. There is the implementation of a hierarchical structure with AND/OR addition rules in the orthogonal sum instead of only an AND rule as in the standard DS method. If one looks at the taxonomy tree in the right-hand side of Figure 1 an incoming report "*Air*" would not only fill the bits corresponding to all air platforms but would also fill the bits corresponding to Fighter and Bomber (they have a member relationship with the report "Air") and the bit corresponding to "*Air*". This is useful because it allows a better treatment of conflicting reports. In the standard DS method, addition of information is done through the use of a logical AND applied only on the platform elements.

Algorithmically, each level in the taxonomy tree has a section in the bitstream: section[I] with I=1,2,3. Section[1] corresponds to Level 1, containing the section Air, Land, Surface; Section[2] to Level 2, containing the portion Fighter, Bomber; and the last level, Section[3], corresponds to the platform elements portion. The algorithm, referred to as HOS, goes as indicated in the following text box. In this box the operator "==" refers to the standard equality comparison operator, $\cup$ and $\cap$ refer to the logical OR and AND operations applied on bitstream, and the $\oplus$ symbol stands for the concatenation procedure applied on a section of bitstream.

Given two propositions described by the bitstreams A and B:

Add: A.section[3]$\cap$B.section[3]==$\emptyset$
 No: C=A.section[1,2,3]$\cap$B.section[1,2,3]
 Yes: A.section[2]$\cap$B.section[2]==$\emptyset$
  No:
   C={(A.section[3]$\cup$B.section[3])}$\oplus${A.section[1,2]$\cap$B.section[1,2]}
  Yes: A.section[1]$\cap$B.section[1]==$\emptyset$
   No:
   C={A.section[2,3]$\cup$B.section[2,3]$\oplus${A.section[1]$\cap$B.section[1]}
   Yes: C=Ignorance.

This algorithm allows us to move up the taxonomy tree to the last non-conflicting information level ("Air" in our case) and provides a better treatment of conflicts. At the fusion step we use the whole taxonomy. When calculating a score through the association step, one however compares only based on the platform bits. Thus in the last case, the conflict between the two reports would be reported fully. This is necessary because prior to fusing a new proposition we need to gate the incoming information with the one available. This also allows us to better report information when there is conflicting information: if in our previous example, the conflicting report had a very strong belief value, then with the standard DS method, the statement "Air" would have been lost until other reports appeared. With the HOS method, we kept the information available at all times.

Suppose that for the same target track, we receive two input attributes: one suggesting CF-18 and the other Bomber. Then, the information conflicts (a CF-18 is not a Bomber) in the platform section but in the taxonomy part it has the elements "*Air*" in common. By adding the taxonomy bits though, when a logical AND rule is performed, we observe that although the addition of the two reports is conflicting, the statement "Air" is common to both and the common information is kept, if we assume both reports to be valid (non-negligible confidence value). The way to handle conflicting inputs is to use the redundant information contained in the elongated bitstreams.

It should be noted that strong conflicts are expected to be rare, so that the previous algorithm remains very fast. In fact, on average, it is almost transparent compared to the previous approach. On top of this we use a truncation mechanism to keep the number of combined propositions in a track manageable in terms of Central Processing Unit (CPU) time. This truncation mechanism has been explained previously in detail.

8.2.3 An Example of the HOS

Consider an AIR target for which the three levels of taxonomy shown in Figure 1 apply [1]. In order to provide conflict (as would be the case in countermeasures), one adds a dummy emitter that, is only found on bomber planes in our database. Thus the ESM will report fighter emitters most of the time, but once in a while will report a bomber emitter for that same plane. The other sensors simulated here do not report any conflicting information. Figure 9 schematizes the results of the best ID proposition, the one with the largest BPA, found for this target in one simulation case.

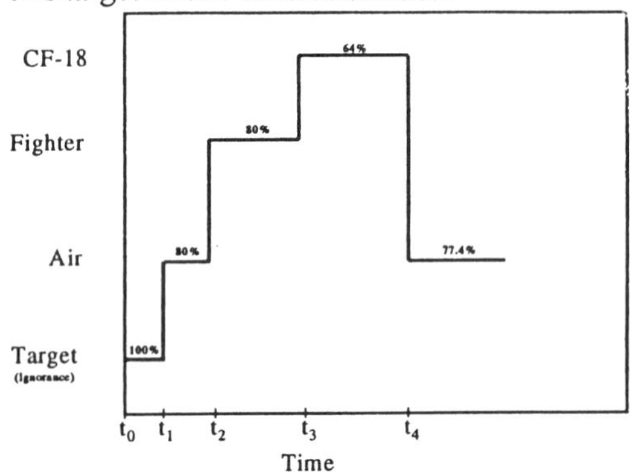

Figure 9. Taxonomy Level of the Target Best ID Proposition as a Function of the Time

The results are as follows. At time t_1 a report is received from the radar sensor reporting only kinematics. Using a fuzzy logic test, this leads to the statement Air with a BPA of 80%. At time t_2, the first ESM report is received. It reports one of the emitters for a fighter (with a BPA of 80%), which we'll call E_1. At this point this emitter is sufficient to correlate the information with the database and we know that this emitter is on a fighter without knowing which one. At time t_3 another fighter emitter, E_2, is reported giving the full identification of the CF-18 which is the only one with both emitters on board. Then, at time t_4, the conflicting report is received reporting an emitter E_3 belonging to bombers only. We see that, at this point, there is some identity degradation once the report is fused with the previous information. Table 9 summarizes the results for all identities by showing the BPA of each proposition as a function of the time of fusion:

TABLE 9. BPA for Scenario with Conflicting Evidence Resolved by the HOS

Proposition	t_1	t_2	t_3	t_4
CF-18	0%	0%	64%	12.8%
Air-Fighter	0%	80%	32%	6.4%
Air	80%	16%	3.2%	77.4%
Bomber	0%	0%	0%	3.2%
Ignorance	20%	4%	0.8%	0.2%

Even in the presence of conflict, the identity of the platform after the conflicting report remained partly correct. We observe that the belief in the *Air* statement (sum of all BPAs of propositions that are sub-sets of air) remained after t_4 with 99.8%. This is in contrast to the previous method that would have sent the conflicting report to the ignorance propositions, which would have reduced the BPA of the Air proposition to approximately 24%. Also, the probability of the platform being a Fighter diminished and a new Bomber statement appeared with lower confidence. The next ESM report, if not conflicting, would have restored the correct identity. This treatment of conflict is, we believe, more adapted to what an operator would like to see in the presence of conflicting evidence.

8.3 TRUNCATED DEMPSTER-SHAFER (DS)

The DS theory of evidence, because it assigns BPA to subsets of a PDB and not to its individual elements, can generate an exponential number of propositions, which is not very useful if used in real time applications. This is where the truncation algorithm comes in. The Truncated DS (TDS) scheme retains propositions according to the rules below (or equivalent ones):

(1) All combined propositions with BPA > MAX_BPM are kept
(2) All combined propositions with BPA < MIN_BPM are discarded
(3) If the number of retained propositions in step 1 is smaller than MAX_NUM, it retains, by decreasing BPA, propositions of length 1
(4) If the number of retained propositions in step 3 is smaller than MAX_NUM, it does the same things with propositions of length 2
(5) Repeat a similar procedure for propositions of length 3
(6) If the number of retained propositions is still smaller than MAX_NUM, it retains propositions by decreasing BPA regardless of length.

The propositions that are discarded are either returned to the ignorance (in most applications) or split between the ignorance and the remaining propositions (in some applications). Optimization and benchmarking [2] on complex realistic scenarios have catalogued the possible values of the three parameters MAX_BPM, MIN_BPM, and MAX_NUM, as well as their interrelationship and their dependence on PDB size.
Measures of Performance (MOP)

$$\left(\frac{BPAM}{100}\right)^2 = 0.5\left(1+\sum_{i,j=1}^{s} m_i m_j \frac{N(P_i \cap P_j)}{N(P_i)N(P_j)}\right) - \sum_{i=1}^{s} m_i \frac{N(P_i \cap E_g)}{N(P_i)}$$

The purpose of a classification MOP is to score how well an MSDF algorithm identifies targets. The Bayesian Percent Attribute Miss (BPAM) MOP is defined below. If one denotes by N(P) the length of a proposition P (i.e., the number of platforms in P) and by E_g the ground truth ID for the target studied, then the square of the BPAM is a number between 0% and 100%, given by Knowledge of ground truth is available when a simulator is used with a known scenario and known targets, which is usually the case. A BPAM value of 0% is attained if the correct singleton is identified, while a value of 100%

corresponds to certainty in an incorrect singleton. The BPAM thus measures how close one is to identifying the ground truth.

9. Conclusions

This lecture has provided an exhaustive list of platform attributes needed for evidential reasoning techniques that aim to provide an ID within a well-defined taxonomy tree (several of which were presented others can be found in [10,11,13,17]). It also has presented the formalism and real-world examples for four well-known reasoning frameworks, mostly applicable to Level 1 and 2 fusion, a partial list being:

(1) fuzzy logic, in particular its use in fuzzification (pre-processing) for other means of reasoning, but also with an ESM application utilizing fuzzy combination rules and defuzzification
(2) NNs, particularly useful for large sets of data, such as imagery datasets, which are easily decomposable into training, validation and test sets, with FLIR and SAR examples detailed
(3) Bayesian approach (with a priori information) with an application to classifiers where an attribute stands out as a discriminator 9, such as line ship length for a SAR classifier)
(4) DS approach with its variants: the original orthogonal sum and its renormalization though conflict, followed by the HOS and its novel conflict resolution through the taxonomy tree, and finally generic considerations about truncating the exponentiation of proposition (NP-hard aspect of the problem).

In addition, examples were given where all of the above methods form crucial parts of a larger more versatile classifier.

10. References

Canada, L.M., Journal articles, where more references can be found (in reverse chronology):

1. Simard, M.-A, Valin, P., and Lesage, F. (2000) The ID box: a Multisource Attribute Data Fusion Function for Target Identification, Proc. SPIE Aerosense Conf. 4051 on Sensor Fusion: Architectures, Algorithms and Applications IV, Orlando, April 24-28, pp. 180-190.
2. Valin, P., and Boily, D. (2000) Truncated Dempster-Shafer Optimization and Benchmarking, Proc. SPIE Aerosense Conf. 4051 on Sensor Fusion: Architectures, Algorithms and Applications IV, Orlando, April 24-28, pp. 237-246.
3. Valin, P., Tessier, Y., and Jouan, A. (1999) Hierarchical Ship Classifier for Airborne Synthetic Aperture Radar (SAR) Images, Proc. 33rd Asilomar Conference on Signal, Systems and Computer, 24-27 October, Pacific Grove, pp. 1230-1234.
4. Jouan, A., Valin, P., and Bossé, E. (1999) Testbed for Fusion of Imaging and Non-Imaging Sensor Attributes in Airborne Surveillance Missions, FUSION 99, Sunnyvale, CA, Vol. 2, pp. 823-830.
5. Valin, P., Jouan, A., and Bossé, E. (1999) Fusion of imaging and non-imaging sensor information for airborne surveillance, Proc. SPIE Aerosense Conf. 3719 on Sensor Fusion: Architectures, Algorithms, and Application III, Orlando, 7-9 April, pp. 126-138.
6. Jouan, A., Gagnon, L., Shahbazian, E., and Valin P. (1998) Fusion of Imagery Attributes with Non-Imaging Sensor Reports by Truncated Demspter-Shafer Evidential Reasoning, in FUSION 98, Las Vegas, 6-9 July, Vol. II, pp. 549-556.
7. Simard, M.-A., and Jouan, A. (1997) Platform Database and Evidential Combination Method to Fuse ESM Detections with the Attribute Data From Other Sensors, NATO AGARD Symposium on EW Integration for Ships, Aircraft, and Land Vehicles, May, Ottawa, pp. 2-1 to 2-16.

8. Bossé, E., and Simard, M.A. (1997) Identity and Attribute Information Fusion Using Evidential Reasoning, Proc. SPIE Aerosense Conf. 3067 on Sensor Fusion: Architecture, Algorithms, and Applications, Orlando, 20-25 April, pp. 38-49.
9. Shahbazian, E., Gagnon, L., Duquet, J.-R., Macieszczak, M., and Valin, P. (1997) Fusion of Imaging and Non-Imaging Data for Surveillance Aircraft, Proc. SPIE Aerosense Conf. 3067 on Sensor Fusion: Architecture, Algorithms, and Applications, Orlando, 20-25 April, pp. 179-189.
10. Valin, P., Couture, J., and Simard, M.A. (1996) Position and Attribute Fusion of Radar, ESM, IFF and Data Link for AAW missions of the Canadian Patrol Frigate, Proc. of Multisensor Fusion and Integration for Intelligent Systems (MFI 96), Washington, Dec. 8-11, pp. 63-71.
11. Couture, J., Boily, E. and Simard, M.A. (1996) Sensor Data Fusion of Radar, ESM, IFF and Data Link of the Canadian Patrol Frigate and the Data Alignment Issues, in Proc. SPIE' Aerosense Conf 2759 on Signal and Data Processing of Small Targets 1996, Oliver E. Drummond, ed., 8-12 April, Orlando, pp. 361-372.
12. Simard, M.A., Bossé, E., and Couture, J. (1996) Data Fusion of Multiple Sensors Attribute Information for Target Identity Estimation Using a Dempster-Shafer Evidential Combination Algorithm, Proc. SPIE Aerosense Conf. 2759, on Signal and Data Processing of Small Targets 1996, Oliver E. Drummond, ed., 8-12 April, Orlando, pp. 577-588.
13. Bégin, F., Kamoun, S., and Valin, P. (1994) On the Implementation of AAW Sensors Fusion on the Canadian Patrol Frigate, Proc. SPIE Aerosense Conf. 2235, on Signal and Data Processing of Small Targets 1996, Oliver E. Drummond, ed., 4-8 April, Orlando, pp. 519-528.
14. Simard, M.A., Valin, P., and Shahbazian E. (1993) Fusion of ESM, Radar, IFF and other Attribute Information for Target Identity Estimation and a Potential Application to the Canadian Patrol Frigate, AGARD 66th Symposium on Challenge of Future EW System Design, 18-21 October Ankara AGARD-CP-546, pp. 14.1-14.18.

Journal and Review articles:

15. Bossé, E. (2000) Possibilistic and Evidential Reasoning for Identity Information Fusion, NATO ASI 2000 on Multisensor and Sensor Data Fusion, these proceedings.
16. Valin, P. (2000) Random sets and unification, NATO ASI 2000, on Multisensor and Sensor Data Fusion, Pitlochry, Scotland, 25 June – 7 July, these proceedings.
17. Bossé, E., and Simard, M.A. (1998) Managing Evidential Reasoning for Identity Information Fusion Optical Engineering, Vol. 37, no. 2, pp. 391-400.

RANDOM SETS AND UNIFICATION

P. VALIN
Advisory Member of R&D
Lockheed Martin Canada
6111 Royalmount ave
Montréal, QC, H4P 1K6
Canada

1. Introduction

Until recently, data fusion problems have been solved heuristically using fuzzy logic, rule-based inference, the Dempster-Shafer (DS) theory of evidence, etc. Beginning in the late 1970s, however, researchers have shown how data fusion can be placed under a purely probabilistic and theoretically rigorous paradigm based on the theory of (closed) random sets. The purpose of this talk is to provide a brief "mathematician's overview" of this work, especially Finite Set Statistics (FISST). Random sets provide a natural setting for data fusion in two respects:

(a) as a natural way of formulating Multi-Sensor (MS) Multi-Target (MT) problems
(b) as a means of modelling ambiguous evidence.

Suppose, on the one hand, that one or more sensors collect observations from one or more targets. Then the total observation (taken over all sensors) consists of a finite set of (conventional) measurements that varies randomly (with respect to cardinality as well as individual measurements). Likewise, our total estimate of reality consists of a finite set of (conventional) estimates of individual targets that varies randomly.

On the other hand, consider an English-language statement such as "The helicopter is near the ownship." The concept "near" could be interpreted as a single closed circular region surrounding the ownship. More generally, it could be modelled as a family of such circular regions, each assigned the subjective probability that it is the most likely interpretation of "near". Such a family is a random set that models the concept "near". One easily sees, accordingly, that a "natural" mathematical universe for data fusion is a space whose elements are sets consisting of a finite number of (closed) random sets. However, such a framework is "natural" in an engineering sense only if it can be formulated in a manner that facilitates practical application. This lecture will outline such a formulation.

Random set theory and FISST applications in Multi-Sensor Data Fusion (MSDF) for MS-MT tracking and identification (ID), have been spearheaded since the mid 1990 by researchers at Lockheed Martin NE&SS in Eagan, particularly by Ronald Mahler. This lecture attempts to summarize the approach of his book [1] "Mathematics of Data

A. K. Hyder et al. (eds.), Multisensor Fusion, 247–266.

Fusion". In order not to burden this presentation with many references, all bibliographical references can be obtained from that book.

2. Conventional Approach, Known Problems And Potential Solution

2.1 THE BAYESIAN APPROACH

Recent years have seen the emergence of a "cookbook Bayesian" viewpoint based on Bayes rule,

$$f_{posterior}(x \mid z) = \frac{f(z \mid x) f_{prior}(x)}{f(z)}$$

where x denotes the unknown quantities of interest (kinematic and ID), the prior encapsulates our previous knowledge about x, z is the new data, the likelihood function $f(z|x)$ describes the generation of data and the posterior encapsulates our calculated current knowledge about x. Finally

$$f(z) = \int f(z \mid x) f_{prior}(x) dx$$

is a normalization constant. If time has expired before collection of new information z, then the prior has to be extrapolated to a new prior that reflects the uncertainties caused by possible interim target motion, which is usually done by evaluating the Markov time-prediction integral

$$f^{+}_{prior}(x) = \int f^{+}(x \mid y) f_{prior}(y) dy$$

where $f^{+}(x|y)$ is the Markov transition density describing the likelihood of the target having state x if it previously had state y. Translation of the calculated posterior into operator language is obtained through Bayes-optimal state estimators such as the Maximum A Posteriori (MAP) or Expected A Posteriori (EAP):

However, it is merely the visible tip of a conceptual iceberg, the existence of which

$$\hat{X}^{MAP} = \arg\max_{x} f_{posterior}(x \mid z) \qquad \hat{X}^{EAP} = \int x f_{posterior}(x \mid z) dx$$

tends to be forgotten precisely because the rest of the iceberg is taken for granted. Both the optimality and simplicity of the Bayesian framework can be taken for granted only within the confines of standard applications addressed by standard textbooks. When one ventures out of these confines, one must exercise proper engineering prudence which includes verifying that standard textbook assumptions still apply. The Bayesian "iceberg" has many facets:

(a) Sensor Models: Bayes' rule exploits to the best possible advantage the high-fidelity knowledge about the sensor contained in the likelihood function $f(z|x)$. Many forms of data, e.g., generated by tracking radars, are adequately characterized that $f(z|x)$ can be constructed with sufficient fidelity. Other kinds of data, e.g., Synthetic Aperture Radar (SAR), are proving to be so difficult to simulate that it is unclear whether sufficiently high fidelity will ever be achieved. There is also the kind of data, features extracted from signatures, English-language statements received over datalink, rules drawn from knowledge bases, etc., which is statistically so poorly understood, that probabilistic approaches are not even obviously applicable.

(b) Target Models: Much of what has been said about likelihoods $f(z|x)$ applies with equal force to Markov densities $f^{+}(x|y)$. The more accurately that $f^{+}(xly)$ models target motion, the more effectively Bayes' rule will do its job. In real-world scenarios targets can appear (e.g., MIRVs and decoys emerging from a ballistic missile re-entry vehicle) or disappear (e.g., aircraft that drop beneath radar coverage) in correlated ways. Consequently, MT filters that assume uncorrelated motion and/or constant target number may perform poorly against dynamic MT environments, for the same reason that single-target trackers that assume straight-line motion perform poorly against manoeuvring targets.

(c) Output Generation: When we are faced with the problem of extracting an "answer" from the posterior distribution, complacency may encourage us to blindly copy state estimators from textbooks, e.g., the MAP and EAP estimators of equations, or else define ad hoc ones. Great care must be exercised in the selection of a state estimator, however. For example, the EAP estimator plays an important role in theory but often produces erratic and inaccurate solutions when the posterior is multimodal. In the MT case, the dangers of taking state estimation for granted become even more acute. For example, we may fail to notice that the MT versions of the standard MAP and EAP estimators are not even defined, let alone provably optimal.

(d) Formal Optimality: The failure of the standard Bayes-optimal state estimators in the MT case has far-reaching consequences for optimality. Because the standard Bayes-optimal state estimators fail in the MT case, we must construct new MT state estimators and prove that they are well behaved.

(e) Computability: The prediction integral and Bayes normalization constant must be computed using numerical integration and, since an infinite number of parameters are required to characterize $f_{posterior}(x|z)$, in general, approximation is unavoidable.

2.2 WHY MULTI-EVIDENCE PROBLEMS ARE TRICKY

Given the technical community's increased understanding of the actual complexity of real signature and other kinds of data, it is no longer credible to invoke Bayesian filtering and estimation as a cookbook panacea. Any such approach that does not also include a robustness strategy, i.e., a means of dealing with likelihood functions that cannot be specified with sufficient fidelity or with data that contains inherently difficult-to-characterize uncertainties, is simply evading the real issues. One needs systematic and fully probabilistic methodologies for:

- modelling uncertainty in poorly characterized likelihoods
- modelling ambiguous data and likelihoods for such data

- constructing MS likelihoods for ambiguous data
- efficiently fusing data from all sources (ambiguous or otherwise)

2.3 WHY MULTI-TARGET PROBLEMS ARE TRICKY

Likewise, given the technical community's increased understanding of the actual complexity of MT problems, it is not credible to propose yet another *ad hoc* MT tracking approach. When one tries to apply the standard statistical thinking just described to the MT case with an unknown number of targets, one quickly discovers that Bayes-optimal MT estimation and filtering encounters fundamental conceptual and practical difficulties. One needs a systematic means of constructing:

- provably true *(as opposed to* heuristic) MS-MT likelihood functions from the underlying sensor models of the sensors
- provably true *(as opposed to* heuristic) MT Markov densities from the underlying motion models of the targets, which account for target motion correlations and changes in target number
- stable, efficient, and provably optimal MT state estimators that address the failure of the classical Bayes-optimal estimators and, in particular, simultaneously determine target number, target kinematics, and target identity without resort to operator intervention or optimal report-to-track association

It should be understood that many MS-MT tracking and association problems for positional and ID fusion seen in real-life situations may not require all the complexity espoused here. As usual, "one uses what one can get away with".

2.4 FINITE-SET STATISTICS (FISST)

One of the major goals of FISST is to address the "Bayesian iceberg" issues described previously. FISST deals with imperfectly characterized data and/or measurement models by extending Bayesian approaches in such a way that they are robust with respect to these ambiguities. FISST deals with the difficulties associated with MS and/or MT problems by directly extending engineering-friendly single-sensor, single-target statistical calculus to the MS-MT realm. Finally, FISST provides mathematical tools that may help address the formidable computational difficulties associated with MS-MT filtering (whether optimal or robust).

The basic approach is as follows. A suite of known sensors transmits, to a central data fusion site, the observations they collect regarding targets whose number, positions, velocities, identities, threat states, etc. are unknown. Then:

(a) reconceptualize the sensor suite as a single sensor (a "global sensor")
(b) reconceptualize the target set as a single target (a "global target") with MT state $X=\{x_1,...,x_n\}$ (a "global state", note the use of a capital letter)
(c) reconceptualize the set $Z=\{z_1,...,z_n\}$ of observations, collected by the sensor suite at approximately the same time, as a single measurement (a "global

measurement", note the use of a capital letter) of the "global target" observed by the "global sensor"

(d) represent statistically ill-characterized ("ambiguous") data as random closed subsets Θ of (MS) observation space. Thus, in general, $Z = \{z_1,\ldots,z_n, \Theta_1,\ldots \Theta_{m'}\}$

(e) just as single-sensor, single-target data can be modelled using a measurement model $Z=h(x, W)$, model MT MS data using an MS MT measurement model - a randomly varying finite set $\Sigma=T(X)\cup C(X)$

(f) just as single-target motion can be modelled using a motion model $X_{k+1}=\Phi(x_k, V_k)$, model the motion of MT systems using an MT motion model - a randomly varying finite set $\Gamma_{k+1}=\Phi(X_k, V_k)\cup B_k(V_k)$.

Given this, we can reformulate MS-MT estimation problems as single-sensor, single-target problems. The basis of this reformulation is the concept of belief-mass β. Belief-mass functions are non-additive generalizations of probability-mass functions. The FISST MS-MT differential and integral calculus is what transforms these mathematical abstractions into a form that can be used in practice. The following general methodology, nicknamed Almost-Parallel Worlds Principle (APWOP), for attacking MS-MT data fusion problems works, modulo certain remarks. Nearly any concept or algorithm phrased in random-vector language can, in principle, be directly translated into a corresponding concept or algorithm in the random-set language. One says "almost-parallel" because, as with any translation process, the correspondence between dictionaries is not precisely one-to-one (for example, vectors can be added and subtracted whereas finite sets cannot). Nevertheless, the parallelism is complete enough that, provided one exercises some care, a hundred years of accumulated knowledge about single-sensor, single-target statistics can be directly brought to bear on MS-MT problems. This is summarized in the following Table 1:

TABLE 1. Mathematical parallels between single sensor, single target and MS-MT problems

Random Vector, Z	Finite Random Set Σ
Sensor	global sensor
Target	global target
observation, z	observation-set, Z
parameter, x	parameter-set, X
sensor model, z = h (x,W)	MT s. m., Z = T(X) U C(X)
motion model, $x_{k+1}=\Phi(x_k,V_k)$	MT m. m., $\Gamma_{k+1}=\Phi(X_k,V_k)\cup B_k(V_k)$
differentiation, dp/dz	set differentiation, δβ/δZ
integration, $\int_S f(z\|x)dz$	set integration, $\int_S f(Z\|X)\delta Z$
probability-mass function, p(S|x)	belief-mass function, β
likelihood function, f(z|x)	MT l. f., f(Z|X)
posterior density	MT posterior
Markov densities	MT Markov densities

The performance of an MT data fusion algorithm can be measured by constructing information-based measures of effectiveness, e.g., the following MT generalization of the Kullback-Leibler discrimination in terms of the densities f and g:

$$K(f,g)\underline{\Delta}\int f(x)\log\left(\frac{f(x)}{g(x)}\right)dx \Rightarrow K(f,g)\underline{\Delta}\int f(X)\log\left(\frac{f(X)}{g(X)}\right)\delta x$$

One should realize that such a correspondance table occurs in many other scientific fields, for example in physics. There, we have the gradual increase in complexity from large scale linear problems dealt with classical methods (e.g. vector mechanics, or tensor formulations of electromagnetism), to the small-scale low-energy problems (with actions of the order of Plank's constant h) treated by quantum theory, to ultimately, the high-energy small-scale realm of quantum field theory, where even the vacuum is a complicated concept. The same can be said of everyday engineering thermodynamics being underpinned by statistical mechanics, itself a subset of quantum statistical theory. Table 2 parallels Table 1.

TABLE 2. Mathematical correspondances between physical phenomena at various scales

Classical theory	Quantum theory	Quantum field theory
Fixed no. of large particles	Fixed no. of small particles	Variable no. of small particles
3D Momentum variable	3D Momentum operator	4D Momentum operator
State vector	Wavefunction concept	Field concept
Pos. & mom. both defined	Pos. & mom. operators do not commute	Fields commute(bosons) or anti-commute (fermions)

3. Basic Statistics For Tracking & ID

The foundation of applied tracking and identification, namely the recursive Bayesian nonlinear filtering equations, is described in Section 1. The procedure for constructing provably true sensor likelihood functions from sensor models, and provably true Markov transition densities from target motion models, is described in Sections 2 and 3, respectively. The basis for fully automated tracking and identification, Bayes-optimal state estimators, is reviewed in Section 4. Section 5 provides a very brief survey of some of the major computational issues and approaches in real-time nonlinear filtering.

3.1 BAYES RECURSIVE FILTERING

Most signal processing engineers are familiar with the Kalman filtering equations. Suppose that we are given a linear sensor measurement model $Z=Hx+W$ and a linear target motion model $X_{k+1}=\Phi_k(X_k)+V_k$ where H, Φ_k are matrices, while W and the set V_k are independent, zero-mean Gaussian noise vectors. Then the Kalman time-update and information-update equations are:

$$\hat{x}_{k+1|k} = \Phi_k \hat{x}_{k|k} \qquad \hat{P}_{k+1|k} = \Phi_k \hat{P}_{k|k} \Phi_k^T + Q_k$$

$$\hat{x}_{k+1|k+1} = \hat{x}_{k+1|k} + K_{k+1}\left(z_{k+1} - H \hat{x}_{k+1|k} \right) \qquad \hat{P}_{k+1|k+1} = \left(I - K_{k+1}H\right) P_{k+1|k}$$

The Kalman filter is a special case of the Bayesian discrete-time recursive nonlinear filter. This more general filter is nothing more than the equations in Paragraph 1.1 applied recursively, namely:

$$f_{k+1|k}(x_{k+1} \mid Z^k) = \int f_{k+1|k}(x_{k+1} \mid x_k) f_{k|k}(x_k \mid Z^k) dx_k$$

$$f_{k+1|k+1}(x_{k+1} \mid Z^{k+1}) = \frac{f(z_{k+1} \mid x_{k+1}) f_{k+1|k}(x_{k+1} \mid Z^k)}{f(z_{k+1} \mid Z^k)}$$

The Bayes nonlinear filter equations assume certain conditional independence properties that need not be discussed here. The practical success of these equations depends upon our ability to effectively construct the likelihood function $f(z|x)$ and the Markov transition density $f_{k+1|k}(x_{k+1}|x_k)$. Though likelihood functions are sometimes constructed via direct statistical analysis of data, more typically they are constructed from sensor measurement models. Markov densities are typically constructed from target motion models.

The usual choices for the "true" likelihood function is $f(z|x)=f_w(z-h(x))$ if the measurement model is $Z = h(x)+W$, and for the true Markov density $f_{k+1|k}(x_{k+1}|x_k)=f_{Vk}(x_{k+1}-\Phi_k(x_k))$ if the motion model is $X_k=\Phi_k(x_k)+V_k$, where both W and V_k are again independent, zero-mean Gaussian noise vectors. How one gets to such expressions is detailed in the next two sections. Finally the most common "good" Bayes state estimators are just the previously defined MAP and EAP evaluated at time k.

3.2 CONSTRUCTING LIKELIHOODS FROM SENSOR MODELS

Suppose that a target with randomly varying state X is interrogated by a sensor which generates observations of the form $Z=Z_{X=x}=h(x)+W$, where W is a zero-mean random noise vector with density $f_w(w)$, but which does not generate missed detections or false alarms. The statistical behaviour of Z is characterized by its likelihood function $f(z|x)$ which describes the likelihood of the sensor collecting measurement z given that the target has state x. How do we compute this likelihood function? We begin with the probability mass function (a.k.a. probability measure) of the sensor model:

$$p(S\,|\,x) \triangleq p_{z|x}(S\,|\,x) \triangleq \Pr(Z \in S)$$

This is just the total probability that the random observation Z will be found in any given region S if the target has state x. The total probability mass $p(S|x)$ in a region S is just the sum of all the likelihoods in that region:

$$p(E_z\,|\,x) = \int_{E_z} f(y\,|\,x)dy \cong f(z\,|\,x)v(E_z)$$

where $v(E_Z)$ is a small (hyper)volume about the point z. In the limit of infinitely small $v(E_Z)$, division of the above equation by the infinitesimal volume defines the Radon-Nikodyn derivative

$$f(z\,|\,x) = \frac{\delta p}{\delta z} \triangleq \lim_{v(E_z)\to 0} \frac{p(E_z\,|\,x)}{v(E_z)}$$

Thus we have found the provably true likelihood function, i.e., the density function that faithfully describes the measurement model $Z=h(x,W)$. For the additive version of the model $Z=Z_{X=x}=h(x)+W$, it can easily be shown that this leads to $f(z|x)=f_W(z-h(x))$, namely a "noisy" version about the expected observation.

3.3 CONSTRUCTING MARKOV DENSITIES FROM MOTION MODELS

Suppose that, between the kth and (k+1)[st] measurement collection times, the motion of the target is best modelled by an equation of the form $X_{k+1}=\Phi_k(X_k)+V_k$. That is, if the target had state X_k at time-step k then it will have state $\Phi_k(X_k)$ at time-step k+l, except that possible error in this belief is accounted for by appending the random variation V_k. How do we construct $f_{k+1|k}(x_{k+1}|x_k)$? It is clear that this situation is completely parallel to the previous section. Specifically, the probability mass function $p_{k+1}(S|x_k)=Pr(X_{k+1}\in S)$ is the total probability that the target will be found in region S at time-step k+1, given that it had state X_k at time-step k, namely

$$f_{k+1|k}(x_{k+1}\mid x_k)\triangleq\frac{\delta p_{k+1|k}}{\delta x_{k+1}}=\lim_{v(E_{x_{k+1}})\to 0}\frac{p(E_{x_{k+1}}\mid x_k)}{v(E_{x_{k+1}})}$$

is the "true" Markov density associated with the motion model. For the additive version of the model, it can easily be shown that this leads to

$$f_{k+1|k}(x_{k+1}\mid x_k)=f_{V_k}(x_{k+1}-\Phi_k(x_k))$$

i.e. a "noisy" version about the expected position from the motion model.

3.4 COMPUTABILITY ISSUES IN SINGLE-TARGET FILTERING

The following is a very cursory survey of some of the major computational strategies.

(a) Approximate Filters: approximate filters achieve computational tractability by replacing likelihoods and posteriors by their linear, quadratic, higher-order, or other kinds of approximations. The best known approximate filters are the Extended Kalman Filter (EKF), the Iterated EKF (IEKF), and quadratic and higher-order filters. A major problem with approximate filters is the fact that, over time, approximation error tends to accumulate and can cause poor performance or even divergence.

(b) Exact Filters: an exact filter is one that achieves computational tractability by assuming that posterior distributions belong to some family of densities, e.g., Gaussians or generalized exponential. Exact filters will perform badly when the real-world posteriors deviate significantly from this form. Besides the Kalman and Kalman-Bucy filters, the best known examples of exact filters are those of Benes and Daum.

(c) Infinite-Dimensional Exact Filter: Kouritzin's convolution filter is apparently the only existing infinite-dimensional exact filtering technique. Assume that the old posterior and its time-update must be related by some (unknown) convolution kernel K. Once K has been determined off-line, one can compute $f_{k+1|k}$ in real time using Fast Fourier Transforms.

(c) Spectral Separation Filter: the Spectral Separation Scheme (S^3) filter of Rozovskii, Lototsky, et. al. is also an off-line technique. It is based on the fact that the unnormalized posterior can be approximated as a truncation of a rapidly convergent stochastic infinite series. The S^3 filter can accommodate quite general continuous-time motion models, but observation noise is assumed to be additive and Gaussian. It is therefore typically used in conjunction with non-parametric estimators to determine and pre-process the actual noise backgrounds.

4. MT Sensor And Motion Models

MT measurement models have been constructed for the following successively more realistic situations:

(a) MT measurement models with no missed detections and no false alarms/clutter
(b) MT measurement models with missed detections
(c) MT measurement models with missed detections and false alarms or clutter
(d) MT measurement models for clutter caused by targets equipped with Electronic Counter-Measures (ECM)
(e) MT measurement models for the multiple-sensor case
(f) single-target and MT measurement models for data that is imperfectly characterized or otherwise" ambiguous".

MT motion models have also been constructed for the following successively more realistic situations:

(a) MT motion models assuming that target number does not change.
(b) MT motion models assuming that target number can decrease. These models are analogous to MT measurement models with missed detections.
(c) MT motion models assuming that target number can decrease or increase. These models are analogous to MT measurement models with missed detections/clutter.

5. Unification Through The FISST MS-MT Calculus And Statistics

This chapter introduces the mathematical core of FISST: the FISST MT integral and differential calculus. The basic purpose is to establish the parallelisms between single-target problems and MS-MT shown in Table 3:

TABLE 3. Parallelisms in calculus and statistics

probability-mass function, $p(S\|x)$	belief-mass function, β
differentiation, dp/dz	set differentiation, $\delta\beta/\delta Z$
integration, $\int_S f(z\|x)dz$	set integration, $\int_S f(Z\|X)\delta Z$

This will allow true MS-MT likelihood functions to be constructed from the measurement models of the individual sensors and true MT Markov transition densities to be constructed from the motion models of the individual targets. Thus FISST provides a formalism that unifies a multitude of separate single target algorithms into a single coherent picture of MS-MT.

5.1 THE BELIEF-MASS FUNCTION OF A SENSOR

Just as the statistical behaviour of a random observation vector Z is characterized by its probability mass function $p(S|x) = Pr(Z \subset S)$, so the statistical behaviour of the random observation-set Σ is characterized by its belief-mass function (a.k.a. belief measure):

where Γ is the random MT state. The belief mass is just the total probability that all

$$\beta(S \mid X) \triangleq \beta_{\Sigma|\Gamma}(S \mid X) \triangleq \Pr(\Sigma \subseteq S)$$

observations in a sensor (or multi-sensor) scan will be found in any given region S, if targets have MT state X. For example, if $X=\{x\}$ and $\Sigma=\{Z\}$ where Z is a random vector then

$$\beta(S \mid X) = \Pr(\Sigma \subseteq S) = \Pr(Z \in S) = p(S \mid x)$$

i.e., the belief mass of a random vector is just its probability mass.

5.2 THE BELIEF-MASS FUNCTION OF A MOTION MODEL

In single-target problems, the statistics of a motion model $X_{k+1}=\Phi(x_k, V_k)$ are described by the probability-mass function $p_{k+1}(S|x_k)=Pr(X_{k+1} \in S)$, which is the probability that the target-state will be found in the region S if it previously had state $\mathbf{X}_k$. In like manner, suppose that $\Gamma_{k+1}=\Phi(X_k, V_k)$ is an MT motion model, then the statistics of the finitely varying random state-set Γ_{k+1} is described by its belief-mass function

$$\beta_{\Gamma_{k+1}}(S / X_k) \triangleq \Pr(\Gamma_{k+1} \subseteq S)$$

which is the total probability of finding all targets in region S at time-step k+1 if, in time-step k, they had MT state $X_k=\{x_{k,1},\ldots,x_{k\,n(k)}\}$.

5.3 FISST, THE SET INTEGRAL AND SET DERIVATIVE

The conventional probability summation equation generalizes to a set integral

$$p(S \mid x) = \int_S f(z \mid x) dz \Rightarrow \beta(S \mid x) = \int_S f(Z \mid X) \delta Z$$

where the meaning of any set integral is the following:

$$\int_S F(Y)\delta Y \triangleq F(0) + \sum_{j=1}^{\infty} \frac{1}{j!} \int_{\underbrace{S \times \ldots \times S}_{j \text{ times}}} F(\{y_1, \ldots y_j\}) dy_1 \ldots dy_j$$

Note that the set notation (curly brackets) and the regular ordering notation (round brackets) are just related by $F(\{y_1,\ldots y_j\})=j!F(y_1,\ldots y_j)$, so that the factorial in the above equation does not carry any deep significance. In the above, F could be an MS-MT likelihood $F(Z)=f(Z|X)$, an MT Markov density $F(X)=f_{k+1|k}(X|X_k)$ or an MT prior or posterior $F(X)=f_{k|k}(X|Z_k)$. Some care must be exercised with set notation and integration when the variable y is completely discrete, but this is outside the scope of this presentation.

The inverse function is the set derivative which, for Z containing distinct entries, reads

$$\frac{\delta\beta}{\delta Z}(S) \triangleq \frac{\delta^m \beta}{\delta z_1 \cdots \delta z_m}(S) \triangleq \frac{\delta}{\delta z_1} \cdots \frac{\delta}{\delta z_m} \beta(S)$$

With this definition the following standard rules are well defined:

(a) Sum rules (linear combination distributivity)
(b) Product rules (differentiate each term leaving others in product constant, then sum)
(c) Constant rule (differentiation gives zero)
(d) Chain rule (for composite functions)
(e) Power rule (with appropriate factorials).

With these tools, it is a straightforward but tedious task to show that the remaining correspondences of the introduction follow naturally, as shown in Table 4:

TABLE 4. Parallelisms for likelihoods and densities

likelihood function, $f(z\|x)$	MT likelihood $f(Z\|X)$
posterior density $f_{k\|k}(x\|Z^k)$	MT posterior $f_{k\|k}(X\|Z^k)$
likelihood function, $f(z\|x)$	MT likelihood $f(Z\|X)$
Markov densities $f_{k+1\|k}(x_{k+1}\|x_k)$	MT Markov densities $f_{k+1\|k}(X_{k+1}\|X_k)$

6. Bayes Fusion, Tracking & ID

6.1 OPTIMAL BAYES FUSION, TRACKING & ID

Bayesian MT filtering is inherently nonlinear. The easiest way to see this is to look at the simplest true MT problem: two targets observed by a single Gaussian sensor with noise model $N_Q(z\text{-}Hx)$, perfect detections, and no false alarms. In this case

$$f(\{z_1, z_2\} \mid \{x_1, x_2\}) = N_Q(z_1 - Hx_1)N_Q(z_2 - Hx_2) + N_Q(z_2 - Hx_1)N_Q(z_1 - Hx_2)$$

When the targets are greatly separated from each other, one of the two terms dominates and this distribution is essentially Gaussian. If they are not greatly separated it becomes a Gaussian mixture. This particular Gaussian mixture is highly non-Gaussian when the two targets have intermediate separation, i.e., not too far from each other but also not too near. In other words, multi-target nonlinear filtering is unavoidable if our goal is optimal-Bayes tracking of multiple, closely-spaced targets. Using FISST, the nonlinear filtering equations can be generalized to MS-MT problems.

The single-sensor, single-target Bayesian nonlinear filtering equations of Section 2 are already computationally demanding, and computational difficulties can get only worse when we attempt to implement the MT nonlinear filtering equations. Instead of the single numerical multivariate integrals of the single-target case, we are now faced with numerical set integrals. For example, in a scenario with as many as M targets, numerical evaluation of a set integral will require as many as $M^2/2$ single-target approximate multivariate integrals. Likewise, the already difficult problem of compressing a single-target posterior into a representation with a small number N of parameters expands into the much more difficult problem of compressing a multi-target posterior into a representation that requires on the order of N^M parameters. Clearly, if multi-target optimal Bayesian filtering is ever to become practical, drastic approximations will be required. The usual approximations involve the Gaussian approximation since it is well known that a convolution of Gaussians will yield another Gaussian.

6.2 ROBUST-BAYES FUSION, TRACKING & ID

Up until this point one has assumed that observations *z* are precisely defined and that their corresponding likelihood functions *f(z|x)* are known with sufficient fidelity to support optimal-Bayes MS-MT fusion, detection, tracking, and identification. However, as noted previously, it is unclear whether real-time, sufficiently high-fidelity likelihoods will ever be achieved for certain kinds of data, e.g. SAR images or HRR range profiles. In such situations optimal-Bayes algorithms can be non-robust, i.e., perform more poorly than anticipated because of the discrepancies between the modelled and actual likelihoods. Similarly, other kinds of data, such as features extracted from signatures, English-language statements received over link, etc., are so "ambiguous" (poorly understood from a statistical point of view) that probabilistic approaches are not even obviously applicable. The purpose of this chapter is to address the following question: How does one extend Bayesian (or other forms of probabilistic) inference to situations in which likelihood functions and/or data are imperfectly understood? This section presumes working familiarity with several reasoning frameworks outlined in my previous lecture, such as fuzzy logic and the Dempster-Shafer theory of evidence.

6.3 RANDOM SET MODELS OF "AMBIGUOUS" DATA

The FISST approach to data that is difficult to statistically characterize is based on the following key point: ambiguous data can be probabilistically represented as random *closed subsets* of (MS) measurement space.

Consider a simple example. Suppose that a near-sighted observer (like myself) examines a distant object that is known to belong to the set of objects

$$V = \{\text{redball, greenball, redcube, greencube, bluecube}\}$$

Suppose that the observations available to this observer are

$$U = \{\text{RED, GREEN, BLUE, ROUND, SQUARE, UNCERTAIN}\}.$$

These observations U can be associated with certain subsets of V as shown in Table 5.

TABLE 5. Observations and their associated subsets

Red	{redball, redcube}
Green	{greenball, greencube}
Blue	{bluecube}
Round	{redball, greenball}
Square	{redcube, greencube, bluecube}
Uncertain	V

In other words, any particular observation implies only a constraint on what the identity of the target could be. Suppose, however, that the observation is equivocal in the sense that the observer can at best specify only a range of hypotheses about what he or she actually saw. That is, the observer may specify a list $T_1 \ldots T_N$, of subsets of V, believed to be valid with respective degrees of confidence m_I where each T_i is drawn from the six subsets of V listed above. This sort of ambiguous evidence can be represented as a random subset Θ with probability distribution

$$Pr(\Theta = T_i) = m_i \qquad (i = 1 \ldots, N)$$

Because Θ is a random subset of the state space V, ambiguous observations constrain the state of the target directly. In general, however, many state variables may be hidden from the direct view of the observer. Accordingly, more general ambiguous observations will be random subsets of observation space U. In this case it is the data that is directly constrained by the ambiguous evidence rather than the state. Nevertheless, the state will still be constrained indirectly since a constraint on data implies a constraint on the state.

6.4 UNIFICATION OF TECHNIQUES FOR "AMBIGUOUS" DATA

It is one thing to recognize that random sets provide a common probabilistic foundation for various kinds of statistically ill-characterized data. It is quite another to construct practical random set representations of such data. Certain aspects of expert-systems theory, such as fuzzy logic, the Dempster-Shafer theory of evidence, and rule-based inference, provide well-known methodologies for mathematically encoding many kinds of ambiguous evidence. As noted in the last section, data that exhibits a great deal of imprecision can be represented as lists of evidential hypotheses, more commonly known as Dempster-Shafer bodies of evidence or "mass assignments." Likewise, "vague" concepts such as the "near" in the English-language statement "The helicopter is near the ownship" can be represented as fuzzy sets. The purpose of this section is to show how two kinds of ambiguous data, namely imprecise and vague data, can be represented probabilistically by random sets (it can be shown that contingent, and constrained-probabilistic data can also be put in random set form). This will provide us with the following recipe for processing ambiguous data:

(a) represent data using expert-system models
(b) transform these models to probabilistic random set mathematics
(c) compute probabilistic formulas.

6.4.1 *Vague Data: Fuzzy Logic*

A fuzzy membership function on some (finite or infinite) universe U is a function that assigns a number *f(u)* between zero and one to each member u of U. The value of *f(u)* is interpreted as the "degree of membership" of u in the fuzzy set represented by f, with f(u)=1 indicating unequivocal membership and f(u)=0 unequivocal non-membership. Fuzzy evidence, as well as the operations of fuzzy logic, can be represented in random set form as follows. Let Σ be a random subset of some (finite or infinite) universe. The one-point covering function μ_Σ of Σ is the fuzzy membership function defined as $\mu_\Sigma(u)=p(u\in\Sigma)$. Conversely, let $f\colon U \to [0,1]$ be a fuzzy membership function of the universe and let A be a uniformly distributed random number on the unit interval [0,1]. Then the random subset $\Sigma_A(f)$, called the canonical random set representation of the fuzzy subset *f*, is

$$\Sigma_A(f) \triangleq \{u \in U \mid A \leq f(u)\}$$

That is, the random set $\Sigma_A(f)$ can be interpreted as a faithful probabilistic representation of the fuzzy membership function *f*. The conventional fuzzy logic complementation operator can be regarded as the result of forcing algebraic closure on the canonical random set representation of the complement. Thus the celebrated non-Boolean behaviour of fuzzy logic can be interpreted as the result of ignoring the statistical correlations between fuzzy subsets. Also, the fact that there are an infinite number of possible fuzzy logics can be interpreted to mean that many fuzzy logics correspond to different assumptions about the statistical correlations between fuzzy sets.

6.4.2 *Imprecise Data: Dempster-Shafer Bodies of Evidence*

A Dempster-Shafer body of evidence B on some space U consists of nonempty subsets $B{:}B_1,\ldots,B_b$ of U and nonnegative weights $m_1,\ldots, m_b$ that sum to one. Each B is interpreted as a hypothesis "$u\in B_i$" constraining the possible value of the unknown quantity u; and the quantity m_i is the amount of mass that accrues to the hypothesis B_i alone and not to any strictly smaller subset of B_i. Every body of evidence B has an associated belief function and plausibility function defined, as in my previous lecture by

If one defines the random subset Σ of U by $p(\Sigma=B_I)=m_I$, for i=1,…b, Σ is said to be

$$Bel_B(S) \triangleq \sum_{i \ni B_i \subseteq S} m_i \qquad Pl_B(S) \triangleq \sum_{i \ni B_i \cap S \neq \phi} m_i$$

the random set representation of B and one writes $B=B^\Sigma$. The mass function of the random set Σ is the same thing as the belief function of B^Σ.

The usual Dempster rule of combination that provides a means of combining two independent bodies of evidence B, C into a single body of evidence B*C can be interpreted as an algebraic operation that mimics set-theoretic intersection of the nonempty, independent random subsets of U. Let us recall from the previous lecture that, for each nonempty hypothesis D_k obtained by this set intersection, the combination rule (a.k.a. the orthogonal sum) assigns the mass $(B*C)_k=d_k/(1-K)$ where d_k is the sum of all b_ic_j such that $B_i\cap C_j=D_k$ and where K is the sum of all b_ic_j such that $B_i\cap C_j=0$.

7. Some Recent Applications Of FISST

The success of any approach in engineering mathematics is measured by the applications that it enables. The purpose of this section is to summarize some recent applications of FISST to practical engineering R&D problems. This work is being conducted cooperatively by Lockheed Martin, Eagan (MN), Scientific Systems Company, Inc. (SSCI), and Summit Research Corporation (SRC). The applications to be described in the following subparagraphs are:

(a) MS datalink fusion and Non-Cooperative Target Identification (NCTI)
(b) naval passive-acoustic antisubmarine fusion and target identification
(c) air-to-ground Automatic Target Recognition (ATR) using SAR
(d) scientific performance evaluation of MS-MT data fusion algorithms
(e) unified detection and tracking using true MT likelihood functions and approximate nonlinear filters
(f) joint tracking, pose estimation, and target identification using High Range Resolution Radar (HRRR).

7.1 OFFBOARD MULTISOURCE ATTRIBUTE FUSION AND NCTI

The goal of this work is to develop a concept-feasibility prototype of an optimally self-reconfiguring adaptive and robust NCTI algorithm based on fusion of multi-source datalink features. On many surveillance aircraft, workstation operators examine various forms of data such as images, extract "features" from this data (e.g., number of hubs or wheels, presence or absence of a gun, etc.), and then transmit these features onto a datalink in a standard alphanumeric format. Features of this kind are corrupted by certain irreducible uncertainties, such as "fat-fingering" (i.e., the operator accidentally hit the wrong key), degradations in operator effectiveness due to fatigue and/or data overload, differences in training, or differences in individual ability. In principle, any individual operator could be "calibrated" to statistically characterize these effects. Because this is not possible in reality, one can never know any given operator's likelihood function $f_{operator}(feature|state)$ with any great certainty.

Test data for this problem came from a simulated Persian Gulf scenario called "LONEWOLF-98". It consists of features parsed from datalink message lines derived from Synthetic Aperture Radar (SAR), Long Range Optical (LRO), Moving Target Indicator (MTI) radar, Electronic Intelligence (ELINT), and Signals Intelligence (SIGINT). The possible target types are tanks, self-propelled guns, missile launchers, and personnel carriers.

A FISST robust-Bayes classifier was constructed based on a knowledge base of model features for the nine target types. A FISST Bayes-rule classifier was tested against two types of data: (1) low-quality synthetic "pseudo-data"; and (2) "as-real-as-possible-but-unclassified" data from LONEWOLF-98. The pseudo-data was highly corrupted by erroneous clutter features. The FISST robust-Bayes classifier correctly identified the correct target with high confidence (greater than 98%) in all nine instances. The FISST classifier was also blind-tested against sixteen unknown targets in LONEWOLF-98. Data for LONEWOLF-98 consisted of nearly a hundred datalink messages containing a total of 145 message lines containing feature information. The FISST classifier identified the correct target in all instances with very high confidence (98% or greater in ten cases and 93% or greater in the remaining seven).

7.2 ROBUST NAVAL ASW / ASUW TARGET IDENTIFICATION

The system being developed under the SPAWARSYSCOM Passive Acoustic Classification System (PACS) program is a prototype MT detection, tracking, and identification system for Anti-Submarine Warfare (ASW) and Anti-Surface Warfare (ASuW). It uses the following data sources: time-frequency passive-acoustic; and datalink attributes from passive-acoustic, magnetic-field, electric-field, and ELINT. The system includes the following major subsystems:

(a) a robust-Bayes classifier algorithm
(b) a supporting knowledge base of robust target signature models, which mimics the "hierarchical" decision-making process used by master-level INTELL (in particular, acoustics intelligence) analysts
(c) a multi-hypothesis correlator-tracker
(d) a semi-automated passive-acoustic operator-assist interface
(e) a naval contact manager.

The automated classifier is based on the "Fuzzy Conditioned Dempster-Shafer (FCDS)" approach that LM at Eagan has used in practical classifiers. The general idea is as follows. A typical passive-acoustic feature z consists of a time-varying random centre frequency surrounded by a band with time-varying random width. There are thousands of distinct acoustic features, most of which are extremely difficult to characterize statistically using a standard likelihood function $f(z|v)$, where v denotes target type. Instead, we use the concept of model-matching using generalized likelihood functions. When used in conjunction with the SRC acoustic model-signature knowledge base, classifiers based on both approaches have proved effective in identifying subsurface targets using passive-acoustic data in either datalink-feature form or time-frequency form. They have been blind-tested on real acoustic data with an approximate 85% correct-identification rate.

7.3 ROBUST ATR FOR SAR

One familiar approach to ATR of ground targets using sensors such as SAR is to compute a Maximum Likelihood Estimate (MLE) on the image vector. That is, let z_i be the intensity in the ith pixel of the image, let $f_i(z|\theta,v)$ be the likelihood function for the ith pixel given pose θ and target class v and, for simplicity, assume pixel-to-pixel independence. Then the MLE estimate of the target type is that v which maximizes the value of the joint likelihood function $f(z|\theta,v)=f_1(z_1|\theta,v)...f_M(z_M|\theta,v)$ of the entire image-vector $z=(z_1,..., z_M)$. However, real-world SAR images are corrupted by irreducible uncertainties that affect target signatures-e.g. dents, wet mud, irregular placement of surface equipment such as grenade launchers, etc. Such uncharacterizable statistical irregularities make it impossible to specify the likelihood functions $f_i(z_i|\theta,v)$ with enough fidelity to ensure proper performance of MLE (or related Bayesian) techniques. FISST provides a means of modelling these kinds of uncertainties, potentially resulting in algorithms more robust with respect to model-mismatch errors. Preliminary tests using simulated SAR data show that the MLE technique based on this approach is less likely to mis-identify a target than the conventional MLE, i.e., it is more robust, but it is also less confident than the conventional MLE about correct identifications.

7.4 Scientific Performance Evaluation for Data Fusion

The ability to meaningfully assess the competence of algorithms is a crucial part of developing and comparing practical MS-MT data fusion systems, whether this be MS integration (Level 1), threat and situation refinement (Levels 2 and 3), or sensor/resource management (Level 4). However, probably no other vital aspect of MS data fusion has been more heuristic, more poorly understood, and less deeply examined than metrology. Whereas considerable effort has been put into developing fusion algorithms that adhere to sound standards of statistical reasoning, fusion Measures of Effectiveness (MoEs) tend to be cobbled together as afterthoughts of the development of particular fusion algorithms. This situation is true even in Level 1 fusion where, despite the existence of a plethora of metrics and a vast infrastructure of statistical and engineering insight, ad hoc approaches to metrology lead to serious difficulties. In Level 1 fusion, the traditional approach is to use MoEs that measure some particular ("local") aspect of algorithm competence, e.g., localization miss distance, track purity, etc. Such MoEs often produce more confusion than clarity because optimization of an algorithm with respect to one particular local MoE (e.g., target localization accuracy) will not infrequently result in algorithm degradation as measured by another local MoE (e.g., target identity accuracy).

In Levels 2, 3, and 4 fusion the situation is even worse. There are few MoEs of any kind, *ad hoc* or otherwise. This section describes ongoing work whose purpose is to develop a systematic and scientifically defensible and practical approach to data fusion metrology at all Levels. The basic concept is that of directly generalizing information theory to the MS-MT realm.

Mathematical information is constructed and computed using information MoEs.

$$\Delta I_{SM} = I_{DF+SM} - I_{DF}$$

Suppose, for example, that a Kalman tracker algorithm is tracking a single target whose position at any time we know. At any instant, how much information is the tracker producing about the target, compared to ground truth? One useful approach is to compute the Kullback-Leibler cross-entropy or discrimination

$$K(f_{Grnd}; f) = \int f_{Grnd}(x) \log\left(\frac{f_{Grnd}(x)}{f(x)}\right) dx$$

where f denotes the probability distribution associated with the Kalman filter output, and f_{Grnd} is a probability distribution that corresponds to perfect knowledge of ground truth.

This metric is always nonnegative and vanishes if and only if $f = f_{Grnd}$. It measures the "directed information distance" between the tracker's estimate and ground truth in the sense that the larger the value of $K(f_{Grnd}; f)$, the less useful information the tracker is producing.

While overall information (whether with respect to complete knowledge or complete ignorance) is desirable, end users may be more interested in performance with respect to some specific function, e.g., target identification. If desired, the amount of information that is attributable to some specific fusion function (e.g., competence in target identification) can be parsed out of this total information score as a percentage. This quantity can be computed by altering the output of the data fusion algorithm to disable the function of interest, and then computing the difference in the information produced by the original and disabled algorithms:

This reasoning is easily extended to Level 4 fusion. The reason is that sensor management is a support function, i.e., its purpose is not to estimate parameters of interest but rather re-allocate sensor dwells in order to collect higher-quality input data for the core fusion algorithm(s). In other words, sensor management subsystem is performing well if the core fusion system generates more information than would be the case if the sensor management subsystem were absent. The following formula encapsulates this idea into an implementable mathematical form:

$$\Delta I = I_{enabled} - I_{disabled}$$

where I_{DF+SM} is the information produced by the data fusion algorithm with the assistance of sensor management and where I_{DF} is the information produced by the algorithm without such assistance. Through suitable choice of the underlying information MoE, it can be computed relative to (1) perfect knowledge of ground truth, (2) imperfect knowledge of ground truth, or (3) complete ignorance of ground truth. Since information provides a common language for all performance measurements, it can be expressed as a percent increase *above the* baseline provided by the core data fusion algorithm (without sensor management). As before, the effectiveness of the sensor management system in aiding specific fusion functions (e.g., target identification) can be specified as numerical percentages of ΔI_{SM}. Current results appear to verify the potential utility of the approach against simple scenarios.

7.5 TRACK-BEFORE-DETECT FILTERING IN CLUTTER

LM at Eagan has designed and implemented an approximate nonlinear filter with the following properties:

(a) its computational complexity is O(logN) or O(N), depending on the computational strategy

(b) it is theoretically guaranteed to not diverge in the non-moving target case (a result which carries over to moving-target scenarios in which the data rate is high enough).

This filter has been tested successfully in several simple one-dimensional model problems. A generalization of this filter, intended to jointly detect and track more than one target in clutter, is currently under development.

7.6 ROBUST JOINT TRACKING AND IDENTIFICATION FOR HRRR

A HRRR signature is a plot of an RF intensity vector I_i versus range bin i, representing the time-sequence of RF returns from a target as the HRRR plane wave passes through the target extent. This signature is highly aspect-dependent (pose θ) and is rich with features that can aid identification of target type T. The goal of this project is to prove the feasibility of an HRRR joint tracking and identification algorithm that:

(a) is robust with respect to uncertainties in the likelihood $f(I_i|T,\theta)$
(b) is robust with respect to novel target types
(c) optimally estimates pose θ, target type T, and position/velocity
(d) will potentially offer real-time operation when hosted on those processors expected to be resident on future generations of fighter aircraft.

A real-time kinematic nonlinear filter was integrated with an existing HRRR target identification algorithm from Airforce Rome Lab, called STaF, which was designed to overcome the lack of robustness of conventional Bayesian approaches to ATR for HRRR. Let i denote the ith range bin of an observed HRRR profile and let $I_i=a(i)$ be its corresponding intensity amplitude. STaF uses the following procedure:

(a) extract features, i.e., intensity peaks in the range-profile that are attributable to scatterers on the target surface rather than to random noise
(b) from each extracted feature, construct modified posterior distributions that hedge against the possibility of novel or unknown target types; and
(c) use Dempster's rule to combine the modified posteriors into n fused posteriors, each corresponding to a different target-identity hypothesis.

A concept-feasibility prototype algorithm based on this approach has been implemented in a restricted two-dimensional problem. Aircraft are assumed to move at constant altitude and to manoeuvre using coordinates turns (i.e., turns in which the length of the aircraft body is tangential to the flight path). Under such assumptions, the complete target state reduces to (x, y, v_x, v_y, w, T), where w denotes planar angular velocity, and between-measurements motion can be modelled using a nonlinear Markov state transition based on a small-angle approximation for each type T:

$$x^{k+1|k} = x^{k|k} + \Delta t \cdot v_x^{k+1|k} \qquad y^{k+1|k} = y^{k|k} + \Delta t \cdot v_y^{k+1|k}$$

$$v_x^{k+1|k} = v_x^{k|k} + \omega^{k|k} \Delta t \cdot v_y^{k|k} \qquad v_y^{k+1|k} = -\omega^{k|k} \Delta t \cdot v_x^{k+1|k} + v_y^{k|k}$$

$$\omega^{k+1|k} = \omega^{k|k}$$

The approximate nonlinear filter of the previous section was modified to make use of this motion model and the FISST likelihood function, and integrated with an SSCI-developed implementation of the STaF algorithm. The integrated algorithm was tested on real, simultaneously collected data *I, z* for three target types Tl, T2, and T3 with the following results. STaF was trained to recognize Tl and T2 but not T3. The integrated classifier-tracker algorithm was tested on data for Tl, T2, and T3 that had not been used in the training. When fed data from Tl or T2, the integrated algorithm decisively chose the correct target type. When fed data from T3, on the other hand, the algorithm was unable to make a determination between type Tl or type T2.

8. Summary And Conclusions

FISST has been created in part to address the issues in probabilistic inference that the "cookbook Bayesian" viewpoint encourages us to ignore or even be unaware of. These issues include:

- dealing with poorly characterized sensor likelihoods
- dealing with "ambiguous" data
- constructing likelihoods for "ambiguous" data
- constructing true MT likelihoods and Markov transition densities
- dealing with the "curse of dimensionality" in MT problems
- providing a single, fully probabilistic, systematic, and unified foundation for MS-MT detection, tracking, ID, data fusion, sensor management, performance estimation, and threat estimation and prediction, while
- accomplishing all of this within the framework of a direct, relatively simple generalization of standard statistics and undergraduate calculus

During the last two years FISST has emerged from the realm of basic research to a range of practical engineering research applications. The purpose of this lecture has been to summarize the FISST approach and its use in such applications. The main challenges ahead are to increase the calculability of MT filtering, in general, beyond the Gaussian approximation.

9. Reference

1. Goodman, I.R., Mahler, R.P.S., and Nguyen, H.T., (1997) *Mathematics of Data Fusion*, Kluwer Academic

FUSION OF INFORMATION UNDER IMPRECISION AND UNCERTAINTY, NUMERICAL METHODS, AND IMAGE INFORMATION FUSION[1]

I. BLOCH
Ecole Nationale Supérieure des Télécommunications
TSI - CNRS URA 820
46 rue Barrault, 75013 Paris, France
Email: Isabelle.Bloch@enst.fr

1. Introduction

The increased research in data fusion in several domains is motivated by the multiplication of sources of knowledge and data, and of techniques for their acquisition. The huge volume of data to be processed and the complexity of the problems that are addressed induce a real need for the development of data fusion techniques.

In particular, data (or information) fusion appeared in the domain of image processing and computer vision a few years ago, as a necessary stage for several applications like medical imaging, aerial and satellite imaging, quality control, robot vision, vehicle or robot guidance. The need for information fusion in these domains originates from the increasing number of information sources, mostly images issued from different sensors, and from other types of information as well (maps, atlases, expert knowledge, etc.). Imprecision and fuzziness being inherent to these pieces of information to merge, data fusion methods have to take this diversity of images into account, in order to improve the combined result.

The aim of this chapter is to underline the major characteristics of information fusion and the ways to explore the implications of fusion in image interpretation, without providing a deep study of all these points. The aim of this chapter is thus to provide an overview of general characteristics of fusion problems, and to highlight their specificities in image information fusion. We restrict the presentation to the problem of information fusion under imprecision and uncertainty, and to numerical methods to account for these imperfections in the fusion process.

This chapter is organized as follows. In Section 2, some definitions of fusion are given, in particular with respect to related problems, not directly addressed here. In Section 3, we present some general characteristics of information that have to be taken into account in a fusion process. In Section 4 we briefly describe the main schemes for information fusion. In Section 5, we present problems in image fusion according to several criteria and points of view, and underline their specificities with respect to other fusion problems. In Section 6, we shortly mention how probabilistic and Bayesian approaches are used in information fusion. In our opinion, fuzzy methods are very appropriate to deal with imprecision in fusion (in particular image fusion), and this is addressed in Sections 7 (for fuzzy and possibilistic approaches) and 8 (for approaches based on belief functions).

[1]This course is based mainly on previously published papers [22],[20],[21],[23],[4].

A. K. Hyder et al. (eds.), Multisensor Fusion, 267–293.

2. Definitions

2.1 DEFINITION OF INFORMATION FUSION

We consider here the term "information" in a broad sense. In particular it covers data (for instance measurements, observations, images, etc.) and pieces of knowledge (on the data, on the domain, on constraints, etc.), which can be either generic or specific.

The definition that we will use in this paper is the following [18,22]:

Definition 1 - *Information fusion consists in combining pieces of information issued from several sources in order to improve the decision making step.* This definition has the interest to focus on the combination and decision steps, which can take different forms depending on the problem at hand and on the considered application. For each type of problem and application, this definition can be made more specific, by answering a number of questions: What is the aim of fusion? How is the decision expressed? What are the pieces of information to be combined? Where are they originating from? What are their characteristics? Which methodology should be chosen? How to evaluate and validate the method and the results? Which are the main difficulties, limitations? etc.

Let us compare this definition to a few others proposed by some working groups that tried to unify terminology and concepts in the fusion community.

Definition 1 is slightly more specific than the one proposed by the European working group FUSION [4], which worked on fusion in different domains from 1996 to 1999. Resulting from discussions in the working group, the proposed informal definition is as follows: *fusion consists in conjoining or merging information that stems from several sources and exploiting that conjoined or merged information in various tasks such as answering questions, making decisions, numerical estimation, etc.* In this definition, which is also focused on combination and aims, the aims are more general. Here we restrict to the amelioration of knowledge about how the world is, and do not refer to how we would like the world be, a problem addressed for instance in fusion of preferences.

A unified terminology has been proposed by the Joint Directors of Laboratories Data Fusion Working Group. This group was created in 1986 in the US and tried to codify the data fusion terminology, producing a Data Fusion Lexicon [1]. The proposed model is dedicated mainly to military applications (such as target tracking, automatic target recognition and identification) and is functionally oriented, identifying processes, functions and techniques. It focuses mainly on a description of a hierarchy of processing steps in a system. Here we take another point of view which is adapted to fusion problems in several domains, including engineering (e.g. in signal and image processing for pattern recognition, diagnosis and decision making in non military applications). Instead of focusing on the system, we define data fusion as the process that combines information issued from different sources in order to take a decision, and therefore we focus on techniques for combining information towards a specific goal.

Another European working group, of EARSeL (Association of Remote Sensing Laboratories), has extended the JDL definition to the more general domain of satellite imaging [100]: data fusion constitutes a formal framework in which data issued from different sources are expressed; it aims at improving information quality; a more precise definition of quality depends on the application. This definition includes several definitions proposed in satellite imaging, summarized in [100]. **Definition 1** goes farther, since it includes the decision step as ultimate goal.

2.2 FUSION VS. RELATED CONCEPTS

The term data fusion being rather vague and understood in a more or less broad sense, we restrict it to a more precise domain, and distinguish it from related concepts.

- **Fusion and estimation** On the one hand, estimation aims at combining several values of a parameter or a distribution to deduce a possible value of it. This includes combination and decision which are the two main ingredients of **Definition 1**. On the other hand, numerical fusion methods often include an estimation step, as will be seen in the following, which therefore appears as one step in a fusion process.
- **Fusion and revision or update** Revision and update consist of augmenting or modifying information in light of new pieces of information. They can be considered as fusion problems. Sometimes, fusion is considered in a more restricted sense, where the combination is symmetrical, while for revision the combination is usually not symmetrical.
- **Fusion and association** Data association is the action which determines from many different signals issued from two or more sources, those which are emitted by the same object (source or target). According to Bar-Shalom and Fortmann [6], data association is the most difficult stage in multi-target tracking. It consists of detecting and associating some noise corrupted measurements, the origins of which are uncertain due to many factors, including random false alarms in the detection, clutter, interfering targets, decoys or other counter-measures. Although uncertainty has to be understood here in the most general sense, possibly including imprecision and fuzziness, most of the methods developed for data association and tracking are mainly dedicated to the processing of probabilistic uncertainty. The main models used in this field are either deterministic (involving classical hypothesis testing) or probabilistic (relying mainly on the Bayesian framework) [6],[81],[61]. The most common method [6] relies on the Kalman filter under Gaussian hypothesis. Therefore, the problems addressed in association are quite different from the ones covered by **Definition 1**.
- **Fusion and data mining** Data Mining consists in extracting interesting parts of information and data, which can be, for instance, data satisfying some particular property, or rare data. It can be opposed to fusion problems where explications or general tendencies are searched, but that are not considered here.

2.3 A MULTIDISCIPLINARY FIELD

A common way to introduce and present pattern recognition is to make a parallel between the way human beings process information for recognizing patterns on the one hand, and the attempts to perform similar tasks by automatic machine processing on the other hand.

A similar parallel can be drawn for data fusion, since human perception consists essentially in integrating multi-sensory information for understanding environment, scenes, situations, and taking a decision for interpretation, choosing a behavior, etc. This has been underlined by several authors, e.g. in [43]. A need occurs for developing computer systems emulating such complex information fusion processing for various tasks and various applications. This parallel has been stimulating researchers in science and engineering to develop the field of data fusion.

Although the concept of fusing information may be understood in a simple and intuitive way, building methods and systems for data fusion appears to be a very difficult and complex task, that did not find until now a consensus, even concerning formal definitions. This task calls for multidisciplinary research, including engineering, computer science, artificial intelligence. More specifically, methods developed in fields as different as process control and monitoring, machine engineering, signal and image processing, logics, statistics, data base management, system theory, diagnosis, tracking, robotics, etc. have been exploited and gathered for developing data fusion systems and methodologies.

3. General characteristics of information

In this Section, we describe the general characteristics of information that have to be taken into account in a fusion process.

3.1 TYPES OF INFORMATION

A first characteristic concerns the type of information to be combined. It can consists of direct observations, of processing results on these observations, of more or less generic pieces of knowledge, of expert opinions, etc. These pieces of information can be expressed in numerical or symbolic form. A particular attention has to be paid to the scale used for representation. It does not necessarily has to have an absolute meaning, but at least some commensurability or normalization has to be guaranteed. A more detailed description of different types of information and their representation can be found in [4].

The level of information is also an important aspect. Depending if the information is considered at low level (typically the original data), or at higher level, which often calls for preliminary processing, constraints and difficulties that arise are not the same.

Other distinctions can be underlined, since they lead to different types of modeling and processing, for instance distinction between frequent and rare data, between static and dynamic (evolving in time) information, between generic or factual information (the second one being often more specific than the first one). Typically, generic information can be ignored if it contradicts a reliable factual (observed) information. On the contrary if no observation can be made, or if it is not reliable, then generic information has to be used, and is then considered as a "default". Information involved in a fusion process consists on the one hand of pieces of information to be combined, and on the other hand of additional knowledge, which serves to guide or help the combination. It can be information about the information to be combined as information about sources, their dependencies, their reliability, about preferences, etc. It can also be contextual information, on the domain. This additional knowledge is generally not expressed in the same formalism as information to be combined.

3.2 IMPERFECTION OF INFORMATION

One important characteristic of information in fusion is its imperfection. It is always present (otherwise fusion would not be necessary). It can take different forms, that are briefly described in this Section. Note that there is no real consensus about these definitions in the literature. The ones that are given here are certainly not universal, but are well adapted to fusion problems.

- **Uncertainty** Uncertainty is related to the truth of some information, characterizing its adequation to reality [40]. It refers to the nature of the considered object, to its quality, or to its occurrence.
- **Imprecision** Imprecision concerns the content of information and describes a quantitative defect of knowledge or measure [40]. It concerns the lack of precision in quantity. This notion is often included in a larger sense of uncertainty, but it may be useful to distinguish between them. More refined classes of uncertainty have also been proposed (e.g. [55]).
- **Incompleteness** Incompleteness of information issued from each source is the main reason that motivated the fusion. Information provided by a source is generally partial, and gives only one point of view or one aspect of the observed phenomenon.
- **Ambiguity** Ambiguity of information extracted from each source has also to be taken into account. For instance an imprecise measurement can prevent to distinguish between two different situations, or incomplete information can induce confusions between objects or situations. One of the objective of fusion is therefore to solve ambiguities using additional sources of information and knowledge.
- **Conflict** Conflictual situations occur often in fusion problems, and are usually difficult to solve. First, detecting conflicts is not easy. They can be confused with other types of imperfection, or even with complementarity between sources. Then their identification and typology is a question that often arises, but differently depending on the domain. At last, solving conflict can take different forms. It can rely on the elimination of non reliable sources, on introducing additional information, etc. In some cases, it is preferable to delay combination and wait for more information, or even not to fusion at all (for instance if conflict appears because the sources actually do not speak about the same phenomenon). These issues are addressed in detail in [4].

3.3 REDUNDANCY AND COMPLEMENTARITY

Other characteristics of information are more positive, and can be exploited in order to limit or reduce imperfections.

- **Redundancy** Redundancy between sources comes from the fact the they give information about the same phenomenon. Ideally, redundancy should be exploited in order to reduce uncertainties and imprecisions.
- **Complementarity** Complementarity between sources comes from the fact that they usually look at different aspects of the observed phenomenon. Ideally, complementarity should be exploited to get a more complete information and to solve ambiguities.

3.4 SYMBOLIC VS. NUMERICAL

There have been large discussions in the fusion community about the duality between numerical and symbolic fusion. Our aim here is not to go deeply in this discussion but rather to present the different levels at which this question may be addressed. Since most of the discussions originate from the fact that the concepts are confusing if this level is not specified, this type of presentation may help to clarify these concepts. The three levels at which we would like to make a distinction between numerical and symbolic are (i) the type of data to be treated, (ii) the type of processing applied to these data, (iii) the role of representations. Examples are taken mainly from the image processing and computer vision domain.

3.4.1 Data and information

By numerical information, we mean data directly given as numbers. These numbers may represent various features, typically physical measures, grey levels, response to an image processing operator, etc. They may be directly read from the sources to be fused, or attached to the domain or contextual knowledge (e.g. wave lengths in satellite imaging, acquisition times in medical imaging, etc.).

By symbolic information, we mean all information given as symbols, propositions, rules, etc. Such information can be related to the data to be combined (e.g. graphical information in a map or in an anatomical atlas, attributes computed on data or objects previously extracted from the images) or related to the domain knowledge (e.g. propositions about the properties of the problem at hand, structural information stating for instance that a road network can be represented as a graph using roads and cross-roads, propositional knowledge stating general rules about the scene like "the ventricles are always inside the white matter", etc.).

The classification of data and information in symbolic and numerical classes cannot always be done in a crisp way. We may have to deal with "hybrid" kind of information, where numbers are used for coding information that is not necessarily of numerical nature. This is typically the case for the evaluation of some data or treatment, for the quantification of imprecision or uncertainty. In such cases, the absolute values of these numbers are not important, this is rather the ranking which plays an important role. These numbers may be attached to symbolic information as well as to numerical information. In image processing, examples can be found for quantifying the quality of a detector, the evaluation of some symbolic data, of source reliability, of confidence in some measurement or numerical value, etc.

3.4.2 Processing

As far as processing of information is concerned, we mean by numerical treatment any computation on numbers. In data fusion, it concerns approaches that combine numbers by some formal calculus. Note that such kind of treatment does not make any assumption on what kind of data is represented by numbers. Data may be originally of numerical as well as of symbolic nature.

Symbolic types of treatment include formal computation on propositions (logic is but one example), possibly taking into account numerical knowledge. Structural approaches, like graph-based approaches often used in structural pattern recognition, can be considered as belonging to this class.

We consider as hybrid types of treatment the methods where prior knowledge is used in a symbolic way to control numerical treatments, for instance by stating some propositional rules that suggest/allow/prevent specific numerical operations. Typically, a proposition stating in which cases sources A and B are independent can be used in the way probabilities are combined, or knowing that the recognition depends only on some local or contextual knowledge may lead to an appropriate modeling of the scene as a Markovian field. Such kind of hybrid processing is widely used in image processing and image fusion.

3.4.3 Representations

As it appears from the two previous subsections, the representations and their type may play very different roles. Numerical representations can be used for intrinsically numerical data as well as for evaluation and quantification of symbolic data. An important use of numerical representations in data fusion is for quantifying imprecision, uncertainty or reliability of the information (this information may be of numerical as well as of symbolic nature), therefore representing rather information about the information than the data themselves. We will focus on such kind of representations in the description of the main numerical approaches for image fusion. Numerical representations are also often used for degrees of belief attached to numerical and/or symbolic knowledge, and for degrees of consistency or inconsistency in a database. Note that the same (numerical) formalism can be used for representing very different kinds of data or knowledge: the most obvious example is the use of probabilities for representing data as different as frequencies, subjective beliefs, etc. [11].

Symbolic representations can be used in logical systems, or knowledge-based systems, but also as prior knowledge for guiding numerical treatment, as a structural support for image fusion (see e.g. [65]), and of course as semantics attached to the manipulated objects.

In several examples, a strong duality can be observed between the roles of numerical and symbolic representations. An example will be given in Section 5.

4. Fusion schemes

A general fusion problem can be stated in the following terms: given l sources S_j representing heterogeneous data on the observed phenomenon, take a decision D_i on an element x, where x can be for instance a pixel in image processing or any other object, and D_i belongs to a decision space $D = \{D_1, ..., D_n\}$ (or set of hypotheses). As opposed to symbolic fusion methods (see e.g. [91],[69],[44]), in numerical fusion methods, the information relating x to each possible decision D_i according to each source S_j is represented as a number M_i^j having different properties and different meanings depending on the mathematical fusion framework. The global fusion scheme consists in taking a decision on x as a function of all M_i^j, $1 \leq i \leq n$, $1 \leq j \leq l$. Typically, M_i^j evaluates how much source j supports decision i. In this scheme, all information is taken into account, but is complex and hardly tractable. Therefore, two degraded systems are often defined, that will be be referred as centralized and distributed schemes in the following.

In the distributed fusion scheme, decision is taken locally on each source S_j separately, from M_i^j, $i=1,..,n$. Then the local decisions are merged in a global one. Many distributed systems have been developed for real-time and military applications, since partial decision are taken as soon as information is available. This scheme is also called decision fusion [96],[33]. Methods in distributed decision often assume 2 possible decisions only D_0 or D_1 (or sometimes 3, the third one representing no decision). In this case, the problem amounts to find an application f from $\{D_0, D_1\}^l$ into $\{D_0, D_1\}$ [80]. The main methods rely on probabilistic models, empirical estimations based on samples, nearest neighbor rules, etc. In this scheme, imprecision can be taken into account for each sensor or each information source separately. Models are then designed adaptively for each type of information, and the addition of a new source of information is easily done. However, no process in such schemes leads to a cooperation between sensors for reducing imprecision or ambiguity provided by one source by using information provided by another one, except at the decision level, which is quite rough for this task. Moreover, conflicts between local decisions are difficult to solve, since all numerical information on information imperfections is lost[2].

In the centralized fusion scheme, the measures related to each possible decision i and provided by all sources are combined in a global (still numerical) evaluation of this decision, taking the form, for each i : $M_i = F[M_i^1, M_i^2, ..., M_i^l]$, where F is a fusion operator. Then a decision is made from the set of M_i, $i=1,..,n$. In this scheme, the final (binary) decision is issued at the end of the processing chain. Therefore we avoid to made decisions at intermediate steps with partial information only, and thus we diminish contradictions and conflicts, which usually require a difficult control or arbitration step. In centralized schemes, imprecision and more generally imperfection of information is combined at a numerical level, and therefore fusion between sources can be better exploited to reduce these imperfections before the final decision is made.

What follows in this paper relies mainly on the centralized scheme, which is for us the most efficient when real time decision is not mandatory[3]. The main steps of fusion can then be described as [18],[22]: (i) modeling information and its imperfections, (ii) estimation of the M_i^j, $1 \leq i \leq n$ according to the chosen mathematical framework, (iii) combination, i.e. choice of an appropriate fusion operator F [12], (iv) decision.

Although the final aim is decision, the previous steps are very important and should be designed properly in order to exploit as well as possible the specific properties of the data,

[2] One possible solution consists in keeping several possible individual decisions for each source, and in considering all of them for the fusion.

[3] A scheme that can be seen as intermediary between centralized and distributed decision consists in choosing the necessary information among the M_i^j , i=1,..,n, j=1,..,l in order to solve some specific decision problems. Such hybrid schemes often involve expert knowledge and are mainly used in rule-based and multi-agent systems.

their complementarity and redundancy. Different choices in the modeling and combination steps may lead to drastically different decisions.

5. Fusion in image processing

In this Section, we highlight the characteristics of image information fusion, with respect to other domains.

5.1 DATA AND KNOWLEDGE CHARACTERISTICS

In this Section, we summarize some aspects of image information that have to be taken into account in image fusion. Some of them are very specific to image processing, and have to be introduced in fusion schemes and mathematical frameworks, which are generally derived from other domains.

First, the information to handle in image fusion is often **heterogeneous**. In several applications, several imaging techniques have to be combined to answer a specific question. They provide different aspects and different points of view on the problem by exploiting different physical properties. For instance, when planning some surgical operation in medical imaging, necessary images can be as heterogeneous as anatomical images (provided by MRI or CT), angiographic images (MRA, spiral CT, etc.), functional images (PET, functional MRI). These images are not informative about the features they are not dedicated to. Similar examples can be found in other domains, in particular in satellite or aerial imaging. An additional cause of heterogeneity comes from the fact that image information needs to be combined with external information to make sense. This can be information on acquisition conditions, on the observed phenomenon, or more generally expert knowledge related to the problem at hand. This knowledge can be expressed either in "iconic" terms, under the form of atlases or maps, or in "propositional" terms, under the form of linguistic expressions or rules. This is an aspect that is quite specific to image processing. Indeed, unlike in other domains where data fusion is used, especially in artificial intelligence, symbolic information is not only the information injected in the process by the expert in a propositional way. It may also be derived from graphical documents since such representations carry a rather abstract interpretation of the objects contained in the scene. Such information is considered as symbolic, as opposed to the numerical information constituted by location, intensity, etc. that may be found in images.

Duality between numerical and symbolic information emerges at this level. Let us take the example of a map and an aerial image of the same area [65]. The numerical information carried by the image provides a quite accurate description of the scene, but the interpretation attached to it is hard to derive. For instance, it is generally difficult to assess the type of a building, although its drawing on the image is accurate. On the contrary, the map carries symbolic information as a semantic meaning of the objects represented on the map but its shape is often sketchy. Here lies also an example of duality between imprecision and uncertainty (imprecision on location and uncertainty on the nature of the objects). This example illustrates how imprecision may be quite different depending on the level of representation (pixel in an image, object in a map for instance).

The situation is even more complex if we include in this scheme image processing algorithms used to extract information about objects, and the attributes describing the objects. These attributes can be numerical or symbolic, and they contribute to the nature of the objects. Therefore, the information we have to deal with is embedded in a complex net, where images, objects, algorithms, measures, interpretations are closely related to each other. Imprecision can be attached to any of these elements. One of the problems to be solved is therefore to understand and model the influence, on the nature of one element, of the imprecision associated with the others. In particular, in order to define accurately these imprecisions and their relationships, we found it difficult to provide some reasonable constructive intermediate between generic definitions and naive instances.

Numerical-symbolic fusion and it specificities when applied to image information under imprecision is a very complex problem, that still did not find a definite solution and that remains a very active field of research.

However, a first step towards understanding and management of imprecisions and their relationships can be performed by studying the different **causes of imprecision**. Indeed, imprecision in image information may be due to several factors, ranging from the observed phenomenon to the algorithms artifacts. In biology, a soft transition between tissues (e.g. healthy and pathological tissues) is surely a cause of imprecision inherent to the nature of the observed objects. Also, if tissues have similar characteristics, images that represent these characteristics will poorly discriminate these tissues. This will result in an uncertainty on the belonging of a pixel to one or the other tissue. Another cause of imprecision is due to the discrete nature of numerical images, resulting in a delocalization of information contained in a small volume at only one point. The partial volume effect (the presence of several tissues in one pixel or voxel) belongs also to this type of spatial imprecision. Other image imperfections can be caused by numerical reconstruction algorithms in computed imaging. One example is the Gibbs effect that may appear in MRI around sharp transitions. At the processing level, imprecision is often induced by the chosen algorithms (e.g. filtering, contour detection, registration between images, etc.), another place where the duality between imprecision and uncertainty can be found [87].

Another aspect that is quite specific to image processing, as compared to other application domains of data fusion, is the **complexity of the information**. This is partly due to the previously mentioned characteristics but also to the increasing number of acquisition techniques and to the huge data sets that have to be dealt with. Typically, one MRI brain image contains $256 \times 256 \times 128$ voxels, one satellite image contains 6000×6000 pixels, and several images of this size have to be combined in a fusion process. The large data volumes, and the statistical measures that are therefore made possible, may explain the use of statistical approaches in most image fusion schemes. The complexity of the fusion process also comes from the simultaneous redundancy and complementarity between images, closely related to the heterogeneity aspects. One of the main tasks of image fusion is to exploit redundancy, in order to increase the global information, and complementarity, to improve certainty and precision. The decision is thus improved by the fusion in terms of both quantity and quality.

The main information in image processing, which is specific to this domain, is the **spatial information**. The previous examples have already shown how it appears at all levels, and the types of imprecision attached to it. When using a fusion method, often issued from another domain, we thus have to incorporate spatial information in the process. This point is discussed in the conclusion.

One of the main problems in image fusion (like in other fields of image processing) concerns the **validation** of methods. If the truth generally more or less exists, it is often difficult or impossible to access[4].

5.2 FUSION LEVELS

In image fusion, like in most image processing domains, several levels of information representation are distinguished. At the lowest level, information carried by each pixel is considered. Fusion schemes at this level aim at taking a decision for each pixel, e.g. attributing each pixel to a class of interest. All raw image information is then used, but such schemes are lacking in structural information and spatial coherence [29]. A precise registration between all images to be combined is generally considered as unavoidable at this level. At an intermediate level, the fusion is performed on features extracted from each image using some image processing tools. These features are typically segments, contours, or regions. Precise registration is then crucial. At the highest level, objects extracted from images are combined. This involves a preliminary recognition step and is therefore related to decision fusion. Structural information can be easily introduced at this level, and therefore makes registration less mandatory. However, the richness of the basic numerical information is lost, and conflicts may be difficult to solve, unless some numerical evaluation is attached to the objects and structures extracted from each image. Applications where segmentation results are fused (e.g. [31],[28]) can be considered as intermediate level fusion if the segmented regions or contours have no semantic interpretation (i.e. no real decision is taken on each image), or as high level fusion otherwise.

5.3 SOME EXAMPLES OF DECISION PROBLEMS IN IMAGE FUSION

Decision can take various forms, ranging from hypothesis evaluation, classification to object recognition, and to scene interpretation and understanding.

Several decision tasks can benefit from fusion of multi-source data, including estimation, hypothesis testing, or model validation. Decision can also take the form of detecting the presence of some phenomenon (e.g. a particular form of an electrocardiogram, or some rupture in a statistical model).

A field where data fusion has been widely used concerns tracking, i.e. the decision consists in assessing the movement of some objects. This task may include several partial decisions like detection of the objects of interest, localization, and possibly recognition.

An important field concerns diagnosis, classification, and pattern recognition. In such problems, the decision consists in labeling objects in a scene, in attributing each pixel or object in an image to a class, in detecting unexpected features, etc. In image fusion, classification and recognition is probably one of the most typical type of decision we may find in the literature. This is a problem closed to multi-criteria aggregation problems. These

[4] In that sense, aerial or satellite imaging is probably a better field of investigation than medical imaging. At least, it allows an easier evaluation.

tasks are part of the more general problem of scene interpretation. The aim of fusion is then to get a global point of view on a scene, by gathering the different points of view provided by the different sources (the sources are then considered as experts or observers, each of them looking at the scene under a different aspect, using its own criteria and specificities). The sources can be different images but also different features extracted from the same image using image processing tools dedicated to specific aspects of the image. Under this form, signal and image fusion can be compared to expert opinion pooling problems. Another part of scene interpretation concerns updating, which aims at improving the information using additional sources, i.e. at getting more precise and more reliable information about the scene. In the case of dynamic scenes, evolving in time, updating also involves processes able to introduce new information or changes in the objects in the global interpretation.

5.4 IMAGE FUSION AND OTHER DOMAINS

To conclude this Section, let us briefly illustrate another reason why image fusion is so different from other application domains of data fusion (see **[20] for** a description of fusion problems in several domains). Of course, the nature of information is an important point, and the spatial aspects as well. But, in our opinion, with respect to multi-criteria aggregation and optimization, the main difference relies on the fact that in that field, the aim is to try to find a solution that satisfies as well as possible some flexible constraints. In image fusion, the sensors do provide (more or less explicitly) a degree of satisfaction (of membership to classes for instance) and the decision amounts rather to choose the best one. With respect to voting problems, the difference is that in those problems, no truth exists, and that probably subjectivity has more to be taken into account, but according to some "ethic" rule. Subjectivity is also part of combination of expert opinions. In that domain, information is moreover usually more sparse, which makes learning perhaps more difficult than in image processing. However, this last problem is probably the closest one to our.

6. Probabilistic and Bayesian approaches

The most used framework in numerical fusion, and in particular in image fusion, is undoubtedly the probabilistic framework, specifically Bayesian approaches.

6.1 MODELING AND ESTIMATION

In the Bayesian framework, the M_i^j's represent conditional probabilities: $p(x \in D_i \mid I_j)$.

These probabilities are computed from characteristics extracted from the images, typically grey-levels or texture indices at low level, or other features and object properties at higher level. They represent mainly the probabilistic uncertainty attached to image information. One of the advantages of Bayesian fusion relies on the large experience on learning that allows the user to perform estimation of these conditional probabilities.

Uncertainty may be expressed in three different ways depending on our a priori knowledge of the system:

- as extreme values which limit the possible excursion of parameters,
- as confidence intervals where the parameter is supposed to have a uniform probability,
- as an exact distribution law (often reduced to a normal distribution).

Another advantage of probability originates from the notion of entropy, of conditional entropy and of mutual information [52],[58],[64]. They can be efficiently used in order to assess the complementarity and redundancy between images at any level of representation. Let us just mention one example, where information extracted at several levels and different spatial scales are fused using mutual information between scales in order to provide an interpretation of aerial images [10]. Although similar entropy concepts have been defined for the types of information represented by fuzzy sets or belief functions (see e.g. [63],[25],[92]), they are still not used in image fusion.

Because of the very strict dependencies between the many propositions of a probabilistic modeling, too complex systems cannot be developed without care in this framework. Arbitrary probabilities, inconsistent propositions, and logical dead locks are the most probable pitfalls of very large probabilistic systems. To avoid these limitations Bayesian networks have been proposed [74] .

6.2 COMBINATION

In the Bayesian framework, the combination of image information as well as of prior probabilities is performed through the Bayes theorem. In image processing, due to the difficulty to learn joint distributions[5], independence between sources is often assumed. However, statistical tests show that this is seldom the case, and that better results can be obtained without independence assumption [29]. Then the combination step is included in the estimation step.

Bayesian methods are probably the most widely used in probabilistic image fusion. They lead mainly to conjunctive data fusion (under the independence hypothesis, the combination operator is a product). Other probabilistic techniques have been proposed, that are able to model different kinds of fusion logic (e.g. [77]).

6.3 DECISION

The most used decision rule in Bayesian decision is the maximum a posteriori. However, many other criteria have been developed by probabilists and statisticians, including maximum posterior marginal, maximum likelihood, maximum entropy, minimum expected risk, etc. Each of them is answering a specific problem which better fits the user's demand or the user's constraints. Quality of decision can be estimated using confidence levels, and rules leading to ambiguity and distance rejection have been proposed. Such rules are widely

[5] When combining n images with 256 grey levels, probability tables of dimension 256^n are needed.

used in image pattern recognition, either as tables or criteria to validate the decision. For instance using Markov random fields, probabilistic approaches may allow not only to decide on the local best solution, but also on the global one for the complete images to be processed.

7. Fuzzy approaches

7.1 MODELING AND INFORMATION REPRESENTATION

In the framework of fuzzy sets and possibility theory [104],[105],[38], the M_i^j 's represent membership degrees to a fuzzy set or possibility distributions. For instance for classification problems, one possible model consists in setting, for each element x, $M_i^j(x) = \mu_i^j(x)$, where $\mu_i^j(x)$ denotes the membership degree of x to the class i according to source j. A global evaluation of the membership of x to class i is then provided by the combination of $\mu_i^j(x)$, $j=1,..,l$. Another model consists in interpreting $M_i^j(x)$ as a possibility degree that x belongs to class i, and therefore setting $M_i^j(x) = \pi_x^j(i)$, where π_x^j is a possibility distribution according to source j [6]. Such models explicitly represent imprecision in the information, as well as possible ambiguity between classes or decisions (for instance, it is possible to evaluate the possibility and necessity of any subset A of D : $\Pi(A) = \sup_{d \in A} \pi(d)$.

Fuzzy sets have powerful features when used specifically for image information processing [56],[10], in particular for three points:

- the ability of fuzzy sets to represent spatial information in images along with its imprecision,
- the adequacy of operations recently generalized to fuzzy sets in order to manage spatial information,
- and the flexibility of information fusion using fuzzy combination operators. The two first points will be detailed in this Section, and the third point in the next one.

Some examples of applications using fuzzy sets have already shown excellent results, even if applications of fuzzy logic and fuzzy sets to image processing are quite recent, compared with other fields like control theory. Fuzzy set techniques give rise now to a large improvements in image processing. This may be explained by several factors, and in particular by the ability of fuzzy sets to represent and to manage imprecise spatial information, which is essential for image fusion and decision. Another factor is that this development can take advantage of the degree of maturity of both image processing and fuzzy logic fields.

Fuzzy sets have several advantages for representing imprecision inherent to image processing. First, they are able to represent several types of imprecision in images, as, for instance, imprecision in spatial location of objects, or imprecision in membership of an object to a class. In the case of partial volume effect, which occurs frequently in medical imaging, a consistent representation is given with fuzzy sets (membership degrees of a

[6] Approaches based on fuzzy sets consider classes as intrinsically fuzzy, while approaches based on possibility theory consider them as crisp.

voxel to tissues or classes directly represent partial membership to the different tissues mixed up in this voxel, leading to a modeling very close to reality). Second, image information can be represented at different levels with fuzzy sets (local, regional, or global), as well as under different forms (numerical, or symbolic). For instance, classification based only on grey levels involves very local information (at the pixel level); introducing spatial coherence in the classification [26] or relationships between features like parallelism [54] involves regional information; and introducing relationships between objects or regions for scene interpretation involves more global information [79]. Third, the fuzzy set framework allows for the representation of very heterogeneous information, and is able to deal with information extracted directly from the images, as well as with information derived from some external knowledge, like expert knowledge for instance. This is exploited in particular in model-based pattern recognition (see e.g. [79],[35]), where fuzzy information extracted from images is compared and matched to a model representing knowledge expressed in fuzzy terms.

The representation of image information using fuzzy sets can be seen from two different points of view: one leads to a symbolic approach and the other to a numerical one. The first one already gave rise to numerous developments, whereas the second one is worth to be amplified and extended to other problems than the traditional classification problem. In the first approach, knowledge, reasoning, and rules are represented using concepts of fuzzy logic, but the basic spatial information in image is represented in a traditional way [35]. This approach is used for reasoning about image information by means of fuzzy rules and fuzzy external knowledge, for instance for scene interpretation tasks by fusing several sources of information. However, the specificity of image information, i.e. spatial information, is not explicitly taken into account using this approach. In the second approach, fuzzy sets are used to represent directly spatial structures in images (regions, classes, contours). The modeling consists in assigning, to each image point, a membership degree to a class of interest. This approach was adopted mostly for fuzzy classification or clustering for pattern recognition ([9]and the widely used fuzzy C-means algorithm,[75]), and for measures on fuzzy sets (set operations, topological or geometrical characteristics, issued from the works of Zadeh [104] and Rosenfeld [83]). Both approaches have advantages the combination of which can be very fruitful: the first, symbolic, approach can take advantage of the granularity concept that allows to represent symbolic information by linguistic variables for instance taking only a few values. For example, considering distance between objects, it can be evaluated using a few fuzzy sets representing "far of", "near to" (e.g. [35]), the same for color description [8]. The second, numerical, approach uses fuzzy sets through their spatial interpretation, and is therefore very close to the basic information in image processing. Again considering distances with this approach, the problem amounts to defining a distance between two fuzzy sets, where each fuzzy set defined on the image represents an object or a region. Distance is an example of useful relationships in image processing which gains from being expressed with fuzzy sets. Since the introduction of fuzzy sets, several measures or operators have been defined on fuzzy sets. They are able to take into account several aspects of the information, that are widely used in image processing, like set theoretical, geometrical, topological, morphological characteristics [83],[73],[57],[107],[19],[]. They have helped in several applications in particular for image segmentation (e.g. [73],[34]). We distinguish two kinds of spatial operations. Those well defined if the objects are crisp, like adjacency, inclusion or other set relationships may benefit from fuzzifying these concepts when dealing with imprecise spatial image information, since they are highly sensitive to errors or imprecision in segmentation. Other relationships are inherently vague concepts, like relative position. Fuzzy definitions of such relationships are more consistent than crisp ones [24],[15],[16].

7.2 COMBINATION

For the combination step in the fusion process, the advantages of fuzzy sets and possibilities rely in the variety of combination operators, which may deal with heterogeneous information [39],[102],[42]. We proposed a classification of these operators with respect to their behavior (in terms of conjunctive, disjunctive, compromise [39]), the possible control of this behavior, their properties and their decisiveness, which proved to be useful for several applications in image processing [12]. It is of particular interest to note that, unlike other data fusion theories (like Bayesian or Dempster-Shafer combination), fuzzy sets provide a great flexibility in the choice of the operator, that can be adapted to any situation at hand. Indeed, image fusion has often to deal with situations where an image is reliable only for some classes, or does not provide any information about some class, or is not able to discriminate between two classes while another does. In this context, some operators are particularly powerful, like operators that behave differently depending if the values to be combined are of the same order of magnitude or not, if they are small or high, and operators that depend on some global knowledge about source reliability about classes, or conflict between images (global or related to one particular class) [42]. Again, the combination process can be done at several levels of information representation, from pixel level to higher level. Whatever the level of representation, such a process corresponds to the numerical approach described above. A noticeable advantage of this approach is that it is able to combine heterogeneous information, like it is usually the case in multi-image fusion, and to avoid to define a more or less arbitrary and questionable metric between pieces of information issued from these images. Another scheme relies on the symbolic approach and consists in combining information described from images but not directly attached to pixels or regions. This is typically used for combining rules where the rule elements involve fuzzy attributes, fuzzy measures or fuzzy relationships. Some authors use a neural architecture for such schemes, that allows for learning and several combination operations (see e.g. [67]).

7.3 DECISION

Decision is usually based on the maximum of membership values after the combination step. Constraints can be added to this decision, typically for checking for the reliability of the decision (is the obtained value high enough?) or for the discrimination power of the fusion (is the difference between the two highest values high enough?). In image fusion, very few works addressed this decision problem, unlike in probabilistic approaches, and very few tools allow to take refined decisions.

7.4 APPLICATIONS OF FUZZY IMAGE FUSION

The concept of data fusion (combination of several images as well as combination of several measures extracted from an image) is, in our opinion, a central point in the application of fuzzy sets to image processing, and it is found in almost all applications. For preprocessing, it is typically possible to make use of fuzzy fusion operators or fuzzy rules for combining local or regional measures of homogeneity and measures of contrast in order to smooth homogeneous regions while preserving or even sharpening edges (see e.g. [84]). Fuzzy classification is up to now the widest developed application of fuzzy sets for image processing. Most applications make use of the variety of information that can be extracted from one or several images, and therefore, such applications are related to data fusion. At pixel or intermediate level, applications can be found for color image classification [49], for medical imaging [35],[26], satellite imaging [29],[67], etc. Low level information, represented as membership degrees to a class, can be combined with some more global information like uncertainty due to feature acquisition, and credibility of a set of features to provide reliable information concerning each class. This has been done using possibility integrals e.g. in [103],[95]. When several levels of representation have to be combined, the most widely developed approach consists in using fuzzy rules (see e.g. [47] for medical image classification, [53] for combining numerical information with model-based and knowledge-based information, [67] for multi-spectral satellite image classification). Model-based recognition is a more recent application of fuzzy sets to image processing, that is likely to lead to a large development in a near future. Indeed, it allows to combine low and intermediate level spatial information with a more structural information, for instance provided by spatial relationships like described in the previous section. The problem amounts to the (fuzzy) matching between image information and models. Structural descriptions like graphs and fuzzy graphs are often used for this task [36],[79],[76]. The matching requires the definition of a similarity measure combining the different features of interest using different levels of representation. Let us mention applications for recognition of anatomical structures in medical imaging [65],[45], recognition of facial expressions [78], character recognition [66], partial shape matching [72], space shuttle simulator panel [79], flexible interrogation of digital angiography database [51], all of them involving aggregation of fuzzy information.

8. Evidential approaches

Dempster-Shafer evidence theory (DS) has been already widely used in satellite image processing (see e.g. [60],[99],[32],[106],[68],[48],[97]). On the contrary, when considering medical applications, it seems that DS has been applied mainly for topics derived from artificial intelligence (DS actually issued from this domain, for reasoning under uncertainty), and thus addresses the medical diagnosis, where propositional representations of evidences and knowledge are set in the DS framework [46],[5]. Evidences issued from the image signal (of "iconic" type rather than propositional) are seldom considered. To our knowledge, only a few papers report on image fusion by DS in medical imaging [59],[30] (both dealing with brain MRI), [94] (applied for MRI left ventricle). However, DS offering a number of advantages for image fusion deserves to be applied, also in medical imaging [13].

8.1 MODELING AND INFORMATION REPRESENTATION

DS, as possibility theory, allows to represent both imprecision and uncertainty, using plausibility and belief functions derived from a mass function defined on 2^D rather than on D only [86],[88]. This is one of the main advantages of the DS approach. Indeed, it leads to a very flexible and rich modeling, able to fit a very large class of situations, occurring in particular in image fusion. A few examples of situations where DS theory may be successfully used are:

- in ideal cases where all information relevant to the problem is known, i.e. cases where Bayesian fusion applies;
- when a source provides information concerning only a few of several classes: for instance brain PET images under some conditions allow for the detection of the brain surface but not of the head surface;
- when a source differentiates two classes and another does not: DS allows to deal with hesitation or ambiguity between these two classes;
- in the case of partial volume effects (often at the border of classes): it can also be taken into account by assigning masses to the union of the two classes mixed in the considered area;
- in cases where global source reliability has to be taken into account: this may be done by weakening all masses and reinforcing $m(D)$, using a discounting process;
- in cases where knowledge of source reliability is available only for some classes: it can be taken into account by modifying accordingly the masses assigned to these classes and by introducing ignorance (for instance, PET images are not very reliable concerning anatomical information whereas MRI images are more reliable);
- in cases where a priori information has to be introduced: even if it is not represented in a probabilistic manner, it can be taken into account if it induces a way to assign masses, in particular to compound hypotheses; for instance if we know that a source is not able to distinguish between two classes, then it is not worth trying to estimate masses for these two classes separately, but the estimation has to be made on the union of the classes and mass is then assigned to the corresponding compound hypothesis.

In our opinion, one of the main differences between DS and fuzzy sets lies at the level of flexibility. The situations described above are easily taken into account in the DS framework at the modeling level, while in the fuzzy set framework this flexibility relies on the wide range of available combination operators.

8.2 ESTIMATION

The definition of mass functions remains a largely unsolved problem, which did not yet find a general answer. In image processing, they may be derived at three different levels. At the highest, most abstract level, information representation is used in a way similar to that in artificial intelligence and masses are assigned to propositions, often provided by experts [71],[5],[46]. Up to now, this kind of information is usually not derived from measures on the images. At an intermediate level, masses are computed from attributes, and may involve simple geometrical models [30],[99],[32],[2]. This is well adapted to model-based pattern recognition but it is difficult to use for image fusion classification of complex structures without a model. At the pixel level, mass assignment is inspired from statistical pattern recognition. The most widely used approach is as follows: masses on simple hypotheses are computed from probabilities or from the distance to a class center [60],[93],[94],[59],[62],[106],[3],[50]. Then a global ignorance $m(D)$ is introduced as a discounting factor, often as a constant on all pixels (e.g. [59],[106]). In most cases no other compound hypothesis is considered, and this drastically under-exploits the power of DS. One way to overcome this problem consists in using the decomposability property on the masses and the Barnett approach (only masses on singletons and on their complements are considered) [7],[59],[30],[46]. Although this approach may be interesting for pattern recognition purposes where each class has to be tested against all the others, it may however be questionable when a source provides information on a compound hypothesis which is neither a singleton nor the complement of a singleton. In [60], the use of mixed pixels in a probabilistic like approach is suggested. In [94], nested focal elements are used, by sorting probabilities. In our opinion, these approaches are quite promising but are also restrictive since they do not allow to take into account all occurring situations. The way we assigned masses in the example presented in [13] is based on a reasoning approach where knowledge about the information provided by each image is used to choose the focal elements. For instance, we just consider a disjunction of two classes for images that are not able to discriminate between these classes. Partial volume effect is modeled in a similar way. However, in case of large numbers of classes, this process would become too tedious, and unsupervised methods are needed, for instance like the one proposed in [] for SAR imaging or in [70] for fusion of several classifiers.

8.3 COMBINATION

In the DS framework, masses are combined by the orthogonal rule of Dempster [86]. For m_j being the mass function associated with source j (j=1...l), this rule is written, for all non-empty subset A of D :

$$m(A) = (m_1 \oplus m_2 \oplus \ldots \oplus m_l)(A) = \sum_{B_1 \cap \ldots \cap B_l = A} m_1(B_1) m_2(B_2) \ldots m_l(B_l). \qquad (1)$$

Similar equations can be derived for directly combining belief or plausibility functions.

In this unnormalized form [89], $m(\emptyset) = \Sigma_{B_1 \cap \ldots \cap B_l = \emptyset}\, m_1(B_1)\, m_2(B_2)\ldots m_l(B_l)$ can be interpreted as a measure of conflict between the sources. It can be directly taken into account in the combination as a normalization factor. It is very important to consider this value for evaluating the quality of the combination: when it is high (in case of strong conflict), the normalized combination may not make sense and may lead to questionable decisions (moreover, the normalized combination rule is not continuous if the conflict is very close to Equation 1 [4]). Several authors suggest not to normalize the combination result (see e.g. [89]), which corresponds to Equation 1.

This fusion operator has a conjunctive behavior[7]. We claim that this means that all imprecision on the data has to be introduced explicitly at the modeling level, in particular in the choice of the focal elements. For instance, ambiguity between two classes in one image has to be modeled using a disjunction of hypotheses, so that conflict with other images can be limited and ambiguity can be possibly solved during the combination.

8.4 DECISION

After the combination, the final decision is usually taken in favor of a simple hypothesis using one of several rules [37]: for instance, the maximum of plausibility (generally over simple hypotheses), the maximum of belief, the maximum of belief without overlapping of belief intervals, i.e. in favor of d *in* D such that $Bel(d) \geq \max_{d' \in D, d' \neq d} Pls(d')$ (a very strict condition), the pignistic decision rule [90], or rules using expected utility [93].

In [13], the decisions have been taken according to two rules: a classical one, where a decision is always taken in favor of a simple hypothesis, and a second, original one, where it can also be decided in favor of a compound hypothesis. This second rule has several advantages: it is robust with respect to evidence weighting, as seen when testing different weights on mass functions and on partial volume, and it fits reality by highlighting for instance regions with partial volume effect. Moreover it is adapted to the expert's way of reasoning.

9. Conclusion

This papers briefly summarized several aspects that have been taken into account in fusion problems. We described more specifically aspects related to image fusion. We have seen the large variety of numerical techniques used in fusion under imprecision and uncertainty. This proliferation is due to the many different tasks which contribute to the decision from the management of different sources of information. Probabilistic approaches still remain the most widely used methods, in particular in image fusion, because of the exceptional achievement of the tools developed under its umbrella. They benefit from a very long experience and may drive the user from the very elementary modeling stage up to decision. At this level (decision making) probability theory reveals as the most efficient tool, the modeling one being on the contrary rather questionable. Fuzzy set theory proposes a quite intuitive modeling well adapted to ill-posed problems. Its decision stage, as far as image processing applications are concerned, is unfortunately still rather coarse and most of the user's knowledge has to be put at the fusion stage, to model with adapted operators its

[7] Other combination rules have been proposed in the DS framework, in particular disjunctive rules [CITE:SMET-90].

knowledge on the problem. DS evidence theory has the most powerful modeling tools with capacities to handle uncertainty and imprecision as well. It is rather rigid in its way to combine information (when restricted to Dempster rule) and allows up to now only few different strategies at the decision level.

The different numerical approaches described above are all dedicated to specific types of imperfection of data and to specific combination behaviors. None of them covers all types of questions raised from any information fusion. Therefore an open field for research concerns the fusion of methods, i.e. the joint use of several formalisms in order to provide better performing algorithms. A promising direction consists in combining imprecision represented as fuzzy sets with uncertainty represented as probabilities. For instance, in [27] and [85], fuzzy sets are used together with a probabilistic Markovian approach, allowing to take into account fuzzy classes and mixed pixels as well as contextual spatial information. Another example concerns the use of fuzzy morphology [19] in the framework of Dempster-Shafer theory for taking into account imprecision in the representation of information and for defining mass functions on compound hypotheses[14]. Fuzzy mathematical morphology is also involved in the definition of several spatial relationships, both at low level and at structural level (e.g. [24],[15],[16]). Such relationships are more and more involved in image fusion problems, particularly at higher level, in order to guaranty a global coherence to the interpretation of the scene. These remarks can be extended to non-numerical formalisms. An active field of research concerns for instance the combination of numerical and symbolic methods (e.g. logics or rule-based approaches and numerical representations of uncertainty).

As stated before, spatial information is an essential aspect of image information, which suffers from several types of imprecision. One of the main stakes is to guarantee that the resulting decision is spatially consistent. There are several ways to introduce spatial information in a fusion process depending on the level of representation. At pixel level, spatial context consists of a few pixels in the neighborhood of the considered point. At feature level (e.g. segments, contours, regions), spatial context involves relationships between features, and contains more structural information, like parallelism, similarities in directions, continuity, etc. (see e.g. [98]). At object level, spatial context involves topological and geometrical relationships between objects. Spatial context can also be introduced at all steps of the fusion process. At the first modeling and estimation step, spatial context can help in validating the measures describing the belonging of an element to a class. This has been done especially at pixel level, for instance using statistical approaches, using measures performed not only at a point, but also in a neighborhood of this point. At the next combination step, an interpretation of spatial context as additional source of information [22] can be particularly useful, since it can be combined with image information using the fusion operator. However, this aspect has not been widely developed until now, except in probabilistic methods. For the other methods, it is still an open field for research. At the final decision step, the most common way to introduce spatial context relies on some rules. Finally, spatial information can be introduced in different mathematical frameworks, and therefore under different forms. Fuzzy mathematical morphology [19] appears as an interesting tool to introduce spatial information at the estimation level [14], or at the combination step through the representation of structural information expressed as spatial relationships [17].

Several questions have to be answered when designing a fusion method. The first one concerns the **data**: the data and knowledge have to be represented in a way they include the specificities of the source they are originating from, the quality of these sources, the types of imperfection attached to the information. Then **relationships between sources** have to be taken into account. Important questions related to these relationships concern independence between sources and conditioning (especially in sequential fusion for instance for updating problems, where the process is dynamic and no longer static): how to assess or verify these properties, how to deal with them? For instance, the probabilistic framework provides methods for testing statistical independence. This is often considered as to constraining and other frameworks like Dempster-Shafer theory prefer to deal with cognitive dependence, which is related to the way knowledge and data are acquired rather than to their nature (typically two different points of view on the same phenomenon can be considered as independent in this sense although they may be statistically dependent) [86],[88]. A very difficult question, that is still not completely solved, concerns **conflict**. When combining different sources of information, conflict may occur due to several reasons. These reasons have to be made as explicit as possible in order to avoid inconsistencies in the final decision process. We distinguish here two types of conflict. The first type concern only "apparent" conflict and is in fact rather an aspect of complementarity: if a source systematically includes class B in class A, while another one provides a good discrimination between these classes, then apparently these two sources are conflicting. If we have some prior or learned knowledge about the fact that the first source has no discrimination power for these classes, then this conflict reduces simply to complementarity and is solved by priviledging information issued from the second source for the concerned classes. A second type of conflict comes from source inconsistencies that may originate e.g. from temporal changes between data acquisition using each source, or to "mistakes" made by one source (or by both of them). This type of conflict is much more difficult to solve. It is even not clear if it has to be solved at all. Conflict resolution methods involve choice, compromise, decision delaying, hypothesis generation and testing, etc. Another question concerns **algorithms**. This includes the choice of an adequate processing for each source of information, and, more important, the choice of a combination method adapted to the application at hand. Again, no general solution exists until now. We may even find in the literature contradicting conclusions concerning the comparison of different methods applied to similar applications. Two main reasons can be advocated: first, the way different formalisms are used strongly influences the results. For instance a specialist of probabilistic approaches will probably always have better results with the method he masters the best. The second reason is that the evaluation of performances and accuracy of methods is a very difficult task, in particular in signal and image processing, where the truth is not necessarily available.

10. References

1. White, F.F., (1991) Data Fusion Lexicon, *Data Fusion Subpanel of the Joint Directors of Laboratories Technical Panel for C^3.*, Code 4202, NOSC, San Diego, CA.
2. Andress, K.M., and Kak, A.C., (1988) Evidence Accumulation and Flow Control in a Hierarchical Spatial Reasoning System. *AI Magazine,* pages 75 94.
3. Appriou, A., (1993) Formulation et traitement de l'incertain en analyse multi-senseurs. In *Quatorzième Colloque* GRETSI, pages 951-954, Juan les Pins.

4. Apprion, A., Ayoun, A., Benferhat, S., Besnard, P., Bloch, I., Cholvy, L., Cooke, R., Cuppens, F., Dubois, D., Fargier, H., Grabisch, M., Hunter, A., Krase, R., Lang, J., Moral, S., Prade, H., Saffiotti, A., Smets, P., and Sossai, C., (2000) Fusion: GeneralConcepts and Characteristics (to appear). *International Journal of Intelligent Systems.*
5. Baidwin, J.F., (1992) Inference for Information Systems Containing Probabilistic and Fuzzy Uncertainties. in L. Zadeh and J. Kacprzyk, editors, *Fuzzy Logic and the Management of Uncertainty*, pages 353 375. J. Wiley, New York.
6. Bar-Shalom, Y., and Fortmann, T.B., (1988) Tracking and Data Association.
7. Barnett, J.A. (1981), Computational Methods for a Mathematical Theory of Evidence. In *Proc. of 7th IJCAI,* pages 868 875, Vancouver.
8. Benoit, E., and Foulloy, L., (1993), Capteurs flous multi-composantes: applications à la reconnaissance des couleurs. In *Les Applications des Ensemo'ies Flous,* pages 167 176, Nimes, France.
9. Bezdek, J.C., (1981), Pattern Recognition with Fuzzy Objective Function Algorithms. Plenum, New York.
10. Bloch, I., (1996), Image Information Processing using Fuzzy Sets. In World Automation Congress, Soft Computing with Industrial Applications, pages 79 84, Montpellier, France.
11. Bloch, I., (1996), Incertitude imprécision et additivite en fusion de données: point de vue historique. *Traitement du Signal,* 13(4):267-288.
12. Bloch, I., (1996), Information Combination Operators for Data Fusion: A Comparative Review with Classification. *IEEE Transactions on Systems, Man, and Cybernetics,* 26(1): 52-67..
13. Bloch, I., (1996), Some Aspects of Dempster-Shafer Evidence Theory for Classification of Multi-Modality Medical Images Taking Partial Volume Effect into Account. *Pattern Recognition Letters,* 17(8):905-919.
14. Bloch, I., (1997), Using Fuzzy Mathematical Morphology in the Dempster-Shafer Framework for Image Fusion under Imprecision. In *IFSA '97*; pages 209-214, Prague.
15. Bloch, I., (1999), Fuzzy Relative Position between Objects in Image Processing: a Morphological Approach. *IEEE Transactions on Pattern Analysis and Machine Intelligence,* 21(7):657-664.
16. Bloch, I., (1999), On Fuzzy Distances and their Use in Image Processing under Imprecision. *Pattern Recognition,* 32(11):1873-1895.
17. Bloch, I., (2000), Fusion of Numerical and Structural Image Information in Medical Imaging in the Framework of Fuzzy Sets. In P. Szczepaniak et al., editor, *Fuzzy Systems in Medicine,* Series Studies in Fuzziness and Soft Computing, pages 429-447.Springer Verlag.
18. Bloch, I., and Maître, H., (1994), Fusion de données en traitement d'images: modèles d'information et décisions. *Traitement du Signal,* 11(6):435-446.
19. Bloch, I., and Maître, H., (1995), Fuzzy Mathematical Morphologies: A Comparative Study. *Pattern Recognition,* 28(9):1341-1387
20. Bloch, I., and Maitre, H., (1997), Data Fusion for Decision Making and Diagnostics in Signal and Image Processing (invited conference). In *IEEE International Symposium on Diagnosis for Electrical Machines,* pages 148-155, Carry-Le-Rouet, France.
21. Bloch, I., and Maître, H., (1997), Data Fusion in 2D and 3D Image Processing: An Overview (invited conference). In X *Brazilian Symposium on Computer Graphics and Image Processing, SIBGRAPI,* pages 127-134, Campos do Jordao, Brazil.
22. Bloch, I., and Maître, H., (1997), Fusion of Image information under Imprecision. In B. Bouchon-Meunier, editor, *Aggregation and Fusion of Imperfect Information,* Series Studies in Fuzziness, pages 189-213. Physica Verlag, Springer.
23. Bloch, I., and Maître, H., (1998), On Some Features of Fuzzy Set Theory for Data Fusion (invited conference). In *9th Symposium on Information Control in Manufacturing INCOM98,* Nancy-Metz, France, June.
24. Bloch, I., Maître, H., and Anvari, M., (1997), Fuzzy Adjacency between Image Objects. *International Journal of Uncertainty, Fuzziness and Knowledge-Based Systems.* 5(6):615-653.
25. Bouchon-Meunier, B., and Yager, R.R., (1993), Entropy of Similarity Relations in Questionnaires and Decision Trees. In *Second IEEE Int. Conf on Fuzzy Systems,* pages 1225-1230, San Francisco, California
26. Boujemaa, N., Stamon, G., Lemoine, J., and Petit, E., (1992), Fuzzy Ventricular Endocardium Detection with Gradual Focusing Decision. In *14th IEEE EMBS Conference,* pages 1893-1894, Paris, France.
27. Caillol, H., Hillion, A., and Pieczynski, N., (1993), Fuzzy Random Fields and Unsupervised Image Segmentation. *IEEE Trans. on Geoscience and Remote Sensing,* 31(4):801-810.

28. Charroux, B., (1996), Image Analysis; Interpretation- Guided Cooperation between Segmentation Operators (in French). PhD thesis, University Paris XI.
29. Chauvin, S., (1995), Evaluation of Decision Theories Applied to Data Fusion in Satellite Imaging (in French). PhD thesis, ENST and Nantes University.
30. Chen, S.Y., Lin, W.C., and Chen, C.T., (1993), Evidential Reasoning based on Dempster-Shafer Theory and Its Application to Medical Image Analysis. In *SPIE*, volume 2032, pages 35-46.
31. Chu, C.C., and Aggarwal, J.K., (1993), The Integration of Image Segmentation Maps using Region and Edge Information. *IEEE Trans. on Pattern Analysis and Machine Intelligence*, 15(12):1241-1252.
32. Cucka, P., and Rosenfeld, A., (1992), Evidence-based Pattern Matching Relaxation. Technical Report CAR-TR-623, Center of Automation Research, University of Maryland.
33. Dasarathy, B.V., (1996), Fusion Strategies for Enhancing Decision Reliability in Multi-Sensor Environments. *Optical Engineering*, 35(3):603-616.
34. Dellepiane, S., Fontana, F., and Vernazza, G., (1994), A Robust Non-Iterative Method for Image Labeling using Context. In *IEEE Int. Conf on lmage Processing*, volume II, pages 207-211, Austin, Texas.
35. Dellepiane, S., Venturi, G., and Vernazza, G., (1992), Model Generation and Model Matching of Real Images by a Fuzzy Approach. *Pattern Recognition*, 25(2):115-137.
36. Demko, C., Loonis, P., and Zahzah, E.H., (1995), Isomorphism of Fuzzy Structures: a New Method for Image Classification. In *9th Scandinavian Conference on Image Analysis*, pages 297-304, Uppsala, Sweden.
37. Denoeux, T., (1995), A k-nearest Neighbor Classification Rule based on Dempster-Shafer Theory. *IEEE Transactions on Systems, Man and Cybernetics*, 25(5):804-813.
38. Dubois, D., and Prade, H., (1980), *Fuzzy Sets and Systems: Theory and Applications*. Academic: Press, New-York.
39. Dubois, D., and Prade, H., (1988), A Review of Fuzzy Set Aggregation Connectives. *Information Sciences*, 36:85-121.1985.
40. Dubois, D., and Prade, H., *Possibility Theory*. Plenum Press, New-York.
41. Dubois, D., and Prade, H., (1988), Representation and Combination of Uncertainty with Belief Functions and Possibility Measures. *Compu. Intell.*, 4:244-264.
42. Dubois, D., and Prade, H., (1997), Combination of Information in the Framework of Possibility Theory. In M. Al Abid et al., editor, *Data Fusion in Robotics and Machine Intelligence*. Academic Press, 1992.
43. Varshney, P.K., (Ed.). Special Issue on Date Fusion. *Proceedings of the IEEE*, 85(1).
44. Gee, S.S., and Newman, A.M., (1993), RADIUS: Automating Image Analysis Through Model-Supported Exploitation. In *Image Understanding Workshop*, pages 185-196, Washington D.C.
45. Géraud, T., Bloch, I., and Maître, H., (1999), Atlas-guided Recognition of Cerebral Structures in MR1 using Fusion of Fuzzy Structural Information. In *CIMAF'99 Symposium on Artificial Intelligence*, pages 99-106, La Havana, Cuba.
46. Gordon, S., and Shortliffe, E.H., (1985), A Method for Managing Evidential Reasoning in a Hierarchical Hypothesis Space. *Artificial Intelligence*, 26:323-357.
47. Hall, L.O., Machrzak, T.L., and Silbiger, M.S., (1994), Obtaining Fuzzy Classification Rules in Segmentation. In *IPM U*, pages 619-624, Paris, France.
48. Le Hégarat-Mascle, S., Bloch, I., and Vidal-Madjar, D., (1998), Introduction of neighborhood Information in Evidence Theory and Application to Data Fusion of Radar and Optical Images with Partial Cloud Cover. *Pattern Recognition*, 31(11):1811-1823.

49. Huntsberger, T.L., Rangarajan, C., and Jayararamamurthy, S., (1986), Representation of Uncertainty in Computer Vision using Fuzzy Sets. *lEEE Trans. on Computers,* C-35(2):145-156.
50. Ip, H.H.S., and Ng, J.M.C., (1994), Human Face Recognition using Dempster-Shafer Theory. In *ICIP,* volume II, pages 292-295, Austin, Texas.
51. Jaulent, M.C., and Yang, A., (1994), Application of Fuzzy Pattern Matching to the Flexible Interrogation of a Digital Angiographics Database. In *IPMU,* pages 904-909, Paris, France.
52. Jaynes, E.T., (1957), Information Theory and Statistical Mechanics. *Physical Review,* 106(4):620-630.
53. Kandel, A., Schneider, M., and Langholz, O., (1994), Autonomous Fuzzy Intelligent Systems for Image Processing. In *IPMU,* pages 613-618, Paris, France.
54. Kang, H.B., and Walker, E.L., (1994), Characterizing and Controlling Approximation in Hierarchical Perceptual Grouping. *Fuzzy Sets and Systems,* 65:187-223.
55. Klir, G.J., and Folger, T.A., (1998), *Fuzzy Sets, Uncertainty, and Information.* Englewood Cliffs.
56. Krishnapuram, R., and Keller, J.M., (1992), Fuzzy Set Theoretic Approach to Computer Vision: an Overview. In *IEEE Int. Conf. on Fnzzp Syst ems)* pages 135 142, San Diego, CA).
57. Krishnapuram, R., Keller, J.M., and Ma, Y., (1993), Quantitative Analys s of Properties and Spatial Relations of Fuzzy Image Regions. *lEEE Transactions on Fuzzy Systems)* 1(3):222-233.
58. Kuliback, S., (1959) *Information Theory and Statistics,* Wiley, New York.
59. Lee, R.H., and Leahy, R., (1990) Multi-Spectral Classiflcat on of MR Images Using Sensor Fusion Approaches. In *SPIE Medical Imaging IV: Image Processing)* volume 1233, pages 149-157.
60. Lee, T., Richards, J.A., and Swain, P.H., (1987), Probabilistic and Evidential AQ preaches for Multisource Data Analysis. *IEEE Transactions on Geoscience and Remote Sensing,* GE-25(3):283-293.
61. Leung, H., (1996), Neural Networks Data Association with Application to Multiple-Target Tracking. *Optical Engineering,* .35(3):693-700.
62. Lowrance, J.D., Strat, T.M., Wesley, L.P., Carvey, T.D., Ruspini, E.H., and Wilkins, D.B., (1991), The Theory, Implementation and Practice of Evidential Reasoning. SRI project 5701 final report, SRI, Palo Alto.
63. De Luca, A., and Termini, S., (1972), A Definition of Non-Probabilist c Entropy in the Setting of Fuzzy Set Theory. *Information and Control,* 20:301 312.
64. Maître, H., (1996), Entropy, Information and Image. In H. Maitre and J. Zinn-Justin, editors, *Progress in P cture Processing, Les Houches Session LVIII,* pages 881-1115. Springer Veriag.
65. Maître, H., Bloch, I., Moissinac, H., and Coninaud, C., (1995), Cooperative Use of Aerial Images and Maps for the Interpretation of Urban Scenes (invited conference). In *Ascona Workshop 'Automatic Estraction* Qf Man-Made *Objects from Aerial and Space Images",* pages 297 306, Ascona, Suisse.
66. Man, G.M.T., and Peon, J.C.H., (1992), A new Similarity Measurement Method for Fuzzy-Attribute Graph Matching and its Application to Handwr tten Character Recognition. In *Int. Carnahan Couf. on Security Technology,* pages 46 49, Lexington, KY.
67. Mascarlila, L., (1994), Rule Extraction based on Neural Networks for Satellite Image interpretation. In *SPIE Image and Signal Processing for Remote Sensing,* volume 2315. pages 657 668, Rome, Italy.
68. Mascle, S., Bloch, I., and Vidal-Madjar, D., (1997), Application of Dempster-Shafer Evidence Theory to Unsupervised Classification in Multisource Remote Sensing. *lEEB Transactions on Geoscience and Remote Sensing,* 35(4):1018-1031.

69. McKeown, D., Harvey, W.A., and McDermott, J., (1985), Rule-Based Interpretation of Aerial imagery. *IEEE Trans, on Pattern Analysis and Machine Intelligence,* 7(5):57W585.
70. Ménard, M., Zahzah, B.H., and Shahin, A., (1996), Mass Function Assessment: Case of Multiple Hypotheses for the Evidential Approach. In *Europto Conf on Image and Signal Processing for Remote Sensing,* Taormina, Italy.
71. Neapolitan, R.B., (1992) A Survey of Uncertain and Approximate inference. In L. Zadeli and J. Kaprzyk, editors, *Fuzzy Logic for the Management of Uncertainty,* pages 55-82. J. Wiley, New York.
72. Ogawa, H., (1994), A Fuzzy Relaxation Technique for Partial Shape Matching. *Pattern Recognition Letters,* 15:349-355.
73. Pal, S.K., (1992), Fuzzy Set Theoretic Measures for Automatic Feature Evaluation. *Information Science,* 64:165-179.
74. Pearl, J., (1986), Fusion, Propagation, and Structuring in Belief Networks. *Artihcial Intelligence,* 29:241-288.
75. Pedrycz, W., (1990), Fuzzy Sets in Pattern Recognition: Methodology and Methods. *Pattern Recognition,* 23(1/2): 121-146.
76. Perchant, A., Boeres, C., Bloch, I., Roux, M., and Ribeiro, C., (1999), Model-based Scene RecognItion Using Graph Fuzzy Homomorphism Solved by Genetic Algorithm. In *GbR '99 2nd International Workshop on Graph-Based Representations in Pattern Recognition,* pages 61-70, Castle of Haindort Austria.
77. Piat, E., (1996), Fusion de croyances dans le cadre combiné de la logique des propositions et de la théorie des probabilités, application à la reconstruction de scène en robotique mobile. PhD thesis, Universite de Technologie de Compiègne.
78. Ralescu, A., and Hartani, R., (1994), Modeling the Perception of Facial Expressions from Face Photographs. In 10th Fuzzy System Symposium, pages 405-408, Ozaka, Japan.
79. Ranganath, H.S., and Chipman, L.C., (1992), Fuzzy Relaxation Approach for Inexact Scene Matching. *Image and Vision Computing,* 10(9):631-640.
80. Rao, N.S.V., and Iyengar, S.S., (1996), Distributed Decision Fusion under Unknown Distributions, *Optical Engineering,* 35(3) :617-624.
81. Romine, J.B., and Kamen, E.W., (1996), Modeling and Fusion of Radar and imaging Sensor Data for Target Tracking. *Optical Engineering,* 35(3):659-673.
82. Rosenfeld, A., (1979), Fuzzy Digital Topology. *Information and Control,* 40:76-87.
83. Rosenfeld, A., (1984), The Fuzzy Geometry of Image Subsets. *Pattern Recognition Letters,* 2:311-317.
84. Russo, F., and Ramponi, C., (1995), An Image Enhancement Technique based on the FIRE Operator. In *IEEE mt. Couf. on Image Processing,* volume I, pages 155-158, Washington DC.
85. Salzenstein, F., and Pieczynski, W., (1995), UnsupervIsed Bayesian Segmentation using HIdden Fuzzy Markor Fields. In *IEEE mt. Conf. on Acoustics, Speech and Signal Procesing,* Detroit, Michigan.
86. Shafer, C., (1996), *A Mathematical Theory of Evidence.* Princeton University Press.
87. Simon, J.C., (1989), *from Pixels to Features.* V-X, North Holland, Amsterdam.
88. Smets, P., (1990), The Combination of Evidence in the Transferable Belief Model. *IEEE Transactions on Pattern Analysis and Machine intelligence,* 12(5):447-458.
89. Smets, P., (1993), Belief Functions: The DisjunctIve Rule of Combination and the Generalized Bayesian Theorem. *International Journal of Approximate Reasoning,* 9:1-35.
90. Smets, P., (1995), The Transferable Belief Model for Uncertainty Representation. Technical Report TR/IRIDIA/95-23, IRIDIA, UniversitA Libre de Bruxelles, Bruxelles, Belgium.
91. Léa Sombé (1989), *Raisonnements sur des inform ations incompletes en intelligence artificielle.* Teknea., Marseille.

92. Stephanou, H.B., and Lu, S.Y., (1984), Measuring Consensus Effectiveness by a Generalized Entropy Criterion, In *First Conference on Artificial Intelligence AppUcations,* pages 518-523, Denver.
93. Strat, T.M., (1989), Decision Analys s using Belief Functions. Technical Note 472, SRI.
94. Suh, D.Y., R.M., Merserean, R.L. Eisner, and R.I. Pettigrew (1990), Automatic Boundary Detect on on Card ac Magnetic Resonance Image Sequences for Four Dimensional Visualization of the Left V'entricle. In *First Conference on Visualization in Biomedical Computing,* pages 149-156, Atlanta GE.
95. Tahani, H., and Keller, J.M., (1990), Information Fusion in Computer VisIon Using the Fuzzy Integral. *IEEE Transactions on System, Man and Cybernetics,* 20(3 '733 741.
96. Thomopoulos, S.C.A., (1990), Sensor Integration and Data Fusion. *Journal of Robotics Systems,* 7(3):337-372.
97. Tupin, F., Bloch, I., and Maitre, H., (1999), A First Step Towards Automatic Interpretation of SAR Images using Evidential Fusion of Several Structure Detectors. *IEEE Transactions on Geoscience and Remote Sensing,* 37(3):1327-1343.
98. Tupin, F., Maitre, H., Mangin, J-F., Nicolas, J-M., and Pechersky, E., (1996), LInear Feature Detection on SAR Images: Application to the Road Network. Technical report, Ecole Nationale Supérieure des Télécommunications (96D006).
99. J. van Cleynenbreugel, Osinga, S.A., Fierens, F., Suetens, P., and Oosterlinck, A., (1991), Road Extract on *from* Multi-temporal Satellite Images by an Evidential Reasoning Approach. *Pattern Recognition Letters,* 12:371-380.
100. Wald, L., (1999), Some Terms of Reference in Data Fusion. *IEEE Transactions on Geoscience and Remote Sensing,* 37(3):1190-1193.
101. Winter, A., Maître, H., Cambou, N., and Legrand, E., (1996), Object Detection Using a Muitiscale Probability Mo del. In *IEEE Int. Conf. on Image Processing ICL 'gc,* volume I, pages 269-272, Lausanne.
102. Yager, R.R., (1991), Connectives and Quantifiers in Fuzzy Sets. *Fuzzy Sets and Systems,* 40:39-75.
103. Yan, B., (1993), Semiconormed Possibility Integrals and Multi-Feature Pattern Classification, *Pattern Recognition,* 26(12):1855-1862.
104. Zadeh, L.A., (1965), Fuzzy Sets. *information and Contro4* 8:338-353.
105. Zadeh, L.A., (1978), Fuzzy Sets as a Basis for a Theory of Possibility. Fuzzy Sets and Systems, 1:3-28.
106. Zalizab, E., (1992), Contribution à la représentation des connaissances et à leur utilisation pour l'interprétation automatique des images satellites. Thèse de doctorat, Université Paul Sabatier, Toulouse.
107. Zwick, R., Carktein, E., and Budescu, D.V., (1987), Measures of Simflarity Amonr Fuzzy Concepts: A Comparative Analysis. *International Journal of Approximate Reasoning)* 1:221-242.

MULTISENSOR FUSION UNDER UNKNOWN DISTRIBUTIONS FINITE-SAMPLE PERFORMANCE GUARANTEES

N.S.V. RAO
Center for Engineering Science
Advanced Research Computer Science and Mathematics Division
Oak Ridge National Laboratory
Oak Ridge, TN 37831-6355
Raons@ornl.gov

1. Introduction

Fusion or combination of information from multiple sources to achieve performances exceeding those of individual sources has been recognized for centuries in diverse areas such as political economy models [22], and composite methods [17]; a brief overview of these works can be found in [31]. In the twentieth century, such fusion methods continued to be applied in a wide spectrum of areas such as reliability [76], forecasting [29], pattern recognition [12], neural networks [24], decision fusion [74], and statistical estimation [11,16].

In engineering systems, the fusion methods have been proven to be particularly important since they can provide multiple sensor capabilities, which are significantly beyond those of single sensor systems. In particular, many researchers realized the fundamental limitations of single sensor systems in a number of application areas such as robotics [10,1], and tracking [7], thereby motivating the deployment of multiple sensors. Information fusion is very important in such multiple sensor systems, and is often referred to as *multisensor fusion* or simply *sensor fusion*. The overall objective is the same across the disciplines, namely to "fuse" the information from many different sensors to overcome the limitations of a single sensor. Although one would expect a fuser to perform better than any of the sensors, it is not as clear as to how to design such fuser. Thus, systematic approaches to fuser design are very critical to the overall performance, for an inappropriate fuser can render the system worse than the worst individual sensor.

In engineering systems, the sensor fusion problems present technical challenges in ways unprecedented in other disciplines. Early information fusion methods require either independence of sensor errors or closed-form analytical expressions for sensor distributions. In the first case, a general majority rule suffices, while in the second a fusion rule can be computed using the Bayesian methods. Several popular distributed decision fusion methods belong to the latter class [16,74]. In engineering systems, however, independence can seldom be assured and, in fact, may not be satisfied. Bayesian methods are not particularly conducive either from an information or from a cost perspective. Typically, the fusion rules are "selected" from a specific function class chosen by the designer to ensure the convergence of the fuser computation. On the other hand, the sensor distributions are not within the hands of the user, and can be arbitrarily complicated in complex engineering

A. K. Hyder et al. (eds.), Multisensor Fusion, 295–329.

systems. As a result, the problem of obtaining the sensor distributions required by the Bayesian methods is more difficult, in an information-theoretic sense, than the fusion problem itself [72]. In addition, deriving closed form expressions for sensor distributions is a very difficult and expensive task in these systems since it requires the knowledge of a variety of areas such as device physics, electrical engineering, and statistical modeling.

The fusion problems arising in operational engineering and robotic systems, however, have a positive side: it is easy to collect "data" by sensing objects and environments with known parameters. Thus, practical solutions to these sensor fusion problems must exploit the empirical data available from the observation and/or experimentation, which is the main topic of this paper. We present a summary of methods for fusion rule estimation from empirical data, which has became possible largely due to the developments in empirical process theory and computational learning theory. Our main focus is on methods that provide *performance guarantees based on finite samples* from a statistical perspective. In particular, we do not cover adhoc fusion rules with no performance bounds or results based on asymptotic guarantees that are valid *only* as the sample size approaches infinity. Our approach is based on a statistical formulation of the problem for the most part with the exception of physical systems in Section 7.1. In this respect, our solutions do not fully capture the non-statistical aspects such as calibration and registration, which can be incorporated into a suitable cost function. But, our results provide an analytical justification of sample-based approaches to the sensor fusion problem, and establish the basic tractability of the solutions. We believe that our performance bounds can be improved in specific cases by suitably incorporating the application specific details.

The organization of this paper is as follows. We present the problem formulation in Section 2. In Section 3, we present several solutions based on the empirical risk minimization methods. In Section 4, we present solutions based on non-linear statistical estimators. We describe applications of these methods in Section 5. In Section 6, we address the issues of relative performance of the composite system and the individual sensors. We describe the metafusers in Section 7. Then, in Section 8, we discuss a special class of the fusion problem which can be efficiently solved by utilizing the physical laws. In Section 9, we present an approach to incorporating known conditional distributions into the fuser computation. Our presentation is tutorial in nature in that we describe only the main results, and the specific details can be found in the references.

2. Problem Formulation

In a *generic sensor system* of N sensors, the sensor S_i, $i=1, 2,\ldots, N$, outputs $Y^{(i)} \in \Re^d$ corresponding to input $X \in \Re^d$ according to the distribution $P\ Y^{(i)} \mid X$. The input X is the quantity that needs to be "estimated" or "measured" by the sensors, such as the presence of a target or a value of the feature vector. The *expected error* of the sensor S_i is given by

$$I(S_i) = \int C(X, Y^{(i)})\ dP_{Y^{(i)},X},$$

where $C: \Re^d x\ \Re^d \rightarrow \Re^d$ is a cost function. $I(S_i)$ is a measure of how good the sensor S_i is in "sensing" the input feature X. For example, if S_i is a target detector or classifier, a choice could be X *in* {0, 1} and $Y^{(i)} \in (0, 1\}$, where X=1 (0) corresponds to the presence (absence) of a target. Then

$$I(S_i) = \int \left[X \oplus Y^{(i)}\right] dP_{Y^{(i)},X}$$

corresponds to the probability of misclassification (false alarm or missed detection) of S_i, where $\oplus$ is the exclusive-OR operation.

There are two types of errors that a sensor can make. The *measurement error* corresponds to the randomness involved in measuring a particular value of the feature X, which is distributed according to $P\, Y^{(i)} | X$. The *systematic error* at X corresponds to $E\,[C\,(X,Y^{(i)})\,|X]$ which must be 0 in the case of a perfect sensor.

Example 2.1: Consider a sensor system consisting of two sensors. For the first sensor, we have $Y^{(1)} = a_1 X + Z$, where Z is normally distributed with zero mean, and is independent of X. That is, first sensor has a scaling error and a random additive error. For the second sensor, we have $Y^{(2)} = a_2 X + b_2$, which has a scaling and bias error. Let X be uniformly distributed over [0,1], and $C[X, Y]=(X - Y)^2$. Then, we have $I(S_1) = (1 - a_1)^2$ and $I(S_2) = (1 - a_2 - b_2)^2$, which are non zero in general.

We consider a fuser $f: \Re^d \to \Re^d$ that combines the outputs of sensors $Y = (Y^{(1)}, Y^{(2)}, Y^{(N)})$ fused output $f(Y)$. The *expected error* of the fuser f is given by

$$I_F(f) = \int C(X, f(Y))\, dP_{Y,X}$$

where $Y = (Y^{(1)}, Y^{(2)}, \ldots, Y^{(N)})$. The objective of fusion is to ensure that $I_F(f)$ is as small as possible. Note that a fuser must account for both the systematic and measurement errors of the sensors in order to achieve a low expected error. The fuser can be chosen from a family of fusion rules $F = \{f: \Re^N \to \Re\}$and the *expected best* fusion rule f^* minimizes I_F (.) over F, i.e.

$$I_F(f^*) = \min_{f \in F} I_F(f).$$

For example, F could be the set of sigmoidal neural networks obtained by varying the weight vector for a fixed architecture. In this case $f^* = f_{w^*}$ corresponding to the weight vector w^* that minimizes I_F(.) over all the weight vectors.

Example 2.1: (Continued) Consider the fuser

$$f(Y^{(1)}, Y^{(2)}) = \frac{Y^{(1)}}{2a_1} + \frac{1}{2a_2}(Y^{(2)} - b).$$

For this fuser, we have $I_F(f) = 0$, since the bias b is subtracted from $Y^{(2)}$ and the multipliers cancel the scaling error. In practice, however, such fuser can be designed only when the sensor distributions are known.

In our formulation, since I_F (.) depends on the *unknown* error distribution $P_{Y,X}$, f^* cannot be computed even in principle. We consider that only an independently and identically distributed (iid) l -sample

$$(X_1, Y_1), (X_2, Y_2), \ldots, (X_l, Y_l)$$

is given, where $Y_i = (Y_i^{(1)}, Y_i^{(2)}, \ldots, Y_i^{(N)})$ and $Y_i^{(j)}$ is the output of S_j in response to input X_i. Now the question is what type of performance guarantees are reasonable to expect under this formulation? The answer can be found in the area of Probably Approximately Correct (PAC) learning [72,70]. We consider methods to complete an estimator $\hat{f}$, based only on a sufficiently large sample, such that

$$P^l_{Y,X}\left[I_F(\hat{f}) - I_F(f^*) > \varepsilon\right] < \delta \qquad (1)$$

where $\varepsilon > 0$ and $0 < \delta < 1$, and $P^l_{Y,X}$ is the distribution of iid l -samples. For simplicity, we denote P^l_{YX} by P. Informally, this condition states that the "error" of $\hat{f}$ is within ε of the optimal error (of f^*) with an arbitrary high probability $1 - \delta$, *irrespective* of the underlying sensor distributions. Intuitively, it is a reasonable criterion since $\hat{f}$ is to be "choosen" from an infinite set, namely f, based only on a finite sample. More concretely, conditions that are strictly stronger than Eq [1] are generally not possible. For example, consider the condition $P^l_{Y,X}[I_F(\hat{f}) > \epsilon] < \delta$ for the case $F = \{f: [0,1]^N \rightarrow \{0,1\}\}$. This condition cannot be satisfied, since for any $f \in F$, there exists a distribution for which $I_F(f) > 1/2 - \rho$ for any $\rho \in [0,1]$; see Theorem 7.1 of [18] for details.

Example 2.1: (Continued) To illustrate the effects of finite samples, consider that we generate three values for X given by $\{0.1, 0.5, 0.9\}$ with corresponding Z values given by $\{0.1, -0.1, -0.3\}$. The corresponding values for $Y^{(1)}$ and $Y^{(2)}$ are given by $\{0.1a_1 + 0.1, 0.5a_1 - 0.1, 0.9a_1 - 0.3\}$ and $\{0.1a_2 + b_2, 0.5a_2 + b_2, 0.9a_1 + b_2\}$ respectively. Consider the class of linear fusers such as $f(Y^{(1)}, Y^{(2)}) = w_1 Y^{(1)} + w_2 Y^{(2)} + w_3$. Based on the measurements, the following weights enable the fuser outputs to exactly match X values for each of the measurements:

$$w_1 = \frac{1}{0.2 - 0.4a_1}, \quad w_2 = \frac{1}{0.4a_2} \quad \text{and} \quad w_3 = \frac{0.1a_1 + 0.1}{0.4a_1 + 0.1} - \frac{0.1a_2 + b}{0.4a_2}$$

While the fuser with these weights achieves zero error on the measurements it does not achieve zero value for I_F. Note that a fuser with zero expected error exists, and can be computed if the sensor distributions are given. The idea behind the criterion in Eq [1] is to achieve performances close to optimal fuser using only a sample. To meet this criterion one needs to select a suitable F, and then achieve small error on a sufficiently large sample, as will be illustrated subsequently.

The generic sensor fusion problem formulated here is fairly general and requires very little information about the sensors: a sensor could be a hardware device, a software module or a combination. We now briefly describe some concrete examples, which are described in detail in Section 5. In the door detection example [43], S_i is an ultrasonic or infrared detector, and the objective is to detect doors that are wide enough for a robot to pass through. In the example of function estimation [42], S_i is a software module that predicts a function value with certain errors, and the objective is to fuse a number of such modules to improve the prediction accuracy. In both examples, the sensors are available so that the training sample can be collected by experimentation.

2.1 RELATED WORKS

Due to the generic nature of the sensor fusion problem described here, it is related to a number of similar problems applied in a wide variety of areas. Here we briefly show its relationship to some of the well-known methods in engineering areas. If the sensor error distributions are known, several fusion rule estimation problems have been solved by methods not requiring the samples. Some of the earlier work in this direction is due to [12]. This problem is also related to the group decision models studied extensively in political economy; for example see [22]. Indeed, early majority methods of combining the outputs of probabilistic Boolean elements date back to 1786 under the name of Condorcet jury models. The distributed detection problem based on probabilistic formulations has been extensively studied [74]. These problems are special cases of the generic fusion problem such that X *in* $\{0, 1\}$ and Y *in* $\{0,1\}^N$, but the difference is that they assume that various probabilities are available. In the systems where only measurements are available the existing methods are not useful, but the solutions to the generic sensor fusion problem are applicable. In general, however, much tighter performance bounds are possible since distribution detection is a special (namely Boolean) case of the generic sensor fusion problem [55]. Also, in many cases the solutions based on known distributions case can be converted to sample-based ones [41].

Many of the existing information integration techniques are based on maximizing a posteriori probabilities of hypotheses under a suitable probabilistic model. However, in situations where the probability densities are unknown (or difficult to estimate) such methods are ineffective. One alternative is to estimate the density based on a sample. But, as illustrated in general by [73], the density estimation is more difficult than the subsequent problem of estimating a function chosen from a family with bounded capacity. This property holds for several pattern recognition and regression estimation problems [73]. In the context of feedforward neural networks that are employed to "learn" a function based on sample, the problem is to identify the weights of a network of chosen architecture. The choice of weights picks a particular network $\hat{f}$ from a family F of neural networks of a particular architecture. This family F satisfies bounded capacity [4] and Lipshitz property [67]. Both these properties are conducive for the statistical estimation of $\hat{f}$ as explained in the next section. On the other hand, no such information is available about the class from which the unknown density is chosen, which makes it difficult to estimate the density.

3. Empirical Risk Minimization

In this section we present fusion solutions based on the empirical risk minimization methods [72]. Consider that the empirical estimate

$$I_{emp}(f) = \frac{1}{l}\sum_{i=1}^{l}\left[X_i - f(Y_i^{(1)}, Y_i^{(2)}, \ldots, Y_i^{(N)})\right]^2$$

is minimized by $\hat{f} \in F$. Using Vapnik's empirical risk minimization method [72], for example, we can show [40] that if F has finite capacity, then under bounded error, or bounded relative error for sufficiently large sample

$$P_{Y,X}^{l}\left[I_F(\hat{f}) - I_F(f^*) > \varepsilon\right] < \delta$$

for arbitrarily specified $\varepsilon > 0$ and δ, $0 < \delta < 1$. Typically, the required sample size is expressed in terms of ε and δ and the parameters of F.

In this section, we describe several different classes of F and their sample size estimators. We first describe the most general condition available on F under which the performance guarantee in Eq [1] can be assured. The sample sizes in this case are based on the scale-sensitive dimension [3]. Then we present specific cases of feedforward networks and vector spaces that will also provide the performance guarantees. We apply the specific properties of these function classes to derive the sample size estimates, which are in general tighter than the general bounds.

3.1 SCALE-SENSITIVE DIMENSION

We first present the definition of scale-sensitive dimension [3]; also see [5] for a more detailed discussion. Let F be a class of [0,1] -valued functions on some domain set D and let ρ be a positive real number. We say that F P_ρ *- shatters* a subset A *of* D if there exists a function $s{:}A \to [0,1]$ such that for every subset E of A there exists some $f_E \in F$ satisfying: for every x *in* $A - E$, $f_E(x) \le s(x) - \rho$, and for every x in E, $f_E(x) \ge s(x) + \rho$. Let the P_ρ *- dimension* of F, denoted by P_ρ - dim (F), be the maximal cardinality of a set A in D that is P_ρ -shattered by F.

The following theorem presents an estimate of the sample size to ensure the condition in Eq [1] when F has finite a scale-sensitive dimension [3] and X in [0,1].

Theorem 1 *[48] Let f^* and $\hat{f}$ denote the expected best and empirical best fusion rules chosen from a function class F with range [0,1]. Given a sample of size*

$$\frac{040}{\varepsilon^2}\max\left\{d_S \ln^2\frac{50\, d_S}{\varepsilon}, \ln\frac{48}{\delta}\right\}$$

where $d_s = P_{\varepsilon/4} - \dim(F)$, we have $P\,[I_F(\hat{f}) - I_F(f^*) > \varepsilon] < \delta$.

Theorem 1 provides us the sufficient sample size as a function of ε, δ and d_S. One needs to simply compute $\hat{f}$ that minimizes the empirical error on a sample of sufficient size to ensure the performance condition. The scale-sensitive dimension is known for several classes such as neural networks and linear combinations [5]. While the above bound is not tight (due to its general applicability), it does establish that the fuser rule estimation is a tractable problem from a statistical standpoint. The result of Theorem 1 is more general than that in [39,40] which is based on capacity of F [73] in that finiteness of capacity implies that of scale-sensitive dimension but not vice versa.

This theorem can be generalized in a straight forward manner to handle the cases: (a) $Y^{(i)}$ is a multi-dimensional vector from $\Re^d$ and/or X in $[0, \tau]$, $\tau > 0$. The cost function can also be generalized to Lipschitz cost functions with an appropriate change in the sample size [59].

The sample bound is based on uniform convergence of empirical means to their expectations for function classes, which are available from the empirical process theory [35,71,66] and its applications to machine learning [73,25]. Results of this kind are available based on a number of characterizations of F such as pseudo-dimension [35], fat VC-dimension [37], scale-sensitive dimension [3], graph dimension [19], and Euclidean parameters [66,34], which can be used to obtain sample size estimates along the lines of Theorem 1. Finiteness of these parameters is only sufficient for the "learnability" of bounded functions, while that of the scale sensitive dimension is both necessary and sufficient [3]. Moreover, the latter is only such deterministic quantity known to us, while other similar quantities are based on expected capacity or entropy [73,71].

The empirical risk minimization simply requires that $\hat{f}$ minimize the empirical error, and does not specify methods to *compute* it. In general computation of $\hat{f}$ in this general framework is intractable. For example, in the special case that F is the set of feedforward neural networks with threshold hidden units, this problem is NP-complete even for simple architectures [9]. Also, since no restrictions are placed on the information of the sensors, even if there is no probabilistic error in the sensors, several multisensor fusion problems are NP-complete [69,38,54]. For the more restrictive cases where F is chosen to be a special class, the computational problems is not always easier. Indeed very few subclasses of the fusion estimation problems are known to be inherently polynomial-time solvable. In linearly separable systems [39], the associated computational problem can be solved (exactly) as a quadratic programming problem, which be solved in polynomial time.

In the next two sections, we describe two practical solutions to the sensor fusion problem. The first method is based on feedforward neural networks. It is applied in a number of practical cases with good results although the underlying computational problem is not known to be polynomial-time computable. The second method is based on the vector space method which includes linear fusers as a special case. This method is widely used in practice and has the advantage of being polynomial-time solvable. Thus, the choice of F has a significant effect on both the sample complexity as well as computational complexity.

3.2 FEEDFORWARD SIGMOIDAL NETWORKS

We consider a feedforward network with a single hidden layer of k nodes and a single output node. The output of the j th hidden node is $\sigma(b_j^T y + t_j)$, where y in $[-B, B]^N$, $b_j \in \Re^N$, $t_j \in \Re$, and the nondecreasing $\sigma: \Re \mapsto [-1,+1]$ is called the *activation function*. The output of the network corresponding to input y is given by

$$f_w(y) = \sum_{j=1}^{k} a_j \sigma(b_j^T y + t_j)$$

where $w = (w_1, w_2, \ldots, w_{k(d+2)})$ is the *weight vector* of the network consisting of $a_1, a_2, \ldots, a_k, b_{11}, b_{12}, \ldots, b_{1d}, \ldots, b_{k1}, \ldots, b_{kd}$, and $t_1, t_2, \ldots, t_k$. Let the set of *sigmoidal feedforward networks* with *bounded weights* be denoted by

$$F_W^{\gamma} = \{ f_w : w \in [-W, W]^{k(d+2)} \}$$

where $0 < \gamma < \infty$ and $\sigma(z) = \tanh^{-1}(\gamma z)$, $0 < W < \infty$.

The following theorem provides several sample size estimates for the fusion rule estimation problem based on the different properties of neural networks.

Theorem 2. *[47] Consider the class of feedforward neural networw F_W^{γ}. Let X in $[-A, A]$ and $R = 8(A+kW)^2$. Given a sample of size at least*

$$\frac{16R}{\varepsilon^2}\left(\ln(18/\delta) + 2\ln(8R/\varepsilon^2) + \ln(2\gamma^2 W^2 kR/\varepsilon) \right. \qquad (2)$$

$$\left. + \frac{\gamma W^2 kR}{\varepsilon}\left[\left(\frac{\gamma W^2 kR}{\varepsilon} - 1\right)^{N-1} + 1\right]\right), \qquad (3)$$

the empirical best neural network $\hat{f}_w$ in F_W^{γ} approximates the expected best f^*_w in F_W^{γ} such that

$$P\left[I_F(\hat{f}_w) - I_F(f_w^*) > \varepsilon \right] < \delta.$$

The same condition can also be ensured under the sample size

$$\frac{16R}{\varepsilon^2}\left(\ln(18/\delta) + 2\ln(8R/\varepsilon^2) + k(d+2)\ln(L_w R/\varepsilon) \right])$$

where $L_w = \max(1, WB\gamma^2/4, W\gamma^2/4)$, or, for $\gamma = 1$,

$$\frac{28R}{\varepsilon^2}\max\left\{ \ln\left(\frac{8}{\delta}\right), \ln\left(\frac{16e(k+1)R}{\varepsilon}\right) \right\}.$$

These sample sizes are based on three qualitatively different parameters of F, namely,

(a) Lipschitz property of $f(y) \in F$ with respect to input y [39],
(b) compactness of weight set and smoothness of $f \in F$ with respect to weights, and
(c) VC-dimension of translates of sigmoid units [30].

The three sample estimates provide three different means for controlling the sample size depending on the available information and intrinsic characteristics of the neural network class F_W^{γ}. The sample sizes in the first and second bounds can be modified by changing the parameter γ. For example, by choosing $\gamma = \dfrac{\varepsilon}{W^2 k R}$ the first sample size can be reduced to a simpler form

$$\frac{6R}{\varepsilon^2}\left[\ln\,(18/\delta)\;+\ln\left(\frac{128R}{W^2 k\,\varepsilon^2}\right)\right].$$

Also, by choosing $\gamma^2 = \dfrac{4}{W \max(1, B)}$, we have a simpler form of the second sample size estimate

$$\frac{16R}{\varepsilon^2}\left[\ln\left(\frac{1152}{\delta}\right) + k(d+2)\ln\,(R/\varepsilon)\right],$$

for $R \geq 1$. In practice, it could be useful to compute all three bounds and choose the smallest one.

These three sample bounds are based on utilizing the boundedness of weights of neural network. If boundedness is not satisfied, one utilizes either VC-bounds [28] or scale-sensitive dimension [4] to estimate the sample size, which in general results in larger sample bounds; see [51] for details. We believe, however, in practical implementations the weights are always bounded.

The problem of computing the empirical best neural network $\hat{f}w$ is still difficult, and is NP-complete for very general subclass of $F^{\gamma}W$ [65]. However, in several practical cases very good results have been obtained using the backpropagation algorithm which provides an approximation to $\hat{f}w$. This algorithm is very easy to implement, and is also available in a number of commercial neural network packages. For the vector space method in the next section, the computation problem is polynomial-time solvable.

3.3 VECTOR SPACE METHODS

We now present a method that is attractive from an analytical point of view. Consider that F forms a finite dimensional vector space. In this case: (a) sample size is a simple function of the dimensionality of F, (b) $\hat{f}$ can be easily computed by well-known least square methods in polynomial time, and (c) no smoothness conditions are required on the functions or distributions.

Theorem 3. [46] *Let f^* and $\hat{f}$ denote the expected best and empirical best fusion functions chosen from a vector space of dimension d_V and range [0,1]. Given an iid sample of size*

$$\frac{512}{\varepsilon^2}\left[d_V \ln\left(\frac{64e}{\varepsilon} + \ln\frac{64e}{\varepsilon}\right) + \ln\,(8/\delta)\right],$$

we have

$$P\left[I_F(\hat{f}) - I_F(f^*) > \varepsilon\right] <$$

If $\{f_1, f_2, \ldots, f_{d_V}\}$ is a basis of $F, f \in F$ can be written as $f(y) =$

for $a_i \in \Re$. Then consider

$$\hat{f} = \sum_{i=1}^{d_V} \hat{a}_i f_i(y)$$

such that $\hat{a} = (\hat{a}_1, \hat{a}_2, \ldots, \hat{a}_{d_V})$ minimizes the cost expressed as

$$I_{emp}(a) = \frac{1}{l} \sum_{k=1}^{l} \left(X_k - \sum_{i=1}^{d_V} a_i f_i(Y_k) \right)^2,$$

where $a = (a_1, a_2, \ldots, a_{d_V})$. Now $I_{emp}(a)$ can be written in the quadratic form $a^T Ca + a^T D$, where $C=[c_{ij}]$ is a positive definite symmetric matrix, and D is a vector. This problem can be solved in polynomial-time using quadratic programming methods [75].

This method subsumes two very important cases:

- *Potential Functions:* The potential functions of Aizerman *et al.* [2], where $f_i(y)$ is of the form $exp((y-\alpha)^2/\beta)$ for suitably chosen constants α and β, constitute an example of the vector space methods. An incremental algorithm was originally proposed for the computation of the coefficient vector a, for which finite sample results have been derived recently [59] under certain conditions. The sample size estimate of Theorem 3. is simpler compared to the existing finite sample results. Note that the above sample size is valid only for the method that minimizes $I_{emp}(.)$ and is not valid for the original incremental algorithm of the potential functions.
- *Special Neural Networks:* In two-layer sigmoidal networks of [29], the unknown weights are only in the output layer. The specific form of these networks enables us to express each network in the form

$$\sum_{k=1}^{d_V} a_i \eta_i(y)$$

where $\eta_i(.)$'s are universal. These networks have been shown to approximate classes of the continuous functions with arbitrarily specified precision in a manner similar to the general single layer sigmoidal networks [14].

3.4 ASYMPTOTIC CONSISTENCY

In statistics literature, the asymptotic consistency results are more common. The finite sample guarantees in Eq 1 often lead to asymptotic results by a direct application of the Borel-Cantelli Lemma [8] as will be shown here. The estimate $\hat{f}$ is *consistent* if $I_F(\hat{f}) \to I_F(f^*)$ with probability one as $l \to \infty$. In several cases, δ can be rewritten as a function of the sample size, ε, and parameters of F. We consider the general case in Section 3.1. In this case we have [48]

$$\delta = 48l\left(\frac{4608l}{\varepsilon^2}\right)^{d_S \log_2(96el/(d\varepsilon))} e^{-\varepsilon^2 l/144}$$

where $d_S = P_{\varepsilon/4}$ -dim (F). Under the finiteness of d_S, the consistency result follows from the Borel-Cantelli if

$$\sum_{l=1}^{\infty} 48l\left(\frac{4608l}{\varepsilon^2}\right)^{d \log_2(96el/(d_S\varepsilon))} e^{-\varepsilon^2 l/144} \leq \infty$$

for every $\varepsilon > 0$, which can be easily shown. Similar results can be shown for various conditions used for sigmoid neural networks, and vector space methods.

4. Statistical Estimators

The fusion rule estimation problem is very similar to the regression estimation problem. In this section we present a polynomial-time (in sample size l) computable estimator which guarantees the criterion in Eq 1 under additional smoothness conditions. Our presentation is based on Nadaraya-Watson estimator applied to the sensor fusion problem. As is the case with most available statistical estimators, their original performance guarantees are asymptotic. In general finite sample results had to be derived for these estimators to be applied the sensor fusion problem under criterion in Eq 1.

We first present some preliminaries needed for the main result. Let Q denote the unit cube $[0,1]^N$ and $C(Q)$ denote the set of all continuous functions defined on Q. The modulus of smoothness of $f \in C(Q)$ is defined as

$$\omega_\infty(f; r) = \sup_{\|y-z\|_\infty < r,\, y,z \in Q} |f(y) - f(z)|$$

where $\|y - z\|_\infty = \max_{i=1}^{M} |y_i - z_i|$. For $m = 0, 1, \ldots$, let Q_m denote a family of diadic cubes (Haar system) such that $Q = \bigcup_{J \in Q_m} J$, J, $J \cap J'$ for $J \neq J'$, and the

N -dimensional volume of J, denoted by $|J|$, is 2^{-Nm}. Let $1_J(y)$ denote the indicator function of $J \in Q_m$: $1_J(y) = 1$ if $y \in J$, and $1_J(y) = 0$ otherwise. For given m, we define the map P_m on $C(Q)$ as follows: for $f \in C(Q)$, we have $P_m(f) = P_m f$ defined by

$$P_m f(y) = \frac{1}{|J|} \int_J f(z)\,dz$$

for $y \in J$ and $J \in Q_m$ [13]. Note that $P_m f\ Q \rightarrow [0,1]$ is a discontinuous (in general) function which takes constant values on each $J \in Q_m$. The *Haar kernel* is given by

$$P_m(y,z) = \frac{1}{|J|} \sum_{J \in Q_m} 1_J(y)\, 1_J(z)$$

for $y, z \in Q$.

Given l -sample, the Nadaraya-Watson estimator based on Haar kernels is defined by

$$\hat{f}_{m,l}(y) = \frac{\sum_{j=1}^{l} X_j P_m(y, Y_j)}{\sum_{j=1}^{l} P_m(y, Y_j)} = \frac{\sum_{Y_j \in J} X_j}{\sum_{Y_j \in J} 1_J(Y_j)}$$

for $y \in J$ [37,20]. The second expression indicates that $\hat{f}_{m,l}(y)$ is the mean of the function values corresponding to Y_j's in J that contains y. This property is the key to efficient computation of the estimate [59].

The Nadaraya-Watson estimator based on more general kernels is well-known in statistics literature [33]. This estimator was found to be very effective in a number of applications involving nonlinear regression estimation. The typical results about the performance of this estimator are in terms of asymptotic results, and are not particularly targeted towards fast computation. The above computationally efficient version based on Haar kernels is due to [20], which was subsequently shown to yield finite sample guarantees in [59]. The result of [59] requires finiteness of capacity of T in addition to smoothness, and the following theorem specifies the sample size based only on smoothness.

Theorem 4 [42]. *Consider a family of functions $F \subseteq C(Q)$ with range $[0,1]$ such that $w_\infty(f; r) \le kr$ for some $0 < k < \infty$. We assume that: (i) there exists a family of densities $P \subseteq C(Q)$; (ii) for each $p \in P$, $w_\infty(p; r) \le kr$; and (iii) there exists $\mu > 0$ such that for each $p \in P$, $p(y) > \mu$ for all $y \in [0, 1]^N$. Suppose that the sample size, l, is larger than*

$$\frac{2^{2m+4}}{\varepsilon_1^2}\left[\left(\frac{k\,2^m}{\varepsilon_1}\left[\left(\frac{k\,2^m}{\varepsilon_1} - 1\right)^{N-1} + 1\right] + m\right) \ln(2^{m+1}k/\varepsilon_1)\right.$$

$$\left. + \ln\left(\frac{2^{2m+6}}{(\delta-\lambda)\varepsilon_1^4}\right)\right]$$

where $\varepsilon_1 = \varepsilon\,(\mu - \varepsilon)/4$, $0 < \beta < \dfrac{N}{2(N+1)}$, $m = \left\lceil \dfrac{\log l\beta}{N} \right\rceil$ and

$$\lambda = b\left(\frac{2}{\varepsilon}\right)^{1/N+1\;-\;1/2\beta} + b\left(\frac{2}{\varepsilon_1}\right)^{1/N+1\;-\;1/2\beta}$$

Then for any $f \in F$, we have $P\left[\,I_F(\hat{f}_{m,l}) - I_F(f^*)| > \varepsilon\right] < \delta$.

The value of $\hat{f}_{m,l}(y)$at a given y is the ratio of local sum of X_i 's to the number of Y_i 's in J that contains y. The range-tree [36] can be constructed to store the cells J that contain at least one Y_i ; with each such cell, we store the number of the Y_i 's that are contained in J and the sum of the corresponding X_i 's. The time complexity of this construction is $O(l(\log l)^{N-1})$ [36]. Using the range tree, the values of J containing y can be retrieved in $O((\log l)^N)$ time [59].

The smoothness conditions required in Theorem [4] are not very easy to verify in practice. However, this estimator is found to perform well in a number of applications including those that do not have smoothness properties (see next section). Several other statistical estimators can also be used for fusion rule estimation, but finite sample results must be derived to ensure the condition in Eq 1. Such finite sample results are available for adapted nearest-neighbor rules and regressograms [59] which can also be applied for the fuser estimation problem.

5. Applications

We present three concrete applications to illustrate the performance of methods described in the previous sections -- the first two are simulation examples and the third one is an experimental system. In addition, the first two examples also provide results obtained with the nearest neighbor rule, which is analyzed elsewhere [39]. In the second example, we also consider another estimate, namely, the empirical decision rule described in [53]. Pseudo random number generators are used in both the simulation examples.

Example 5.1: *Fusion of Noisy Function Estimators:* [42] Consider five estimators of a fuction g: $[0,1] \rightarrow [0,1]$ such that i th estimator outputs a corrupted value $Y^{(I)} = g_i(X)$ of $g(X)$ when presented with input $X \in [0,1]^d$. The fused estimate $f(g_1(X), \ldots, g_5(X))$ must closely approximate $g(X)$. Here g is realized by a feedforward neural network, and, for i=1, 2,..., 5, $g_i(X) = g(X)(1/2 + iZ/10)$ where Z is uniformly distributed over [-1,1]. Thus we have $1/2 - i/10 \le g_i(X)/g(X) \le 1/2 + i/10$. Table 1 corresponds to the mean square error in the estimation of f for d=3 and d=5, respectively, using the Nadaraya-Watson estimator, nearest neighbor rule, and a feedforward neural network with backpropagation learning algorithm. Note the superior performance of the Nadaraya-Watson estimator.

TABLE 1. Fusion of function estimators: mean square error over test set.

Training Set	Testing Set	Nadaraya-Watson	Nearest Neighbor	Neural Network
100	10	0.000902	0.002430	0.048654
1000	100	0.001955	0.003538	0.049281
10000	1000	0.001948	0.003743	0.050942

(a) d=3

Training Set	Testing Set	Nadaraya-Watson	Nearest Neighbor	Neural Network
100	10	0.004421	0.014400	0.018042
1000	100	0.002944	0.003737	0.021447
10000	1000	0.001949	0.003490	0.023953

(b) d=5

Example 5.2: *Decision Fusion:* [53,47] We consider 5 sensors such that $Y \in \{H_0, H_1\}^5$ such that $X \in \{H_0, H_1\}$ corresponds to "correct" decision, which is generated with equal probabilities, i. e., $P(X=H_0) = P(X=H_1) = 1/2$. The error of sensor S_i, i=1,2, ..., 5, is described as follows: the output $Y^{(i)}$ is correct decision with probability of 1-i/10, and is the opposite with probability i/10. The task is to combine the outputs of the sensors to predict the correct decision. The percentage error of the individual detectors and the fused system based on the Nadaraya-Watson estimator is presented in Table 2. Note that the fuser is consistently better than the best sensor S_1 beyond the sample sizes of the order of 1000. The performance results of the Nadaraya-Watson estimator, empirical decision rule, nearest neighbor rule, and the Bayesian rule based on the analytical formulae are presented in Table 3. The Bayesian rule is computed based on the formulae used in the data generation and is provided for comparison only.

TABLE 2. Performance of Nadaraya-Watson estimator for decision fusion.

Sample Size	Test set	S_1	S_2	S_3	S_4	S_5	Nadaraya-Watson
100	100	7.0	20.0	33.0	35.0	55.0	12.0
1000	1000	11.3	18.5	29.8	38.7	51.6	10.6
10000	10000	9.5	20.1	30.3	39.8	49.6	8.58
50000	50000	10.0	20.1	29.8	39.9	50.1	8.860

TABLE 3. Comparative Performance.

Sample Size	Test Size	Bayesian Fuser	Empirical Decision	Nearest Neighbor	Nadaraya-Watson
100	100	91.91	23.00	82.83	88.00
1000	1000	91.99	82.58	90.39	89.40
10000	10000	91.11	90.15	90.81	91.42
50000	50000	91.19	90.99	91.13	91.14

Example 5.4: *Door Detection Using Ultrasonic and Infrared Sensors:* Consider the problem of recognizing a door (an opening) wide enough for a mobile robot to move through. The mobile robot (TRC Labmate) is equipped with an array of four ultrasonic and four infrared Boolean sensors on each of four sides as shown in Figure 1. We address only the problem of detecting a wide enough door when the sensor array of any side is facing it. The ultrasonic sensors return a measurement corresponding to distance to an object within a certain cone as illustrated in Figure 1. The infrared sensors return Boolean values based on the light reflected by an object in the line-of-sight of the sensor; white smooth objects are detected due to high reflectivity, while objects with black or rough surface are generally not detected. In practice, both ultrasonic and infrared sensors are unreliable, and it is very difficult to obtain accurate error distributions of these sensors. The ultrasonic sensors are susceptible to multiple reflections and the profiles of the edges of the door. The infrared sensors are susceptible to surface texture and color of the wall and edges of the door. Accurate derivation of probabilistic models for these sensors requires a detailed knowledge of the physics and engineering of the device as well as a priori statistical information. Consequently, a Bayesian solution to this problem is very hard to implement. On the other hand, it is relatively easy to collect experimental data by presenting to the robot doors that are wide enough as well as those that are narrower than the robot. We employ the Nadaraya-Watson estimator to derive a non-linear relationship between the width of the door and the sensor readings. Here the training sample is generated by actually recording the measurements while the sensor system is facing the door. Positive examples are generated if the door is wide enough for the robot, and the sensory system is facing the door. Negative examples are generated when the door is not wide enough or the sensory system is not correctly facing a door (wide enough or not). The robot is manually located in various positions to generate the data. Consider the sensor array of a particular side of the mobile robot. Here $Y^{(1)}$, $Y^{(2)}$, $Y^{(3)}$, $Y^{(4)}$ correspond to the normalized distance measurements from the four ultrasonic sensors, and $Y^{(5)}$, $Y^{(6)}$, $Y^{(7)}$, $Y^{(8)}$ correspond to the Boolean measurements of the infrared sensors. X is 1 if the sensor system is correctly facing a wide enough door, and is 0 otherwise. The training data included 6 positive examples and 12 negative examples. The test data included 3 positive examples and 7 negative examples. The Nadaraya-Watson estimator predicted the correct output in all examples of test data.

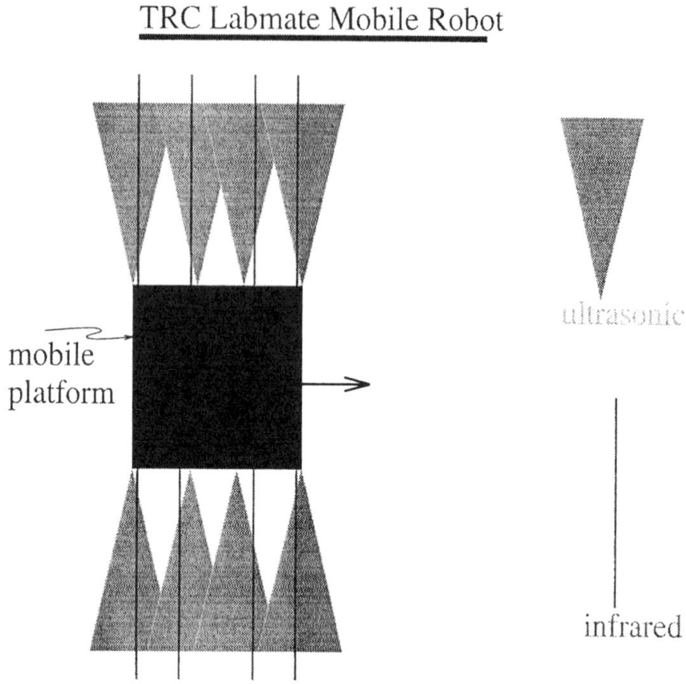

Figure 1. Schematic of sensory system (only the side sensor arrays are shown for simplicity)

6. Performance of Fused System

In the empirical risk minization methods IF $(\hat{f})$ is shown to be close to $I_F(f^*)$, which depends on F. In general $I_F(f^*)$ could be very large for particular fuser classes. Note that one cannot simply choose an arbitrary large F : if so, the performance guarantees of the type in Eq 2 cannot be guaranteed. If $I_F(f^*) > I(S_i)$, then fusion is not useful, since one is better off just using S_i. In practice, however, such condition cannot be verified if the distributions are not known. In this section, we address the issue of the relative performance of the composite system, composed of the fuser and $S_1, S_2, \dots, S_N$, and the individual sensors or sensor subsets. We obtain sufficiency conditions under which the composite system can be shown to be at least as good as the best sensor or best subset of sensors.

For simplicity, we consider a system of N sensors such that X in [0,1], $Y^{(I)}$ in [0,1] and the *expected square error* is given by

$$I_S(S_i) = \int [X - Y^{(i)}]^2 dP_{Y^{(i)},X}.$$

The *expected square error* of the fuser f is given by

$$I_F(f) = \int [X - f(Y)]^2 dP_{Y,X}$$

respectively, where.

$$Y = (Y^{(1)}, Y^{(2)}, \ldots, Y^{(N)})$$

6.1 ISOLATION FUSERS

If the distributions are known, one can derive the best sensor S_{i^*} such that

$$I_S(S_{i^*}) = \min_{i=1}^{N} I_S(S_i).$$

In the present formulation, the availability of *only* a sample makes the selection (with probability 1) of the best sensor infeasible, even in the special case of the target detection problem [18]. In this section, we present a method that circumvents this difficulty by fusing the sensors such that the performance of best sensor is achieved as a minimum. The method is fully sample-based in that no comparative performance of the sensors is needed -- in particular, the best sensor may be unknown.

A function class $F = \{f: [0,1]^k \mapsto [0,1]\}$ has the *isolation property* if it contains the functions $f^i(y_1, y_2, \ldots, y_k) = y_i$ for all $i=1, 2, \ldots, k$. If F has the isolation property, we have

$$I_F(f^*) = \min_{f \in F} \int (X - f(Y))^2 dP_{Y,X} \leq \int (X - f^i(Y))^2 dP_{Y,X} \quad (1)$$

$$= \int (X - Y^{(i)})^2 dP_{Y,X} = I_S(S_i), \quad (2)$$

which implies $I_F(f^*) = \min_{i=1}^{N} I_S(S_i) - \Delta$ for some Δ in $[0, \infty)$. Due to the isolation property, we have $\Delta \geq 0$, which implies that the error of f^* is no higher than $I(S_{i^*})$, but can be significantly smaller. The precise value of Δ depends on F, but the isolation property guarantees that $I_F(f^*) = \min_{i=1}^{N} I_S(S_i)$ as a minimum.

Let the set S be equipped with a pseudodometric ρ. The *covering number* $N_C(\varepsilon, \rho, S)$ under metric ρ is defined as the smallest number of closed balls of radius ε, and centers in S, whose union covers S. For a set of function $G = \{g : \Re^M\ [0,1]\}$, we consider two metrics defined as follows: for $g_1, g_2 \in G$ *we* have

$$d_P(g_1, g_2) = \int_{z \in \Re^M} |g_1(z) - g_2(z)|\, dP,$$

for the probability distribution P defined on $\Re^M \mapsto [0,1]\}$ and

$$d_\infty(g_1, g_2) = \sup_{z \in \Re^M} |g_1(z) - g_2(z)|.$$

This definition is applied to functions defined on

This definition is applied to functions defined on $A \subseteq \Re^M$ by extending them to take value 0 on $\Re^M \setminus A$.

Theorem 5 [44]. *Consider a fuser class* $F = \{f: [0,1]^N \mapsto [0,1]\}$, *such that* $I_F(f^*) = \min_{f \in F} I_F(f)$ and $\hat{I}_F(\hat{f}) = \min_{f \in F} I_F(f)$. *If F has the isolation property, we have*

$$I_F(f^*) = \min_{i=1}^{N} I_S(S_i) - \Delta,$$

for $\Delta \in [0, \infty)$, and

$$P^l_{Y,X}\left[I_F(\hat{f}) - \min_{i=1}^{N} I_S(S_i) + \Delta > \varepsilon\right] <$$

given the sample size l of at least

$$\frac{2048}{\epsilon^2}[\ln N_C(\epsilon/64, \mathcal{F}) + \ln(4/\delta)]$$

for cases: (i) $N_C(\varepsilon, F) = N_C(\varepsilon, d_\infty, F)$ and (ii) $N_C(\varepsilon, F) = N_C(\varepsilon, dp, F)$ *for all distributions P.*

If F has the isolation property, the fuser is guaranteed to perform at least as good as the best sensor in PAC sense. No information other than the iid sample is needed to ensure this result. Since $\Delta \geq 0$, under the sample size of Theorem 5, we trivially have

$$P\left[I_F(\hat{f}) - \min_{i=1}^{N} I_S(S_i) > \varepsilon\right] < \delta.$$

The sample size needed is expressed in terms of d_∞ or distribution-free covers for F. Note that for smooth fusers such as sigmiod neural networks, we have simple d_∞ cover bounds. The pseudo-dimension and scale-sensitive dimension of F provide the distribution-free cover bounds needed in Theorem 5 when smoothness conditions may not be satisfied.

The isolation property was first proposed in [55,39] for concept and sensor fusion problems. For linear combinations, i.e.

$$f(y_1, y_2, \ldots, y_k) = w_1 y_1 + w_2 y_2 + \ldots + w_k y_k,$$

for $w_i \in$. this property is trivially satisfied. For several well-known fuction classes such as potential functions [2] and feedforward sigmoid networks [64], this property is not satisfied in general; see [52] for a more detailed discussion on the isolation property and various function classes that have this property.

Consider the special case where S_i's are classifiers obtained using different methods as in [22]. For Boolean functions, the isolation property is satisfied if F contains all Boolean functions on k variables. S_i computed based an iid l-sample is *consistent* if $I_S(S_i) \rightarrow I_S(S^*)$, where S^* is the Bayes classifier. By the isolation property, if one of the classifiers is consistent, the fused classifier system (trained by l-sample independent from n-sample used by the classifiers) can be seen to be consistent. Such result was obtained in [32] for linear combinations (for which $N_C(\varepsilon, F)$ is finite). Our result does not require the linearity, but pinpoints the essential property, namely the isolation. More generally, linear combinations have been extensively used as fusers in various applications such as combining neural network estimators [23], regression estimators [11,68], and classifiers [32]. Since the linear combinations possess the isolation property, Theorem 5 can be viewed as providing some analytical justification for these methods. A more detailed treatment of Boolean problems, namely for classifier problems, can be found in [45].

6.2 PROJECTIVE FUSERS

A *projective fuser [49]*, f_P, corresponding to a *partition* $P = \{\pi_1, \pi_2, ..., \pi_k\}$, $k \leq N$, of input space $\Re^d$ ($\pi_i \subseteq [0,1]^d$, $\bigcup_{i=1}^{k} \pi_i = \Re^d$, and $\pi_i \cap \pi_j = \phi$ for $i \neq j$), assigns each block π_i to a sensor S_j such that

$$f_P(Y) = Y^{(j)}$$

for all $X \in \pi_i$, i.e. the fuser simply transfers the output of the sensor S_j for every point in π_i. An *optimal projective fuser*, denoted by $f_P{}^*$, minimizes I (.) over all projective fusers corresponding to all partitions of $\Re^d$ and assignments of blocks to sensors $S_1, S_2, ... S_N$.

We define the *error regression* of the sensor S_i and fuser f_F as

$$\mathcal{E}(X, S_i) = \int C(X, Y^{(i)}) dP_{Y|X}$$

and

$$\mathcal{E}(X, f_P) = \int C(X, f_P(Y)) dP_{Y|X},$$

respectively. The *projective fuser* based on the lower envelope of error regressions of sensors is defined by

$$f_{LE}(Y) = Y^{(i_{LE}(X))}$$

where

$$i_{LE}(X) = \arg \min_{i=1,2,...,N} \mathcal{E}(X, S_i).$$

We have $E(X, f_{LE}) = \min_{i=1,\ldots,N} E(X, S_i)$, or equivalently the error regression of f_{LE} is the lower envelope with respect to X of the set of error regressions of sensors given by $\{E(X, S_1), \ldots, E(X, S_N)\}$.

Example 6.1: [49] Consider that X is uniformly distributed over [0,1], which is measured by two sensors S_1 and S_2. Let $C(X, Y^{(I)}) = (X - Y^{(I)})^2$. Consider $Y^{(1)} = X + |X - ½| + U$ and $Y^{(2)} = X + 1/[4(1+|X - ½|)] + U$, where U is an independent random variable with zero mean. Thus, for both sensors the measurement error at any X is represented by U. Note that

$$E[Y^{(1)} - X] = |X - ½|$$

$$E[Y^{(2)} - X] = 1/[4(1+|X - ½|)].$$

Thus, S_1 achieves a low error in the middle of the range [0,1], and S_2 achieves a low error towards the end point of the range [0,1]. The error regressions of the sensors are given by

$$E(X, S_1) = (X - ½)^2 + E[U^2]$$

$$E(X, S_2) = 1/[16(1 + |X - ½|)^2] + E[U^2]$$

We have

$$I(S_1) = 0.0833 + E[U^2] \text{ and } I(S_2) = 0.125 + E[U^2],$$

which indicates that S_1 is the better of the two sensors. Now consider the projective fuser f_{LE} specified as follows, which corresponds to the lower envelope of $E(X, S_1)$ and $E(X, S_2)$.

range for X	sensor to be projected
[0,0.134]	S_2
[0.134,0.866]	S_1
[0.866,1]	S_2

Then, we have $I(f_{LE}) = 0.0828 + E[U^2]$, which is lower than that of the best sensor.

Example 6.2: [49]. We consider a classification example such that $X \in [0,1] \times \{0, 1\}$ is specified by a function $f_X = 1_{[1/4,3/4]}$, where $1_A(z)$ is the *indicator function* (which has a value 1 if and only if $z \in A$ and has value 0 otherwise). The value of X is generated as follows: a random variable Z is generated uniformly in the interval [0,1] as the first component, and then $f_X(Z)$ forms the second component, i.e. $X = (Z, f_X(Z))$. In the context of the detection problem, the second component of X corresponds to the presence ($f_X(Z) = 1$) or absence ($f_X(Z) = 0$) of a target, which is represented by a feature Z taking a value in the interval [1/4,3/4]. Each sensor consists of a device to measure the first component of X and an algorithm to compute the second component. We consider that S_1 and S_2 have ideal devices that measure Z without an error, but make errors in utilizing the measured features. Consider that $Y^{(1)} = (Z, 1_{[1/4 - \epsilon_1, ¾]}(Z))$ and $Y^{(2)} = (Z, 1_{[1/4 - \epsilon_1, 3/4]}(Z))$ for some $0 < \epsilon_1, \epsilon_2 < ¼$ (see Figure 2). In other words, there is no

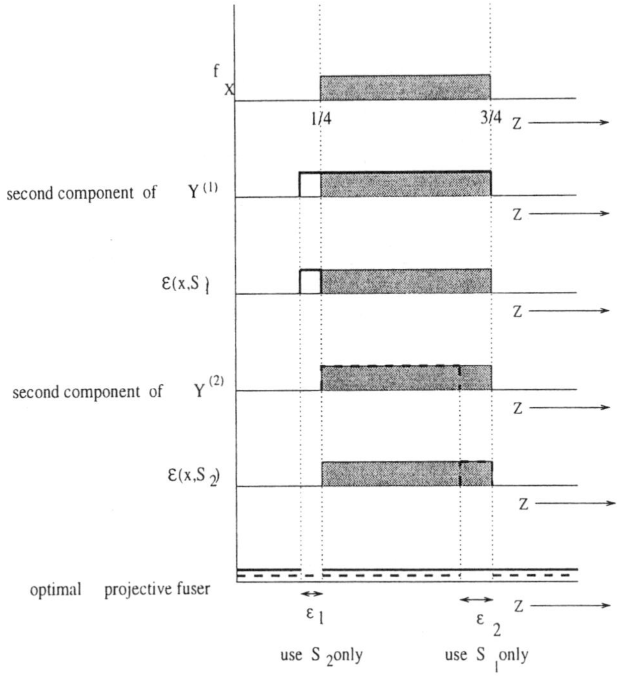

Figure 2. **Illustration for example**

measurement noise in the sensors but just a systematic error due to how the feature value is utilized; addition of independent measurement noise as in Example 6.1 does not change the basic conclusions of the example. Now consider the quadratic cost function $C\,(X,Y^{(i)}) = (X-Y^{(i)})^T(X-Y^{(i)})$. The error regressions are given by $E\,(X, S_1) = 1_{[1/4 - \epsilon 1\,,\,1/4]}\,(Z\,)$ and $E\,(X, S_2) = 1_{[3/4 - \epsilon 1\,,\,3/4]}\,(Z\,)$ which corresponds to disjoint intervals of Z as shown in Figure 3. The lower envelope of the two regressions is the zero function hence

$I\,(f_{LE}) = 0$, where as both $I\,(S_1)$ and $I\,(S_2)$ are positive. The profile of f_{LE} is shown at the bottom of Figure 1, wherein S_1 and S_2 are projected based on the first component of X in the intervals $[3/4 - \epsilon_2, 3/4]$ and $[1/4 - \epsilon_1, 1/4]$, respectively, and in other regions either sensor can be projected.

The projective fuser based on error regressions is optimal as in the following theorem.

Theorem 6. [49]. *The projective fuser based on the lower envelope of error regressions is optimal among all projective fusers.*

A special case of this theorem for function estimation can be found in [49], and for classifiers can be found in [50]. A sample-based version of projective fuser is discussed in [49].

We close this section by emphasizing that f_{LE} may not be optimal in a larger class of fusers where some function of the sensor output (as opposed to just the output) can be projected.

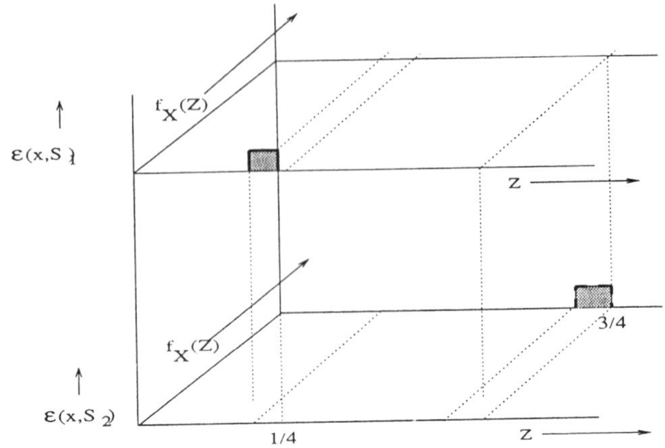

Figure 3. Illustration of error regressions

Example 6.3: [49]. In Example 2.2, consider $f_X = 1_{[1/4,3/4]}$,

$$Y^{(1)}(X) = (Z, 1_{[1/4 - \epsilon 1, 3/4]}(Z))$$

$$Y^{(2)}(X) = (Z, 1_{[1/4, 3/4 - \epsilon 1]}(Z))$$

for some $0 < \epsilon_1 < \epsilon_2 < 1/8$, and $\epsilon_1 < \epsilon_2$. Thus, we have $E(X, S_1) = 1_{[1/4 - \epsilon 1, 1/4]}(Z)$ and $E(X, S_2) = 1_{[3/4 - \epsilon 2, 3/4]}(Z)$, whose lower envelope is not the zero function. Thus, we have $E(X, f_{LE}) = 1_{[3/4 - \epsilon 2, 3/4 - \epsilon 1]}(Z)$, and $I(f_{LE}) = \int_{[3/4-\epsilon_2, 3/4-\epsilon_1]} dP_z$. By changing the assignment $Y^{(1)}$ of f_{LE} to $1 - Y^{(1)}$ for Z in $[3/4 - \varepsilon_2, 3/4 - \varepsilon_1]$, one can easily achieve zero error.

7. Metafusers

In this section, we first compare projective fusers to linear fusers and show that they have complementary performances. Thus it is natural to combine the fusers -- much the same ways as sensors -- to exploit the relative merits of individual fusers. Such approach leads to the idea of *metafusers*. There are many possible ways of designing metafusers, and here we describe how isolation property can be utilized to develop systems that exploit the complementary performances sensors as well as fusers.

The output of linear fuser corresponding to input X and sensor output $Y = (Y^{(1)}, \ldots, Y^{(N)})$ is defined as

$$f_L(Y) = \sum_{i=1}^{N} \alpha_i Y^{(i)}$$

where α_i is a d x d matrix. For simplicity, we consider the case d=1 such that $(\alpha_1, \ldots, \alpha_2) \in \Re^N$ An *optimal linear combination fuser*, denoted by f_{L*}, minimizes I (.) over all linear combinations. In terms of relative performance, f_{LE} is better than f_{L*} if the individual sensors perform better in certain localized regions of $\Re^d$. On the other hand, if the sensors are equally distributed around certain values in global sense, f_L performs better as illustrated follows.

Example 6.4: [49]. In the Example 6.2, for $f_L = \alpha_1 Y^{(1)} + \alpha_2 Y^{(N)}$, we have

$$I(f_L) = \alpha_1^2 \int_{[1/4-\epsilon_1,\ 1/4)} dP_Z \tag{3}$$

$$+ (1-\alpha_1-\alpha_2)^2 \int_{[1/4,\ 3/4-\epsilon_2)} dP_Z \tag{4}$$

$$+ (1-\alpha_1)^2 \int_{[3/4-\epsilon_2\ 3/4]} dP_Z \tag{5}$$

which is non-zero no matter what the coefficient are. The error regressions of S_1 and S_2 take non-zero values in the intervals $[1/4 - \epsilon_1, 1/4]$ and $[3/4 - \epsilon_2, 3/4]$ of Z, respectively. Since, these intervals are disjoint, there is no possibility of the error of one sensor being canceled by a scaler multiplier of the other.

In general, the argument of Example 6.4 is true: if the error regressions of the sensors take non-zero values on disjoint intervals, then any linear fuser will have non-zero error. On the other hand, the disjointness yields $E(X, f_{LE}) = 0$, for all X, and hence $I(f_{LE}) = 0$. We now present an example where a linear fuser outperforms f_{LE}.

Example 6.5: Consider that in Example 6.2, $f_X = 1$ for $Z \in [0,1]$,

$$Y^{(1)}(X) = (Z, \epsilon Z + 1 - \epsilon)$$

$$Y^{(2)}(X) = (Z, \epsilon Z + 1 + \epsilon),$$

for $0 < \epsilon < 1$. The optimal linear fuser is given by $F_L^*(Y) = ½(Y)^{(1)} + (Y)^{(2)} = 1$, and $I(f_{L*}) = 0$. At every $X \in [0,1]$, we have

$$E(X, S_1) = E(X, S_2) = \epsilon^2 (1 - Z)^2 = E(X, f_{LE}).$$

Thus, $I\,(f_{LE}\,) = \epsilon^2 \int_{[0,1]} (1 - Z\,)^2\, dP_Z > 0$, whereas $I\,(f_{L^*}\,) = 0$.

In summary, the performance of the optimal linear and projective fusers are complementary in general. We now combine linear and projective fusers to realize various metafusers that are guaranteed to be at least as good as the best sensor as well as best sensor. By including the optimal linear combination as S_{N+1}, we can guarantee that , $I\,(\,f_{LE}\,) \leq I\,(\,f_{L^*}\,)$ by the isolation property of projective fusers [49]. Since linear combinations also satisfy the isolation property, we in turn have $I\,(f_{L^*}\,) \leq.\ \min_{i=1,\ldots,N} I\,(S_i\,)$.

The roles of f_{L^*} and f_{LE} can be switched -- by including f_{LE} as one of the components of f_{L^*} - to show that

$$\mathrm{I}\,(f_{L^*}\,) \leq I\,(\,f_{LE}\,) \;\leq\; \min_{i=1,\ldots,N} I\,(S_i\,)\,.$$

One can design a metafuser by utilizing the available sensors which are combined using a number of fusers including a fuser based on isolation property (for example, a linear combination). Consider that we employ a metafuser based on a linear combination of the fusers. Then the fused system is guaranteed to be at least as good as the best of the fusers as well as the best sensor. If at a latter point, a new sensor or a fuser is developed, it can be easily integrated into the system by retraining the fuser and/or metafuser as needed. As a result, we have a system guaranteed (in PAC sense) to perform at least as good as the best available sensor and fuser at all times. Also, the computational problem of updating the fuser and/or metafuser is a simple least squares estimation that can be solved using a number of available methods.

8. Fusers For Physical Systems

In this section, we consider a multisensor fusion problem that does not exactly fit into the classical formulation of the fusion problem. Consider a physical system described by a set of parameters, such that each parameter is either measured by a number of sensors or estimated by a set of computer programs using sensor measurements. As a result, the resultant parameter values could be widely varying. In comparison with the traditional fusion problems, there is no training set that provides the actual parameter values. Furthermore, since every parameter is measured or estimated, there are no parameters whose actual values are known. In this section, we describe a fuser based on the least violation of the physical laws that relate the parameters. Under very general conditions on the physical law, we derive distribution-free performance bounds based on finite samples. We illustrate the effectiveness of this method for a practical problem of fusing well-log data in methane hydrate exploration.

8.1 PHYSICAL LAWS

We consider a physical system specified by the parameters

$$P(z) = (p_1(z), p_2(z), \ldots, p_n(z))$$

with $p_i(z) \in \Re$ where z is one-dimensional variable such as time or position. Each parameter p_i is measured by a_i instruments and estimated by b_i estimators ($a_i \geq 0$, $b_i \geq 0$, and $a_i + b_i \geq 1$). The measurements corresponding to $p_i(z)$ are denoted by

$$m_i(z) = \{ m_{i,1}(z), m_{i,2}(z), \ldots, m_{i,a_i}(z) \}$$

and the corresponding estimators are denoted by

$$e_i(z) = \{ e_{i,1}(z), e_{i,2}(z), \ldots, e_{i,b_i}(z) \}.$$

The measurements are noisy in that repeated measurements by a sensor of $p_i(z) = x$ for a fixed value are distributed independently according to the distribution $P_{m_{i,j}|x}$, which is denoted by $P_{m_{i,j}|p_i(z)}$. Thus, $m_{i,j}$ is a random variable. The estimator $e_{i,j}$ is a (deterministic) function of the measurements, and hence is also a random variable. The joint distribution of the measurements is denoted by $P_{m1, m2, \ldots, mn | p1, p2, \ldots, pn}$. Note that there are $a_i + b_i$ competing values for each parameter, and in general we do not know which one is more accurate.

There is a physical law

$$L[p_1(z), p_2(z), \ldots p_n(z)] = 0$$

which relates the actual parameters corresponding to z. We assume that $L[.]$ satisfies the reasonable *monotonicity* condition: for any y_1, y_2, $|y_1| \leq |y_2|$, we have

$$|L[p_1(z), \ldots, p_i(z) + y_1, \ldots, p_n(z)]| \leq |L[p_1(z), \ldots, p_i(z) + y_2, \ldots, p_n(z)]|.$$

This condition means that accurate parameter estimators yield no lesser "magnitude" of violation of the law compared to less accurate estimators.

Let us choose a single estimator or measurement $\hat{p}_i$ for the parameter. The closeness of $L[\hat{p}_1(z), \hat{p}_2(z), \ldots, \hat{p}_n(z)$ to 0 determines how closely the law is satisfied. Let a *basic set*, denoted by S, be a set of measurements and estimators such that for each parameter we choose precisely one measurement or estimator (but not both). The total error due to S is given by

$$\hat{E}(S) = \sum_z L[\hat{p}_1(z), \hat{p}_2(z), \ldots, \hat{p}_n(z)].$$

Note that in all there are $\prod_{i=1}^{n}(a_i + b_i)$ possible basic sets, and $\widehat{S}$ be the one with least error such that $\widehat{E}(\widehat{S}) = \min_{S} \widehat{E}(S)$. The expected error of S is denoted by

$$E(S) = \sum_{z} \int L\left[\hat{p}_1(z), \hat{p}_2(z), \ldots, \hat{p}_n(z)\right] P_{m_1,\ldots,m_n|p_1,\ldots,p_n},$$

and let S^* be the one with the least expected error such that $E(S^*) = \min_{S} E(S^*)$.

Note that S^* minimizes the expected error but $\hat{S}$ in general does not. More detailed discussion of the physical laws can be found in [62].

Example 8.1: [62]. We consider a simple illustrative example of known mass m subjected to a constant force f in a friction-free environment. In this case the physical law is $f = ma$, which can be rewritten as $L[f,a]=(f-ma)^2=0$, where a is acceleration and f is force. Let $p_1(z) = f$ and $p_2(z) = a$. We are given a sensor that measures force and two sensors that measure acceleration. The force measurements have simple bias error such that $m_{1,1}(z)=f+\varepsilon$, for some deterministic ε. The acceleration measurements are given by $m_{2,1}=a+\delta$, $m_{2,2}=0.7a$, where δ is a small normally distributed error. Then, we have

$$L[m_{1,1}, m_{2,1}] = (\varepsilon - m\delta)^2$$

$$L[m_{1,1}, m_{2,2}] = (\varepsilon + 0.3ma)$$

Consider that $a > 0$, and ε□$m\,\delta$. If $|\delta| \leq |0.3a$, we have

$$L[m_{1,1}, m_{2,1}] \leq L[m_{1,1}, m_{2,2}],$$

i.e. for large values of a, the better choice is $m_{2,1}$, otherwise $m_{2,2}$ is a better choice.

8.2 FUSER COMPUTATION

The *fusion function* $f_i \in F_i$ for parameter p_i combines the measurements and estimators such that $f_i(m_i(z), e_i(z))$ is an estimate of $p_i(z)$. Let $f = (f_1, \ldots, f_n)$ denote the *fuser* for all parameters. The expected error due to the fused estimate is

$$E(f) = \sum_{z} \int L\left[f_1(m_1(z), e_1(z)), \ldots, f_n(m_n(z), e_n(z))\right] dP_{m_1,\ldots,m_n|p_1,\ldots,p_n},$$

and let $f^* \in F_1 \times \ldots \times F_n$ be the one with the least expected error. As before $E(f)$ cannot be computed if the error distributions are not known, and hence f^* is not computable. In stead, we compute $\hat{f}$ that minimizes the empirical cost given by

$$\hat{E}(f) = \sum_z L\left[f_1(m_1(z), e_1(z)), \ldots, f_n(m_n(z), e_n(z))\right],$$

based on a set of iid measurements (also called the sample)

$$\{<(m_1(z), e_1(z)), \ldots, (m_n(z), e_n(z))>: z = 1, \ldots, s\}$$

We now describe methods that ensure $E(f^*) < E(S^*)$, and more importantly based on a computable $\hat{f}$ that

$$E(\hat{f}) < E(S^*),$$

with a specified probability based entirely on the measurements and without any knowledge of the underlying distributions.

If each F_i satisfies the isolation property, then the following conditions are directly satisfied.

$$E(f^*) \leq E(S^*) \text{ and } \hat{E}\ \hat{f} \leq \hat{E}(\hat{S})$$

The first condition is useful only if f^* can be computed, which in turn requires the knowledge of the distributions. If the distributions are not known, then $\hat{f}$ can be used as an approximation. We subsequently show finite samples guarantees for such fuser.

8.3 SMOOTH PHYSICAL LAWS

For any function $g : [-A, A]^d \mapsto \Re$, let

$$\|g(r)\|_\infty = \sup_{r \in [-A,A]^d} |g(r)|.$$

A function: $g(y) : [-A, A]^d \mapsto \Re$ is *Lipschitz* with constant k_g if for all $y_1, y_2 \in [-A, A]^d$, we have

$$\|g(y_1) - g(y_2)\|_\infty \leq k_g \|y_1 - y_2\|_\infty.$$

For example, the sigmoid neural networks are Lipschitz with constant specified by the parameters of the network [51].

Theorem 7 [62] *Consider that the physical law is Lipschitz with constant k_L, and parameters, estimators and measurements are bounded such that $p \in [-C,C]^n$, $m_i \in [A, A]^{ai}$,* *and* $e_i \in [B, B]^{b_i}$. *Let each fuser class* F_i *be Lipschitz with constant* k_{fi}. Let $d = \sum_{i=1}^{n} (a_i + b_i)$ and $k = k_L \max(k_{f1}, k_{f2}, \ldots, k_{fn})$. *Then given a sample size*

$$s = \frac{512k(A+B)}{\varepsilon^2}\left[d \ln\left(\frac{32k(A+B)}{\varepsilon}\right) + \ln(8/\delta)\right],$$

we have

$$P\left[E(\hat{f}) - E(f^*) > \varepsilon\right] \le \delta,$$

irrespective of the sensor distributions. Furthermore, $E(\hat{f}) \to E(f^*)$, *as* $s \to \infty$ *theorem.*

This theorem provides a distribution-free finite sample result: given a sufficiently large sample size, with a probability 1-δ, the cost of the sample-based solution is within ε of the lowest achievable cost (which can only be computed if all error distributions are known). Results similar to the asymptotic result shown in the above theorem are more common in the statistics literature [37]. The finite sample result, however, is stronger in that it implies the asymptotic result, and also establishes that the method is justified even for small sample sizes.

The smoothness conditions required in this theorem are quite reasonable. The Lipschitz condition is satisfied for a number of physical laws, although not always guaranteed. Similar condition was used in converting the decision fusion rules designed for known distributions to sample-based ones in [41]. The isolation and Lipschitz properties required of the fusers are satisfied in a number of cases such as linear combinations with bounded coefficients and piecewise linear feedforward networks [44,52].

Example 8.2: [62] Consider the scenario in Example 8.1, and

$$f_2(m_{2,1}, m_{2,2}) = w_1 m_{2,1} + w_2 m_{2,2}.$$

For the choice $w_1 = 0.3$ and $w_2 = 1$, we have $f_2(m_{2,1}, m_{2,2}) = a + 0.3\ \delta$, and

$$L[m_{1,1}, f_2(m_{2,1}, m_{2,2})] = (\varepsilon - 0.3m\,\delta)^2,$$

which is always smaller than $L[m_{1,1}, m_{2,1}]$, and is smaller than $L[m_{1,1}, m_{2,2}]$ for small $\delta \le a$ which is true with probability one since δ is a zero-mean random variable.

8.4 NON-SMOOTH PHYSICAL LAWS

In general physical laws may not be Lipschitz, especially if they involve discrete components or discontinuities. For example, consider the simple case of H_2O heated in a container, where p_1 denotes the temperature and $p_2 \in \{0,1\}$ is the state, i. e. $p_2 = 0$ denotes liquid and $p_2 = 1$ denotes steam. Let T_0 denote the boiling temperature under this condition. Then, one of the physical laws is: $p_2 = 0$ if $p_1 < T_0$ and $p_2 = 1$ otherwise. This law can be represented as

$$L[p_1, p_2] = p_2 1_{\{p_1 < T_0\}} + (p_2 - 1) 1_{\{p_1 \geq T_0\}} = 0,$$

where the indicator function 1_C is 1 if condition C is true and is 0 otherwise. Here, $L[.]$ is not Lipschitz. The class of functions with bounded variation [61], allow for discontinuities and discrete values, and include Lipschitz functions as a subclass. We now describe a generalization of the results of last section, which enable the utilization of non-smooth physical laws.

Consider a function one-dimensional function $h : [-A, A] \mapsto \Re$. For $A < \infty$, a set of points $P = \{x_0, x_1, \ldots, x_n\}$ such that $-A = x_0 < x_1 < \ldots < x_n = A$ is called a *partition* of $[-A, A]$. The collection of all possible partitions of $[-A, A]$ is denoted by $P[-A, A]$. A function $g : [-A, A] \mapsto \Re$ is of *bounded variation*, if there exists M such that for any partition $P = \{x_0, x_1, \ldots, x_n\}$, we have $\sum(P) = \sum_{k=1}^{n} |f(x_k) - f(x_{k-1})| \leq M$. A multivariate function $g : [-A, A]^d \mapsto \Re$ is of bounded variation if it is so in each of its input variable for every value of the other input variables.

The following are facts about the functions of bounded variation: (i) not all continuous functions are of bounded variation, e.g. $g(x) = x \cos(\pi / (2x))$ for $x \neq 0$ and $g(0)=0$; (ii) differentiable functions on compact domains are of bounded variation; and (iii) absolutely continuous functions, which include Lipschitz functions, are of bounded variation.

For the case of non-smooth physical laws, we utilize the fuser classes with finite pseudo-dimension [5], which is described as follows. Let G be a set of functions mapping from a domain X to $\Re$ and suppose that $S = \{x_1, x_2, \ldots, x_m\} \subseteq X$. Then S is *pseudo-shattered* by F if there are real numbers $r_1, r_2, \ldots, r_m$ such that for each $b \in \{0, 1\}^m$ there is a function g_0 in G with $\text{sgn}(f_b(x_i) - r_i) = b_i$ for $1 \leq i \leq m$. Then G has the *pseudo-dimension* d if d is the maximum cardinality of a subset S of X that is pseudo-shattered by G. If no such maximum exists, we say that G has infinite pseudo-dimension. The pseudo-dimension of G is denoted Pdim (G). Pseudo-dimensions are known for several classes such as sigmoid neural networks, vector spaces, and linear combinations [5].

Theorem 8 [63]. *Consider that the physical law is of bounded variation such that* $|L(p)| \leq M_L$ for all p. *Let parameters, estimators and measurements are bounded. Let each fuser class* F_i *have finite pseudo-dimension* d_i, *and each fuser function g be bounded such that* $|g(.)| \leq M$ *for all parameters. Let* $d = \sum_{i=1}^{n} d_i$. *Then given a sample of size*

$$s = \frac{256 M_L^2}{\varepsilon^2} \left[4d \ln\left(\frac{128\, e M}{\varepsilon} \right) + (n+1) \ln(4/\delta) \right],$$

we have

$$\mathbf{P}[E(\hat{f})-E(f^*)>\epsilon]\leq\delta,$$

irrespective of the sensor distributions. Furthermore, $E(\hat{f})\rightarrow E(f^*)$, as $s\rightarrow\infty$.

The following corollary is a weaker version of Theorem 8 since $E(f^*)\leq E(S^*)\leq E(\hat{S})$.

Corollary 1 Let F_i *satisfy the isolation property for all* $i = 1, 2, \ldots, n$. *Under the same conditions as Theorem 8, we have following conditions satisfied.*

$$\mathbf{P}[E(\hat{f})-E(S^*)>\epsilon]\leq\delta \text{ and } \mathbf{P}[E(\hat{f})-E(\hat{S})>\epsilon]\leq\delta.$$

Informally speaking, this corrollary shows that the error of the computed fuser $\hat{f}$ is not likely to be much higher than that of the best basic set, and could be much smaller. Theorem 8 offers a stronger result: $\hat{f}$ will be closer to f^* which can have much smaller error than S^*.

8.5 METHANE HYDRATES WELL LOGS

We now briefly describe a practical application of the proposed fusion method based on physical laws. Gas hydrates are crystalline substances composed of water and gas, in which gas molecules are contained in cage-like lattices formed by solid water. A challenge is to predict the prescence of hydrates using measurements collected at wells located in certain locations such as off the US coast in mid-Atlantic and Mackenzie Delta in Northwest Canada. A number of measurements are collected at each well using a suite of sensors. These measurements include density, neutron porosity, acoustic transit-time, and electric resistivity, collected at various depths in the well [15]. We consider only the estimation of the *porosity* at various depths. Our data consists of 3045 sets of measurements each collected at different depths in a single well. There are a variety of methods to estimate porosity based on different principles and utilizing different measurements. We employed six known methods for estimating the porosity based on neutron measurements ($\hat{\phi}_1$), density measurements ($\hat{\phi}_2$), fluid velocity equation ($\hat{\phi}_3$), acoustic travel time based on S-wave ($\hat{\phi}_4$), time-average equation based on P-wave ($\hat{\phi}_5$), and Wood's equation ($\hat{\phi}_6$).

We consider one of the well-established physical laws relates the parameters of porosity (ϕ), density (ρ), and hydrate concentration (ψ), as follows

$$L[\phi, \psi, \rho] = (\phi[\rho_m - (1-\psi)\rho_w + \psi\rho_h] - \rho + \rho_m)^2 = 0,$$

where ρ_m, ρ_w, and ρ_h are known constants. In this equation, we use the only one measurement for density $\hat{\rho}$ and and a single estimator $\hat{\psi}$ for the hydrate concentration using the Archie's equation. We consider a fuser based on the linear combination of the estimators

$$\hat{\phi}_F = w_7 + \sum_{i=1}^{6} w_i \hat{\phi}_i ,$$

where $(w_1, \ldots, w_7) \in \Re^7$ is the weight vector that minimizes the error based on measurements. The error achieved by $\hat{\phi}_F$ is about 20 times better than that of the best estimator $\hat{\phi}_4$ (details can be found in [62]). Note that $L[.]$ and the fusers employed here satisfy the conditions of Corollary 1. Incidentally, they also satisfy the smoothness conditions of [62].

9. Known Distributions

It is not necessary that all sensor distributions be unknown to apply the main results of this paper. Consider that the joint conditional distribution $P_{Y(1), \ldots, Y(M) | Y(M+1, \ldots, Y(N)}$ of the sensors $S_1, \ldots, S_M$, for $M < N$, is known. Then we can rewrite

$$I_F(f) = \int \Phi(X, f(Y)) \, dP_{Y^{(M+1)}, \ldots, Y^{(N)}, X}$$

where $\Phi(.)$ is suitably derived from the original cost function $C(.)$ and the known part of the conditional distribution. Then various theorems can be applied to this new cost function with only a minor modification. Since the number of variables with unknown distribution is reduced now, the statistical estimation process is easier. It is important note that it is not sufficient to know the individual distributions of the sensors, but the *joint* conditional distributions are required. In the special case of statistical independence of sensors the joint distribution is just the product, which makes the transformation easier. In general for sensor fusion problems, however, the interdependence between the sensors is a main feature to be exploited to overcome the limitations of single sensors.

10. Conclusion

In a multiple sensor system, we considered that for each sensor the outputs are related to the actual feature values according to a certain probability distribution. We presented an overview of informational and computational aspects of a fuser that combines the sensor outputs to more accurately predict the feature, when the sensor distributions are unknown but iid measurements are given. Our performance criterion is the probabilistic guarantees in terms of distribution-free sample bounds based entirely on a finite sample. We first discussed a number of methods based on the empirical risk minimization approach, which yield a fuser that is guaranteed, with a high probability, to be close to an optimal fuser. Note that the optimal fuser is computable only under a complete knowledge of sensor distributions. Then we described the isolation fusers that are guaranteed to perform at least as good as the best sensor. We then discussed the projective fusers that are guaranteed to perform at least as good as the best subset of sensors. We briefly discussed the notion of metafusers that can combine fusers of different types. We considered physical systems wherein the training data consisting of actual physical values is not available. For this case, we discussed methods that utilize the physical laws to obtain a suitable fuser.

The overall focus of this paper is very limited: we only considered sample-based fuser methods that provide finite sample performance guarantees. Even then, there are a number of important issues for the fuser rule computation that have not been covered in this paper. We did not discuss stochastic algorithms for fuser computation [39,55]. Another important aspect is the utilization of fusers that have been designed for known distribution cases for the sample-based case. In many important cases, the fuser formulas expressed in terms of probabilities can be converted into sample-based ones by utilizing suitable estimators [41]. For the most part, we only considered stationary systems, and it would of future interest to study sample-based fusers for time-varying systems.

11. Acknowledgments

This research is sponsored by the Engineering Research Program of the Office of Basic Energy Sciences, U.S. Department of Energy, under Contract No. DE-AC05-00OR22725 with UT-Battelle, LLC, Office of Naval Research under order No. N00014-96-F-0415, and Ballistic Missile Defense Organization under MIPR No. 0100568954.

12. References

1. Abid, M.A. and Gonzalez, R.C., editors (1992), *Data Fusion in Robotics and Machine Intelligence.* Academic Press, New York.
2. Aizerman, M.A., Braverman, E.M., and Rozonoer, L.I. (1970), *Extrapolative problems in automatic control and method of potential functions*, Volume 87 of American Mathematical Society Translations, p. 281-303.
3. Alon, N., Ben-David, S., Cesa-Bianchi, N., and Hausler, D., (1993), Scale-sensitive dimensions, uniform convergence, and learnability. *In Proc of 1993 IEEE Symp. on Foundations of Computer Science.*
4. Anthony, M. (1994), Probabilistic analysis of learning in artificial neural networks: The PA model and its variants. Neuro COLT Technical Report Series NC-TR-94-3, Royal Holloway, University of London.
5. Anthony, M., and Bartlett, P.L., (1999). *Neural Network Learning: Theoretical Foundations.* Cambridge University Press.
6. Apostol, T.M. (1974), *Mathematical Analysis.* Addison-Wesley Pub. Co.
7. Bar-Shalom, Y., and Li, X.R. (1995), *Multitarget-Multisensor Tracking: Principles and Techniques.* YBS Publishing.
8. Billingsley, P. (1986), *Probability and Measure.* John Wiley and Sons, New York, second edition.
9. Blum, A.L., and Rivest, R.L., (1992), Training a 3-node neural network is NP-complete. *Neural Networks*, 5:117-127.
10. Brady, J.M., (1988). Forword. *Int. J. Robotics Research*, 8(5):2-4. Special Issue on Sensor Data Fusion.
11. Brieman, L. (1996), Stacked regressions. Machine Learning, 24(1): 49-64.
12. Chow, C.K., (1965), Statistical independence and threshold functions. *IEEE Trans. Electronic Computer*, EC-16:66-68.
13. Ciesielski, Z., (1988). Harr system and nonparametric density estimation in several variables. *Probability and Mathematical Statistics*, 9:1-11.
14. Cybenko, G. (1989), Approximation by superpositions of a sigmoidal function. *Mathematics of Controls, Signal, and Systems*, 2:303-314.

15. Dallimore, S.R., Uchida, T., and Collett, T.S., editors (1999), *Scientific Results from JAPEX/JNOC/GSC/ Mallik 2L-38 Gas Hydrate Research Well, Mackenzie Delta: Geological Survey of Canada Bulletin 544.* Geological Survey of Canada, Bulletin 544.

16. Dasarathy, B.V., (1994), *Decision Fusion.* IEEE Computer Society Press, Los Alamitos, California.

17. deLaplace, P.S., (1818). Deeuxième supplement à la théorie analytique des probabilitiés. Reprinted (1847) in *Oeuvres Complètes de Laplace*, vol. 7 (Paris, Gauthier-Villars) 531-580.

18. Devroye, L., Gyorfi, L., and Lugosi, G., (1996), *A Probabilistic Therory of Pattern Recognition.* Springer-Verlag, New York.

19. Dudley, R. (1987). Universal Donsker classes and metric entropy. *Annals of Probability*, 15:1306-1326.

20. Engel, J. (1994). A simple wavelet approach to nonparametric regression from recursive partitioning schemes. *Journal of Multivariate Analysis*, 49:242-254.

21. Granger, C.W.J. (1989), Combining forecasts – twenty years later. *J. of Forecasting*, 8:167-173.

22. Grofman, B. and Owen, G. editors (1986). *Information Pooling and Group Decision Making.* Jai Press, Inc., Greenwich, Connecticut.

23. Hashem, S. (1997), Optimal linear combination of neural networks. *Neural Networks*, 10⊗4):599-614.

24. Hashem, S., Schmeiser, B., and Yih, Y. (1994). Optimal linear combinations of neural networks: An overview. In *Proceedings of 1994 IEEE Conf. On Neural Networks*, pages 1507-1512.

25. Haussler, D. (1992). Decision theoretic generalizations of the PAC model for neural net and other learning applications. *Information and Computation*, 100:78-150.

26. Juditsky, A. and Nemirovski, A. (1996). Functional aggregation for nonparametric estimation. *Publication Interne*, IRISA, (993).

27. Kearns, M.J. and Schapire, R.E. (1994). Efficient distribution-free learning of probabilistic concepts. *Journal Computer and Systems Sciences*, 48(3):464-.

28. Koiran, P. and Sontag, E.D. (1997). Neural networkswith quadratic VC dimension, *Journal of Computer and System Sciences.* 54:190-198.

29. Kurkova, V. (1992). Kolmogorov's theorem and multilayer neural networks. *Neural Networks*, 5:501-506.

30. Lugosi, G. and Zeger, K. (1995). Nonparametric estimation viaa empirical risk minimization. *IEEE Transactons on Information Theory*, 41(3):677-687.

31. Madan, R.N. and Rao, N.S.V. (1999). Guest editorial on information/decision fusion with engineering applications. *Journal of Franklin Institute*, 336B(2).199-204.

32. Mojirsheibani, M. (1997). A consistent combined classification rule. *Statistics and Probability Letter*, 36:43-47.

33. Nadaraya, E.A. (1989). *Nonparametric Estimation of Probability Densities and Regression Curves.* Kluwer Academic Publisher, Dordrecht.

34. Nolan, D. and Pollard, D. (1987), U-Processes: Rates of convergence. *Annals of Statistics*, 15(2):780-799.

35. Pollard, D. (1990). *Empirical Processes: Theory and Applications.* Institute of Mathematical Statistics, Haywood, California.

36. Preparata, F.P. and Shamos, I.A. (1985). *Computational Geometry: An Introduction, Springer-Verlag*, NewYork.

37. Rao, B.L.S.P. (1983). *Nonparametric Functional Estimation.* Academic Press, New York.

38. Ras, N.S.V. (1991). Computational complexity issues in synthesis of simple distributed detection networks. *IEEE Transactions on Systems, Man and Cybernetics*, 21(5):1071-1081.

39. Rao, N.S.V. (1994). Fusion methods for multiple sensor systems with unknown error densities. *Journal of Franklin Institute*, 331B(5):509-530.

40. Rao, N.S.V. (1995). Fusion rule estimation in multiple sensor systems using training. In Bunke, H., Kanade, T., and Noltemeier, H.m editors, *Modelling and Planning for Sensor Based Intelligent Robot Systems.* Pages 1799-190. World Scientific Pub.

41. Rao, N.S.V. (1996). Distributed decidion fusion using empirical estimation. *IEEE Transactions on Aerospace and Electronic Systems*, 33(4):1106-1114.

42. Rao, N.S.V. (1997b). Nadaraya-Watson estimator for sensor fusion. Optical Engineering, 36(3):642-647.

43. Rao, N.S.V. (1997b). Nadaraya-Watson estimator for sensor fusion problems. *In Proceedings of IEEE International Conference on Robotics and Automation*, pages 2069-2074.

44. Rao, N.S.V. (1998a). A fusion method that performs better than best sensor. *In First International Conference on Multisource-Multisensor Data Fusion*, 19-26.
45. Rao, N.S.V. (1998b). To fuse or not to fuse: Fuser versus best classifier. *In SPIE Conference on Sensor Fusion: Architectures, Algorithms, and Applications 11*, pages 25-34.
46. Rao, N.S.V.(1998c). Vector space methods for sensor fusion problems. *Optical Engineering*, 37(2):499-504.
47. Rao, N.S.V. (1999a). Fusion methods in multiple sensor systems using feedforward neural networks. *Intelligent Automation and Soft Computing*, 5(1):21-30.
48. Rao, N.S.V.(1999b). Multiple sensor fusion under unknown distributions. *Journal of Franklin Institute*, 336(2)289-299.
49. Rao, N.S.V. (1999c). On optimal projective fusers for function estimators. In *IEEE/SICE/RSJ International Conference on Multisensor Fusion and Integration for Intelligent Systems*, pages 1-6.
50. Rao, N.S.V. (1999d). On optimal projective fusers for function estimators. In *Second International Conference on Information Fusion*, pages 296-301.
51. Rao, N.S.V. (1999e). Simple Sample bound for feedforward sigmoid networks with bounded weights. *Neurocomputing*. 29:115-44.
52. Rao, N.S.V. (2000). Finite sample performance guarantees of fusers for function estimators. *Information Fusion*, 1(1):35-44.
53. Rao, N.S.V. and Iyengar, S.S. (1996). Distributed decision fusion under unknown distributions. *Optical Engineering*, 35(3):617-624.
54. Rao, N.S.V., Iyengar, S.S., and Kashyap, R.L. (1993). Computational complexity of distributed detection problems with information constraints. *Computers and Electrical Engineering*, 19(6):445-451.
55. Rao, N.S.V. and Oblow, E.M. (1994). Majority and location-based fusers for PAC concept learners. *IEEE Trans. On Syst., Man and Cybernetics*, 24(5):713-727.
56. Rao, N.S.V. and Oblow, E.M. (1997). N-learners problem: System of PAC learners. In *Computational Learning Theory and Natural Learning Systems, Vol IV: Making Learning Practical*, pages 189-210. MIT Press.
57. Rao, N.S.V., Oblow, E.M. and Glover, C.W. (1994a), Learning separations by Boolean combinations of hyperplanes. *IEEE Transactions on Pattern Analysis and Machine Intelligence*, 16(7): 765-768.
58. Rao, N.S.V., Oblow, E.M., Glover, C.W., and Liepins, G.E. (1994b). N-learners problem: Fusion of concepts. *IEEE Transactions on Systems, Man and Cybernetics*, 24(2):319-327.
59. Rao, N.S.V. andProtopopescu, V. (1996). On PAC learning of functions with smoothness properties using feedforward sigmoidal networks. *Proceedings of the IEEE*, 84(10):1562-1569.
60. Rao, N.S.V. and Protopopescu, V. (1998). Function estimation by feedforward sigmoidal networks with bounded weights. *Neural Processing Letters*, 7:125-131.
61. Rao, N.S.V., Protopopescu, V., Mann, R.C. Oblow, E.M., and Iyengar, S.S. (1996). Learning algorithms for feedforward networks based on finite samples. *IEEE Trans. On Neural Networks*, 7(4):926-940.
62. Rao, N.S.V., Reister, D.B., and Barhen, J. (2000a). Fusion method for physical systems based on physical laws. *In Proceedings fo 3rd International Conference on Information Fusion.*
63. Rao, N.S.V., Reister, D.B., and Barhen, J. (2000b). Information fusion in physical systems using physical laws. *In Proceedings of Eighteenth Symposium on Energy Engineering Sciences.*
64. Roychowdhury, V., Siu, K., and Orlitsky, A., editors (1994). *Theoretical Advaces in Neural Computation and Learning*. Kluwer Academic Pub., Boston.
65. Sima, J. (1996). Back-propagation is not efficient. *Neural Networks*, 9(6):1017-1023.
66. Talagrand, M. (1994). Sharper bounds for Gaussian and empirical processes. *Annals of Probability*, 22(1):28-76.
67. Tang, Z., and Koehler, G.J. (1994). Lipschitz properties of feedforward neural networks. Technical report.
68. Taniguchi, M. and Tresp. V. (1997). Averaging regularized estimators. *Neural Computation*, 9:1163-1178.
69. Tsitsiklis, J.N. and Athans, M. (1985). On the complexity of decentralized decision making and detection problems. *IEEE Transactions on Automatic Control*, AC-30(5):440-446.
70. Valiant, L.G. (1984). A theory of the learnable. *Communications of the ACM*, 27(11):1134-1142.
71. van der Vaart, A.W. and Wellner, J.A. (1996). *Weak Convergences and Empirical Processes*. Springer-Verlag, New York.

72. Vapnik, V. (1982). *Estimation of Dependences Based on Empirical Data.* Springer-Verlag, New York.
73. Vapnik, V.N. (1995). *The Nature of Statistical Learning Theory.* Springer-Verlag, New York.
74. Varshney, P.K. (1996). *Distributed Detection and Data Fusion.* Springer-Verlag.
75. Vavasis, S.A., (1991). *Nonlinear Optimization.* Oxford University Press, New York.
76. von Neumann, J. (1956). Probabilistic logics and the synthesis of reliable organisms from unreliable components. In Shannon, C.E. and McCarthy, J., editors, *Automata Studies*, pages 43-98. Princeton University Press.

DATA ASSOCIATION AND MULTITARGET TRACKING

J.-P. LE CADRE
IRISA/CNRS
Camput de Beaulieu
35 042, Rennes, France
email:lecadre@irisa.fr

1. Introduction

Multitarget tracking (MTT here after) deals with state estimation of an unknown number of (moving) targets. Available measurements may have originated from the targets if they are detected or of a special model called "clutter". Clutter is generally considered as a model describing false alarms. Its (spatio-temporal) statistical properties are quite different from target ones which render possible the separation of target tracks on the one hand and clutter model on the second. To perform multitarget tracking the observer has at its disposal a huge amount of data, possibly collected on multiple sensors [2]. Elementary measurements are receiver outputs; e.g. bearings, ranges, time-delays, dopplers, etc. But the main difficulty comes from the assignment of a given measurement to a target model. For critical situations, these assignments are unknown, as are the true target models. This is a neat departure from classical estimation problems. Two distinct problems have to be solved *jointly* : data association and estimation.

Since the mid-sixties, this subject has attracted considerable attention. This interest is due to the challenging difficulty of the problem as well as to the operational requirements and computation possibilities. The simpler approach is probably the Nearest Neighbor approach. However, despite its simplicity, it has fundamental limitations. A natural framework is to face the combinatorial complexity of the data association problem. Pioneering work in this way has been made by Morefield [11], using integer programming. More recently, this approach has been greatly improved by means of approximate methods (Lagrangian duality) for solving the combinatorial assignment problem [12,13]. Other approaches are more relevant to the general class called Multiple Hypotheses Tracker (MHT), in which measurements received at a scan are assigned to intialized tracks, new targets or false alarms [15]. Pruning and gating techniques are used to retain the most likely hypotheses and limit their number. By this way, combinatorial explosion may be strictly contained inside reasonable bounds. However, there is a strong risk to eliminate correct assignments.

A fundamentally different approach is to examine the data association problem from a probabilistic viewpoint and more precisely to consider the measurement-model asignment as missing data. This way has been pioneered by Avitzour [1]. In their seminal papers, Streit and Luginbuhl gave a more rigorous application of the EM algorithm, the PMHT algorithm. This probabilistic approach is original and avoids enumeration of the measurement to target assignment. Following these guidelines, we give a general presentation of the PMHT. The

A. K. Hyder et al. (eds.), Multisensor Fusion, 331–349.

general framework is valid either for a linear measurement model or a non-linear one, deterministic states or random ones. However, despite its generality and versatility, it may suffer from convergence problems.

This is especially true for non-linear measurements (e.g., passive sonar measurements). Actually, in this case, the likelihood functional is (generally) not convex and numerous local extrema exist. Therefore, it seems impossible to completely avoid combinatorial problems. These problems may be seen from various wiewpoints, ranging from performing heuristics (e.g., gating, pruning) to more rigorous approaches like Lagrangian relaxation. In all the cases, it is more than likely than efficient multitarget tracking should require combination of various approaches.

2. A general formulation of the multitarget tracking problem

Let us first present the general context of MTT as well as the basic definitions, which is very roughly:

- Associate the measurements obtained at various instants to form tracks.
- A track is a measurement subset corresponding to a common hypothesis (e.g. trajectory).
- Determination of the track parameters.

So, we must to solve **two** problems, completely entangled:

- A data association problem.
- An estimation problem. Assume that we dispose of (spatio-temporal) measurements collected on T instants, denoted Y. We thus have :

$$Y = \{Y(1), \cdots, Y(T)\} \ .$$

Each measurement vector $Y(t)$ is made of all the elementary measurements (e.g., estimated bearings) received at the time (or scan) t. Each measurement is the result of an elementary detection process (thresholding). At each time, a new measurement vector is received, i.e.:

$$Y(t) = (y_1(t), \cdots, y_{m_t}(t)) \ ,$$

where m_t is the (variable) number of measurements received at time t. Variations in measurement number are due to numerous factors: non-detection, false alarms, track merging, etc. Associated with this measurement set, there is a vector set

$$X \ (X = \{X(1), \ldots, X(T)\} \ )$$

Target motions are generally modelled by Markov process, i.e.:

$$x_s(t) = f_s[x_s(t-1), v_s(t)] \ .$$

A general assumption is that the function f_s is linear and v is w.g.n (white gaussian noise). The associated measurement equation is :

$$y(t) = h_s[x_s(t), n_s(t)] \ , \tag{1}$$

where $n_s(t)$ is the measurement noise and $y(t)$ the measurement originating from model s. In order to take into account the uncertainty in the measurement origin, the assignment vector is introduced, i.e.

$$\begin{cases} K = \{K(1), \ldots, K(T)\} \ , \\ K(T) = (k_1(t), \ldots, km_t(t)) \ , \\ k_j(t) = m \Leftrightarrow j-th\ measurement\ is\ associated\ with\ model\ m \ . \end{cases} \tag{2}$$

A track T_i is defined as a set of measurements associated with the same model i:

$$T_i = \{y_j(t) \mid k_j(t) = i,\ 1 \le t \le T\},\ 1 \le i \le M \tag{3}$$

In this definition, a track may contain missed detections. A partition of the measurement set Y into tracks is our general aim, and a general definition of our problem is:

Estimate X, from measurements Y, when K is not available.

A (very) brief history of MTT ;

- "Exhaustive" (but practically pruning and gating) of the set of hypotheses: MHT (Multiple Hypothesis Tracker) [15].
- Combinatorial approaches [11,12].
- Probabilistic approaches:

Jpdaf (Joint Probabilistic Data Association Filter [2], PMHT (Prob. Multi - Hpotheses Tracking) [18]. (4)

Concerning probabilistic approaches, classical constraints on track formation are :

(a) A measurement is either associated with a target track or a false alarm.

(b) A target can generate *at most* one measurement. If we denote P_n a partition of the observation space in tracks at time n, the probability of such a partition is computed recursively ($Y^n=(Y_1,\ldots,Y_n)$):

$$p(P_n|Y^n) = c\ p(Y(n)|P_{n-1}, K(n), Y^{n-1}) \ , \tag{5}$$

$$\times p(K(n)|P_{n-1}, Y^{n-1})\ p(P_{n-1}|Y^{n-1}) \ .$$

In MHT, when a new set of measurements is received, the probability $p(P_n \mid Y^n)$ is calculated recursively by means of the previous formula, while jpdaf deals with the two first terms of the previous equlaity.

Conversely in the PMHT [17], the basic idea is to introduce data association in the estimation. More precisely, the assignment vector is considered as a random variable. Let Π the vector of the assignment probabilities of a measurement to a model, its components are defined by:

$$\pi_m(t) = p[k_j(t) = m], \quad \forall j = 1, \ldots m_t . \tag{6}$$

We note that there is no ordering in the measurements and that assignment variables are assumed independent. The probability $\pi_m(t)$ is thus independent of a particular measurement. Under these assumptions, a general expression of the likelihood functional is:

$$p(Z|\Phi) = \prod_{t=1}^{T} \prod_{i=1}^{m} \left\{ \sum_{m=0}^{M} p(z_i | x_m)\, \pi_m(t) \right\}$$

The density of each measurement is a mixture of M+ elementary densities. The problem is then to estimate the vector O = (X,Π).

To that aim, the E.M. algorithm will be instrumental.

3. Multitarget tracking as an incomplete problem

We shall deal, in a first time, with the estimation of the component proportions (and parameters) of a normal mixture. In this framework, the calculations are relatively simple but they constitute the basis of PMHT and PMHT-like methods [14,6]. However, even if these methods are statistically sounded they may suffer from serious drawbacks, mainly related to their initialization. To remedy this problem, the combinatorial approach appears to be a feasible and efficient way.

3.1 THE EM METHOD

We briefly present here the two steps of the Expectation Maximization (EM) method [10] Let $Y=\{y\}$ be the observations (sensor outputs), with an associated probability density function (PDF) $g(y;\Psi)$. Furthermore, we denote the $gc(y;\Psi)$ PDF of *complete* observations x. and $\ln L_c(\Psi) \equiv \ln g_c(x:\Psi)$, its associated likelihood functional. We then have:

$$g(y, \Psi) = \int_{x(y)} g_c(x; \Psi)\, dx, \quad \text{where:} \quad X(y) = \{x | y(x) = y\} . \tag{7}$$

The E.M algorithm is then made of two steps :

(a) E-step: calculation of $Q(\Psi, \Psi_k) = \mathbf{E}_{\Psi_k}[\ln(g_c(x; \Psi)|y]$,

(b) M-step: $Q(\Psi_{k+1}, \Psi_k) \geq Q(\Psi, \Psi_k), \forall \Psi \in \Omega$.

Then, the following results hold:

$(L(\Psi) \equiv g(y, \Psi)) \rightarrow L(\Psi_{k+1}) \geq L(\Psi_k)$,
$L(\Psi_k) \rightarrow L(\Psi_*)$, where Ψ_* is a stationary point of the E.M. algorithm.

3.2 ESTIMATION OF THE PROPORTIONS OF A DENSITY MIXTURE [10]

Consider the following density mixture defined by

$$f(y;\Psi) = \sum_{i=1}^{g} \pi_i \, f_i(y_i) \, . \tag{8}$$

We define the parameter vector Ψ as:

$$\Psi = (\pi_1, \ldots, \pi_g - 1)^t \ , \quad \text{subject to:} \quad \sum_{i=1}^{g} \pi_i = 1. \tag{9}$$

$Y = (y_1, \ldots, yn)$ are the observations. Then the incomplete log-likelihood functional $L(\Psi)$ is defined by:

$$\ln L(\Psi) = \sum_{j=1}^{n} \ln\left(\sum_{i=1}^{g} \pi_i \, f_i(y_j) \right) . \tag{10}$$

Let us introduce now unobservable variables (assignment variables):

$$Z = (z_1, \ldots, z_n), \quad \text{where:} \quad (z_j)_i \equiv z_{ij} \in \{0,1\} \text{ is the } i-th \text{ component of the } j-th \text{ assignment vector.} \tag{11}$$

We shall now detail the expectation and maximization steps of the EM algorithm[3].

3.2.1 E-Step

Consider now the complete observation $X=(Y,Z)$ and its associated functional $L_c(\Psi)$:

$$\ln L_c(\Psi) = \sum_{i=1}^{g}\sum_{j=1}^{n} z_{ij} \ln \pi_i + \sum_{i=1}^{g}\sum_{j=1}^{n} z_{ij}(f_i(y_j)) \ , \tag{12}$$

and let $\tilde{z}_{ij}$ therandom variable associated with z_{ij}. We must calculate $E_{\Psi k}(\tilde{z}_{ij}|Y) = pr_{\Psi k}(\tilde{z}_{ij}|Y) \equiv z_{ij}^k$.

This is achieved by using Bayes theorem, yielding:

$$z_{ij}^{k} = \frac{p(Y|z_{ij}=1;\Psi_k)\ p(z_{ij}=1;\Psi_k)}{p(Y;\Psi_k)}, \qquad (13)$$
$$= \frac{\pi_i^k f_i(y_j)}{\sum_{i=1}^{g} \pi_i^k f_i(y_j)}.$$

In the above formula $z_{ij}^{k} = \tau_i(y_j;\Psi^k)$ represents an *a posteriori* density.

3.2.2 M-Step

We then have to deal with the following optimization problem:

$$\max_{\pi_i} \ln\ L_c(\Psi) = \sum_{i=1}^{g}\sum_{j=1}^{n} z_{ij}^k \ln \pi_i + \text{cst.}\ ,$$
$$\text{subject to :} \qquad (14)$$
$$\sum_{i=1}^{g} \pi_i = 1\ .$$

Since the optimization variables (π_i) are separated both in the cost functional and the (equality) constraint, this an elementary optimization problem. Forming the associated Lagrangian and solving sationarity conditions, we obtain straightforwardly:

$$\begin{cases} \pi_i^{k+1} = \frac{1}{n}\left(\sum_{j=1}^{n} z_{ij}^k\right), \\ \\ z_{ij}^k = \frac{\pi_i^k f_i(y_j)}{\sum_{i=1}^{g} \pi_i^k f_i(y_j)}. \end{cases} \qquad (15)$$

3.3 ESTIMATION OF THE COMPONENTS OF A MIXTURE OF DENSITIES [14]

As previously, we denote X the complete data ($X=(Y,K)$); Y the incomplete data (observations) and K the assignment vector. We then have:

$$\begin{cases} X = ((y_1, k_1), \ldots, \ldots, (y_n, k_n)), \\ Y = (y_1, \ldots, y_n), \\ K = (k_1, \ldots, k_n), p(y_j | \Phi) = \Sigma_i \pi_i p(y_j | \Phi). \end{cases} \tag{16}$$

3.3.1 E-step

For the expectation step, we must calculate the quantity $Q(\Phi,\Phi_i)$:

$$\begin{aligned} Q(\Phi|\Phi_i) &= \boldsymbol{E}_{\Phi_i}\left[\ln p(X|\Phi) \; |Y\right], \\ &= \boldsymbol{E}\left[\ln p(X|\Phi) \; |Y, \Phi_i\right], \\ &= \sum_{K,Y} \ln p(X|\Phi) \; p(K|Y, \Phi_i) \; . \end{aligned} \tag{17}$$

To that aim, calculation of the elementary terms of $Q(\Phi,\Phi_i)$ is needed and is detailed below.

(a) Calculation of $p(X|\Phi$

$$\begin{aligned} p(X|\Phi) &= p(Y, K|\Phi) = p(Y|K, \Phi) \; p(K|\Phi) \\ &= \prod_{j=1}^{n} p(y_j|\phi_{k_j}) \; \pi_{k_j} \; . \end{aligned} \tag{18}$$

(b) Calculation of $p(Z|Y,\Phi_i)$

$$p(K|Y, \Phi_i) = \frac{p(K, Y|\Phi_i)}{p(Y|\Phi_i)} = \prod_{j=1}^{n} \frac{p(y_j|\phi_{k_j}) \; \pi_{k_j}}{p(y_j|\Phi)} \; . \tag{19}$$

(c) Calculation of $Q(\Phi,\Phi_i)$

Collecting the preceding results, we obtain:

$$\begin{aligned} Q(\Phi, \Phi_i) &= \sum_{k_1=1}^{m} \cdots \sum_{k_n=1}^{m} \left\{ \sum_{j=1}^{n} \ln\left(\pi_{k_j} p(y_j|\phi_{k_j})\right) \prod_{j'=1}^{n} \frac{p(y_{j'}|\phi_{k^i_{j'}}) \; \pi_{k^i_{j'}}}{p(y_{j'}|\Phi)} \right\}, \\ &= \sum_{j=1}^{n} \left[\sum_{k_1=1}^{m} \cdots \sum_{k_n=1}^{m} \ln\left(\pi_{k_j} p(y_j|\phi_{k_j})\right) \prod_{j'=1}^{n} \frac{p(y_{j'}|\phi_{k^i_{j'}}) \; \pi_{k^i_{j'}}}{p(y_{j'}|\Phi)} \right] . \end{aligned} \tag{20}$$

It remains to simplify this expression. Consider then a term of this sum and denote j the corresponding index. There is an index j' of the product equal to j, so that:

$$\sum_{k_1=1}^{m} \cdots \sum_{k_n=1}^{m} \ln\left(\pi_{k_j} p(y_j|\phi_{k_j}\right) \left\{ \frac{p(y_{j'}|\phi_{k^i_{j'}})\ \pi_{k^i_{j'}}}{p(y_{j'}|\Phi)} \right\}$$

$$= \sum_{k_j=1}^{m} \ln\left(\pi_{k_j} p(y_j|\phi_{k_j}\right) \frac{p(y_j|\phi_{k^i_j})\ \pi_{k^i_j}}{p(y_j|\Phi)} \times \left(\sum_{k_1=1}^{m} \cdots \sum_{k_n=1}^{m} \prod_{j' \neq j} \frac{p(y_{j'}|\phi_{k^i_{j'}})\ \pi_{k^i_{j'}}}{p(y_{j'}|\Phi)} \right) . \tag{21}$$

Let S_j be this sum. The following (elementary) remark is then *instrumental*:

$$\left(\sum_{k_1=1}^{m} \cdots \sum_{k_n=1}^{m} \prod_{j' \neq j} \frac{p(y_{j'}|\phi_{k^i_{j'}})\ \pi_{k^i_{j'}}}{p(y_{j'}|\Phi)} \right),$$

$$= \sum_{k_1} \cdots \sum_{k_{j-1}} \sum_{k_{j+1}} \cdots \sum_{k_n} p\left(k_1, \cdots, k_{j-1}, k_{j+1}, \cdots, k_n | y, \Phi_i\right) = 1 . \tag{22}$$

We then have:

$$S_j = \sum_{k_1=1}^{m} \ln\left(\pi_{k_j} p(y_j|\phi_{k_j}\right) \frac{p(y_j|\phi_{k^i_j})\ \pi_{k^i_j}}{p(y_j|\Phi)} ,$$

so that, finally we have obtained:

$$\begin{aligned} Q(\Phi, \Phi_i) &= \sum_{k=1}^{m} \left[\sum_{j=1}^{n} \frac{p(y_j|\phi^i_k)\ \pi^i_k}{p(y_j|\Phi^i)} \right] \ln \pi_k , \\ &\quad + \sum_{k=1}^{m} \sum_{j=1}^{n} \ln\left(p(y_j|\phi_k\right) \frac{p(y_j|\phi^i_k)\ \pi^i_k}{p(y_j|\Phi^i)} , \end{aligned} \tag{23}$$

which ends the "E" step.

3.3.2 M-Step

Let us consider now the "M" step. We have to solve the following optimization problem:

$$\begin{cases} \max_\pi g(\pi) = \sum_{k=1}^{m} c_k \ln \pi_k \,, \\ \quad where: \\ \quad c_k = \dfrac{p(y_j|\phi_k^i)\ \pi_k^i}{p(y_j|\Phi^i)} \,, \\ subject\ to: \sum_{k=1}^{m} \pi_k = 1 \,. \end{cases} \tag{24}$$

Again, this is an elementary optimization problem, with an equality constraint. Forming the Lagrangian and solving stationarity conditions, we obtain (i:iteration index):

$$\pi_k = \frac{c_k}{n} = \frac{1}{n} \sum_{j=1}^{n} \frac{\pi_k^i p(y_j|\phi_k^i)}{p(y_j|\Phi^i)} \,. \tag{25}$$

3.3.3 The algorithm[14], [10]

Summarizing previous results, we can give now the general form of the E.M. algorithm for estimating the mixture components for unlabelled data:

E-step : updating π_k, k=1,..., m, (i+1) -th iteration:

$$\pi_k^{i+1} = \frac{1}{n} \sum_{j=1}^{n} w_{j,k}^{i+1} \,,$$

$$where\ the\ weights\ w_{j,k}^{i+1}\ are\ given\ by: \tag{26}$$

$$w_{j,k}^{i+1} = \frac{\pi_k^i\ p(y_j|\phi_k^i)}{\sum_{k=1}^{m} \pi_k^i\ p(y_j|\phi_k^i)} \,,$$

M-step : updating ϕ_k, k=1,...,m, (i+1)-th iteration :

$$\phi_k^{i+1} \in \boldsymbol{arg}\ \max_{\phi_k} \sum_{j=1}^{n} w_{j.k}^{i+1} \ln\left[p(y_j|\phi_k\right] \,. \tag{27}$$

4. Formulation of a Multitarget Association Problem [16,6]

The problem parameters are denoted O (O=X,Π)), where χ is the set of the target state vectors and Π the vector of assignment probabilities. T scans are available and each scan is made of m_T measurements. The number of target models is denoted M. Again X represents complete data (X=(Y,K)), Y the observations. Assuming temporal and spatial independence of the components of the assignment vectors (K), we have:

$$p(X|O) = \prod_{t=1}^{T}\prod_{j=1}^{m_t} p\left(y_j(t)|x_{k_j(t)}\right)\pi_{k_j}(t)\ ,$$
$$p(K|Y,O) = \prod_{t=1}^{T}\prod_{j=1}^{m_t} p(k_j(t)|y_j(t), O(t))\ , \tag{28}$$

whence:

$$Q(O|O^*) = E\ \{\ln p(Y,K|O)|Y, O^*\}\ , \tag{29}$$
$$= \sum_K \ln\left[p(Y,K|O)\right]p(K|Y,O^*)\ ,$$
$$= \sum_K\left\{\sum_{t=1}^{T}\sum_{j=1}^{m_t}\ln\left[p(y_j(t)|x_{k_j}(t)\right]\pi_{y_j}(t)\right\}\times\prod_{t=1}^{T}\prod_{j=1}^{m_t}p(k_j(t)|y_j(t), O^*(t))\ .$$

Using the same reasoning as previously, we have:

$$Q(O|O^*) = \sum_{t=1}^{T}\sum_{j=1}^{m_t}\sum_{k_j=1}^{M}\ln\left[\pi_{k_j}(t)\right]w_{j,k_j}(t) \tag{30}$$
$$+\sum_{t=1}^{T}\sum_{j=1}^{m_t}\sum_{k_j=1}^{M}\ln\left[p(y_j(t)|x_k(t)\right]w_{j,k_j}(t)\ .$$

where:

$$w_{j,k}(t) = \frac{\pi_k^*\, p(y_j(t)|x_k^*(t))}{p(y_j(t)|O^*(t))}\ ,$$

which ends the E-step.

Again, the M-step reduces to:

$$
\begin{aligned}
&\forall t \ , \max_{\Pi} \ g(\Pi(t)) = \sum_{k=1}^{M} \sum_{j=1}^{m_t} \ln[\pi_k(t)] \, w_{j,k}(t) \ , \\
&\text{subject to :} \\
&\textstyle\sum_{k=1}^{M} \pi_k(t) = 1 \ ,
\end{aligned} \tag{31}
$$

So that:

$$
\begin{aligned}
\pi_k(t) &= \frac{1}{m_t} \sum_{j=1}^{m_t} w_{j,k}(t) \ . \\
x_k(t) &\in \arg\max_{x_k} \sum_{j=1}^{m_t} \ln\left[p(y_j(t)|x_k(t)\right] \ w_{j,k}(t) \ .
\end{aligned} \tag{32}
$$

We can now give the general form of the PMHT. Notice that it was first derived for a linear modelling of the observation, which is only a special case.

General form of the PMHT [16,17,18] : (i: iteration index.)

(a) Update of the π_k :

$$
\begin{aligned}
\pi_k^{i+1}(t) &= \frac{1}{m_t} \sum_{j=1}^{m_t} w_{jk}^{i+1}(t) \ , \\
\text{where :} & \\
w_{jk}^{i+1}(t) &= \frac{\pi_k^i p(z_j(t)|x_k^i(t))}{p(z_j(t)|\mathcal{O}^i(t))} \ ,
\end{aligned} \tag{33}
$$

for t=1,...., T and k=1, ..., M $(p(z_j(t) \mid O^i(t)) = \Sigma_k \, p(z_j(t) \mid {}^i(t)))$.

(b) State vectors updating:

(1) For the Maximum Likelihood (deterministic target motions):

$$
M.L. : x_{k_0} \in \max \sum_{t=1}^{T} \sum_{j=1}^{m_t} \ln\left[p(y_j(t)|x_{k_0}\right] \, w_{j,k}(t), \quad k_0 = 1, \cdots, M \ . \tag{34}
$$

(2) EM for the map (k=1,...,T):

$$
(x_k^0, \cdots, x_k^T) \in ArgMax_{\chi_k} \left\{ \sum_{t=1}^{T} \sum_{j=1}^{m_t} \ln\left[p(y_j(t)|x_k(t)\right] w_{j,k}(t) + \sum_{t=0}^{T} \ln\left[p(x_k(t)|x_k(t-1)\right] \right\} . \tag{35}
$$

Calculation of the $\pi_k^{i+1}(t)$ is identical to M.L. In the case of a linear model for the observations and Markovian states, we have:

$$\prod_i^{m_t}\left[p(y_i(t)|x_k(t)\right]^{w_{j,k}^{i+1}(t)} \propto \prod_i^{m_t} N(\bar{y}_k(t)|H_k x_k(t),(m_t\pi_k^{i+1})^{-1}R_k) \ , \tag{36}$$

where:

$$\bar{y}_k(t) = \frac{1}{m_t\pi_k^{i+1}(t)} \sum_{j=1}^{m_t} w_{j,k}^{i+1}(t)\, y_j(t)$$

It has thus been possible to derive Kalman-filtering version of the PMHT where the measurements are replaced by their centroids [18] $\bar{y}_k(t)$, with covariance matrix $\tilde{R}_k$,

$$\tilde{R}_k = \frac{1}{m_t\pi_k^{i+1}(t)} R_k$$

5. PMHT for the Passive Sonar

In this context, the measurement equation is given by:

$$\begin{cases} y(t) = \beta(t) + n(t) , \\ \beta(t) = \tan^{-1}(r_x(t) / r_y(t)). \end{cases} \tag{37}$$

For deterministic target motions, the state equations are:

$$x(t+1) = Fx(t) + u(t), \ \ u\text{: } \textit{observer controls}. \tag{38}$$

In this case, state etimation reduces to the determination of the state vector at a reference time, i.e. $x_k(0)$. This can be achieved by means of various optimization methods (Gauss-Newton, min). For Markovian targets this optimization step (see the map version of the PMHT) is more demanding. Methods like HMM tracking (Viterbi), Extended Kalmam Filter (EKF) or particle filtering are then appropriate. In all the cases we stress that, *for multiple target tracking*, convergence toward the true maximum of the likelihood functional is not guaranteed. This is due to interactions between tracks (see Appendix) [8,9].

Especially in this context (non-linear measurement equation), initialization is the main difficulty. To a large extend, it is basically a combinatorial optimization problem for which efficient methods exist (see section 6).

The PMHT framework is quite versatile and permits to take into account more realistic modellings (non-detection, clutter, etc.). For example, a false alarm model [6] may be added so that:

$$p(y_i(t)|x_{k,0}(t)) = \begin{cases} c\ \exp\left\{-\dfrac{1}{2\sigma_k^2(y_i(t-\beta(x_{k,0}(t)))}2\right\}, & 1\le k\le M , \\ 1/U_\beta, & k=M+1 . \end{cases} \tag{39}$$

For simulation results, we refer to [6].

6. Combinatorial Approaches to Data Association [12]

As we have seen previously initialization may be a very critical step for MTT. In fact, it is not possible to completely avoid combinatorial problems.

6.1 THE COMBINATORIAL OPTIMIZATION PROBLEM

We consider a spatio-temporal measurement set Y. This set is made of N scans, the following notations are then used :

- y_{i1} is the i_1 -th measurement of the 1 -st scan, y_0 denotes the "dummy" measurement (non-detection, false alarms).
- $Y_{i1,i2,\ldots,iN} \Delta \{y_{i1}, y_{i2},\ldots,y_{iN}\}$, where $i_1 \in\{0, 1, \ldots, n_1\}$, …, $i_N \in \{0, 1. \ldots, n_N\}$; n_j is the number of measurements in the j -th scan. $Y_{i1,i2,\ldots, iN}$ represents a spatio-temporal association (a track).
- $P = \{Y_{i1,i2,\ldots, iN}\}$ is a feasible partition o y, satisying the following constraints :

$$\begin{aligned} &\text{for } i_j, i'_j = 1, \ldots, n_j \text{ and } i_j \neq i'_j, j = 1, \ldots, N, \\ &Y_{i1,i2,\ldots,iN} \bigcap Y_{i'1,i'2,\ldots,i'N} = 0, \\ &\bigcup Y_{i1,i2,\ldots,iN} = y. \end{aligned} \tag{40}$$

- $\Gamma = \{P\}$ is the set of all feasible partitions, P is the partition only made of false alarms.
- H_P is the hypothesis: P is the true partition, $H = \{HP \mid P \in \Gamma$ is the set of all feasible events. The problem is to determine the best partition P, say the partition which maximizes the ratio [12,7]

$$\frac{\max_X p(H_P | Y, X)}{\max_X p(H_{P_0} | Y, X)} . \tag{41}$$

- The cost of the association $Y_{i_1,i-2,\ldots, in}$ is denoted $c_{i_1,i-2,\ldots, in}$. Its calculations requires the computation of $p(y_{i_1,\ldots,i_N} | xm)$, itself defined by:

$$p\left(y_{i_1,\cdots,i_N}|x_m\right) = \prod_{n=1}^{N} \left(P_D(n)\,\mathcal{N}\left[y_{i_n}, \beta(x_m,n)\right]\right)^{1-\delta_{n,i_n}} .[1-P_D]^{\delta_{n,i_n}} ,$$

where :

$$\delta_{n,i_n} = 1 \text{ if : } i_n = 0 \text{ , } 0 \text{ else :} \tag{42}$$

and :

$$c_{i_1,i_2,\cdots,i_n} = -\ln\left(\frac{\max_{x_m} p(Y_{i_1,i-2,\cdots,i_n}|x_m)}{p(Y_{i_1,i-2,\cdots,i_n}|x=\emptyset)}\right) .$$

For any feasible association $\{ {}_{i_1,\ldots,i_n}$, the cost $c_{i_1,i-2,\ldots,in}$ must be calculated. The partitioning problem then becomes:

$$find\Phi_N = \max_{\rho_{i_1,\ldots,i_N}} \sum_{i_1=0}^{n_1} \cdots \sum_{i_N=0}^{n_N} c_{i_1,i_2\ldots,i_N}\, \rho_{i_1,i_2\ldots,i_N} ,$$

$$subject\ to:$$

$$\begin{cases} \sum_{i_2=0}^{n_2} \cdots \sum_{i_N=0}^{n_N} \rho_{i_1,i_2\ldots,i_N} = 1 , \quad i_1 = 1,\cdots,n_1 \\ \vdots \\ \sum_{i_1=0}^{n_1} \cdots \sum_{i_{N-1}=0}^{n_{N-1}} \rho_{i_1,i_2\ldots,i_N} = 1 , \quad i_1 = 1,\cdots,n_1 , \\ \rho_{i_1,i_2\ldots,i_N} \in \{0,1\} . \end{cases} \tag{43}$$

The problem is then a N-dimensional assignment one. This is a classical problem of combinatorial optimization. A Lagrangian relaxation of the integer constraints seems to be the best way to give an *approximate* solution[12].

6.2 LAGRANGIAN RELAXATION

The basic idea is to relax set of constraints in order to obtain a two-dimemsional assignment problem for which polynomial algorithms exist (e.g., the Auction algorithm [4]). This idea requires us to treat with specific care dummy measurements. Suppose the last N-2 sets of constraints are relaxed and denote $u_{3,i_3},\ldots,u_{N,i_N}$ the corresponding Lagrange multipliers, then we have to consider the following 2 -dimensional assignment problem [7] :

$$\Phi_u^2(u_3,\cdots,u_n) = \max_{w_{i_1,i_2}} \sum_{i_1=0}^{m_1} \sum_{i_2=0}^{m_2} d_{i_1,i_2} w_{i_1,i_2} + \sum_{i_3=0}^{m_3} u_{3,i_3} + \cdots + \sum_{i_N=0}^{m_N} u_{N,i_N} , \tag{44}$$

Subject to:

$$\begin{cases} \sum_{i_2=0}^{m_2} w_{i_1,i_2} = 1 \quad , i_1 = 1,\cdots,m_1 , \\ \sum_{i_1=0}^{m_1} w_{i_1,i_2} = 1 \quad , i_2 = 1,\cdots,m_2 , \end{cases} \tag{45}$$

and $d_{i_1,i}$, $w_{i_1,i}$ defined by:

$$d_{i_1,i_2} = \begin{cases} \min_{i_3}\cdots\min_{i_N}\left(c_{i_1,i_2\ldots,i_N} - u_{3,i_3} - \cdots - u_{N,i_N} \right) , & \forall (i_1,i_2) \neq (0,0) , \\ 0 & (i_1,i_2) = (0,0) , \end{cases} \tag{46}$$

$$w_{i_1,i_2} \in \{0,1\} .$$

The interest of Lagrangian relaxation methods comes from the following set of inequalities[12] :

$$\Phi_N \leq \Phi^2_* \triangleq \min_{u_1,\ldots,u_N} \Phi^2_u \leq \Phi^2_u \ . \tag{47}$$

These inequalities are classical in duality [19]. The difference $(\Phi^2_* - \Phi_N)$ is called the duality gap. It remains now to minimize the dual function Φ_u^2 with respect to $\{u_{3,i3},\ldots,u_{N,,iN}\}$. It is worth noting that the dual functional is piecewise linear and *convex*. So, subgradient algorithms perform satisfactorily with an ensured convergence toward the global extremum. This algorithm takes the following form (k iteration index):

$$u^{(k+1)}_{j,i_j} = u^{(k)}_{j,i_j} + \lambda^{(k)} \ g^{(k)}_{j,i_j} \ , \ j-3,\cdots,N \ . \tag{48}$$

Practically, the determination of convenient values for the sequence of stepsizes $\{\lambda^{(k)}\}_k$ remains an important issue [3]. The subgradient itself is easily calculated, i.e.:
Where $\rho_{i_1 \ldots i_j \ldots i_N}$ is the optimal solution of the Lagrangian relaxation.

7. Conclusion

$$\forall j = 3,\cdots, N, \quad , i_j = 1,\cdots, n_j \ ,$$

$$g_{j,i_j}^{(k)} = 1 - \sum_{i_1=0}^{n_1} \cdots \sum_{i_{j-1}=0}^{n_{j-1}} \sum_{i_{j+1}=0}^{n_{j+1}} \sum_{i_N=0}^{n_N} \rho^{*}_{i_1 \cdots i_j \cdots i_N} \ , \tag{49}$$

The EM method is a powerful approach in the MTT context. It is at the origin of PMHT and of its numerous and important extensions. For instance we may mention studies in maneuvering target tracking, multisensor-multitarget tracking and classification. On the other hand, combinatorial optimization problems are omnipresent in multisensor multitarget tracking. Relaxation algorithms permit us to handle a great variety of problems within a unique framework. However, their computation and memory requirements limit their applicability to reasonable dimensions. So, it seems necessary to consider cooperations between data association methods.

APPENDIX A

On the convergence of Iterative Methods for MTT

Let us now consider the case of multiple sources in the passive sonar context [9]. First, for the sake of brevity, our analysis will be restricted to the case of two sources and two independent mesurements. The state vector is made of the source state vectors (x_1 and x_2) and the mixture parameters (π_1 and $\pi_2=1-\pi_1$). At each scan, two measurement models are observed; with an assignment probability π_1, resp. π_2. The likelihood functional $L(p)$ (p scans) then takes the following form [16, 6] :

$$
\begin{aligned}
L(p) &= \prod_{t=1}^{p} \{ \prod_{j=1}^{2} (\pi_1 p(\tilde{y}_{j,t}|x_1) + \pi_2 p(\tilde{y}_{j,t}|x_2)) \} \ , \\
&= \prod_{t=1}^{p} L_t \ ,
\end{aligned}
\tag{50}
$$

where $\tilde{y}_{1,t}$ and $\tilde{y}_{2,t}$ are the measurements for scan t.

Let us now briefly detail the term L_t. Omitting the time index (in L_t), we have:

$$
\begin{aligned}
L_t &= (\pi_1 p(\tilde{y}_1|x_1) + \pi_2 p(\tilde{y}_1|x_2)) \, (\pi_1 p(\tilde{y}_2|x_1) + \pi_2 p(\tilde{y}_2|x_2)) \ , \\
&\triangleq p(\tilde{y}_1, x_1, x_2) p(\tilde{y}_2, x_1, x_2) \ .
\end{aligned}
\tag{51}
$$

Quite analogously to the unique source case, we obtain $p'(\tilde{y}_j|x_1) = \frac{1}{\sigma} p(\tilde{y}_j|x_1)$

$$
\begin{aligned}
r_1^2 \nabla_{x_1} L_t &= p(\tilde{y}_2, x_1, x_2) p'(\tilde{y}_1|x_1) \pi_1 [\tilde{y}_1 - y(x_1)] \, W_1 \\
&\quad + p(\tilde{y}_1, x_1, x_2) p'(\tilde{y}_2|x_1) \pi_1 [\tilde{y}_2 - y(x_1)] \, W_1 \ ,
\end{aligned}
\tag{52}
$$

where:

$$
W_1 = (r_{y,1}, -r_{x,1}, t r_{y,1}, -t r_{x,1})^* \ .
$$

From props. 1-2 of [9], the following equalities are straightforwardly deduced :

$$
\begin{aligned}
(\sigma^2 \pi_1^{-1} \frac{\hat{r}_1}{r_1}) (\hat{x}_1 - x_1)^* \nabla_{\hat{x}_1} L_t &= p(\tilde{y}_2, \hat{x}_1, \hat{x}_2) p(\tilde{y}_1|\hat{x}_1) (\tilde{y}_1 - \hat{y}_1) \sin (\tilde{y}_1 - \hat{y}_1) \\
&\quad + p(\tilde{y}_1, \hat{x}_1, \hat{x}_2) p(\tilde{y}_2|\hat{x}_1) (\tilde{y}_2 - \hat{y}_1) \sin (\tilde{y}_1 - \hat{y}_1) \ ,
\end{aligned}
\tag{53}
$$

and, similarly:

$$(\sigma^2\pi_2^{-1}\frac{\hat{r}_2}{r_2})\,(\hat{x}_2-x_2)^{\,*}\,\nabla_{\hat{x}_2}L_t = p(\tilde{y}_2,\hat{x}_1,\hat{x}_2)\,p(\tilde{y}_1|\hat{x}_2)\,(\tilde{y}_1-\hat{y}_2)\,\sin\,(\tilde{y}_2-\hat{y}_2) \qquad (54)$$
$$+\; p(\tilde{y}_1,\hat{x}_1,\hat{x}_2)\,p(\tilde{y}_2|\hat{x}_2)\,(\tilde{y}_2-\hat{y}_2)\,\sin\,(\tilde{y}_2-\hat{y}_2)\;.$$

We thus have finally :

$$(\hat{x}_1-x_1)^{\,*}\nabla_{\hat{x}_1}L(T) = \sum_t L_1\cdots L_{t-1}.(\hat{x}_1-x_1)^{\,*}\nabla_{\hat{x}_1}L(t)\,.L_{t+1}\cdots L_T\,, \qquad (55)$$

where the term $(\hat{x}_1-x_1)^{\,*}\nabla_{\hat{x}_1}L_t$ is given by (Eq. 54). Eq. 54 extends the main results to the case of a multiple source mixture. However, the conclusions are much less affirmative. In fact, let us consider the sign of the elementary term $(\hat{x}_1-x_1)^{\,*}\nabla_{\hat{x}_1}L_t$.

Analogously to the unique source case, the term $(\tilde{y}_1-\hat{y}_1)\,\sin\,(\tilde{y}_1-\hat{y}_1)$ is positive but the trouble comes from the term $(\tilde{y}_2-\hat{y}_1)\,\sin\,(\tilde{y}_1-\hat{y}_1)$ which represents an interaction term between the two components of the mixture. The analysis of the term $(\hat{x}_2-x_2)^{\,*}\nabla_{\hat{x}_2}L_t$ leads to an identical conclusion. In fact, this is due to the combinatorial nature of the data association problem in the multiple source case.

Summarizing the previous steps, the general guidelines of the relaxation algorithm are given below :

- **Relaxation algorithm :**
- Compute all the costs $c_{i_1,\ldots,i_N}$ for all the N-tuple $c_{i_1,\ldots,i_N}$.
- Initialize Lagrangian multipliers.
- Repeat until the stop condition be satisfied :
  - Form the relaxed problem for the updated multipliers.
  - Solve the associated 2-*D* assignment problem by the Auction algorithm.
  - Update the subgradient and the Lagrange multipliers.
  - Compute a feasible solution to the primal problem and give an upper bound of the duality gap.

Bibliography

(1) Avitzour, S., (1992) A Maximum Likelihood Approach to Data Association. *ieee Trans. on Aerospace and Elect. Systems*, vol. 28, No. 2, pp. 560--565.

(2) Bar-Shalom, Y., and Fortmann, T.E. (1988) *Tracking and Data Association*. Academic Press.

(3) Bazaraa, M.S., and Sherali, H.D. (1991) On the choice of the step size in subgradient algorithms. European Journal of Operational Research, vol. 7, pp. 380-388.

(4) Bertzekas, D.P. (1988) The auction algorithm: a distributed relaxation method for the assignment problem. Annals of Operations Research, vol. 14, pp. 105-123.

(5) Fortmann, T.E., Bar-Shalom, Y., and Scheffe, M., (1983) Sonar tracking of multiple targets using joint probabilistic data association. *ieee Journal of Oceanic Eng,*, vol. 8, No. 3, pp. 173-184.

(6) Gauvrit, H., Le Cadre, J.P., and Jauffret, C. (1997) A formulation of multitarget tracking as an incomplete data problem. *IEEE Trans. on Aerospace and Elect. Systems*, vol. 33, No. 4, pp. 1242-1257.

(7) Gauvrit, H., Jauffret C., and Le Cadre, J.P., (1996) A N-dimensional assignment algorithm to solve multitarget tracking problems. In *Proc. of the first Australian Data Fusion Symp.*, Adelaide, S. A., pp. 172-177.

(8) Le Cadre, J.P., Gauvrit, H., and Trarieux, F., (2000) Approximations of the Cramr-Rao bound for multiple-target motion analysis. *IEEE Proc.-Radar, Sonar Navig.*, vol. 147, No. 3.

(9) Le Cadre, J.P.and Jauffret, C., (1999) On the convergence of iterative methods for bearings-only tracking. *ieee Trans. on Aerospace and Elect. Systems*, vol. 35, No. 3, pp. 801-817.

(10) Mc Lachlan G.J. and Krishnan, T., (1997) *The EM algorithm and extensions*. Wiley series in probability and statistics, Wiley, N.Y.

(11) Morefield, C.L., (1977) Application of 0-1 integer programming to multitarget tracking problems. *IEEE Trans. on Automatic Control*, vol. 22, No. 3, pp. 302-311.

(12) Pattipati, K.R., deb, S., Bar-Shalom, Y., and Washburn, R.B., (1992) A new relaxation and passive sensor data association. *ieee Trans. on Automatic Control*, vol. 37, No. 2, pp. 198-213.

(13) Poore, A.B., and Rijavec, N., (1991) Multi-target tracking and multi-dimensional assignment problems. In *Proc. SPIE symp. Signal and Data Proc. of Small Targets*, vol. 1481, Orlando, FL.

(14) Redner, R.A., and Walker, H.F., (1984) Mixture densities maximum likelihood and the EM algorithm. siam review, vol. 26, No. 2, pp. 195-239.

(15) Reid, D.B., (1979) An algorithm for tracking multiple targets. *IEEE Trans. on Automatic Control*, vol. 24, No. 6, pp. 843-854.

(16) Streit, R.L. and Luginbuhl, T.E., (1993) A probabilistic multi-hypotheses tracking algorithm without enumeration and pruning. In *Proc. of the Sixth Joint Service Data Fusion Symposium*, Laurel, MD, June 14-18, pp. 1015-1024.

(17) Streit R.L. and Luginbuhl, T.E., (1994) Maximum likelihood method for probabilistic multi-hypothesis tracking. In *Proc. SPIE symp. Signal and Data Proc. of Small Targets*, vol. 2335-24, Orlando, FL.

(18) Streit R.L. and Luginbuhl, T.E., (1995) Probabilistic multi-hypothesis tracking. Technical report 10428, Naval Undersea Warfare Center, Newport, RI.

(19) Wolsey, L.A., (1998) *Integer Programming*. Wiley-Interscience series in discrete mathematics and optimization, Wiley.

WAVELETS FOR MODELING AND DATA FUSION IN REMOTE SENSING

T. RANCHIN
Ecole des Mines de Paris
Centre d'Energétique, Groupe Télédétection & Modélisation
BP 207, 06904 Sophia Antipolis cedex
France

1. Introduction

Modeling is a very important topic whatever the research domain. In data fusion, many models were developed to classify the different possible types of fusion (see e.g. the JDL Model described in several other chapters in this text). Modeling has different objectives. The main ones are:

- mathematical description of the reality,
- mathematical description of interactions between the physical phenomena of interest,
- mathematical description of the behavior of phenomena,
- prediction of phenomena, behaviors and interactions,
- definition of sensor(s),
- helping us to think [1]

The various architectures and algorithms presented in this text demonstrate that modeling and data fusion are strongly linked and profit from each other.

In this chapter, the benefits of using wavelet transform (WT) and multiresolution analysis (MRA) for modeling complex multiscale structures are described. The description of these tools puts the emphasis on the hierarchical description of phenomena obtained. An example of modeling applied to the extraction of streets from satellite images of urban areas is described. A summary of some winning applications of data fusion in the field of remote sensing, using these tools is also provided.

These examples show the potentialities and the benefits of WT and MRA for modeling and data fusion in the field of remotely sensed images.

2. Wavelets and Multiresolution Analysis

In the begin of the 1980's the wavelet transform (WT) appeared at the junction of signal processing, applied mathematics and quantum mechanics and provided an efficient mathematical framework for the study of non-stationary signals. Combined with the multiresolution analysis (MRA), introduced in 1989 by Mallat [2], the WT allows the analysis and the processing of images.

A. K. Hyder et al. (eds.), Multisensor Fusion, 351–363.

In the field of Earth Observation, the first applications of these two mathematical tools appear in the early 1990's with the analysis of remotely sensed images in different fields (geology, urban areas, oceanography, etc.), the processing of SAR imagery and data fusion for the improvement of spatial resolution of images. After these promising beginnings in analysis and processing, use of WT and MRA was developed for modeling and data fusion in the field of remote sensing.

2.1. WAVELET TRANSFORM

WT will be presented in its continuous version and in the mono-dimensional case. The main property of WT is to adapt the analysis window to the phenomena under study, providing a local information. WT leads to a time-frequency representation. In the case of images, WT leads to a scale-space representation.

As the Fourier transform, WT is equivalent to a decomposition of the signal in a base of elementary functions: the wavelets. The base is generated by dilatation and translation of a single function called the mother wavelet:

$$\psi_{a,b} = |a|^{\frac{-1}{2}} \psi\left(\frac{x-b}{a}\right) \tag{1}$$

where $a, b \in \Re$ and $a \neq 0$. a is called the dilation step and b the translation step.

Many mother wavelets exist. They are all oscillating functions, which are well localised both in time and frequency. All the wavelets have common properties such as regularity, oscillation and localisation, and satisfy an admissibility condition. For more details about the properties of the wavelets, one may refer to Meyer [3] or Daubechies [4]. Even if they have common properties, each of them brings to a single decomposition of the signal related to the used mother wavelet. In the one dimension case, the continuous WT of a function f(x) is:

$$WT_f(a,b) = \langle f, \psi_{a,b} \rangle = \frac{1}{\sqrt{|a|}} \int_{-\infty}^{+\infty} f(x) \overline{\psi(\frac{x-b}{a})} dx \tag{2}$$

where $\psi_{a,b}$ is defined as in Eq. 1 and $\overline{\psi(\frac{x-b}{a})}$ is the complex conjugated of ψ. $WT_f(a,b)$ represents the information content of $f(x)$ at scale a and location b. For fixed a and b, $WT_f(a,b)$ is called the wavelet coefficient. The computation of WT for each scale and each location of a signal provides a local representation of this signal. The process can be reversed and the original signal reconstructed exactly (without any loss) from the wavelet coefficients by:

$$f(x) = \frac{1}{C_\psi} \int_{-\infty}^{+\infty} \int_{-\infty}^{+\infty} WT_f(a,b) \psi_{a,b}(x) \frac{da\, db}{a^2} \tag{3}$$

where C_ψ is the admissibility condition of the mother wavelet.

This last equation can be interpreted in two ways:

- *f(x)* can be reconstructed exactly if one knows its wavelet transform,
- *f(x)* is a superimposition of wavelets

These two points of view lead to different applications of WT. In the first case, the processing of the signals and in the second, their analysis.

2.2. MULTIRESOLUTION ANALYSIS

The concept of MRA introduced by Mallat [2] derives from the Laplacian pyramids [5]. In this approach, the size of a pixel is defined as a resolution of reference to allow a measure of local variations in the image. Note that the resolution is related to the inverse of the scale used by cartographers. Hence, the larger the resolution of an image, the smaller the size (or the characteristic length or characteristic scale) of the smallest visible object.

Figure 1 is a very convenient description of MRA and more generally of pyramidal algorithms. MRA computes successive, coarser and coarser approximations of the same original image. The basis of the pyramid corresponds to the original image. Climbing the pyramid, the different floors represent the successive approximations of the image. The theoretical limit of these algorithms is the top of the pyramid that corresponds to a unique pixel. The difference of information existing between two successive approximations of the same image is described by the wavelet coefficients.

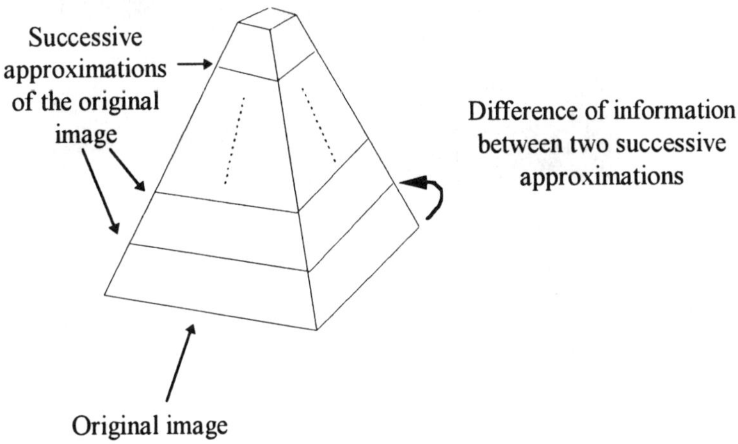

Figure 1. Representation of the successive approximations of an image by the means of a multiresolution algorithm

The application of the Mallat's algorithm to images is well known within the wavelet community. Another algorithm of interest is the one proposed by Dutilleux [6], the so-called "à trous" algorithm. In this algorithm, only a scale function is used. The approximation of the original image is obtained by filtering the original image and the wavelet coefficient image is obtained by subtracting the approximation to the original image in a pixel-to-pixel basis. This algorithm provides at each step one context and one non-directional wavelet coefficient image. To perform a dilation of the scale function, one adds zero between each coefficient of the filter. Hence no subsampling of the image is performed and all images have the same size. The reconstruction is done by summation of the last context and the wavelet coefficient images computed.

2.3. WAVELET COEFFICIENTS IMAGE STATISTICS

The wavelet coefficients images always have histograms centered at zero and present a very sharp distribution. Figure 2 is a typical example of the histogram of the wavelet coefficients image for a natural image. This histogram which is a representation of the probability density function (pdf) of the image can be adjusted to a theoretical pdf obeying a Laplacian or a Gaussian law.

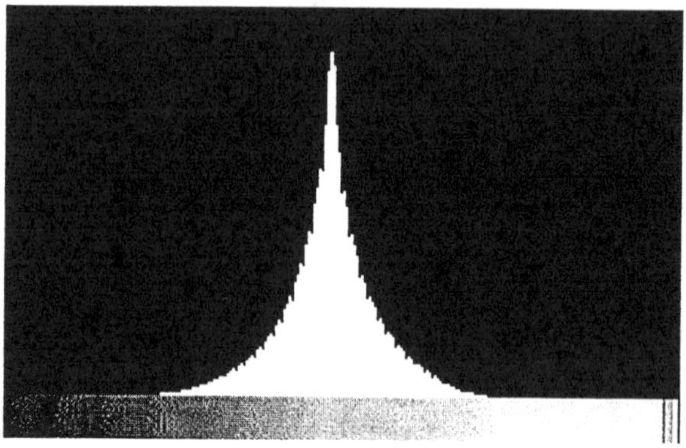

Figure 2. Typical histogram of the wavelet coefficients image representing the information at a given scale (here between one and two meters) for a natural image

Following Antonini *et al.* [7], the best results were achieved by fitting the histogram to a generalized Gaussian law:

$$p(x) = ae^{-|bx|^{\alpha}}$$

$$\text{where} \quad a = \frac{b\alpha}{2\Gamma(1/\alpha)} \quad \text{and} \quad b = \frac{1}{\sigma}\sqrt{\frac{\Gamma(3/\alpha)}{\Gamma(1/\alpha)}}$$

σ is the standard deviation of the distribution of the wavelet coefficients image, Γ is the gamma function, and α is a coefficient estimated through a t_{χ^2} test given by:

$$t_{\chi^2} = \sum_{i=1}^{n} \frac{(P_x(x_i) - P(x_i))^2}{P(x_i)}$$

in order to fit the wavelet coefficient image distribution to the model. If α is equal to 1, the pdf is equivalent to a Laplacian distribution, if α is equal to 2 the pdf is equivalent to a Gaussian distribution. Most often, α is comprised between 0.5 and 1.

The contents of the histogram can be interpreted as follows:

- the values close to zero (central peak of the histogram) represent noise or very small variation of the image,
- the high values of the histogram (left and right parts of the histogram) represent the strong variations of the image *i.e.* well-marked structures as borders between different elements.

3. Modeling Of Streets In Urban Areas

The example illustrating the use of WT and MRA for modeling in remote sensing, deals with urban mapping and proposes a modeling of streets in the cities. Couloigner and Ranchin [8] described a semi-automatic method for the extraction of streets from an image acquired over a city. An explicit and generic model of «streets» was developed according to MRA of images associated with a wavelet transform. The method consists of two steps of processing: the multiresolution extraction of sides of streets, and, if existing, the multiresolution extraction of strips of streets. The extraction of sides is achieved by the use of a multiscale representation of images, and the extraction of strips is done by a modeling of distinct substructures of streets at different characteristic scales achieved by the application of the associated WT.

Different methods to extract road network from remotely sensed images exist. All methods are based on very simple models of roads that can be only applied, in our study, to simple streets. The main properties of these models are common for all authors, *i.e.* the radiometry along one road is relatively homogeneous and when compared to its background could be highly or poorly contrasted according to the surroundings of the road under consideration. Moreover there are different widths and different types of roads, that provide different profiles. The example proposed is applied to an image that was acquired by the Russian analog sensor KVR-1000 with a spectral range of 0.51 to 0.71 μm and digitally scanned with a 2 m spatial resolution (ground sampling distance) over the city of Jeddah in Saudi Arabia. This kind of image, which is commercially available, is representative of the future very high spatial resolution of spaceborne images. The road network of Jeddah consists of four classes of streets: the primary streets with a 68 m width and a median and two secondary strips (class 1), the secondary streets with a median strip and two possible widths, 54 (class 2) or 34 m (class 3), and the tertiary streets with a 15 m width (class 4).

The extraction of cross-sections (transversal profiles) of different streets allows a better understanding of the evolution of the representation of streets when the scale of approximations (Figure 3a) and of wavelet coefficients images (Figure 3b) decreases. These cross-sections are similar along one class of streets according to the spatial resolution of the images under consideration. One cross-section of one street depends both on its class and on the observed scale.

Figure 3a displays the evolution of the cross-section of one primary street with the spatial resolution of approximations. In this Figure, the secondary strips disappear for a spatial resolution greater or equal to 8 m and the median strip for a spatial resolution greater or equal to 32 m.

Moreover, on these two figures, two typical points emerge. They are the points of intersection of each cross-section at different scales. These two points represent the positions of the two sides of the street. Indeed the gradient of the transition, that locates a side of road for example, tones down when it is smoothed. Hence, the location of the transition can be found by the intersection of the cross-section, including the side, computed with different spatial resolution.

Figure 3b brings up the positions of the street strips. In this Figure, a cross-section acquired from three successive wavelet coefficient images could be observed. At the 32-64 m scale (see the line with triangles in Figure 3b), only the position of the median strip of the considered primary street is marked. It corresponds to the minimum of the cross-section included between the two sides of the street. At the 16-32 m scale (see the line with crosses in Figure 3b), this position is characterized by a maximum and one observes the positions of the two secondary strips marked by two minima. At the 8-16 m scale (see the line with squares in Figure 3b), these two minima become two maxima of the cross-section included between one side of the street and its median strip.

According to the state-of-the-art and the multiscale analysis of data, a model of «streets» is defined to extract the pattern under consideration from images and to elaborate appropriate processes. The model is based on the three following properties:

- geometry: streets are linear and the range of their width is known. Streets can be represented by a quadrilateral.
- radiometry: this is homogeneous along street axis and exhibits contrast in the perpendicular direction.
- topology: depends on the class of street of interest and on the spatial resolution (Figures 3 and 4).

This model is generic and explicit. The method described in [8] is based on this model and a comparison with another algorithm of street extraction.

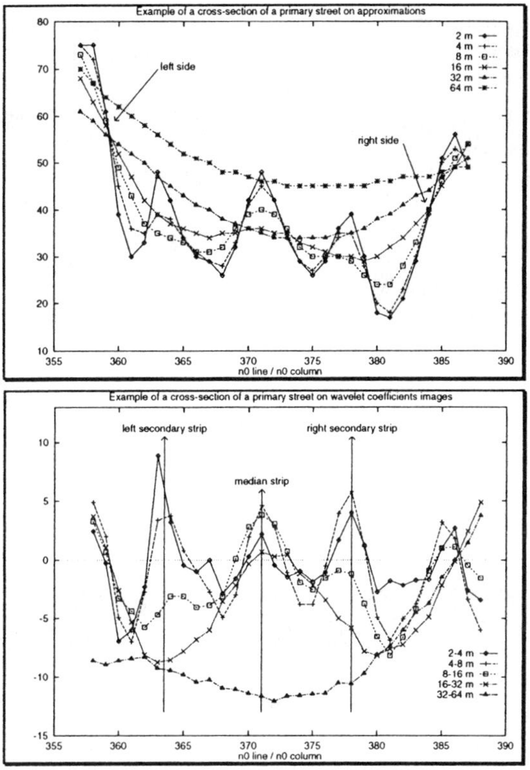

Figure 3. Example of a cross-section of a primary street at different spatial resolution on the approximations (a) and on the wavelet coefficients images (b).

4. Data Fusion Using WT And MRA In Remote Sensing

The first example of Data fusion based on the use of WT and MRA concerns the combination of information extracted from a SAR image with an optical contextual image. Figure 4 presents an example of a SPOT electro-optical image and an ERS-1 SAR image of the airport of Marignane, in southern part of France.

A multiresolution analysis combined with a wavelet transform was applied to Figure 4b. A simple thresholding applied to the first wavelet coefficient image selects the brightest points. Because the two images are superimposable, these bright points have been introduced on the original SPOT image. Figure 5a presents an enlargement of the original SPOT image and Figure 5b the results of the merging of the original SPOT image and of the information extracted from Figure 4b.

The bright points introduced in Figure 5b have been verified by a ground-truth travel. This travel allows us to verify if the points we have detected are corresponding to sub-pixel features or to artifacts. The main part of these points where identified as man-made features. A 3 m x 3 m metal house, a car which is used for frightening birds in the airport, and an antenna, were detected. The border of the concrete buildings were also underlined by the bright points.

(a) (b)

Figure 4. (a) Original SPOT image acquired in 1992 over the airport of Marignane. The spatial resolution is 10 m. Copyright CNES 1992. (b) Original ERS-1 image acquired in 1992 over the airport of Marignane. The original spatial resolution is 12.5 m. The geometrical correction of the image leads to the resampling of the image at 10 m spatial resolution. Copyright ESA 1992.

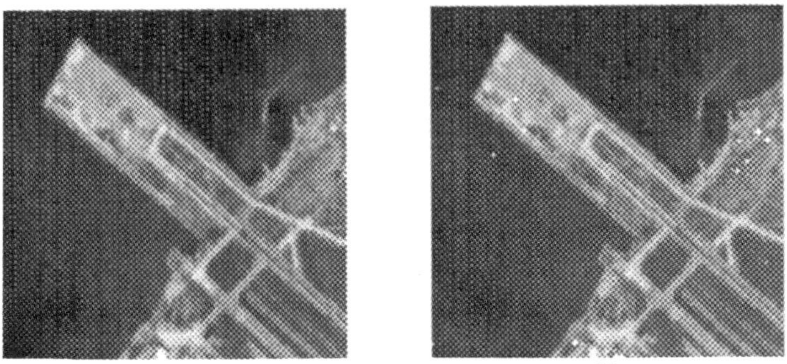

(a) (b)

Figure 5. (a) Enlargement of Figure 4a. (b) Resulting image of the merging of SPOT and bright points extract from Figure 4b.

The second example illustrating the use of WT and MRA in data fusion deals with automatic vine detection. The automatic characterization of the heterogeneous environments with regular structures are of interest for cultures such as orchards, vines, market gardening, etc. It is of tremendous importance for the establishment and the survey of national or European registers. But it is a very complex process because of the spatial structure of the objects that composed these environments: plants, soil, shadow, grass, stones, organic residues, etc. The automatic recognition of the parameters of the vines can only be achieved when the vine parcels detection is effective. If only the spectral information is considered, the observed parcels can be mixed up with non-cultivated soils or lying fallow. Depending on the cultivation technique of the vineyard, confusion is possible with orchards or asparagus, if only the textural information is considered. Hence, approaches using separately only the texture or the structure or the spectral or the temporal information can not reach this detection. The vineyard is a patchwork of objects evolving with the sun position, the season, the orientation of the rows of vines, the type of vine, the growing method, etc. For a spectral and temporal study, and for characterization of the vines, detection of vine parcels is essential.

Ranchin *et al.* [9] proposed a method based on a multiresolution model for automatic method for detection of vine parcels. The algorithm used in this application is the "à trous" algorithm [6]. It is proposed to separate the structural and the spectral information of an IRC airborne image over vineyards through MRA. Figure 6 presents the strategy applied.

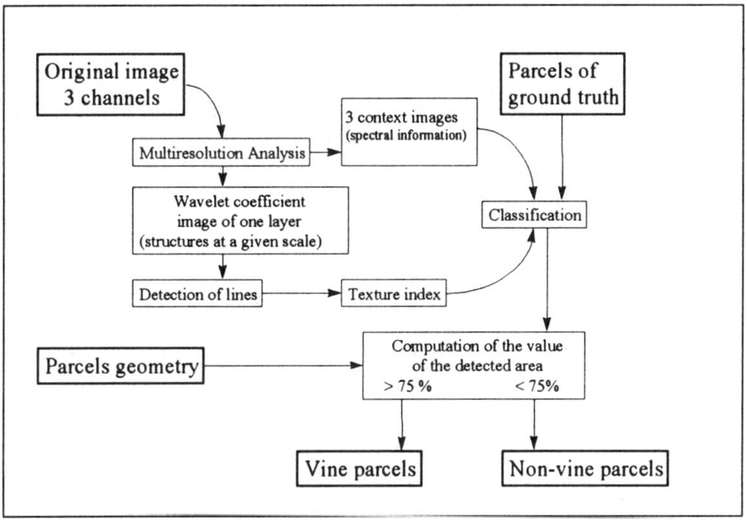

Figure 6. Strategy for automatic vine detection

The strategy for automatic vine detection is applied through the following steps:

- The original IRC image is decomposed in conventional red, green, blue layers. On one layer, a MRA using WT is applied,
- the line detection is applied on the wavelet coefficients image representing the structures to be detected. The choice of this parameter depends on an a priori knowledge of the distances between the rows of vines. In the following case, the main part of the vineyard is growing on lines spaced of between 1.5 and meters. It can easily be tuned
- from the results of this lines detection, a textural index is derived,
- the approximations of the original data computed through MRA and the textural index are used to derive a supervised classification of the whole image,

- in order to give the results in a parcel basis, the results of the classification is analyzed on a GIS environment. Then a combination of the parcel shapes and the results of the classification is performed. This allows obtaining the results in a parcel by parcel representation. To determine if the parcel is a vine parcel, a threshold is applied. All the parcels with a value in surface detected greater than 75 % of the reference surface are considered as vine. This threshold was determined empirically from the individual parcel results.

The high spatial resolution imagery allows us to access the texture of the vine areas. But if we process the original images and compute a classification in order to determine vine versus non vine areas, all the textures coming from all the details in the area will be included. Hence, the use of smooth approximations of the original channel allows to mix up the shadows, the leaves, the ground, and even the grass around the stocks of vines and to obtain a pixel representing the spectral signature of vines.

The line detection is applied on the wavelet coefficients image representing the structures with a spatial resolution comprised between one and two meters. The method of line detection is based on this observation. Hence, a thresholding of the wavelet coefficients image of the structures to detect, will enhance the well-marked structures and shows the lines of vines. The threshold was determined empirically to be $0.7\ \sigma$ where σ is the root mean square of the wavelet coefficients image between one and two meters of the image. Mainly, this threshold depends on the application and on the size of the structures we need to detect. Some tests are always needed to fit this value to the aim.

The area of interest is situated on the district of Villeneuve les Maguelone (Hérault). It comprises the Viticultural Experiment Station of INRA called "Domaine du Chapitre", surrounded by a representative and exhaustive samples of the languedocian vineyard. In 1984, (date of acquisition of the image), this Station presents structures of plantations that are presently used (*i.e.* 2 m in row with a distance of 1.2 m between stocks, 2 m x 1.3 m and 2.5 m x 1.2 m) in bush vines (goblet-pruned vines) or lateral cordon-constrained and trellised vines with numerous different varieties of vines. Around the Experiment Station, the parcels are not homogeneous. A lot is planted in bush-vines and with a square system of planting (1.5 m x 1.5 m or less). In these last ones a path exists every 3 or 4 rows, for the treatment engines. The image was acquired using an IRC (InfraRed Color) emulsion, at the scale of 1/5000 in October, 16, 1984 and is composed of urban area (more or less diffuse), vineyard (the "Domaine du Châpitre" represents 20 hectares of the vineyard), lying fallow, asparagus and orchards.

Figure 7 presents an example of application of the complete strategy. The inter-rows distance of vines is between one and two meters. Then, the line detection algorithm has been applied on the wavelet coefficients image describing the information between one and two meters. The spatial resolution of the approximations used for the spectral description of the area is 8 meters in this case. The masking was applied on the results of the classification. The surfaces of the resulting parcels are then compared to the reference surfaces. If the individual difference for each parcel is lower than 25 % in surface, the parcel is classified as vine. If the difference is greater than 25 % the parcel is classified as non-vine. An evaluation of the results is presented in [9] demonstrating the contribution of the multiresolution modeling.

Figure 7. Example of the strategy for automatic vine detection.
The upper left corner is an extract of the reference image. The pixel size is 0.25 m and the image was acquired using an InfraRed color emulsion at the scale of 1/5000. The upper right corner is the result of the line detection algorithm using the wavelet transform. The lower left corner is the superposition of the original image and the parcel information. The lower right corner is the combination of the results of the classification scheme described and the parcels information.

The last application proposed for illustrating data fusion using wavelets in remote sensing deals with the improvement of the spatial resolution of satellite images. Ranchin and Wald [10] proposed the description of a concept in this objective. From a set of remotely sensed images with different spatial and spectral resolutions, the aim is to synthesize images with the highest spatial resolution available in the set and with an appropriate spectral content. Several sensor fusion methods exist; most of them improve the spatial resolution but with a poor quality of the spectral content of the resulting image. Based on a multiresolution modeling of the information, the ARSIS concept (from its French name "Amélioration de la Résolution Spatiale par Injection de Structures") was designed in the aim of improving the spatial resolution together with a high quality in the spectral content of the synthesized images.

Let A_h be the high spatial resolution image, B_l the multispectral image, h the original spatial resolution of A and l the original spatial resolution of B ($l < h$). The MRA is applied to the two images A and B. A scale by scale description of the information content of both images is obtained. The missing spatial information (high frequency) between A_h

and B_l, represented by the wavelet coefficients, is extracted and used for the construction of the synthetic image B^*_h. Figure 8 illustrates this approach in the case of a pyramidal description. The missing information to be injected in the pyramid B from image A is located in the missing bottom of the pyramid B (dotted line). Only this part is needed to improve the spatial resolution of image B. But if the missing information is set equal to that provided by the image A, the synthesized image B^*_h will not be equivalent to "what would be seen by the sensor B if it has the spatial resolution of the sensor A". Hence, in order to improve the quality of the synthesized image, a transformation should be applied to convert the information provided by the multiscale representation of image A into the information needed for the synthesis of image B.

A more complete description of this application is proposed in [11].

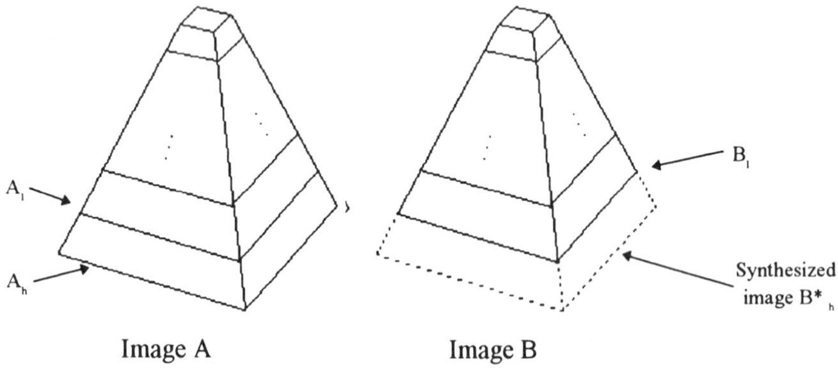

Figure 8. The use of the pyramidal approach for the fusion of high spatial and s pectral resolution images in the ARSIS concept.

5. Conclusions

The examples provided in this paper enhance the capability and the benefits of the hierarchical description of the information provided but MRA and WT for data fusion. In each case, the results achieved where evaluated and compared to methods already in use. These comparisons demonstrate the benefits of the multiresolution models derived from the MRA representation. The models derived from this representation are of various complexity and are based on the statistics of the wavelet coefficients images.

Some other classes of models have been defined. Basseville *et al.* [12, 13] have linked the multiscale representation of information in an image with stochastic processes through the definition of multiscale autoregressive models. As an example in the field of RS, Fosgate *et al.* [14] have demonstrated the efficiency of these models for the segmentation of SAR images and the detection of anomalies in these images. New classes of models are under definition and some links between the geostatistics field and the wavelet are emerging.

More and more applications in the field of RS take advantage of this hierarchical description of the information provided by MRA and WT. The development of the modeling based on WT in the field of RS will lead to a better understanding of the natural processes observed from satellites.

6. References

1. Bedworth M., Verlinde P., Litman A., Lowe D., James C. and Ranchin T. (2000) Personal communication, NATO Advanced Studies Institute on Multisensor Data Fusion, Pitlochry, Perthshire, Scotland.
2. Mallat, S. G., (1989) A theory for multiresolution signal decomposition: the wavelet representation. *IEEE Transactions on Pattern Analysis and Machine Intelligence*, 11, 7, pp. 674-693.
3. Meyer, Y., (1990) *Ondelettes et opérateurs 1: Ondelettes.* Hermann, Paris, France, 215 p.
4. Daubechies, I. , (1992) *Ten lectures on wavelets.* CBMS-NSF regional conference series in applied mathematics 61, SIAM, Philadelphia, USA, 357 p.
5. Burt, P. J. and E. H. Adelson, (1983) The Laplacian pyramid as a compact image code. *IEEE Transactions on Communications*, 31, 4, pp. 532-540.
6. Dutilleux, P. (1987) An implementation of the "algorithme à trous" to compute the wavelet transform. In *Proceedings du congrès ondelettes et méthodes temps-fréquence et espace des phases*, Marseille, Springer Verlag Editors, pp.298-304.
7. Antonini, M., Barlaud, M., Mathieu, P. and Daubechies, I., (1992) Image coding using wavelet transform. *IEEE Transactions on Image Processing*, 1, 2, 205-220.
8. Couloigner, I. and Ranchin, T., (2000) Mapping of urban areas: a multiresolution modeling approach for semi-automatic extraction of streets. *Photogrammetric Engineering & Remote Sensing*, 66, 7; pp. 867-874.
9. Ranchin, T., B. Naert, M. Albuisson, G. Boyer and P. Astrand, (2000). An automatic method for vine detection in airborne imagery using wavelet transform and multiresolution analysis. To appear in *Photogrammetric Engineering & Remote Sensing.*
10. Ranchin, T. and Wald, L. (2000) Fusion of high spatial and spectral resolution images: the ARSIS concept and its implementation. *Photogrammetric Engineering & Remote Sensing*, 66, 1, pp. 49-61.
11. Ranchin, T. (2000) Data Fusion in remote sensing and improvement of the spatial resolution of satellite images. NATO Advanced Studies Institute on Multisensor Data Fusion, Pitlochry, Perthshire, Scotland.
12. Basseville, M., Benveniste, A.B. and Willsky, A.S., (1992a) Multiscale Autoregressive Processes, Part I: Schur-Levinson Parametrizations. *IEEE Transactions on Signal Processing*, Vol. 40, No. 8, pp.1915-1934.
13. Basseville, M., Benveniste, A.B. and Willsky, A.S., (1992b) Multiscale Autoregressive Processes, Part II: Lattice Structures for Whitening and Modeling. *IEEE Transactions on Signal Processing*, Vol. 40, No. 8, pp.1935-1954.
14. Fosgate, C.H., Irving, W.W., Krim, H., Willsky, A.S., Karl, W.C. and Chaney, R.D., (1997) Multiscale Segmentation and Anomaly Enhancement of SAR Imagery. Special issue of the *IEEE Transactions on Image Processing on Automatic.*

SOFT SENSOR MANAGEMENT FOR MULTISENSOR TRACKING ALGORITHM

V. NIMIER
ONERA, BP 72, 92322
Chatillon Cedex, France
E-mail: nimier@onera.fr

1. Introduction

This chapter describes a method and an algorithm for combining symbolic and numerical information for data fusion in a tracking application. In recent years, the synergetic use of multiple sensors has received a great deal of attention. The major benefits of such use are an increase in the ability to analyse complex situations and an improved robustness of the fusion process in a cluttered environment. Military command and control, battlefields management, mobile robot navigation, and multitarget tracking are all typical applications that can benefit from the use of multiple sensors. Therefore, both the integration and the fusion of information, provided by multiple sensors, appear to be fields of growing interest.

The conventional techniques of signal processing using the probability theory have been extended to data fusion. Nevertheless, new problems appear in the use of multiple sensors that need new models for the processing of the information. To deal with these problems, Fuzzy Logic [12], [13] appears to be an alternative to probability theory and a duality exists between these two theories. The probability is devoted to stochastic phenomena, whereas the Fuzzy logic deals with the representation of human knowledge. Measurements provided by sensors and human operators are both available in the multisensor system. Therefore, these two theories for the representation of the information can be used in data fusion to obtain an overall system with better performances.

In the sensor fusion field, the designation ("quotation") of sensor reliability and information credibility is very important. This notion is relative to the quotation associated with the information uncertainty. This notion is well defined in an operational context: A NATO Standardization Agreement (STANAG 2022) addresses this issue. Yet, in data fusion algorithms this quotation notion is not fully addressed. And there is still no compatibility between the quotation used in data fusion and the operational one.

In this chapter quotation rules are presented from an operational point of view. Then a data fusion algorithm, using similar quotation rules is presented. We conclude by a simulation which illustrate the efficiency of the method.

A. K. Hyder et al. (eds.), Multisensor Fusion, 365–379.

2. Quotation Of Information

In STANAG 2022, the evaluation of the degree of confidence placed in any item or information is achieved by adopting an alphanumeric system of rating which combines a measurement of the reliability of the source of the information with a measurement of the credibility of that information when examined in the light of existing knowledge.

The reliability of the source is designed by a letter between A and F signifying various degrees of confidence as indicated below :

(a) Completely reliable
(b) Usually reliable
(c) Fairly reliable
(d) Not usually reliable
(e) Unreliable
(f) Reliability cannot be judged

The credibility of information is designed by a numeral between 1 and 6 signifying varying degrees of confidence as indicated below :

(a) Confirmed by other (information)
(b) Probably true
(c) Possibly true
(d) Doubtful
(e) Improbable
(f) Truth cannot be judged

These notions are explained in the following example (Figure 1):

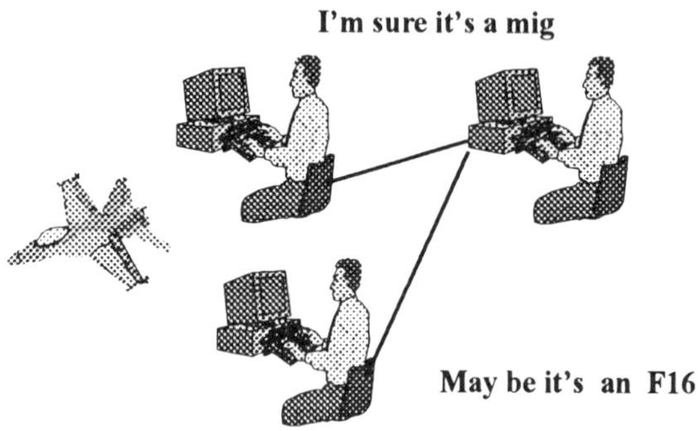

Figure 1. Credibility of information

An experienced anti-aircraft system manager has to identify a target in his volume of interest based on the information obtained from two operators. Each operator has in charge one or many sensors covering the same volume. After detection the manager asks the operators on the identity of the target. The manager knows that operator A is young and inexperienced. He judges that A is not usually reliable. The manager has a long experience with operator B and knows that he is completely reliable. Operator A answer to the question is "I'm sure it's a Mig". Operator B answer is "may be it's an F16". With these elements the manager makes the decision "it's an F16". From this example it is clear that if the manager has only the two operators' answers, his own decision would be different from the one considering the reliability of the operators.

This notion of reliability is very important in the fusion process. It is crucial to have, in addition to the credibility (degree of confidence) of the information, the reliability of the source to avoid corrupting good quality information with information that has a good credibility but lacks reliability of the source.

The question is how to evaluate these two degrees of confidence. As it is mentioned in STANAG 2022, the reliability and the credibility must be considered independently. In our example the reliability of the source is evaluated by the manager. Based on his experience, a learning process, the manager is able to assign a degree of reliability to each of the operator. However, this degree of reliability can change depending on the situation. For example, let's consider that the crew member B is finishing his work shift of 14 hours and A is just starting his shift, then the manager will assign to them a different degree of reliability . We propose here the same reasoning for sensor fusion but with a higher degree of automation. As the reliability and credibility must be considered independently, we consider that the credibility is given by the likelihood of the measurement which is a function of the sensor's characteristics, and the reliability is given by other parameters that we call contextual parameters. These parameters allow to evaluate the use of the sensor given the scenario and the environment.

3. Contextual Space

3.1 DEFINITION

Here we consider a system with n sensors, the set of all the sensors is noted N={1, ..., n}. A context is described by p contextual variables z_j, with $j \in \{1, ..., p\}$. The origin and the nature of these variables are very diverse. For instance, they can be represented by the measurements of luminosity or pluviometry of an annex sensor. They can also be parameters resulting from signal processing, for example: the signal-to-noise ratio, or the width of a pick of correlation. Parameters that characterize the performances of the sensors under normal conditions can also be useful in practical situations. All these variables help us to evaluate the reliability of the sensors and the quality of the measurements they provide.

The set of all contextual variables is noted Z. In this space we define the domain of validity of each sensor as it is presented in Figure 2. This domain is a subset of the contextual variables for which each sensor is reliable and provides correct measurements of the target state.

Here, we distinguish two kinds of validity domains. First, we note C_i the inclusive validity domain which represents the subset of contextual variables in which the sensor i, with $i \in N$, is reliable; there is no assumption on the other sensors, they can or not be reliable. Next, we note c_i the exclusive validity domain where, excluding the other sensors, only sensor i is valid.

We can write the exclusive validity domain as a function of inclusive validity domain. For example, for a multisensor system with 3 sensors, the exclusive validity domain of sensor 1 is of the form: $c_1 = C_1 \cap \overline{C}_2 \cap \overline{C}_3$. For the subset {1,2}, the exclusive validity domain is defined as: $c_{\{1,2\}} = C_1 \cap C_2 \cap \overline{C}_3$. More generally, we can extend this notion to all the subsets J of N. The relation between the two domains is of the following form: $c_J = \bigcap_{j \in J} C_j \bigcap_{i \in \bar{J}} \overline{C}_i$. We can define the set A of exclusive validity domains with $A = \{c_\emptyset, c_1, ..., c_{\{1,2\}}, ..., c_N\}$. This set is composed of exclusive subset of Z and form a partition of Z.

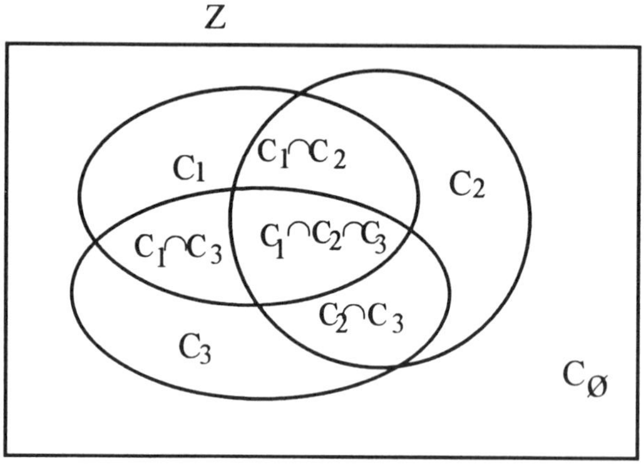

Figure 2. Domains of validity

3.2 PROBABILITY OF VALIDITY OF A SENSOR

The previous definition of a domain is based on binary logic: an element belongs or not to the domain. Because the definition of the validity domain of a sensor results from human knowledge, it is more convenient to define this domain by the membership function of a fuzzy subset. This membership function takes its values in the interval [0,1] allowing a better characterisation of the incertitude intrinsic to this form of information. Let $\mu_{ij}(z_j)$ the elementary membership function of the domain of sensor i on the contextual variable j. The global membership function $\mu_i(z)$ is obtained by conjunction of the elementary functions following

$$\mu_i(z)=\mu_{i1}(z_1)\wedge\mu_{i2}(z_2)\wedge\ldots\mu_{ip}(z_3)$$

where $\wedge$ is an operator of conjunction as, for instance, the Min operator. By convention, if the contextual variable z_j does not affect the validity of sensor i, we write: $\mu_{ij}(z_j)=1$, for all values of z_j.

A sensor confidence coefficient is defined as a function of the probability of validity of this sensor. This probability is defined by the following relation

$$P(C_i / z^m) = \int \mu_i(z)\, p(z / z^m)\, dz \tag{1}$$

Knowing the measurement z^m, then $P(C_i/ z^m)$ is the probability that the value of z belong to the subset C_i. Because C_i is a fuzzy subset, $P(C_i/ z^m)$ is the probability of a fuzzy event as defined by Zadeh [13].

3.3 PROBABILITY OF VALIDITY FOR AN ASSOCIATION OF SENSORS

For an association or a subset J of the inclusive domain, the probability of validity is given by the relation:

$$P(\cap_{i\in J} C_i/z^m) = \int \underset{i\in J}{\mathrm{Min}}(\mu_i(z))\; p(z/z^m)dz$$

Then, the expression of the exclusive domain probability is given by

$$\beta_J = p(c_J) = \sum_{\{I\subseteq N/\, J\subseteq I\}} (-1)^{|I-J|} P(\cap_{i\in I} C_i)$$

and $$\beta_\varnothing = P(c_\varnothing) = P(\cap_{j\in N} \overline{C}_j) \tag{2}$$

where I and J are two subset of N, and $|I-J|$ is the cardinal of the subset I-J. For the sake of simplicity we have omitted to write the conditional variable z^m in the above expression. There are 2^n possible associations of sensors corresponding to the subset of N and therefore 2^n probabilities β_J that satisfy the expression

$$\sum_{J\subseteq N} \beta_J = 1$$

which is the normalisation condition.

4. Supervision of the Fusion Process by Contextual Analysis

In this section, the supervision algorithm for the multisensor tracking algorithm which takes into account the contextual information is presented. Each association J of sensors of the system provides measurements. A confidence coefficient of such association is evaluated based on the probability β_J. First we consider a static formulation of the problem, then we develop the filter equations in a dynamic context.

4.1 STATIC ESTIMATION

4.1.1 *Problem formulation*

Consider a multisensor system with n sensors. Each sensor provides an observation noted y^i, with $i \in N$. The measurement equation is given below

$$y^i = H_i x + b^i \qquad \text{with } i \in N$$

where H_i is the observation matrix and x is the vector of parameters to be estimated. The process noise b^i is a zero mean gaussian with a covariance matrix $E(b^i b^{i^T}) = R_i$. The set of measurements is noted $Y = \{y^1, \ldots, y^n\}$.

4.1.2 *Equations for the static estimation problem*

The mean square estimation of x given Y is written as

$$\hat{x} = E(x/Y) = \int x \ P(x/Y) \ dx \tag{3}$$

Using the law of total probability, the probability density $P(x/Y)$ can be written as

$$P(x/Y) = \sum_{J \subseteq N} P(x/Y, c_J) P(c_J) \tag{4}$$

Considering the contextual information in the estimation process, we substitute the probability density $P(x/Y)$ in (3) by its value in (4). It follows that

$$\hat{x} = P(c_\varnothing) \int x \ P(x/Y, c_\varnothing) \ dx + \sum_{J \subseteq N - \{\varnothing\}} P(c_J) \int x \ P(x/Y, c_J) \ dx \tag{5}$$

Considering c_J in the definition of the conditional probability density $P(x/Y)$ means that only the sensors for which the subscript belongs to the subset J are valid. Therefore only their measurements are considered.

$Y_J = \{y_i\}_{i \in J}$ defines this subset of measurements. The expression (5) can then be written as

$$\hat{x} = \beta_{ø} x_0 + \sum_{J \subseteq N - \{ø\}} \beta_J E(x/Y_J) \tag{6}$$

For convenience of notation, the probabilities $P(c_J)$ are replaced by the coefficients β_J which are defined in equation (2). The variable x_0 is given a priori when no sensor is valid.

4.1.3 *Example*

We consider a system with two sensors : the first one is a radar and the second one an optical sensor. Each sensor has one variable for evaluating contextual environment, for example : z_1 is an indicator of the smooth working of the radar, and z_2 is a parameter which determinates the luminosity of the observed scene. The membership functions that bound the two domains of validity C_1 and C_2 of each sensor are noted $\mu_1(z_1)$ and $\mu_2(z_2)$.

For this system, formula (2) gives the following exclusive probability of validity for each association of sensors

$$\begin{aligned}
\beta_{\{ø\}} &= P(\overline{C}_1 \cap \overline{C}_2) \\
\beta_{\{1\}} &= P(C_1) - P(C_1 \cap C_2) \\
\beta_{\{2\}} &= P(C_2) - P(C_1 \cap C_2) \\
\beta_{\{1,2\}} &= P(C_1 \cap C_2)
\end{aligned} \tag{7}$$

The sensors provide observations y_1 and y_2 which are independent, gaussian, with mean x and variance σ_1^2 and σ_2^2 respectively. We assume a uniform a priori distribution of the variable x. Given these assumptions the estimates of each association of sensor are the following

$$\begin{aligned}
E(x/\{ø\}) &= x_0 \\
E(x/\{y_1\}) &= y_1 \\
E(x/\{y_2\}) &= y_2
\end{aligned}$$

$$E(x/\{y_1, y_2\}) = \frac{y_1 \sigma_2^2}{\sigma_1^2 + \sigma_2^2} + \frac{y_2 \sigma_1^2}{\sigma_1^2 + \sigma_2^2} \tag{8}$$

Using equations (6), (7), and (6) the estimate expression is then

$$\begin{aligned}
\hat{x} = P(c_{\{ø\}}) x_0 &+ y_1 [P(C_1) - \frac{\sigma_1^2}{\sigma_1^2 + \sigma_2^2} P(C_1 \cap C_2)] \\
&+ y_2 [P(C_2) - \frac{\sigma_2^2}{\sigma_1^2 + \sigma_2^2} P(C_1 \cap C_2)]
\end{aligned} \tag{9}$$

Assuming that the measurement errors of the contextual variable are negligible and the probability densities in formula (1) are dirac function, then the probabilities are directly given by the membership function of the fuzzy subset. The inclusive domain probabilities are given by

$$P(C_{\{\emptyset\}}) = \mathrm{Min}[1-\mu_1(z_1), 1-\mu_2(z_2)]$$
$$P(C_{\{1\}}) = \mu_1(z_1)$$
$$P(C_{\{2\}}) = \mu_2(z_2)$$
$$P(C_{\{1,2\}}) = \mathrm{Min}[\mu_1(z_1), \mu_2(z_2)]$$

Hereafter we study some special cases.

Case 1 - $\mu_1(z_1)=\mu_2(z_2)=1$

All sensors are reliable. The estimate is given by

$$\hat{x} = \frac{y_1\sigma_2^2}{\sigma_1^2+\sigma_2^2} + \frac{y_2\sigma_1^2}{\sigma_1^2+\sigma_2^2}$$

This estimate can also be obtained by a standard (probabilistic) approach to the fusion: Maximum likelihood.

Case 2 - $\mu_1(z_1)=0$ and $\mu_2(z_2)=1$

Sensor 2 is reliable but not sensor 1. The estimate is then

$$\hat{x} = y_2$$

It follows that only the measurement of the second sensor is taken into account.

Case 3 - $\mu_1(z_1)=0.5$ and $\mu_2(z_2)=1$

This case is an intermediate situation between case 1 and case 2. We obtain

$$\hat{x} = y_1\frac{0.5\,\sigma_2^2}{\sigma_1^2+\sigma_2^2} + y_2\frac{\sigma_1^2+0.5\,\sigma_2^2}{\sigma_1^2+\sigma_2^2}$$

The measurement y_2 has the greatest importance in the weighted mean compared to case 1 which assumes the two sensors are valid.

Case 4 - $\mu_1(z_1)=0$ and $\mu_2(z_2)=0$

We obtain

$$\hat{x} = x_0$$

It is a limit case where no sensor is valid. The estimate is identical to the *a priori* value.

4.1.4 *Remark*

The estimate $E(x/\{y_1, y_2\})$ given by the formula (8) is a classical result obtained for data fusion when using the maximum likelihood approach. It is a weighted mean that depends only on the variances of the observation noise. The variance of the estimate is defined by

$$\sigma_{\hat{x}}^2 = \frac{\sigma_1^2 \sigma_2^2}{\sigma_1^2 + \sigma_2^2}$$

which is always inferior to the lowest variance of each sensor.

The first advantage of data fusion is to improve the accuracy of the estimate. Nevertheless, the maximum gain is obtained when the two sensors have the same variance σ^2. It follows that the variance of the estimate is $\sigma_{\hat{x}}^2 = \frac{\sigma^2}{2}$. It is a modest gain compared to the complexity of the implementation of a multisensor system. Moreover, sensors often have different resolutions therefore, the variance of the estimate can only converge to the variance of the best resolution sensor.

The major benefit of a multisensor system is its robustness in a cluttered environment. Because the sensors use different physical phenomena's, they are rarely obstructed simultaneously. The reliability of each sensor and the context in which it works must be evaluated to calculate confidence coefficients used to supervise the fusion process. The objective is to discard the measurements of the obstructed sensors to avoid the corruption of measurements provided by the reliable sensors. In this case the result of the fusion process is always better than the one obtained by the *best* of the sensors.

4.2 DYNAMIC ESTIMATION

4.2.1 *Mathematical Model*

The kinematic model of the target is described by the following equation

$$x_k = F x_{k-1} + v_k$$

where x_k is the target state vector. The input noise is described by a white gaussian noise, zero mean, with covariance matrix

$$E(v_k v_l^T) = Q\delta(k,l)$$

where δ is the Kronecker symbol.

The measurements of the sensors are defined by

$$y_k^i = H_i\, x_k + b_k^i \quad \text{with } i \in \{1, ..., n\}$$

where the measurement noise is assumed to be white gaussian, zero mean, independent from sensor to sensor. The covariance matrix is given by

$$E(b_k^i b_l^{j^T}) = R\, \delta(k,l)\delta(i,j)$$

The set of measurements of sensor i at time k is defined by

$$Y_k^i = \{y_l^i\}_{l=1}^{l=k}$$

For a subset of sensors $J \subseteq N$, the cumulative measurements are denoted

$$Y_k^J = \{Y_k^i\}_{i \in J}$$

and the cumulative measurements of all sensors are defined by

$$Y_k = \{Y_k^i\}_{i=1}^{i=n}$$

4.2.2 *Filter equations for an association of sensors*

For an association of sensors J, the minimum variance estimate of the state is defined by

$$\hat{x}_{k/k}^J = E(x_k / Y_k^J)$$

then the update equation of the state is of the form

$$\hat{x}_{k/k}^J = \hat{x}_{k/k-1}^J + \sum_{j \in J} K_J^j(k)(y_k^j - H_j \hat{x}_{k/k-1}^J) \qquad (10)$$

where K_J^j is the Kalman gain defined in the information form as

$$K_J^j = P_J(k/k) H_i^T R_i^{-1}$$

The covariance matrix of the state is given by

$$P_J^{-1}(k/k) = P_J^{-1}(k/k-1) + \sum_{j \in J} H_j^T R^{-1} H_j$$

4.2.3 *Filter equations for the global system*

If we consider the overall set of sensors, the minimum variance state estimate is defined by

$$\hat{x}_{k/k} = E(x_k / Y_k)$$

The context and the reliability of the sensor are taken into account according to the total probability law

$$\hat{x}_{k/k} = \sum_{J \subseteq N} E(x_k / Y_k, c_J) P(c_J) \tag{11}$$

Using equations (2) and (3), one can write

$$\hat{x}_{k/k} = \beta_{\emptyset} E(x_k / Y_k^{\emptyset}) + \sum_{J \subseteq N - \{\emptyset\}} \beta_J E(x_k / Y_k^J) \tag{12}$$

$E(x_k / Y_k^{\emptyset})$ is the state estimate obtained in the absence of measurements; it can be substituted by the prediction of the state $\hat{x}_{k/k-1}$. Each estimate $E(x_k / Y_k^J)$ is replaced in (12) by its expression in (10) where the prediction $\hat{x}^J_{k/k-1}$ is replaced by the global prediction $\hat{x}_{k/k-1}$. Relation (12) becomes

$$\hat{x}_{k/k} = \beta_{\emptyset} \hat{x}_{k/k-1} + \sum_{J \subseteq N - \{\emptyset\}} \sum_{j \in J} \beta_J K^j_J (y^j_k - H_j \hat{x}_{k/k-1}) \tag{13}$$

A simpler form of this last equation is obtained by considering the Kalman gain

$$K_j(k) = \sum_{\{J / j \in J\}} \beta_J(k) K^j_J(k)$$

where $\{J / j \in J\}$ defines the set of all subsets of sensors which contain the sensor j. Thus using the gain expression, the estimate equation is given by

$$\hat{x}^J_{k/k} = \hat{x}_{k/k-1} + \sum_{j \in N} K_j(k)(y^j_k - H_j \hat{x}_{k/k-1})$$

According to expression (12) the covariance matrix is given by [2]:

$$P(k/k) = \sum_{J \in N} \beta_J(k)[\, P_J(k/k) + (\hat{x}_{k/k} - \hat{x}^J_{k/k})(\hat{x}_{k/k} - \hat{x}^J_{k/k})^T]$$

Finally, the predicted state and its covariance matrix are given by the classical equations

$$\hat{x}_{k/k-1} = F\ \hat{x}_{k-1/k-1} \text{ and } P_{k/k-1} = F\ P_{k-1/k-1} F^T + Q_k$$

5. Simulation

In this section, we present simulation results of the proposed algorithm in the case of measurements provided by three sensors. For the sake of simplicity, a constant velocity on azimuth defines the dynamics of the target. The simulated data are presented in Figure 3. The first sensor is a search radar, under normal conditions with measurement y3. The second sensor is a tracking radar with measurement y2. This sensor is obstructed by an important noise between time 80 and time 140. The third sensor is an I.R camera with measurement y1; the signal is corrupted by an impulse noise throughout the tracking.

The time variations of the contextual parameters, shown in Figure 4, are supposed to be known. They are provided by annexe treatments. The variable z2 of sensor 2 indicates the presence of a noise but not its amplitude. The variable z1 indicates the presence of a perturbation on sensor 1.

The result obtained by the classical multisensor algorithm xf, and the one obtained by the proposed algorithm xc, are presented in Figure 5. When the context is not taken into account, it is obvious that a perturbation on one sensor affects the result of the overall fusion process. In this case, the result of the fusion can be worse than the one obtained with a reliable sensor having the worst resolution in the set of used sensors: the search radar on the above example.

With the proposed algorithm, the perturbation effect is very limited and a good quality track is maintained. Figure 6 presents the mean square error and shows a clear difference between the performances given by the two algorithms. It can be observed that the two algorithms present similar performances in the absence of perturbations.

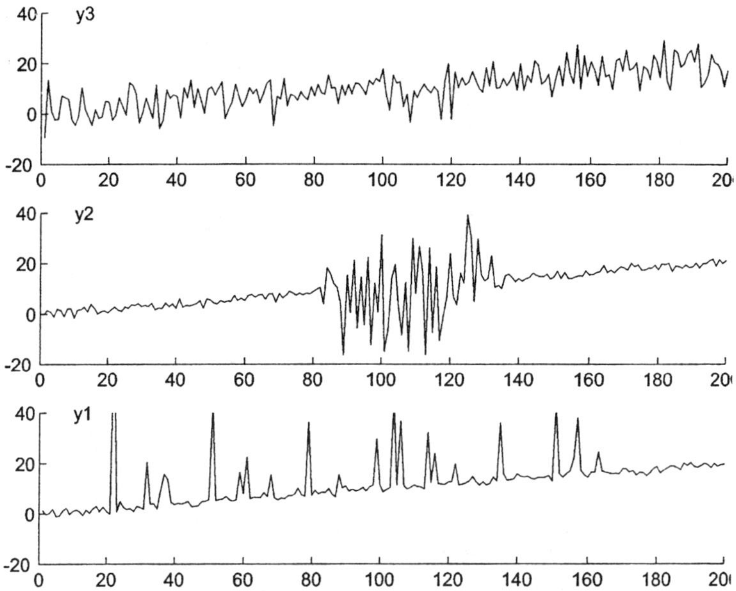

Figure 3. Measurements provided by sensors.

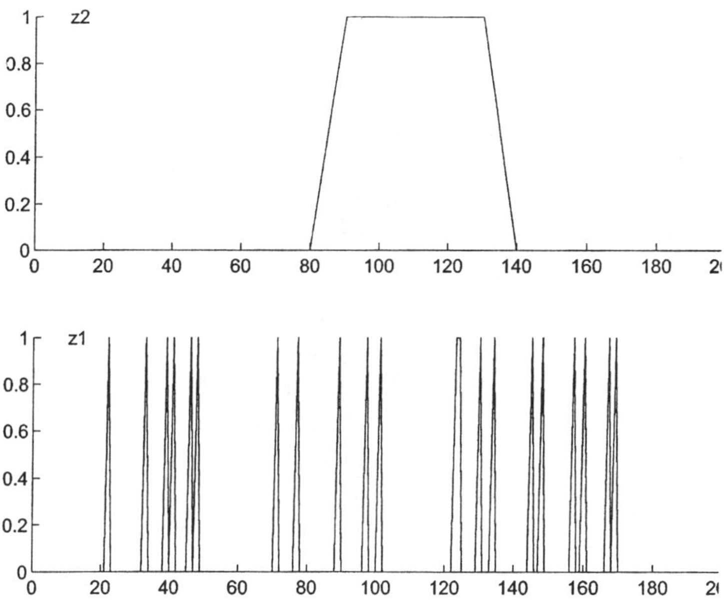

Figure 4. Contextual variables.

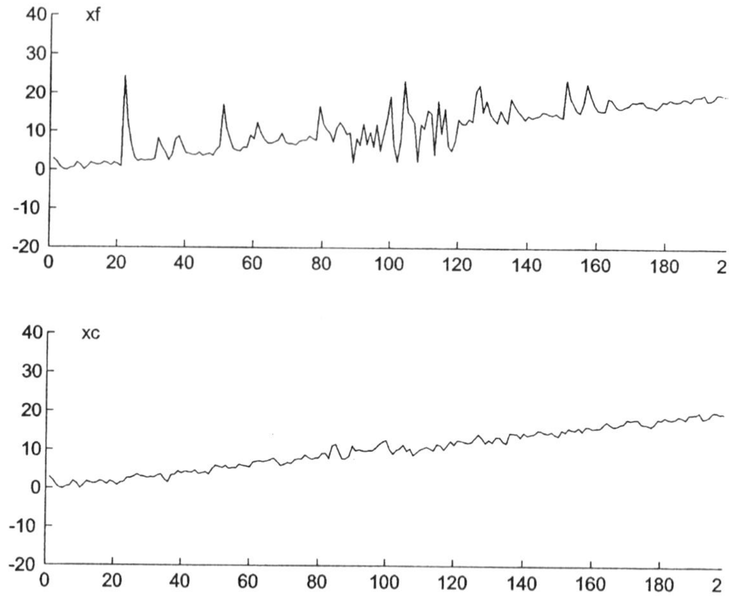

Figure 5. Top-result from classical multisensor tracking algorithm. bottom-result from the proposed algorithm.

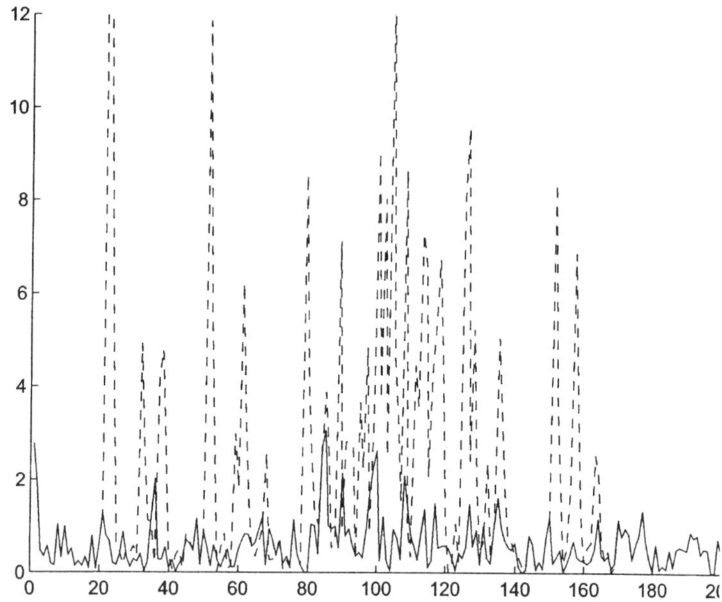

Figure 6. Root mean square error. Dashed line: classical multisensor tracking algorithm. Solid line: proposed algorithm.

6. Conclusion

An algorithm capable of reliably tracking a target using measurements of diverse sensors in the presence of perturbations has been presented. This algorithm is based on the analysis of the context which enable the evaluation of validity probabilities. These probabilities allow to increase the importance of measurements provided by reliable sensors and to decrease the importance of the corrupted ones in the update equations of a multisensor Kalman filter. Fuzzy logic is used for the representation of the human knowledge that describes the validity domain of the sensors. This theory is consistent with the probability theory usually used in modelling stochastic signals. Simulations results show a significant gain in performance. This gain in performance is mainly characterized by a strong robustness to natural and artificial perturbations in an operational scenario.

7. References

1. Appriou A. (1990) "Perspectives liées à la fusion de données", Science et défense 90, Dunod, Paris.
2. Bar-Shalom, Y., Fortman, T.E., (1988) *Tracking and data association*, New York: Academic Press.
3. Bar-Shalom Y. (1992) "Multitarget - Multisensor Tracking: Application and Advances", Artech House.
4. Bloch I., (1996) "Information Combination Operators for Data Fusion", IEEE Trans. on Systems, Man, and Cybernetics, Vol. 26, n° 1, pp 52-56.
5. Chong C.Y., (1979) "Hierarchical Estimation", Proc. 2nd MIT/ONR Workshop on Distributed Communication and Decision Problems, Monterey, CA.
6. Chong C.Y., Mori S.,Tse E, Whisner R.P. (1979) "Distributed Estimation in Distributed Sensor Networks", Proc IEEE American Control Conference, Arlington, CA.
7. Haimovich A.M., "Fusion of Sensors with dissimilar Measurement/tracking accuracies", IEEE

INTELLECTUALIZATION OF MULTI-SENSOR SYSTEMS FOR DECISION MAKING PROCESS

T.E. BEKMURATOV
70 Acade
Tashkent, 70047
Uzbekistan

The complex objects of control, diagnostics, recognition and identification are characterized by a lump of parameters. The information on them is formed accordingly by different independent sources (sensors). Therefore such objects attribute to the class multisensors.

As a rule, much data about physical parameters of object and their interdependencies (quantitative and qualitative) are uncertain, i.e., random, fuzzy, ambiguous and doubtful. Therefore decision making with the purpose of diagnostics, recognition, identification and control of such objects happens in the conditions of uncertainty.

The methods, used for decision making, of the statistical solutions, adaptive control, multifactor analysis, planning of an experiment are effective only if there is sufficient information on the object. In the real cases, as a rule, this condition is defaulted.

Therefore for decision making in conditions of uncertainty, the broad application is received by the methods of an artificial intelligence: heuristic methods, the theory of fuzzy sets, procedures of the logical inference with usage of fuzzy knowledge, fuzzy rules and algorithms, methods of removal of uncertainties.

In the given work, the methods of representation of fuzzy relations of initial data, verification and estimation of their accuracy are considered with the purpose of selection of the most reliable, and representing them for decision making.

The problem is researched, when the initial set of fuzzy relations is set by the way:; $E = \{g_1, g_2, \ldots g_n\}$, where the relations are differentiated by the type on unconditional and fuzzy relations $g_i = <x \text{ is } H_i>$ and conditional fuzzy relations $g_i = <\text{if } X \text{ is } G_i, Y \text{ is } H_i>$.

The variables X and Y receive values on universal sets U and V accordingly. The fuzzy subsets H_i and G_i are part of sets U and V accordingly and have membership functions $\mu_{Hi}(u)$, $u \varepsilon U$ and $\mu_{Gi(v)}$, $v \varepsilon V$.

In this problem, it is required to analyze initial set of fuzzy relations with the purpose verification, matching and estimation of a degree of their accuracy with the purpose of selecting the most acceptable for the correct decision-making.

Let's consider a case, when the information fuzzy relations gi look like

$$g_i = <\text{if } Y = H_i, X = G_i>, \quad (1)$$

where H_i, G_i – fuzzy parameter values Y and parameter X with membership functions μ_{Hi}, μ_{Gi}, accordingly. The degree of internal incompleteness of the combinations of these relations is expressed by the value

A. K. Hyder et al. (eds.), Multisensor Fusion, 381–385.

$$\text{in } E = \sup_{y} \max \{0, 1 - \sum_{i} \mu_{Hi}(y)\} \tag{2}$$

Accordingly, the degree of an internal redundancy of the same set of the expressions can be determined by the value:

$$\text{red } E = \sup_{y} \max \{O, \sum_{i} \mu_{Hi}(y) - 1\}. \tag{3}$$

If the combination $E = \{g_1, g_2,\ldots g_n\}$ describes relations of the parameter X from vectors $Y = Y_1\ldots, Y_m)$ then the conditional expressions are made by the following way.

$$G_i = \langle \text{If } Y_1 = H_{1i}, \ldots, Y_i = H_{ij} \text{ that } X = G_i \rangle, \tag{4}$$

In this case the degree of internal incompleteness and degree of internal redundancy of a set E can be calculated similarly:

$$\text{in } E = \sup_{Y1,\ldots,Ym} \max \{0, 1 - \sum_{i} \min_{j} \mu_{Hi}(y_j)\} \tag{5}$$

$$\text{red } E = \sup_{Y1,\ldots,Ym} \max \{0, \sum_{i} \min_{j} \mu_{Hi}(y_j) - 1\}. \tag{6}$$

Inconsistency of the information concluded in a set of the expressions E, appears when on the basis of this set of the expressions can be made different conclusions about the value of some criterion of coordination of information. If the value of criteria of coordination of the information, calculated for some set of the expressions, indicate excessive redundancy then it is necessary to reveal a cause, and if necessary with the help of the person who takes decision, to correct the parametric description of alternatives. It is necessary, in particular, to test and correct the information inclusive in the expressions about parametric relation and in the expressions about probability of values of parameters.

Thereby, there is an attempt to bring the information in a condition, which was matched with representation of the person taking decision.

The verification of the inconsistency of a set of the conditional expressions a little other, when except for a set of the expressions, to other information on character of relation of criteria from parameters is not preset and when, contrary, is known that has some properties (for example, when the person who is taking a decision considers that relation monotone or convex). If such information is not present, it is necessary, that the only requirement was executed: the expressions linking similarities (i.e. having common basic subsets) value of parameters, should correspond to no less similar values of criteria.

$$(\forall\, t_1, t_2, = 1, n)\; G\,(G_{t1}, G_{t2}) \geq C\,``(H_{t1}, H_{t2}), \tag{7}$$

Where C – coordination of fuzzy values of criterion and parameters defined as follows. The coordination of two fuzzy values of criterion G_{t1}, and G_{t2} are expressed as:

$$C\,(G_{t1}, G_{t2}) = \sup_{x} \min \{\mu_{Gt1}(x), \mu_{Gt2}(x)\} \tag{8}$$

In turn, the coordination of two vectorial parameter values H_{t1} и H_{t2} can be calculated with the help of

$$C\,(H_{t1}, H_{t2}) = \min_{j} \sup_{yj} \min \{\mu_{Hjt1}(y_j), \mu_{Hjt2}(y)\} \tag{9}$$

Thus expression (7) means, that the coordination of values of criterion for each pair of expressions such as (1) and (4) should be not less than the coordination of vectorial value of parameters in the same expressions.

The estimation of an accuracy of fuzzy relations implements on the basis of calculation of criterion estimations of alternatives, i.e. the valued information. For this purpose it is necessary:

-to determine criteria of an estimation of the alternatives;

-to determine parameters influential on criteria estimations of alternatives;

-to determine subsets of parameters influential on estimations of alternatives by different criteria;

-to select in each set of parameters influential on estimations of alternatives by one criterion, independent subsets;

-to determine types of expressions about the relation of criteria to each subset of parameters;

-to form the expressions, which compare different values of criteria and parameters;

-to correct the relations with the help of verification of the gated expressions by criteria of consistency, entirely without redundancy.

In case of the determined parameters for calculation of fuzzy criteria estimations it is enough to have the fuzzy descriptions of relations of criteria from separate parameters with the help of the expressions such as (4).

There are at least two ways to approach the calculation of fuzzy criteria estimations on the basis of sets of the expressions such as (4): application or composite production rule, or fuzzy interpretations. In first of them it is supposed, that the set of the expressions E sets indistinct relation R between parameters and criteria, and $\mu_R(y,x)$ by number expresses a degree, from which the values of parameters $y_j \in Y_j$, where $j=\overline{1,m}$ correspond to criteria estimations $x \in X$.

The separate expression g_i such as (4) sets a fragment R_{ij} of fuzzy relation between the parameter values Y_j and criterion, dependent on it X by the way of fuzzy parameters the Cartesian product $Y_j \times X$ with a membership function $\mu_{Rji}(y_j,x) = \min\{\mu_{Hji}(y_j), \mu_{Gji}(x)\}$.

The combination of the expressions such as (4) sets on $Y_j \times X$ relation $R_j = \underset{i}{V} R_{ji}$ by a membership function.

$$\mu_{R(y,x)} = \underset{i}{\text{man}} \min \{\mu_{Hji}(y_j), \mu_{Gji}(x)\}$$

Thus, the fuzzy description of relations of criteria from parameters set fuzzy relations R with a membership function:

$$\mu_{R(y,x)} = \min_{i} \mu_{c(Rj)}, (x,y), y \in Y, x \in X.$$

If now in space Y to set vector of fuzzy values of parameters $H=(H_i,\ldots,H_m)$, then on the basis of the composite rules it is possible to determine an estimation appropriate to fuzzy criterion, G by membership a function $\mu_G(x)=\sup_y \otimes [\ \mu_R(y,x), \mu_H(y)]$, where the operation of limited product $\otimes$ is set by expression.

$$\otimes[\ \mu_R(y,x), \mu_H(y)]=\max\{0, \mu_R(y,x) + \mu_H(y) - 1\},$$

and the value $\mu_H(y)$ is evaluated under the formula: $\mu_H(y)=\min_J \mu_{Hj}(y_j)$.

Fuzzy criteria estimations X we shall call deductive from fuzzy values of parameters Y by the way of fuzzy relation of E such as (6), if there is a rule permitting to each feasible vector of fuzzy values of parameters E to put in conformity some fuzzy criterion an estimation G. The fact of deductibility we shall write by the following way $X=\Phi_E(Y)$.

In particular, the fuzzy relation E can contain the description of dependence of criterion X only from one of m parameters forming space Y.

The fuzzy criteria estimations G_1 and G_2 deduced on the basis of two different fuzzy relations E_1 and E_2 from the same vector of fuzzy value of parameters H, i.e. $G_1 = \Phi_1(H)$, $G_2 = \Phi_2(H)$, can differ, if the relations E_1 and E_2 are not completely compatible. In this case, it is necessary to develop the methods of combination deduced from criteria estimations G_1 and G_2, permitting to smoothen inconsistency between E_1 and E_2.

The analysis of different ways of combination allows to propose the following generalized formula.

Let E_R-fuzzy relation for which is known or can be calculated the values of parameters and which usage allows to calculate value of criterion X. Let also L – set of numbers of such relations with known confidence figures α_R, $R\in L$, G_R-fuzzy criteria of estimation, deduced on the basis of fuzzy relation ER from vector of fuzzy values of parameters H, i.e. $G_R = \Phi_R{}^{\alpha}{}_R(H)$. In this case the membership the function of a combined fuzzy criteria estimation G is determined on the basis of a membership $\mu_{GR}(x)$, $R\in L$, as follows:
$\mu_G(x)= \otimes(1-\max \alpha_t) \min\{\mu_{GR}(x), \min[1-\mu_{Gt}(x)]\}$, where $a \otimes B=\min\{1, a+b\}$. The program complex intended to implement of a described procedure of decision making, should provide in the dialog mode (including experts): input and output fuzzy expressions, formation of the answers on questions of the user, an estimation of accuracy of fuzzy expressions, fulfilment of fuzzy arithmetic and set-theoretical operations over the fuzzy multiple expressions, and also updating of the fuzzy data and correcting of the fuzzy rules.

The program complex contains the following modules.

The leading control module, which operates the process of decision-making, allows the user to select the necessary procedure.

The inputting fuzzy data module provides replenishment of the dynamic database (DDB) with the new information.

The deletion fuzzy data module executes deletion of the out-of-date data from DDB.

The adjusting fuzzy rules module provides dynamic change, input and deletion of the fuzzy rules from the knowledge base.

The formatting the answers on the questions of user module executes output of information and the rules necessary for the user in estimating the situation and decision-making.

The syntactic analysis and interpretation of arithmetic-logic operations module. This module executes an estimation of a degree of an accuracy of fuzzy expressions, executes fuzzy arithmetic, and set-theoretical operations.

Literate

1. Zadeh, L.A., (1979) Approximate reasoning in fuzzy logic. Proc/ Int. Conf/ on Artificial Intelligence. Tokyo.
2. Bekmuratov, T.F., (1999) Khodjimatova G.M. Systems of Decision Making. Problems of Information and Energy. (Problemi Informatiki I Energetiki). Tashkent, N1.

IMPLEMENTING DATA FUSION SYSTEMS

R. REYNAUD
IEF Université Paris 11
F91405 Orsay Cedex
Email: Roger.Reynaud@ief.u-psud.fr

1. Data Fusion Examples

1.1 NAVIGATION AND LOCALISATION

Navigation is a fundamental problem in the domain of mobile robotics. To reach the goals of a lot of applications, it is necessary for the onboard computer to know its relative or absolute localisation [1]. It is well known today that the solution [41] results from the competition of two sub-systems: a dead reckoning navigation based on data issued from various proprioceptive[1] sensors; and a re-calibration at regular intervals based on data issued from exteroceptive [2] sensors giving absolute or relative estimates of obstacles or landmarks.

As described in Figure 1, the dead-reckoning system uses sensors such as gyroscopic sensor, acceleration sensor, steering angle measurement, pedometer, odometer (angular encoders attached to the wheels). The sole use of dead-reckoning navigation provides continuous information with no delay by integration of the data, but cannot avoid the accumulation of errors and the introduction of a drift between the estimated and the real trajectory.

The observation of the real world is realised through the use of sensors such as CCD cameras, range finder, acoustic belt, optical barrier, radar, doppler, lydar, global positioning system,... All these sensors are relatively expensive to implement all together on a prototype because they don't sense the same features of the obstacles or the landmarks present in the scene. Moreover, they provide precise absolute locations at discrete time instants, with a delay that can be too large for real-time applications if they are the sole sources of data[40].

A simple scheme of fusion is to take some selected pieces of data from the two kinds of sensors and to combine them by an Extended Kalman Filter [21]. A future state can be computed in 3 steps from an initial state: the first step is a short-term prediction of the location, based on a dead reckoning navigation, and taking into account as many pieces of internal data as necessary; followed when an external piece of data is available by a step of self localisation technique given the measured location of the vehicle; the final step is then an estimation or update comparable to a weighted interpolation between the prediction and the position computed from the external observation. The use of an appropriate Kalman filter can thus supply a statistically optimal estimate of the internal state of the vehicle in the theory of stochastic dynamic systems [27,28]. Both internal state and measured location are

[1] Proprioceptive: measuring some features of its own state.

[2] Exteroceptive: measuring the variations of the features of external objects present in the scene.

A. K. Hyder et al. (eds.), Multisensor Fusion, 387–418.

represented by two stochastic gaussian processes, and are characterised by their mean and standard deviation. Figure 1 shows the final step that interpolates the measured location and the predicted location.

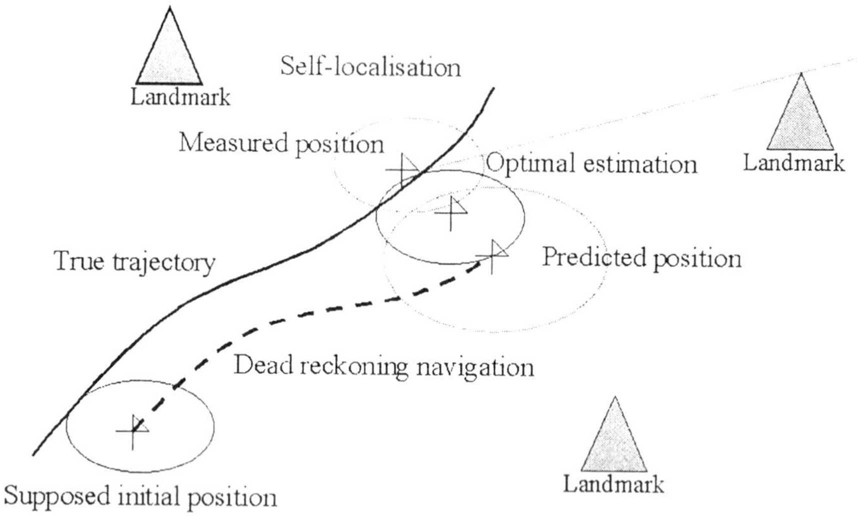

Figure 1. Navigation and localisation

The main problem lies in finding a self location mechanism through the use of suited sensors and associated algorithms. It depends greatly on the environment and on the desired accuracy. The environment can be fully structured (full of landmarks that lead to easy reconstruction of a measured location), partially structured (there exists some landmarks which can allow a regular re-calibration, the difficulty lies then in finding the landmarks and associating them with possible locations) or not structured. As a matter of fact, precision depends on the number of landmarks and on their configuration [10].

1.1.1 Example of autoguided missile

Before the availability of the Global Positioning System, the control of an autoguided missile was a good example of navigation system. In the ninety's, the technology led to build autoguided missile following the model [31] depicted in Figure 2.

Just before the launching of the missile, the pieces of information in the right of Figure 2 are loaded in the onboard computer.

The system can be considered as composed of three stages of tasks:

- The first stage realises different mechanisms of continuous search and extraction of the features of the landmarks to be identified.
- The intermediate stage is a centralised detection block which searches the association between found landmarks and marks compiled in the pre-loaded map. When there is a validated detection, often with a slow rate, then the system accepts

this position issued from a self localisation and takes it into account by an extended Kalman Filter in the final stage.

- The last one is a dead reckoning navigation taking into account frequent updated observations of the proprioceptive sensors.

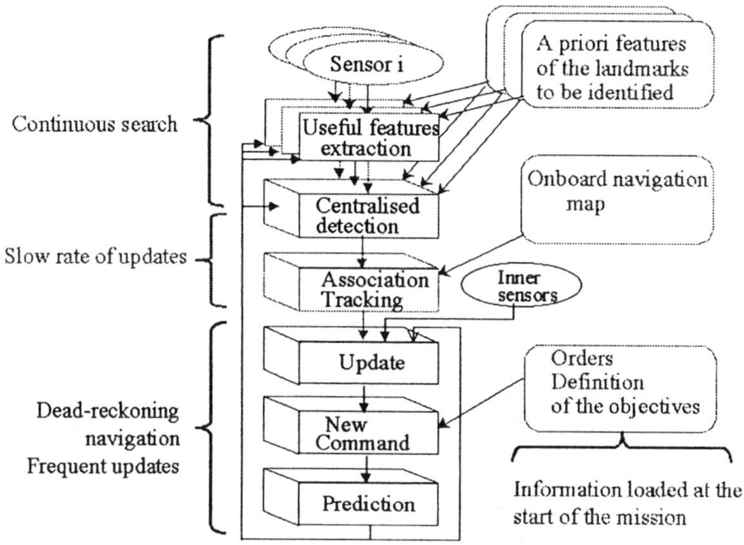

Figure 2. Autoguided Missile

1.1.2 Mechanical coupling of two sensors

The fusion can be seen as a problem close to the sensor. A basic example is to pair two sensors on the same site such as a radar and an infrared 2D retina.

The strength of this coupling is to have complementary functions. The radar holds the two sensors in the direction of the target thanks to a tracking system. The 2D vision system is mechanically coupled and its function is to allow an early detection of a change in the acceleration of the target. Without the coupling, the tracking system is also able to detect such a change, but it must wait that the variance of some selected variables reaches a threshold to assume this change in direction. Globally, the system is more efficient, because the early detection of change makes the system more robust.

Other examples of mechanical or geometrical pairing can be found in the literature such as: Camera + scanning telemeter [30], Camera + radar [29], two linear cameras [16], etc.

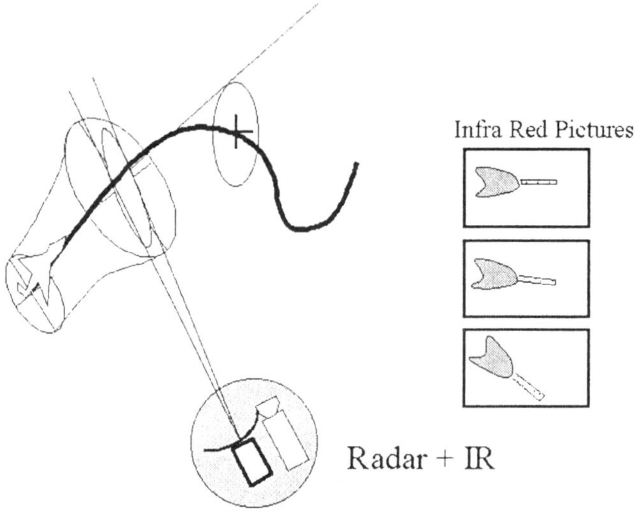

Figure 3. Coupled sensors

1.1.3 Triangulation and range finder

Range measurement is very useful because the simpler state used in Kalman filtering is to have the position and the speed of the tracked objects.

But it is not obligatory, because a pseudo range finder can be created by a mechanism of triangulation from the knowledge of at minimum two angular measurements. If we use only two angular measurements, the precision of the result provided by this new virtual sensor is good near the sensors and is deteriorated as the distance increases. With three or more sensors, the precision increases greatly.

The availability of the technology of the Global Positioning System simplifies greatly the absolute self-localisation for outdoor vehicles when the precision is around the meter. This includes all the applications of vehicle path planning in the management of vehicles fleets, but needs some additional sensors capabilities when the precision must reach one to ten centimetres.

1.2 TRACKING

In a tracking system [3,4,5], it is not possible to implement a dead reckoning navigation of the evolving targets, because we have no access to the corresponding proprioceptive data. This sub-system is replaced by a model based prediction that allows to estimate the target position between two localisations.

Apart from this sub-system, the remaining system is identical. When a target is observed by a sensor, this observation is translated into a centralised coordinate space, and is used to readjust the state (position, speed, acceleration) of the track through a version of an Extended Kalman Filter.

Each target is autonomous and is not necessarily cooperative, so the remaining difficulty is to choose an appropriate model.

Another difficulty comes from systems that have sensors in different locations. Their data are not synchronous (we will discuss this point later) and we must imagine the mechanisms of fusion. The most classical way to do this is to define a granule of time. If the data coming from all the sensors belong to the same granule, they are processed in one batch in a predetermined sequence.

The first place of fusion is then to chain the processing. The output variables of the processing stage associated to the first sensor are taken as input variable of the following stage associated to the second sensor and so on.

The second place of fusion is constituted by a mechanism global to the whole processing stages and is updated once per iteration. In an Interacting Multiple Model, the probability of each model is estimated from the initial probabilities, from the likelihood of belonging to each model at each processing stage, and from a modelling of the behaviour of the object being tracked.

The system must remain operational when an incoming data is missing. In this case, the corresponding stage is by-passing (the input variables are presented as output), and the corresponding likelihoods of belonging to a model are equal to each other, so that they don't interfere on the results.

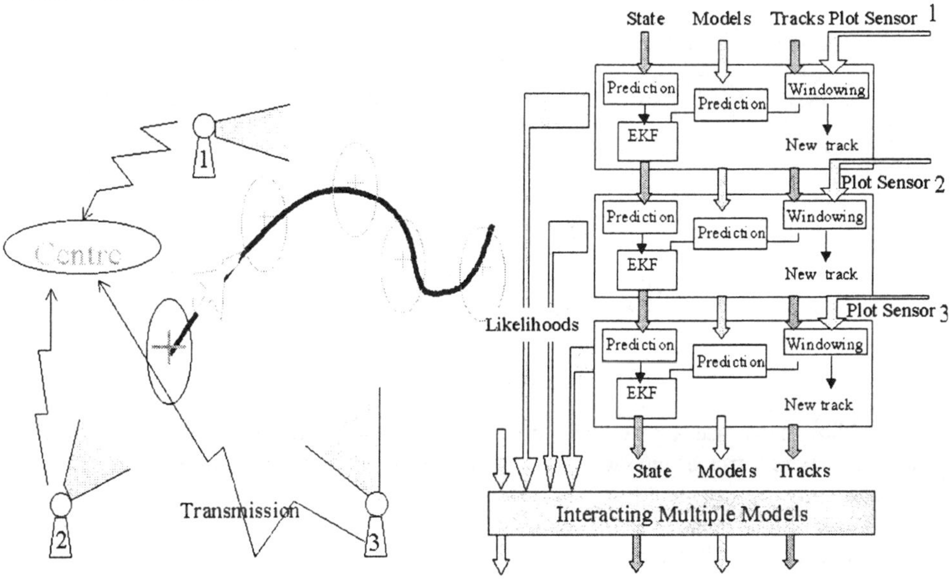

Figure **4. Multi-sensor IMM tracker**

1.3 MULTI-AGENTS SYSTEMS

Another approach can be used to build an autonomous system with sensorial modalities [15,25]. The application is split into tasks named agents. Each agent has a level of knowledge, a level of autonomy and some possibilities of interacting with other agents. To make such a system efficient, it is important to understand the mechanisms of interaction between agents and with the environment.

In the framework of the research group ISIS[3] of the CNRS[4]

We can see that the fusion corresponds only to a small part of this classification. There are two reasons: the notion of fusion is restricted to the mechanism of combination of information in the mind of the members of this community; but it also means that, if we want to implement original and efficient mechanisms of reasoning, then the system must not only include fusion mechanisms but also some others forms of reasoning.

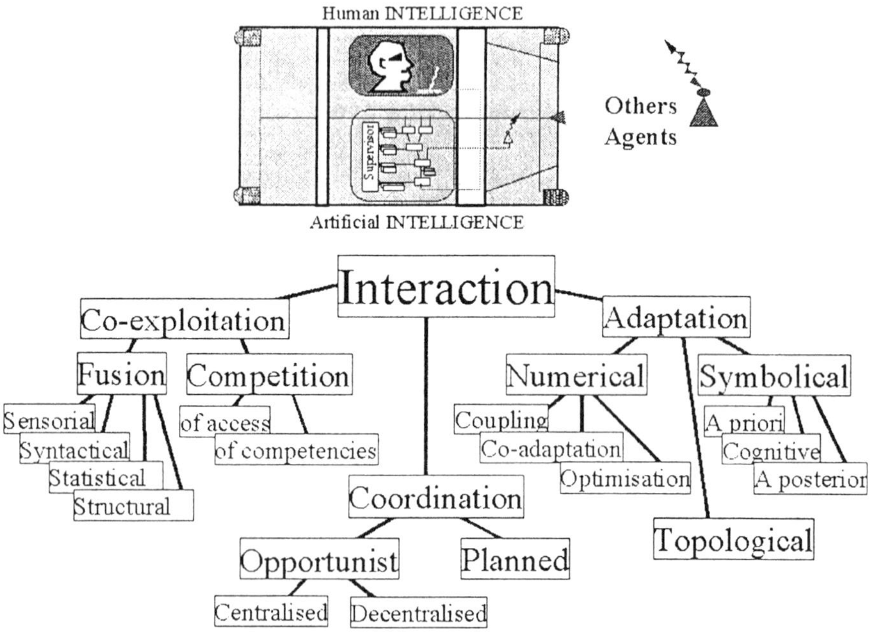

Figure 5. Taxonomy of possible interactions

So, although we use a processing line as we describe it further, we must be able to implement others forms of interacting, so that the assistance system becomes more efficient by using finer reasoning [12,13].

2. Onboard Obstacle Detection

2.1 THE APPLICATION

This application has been demonstrated in ninety-five during a demonstration day of the PROCHIP[5] project [38] in the framework of an European program with the label PROMETHEUS[6].

[3] ISIS
[4] CNRS: National Centre of Scientific Research in France
[5] PROCHIP: Acronym of PROgram for Custom Hardware for Intelligent Processing

We don't insist on the choice of the sensors. It is sufficient to know that the basic modality consist of two scanning infrared lidars that provide range and angular measurement after a step of triangulation. This modality brings information up to forty meters in front and in back of the smart car. An additional radar takes over from the lidar in front of the smart car for ranges from thirty meters up to one hundred and eighty meters. The smart car interacts with external agents by means of beacons. The environment is hence supposed cooperative and relatively highly structured. All lanes are visible and each obstacle is suitably equipped so that the spatial hypotheses are respected.

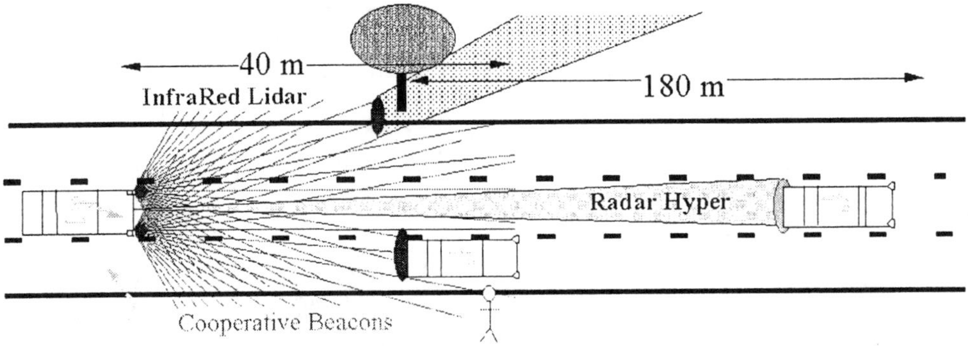

Figure 6. Typical application: Highway driving

At the question what to observe, the reply is application dependent. At the question how to observe, the reply is sensor dependent. At the question where to observe, the reply is context dependent.

At the questions how, what and where to fusion, to process or to decide, the reply is implementation dependent and each designer has several options to implement his version of the system. We have opted for a data flow processing line that is continuously regulated by a competition among models which control all the stages of the processing line.

2.2 ASSISTANCE TO DECISION

This leads us to propose a system corresponding to the following schema of assistance to decision:

(a) Each modality can be pre-processed to provide data in the form of [relative position, relative speed] which is the state variable allowing to reconstruct a map of the environment.

(b) These data are combined by Extended Kalman Filtering to provide a reconstructed map of the local neighbourhood with the smart car localised in the centre of the 2D space. The obstacles are positioned relatively to the smart car.

[6] PROMETHEUS: Acronym of PROgram for an European Traffic with Highest Efficiency and Unprecedented Safety

(c) From this map, and according to the contents of the database, different interpretations can be done, by co-exploiting the same inputs in a form of competition of competence.
(d) The decisions are fused and presented to the driver
(e) The supervisor allows to implement several forms of adaptation to make the system more precise and more robust. All the no determinism of the application is gathered in this module.

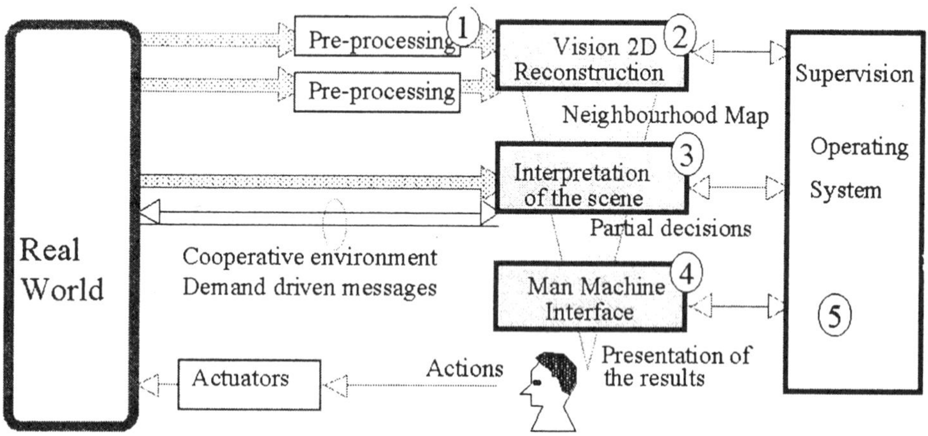

Figure 7. Assistance to decision

2.3 THE FUSION PROCESSING LINE

The first generic goal of a data fusion system is to decrease the amount of data issued from the observation and, in the same time, to increase the semantic level of these data and their robustness towards imprecision and uncertainty.

To realise these two objectives, the processing is organised along a vertical data flow processing line consisting of sequential stages [39]. Each stage consumes more data as input than it provides data as output, and it accompanies this circulation of data with a refinement of their semantic level. Alone, such an organisation is unable to process in a correct manner the data in order to reach a sufficient semantic level of decision. But it plays correctly its part of concentration of the information due to the systematic fetch and consumption of the data, mechanism which can be easily implemented at the system level.

To obtain an efficient system, this processing line is supervised thanks to the choice of different models at each processing stage: Models in competition at the preprocessing and intermediate fusion stage; Context dependent conversion at the numeric to symbolic stage; Selection of different rules sub-sets at the symbolic processing level.

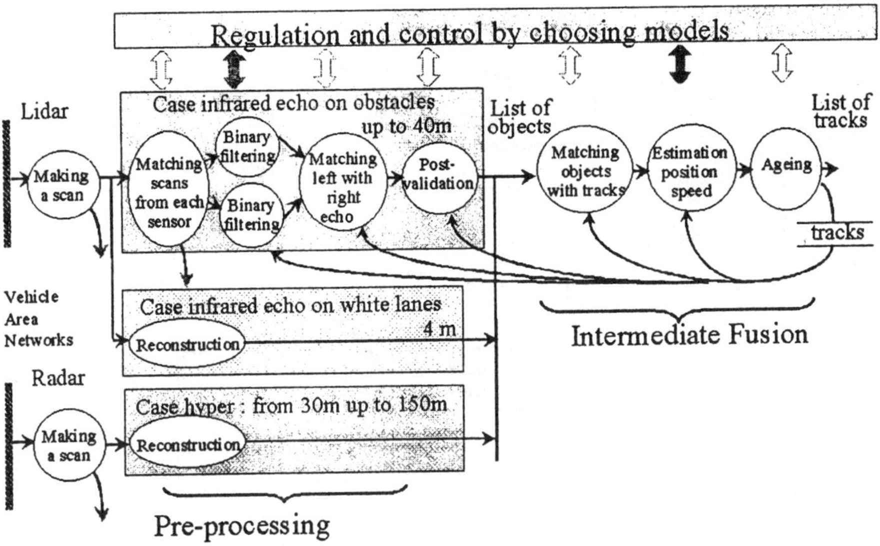

Figure 8. The fusion processing line

2.3.1 Signal fusion

With the example of the preprocessing stage and the intermediate fusion filtering, we treat of the notion of model switching and regulation of the processing line by competition among different models [37,20].

In Figure 9 the processing line is horizontal. All the stages of the processing line can be regulated through the choice of different models. We present two examples of this regulation.

Tuning a Kalman Filter

A well-known mechanism of regulation of a Kalman filter is the set up of the signal to noise ratio that controls the classic trade-off between the relative weight that we grant to the observation and the a priori. When this ration decreases, we have more confidence in the knowledge of the current model and so it corresponds to a refinement of our knowledge. When this ratio increases, we give more weight to the observation than to the a priori. We have less confidence in the knowledge of the model, so it corresponds to a deterioration mode.

To be operational, the context of application of all models must be nested with each other. They are thus organised in a structure of a hierarchical tree. At the base, the less informative model must include the other finer models which are more and more informative in proportion of the progress in the tree. When the regulation switches from a model to another, the transition are not all authorised to secure the functioning of the regulation. In all the cases, the idea is to find, as soon as possible, that the current model is out-of-date and is no longer representative of the reality of the behaviour of the phenomena, and to switch to a model becoming coarser but consistent with the reality as soon as possible [2,15].

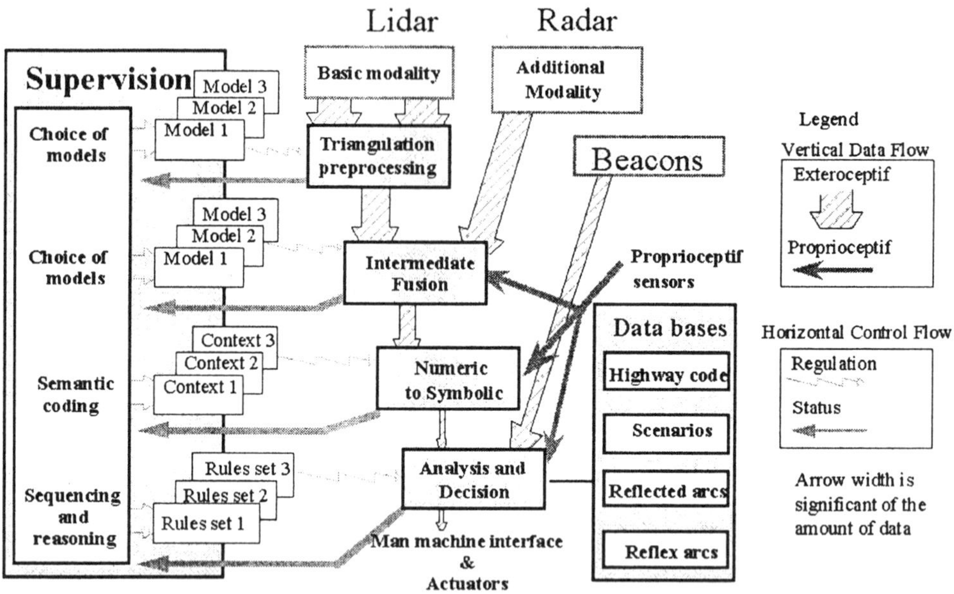

Figure 9. Pre-processing and intermediate fusion

Tuning a binary filter in the pre-processing step

We present here another choice of control of a processing step thanks to the choice of a model among a list of available models [18,19].

For a lidar sensor, a vehicle is seen as a packet of contiguous bits setting to one, and having a given angular width. In the scenario of Figure 10, the smart car cannot see the vehicle V2 which is masked by the vehicle V1. At the end of the scene, the two vehicles are observable.

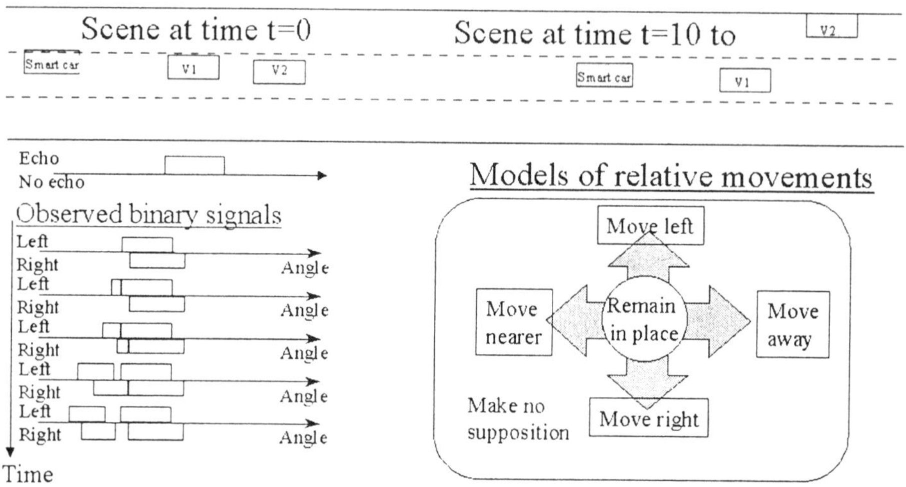

Figure 10. Tuning a binary filter

If the supervisor has a sufficient confidence in the knowledge of the movement of the obstacle V1, then it can propose to switch to refiner models of relative movement such as: remain in place, move left, move away, move right and move nearer. The binary filtering process can then favour selected patterns opposite to others patterns, and thus, it helps to early find the coming of the new vehicle in the scene.

With this way of thinking, a deteriorated model must also be implemented when we have no sufficient information. The by-passing of the binary filtering task means that we have no favour for any model of relative movement, and brings a prejudice to the early detection of a new obstacle.

2.3.2 Symbolic fusion

Timing constraints of the decision making system excludes to make an exhaustive analysis of the whole set of rules at each recursion. Therefore, the knowledge base is structured in rules sub-sets. Each specific sub-set can be designed to solve a particular problem, or dedicated to a given scenario, or to a way of reasoning such as reflex, reflected or diagnostic arcs. There are also meta-rules that chain a train of sub-sets to obtain coherent actions.

It allows coarse or fine reasoning to be introduced. The completeness and the coherence can be tested, module by module. Resulting code is as efficient as if it was written in C language. Using the information represented by rules and processing it by an expert system are not detrimental to the efficiency of the implementation of the resulting code.

It is always unrealistic to try all expected combinations and the supervisor must regularly decide which rule sets have to be executed at one time to secure a reaction in a given delay. Thus we have two kinds of rules. The ones in the first part can all be compiled.

Some of the rules play a role of regulation of the processing, or a role of planning of the other sub-sets of rules. Other standard rules are directly concerned by the answer of the initial issue: is driving situation dangerous? All these last rules compose the second part which is called supervision.

Moreover and considering the emergency of a situation, the supervisor is then able to activate incomplete but very fast sets of rules in emergency cases (reflex arcs), whereas a more detailed reasoning is considered otherwise (reflected of more complete reasoning on driver's intentions when an indicator is put on). The lack of determinism, inherent to all decision systems based on expert systems, is here located only in the supervisor where it can more easily be controlled.

Context dependant conversion

We present also an example of a context dependent conversion from numeric to symbolic. The problem considered in [8] is to set the variable Distance_To_Obstacle to one of the three values: little; medium; big. This conversion is obviously dependent on the relative speed between the smart car and the obstacle. It can be given for example by two charts symbolising the two frontiers between the crisp symbols. The objective is, at this level, to reduce the number of rules by creating a kind of abstraction that allows to gather likely rules.

Examples of rules

In a fusion processing line, the data sources are incrementally taken into account. The basic modality provides a first map, the additional modalities allow precision, certainty and processing safety to be increased, either by systematically processing data from sensors, or by requesting data to make the reasoning finer.

The scenario is the following: the driver of the smart car has decided to overtake another car in front of it.

Rule overtaking 1

If the driver sets the left indicator

If a sensor has detected the presence of an obstacle at the left stern

If the distance of this obstacle becomes dangerous

then the driver's behaviour is dangerous

then the situation becomes urgent

then the action is to warn the driver

Rule overtaking 2

If the driver sets the left indicator

then the supervisor allocates immediately sensor and computing resources to detect the presence of an obstacle at the left stern

When the result is ready then the first rule of Overtaking 1 is evaluated.

Because of the no determinism inherent to a data request protocol, this second rule must be managed at the supervisor level. In counterpart, the first rule can be implemented in the fusion processing line insofar as it supposes that the data are not out-of-date and that a systematic prefetch of the data guarantees the determinism of this action.

The two rules manage differently the confidence and the perennialty of a previous detection. The questions are: how long since a detection has been made? Are the contents of the symbol indicating that an obstacle have been detected still valid? Therefore, according to the context (short or long delay), and thanks to the use of alternate subsets of rules, it is possible to modify the protocols of fetching data and of validating the evolving content of the blackboard.

2.4 ARCHITECTURE

In order to make the prototyping of a system, two classical design approaches exist: top-down and bottom-up.

In a top-down approach, we can hope a perfect respect of the initial specification in a formal theoretical frame with the possibility to use tools of verification and optimisation. We will see such an approach based on the use of the Signal language.

On the other side, a goal is to obtain a rapid prototyping of the system, with fast feedback, necessary when the application is evolving with the time. The goal is then to minimise the resource necessary to bring the demonstration to a successful conclusion. This context is thus in favour of modularity and in favour of a maximal re-use of already verified blocks.

None of the two approaches allows to obtain an operational system. The designer finds himself in the necessity to validate a method or a methodology to join the two approaches at an intermediate level.

For our part [11], we start from the notion of an idealised architecture. All the dedicated processing blocks manage a private network by which they collect the data to be pre-processed. For the symbolic fusion task, the network collects data from the proprioceptive sensors, but also controls the actuators with the same network. These blocks are linked thanks to an internal network that can be generic, but that is also very often dedicated to the application.

For a rapid prototyping, the keywords are: open system; symmetrical world; low parallelism; specific sensor networks; centralised control; concentrating information.

2.4.1 Hardware

This is a proposal of hardware, approximating the idealised architecture. There are five levels: sensor level; vehicle area network; processing level; loading and monitoring level; and off-line testing level to record a lot of pieces of information and replay the same scenario in order to debug a difficulty.

The hardware of the PROCHIP demonstrator was based on TMSC40 processors in Tim40 modules of Texas Instrument and on a PC host computer for loading and diagnostic tasks. The network is dual: two asynchronous vehicle area networks for sensors; some parallel links between processors and towards PC.

The following static allocation is used: a processor, connected to a private Vehicle Area Network, maintains the data input, and builds the map of the environment. A second processor performs the interpretation of the scene by symbolic heuristics. A third processor supervises the whole and controls the links with the PC host and with a VAN dedicated to inner sensors, actuators and beacons.

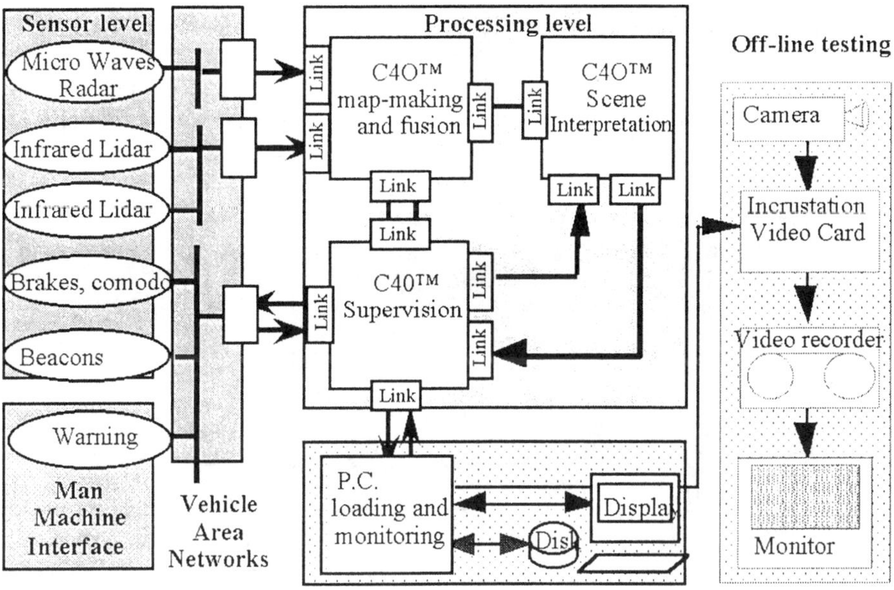

Figure 11. Hardware structure

2.4.2 The system

The system must be able to efficiently manage a parallelism of four processors with a few micro-controllers on the communication interfaces [39].

In this kind of application, timing constraints are difficult to respect. We will see in the next part that a synchronous methodology can be used to specify and validate a great part of the time control. But the remaining part is made by the supervisor at the applicative level. The time control is explicitly coded by using the basic primitives of a real-time kernel.

The control paths are used to regulate the other tasks from the supervisor. This flow is less regular in topology and in frequency. It can be useful to implement a data request protocols. The complexity is then greater but remains limited by the lower amount of transmitted control.

2.5 CONCLUSION

It is possible to implement an onboard system of vehicle avoidance from two ideas. The first idea is to build the functional and hardware structures from a data flow processing line. That means that we favour firstly the aspect of multisensor data fusion. The second idea is to understand that it is illusory to believe that a complex and anthropomorphic system could be regulated by the sole mean of standard generic automates.

A part of the control is application specific and must be coded at the supervisor's level.

Therefore we find in the functional structure the use of the following logical principles of interactions of Figure 5: coexploiting with fusion and competition; coordinating with the notion of planned vision and opportunist approach; adapting at the numeric, symbolic and topologic level.

In Figure 12, we find again on the hardware structure the data flow line as a concentrator of data and the control flow paths as arrows. A great part of the control flow is present only to allow to implement alternate forms of reasoning.

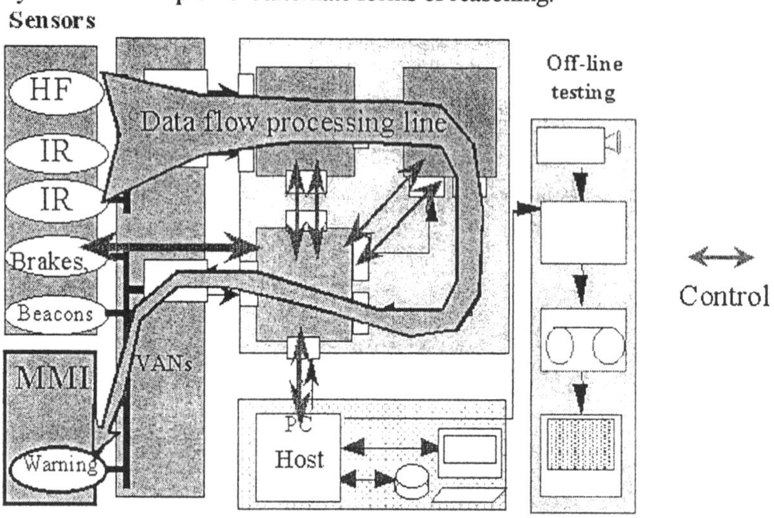

Figure 12. Hardware structure with data and control flows

But the main effect of the supervisor related to the ability of reasoning remains the sequencing and the chaining of the rules sub-sets, and its ability to control the fusion processing line by the choice of a model of regulation at each stage of the processing. The models in competition must be organised in a hierarchical structure allowing to switch to a lower or greater refinement on the knowledge of the behaviour of the object being processed.

3. Timing specifications

The second part deals more particularly with the timing specifications. Two approaches has been used in the PROCHIP project. The first one is an asynchronous approach with its well-known difficulty to guarantee the determinism of its functioning. Specific of the French community, another way is to use a synchronous approach in parallel with tools of validation and generation of executive. We show with an example of timing protocol how to naturally take into account asynchronous data in a synchronous framework.

On the right side, the other presented point tries to reply to the question: When a piece of data has too long a delay, but becomes available in the fusion centre, then what should we do with this piece of data? Integration in the fusion process or withdrawal of this data, if we think that it contains no innovative information.

3.1 MANAGING THE TIME

We propose an inventory of the problems that occur when we want to manage more precisely the time.

The sensors that provide the data meant for the fusion have different reply times, with an asynchronism of acquisitions and the presence of temporal drift. So the frequencies of arrival of information, and of processing update are also different. A first idea in a fusion system is to simultaneously exploit all the incoming data at the same instant. A set of valid data is created corresponding to the present time, or more often to a common time with a known delay. This set is considered either as a vector of observations, or as a set of observations sequentially processed as in Figure 4. The difficulty lies then in the estimation of the components of this vector when they are not synchronous. If the evolving models of the observations and the dynamic models of the vehicles or obstacles present in the scene are available, then either the oldest measurement are modified by prediction, then fused with the other data, or are fused then the result of the fusion is propagated to the present time. But the model validity during the time is not guaranteed during the step of prediction.

This method of temporal calibration supposes a technique of dating of the measurements that presents other difficulties. The delays of transmission and the latency of the processing are not necessarily constant even in a centralised system. It remains some uncertainties on the dates and the delays in a distributed system with the possible existence of temporal drift among all the physical clocks. A time stamps is used to re-synchronise the clocks, but it remains some physical mechanisms which are not deterministic in time such as the protocols on a vehicle area network.

Given a particular attention to the problem of temporal marking, synchronism and the notion of logical time, and an associated physical notion of granule of time. An important point in a dynamic system is to succeed in defining, for each physical modality, the corresponding granule of time. When we accept that two data belong to the same granule, then we can say that they are synchronous in the sense of the Shannon theorem. Their dynamic evolution is low compared to this granule of time. The specification can become simpler because we are making an approximation of the reality. Working in a synchronous world is the necessary in order to make temporal verification and automatic proof of programs. This leads to the use of synchronous and asynchronous languages and specifications.

The real time systems need also to have a determinism of working [17]. It goes through the reduction of the latency of the tasks as in a pre-compilation of the decisions trees in an expert system for example. It goes also through the use of synchronous specifications concerning the communications among the tasks and a step of automatic verification of the temporal part of the specification. This leads to the use of automatic generation of distributed executives that respect the initial specification.

Other problems occur when we want to decide the integration or the withdrawal of a delayed data. There are two cases: either the management of the obsolescence of out-of-date or no longer valid data; or the use of valid data that contains no more interesting part of information for the fusion processing. All the policies of anticipation to fetch the data can be associated to these notions as: automatic prefetch of the data to be processed, or demand driven request of data, when there is an access on an out-of-date data.

We have evaluated two approaches within the PROCHIP project: A synchronous approach and an asynchronous approach. The asynchronous approach [14] , that will be quickly presented, is the more easy to implement, because it corresponds to the physical world, but it doesn't guarantee the determinism and the temporal validation of the working of the system. In a synchronous approach [6], because of the no-use of the tasks duration in a first step, a formal verification of the temporal specifications is possible. But it is then necessary to take into account the time in a second step whose goal is to optimise the placement of the functional tasks on the hardware structures, and to generate, when it is possible, the corresponding dedicated system which verifies automatically the specifications.

Finally the synchronous distributed executive, as in an asynchronous approach, is based on some basic and native primitives of a real-time kernel.

3.1.1 Asynchronous approach

What does asynchronism mean? It means that it is possible that no temporal relation exists among some considered events.

The principles are: tasks or modules can operate in parallel or in sequence, their code is executable in finite time, with no deadlock, it can be in three states, {ready, interrupted, not operating}, the kernel have to schedule the activity of the tasks by using a policy reacting to asynchronous events, giving the rhythm {activating, interrupting} of the modules. Each event is in the following state {active, stored, not operating}. Usually the kernel manages the pre-emption of the processing resource with priorities that have to be defined by the user and that depend on the execution time that is supposed to represent the real world and so to be different from zero.

3.1.2 Synchronous approach

Although it doesn't correspond to the reality of the physical world, it is often, and perhaps always interesting to sue a synchronous top-down approach (top-down means here from specification to dedicated executive on the hardware structure). What does synchronism mean? It means that it always exists a temporal relation between each variable. These approaches operate in two steps: For the step of specification in the language Signal [6], the durations are forgotten and the hardware temporal constraints are not taken into account. The logical time (instant when a piece of data is valid) replaces the physical time. This tool deals with the class of the systems reacting infinitely fast on all stimulus. This transformation is the result of a deep study of the physics of the sensors, which leads to approximate the reality and allows to identify when data can be reasonably considered as synchronous. This time consuming work is very interesting insofar as it allows in fact the extraction of the potential parallelism induced by the specification.

Language Signal

The language Signal [7] allows the specification of a program either in textual form or as a Conditioned Data Flow Graph. The basic elements are: the logical clocks, a set of logical instants when a piece of data is valid; the variables of kind Signal, a set composed of a clock and the values the data take when the clock is valid. The following time primitives allows to build a description of an algorithm: simultaneous functional expressions, the delay, the condition when, the mixing of two signals with priority named default, and the clock synchronisation. A step of compilation tests the coherence of the temporal specification, that is to say: only the expression with synchronous signals are possible. At this point, all the relations between clocks are known trough a hierarchical conditioned graph, so benchmarks in logical time can be tested to verify the similarity of results between the wanted specifications and the resulting program coded in Signal language.

Tool SynDEx

The second tool is involved in the placement of the tasks on the processors and in the generation of the corresponding dedicated executive. We have used the tool SynDEx [42] of INRIA. This tool has two inputs: a description of the algorithm in the form of a Conditioned Data Flow Graph as it is generated by the Signal tool, with its known potential parallelism, as its partial order corresponding to the clocks relations; and a description of the target hardware structure in the form of a hyper-graph representing the network of processors and so the really available parallelism, with the knowledge of the duration of all the tasks on each kind of processor.

At the question: is the algorithm able to run on the target hardware structure by respecting the timing constraints? The reply can be positive or negative. When the reply is positive, that means that there is a compatibility or adequation between the specification of the algorithm and the architecture under the timing constraints [9].

In this case, the tool provides an estimation of the ratio of acceleration due to the use of parallel processing and an estimation of the latency of the reply to each stimulus. For some reasons of efficiency, the generation of the executive kernel is generic, written in macro-instructions code resorting to basic primitives of a real-time kernel, and is translated in operational code according to the target processor. It is an optimisation tool, so the optimising criterion can give greater importance to the amount of processors, or the minimisation of the latency, or other aspects of the implementation.

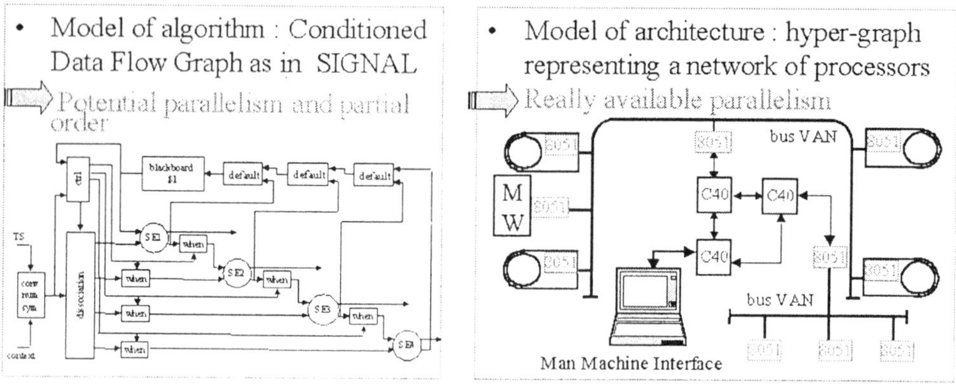

Figure 13. The inputs of the tool SynDEx

Synchronous hypothesis and real world

The positive result of the tool is the generation of a synchronous distributed executive, ensuring the respect of the initial specifications in a synchronous language.

At the functional level, the designer can then consider that, as in the case of the compilation of the rules set in an expert system under the form of logical equations, the control is pre-compiled under the form of automata of control of activation and communication of tasks. The resulting advantage is that the no determinism is pushed away more uphill in the functional description of the algorithm of control and the hope is to have to deal with longer time constraints and with tasks whose behaviour is deterministic (the worst case is known and not too long), so that the remaining problem could be considered as a resource management problem, essentially a resource allocation under timing constraints. The tracking of the obstacles and the refinement of the reasoning are part of a strategy depending from the threat of the situation, of dynamic behaviour of the obstacle and of the timing constraints and delays to secure the smart car from a dangerous situation.

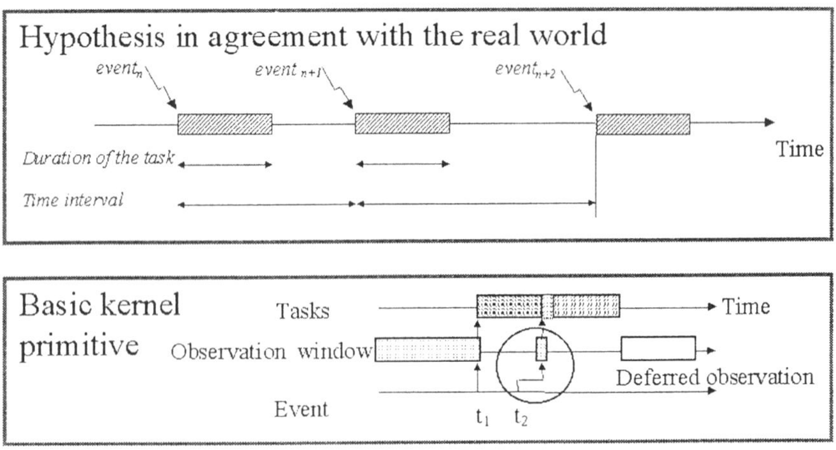

Figure 14. Synchronous hypothesis and real world

At the hardware level, the basic synchronous hypothesis is that the duration of a task must be slower than the time interval between two consecutive events that activate this task. The mechanism of taking into account an event and of activating a task can be implemented under several ways by the real-time kernel, and these details don't matter to the designer insofar as the executive is assured of respecting the initial synchronous specification.

3.2 LOGIC TIME AND REAL TIME

We present now an example of what it is necessary to do to pass from the physical world to the logical time [20]. On the top graph of the following Figure, the little vertical lines are set on the horizontal axis which represents the physical time, at the exact date of acquisition. Each turn lasts for 100 milliseconds. Each sensor scans a third of turn, that is 120 degrees. These data are sending to the central processing unit by means of frames on a Vehicle Area Network as soon as a packet is ready. But we don't know the pure delay, that is variable, due to the communication on the network. Therefore the arrival time in the processing unit has not any relation with the real time. So the date of acquisition must be transported from the sensor as a data in the message. The question is to precise if it is necessary to code this date for each scanning echo or if it is possible to consider that these data are synchronous with a granule of time equal to 100 milliseconds. In this application, considering the dynamic behaviour of the obstacles, it is sufficient to have only a temporal mark at the end of a train of each scan.

We know moreover that there is a variable drift between the two left and right scanning sensors.

The simulation can be realised by the use of a specification in Signal language, because we have only logical marks that have no relation with the real processing time. This first step is followed by a second step that takes into account the timing constraints. Several protocols have been tested. The idea is to code a specification that implements the following protocol.

- We consider only the temporal marks set at the end of a train coming from the left and right side sensors, although we could estimate a more precise time of acquisition, knowing that the speed is quite constant.
- We want to pair two marks separated by the minimum interval of time. These data are the nearest in time and correspond the best to the notion of synchronous data.
- We want to maximise the real-time interval corresponding to the activation of the successive pairing tasks allowing to increase the mean duration of the task that is activated. So we accept to lose a turn. By doing like this, we want to facilitate the work of the SynDEx tool constituting of the second step of the implementation.

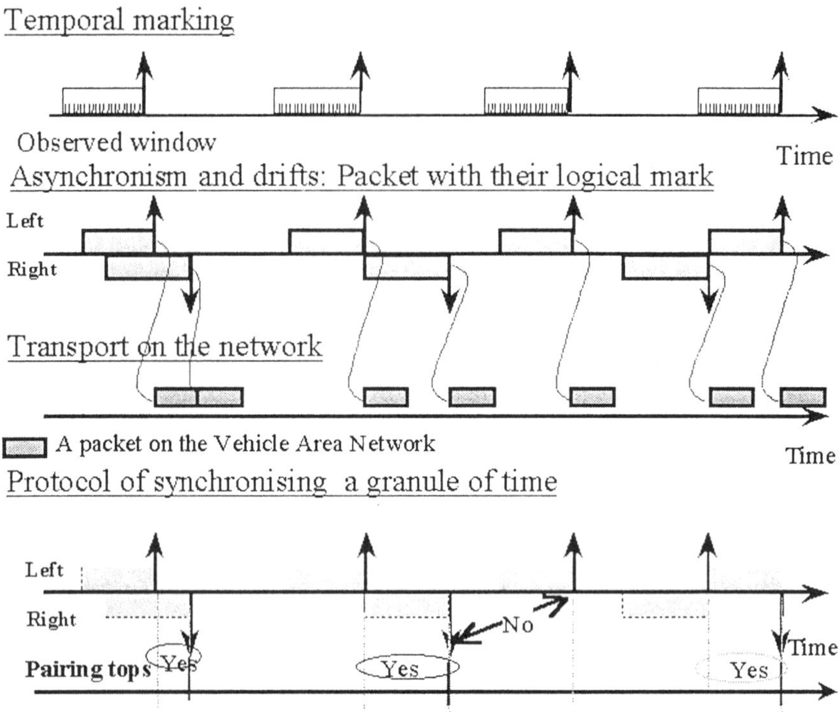

Figure 15. Temporal marks and pairing protocol

3.3 CONCLUSION ON SYNCHRONOUS SPECIFICATION

Coming from the system engineering community, we are convinced of the necessity to use synchronous specifications to obtain a formal verification of all the timing and control protocols.

The main negative point is that a synchronous specification doesn't correspond to the reality of the notion of physical event. So a certain effort is necessary to model the real world with a synchronous specification, all the more so since it is expressed in logical time with no notion of timing constraints, that are taken into account in a second step. But if we want that the second step could succeed, the proposed protocols of data pairing have also to take into account the fact to maximise the expected mean time interval between successive activation of the same task. By doing so, we decrease the constraint on the cadence of the processing.

In counterpart, the work induced by passing in logical time can be seen as positive. In the multi-sensor data fusion applications, a difficulty comes from the asynchronism of the data acquisition on each site. Different methods of re-synchronisation can be implemented either at the hardware level or at the prediction level by the use of models of behaviour, one model or a set of models for each obstacle. Another way is to study in which extent some real events can be gathered in the same logical instant, that we name granule of time. This approximation avoids the use of association-prediction task, it has also the important advantage to reduce the complexity of the verification, or it simply makes a formal verification possible.

All what concerns the hierarchy of the clocks, the dependence between the clocks, the sub-sampling and the verification of the clock trees is formally done.

The final point of interest is to consider a synchronous distributed executive provided by the second step of the implementation tools. That is to say that it is possible to obtain a static pre-placement of the tasks on the processors and a pre-compilation of the communications management between the tasks. This anticipated and static schedule simplifies greatly the systematic pre-fetch of all the corresponding data. It remains a part of the control that is specific to the application, this part is managed by a dedicated module named the supervisor. Its function is to reply to the questions: is the current driving situation dangerous? And how to switch between all the combinations to secure a good functioning according to task duration and available resources?

As in the case of the compilation of expert systems, the lack of determinism, inherent to all versatile decision systems, is located only in the supervisor, where, we hope, it can more easily be controlled.

4. Multiple hypotheses testing with fuzzy temporal logic for the classification of behaviour

This work has been done in the framework of a convention between Aérospatiale & IEF University of ORSAY during the framing of the thesis [32] of Alain Nifle, who is actually at ISF Thomson-CSF.

4.1 FRAMEWORK

The aim of this particular work is to design a method leading to the identification of a ballistic missile thanks to the infra-red observation of its attributes and of the occurring of events during the propelled phase.

The global method needs to use heterogeneous pieces of information. For the missile identification part, the observations stem from image processing of low or high level, and from the dynamics of the trajectory. But other pieces of information characterising the missile or concerning the potential mission of this missile are provided from databases. The nature of these pieces of information can be:

- Tactical and strategic information,
- Imperfect and probabilistic information,
- Probabilistic well-known distributions about the sensors.

The global method is based on data fusion techniques, to take advantage of all the piece of information, of their redundancy and of their complementarity.

A previous work [33] has elaborated the temporal models providing from the real observations a distribution of occurring possibility of the events. Therefore the missile identification part is designed in the framework of the theory of possibility. The result of this part is to obtain a first step of discrimination among the missiles stored in the corresponding database. It provides a measure of compatibility between the imprecise but reliable behaviour of a missile and the whole observations that are also imprecise but not reliable.

The remaining parts deal with the hybrid combination of a double representation of information, probabilistic and possibilistic [36] or the combination of other heterogeneous pieces of information in preparation.

We have named this part Fuzzy Temporal Multiple Hypotheses Testing and the outline is the following: presentation of the algorithm.

4.2 ALGORITHM FTMHT FUZZY TEMPORAL MULTIPLE HYPOTHESES TESTING

Evaluating the threat represented by detected ballistic missiles implies the identification of the missiles and the estimation of their ballistic trajectory. To reach these two objectives supposes that the method is jointly suited to the temporal feature of the problem and to the imperfections of the sources of information [34,35]. The three main aspects constituting the problematic are described now.

To benefit from the observation of dated events and from the a priori knowledge of the time intervals between events

The necessary use of the whole set of observations distributed in the time forces to manage a mechanism of prediction--readjustment concerning the time itself. As a matter of fact, the observations provide the absolute dates of the occurring of the events, and the behaviours to be found provide relative delays through the a priori knowledge of the time intervals between the events. The fact of mixing relative & absolute pieces of information is realised by a simple addition of the distributions of possibility [22].

To take into account the uncertainties

The description of the behaviour is supposed to be fully reliable. In counterpart, the sources of observations are not fully reliable. The observations present some False Alarms, No Detection, out-of-margin errors as in all system of information in signal processing applications. Such an adaptive method is soon proposed by Dubois and Prades in [24] and this work can be seen as an extension of their work to the case of problems with taking into account the time. In order to be robust to these uncertainties, the algorithm starts by enumerating the hypotheses concerning the uncertainties which have to be taken into account.

To associate a belief coefficient to each kind of behaviour in order to quantify its compatibility with the observations

Using direct or conditional distributions is not sufficient to discriminate among the behaviours. We need as in [24] to gather the hypotheses in sub-sets or to propose a résumé that allows to exhibit a degree of belief related to each behaviour, corresponding to all the observations, and robust to all the tested hypotheses of uncertainties.

4.2.1 General outline of algorithm FTMHT

In the top of Figure 16, the three groups of boxes with round corners represent the tree inputs of the algorithm. Each left top box is the behaviour of a missile. The middle top box represents the whole set of the dated observations. The right top box represents all the basic imperfections that we accept to consider in this application.

The first step consists in enumerating all the tested associations. As in a Probabilistic Data Association Filter, each observed date is associated with a potential event. The result is a list of possible hypotheses of association.

Then, from each behaviour and each association, two functions are evaluated: the consensus and the cost.

Finally, for a given behaviour, the graph of the points formed by the consensus and the cost for an tested association is drawn. A belief function is extracted from this graph as a résumé of the information that we dispose.

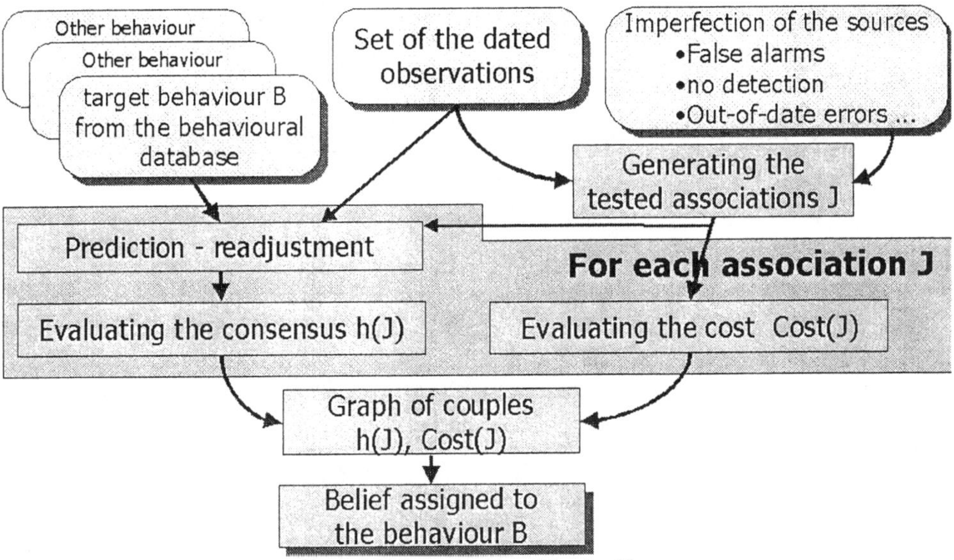

Figure 16. General outline of the algorithm

4.2.2 Definition of a Behaviour

A behaviour is characteristic of a missile. The events that we consider in this application are the following:

(a) Launching
(b) A fast variation of luminance due to combustion effect characteristic of the altitude
(c) The first stage separation,
(d) The second stage separation corresponding to the end of the propelled phase and named injection date.

So a behaviour is defined by giving the 3 distributions of possibility of the time intervals between 4 consecutive events. These behavioural models are supposed reliable. They can be close to each other. The modelling is possibilistic, it represents fuzzy possible time-intervals [23]. It is issued from the knowledge of experts more than from the observation of frequent experiments.

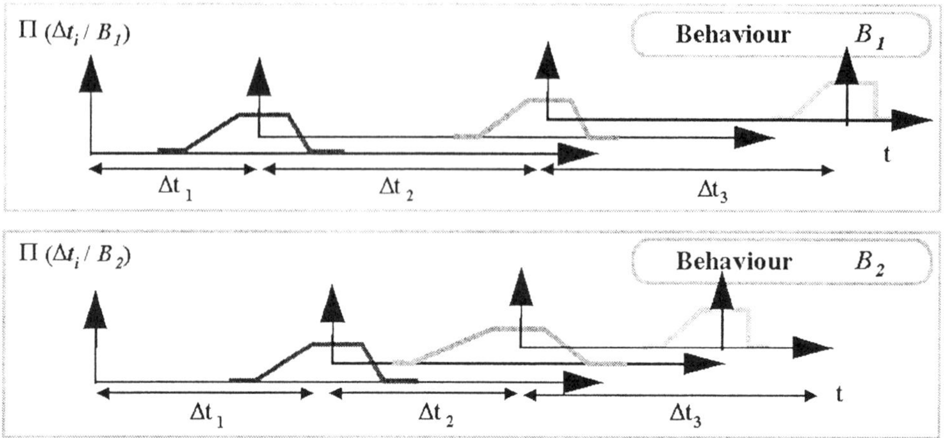

Figure 17. Definition of two behaviours

4.2.3 *Example of Observations : 4 observations for 3 events*

The mechanism of dating is imprecise due to the pre-processing step that converts the native observation in a date of event. The events are less than ten. The great difference with the modelling of a behaviour is in the fact that the observations are not reliable. Because of the presence of several sensors, it is possible to have several observations of the same event.

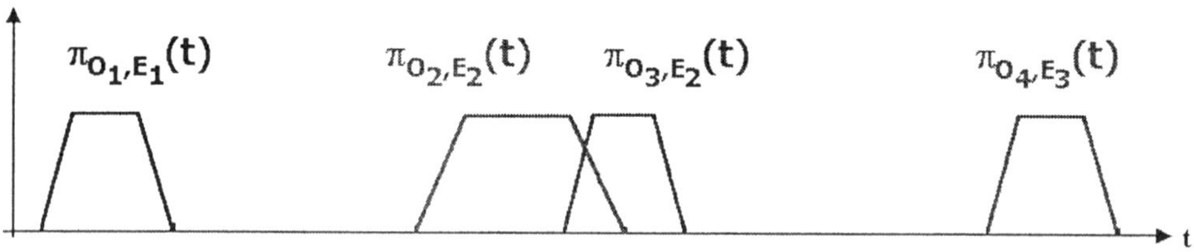

Figure 18. Example of a set of observations

4.2.4 *Generating the authorised associations*

As in a classic association mechanism, it is useful to build a matrix of association. In the following Figures, when a number 1 is present in a case, that means that the corresponding date of an observation is associated with an event. In the example of 3 observations for 3 events, the matrix has 3 rows for 3 observations and four columns because of three events and a column for the false alarm. A standard association is for example

TABLE 1. Standard Association

	Event 1	Event 2	Event 3	False Alarm
Observed date 1	1			
Observed date 2		1		

The three other examples represent the supposition of the presence of a basic error.

TABLE 2. No detection

	Event 1	Event 2	Event 3	False Alarm
Observed date 1	1			
Observed date 2			1	

TABLE 3. False Alarm

	Event 1	Event 2	Event 3	False Alarm
Observed date 1	1			
Observed date 2				1

TABLE 4. Dating error out of imprecision margins

	Event 1	Event 2	Event 3	False Alarm
Observed date 1	1			
Observed date 2				1

This last basic error is not the simple concatenation of a no detection and a false alarm. Because this kind of uncertainty must have a specific cost, it must be processed differently.

The lack of reliability of the observations is an important point. When we have up to 10 observations, 3 or 4 basic errors can be simultaneously accepted. All the corresponding associations are then generated. It is then obvious that the complexity, and above all, the robustness of the method depend of the enumerated cases which have been considered.

4.2.5 Prediction, readjustment & consensus

For a triplet {one behaviour, a set of observations, one association}, it is possible to propose a mechanism of prediction -- conjunction that gives a possibilistic distribution related to the last event. We propose an iterative mechanism of prediction--conjunction that exhibits the degree of conflict as soon as possible in order to limit the loss of information during the step of prediction.

It is like a Kalman Filtering with:

- A dynamic model of prediction, a behaviour, given by the knowledge of time intervals between the events;
- A step of prediction [22] that is made by a mechanism of fuzzy addition [43] that adds two distributions of possibility characterising the uncertainty on the estimation concerning the date an the uncertainty on the model concerning the delay to add. As in the case of the addition of two probabilistic distributions, the uncertainty increases during the step of the prediction.

$$\pi_{\text{predicted}} = \pi_{i-1} \oplus \pi(\Delta t_i) \quad (1)$$

- A step of possibilistic conjunction without normalisation so as to preserve the pertinent information about the conflict. At this step, in presence of several observations associated to the same event, the distributions are simultaneous combined. The operator of conjunction is the minimum of the two distributions at each time.

$$\pi_i = \pi_{predicted} \wedge \pi(Observed.date) \qquad (2)$$

The mechanism is initialised by the distribution concerning the first observed date. The definition of the consensus is classical as in Dubois92. It is the maximum of the final distribution for each time. The maximum of the final distribution is representative of the compatibility between the set of observations, the behaviour and the current hypothesis of an association.

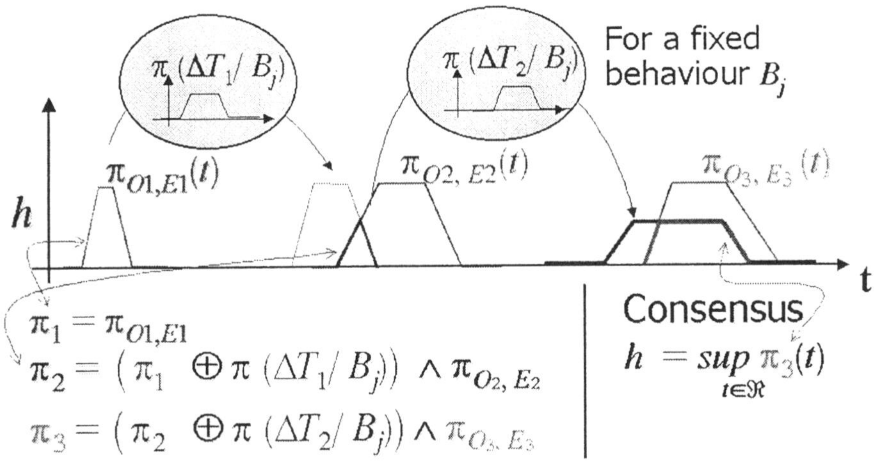

Figure 19. Example of the evaluation of the consensus

4.2.6 Examples of behaviours

The first behaviour, given in Figure 20, is little compatible with the observation. The second behaviour, given in Figure 21, is fully compatible with the observations.

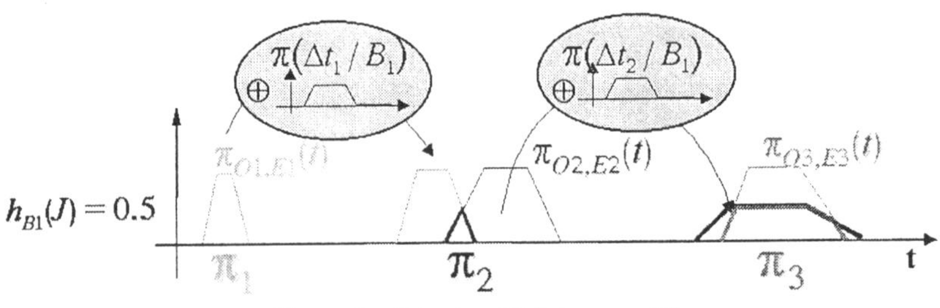

Figure 20. Behaviour B1 little compatible

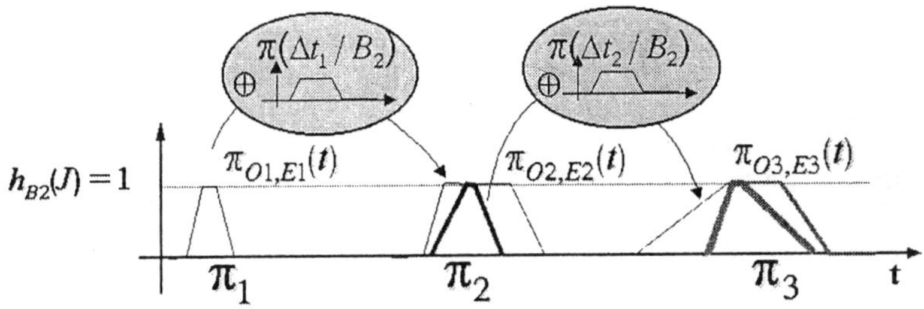

Figure 21. Behaviour B2 fully compatible

4.3 INFORMATIVE RÉSUMÉ

We dispose now of fifty to eighty associations for each behaviour according to the number of associations that we accept to consider. We have the same number of consensus values. And we want to turn these intermediate degrees of possibility in good account to extract an informative résumé, expressing a belief degree, and robust according to the tested uncertainties.

The first idea is to define a cost function for each association, for example the number of sources that has been simultaneously combined to obtain the consensus as in [24].

To produce an informative résumé, the graph of the points {cost, consensus} is drawn with all the impacts of the points corresponding to the accepted associations. What we propose here, supposes naturally a loss of information inherent to the notion of résumé. But it can be envisaged to keep all the associations and their degree of compatibility, the consensus, to delay the decisions, hoping that new pieces of observations allow to weaken some hypotheses. But we suppose that we are in the case where we must provide a belief degree to decision system.

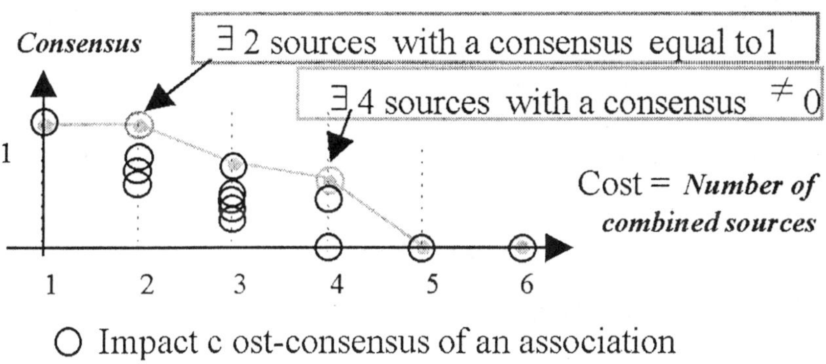

Figure 22. Cost-Consensus Graph version 1

4.3.1 Costs evaluation

The previous cost isn't suitable. It is an additive cost; this kind of cost is not sufficient to handicap the associations having a great number of errors, knowing that, in this case, it is easy to exhibit an association giving a high consensus in this condition. Therefore we propose a new definition of the cost.

For each association J, a cost is computed taking into account: the number of false alarms and the cost of a false alarm; the number of no detection and the cost of a no detection; the number of rejected sources without shifting the numbering (case of dating errors) and the associated cost.

For several errors present in the same association, we propose a multiplying cost function. For the example of an association with 3 basic elements of equivalent error cost CostError, the global corresponding cost is:

$$\text{Cost (association)} = 1 - (1 - \text{CostError})^3 \qquad (3)$$

This confirms the choice of an ad-hoc probabilistic modelling of the cost.

4.3.2 Graph cost/consensus and belief definition

We dispose now of a graph cost-consensus using the new definition of the cost. We define an intermediate multiplying function (1 - cost)consensus for each association. The degree of belief is defined as the maximum of this function for all the tested associations. For each value of belief, it exists an iso-belief curve. A result is that at least one winning association is exhibited corresponding to the maximum of the belief.

The definition of the belief is given by the following equation:

$$\text{Belief}(B) = \text{Max}((1 - \text{Cost}(J)) \times h_B(J)) \qquad (4)$$

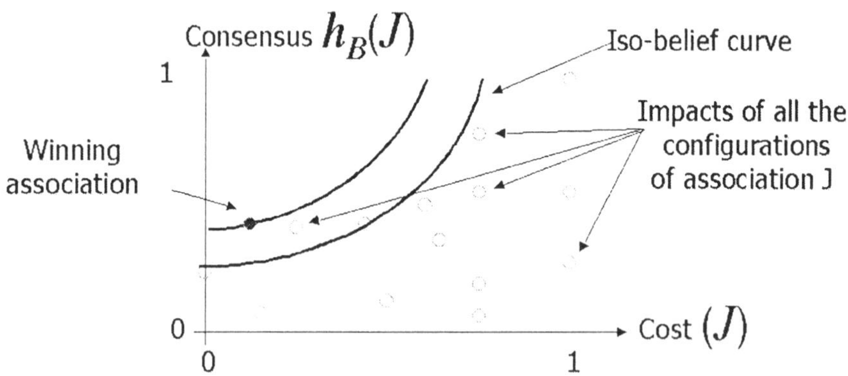

Figure 23. Cost-Consensus Graph version 2

4.4 SIMULATIONS

We employ a simulator built with the Matlab tool. We can set a behaviour, a set of observations, the cost of each basic error and the maximum number of errors accepted in the simulation. For the following simulation, the basic costs are equal, the maximum number of error is 2 for 4 observations, and three time intervals to describe a behaviour. The simulator generates 86 associations to be tested. It computes the cost and the consensus for each association, it draws the graph cost-consensus, and highlights the winning association. The use of this interface for several behaviours allows finally to obtain a discrete distribution of belief degree, that is not normalised to one, but which is a measure of the discrimination of the behaviours.

The Figure presents a simulation with 4 behaviours in competition. The two first behaviours are fully compatible, the third behaviour is partially compatible, the last one is nearly incompatible.

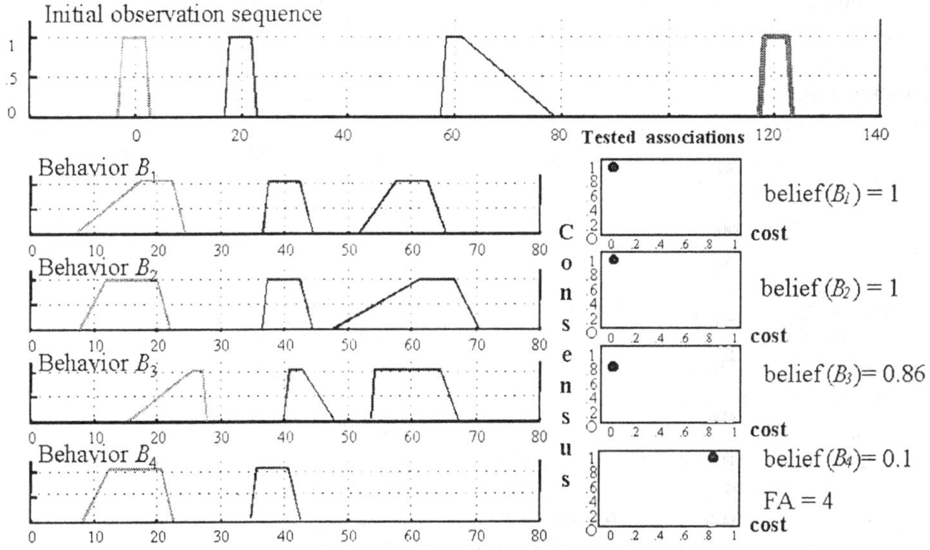

Figure 24. Simulations

4.5 CONCLUSION ON THE FTMHT

The proposed method shows that the possibility theory is well suited to the problem of the management of the possibilities of occurrence of time events. The representation of the time information is correctly translated by continuous distributions of possibility. The prediction is realised by the use of a fuzzy addition and the combination of the information by a conjunction operator.

The aspect " test of discrete possibilities by enumerating the imperfections to be considered" is another important point that we find in all the efficient mechanisms to take into account a unreliable information. It allows to have a robustness towards errors (No Detection, False Alarm, out-of-margin error) by enumerating the tested cases, and by considering the multiple hypotheses in competition.

The computational complexity is then directly dependent on the number of the considered configurations.

But the objective remains the discrimination of behaviour. Therefore a belief degree has been elaborated issued from the compatibility with the observations, and robust towards the considered errors. This degree comes from a graph cost-consensus, based on an ad-hoc modelling; with the definition of a multiplying cost function which is highly penalising in the presence of several errors; and the definition of a possibilistic consensus.

5. Conclusion

We have presented our experience in the implementation of data fusion systems. We have build a philosophy concerning the way to implement operational systems.

The first point is to be able to implement the classical mechanisms of fusion and combination of the information but also a great number of other forms of reasoning in order to obtain a sufficiently fine decision mechanism. The competition among models that supervise different stages of a data flow processing line remains the most efficient mean to implement operational executive with a sufficient throughput. To have a stable system, the models must be chosen so that they present a hierarchical structure and in the case of the use of a mechanism of model switching, the transitions among the models must represent either a refinement of the mode or, in counterpart, a deterioration as if the models could be represented as a set of nested areas, the finest models having the lower surface. In this topology, the transitions are only allowed between adjacent area. This restricted scheme of control applied to a fusion processing line allows to implement different forms of reasoning while keeping its efficiency according to the compression of the amount of data.

The management of the time is a problem which becomes the most dominant in the implementation of data fusion systems. Three aspects have been presented:

The first aspect is the classical management of the parallelism inherent to all distributed systems, in particular to the multisensor systems. Several policies can be implemented, in particular a data flow processing line is an efficient structure that operates a systematic prefetch on the data. But the data can become out-of-date or they can have different clock frequencies.

The temporal specification of the system can be done in a synchronous language allowing to operate a formal verification of this specification. Associated with a tool of generation of distributed executive, this way to do contributes to push the no determinism back in a supervision module. The no determinism is then localised only in this module, it is managed by an application specific time control module, and with a granule of time sufficient to be regulated.

The third aspect is to consider that the time is a physical variable as the others. So the measurement done on this variable can present some imprecision or uncertainty as for any variable. To solve such a problem, the time becomes itself a state variable, the mechanisms of prediction, re-adjustment or combination of information must be introduced in this system of coordinate. In the example of the identification of behaviours, the modelling of the possibility of occurrence of an event by a possibilistic distribution leads to discriminate the behaviours by providing a degree of compatibility between the set of observed data, and each behaviour, degree robust to the no reliability of the sources of observation.

6. References

1. Abidi, M. and Gonzales, C. (1992) Data fusion in robotics and machine intelligence, Academic Press, New York.
2. Albus, J.S. (1991) Outline for a Theory of Intelligence, IEEE Trans., SMC-2. p 473.
3. Appriou, A. (1998) " Uncertain data aggregation in classification and tracking processes", in Aggregation and fusion of imperfect information, B. Bouchon_Meunier, pp 231-260.
4. Bar-Shalom, Y. and Fortmann, T.E., (1988) Tracking and Data Association, Academic Press.

5. Bar-Shalom, Y. and Li, X.R., (1993) Estimation and Tracking Principles, Techniques and Software, Artech House, Norwood.
6. Benveniste, A., Guernic, P.Le, Sorel, Y. and Sorine, M., (1992) "A denotational theory of synchronous communication system", Information on Computation, vol. 99 n° 2, pp 192-230.
7. Benveniste, A., Guernic, P.Le., and Jacquemot, C. (1991) "Synchronous programming with events and relations. The SIGNAL language and its semantics", Sciences of computer programming, vol. 16 n □ 2, pp103-149.
8. Berschandy, D. (1993) "Comparaison des approches classiques et neuronales dans l'architecture d'un système d'intelligence artificielle embarqué. Application à la détection temps réel de danger automobile", Thèse de doctorat, Université Paris XI, Orsay.
9. Bizouerne, F. (1998) " Specification et Implantation d'Algorithmes sur une architecture distribuée hétérogène avec l'approche des langages synchrones", These Université Paris Sud.
10. Bonnifait, P., and Garcia, G. (1996) "Continuous Localisation of Outdor Mobile Robot using Odometry and Relative Azimuth Angles of known Landmarks", (1996) Proc. CESA'96, Symposium on Robotics and Cybernetics, p 30.
11. Bouaziz, S., Reynaud, R., and Maurin, T. (1993) "Parallel architecture for an embedded real-time application", ICSPAT'93, SantaClara, Vol 1, pp151-155.
12. Bouchefra K., Reynaud R., Maurin T, (1995)"IVHS Viewed from a Multiagent Point of View", Proc IEEE Intelligent Vehicles '95. pp 113-117.
13. Bouchefra K., Reynaud R., Maurin T, (1995) "Towards a Robust Algorithm for Dynamic Obstacle Tracking". Proc. of the Workshop on Intelligent Components for Autonomous and Semi-Autonomous Vehicles, IFAC-ICASAV'95, pp 111-116.
14. Bouchefra K., Augé P., Maurin T., Rozoy B., and Reynaud R., (1998) Multi-agent based architecture specification and verification, KAW'98. Agent-oriented Approaches to Knowledge Engineering.
15. Brooks, R.A. (1986) A Robust layered Control System for a Mobile Robot, IEEE Journal of Robotics and Automation, RA2(1), pp 14-23.
16. Burie, J.C., and Bruyelle, J.–L., (1996) Detection and Tracking of Obstacles in front of a Moving Car with a Linear Stereo Vision System, CESA'96, Symp. On Robotics and Cybernetics, pp 443-448.
17. Calvez, J.P., (1993) "Embedded Real-Time Systems, a Specification and Design Methodology", Wiley series in Sofware Engineering Practice..
18. Chebira, A., Reynaud, R., Maurin, T., and Berschandy, D., (1991) "On Board data fusion and decision system used for obstacle detection : a network and a real time approach", Euromicro'91 workshop on real-time systems, p193-200, Paris June 12-14.
19. Chebira, A., Reynaud, and R., Maurin, T., (1993) "Reconstruction of 2Dmar Parameters using Infra-Red Sensors". SPIE, Int Symp. On Optics, Imaging and Instrumentation, San Diego.
20. Chebira, A., (1994) " Détection d'obstacles par reconstruction de l'environnement à partir de capteur infrarouges", Thése n° 3088 Université Paris 11.
21. Chuy, C.K., and Chen, G., (1991) Kalman Filtering with Real Time Application, 2nd Ed, Springer-Verlag.
22. Dubois, D., and Prade, H., (1988) Representation and combination of uncertainty with belief functions and possibility measures. Comput. Intelligence, 4, pp 244-264.
23. Dubois, D., and Prade, H., (1989) "Processing Fuzzy Temporal Knowledge". IEEE Transactions on Systems, Man and Cybernetics. Volume 19 No 4.
24. Dubois, D., and Prade, H., (1992) "Combination of Fuzzy information In the Framework of Possibility Theory" in Abidi [1] p 481-505.
25. Durant-Whyte, H.F., (1988) "Integration, coordination and control of multi-sensor robot systems", Ed. Kluwer, ISBN 0-89838-247-5.
26. Garbay, C., (1997) "Interaction Systeme/Environement pour l'interpretation des Signaux et des Images", Synthetic Report of GdR ISIS. http://www-isis.enst.fr/.
27. Gelb, A., (1984) "Applied optimal estimation", Cambridge, MIT press.
28. Gopal, M., (1993) "Modern Control System Theory", New Age Intern. Publishers.
29. Kester, L., and Theil, A., (2000) "Fusion of Radar and EO-sensors for surveillance. (2000). Proc of 3rd Int. Conf. On Information Fusion, (2000). Paris.
30. Moutarlier, P., Grandjean, P., and Chatila, R., "Multi-sensory Data Fusion for Mobile Robot Location and 3D Modelling", IARP Workshop on multi-sensor fusion and environment Modelling, Toulouse 89.
31. Nasr, A., Bahnu, F., and Shaffer, G., (1987) "Guiding an Autonomous Land Vehicle using Knowledge-based Landmark Recognition". Proc. Of Image Understanding Workshop, p 432.

32. Nifle, A., (1988) Modélisation comportementale en fusion de données. Application à la détection et l'identification d'objets ou de situations, Thèse de doctorat de l'Université d'ORSAY N° 5403
33. Nifle A., Reynaud R., (1997), "Classification comportementale fondée sur l'occurrence d'évé nements en théorie des possibilités", Revue Traitement du Signal Volume 14 - n° 5, pages 523-534.
34. Nifle A., Reynaud R, (1998) "A FTMHT - Fuzzy Temporal Multiple Hypothesis Testing - for Behaviour classification", IPMU'98 Poster Proc. Conference on Information Processing and Management of Uncertainty in Knowledge-Based Systems, pp 1866-1867.
35. Nifle A., Reynaud R, (1999) "Fuzzy Temporal Multiple Hypothesis Testing for Behaviour classification", Rencontre francophone sur la Logique Floue et ses Application. CEPAD p 303.
36. Nifle, A. and Reynaud, R., (2000) "Double Representation of Information and Hybrid Combination for Identification Systems", Fusion2000, Proc. of the 3rd Int. Conf. On Information Fusion.
37. Reynaud, R., Sorel, Y., and Lavarenne, C., (1993) "Spécification et validation à l'aide d'un langage synchrone d'un protocole d'appriement de données asynchrones", Proc. GRETSI 93, Juan-les-Pins, France.
38. Reynaud, R., and Maurin, T., (1994) "On Board Data Fusion and Decision System used for Obstacle Detection : 2D vision versus 1D sensor Fusion", IEEE Intelligent Vehicles '94, pp 243-248.
39. Reynaud, R., and Maurin, T., (1996) "Control and supervision of a fusion processing line by competition among models", Proc. CESA'96 IMACS, Symposium on Robotics and Cybernetics, pp 494-499.
40. Rombauts, M., and et. al. (1998) "La perception et la fusion de données pour véhicules intelligents", Synthetic Report of GDR ISIS. http://www-isis.enst.fr/
41. Stelle, E., and et. al., (1995) "Position estimation for a mobile robot using Data Fusion". IEEE Int. Symp on Intelligent Control. p 565.
42. Sorel, Y., (1994) "Massively parallel computing system with real time constraints. The algorithm Architecture Adequation methodology", Proc Massively parallel computing sytems, the challenge of general-purpose and special-purpose computing.
43. Ziqiang, C., (1992). "Uncertain Temporal Knowledge Management". 4th Proc. of International Conference on Software Engineering and Knowledge Engineering.

THE IMPLEMENTATION OF DATA FUSION SYSTEMS

D.L HALL
Associate Director
The Pennsylvania State University
Applied Research Laboratory
P. O. Box 30
State College, Pennsylvania USA 16801-0030

1. Introduction

Data fusion systems seek to combine information from multiple sensors and sources to achieve improved inferences than those achieved from a single sensor or source. Applications of data fusion related to the Department of Defense (DoD) span a number of areas including automatic target recognition (ATR), identification-friend-foe-neutral (IFFN), *smart* weapons, battlefield surveillance systems, threat warning systems (TWS), and systems to support precision guided weapons. Waltz and Llinas [1], Hall [2], and Hall and Llinas [3] provide a general introduction to multisensor data fusion. Additional information can be obtained from the texts by Blackman [4], Antony [5], and Hall [6]. Data fusion systems typically use a variety of algorithms and techniques to transform the sensor data (e.g., radar returns, and infrared spectra) to detect, locate, characterize, and identify entities such as aircraft and ground-based vehicles. These techniques include signal and image processing, statistical estimation, pattern recognition, and many others (see Hall and Linn [7]). In addition, the fusion systems may use automated reasoning techniques to understand the context in which the entities are observed (i.e., situation assessment) and to understand the intent and possible threat posed by the observed entities (i.e., threat assessment).

Over the past two decades, significant DoD funding has been applied to the problem of data fusion systems, and a large number of prototype systems have been implemented (Hall, Linn, and Llinas [8]). The data fusion community has developed a data fusion process model [9], a data fusion lexicon [10], and engineering guidelines for system development [11]. While a significant amount of progress has been made (Hall and Llinas [12,3]), much work remains to be done. Hall and Garga [13], for example, identified a number of pitfalls or problem areas in implementing data fusion systems. Hall and Llinas [14] described some shortcomings in the use of data fusion systems to support individual soldiers, and M. J. Hall, S. A. Hall and Tate [15] discuss issues related to the effectiveness of human-computer interfaces for data fusion systems.

This paper provides a summary of current progress in multisensor data fusion and identifies areas in which additional research is needed. In addition, the paper describes some issues in the current state of practice of data fusion systems.

A. K. Hyder et al. (eds.), Multisensor Fusion, 419–433.

2. The JDL Data Fusion Process Model

In order to make this paper self-contained, we provide here a brief summary of the Joint Directors of Laboratories (JDL) data fusion process model [9,2,3]. A top-level view of the model is illustrated in Figure 1, and a summary of the processes is shown in Figure 2. This model is commonly used in the data fusion community to assist communications concerning data fusion algorithms, systems, and research issues. It will be used here for the same purpose.

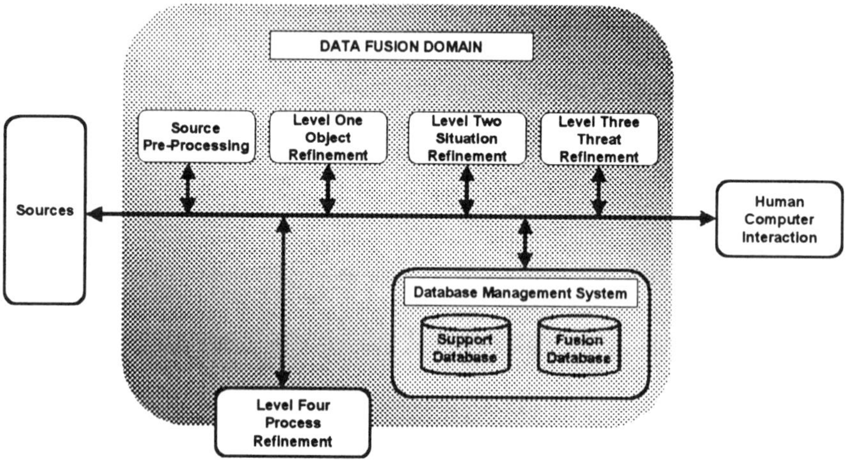

Figure 1. Top-Level View of the JDL Data Fusion Process Model

The JDL model is a two layer hierarchical model that identifies fusion processes, processing functions and processing techniques to accomplish the functions. The model was intended for communications among data fusion researchers and implementation engineers, rather than a prescription for implementing a fusion system or an exhaustive enumeration of fusion functions and techniques. The model has evolved since its original exposition to the data fusion community. Steinberg and Bowman [17], for example, have recommended the inclusion of a new *Level zero* processing to account for processing such as pre-detection fusion and coherent signal processing of multi-sensor data. In addition, they suggest a re-naming and re-interpretation of the Level 2 and Level 3 processes to focus on understanding the external world environment (rather than a military-oriented situation and threat focus). C. Morefield [18] has suggested that the distinction between Level 2 and Level 3 is artificial, and that these processes should be considered as a single process. Bowman has suggested that the JDL model can be detrimental to communications if systems engineers focus on the model rather than a systematic architecture analysis and decomposition approach. Many of these comments have merit. However, for the purpose of this paper we will utilize the JDL model for describing the current state of practice and limitations.

Process components	Process Description	Functions
Sources of information	Local and remote sensors accessible to the data fusion system; information from reference systems and human inputs	Local and distributed sensors; external data sources; human inputs
Human Computer Interface (HCI)	Provides an interface to allow a human to interact with the fusion system	Graphical displays; natural language processing
Source Preprocessing	Processing of individual sensor data to extract information, improve signal to noise, and prepare the data for subsequent fusion processing.	Signal and image processing; canonical transformations; feature extraction and data modeling
Level 1 Processing: Object Refinement	Association, correlation, and combination of information to detect, characterize, locate, track, and identify objects (e.g., tanks, aircraft, emitters)	Data alignment; correlation; position, kinematic, attribute estimation; object identity estimation
Level 2 Processing: Situation Refinement	Development of a description of the current relationships among objects and events in the context of their environment	Object aggregation; event and activity interpretation; context-based reasoning
Level 3 Processing: Threat Refinement	Projection of the current situation into the future to draw inferences about enemy threats, friendly and enemy vulnerabilities, and opportunities for operations	Aggregate force estimation; intent prediction; multi-perspective analysis; temporal projections
Level 4 Processing: Process Refinement	A meta-process that seeks to optimize the on-going data fusion process (e.g., to im-prove accuracy of inferences, utilization of communication and computer resources)	Performance evaluation; process control; source requirement determination; mission management
Data Management	Provide access to, and management of, dynamic data fusion data including; sensor data, target state vectors, environmental information, doctrine, physical models, etc.	Data storage and retrieval; data mining; archiving; compression; relational queries and updates

Figure 2. Summary of JDL Processes and Functions.

3. Current Practices in Data Fusion

A summary of the current state and limitations of data fusion is provided in Figure 3. This is an update of a similar figure originally introduced by Hall and Llinas in 1993 [12] and updated by Hall and Llinas in 1997 [3]. For each of the key components of the JDL process, the figure provides a summary of the current practices and limitations. These are summarized below.

JDL Process	CURRENT PRACTICES	LIMITATIONS & CHALLENGES
Level 1: Object refinement	■ Sensor preprocessing using standard signal and image processing methods ■ Explicit separation of correlation and estimation problem ■ Multiple target tracking using MHT [4], JPDA [21], etc. ■ Use of ad hoc maneuver models ■ Object ID dominated by feature based methods [29] ■ Pattern recognition using ANN [27] ■ Emerging guidelines for selection of correlation algorithms [11,41] ■ Promising integrated solutions by Poore [18], Mahler [19], Barlow, et al [40]	■ Dense target environments ■ Rapidly maneuvering targets ■ Complex signal propagation ■ Co-dependent sensor observations ■ Background clutter ■ Context-based reasoning ■ Integration of identity and kinematic data ■ Lack of available ANN training data (for target identification) [27] ■ No true fusion of image and non-image data (at the data level)
Level 2: Situation Refinement	■ Numerous prototype Systems [8] ■ Dominance of rule-based KBS ■ Variations include blackboard systems [23], logical templating [22], and case-based reasoning [24] ■ Emerging use of fuzzy-logic [25] and agent-based systems [26]	■ Very limited operational systems ■ No experience in scaling up prototypes to operational systems ■ Very limited cognitive models [15] ■ Perfunctory test and evaluation against toy problems [8] ■ No proven technique for knowledge engineering [2]
Level 3: Threat Refinement	■ Same as Level 2 Processing ■ Limited advisory status ■ Limited deployment experience ■ Dominated by ad hoc methods ■ Doctrine-specific, fragile implementations	■ Same as level 2 ■ Difficulty to quantify intent [6] ■ Models require established enemy doctrine ■ Difficult to model rapidly evolving situations
Level 4: Process Refinement	■ Robust methods for single-sensor systems ■ Formulations based on operations research [2] ■ Limited context-based reasoning ■ Focus on measures of performance (MOP) versus measures of effectiveness (MOE) [1]	■ Difficult to incorporate mission constraints ■ Scaling problem when many sensors (10^N) and adaptive systems [36] ■ Difficult to optimally use non-commensurate sensors ■ Very difficult to link human information needs to sensor control [28]
Human Computer Interface (HCI)	■ HCI dominated by the technology of the week ■ Focus on ergonomic versus cognitive-based design ■ Numerous graphics-based displays and systems [30,31] ■ Advanced, 3-D full immersion HCI available [32] and haptic interfaces [33,43]	■ Very little research has been performed to understand how human analysts' process data and make accurate inferences. ■ Creative HCI is needed to adapt to individual users and to provide mitigation of known cognitive biases and illusions [15,35]
Data Base Management	■ Extensive use of 4th and 5th generation COTS DBMS ■ DBMS individually optimized for text, signal data, imagery, or symbolic infor-mation (but not the intersection of any two) ■ DBMS requires extensive tailoring for individual data fusion systems	■ Need a generalized DBMS capability for text, signal data, images, and symbolic information ■ Need a software solution to multi-level security

Figure 3. Summary of Current State of Multisensor Data Fusion.

Level 1: Object Refinement: Level One processing seeks to combine information about the location and attributes of entities (such as tanks or aircraft) to detect, locate, characterize, track, and identify the entities. Level one processing involves data assignment/correlation, estimation of the state of an entity, and an estimate of the entity's identity. The typical data fusion system partitions the object refinement problem into three basic components; (1) data assignment/correlation, (2) estimate of a state vector (e.g., for target tracking), and (3) estimation of a target's identity. Object refinement is relatively easy when there are a relatively few, widely separated targets moving in predictable paths. Target identity classification can generally be performed when there are observable target attributes (e.g., size, shape, and spectral signature) that can be uniquely mapped to target class or identity. This requires either an accurate model to link attributes with target identity, or a very large set of training data to train a pattern classification algorithm.

When these observing conditions are violated, the problem becomes much more challenging. Closely spaced, rapidly maneuvering targets, for example, are difficult to track because we cannot easily associate the sensor measurements to the appropriate targets. In addition, since acceleration cannot be observed directly, maneuvering targets cause a potential loss of track because we cannot accurately predict the future position of the targets. Complex observing environments, involving multi-path signal propagation clutter, dispersion, or other effects on signal to noise, can cause difficulties in data association and state estimation (because we may lack an accurate model to link the value of a target state vector to predicted observations). It is difficult to combine data from sensors that are co-dependent (viz., for which the sensor data are not statistically independent). Finally, complex targets without distinguishing attributes are difficult to classify or identify.

Current Level One processing is dominated by estimation techniques such as Kalman filters [4], multiple hypothesis tracking (MHT) [4], joint probabilistic data association (JPDA) filters [21], or related techniques. The problem of identity declaration is generally performed using a feature-based, pattern recognition approach [2,29]. This involves representing the sensor data using extracted features (e.g., spectral peaks in a radar cross section observation) and mapping the feature vector to a location in feature-space that can be uniquely identified with a target class or identity. Typical techniques include artificial neural networks (ANN) or cluster algorithms [29]. This identification process works well when there is a unique map between the observed features and the target class, but requires a significant amount of training data. However, the methods fail when training data is lacking [27], or there is ambiguity in the feature-to-target class mapping. Emerging methods include both model-based techniques and syntactic methods that develop descriptions of the makeup of a target in terms of elementary components.

Aubrey Poore [19] and R. Mahler [20] developed two promising methods in Level One fusion. Poore revisited the approach of separating the problems of object correlation, target tracking, and identity estimation. Poore re-links these problems into one single optimization problem having multiple constraints (viz., find the set of state vectors (including the association between observations and tracks) that best fits the observational data). While this larger problem is even more difficult than the original sub-problems,

Poore has developed approximation methods to improve the computational feasibility. By contrast, Mahler has developed applications of random set theory to address the joint problem of data association and state estimation. A unified method based on Bayesian inference has been used by Barlow, Stone and Finn [40] to simultaneously estimate target state, identity, and association of the data. Finally, an extensive survey of methods for data correlation has been performed by Llinas *et al* [41].

Level 2: Situation Refinement: Level Two processing seeks to understand the relationships among observed entities and their relationship to the environment. This process involves recognition of patterns, context-based reasoning, and understanding of spatial, temporal, causal, and functional relationships. In general this is a difficult problem. There are numerous prototype systems that have been developed for DoD applications [8]. The predominance of the methods involves knowledge-based systems utilizing production rules [2], fuzzy logic [25], logical templates [22], or case-based reasoning [24]. Emerging systems are beginning to utilize agent-based approaches [26] and blackboard architectures [23].

While this is a very active area of research, the results to date are relatively disappointing. Very few operational systems have been deployed. Many of the prototype systems have addressed limited or *toy* problems with little or no test and evaluation. There is little experience on how to scale these small prototype systems to larger scale operational systems. A key problem for Level Two processing (as well as for Level Three) is the lack of cognitive models for how to perform situation assessment. Current cognitive models can be described by some as *pathetic*. We simply do not know how to model the reasoning process to perform a *gestalt* type of situation assessment. Numerous *ad hoc* methods (e.g., rules, frames, fuzzy logic, decision trees, scripts, templates, etc.) have been applied. One difficulty involves how to perform the knowledge engineering to identify the key information, inter-relationships, and the associated uncertainty information. Here again, Mahler's random set theory [20] provides a basis for a unified calculus of uncertainty. However, the application to realistic problems is far from routine. A general implementation approach has not yet been developed.

It is likely that improvements to Level Two processing will emerge from an improved understanding of how to select and use existing methods for knowledge representation (e.g., rules, frames, scripts, fuzzy logic), coupled with a better understanding of the strengths and weaknesses of human cognition for these types of tasks. One example would be the incorporation of so-called negative information in reasoning. Negative information involves reasoning about information that has not been observed (but would be expected to for a hypothesized situation). The use of negative reasoning appears to be a key element of successful diagnosis and inference in many areas such as medical diagnosis or diagnosis of mechanical faults [34]. Another promising area for research involves the development of aids for analysts that would address known cognitive biases and shortcomings (e.g., confirmation bias in which humans seek information that confirms a proposed hypothesis rather than evidence that refutes the hypothesis, miss-use of probability, and other biases [15,35]). The original research by J. Wohl [42] and his associates to develop tools for assisting an antisubmarine warfare (ASW) analyst is particularly intriguing. The research suggests that some fairly simple cognitive aids could be developed to significantly improve the data fusion/analysis process.

Level 3: Threat Refinement: Level Three processing involves an interpretation of the situation from a consequences point of view. That is, what is the meaning of the situation in terms of potential opportunities and threats? Alternative hypotheses are generated and projected into the future to determine what are the likely courses of action for engagements, and the consequences of those courses of action. The state of Level Three processing is similar to that of Level Two. There are a number of prototype systems that have been developed, but few deployed systems. The main focus of Level Three processing has been the application of automated reasoning systems and techniques from the discipline of artificial intelligence. A special challenge for Level Three processing is the determination of enemy intent. Conceptually, the determination of an enemy's intent involves a mind-reading exercise; what will the enemy do, under what circumstances, and with what motivation? When a well-known enemy doctrine exists, this can be modeled using a variety of techniques. However, in modern conflict situations this doctrine is often unknown. Hence, it is challenging to automate the process of threat refinement. Another problem for threat refinement is the role of adaptive intelligence opponents. How can engagements be modeled in which an opponent adapts to the actions of a protagonist? Much research has been performed in game theory to address this issue, but there is limited success in applying this work to realistic tactical situations.

Level 4: Process Refinement: The Level Four process is a meta-process; it is a process that monitors the overall data fusion process and seeks to optimize the data fusion within operational and physical constraints [1,2]. Types of functions within Level Four processing include generation of sensor *look angles* (to indicate where to point the sensors to track targets), computation of measures of performance (MOP) and measures of effectiveness (MOE), determination of information needs and sources, and process optimization. Level Four processing is relatively mature for single sensor environments. For single sensors, or a small number of commensurate sensors, Level Four processing becomes a routine problem in multi-objective optimization. This is an area that has received an extensive amount of research, e.g., for applications such as industrial process control.

The Level Four process becomes more challenging under a number of circumstances. These include use of a large number of sensors, use of co-dependent sensors, utilization of non-commensurate sensors (e.g., measuring very diverse physical phenomena on greatly different time scales), and use of sensors in a geographically distributed environment. Modern data fusion systems often involve geographically distributed collection and processing with adaptive systems that self-adjust for system failures and other problems [36]. Under these circumstances it is difficult to develop global MOE and MOP models and to optimize the overall system performance. Another challenge involves modeling sensor performance in realistic data collection environments. Finally, the most effective Level Four process would link the information needs of a human decision-maker to the sensor and source tasking in real-time.

Much research remains to be performed in the Level Four area. However, the improved intelligence and agility of modern sensors makes this an area in which major improvements can be obtained with relatively modest effort.

Human-Computer Interface: The human-computer interface (HCI) area in data fusion is one that appears to be technology rich and theory poor. M. J. Hall, S. A. Hall, and Tate [15] point out that there is a rapidly evolving capability in HCI technology to provide interfaces such as full-immersion, three dimensional displays [32], haptic interfaces [33,43], three dimensional sound, and other types of interfaces to access and analyze data. However, they note that these interfaces smack of the *technology du jour* and have not been applied with a solid theoretical understanding of how humans access and respond to information displays. Many of the existing HCI for data fusion systems involve Geographical Information System (GIS) type displays and data access [30,31]. While these are useful, it is not clear that these interfaces truly assist the understanding of information available from a data fusion system, or whether they may actually impede the inference process. B. Feran [38] has argued that the HCI for intelligence systems can actually act as a bottleneck that limits the ability of a user to access and analyze data. Other studies have investigated the issue of trust in decision support systems (Llinas *et. Al* [39]), and how the HCI affects the extent to which a user believes and trusts the results.

Data Base Management: The final area to be addressed involves data base management for data fusion systems. This is an important area for data fusion systems for several reasons. First, data base management software constitutes the single major amount of software to be developed for a data fusion system (even if one utilizes sophisticated commercial-off-the-shelf (COTS) DBMS packages) [2]. Data required for fusion systems ranges from sensor data (e.g., scalars, vectors, time series, and images), to information input by human users, environmental data, textual information, and knowledge such as doctrine. The data base management for a data fusion system must simultaneously accept data at the rate provided by the contributing sensors, and also allow algorithms and users to rapidly retrieve large amounts of data using general Boolean queries. The combination of the complexity of the data sets and need for real-time data storage and retrieval complicates the data base management for data fusion. In addition, the data associated with fusion systems often involves multiple levels of security. Handling multi-level security is currently difficult to do via a software approach. For all of these reasons, extensive special software must be implemented for data fusion systems.

4. Implementation of Data Fusion Systems

Several of the most critical issues related to the implementation of data fusion systems are addressed in this section. While each data fusion system constitutes a separate problem in systems engineering, there are some general remarks that can be made about requirement analysis, sensor selection, architecture selection, algorithm selection, software implementation, and test and evaluation. These are summarized below.

Requirement Analysis: The initial step for the implementation of any complex system is to perform a requirement analysis. In a typical structured system development approach, formal analyses are performed, culminating in a requirement review and associated documentation. This approach is also recommended for the development of data fusion systems. It should be remembered that data fusion is a derived requirement. Strictly speaking there is no such thing as a data fusion system. Instead there are systems designed to address specific applications or to achieve specified missions. These systems may utilize data fusion algorithms to achieve the application or mission goals. Under this perspective, the system designer must understand what the mission goals or application goals are. What targets or entities must be detected, characterized, or identified? Who are the intended users of the system and what decisions must these users make (to achieve a mission)? This approach focuses on the end-user of the data fusion system rather than on the sensors. All too often, system designers treat a data fusion system as a *bucket* used to collect or process the sensor data, rather than a system to support an end-user. The requirement analysis must consider the effects of the observing environment, the end-user, the platform (on which the sensors are located), communication constraints, and computing limitations. Key issues include the observing and decision timeline and required level of specificity and accuracy. An excellent discussion of requirement analysis for data fusion system is available in Waltz and Llinas [1].

Sensor selection and analysis: Often, the sensors associated with a data fusion system are specified *a priori*. This is frequently the case for existing platforms such as aircraft or ships for which a data fusion system must be developed. However, even if the sensors are specified *a priori*, it is necessary to perform an analysis of the sensors. In general, there is no single sensor that can detect, locate, characterize, and identify the targets of interest under all circumstances (i.e., there is no perfect sensor). Indeed, the lack of a perfect sensor is one of the motivations for developing a data fusion system. A key component of the system development process is to understand how the sensors perform, both individually and in concert, to contribute to inferences sought by the data fusion system. Sensor selection and analysis must determine what can be observed, and how these observable quantities relate to the targets or inferences of interest. Steinberg [45] provides an excellent example of this type of analysis for a tactical aircraft. The system designer should develop or utilize high fidelity models that predict how sensors will perform in realistic environments. These models should include the effects of the target, the signal propagation environment, the location of the sensor antenna on an observing platform, and the internal sensor processing. In addition, real data should be collected and analyzed. It is important to understand and accurately model sensor performance. These estimates should be used to help weight the sensor data in the data fusion process. If the performance estimates are not accurate, then it can corrupt all of the down-stream fusion processing.

Architecture selection: Modern data fusion systems often involve distributed sensors and processing systems. Examples include systems on-board a single platform (e.g., an aircraft) linked via a computer network such as Ethernet and unattended ground sensor systems in which multiple sensors are linked using wireless communications. In these systems processing is performed both at the sensors and at a centralized processing computer. A basic question for the system designer is to determine where in the processing flow should the fusion actually be performed? Conceptually, there are three basic types of architectures for data fusion [2]; (1) centralized fusion, (2) distributed fusion, and (3) hybrid fusion. The centralized fusion architecture involves passing raw sensor data (e.g., time series, images, and vectors) from the sensors to a central fusion processing function. In this approach the raw sensor data must be associated, correlated, and fused. The centralized fusion architecture is potentially the most accurate approach for fusion of information (since the raw data are available for processing). However, the centralized approach may require significant communication resources to send the data from the sensors to the centralized fusion process. In addition, the centralized approach may be difficult to actually achieve for non-commensurate sensors.

In the distributed fusion approach, sensor data are processed to result in state vector estimates for observed targets. These estimates may contain estimates of the target's position, velocity, attributes, and identities. In the distributed fusion architecture the processing is effectively distributed among the sensors and other fusion processing computers. Distributed fusion reduces the amount of information that must be communicated between the sensors and the fusion processor (i.e., instead of raw data sent from the sensors to the fusion processor, only state vectors or feature vectors are sent from the sensors to the fusion processor). In addition, distributed fusion is convenient for processing information from diverse types of sensors. A potential problem with distributed fusion is a loss of accuracy because the raw data are not available for fusion.

Finally, hybrid fusion architectures combine both a centralized approach and a distributed approach. This allows the system to limit communication resource utilization (using the distributed mode of operation), or to improve accuracy by processing the raw sensor data. Hybrid fusion provides the most flexible type of architecture at the expense of added computational overhead to address the decision processes associated with operating the hybrid processing.

How does one select an appropriate architecture for a fusion system? This is a basic problem in system engineering. Steinberg and Bowman[17] provide basic guidelines for system design and architecture selection. Bowman in particular has developed an approach that involves a hierarchical decomposition of fusion system into nodes with an allocation of functions to nodes based on the characteristics of the sensors, communications, and processing systems. The system designer should explicitly consider alternative architectures for the fusion system to be designed. Each alternative can be modeled to evaluate the system effectiveness and the demands on system resources such as computing and communications.

Algorithm selection – Perhaps the most controversial issue in data fusion is the selection of algorithms. There are many algorithms that can be applied to the different processes within the data fusion process. For example, many different algorithms have been developed for target tracking and target identification. Even for a function as basic as data correlation, a wide variety of techniques can be applied. Llinas *et al* [41] identified over 50 different techniques used for data correlation and association. The methods involve different assumptions concerning the nature of the input data, understanding of the uncertainty of the input data, available computing resources, and other factors.

One challenge in the implementation of a data fusion system is how to select from these different techniques. Hall and Linn [7] provide some general guidelines for the selection of data fusion algorithms. However, there is no prescription that provides a unique specification of which algorithms to use. Unfortunately, many proponents of specific techniques (e.g., Bayesian inference, Dempster-Shafer's method, multiple-hypothesis tracking, etc.) argue that their method is the only method to be used for every application. The choice of a set of algorithms for a fusion system must be based on a system's engineering approach. The designer must have a clear understanding of the algorithms (including the underlying assumptions, required a priori data, etc.), the processing constraints of the fusion system, and the limitations in the observing environment. In some cases, sophisticated algorithms may be mathematically appealing, but cannot be effectively used because the requisite a priori data is not available. Hall and Linn [46] provide an example of how easily complex algorithms can be corrupted by input of incorrect data. Special care must be taken to understand how fragile an algorithm is to variations in the input data. A recommended approach is to deliberately identify and tradeoff candidate algorithms prior to the selection of an algorithm suite for a data fusion system.

Software implementation: Implementation of a data fusion system generally involves development of an extensive set of software. Indeed, an analysis of 35 data fusion systems by Hall [2] showed that for the average data fusion system, only 12 per cent of the software are actually concerned with data fusion. The remaining software involves infrastructure functions such as data base management, human-computer interface, message processing, sensor control, and other functions. The extensive development of this infrastructure software is one of the reasons that many data fusion development efforts fail. The system resources are consumed by the infrastructure. Hence, Hall and Linn [47] recommend the use of commercial off the shelf (COTS) software where possible. However, care must be taken to distinguish between usable general-purpose software and limited *point solutions*. Many of the existing data fusion systems cannot be readily used for other applications. Unfortunately, at this time there does not exist the equivalent of a numerical methods library for data fusion (although there are some useful tool kits such as *Khoros* [48] and the data fusion tool kit developed at the Pennsylvania State University Applied Research Laboratory [49]).

Test and evaluation: The test and evaluation of a data fusion system involves the traditional problems associated with testing and evaluating any large-scale hardware and software system. Structured *waterfall* software development methods prescribe an approach that involves a sequence of steps; requirements analysis, preliminary design, detailed design, software code and unit test, integration, and system testing. Each of these steps is conducted in accordance with strict standards and procedures. Formal reviews are conducted at the end of each step. These reviews act as a gate to the subsequent development steps.

One of the particular challenges for testing data fusion systems involves the automated reasoning associated with Level 2 and Level 3 processing. In many cases these processing functions are implemented using an expert system. Expert systems are notoriously difficult to test and evaluate (see Hall, Heinze, and Llinas [50]). Challenges include the difficulty of obtaining an appropriate standard, the large magnitude of the search space, and the difficulty of defining acceptable transitions in the search space. Hall, Heinze, and Llinas provide a discussion of the challenges of testing expert systems and a survey of methods for test and evaluation. For situation assessment and threat assessment, one must obtain examples of tactical or strategic situations (e.g., sensor data associated with emitters, platforms, environmental conditions, and terrain). Each single test case thus involves a large set of data that must be gathered or modeled. In general, there are only limited test cases available. As a result, it is very difficult to determine how well a data fusion system would perform. The test and evaluation issue remains one of the most challenging problems for implementing effective data fusion systems.

5. Pitfalls in Data Fusion

The previous part of this paper has provided a broad overview of the state of data fusion technology and identification of potential implementation issues. A practitioner might well ask the question; so what do I do tomorrow to implement a system? What are some problems and challenges that need to be addressed? It is well beyond the scope of this paper to provide a prescription for the implementation of data fusion systems. However, there are several areas worth noting. First, Bowman and Steinberg [11] provide an overview of the general systems engineering approach for implementation of data fusion systems. Engineering guidelines for selection of correlation algorithms are described by Llinas *et al* [41]. Several texts, such as those of Hall [2] and Waltz and Llinas [3] provide detailed information on data fusion algorithms. R. Antony [5] describes issues in data base management systems, and texts are available on specific applications to target tracking (e.g., Blackman [4]) and signal processing techniques [44].

Hall and Garga [13] have discussed the problem of implementing data fusion systems and identified a number of problems or pitfalls. These include the following dictums.

- *There is no substitute for a good sensor:* No amount of data fusion can substitute for a single accurate sensor that measures the phenomena that you want to observe.
- *Downstream processing cannot make up for errors (or failures) in upstream processing:* Data fusion processing cannot correct for errors in processing (or lack of pre-processing) of individual sensor data.
- *Sensor fusion can result in poor performance if incorrect information about sensor performance is used:* A common failure in data fusion is to characterize the sensor performance in an *ad hoc* or convenient way. Failure to accurately model sensor performance will result in corruption of the fused results.
- *There is no such thing as a magic or golden data fusion algorithm:* Despite claims to the contrary, there is no perfect algorithm that is optimal under all conditions. Often real applications do not meet the underlying assumptions required by data fusion algorithms (e.g., available prior probabilities or statistically independent sources).
- *There will never be enough training data:* In general there will never be sufficient training data for pattern recognition algorithms used for automatic target recognition or IFFN. Hence, hybrid methods must be used (e.g., model-based methods, syntax representations, or combinations of methods).
- *It is difficult to quantify the value of a data fusion system:* A challenge in data fusion systems is to quantify the utility of the system at a mission level. While measure of performance can be obtained for sensors or processing algorithms, measures of mission effectiveness are difficult to define.

- *Fusion is not a static process:* The data fusion process is not static, but rather an iterative dynamic process that seeks to continually refine the estimates about an observed situation or threat environment.

We note that these issues must be addressed for implementation of an effective data fusion system.

6. Summary

The technology of multisensor data fusion has made major strides in the past two decades. Extensive research has been performed on data fusion algorithms, distributed architectures, automated reasoning techniques, and new resource allocation and optimization techniques. There is an emerging consensus in the data fusion community concerning basic terminology and engineering guidelines. Recent activities by the US DoD to initiate Data Fusion Information Analysis Center (FUSIAC) promise to accelerate the development of data fusion technology by increasing the communications among researchers and system implementers. Despite these rapid advances, however, much research remains to be done. This paper has presented a perspective on the current limitations and challenges in data fusion, and identified challenges for system implementation.

7. References

1. Waltz, E. and Llinas, J. (1990) *Multisensor Data Fusion*, Artech House, Inc., Norwood, MA.
2. Hall, D.L. (1992) *Mathematical Techniques in Multisensor Data Fusion*, Artech House, Inc., Norwood, MA.
3. Hall, D.L. and Llinas, J. (1997) *An introduction to multisensor data fusion*, Proceedings of the IEEE, vol. 85, No. 1, pp. 6-23.
4. Blackman, S.S. (1987) *Multiple-Target Tracking with Radar Applications*, Artech House, Inc., Norwood, MA.
5. Antony, R.T. (1995) *Principles of Data Fusion Automation*, Artech House, Inc., Norwood, MA.
6. Hall, D.L. (2000) *Lectures in Multisensor Data Fusion*, Artech House, Inc., Norwood, MA.
7. Hall, D.L. and Linn, R.J. (1987) *Algorithm selection for data fusion systems*, Proceedings of the 1987 Tri-Service Data Fusion Symposium, APL Johns Hopkins University, Laurel, MD, Vol. 1, pp. 100-110.
8. Hall, D.L., Linn, R.J. and Llinas, J. (1991) *A survey of data fusion systems*, Proceedings of the SPIE Conference on Data Structures and Target Classification, Vol. 1470, Orlando, FL, pp. 13-36.
9. Kessler, O. et al, (1992) *Functional Description of the Data Fusion Process*, Technical Report, Office of Naval Technology, Naval Air Development Center, Warminster, PA.
10. F. E. White, *Data Fusion Lexicon*, Data Fusion Sub-Panel of the Joint Directors of Laboratories Technical Panel for C^3 NOSC, San Diego, CA, 1991.
11. Steinberg, A.N. and Bowman, C.L. (1998) *Development and application of data fusion system engineering guidelines*, Proceedings of the National Symposium on Sensor and Data Fusion (NSSDF), Lexington, MA.
12. Hall, D.L. and Llinas, J. (1994) *A challenge for the data fusion community I: research imperatives for improved processing*, Proceedings of the 7th National Symposium on Sensor Fusion, Albuquerque, NM.
13. Hall, D.L. and Garga, A.K. (1999) *Pitfalls in data fusion (and how to avoid them)*, Proceedings of the 2nd International Conference on Information Fusion – Fusion '99, vol. 1, pp. 429-436, Sunnyvale, CA.
14. Hall, D.L. and Llinas, J. (1998) *From GI Joe to starship trooper: the evolution of information support for individual soldiers*, Proceedings of the International Conference on Circuits and Systems, Monterey, CA.
15. Hall, M.J., Hall, S.A. and Tate, T. (2000) *Removing the HCI bottleneck: how the human computer interface (HCI) affects the performance of data fusion systems*, Proceedings of the 2000 Meeting of the MSS, National Symposium on Sensor and Data Fusion, San Antonio, Texas.

16. Steinberg, A. (1999) *Standardization in data fusion: new developments*, Proceedings of EuroFusion99: International Conference on Data Fusion, Stratford-upon-Avon, UK, pp. 269-278.
17. Steinberg, A.N. and Bowman, C. (1998) *Revisions to the JDL data fusion process model*, Proceedings of the National Symposium on Sensor and Data Fusion (NSSDF), Lexington, MA.
18. Morefield, C. (2000) *AlphaTech Corporation, private communication to D. L. Hall and A. N. Steinberg.*
19. Poore, A. (1994) *Multi-dimensional assignment formulation of data association problems arising from multi-target and multi-sensor tracking*, Computational Optimization Applications, vol. 3, pp. 27-57.
20. Mahler, R. (1987) *A unified foundation for data fusion*, in Proceedings of the 1994 Data Fusion Systems Conference, Applied Physics Laboratory, Johns Hopkins University.
21. Bar-Shalom, Y., and Blare, W. (ed.) (1998) *Multi-Target, Multi-Sensor Tracking: Applications and Advances Vol III*, Artech House, Norwood, MA.
22. Hall, D.L. and Linn, R.J. (1989) *Comments on the use of templating for multi-sensor data fusion*, Proceedings of the 1989 Tri-Service Data Fusion Symposium, Vol. 1, pp. 345-354.
23. Llinas, J. and Antony, R. (1993) *Blackboard concepts for data fusion and command and control applications*, International Journal on Pattern Recognition and Artificial Intelligence, vol. 7, no. 2.
24. Watson, I. and Marir, F. (1994) *Case-based reasoning: a review*, The Knowledge Engineering Review, Vol. 9, No. 4, (http://www.surveying.salford.ac.uk...br-mirror/classroom/cbt-review.htm).
25. Gibson, R., Hall,a D.L. and Stover, J. (1994) *An autonomous fuzzy logic architecture for multi-sensor data fusion*, Proceedings of the 1994 Conference on Multi-Sensor Fusion and Integration for Intelligent Systems, Las Vegas, NV. pp. 143-150.
26. Russell, S. and Norvig, P. (1995) *Artificial Intelligence: A Modern Approach*, Prentice Hall Series in Artificial Intelligence, Prentice Hall, New Jersey.
27. Mush, D. and Horne, B. (1993) *Progress in supervised neural networks: what's new since Lippman*? IEEE Signal Processing Magazine, pp. 8-39.
28. Hall, D.L. and Garga, A.K. (1999) *New perspectives on level four processing in data fusion systems*, Proceedings of the SPIE AeroSense'99 Conference: Digitization of the Battlefield IV, Orlando, FL.
29. Fukanaga, K. (1990) *Introduction to Statistical Pattern Recognition*, 2nd, ed., Academic Press, NY.
30. Brendle, B.E., Jr. (1997) *Crewman's associate: interfacing to the digitized battlefield*, Proceedings of the SPIE: Digitization of the Battlefield II, Vol, 3080, Orlando, Florida, 22-24, pp. 195-202.
31. Steele, A., Marzen, V. and Corona, B. (1999) *Army Research Laboratory advanced displays and interactive displays, Fedlab technology transitions*, Proceedings of the SPIE: Digitization of the Battlespace IV, Vol. 3709, Orlando, Florida, pp. 205-212.
32. The CAVE at NCSA, http://www.ncsa.uiuc.edu/VEG/ncsaCAVE.html.
33. Ellis, R.E., Ismaeil, O.M. and Lipsett, M. (1996) *Design and evaluation of a high-performance haptic interface*, Robotica, vol. 14, pp. 321-327.
34. Hall, D.L., Hansen, R.J. and Lang, D.C. (1997) *The negative information problem in mechanical diagnostics*, Transactions of the ASME, vol. 119, pp. 370-377.
35. Piattelli-Palmarini, M. (1994) *Inevitable Illusions: How Mistakes of Reason Rule over Minds*, John Wiley and Sons, New York.
36. Hall, D.L., Sexton, P., Warren, M. and Zmyslo, J. (2000) Dynamo*: A tool for modeling integrated air defense systems*, Proceedings of the 2000 Meeting of the MSS, National Symposium on Sensor and Data Fusion, San Antonio, Texas.
37. Nixon, M. (2000) *Application of economic auction methods to resource allocation and optimization*, presented to a Data Fusion Workshop, Washington, DC.
38. Ferran, B. (1999) *presentation to the National Reconnaissance Office (NRO)*, spring, (available on video from the NRO office in Chantilly, Virginia).
39. Llinas, J., Drury, C., Bialas, W., and Chen, An-che, (1997) *Studies and Analyses of Vulnerabilities in Aided Adversarial Decision-Making*, Technical Report, State University of New York at Buffalo, Dept. of Industrial Engineering.
40. Barbar, C.A., Stone, L.D. and Hun, M.V. (1996) *Unified data fusion*, Proceedings of the 9th National Symposium on Sensor Fusion, Vol. 1, 12-14, pp. 321-330.
41. Llinas, J., Neuenfeldt, B., McConnell, L. Bohney, D., Bowman, C., Hall, D., Lochacki, J. and Applegate, P. (1996) *Studies and analyses within project Correlation: an in-depth assessment of correlation problems and solution techniques*, Proceedings of the 9th National Symposium on Sensor Fusion, Vol. 1, 12-14, pp. 171-188.

42. Wohl, J.G., Entin, E.E., Serfaty, D., James, R.M. and Deckert, J.C. (1987) *Human cognitive performance in ASW data fusion*, Proceedings of the 1987 Tri-Service Data Fusion Symposium, Johns Hopkins University, Applied Physics Laboratory, Laurel MD, pp. 465-479.
43. http://haptic.mech.nwu.edu
44. Swanson, D.C. (2000) *Signal Processing for Intelligent Sensing Systems*, Marcel Dekker, Inc.
45. Steinberg, A. (1987) *Threat Management Systems for Combat Aircraft*, Proceedings of the 1987 Tri-Service Data Fusion Symposium, vol. 1, Johns Hopkins University, Applied Physics Laboratory, Laurel, MD, pp. 528-543.
46. Hall, D.L. and Linn, R.J. (1991) *A Comparison of Association Metrics for Emitter Classification*, Technical Proceedings of the 1991 Data Fusion Systems Conference, DFS-91, Johns Hopkins University, Applied Physics Laboratory, Laurel, MD, pp. 447-462.
47. Hall, D.L. and Linn, R.J. (1993) *Survey of Commercial Software for Multisensor Data Fusion*, Proceedings of the SPIE Conference on Sensor Fusion and Aerospace Applications, Orlando, FL.
48. Khoros Software, see Khoral Research Inc., 6001 Indian School Rd. NE Suite 200, Albuquerque, NM 87110, USA, http://www.khoral.com/
49. Hall, D.L. and Kasmala, G.F. (1996) *Visual Programming Environment for Multisensor Data Fusion*, Proceedings of the SPIE Aerosense 1996 Symposium, Orlando, FL, Vol. 2764, pp. 181-187.
50. Hall, D.L., Heinze, S.D. and Llinas, J. (1989) *Test and Evaluation of Expert Systems*, in AI in Manufacturing: Theory and Practice, S.T. Kumara, R. Kashyap, and A.L. Soyster (eds.), Institute of Industrial Engineering, pp. 121-140.

MULTISENSOR DATA FUSION USING KALMAN FILTERS BASED ON NEURO-FUZZY LINEARISATION

C. HARRIS AND Q. GAN
Image, Speech and Intelligent Systems Research Group
Department of Electronics and Computer Science
University of Southampton
Southampton SO17 1BJ, UK
E-mail: cjh@ecs.soton.ac.uk,qg@ecs.soton.ac.uk

1. Introduction

Target tracking is an important issue in automatic control, such as in military surveillance systems and civilian traffic control systems. The main purpose of a tracking system is to obtain a target trajectory estimate, which is more accurate than the trajectory observation, and to predict the target behaviour in the near future. Most target tracking methods are model-based, with the following linear discrete-time state-space model describing the target dynamics:

$$\mathbf{x}(t) = A(t)\mathbf{x}(t-1) + B(t)\mathbf{u}(t) + \mathbf{v}(t) \tag{1}$$

$$y(t) = C(t)x(t) + w(t) \tag{2}$$

where $x(t) = [x_1(t) \dots x_n(t)]^T$ is a state vector, $u(t) = [u_1(t) \dots u_m(t)]^T$ is an input vector representing maneuvers, $y(t) = [y_1(t) \dots y_L(t)]^T$ is an observation vector, $v(t)$ and $w(t)$ represent additive Gaussian noise, satisfying $E[v(t)] = E[w(t)] = \mathbf{0}$, $E[v(i)v^T(j)] = \delta_{ij}Q(i)$, $E[w(i)w^T(j)] = \delta_{ij}R(i)$, and $E[v(i)w^T(j)] = \mathbf{0}$.

The Kalman filter (Balakrishnan, 1987; Chui and Chen, 1987) [2,19] is the most popular approach to the solution of the above linear target tracking problem, which produces an unbiased and optimal estimate of the state vector in the sense of minimum error variance if the model described by (1) and (2) is completely known. The Kalman filter algorithm can be described by the following equations (Balakrishnan, 1987; Chui and Chen, 1987; Gan and Harris, 1999e; Manyika and Durrant-Whyte, 1994) [2,19,26,38]:

Prediction:

$$\hat{x}(t|t-1) = A(t)\hat{x}(t-1|t-1) + B(t)u(t) \tag{3}$$

$$P(t|t-1) = A(t)P(t-1|t-1)A^T(t) + Q(t) \tag{4}$$

A. K. Hyder et al. (eds.), Multisensor Fusion, 435–456.

Estimate (Correction):

$$K(t) = P(t\,|\,t-1)C^T(t)[C(t)P(t\,|\,t-1)C^T(t)+R(t)]^{-1} \quad (5)$$
$$\hat{x}(t\,|\,t) = \hat{x}(t\,|\,t-1)+K(t)[y(t)-C(t)\hat{x}(t\,|\,t-1)] \quad (6)$$
$$P(t\,|\,t) = [I-K(t)C(t)]P(t\,|\,t-1) \quad (7)$$

where $\hat{x}(t\,|\,t)$ represents the estimate of the state vector $x(t)$, $P(t|t)$ is the state estimate covariance matrix, $K(t)$ the Kalman gain matrix, $Q(t)$ the process noise variance matrix, and $R(t)$ the observation noise variance matrix. Usually, $[y(t)-C(t)\hat{x}(t\,|\,t-1)]$ is called measurement residual or innovations, and $[C(t)P(t\,|\,t-1)C^T(t)+R(t)]$ is the innovations covariance matrix, which provide very useful information for adaptive filtering.

For a time-invariant kinematic process, the state estimate covariance $P(t|t)$ will converge to a steady-state value if the following conditions are satisfied (Balakrishnan, 1987; Bar-Shalom and Fortmann, 1988): (i) The noise is stationary, *i.e.*, the covariance matrices Q and R are constant; (ii) $\{A, C\}$ is completely observable; (iii) $\{A, D\}$ is completely controllable, where D is defined by $DD^T=Q$. In this case, the Kalman filter can be simplified into its steady-state form, for instance, the $\alpha-\beta$ filter for two-dimensional state vector estimation, which is formed by simplifying the estimate phase (5)-(7) into the following (Bar-Shalom and Fortmann, 1988):

$$\hat{x}(t\,|\,t) = \hat{x}(t\,|\,t-1)+\begin{bmatrix}\alpha\\ \beta\end{bmatrix}[y(t)-C(t)\hat{x}(t\,|\,t-1)] \quad (8)$$

where α and β are constant and determined by the noise covariance Q and R. Despite its simplicity, the steady-state filter is by far the most successful and important application of the Kalman filter.

For simple scenarios, the Kalman filter is very effective. However, many problems arise in practical applications where there exist nonlinearities and uncertainties. To cope with nonlinearities, the conventional approach is the extended Kalman filter using a linearisation technique based on the first-order Taylor series expansion (Balakrishnan, 1987; Chui and Chen, 1987) [2,19]. New linearisation techniques in state estimation include local linearisation (Gan and Harris, 1999b; Gan and Harris, 1999c; Gan and Harris; 1999f; Gan and Harris, 1999g; McGinnity and Irwin, 1996; Wu and Harris, 1997) [23,26-28,43,49] and feedback linearisation (Gan and Harris, 1999d; Marino and Tomei, 1995; Marino and Tomei, 1992). Due to the model error introduced by the linearisation process, the filtered results are no longer unbiased and optimal, and potentially worse the filter algorithm may diverge. Uncertainties exist in unknown state-transition functions, unpredictable maneuvers, and nonstationary noise. The techniques for tackling uncertainties can be divided into two main categories: the residual based adaptive technologies (Amoozegar, 1998; Bar-Shalom et al., 1989; Bogler, 1987; Chan et al., 1982; Chan et al., 1979) [1,6,12,14,13] and multiple model approaches (Blom and Bar-Shalom, 1988; Lerro and Bar-Shalom, 1993; Li and Bar-Shalom, 1993a; Li and Bar-Shalom, 1993b; Mazor et al., 1998; McGinnity and Irwin, 1998) [11,35-37,41,43]. The key issue in these methods is the detection and estimation of abrupt changes in the target and measurement dynamic processes. Multiple model approaches, which have drawn much attention in recent years, make use of *a priori* knowledge about targets and sensors, and are generally more effective than single model methods. It is noteworthy that the adaptive techniques for tackling uncertainties can also be used for coping with nonlinearities, since linearisation errors can be viewed as nonstationary noise, a sort of uncertainties.

In many engineering applications, there exist even more complicated situations. For example, the nonlinearities and uncertainties in multitarget multisensor tracking problems are more severe than those in single-target single-sensor tracking problems (Bar-Shalom and Li, 1995; Bar-Shalom and Li, 1993; Bar-Shalom, 1992; Bar-Shalom and Fortmann, 1988; Shams, 1996) [9,8,4,7,48]. However, there are more data sources available for developing new techniques such as data fusion in multitarget multisensor tracking. The technologies for multitarget multisensor tracking problems mainly include multitarget data association (Amoozegar, 1998; Bar-Shalom and Fortmann, 1988; Chang and Bar-Shalom, 1989; Chang et al., 1986; Sengupta and Iltis, 1989) [1,7,15,47] and multisensor data fusion (Bar-Shalom and Fortmann, 1988; Bar-Shalom and Campo, 1986; Bar-Shalom, 1981; Blair and Bar-Shalom, 1996; Chang et al., 1998; Chang et al., 1997; Doyle and Harris, 1996; Gan and Harris, 1999a; Gan and Harris, 1999e; Grime and Durrant-Whyte, 1994; Hall and Llinas, 1997; Harris et al., 1998; Harris et al., 1997; Manyika and Durrant-Whyte, 1994; Roecker and McGillem, 1988; Saha and Chang, 1998; Saha, 1996). [7,5,3,10,18,17,20,22,24,29-31,34,38,44,46,45] In the following sections, this paper will focus on neuro-fuzzy linearisation techniques for process modelling and state estimation, and multisensor data fusion methods using Kalman filters based on neuro-fuzzy linearisation models.

2. Neuro-Fuzzy Linearisation Techniques For Modelling And State Estimation

In general, a dynamic process can be described by

$$x(t) = f[x(t-1), u(t)] + v(t) \tag{9}$$

$$y(t) = h[x(t)] + w(t) \tag{10}$$

where *f[.]* is a smooth vector function describing the state transition process, *h[.]* is a smooth vector function describing the measurement process. In many applications, *f[.]* and *h[.]* can be derived from physical laws such as ship motion (Fossen, 1994), [21] but in some practical applications they are even totally unknown.

If *f[.]* and *h[.]* are nonlinear but known *a priori*, the popular method for nonlinear state estimation is the extended Kalman filter which is defined by the following equations:

Prediction:

$$\hat{x}(t|t-1) = f[\hat{x}(t-1|t-1), u(t)] \tag{11}$$

$$P(t|t-1) = \nabla f_x(t) P(t-1|t-1) \nabla f_x^T(t) + Q(t) \tag{12}$$

Estimate (Correction):

$$K(t) = P(t|t-1)\nabla \boldsymbol{h}_x^{\ T}(t)[\nabla \boldsymbol{h}_x(t)P(t|t-1)\nabla \boldsymbol{h}_x^{\ T}(t) + R(t)]^{-1} \tag{13}$$

$$\hat{\boldsymbol{x}}(t|t) = \hat{\boldsymbol{x}}(t|t-1) + K(t)[\boldsymbol{y}(t) - \boldsymbol{h}(\hat{\boldsymbol{x}}(t|t-1))] \tag{14}$$

$$P(t|t) = [I - K(t)\nabla \boldsymbol{h}_x(t)]P(t|t-1) \tag{15}$$

where $\nabla \boldsymbol{f}_x(t)$ and $\nabla \boldsymbol{h}_x(t)$ are the Jacobian matrices of $\boldsymbol{f}[.]$ and $\boldsymbol{h}[.]$, respectively. The extended Kalman filter is based on the first-order Taylor series expansion approximation which results in a biased and suboptimal estimation and may lead to divergence. When the nonlinear functions $\boldsymbol{f}[.]$ and $\boldsymbol{h}[.]$ are *a priori* unknown, the error in the Jacobian matrix calculation, which is based on approximated models, may be large, even if the model approximation error is small. This problem can be avoided by using state estimators based on neuro-fuzzy local linearisation (NFLL) or neuro-fuzzy feedback linearisation (NFFL) models (Gan and Harris, 1999d; Gan and Harris, 1999f; Gan and Harris, 1999g; Harris et al., 2000; McGinnity and Irwin, 1996; Wu and Harris, 1997) [25,27-28,33,42,49].

Now let us introduce the NFLL modelling method. For simplicity, consider first a scalar nonlinear function described by the input-output relation: $y(t)=f[\boldsymbol{x}(t)]$, where $f[.]$ represents a nonlinear function and $\boldsymbol{x}(t) = [x_1(t)\ x_2(t)\ \ldots\ x_n(t)]^T$ represents the input vector. If the nonlinear function is dynamical, the input vector will include both observed external inputs and previous output feedback, for example, $\boldsymbol{x}(t) = [y(t-1)\ \ldots\ y(t-d_y)\ u(t-1)\ \ldots\ u(t-d_u)]^T$. It is easy to generalise the scalar function case to the vector function case.

One of the most popular methods widely used in function approximation or modelling is the following B-spline basis function expansion (Harris et al., 1996) [32]:

$$\hat{f}(\boldsymbol{x}) = \sum_{k=1}^{K} N_k(\boldsymbol{x})a_k \tag{16}$$

with $f(\boldsymbol{x}) = \hat{f}(\boldsymbol{x}) + \Delta f(\boldsymbol{x})$, where a_k represent expansion coefficients, $N_k(\boldsymbol{x})$ are B-splines which can be interpreted as membership functions in a neuro-fuzzy system, and $\Delta f(\boldsymbol{x})$ represents the model mismatch. If the constant coefficients a_k in (16) are replaced by linear functions of the state $\boldsymbol{x}$, then (16) is converted into a NFLL model:

$$\hat{f}(\boldsymbol{x}) = \sum_{k=1}^{K} N_k(\boldsymbol{x})(a_{k0} + \boldsymbol{a}_k^T \boldsymbol{x}) \tag{17}$$

Because multi-dimensional B-splines $N_k(\boldsymbol{x})$ are usually formed by tensor production, the curse-of-dimensionality problem will be severe for B-spline basis function expansion when the dimension of the input space is high (say, $n>4$). In order to resolve this problem, the analysis-of-variance (ANOVA) decomposition is often adopted to decompose a high dimensional model into a sum of low dimensional submodels (Gan and

Harris, 1999b; Gan and Harris, 1999g). By using the ANOVA decomposition in the NFLL model, Eq. (17) is reformed as follows:

$$\hat{f}(\boldsymbol{x}) = \sum_{s=1}^{S} f_s(\boldsymbol{x}_s) = \sum_{s=1}^{S}\sum_{k=1}^{K_s} N_k^{(s)}(\boldsymbol{x}_s)(a_{k0}^{(s)} + \boldsymbol{a}_k^{(s)T}\boldsymbol{x}_s) \tag{18}$$

where s is the index of B-spline submodels, $\boldsymbol{x}_s = [x_{s1}\ x_{s2} \dots x_{sn_s}]^T$ represent low-dimensional subspaces of the input space, $\boldsymbol{a}_0^{(s)} = [a_{10}^{(s)}\ a_{20}^{(s)} \dots a_{K_s0}^{(s)}]^T$ and $A^{(s)} = [a_{ki}^{(s)}]_{K_s \times n_s}$ are local model parameters, and $N_k^{(s)}(\boldsymbol{x}_s)$ are B-splines used as membership functions of fuzzy sets that partition the subspaces.

The model described by (18) is linear in parameters $a_{ki}^{(s)}$, and can be reformulated as follows:

$$\hat{f}(\boldsymbol{x}) = \sum_{s=1}^{S}\sum_{k=1}^{K_s}\sum_{i=0}^{n_s} [N_k^{(s)}(\boldsymbol{x}_s)\cdot x_{si}]a_{ki}^{(s)} = \sum_{l=0}^{K'} N'_l(\boldsymbol{x})\theta_l \tag{19}$$

where, $x_{s0} = 1$, $K' = \sum_{s=1}^{S} K_s \times (n_s + 1)$, $\theta_l = a_{ki}^{(s)}$, $l = i + (k-1)\times(n_s+1) + \sum_{j=1}^{s-1} K_j \times (n_j + 1)$. With $\{x(t), y(t)\}_{t=1}^{M}$ as training data pairs, by setting $\hat{y} = [\hat{f}(x(1)) \dots \hat{f}(x(M))]^T$, Eq. (19) can be represented in a matrix form $\hat{y} = \boldsymbol{N}'\boldsymbol{\theta}$. This is a typical equation which is directly suitable for the least squares algorithm to identify the free parameters θ_l. Together with the parameter identification, the B-spline structures in (19) can also be identified automatically by a modified adaptive spline modelling (MASMOD) algorithm (Gan and Harris, 1999b) [23], which is based on a structural risk minimisation (*SRM*) criterion.

The *SRM* index is a function of the root-mean-squares (*RMS*) of model errors, the degrees of freedom (*dof*) of the NFLL model, and the number of training samples (*M*), defined as follows:

$$SRM = \begin{cases} \dfrac{RMS}{\sqrt{1-PN}} & if\ \ PN < 1 \\ \infty & otherwise \end{cases} \tag{20}$$

with

$$PN = \beta_1\sqrt{(dof \cdot log(2M) - log(dof!) - \beta_2)/M} \tag{21}$$

where, $RMS\sqrt{\sum_{t=1}^{M}\{y(t) - \hat{f}[x(t)]\}^2 / M}$, *dof* is defined as the rank of the matrix $\boldsymbol{N}'$, and β_1 and β_2 are constants.

The MASMOD algorithm using the *SRM* criterion consists of the following steps:

(a) Initialise the model, *i.e.*, set up an empty model or a minimum-sized model, and set the initial order of B-splines and initial values of all the free parameters.
(b) Update parameter vector $\boldsymbol{a}_0^{(s)}$ and matrix $A^{(s)}$ using the least-squares (LS) or least-mean-squares (LMS) algorithm, based on the training input-output data pairs.
(c) Compute the *SRM* criterion. If it is satisfied, exit.
(d) Refine the model by constructing the B-splines. The refinement includes adding new univariate submodels, inserting new knots in the existing submodels, forming tensor product submodels, pruning knots or reducing the order of B-splines.
(e) Go to Step (b).

The above algorithm can be used to automatically determine the number of submodels, the number of B-splines in each submodel, and their orders and knot positions, in such a way that the *SRM* index is minimised. The most important part in the MASMOD algorithm is the model structural learning. Basically, the MASMOD algorithm should be run off-line. After the model has been constructed, however, its parameters can be adapted on-line in the application phase, with its structure fixed. In the application phase, there is no training data to obtain direct model errors, hence error signals has to be calculated in a somewhat indirect way, such as based on the measurement residual in state estimation applications.

The MASMOD algorithm can also be used for developing NFFL models (Gan and Harris, 1999d) [25], which are recurrent networks and very useful for state-space modelling if the problem of convergence can be successfully resolved. Both NFLL and NFFL can be used for state-space linearisation modelling, based on which the standard Kalman filter is directly applicable to nonlinear state estimation (no need to calculate Jacobian matrix).

For instance, by reformulating the NFLL models, a nonlinear state-space model can be transformed into a time-varying linear model. Let $\boldsymbol{x}'(t)=[\boldsymbol{x}^T(t-1)\ \ \boldsymbol{u}^T(t)]^T$ and $\hat{\boldsymbol{f}}[\boldsymbol{x}(t-1),\ \boldsymbol{u}(t)]=\hat{\boldsymbol{f}}[\boldsymbol{x}'(t)]=[\hat{f}_1(\boldsymbol{x}')\quad \hat{f}_2(\boldsymbol{x}')\cdots\hat{f}_n(\boldsymbol{x}')]^T$. Based on the NFLL modelling formulation, $\hat{f}_l(\boldsymbol{x}')$, $l=1,\ 2,\ \ldots,\ n$, can be modelled as follows:

$$
\begin{aligned}
\hat{f}_l(\boldsymbol{x}') &= \sum_{k=1}^{K} N_k(\boldsymbol{x}')[a_{kl1}x_1(t-1)+\cdots+a_{kln}x_n(t-1) \\
&\qquad + b_{kl1}u_1(t)+\cdots+b_{klm}u_m(t)] \\
&= \alpha_{l1}(t)x_1(t-1)+\cdots+\alpha_{ln}(t)x_n(t-1) \\
&\qquad + \beta_{l1}(t)u_1(t)+\cdots+\beta_{lm}(t)u_m(t)
\end{aligned}
\tag{22}
$$

with

$$
\alpha_{li}(t)=\sum_{k=1}^{K} N_k(\boldsymbol{x}')\,a_{kli},\qquad i=1,2,\ldots,n \tag{23}
$$

$$\beta_{lj}(t) = \sum_{k=1}^{K} N_k(\mathbf{x}')\, b_{klj}, \qquad j = 1,\ldots,m \tag{24}$$

$$N_k(\mathbf{x}') = \prod_{i=1}^{n} B_{i,k_i}(x_i(t-1)) \bullet \prod_{j=1}^{m} B_{j,k_{n+j}}(u_j(t)) \tag{25}$$

where k corresponds to an ordered sequence $\{k_1,\cdots,k_n,k_{n+1},\cdots,k_{n+m}\}$, and K represents the total number of the possible sequences. Based on (22)-(25), Eq. (9) can be reformed as

$$\mathbf{x}(t) = A(t)\mathbf{x}(t-1) + B(t)\mathbf{u}(t) + \mathbf{v}'(t) \tag{26}$$

where, $A(t) = [\alpha_{li}(t)]_{n\times n}$, $B(t) = [\beta_{lj}(t)]_{n\times m}$, and $\mathbf{v}'(t)$ represents the original process noise plus the error introduced by model mismatch. In a similar way, Eq. (10) can be reformed as follows:

$$\mathbf{y}(t) = C(t)\mathbf{x}(t) + \mathbf{w}'(t) \tag{27}$$

If the underlying nonlinear vector function $\mathbf{f}(\mathbf{x}, \mathbf{u})$ is approximately linear over part of a process operating regime (as is frequently the case in practice), then the coefficients in (26) and (27) will change slowly in the operating regime and the standard Kalman filter is directly applicable to (26) and (27) for target tracking. In addition, in order to compensate model errors, model parameters, such as process noise covariance, can be adapted on-line based on the measurement residual in the tracking process.

3. Kalman-Filter-Based Multisensor Data Fusion Techniques

To improve target tracking accuracy and robustness, multiple data sources are integrated (fused) to obtain a better estimate. Usually data association techniques (Bar-Shalom and Fortmann, 1988; Chang et al., 1989; Chang et al., 1986; Sengupta and Iltis, 1989) [7,15-16,47] are required before data fusion to ensure that the available data is from the same target. This section introduces Kalman-filter-based multisensor data fusion methods, assuming that the measurements are taken from the same target and that the measurement models are given by

$$\mathbf{y}_j(t) = C_j(t)\mathbf{x}(t) + \mathbf{w}_j(t), \qquad j = 1,2,\ldots,N \tag{28}$$

where N is the number of sensors, and $\mathbf{w}_j(t) \sim N(\mathbf{0}, R_j(t))$. To some extent, multisensor data fusion can be viewed as a multiple model approach.

In multisensor data fusion applications, the information form of the Kalman filter (or information filter) is usually more amiable than the standard Kalman filter. If we define a so-called information state vector: $\hat{\mathbf{z}}(t_1 \mid t_2) \equiv P^{-1}(t_1 \mid t_2)\hat{\mathbf{x}}(t_1 \mid t_2)$, where $P^{-1}(t_1 \mid t_2)$ is called the information matrix, the standard Kalman filter can be transformed into the following information form (Manyika and Durrant-Whyte, 1994; Gan and Harris, 1999e) [38,26]:

Prediction:

$$\begin{aligned}\hat{z}(t|t-1) = P^{-1}(t|t-1)A(t)P(t-1|t-1)\hat{z}(t-1|t-1) \\ + P^{-1}(t|t-1)B(t)\mathbf{u}(t)\end{aligned} \tag{29}$$

$$P(t|t-1) = A(t)P(t-1|t-1)A^T(t) + Q(t) \tag{30}$$

Estimate (Correction):

$$\hat{z}(t|t) = \hat{z}(t|t-1) + C^T(t)R^{-1}(t)\mathbf{y}(t) \tag{31}$$

$$P^{-1}(t|t) = P^{-1}(t|t-1) + C^T(t)R^{-1}(t)C(t) \tag{32}$$

where $C(t)$ and $\mathbf{y}(t)$ can be from a particular measurement process or some kind of combination of multiple sensor measurement processes.

The information filter is functionally equivalent to the standard Kalman filter, but has some computational advantages, especially in multisensor data fusion applications where the innovations covariance matrix $[C(t)P(t|t-1)C^T(t) + R(t)]$ in the Kalman filter is usually highly dimensional and non-diagonal. The major difference between the standard Kalman filter and the information filter lies in their estimate phase where data fusion takes place. The estimate phase of the information filter is simpler because the gain matrix $K(t)$ in the standard Kalman filter is much more complicated than the term $C^T(t)R^{-1}(t)$ in the information filter, especially in multisensor data fusion applications where the term $[C(t)P(t|t-1)C^T(t) + R(t)]^{-1}$ in $K(t)$ may become computationally prohibitive.

Methods for Kalman-filter-based data fusion includes measurement fusion and state-vector fusion. Measurement fusion methods directly fuse sensor measurements, which then drives a single Kalman filter. Whereas state-vector fusion methods use a group of Kalman filters to obtain individual sensor based state estimates which are then fused to obtain an improved joint state estimate.

3.1 MEASUREMENT FUSION

There are two commonly used measurement fusion methods which assume the measurement noise is independent. The first method (MF-I) integrates the measurement information as follows (Manyika and Durrant-Whyte, 1994; Gan and Harris, 1999e; Doyle and Harris, 1996) [38,26,20]:

$$\mathbf{y}(t) = \mathbf{y}^{(1)}(t) = [\mathbf{y}_1(t) \ \cdots \ \mathbf{y}_N(t)]^T \tag{33}$$

$$C(t) = C^{(1)}(t) = [C_1(t) \ \cdots \ C_N(t)]^T \tag{34}$$

$$R(t) = R^{(1)}(t) = diag[R_1(t) \ \cdots \ R_N(t)] \tag{35}$$

The second method (MF-II) obtains the fused measurement information in the following way (Gan and Harris, 1999e; Roecker and McGillem, 1988) 26,44]:

$$y(t) = y^{(II)}(t) = [\sum_{j=1}^{N} R_j^{-1}(t)]^{-1} \sum_{j=1}^{N} R_j^{-1}(t)\, y_j(t) \tag{36}$$

$$C(t) = C^{(II)}(t) = [\sum_{j=1}^{N} R_j^{-1}(t)]^{-1} \sum_{j=1}^{N} R_j^{-1}(t)\, C_j(t) \tag{37}$$

$$R(t) = R^{(II)}(t) = [\sum_{j=1}^{N} R_j^{-1}(t)]^{-1} \tag{38}$$

Comparing (33)-(35) with (36)-(38), we note that the treatment in the two measurement fusion methods are quite different. For example, the dimension of the measurement information ($y(t)$, $C(t)$, $R(t)$) in MF-I is N times as large as that in MF-II. However, it has been demonstrated that there exists a certain form of functional equivalence between the two methods, contained in the following theorem (Gan and Harris, 1999a; Gan and Harris, 1999e) [22,26]:

THEOREM *If the sensors used for data fusion, with different and independent noise characteristics, have identical measurement matrices, i.e.,* $C_1(t) = C_2(t) = \ldots = C_N(t)$, *then MF-I is functionally equivalent to MF-II.*

PROOF The following formula in linear algebra will be used to cope with the inversion of matrices:

$$\begin{bmatrix} A_1 & A_2 \\ A_3 & A_4 \end{bmatrix}^{-1} = \begin{bmatrix} B_1 & B_2 \\ B_3 & B_4 \end{bmatrix} \tag{39}$$

$$(A + HBH^T)^{-1} = A^{-1} - A^{-1}H(B^{-1} + H^T A^{-1} H)^{-1} H^T A^{-1} \tag{40}$$

where $B_1 = (A_1 - A_2 A_4^{-1} A_3)^{-1}$, $B_2 = -B_1 A_2 A_4^{-1}$, $B_3 = -A_4^{-1} A_3 B_1$, and $B_4 = A_4^{-1} + A_4^{-1} A_3 B_1 A_2 A_4^{-1}$.

If the standard Kalman filter is used, in order to demonstrate the functional equivalence of the two measurement fusion methods we only need to check whether the terms $K(t)C(t)$ and $K(t)y(t)$ in MF-I are functionally equivalent to those in MF-II. Alternatively, if the information filter is used, then we need to check the functional equivalence between terms $C^T(t)R^{-1}(t)C(t)$ and $C^T(t)R^{-1}(t)y(t)$ of both methods. To avoid confusion, the variables in MF-I are indexed by $^{(I)}$ and the variables in MF-II indexed by $^{(II)}$. For simplicity, the time argument in $C_j(t)$ and $R_j(t)$ will be omitted in the following derivation.

First, consider the case when the standard Kalman filter is applied. Because of the complexity of the innovations covariance matrix, direct derivation via this route is complex. Let us begin with sensor-to-sensor fusion, i.e. $N=2$. Letting $C_1 = C_2 = C_0$ and $\Omega^{(I)} = C_0 P^{(I)}(t|t-1) C_0^T$, we obtain the Kalman gain in MF-I as follows:

$$\begin{aligned} K^{(I)}(t) &= P^{(I)}(t|t-1)C_0^T [I \ \ I] \begin{bmatrix} R_1 + \Omega^{(I)} & \Omega^{(I)} \\ \Omega^{(I)} & R_2 + \Omega^{(I)} \end{bmatrix}^{-1} \\ &= P^{(I)}(t|t-1)C_0^T [(R_2 + \Omega^{(I)})^{-1} R_2 \cdot \\ &\quad [R_1 + \Omega^{(I)} - \Omega^{(I)}(R_2 + \Omega^{(I)})^{-1}\Omega^{(I)}]^{-1}, \\ &\quad (R_2 + \Omega^{(I)})^{-1} - (R_2 + \Omega^{(I)})^{-1} R_2 \cdot \\ &\quad [R_1 + \Omega^{(I)} - \Omega^{(I)}(R_2 + \Omega^{(I)})^{-1}\Omega^{(I)}]^{-1}\Omega^{(I)}(R_2 + \Omega^{(I)})^{-1}] \end{aligned} \tag{41}$$

In the Appendix, the following equalities will be proved:

$$\begin{aligned} &(R_2 + \Omega^{(I)})^{-1} R_2 [R_1 + \Omega^{(I)} - \Omega^{(I)}(R_2 + \Omega^{(I)})^{-1}\Omega^{(I)}]^{-1} \\ &= [\Omega^{(I)} + R_1(R_1 + R_2)^{-1} R_2]^{-1} R_2 (R_1 + R_2)^{-1} \end{aligned} \tag{42}$$

and

$$\begin{aligned} &(R_2 + \Omega^{(I)})^{-1} - (R_2 + \Omega^{(I)})^{-1} R_2 \cdot \\ &[R_1 + \Omega^{(I)} - \Omega^{(I)}(R_2 + \Omega^{(I)})^{-1}\Omega^{(I)}]^{-1}\Omega^{(I)}(R_2 + \Omega^{(I)})^{-1} \\ &= [\Omega^{(I)} + R_1(R_1 + R_2)^{-1} R_2]^{-1} R_1 (R_1 + R_2)^{-1} \end{aligned} \tag{43}$$

Based on (41) - (43), we have

$$\begin{aligned} K^{(I)}(t) = P^{(I)}(t|t-1)C_0^T &[C_0 P^{(I)}(t|t-1)C_0^T + R_1(R_1 + R_2)^{-1} R_2]^{-1} \cdot \\ &[R_2(R_1 + R_2)^{-1}, \ R_1(R_1 + R_2)^{-1}] \end{aligned} \tag{44}$$

$$\begin{aligned} &K^{(I)}(t) C^{(I)}(t) \\ &= P^{(I)}(t|t-1)C_0^T [C_0 P^{(I)}(t|t-1)C_0^T + R_1(R_1 + R_2)^{-1} R_2]^{-1} C_0 \end{aligned} \tag{45}$$

$$\begin{aligned} &K^{(I)}(t) y^{(I)}(t) \\ &= P^{(I)}(t|t-1)C_0^T [C_0 P^{(I)}(t|t-1)C_0^T + R_1(R_1 + R_2)^{-1} R_2]^{-1} \cdot \\ &\quad [R_2(R_1 + R_2)^{-1} y_1(t) + R_1(R_1 + R_2)^{-1} y_2(t)] \end{aligned} \tag{46}$$

If $C_1 = C_2 = C_0$, then $C^{(II)} = C_0$, $R^{(II)} = R_1(R_1 + R_2)^{-1} R_2$, and we obtain the Kalman gain in MF-II as follows:

$$K^{(II)}(t) = P^{(II)}(t|t-1)C_0^T[C_0P^{(II)}(t|t-1)C_0^T + R_1(R_1+R_2)^{-1}R_2]^{-1} \tag{47}$$

$$\begin{aligned} &K^{(II)}(t)C^{(II)}(t) \\ &= P^{(II)}(t|t-1)C_0^T[C_0P^{(II)}(t|t-1)C_0^T + R_1(R_1+R_2)^{-1}R_2]^{-1}C_0 \end{aligned} \tag{48}$$

$$\begin{aligned} &K^{(II)}(t)y^{(II)}(t) \\ &= P^{(II)}(t|t-1)C_0^T[C_0P^{(II)}(t|t-1)C_0^T + R_1(R_1+R_2)^{-1}R_2]^{-1} \cdot \\ &\quad [R_2(R_1+R_2)^{-1}y_1(t) + R_1(R_1+R_2)^{-1}y_2(t)] \end{aligned} \tag{49}$$

Note that (45) and (48) are in the same form and that (46) and (49) are also in the same form. Therefore, with the same initial conditions, *i.e.*, $P^{(I)}(0|0) = P^{(II)}(0|0)$ and $\hat{x}^{(I)}(0|0) = \hat{x}^{(II)}(0|0)$, the Kalman filters based on observations described by (33)-(35) and observations by (36)-(38) will result in the same state estimate $\hat{x}(t|t)$. This means that the two measurement fusion methods are functionally equivalent in the sensor-to-sensor case. This result can be generalised for N sensors. More conveniently, the generalised result can also be obtained based on the following proof with the information filter and the functional equivalence between the standard Kalman filter and information filter.

Now, consider the case when the information filter is applied. From (33)-(38), it is easy to prove the following equalities:

$$[C^{(I)}(t)]^T[R^{(I)}(t)]^{-1}C^{(I)}(t) = \sum_{j=1}^{N} C_j^T R_j^{-1} C_j \tag{50}$$

$$[C^{(I)}(t)]^T[R^{(I)}(t)]^{-1}y^{(I)}(t) = \sum_{j=1}^{N} C_j^T R_j^{-1} y_j \tag{51}$$

$$[C^{(II)}(t)]^T[R^{(II)}(t)]^{-1}C^{(II)}(t) = [(\sum_{j=1}^{N} R_j^{-1})^{-1}\sum_{j=1}^{N} R_j^{-1}C_j]^T \sum_{j=1}^{N} R_j^{-1}C_j \tag{52}$$

$$[C^{(II)}(t)]^T[R^{(II)}(t)]^{-1}y^{(II)}(t) = [(\sum_{j=1}^{N} R_j^{-1})^{-1}\sum_{j=1}^{N} R_j^{-1}C_j]^T \sum_{j=1}^{N} R_j^{-1}y_j \tag{53}$$

If $C_j = C_0$, $j = 1,2,\ldots,N$, then we have

$$[C^{(I)}(t)]^T[R^{(I)}(t)]^{-1}C^{(I)}(t) = [C^{(II)}(t)]^T[R^{(II)}(t)]^{-1}C^{(II)}(t) \tag{54}$$

$$[C^{(I)}(t)]^T [R^{(I)}(t)]^{-1} y^{(I)}(t) = [C^{(II)}(t)]^T [R^{(II)}(t)]^{-1} y^{(II)}(t) \tag{55}$$

From the information filter formulation (29)-(32), it is obvious that there exists a functional equivalence between the two measurement fusion methods.

The functional equivalence of the two measurement fusion methods is limited to the case when $C_1 = C_2 = ... = C_N$. If $C_1 \neq C_2 \neq ... \neq C_N$, then the functional equivalence will no longer hold. Because MF-I makes use of all the raw measurement information, it most likely provides a better estimate. Additionally, if the sensors are dissimilar, MF-II may not work at all (*e.g.*, if C_i and C_j are of different sizes). Based on the above theorem, we should choose MF-II if $C_1 = C_2 = ... = C_N$ is satisfied, otherwise MF-I is a better choice. These results are very useful in guiding the design of practical multisensor data fusion systems.

3.2 STATE-VECTOR FUSION

State-vector fusion integrates local state estimates to obtain an improved joint state estimate. In nonlinear situations, it is difficult to optimally combine the local state estimates because they are usually dependent on each other and sometimes it is difficult to obtain the correct state estimate covariance matrices. For the sake of comparison, track-to-track fusion and state-vector assimilation fusion are introduced in this section.

3.2.1 Track-to-Track Fusion (TTF)

The track-to-track fusion (*i.e.*, *N*=2) based on the standard Kalman filter is a widely studied optimal approach, which is formulated as follows (Bar-Shalom and Fortmann, 1988; Bar-Shalom and Campo, 1986; Bar-Shalom, 1981; Chang et al., 1998; Chang et al., 1997; Saha and Chang, 1998; Saha, 1996) [7,5,3,18,17,46,45]:

$$\hat{x}(t|t) = \hat{x}_1(t|t) + P_{xy} P_{yy}^{-1} [\hat{x}_2(t|t) - \hat{x}_1(t|t)] \tag{56}$$

$$P(t|t) = P_1(t|t) - P_{xy} P_{yy}^{-1} [P_1(t|t) - P_{21}(t|t)] \tag{57}$$

$$P_{xy} = P_1(t|t) - P_{12}(t|t) \tag{58}$$

$$P_{yy} = P_1(t|t) + P_2(t|t) - P_{12}(t|t) - P_{21}(t|t) \tag{59}$$

where $\hat{x}_1(t|t)$ and $\hat{x}_2(t|t)$ are the state estimates from local Kalman filters, $P_1(t|t)$ and $P_2(t|t)$ are covariance matrices of $\hat{x}_1(t|t)$ and $\hat{x}_2(t|t)$, respectively, $P_{12}(t|t)$ and $P_{21}(t|t)$ are cross-covariance between $\hat{x}_1(t|t)$ and $\hat{x}_2(t|t)$, $\hat{x}(t|t)$ and $P(t|t)$ are the fused state estimate and the corresponding covariance. The cross-covariance is very important to the accuracy of the fused state estimate, but its computation is very complex. There have been a plenty of work on the efficient algorithms and performance evaluation for the track-to-track fusion (Chang et al., 1998; Chang et al., 1997; Saha and Chang, 1998; Saha, 1996) [18,17,46,45]. In spite of lots of

3.2.2 State-Vector Assimilation Fusion (SVAF)

SVAF, without making use of the statistical properties of the data, is not an optimal method for data combination. It is simply based on the assimilation of local state estimates. Both the standard Kalman filter and information filter can be used in SVAF. The formulation of SVAF given below is based on information filter (Grime and Durrant-Whyte, 1994):

Local estimate:

$$\hat{z}_j(t|t) = \hat{z}_j(t|t-1) + C_j^T(t)R_j^{-1}(t)y_j(t) \tag{60}$$

$$P_j^{-1}(t|t) = P_j^{-1}(t|t-1) + C_j^T(t)R_j^{-1}(t)C_j(t) \tag{61}$$

Assimilation fusion:

$$\hat{z}(t|t) = \hat{z}(t|t-1) + \sum_{j=1}^{N}[\hat{z}_j(t|t) - \hat{z}(t|t-1)] \tag{62}$$

$$P^{-1}(t|t) = P^{-1}(t|t-1) + \sum_{j=1}^{N}[P_j^{-1}(t|t) - P^{-1}(t|t-1)] \tag{63}$$

where $\hat{z}_j(t|t)$ and $P_j^{-1}(t|t)$, $j=1,2,\ldots,N$, are local estimates of the information state vector and information matrix, $\hat{z}(t|t)$ and $P^{-1}(t|t)$ are the fused information state vector and information matrix. SVAF is a simple method with a low computational cost. It is very suitable for decentralised data fusion implementation, but does not provide optimal fusion results.

Most data fusion methods make use of signal statistical properties, such as covariances, of various data sources to obtain more accurate fused data or information. Measurement fusion methods utilise the covariance matrices of measurement noise. In state-vector fusion, the state estimate covariance matrices should be used. However, the state estimates of the same target from different estimators are often dependent on each other, even if the measurement noise of different sensors is independent. Sometimes, the state estimates are inconsistent and the estimate covariance matrices can not reflect real estimation errors. These problems make the design of state-vector fusion more difficult than measurement fusion. In general, measurement fusion methods provide better overall estimation performance, while state-vector fusion methods have a lower computation and communication cost and have the advantages of parallel implementation and fault-tolerance (Chang et al., 1998; Chang et al., 1997; Roecker and McGillem, 1988) [18,17,44]. In addition, a lot of work ought to be done before state-vector fusion methods can be successfully applied in nonlinear target tracking. For example, adaptive techniques may be helpful in this aspect.

3.3 HIERARCHICAL MULTISENSOR DATA FUSION

Centralised and decentralised data fusion architectures have their own advantages and disadvantages. Centralised architectures are less complex, and easy in managing sensor resources, but are inflexible, and vulnerable to system failure, especially the central processor on which severe computational load is imposed. Alternatively, decentralised architectures have some complementary properties. As a compromise, a hierarchical multisensor data fusion architecture, as shown in Figure 1, is developed here for target tracking. There are local fusion nodes, each having a set of sensors and being able to receive globally fused track feedback, and also a global fusion centre which may have its own sensors and can receive local track estimates or local sensor measurements with related information such as covariance matrices. The globally fused track feedback is utilised at local fusion nodes to improve local estimates.

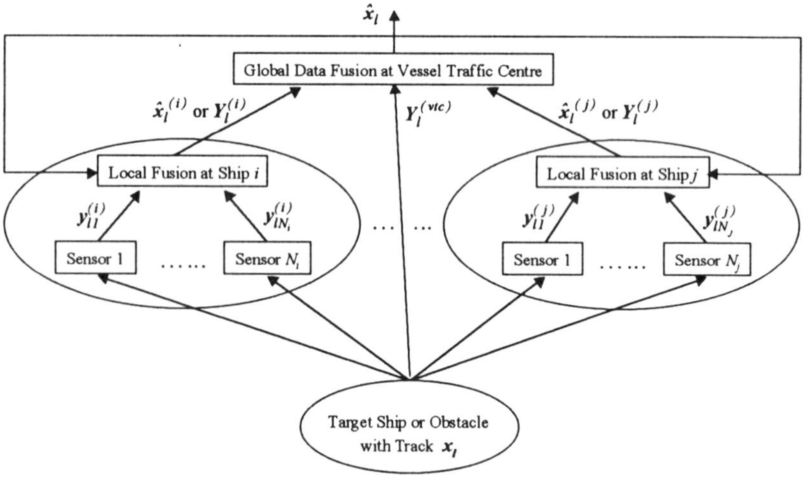

Figure 1. A hierarchical architecture for multisensor data fusion

At the global level, Kalman-filter-based multisensor data fusion algorithms introduced in the previous sections can be applied directly. At local levels, there exist not only local sensor measurements but also globally fused track feedback. As shown in Figure 1, for a particular target with track x_l, at local fusion node j, local sensor measurement vector $Y_l^{(j)}$, previous local track estimate $\hat{x}_l^{(j)}(t-1|t-1)$, and globally fused track estimate $\hat{x}_l(t-d|t-d)$ (d represents a track feedback delay which depends on feedback rate.), and the corresponding covariance matrices are available for updating the local track estimate. If track-to-track fusion methods are directly used at local levels, *i.e.*, local state estimates are fused with the globally fused track feedback, problems such as timing, recounting and cross-correlation have to be carefully considered. To avoid these problems, if $d=1$, the global track feedback is used to replace the previous local track estimate (including the covariance matrix) for track prediction, which is then corrected by the fused local sensor measurement. If $d>1$ (This happens when the feedback rate is less than the measurement rate.), the previous local track estimate will be used for track prediction until a new global track feedback is received.

4. A Simulation Example

A simulation example of nonlinear trajectory estimation, which integrates some of the techniques introduced in the previous sections, is given in this section. The hierarchical multisensor data fusion architecture shown in Figure 1, is utilised to cope with multiple data sources and the NFLL technique is used for unknown process modelling. The target state vector is defined by $x(t)=[r(t),\ \theta(t),\ \dot{r}(t),\ \dot{\theta}(t)]^T$, with $r(t),\ \theta(t),\ \dot{r}(t)$, and $\dot{\theta}(t)$ representing the range, bearing, range velocity, and bearing velocity of the target, respectively. The target trajectory data for training NFLL models are generated by a dynamic process model consisting of four nonlinear state transition equations and sensor models (Gan and Harris, 1999g) [28]. The target dynamics is *a priori* unknown to the estimators. It is modelled directly from the training data by neuro-fuzzy linearisation modelling techniques. For simplicity, we only consider two local nodes, each having one sensor (Sensor 1 and Sensor 2, respectively) to detect the same nonlinear dynamic target and being able to receive global track feedback. At the global fusion centre, measurements from Sensor 1 and Sensor 2 can be received and fused to obtain a global target track estimate. This is the simplest case described by Figure 1.

Sensor 1 and Sensor 2 only measure the range and bearing of the target. The measurement matrices are C_1 and C_2 respectively, with the size of 2×4. The measurement noise covariance matrices are $R_1 = diag\{R_{11}, R_{12}\}$ and $R_2 = diag\{R_{21}, R_{22}\}$. A typical trajectory and its noisy observations by the sensors with different noise levels are shown in Figure 2 (observed by Sensor 1: $R_{11} = 0.25, R_{12} = 0.0025$) and Figure 3 (observed by Sensor 2: $R_{11} = 0.01, R_{12} = 0.0001$), respectively. The units of range and bearing in the simulation are kilo-meter and radian, respectively.

The trajectory estimation result using measurement fusion at the global fusion centre is given in Figure 4, with the solid line and dashed line denoting the real trajectory and estimated trajectory, respectively. The trajectory estimation results at the local node with Sensor 1, which are based on the noisy measurement from Sensor 1 and the globally fused track feedback, are shown in Figures 5-7, with variation of the track feedback rate. The ratios of measurement rate to track feedback rate (*m/f* ratio) in Figures 5 and 6 are 1 and 5, respectively. In Figure 7, no track feedback is utilised and the trajectory estimation is based on the measurement from Sensor 1 only, which naturally obtains the worst performance. From Figures 4-7 we can see that the global measurement fusion achieves the best performance, and that the globally fused track feedback is obviously useful to improve the local state estimates. At the local node with Sensor 2, which has smaller noise, the estimation performance is also improved by sharing the measurement information from Sensor 1 (with larger noise) via global track feedback, although it is not so obvious as at the local node with Sensor 1. However, the major benefit from data fusion at node Sensor 2 is the robustness in the trajectory estimation. Finally, for a numerical comparison, the root mean square errors (RMSEs) of the trajectory estimation results shown in Figures 4-7 are given in Table 1.

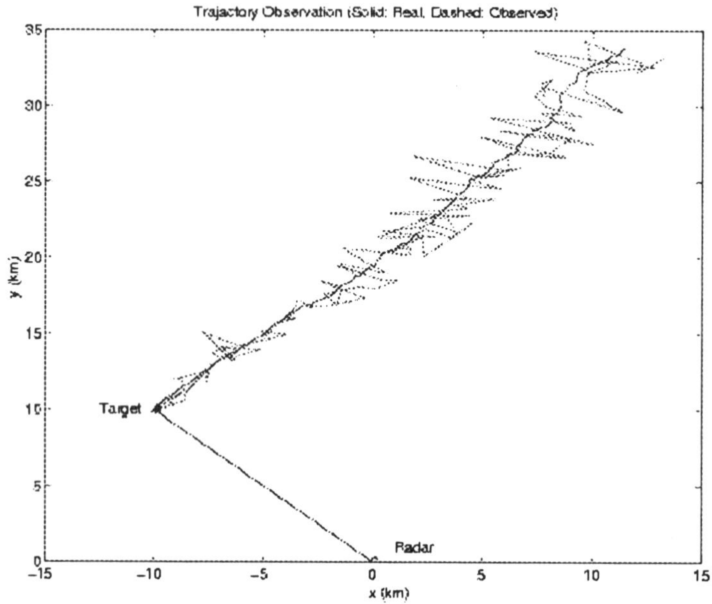

Figure 2. A target trajectory and its observation at Sensor 1

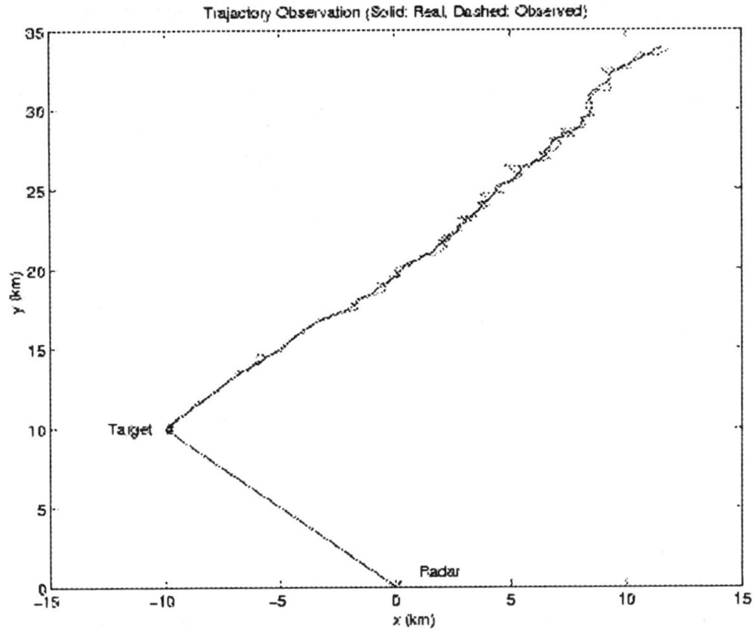

Figure 3. A target trajectory and its observation at Sensor 2

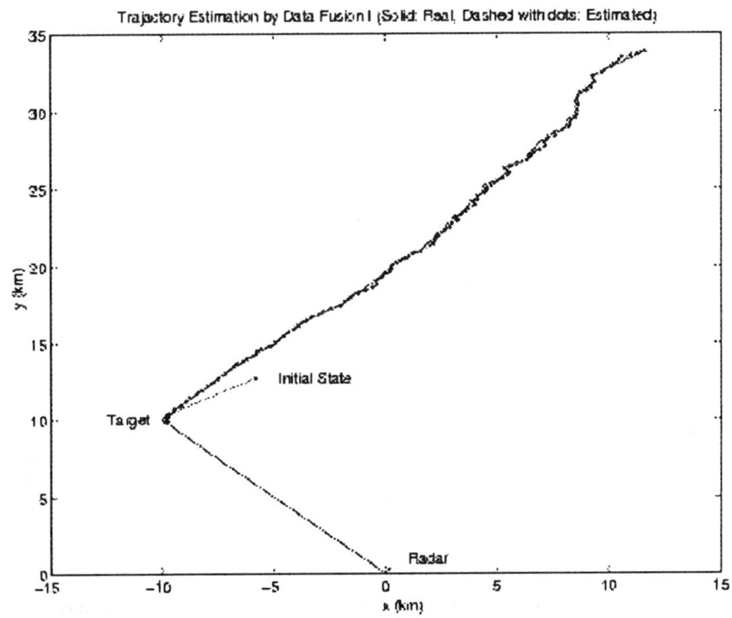

Figure 4. Global trajectory estimation by fusion of data from the two sensors

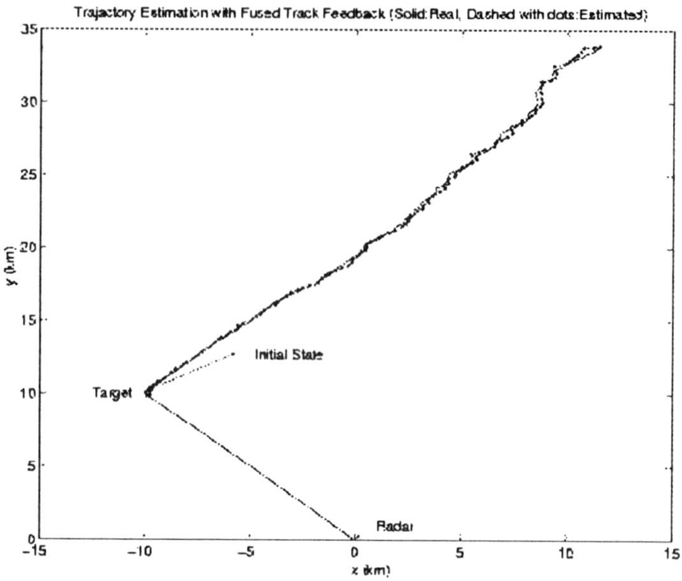

Figure 5. Local trajectory estimation at Sensor 1 with track feedback, m/f ratio=1

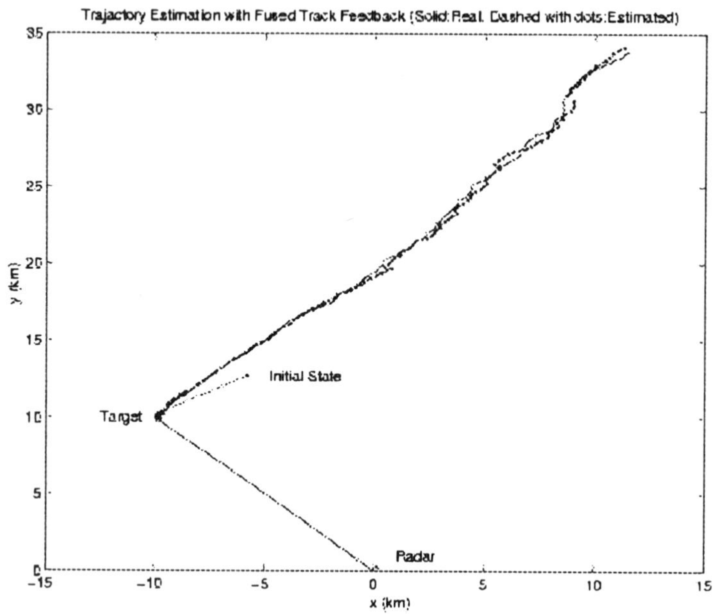

Figure 6. Local trajectory estimation at Sensor 1 with track feedback, m/f ratio=5

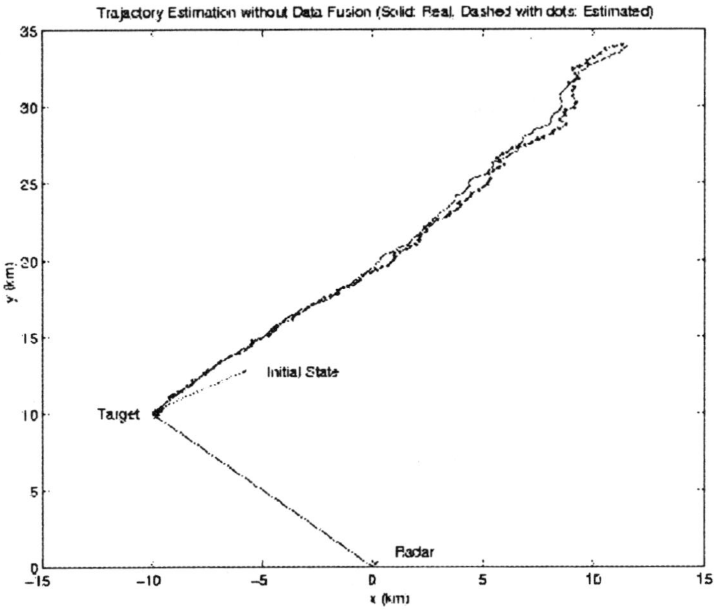

Figure 7. Local trajectory estimation at Sensor 1 without track feedback

TABLE 1. Comparison of RMSEs of trajectory estimation results

Methods	RMSEs
Global measurement fusion	0.1233
Local estimation, m/f ratio=1	0.1732
Local estimation, m/f ratio=5	0.2662
Local estimation without track feedback	0.3588

5. Conclusion

Nonlinearities and uncertainties exist naturally in many practical target tracking tasks and have posed a lot of challenging problems. The well-developed adaptive techniques have been widely applied to tackle these problems. Multisensor data fusion has provided another way to make use of multiple data sources for accurate and robust target tracking. This paper presents neuro-fuzzy linearisation modelling techniques to cope with nonlinear and/or unknown target dynamics and the Kalman-filter based multisensor data fusion techniques to deal with uncertainties.

It has been widely realised that most target tracking tasks are too complex to be tackled by any single tracking technique. In order to obtain accurate and robust tracking performance, it is required to utilise multiple data sources and multiple experts (models) in target tracking. The major challenges here include how to design individual data sources and experts, and how to integrate them in an efficient way. Also, it should be noted that there still exist some limitations of multisensor data fusion methods. For example, basically multisensor data fusion cannot resolve the problems caused by nonlinearities (some multisensor data fusion methods are effective only for linear target tracking) and unpredictable maneuvers. Additionally, theoretical work is in urgent need to guide the design of multisensor data fusion systems.

6. Appendix

Based on the matrix inversion lemma (40), Eq. (42) can be proved as follows:

$$
\begin{aligned}
&(R_2+\Omega^{(1)})^{-1}R_2[R_1+\Omega^{(1)}-\Omega^{(1)}(R_2+\Omega^{(1)})^{-1}\Omega^{(1)}]^{-1} \\
&=(R_2+\Omega^{(1)})^{-1}R_2[R_1+\Omega^{(1)}(R_2+\Omega^{(1)})^{-1}R_2]^{-1} \\
&=(R_2+\Omega^{(1)})^{-1}R_2[(R_1+R_2)-R_2(R_2+\Omega^{(1)})^{-1}R_2]^{-1} \\
&=(R_2+\Omega^{(1)})^{-1}R_2[(R_1+R_2)-R_2(R_2+\Omega^{(1)})^{-1}R_2]^{-1} \\
&\quad\cdot R_2(R_2+\Omega^{(1)})^{-1}(R_2+\Omega^{(1)})R_2^{-1} \\
&=\{[(R_2+\Omega^{(1)})-R_2(R_1+R_2)^{-1}R_2]^{-1}-(R_2+\Omega^{(1)})^{-1}\}\cdot(R_2+\Omega^{(1)})R_2^{-1} \\
&=\{[\Omega^{(1)}+R_1(R_1+R_2)^{-1}R_2]^{-1}-(R_2+\Omega^{(1)})^{-1}\}\cdot(R_2+\Omega^{(1)})R_2^{-1} \\
&=[\Omega^{(1)}+R_1(R_1+R_2)^{-1}R_2]^{-1}(R_2+\Omega^{(1)})R_2^{-1}-R_2^{-1} \\
&=[\Omega^{(1)}+R_1(R_1+R_2)^{-1}R_2]^{-1}\{(R_2+\Omega^{(1)})-[\Omega^{(1)}+R_1(R_1+R_2)^{-1}R_2]\}R_2^{-1} \\
&=[\Omega^{(1)}+R_1(R_1+R_2)^{-1}R_2]^{-1}R_2(R_1+R_2)^{-1}
\end{aligned}
\tag{A.1}
$$

Making use of the above result, we can prove (43) as follows:

$$
\begin{aligned}
&(R_2+\Omega^{(1)})^{-1}-(R_2+\Omega^{(1)})^{-1}R_2[R_1+\Omega^{(1)}-\Omega^{(1)}(R_2+\Omega^{(1)})^{-1}\Omega^{(1)}]^{-1} \\
&\cdot\Omega^{(1)}(R_2+\Omega^{(1)})^{-1} \\
&=(R_2+\Omega^{(1)})^{-1}-[\Omega^{(1)}+R_1(R_1+R_2)^{-1}R_2]^{-1}R_2(R_1+R_2)^{-1} \\
&\quad\cdot\Omega^{(1)}(R_2+\Omega^{(1)})^{-1} \\
&=[\Omega^{(1)}+R_1(R_1+R_2)^{-1}R_2]^{-1}\cdot\{[\Omega^{(1)}+R_1(R_1+R_2)^{-1}R_2] \\
&\quad-R_2(R_1+R_2)^{-1}\Omega^{(1)}\}\cdot(R_2+\Omega^{(1)})^{-1} \\
&=[\Omega^{(1)}+R_1(R_1+R_2)^{-1}R_2]^{-1}\cdot[R_1(R_1+R_2)^{-1}R_2+R_1(R_1+R_2)^{-1}\Omega^{(1)}] \\
&\quad\cdot(R_2+\Omega^{(1)})^{-1} \\
&=[\Omega^{(1)}+R_1(R_1+R_2)^{-1}R_2]^{-1}\cdot R_1(R_1+R_2)^{-1}\cdot(R_2+\Omega^{(1)})(R_2+\Omega^{(1)})^{-1} \\
&=[\Omega^{(1)}+R_1(R_1+R_2)^{-1}R_2]^{-1}R_1(R_1+R_2)^{-1}
\end{aligned}
\tag{A.2}
$$

7. References

1. Amoozegar, F. (1998), Neural-network-based target tracking state-of-the-art survey. *Optical Engineering*, 37(3): 836-846.
2. Balakrishnan, A. V. (1987). *Kalman Filtering Theory*. Optimization Software, Inc., Publication Division.
3. Bar-Shalom, Y. (1981). On the track-to-track correlation problem. *IEEE Trans. on Automatic Control*, 26(2): 571-572.
4. Bar-Shalom, Y. (1992). *Multitarget-Multisensor Tracking: Applications and Advances*. Norwood, MA: Artech House.
5. Bar-Shalom, Y. and Campo, L. (1986). The effect of the common process noise on the two-sensor fused-track covariance. *IEEE Trans. on Aerospace and Electronic Systems*, 22(6): 803-805.
6. Bar-Shalom, Y., Chang, K. C., and Blom, H. A. P. (1989). Tracking a maneuvering target using input estimation versus the interacting multiple model algorithm. *IEEE Transactions on Aerospace and Electronic System*, 25(2): 296-300.
7. Bar-Shalom, Y. and Fortmann, T. E. (1988). *Tracking and Data Association*. New York: Academic Press.
8. Bar-Shalom, Y. and Li, X. R. (1993). *Estimation and Tracking: Principles, Techniques, and Software*. Norwood, MA: Artech House.
9. Bar-Shalom, Y. and Li, X. R. (1995). *Multitarget-Multisensor Tracking: Principles and Techniques*. Storrs, CT: YBS Publishing.
10. Blair, W. D. and Bar-Shalom, Y. (1996). Tracking maneuvering targets with multiple sensors: Does more data always mean better estimates? *IEEE Trans. on Aerospace and Electronic Systems*, 32(1): 450-455.
11. Blom, H. A. P. and Bar-Shalom, Y. (1988). The interacting multiple model algorithm for systems with Markovian switching coefficients. *IEEE Trans. on Automatic Control*, 33(8): 780-783.
12. Bogler, P. L. (1987). Tracking a maneuvering target using input estimation. *IEEE Trans. on Aerospace and Electronic Systems*, 23(3): 298-310.
13. Chan, Y. T., Hu, A. G., and Plant, J. B. (1979). A Kalman filter based tracking scheme with input estimation. *IEEE Trans. on Aerospace and Electronic Systems*, pages 237-244.
14. Chan, Y. T., Plant, J. B., and Bottomley, J. R. T. (1982). A Kalman tracker with a simple input estimator. *IEEE Trans. on Aerospace and Electronic Systems*, pages 237-244.
15. Chang, K. C. and Bar-Shalom, Y. (1989). Distributed adaptive estimation with probabilistic data association. *Automatica*, 25(3): 359-369.
16. Chang, K. C., Chong, C. Y., and Bar-Shalom, Y. (1986). Joint probabilistic data association in distributed sensor networks. *IEEE Trans. on Automatic Control*, 31(10): 889-897.
17. Chang, K. C., Saha, R. K., and Bar-Shalom, Y. (1997). On optimal track-to-track fusion. *IEEE Trans. on Aerospace and Electronic Systems*, 33(4): 1271-1276.
18. Chang, K. C., Tian, Z., and Saha, R. K. (1998). Performance evaluation of track fusion with information filter. In *Proc. of Int. Conf. on Multisource-Multisensor Information Fusion*, pages 648-655, USA.
19. Chui, C. K. and Chen, G. (1987). *Kalman Filtering: with Real-Time Applications*. Springer-Verlag.
20. Doyle, R. S. and Harris, C. J. (1996). Multi-sensor data fusion for helicopter guidance using neuro-fuzzy estimation algorithms. *The Royal Aeronautical Society Journal*, pages 241-251.
21. Fossen, T. I. (1994). *Guidance and Control of Ocean Vehicles*. John Wiley & Sons.
22. Gan, Q. and Harris, C. J. (1999a). Comparison of two measurement fusion methods for Kalman-filter-based multisensor data fusion. *Technical report*, Dept. of Electronics and Computer Science, University of Southampton.
23. Gan, Q. and Harris, C. J. (1999b). Fuzzy local linearisation and local basis function expansion in nonlinear system modelling. *IEEE Trans. on Systems, Man, and Cybernetics*, 29: Part B(4): 559-565.
24. Gan, Q. and Harris, C. J. (1999c). A learning scheme combining EM and MASMOD algorithms for nonlinear system modelling. *Technical report*, Dept. of Electronics and Computer Science, University of Southampton.
25. Gan, Q. and Harris, C. J. (1999d). Linearisation and state estimation of unknown discrete-time nonlinear dynamic systems using recurrent neurofuzzy networks. *IEEE Trans. on Systems, Man, and Cybernetics*, 29: Part B(6): 802-817.
26. Gan, Q. and Harris, C. J. (1999e). Multisensor data fusion using Kalman filters based on linearised process models. In *Proc. of International Conference on Data Fusion, EuroFusion99*, pages 105-112, UK.

27. Gan, Q. and Harris, C. J. (1999f). Neurofuzzy local linearisation modelling for nonlinear state estimation. In Mohammadian, M., editor, *New Frontier in Computational Intelligence and its Applications*, pages 209-217. IOS Press.
28. Gan, Q. and Harris, C. J. (1999g). Neurofuzzy state estimators using a modified ASMOD and Kalman filter algorithm. In *Proc. of the International Conference on Computational Intelligence for Modelling Control and Automation*, volume 1, pages 214-219, Vienna, Austria.
29. Grime, S. and Durrant-Whyte, H. (1994). Data fusion in decentralised sensor networks. *Control Engineering Practice*, 2(5): 849-863.
30. Hall, D. L. and Llinas, J. (1997). Introduction to multisensor data fusion. *Proceedings of the IEEE*, 85(1): 6-23.
31. Harris, C. J., Bailey, A., and Dodd, T. J. (1998). Multi-sensor data fusion in defence and aerospace. *The Aeronautical Journal*, 102(1015): 229-244.
32. Harris, C. J., Brown, M., Bossley, K. M., Mills, D. J., and Feng, M. (1996). Advances in neurofuzzy algorithms for real-time modeling and control. *Engineering Applications of Artificial Intelligence*, 9(1): 1-16.
33. Harris, C. J., Hong, X., and Gan, Q. (2000). Neurofuzzy state estimators. In N. K. Sinha and Gupta, M. M., editors, *Soft Computing and Intelligent Systems: Theory and Applications*, pages 377-402. Academic Press.
34. Harris, C. J., Wu, Z. Q., Bossley, K. M., and Brown, M. (1997). Intelligent neurofuzzy estimators and multisensor data fusion. In Tzafestas, S. G., editor, *Methods and Applications of Intelligent Control*, pages 283-303. Kluwer Academic.
35. Lerro, D. and Bar-Shalom, Y. (1993). Interacting multiple model tracking with target amplitude feature. *IEEE Trans. on Aerospace and Electronic Systems*, 29(2): 494-508.
36. Li, X. R. and Bar-Shalom, Y. (1993a). Design of an interacting multiple model algorithm for air traffic control tracking. *IEEE Trans. on Control Systems Technology*, 1(3): 186-194.
37. Li, X. R. and Bar-Shalom, Y. (1993b). Performance prediction of the interacting multiple model algorithm. *IEEE Transactions on Aerospace and Electronic Systems*, 29(3): 755-771.
38. Manyika, J. and Durrant-Whyte, H. (1994). *Data Fusion and Sensor Management: A Decentralized Information-Theoretic Approach*. New York: Ellis Horwood.
39. Marino, R. and Tomei, P. (1992). Global adaptive observers for nonlinear systems via filtered transformations. *IEEE Trans. on Automatic Control*, 37(8): 1239-1245.
40. Marino, R. and Tomei, P. (1995). Adaptive observers with arbitrary exponential rate of convergence for nonlinear systems. *IEEE Trans. on Automatic Control*, 40(7): 1300-1304.
41. Mazor, E., Averbuch, A., BarShalom, Y., and Dayan, J. (1998). Interacting multiple model methods in target tracking: A survey. *IEEE Transactions on Aerospace and Electronic Systems*, 34(1): 103-123.
42. McGinnity, S. and Irwin, G. W. (1996). Nonlinear state estimation using fuzzy local linear models. *International Journal of Systems Science*, 28(7): 643-656.
43. McGinnity, S. and Irwin, G. W. (1998). Fuzzy logic approach to maneuvering target tracking. *IEE Proc. - Radar, Sonar and Navigation*, 145(6): 337-341.
44. Roecker, J. A. and McGillem, C. D. (1988). Comparison of two-sensor tracking methods based on state vector fusion and measurement fusion. *IEEE Trans. on Aerospace and Electronic Systems*, 24(4): 447-449.
45. Saha, R. K. (1996). Track-to-track fusion with dissimilar sensors. *IEEE Trans. on Aerospace and Electronic Systems*, 32(3): 1021-1029.
46. Saha, R. K. and Chang, K. C. (1998). An efficient algorithm for multisensor track fusion. *IEEE Trans. on Aerospace and Electronic Systems*, 34(1): 200-210.
47. Sengupta, D. and Iltis, R. (1989). Neural solution to the multi-target tracking data association problem. *IEEE Trans. on Aerospace and Electronic Systems*, (1).
48. Shams, S. (1996). Neural networks optimization for multi-target multi-sensor passive tracking. *Proc. of the IEEE*, 84(10).
49. Wu, Z.-Q. and Harris, C. J. (1997). A neurofuzzy network structure for modelling and state estimation of unknown nonlinear systems. *International Journal of Systems Science*, 28: 335-345.

IDENTIFY INFORMATION FUSION THROUGH EVIDENTIAL REASONING

É. BOSSÉ
Defense Research Establishment Valcartier
2459 Pie XI Blvd. North
P.O. 8800, Val-Bélair, G3J 1X5
Canada

1. Introduction

Modern military operations take place within an enormously complex environment to accomplish missions across the spectrum of conflict from humanitarian assistance to high intensity combat. In the past several decades, the battlespace has expanded enormously in the face of increasingly potent and accurate weapons capable of being launched at progressively further ranges from their targets. In response to these challenges, powerful new sensors have been deployed at sea, ashore and in space, while the capacity of communications systems has multiplied to make available huge volumes of data and information to commanders and their staffs. In short, technological improvements in mobility, range, lethality and information acquisition continue to compress time and space, forcing higher operating tempos and creating greater demands on getting situation awareness for better decision-making.

Situation awareness is for commanders/deciders to understand the current situation and to fully grasp all of the implications associated with each of several possible courses of action. Achieving this is an inherently complex activity. As a process, it begins by rendering, by both automated and manual methods, huge amounts of raw data into information and displaying it by various means for the decider's use. A key element is data/information fusion, which offers the possibility of greatly improved situational awareness, particularly through exploitation of various sensors/sources such as sonar, satellites (both military and commercial), Electro Optic (EO) sensors, or radars, as well as non-organic information from off-board sensors and intelligence. There is clearly a need for improved data and information fusion capability to make best use of available information to reduce the workload and speed up the response of the command team.

This talk presents a multisource data fusion system with the emphasis on the identity information fusion process. With the increasing use of the remote sources and the increasing use of information that has already been fused, there is a requirement to define an identity data fusion function that can handle organic and non-organic, local and remote types of information characterized by different accuracy and timeliness. The discussion is specifically on the fusion of identity information emanating from different sources through the use of statistical analysis rooted in the Dempster-Shafer theory of evidence. Two identity information fusion architectures are discussed here. The first is concerned with the fusion of identity declarations where the sources are expected to provide only

A. K. Hyder et al. (eds.), Multisensor Fusion, 457–480.

useful and complete results such as an identity declaration. The second is concerned with the fusion of attribute information using a modified version of the Dempster-Shafer evidential combination algorithm.

The first identity information fusion architecture is concerned with the fusion of identity declarations which can be represented in a hierarchical manner. The Dempster-Shafer evidential theory is used with an emphasis on the Shafer-Logan algorithm for fast computation when the information is hierarchically structured. The second architecture is concerned with the fusion of attribute information to declare an identity using a modified version of the Dempster-Shafer evidential combination algorithm. Here the sources of information or the sensors can only provide an incomplete set of data or attributes to identify objects. Two examples are presented to illustrate the use of both architectures for fusing identity information.

2. The Fusion Of Identity Information

2.1 THE KIND OF IDENTITY INFORMATION

In a military environment (e.g. maritime), various surveillance systems, electronic intelligence and human observations are examples of information sources available to the commander. Output from these sources is partitioned according to the type of information they provide; output data may be characterized as either positional or identity information. *Positional information* represents the dynamic parameters describing the movement associated with an object (contact). This generally includes position, velocity and acceleration. *Identity information* can be defined as declarations, propositions or statements that contribute to establish the identity of an object. Equivalently, identity information may be seen as information from various sources that help in distinguishing one object from another. Possible values for identity information can span the range from *sensor signals*, to *attributes*, to *identity declarations*.

The sensor signals represent some characteristics of the energy sensed. Attributes such as size, shape, degree of symmetry, emitter type, etc. are inferred from these characteristics. Identity declarations specify the detected object; in the Canadian Navy, for example, they can consist of a general classification of which the observed object is a member (surface combatant), a specific type of ship (frigate), a specific class (City Class) or a unique identity (Ville de Québec). Therefore surface combatant, frigate, City Class and Ville de Québec are all examples of identity declarations. Identity declarations can also include information concerning the threat designation of an object: pending, unknown, assumed friend, suspect, friend, neutral or hostile.

In addition, the identity information can be characterized by being local, remote, organic and non-organic. The notion of local and remote are self-explaining but the organic and non-organic require to be defined. Data collected and managed by assets under Commanding Officer's (CO) direct control can be defined or characterized as organic information. Organic information must be sufficiently timely and accurate to be used in real-time, responsive systems. Consequently, it can be used to produce a local tactical picture related to the situation. Note that remote sources can be organic and local sources can be non-organic. Here are examples of organic information: radars, datalink (e.g. Link-11), ESM, weather, ...etc.

Data collected by agents not under CO direct control, is referred as non-organic information. Non-organic information is less timely, reduced in accuracy, differently structured and has differing identification confidence levels. Transient information such as intelligence reports or Participating Unit information is non-organic as means to support the various processes providing the commander at sea to gain a level of situational awareness.

Another source of data is the human contribution to the identification process through a Human Computer Interface (HCI). Human can participate actively by providing inputs, based on his experience, his training and his awareness of the situation within the identification process. The precise nature of this human-computer interaction remains to be determined.

A Priori Knowledge contains static information as means to support the various processes providing the decider to gain a level of situational awareness. A Priori Knowledge is used to refine identity information. Here are some examples of a priori knowledge source: social and political, geographical, platforms characteristics, mission guidelines, weapons characteristics, corridor and flight paths, EM characteristics, lethality, emitter characteristics, doctrines, ...etc.

2.2 IDENTITY INFORMATION FUSION ARCHITECTURE

One of the key issues in developing a MultiSource Data Fusion (MSDF) system is the question of where in the data flow to actually combine or fuse the data. The MSDF architecture is an important issue since the benefits are different depending on the way the sensor or other source data are combined. There are four broad alternatives to fusing positional information: at the signal level, at the contact (plot) level, at the track level and finally a hybrid approach which allows fusion of either contact or track data. The benefits of theses architectures for positional fusion have been extensively analyzed in [5]. For identity fusion, there are several types of architecture which can be used: signal level fusion, feature level fusion and decision level fusion.

There is no universal architecture, which is applicable to all situations or applications. The selection of the MSDF architecture type should be aimed at optimizing the target detection, tracking and identification performance required for a specific platform given its missions. However, the selection is also constrained by the technological capabilities (both hardware and software). It depends on the quality of the sensors being fused, the availability of computer processing power, the bandwidth of the available data transmission paths, and the degree to which operator intervention is required or desired.

2.3 THE DEMPSTER-SHAFER THEORY

If *a priori* probability distributions and conditional probability matrices are available and if probabilities can be distributed to single elements, then the Bayesian approach should be used to fuse uncertain information. However, the Bayesian approach suffers from major deficiencies in a hierarchical context, when fully specified likelihoods are not available. Pearl's rules [3] can be applied but may not always lead to an appropriate estimation of the posterior probabilities associated with certain nodes of a hierarchy. Other problems associated with the Bayesian approach are the coding of ignorance and the strict requirements on the belief of a hypothesis and its negation.

The Dempster-Shafer technique [6] does not require prior probabilities nor does it need to know the capability of each source. Evidence is not committed to any specific event or set of events until evidence is gained. Also, it is not required that belief not committed to a given proposition should be committed to its negation, nor that belief committed to a given proposition should be committed more specifically. The Dempster-Shafer technique has the capability of expressing ignorance explicitly. The technique actually focuses on the probability of a collection of points belonging to the sample space, whereas the classical probability theory is interested in the probability of the individual points.

In the Dempster-Shafer evidential theory, the terminology is slightly different from that used in Probability theory. The *frame of discernment* Θ is defined as an exhaustive set of mutually exclusive events or propositions of a particular experiment. We denote 2^{Θ} the set of all subsets of Θ. The elements of 2^{Θ}, or equivalently the subsets of Θ, are the class of general propositions and include the empty set $\varnothing$, which corresponds to a proposition that is known to be false, and the whole set Θ, which corresponds to a proposition that is known to be true. We will assume throughout our discussion that the frame of discernment is finite.

Let Let A be a subset of Θ. The evidential theory differentiates between the measure of belief committed exactly to A and the total belief committed to A. The first is characterized by the basic probability assignment and the latter by the belief function. A function m is a *basic probability assignment* if it assigns a number m(A) to each subset $A \in 2^{\Theta}$, in such a way that: $m(\varnothing) = 0$, and

$$\sum_{A \subseteq \Theta} m(A) = 1.$$

The quantity m(A) is called A's *basic probability number* and represents the exact belief in the proposition depicted by A. A function Bel is a *belief function* if it assigns a number Bel(A) to each subset $A \in 2^{\Theta}$ in such a way that: $Bel(\varnothing) = 0$, $Bel(\Theta) = 1$ and for every positive integer n and every collection, $A_1, A_2, \ldots, A_n$ of subsets of Θ,

$$Bel(A_1 \cup \ldots \cup A_n) \geq \sum_i Bel(A_i) - \sum_{i<j} Bel(A_i \cap A_j) + \ldots + (-1)^{n+1} Bel(A_1 \cap \ldots \cap A_n)$$

$$\geq \sum_{\substack{I \subset \{1 \ldots n\} \\ I \neq \varnothing}} (-1)^{|I|+1} Bel\Big(\bigcap_{i \in I} A_i\Big)..$$

The quantity Bel(A) is called the *degree of belief* about proposition A. A belief function assigns to each subset of Θ a measure of the total belief in the proposition represented by the subset. Here |I| stands for the cardinality of the set I. The simplest belief function is the one where $Bel(\Theta) = 1$ with $Bel(A) = 0$ for all $A \neq \Theta$; this function is called the *vacuous belief function.* A belief function can be obtained from the basic probability assignment:

$$Bel(A) = \sum_{B \subseteq A} m(B), \text{ for all } A \subseteq \Theta.$$

Another function conveys the same information as the belief function; it is called the plausibility function. Let Bel be a belief function over a frame Θ a function Pl is a *plausibility function* if it assigns a number Pl(A) to each subset $A \in 2^{\Theta}$ Θ in such a way that: $Pl(A) = 1 - Bel(\overline{A})$ or

$$Pl(A) = \sum_{A \cap B \neq \varnothing} m(B), \text{ for all } A \subseteq \Theta$$

The quantity Pl(A) is known as the *degree of plausibility* of A and expresses the extent to which one fails to doubt A. Pl(A) will be zero when the evidence refutes A, and one when there is no evidence against A. The degree of belief and degree of plausibility summarize the impact of the evidence on a particular proposition A in the following manner: the first shows how well the evidence supports proposition A and the second reports on how well its negation is supported. This information can be expressed in the form of an interval called *evidential interval* whose length Pl(A) - Bel(A), can be referred to as the *ignorance* remaining about subset A. If ignorance about subset A is zero, then the Dempster-Shafer process is identical to the Bayesian approach such that $Bel(A) = P(A) = Pl(A)$; if ignorance about A is equal to one, then no knowledge at all is available concerning subset A. More generally, the relationship between the Bayesian and Evidential theories can be described by the following equation: $Bel(A) \leq P(A) \leq Pl(A)$. Thus, the degree of belief and degree of plausibility on a hypothesis can be seen as the lower bound and upper bound on the probability of that hypothesis.

The evidential theory proposes a combination rule, called Dempster's rule of combination, which synthesizes basic probability assignments and yields a new basic probability assignment representing the fused information. The combination rule is known as the *orthogonal sum* and is denoted by $\oplus$. Basically, this rule corresponds to the pooling of evidence: if the belief functions being combined are based on entirely distinct bodies of evidence and the set Θ discerns the relevant interaction between those bodies of evidence, then the orthogonal sum gives degrees of belief that are appropriate on the basis of the combined evidence.

Let m_1 and m_2 be basic probability assignments, on the same frame of discernment Θ, for belief functions Bel_1 and Bel_2 respectively. If Bel_1 focal elements are $B_1, \ldots, B_k$, and Bel_2's focal elements are $C_1, \ldots, C_n$, the total portion of belief exactly committed to A ($A \neq \varnothing$) is given by the orthogonal sum $m = m_1 \oplus m_2$:

$$m(A) = K \sum_{\substack{i,j \\ B_i \cap C_j = A}} m_1(B_i) \cdot m_2(C_j),$$

where $1/K = 1 - \sum_{\substack{i,j \\ B_i \cap C_j = \varnothing}} m_1(B_i) \cdot m_2(C_j) = \sum_{\substack{i,j \\ B_i \cap C_j \neq \varnothing}} m_1(B_i) \cdot m_2(C_j)$.

The scalar K is a normalizing constant. It normalizes to one the total portion of belief exactly committed to A. If m_i are a collection of basic probability assignments with focal elements A_i, B_j, C_k, D_l, ... respectively, then

$$m(A) = K \sum_{\substack{A_i, B_j, C_k, D_m, \dots \\ A_i \cap B_j \cap C_k \cap D_m \cap \dots = A}} [m_1(A_i)\ m_2(B_j)\ m_3(C_k)\ m_4(D_m) \cdots]$$

$$m(\varnothing) = 0$$

$$1/K = 1 - \sum_{A_i \cap B_j \cap C_k \cap D_m \cap \dots = \varnothing} [m_1(A_i)\ m_2(B_j)\ m_3(C_k)\ m_4(D_m) \cdots]$$

3. A Multi-Source Data Fusion System

The Multi-Source Data Fusion (MSDF) system [4] deals with the alignment, association or correlation, filtering and combination of data and information obtained from multiple sources to achieve refined state and identity estimation for each target present in the operational environment. As shown in Figure 1, an MSDF system can be broken into several pieces: input data preparation, system track selection, data alignment, data association, cluster management, kinematics information fusion, identity information fusion, track management, configuration monitoring and control, and internal system track data store (ISTDS). These pieces operate in a well-orchestrated manner to map the source data onto the ISTDS.

In this talk, the emphasis is on the target identity information fusion process that is an important process to achieve situation awareness. Figure 2 details the identity information fusion process involving the items indicated by a check mark in Figure 1. The first task of the identity information fusion process (Figure 2) is to compare the attribute information and identity declarations obtained from various sources with a Platform Data Base (PDB) ideally containing all the possible identity values that the potential target may take. Each record of this database contains information related to the measured sensor attributes and identity declarations.

The second task of the identity information fusion function is the data association process. This process determines to which MSDF track the received sensor information belongs. This is a critical process of any fusion function. Following the conclusions of the data association process, the identity proposition construction should be completed with the attributes derived from the positional fusion (speed, acceleration, etc.).

The fusion rules are then applied to fuse the MSDF target track identity propositions list with the input propositions. Decision rules are then applied for pruning the "unessential" propositions, selecting the "best" identity propositions. A set of identity proposition is then stored in the store ISTD. The fusion and decision rules for both the attribute identity information and the identity declaration information fusion will be discussed in details in Section 4 and 5 respectively.

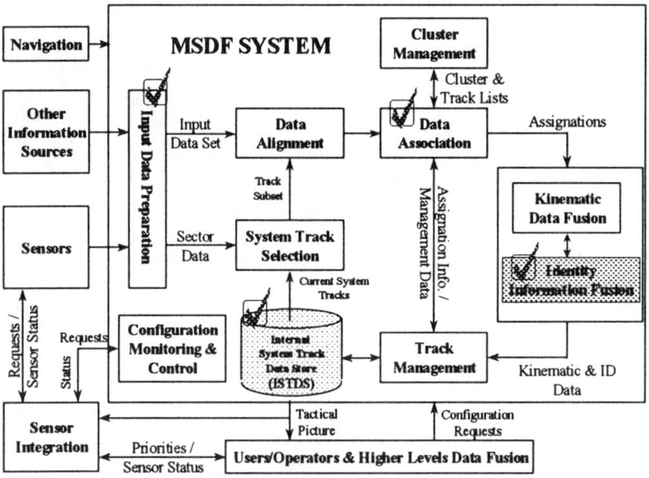

Figure 1. A Multi-Source Data Fusion System (MSDF)

The last step in the identity information fusion process is the refinement of the identity propositions.Refinement of the identity propositions can be achieved by reasoning in terms of groups or organizations of objects. This is done in the perception refinement process of the higher levels of data fusion [2] and illustrated as part of the integration process of Figure 2.

3.1 INPUT DATA PREPARATION/PROPOSITION CONSTRUCTION

The input data preparation and proposition construction use data from the sources on a continuous basis. This data is given in the format of the sources. It can be sensor signals, attribute information or identity declarations. When sensor signals are presented at the input, the proposition construction infers the attributes and transforms the attribute information into a subset of the Platform Data Base. When more than one piece of data is produced by a given source at a given time, these pieces of data must be combined to create one single identity proposition for that source and the Basic Probability Assignment (BPA), m, evaluated for each identity proposition. We assume that all sources capable of providing identity declarations will do so by attaching to each declaration a quantitative measure of uncertainty.

The attribute data is statistically different from the positional data. The attributes usually do not belong to a continuous space. In most cases, they pertain to a space of discrete values to which the concept of standard deviation is not applicable. A declaration, from a radar for example, can be of the form "it is a small target". The word small can be visualized as an indication of the Radar Cross Section (RCS). In spite of the capability of some radars to measure accurately the RCS, in general the associated declaration will be considered as a "fuzzy" statement where the word small refers to an order of magnitude. Table I shows an example of the RCS attribute.

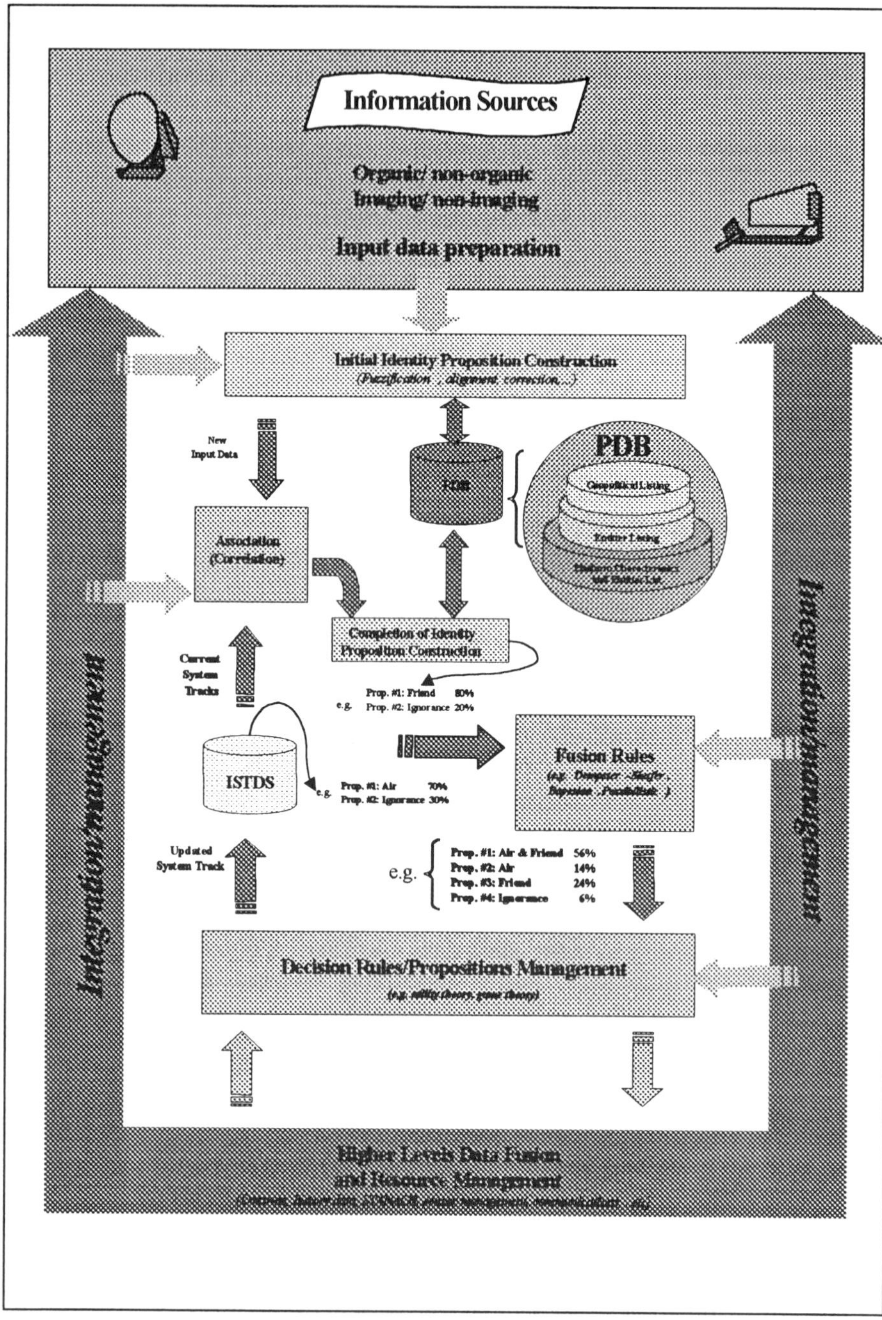

Figure 2. An identity information fusion process

The application of the evidential formalism implies a complete and exhaustive frame of discernment Θ (or the PDB), which implies that all the fused sensors provide data under the form of a complete list of subsets of PDB and implies that the BPA of these subsets be known. In practical implementations, it is possible to build a complete and exhaustive PDB but it is usually impossible to claim that the sources will provide data in the right format, that they are complete, and even that the BPA will be known. As in any estimation problem one has to rely on an incomplete solution and on an estimated confidence level. To apply the data fusion rules the attribute information proposition construction processes attributes as fuzzy logic variables and uses a fuzzification mechanism to build the basic probability assignment, m.

TABLE 1. Target size identity information based on the RCS measurement

Radar RCS Attribute	Interval, AIR	interval, SURFACE
very large	$RCS \geq 5.0\ m^2$	$RCS \geq 50.0\ m^2$
large	$2.0\ m^2 \leq RCS < 5.0\ m^2$	$20.0\ m^2 \leq RCS < 50.0\ m^2$
medium	$0.5\ m^2 \leq RCS < 2.0\ m^2$	$5.0\ m^2 \leq RCS < 20.0\ m^2$
small	$0.2\ m^2 \leq RCS < 0.5\ m^2$	$2.0\ m^2 \leq RCS < 5.0\ m^2$
very small	$RCS < 0.2\ m^2$	$RCS < 2.0\ m^2$

Table 2 below outlines a typical output of the sensor declaration routine applicable to a surveillance radar.

TABLE 2. Example of identity proposition construction from a surveillance radar.

Radar attribute	Measured value	Fuzzyfied attribute	Associated subsets of Θ	m %	m(Θ) %
Range	50 km	air type	{1,2,3,..,19}	60	10
		very large surface type	{20,21,22}	30	
RCS	$10\ m^2$	very large air + very small surface	{7,8,9,26,27}	60	10
Course - bearing	175 °	very very large air + small surface	{15,16, 32,33,34}	15	
		large air + very very small surface	{10,11,12, 35,36}	15	
Speed	60 km/s	helo + very very slow air type + very very fast surface	{8,1,2, 32,20,21}	75	25

The proposition construction routine applies a fuzzification mechanism to appraise the basic probability assignment, m, of the sensor declarations. Fuzzy logic is used because little is known about how probabilistically correct are the measured attributes. The identification problem based on vague information is not a purely measurable mechanism. The computation of the basic probability assignments relies on the declaring sensor and is a function of many parameters. If the declaration is from a radar and concerns the RCS, the basic probability assignments may be a function of the signal-to-noise ratio of the contacts, the knowledge of the aspect from the course value, the radar environmental conditions, the target density, the range of the contacts, etc. When dealing with the fusion of attribute information, one has to consider that most of the time the declarations are subjective and consequently their confidence level are inherently imprecise. It is thus irrelevant to develop rigorous expressions; the fusion of attribute information may be acceptably performed with basic probability assignments which are, for each attribute, phenomenologically in agreement with the expected behavior. The "m" value shown in the fifth and sixth column of Table 2 are dictated by common sense and are indicative of what have been implemented for proof-of-concept purpose.

3.2 PLATFORM DATA BASE STRUCTURE (PDB)

This sub-section presents the structure of a PDB that is used by a identity information fusion process to fuse data from dissimilar sources. The PDB is a listing of all the surface, sub-surface, air and land platforms which can be potentially the identity of the detected and reported target by the sensors or the information sources of a military surveillance system. Here the word surveillance takes a general sense which include also tracking, detecting, monitoring, etc. The platforms are classified according to STANAG 4420 [1] and described in term of the parameters which are measured and seen by the sensor and other sources. Properly speaking, the PDB is composed of three objects since two other secondary listing are linked to some field of the main PDB. There is the linked Geo-Political Listing (GPL) which lists the attribute data that are assessments by the CIS for each country or organization of the world. The other linked list is an Emitter Name Listing (ENL) which includes the name and class of all the radio emitter source that can be detected by the ESM sensor. Table 3 shows an example of a PDB.

TABLE 3. Example of a Platform Data Base (one element)

Platform Data Base (PDB)

Platform index	Platform name	Platform type	Sub-classification
000002	HALIFAX-CPF	Surface military	Frigate

Lethality	Country acronym	Minimum speed	Maximum speed	Acceleration	Maximum altitude
Medium	CANA	0	35	999	0

Platform length	Platform height	Platform width	Gaz turbine / diesel engine	Number of cylinders	RCS side view	RCS front view	RCS top view
130	5	16	3	20	65000	8000	208000

Type of Infrared Signatures	Rotating parts	Number of blades/rotating parts	Number of emitters	List of emitters
B	2	44	9	1 2 3 4 5 6 7 8

Emitter Name List (ENL)

Emitter index	Emitter name	Emitter type and its usage
00456	AN/SPS-49-V1	NAV2DSU (Naval 2D Surveillance radar)

Geo-Political Listing (GPL)

Country acronym	Country name	Allegiance	Number of languages	Language list
INDI	INDIA	NEUTRAL	5	English Indi Gujrati Benghali Punjabi

The PDB contains all the attribute information related to each platform where the identity is declared according to STANAG 4420. The PDB elements must be described according to the sensor suites utilized and fused. If a sensor is capable of providing an attribute such as the image moment [as obtained from an Infra Red (IR) camera], the PDB must provide the range of values for this attribute for each PDB element. Otherwise this attribute can not be fused or some PDB elements will be excluded as potential identifications. The database should include all the types of potential objects that can be seen by the sensors. This includes the various commercial and civil platforms of all types, the beacons, buoy, etc... *In fact the quality and performance of the identification process is much more a function of the completeness of the Platform Database than anything else.*

3.3 DATA ASSOCIATION

The association process is the cornerstone of the fusion function. It performs the task of selecting the various ways the received source information may be combined with the existing tracks. Identity of an object is meaningless unless it can be associated with a position. Therefore, identity declarations provided by various independent sources must be partitioned into groups representing observations belonging to the same observed object. Algorithms using positional and identity information are applied to associate declarations pertaining to the same object but originating from different sources.

Classical tracking systems perform association on the basis of the positional information only by comparing the statistical distance between a contact and a list of tracks. When many contacts have to be associated to many tracks an optimal assignment algorithm has to be used. We assign to the statistical distances a Bayesian probability using the appropriate chi-squared distribution. These Bayesian probabilities are then multiplied by the basic probability assignment associated to the *conflict*. If the conflict is equal to zero, the association is performed on the basis of the positional information only. If the conflict is large, the association probability is strongly reduced even if positional information favor a correlation (this happens in high target density or in splitting track situations).

4. Attribute Identity Information Fusion

The fusion of attribute information is performed when the sources of information or the sensors can only provide an incomplete set of data or attributes to identify objects. This section presents briefly the fusion rules and a complete description of the decision rules to control the number of generated propositions when applying the Dempster-Shafer theory. An example is provided to illustrate the concepts.

4.1 FUSION RULES

One of the first rules for fusion is to extract identity information from the positional fusion (state vector) and combine this information with the propositions generated under the attribute information proposition construction. The fusion of the identity propositions is done using evidential reasoning. Evidential theory proposes a combination rule, called Dempster's rule of combination, which synthesizes basic probability assignments and yields a new basic probability assignment representing the fused information. The combination rule is known as the *orthogonal sum* (see Section 2.3).

One drawback of the Dempster-Shafer Evidential Theory is the long calculation time required by its high computational complexity. Because the combination rule produces basic probability numbers on the subsets of Θ, the calculations are time exponential. In comparison, the Bayesian approach provides probability statements on the elements of Θ. Therefore, if Θ consists of 4 possible points/outcomes, the definition of a probability function on Θ requires the assignment of probability to 4 points, whereas the definition of the Dempster-Shafer basic probability assignment requires the definition of m(A) for $2^4=16$ subsets A of Θ.

Here we discuss the problem of attribute information with non-hierarchical structure. As various evidences are combined over time, Dempster-Shafer combination rules will have a tendency to generate more and more propositions which in turn will have to be combined with new input evidences. Since this problem increases exponentially, the number of retained solutions must be limited. This pruning mechanism is achieved through a set of decision rules for the attribute information fusion problem.

4.2 DECISION RULES

The Dempster-Shafer theory of evidence performs the conventional combination rules but our specific truncation (primary) approach [9] retains the final solution proposition according to the following criteria:

(a) All combined propositions which have a basic probability assignment higher than MaxBe are retained.
(b) All combined propositions which have a basic probability assignment lower than MinBe are eliminated.
(c) If the number of retained propositions in step "a" is smaller than MaxNum, the subroutine will retain, by decreasing basic probability assignment, the propositions consisting of one element (singleton) until MaxNum is reached. If MaxNum is not reached, the subroutine retains, by decreasing basic probability assignment, the propositions consisting of two elements. The process is repeated until MaxNum is reached.

The computation of the basic probability assignment in the ignorance or frame of discernment Θ must be modified. The parameters MinBe and MaxNum have been used to eliminate a portion of the basic probability assignment which belongs to some dropped propositions. The most critical approximation of our proposed Dempster-Shafer algorithm is made here. The basic probability assignment of the dropped propositions will be assigned to the ignorance, such that:

$$m(\Theta) = Maximum\left[(1 - \sum_{i=1}^{MaxNum} m(A_i)) \, , \, MinIgno \right]$$

where the summation in the first term of the square bracket is performed only over the retained propositions. If the BPA of the dropped propositions is too small, m(Θ) will receive a small value "MinIgno". This last condition is to ensure that the ignorance will never approach zero. As explained below this is required in order to remain adaptive. If MinIgno has been selected (a value of 0.01 has been used), the retained propositions have to be renormalized according to:

$$new\_m(A_i) = MinIgno \cdot \left(\sum_{i=1}^{MaxNum} m(A_i) \right)^{-1} \cdot m(A_i)$$

The two approximations associated with the parameters MinBe and MaxNum lead to two great advantages. They control the population of the processed propositions and ensure adaptation in case of contradictions between the new set of sensor declarations and the previous retained solution. In highly conflicting sensor declarations, for example in adverse weather conditions or when counter measures are deployed, this adaptation is necessary. If m(Θ) is allowed to approach zero, the combination rules spoil the chances of new focal propositions to surviving the first two truncation criteria.

4.3 A CASE STUDY: ATTRIBUTE IDENTITY INFORMATION FUSION

The following example is provided to illustrate the case of attribute information fusion.

4.3.1 *The Test Scenario*

There are 4 aircraft targets:

(a) a very very fast friendly aircraft (MSDF-1, USAF SR-71) with a known friendly emitter and an IFF mode-4 response;
(b) a hostile very fast aircraft (MSDF-2, TU-27 Backfire) which is somewhat slower and has a Closest Point of Approach (CPA) larger than the three others;

(c) a very very fast hostile aircraft (MSDF-3, MIG-31 Fulcrum); and
(d) a very very fast commercial aircraft (target 4, France Concorde) with commercial emitter and IFF in mode-3 and C response.

The test scenario is shown in Figure 3 and Table 4 presents some parameters of the TU-27 track. This target was specially used to challenge our Dempster-Shafer approach. The four aircraft are members of the PDB and the six mentioned emitters are members of the emitter database. The emitters, platforms and allegiance have been arbitrarily given only for the purpose of this study and are not related to any current political or military situation.

During this simulation, the IFF is interrogating, in mode 3 from 00.00.20 to 00.00.40 in sector 260 to 280 degrees and in mode 4 from 00.01.20 to 00.01.40 in all sectors. ESM reports a new message each time a bearing variation of few degrees is detected for the same electromagnetic emission. Except for the TU-27, the three aircraft approach CPF from about 120 DM with nearly the same velocity. The characteristics are chosen such that these targets will eventually satisfy the automatic engagement mode criteria. The hostile aircraft TU-27 is somewhat slower than the others, but all its dynamic parameters satisfy automatic engagement mode criteria.

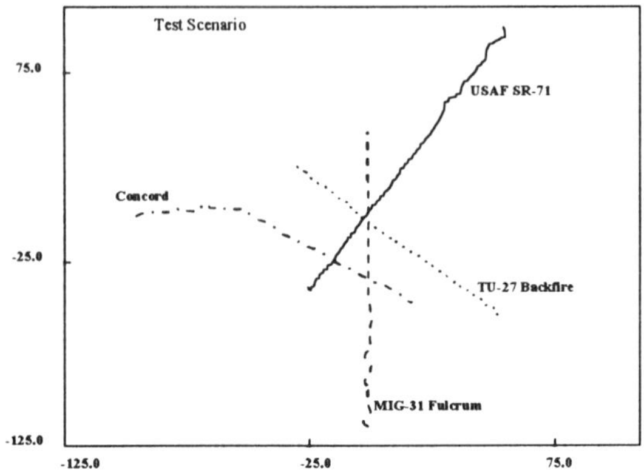

Figure 3. Scenario used to test the Dempster-Shafer algorithm

TABLE 4. Parameters of the scenario

Parameters	TU-27
Creation time	00.00.01.0
Identity	TU-27 Backfire
Allegiance	hostile
Range	80 DM
Azimuth	135 degrees
Speed	1013.4 knots
Heading	314.5 degrees
Altitude	10,000 ft
IFF response	No response
Emitting report time and name	00.01.35.0 DOWNBEAT 00.03.10.0 SKIP-SPIN 00.03.30.0 APQ-55 00.03.45.0 DOWNBEAT

4.3.2 The Simulation Results

The track TU-27 has been specifically designed to show the subtlety of the proposed decision rules with the Dempster-Shafer algorithm. This target is considered as a foe in the PDB and the scenario has imitated a situation where this target was trying Electronic Counter Measure (ECM) techniques which let the CPF ESM sensor declare the presence of an APQ-55 emitter. In the PDB, this target does not carry such a radar. The time evolution of the three following identity propositions is shown in Figures 4-5. 1) the main proposition (the largest BPA, Figure 4); and 2) the correct identity proposition (TU-27) (Figure 5).

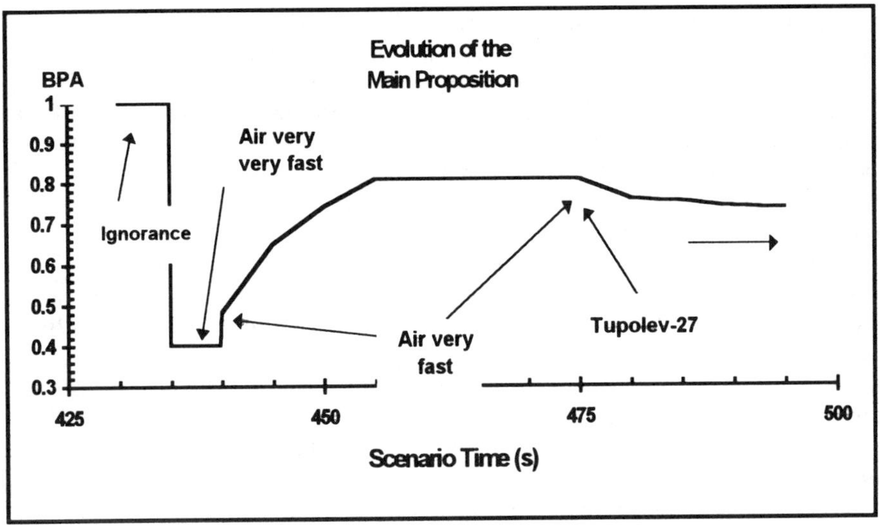

Figure 4. Time evolution of the main identity proposition

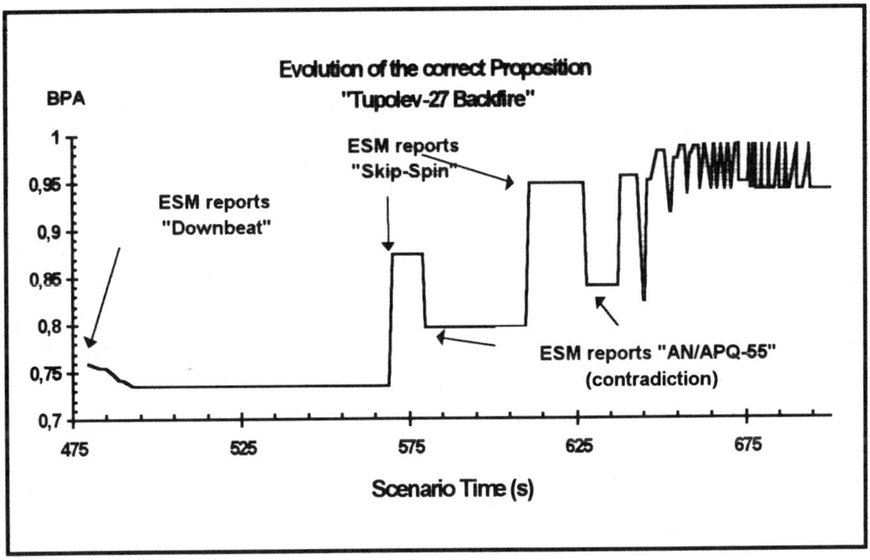

Figure 5. Time evolution of the correct identity proposition.

At the beginning, after the speed has been computed by the positional fusion function, only three propositions are estimated by Dempster-Shafer: the very fast and very very fast air objects of the PDB and of course, the ignorance. The ignorance minimal (IgnoMin) is 1%. After few radar scans the estimated speed stabilizes. The air very very fast ID proposition and soon later the air very fast ID proposition reaches the highest value while the ignorance reaches a minimal BPA value. This can be seen in Fig. 4 within the first 30 seconds of the scenario. In fact the target speed attributes are fused every 5 seconds, each time a new SPS-49 radar scan has been associated with the track. Later the fusion module receives an ESM report associated with this track, 45 seconds from the beginning of the scenario, indicating that this target operates a radar emitter named Downbeat. Since the BPA of this ESM report is 80%, the BPA of the air very fast ID proposition drops to 17% and a new singleton proposition, made of a single PDB element "TU-27 Backfire," which is the correct answer, is proposed.

After 140 seconds from the beginning of the scenario, another ESM contact reports the presence of a Skip-Spin emitter which according to the emitter database is carried by many aircraft of the PDB. The total BPA of the proposition "TU-27, Backfire" has increased since this attribute information supports this proposition. Ten seconds later a new ESM contact declares the presence of the APQ-55, a contradictory information because this radar is not on the "TU-27, Backfire" platform. The fusion of this new contact contributes to generate more propositions from the Dempster-Shafer combination rules. This explains why the ignorance proposition Θ gets a sudden increase and some new propositions are suggested with BPA which has been mainly taken from the "TU-27, Backfire" proposition. One of these new propositions is of generic nature and states: the air platforms that include an AN/APQ-55 radar which steals some BPA from the air very very fast and the air very fast ID proposition.

The contradictory sensor declaration has been voluntarily included in the scenario. The simulated ECM action produces declarations which were not compatible with the simulated target platform (TU-27). Early in the scenario the contradiction would have produced no difference between the results obtained from the conventional Dempster-Shafer (D-S) and our modified version because the MaxNum criteria would have not been reached. But later during the scenario, when the truncation criteria has been used many times some platform and the ignorance Θ got more belief than would have been allowed by the conventional D-S approach. These contradictions were included for the purpose of testing the estimator robustness and ability to recover on the right identity.

The possibility of an increase in the belief of Θ is useful in some circumstances. The case where many wrong attribute data followed by "correct" ones are associated to a track (this happens with non-cooperative targets due to counter-measure or systematic errors) is an example. In such a case, the combination of evidence converges to a wrong identification but has the capability to reorient the belief to some more compatible solutions if correct attribute data are obtained (due to adequate corrective action to counter-measure effect). The recovering would have been more complicated for the conventional Dempster-Shafer. Usually, when a Dempster-Shafer algorithm puts a large belief on a wrong proposition and the belief in the ignorance is very small (this happens after many executions of the identification algorithm), it will take a lot of declarations to reorient the algorithm to the correct proposition. Decision rules such as those proposed in this section have the great advantage of keeping the ignorance large enough to make the recovery easier and faster.

5. Identity Declaration Information Fusion

This section is concerned with the fusion of identity declarations, which can be represented in a hierarchical manner. The Bayesian approach is found to suffer from major deficiencies in a hierarchical context, when fully specified likelihoods are not available. Other problems associated with this approach are the coding of ignorance and the strict requirements on the belief of a hypothesis and its negation. One drawback of the Dempster-Shafer evidential theory is the long calculation time required by its high computational complexity. Due to the hierarchical nature of the evidence, an algorithm proposed by Shafer & Logan [8] is implemented which reduces the calculations from exponential to linear time proportional to the number of nodes in the tree.

5.1 FUSION RULES

An interesting approach, which also assumes that evidence is hierarchically structured, was proposed by Shafer [7]. He suggested that belief functions being combined should be carried by a partition *P* of Θ, which has fewer elements than Θ. This is done by separating the frame of discernment into smaller more manageable frames, one for each set of mutually exclusive hypotheses. Of course, once the elements have been separated into multiple frames, items from different frames can no longer be compared since they then pertain to different belief functions. The Shafer-Logan approach combines simple support functions focused on a subset of Θ or its complement (Shafer & Logan, [8]. This approach is detailed below.

Shafer & Logan [8] have introduced a new terminology and notation that will greatly facilitate the understanding of the combination process. Let $\mathfrak{R}$ be the collection of all nodes below Θ. As illustrated in Figure 6, $\mathfrak{R}$ is represented by {C, D, E, F, G, H, I, J, K}. The node D is said to be the child of C, and C is D's parent. The set of nodes that consists of all thc children of a given non terminal node is called a *sib*; the sib ℓ_C consists of the children of C.

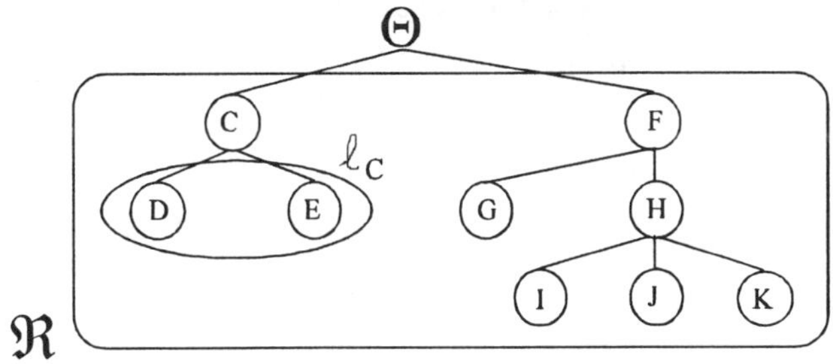

Figure 6. Hierarchical Evidence

It is assumed that for each node A in $\mathfrak{R}$, there is a single dichotomous belief function Bel_A with dichotomy $\{A, \overline{A}\}$. Furthermore, $Bel_A(A)$ and $Bel_A(\overline{A})$ are both strictly less than one, but either, or both, can be zero. For each node A in the tree, $Bel_A^{\downarrow}$ denotes the orthogonal sum of $\mathrm{Bel}_{\mathrm{B}}$ for all nodes B that are strictly below A. For each node in $\Re$, $\mathrm{Bel}_{\mathrm{A}}^{\lozenge}$ denotes the orthogonal sum of $\mathrm{Bel}_{\mathrm{B}}$ for all nodes B in $\Re$ that are neither below A nor equal to A. For example, from Figure 6.

$$\mathrm{Bel}_{\mathrm{F}}^{\downarrow} = \mathrm{Bel}_{\mathrm{G}} \oplus \mathrm{Bel}_{\mathrm{H}} \oplus \mathrm{Bel}_{\mathrm{H}}^{\downarrow}$$

$$= \mathrm{Bel}_{\mathrm{G}} \oplus \mathrm{Bel}_{\mathrm{H}} \oplus \mathrm{Bel}_{\mathrm{I}} \oplus \mathrm{Bel}_{\mathrm{J}} \oplus \mathrm{Bel}_{\mathrm{K}}$$

and

$$\mathrm{Bel}_{\mathrm{C}}^{\lozenge} = \mathrm{Bel}_{\mathrm{F}} \oplus \mathrm{Bel}_{\mathrm{F}}^{\downarrow}$$

In terms of this nomenclature, the aim of Shafer & Logan's approach amounts to calculating the values of all the nodes $\mathrm{A} \in \Re$:

$$\mathrm{Bel}_{\Theta}^{\downarrow} = \oplus\{\mathrm{Bel}_{\mathrm{A}} | \mathrm{A} \in \Re\},$$

or equivalently

$$\mathrm{Bel}_{\Theta}^{\downarrow} = \mathrm{Bel}_{\mathrm{A}}^{\downarrow} \oplus \mathrm{Bel}_{\mathrm{A}} \oplus \mathrm{Bel}_{\mathrm{A}}^{\lozenge}.$$

Using the definitions of partition of a frame of discernment and the concept of discernment of evidence, Shafer & Logan have shown that:

(a) if P is a partition of Θ and $\mathrm{P} \in \Re \cap P$, then $(\mathrm{Bel}_{\mathrm{P}}^{\downarrow})_{P} = (\mathrm{Bel}_{\mathrm{P}}^{\downarrow})_{\{\mathrm{P},\overline{\mathrm{P}}\}}$,

(b) if P is a partition of Θ, $\mathrm{A} \in \Re$ and $\overline{\mathrm{A}} \in P$ then $(\mathrm{Bel}_{\mathrm{A}}^{\lozenge})_{P} = (\mathrm{Bel}_{\mathrm{A}}^{\lozenge})_{\{\mathrm{A},\overline{\mathrm{A}}\}}$,

(c) if P is a partition of Θ, then P discerns the interaction relevant to itself among the belief functions $\{\mathrm{Bel}_{\mathrm{P}}^{\downarrow} | \mathrm{P} \in \Re \cap P\}$ and

$$\{\mathrm{Bel}_{\mathrm{P}} | \mathrm{P} \in \Re \cap P\}.$$

From the characteristics of a partition of a frame of discernment and the definition of $\mathrm{Bel}_{\mathrm{A}}^{\downarrow}$, we obtain, say, for a partition $\ell_{\mathrm{A}} \cup \{\overline{\mathrm{A}}\}$:

$$(\mathrm{Bel}_{\mathrm{A}}^{\downarrow})_{\ell_{\mathrm{A}} \cup \{\overline{\mathrm{A}}\}} = \oplus\{(\mathrm{Bel}_{\mathrm{B}})_{\ell_{\mathrm{A}} \cup \{\overline{\mathrm{A}}\}} \oplus (\mathrm{Bel}_{\mathrm{B}}^{\downarrow})_{\ell_{\mathrm{A}} \cup \{\overline{\mathrm{A}}\}} | \mathrm{B} \in \ell_{\mathrm{A}}\}.$$

By (a) above, this is equivalent to

$$(\mathrm{Bel}_{\mathrm{A}}^{\downarrow})_{\ell_{\mathrm{A}} \cup \{\overline{\mathrm{A}}\}} = \oplus\{\mathrm{Bel}_{\mathrm{B}} \oplus (\mathrm{Bel}_{\mathrm{B}}^{\downarrow})_{\{\mathrm{B},\overline{\mathrm{B}}\}} | \mathrm{B} \in \ell_{\mathrm{A}}\}$$

Note that if the element B of ℓ_{A} is a terminal node, then $\mathrm{Bel}_{\mathrm{B}}^{\downarrow}$ is vacuous and the orthogonal sum reduces to $\mathrm{Bel}_{\mathrm{B}}$.

The mechanism for passing up belief values between different levels of hierarchy satisfies the intuitive argument to the effect that any evidence confirming a node should also provide evidence confirming its parent. Generalizing for Θ, we obtain:

$$(\mathrm{Bel}_{\Theta}^{\downarrow})_{\ell_{\Theta}} = \oplus\{\mathrm{Bel}_{\mathrm{B}} \oplus (\mathrm{Bel}_{\mathrm{B}}^{\downarrow})_{\{\mathrm{B},\overline{\mathrm{B}}\}} | \mathrm{B} \in \ell_{\Theta}\}$$

The partition becomes ℓ_Θ instead of $\ell_\Theta \cup \{\overline{\Theta}\}$, since $\ell_\Theta \cup \{\overline{\Theta}\} \subset \ell_\Theta$. A statement even stronger than statement *c.* above is: if A is a non terminal element of $\mathfrak{R}$, then the partition $\ell_A \cup \{\overline{A}\}$ discerns the interaction relevant to itself among $\mathrm{Bel}_A^{\downarrow}$, Bel_A and $\mathrm{Bel}_A^{\Diamond}$. Therefore, we obtain:

$$(\mathrm{Bel}_\Theta^{\downarrow})_{\ell_A \cup \{\overline{A}\}} = \{(\mathrm{Bel}_A^{\downarrow})_{\ell_A \cup \{\overline{A}\}} \oplus (\mathrm{Bel}_A)_{\ell_A \cup \{\overline{A}\}} \oplus (\mathrm{Bel}_A^{\Diamond})_{\ell_A \cup \{\overline{A}\}}\}$$

or, equivalently:

$$(\mathrm{Bel}_\Theta^{\downarrow})_{\ell_A \cup \{\overline{A}\}} = \{(\mathrm{Bel}_A^{\downarrow})_{\ell_A \cup \{\overline{A}\}} \oplus \mathrm{Bel}_A \oplus (\mathrm{Bel}_A^{\Diamond})_{\{A, \overline{A}\}}\}$$

since Bel_A is carried by $\ell_A \cup \{\overline{A}\}$ and $(\mathrm{Bel}_A^{\Diamond})_{\ell_A \cup \{\overline{A}\}} = (\mathrm{Bel}_A^{\Diamond})_{\{A, \overline{A}\}}$ from statement *b*.

The evidence from above A and down other branches affects the degree of belief of the children of A only if the degree of belief is for or against A itself. This is the mechanism for passing, confirming and disconfirming evidence to the lower levels of the hierarchy. Using the algorithm proposed by Shafer & Logan, the combination of hierarchical evidence is accomplished in four stages.

At stage 0, evidence is received for a specific node in the form of a simple support function or dichotomous function. This evidence is first combined with the existing belief value associated with the node using Dempster's rule of combination. In this case the combination is easy to calculate, since the belief functions being combined are simple support functions or dichotomous functions focused on the same node. The result of this simple combination is a dichotomous belief function which is then propagated through the hierarchical tree using the Shafer-Logan algorithm stages 1 to 3 below.

At stage 1, the dichotomous belief functions attached to terminal nodes are combined to find degrees of belief for and against their parents; the same is done to the parents' parents and so on, until there is a dichotomous belief function attached to each child of A to obtain the values of $(\mathrm{Bel}_A^{\downarrow})_{\ell_A \cup \{\overline{A}\}}$.

At stage 2, we obtain the values of each (direct) child of Θ : $(\mathrm{Bel}_\Theta^{\downarrow})_{\ell_\Theta}$.

At the last stage, the degree of belief of each node is reevaluated to take into account the influence of other nodes using; this process calculates the degrees of belief by moving back down the tree.

The amount of arithmetic involving a particular node depends linearly on the number of daughters of the node. Furthermore, the computational complexity of the algorithm is linear in the number of nodes in the tree. This is the case because the belief functions being combined are simple support functions focused on nodes or their complements. However, if we were to combine belief functions carried by the field of subsets generated by the children of a node, more precisely by a partition $\ell_A \cup \{\overline{A}\}$ (such as suggested by Shafer, 1985), then the amount of arithmetic would become exponential in the sib size but remain proportional to the number of sibs.

Note that a special case occurs when the belief functions are Bayesian, that is, when the belief function Bel_A carried by $\ell_A \cup \{\overline{A}\}$ satisfies:

$$\mathrm{Bel}_A(B|A) + \mathrm{Bel}_A(\overline{B}|A) = 1$$

and

$$\mathrm{Bel}_A(B|\overline{A}) + \mathrm{Bel}_A(\overline{B}|\overline{A}) = 1$$

for every element B of the field $(\ell_A \cup \{\overline{A}\})^*$; the arithmetic is then linear in the sib size. This case is specifically the one devised by J. Pearl.

5.2 DECISION RULES

An important element to take into consideration in the design of an identity information fusion function is the decision making process required to select the identity declaration which best supports the combined declarations. Statistical decision making is necessary since, after fusion, the result may contain many identity declarations with a non-null confidence value. Because decisions are subjective, the decision maker will undoubtedly depend on his or her own judgement as well as information collected from various sources. The fusion function should still suggest to the decision maker the "best" candidate or candidates according to predetermined decision rules. A number of factors are involved in the choice of the decision rules to be used :

- rule complexity (if the computational time required to make a decision is too high, the process becomes useless),
- confidence level (decision rules are made on a probabilistic basis),
- fusion technique used to combine uncertain information (for example the Dempster-Shafer approach produces belief and plausibility values instead of single probability values),
- type of application (the application environment can dictate decision rules).

When dealing with a hierarchical structure, the decision making techniques for an evidential theory approach cannot rely on the maximum value of belief because the belief of parent will always be equal to, or higher than, the belief of each of his children. Also, the plausibility of a parent will always be equal to, or higher than, the plausibility of each of his children. Furthermore, an important aspect that should not be forgotten in military applications is the fact that decision making is scenario and mission dependent. What we propose, therefore, is a semi-automated approach based on the belief and plausibility values. It is called semi-automated because threshold values will be applied to the belief and plausibility measures. However, the final decision will be taken by the decision maker, because he/she remains an important part of the process and because the choice of the final identity is typically scenario and mission dependent. The decision making approach proposed is as follows:

(a) select all alternatives with a plausibility value greater than a certain threshold T_{Pl};

(b) plot the chosen alternatives according to hierarchical structure and indicate for each node its belief value;

(c) add to the graph all the nodes directly below Θ with their belief values.

Based on the graph constructed by this decision approach, the decision maker can select the best alternative according to the highest belief value, or the hierarchical level of interest or any other criteria. The plausibility value is not included in the graph as it is not deemed to be essential to the decision making process.

5.3 A CASE STUDY: IDENTITY DECLARATION INFORMATION FUSION

The following example illustrates the use of the Shafer-Logan algorithm for the specific problem of combining identity declarations using two frames of discernment. The scenario is as follows: the commander of a Canadian Patrol Frigate-type ship receives a series of identity declarations concerning one object; he/she must determine the identity of the object and take action. For example, the types of information received can consist of a general classification (surface combatant), a specific type of ship (frigate), a specific class (City class) or a unique identity (Ville de Québec). The belief and plausibility values at each node of the two hierarchical trees are zero. The evidences received from various information sources are given in Table 5 (assuming that the order of the received evidences is unimportant). The evidences are the only ones received concerning the object; a decision could be taken after each evidence if the plausibility of at least one node is greater than T_{Pl}. However, we have chosen to combine all the evidences before the decision making process. In this example, T_{Pl} will be set to .5. Evidences m({air}) = .5 and m({ $\overline{\text{air}}$ }) = .1 are contradictory but not enough for the Dempster's combination rule to produce irregular results. Figures 7 and 8 show the belief and plausibility measures for each node after the combination of evidence for the two frames of discernment respectively. Figures 9-10 give the graphs which are available to the decision maker. It seems that the object is airborne and hostile; there is a fairly good chance that it is a fixed wing fighter. According to the mission, the decision maker will choose the best alternative and take appropriate action if necessary.

TABLE 5. Evidences received from various sources

Frame of discernment Θ_1	Frame of discernment Θ_2
m({fighter}) = 0.4	m({unknown}) = 0.2
m(Θ) = 0.6	m(Θ) = 0.8
m({ $\overline{\text{carrier}}$ }) = 0.7	m({ $\overline{\text{friend}}$ }) = 0.6
m(Θ) = 0.3	m(Θ) = 0.4
m({fixed-wing}) = 0.3	m({hostile}) = 0.7
m(Θ) = 0.7	m(Θ) = 0.3
m({ $\overline{\text{non - combatant}}$ }) = 0.8	m({neutral}) = 0.3
m(Θ) = 0.2	m(Θ) = 0.7
m({air}) = 0.5	m({hostile}) = 0.8
m(Θ) = 0.5	m(Θ) = 0.2
m({ $\overline{\text{air}}$ }) = 0.1	
m(Θ) = 0.9	
m({MiG-25}) = 0.1	
m(Θ) = 0.9	
m({helicopter}) = 0.5	
m(Θ) = 0.5	
m({MiG-19}) = 0.6	
m(Θ) = 0.4	
m({ $\overline{\text{missile}}$ }) = 0.7	
m(Θ) = 0.3	
m({MiG-25}) = 0.4	
m(Θ) = 0.6	

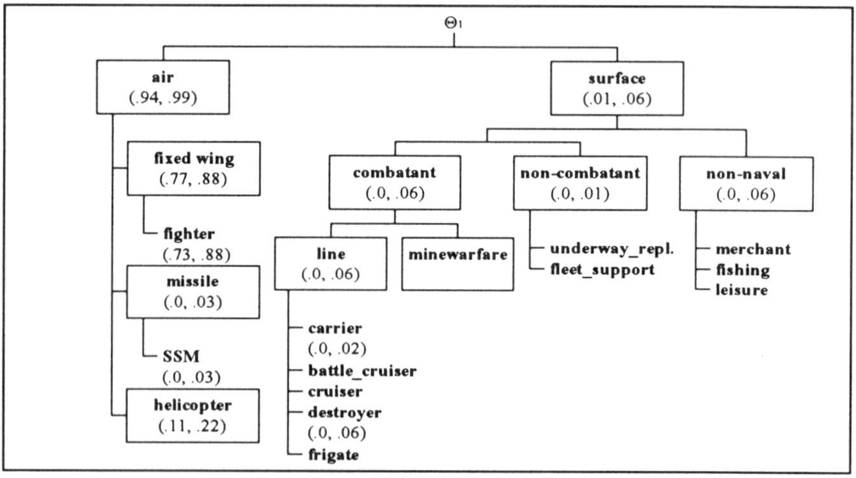

Figure 7. Results From Combining Evidences Using Θ_1

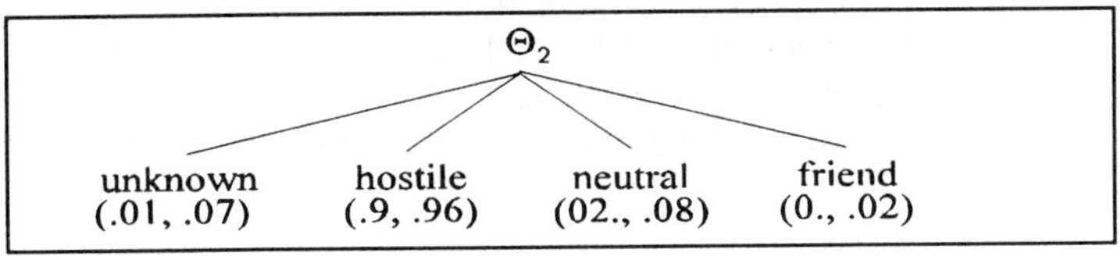

Figure 8. Results From Combining Evidences Using Θ_2

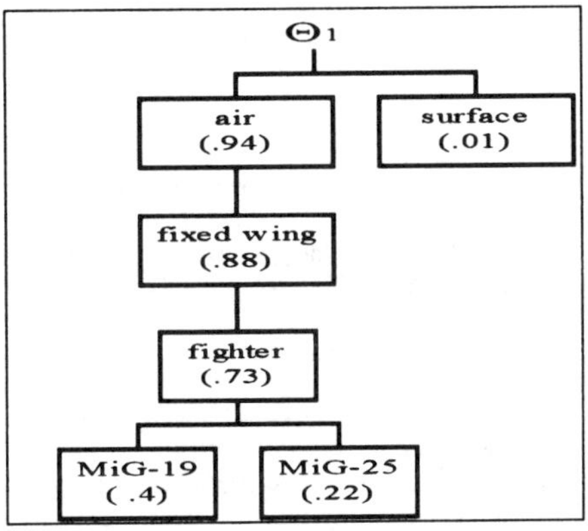

Figure 9. Decision Graph Representing Best Alternatives Using Θ_1

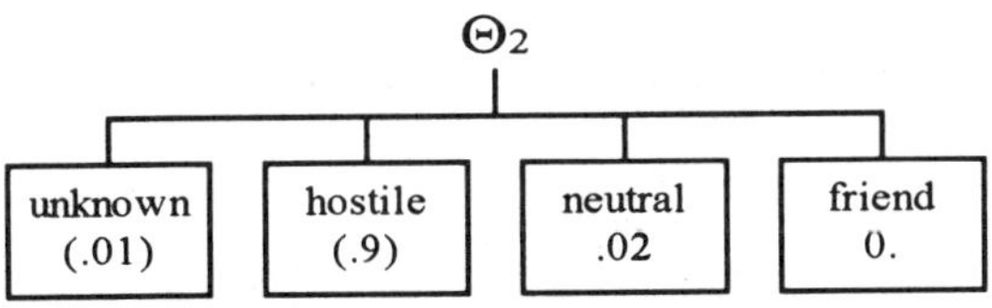

Figure 10. Decision Graph Representing Best Alternatives Using Θ_2

6. Summary And Conclusions

This talk discussed the use of the Dempster-Shafer theory of evidence for the fusion of identity declarations and attribute information within a military environment. It proposed to hierarchically structure the identity declarations according to NATO's STANAG 4420 charts, which provide a better base for achieving interoperability in information exchange between nations than uncontrolled alternatives. One drawback of the Dempster-Shafer evidential theory is the long calculation time required by its high computational complexity. Due to the hierarchical nature of the evidence, an algorithm proposed by Shafer & Logan was implemented which reduces the calculations from exponential to linear time proportional to the number of nodes in the tree. A semi-automated decision making technique, based on belief and plausibility values, was then described to select alternatives which best support the combined identity declarations. The second architecture was concerned with the fusion of attribute information to declare an identity using a modified version of the Dempster-Shafer evidential combination algorithm. We used fuzzy logic to make possible the fusion of apparently incomplete attribute information.

7. Acknowledgements

The author acknowledges the contribution of P. Valin, A. Jouan and M.A. Simard from Lockheed Martin Canada in several aspects of this work.

8. References

1. NATO. STANAG 4420. (1989) Display Symbology and Colours for NATO Maritime Units. Edition 1. NATO.
2. Paradis, S., Roy, J. and Treurniet, W., (1998) Integration of All Data Fusion Levels Using a Blackboard Architecture. In *Proceedings of Eurofusion98: International conference on Data Fusion.* Great Malvern, UK.
3. Pearl, J. (1988) *Probabilistic Reasoning in Intelligent Systems: Networks of Plausible Inference.* Morgan Kaufmann Publishers, San Mateo, CA.
4. Roy, J. and Bossé, É., (1998) A Generic Multi-Source Data Fusion System. *Defence Research Establishment Valcartier. Technical Report.* DREV-R-9719.
5. Roy, J.M.J, Bossé, É. and Duclos-Hindié, N. (1998) Performance comparison of contact-level and track-level sensor fusion architectures. *Optical Engineering, special section on sensor fusion.*
6. Shafer, G. (1976) *A Mathematical Theory of Evidence.* Princeton University Press.
7. Shafer, G. (1985) Hierarchical evidence. In *Proceedings of the Second Conference on Artificial Intelligence Applications*, pp.16-21.
8. Shafer, G. & R. Logan, (1987) Implementing Dempster's rule for hierarchical evidence. *Artificial Intelligence*, **33**, 271-298.
9. Bossé, E., and Simard, M.A., (1998) Managing Evidential Reasoning for Identity Information Fusion Optical Engineering, Vol 37, no. 2, pp. 291-00.

ASSOCIATION TECHNIQUES USED IN MULTISENSOR DATA FUSION

E. SEMERDJIEV
Central Laboratory for Parallel Processing
Bulgarian Academy of Sciences
Department "Mathematical Methods for Sensor Data Processing"
25-A Acad. G. Bonchev Str., 1113 Sofia, Bulgaria
Email: signal@bas.bg

1. Introduction

This chapter provides a brief overview of the problems of associating observations and tracking of targets in the presence of random interference. It emphasizes the application of the Kalman filter, as a basic estimation technique used to develop different association mechanisms. Five basic association approaches are considered in this chapter: the Nearest Neighbor approach, the Strongest Neighbor approach, the Probabilistic and Joint Probabilistic Data Association Filters, as well as the Multiple Hypothesis Tracking algorithm. All are applied in sonar, radar and other surveillance systems applications.

2. The Kalman filter – the keystone of the tracking algorithm

The *Kalman filter* (KF) is a mathematical technique for state estimation and prediction of continuous and discrete stochastic dynamic systems observed using sequences of noisy measurements. Because of the usually discrete nature of the measurement process, the discrete KF is the keystone element of the discussed measurement (observation) association algorithms. Its standard form is derived for linear stochastic dynamic systems and sensor models, in the form described below:

$$x(k) = \Phi x(k-1) + q(k-1) + Gu(k-1);$$

$$y(k) = Hx(k) + v(k).$$

The main assumption, concerning the above equations is that the two vectors (the n -dimensional vector x, that includes the target states which have to be estimated and the m-dimensional ($0 < m <= n$) measurement vector y), are both normally distributed. The p-dimensional control vector u is a known deterministic control input, which causes changes in the object's relative position, acceleration and deceleration. It is set here u=0. The n x n system (target) transition matrix Φ, the n x p control matrix G, as well as the m x n measurement matrix H are known. The second important assumption considers the system and the measurement noises q and v. They are assumed to be mutually independent, zero-mean, white, Gaussian noises, with known covariance Q and R.

A. K. Hyder et al. (eds.), Multisensor Fusion, 481–495.

The above assumptions have an important practical sense. The vector of mathematical expectations (usually named estimations) and the respective covariance matrix completely describe the statistical properties of any normally distributed stochastic vector. Such a vector, passed trough linear transformations remains normally distributed, and his new statistical parameters may be computed upon the base of the previous ones. The above assumptions are invalid when one or both models are non-linear or when the noises are not Gaussian nor independent. In this case KF approach still could be used by applying additional transformations to the real non-linear models, which are linearized and the colored noises are converted to white ones. Theoretically, in this way the best possible approximation of the optimal results could be obtained. Generally we can say, that the KF and its simplified or more sophisticated modifications represent about all of the knowledge needed for the development of successful tracking algorithms.

The KF algorithm consists of two important steps: prediction and update (filtering).

The prediction step is based on extrapolation of the vectors of the state and measurement estimations, from the current k-1-st observation frame (k-1-st sensor scan, corresponding to moment $t=(k-1)T$, where T is sampling interval), to the next scan k, in order to predict the expected development of the observed stochastic process. The target state and measurement predictions as well as the covariances of the corresponding predicted state and measurement errors for scan k are simply:

$$\hat{x}(k/k-1) = \Phi\hat{x}(k-1/k-1); \tag{1}$$

$$\hat{y}(k/k-1) = H\hat{x}(k/k-1), \tag{2}$$

$$P(k/k-1) = \Phi P(k-1/k-1)\,\Phi^T + Q. \tag{3}$$

$$S(k) = HP(k/k-1)H^T + R. \tag{4}$$

The predicted estimate $\hat{x}(k/k-1)$ and its error covariance $P(k/k-1)$ define the expected state vector distribution, based on a *priori* information:

$$p[x(k)] = N[x(k);\hat{x}(k/k-1),P(k/k-1)]$$

The next filtering step is used to correct a *priori* made prediction, when the observation has been already obtained. This is made on the base of the next equation named *state update*:

$$\hat{x}(k/k) = \hat{x}(k/k-1) + K(k)\tilde{y}(k) \tag{5}$$

where the residual vector:

$$\tilde{y}(k) = y(k) - \hat{y}(k/k-1), \tag{6}$$

represents the differences between the measured and predicted values (the measurement prediction error), which are normally distributed, too:

$$p[\tilde{y}(k)] = p[y(k)|\hat{x}(k/k-1)] = N[\tilde{y},0,S(k)]$$

The state update operation is also named filtering, due to the presence of the KF gain $K(k)$, which plays the role of a scaling factor:

$$K(k) = P(k/k-1)H^T S^{-1}(k). \tag{7}$$

It varies directly proportional to the prediction errors. If there is no prediction error $K(k) = P(k/k\text{-}1)=0$ and then $\hat{x}(k/k) = \hat{x}(k/k-1)$. The higher the measurement errors R, the lower the gain $K(k)$ is. For example, the significance of the correction term $K(k)\tilde{y}(k)$ directly depends on the accuracy of the information brought by the sensor measurement.

Finally, the state error covariance at scan k is corrected:

$$P(k/k) = P(k/k-1) + K(k)S^{-1}(k)K^T(k). \tag{8}$$

The filtered estimate $\hat{x}(k/k)$ and its covariance $P(k/k)$ define the Gaussian conditional posterior state vector distribution. Note, that in the KF filter there is only one always available measurement:

$$p[x(k)|y(k)] = N[x(k), \hat{x}(k/k), P(k/k)].$$

Sometimes the target is not detected (usually the sensor detection probability P_D is lower than 1). This case is not considered in the KF at all. It is easy to note, that when there is no measurement, the posterior state vector distribution does not exist. That's why in all tracking algorithms it is usually set:

$$\hat{x}(k/k) = \hat{x}(k/k-1),$$
$$P(k/k) = P(k/k-1).$$

In the multitarget case, a separate KF is allocated for every target. The other targets and the false alarms are not processed in it. A respective number of KF running in parallel may be applied in multiple target case. This approach is appropriate when observed targets are sufficiently far from one another. A wrong state (track) update can be obtained if we use incorrect observations originating from other targets. This can be avoided by introducing a specific correlation mechanism, which defines a normalized distance d^2 between each arriving observation and the predicted target state:

$$d^2 = \tilde{y}(k)\, S^{-1}(k)\tilde{y}(k).$$

A region in the space where the observation will be found with given probability could be defined, assuming that the true observation at scan k is normally distributed and the predicted state vector is given:

$$d^2 \leq C,$$

where C is constant. This region is called *validation region or gate*, whose volume is denoted by V_g. It is the ellipse (or ellipsoid) of probability concentration - the region of minimum volume that contains a given probability mass under the Gaussian assumption [5].

Observations that lie inside the gate are considered valid; those outside are discarded. For example, in radar tracking problems, the return from the target of interest is sought within a time interval determined by the delay corresponding to the anticipated range of the target when it reflects the energy transmitted by the radar. In general, the measurement vector consists of multiple components - range, bearing, elevation and Doppler frequency, for example. So a multidimensional gate should be set up for detecting the return from the target.

The parameter C is obtained from tables of the chi-square distribution, since the weighted norm of the innovation that defines the validation region is chi-square distributed with number of degrees of freedom equal to the dimension m of the measurement. Table 1 gives the probability mass ([5], p.151):

$$P_G = Pr\,[y(k) \in V_g] = Pr\,[d^2 \le C].$$

which is the probability that the true observation will fall in the gate, for various values of C and dimensions m of the measurement. The square root $\sqrt{C}$ is sometimes referred to as "number of sigmas" (standard deviations) of the gate. The gate size is usually selected beforehand and kept constant for any given application.

Finally, accepting the underlying assumption associated with false alarm (FA) that they are randomly distributed in space, time, and intensity, it can happen that FA will be found in the validation region. Other targets may also move in the vicinity of the center of the region. That is why in all tracking applications there is a procedure (whose place is between the KF prediction and filtering steps) that selects the observation to be incorporated into the state estimator among several candidates.

TABLE 1. Gating thresholds and values of probability mass in the gate

C $\sqrt{C}$	1 1	4 2	9 3	16 4	25 5	6.6 2.57	9.2 3.03	11.4 3.38
m=1	0.683	0.954	0.997	0.99994	1.0	0.99		
m=2	0.393	0.865	0.989	0.9997	1.0		0.99	
m=3	0.199	0.739	0.971	0.9989	0.99998			0.99

The simplest methods for resolving the ambiguity about the observations' origins is to chose the observation (O_K) nearest to the predicted ($\hat{O}_K$) observation position (the *"nearest neighbor"* -- Figure 1). In other tracking applications the true measurement always has the highest signal strength (the *"strongest neighbour"* -- Figure 2). Unfortunately, these techniques are based on assumptions, which are not valid in most of the real-life tracking implementations. That is why two other approaches are usually applied.

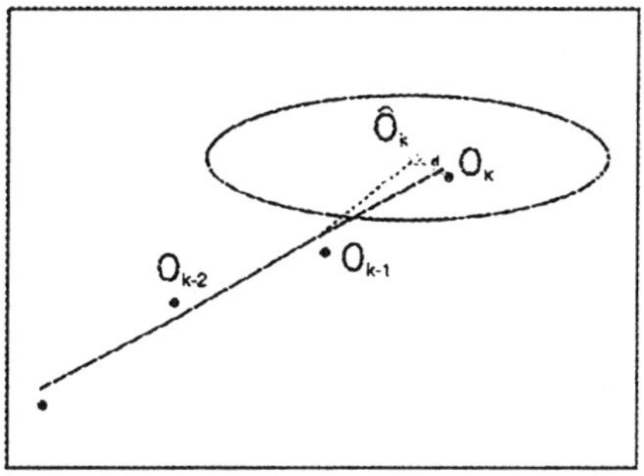

Figure 1. Nearest Neigh bor (O_K) Association

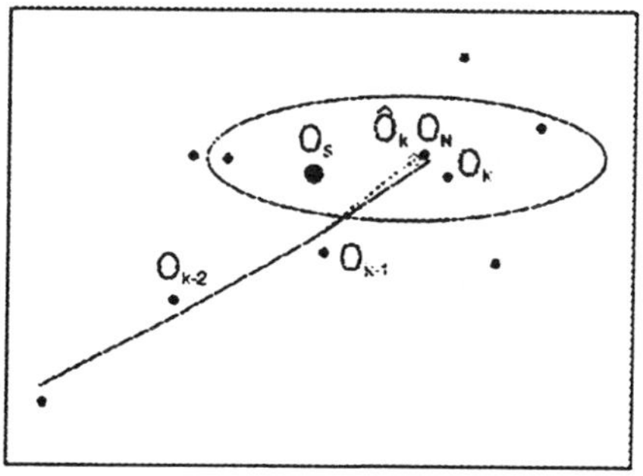

Figure 2. Strangest Neighbor (O_S) Association

3. The PDA and JPDA filters

The *Probabilistic Data Association* (PDA) algorithm and its multitarget version - the *Joint Probabilistic Data Association* (JPDA) algorithm are *"all neighbors"* association techniques, based on appropriately modified KF called PDAF and JPDAF. These "track-oriented" algorithms always associate arriving observations with previously established track(s) -- the hypothesis suggesting that the measurement is from a "newly born" track is not considered at all. It should be also noted that PDA/JPDA tracking algorithms have no explicit provision for track initiation. Thus, a separate track initiation function, must be defined. Excellent description of these methods is given in [1], [5],[6].

3.1 THE PDA METHOD

The PDA algorithm, first presented by Bar-Shalom and Tse [6], considers the case when all target trajectories are sufficiently far from one another to be considered independently. Let N measurements be validated in a certain track gate after the standard KF prediction step including equations (1) through (4). No observations, or maximum one observation of them may originate from the true target and, evidently, all other observations are FA. On the other hand, it is not known which one is the true observation. So, N suggestions (named hypotheses H_j, $j = 1, \ldots, N$) consecutively considering each validated observation as the true one may be built. The $(N+1)$-st hypotheses H_0 considers the case when the true measurement hasn't been detected. If we assume that the extraneous return is with Poisson density, with $\beta=\beta_{NT}+\beta_{FA}$, including new targets and FA densities, then the probability of the hypotheses H_0 is proportional to p ([1], p.300), where

$$p_0' = \beta^N (1 - P_D P_G).$$

Similarly, the probability of the hypothesis H_j, $j = 1, \ldots, N$ that observation j is valid return is proportional to

$$p_j' = \frac{\beta^{N-1} P_D P_G e^{-0.5 d_j^2}}{(2\pi)^{0.5m} \sqrt{det(S)}}, \quad j = 1, \ldots, N$$

Finally, the probabilities p_j associated with the above $N+1$ hypotheses are computed through the normalization equation:

$$p_j = \frac{p_j'}{\sum_{i=0}^{N} p_i'}, \quad j = 1, \ldots, N$$

Therefore, a convenient simplified form becomes

$$p_j = \begin{cases} \dfrac{b}{b + \sum_{i=1}^{N} \alpha_j}, & j = 0 \quad (no\ valid\ observation) \\[2ex] \dfrac{\alpha_j}{b + \sum_{i=1}^{N} \alpha_j}, & 1 \le j \le N \end{cases}$$

where ([1], p.301):

$$b = (1 - P_D P_G)\, \beta (2\pi)^{0.5m} \sqrt{det(S)},$$

$$\alpha_j = P_D P_G e^{-0.5 d_j^2}$$

The next step is to merge the received observations, respectively merging their hypotheses. This is made through generation of the super residual:

$$\tilde{y}(k) = \sum_{j=1}^{N} p_j \tilde{y}(k) = \sum_{j=1}^{N} p_j [y_j(k) - H\hat{x}(k/k-1)],$$

which is used in the PDAF instead of the residual vector described by equation (5), and during the filtering step in the standard KF update equation (6).

The covariance, at scan k, is modified according to the equation:

$$P(k/k) = P^0(k/k) + dP(k),$$

where $P^0(k/k)$ is the standard Kalman covariance that would be computed if a single return were present, and $dP(k)$ is an increment taking into account the effect of uncertain correlation:

$$dP(k) = K(k)\left[\, \sum_{j=1}^{N} p_j \tilde{y}_j(k) \tilde{y}_j^T(k) - \tilde{y}(k)\tilde{y}^T(k) \,\right] K^T$$

$$P^0(k/k) = p_0 P(k/k-1) + (1-p_0)P^*$$

The standard KF covariance defined by equation (8) is denoted above trough $P^*(k/k)$. $P^0(k/k) = P(k/k-1)$, when no observations are validated (i.e. $p_0=1$). The term $dP(k)$ increases the covariance according to both the *a posteriori* probabilities and the spread of the observations found within the track gate. Finally, it should be noted that the PDAF tracking a single target in clutter shows a significant decrease in the number of lost tracks, when compared to a standard KF using nearest neighbor correlation method [1]. However, an extension, discussed next, is required if multiple target tracks are considered.

3.2 EXTENSION TO JPDA

The JPDA method is identical to the PDA except for the changed assumption about the number of true targets in any of the considered two or more tracks' gates. The association probabilities are now computed using all observations and all tracks. The state estimation gain and covariance are computed, as before, just the probability computation must be extended, to include multiple tracks ([1], p.302).

An example similar to those given in [1] is presented below. It considers a typical conflict situation (Figure 3), in which multiple observations fall within the validation gates of multiple target tracks. In this example, there are two observations (O_1 andO_2) within the gate of the predicted measurement position ($\hat{O}_1$) of the first track, while O_2 and O_3 are within the gate of the predicted measurement position ($\hat{O}_2$) of the second track.

The JPDA computes weighted residual for the first track based upon the two observations, but the weights for O_2 will be reduced to account for its presence within the gate of the second track. The residual for the second track will be formed using O_2. Table 2 gives the hypothesis matrix and the associated hypothesis probabilities for this example. The numbers represent the observations assigned to the tracks, and 0 represents the assignment of no observation to a given track. Also, as previously defined, α_{ij} is the Gaussian likelihood function associated with the assignment of the j -th observation to track i. Table 2 shows the structure for computing the hypothesis probabilities Pr . Defining N_O and N_T to be the numbers of observations and tracks, respectively, note that certain common factors may appear in all Pr. If $N_0 > N_T$, as in our example, the common factor is $\beta^{N_0 - N_T}$ while if $N_T > N_0$, the common factor will be $(1 - P_D)^{N_T - N}$. The same probability of detection P_D has been assumed for all tracks, but the extension to track dependent P_D is direct. If an observation does not fall within the gate of a given track, the associated probability is, of course, taken to be zero. Finally, the normalised probabilities Pr are computed in the standard manner (where N_H is the total number of hypotheses):

$$Pr(H_l) = \frac{Pr'(H_l)}{\sum_{j=0}^{N_H} Pr'(H_j)}$$

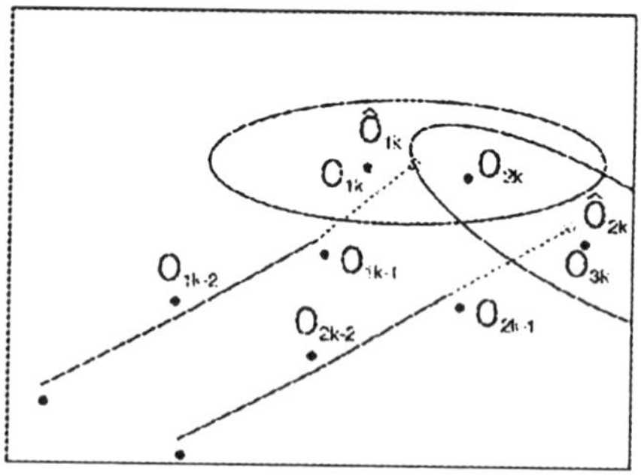

Figure 3. Multiple observation within two overlapped validation gates

TABLE 2. Target oriented hypothesis matrix for example of Figure 3

Hypothesis #	Track 1	Track 2	Hypothesis Probability
1	0	0	$(1-P_D)^2\beta^3$
2	1	0	$\alpha_{11}P_D(1-P_D)\beta^2$
3	2	0	$\alpha_{12}P_D(1-P_D)\beta^2$
4	0	2	$\alpha_{22}P_D(1-P_D)\beta^2$
5	1	2	$\alpha_{11}\alpha_{22}P_D^2\beta$
6	0	3	$\alpha_{23}P_D(1-P_D)\beta^2$
7	1	3	$\alpha_{11}\alpha_{23}P_D^2\beta$
8	2	3	$\alpha_{12}\alpha_{23}P_D^2\beta$

To compute the probability p_{ij} that observation j should be assigned to track i, a sum is taken over the probabilities from those hypotheses in which this assignment occurs. For example, for the first track (noting that j=0 refers to no assignment), the probabilities are:

$$p_{10} = P(H_1) + P(H_4) + P(H_6)$$
$$p_{11} = P(H_2) + P(H_5) + P(H_7)$$
$$p_{12} = P(H_3) + P(H_8)$$
$$p_{13} = 0.$$

Similarly, for the second track:

$$p_{20} = P(H_1) + P(H_2) + P(H_3)$$
$$p_{21} = 0$$
$$p_{22} = P(H_4) + P(H_5)$$
$$p_{23} = P(H_6) + P(H_7) + P(H_8)\,.$$

Thus, as expected, the most heavily weighted events are computed to be the assignment of O_2 to the first track and either O_3 to the second one.

4. Multiple Hypothesis Tracking

The *Multiple Hypothesis Tracking* (MHT) is a term applied to a class of algorithms that are able to entertain many "consistent" hypotheses about the state of a target, with certain hypotheses emerging as "likely" and others as "unlikely" as observations are received over time [7]. For reduction of the algorithm complexity it is often assumed that two or more tracks in a hypothesis cannot share observation, that satisfies the state predictions of the track in question.

To describe the MHT algorithm let us assume that a set of tracks has been always initiated and separated in independent subsets (named clusters), which do not share observations among them (directly or via other interacting tracks). Within each cluster, multiple hypotheses are generated each of which constitutes a consistent interpretation of the received observations in the region. The tracks within a cluster compulsory share directly or via other tracks common observations in their track history. The track history is a list of observations associated with the track, whose length theoretically may vary from 2 to k scans. The main benefit of maintaining numerous interpretations of the received measurement data in parallel is to have the ability to go back to the right chain of association decisions, if one or more wrong association decisions has been made in the past. The longer the track history is, the farther one can go back trying to "repair" his chain of association decisions and the higher the computational requirements will be.

The MHT processing chain starts with propagation of all tracks, from scan k-1 to the next scan time k. The respective track gates are used to correlate these tracks with the arrived observations. According to the applied association logic, two or more track clusters are immediately merged if an observation is simultaneously validated by two or more tracks belonging to different track clusters.

Once the clusters are formed, hypotheses are generated for the observations within each cluster. Each track can be associated with only one observation per scan, which falls in the respective track's gate. A set of mutually non-contradicting associations forms candidate tracks depicted by a separate hypothesis. If all observations are either: (a) associated with only one track belonging to the hypothesis or (b) declared to be a FA (under the given hypothesis) or (c) considered as a new track, a complete hypothesis set is generated for a cluster using the following steps [7]:

(a) Start with an empty hypothesis (empty track set).
(b) List all possible tracks formed within the cluster.
(c) Select a track from within the cluster. Add this to the hypothesis.
(d) Remove from the list all tracks that are inconsistent with the track selected.
(e) Select another track from those remaining.
(f) Go to 4.

This complicated association mechanism generates multiple branches of already existing tracks as well as multiple new tracks. Numerous tracks are built even when a single target is observed. In case (a), the track whose gate is considered will be updated with validated observation. In all other cases it is suggested that the observation from the considered track is not detected, so the track is propagated using just a priori data. The measurement orientation of the MHT algorithm is based on the constant presence of the hypothesis corresponding to case (c). In order to provide timely track split detection, it generates new tracks on the base of each validated measurement. In most cases this hypothesis is inconsistent and generates new tracks, which are copies of already existing tracks, but in military applications the new target detection is especially important system requirement. In more complicated cases, when one or more observations are simultaneously validated in two or more gates, each validated observation is used to update consecutively each track, in whose gate it is validated, suggesting that all remaining observations in this gate are FA or new targets. Figure 4 shows the track initiation, updating and branching possibilities resulting from the new sets of observations arriving in two consecutive scans k-1 and k. Evidently, the number of generated tracks grows exponentially, leading to "combinatorial explosion" of the hypothesis tree. This feature makes the MHT algorithm the most difficult algorithm for practical implementation.

Once the tracks are branched and/or initiated, MHT evaluates the likelihood that each track (group of detections satisfying track gates) represents a true target, as opposed to the possibility that all detections are FA. The confirmation and deletion of tracks is based on a track scoring function. This score function utilizes the previous score of a track and residual information associated with it to update scores for each track at each update. Based on these scores, a sequential probability ratio test is used to determine which tracks should be deleted and which tracks should be declared as valid [7].

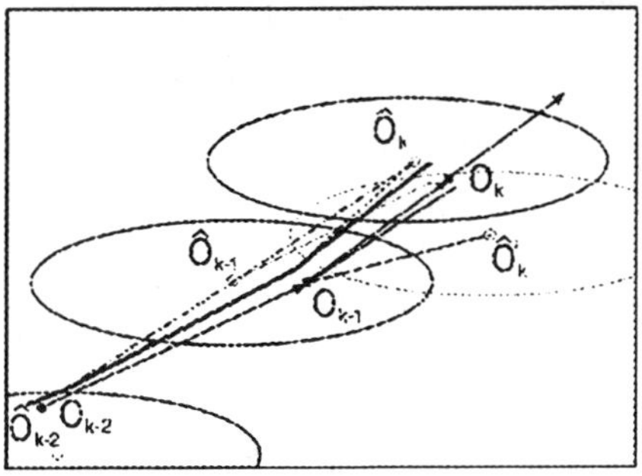

Figure 4. Multiple hypotheses resulting from two consecutive scans

The a posteriori probability of track validity given sequence of observation data $O^k=\{O_j\}^k_{j=1}$ is obtained:

$$Pr(T/O^k) = \frac{Pr(O_k/T)\,Pr(T/O^{k-1})}{Pr(O_k)}$$

where: $Pr(T/O^k)$ is the a posteriori probability that the sequence of returns $O^k=\{O_k,O^{k-1}\}$ used to form track T at scan k represents a sequence of true target observations voting in the track gate; $Pr(O_k/T)$ is the likelihood function-- the probability of receiving the observation O_k given that a true target is present; $Pr(T/O^{k-1})$ is the a priori probability of a true target being present in the track which has associated the observation sequence O^k and, finally, $Pr(O_k)$ is the probability of receiving the observation O_k. The probability $Pr(T/O^k)$ can be also defined in terms of likelihood ratios as:

$$Pr(T/O^k) = \frac{L_k Pr(T/O^{k-1})}{L_k Pr(T/O^{k-1}) + 1 - Pr(T/O^{k-1})}, \qquad (9)$$

where L_k is the likelihood ratio for the observation O_k defined as:

$$L_k = L(O_k) = \frac{Pr(O_k/T)}{Pr(O_k/FA)}$$

and $Pr(O_k/FA)$ is defined for FA in the same manner that $Pr(O_k/T)$ is defined for true targets. The initial track probability $Pr(T/O^0)=Pr(T)$ is based on estimated spatial densities β_{NT} and β_{FA} :

$$Pr(T) = \frac{\beta_{NT}}{\beta_{NT}+\beta_{FA}}.$$

In order to calculate L_k, the probabilities $Pr(O_k/T)$ and $Pr(O_k/FA)$ are first computed:

$$Pr(O_k/T) = P_D\alpha = P_D\frac{e^{-0.5d^2}}{(2\pi)^{0.5m}\sqrt{det(S)}}, \quad and \quad Pr(O_k/FA) = \frac{P_{FA}}{V_g}$$

Since $P_{FA}=\beta_{FA}V_g$, where β_{FA} is the FA or clutter spatial density, L_k may be expressed as:

$$L_k = P_D\frac{e^{-0.5d^2}}{\beta_{FA}(2\pi)^{0.5m}\sqrt{det(S)}}, \qquad (10)$$

if a detection is observed, and similarly, if no detection is observed:

$$L_k = \frac{1-P_D}{1-P_{FA}}. \qquad (11)$$

Once the above likelihood are calculated and fed into equation (8), more convenient log-likelihood ratios (or "scores") are usually computed as

$$L_{LR}(k) = lg\left[\frac{Pr(T/O^k)}{1-Pr(T/O^k)}\right] = lg\left[L_k\frac{Pr(T/O^{k-1})}{1-Pr(T/O^{k-1})}\right] =$$

$$= lg\left[\prod_{j=0}^{k} L_j\right] = \sum_{j=0}^{k} lg L_j, \qquad (12)$$

where the initial score is given by

$$L_0 = \frac{Pr(T)}{1-Pr(T)} = \frac{\beta_N}{\beta_{FA}}. \qquad (13)$$

$lg L_j <$

In summary, MHT track scores are initiated using equation (13), and further incremented by the $_j$ of either equation (10) or equation (11), whichever is appropriate. Looking at equation (9), we see that the $_j$ can be either negative or positive, depending on if detection was observed, and if so, where that detection fell within the correlation gates. Evidently, the tract score will be decremented for every missed updated, since usually

$L_j = \frac{1 - P_D}{1 - P_{FA}} < 1$, so $\lg L_j < 0$, and it will grows in the opposite case. The computed track scores are used to perform a sequential probability ratio test that compares the scores with two thresholds (upper T_U, for track confirmation and lower T_L, for track cancellation). The following thresholding decision logic is then used [7]:

(a) $L_{LR} <= T_L$, drop track;
(b) $L_{LR} >= T_U$, confirm track;
(c) $T_L < L_{LR} < T_U$, wait for more data.

The processor load is controlled by controlling the number of track processed during a given period of time. Tracks, whose track score falls below computed deletion threshold T_{DR}, are terminated. Merging tracks with identical histories or kinematically close tracks [1] over the last s frames ($2<s<6$, for example) limits redundant tracks on the same target and excludes the track-copies generated by the hypothesis considering each observation as a new target.

On the other hand, the score of any hypothesis within a cluster is simply the sum of the scores of the tracks contained within the hypothesis. To reduce the computational load hypotheses containing one and the same tracks are merged, too [1], and their scores are summarized. Sorting the remaining hypotheses it becomes possible to prune these whose score is lower than given threshold.

Finally, the tracks belonging to remaining hypotheses undergo Kalman filtering and a cluster is split (if it is possible) to reduce the processor throughput. This operation leads to computational savings since the processing requirements for several functions within MHT are roughly proportional to n^3 [7], where n is the number of tracks considered together (in a cluster). By dividing n into, for example, 10 clusters of equal size the processing required decreases roughly as $10\left(\frac{n}{10}\right)^3 = \frac{n^3}{10}$, two orders of magnitude reduction. At the next scan the MHT processing chain begins again with tracks propagation.

The general MHT algorithm has the following form [4]:

First scan: *Read basic input data & first scan data;*
For all next scans: *Read the next scan data*
-For each track of each cluster:
--For each observation:
----If observation falls in a track gate of this cluster & in a track gate from another one:
----Make super cluster from these clusters.
---If an observation is out of all gates, then make new cluster.
-For each old cluster:
--For each new observation associated with track from this cluster:

---For each hypothesis considering tracks from this cluster:
----For each track from this hypothesis:
-----If the observation falls in the track gate:
------If this track-observation pair is not encountered yet:
------Create a new track from this pair, continuing the old one, and
------Add a new hypothesis with a new track in the cluster.
------Make a new track from the current observation, and
------Add the new track to the above hypothesis;
----Leave first M1 hypotheses & prune others;
--Combine closely spaced tracks or tracks made up of one and the same observations;
--Combine hypotheses made up of one and the same tracks;
--Leave best M2 hypotheses & prune others;
--Filter all tracks in the cluster;
--Split clusters if possible;
-Delete clusters with no tracks in them.

5. Conclusions

The Kalman filter and all its simplified or sophisticated versions are usually the basic estimation technique used to develop different association mechanisms. Based on the predicted target state it becomes possible to estimate the expected observation position and to correlate the arriving target returns and false measurements with tracks. Five basic association techniques are considered in this chapter: the Nearest and Strongest Neighbor approaches (effective just for sparse environments), the Probabilistic and Joint Probabilistic Data Association Filters, and finally, the Multiple Hypothesis Tracking algorithm. The PDA/ JPDA algorithms are considered in [1] as a special case of MHT. Their calculations are similar to those, required for MHT. According to it, the advantage of the JPDA is that it is a relatively simple recursive method, which does not require the storage of past observation data nor multiple hypotheses. An apparent disadvantage associated with the PDA/JPDA is the lack of an explicit mechanism for track initiation. As discussed in [1], however, it is suitable to employ other batch- type algorithms for track initiation and then to use the PDA/JPDA for track maintenance. It is also noted [1] that the most important factor in the choice of MHT versus JPDA methods is probably the FA density. For high FA densities, such as in sonar or radar air-to-ground tracking applications, MHT is considered not feasible, and the JPDA is favored. The MHT is considered feasible for the lower false target densities associated with the radar air-to-air tracking problem. In contrast to this opinion, it is shown in [4], that introducing batch-processing track initiation procedure based on application of Hough Transform it becomes possible to decrease significantly the required computational load, generally improving the MHT algorithm performance.

6. References

1. Blackman S.S. (1986). *Multiple Target Tracking with Radar Applications*, Artech House.
2. Duda P.O., Hart, P.E. (1973). *Pattern Classification and Scene* Analysis, Willey.

3. Reid, D.B. (1979). "An Algorithm for Tracking Multiple Targets", IEEE Trans. on Automatic Control, AC-24, pp.843-854.
4. Semerdjiev E.A., Alexiev,K.M., Djerassy, E.N., Constantinova, P.D. (1999). "Multiple Hypothesis Tracking Using Hough Transform Track Detector". INFORMATION & SECURITY, An International Journal, Vol. 2, Sofia, Bulgaria, pp. 113-122.
5. Bar-Shalom Y., Fortman, T.E. (1988). *Tracking and data association*. Academic Press.
6. Bar-Shalom Y., Tse, E. (1975). "Tracking in cluttered environment with Probabilistic Data Association", Automatica, Vol. 11, pp. 451-460.
7. Werthmann (1992). "A step by step description of a computationally efficient version of multiple hypothesis tracking", SPIE vol.1698 Signal and Data Processing of Small Targets pp. 0-8194-0863-8/92.

CHANGE DETECTION THROUGH MULTISENSOR PROCESSING

E. SEMERDJIEV
Central Laboratory for Parallel Processing
Bulgarian Academy of Sciences
Department "Mathematical Methods for Sensor Data Processing"
25-A Acad. G. Bonchev Str., 1113 Sofia, Bulgaria
Email: signal@bas.bg

1. Introduction

The problem of detecting changes in stochastic dynamic systems arises in many important areas of human activity, such as geophysics, medicine, technical diagnostics, transportation, control of technological processes and etc. E. S. Page formulated it as a mathematical statistics problem during the 1950s, in his papers [1,2]. At the end of the 1970s and at the beginning of the 1980s the problem of abrupt change detection was recognised as the most difficult problem of its kind. The scientific efforts in this area supported the intensive research in the fields of fault detection and localisation, image and signal processing and moving objects tracking. In [3] A.S Willsky, M.A. Bassewille and B. Benveniste described the nature of this problem, stating that the Multiple Model (MM) approach is the most effective solution. Describing the abrupt changes in the behaviour of stochastic systems by a set of appropriate models it becomes possible to design efficient methods for their detection. Processing sequences of measurements obtained from the observed system, it becomes possible to acquire indications for change, to accumulate respective statistics and to apply rules for timely and reliable change detection. Finally, H. A. P. Blom and Y. Bar-Shalom [4,5] have proposed the most effective algorithm for abrupt change detection - the Interacting MM (IMM) algorithm, described in numerous publications. The MM change detection mechanism manages a set of mathematical models, created according to given set of hypotheses about the stochastic system behaviour. In practice, every model from the above set is "embedded" in a respective Kalman Filter (KF). This allows assessing of the validity of each hypothesis through investigation of the KF statistical properties. That is why a brief description of the KF and the basic techniques for change detection is given in Section 1. The next Sections 2, 3 and 4 are devoted to the MM approach and its simplified applications - the General Pseudo-Bayesian (GPB) and IMM approaches [6], which are the two most popular MM change detection algorithms. The multiple sensor change detection case is considered in Section 5. Finally, an IMM algorithm implementation for the purposes of change detection in adaptive control systems is given, in Section 6.

A. K. Hyder et al. (eds.), Multisensor Fusion, 497–510.

1.1 THE KALMAN FILTER - KEY ELEMENT OF THE CHANGE DETECTION ALGORITHMS

The Kalman filter (KF) is a popular mathematical technique for state estimation and prediction of stochastic dynamic systems, observed upon the base of sequence of noisy measurements. Its discrete standard form is derived for linear stochastic dynamic systems and linear sensor models, as follows:

$$\begin{aligned} x(k) &= F[M(k-1)]\,x(k-1) + q[k-1, M(k-1)] + \\ &+ G[M(k-1)]\,u[k-1, M(k-1)] \\ y(k) &= H[M(k)]\,x(k) + v[k, M(k)]\,. \end{aligned} \tag{1}$$

The two vectors - the n -dimensional vector x, that includes the target states to be estimated and the m -dimensional ($0<m<n$) measurement vector y, are considered Gaussian. The p -dimensional control vector u may be unknown control input. The $n \times n$ system transition matrix ϕ, the $n \times p$ control matrix G, as well as the $m \times n$ measurement matrix H are assumed known. The system and the measurement noises q and v are assumed mutually independent, zero-mean, white, Gaussian noises with known covariances Q and R.

The presence of the system mode $M(k)$ in equations (1) and (2) denotes the fact, that system model may change in moment k, i.e. during the interval $(k-1;k)$. It is also suggested that the system can change its mode in discrete moments of time 1,2,…,k. The models may differ by their parameters, state vectors length, noise levels, or by deterministic inputs. The number of possible system modes r is usually considered finite, constant and known:

$$M(k) \in \{M_j\}_{j=1}^{r}. \tag{2}$$

Because of the presence of both continuous (noise) and discrete (model) uncertainties, such systems are often called *hybrid* [6].

The most of change detection algorithms utilise the Kalman Filter (KF), which is a powerful tool for real-time estimation of unknown parameters. It consists of two steps. On the first step the algorithm predicts the state and measurement vectors from the current moment k-1 to the next moment k and calculate their error covariances:

$$\begin{aligned} \hat{x}(k/k-1) &= \Phi[M(k-1)]\,\hat{x}(k-1/k-1) + \\ &+ G[M(k-1)]\,u[k-1, M(k-1)], \end{aligned} \tag{3}$$

$$\hat{y}(k/k-1) = H[M(k-1)]\,\hat{x}(k/k-1), \tag{4}$$

$$\begin{aligned} P(k/k-1) &= \Phi[M(k-1)]\,P(k-1/k-1)\,\Phi^T[M(k-1)] + \\ &+ Q[M(k-1)], \end{aligned} \tag{5}$$

$$S(k) = H[M(k-1)]\,P(k/k-1)\,H^T[M(k-1)] + R[M(k-1)], \tag{6}$$

The predicted estimate $\hat{x}(k/k-1)$ and covariance $P(k/k\text{-}1)$ exhaustively describe the expected state vector distribution, based on *a priori* information:

$$p[x(k)] = N[x(k); \hat{x}(k/k-1), P(k/k-1)]\,. \tag{7}$$

On the second step *a posteriori* state vector distribution, which is conditioned on the received measurement (observation), is defined as

$$p[x(k)|y(k)] = N[x(k);\hat{x}(k/k),P(k/k)]. \quad (8)$$

The vector of filtered estimates $\hat{x}(k/k)$ and the corresponding error covariance matrix $P(k/k)$ are:

$$\hat{x}(k/k) = \hat{x}(k/k-1) + K(k)\tilde{y}(k), \quad (9)$$

$$P(k/k) = P(k/k-1) + K(k)S^{-1}(k)K^{T}(k). \quad (10)$$

The KF gain $K(k)$ and the measurement prediction errors $\tilde{y}(k)$ (named also *residual* vector) are defined as

$$K(k) = P(k/k-1)H^{T}[M(k)]S^{-1}(k), \quad (11)$$

$$\tilde{y}(k) = y(k) - \hat{y}(k/k-1) = H[M(k)][x(k) - \quad (12)$$

$$-\hat{x}(k/k-1) + v[M(k)].$$

The residual vector is zero-mean. Any sequence of N such vectors generated by KF is white, if the filter model exactly matches the true system dynamics, i.e.:

$$E[\tilde{y}(i)] = 0, \; i = 1,\dots,k \quad (13)$$

$$E[\tilde{y}^{T}(i)\tilde{y}(j)] = 0, \; if \; i \neq j \quad (14)$$

where $E[.]$ is the mathematical expectation operator performed over the value enclosed in the brackets. The above statements are easy to be proven through substitution of equations (1)--(4) in equation (7) and taking into account the statistical properties of the system and measurement noises. When the KF model coincides with the true system model (for example, if u [.] is known, or if $u(k)$ =0 in both models), the *residual* vector is normally distributed:

$$p[\tilde{y}(k)] = p[y(k)|\hat{x}(k/k-1)] = \frac{e^{-0.5d^2(k)}}{(2\pi)^{0.5m}\sqrt{det[S(k)]}} \quad (15)$$

The normalised distance squared d^2 is a stochastic vector, which obeys the chi-square distribution with m, the dimension of the measurement, degrees of freedom [6], [10]

$$d^2(k) = \tilde{y}^{T}(k)S^{-1}(k) \sim \chi^2_m \quad (16)$$

Using the standard tables describing the chi-square distribution it may be stated with given tail probability α that d^2 won't exceed a preliminary defined threshold d^2_{max} :

$$Pr\{d^2(k) < d^2_{max}\} = 1 - \alpha. \quad (17)$$

The bias $H[M(k)]G[M(k)]u[k-1,M(k)]$ appears when the KF system model does not match reality. In this case the statistical properties of the normalised distance squared d^2 will change, exceeding the threshold d^2_{max}. This allows development of a single test change detection technique. A time average over a sum of N consecutive normalised distances squared:

$$\sum_{j=k-N+1}^{k} d^2(j) \quad (18)$$

is often used instead the above statistic. It obeys the chi-square distribution with Nm degrees of freedom. This sum is collected through a "sliding window" incorporating each newly computed distance $d^2(k)$ and discarding the "oldest" one $d^2(k-N)$. This mechanism is used to develop the popular *Input estimation technique*. It considers the residual vector of the mismatched KF as a linear measurement of the unknown input. It is estimated using the least-squared method. An exponentially discounted average $d^2(k)+\phi d^2(k-N)$ (where $0<\phi<1$) may be also used for change detection tests.

2. The Multiple Model Approach

The Multiple Model (MM) approach is the most effective approach for change detection. It considers the behaviour of the observed system dynamics as a sequence of transitions (jumps) between multiple models, composing a model history through time k :

$$M^{k,l} = \{ M^{k-1,s}, M(k) = i \}, \quad l = \overline{1,r^k}, \tag{19}$$

where l and s are the indexes denoting particular model history; i=1,..,r is the current mode number in the mode sequence l corresponding to the last moment k. Each model history represents certain hypothesis uniquely interpreting the already arrived observations. Evidently, the number of possible hypotheses r^k grows up exponentially with the time k. Attempting to reduce their number it is often assumed that the model jump is a Markov process with constant transition probabilities, independent from the system state:

$$\pi_{ij} = Pr\{ M(k) = j | M(k-1) = i \}, \quad i,j = \overline{1,r}. \tag{20}$$

These probabilities are considered known, but actually they are design parameters which have to be selected during the design process.

The probabilities of the posterior state distribution of the observed system are obtained from the prior ones using the total probability theorem with respect to the mutually exclusive and exhaustive set of events describing possible system jumps. The conditional probability of the l -th hypothesis (model history) is evaluated using Bayes' formula:

$$\mu^{k,l} = Pr\{ M^{k,l} | Y^k \} = \frac{1}{c} p\{ y(k) | M^{k,l}, Y^{k-1} \} \pi_{ij} \mu^{k-1,s}, \tag{21}$$

where due to the Markov properties of the model jump, the current model i from the l -th history $M^{k,l}$ depends only on model j - the last model of the s -th history $M^{k-1,s}$ ending at moment k-1 ; c is a normalisation constant.

The likelihood function Λ_i is usually considered Gaussian, and its parameters are obtained from the KF matching model i:

$$\Lambda_i(k) = p\{ y(k) | M^{k,l}, Y^{k-1} \} = p\{ y(k) | M(k) = i \} = N[\tilde{y}_i(k); 0, S_i(k)]. \tag{22}$$

In order to detect the most likely hypothesis the optimal MM approach "runs" separate KF along each possible mode history. Evidently it may be stated that this approach is generally inapplicable. An exponentially increasing number of KF is needed even suggesting that the model-jumps obey Markov chain properties. That is why a number of sub-optimal techniques are developed for detection of possible changes in the target behaviour in practice.

Finally, it may be noted that all MM algorithms are often called algorithms with "soft decision logic" -- since the model switch is made on the basis of last computed mode probabilities, and no change detection decision is needed.

3. The Generalised pseudo-Bayesian approach

3.1 FIRST ORDER GENERALISED PSEUDO-BAYESIAN APPROACH (GPB1)

At the end of the previous moment k-1 the first order generalised pseudo-Bayesian approach (GPB1) [6] combines all r previous one-step system histories in a single one. A single Gaussian distribution approximates the corresponding probabilistically weighted sum of Gaussian distributions. The single estimate $\hat{x}(k-1/k-1)$ and its error covariance $P(k-1/k-1)$ describe it. At the next time k these estimate and error covariance are processed in parallel by r mode-matched KF. The separate filter outputs $\hat{x}^j(k/k)$ and $P^j(k/k)$ are finally combined, thus obtaining the fused input estimate and error covariance for the next sampling interval [6]:

$$\hat{x}(k/k) = \sum_{j=1}^{r} \hat{x}^j(k/k)\mu_j(k), \tag{23}$$

$$P(k/k) = \sum_{j=1}^{r} \mu_j(k)\{P^j(k/k) + [\hat{x}^j(k/k) - \hat{x}(k/k)] \times \tag{24}$$

$$\times [\hat{x}^j(k/k) - \hat{x}(k/k)]^T\}.$$

The model probabilities $x_j(k)$ are used as combination weights. They are computed as

$$\mu_j(k) = \frac{1}{c}\Lambda_j(k)\sum_{i=1}^{r} \pi_{ij}\mu_i(k-1), \tag{25}$$

$$c = \sum_{j=1}^{r} \Lambda_j(k) \sum_{i=1}^{r} \pi_{ij}\mu_i(k-1), \tag{26}$$

$$\Lambda_j(k) = p[y(k)|M(k) = j, Y^{k-1}] = \tag{27}$$

$$= p[y(k)|M(k) = j, \hat{x}(k-1/k-1), P(k-1/k-1)].$$

The respective processing scheme for r=2 models is given in Figure 1.

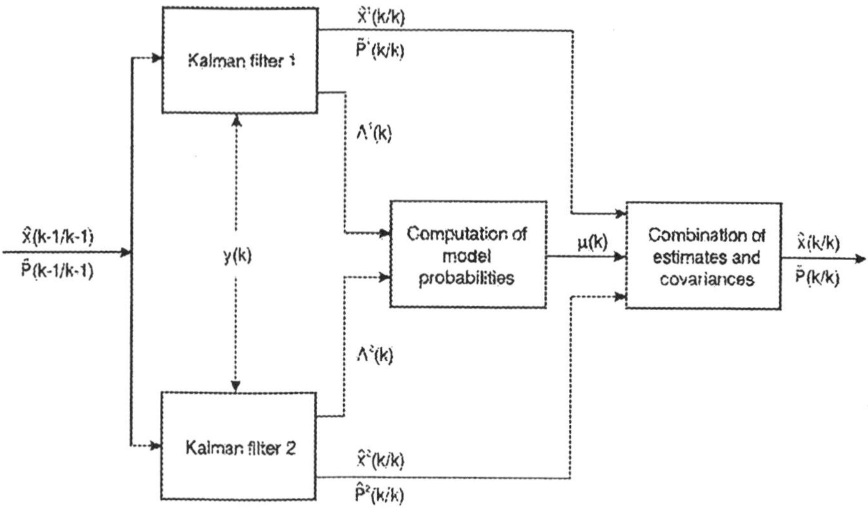

Figure 1. GPB1 for models

3.2 SECOND ORDER GENERALISED PSEUDO-BAYESIAN APPROACH (GPB2)

The second order generalised pseudo-Bayesian approach (GPB2) considers all the possible models included in two-step mode histories. At time k it passes each estimate $\hat{\tilde{x}}^{j}(k-1/k-1)$ ($j=\overline{1,r}$) and error covariance $P^{j}(k\text{-}1/k\text{-}1)$ through each of the r KF generating r^2 hypotheses about possible system transitions. As a result of these r^2 KF applications, r^2 output estimates $\hat{x}^{ij}(k/k)$ and error covariances $P^{ij}(k/k)$ are obtained. Then all estimates $\hat{x}^{ij}(k/k)$ and respective covariances $P^{ij}(k/k)$ corresponding to model histories finishing at certain last model j are merged to obtain the new filter-matched estimate:

$$\hat{\tilde{x}}^{j}(k/k) \tag{28}$$

and estimation errors covariance $P^{j}(k/k)$:

$$\mathrm{x}[x(k/k) - x(k/k)]^{T}\ \}\ \mu_{i|j}\quad (k\text{-}1/k). \tag{29}$$

The combination weights (the model probabilities) are obtained as probabilities that mode i is in effect in moment k-1, if mode j is in effect in moment k, given observation $y(k)$: $(i,j = 1,r)$.

The latest state estimate, covariance and mode probabilities are *only for output only*:

$$\hat{x}(k/k) = \sum_{j=1}^{r} \hat{x}^{j}(k/k)\,\mu_j(k-1/k), \tag{30}$$

$$P(k/k) = \sum_{j=1}^{r} \{P^{j}(k/k) + [\hat{x}^{j}(k/k) - \hat{x}(k/k)] \times \tag{31}$$

$$\times [\hat{x}^{j}(k/k) - \hat{x}(k/k)]^{T}\}\, \mu_j(k-1/k),$$

$$\mu_j(k) = \frac{1}{c}\sum_{i=1}^{r} \Lambda_{ij}(k)\,\pi_{ij}\mu_i(k-1) = \frac{c_j}{\sum_{j=1}^{r} c_j} \quad (i,j = \overline{1,r}). \tag{32}$$

The GPB2 processing scheme is given in Figure 2.

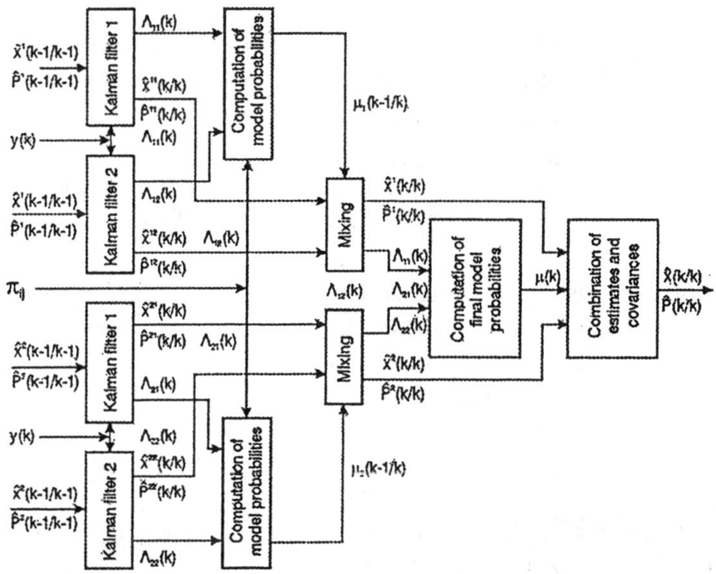

Figure 2. GPB2 for models

4. Interacting multiple-model (IMM) approach

The IMM algorithm is considered as the best compromise between the achieved algorithm performance and the required computational load. It is nearly as good as GPB2 algorithm, and it is practically as simple as the GPB1 algorithm. It provides estimation errors lower than the errors of the rough data during the system transitions, and significantly reduce these errors on the steady state stages. This improvement is achieved taking into account possible change of the system mode "in advance". For this purpose the IMM algorithm makes some initial "preparations" accelerating the change detection procedure. Interacting separate "old" estimates $\hat{x}^j(k-1/k-1)$ and covariances $P^i(k\text{-}1/k\text{-}1)$ of all r model-matched KF, the IMM algorithm prepares r different input mixtures for to them. For example, the initial mixture for the KF matched to model j is obtained combining the above estimates and covariances:

$$\hat{x}^{0j}(k-1/k-1) = \sum_{i=1}^{r} \hat{x}^i(k-1/k-1)\,\mu_{i|j}(k-1/k-1); \tag{33}$$

$$P^{0j}(k-1/k-1) = \sum_{i=1}^{r} \mu_{i|j}(k-1/k-1) \times \tag{34}$$

$$\times \{P^i(k-1/k-1) + [\hat{x}^{0j}(k-1/k-1) - \hat{x}^i(k-1/k-1)] \times$$
$$\times [\hat{x}^{0j}(k-1/k-1) - \hat{x}^i(k-1/k-1)]^T\},$$

where $\mu_{i|j}(k-1)(k-1) \quad (i,j = 1,r)$ is the *a priori* probability of the event that the model i was in effect in the previous moment k-1, given that model j is in effect in the current moment k, and no measurement is made yet:

$$\mu_{i|j}(k-1/k) = Pr\{M(k-1) = i|M(k) = j, Y^{k-1}\} = \frac{1}{\bar{c}_j}\pi_{ij}\mu_i(k-1), \tag{35}$$

$$\bar{c}_j = \sum_{i=1}^{r} \pi_{ij}\mu_i(k-1), \quad i,j = \overline{1,r} \tag{36}$$

In this way, the algorithm takes into account the possible system transition -- the contribution of any of the previous modes is directly proportional to its probability μ_I (k-1) and to the corresponding transition probability. All above operations compose the first algorithm step named *Interaction/Mixing*.

On the next *Mode-matched filtering* step each mode-matched KF propagates and filters the input estimates and covariances $\hat{x}^{0j}(k-1/k-1)$ and $P^{0j}(k\text{-}1/k\text{-}1)$, defining the new separate ones, $\hat{x}^j(k/k)$ and $P^j(k/k)$. The key parameter here is the likelihood function Λ_j :

$$\Lambda_j(k) = N\{\tilde{y}_j(k); 0, S_j(k)\}\;\frac{e^{-0.5d^2}}{(2\pi)^{0.5m}\sqrt{det[S_j(k)]}} \tag{37}$$

It is as low as long is the normalised distance squared d^2. This function is used to update (i.e. to correct) the corresponding mode probability

$$\mu_j(k) = \frac{\Lambda_j(k)\bar{c}_j}{\sum_{i=1}^{r} \Lambda_i(k)\bar{c}_i} \tag{38}$$

This algorithm step is named *Mode-probability update*. Finally, on the *Combination step* the IMM algorithm generates combined estimate and covariance needed *only for output purposes*:

$$\hat{x}(k/k) = \sum_{j=1}^{r} \hat{x}^j(k/k)\,\mu_j(k), \tag{39}$$

$$P(k/k) = \sum_{j=1}^{r} \mu_j(k)\{P^j(k/k) + [\hat{x}(k/k) - \hat{x}^j(k/k)][\hat{x}(k/k) - \hat{x}^j(k/k)]^T\} \tag{40}$$

Figure 3 depicts one step of the IMM algorithm.

The general processing scheme of the MM estimation algorithm is given in Figure 4 [6].

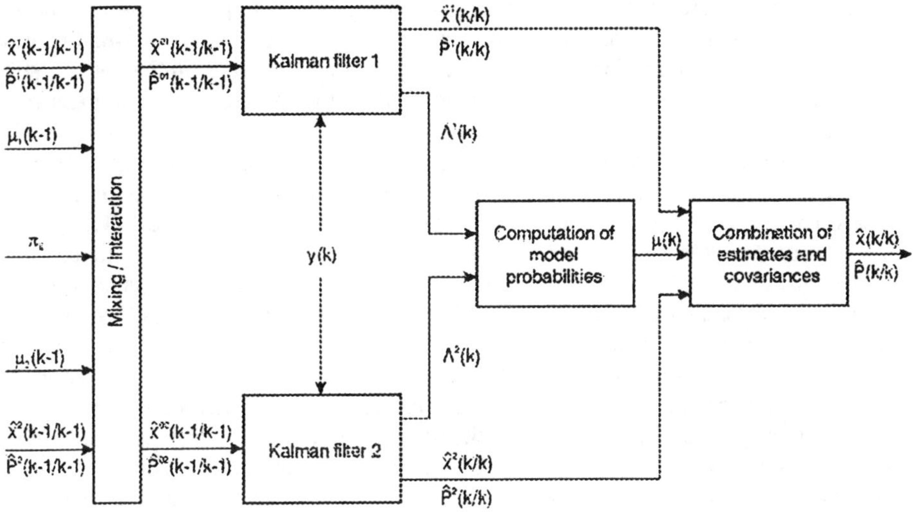

Figure 3. The IMM algorithm

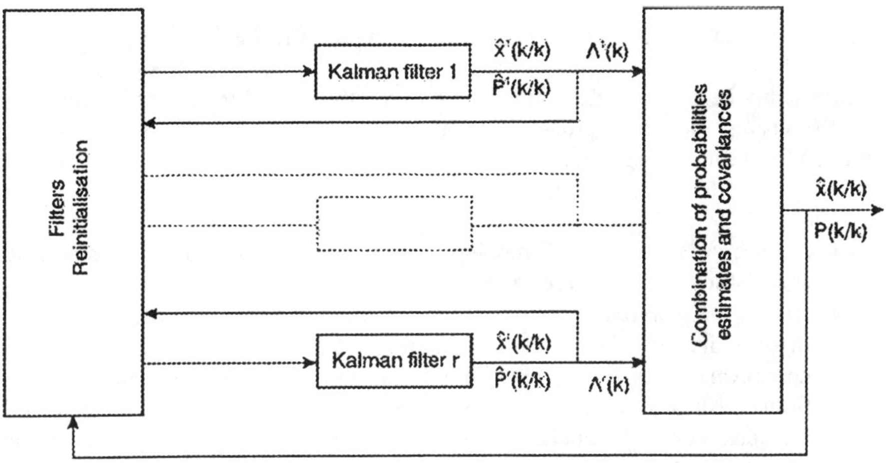

Figure 4. Generalized structure of a MM estimator

5. Multiple sensor change detection

The change detection techniques described above are also directly applicable in the cases when multiple sensors a stochastic system.

For example, in a number of navigation systems different on-board sensors measure different state vector components, thus composing a single joint measurement model. Despite the increased complexity of such models, they do not require any modifications to the applied change detection techniques (except for possible substitution of the standard linear KF with the more sophisticated extended KF, if the composite measurement model is non-linear).

In the case when two or more sensors simultaneously observe one and the same state parameters it becomes necessary to apply some pre-processing techniques on the received measurements, or some post-processing to the estimated state parameters. The purpose of the measurements pre-processing (often called *measurement fusion*) is to combine all measurements made by different sensors, which concern one common system entity (state component, for example). The generated single artificial measurement is then processed in standard the way by a standard change detection algorithm. Because of the usual lack of measurement synchronisation or/and due to the presence of false measurements or measurements concerning other observed stochastic systems, it often becomes very difficult to determine which measurements should be combined. The solution of these *synchronisation* and *assignment problems* may require implementation of sophisticated association logic in the above algorithms, but the core change detection techniques remain the same.

The application of post-processing techniques assumes that the standard change detection procedure is applied in parallel in each separate sensor, and just then the output estimated state vectors are combined (so-called *track fusion*).

Finally, it may be noted that the application of multiple sensors generally improves the overall sensor performance making the change detection process faster and more reliable.

6. An example of adaptive control algorithm using the IMM algorithm

The Interacting Multiple Model estimator was recently applied in [7] for the first time to solve the change detection problems in adaptive control systems. According to the separation theorem, the suggested stochastic control algorithm consists of two parts (Figure 5):

- *Separate estimators* - KF, running in parallel to provide input signals (the partial state estimates) for the regulators;
- *Separate regulators*, generating single-model-based control processes. However, these units are not so easy separable. They interact efficiently, but in a complicated way. The algorithm forms the overall system control $u(k)$ as a probabilistically weighted sum of separate control processes $u_j(k)$ obtained from the corresponding regulators. These regulators are computed as a state feedback minimising the quadratic cost function, where B and C are known positive-definite

$$J = E\{x^T(k)Bx(k) + u^T(k)Cu(k)\}, \tag{41}$$

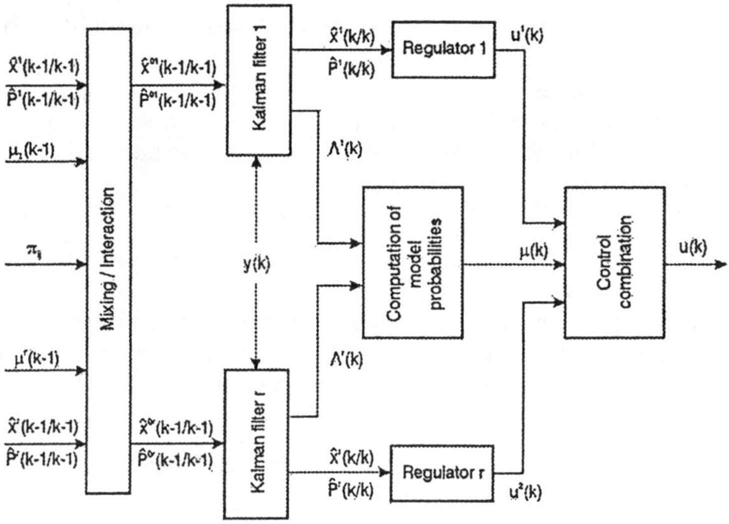

Figure 5. IMM control algorithm

weights. Each regulator is developed for known system model corresponding to a particular system mode. The separate state estimates $\hat{x}^i(k/k)$ generated by the IMM algorithm are based on exactly the same known models. They provide the respective regulatorsÆ inputs. The IMM model probabilities $\mu_j(k)$ represent the weights required for overall control synthesis:

$$u(k) = -\sum_{i=1}^{r} \mu_i(k) u^i(k), \tag{42}$$

where $\hat{x}^i(k/k)$ is the partial state estimate generated by the i -th KF. The partial regulators are linear quadratic Gaussian regulators. The overall state estimate $\hat{x}(k/k)$ and the respective state errors covariance matrix $Px(k/k)$ are not directly used and have the standard form (4) – (6).

Some results illustrating the efficiency of the proposed IMM algorithm for stochastic systems control are given below. The proposed IMM algorithm is applied to a paper machine [7] producing super-thin condenser paper. The state space model of its head-box section [7] has the form described by equation (1), where

$$F = \begin{pmatrix} 0.8667 & 0 & 0 \\ 0 & 0.8667 & 0 \\ 0.1069 & -0.503 & 0.9099 \end{pmatrix}, \quad G_u = \begin{pmatrix} -0.0344 & 0 \\ 0 & -0.6877 \\ 0 & 0 \end{pmatrix}. \tag{43}$$

$G_v = I_3$ and I is the identity matrix. The measurement matrix and the noise covariances are:

$$H = \begin{pmatrix} 1 & 0 & 0 \\ 0 & 0 & 1 \end{pmatrix}, \quad D_v = diag(0.16, 0.16, 1), \quad D_w = diag(0.09, 0.16). \tag{44}$$

Four system models differing by their system matrices cover the uncertainty domain F :

$$F_1 = \begin{pmatrix} 0.1000 & 0 & 0 \\ 0 & 0.1000 & 0 \\ 0.1069 & -0.503 & 0.9099 \end{pmatrix}, \quad F_2 = \begin{pmatrix} 0.3000 & 0 & 0 \\ 0 & 0.3000 & 0 \\ 0.1069 & -0.503 & 0.9099 \end{pmatrix} \tag{45}$$

$$F_3 = \begin{pmatrix} 0.5000 & 0 & 0 \\ 0 & 0.5000 & 0 \\ 0.1069 & -0.503 & 0.9099 \end{pmatrix}, \quad F_4 = \begin{pmatrix} 0.8000 & 0 & 0 \\ 0 & 0.8000 & 0 \\ 0.1069 & -0.503 & 0.9099 \end{pmatrix} \tag{46}$$

The other model matrices in the algorithm coincide with the true ones. The fourth model is the closest model to the true system. The matrices of the quadratic cost function are chosen to provide rapid transient processes of the closed-loop system: $B = 100I_3$, $C = 0.01I$.

The IMM transition probability matrix and the initial mode probability vector are chosen:

$$Pr = \begin{pmatrix} 0.91 & 0.03 & 0.03 & 0.03 \\ 0.03 & 0.91 & 0.03 & 0.03 \\ 0.03 & 0.03 & 0.91 & 0.03 \\ 0.03 & 0.03 & 0.03 & 0.91 \end{pmatrix}, \quad \mu(0) = \begin{pmatrix} 1/4 \\ 1/4 \\ 1/4 \\ 1/4 \end{pmatrix} \tag{47}$$

The following measures of performance are used: the average algorithm mode probabilities (shown in Figure 6), and the recursively-computed value of the cost function (shown in Figure 7). The test scenarios have been artificially complicated to test the algorithm performance in the non-stationary case - in the next scenario abrupt changes arise in some elements of the matrix F:

$$F = \begin{cases} \textit{true matrix } F, & k < 50 \\ F_4, & 50 \le k < 70 \\ F_3, & 70 \le k < 150 \\ F_1, & k \ge 150 \end{cases} \tag{48}$$

The mode probabilities of the algorithm, the optimal regulator and the computed cost function J are presented in Figures 6 - 7. Evidently, the algorithm timely recognises the true models (their probabilities are the greatest).

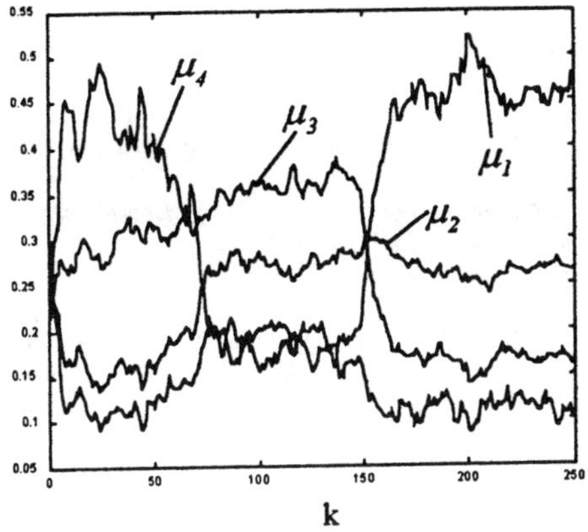

Figure 6. Average Algorithm Mode Probabilities

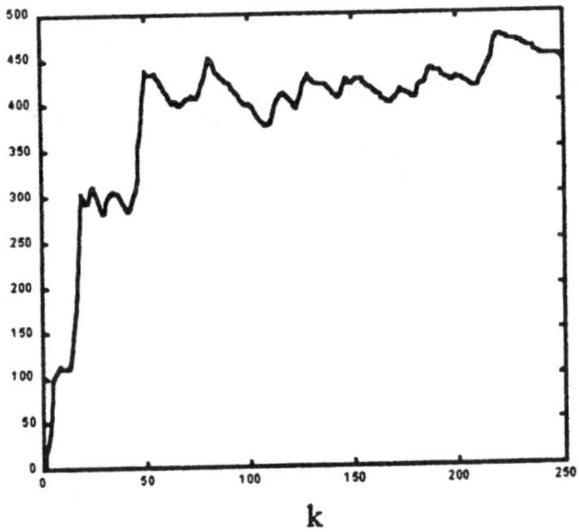

Figure 7. Recursively Computed Cost Function

7. Conclusions

This chapter has provided an introduction to the most popular methods developed to solve the change detection problem. The Kalman Filter is effectively used as an instrument for change detection, by measuring the statistical properties of the innovation process. The optimal Multiple Model approach and its simplified algorithm implementations -- the General Pseudo-Bayesian and the Interacting multiple models algorithms are effective tools for abrupt change detection. The IMM algorithm has proved to be the most cost-effective implementation. Change detection techniques for multisensor change detection may be built on this base. The presented example illustrates its high efficiency in the case of adaptive control. Other more sophisticated IMM algorithm implementations in the fields of fault detection, parameter drift estimation and manoeuvring target tracking can be found in [8],[9].

8. References

1. Page E.S. (1954). "Continuous Inspection Schemes", Biometrika, 41, pp. 100-115.
2. Page E.S. (1955). "A Test for a Change in Parameter Occurring at an Unknown Point", Biometrika, 43, pp. 523- 527.
3. Bassewille M.A., Benveniste, B. (editors) (1986). *Detection of abrupt changes in signal and dynamical systems*, Springer-Verlag, Berlin, Heydelberg, New York, Tokyo.
4. Blom, H.A.P. (1984). "An efficient filter for abruptly changing systems", In Proc. Of 23rd IEEE Conference on Decision and Control.
5. Blom H.A.P., Bar-Shalom, Y. (1987). "The interacting multiple model algorithm for systems with Markovian switching coefficients", IEEE Trans. on Automat. Contr., vol. AES-23, pp. 298-310, May.
6. Bar-Shalom Y., Li, X.R. (1993). *Estimation and Tracking: Principles, Techniques, and Software*. Noorwood, MA,: Artech House.
7. Mihaylova L., Semerdjiev, E. (1999). "An Interacting Multiple Model Algorithm for Stochastic Systems Control", In Proceedings of the European Conference on Multisensor Information Fusion EUROFUSIONÆ99, Stratford-upon-Avon, GB, pp. 49-56.
8. Semerdjiev, E.A., Mihaylova, L.S, and Li, X.R. (1999). "An Adaptive IMM Filter for Aircraft Tracking", 2nd Intern. Conf. On Multisource-Multisensor Information Fusion (FUSION'99), Sunnyvale, USA, Vol. 1, pp.770-777.
9. Mihaylova, L.S, Semerdjiev, E.A, and X.R. Li (1999). "Detection and Localization of Faults in System Dynamics by IMM Estimator", 2nd Intern. Conf. on Multisource-Multisensor Information Fusion (FUSION'99), Sunnyvale, USA, Vol. 1, pp.937-943
10. Bar-Shalom Y., Fortman, T.E. (1988). *Tracking and data association*. Academic Press.

STATISTICAL, LOGIC BASED, AND NEURAL NETWORK BASED METHODS FOR MINING RULES FROM DATA

M. HOLENA
Institute of Computer Science
Academy of Sciences of the Czech Republic
Pod vodárenskou vez' 2
CZ-18200 Praha 8
Czech Republic
martin@cs.cas.cz,
http://www.cs.cas.cz/~martin

1. Introduction

In the seventies, the amount of data typically gathered through collecting or measurements increased so much that the information captured by the data was manageable for humans to utilize it as an aid to decision making. It became obvious that the information needs to be captured in a more succinct and compact way, such as associations, dependencies, class hierarchies, spatial or temporal patterns, in a way more closely reflecting how information is captured in human knowledge. This is particularly true if the data are fused from different sources. As a consequence of that situation, the technology of data mining has developed, which is now experiencing a boom of interest of software producers. Indeed, all main producers of statistical software and several important producers of database systems launched their own data mining tools during the last decade, competed by companies specializing directly in data mining.

Since data mining sometimes needs to be performed in situations when rigorous statistical methods are not applicable, it has encouraged the development of alternative data analysis approaches, relying on artificial neural networks, fuzzy logic, machine learning, etc. On the other hand, approaches with different underlying paradigms can yield different or even contradictory results for the same data. That is an important difficulty pertaining to practical data mining which, as a consequence, can lead to a confusion and to mistrusting the area of data mining as a whole. Another serious difficulty follows from the possibility to capture the same information redundantly in different pieces of knowledge. That possibility entails the danger that an unmanageable amount of data will be eventually replaced by an unmanageable amount of discovered knowledge. Due to both difficulties, integrative data mining frameworks are needed.

In this paper, an integrative framework based on the calculus of observational logic is presented. It has been elaborated already in the seventies, as a theoretical foundation of the method GUHA, one of the first methods for automated knowledge discovery. Originally, observational calculus has been proposed as a means for a unified treatment of different kinds of statistical methods, such as testing and estimation, parametric and nonparametric

A. K. Hyder et al. (eds.), Multisensor Fusion, 511–532.

methods, etc. However, recent research indicates that the framework can be successfully extended also behind statistical methods of knowledge discovery [21,27,28].

In the following section, a survey of the main data mining approaches is given, completed with some practical recommendations to those who want to use them. In addition, two real-world applications are introduced, which are in the sequel used to illustrate the explained concepts and methods. Basic notions pertaining to an observational calculus are presented in section [3] Sections [4] and [5] show how observational calculus can capture the extraction of knowledge from data with several important data mining approaches. The approach considered in section [4] is the extraction of implications by means of statistical hypotheses testing. Finally in section [5], two approaches to the extraction of rules with artificial neural networks are described.

2. Data Mining - An Overview

The term data mining emerged some 20 years ago in statistics. Originally, it was used rather derogatorily, to denote a search for the best fitting model or the most significant hypothesis by trying a large number of models or hypotheses on the same body of data, not worrying about the interpretability and reproducibility of the obtained results [10.36]. Nowadays, data mining belongs to the most prominent information technologies, experiencing a boom of interest both from users and from software producers. During the late 90s, companies in commerce, finance and other sectors started to introduce specialized data mining systems, and many important software producers launched their own systems of that kind (Figure 1). Meanwhile, the meaning of the term has shifted and it is nowadays mostly understood as searching large amounts of data for patterns of interest in some particular representational form or representational forms [14,15]. The representational forms of the sought patterns can be quite various: implications, associations, rules, class hierarchies, clusters, regression functions, dependency structures, and many others.

Figure 1. Logos for six important commercial data mining tools

In connection with data mining, another term, originating in artificial intelligence, is very often used -- knowledge discovery. Differently to data mining, knowledge discovery is usually understood as the overall process of extracting from data what is deemed to be the inherently present knowledge [4,14]. In addition to the data mining step, that process covers a number of further, interactively interleaved steps, most importantly:

- data cleaning;
- data reduction;
- data projection and feature selection;
- incorporating prior knowledge;
- evaluation of the found patterns;
- interpreting the obtained results;
- consolidating the results with previous knowledge.

Needless to say, data mining is the core step of the knowledge discovery process. However, to be able to really discover new valid knowledge, data mining needs to be appropriately combined with the remaining steps. Otherwise, it degrades to its original derogatory meaning.

2.1 BASIC DATA MINING APPROACHES

Traditionally, analysing data and extracting useful knowledge from them has been the domain of statisticians. Therefore, many approaches used in data mining are actually statistical data analysis approaches. Most of them have been known for decades before data mining emerged -- for example, testing statistical hypotheses, estimating parameters of probability distributions, various kinds of regression, discriminant analysis, principal component analysis, clustering and Bayesian classification (see or [12,23] for a survey). Nevertheless, those traditional statistical approaches often have more recent modifications designed specifically to scale to the amount and complexity of data in contemporary databases. In addition, several new statistical approaches with such scaling properties have been elaborated during the last 20 -- 30 years, such as projection pursuit, group method of data handling, or adaptive splines [13,16,17].

Statistical methods underlie also two important approaches outside statistics. One of them, graphical dependency models, forms an independent scientific discipline relying on the one hand on probability theory, on the other hand on combinatorial graph theory [41,54]. They seek probabilistic dependences in data, using as representational form some kind of graph structures. Although still primarily at the research phase, they are of particular interest to data mining since graph structures lend themselves easily to human interpretation.

Also the other approach with a substantial statistical component is attractive due to the simple representational form it uses. It is the extraction of classification rules from data by means of decision trees, an approach belonging primarily to the area of machine learning [5,42]. As long as they are shallow, decision trees are very easy to interpret. In addition, they are also useful in difficult estimation problems where their locality, piecewise constant response surface, and lack of smoothness make them highly robust to outliers in either the predictors or the response variable.

During the last decade, another machine learning approach has been successfully applied in data mining -- inductive logic programming (ILP) [8,39]. It uses the language of first-order logic as a representational form, and consists basically in constructing an intentional definition of a relation from tuples known to belong or not to belong to it, while other relations can be used as background knowledge in the induced definition. Nowadays, a variety of ILP systems exists, for learning single concepts or multiple concepts, in batch mode or in incremental mode, for interactive or non-interactive learning.

The idea of using the language of some logical calculus as a powerful representational form for the knowledge discovered from data is not new to data mining, it has been used already before the advent of ILP. Two kinds of logical calculi should be mentioned here. Observational logic is one of the core topics of this lecture, it will be treated in more detail in the sequel. Fuzzy logic plays a twofold role in connection with data mining. One the one hand, fuzzy rules are one of the basic kinds of rules that can be extracted from data by means of artificial neural networks [2,48]. On the other hand, fuzzy logic serves as a theoretical fundamental for fuzzy generalizations of traditional statistical approaches, such as fuzzy hypotheses testing, fuzzy estimation, fuzzy regression, fuzzy clustering [18,32,35,53]. Those generalizations are useful when either some of the available data are imprecise, or only vague knowledge can be found in them. Finally, in view of the increasing importance that random sets play in data fusion [7], the close connection between fuzzy logic and the theory of random sets should be mentioned [35].

Last but not least, many practically important data mining methods rely on artificial neural networks (ANNs). Most significantly, ANN-based approaches are increasingly successful competitors of traditional statistical approaches to such tasks as classification, clustering, regression, or time series analysis [35,40,50,52]. In addition, one of the main directions of the ANN research has since the last decade been the extraction of symbolic knowledge in the representational form of logical formulae (rules) from trained neural networks. Also this topic will be tackled in some detail later.

Data mining cannot be thought away from various supporting technologies (Figure 2). Probably the most important among them are the database technology, with its specialized descendant -- data warehousing technology, developed specifically for the support of data mining and on-line analytical processing, and several distribute object-oriented technologies, such as CORBA or RMI/JavaBeans [27,51].

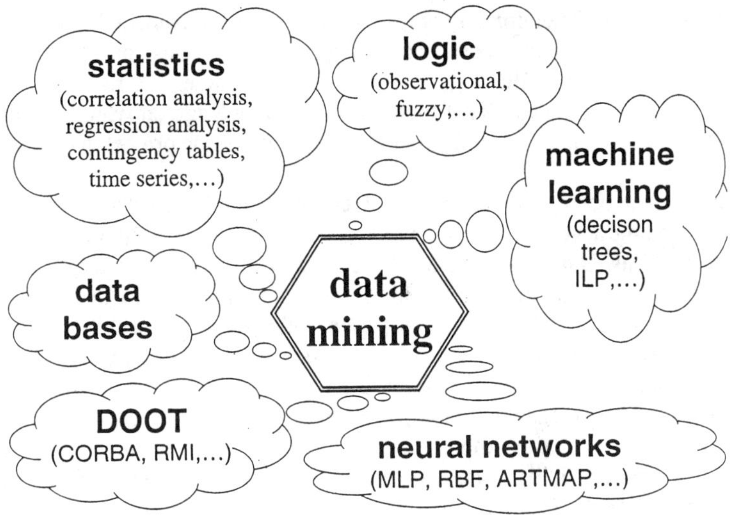

Figure 2. Main data-mining approaches and supporting technologies

2.2 TWO EXAMPLE APPLICATIONS

In this subsection, two real-world applications are sketched that will provide illustrations for the concepts and methods explained in the sequel. One of them will be mentioned only very briefly since it can be only used to illustrate statistical data mining methods. In the other application, on the contrary, all the methods presented in this lecture have been employed, therefore it will be described in more detail.

The first application comes from the clinical research of epilepsy [55]. Here, data mining was applied to search possible relationships between 13 clinical variables commonly used to characterize the disease, and the occurrence of memory disturbances in epileptic patients. The clinical variables concerned mainly the seriousness of the disease (frequency of grand mal seizures, frequency of other epileptic seizures, invalidity, heredity, organic lesions) and its duration, but they included also variables not specific to epilepsy (gender, age, family status, education). A memory disturbance was indicated by a low value of the memory quotient, measured according to the Wechsler Memory Scale test set. The objective of data mining was to find rules associating certain combinations of clinical variables either with a low MQ value or with a high MQ value. It should be emphasized that in this application, like in other typical medical applications, the results of data mining were not the final goal, but rather only a starting point for a subsequent confirmatory analysis (cf. the following subsection [2,3]).

The second application belongs to the area of river ecology [29]. One of the very efficient ways to increase the suitability of rivers for water transport is building groynes. On the other hand, ecologists often fear the changes in the biocoenosis of the river and its banks to which groynes may lead. However, it is a matter of fact that the complex relationships between the biocoenosis and the ecological factors characterizing a groyne field are only poorly understood so far. Therefore, a research project has been launched in Germany 1998, with the objective to investigate those relationships, and to propose an empirically proven model capturing them and allowing to estimate the changes in the biocoenosis that prospective groynes would cause. That model and the knowledge on which it will be based are intended to serve as a basis for constructing, at the end of the project, a decision support system for river ecology.

The project has been founded by the German federal ministry of education and research, and is being accomplished by researchers from four institutions -- the German universities of Cottbus, Darmstadt and Marburg, and the Czech Academy of Sciences. On the important Czech and German river Elbe, five groyne fields have been chosen, and a large amount of empirical data on them has been collected during 1998--1999. The main part of the collected data is formed by nearly 1000 field samples of aquatic fauna and more than 1400 field samples of terrestrial fauna. Each sample includes all animals caught in special traps during some prescribed period of time, ranging from several hours to two days. Simultaneously with collecting those samples, various ecological factors have been measured in the groyne fields, e.g., oxygen concentration, diameter of ground grains, glowable proportion of the ground material, whereas others, such as water level and flow velocity, have been computed using a hydrodynamic simulation model.

The collected data are, first of all, analysed with respect to the species contained in them. Then some preprocessing is performed, and finally data mining methods are applied to the preprocessed data. So far, only a small fraction of the collected data has been analysed with respect to the contained species. A first attempt to apply knowledge extraction methods to those data covered the four most completely documented ecological factors and 13 selected species of aquatic fauna. The order in which the samples of fauna are being analysed with respect to the contained species is determined mainly by the order in which they were collected, and is not independent of the values of the considered ecological factors. Due to this fact, and due to the small fraction of the collected data that is so far available for data mining, the first obtained results have only a limited value from the point of view of the objectives of the whole research project. Nevertheless, they already confirm that data mining can yield interesting knowledge about the relationships between ecological factors and aquatic fauna in groyne fields.

2.3 SOME UTILIZABILITY ASPECTS

Basically, data mining is appropriate whenever a large amount of data concerning many various properties of some real object or phenomenon has been gathered, whereas only poor previous knowledge about possible relationships, dependences etc., between those properties is available. Particularly suitable it may be if we are interested also in relationships and dependences between whole groups or between co-occurring combinations of the considered properties.

A typical example of such a situation can be found in databases of large clinical registers, registering occurrences of some important disease, such as the nation-wide cancer registries [6,33,46]. They contain not only detailed data concerning the registered cases of the disease (precise diagnosis, results of exams, laboratory findings, data about the development of the disease, applied therapies, hospital stays and the like), but also data concerning the overall medical history of the patients, their living conditions, some parts of the medical records of their relatives, demographic data, data about the involved general practitioners etc. Consequently, such registers allow researchers to discover a large number of relationships between various aspects of the considered disease, the medical history, living conditions and genetic dispositions of the patient, and the applied therapy, far behind the previous empirical knowledge, usually limited to relationships between symptoms and diagnostic status.

As another example of a situation suitable for data mining, we can mention information systems of service centres servicing complex and expensive products, such as cars or trucks [43,45]. In the databases of those information systems, not only detailed data about products and their failures are gathered, but also data concerning users, usage conditions, workshops, part suppliers etc. Hence, they offer the possibility to discover many interesting relationships, which can be valuable, e.g., in the area of quality control.

On the other hand, data mining is quite inappropriate if only some particular relationships between properties should be checked in the data, which have already been formulated in advance, and do not need to be discovered first. Caution is also needed if the application in which data mining should be used is concerned with law-like, objectively valid knowledge. For example, the objective of common research studies in medicine and natural sciences is to extend, detail or correct the existing knowledge, typically through confirming or refuting either some in advance postulated conjecture, or some already available piece of knowledge. Data mining, in contrast, is concerned with knowledge about the data in a particular database or dataset. Though that knowledge is also often established through evaluating some hypotheses, with the purpose of confirming or rejecting them, these hypotheses are usually not postulated by humans, but are automatically generated during a search in a predefined hypotheses space.

To summarize, data mining is not able to substitute rigorous data analysis, but it can be very well employed for the often tedious preprocessing of the data, which is needed especially in situations when the data have not been collected with any particular conjecture in mind, but simply as a result of an increased easiness of collecting them. It is this kind of data the amount of which is growing most rapidly. In the context of exploratory data analysis, data mining can play a twofold role. First, some of its results can serve as conjectures for further research. Due to the exhaustive search that data mining methods perform, they can sometimes suggest interesting conjectures that would not be considered otherwise. Second, the high degree of automation inherent to data mining makes the initial exploratory analysis of the available data less time consuming, thus freeing the data analyst to focus on the confirmatory analysis.

Finally, the problem of simultaneously using different data mining approaches should be mentioned. Basically, that problem decomposes into two subproblems. One of them is how to make implementations of several different approaches available to a particular client solving some data mining task. One possible solution is to use large data mining tools, such as Darwin, Enterprise Miner, Intelligent Miner, and the like (see Figure 1). Another, more general solution provide distributed object technologies (DOTs), i.e. middleware technologies allowing clients to pass requests to objects implementing diverse data mining functionality in a distributed environment. Such a situation is schematically depicted in Figure 3. Developers of data mining tools increasingly often realize the need for an

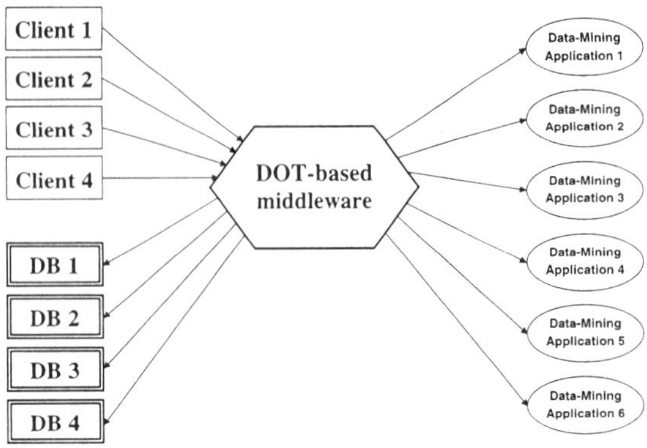

Figure 3. Data Mining Using DOT-based Middleware

underlying DOT, and they start providing their products with corresponding interfaces. Results of a 1998 performed survey of data mining tools supporting some distributed object technology can be found in [99,27].

The second subproblem is how to functionally integrate different data mining approaches -- how to apply them simultaneously to the same data, how to compare obtained results, etc. To this end, integrative frameworks are needed [1,28,34,56]. Such an integrative framework will be described in the next section.

3. Observational Logic And Data Mining

Observational calculus has been developed in the seventies, as a theoretical foundation of the method GUHA (General unary hypotheses automaton), one of the first methods for automated knowledge discovery [22]. GUHA proved to be an extremely vital method. It has been at least four times implemented, from a mainframe implementation 20 years ago, till two recent Windows implementations, one of which has been used also in the groyne fields ecology project. More importantly, it has been always used in practical applications. Several recent applications have been surveyed in [24] and [99].

Formally, a monadic observational predicate calculus is a collection of unary predicates with a single object variable *x*, and of generalized quantifiers. From predicates, open formulae can be built by means of the usual logical connectives (Γ,Λ), whereas generalized quantifiers yield closed formulae when applied to open formulae.

A unary predicate, in general, states a range of values or even one specific value of a variable capturing some single property of an individual object (e.g., a patient, a sample of aquatic fauna), such as in the following predicates

disease duration > 10 years,
sex = male,
grand mal seizures = yes, or more simply, grand mal seizures,
glowable proportion < 10%,
number of trees within 5 m distance = 2,
Tubificidae = yes, or more simply, Tubificidae.

Consequently, an open formula containing logical connectives corresponds to some combined property (combination of single properties), of an individual object, e.g.,

sex = male $\wedge$ *disease duration > 10 years* $\wedge$ *grand mal seizures,*
glowable proportion < 10% $\wedge$ *grain diameter > 1.5mm* $\wedge$ *water level < 0.4m.*

On the other hand, closed formulae state properties characterizing in some way the whole set of objects underlying the considered data. For example, the fact that we deal with data about adult patients can be expressed as (x) age > 15. Similarly, the fact that in all the considered samples of aquatic fauna the family Oligochaeta was present can be expressed as ($\forall$ x) Oligochaeta. In the case of binary quantifiers or other quantifiers of a higher arity, the closed formula $(Q\,x)\,(\varphi_1,\ldots,\varphi_m)$ built from an m-ary generalized quantifier Q and open formulae $\varphi_1,\ldots,\varphi_m$ in general states some relationship between properties corresponding to $\varphi_1,\ldots,\varphi_m$. Examples of such relationships will be given in the next sections.

As models of a monadic observational predicate calculus, data matrices are used, which for the considered objects record the values of variables capturing the considered properties. More precisely, each column records the values of one variable, such as "disease duration" or "glowable proportion". The interpretation in the model of a predicate concerning that variable, e.g., *disease duration > 10 years or glowable proportion < 10 %*, is a $\{0,1\}$ -valued column vector with 1 s on positions corresponding to those rows of the data matrix for which the values of the variable fulfil the predicate. Similarly, the interpretation of an open formula is the column vector that results from combining the interpretations of its constituent predicates in the boolean algebra of $\{0,1\}$ -valued vectors. For example, the interpretation of

sex = male $\wedge$ *disease duration > 10 years* $\wedge$ *grand mal seizures,*
or
glowable proportion < 10% $\wedge$ *grain diameter > 1.5mm* $\wedge$ *water level < 0.4m*

is a vector with ones on positions corresponding to those rows of the data matrix for which the three constituent predicates are simultaneously valid, and with zeros on all remaining positions.

Finally, the interpretation of the closed formula $(Q\,x)\,(\varphi_1,\ldots,\varphi_m)$ is obtained by applying some $\{0,1\}$ -valued function $Tf(Qx)\,(\varphi_1,\ldots,\varphi_m)$ called the truth function of $(Q\,x)\,(\varphi_1,\ldots,\varphi_m)$ to the data matrix serving as the model. The value of the truth function can, in general, depend on the whole data matrix, more precisely on those its columns that are needed to obtain the interpretations of $\varphi_1,\ldots,\varphi_m$. Thus the domain of $Tf(Qx)\,(\varphi_1,\ldots,\varphi_m)$ is the set of all real-valued matrices with an appropriate number of columns. However, in many important cases that dependence reduces to a dependence on the interpretation of ($\varphi_1,\ldots,\varphi_m$). Then the interpretation of $(Qx)(\varphi_1,\ldots,\varphi_m)$ is uniquely determined by a matrix M of m compatible interpretations, no matter what open formulae are actually being interpreted. In such a case, the truth function is a function on the set of all dichotomous matrices with m columns, is denoted simply TfQ and called the truth function of the quantifier Q. For example, the truth functions of the classical unary existential quantifier $\exists$ and universal quantifier $\forall$ are defined $Tf_\exists(M) = 1$ iff $\Sigma M_i > 0$ and $Tf_\forall(M) = 1$ iff M1=M2=...=1, respectively.

If the observational calculus includes at least one of the quantifiers $\forall$ and $\exists$, all tautologies derivable in the classical boolean predicare logic can be derived also in that observational calculus. In addition, a number of other tautologies are derivable also for various classes of generalized quantifiers of a higher arity [22]. Using those tautologies, the validity of some relationships can be deduced from the validity of other relationships, without having to evaluate the closed formula capturing the deduced relationship. Since the evaluation of some quantifiers is quite time-consuming, making use of applicable logical tautologies can substantially increase the efficiency of data mining.

4. Data Mining in Four-fold Tables

Observational calculus has been developed primarily for knowledge extraction by means of statistical hypotheses testing [22]. Hence, they have been developed for situations when the rows of the data matrix are realizations of a random vector, independently generated by some probability distribution P, and the matrix itself is a realization of a random sample from P. Through appropriate constraints on P, it is possible to capture probabilistic relationships between the properties corresponding to the columns of that data matrix. Assuming that P is a priori known to belong to some set $\boldsymbol{D}$ of probability distributions, a statistical test assesses the validity of the constraint H_o: $P \varepsilon D$ for a given $D_o \subset D$ against the alternative H_1: $P \varepsilon D_1 = D \setminus D_o$.

The test is performed using some random variable t, the test statistic, and some borel set C_α, the critical region of the test on a significance level α in (0,1), which are connected to H_0 through the condition $(\forall P \varepsilon D_o)\, P(t \varepsilon C\alpha) \leq \alpha$.

In addition, the test of test often makes use of the fact that if the considered model is a realization of a random sample, then also the matrix M of interpretations of open formulae $\varphi_1,\ldots,\varphi_m$ is a realization of a random sample, more precisely a realization of a multinomial random sample (random sample from a multinomial distribution). That fact allows to choose some test statistic t that is a function of that multinomial random sample, and to define a generalized quantifier ~ corresponding to the constraint H_1 by means of a truth function depending only on M,

$$Tf\sim(M) = 1 \text{ iff } t_M \in C\alpha \tag{1}$$

where t_M is the realization of t provided the realization of the random sample from P yields the matrix of interpretations M.

The existing implementations of GUHA include about a dozen quantifiers defined according to Eqn. 1. All of them are binary, and their truth functions have the special property that the dependence of the truth function on the two-column matrix M of interpretations reduces to its dependence on the four-fold table corresponding to M. Formally, there exists a mapping $\tau : N_0^4 \rightarrow \Re$ such that for each two-column matrix of interpretations M, the following holds

$$t_M = \tau\,(a_M, b_M, c_M, d_M)\,, \tag{2}$$

where a_M, b_M, c_M and d_M are the counts of the pairs (1,1), (1,0), (0,1) and (0,0), respectively, among the rows of M.

The dependence (Eqn. 2) allows to introduce two important classes of generalized quantifiers:

(a) A binary generalized quantifier $\rightarrow$ is called implicational if for each two-column matrices of interpretations M_1, M_2, its truth function $Tf_{\rightarrow}$ fulfills

$$a_{M_2} \geq a_{M_1} \wedge b_{M_2} \leq b_{M_1} \wedge Tf_{\rightarrow}(M_1) = 1 \Rightarrow Tf_{\rightarrow}(M_2) = 1. \tag{3}$$

A closed formula $\varphi \rightarrow \psi$ (standing for the more pedantic $(\sim x)(\varphi, \psi)$) is then called a GUHA-implication.

(b) A binary generalized quantifier $\sim$ is called associational if $Tf_{\sim}$ fulfills

$$a_{M_2} \geq a_{M_1} \wedge b_{M_2} \leq b_{M_1} \wedge c_{M_2} \leq c_{M_1} \wedge d_{M_2} \geq d_{M_1}$$

$$Tf_{\sim}(M_1) = 1 \Rightarrow Tf_{\sim}(M_2) = 1. \tag{4}$$

Observe that each implicational quantifier is also associational. Expectably, a closed formula $\varphi \sim \psi$ (standing for $(\sim x)\,(\varphi\,,\,\psi)$) is called a GUHA association.

Due to (Eqn. 3) and (Eqn. 4), an implicational quantifier $\rightarrow$ is suitable to capture a high probability that the random variable corresponding to the second column of the matrix of interpretations M assumes the value 1 provided the random variable corresponding to the first column of M assumes the value 1, whereas an associational quantifier $\sim$ is suitable to capture a correlation between those random variables. Consequently, a GUHA implication $\varphi \sim \psi$ states a high conditional probability of the value range or combination of value ranges stated by the open formula ψ conditioned on the value range or combination of value ranges stated by φ Similarly, a GUHA association $\varphi \sim \psi$ states a correlation between the value ranges or combinations thereof that are stated by the open formulae φ and ψ. Moreover, the truth function of an associational quantifier can always be defined in such a way that the correlation is positive.

The most fundamental examples of associational quantifiers are the Fisher quantifier, $\sim_\alpha$ corresponding to the one-sided Fisher exact test of independence in four-fold tables with the significance level $\alpha\varepsilon$ in (0,1), and the chi-square quantifier $\sim_\alpha^2$ corresponding to the χ^2 asymptotic test of independence in four-fold tables with the significance level α. The truth functions of their most simple versions are defined as follows:

$$Tf_{\sim F_\alpha}(M) = 1 \quad iff \quad a_M d_M > b_M c_M \wedge \sum_{i=a_M}^{\min(r_M,k_M)} \frac{\binom{k_M}{i}\binom{\ell_M}{r_M-i}}{\binom{n_M}{r_M}} \le \alpha, \qquad (5)$$

$$Tf_{\sim\chi^2_\alpha}(M) = 1 \quad iff \quad a_M d_M > b_M c_M \wedge \frac{n_M(a_M d_M - b_M c_M)^2}{r_M s_M k_M l_M} \ge \chi^2_1(1-2\alpha), \qquad (6)$$

where $\chi^2_m(t)$ for $t > 0$ denotes the t-quantile of the χ^2 distribution with $m \in N$ grades of freedom, $r_M = a_M + b_M$, $s_M = c_M + d_M$, $k_M = a_M + c_M$, $l_M = b_M + d_M$, and $n_M = a_M + b_M + c_M + d_M$, i.e., n_M is the number of rows of the matrix M. The basic example of an implicational quantifier is the lower critical implication $\to^\theta_\alpha$ with the threshold θ in (0,1), which corresponds to the binomial test with a significance level α. The truth function of the most simple version of this quantifier is defined

For example, in the medical application introduced in [2.2], the following relationships have been discovered using the Fisher quantifier:

sex = male $\wedge$ *disease duration > 10 years* $\wedge$
$\wedge$ *only grand mal seizures* $\sim^F$ *memory quotient > 90,*
disease duration > 10 years $\wedge$ *only grand mal seizures* $\wedge$
$\wedge$ *course of the disease = good* $\sim^F$ *memory quotient > 90.*

$$Tf_{\sim^\theta_\alpha}(M) = 1 \quad iff \sum_{i=a_M}^{r_M} \binom{r_M}{i} p^i (1-p)^{r_M-i} \le \alpha . \qquad (7)$$

An example from the other application mentioned in [2.2], the mining of groyne field ecology data, is shown in Table [1]. The table lists the antecedents of formulae found to hold in the data and expressible as

$$(\wedge_{i\in I} E_i \sim^F_{10\%} RD) \wedge (\wedge_{i\in I} E_i \to^{2/3}_{10\%} RD)$$

or, more pedanticly

$$(\sim^F_{10\%} x)(\wedge_{i\in I} E_i, RD) \wedge (\to^{2/3}_{10\%} x)(\wedge_{i\in I} E_i, RD)$$

TABLE 1. Antecedents of found formulae $(\vee_{i \in I} E_i \sim^{F}_{10\%} RD) \wedge (\vee_{i \in I} E_i \rightarrow^{2/3}_{10\%} RD)$

attained significance:	$\sim^{F}_{10\%}$ %	$\rightarrow^{2/3}_{10\%}$ %	attained certainty
90%il of grain diameter < 3mm ∧ flow velocity > 0.2m/s	0.008%	3.6%	71.7%
mean grain diameter < 1.5 mm ∧ flow velocity > 0.2m/s	0.008%	3.6%	66.0%
glowable proportion < 5 % ∧ 90%il of grain diameter < 3mm ∧ flow velocity > 0.2m/s	0.33%	5.5%	71.5%
glowable proportion < 5 % ∧ mean grain diameter < 1.5mm ∧ flow velocity > 0.2m/s	0.33%	5.5%	72.9%

where I is some subset of the covered ecological factors, E_i, for i in I are predicates stating ranges of values of the respective ecological factors, and RD is a predicate stating the occurrence of the species "Robackia demeierei". More precisely, RD states that the number of individuals of "Robackia demeierei" per sample is at least $\frac{1}{10} n_{\max}$, where $n_{\max}$ denotes the maximal number of individuals of this species per sample encountered in the data.

To asses the simultaneous validity of n relationships, generalized quantifiers need to be defined with the significance levels required in the multiple testing method to be applied. Thus for example in the Holm method, the quantifier for the i-th most significant hypothesis must be defined with the significance level $\frac{\alpha}{n+1-i}$, whereas in the Simes method it must be defined with the significance level $\frac{i\alpha}{n}$. In the context of GUHA the methods of Bonferroni, Holm and Simes have been used [7].

5. ANN-Based Extraction of Rules

The extraction of knowledge from data by means of artificial neural networks has received much attention not only in connection with data mining, but also in connection with pattern recognition [2,3,45,47,48]. Actually, already the mapping learned by a neural network incorporates knowledge about the implications that certain values of the input variables have for the values of the output variables. Usually, however, the ANN-based knowledge extraction aims at the more comprehensible representation of those implications as logical formulae, which in the case of artificial neural networks are usually called rules.

A comprehensive survey of ANN-based rule extraction methods has been given in [1,48], where a classification scheme for those methods has been proposed as well. In the context of that classification scheme, two remarks can be made to the possibility to capture the extraction of rules using a generalized quantifier of an observational calculus:

- As far as the primary dimension of the proposed scheme is concerned, the expressive power of the rules, the fact that the truth function of a generalized quantifier is $\{0,1\}$ -valued implies that only boolean rules can be covered. In connection with the other major class of rules, fuzzy rules, ongoing research into fuzzy generalized quantifier should be mentioned [21,22,26,27].
- With respect to the second dimension, the translucency, the extracted rules should concern only network inputs and outputs. The reason is that hidden neurons are not explicitly assigned to properties of real objects, thus they do not correspond to open formulae of an observational calculus.

Even with the choices with respect to those two dimensions fixed, there are innumerable possibilities how to design a rule extraction method. In the following, the definition of a generalized quantifier will be elaborated for two particular methods. Both of them share a common choice of the underlying neural network architecture - a multilayer perceptron (MLP) with one hidden layer. MLPs are the most common kind of neural networks, both in general and in the context of ANN-based rule extraction [1,11,47,49]. If a MLP has m input and n output neurons, then it computes a mapping

$$F = (F_1,\ldots,F_n) : \Re^{m} \rightarrow \Re^{n} \quad (8)$$

In addition, activation functions are assigned to both hidden and output neurons and the same activation function f is assigned to all of them.

Except for these two design decisions, the methods are based on different design choices, therefore also their applicability conditions are different.

5.1 UNCERTAIN DICHOTOMOUS OUTPUTS

In the first method, neural networks with real-valued inputs and uncertain dichotomous outputs are used [26]. In such a network, to each input neuron a real-valued variable is assigned, whereas to each output neuron a dichotomous variable is assigned whose value can sometimes be not quite evident, or it may be ascertained with subjective bias, thus being always assigned some certainty grade from the interval [0,1]. For example in the groyne fields ecology application, some soft animal bodies are damaged or fragmental in some samples of aquatic or terrestrial fauna, not allowing a sure classification of the animal's species. Therefore, the occurrence of a particular species in a particular sample can be connected with some uncertainty. More importantly, when extracting knowledge about the most often encountered species, not only the fact whether an individual belonging to that species has been found in a particular sample is taken into account, but also whether such individuals have been only a few, many, very many, etc. Needless to say, assigning such an abundance label as "few" or "many" to a particular number of occurrences of a particular species in a sample is usually connected with uncertainty. In the software that has been developed for the knowledge extraction, the labels are represented as species-specific fuzzy sets, and the grade of membership of a particular count of occurrences in the fuzzy set representing an abundance label is taken as the certainty grade of that label.

In the case of MLPs with uncertain dichotomous outputs, the functions F_i,, i=1,...,n from (Eqn. 8) fulfill

$$Fi : \Re^m \to [0, 1]. \tag{9}$$

More precisely,

$$F_i \in U_{k=1}^{\infty} MLP_{m:k,f}, \qquad i = 1, \ldots, n, \tag{10}$$

where for each k in N and for each f: R->[0,1], nonconstant and borel-measurable, $MLP_{m.k,f}$ denotes the set of mappings learnable by a multilayer perceptron with the topology ($m.k.f$), whose hidden and output neurons have the activation function f. The function (1.10) is obtained if for some number k of hidden neurons, some learning algorithm L_k is applied to the rows of some data matrix D, whose columns record values of some variables $V_1,\ldots,V_m$, provided a vector c_i of certainty grades corresponding to the rows of D is available for the i -th output:

$$(\exists k) \in F_i = L_k(Dc_i), \tag{11}$$

where

$$L_k: U_{s=1}^{\infty}(\Re^m \times [0,1])^s \to MLP_{m:k,f}. \tag{12}$$

For example, in the groyne fields ecology application, the function F_i,i=1,...3, describes the dependence of the certainty grade of the occurence of the i -th sepcies on the four considered ecological factors. To get an idea about the form of that function, visualization can be used. Naturally, whereas F_i expresses a relationship between 5 variables (certainty grade and 4 ecological factors), only relationships between 3 variables can be visualized. In the groyne fields ecology project, visualization has been used to illustrate relationships between 1--2 ecological factors and the certainty grade of the occurrence of a particular species (see Figure 4, as well as to depict those areas in the product of ranges of 3 ecological factors in which a particular species occurs with a certainty grade above a given threshold, independently of the remaining ecological factor.

The method for extracting rules by means of neural networks with uncertain dichotomous outputs is based on the following discernibility assumption:

There exists a certainty threshold θ in (0,1) above which the dependence of the learning algorithm on particular values v_i of V_i, i=1,...,m, is discernible only with respect to some given finite partition

$$S_i = \{ S_{i,1},\ldots,S_{i,p_i}\}$$

of the value set of V_i..

Defining a congruence ≅ on $\Re^m$ x [0,1] by

$$(v_1,\ldots,v_m,c) \cong (\hat{v}_1,\ldots,\hat{v}_m,c) \ iff$$

$$(\forall i \in \{1,\ldots,m\})(\exists S \in S_i) v_i \in S \wedge \hat{v}_i \in S \wedge (c > \theta \Leftrightarrow \hat{c} > \theta), \tag{13}$$

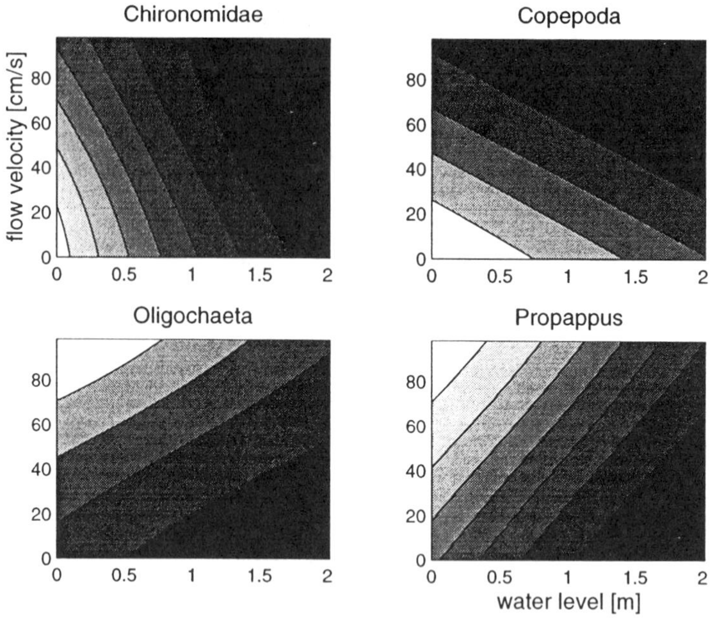

Figure 4. Examples of relationship between two ecological factors and the uncertainty of occurance of one species

this assumption precisely means

$$(\forall r \in N)\,(\forall D,\, \widehat{D} \in \Re^{r,m})\,(\forall c,\, \hat{c} \in \Re^{r})$$
$$[\,(\forall s \in \{1,\ldots,r\})\,(D_{s,1},\ldots,D_{s,m},\, c_s) \cong (\widehat{D}_{s,1},\ldots,\widehat{D}_{s,m},\hat{c}_s)\,] \Rightarrow \tag{14}$$
$$L_k(D_c)^{-1}(\theta,1] = L_k(\widehat{D}\hat{c})^{-1}(\theta,1]\,.$$

Let π_0 be the predicate stating that the certainty grade of the occurrence of the considered property exceeds the certainty threshold, and let for $i=1,\ldots,m$, $j=1,\ldots,p_i$, $\pi_{i,j}$ be the predicate stating that the value of V_i lies in $S_{i,j}$. If the above discernibility assumption holds, then a generalized quantifier $\nabla_{S_1,\ldots,S_m}$ can be defined allowing to state the fact that the occurrence of the considered property is with a high certainty grade related to situations when each variable V_i, $i=1,\ldots,k$, takes values in a given set

$$(\nabla_{S_1,\ldots,S_m} x)\,(\pi_0, \pi_{1,j_1},\ldots,\pi_{m,j_m}),$$
$$Tf_{(\nabla_{S_1,\ldots,S_m} x)(\pi_0,\pi_{1,j_1},\ldots,\pi_{m,j_m})} \tag{15}$$

whose truth function is defined

$$iff\ L_k(Dc)(S_{1,j_1},\dots,S_{m,jm}) \subset (\theta, 1]. \tag{16}$$

$$(\forall r \in N)(\forall D \in R^{r,m})(\forall c \in R^r)\, Tf_{(\nabla_{S_1,\dots,S_m}x)(\pi_0,\pi_{1,j_1},\dots,\pi_{m,jm})}(Dc) = 1$$

Similarly, the fact that the occurrence of the considered property is with a high certainty grade related to situations when only some of the variables, say V_i for $i \in I \subset \{1,\dots,k\}$, $|I| \leq k - 1$ take values in given sets $S_{i,ji} \in S_i$ can be stated by the formula

$$V_{i \in I}\, V_{j_i=1}^{p_i} (\nabla_{S_1,\dots,S_m}x)(\pi_0, \pi_{1,j_1},\dots,\pi_{m,jm}) \tag{17}$$

As an example from the groyne fields ecology application, Table 1 gives the attained certainty grades of the occurrence of "Robackia demeierei" for the listed combinations of value ranges of the ecologicla factors. The partitions of the value sets of the considered ecological factors in (Eqn.. 16) and (Eqn. 17) have been constructed in such a way that they include all the encountered value ranges.

5.2 GENERAL MULTYLAYER PERCEPTRONS

In the second rule extraction method, the variables assigned to the MLP outputs are less restricted, having to fulfil only the requirement of normalizability to the interval [0,1], but a restriction is put on the activation function *f* of the hidden and output neurons. This function is chosen to be a piecewise-linear sigmoidal, i.e.,

$$(\exists q \in N)(\exists a_0, a_1,\dots,a_q \in \Re)\, f((-\infty, a_0]) = 0 \wedge f([a_q, +\infty)) = 1 \wedge$$

$$f \mid [a_{i-1}, a_i]\ is\ linear for i = 1,\dots,q. \tag{18}$$

Observe that if the mapping F in (Eqn. 8) is computed by a MLP with such activation functions, any hyperrectangle (actually, any polyhedron) in the range of F is the map of some finite set of polyhedra. Formally, let P_k and H_k for k in N denote the set of all polyhedra in $\Re^k$ (intersections of finitely many halfspaces in $\Re^k$, any of which may be either open or closed), and the set of all hyperrectangles in $\Re^k$ (Cartesian product of arbitrary intervals), respectively. Then the activation function (Eqn. 18) assures the following to hold:

$$(\forall k \in \{1,\dots,n\})(\forall i_1,\dots,i_k \in \{1,\dots,n\})\, i_1 < \dots < i_k \Rightarrow$$

$$(\forall H \in H_k)(\exists P_1,\dots,P_p \in P_m)\, H = (F_{i_1},\dots,F_{i_k})\,(U_{j=1}^{p} P_j) \tag{19}$$

Observe further that if $F_g : \Re^m \to \Re^n$ is a mapping computable by a MLP to whose hidden and output neurons a continuous sigmoidal activation function g is assigned (such as logistic or arctangent), then given any $\delta > 0$, there exists a piecewise-linear activation function f so close to g that the mapping F computed by a MLP obtained from the original one through replacing g with f fulfills

$$d(F_g, F) < \delta, \tag{20}$$

where d is the distance in an appropriate metric space (such as the L_2 -space, or the space $C(K)$ for a compact K).

Since all known MLP learning algorithms are iterative, the MLP does not learn the mapping F^* that is optimal with respect to a given objective function (e.g., sum of squares, mutual entropy, ..., but only some suboptimal mapping F, sufficiently close to F^* with respect to the distance d. It is worth noticing that F actually does not have to be computed by the same MLP by which F^* would be computed. More precisely, if we fix the topology of the network and denote

$$(\forall f: \Re \rightarrow [0,1]\; C_f = \{\text{F: Fcan be computed by a MLP with the activation function f}\}, \tag{21}$$

then under some assumptions on the objective function, to any continuous sigmoidal activation function g a piecewise-linear sigmoidal f can be found such that for $F^* \in C_f$, a suboptimal mapping F_g can be learned in C_g. Subsequently, (Eqn. 20) can be used to obtain a mapping F in C_f. This is of practical importance if no specific method for learning in a MLP with piecewise-linear activation functions is implemented. However, caution should be taken if computational complexity is an issue, since a MLP with piecewise-linear activation functions has a lower Vapnik-Chervonenkis dimension than a MLP with smooth activation functions [37].

Besides the choice of the activation function (Eqn. 18), there is still another design decision on which this method relies - to replace, if possible, polyhedra through hyperrectangles. From the point of view of an observational calculus, hyperrectangles are particularly important subsets of the input and output value spaces, since each hyperrectangle is the set of those tuples of values for which the conjunction of some appropriate unary predicates holds. Therefore to each polyhedron $P \in P_m$ in (Eqn. 19), the hyperrectangle $H \in H_m$ is sought that is the closest from P with respect to some monotone measure μ_P, possibly depending on P. If such a H exists and the difference between H and P with respect to μ_P is within a given tolerance $\varepsilon > 0$, then it replaces P for the purpose of rule extraction. Formally, a replaceability relation $\approx^{\mu_P}$ between P_m and H_m can be defined by

$$(\forall P \in P_m)(\forall H \in H_m)\; P \approx_{\varepsilon}^{\mu_P} H \quad iff$$
$$\mu_P(P \Delta H) = min\{\mu_P(P \Delta H') : H' \in H_m\} \leq \varepsilon, \tag{22}$$

with ($P \; \Delta \; H$ denoting the set differnece of P and H. More generally, $P \approx^{\mu_P}_{\varepsilon} \oplus_{k \in I} I_k$ for a Cartesian product $\otimes_{k \in I} I_k$ of intervals I_k, $k \in I \subset \{1,\ldots,m\}$, is defined as $P \approx^{\mu_P} H$, where $H \in H_m$ ranges over I_k for coordinates k in I and over the whole $\Re$ for the remaining coordinates. The measures μ_P, $P \in P_m$, on $\Re^m$ can in general be any monotone measures. Practically important are the empirical distribution of the data $\widehat{\mu}$, and its weighted version $\widehat{\mu}_P$, which is for $P \in P_m$ with $\widehat{\mu}(P) \neq 0$ defined

$$(\forall S \subset \Re^m)\, \widehat{\mu}_P(S) = \frac{\widehat{\mu}(S)}{\widehat{\mu}(P)}. \tag{23}$$

For $P \in P_m$ with $\widehat{\mu}(P) =$, the measure $\widehat{\mu}_P$ can be defined arbitrarily, depending on whether we want to admit replaceability of such polyhedra. Notice that, differently to general monotone measures, both for $\widehat{\mu}$ and for the measure defined in (Eqn. 23), the hyperrectangle closest to a given polyhedron always exists.

This rule extraction method allows a straightforward definition of a binary generalized quantifier $\rightarrow^{\mu}_{\varepsilon}$ to be used in closed formulae stating the fact that particular values of some input variables imply particular values of some output variables. Consider a multilayer perceptron with m inputs, n outputs, piecewise-linear activation functions, a fixed number of hidden neurons, and a fixed algorithm to learn a suboptimal mapping F (no matter whether through direct learning or using an auxiliary activation function). For any data matrix D with $m+n$ columns, denote F^D the mapping learned by the considered perceptron with the rows of D used as the training set, the first m columns supplying the inputs, the last n columns supplying the targets. Let further $\mu = (\mu_P)_{P \in P_m}$ be a system of monotone measures on $\Re^m$, and $\varepsilon > 0$. Finally, let $I \subset \{1,\ldots,m\}$, $O \subset \{1\ldots,n\}$ and for each k in I, or k in O, let π_k be a unary predicate stating some interval I_k of values of the k-th input variable, or the k-th output variable, respectively. Then the closed formula $\Lambda_{k \in I} \; \pi_k \rightarrow^{\mu}_{\varepsilon} \Lambda_{k \in I} \; \pi_k$ or more pedanticly $(\rightarrow^{\mu}_{\varepsilon} x)(\wedge_{k \in I} \pi_k, \wedge_{k \in O} \pi_k)$ has the truth function $Tf_{\bigvee_{k \in I} \pi_k \rightarrow^{\mu}_{\varepsilon} \bigvee_{k \in O} \pi_k}$

defined

$$Tf_{\bigvee_{k \in I} \pi_k \rightarrow^{\mu}_{\varepsilon} \bigvee_{k \in O} \pi_k}(d) = 1 \;\; if\; (\exists P_1,\ldots,P_p \in P_m) \overset{N}{}_{k \in O} I_k = (F^D_k)_{k \in O}(U^p_{j=1} P_j)$$
$$\wedge (\exists j \in \{1,\ldots,p\})\, P_j \approx^{\mu_P}_{\varepsilon} \oplus_{k \in I} I_k,$$
$$0 \;\; else.$$

To conclude again with an example, Figure 5 depicts a 2-dimensional projection of the union of polyhedra from (Eqn. 24) that a particular coordinate F^D_k of the mapping F^D learned by a MLP maps to a particular interval I_k. In Figure 5, the considered F^D_k counts the number of individuals of "Robackia demeierei" per sample, and

$I_k = \left\langle \frac{1}{10} n_{\max}, +\infty \right)$. The depicted union of polyhedra is projected to the coordinates corresponding to the ecological factors "glowable proportion" and "90\,%il of grain diameter". Figure 6 shows a projection, to the same coordinates, of those hyperrectangles that replace any polyhedron P from the union depicted in Figure 5 according to the replaceability relation $\approx^{\widehat{\mu}_P}_{10\%}$, where $\widehat{\mu}_P$ is the weighted empirical distribution defined in (Eqn. 26). The projection in Figure 6 reflects the following rules that can be extracted from the data using (Eqn. 24) with $\mu = (\widehat{\mu}_P)_{P \in P_m}$:

glowable proportion < 5.39 % $\rightarrow^{\mu}_{10}$ RD,

glowable proportion < 18.69 % $\wedge$ mean grain diameter < 1.17 mm
$\wedge$ 90\%il of grain diameter < 1.7 mm $\rightarrow^{\mu}_{10}$ RD,

glowable proportion < 9.04 % $\wedge$ mean grain diameter < 1.62 mm
$\wedge$ 90\,%il of grain diameter < 3.06 mm $\rightarrow^{\mu}_{10}$ RD.

Needless to say, it is now very easy to compare such extracted rules with results obtained using statistical hypotheses testing, e.g., with the results in Table 1. This impressively illustrates the usefulness of the employed integrative framework.

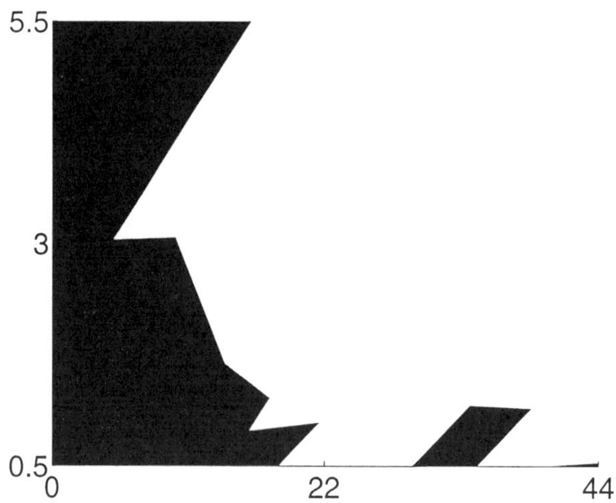

Figure 5. Example of found poly hedra

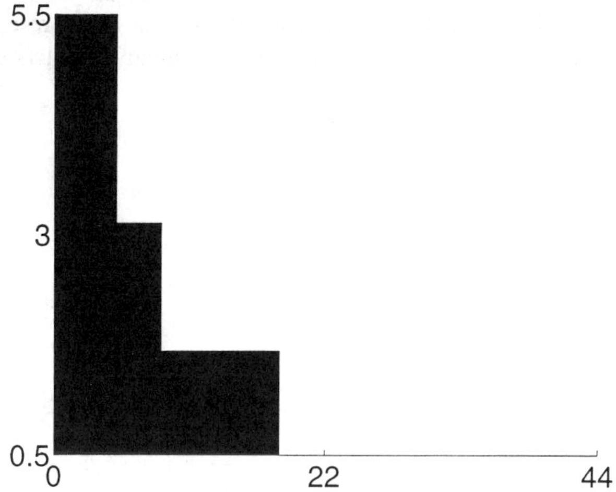

Figure 6. Example of fond hyper rectangles

6. Conclusion

This lecture has given an overall survey of the area of data mining, emphasizing the need for integrative frameworks that allow to apply different data mining methods in a common context and to view them from a common perspective. An integrative framework based on observational logic is described, and its use is explained for two extremely important kinds of data mining methods -- statistical methods based on hypotheses testing, and methods for the extraction of rules from data with artificial neural networks.

The calculus of obsrvational logic has been originally proposed as a means for a unified treatment of different kinds of statisical methods. In this role, it has been more than 25 years used in practice, in the method GUHA, one of the oldest and most vital methods for automated knowledge discovery. The lecture tried to show that the presented framework provides a unique conceptual view of both kinds of considered data mining methods while preserving their specific advantages. Due to the different underlying paradigms and different initial assumptions, artificial neural networks do not necessarily yield the same results as statistical methods. Hence, a coincidence between relationships discovered in the data by both kinds of methods increases the chance that those relationships pertain also to the reality behind the data, to the phenomena that generated them. On the other hand, if a relationship found by means of some quantifier can not be confirmed by means of other quantifiers, including a quantifier based on the other paradigm, then such as relationship is a natural starting point for further, deeper investigations.

7. Acknowledgments

The material presented in sections [3] and [4] relies mainly on the work by Petr Hájek and the late Tomás Havránek. The material in section [5] relies on recent research supported by the grant B2030007 of the Grant Agency of the Czech Academy of Sciences.

MULTI-ASPECT DATA FUSION APPLIED TO ELECTROMAGNETIC TARGET CLASSIFICATION USING ENETIC ALGORITHM

G. TURHAN-SAYAN
Associate Professor
Middle East Technical University
Electrical and Electronics Engineering Department
06531 Ankara, Turkey

1. Introduction

Electromagnetic target detection and classification is an important problem relevant not only to military applications but also to civilian use. In the problem of a breast tumor detection and identification [1], for instance, the main concern is accuracy. In the case of the recognition of a military target such as an aircraft or a ship, on the other hand, speed of classification is as important as the accuracy of the decision as such a decision should be made within a fraction of a second. For the K-pulse target identification technique used in this research, the K-pulse design phase is rather complicated and time consuming but the interactive decision process, that is the only phase of identification to be completed in real-time, is very simple and fast.

The theoretical background of the K-pulse target identification/classification technique was introduced by Kennaugh [2] about two decades ago. His theory was based on the following well-established facts: All the characteristic information about size, shape and electrical properties of a target is implicitely provided by the scattered electromagnetic signals received from the object. However, such data are strongly dependent on polarization and aspect angle of transmitted and received radar signals as well as the measurement band of frequencies. Therefore, the database needed by a target classifier, in general, must contain scattered signatures for each candidate target over a proper frequency bandwidth, at many combinations of aspect angle and/or polarization to provide sufficient characterization. It is obvious that a target classifier relying on the use of such a huge database for real-time data processing would be too slow in making the decision in the order of seconds. In other words, a feasible real-time target classifier must utilize special techniques for data processing, feature extraction and comparison to minimize the time period needed for active target recognition phase, perhaps in the expense of a prolonged and complicated passive recognition phase. The K-pulse technique is a good example in this respect. As originally defined in [2], K-pulse is a target-specific, time-limited excitation signal to produce time-limited scattered responses at *all* possible combinations of aspect angle and polarization. As the K-pulse has a compact support by definition, its Laplace transform must be an entire function of complex frequency having no singularities. The set of zeros of this entire function is required to be the same as the set of complex natural resonance (CNR) frequencies (i.e., system poles) of the related target to produce time-limited target responses at all possible

A. K. Hyder et al. (eds.), Multisensor Fusion, 533–539.

aspects and polarizations. The poles of an electromagnetic scatterer are known to be independent of aspect and polarization of excitation [3], so must be the K-pulse signal. As the accuracy of extracting CNR frequency information from noisy experimental data is quite poor, K-pulse shaping can not be realized as a simple zero-insertion process in the Laplace domain. Instead, the K-pulse synthesis procedure must be based on an optimization problem where the cost function to be minimized is the natural resonance-related late-time energy content of the scatterer response [4, 5, 7 and 8]. The K-pulse optimization process can be realized using the Genetic Algorithm (GA) method which can successfully treat the K-pulse duration as an optimization variable as suggested for the first time by Turhan-Sayan et al. in reference [7] where the K-pulse design problem is formulated to utilize data at an arbitrarily chosen aspect. It is known, however, that a K-pulse target classifier resulting from a single-aspect design has a certain risk of performance degredation due to the fact that some of the target poles may be very weakly excited at the chosen design aspect. Then, such poles can not be represented by the resulting K-pulse.

In the K-Pulse synthesis approach proposed in the present article, on the other hand, the multi-aspect scattered data available at several different aspects are fused by means of the fitness function defined in the GA. As a result of this essential modification, the probability of having most of the target poles represented in the designed K-pulse is considerably increased.

In Section 2, an outline of the K-pulse formulation and multi-aspect data fusion will be given together with the basic definitions of the GA method. Then, in Section 3, the proposed K-pulse synthesis procedure will be demonstrated in its multi-aspect form using two spherical targets which are of the same size but are made of different lossless dielectric materials with different refractive indices. Finally, in Section 4, conlusions and plans for the future work will be presented.

2. Theory

Basic steps of K-pulse formulation, implementation of multi-aspect data fusion and the need for Genetic Algorithm optimization for the solution of this particular electromagnetic target classification problem will be summarized below.

2.1. K-PULSE FORMULATION AND DATA FUSION

Formulation of the K-pulse synthesis problem for a finite-size electromagnetic scatterer can be posed as a linear, time-invariant system problem: The scattered response, $r^i(t)$, of the target to an arbitrary input signal x(t) applied at the ith aspect angle is expressed as

$$r^i(t) = x(t) * h^i(t) \tag{1}$$

where $h^i(t)$ is the scatterer's impulse response and the asterisk " * " denotes the convolution operator. As discussed in [5], to keep the number of optimization variables at the lowest possible number, the continuous part of the K-pulse signal, k(t), can be expanded into a suitably truncated set of modified Legendre polynomials as

$$k(t) = A\, d(t) + \sum_{n=0}^{N} D_n \overline{P}_n(t) \tag{2}$$

where

$$\overline{P}_n(t) = \sum_{m=0}^{n} \alpha_{nm} \left(\frac{2}{T_k} t - 1\right)^m \tag{3}$$

is the nth order modified Legendre polynomial defined over the interval $[0, T_k]$ with T_k being the K-pulse duration and α_{nm}'s are the tabulated construction constants for the original Legendre polynomials which are defined over the interval [-1,1]. D_n's are the expansion coefficients to be optimized. An impulse term may optionally be included in the K-pulse model with a weight factor A. If a target is excited by its own K-pulse signal, the resulting *matched* K-pulse response , $r_k^i(t)$, at the ith combination of aspect and polarization, is expected to satisfy

$$r_k^i(t) = 0 \qquad \text{for} \qquad t \geq T_{L_i} = \frac{L_i}{c} + T_k \tag{4}$$

where L_i is the target dimension along the line of sight at an aspect angle θ_i and c is the speed of light in air. In other words, the span of the target's K-pulse response, T_{L_i} , must be equal to the span of its forced component as the natural response component is expected to be annihilated under the K-pulse excitation. Then, the K-pulse optimization problem can be stated as follows :

$$\textbf{\textit{Minimize}} \quad J(T_k, D_0, D_1, \cdots) = \sum_{i=1}^{I} w_i J_i \tag{5}$$

where

$$J_i = \int_{\frac{L_i}{c}+T_k}^{\infty} [\, r^i(t, T_k, D_0, \cdots)\,]^2 \, dt \tag{6}$$

The cost function **J** (to be minimized with respect to the K-pulse duration, T_k, and the expansion coefficients, D_n's) represents the weighted sum of the natural response-related late-time energy contents, J_i , of the target response at I different aspect angles with w_i terms being the multi-aspect weight factors. Therefore, the multi-aspect data fusion is integrated into the K-pulse design process by this special cost function definition. The GA algorithm is basically a code for *Fitness* maximization. The fitness function in our GA application is chosen to be the inverse of the cost function. Basic steps of the K-pulse design process are shown in the block diagram given in Figure 1.

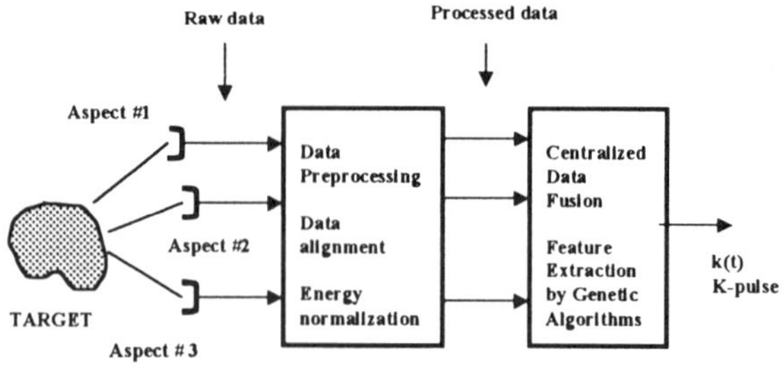

Figure 1. Block diagram for a 3-aspect K-pulse design process.

In establishing a K-pulse target classifier, synthesis of K-pulse waveforms for every target in a given target group forms the passive phase. Then, the active (real-time) target classification phase can be accomplished by convolving the impulse response of an *unknown target* with all these K-pulse signals which are already stored as digital filters in the computer. The K-pulse filter having an almost time-limited output with the shortest time span and with a negligible amount of late-time transient oscillations is chosen to determine the class of the test target. As the cpu time needed to compute a discrete convolution is extremely short, the target classification technique based on K-pulse concept turns out to be very fast.

2.2 IMPLEMENTATION OF THE GENETIC ALGORITHM

Genetic algorithms (GAs) are adaptive parallel search techniques of probabilistic nature and have been inspired by the genetic evolution processes [9]. A GA basically manipulates a fixed-size population of encoded structures, called *chromosomes,* through the application of genetic operators which are *selection, crossover* and *mutation.* A chromosome can be encoded as a string of binary digits and represents a vector of optimization variables in the feasible solution space. Each chromosome is associated with a measure of performance called *fitness*. Chromosomes with higher fitness values have greater chances of reproduction and survival in the population. The most important advantage of the GAs is that they are designed to search for the global optimum without the need for good initial guesses. Furthermore, as GAs do not require gradient information, a lot of flexibility is provided in cost function selection. Due to these facts, complicated optimization problems which can not be solved by traditional optimization techniques can be successfully handled by GA codes as discussed in [6] and demonstrated in [7].

3. Applications

The proposed K-pulse design technique with the use of multi-aspect data fusion is demonstrated for two lossless dielectric spheres of permittivities 4 and 10. In other words, the refractive indices of the first sphere (S1) and the second sphere (S2) are 2 and 3.1623, respectively. The spheres are of the same size of 1 meter radius. The bistatic aspect angles (the angle between the transmission direction and the reception direction of the simulated radar) used for the K-pulse design are zero degrees, 45 degrees and 90 degrees for both targets. The K-pulse signals are designed in 150 iterations with a population which is composed of 100 chromosomes. Each chromosome is made of 7 variables (one of which is the K-pulse duration) where each variable is represented by 8 bits. Elitism (to carry the best chromosome of each iteration to the next iteration) and mutation with a linearly decreasing rate from 0.01 to 0.001 are utilized. The resulting K-pulse waveforms for the dielectric spheres S1 and S2 are shown in Figure 2, with the K-pulse durations of 37 nsec and 49.5 nsec, respectively. These K-pulse waveforms are used to excite both spheres (S1 and S2) at different bistatic aspects including the aspect angles not used in the K-pulse design process. In all of these simulations, the cases for the *matched* K-pulse excitation (i.e. the test target is excited by its own K-pulse) turn out to be creating the time-limited target responses with negligible amount of late-time transient oscillations. For the sake of brevity, results for only one of these simulations will be presented. (The total energy of each response waveform, presented in the following plots, is normalized to unity for fair comparison.) The normalized impulse response and the normalized *mismatched* K-pulse response of the sphere (S1) at 45 degrees bistatic aspect angle are plotted in Figure 3. Next, the normalized impulse response and the normalized *matched* K-pulse response of the same sphere at the same aspect are plotted in Figure 4. It is observed that the matched K-pulse is certainly more successful to minimize the energy of late-time natural oscillations. Finally, the normalized energy curve of the target's impulse response together with the normalized energy curves of the *matched* and *unmatched* K-pulse responses are presented in Figure 5 (part of the curves are shown in an enlarged scale to show details better). As clearly observed in this figure, the normalized energy curve belonging to the *matched* K-pulse response is found to approach the maximum energy level (which is unity) in a shorter time, due to successful target pole annihilation, as expected.

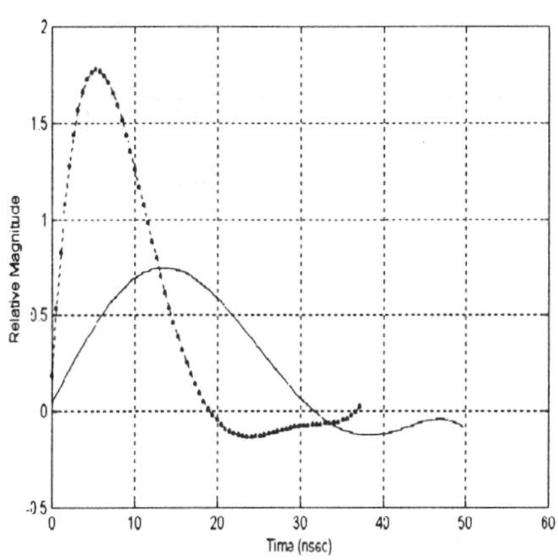

Figure 2. K-pulse waveforms for the targets S1 (. . . . .) and S2 (——) where both waveforms have a unit strength impulse at the initial time (t = 0).

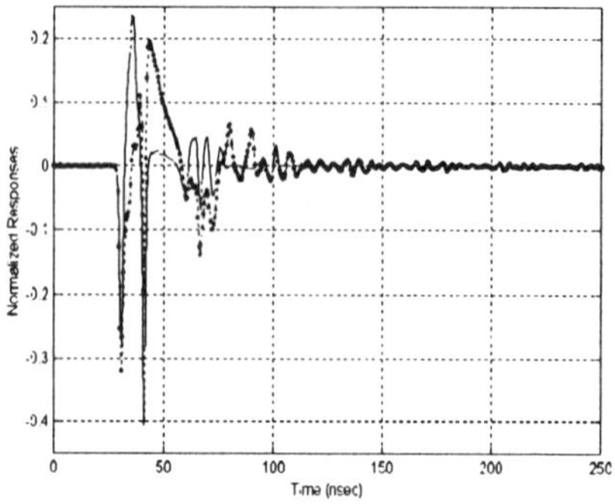

Figure 3. Normalized impulse response (——) and the *mismatched* K-pulse response (. . . . .) for the sphere S1 at 45 degrees bistatic aspect angle.

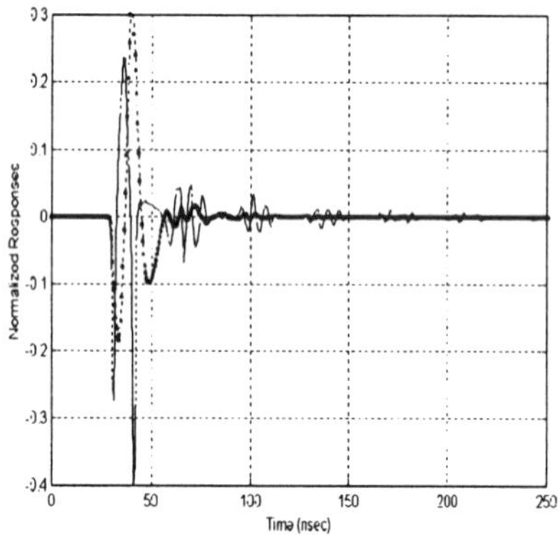

Figure 4. Normalized impulse response (——) and the *matched* K-pulse response (. . . . .) for the sphere S1 at 45 degrees bistatic aspect angle.

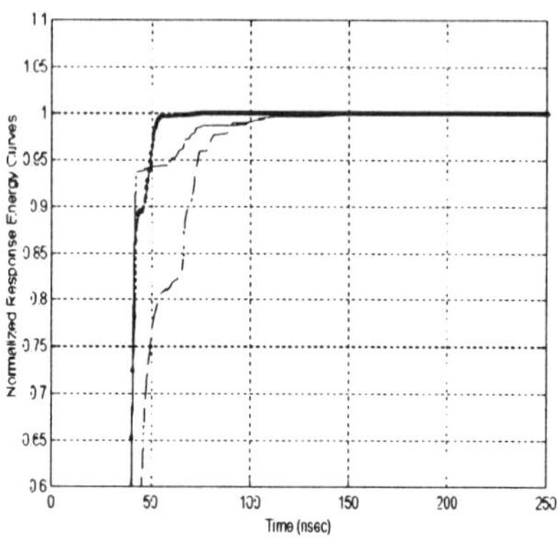

Figure 5. Normalized energy curves for the impulse response (——), *matched* K-pulse response (. . . . .) and *mismatched* K-pulse response (– – –) for the sphere S1 at 45 degrees bistatic aspect angle.

4. Conclusions

For the first time in literature, use of multi-aspect data fusion in the Genetic Algorithm-based K-pulse design problem (i.e. in the passive phase of the proposed target classification process) is formulated and successfully demonstrated for spherical dielectric targets. Target classification tests for two spherical targets with the same size but different refractive indices have resulted consistently correct at all test cases, including the aspects which were not previously used in the K-pulse design process. As a future work, use of decision fusion at the real-time (active) classification phase will be studied to further improve the overall performance of the K-pulse target classification technique.

5. References

Journal articles

1. Hagness, S. C., Taflove, A. and Bridges, J. E. (1998) Two-dimensional FDTD analysis of a pulsed microwave confocal system for breast cancer detection: Fixed-focus and antenna array sensors, *IEEE Trans. Biomed. Eng.* 45, 1470-1479.
2. Kennaugh, E.M. (1981) The K-pulse concept, *IEEE Trans. Antennas Propagat.* 29, 327-331.
3. Marin, L. (1973) Natural-mode representation of transient scattered fields, *IEEE Trans. Antennas Propagat.* 21, 809-818.
4. Turhan-Sayan, G. and Moffatt, D.L. (1989) K-Pulse Estimation and Target Identification of Low-Q Radar Targets, *Wave Motion* 11, 453-461.
5. Turhan-Sayan, G. and Moffatt, D.L. (1990) K-Pulse Estimation Using Legendre Polynomial Expansions and Target Discrimination, *Journal of Electromagnetic Waves and Applications* 4, 113-128.
6. Weile, D. and Michielssen, E. (1997) Genetic Algorithm Optimization Applied to Electromagnetics: A Review, *IEEE Trans. Antennas Propagat.* 45, 343-353.
7. Turhan-Sayan, G., Leblebicioglu, K. and Inan, S. (1998) Input Signal Shaping for Target Identification Using Genetic Algorithms, *Microwave and Optical Technology Letters* 17,128-132.

Book references

8. Turhan-Sayan, G. and Moffatt, D.L. (1991) K-Pulse Estimation and Target Identification for Geometrically Complicated Low-Q Scatterers, in B.W.Noel (ed), *Ultra-Wideband Radar: Proceedings of the First LosAlamos Symposium*, CRC Press, pp. 435-462.
9. Goldberg, D. E. (1989) *Genetic Algorithms in Search, Optimization and Machine Learning*. Reading, MA: Addison-Wesley.

OBJECT DETECTION AND TRACKING IN DISTRIBUTED SURVEILLANCE SYSTEMS USING MULTIPLE CAMERAS

Object Detection and Tracking in Distributed Surveillance Systems Using Multiple Cameras

C.S. REGAZZONI, AND L. MARCENARO
Department of Biophysical and Electronic Engineering
University of Genova
Vial All'Opera Pia 11A, I-16145
Genova - Italy
Carlo@dibe.unige.it

1. Introduction

In the last decade video-surveillance systems have been developed for guarding remote environments in order to detect and prevent dangerous situations. Until few years ago, surveillance was performed entirely by human operators, who interpreted the visual information presented to them on one or more monitors. Sometimes, their work conditions resulted in failing to raise the generation of alarms in case of dangerous situations.

Therefore, recently an effort has been made to develop automatic or semi-automatic video-based surveillance systems [1], able to alert human operators when something unusual occurs within the guarded environments.

Such systems were developed for different tasks, such as object detection [2], vehicle tracking on highways for traffic control purposes [3], analysis of human behavior [4] or people counting [5] in public environments. All the systems must process a scene for detecting dangerous situations or extract significant environmental features (crowd estimation): one example is showed in Figure 1, where an automatic video-surveillance system is able to detect objects moving within the guarded scene, and in Figure 2, where a system for tourist counting is presented [6].

The development of complex surveillance systems is capturing interest of both research and industrial world as there are strong requirements coming from the society in the direction of increasing safety and security in different applications.

Transports, in particular, are the application field from which most pressure comes for innovative research solutions: improving monitoring capabilities on large transport environments is the topic of more interest. Surveillance services of improved capabilities are also more and more requested as a consequence of the increased demand of secure feeling in all-days life coming from people. Finding a trade-off between people security necessity and privacy issues is becoming a worldwide spread political issue.

In order to satisfy a potentially so large market, strong research innovations are required that allow system designers to take advantage of multiple architectural solutions and processing tools developed by researchers. The assessment of widely accepted performance

A. K. Hyder et al. (eds.), Multisensor Fusion, 541–572.

evaluation metrics is also an important step to allow surveillance engineers and end-users to select in a straightforward way architectural choices for different applications.

Figure 1. An example of a video-surveillance system for unattended highway tool-stations

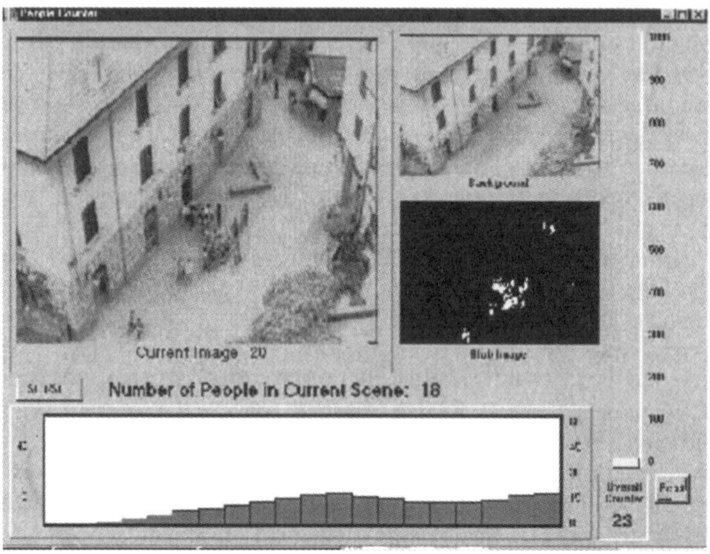

Figure 2. An example of a video-surveillance system implementing tourist counting functionalities

2. Video-based surveillance systems

An automatic surveillance system can be defined as a computer system that, based on input data and provided by scene understanding capabilities, is oriented to either human operators attention focusing or to automatic alarm generation.

Video-surveillance systems have been developed since 60s:

- *First generation (1960-1980):* in these kind of systems there is no information processing; acquired images are presented on one or more monitors, and the scene understanding task is performed by the human operator;
- *Second generation (1990-2000):* digital information processing is completely performed by a centralized system that shows acquired scenes and generates alarms helping the human operator to face potentially dangerous situations;
- *Third generation (2000-):* distributed digital signal processing. Only the information for signaling and describing a dangerous situation is presented to the human operator. A summary of the evolution of the video-surveillance systems functionalities is presented in Table 1 while a generic schema of a standard video-based surveillance system is shown in Figure 3.

TABLE 1. Evolution of surveillance systems

	Sensor	Automatix Processing	Transmission	Information Visualization
1-st Generation	Analog	Absent	Analog dedicated; Digital over wide area network	Cyclic multiplexing
2-nd Generation	Analog-Digital	Centralized	Analog-digital dedicated; Analog-digital over wide area network	Automatic signaling of dangerous siituations
3-rd Generation	Analog-Digital	Centralized-distributed	Digital dedicated; Digital over wide area network	Semantically configurable signaling; Distributed alarms over the network

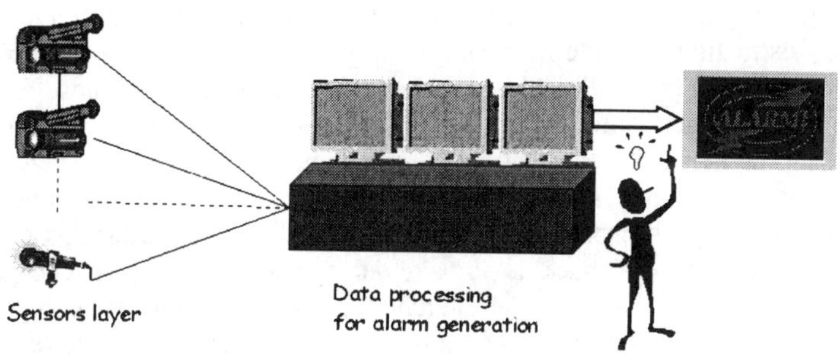

Figure 3. Model for a generic video-surveillance system

Sensors acquiring input data from the outside world to the system can be heterogeneous (TV cameras, microphones, GPS sensors, IR cameras, etc.) and arranged in multisensory configurations. Multisensory nature of surveillance information constitutes an important direction of improvement that is related to improved processing capabilities. Multisensor systems can take advantage from processing either the same type of information acquired from different spatial locations or information acquired by sensors of different type on the same monitored area. Appropriate processing techniques and new sensors providing real-time information related to different scene characteristics can help both to enlarge the size of monitored environments and to improve performances of alarm detection in areas monitored by more sensors.

Scene understanding requirements regarding a video-based surveillance system are severe due to high variability and poor structure of monitored scenes in different surveillance applications. Such variability has several consequences in required processing tools. For example, the use of sophisticated image processing algorithms is necessary for rough signal pre-processing and filtering. Highly variable scene conditions imply the necessity of selecting robust scene description and pattern recognition methods. Automatic learning capabilities are an emerging issue in surveillance systems, as the capability of automatically developing models of scenes to be recognized as potentially dangerous events from a training set of presented examples will be a key issue for improving end-user acceptance of surveillance systems. Real-time and low-cost are two of the most important constraining factors for individuating processing choices acceptable for surveillance systems.

2.1 MULTISENSOR SURVEILLANCE SYSTEMS

A major limitation of current surveillance systems for providing scene interpretation sufficiently robust to operate in outdoor environments (e.g. under environmental changing conditions) is the low level of diversity in the scene interpretation architecture.

Diversity can be obtained for example by increasing the number and type of sensors monitoring the same scene under different point of views or with different resolutions, as well as by extracting multiple features (possibly highly independent each other) from the same scene. Diversity implies also some disadvantages that are most related with the necessity of intelligent information fusion strategies capable to provide consistent and univocal representations and interpretations of monitored scenes. Therefore, real-time multisensory surveillance systems based on diversity concepts are unavoidably more complicate than single sensor systems based on centralized architectures; a natural choice to face this complexity drawback is to take advantage of relatively cheaper processing resources as well as of lower costs of interconnection potentialities for going towards really distributed systems. The advantage of a multisensor system over a single sensor system can be quantified in terms of improvement in the situation assessment ability [7]. There are many factors that contribute to the enhancement of quantifiable system performance and to the improvement of the overall system utility. Some of them are:

- *Improved system reliability and robustness.* Multisensor systems have inherent redundancy. Due to the availability of data from multiple sensors, system performance is improved. When one or more sensors fail or are unable to operate due to interference such as jamming, the system can continue to operate at reduced performance level (graceful degradation).

- *Extended coverage.* Use of multiple sensors allows an increase in coverage, both spatial and temporal. Multiple sensors can observe a region larger than the one observable by a single sensor.
- *Shorter response time.* Since more data are collected by multiple sensors, a prescribed level of performance can be attained in a shorter time.
- *Improved resolution.* Use of sensors with different resolutions can result in an inference with better resolution than any single sensor.

Examples regarding the *coverage extension* improved functionality can be found in [9] where a set of non-overlapping cameras is used for guarding a large and complex environment such as an office building or in [10] where a distributed data fusion system is presented for real-time crowding estimation.

3. Distributed video-based surveillance systems

Modern video-based surveillance systems usually aim to meet some commercial or security requirements in order to provide better performances in terms of some aspects of the surveillance task. For example, deploying more cameras in a monitored environment yields a better coverage of a surveillance area, thus achieving an overall control of the scene contained in the area. There is also a trend towards wireless cameras because they can potentially provide more ubiquitous installation in many places such as natural parks, industrial factories, etc.

Moreover, a real-time processing of events is required if, as usual, temporal bounds to the surveillance application are critical. One way to meet such a requirement is to take advantage of the remote-processing skills of modern computer networks, as shown in Figure 4.

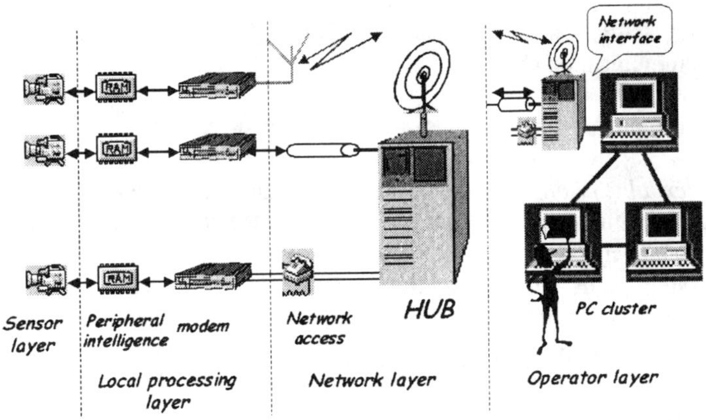

Figure 4. Model for a generic distributed video-surveillance system

In this figure an architecture for a distributed surveillance system is shown: video-cameras constitute the sensor layer of the system, while the peripheral intelligence and the transmission devices form the local processing layer. Sensor and local processing layers can be physically organized together in a so called *intelligent camera*. The principal entity of the network layer is the HUB: the main functionality of the HUB is the fusion between data coming from lower-level layers. Network layer has an interface towards the operator layer that is able of presenting surveillance alarms to the human operators.

Previous analog CCTV-based systems were simple broadcast video applications. Now we are moving towards completely digital systems that utilize wireless or wired networks to provide two-way communication links between surveillance centers and intelligent sensors including cameras.

Distinctive features of a distributed surveillance system with respect to a standard centralized systems are:

- *Multisensor architecture:* distributed surveillance systems are inherently multisensorial with possible heterogeneous sensors;
- *Multiprocessor:* spatially distributed systems use processors or cluster of processors connected through a wide area network. A relevant pressure in the direction of task distribution comes from the fact that we are now in the era of INTERNET; this suggests that it becomes also possible to think to develop surveillance systems based on a mixture of distributed resources going from end user PCs to more powerful processing centers made available by "surveillance services providers". This perspective makes it necessary the study of distributed processing and network management strategies, by providing also interesting market perspectives to the "distributed intelligent surveillance system" concept. As one of the most interesting trends of modern computer networks is represented by the deployment of "intelligence" in the networks themselves, network management strategies should be capable to find a decomposition of surveillance systems in terms of tasks that they have internally to perform and to design distributed solutions capable to allow an optimal allocation of such tasks on available processing resources, depending also on bandwidth characteristics of involved communication channels.

Thus, the main advantages of a distributed video-surveillance system are:

- *Flexibility-configurability:* a distributed surveillance system can be adjusted to different network architectures minimizing the cost of new installation;
- *Modularity:* new capabilities can be easily added to the distributed surveillance system, thus simplifying the work of the surveillance service providers and engineers and promoting the reuse of existing surveillance modules;
- *Robustness:* a distributed system is robust to possible network congestion or peripheral PCs failure: this functionality is achieved by implementing an optimal adaptive task allocation algorithm.

3.1 DESIGN PROBLEMS

The underlying problems of a generic multisensor distributed video-surveillance system are managing and organization of the information flow. The network paradigm is, of course, universal and applicable directly to surveillance networks. Hosts are typically sensors and cameras. HUBs and switches are used to collect information, multiplex it and route it to remote operator in the surveillance center(s).

Starting from these observations, major problems that a video-surveillance system designer has to solve are related with the following points:

(a) **System architecture:** during the realization of a distributed surveillance system, a distinction must be done between physical and logical architecture. Physical architecture is related to the optimal system topology: this affects the following stages of the system implementation, such as network compatibility and multisensor data fusion. Logical architecture involves the design and the subdivision of processing tasks of the system. Finally, a rule for optimal tasks distribution must be carried out;

(b) **Information processing:** methods and algorithms for processing sequences coming from peripheral cameras have to be implemented; processing engine must be efficient and made of units which can be assembled in different ways (modularity);

(c) **Inter-processors telecommunications:** efficient techniques for information transmission must be adopted in order to achieve low error rates and real-time transmission performances;

(d) **Data fusion:** information coming from intelligent cameras must be fused into a common reference system: data fusion can happen at different signal representation levels(images, regions, trajectories, etc.).

(e) **Performance evaluation:** surveillance systems are very complex and their performances rely strongly on values taken by parameters regulating the behavior of the processing algorithms. This makes difficult the comparison between performances in order to choose the most suitable system for the considered problem. Moreover, parameters are set during the installation phase, but a good selection requires time and experience. An evaluation method for characterizing video surveillance systems, in order to provide an easy way for system installation, must be studied.

4. System architecture

The first task in the realization of a distributed video-based surveillance system is the definition of the system architecture. We distinguish here between the physical and logical architecture of distributed surveillance system.

Physical architecture is related with overall system topology: in particular, the design should take into account the features of the network. Links between peripheral cameras, local HUBs, central processing unit and remote control room, have to satisfy certain bandwidth requirements. If the video-surveillance system is to be installed over an existing network, video processing tasks distribution must consider the eventual limitations due to the features of the underlying network architecture.

If a open area network is used, problems regarding potential large scale availability of reserved information must be taken into account. In particular several countries recently promulgated laws concerning the respect of citizens privacy rights: the administrator of a wide area distributed surveillance system must avoid uncontrolled access to personal data or pictures regarding non suspicious people acting or working within the guarded environment.

Figure 5 shows a typical physical network architecture for a distributed surveillance system.

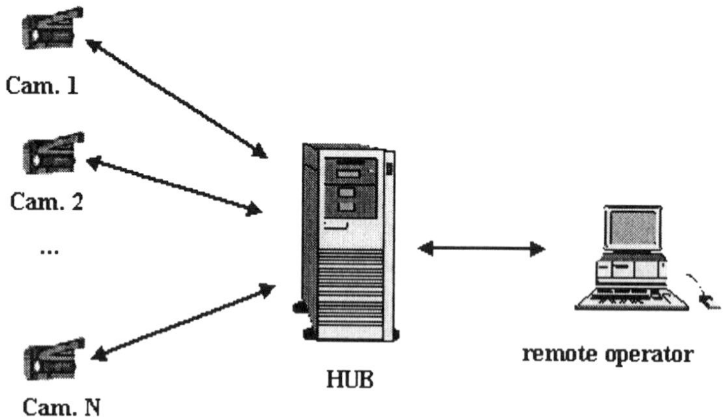

Figure 5. Example of physical architecture of a distributed video-surveillance system

Logical architecture of a distributed surveillance system is connected with tasks subdivision among different system entities. A *task* is a well-defined set of operations that can be executed independently by the different physical entities of the video-surveillance system: such parts receive input data and provide output results consistent with the system. For example image acquisition, change detection, morphology and focus of attention are the basic tasks performed by most video-based surveillance systems, as it will be pointed out in the following section.

4.1 PHYSICAL ARCHITECTURE`

The physical architecture of a distributed surveillance system is completely defined when a graph is given showing the interconnections between sensors, HUBs, and control centers. An example of the topology graph is shown in Figure 6.

The topology of the system should be selected on the basis of the particular installation; one should take into account the capacity of links between different processing units jointly with their processing ability and the possible environmental limitations.

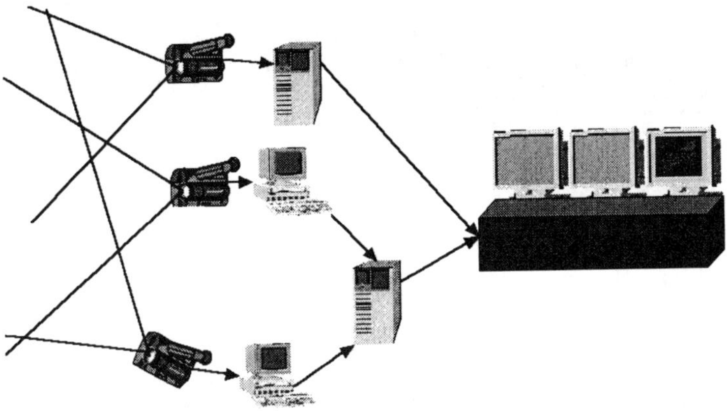

Figure 6. A graph correspondence to a physical architecture of a distributed video-surveillance system

On the other hand, the optimal placement for the cameras with respect to the monitored environment must be found. This strongly depends upon the particular system installation and the environmental obstacles; several cameras installations can be distinguished on the basis of the reciprocal positions of each field of view:

- Overlapped fields of view: having overlapped fields of view can be useful either to improve sensors resolution or to improve the reliability of the video-surveillance system in the overlapped area;
- Non-overlapped fields of view: in this case multiple cameras are used to enlarge the overall field of view of the system.

4.2 LOGICAL ARCHITECTURE

The definition of the logical architecture of the system involves the specification of task subdivision between available processing resources of the system. Figure 7 depicts a logical graph of a distributed surveillance system: each node of the graph corresponds to a task of the surveillance system; different tasks can be grouped together and executed sequentially by the same processing unit. For example, in Figure 7 N_{IC} tasks are allocated to each camera, while N_H and N_{CC} tasks are allocated to the HUB and the control center, respectively. In such a case, grabbing skills of a common camera can be joined with some low-level image processing tasks, and a so called *intelligent camera* is obtained as lower level node of the physical network. An *intelligent camera* is a conventional digital camera with some processing power. Such a camera should be able to execute one or several parts of the surveillance algorithms in order to avoid the transmission (through the HUB) of redundant information to the remote surveillance operator. Moreover, sending only relevant data would make more channel bandwidth available to other cameras, thus fulfilling one of the above mentioned requirements. The price to be paid is the necessity of a computational power to be allocated to the camera.

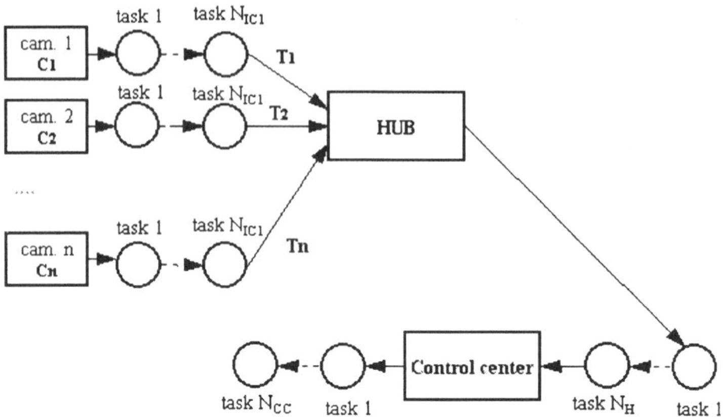

Figure 7. Example of logical architecture of a distributed video-surveillance system

In addition, allocating one part of the surveillance algorithms to the camera and the remaining parts to the HUB (the gathering node of the network), rather than all the parts to the HUB, will allow a distributed implementation of the surveillance program as well as more "clients" will be served by an HUB with a given computational power. If the cameras and the HUB processors are properly synchronized, the parallel execution of the video-processing algorithms will be significantly faster, thus matching the real-time requirement for the execution.

Therefore, the feasibility of subdividing a surveillance algorithms into several, independently executable tasks, becomes a basic issue of remotely distributed processing.

Advanced video cameras with some processing facilities are commercially available and currently used, especially for industrial purposes. For example, a flexible and powerful solution to avoid the overload of the monitoring process is to assign the detection of faulty components to an "on site" checking system using a smart camera with a DSP processor. On the other hand, video-surveillance applications are usually quite particular and different applications involve many variations. The design of an *ad hoc* intelligent cameras should take into account the following requirements:

- Video-surveillance algorithms often process sequences of images rather than single independent images, and many applications require an "off-line" image processing (e.g., retrieving of data associated with alarm events). Thus, as commonly used color images are memory wasting objects, large storing facilities are needed.
- Wireless links between the cameras and the central operating unit (HUB) are advisable for the flexibility of the whole system and for the mobility of some cameras in the surveillance area. The intelligent camera should therefore be

provided with a LAN (Local Area Network) interface to manage the communications with the HUB according to some wired/wireless protocol.

Figure 8 shows the most important components of an intelligent camera. The video digitizer provides a multi-level image that is temporarily stored in the image buffer. One or more ADSP(s) (Advanced Digital Signal Processor) manage the most time-consuming operations, while the actual master processor executes the video-surveillance tasks previously stored in the DRAM (Dynamic RAM) memory. The LAN Interface Controller manages the incoming and outgoing information exchanged with the HUB, like handshake signals, images etc.

Current trends of video-sensor technology suggest a fully on-chip integrated approach to the design of the intelligent camera architecture [3,31]. Task-allocation strategies.

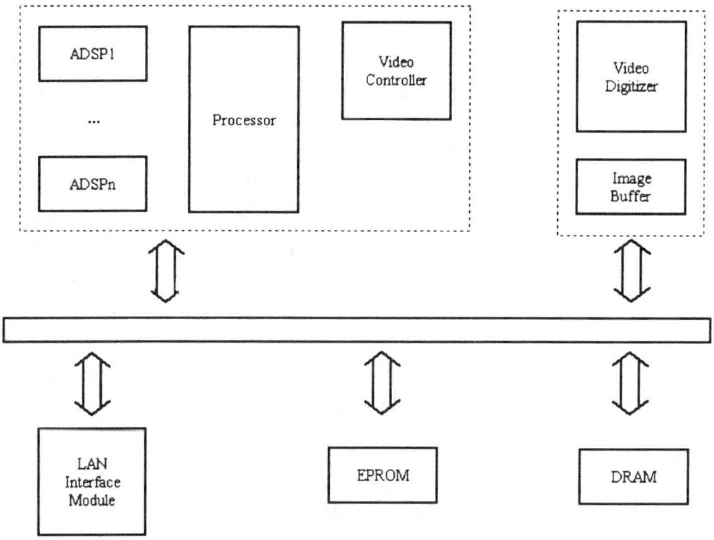

Figure 8. Architecture of an intelligent camera

4.3 TASK ALLOCATION STRATEGIES

Figure 7 shows an example of task allocation to a single surveillance "cell" of a network system. The HUB managing the cell can receive information from N cameras by using a local communication system (e.g. a LAN, possibly wireless), and several tasks run on each camera. Each camera is provided with a given processing power that makes it possible pre-processing of the video stream as well as appropriate coding. The basic processing parameters of the DSP: processor clock frequency, RAM size, cache size, and so on can be considered as parameters bounding the number of tasks that can be allocated to the processor.

The channel between two nodes of the physical architecture (e.g. a camera and the HUB) is characterized by other bounding parameters such as the available bandwidth and the signal-to-noise (S/N) ratio. In the example of Figure 7, where all the cameras share the HUB-processor power and tasks there allocated, the number of tasks running on each camera C_i depends on the processing power of each camera and on the bandwidth characteristics of the link. For example, if links and nodes are characterized by equal parameters, it is reasonable to assume that the same number of tasks will be allocated to each camera. From a practical viewpoint, estimation of the parameters is necessary of a set of tasks and of inter-module links. For example, a benchmark-oriented evaluation system should be used to estimate the computational load of any single task running in each physical node.

An interesting characteristic of any video-surveillance system based on a hierarchical architecture is the progressive *specialization* of the operations performed by each layer (or task) of the flow diagram on incoming information. For example, after performing the *change detection* task on the original images, only the shapes of intruders will be considered by the system.

After the focus of attention step, intruders will be represented by a small-sized data structure collecting information such as coordinates, mean gray levels, inter-frame relationships, and so on. Thanks to the progressive specialization of modules, the amount of information that each task sends to the following task in the logical architecture tends to decrease when the task position in the logical hierarchy becomes higher. The staircase-shaped function of Figure 9 can roughly describe the aforesaid relationship. If a communication channel with a given capacity C (bit/sec) is used to transmit information I (bits), the transmission time $T = I/C$ is required to send images to the HUB. This general simple model can be formalized as described in [8] by denoting as T_m, m=1...M the task sets allocated to the m-th node of a linear physical architecture, P_i. A linear architecture can be defined as an architecture whose nodes have a single input link and a single output link. For example, a surveillance system composed by an intelligent camera, a HUB, and a central control center can be considered as a linear architecture. If a task of the surveillance system is indicated as t_x, one can write:

$$T_m = \{ t_x, \sum_{i=1}^{m-1} N_i < x < \sum_{i=1}^{m} N_i \} \qquad (1)$$

The overall number of tasks in the linear architecture is given by:

$$NT = \sum_{i=1}^{M} N_i \qquad (2)$$

and the global set of tasks can be indicated as:

$$T = \bigcup_{i=1}^{m} T_i$$

The execution time $p(t_i)$ of the i-th task as well as the time $it(t_i)$ necessary to transmit information to the higher level physical node can be considered as the parameters that influence the allocation choice to different processors in the chain. Both parameters depend on the particular physical node to which the task is allocated, i.e., $it(t_i)=it(t_i, m)$; $p(t_i)=p(t_i,m)$. Moreover, it can be assumed that transmission of information between tasks allocated to the same physical module is null, i.e. $it(t_i,m)=0$, if t_i+1 is allocated to node m, too. The task allocation problem can be defined as the problem of optimal partitioning of logical tasks between physical nodes. On the basis of the above considerations, optimality can be defined as finding the task configuration that ensures processing within the minimum possible time. The functional J to be minimized can be indicated as $J=J(T_1, T_2, \ldots, T_{n-1})$, since

$$T_n = \mathrm{T} - \bigcup_{i=1}^{n-1} T_i .$$

The above functional can be considered as given by the sum of processing and information transmission times on all tasks of the surveillance chain. In this way, the resulting problem becomes a simple optimization problem, as tables of processing and information transmission time can be considered as available for a given surveillance problem. However, if a hierarchical multisensor architecture as well as a tree-like structured logical architecture is considered- i.e. multiple paths converge to single nodes, the optimization problem becomes more complicated. New constraints must be considered, for example coming from the level at which data fusion algorithms are available. Fixing such a level, implies that information coming from lower levels must be of the same type, to be appropriately fused. This means that task allocation to lower modules is somehow restricted to separate allocation of disjoint subsets of T, each formed by a linear chain. If multiple fusion levels are possible the number of subsets increases but the problem can be managed in the same way. However, often if multiple fusion methods are available the problem becomes not only the one of allocating tasks, but also the one of choosing between alternative architectures composed by a partially common set of tasks, so making necessary a generalization of the optimization problem. Another complication may arise from different computational power of different physical nodes working at the same level (e.g. intelligent cameras with different computational power). In case of data fusion at a HUB level and links with equal information transmission capacities between cameras and the HUB, for example, a simple heuristic rule of thumb could be to select the task allocation by considering the camera (forming single path with the HUB) whose global processing time (i.e. the sum of processing times $p(ti,1)$) is lower to perform task allocation, and to adopt the same partitioning for the other paths linking the HUB to the other cameras.

5. Image processing levels

Figure 10 shows the main structure of a generic video-surveillance system. Signal processing operations performed by the system, can be categorized low, middle and high level stages. Low-levels perform a first abstraction from the acquired signal: images are processed and objects in the scene are detected. A list of detected objects together with objects features is the input to middle-level stage. Middle-level processing algorithms perform tasks such as blob tracking, blob labeling, etc. Finally, high-levels are responsible of image understanding (behavior analysis) capabilities of the surveillance system.

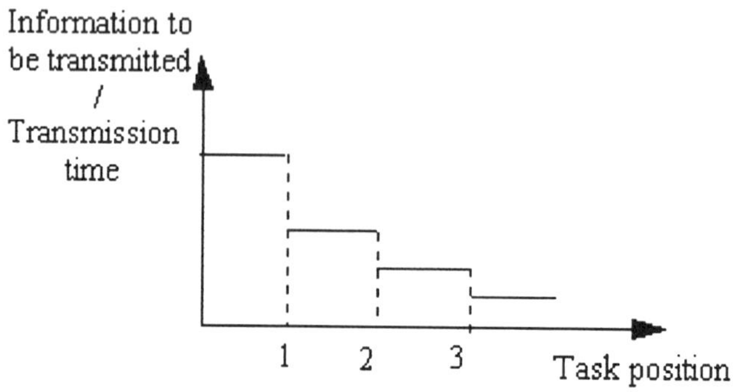

Figure 9. Evaluation of the number of bytes to be sent on the basis of a sample sequence

Below are listed the tasks used by the surveillance system and their main features. Only low-level tasks have been considered, due to their high time-consuming and huge-producing-data features:

- *Image acquisition:* The acquisition of images is performed by a frame grabber: if an analog camera is used, a grabber with A/D capabilities is needed, while if a digital camera is used, a controller board is sufficient (IEEE1394 compliant, for example). A hardware adapter is needed between the camera and the processor, and the acquisition time is depending on the hardware features and image format features (image size, black and white or full color, ...).
- *Change detection:* Change detection operations [12,13] are implemented as a scanning loop that computes, for each pixel of two images, the difference between their grey levels. Such difference is then compared with a threshold, and the image is converted to a two-level image. So, the outgoing result is an image whose pixels grey level can be displayed with just one bit each (1 = white, 0 = black). Given the size and the type of image (black and white, color, ...), the execution time is fixed if we use a conventional change detection algorithm. If we chose a more complex algorithm, such as a light-adaptive one, the time may depend on the number of "intruders" actually perceived inside the room. Anyway, the execution time for change detection algorithms is usually very short.
- *Morphology algorithms:* Morphology algorithms [14] such erosion and dilation will be considered. The execution time depends strongly on the number of intruders inside the guarded room. If no persons or object are detected, then the morphology algorithms are just skipped. When one or more "blobs" are detected, a local 9x9 sized set of operations is performed on each pixel corresponding to the intruder, and if the number of intruders is quite high this might be a time consuming phase. As well as in the change detection phase, the outgoing image is a two level image.

- *Region growing:* The aim of the region growing operation [14] is to identify all pixel-connected regions inside the image received by the previous processing stages. It's an intermediate operation whose function is to provide a compliance set of data to feed in the next phase of "focus of attention". This task is a pixel by pixel and time-consuming operation, but, as in the previous case of morphology, it has to be performed only on those pixels belonging to the shape of an intruder. The output image is no more a two grey levels, but the number of levels depends on the number of intruders that have been detected (one grey level for every intruder).
- *Focus of attention:* Image coming out from the region growing phase provides a "map" of all intruders detected within the monitored room. The focus of attention phase [15] just draws a rectangle around each of them, by computing the horizontal and vertical bounding edges. This is still a pixel-by-pixel operation, but not so high time consuming as the previous ones. The outgoing data is no longer an image, but a set of co-ordinates arranged in a table structure. Actually, the focus of attention module performs some other task, like merging the overlapping rectangles or removing the rectangles whose size is smaller than a given threshold. Hereinafter, the modules of the system following the focus of attention phase can ignore the blob shapes, just managing their co-ordinates.
- *Segmentation:* All modules of the system run independently of the segmentation phase [16]. But for some applications, the segmentation algorithm can be very useful, providing a powerful tool for the "blob tracking" operation performed by the modules in the downstream. For instance, if some person inside a group is very close to other persons, the segmentation can help to find out the borderlines between them, so refining both the focus of attention and blob tracking phase. The main concern with the segmentation operations is the large amount of execution time that they require to be performed, and the need for full color image as input and output. So, if such a set of algorithms is needed, it is advisable to allocate them on the camera side, decreasing the time required to send full color images to the HUB.
- *Shadow correction:* As well as the segmentation, the shadow analysis is not always necessary. However, if shadows are a big problem for the current application, the shadow correction modules can be included in the processing chain without any specific change to the existing framework. Shadow correction needs current image in full color with all intruders inside their own rectangles. As output, the module returns the list of rectangle updated with respect to the new vertical size, computed ignoring the shadow.
- *Background updating:* Background image is periodically updated by a routine that uses algorithm such as Kalman filtering for predicting pixels value of the background image [13]. The aim of the algorithm is to remove from the background all the "interference" that are produced by changes in light conditions or in the environmental structure of the monitored room. However, the use of such algorithm depends on the specific application and for some it could even be counterproductive (remove of abandoned objects from the background).

A detailed graph of the considered image processing modules is shown in Figure 11. Having described the main tasks or algorithms involved in the surveillance system processing chain, a table presenting processing features in terms of output information and computational load of the cited algorithms are presented in Table 2. The computational load of each module depends upon the *Size* of input data of the module itself: this explains the *Size* dependence in the last column of Table 2.

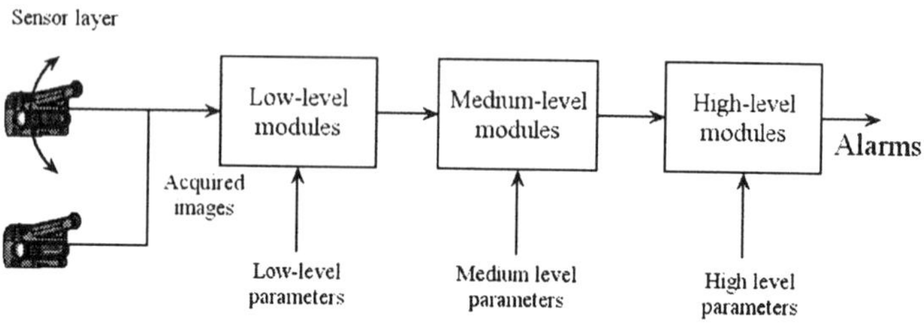

Figure 10. Architecture of a video surveillance system

TABLE 2. Image processing algorithms features

	Output	Computational load
Acquisition	Full color	f_{ac} (Size, Color Depth)
Change detection	Black/White	f_{cd} (Size)
Morphology	Grey-scale	f_{mph} (Size)
Region growing	List of blobs features	f_{rg} (Size)
Focus of attention	Full color	f_{foa} (Size)
Segmentation	List of blobs features	f_{seg} (Size)
Shadow correction		f_{sc} (Size)

6. Inter-processors telecommunication

Arcs in Figure 6 represent transmission channels of the distributed video-surveillance system. In the current section a possible classification of communications links useful for multisensor surveillance systems is proposed. Transmission channels can be first classified on the basis of the physical link type: wireless and wired transmission can be the first rough categorization. Wireless and wired links are both employed in surveillance systems. A further classification can be done depending on the specific characteristics of the transmission channel in terms of waves used to propagate the information (e.g. wireless: satellite, radio-link, WLAN; wired: cable parameters, optical fiber).

From the point of view of the architecture of a distributed multisensor surveillance system (DMSS), choosing a different transmission system can be related to select a different type of link. Among the possible characterization of a link within a distributed surveillance architecture one can select:

- Analog/ digital transmission;
- Long range/short range link;

- Narrowband/wideband signals;
- Source coding method;
- Channel coding method;
- Fixed/variable rate;
- Dedicated/shared channel;
- Mono/Bidirectional

Analog transmission was used in first surveillance systems using closed circuit television, CCTV, while digital transmission is becoming the current standard for DMSSs mainly due to the higher flexibility it allows. Long-range communication systems can be used generally for transmission between the HUB and the control center; in this case long-distance transmission systems such as fiber optics or satellite can be used, while generally the HUB is closer to cameras, so that short range communications as wireless LANs or wired access networks (cable TVs, ADSL lines can be used). Generally, more information to be transmitted in a surveillance system at lower level of the logical architecture: at such levels the information rate can be fixed by considering the acquisition rate (e.g. 2-5 frame per second), the image size (e.g. square 256x256 or 512 x512 images), the image information (full color/ black and white) the sample quantization (e.g. 8bit/pixel/color). Video streams can be coded by using methods that exploit spatial-temporal information (e.g. MPEG, M-JPEG) or that can be applied separately to video frames (e.g. JPEG). Coding can be used to reduce the amount of information to be transmitted by lower levels by allowing compression. Nevertheless, tasks at lower levels of a DMSS are those usually characterized by higher information transmission time requirements. For such tasks, usually wideband signals are generated whenever task allocation made necessary splitting allocation of immediately higher level tasks in the logical hierarchy to be allocated to different physical nodes. Narrowband signals can be generated at higher levels thanks to the specification property of DMSS. In order to establish the information transmission time it is also important to know whether a fixed or variable rate service is granted by the used communication channel. Fixed rate services are usually associated with a wired communication channel between the communicating nodes. Variable rate services can depend either on sharing the communication channel with other users whose number and quantity of transmitted information may vary in time, or on time-varying channel characteristics. Both cases can be present in a DMSS: for example, if two intelligent cameras share the same wireless channel to communicate information to an HUB with other end-users the two situations can occur simultaneously. Robustness is another issue for a communication link as information must be protected for avoiding heavy degradation due to channel characteristics; this can be done either by adding redundancy at the information source, so slowing down information rate or channel coding methods are necessary or by including new tasks at the received physical node that realize restoration techniques. Finally, each node in a lower level of a physical architecture of a DMSS can require not only to transmit information to the immediately higher level node, but also to receive information (e.g. re-configuration software, parameter setting, etc.). All the above aspects should characterize links among nodes in the DMSS, and include further complexity in functions such as task allocation. However, a simple first order model can be summarized by means of a limited set of features to be associated with the representation of each physical link; such features can be:

- Channel capacity;
- Physical link length;
- Channel bandwidth;
- Number of users.

Moreover, a number of communication functionalities must be associated with each task when it is assigned to a given physical node to represent the above characteristics of communication links, such as:

- Source rate;
- Source coding method; Finally, transmission links can be classified on the basis of the adopted transmission techniques.

Table 3 shows a classification of the image processing algorithms described in the previous section from the output rate point of view. Data that are transmitted over physical links are given by the informations coming from the different tasks of the DMSS. Output data from each processing task are transmitted over the distributed network to the following task in the processing chain. Integration tasks accept multiple inputs and perform a data fusion algorithm to obtain a smaller number of outputs. In the following paragraphs, only integration between hierarchically different levels will be considered. The network connectivity within the camera is usually obtained by a communication subsystem (e.g. embedded GSM chipset, IEEE802.11 WLAN, Ethernet or cable modem). Today the technology to integrate communication, networking and image-processing capabilities is available. However, both optimization of performance and cost-effectiveness are still crucial issues, as intelligent networking surveillance cameras are not yet commercially available. If we build a low-cost system with a limited communication bandwidth, this is an extra reason for making a camera capable of image processing. Video encoding is still a relatively complex operation, especially if one uses the MPEG family. Hence, one should choose either a very powerful CPU or dedicated chips to provide compression by encoding. This would result in higher costs and increased power consumption. Sometimes, local processing can be performed at a lower CPU cost than required by encoding.

TABLE 3. Image processing algorithms output rates estimation.

	Output data rate
Acquisition	Very high
Change detection	Medium
Morphology	Medium
Region growing	High
Focus of attention	Low
Segmentation	Very high
Shadow correction	Low

Wireless local area networks (WLANs) are very suitable communication subsystems. Spread-spectrum technology (either DSSS or FHSS) represents a good solution, as it is superior not only from the communication-engineering point of view but also because a large number of low-cost chip-sets and products are available [17,18,19]. The spread-spectrum approach has been studied also for cable-TV network communications, and this leads to interesting applications where the subsystem is very similar for both wireless and wired communications. IEEE 802.11 has a complex MAC with numerous features and various realization options. This gives design space for application providers, but somehow complicates the providers' work. In the MAC design, several issues are addressed. The system allows some level of mobility, as it is able to hand off from one access-point to another access-point. Spread-spectrum technologies and CSMA/CA overcome high-attenuation problems. Finally, power-control features make the system more economic. In conclusion we can say that the IEEE 802.11 is suitable for intelligent cameras. However, it is quite a complex subsystem, and task subdivision for the different algorithms allocated on the microprocessors must be performed very carefully in an intelligent-camera system.

7. Data fusion

Data fusion capabilities can be associated with specific processes in the DMSS: in particular, tasks having the cardinality of the outputs smaller than the cardinality of inputs, are considered as "integration" tasks. As stated in Section 2.1, data fusion allows several advantages such as improved system reliability and robustness, extended coverage and improved resolution. From an architectural design point of view these advantages can be achieved by means of the inclusion in linear architecture of data-fusion tasks. Therefore, the possibility of managing multisensor data implies an increased computational complexity and higher execution times. Nevertheless, the benefits can be high from the point of view of surveillance functionalities.

In the presented DMSS architecture, one can identify data fusion methods referred to both integration at **physical** and **logical** levels. Physical data fusion methods fuse data coming from different physical nodes. In particular, due to the fact that each physical node is devoted to monitor in time a different spatial area, it is natural to associated a spatial-temporal meaning to physical level fusion method. Events detected by different cameras will be fused in an HUB, whereas a control center will be devoted to fuse complex situations and events detected at a HUB level. On the other hand, logical fusion methods concern with the integration of different information useful to characterize and detect events at different levels of the logical architecture. As the representation of an event can be highly structured, logical fusion methods can be quite different in their purpose. Nevertheless, one of the major problem of these methods is to allow cooperation between multiple knowledge sources (i.e. a-priori knowledge) in providing a sufficiently robust event description. Despite the above classification, it is not often simple to identify task functionality that are related to either physical or logical fusion: spatial information can be considered as a specific type of attribute of an event. One example of spatial integration, is finding the absolute position of a detected object, starting from two or more measurements of its position performed by two camera subsystems. If each camera has non-overlapped fields of view, the final task of a possible physical data fusion algorithm could be the handover management between different fields of view of each single subsystem. The integration task should be able to recognize the identity of an object in order to find the correct matching between objects passing from one field of view to another. A further physical level fusion method that will be illustrated in the following can be used when dealing with a mobile-head camera; in this case the data fusion task is performed just beyond the image acquisition process. Integration task accepts a set of partially overlapped images acquired by the mobile-head sensor and fuses the images into a multilayer background image with a mosaicing technique. Examples of logical level integration tasks can be the integration of informations generated from multiple applications of image processing algorithms on data coming from lower levels of the surveillance system. These techniques are addressed to maintain a coherence of the overall information processed by the system with relation to the event to be detected and its attributes.

While physical data fusion techniques aim at find an enhanced physical structure of the guarded scene from a physical point of view (enlargement of the monitored area, increased resolution or precision of object tracking), logical data fusion techniques perform an integration that is functional to the event to be detected by the system. In this way video-surveillance environments based on physical data fusion algorithms, can be seen as systems based on a "a priori" enhanced knowledge of the physical world. If logical data fusion techniques are considered, the enhanced knowledge is an "a posteriori" knowledge that depends upon the type of the event to be detected and it concerns that event only.

In general logical data fusion techniques can be divided into:

- Top-Down techniques: these methods are based for example upon stochastic or symbolic models of the event to be detected by the system (e.g. statistical models);
- Bottom-Up techniques: bottom-up algorithms integrate the logical informations from the image processing levels without using any prior knowledge on the event itself (e.g. correspondences matching);

Depending upon the type of data representing the event different kind of techniques are used. Figures 12 and 13 show respectively two possible system architectures for logical and physical data fusion. The first proposed graph shows a system for crowding evaluation in complex environments[20]. Information acquired by a single sensor is not sufficient to acquire a complete knowledge about the status of the system. For example in complex environments such as a subway station, it is very difficult, mainly due to occlusions, that a single camera is sufficient to monitor the whole scene. In general, it is possible that some areas do not fall in the field of view of any sensor. Further problems are represented by the difficulty of current image processing techniques to provide a good response under different environmental conditions: changes in illuminations and scene structure must be considered as a part of the problem to be solved. At the same time, perspective transformation makes it possible that different 3D-scenes with the same number of people can be perceived in a very different way by the camera. These aspects make it difficult the problem of selecting features to be extracted from an image to perform the mapping: in order to provide a robust mapping, such features should behave in an invariant way under different transformations.

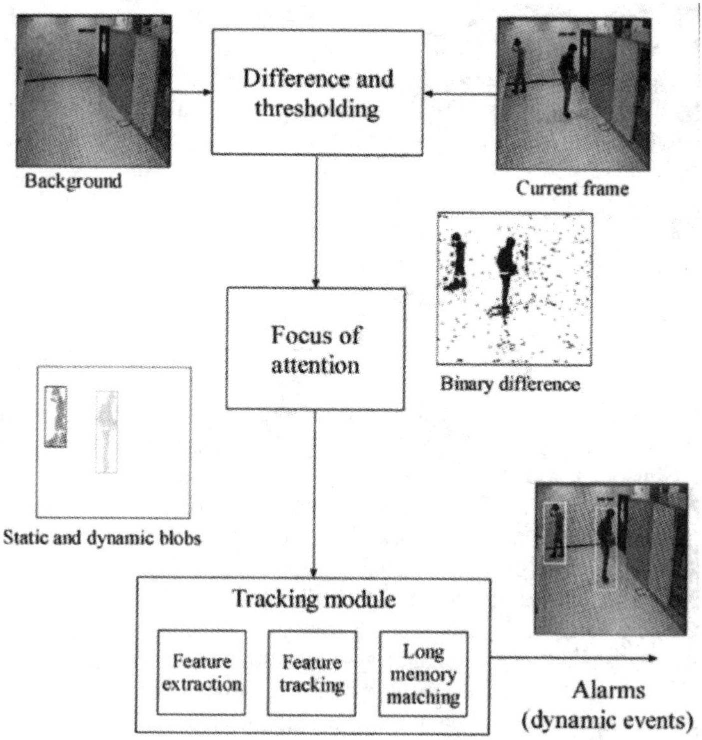

Figure 11. Architecture of a video surveillance system

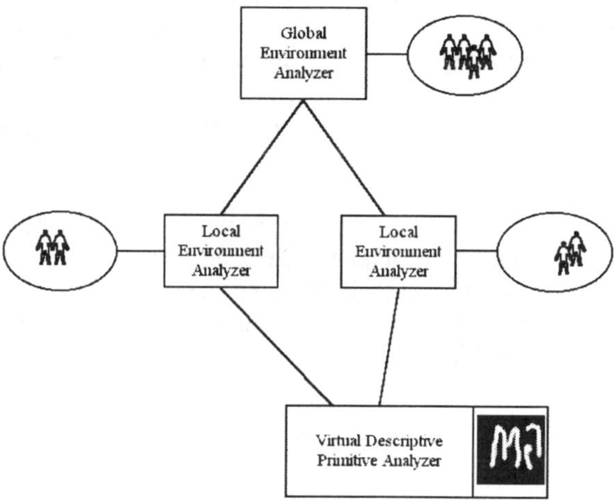

Figure 12. Architecture of a system performing logical data fusion

Architecture shown in Figure 13 has been applied in a car park monitoring system [21]. Events such as a vehicle or person passing through the environment may be simultaneously visible in more than one camera. Moreover the object may move in and out of the view volumes of several cameras in a sequential manner as the object progresses through the car park. An agent-based architecture for performing visual surveillance in a multi-camera environment is used. The loosely coupled nature of the agent approach and the ability to neatly partition responsibilities is ideal for solving such complex issues as *target handover* between cameras, *data fusion* and description of inter-object situations.

8. Performance evaluation

The goal of performance evaluation in DMSSs is twofold: from one side it is necessary to design procedures to define most efficient architectures in terms of use of processing and communication resources as explained in Section 4. On the other hand, it is necessary to assess the goodness of the considered application with respect to the expected surveillance functionalities. As the latter objective of a surveillance system is to detect events and complex situations, a natural choice is to evaluate goodness in terms of false alarm and correct detection rates of events to be detected. The complexity arises to the distributed nature of the system, as well from the fact that each task of the DMSS is usually parametrized, i.e. his behavior changes depending on setting of a given number of parameters. This allows the possibility of the DMSS to be customized to fulfill specific either end-user needs or service provider strategies. Recent studies on performance analysis of real-time video-based surveillance systems allow one to compare on a more theoretical basis different solutions for a given application domain and, at the same time, to select optimal parameters, depending on the specific functionality of a system and on the particular characteristics of the monitored environment.

The definition of a performance evaluation procedure could be helpful in that it allows to select the best objective parameter values and to provide useful guidelines for the installation of a particular system. Nevertheless, performance evaluation of complex systems is an open problem. In the [22] a performance evaluation method has been proposed, that try to attempt a solution to this problem in a surveillance system The method is based on the use of Receiver Operating Characteristic (ROC) curves; these curves are traced for different parameter settings and for different system working conditions.

A ROC curve is generated by computing pairs (P_d, P_f), where P_d is the probability of correct signal detection and P_f is the false alarm probability. These quantities can be related to the event detection performances of a DMSS at some level of the logical hierarchical architecture. Both probabilities depend on the values of the parameters regulating the behavior of the tasks under that level. By varying the parameters, it is possible to derive different curves, so describing the behavior of the system in different conditions (see for example the curves P_1, P_2 in Figure 14.

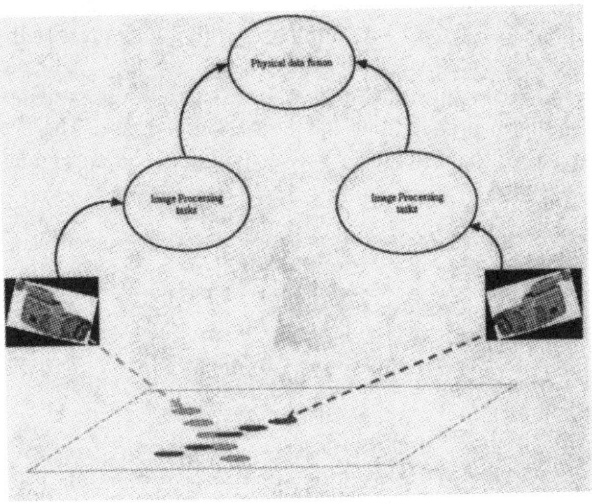

Figure 13. Architecture of a system performing physical data fusion

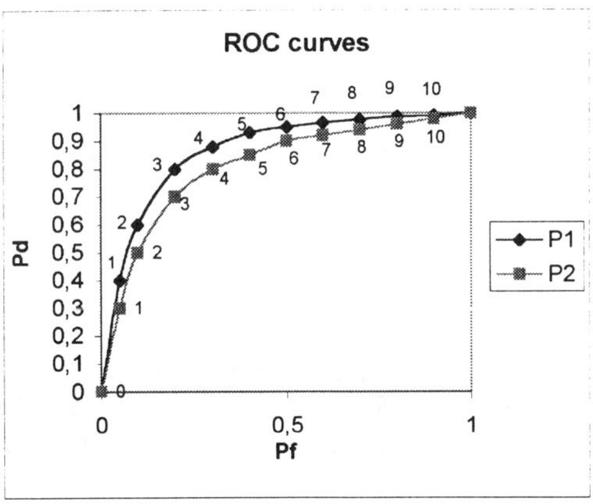

Figure 14. Example of typical ROC curves obtained for different working conditions correspondent to points 1, 2 ,..., 10

Since most of video-based surveillance systems aim at detecting dangerous situations, ROC curves could represent a precise instrument for evaluating video-based surveillance systems, after previous characterization of the subject to be detected.ROC curves can be traced by considering as parameters also different external scenarios in which event detection occurs, as they do not deal in different ways when considering extrinsic or intrinsic parameters.

Such curves can be analyzed in order to allow an operator installing the system to select the parameter set corresponding to an "optimal" working point. Optimality refers to the maximization of the global performances for the possible scenarios that could be presented in the guarded environment. An example of optimality criterion for searching for the optimal working point is related to the minimax criterion as presented in [22] (see Figure 15).

The crucial point in tracing ROC curves for a surveillance system is the estimation of detection and false alarm probabilities for the considered task. These probabilities can be computed on the basis of the differences between outputs of a current task and the expected one.

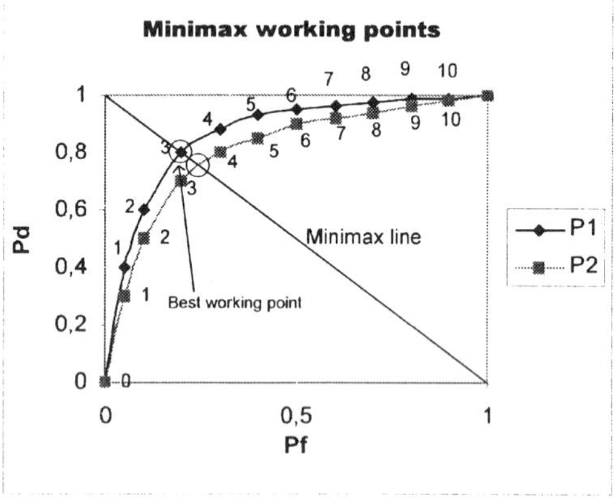

Figure 15. Parameter selection according to minimax criterion

9. Data fusion results

9.1 PHYSICAL DATA FUSION

9.1.1 Agent based multiple camera surveillance system

In this subsection, an agent-based framework for use in scene understanding is presented[21]. The framework was used to implement a car park monitoring prototype system.

Each camera is managed by its own camera agent responsible for detecting and tracking moving objects as they traverse each view volume. For each event detected, the camera agent creates an object agent, which is responsible for progressively identifying the nature of the event and describing the underlying activity. Three-dimensional positional and color observation data of the event are continually streamed to each existing object agent.

The regions of interest in the scene are semantically labelled off-line, and stored in the ground plane agent. (parking areas, entrances, exits and vehicle, bicycle and pedestrian routes). These are stored and used by an object agent to classify its activity. The ground plane agent manages the community of object agents and classifies their activity referencing this semantic labelling. Figuratively speaking, a community of camera agents will continuously detect events and invoke object agents on the ground plane. The resultant population of objects agents model their motion within the semantically labelled ground plane, and their interactions with other objects. A number of important problems can be elegantly handled by requiring that all object agents periodically communicate with each other. Allowing neighboring object agents to compare trajectory and color data facilitates the handover of events between cameras, or the fusion of data from multiple overlapping cameras. When pair of object agents determine that they refer to the same event, they are merged to create a single object agent, which inherits both sets of camera data links. The internal color and trajectory models are updated from both sources of information.

Each camera, which successfully tracks the same event in the 3D scene, creates its own object agent, to which it subsequently streams pixel and 3D positional observations of the event. Consequently agents which correspond to the same 3D event must have some mechanism for first identifying each other and second fusing. This is achieved by requiring all agents to compare trajectory information. Two agents, which have highly similar trajectories are combined to form a new agent. The original data streams from each camera are inherited by the new agent from each of the merged agents. Thus positional information will arrive from multiple cameras. Such a new agent will continue as before to update its trajectory from this combined source using the standard Kalman update equations.

Identical trajectories can be identified using a hypothesis test on the *Mahalanobis* distance between the location of estimates in six dimensional trajectory space. Given two trajectories X and X', and their respective covariances Λ and Λ', the normalized separation $\partial X = (X-X')^T (\Lambda - \Lambda')^{-1} (X-X')$.

Objects whose trajectories are separated by a distance less than some threshold τ_X (where τ_X represents a confidence level of 0.95) are considered identical. In fact merging of agents is only initiated after the trajectories have been repeatedly determined as identical for ΔT seconds (where ΔT is currently set at 5 seconds). Introducing such an inertial characteristic ensures that the costly process of merging agents is not initiated by any occasional gross noise conditions. Figure 15 shows in details the architecture of the agent-based system.

9.1.2 A virtual multiple sensor: mobile head camera surveillance system

A multiple camera environment is able to guard a wider area, but it can be too expensive. The problem could be solved by using a mobile camera: a surveillance system can be realized by using a mobile-head camera able to have pan and tilt movements using motion estimation and optical flow to detect and track the objects in the scene [30]. These techniques are computationally heavy and the realization of a real-time video-surveillance system based on such algorithm on a multi-purpose computer architecture is difficult at the current stage of technology. A method is shown for extending efficient algorithms already developed and tested for fixed cameras to a mobile camera environment [31].

The mobile camera acquires a certain number of images of the empty scene with different position of its head. The integrated pan/tilt unity realizes the movement of the camera. The optical axis of the camera is rotated along two orthogonal axes: the first kind of rotation (equatorial) is done with respect to an axis perpendicular to the ground and allows an angular movement in the range of $\pm 100^o$; the second kind of rotation (vertical) is referred to an axis laying on ground and perpendicular to the optical axis of the camera: the vertical rotation allows an angular movement $\pm 25^o$. The movement of the camera is controlled from a PC with a set of instructions (following Sony VISCA communication protocol [32]) sent to the camera trough the serial port of the host PC. The camera is calibrated by modeling the extrinsic parameters of the sensor by using the technique described in [33]. A set of fifteen images is acquired such that they cover the whole panoramic of the scene: the result of the processing of the acquired images is a global background image used by change detection module.

The result of this phase is the generation of the multi-layer global background: a set of partially overlapped images of the empty scene is acquired (Figures 17 and 18) and the overall background image is generated by using a mosaicing algorithm (Figure 19) able to optimize the change detection phase performances. Because of the movements allowed by the mobile camera the projection that has been used in this case is a spherical one. Multi-layer background allows one to use different change detection techniques.

9.2 LOGICAL DATA FUSION

9.2.1 Crowding estimation with a distributed multiple camera system

In the present section a system described in [20] is illustrated: the considered system addresses the problem of crowding monitoring by following a multisensor data-fusion approach. The goal achieved by the system is the integration of multiple observations acquired in a synchronized way by either a single or multiple visual cameras in order to provide a quantitative estimate of the single sensor and within the overall station. To this end, an architecture (see also Figure 12) composed by two groups of modules is used [35,36,37].

The first subnetwork, constituted by Low-Level Nodes (LLNs), is devoted to provide a quantitative estimate of the number of people within the field of view of a single camera. In this sense, each LLN can be considered as an intelligent sensor whose observations are virtual fused data, i.e., observations of different kinds obtained from the same physical sensor.

The second group of linked nodes, indicated as High Level Nodes (HLNs), has the role of referring estimates (coming from multiple cameras) to the subway-station reference map by using both a-priori knowledge about the spatial displacement of the field of view of each sensor and fusion techniques. Moreover, the highest level module of the HLN has to provide to the surveillance system an estimate about the global number of people within the overall station.

In particular, the lower layer is composed of a set of nodes (Descriptive Primitive Analyzers - DPAs), each performing two sequential main tasks:

(a) extraction of the value of a certain feature type (one for each DPA associated with the same camera) from the analyzed image (e.g. straight edges, points of edge, etc);
(b) mapping of each obtained value into the corresponding Gaussian probability density function (pdf) related to the detected number of people. The Virtual Descriptive Primitive Analyzers (VDPA) is devoted to fusing estimates derived from the lower level into a single, data-dependent pdf.

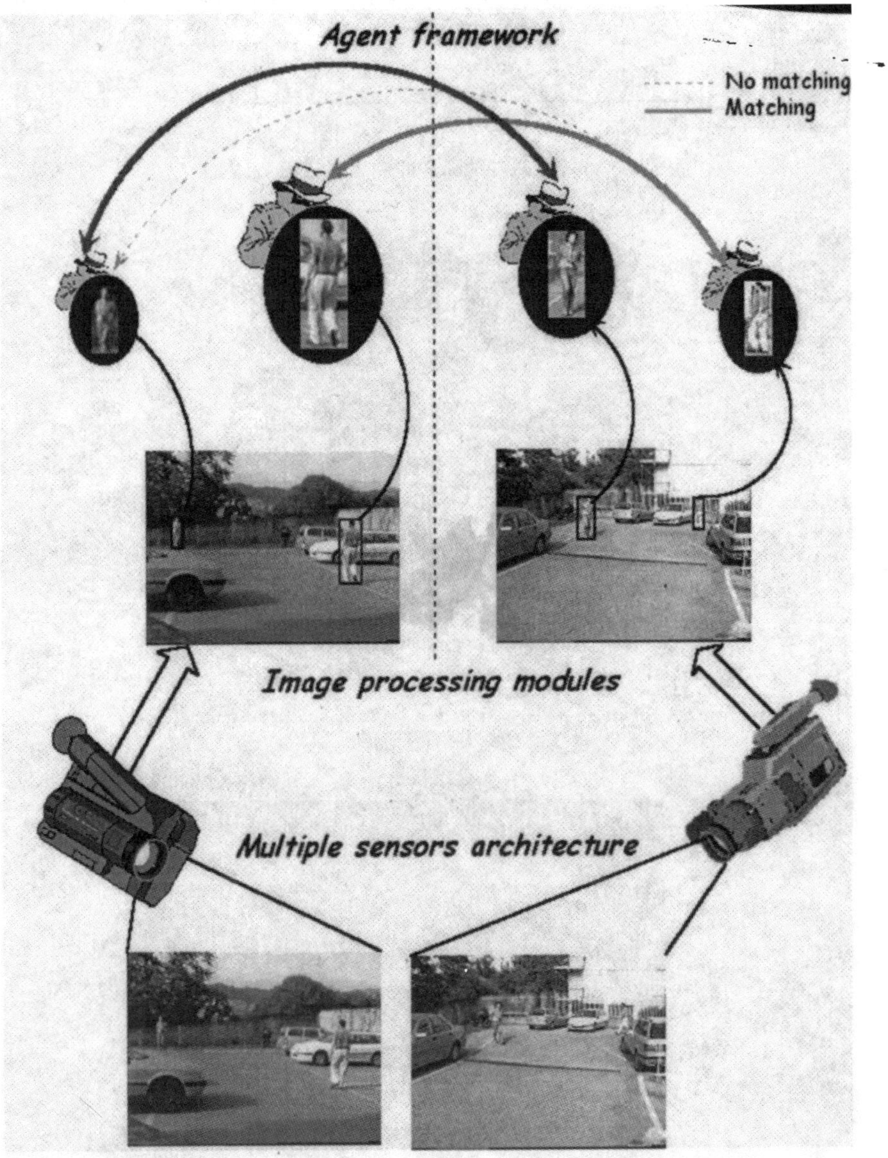

Figure 16. Agent framework applied to a multiple camera system

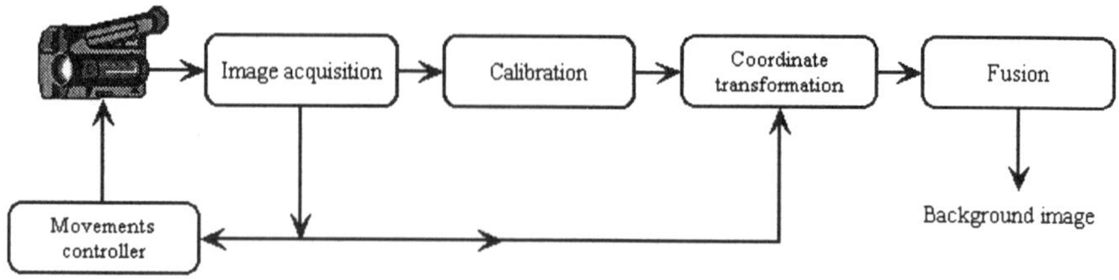

Figure 17. Generation of the multilayer background

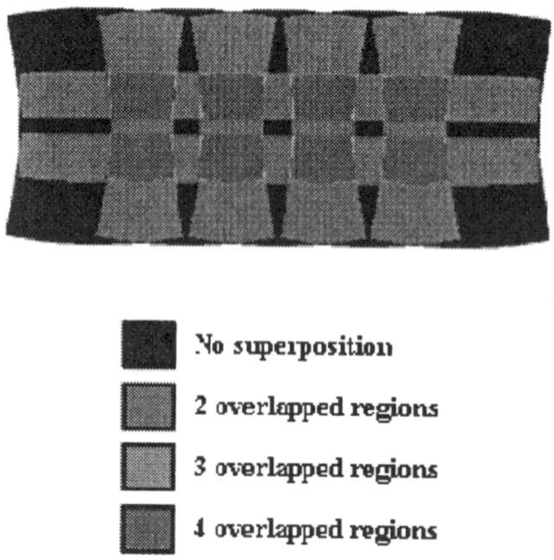

Figure 18. Overlapping regions in multi-layers background

Figure 19. The panoramic background

Any partial estimate coming from the K DPAs referring to the same camera is collected by the common parent VDPA i. Locally, each VDPA fuses all propagated λ_{ij} pdfs, to obtain a unique estimate of crowding for the monitored field of action.

9.2.2 Multiple camera shape analysis

This section describes a multiple-camera extension [38] to the algorithm in[34]: the contour of an object is characterized by a closed B-spline curve that is completely defined by the set of N control points {(x1,y1), (x2,y2), ..., (xn,yn) }. The control points are organized into a 2N dimensional shape vector: $\vec{x} = [x_1 \ldots x_N y_1 \ldots y_N]^T$. A set of s shape vectors $\vec{x_i}$ are aligned with a common reference system; the shape vectors can be seen as a distribution in a 2N dimensional space. By modeling this distribution, one can generate new shape vectors similar to those of the training set, and the new shape vectors can be examined to decide if they represent valid shapes. At this point the Principal Component Analysis (PCA) algorithm is applied over the set in order to characterize the statistical distribution of the training set.

A video surveillance system using M cameras is now considered. If one suppose that a certain object is simultaneously in the field of view of each camera, one can find M shape vectors for each object in a given frame. These shape vectors must be aligned with a reference shape: this kind of processing can be performed separately for each view. As a result, each shape is aligned with its own point of view. The multiple camera model generated through the Principal Component Analysis takes into account the correlations among the different points of view of the same scene. The two main advantages of the proposed method are the data synthesis resulting from the multiple-camera model and an increased robustness in object tracking (see Figure 20). The major disadvantage is an increased computational load, as compared with that of a single-camera system. However, the load is smaller than the one of a system using each camera separately.

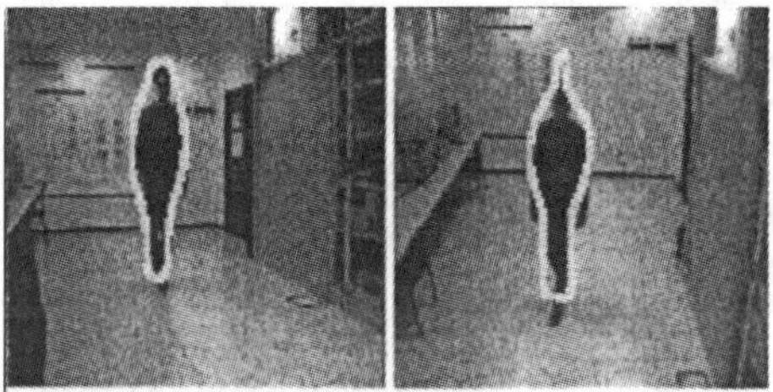

(a)

(b)

Figure 20. A comparison between single (a) and multiple (b) camera model

10. Conclusions

In this paper, the main problems related to the design of the architecture of a Distributed Multisensor Surveillance System have been highlighted. Processing and communication aspects have been considered by providing an insight in several issues and related solutions. The paper has stressed from one side the high variability of possible solutions, while highlighting the possibility of organizing in a general framework the complex set of knowledge and expertise necessary in a multidisciplinary field such as surveillance. It has been shown that, in this context, data fusion methods play a very important role, as they are constitute a key step to allow spatial-temporal fusion occurring at the physical level of the architecture as well as fusion related to attributes of events occurring at the logical level. In this direction, examples have been provided describing new emerging methods and related applications, where either multiple cameras or multiple knowledge sources have been employed. Agent-based fusion has been presented as an example of spatial-temporal information fusion, as well as integration of multiple views to form a mosaicing provided by a physical virtual sensor from views coming from a mobile camera. An example describing fusion of shape information associated with an event detected by more cameras (i.e. a person walking in a scene) has been also presented. A complex system integrating both physical and logical fusion methods in a complex architecture oriented at people number estimation in a wide area scene has been also discussed in the presented framework. The paper is definitively not exhaustive as the global problem is so complex and multidisciplinary that a big amount of research still needs to be carried on: the most interesting issues to be explored in the next years, in the authors' opinion are related to new aspects related to surveillance service providing over open networks, such as internet, as well as to automatic processing tools for ensuring privacy protection issues in DMSSs.

11. Acknowledgments

This work has been partially supported by the University and Scientific Research Ministry (MURST) of the Italian Government under the National Interest Scientific Research Program. The agent logo in Figure 16 is courtesy of Professor Paolo Remagnino, University of Kingston.

12. References

1. Regazzoni, C.S., Vernazza, G. and Fabri G. (eds.) (1999) Advanced Video-Based Surveillance Systems, Kluwer Academic Publishers.
2. Stringa, E. and Regazzoni, C.S. (1998), Content-Based Retrieval and Real-time Detection from Video Sequences Acquired by Surveillance Systems, IEEE International Conference on Image Processing, Chicago, October 4-7, 1998, Vol.3, pp 138-142.
3. Blosseville, J.M. (1999), Image Processing for Traffic Management, in C.S. Regazzoni, G. Vernazza, and G. Fabri (eds.), Advanced Video-Based Surveillance Systems, Kluwer Academic Publishers, Norwell, pp. 67-75.
4. Bogaert, M., Chleq, N., Cornez, P., Regazzoni C.S., Teschioni, A. and Thonnat, M., (1996) The PASSWORDS Project, IEEE International Conference on Image Processing, Lausanne, Vol.III, pp. 675-678.
5. Regazzoni, C.S. and Tesei, A., (1996) Distributed Data Fusion for Real-Time Crowding Estimation, Signal Processing, Vol.53, pp.47-63.

6. Regazzoni, C.S., Sacchi, C. and Dambra, C., (1999) Remote Cable-based Video Surveillance Applications: the AVS-RIO Project, Proceedings of ICIAP99, 27-29th September, Venice, Italy, pp. 1214-1215.
7. Varshney, P.K., (1997) Multisensor data fusion, Electronic and Communication Engineering Journal, Vol 9(6), pp.245-253.
8. Soldatini, F., Mähönen, P., Saaranen, M. and Regazzoni, C.S., (in press) Network Management Within an Architecture for Distributed Hierarchial Digital Surveillance Systems, in C.S. Regazzoni, G.L. Foresti and P. Mähönen (eds.), Multimedia Videobased Surveillance Systems. Requirements, Issues and Solutions, Kluwer Academic Publishers, Norwell.
9. Kettnaker, V. and Zabih, R., (1999) Bayesian Multi-camera Surveillance, Proceedings of IEEE Conference on Computer Vision and Pattern Recognition, pp. 253-259.
10. Regazzoni, C.S. and Tesei, A., (1996) Distributed data fusion for real-time crowding estimation, Signal Processing 53, pp. 47-63.
11. Aranda, J., Amat, J. and Fragola M., (1993) A multitracking system for trajectory analysis of people in a restricted area, Proceedings of the 4th International Workshop on Time-varying image processing and moving object reconigtion.
12. Foresti, G.L. and Regazzoni, C.S., (1994) A change detection method for multiple object localization in real scenes, International Conference on industrial Electronics (IECON 1994), Bologna (I), pp.984-987.
13. Marcenaro, L., Gera, G. and Regazzoni, C.S., (2000) Adaptive change detection approach for object detection in outdoor scenes under variable speed illumination changes, European Signal Processing Conference, Eusipco2000, Tampere, Finland.
14. Serra, J., (1982) Image analisys and mateemathical morphology, Academic Press.
15. Regazzoni, C.S., Teschioni, A., and Stringa, E., (1997) A long memory matching of interacting complex objects from real images sequences, Time Varying Image Processing and Moving Object Recognition, 4, V. Cappellini (eds.), Elsever, pp. 283-288.
16. Haddon, J.F. and Boyce, J., (1990) Image segmentation by unifying region and boundary information, IEEE Trans. on Pattern Analysis and Machine Intelligence, Vol. 12, n. 10, pp. 929-948.
17. Pahlavan, K. and Levesque, A.H., (1994) Wireless Data Communications, Proceedings of the IEEE, Vol. 82, n. 9, pp. 1398-1430.
18. Pahlavan, K. and Levesque, A., (1995) Wireless information networks, John Wiley & Sons, New York.
19. Glisic, S. and Vucetic, B., (1997) Spread Spectrum CDMA Systems For Wireless Communications, Artech House inc., U.S.A..
20. Regazzoni, C.S., Tesei, A. and Vernazza, G., (1997) A Bayesian Network for Automatic Visual Crowding Estimation in Underground Stations, in Jorge L.C. Sanz (eds.), Image Technology, Springer-Verlag.
21. Remagnino, P., Orwell, J., Greenhill, D., Jones, G.A. and Marchesotti, L., (2000) An Agent Society for Scene Interpretation, in C.S. Regazzoni, G.L. Foresti and P.Mähönen (eds.), Multimedia Videobased Surveillance Systems. Requirements, Issues and Solutions, Kluwer Academic Publishers, Norwell.
22. Oberti, F., Stringa, E. and Vernazza, G., (2000) Performance Evaluation Criterion for Characterizing Video Surveillance Systems, Real-Time Imaging Journal.
23. Demigny, D. and Kamlé, T., (1997) A Discrete Expression of Canny's Criteria for Step Edge Detector Performance Evaluation, IEEE Trans. Pattern Analysis and Machine Intelligence, Vol. 19, n. 11, pp. 1199-1211.
24. Peli, T. and Malah, D., (1982) A Study of Edge Detection Algorithms, Computer Graphics Image Processing, Vol. 20, pp. 1-21.
25. Haralick, R.M. and Lee, J.S.J., (1990) Contex Dependent Edge Detection and Evaluation, Pattern Recognition, Vol. 23, n. 1, pp. 1-19.
26. Woods, K. and Bowyer, K.W., (1997) Generating ROC Curves for Artificial Neural Networks, IEEE Trans. on Medical Imaging, Vol.16, n. 3, pp. 329-337.
27. Bone, J.M., Sigillito, V.G. and Shaber, G.S., (1990) Neural Networks in Radiology: an Introduction and Evaluation in a Signal Detection Task, Med. Phys., Vol.17, n. 2, pp. 234-241.
28. Van Trees, H., (1968) Classical Detection and Estimation Theory - Detection, Estimation, and Modulation Theory, John Wiley & Sons, Inc, pp. 19-46.
29. Kanungo, T., Jaisimha, M.Y., Palmer, J., and Haralick, R.M., (1995) A Methodology for Quantitative Performance Evaluation of Detection Algorithms, IEEE Trans. On Image Processing, Vol. 4, n. 12, pp. 1667-1673.
30. Benabdelkader, C., Burlina, P. and Davis, L., (2000) Single Camera Multiplexing for Multi-Target Tracking, in C.S. Regazzoni, G.L. Foresti and P.Mähönen (eds.) Multimedia Videobased Surveillance Systems. Requirements, Issues and Solution, pp. 130-142, Kluwer Academic Publishers, Norwell.

31. Marcenaro, L., Oberti, F. and Regazzoni, C.S., (2000) Change Detection Methods for Automatic Scene Analysis by Using Mobile Surveillance Cameras, International Conference on Image Processing, Vancouver, Canada.
32. Breig, M., Sony EVI D30/31 Pan/Tilt Unit, University of Dortmund.
33. Tsai, R.Y., (1986) An efficient and accurate camera calibration technique for 3D machine vision, Proc. IEEE Comp. Soc. Conf. On CVPR, Miami Beach.
34. Baumberg, A. and Hogg, D., (1994) Learning flexible models from image sequences, Proceedings of European Conference on Computer Vision, vol. 1, pp. 299-308.
35. Ottonello, C., Peri, M., Regazzoni, C.S. and Tesei, A., (1992) Integration of multisensor data for crowding evaluation, Proc. of IEEE International Conference on System, Man and Cybernetics, pp. 791-796.
36. Bozzano, R., Regazzoni, C.S. and Tesei, A., (1993) A Bayesian network for crowding estimation in underground stations, Proc. of 7th International Conference on Image Analysis and Processing.
37. Peri, M., Regazzoni, C.S., Tesei, A. and Vernazza, G., (1993) Crowding estimation in underground stations: a Bayesian probabilistic approach, ESPRIT Workshop on data fusion, Springer Verlag.
38. Marcenaro, L., Regazzoni, C.S., and Vernazza, G., (2000) Automatic Generation of the Statistical Model of a Non-Rigid Object in a Multiple-Camera Environment, International Conference on Pattern Recognition, Barcelona, Spain.

AN ADAPTIVE FILTER FOR TRACKING TARGETS IN CLUTTER

M. EFE
Ankara University
Electronics Engg. Dept.
Ankara, Turkey
Email: efe@science.ankara.edu.tr

1. Introduction

The performance of a tracking algorithm is mainly governed by the performance of the state estimator used. The Kalman filter is the traditional and the most widely used state estimator in target tracking applications. The filter is the general solution to the recursive linear minimum mean square estimation problem. It will minimise the mean square error between the estimates and the actual target dynamics as long as the target dynamics are accurately modelled.

However, if the tracking is attempted in clutter, i.e., where false returns are present, the standard Kalman filter, on its own will not be sufficient to produce reliable estimations. There are several approaches to be used in collaboration with the Kalman filter to address the false return problem. The PDA approach [1] is one of the widely accepted approaches to be used with the Kalman filter in tracking targets in clutter and the resulting filter is called the PDA filter. Although the PDAF constitutes a useful tool to track targets in clutter, its performance can be further improved by modifying the estimator.

This chapter is organised as follows. The next section explains the motivation behind using an adaptive Kalman filter in place of a standard one where section 2 outlines the PDA approach and how to combine the adaptive filter with it. In section 3 some comparative simulation results are presented to highlight the performance improvement by using an adaptive filter. The chapter ends with some concluding remarks.

2. Why Use An Adaptive Filter?

The process of state estimation in the Kalman filter comprises two parallel cycles, namely, *i*) estimation of the state and *ii*) estimation of the state covariance. The final state estimation at the end of each scan is computed from the predicted state, innovation and Kalman gain. The Kalman gain is the 'ratio' of the state covariance to the innovation covariance and can be considered as a correction factor on the final estimate. From a frequency domain viewpoint the magnitude of the Kalman gain determines the bandwidth and response speed of the filter. A small gain value produces a substantial noise reduction when the target is not manoeuvring and a large gain value gives a fast response to changes in the target's dynamics, providing the filter with a larger bandwidth to cover maneuvers. Basically, it can be said that the performance of the Kalman filter is

A. K. Hyder et al. (eds.), Multisensor Fusion, 573–580.

determined by the size of the Kalman gain. However, while the Kalman gain plays an important role in estimating the target's state, it is independent of the measurements taken. In fact, given the process and measurement noise covariances as a function of time, the Kalman gain can be computed off line. This is due to the fact that, the state covariance estimation cycle, in which the Kalman gain is computed, is not affected by the observations.

During the recursive estimation of the state, the Kalman gain reaches a steady-state value determined by the pre-selected and assumed constant process and measurement noise covariances, Q and R respectively. The use of constant values of the covariances imposes a major restriction on the filter's performance. Because if the pre-selected process noise covariance level is not appropriate the correction, by the Kalman gain, on the state prediction will not be suitable and large estimation errors will develop. An intuitive solution to this problem is to adaptively adjust the process noise covariance and thus the Kalman gain, according to changing target dynamics. The simplest method to achieve this is to establish a maneuver detection scheme then modify the process noise covariance following the maneuver detection [2]. The main disadvantage of this approach is the time delay in maneuver detection and changing of the filter. In [3] an adaptive Kalman filtering technique was suggested where the process noise covariance was estimated by means of the difference between the expected prediction error variance and the measurement noise variance. Alternatively, multiple model algorithms can be used, to provide good coverage by using different levels of process noise covariance in each model, at the cost of an increased computational burden and complexity, but even a large number of models does not guarantee a better coverage [4].

The adaptive Kalman filter [5], that will be used in conjunction with the PDA approach, adjusts the gain level of a second order Kalman filter for tracking maneuvering targets. The method introduces a scale factor that represents the current magnitude of process noise covariance, in other words target unpredictability, at time n as estimated from the available data. The aim of the adaptive Kalman filter is to take observations into account while estimating the state covariance, so that the Kalman gain level is adaptively adjusted in accordance with the changing target dynamics. With this feature the filter yields an online gain adjustment without a delay while keeping the computational burden and the complexity at minimum.

3. Adaptive Rescaling Of Process Noise

For the adaptive Kalman filter design, suppose that the target dynamic equation is given by,

$$X(n+1) = \Phi X(n) + \Theta(n)\Gamma W(n) \tag{1}$$

where X(n) is the target state vector, Φ is the known transition matrix, $\Theta(n)$ is the scale factor that represents the current magnitude of the process noise, Γ is the known disturbance transition matrix and $W(t)$ is the unknown zero-mean Gaussian process noise $\sim$ N(0,Q) independent of previous events. The measurements are in the form of linear combinations of the system state variables, corrupted by uncorrelated noise. Thus, the measurement vector is modelled as

$$Z(n) = HX(n) + V(n) \tag{2}$$

where H is the measurement matrix and $V(n)$ is white Gaussian measurement noise ~N(0,R), also independent of previous data. Note that it is assumed that $W(n)$ and $V(n)$ are mutually uncorrelated. At time n, assume that the prior distribution of $X(n)$ for the next transition, using Eqs 1 and 2, is

$$X(n) \sim N(\hat{X}(n), P(n)) \tag{3}$$

This is based on the previous data and depends on the scale factors $\Theta(0), \Theta(1),\ldots, \Theta(n-1)$. For the next stage, $\Theta(n) > 0$ is chosen to specify the model for $X(n+1)$, $Z(n+1)$ as viewed through the sensor.

Note that the normality of Eq 3 and any positive specification of will ensure normality of the posterior distribution

$$X(n+1) \sim N(\hat{X}(n+1), P(n+1))$$

by the standard properties of multivariate normal distributions for a linear Gaussian model. The calculations lead to the following results.

At $n+1$, before observing $Z(n+1)$ we have

$$X(n+1) \sim N(\hat{X}(n+1), M(n+1))$$

where

$$\tilde{X}(n+1) = \Phi\hat{X}(n) \tag{4}$$

$$M(n+1) = \Phi P(n)\Phi^T + \Theta^2(n)\Gamma Q\Gamma^T \tag{5}$$

Thus, the predictive distribution for the new measurement given by Eq 2 is normal with mean vector

$$\tilde{Z}(n+1) = H\tilde{X}(n+1) = H\Phi\hat{X}(n)$$

covariance matrix

$$S(n+1) = HM(n+1)H^T + R \tag{6}$$

And the innovation is

$$v(n+1) = Z(n+1) - H\tilde{X}(n+1) \tag{7}$$

Note that, a-priori, the expectation of the sum of squares $|\nu(n+1)|^2 = \nu(n+1)^T \nu(n+1)$ is the sum of diagonal elements of $S(n+1)$. By Eqs 5 and 6 this expectation contains a term $\propto \Theta^2(n)$. Later, the observed difference $|\nu(n+1)|^2$ - $\Sigma_i[S(n+1)]_{ii}$ will be used to construct the new scale factor $\Theta^2(n+1)$.

Given the observation $Z(n+1)$, the posterior distribution $X(n+1)$ is constructed by standard methods. It is normal with mean vector

$$\hat{X}(n+1) = \tilde{X}(n+1) + K(n+1)\nu(n+1) \tag{8}$$

where

$$K(n+1) = M(n+1)\, H^T S^{-1}(n+1) \tag{9}$$

The new covariance matrix is

$$P(n+1) = \{I - K(n+1)\, H\}\, M(n+1) \tag{10}$$

This completes the transition from one state of information $(\hat{X}(n), P(n))$ to the next $(\hat{X}(n+1), P(n+1))$.

Construction of Θ(n+1): By using Eqs 5 and 6, we have

$$S(n+1) = H\Phi P(n)\Phi^T H^T + R + \Theta^2(n) H\Gamma Q\Gamma^T H^T \tag{11}$$

Assume that, ψ^2, η^2 and δ^2 are the sums of the diagonal elements of $S(n+1)$, $[H\Phi P(n)\Phi^T H^T + R]$ and $[H\Gamma Q\Gamma^T H^T]$ respectively, then we have;

$$\psi^2(n+1) = \eta^2(n+1) + \Theta^2(n)\delta^2(n+1) \tag{12}$$

Roughly speaking, if the observed sum of squares $|\nu(n+1)|^2$ is close to $\psi^2(n+1)$, then the filter is reasonably well focused on the target, so $\Theta(n)$ is already set at an appropriate level. If we let a,b,c be fixed constants, where $a \geq 1$, $b \geq 1$ and $c \geq 1$ and $a+b+c = 1$, then the following sequential relationship offers a possible scheme for constructing $\Theta(n+1)$,

$$\Theta^2(n+1) = \max\{a\Theta^2(0) + b\Theta^2(n) + c(\frac{|\nu(n+1)|^2 - \eta^2(n+1)}{\delta^2(n+1)}),0\} \tag{13}$$

If $|\nu(n+1)|^2 = \psi^2(n+1)$ then

$$\Theta^2(n+1) = \max\{a\Theta^2(0) + (b+c)\,\Theta^2(n),0\} \tag{14}$$

The right hand side of Eq 12 is constrained to remain non-negative. The constant c can be regarded as a sensitivity parameter that can be used to adjust the adaptive behaviour of the filter. The initial value of the scale factor was chosen as $\Theta(0) = 1$ so that

a gives some weight to the original scale factor. In the simulations a small value of c namely 0.05 was used along with the values $a = 0.8$ and $b = 0.15$.

4. The Probabilistic Data Association Filter (PDAF)

The PDAF is simply a Kalman filter that used in conjunction with the PDA approach in order to take the measurement origin uncertainty into account. The filter assumes that the target is detected (perceived) and its track has been initialised. At each sampling interval a validation gate is set up. The measurement originating from the target of interest can be among the possible several validated measurements, hence, the track update is done by taking the weighted some of all observations within the gated region in a probabilistic manner. Measurements outside the validation region are assumed to have originated from false alarms or clutter. The PDAF uses only the latest measurements and the past is summarised approximately by making the following assumption:

$$p[X(n) \mid Z^{n-1}] = N(X(n); \hat{X}, P(n)) \tag{15}$$

which states that the state is assumed to be normally distributed (Gaussian) according to the latest estimate and covariance matrix. Starting from this point, one cycle of the state estimation is described as follows.

Let $Z(k)$ contain measurements from the elliptical validation region for the track at time n

$$Z(k) = \{z_m(n)\}, \qquad m=1,2,\ldots,m_k \tag{16}$$

where m_k is the number of validated measurements in that region. Also, the cumulative set of validated measurements up to scan n is denoted by Z^k. Using the nonparametric version of the PDAF [1] the validated tracks are associated to the track. The combined target state estimate is obtained as

$$\hat{x}(n) = \sum_{m=0}^{m_k} \beta_m(n)\hat{x}_m(n) \tag{17}$$

where β_m is the probability that the mth validated measurement is correct and $\hat{x}(n)$ is the updated state conditioned on that event. The conditionally updated states are given by

$$\hat{x}(n) = \hat{x}(n-1) + K_m(n)v_m(n) \tag{18}$$

where $K_m(n)$ is the filter gain and $v_m(n)$ is the innovation associated with the mth validated measurement. The association event probabilities $\beta_m(n)$ are given by

$$\beta_m = e_m\left(b + \sum_{j=1}^{m_k} e_j\right)^{-1} \qquad m = 1,\ldots,m_k \tag{19}$$

$$\beta_0 = b(b + \sum_{j=1}^{m_k} e_j)^{-1} \qquad m = 0 \tag{20}$$

where

$$e_m = exp\{0.5 \nu^T_i(n+1)S^{-1}(n)\, \nu_i(n+1)\} \tag{21}$$

$$b = m_k (1\text{-}P_D P_G)(P_D V(n))^{-1} \tag{22}$$

and P_D is the probability of detection of a target originated measurement, V is the volume of the validation gate. Then the updated state is

$$\hat{X}(n+1) = \tilde{X}(n+1) + K(n+1)\nu(n+1) \tag{23}$$

where

$$\nu(n+1) = \sum_{i=1}^{m_k} \beta_i(n+1)\nu_i(n+1) \tag{24}$$

is the combined innovation. The error covariance associated with the updated state estimate is

$$P(n+1) = \beta_0(n+1)M(n+1) + [1\text{-}\beta_0(n+1)]P^c(n+1) + \tilde{P}(n+1) \tag{25}$$

Where

$$\tilde{P} = K(n+1)[\sum_{i=1}^{m_k} \beta_i \nu(n+1)\nu^T(n+1) - \nu(n+1)\nu^T(n+1)]K^T(n+1) \tag{26}$$

and

$$P^c(n+1) = [I - K(n+1)H(n+1)]M(n+1) \tag{27}$$

is the covariance of the updated state. Prediction of the state, its covariance, measurement to time n+1 and the innovation covariance are calculated as in the standard Kalman filter.

Although, unlike the standard Kalman filter, being a nonlinear filter the PDAF takes the measurements into account in order to address the measurement origin uncertainty issue while estimating the current state covariance, it is still somehow oblivious to the changes in the target dynamics. In other words the target unpredictability/maneuverabilty is still unaccounted for. This is exactly why the standard Kalman filter in the PDAF is replaced with an adaptive one to establish an adaptive PDAF.

4.1 MAKING THE PDAF ADAPTIVE

As explained in section 2, the scale factor in the adaptive Kalman filter, which accounts for the target maneuverabilty, is introduced in the predicted state covariance. Also, as mentioned above the covariance of the predicted state is calculated in the same way as it is done in the standard Kalman filter i.e., it is independent of the association event probability calculation procedure. Thus, since the adaptiveness of the filter is achieved without breaking the general structure of the standard Kalman filter, the replacement of the Kalman filter, in the PDAF, by the adaptive Kalman filter will also keep the general structure of the PDAF intact. In other words in the adaptive PDAF the standard PDAF structure, in general, is preserved. Moreover, the assumptions, that are valid for the

PDAF, are still valid for the APDAF. On the other hand by the inclusion of the scale factor, the changed target dynamics are taken into account for the next scan resulting in an algorithm for maneuvering target tracking in clutter. In the adaptive PDAF the equations given in section 3 are still valid except for the predicted state covariance which should be replaced by Eq 5. Also, the combined innovation given by Eq 24 should be used to calculate $\Theta^2(n+1)$ in Eq 13.

5. Simulation Results

The simulated target motion is generated in two dimensions (i.e. x-y plane) with a sampling interval of 1.0 second and assumed to last 20 seconds. Measurements are assumed known at the origin of the Cartesian coordinates for the x-y positions of the target with a Gaussian measurement error standard deviation of 100 *m* used for both axes. The process noise standard deviation is taken as 5 *m/s* for each axes, whereas 0.9 and 0.002 values are used for the probability of detection and probability of false alarm respectively. The initial values of the state vector are as follows; $[x(0), y(0), V_x(0), V_y(0)] = [1km, 1km, 0.2km/s, 0.2km/s]$.

A simulated target trajectory is given in the left-hand side of Figure 1 where the rosses indicate the true target positions at each sampling time, the letter *d* denotes a detection and each dot represents a false return.

The right hand figure in Figure 1 depicts the position error produced by the PDAF for the given trajectory. A clear performance degradation can be seen when there is lack of information about the target motion (i.e., no return) between the 9th and the 15th seconds. As it is depicted in the left-hand figure of Figure 2, a similar performance degradation is also present for the APDAF as well. However, the introduction of the scale factor helps the APDAF keep the position error at a lower level. Moreover, by rescaling the process noise covariance level the increase in the position error is confined to a short space of time after which the error is pulled back to a reasonable level. This is achieved by introducing a scale factor whose change in value is given in the right hand figure in Figure 2. Note that the value of scale factor increases as soon as there is an ambiguity about the target motion. Target maneuver can also be accounted for by the scale factor.

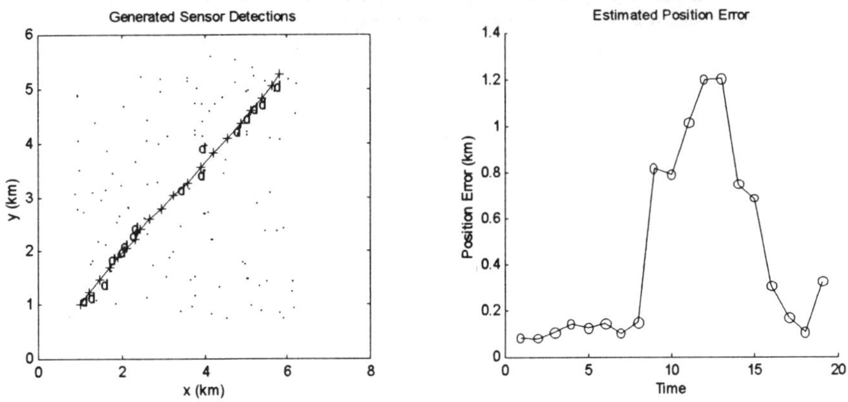

Figure 1. Simulated tracking scenario and position error produced by the PDAF

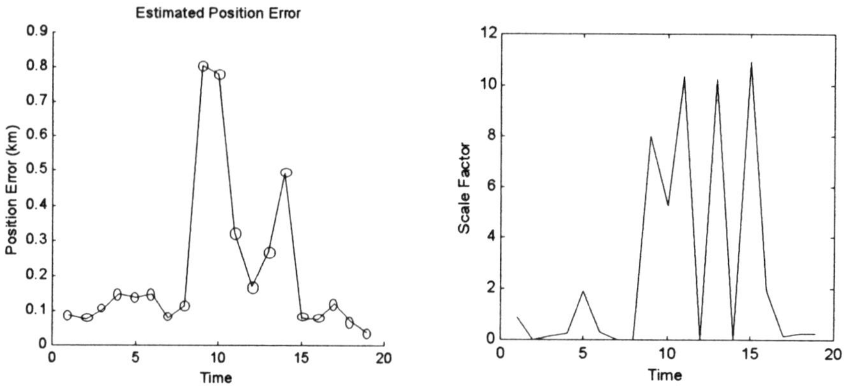

Figure 2. Position error produced by the APDAF and change in the scale factor

6. Conclusions

In this chapter an adaptive filter for tracking targets in clutter is presented. The filter uses an adaptive Kalman filter, in which the level of process noise covariance is adaptively rescaled, in place of standard Kalman filter in the PDA filter. The adaptive PDAF accounts for the target unpredictability/maneuverability as well as the measurement origin uncertainty.

7. References

1. Y. Bar-Shalom and T.E. Fortman, (1988) *Tracking and Data Association*. Academic Press.
2. S.S. Blackman, (1986) *Multiple-Target Tracking With Radar Applications*. Artech House.
3. P.O. Gutman and M. Velger, (1995) Tracking Targets Using Adaptive Kalman Filtering. *IEEE Trans. On Aerospace and Electronic Systems*, Vol. 26, No. 5: 691-698, September.
4. X.R. Li and Y. Bar-Shalom, (1996) Multiple Model Estimation With Variable Structure. *IEEE Trans. On Automatic Control*, Vol. 41, No.4: 478-493, April, 1.
5. M. Efe, J.A. Bather and D.P. Atherton, (1999) An Adaptive Kalman Filter With Sequential Rescaling of Process Noise. *In American Control Conference Proceedings*, 3913-1917.

TRACKING CLOSELY MANEUVERING TARGETS IN CLUTTER WITH AN IMM-JVC ALGORITHM

A. JOUAN*, B. JARRY & H. MICHALSKA
Department of Electrical and Comp. Eng.
McGill University
3480 University Street
Montreal H3A 2A7, Canada
**R&D-Engineering Dept.*
Lockheed Martin Canada
Montreal H4P 1 K6, Canada
Email: michalsk, jarrryb@cim.mcgill.ca
alexandre.jouan@lmco.com

1. Introduction to Fusion, Tracking, Data Association, JVC, IMM

The Interacting Multiple Model state estimators (IMM), [1-5], provides a better tracking accuracy for maneuvering targets than that obtained from other single-scan positional estimators such as the Kalman filter - even with a recursion on the process noise to make it more capable of following a maneuver - or more sophisticated estimators making use of rule-based maneuver detectors [6].

When dealing with closely maneuvering targets, a critical role is played by the chosen contact-to-track associator. Without a good associator, state estimation is at risk. In a previous paper, [7], the performances of two association algorithms were compared, namely the simple Nearest Neighbour (NN) procedure and the Jonker-Volgenant-Castanon (JVC) optimization,[8]. Simulations have shown the superiority of a positional estimator using the JVC optimization for association and an IMM tracker using two motion models (constant velocity CV and a constant acceleration CA) for the estimation. The IMM-JVC algorithm would probably take a very favorable place among the other approaches such as the IMM-JPDAF,[9], or the multi-hypothesis tracking (MHT), [10]. A study aiming at benchmarking the proposed IMM-JVC, the IMMJPDA and an MHT algorithm is currently on-going.

The JVC algorithm optimizes the contact-to-track association by minimizing the cost associated to the assignment matrix. When building this matrix, two processes should be closely monitored since they may alter the performance of the JVC optimization.

The first of these two processes is the bufferization scheme chosen to present the sensor reports to the association module of the Multi Sensor Data Fusion (MSDF) test bed. Most of the time, the characteristics of these buffers (time duration, scanned volume) are intimately related to build-in rules characterizing each sensor. Those rules are based on factors such as scan duration, scanned angular range, estimated target density... Buffers of sensor reports built according those rules are very often used without modification in the contact-to-track association module. There is a requirement for a more

A. K. Hyder et al. (eds.), Multisensor Fusion, 581–592.

sophisticated bufferization scheme to keep the JVC optimization efficient specifically in a dense target environment.

The second of these two processes is the selection from the MSDF track database of some of the tracks that are likely to be updated with a sub-set of the incoming contacts. For MSDF systems having real-time constraints, specific track selection strategies may be required to reduce the processing time.

After a brief description of the mathematical formalism associated to the introduction of the JVC optimization scheme within the IMM positional estimator, this paper will concentrate on the evaluation of the role played by the first of the two processes, namely the sensor report bufferization.

Experimental results are presented for a scenario of two closely maneuvering aircraft. Sensor reports have been contaminated by randomly simulated clutter.

2. The Imm-Jvc Algorithm

The discrete time model used to represent the target dynamic in each mode of the IMM filter is written as:

$$x_{k+1} = F_k x_k + G_k w_k \tag{2.1}$$

$$z_k = H_k^T x_k + v_k \tag{2.2}$$

in which $\{w_k\}$, $\{v_k\}$, are independent, zero mean, gaussian white processes with $E[v_k v_l^T] = R_k \delta_{kl}$, $E[w_k w_l^T] = Q_k \delta_{kl}$, and the initial state x_0 is a gaussian random variable with mean $\overline{x_0}$ and covariance P_0, independent of $\{v_k\}$ and $\{w_k\}$.

The IMM algorithm used in this study mixes the position estimates provided by both a constant velocity and a constant acceleration motion models.

To integrate the JVC optimization scheme within the IMM, it must be integrated in between the state prediction and the actual state update. The detail of the algorithms are given in the next sections.

2.1 THE JVC MODULE

Extensive tests were performed on the JVC algorithm at Lockheed Martin Advanced Technology Laboratories (ATL) for contact-to-track association in the Rotorcraft Pilot's Associate (RPA) program, [11]. JVC proved to provide an excellent compromise between track accuracy and execution time when compared to a Nearest Neighbour (NN) association (fast but lacking of a global optimization of the contact-to-track association problem) or to the Multi-Hypothesis Tracking (MHT) algorithms (much computer intensive).

The contact-to-track assignment matrix is usually rectangular. In our implementation, each matrix element represents a probability of "non-association" between an incoming contact and an existing track. The probability of "non-association" is calculated as

(1-P_G (M)) where P_G (M) represents the probability that a valid observation falls within an M-dimensional gate of size G, [12]. The concept of non-association is used here to build an assignment matrix readily useable by the JVC algorithm since it is coherent with the definition of a cost matrix. The higher the cost, the lower the probability of association.

As described in [13], the JVC algorithm comprises two sequential steps, the first one being similar to an auction algorithm, [14], the second being a modified version for sparse matrices of the Munkres algorithm, [15].

When the number of contacts in the buffer is larger than the number of tracks to be updated with, the contacts left unassigned will contribute to the creation of a new track with a status left as NO-STATUS. If, at a later time, a second contact is assigned to a NO-STATUS track, then the track is promoted as INITIATED. If a third contact is later assigned to an INITIATED track, the track is promoted as TENTATIVE. It is only at the fourth assigned contact, that the status of the track is declared as FIRM.

2.2 THE IMM MODULE

In the multiple model approach, see [2], it is assumed that at any fixed instant of time k the motion of the system can be adequately described by one of a finite number of given models (filter modes) $\{M_j\}_j = 1^r$. For simplicity, we assume that for each $j \leq r$, M_j has a linear representation (2.1)-(2.2) with distinct noise characteristics R_k, Q_k and state space matrices F_k, G_k, H_k.

The interacting multiple model approach then allows for dynamic model switching in which the mode jump process is a Markovian chain with known probabilities $p_{ij} \stackrel{def}{=} P\{M_k = M_j \mid M_{k-1} = M_i\}$ of transition from mode M_i to mode M_j at any time k (the event that mode j is in effect at time k is denoted by $M_k = M_j$). While a rigorous development of an algorithm allowing for model switching would require considering an exponentially increasing number of switching histories, the practical IMM approach computes the state estimate at time k as a mixture of the outputs from all r modes, with each mode using a different combination of the previous model-conditioned estimates - a mixed initial condition. One cycle of such a practical IMM algorithm (for a single, a priori identified track) then consists of the following steps:

Step 1. Calculation of the mixing probabilities

The probability that mode M_i was in effect at time k - 1 given that M_j is in effect at time k conditioned on Z_{k-1} is calculated as

$$\mu_{k-1|k-1}^{i|j} \stackrel{def}{=} P\{M_{k-1} = M_i \mid M_k = M_j, Z_{k-1}\}$$

$$\mu_{k-1|k-1}^{i|j} = \frac{1}{c_j} P\{M_k = M_j \mid M_{k-1} = M_i, Z_{k-1}\}$$

$$P\{M_{k-1} = M_i \mid Z_{k-1}\} = \frac{1}{c_j} p_{ij} \mu_{k-1}^i \tag{2.3}$$

for $i,\ j=1,\ldots,r,$ with the values of

$$\mu_{k-1}^{i} \overset{def}{=} P\{M_{k-1}=M_i \mid Z_{k-1}\} \text{ and}$$
$$c_j = \sum_{i=1}^{r} p_{ij}\mu_{k-1}^{i} \tag{2.4}$$

derived as outputs from the previous cycle of the algorithm (see Step 4.).

Step 2. Mixing of the initial condition

Starting with previous state estimates $\hat{x}_{k-1|k-1}^{i}$ obtained as outputs from the r different

$$\hat{x}_{k-1|k-1}^{0j} = \sum_{i=1}^{r} \hat{x}_{k-1|k-1}^{i}\ \mu_{k-1|k-1}^{i|j}, \qquad j=1,\ldots,r \tag{2.5}$$

Kalman filters (acting as the r different modes), and corresponding covariance matrices $P_{k-1|k-1}^{i}$, a mixed initial condition for the filter M_j at time k is calculated as
and the corresponding covariance as
Step 3. Mode-matched filtering
The mixed initial conditions obtained in Step 2 as well as the new measurement z_k are

$$P_{k-1|k-1}^{0j} = \sum_{i=1}^{r} \mu_{k-1|k-1}^{i|j}\{P_{k-1|k-1}^{i} + [\hat{x}_{k-1|k-1}^{i} - \hat{x}_{k-1|k-1}^{0j}][\hat{x}_{k-1|k-1}^{i} - \hat{x}_{k-1|k-1}^{0j}]^{T}\},\ j=1,\ldots,r \tag{2.6}$$

now used as inputs to each filter M_j at time k producing new modal estimates $\hat{x}_{k|k}^{j}$ and $P_{k|k}^{j}$. Computation of the likelihood functions for the r filters follows next

$$\Lambda_k^j = p[z_k \mid M_k = M_j,\ \hat{x}_{k-1|k-1}^{0j},\ P_{k-1|k-1}^{0j}], \qquad j=1,\ldots,r \tag{2.7}$$

where it is known that the above p.d.f. is gaussian.

Step 4. Mode probability update
The new mode probabilities are calculated as follows

$$\mu_k^j \overset{def}{=} P\{M_k = M_j \mid Z_k\} = \frac{1}{c}\Lambda_k^j \sum_{i=1}^{r} P\{M_k = M_j \mid M_{k-1} = M_i,\ Z_{k-1}\}$$
$$P\{M_{k-1} = M_i \mid Z_{k-1}\} = \frac{1}{c}\Lambda_k^j \sum_{i=1}^{r} p_{ij}\mu_{k-1}^{i} \tag{2.8}$$
$$c = \sum_{j=1}^{r} \Lambda_k^j c_j$$

Step 5. Estimate and covariance combination

The next point along an established track is generated from

$$\hat{x}_{k|k} = \sum_{j=1}^{r} \hat{x}^{j}_{k|k} \, \mu^{j}_{k} \tag{2.9}$$

$$P_{k|k} = \sum_{j=1}^{r} \mu^{j}_{k|k} \{ P^{j}_{k|k} + [\hat{x}^{j}_{k|k} - \hat{x}_{k|k}][\hat{x}^{j}_{k|k} - \hat{x}_{k|k}]^{T} \} \tag{2.10}$$

2.3 INTEGRATION OF THE IMM AND JVC

As explained above, Step 3 of the IMM algorithm employs the next measurement z_k which at this instant needs to be matched with the appropriate track. To avoid matching measurements with all the individual modal estimates

$$\hat{x}^{0j}_{k|k-1} = F_{k-1} \, \hat{x}^{0j}_{k-1|k-1}, \qquad j = 1, \dots, r \tag{2.11}$$

$$P^{0j}_{k|k-1} = F_{k-1} P^{0j}_{k-1|k-1} F^{T}_{k-1} + G_{k-1} Q_{k-1|k-1} G^{T}_{k-1} \tag{2.12}$$

a suitable mixing of the latter is required. This is done by introducing the conditional probabilities $\mu^{j}_{k|k-1} \overset{def}{=} P\{M_k = M_j \mid Z_{k-1}\}$ which are naturally calculated as

$$\begin{aligned} \mu^{j}_{k|k-1} &= \sum_{i=1}^{r} P\{M_k = M_j \mid M_{k-1} = M_i, Z_{k-1}\} \\ P\{M_{k-1} = M_i \mid Z_{k-1}\} &= \sum_{i=1}^{r} p_{ij} \mu^{j}_{k-1} \end{aligned} \tag{2.13}$$

The individual modal estimates are now mixed as follows to produce the "aggregated" state and covariance estimates which identify the tracks for the purpose of measurement to track association.

$$\begin{aligned} \hat{x}^{0}_{k|k-1} &= \sum_{j=1}^{r} \hat{x}^{0j}_{k|k-1} \, \mu^{j}_{k|k-1} \\ P^{0}_{k|k-1} &= \sum_{j=1}^{r} \mu^{j}_{k|k-1} \{ P^{0j}_{k|k-1} + [\hat{x}^{0j}_{k|k-1} - \hat{x}^{0}_{k|k-1}][\hat{x}^{0j}_{k|k-1} - \hat{x}^{0}_{k|k-1}]^{T} \} \end{aligned} \tag{2.14}$$

The latter are readily used as inputs to the JVC module to produce the innovations $\nu_k = z_k - H_k^T \hat{x}^0_{k|k-1}$, and the corresponding probabilistic distances, finally yielding the desired measurement to track associations. The execution of Step 3 of the IMM module is thus enabled, and the two modules: JVC and IMM, are then integrated into one algorithm.

3. Simulation Experiments

The following experimental results were obtained on the agent-based Multi Sensor Data Fusion Testbed developed at Lockheed Martin Canada (Montreal) using a scenario simulator developed by the Canadian Defense Research Establishment at Valcartier (Quebec).

3.1 SCENARIO 1 - AIRCRAFT CLOSELY MANEUVERING WITHOUT CLUTTER

This scenario shows two aircraft flying along straight lines at 100 m/s along a 45° converging path scanned by a radar at a rate of 30 rpm. When the distance between the two aircraft is of 200 m they perform a 90° turn with an acceleration of 3 g during 5 s. The entire 360° angular coverage is divided into 32 equiangular sectors of 11.25° each. The buffers are made with a minimum of one and a maximum of two sensor reports. Figure 1 shows results obtained when using a Nearest-Neighbour (NN) association and the IMM-CVCA for positional estimation. This Figure shows that the aircraft maneuvers are not detected, each aircraft being thought by the MSDF module as pursuing a rectilinear course, which is obviously wrong.

This result is interesting since it also represents the performance we could expect when using the JVC association with buffers of contacts containing only one sensor report. In such a case, the assignment matrix is of dimension *1xm* where *m* is the number of tracks - selected from the track database - which are likely to be updated with the incoming contact. This points out the need for a contact bufferization on order to have a more global approach to the assignment problem.

Figure 2 shows results obtained when the JVC assignment optimization is executed on the buffer of contacts provided by the scenario simulator. Now, the assignment matrix is of size *1xm* or *2xm*. As expected when the track assignment is considered globally by JVC, the maneuvers of the two aircraft are correctly tracked.

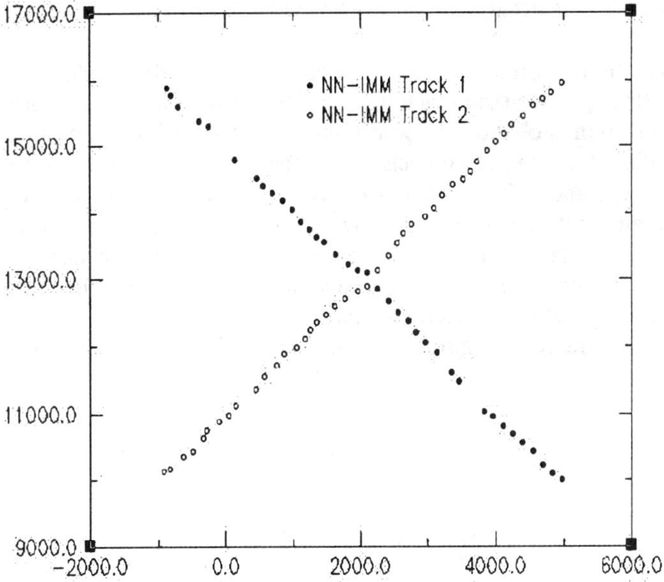

Figure 1. Nearest Neighbor association, tracking with IMM-CVCA

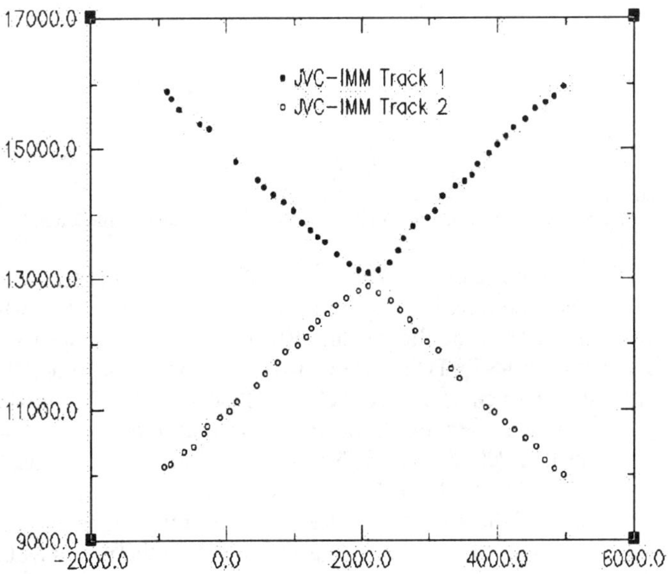

Figure 2. JVC association, IMM-CVCA tracker

3.2 SCENARIO 2 - AIRCRAFT CLOSELY MANEUVERING WITH CLUTTER

In this scenario, the two aircraft of the Scenario 1 are still flying along the same path. After the first 20s of the scenario (each track is FIRM since we are not considering at this point the initialization problem), random clutter is added within each buffer. For each clutter-free buffer of N_t target contacts, a number N_c of clutter contacts is randomly generated between 0 and 3. Then each of the N_c clutter reports is time stamped at a time randomly generated between the time at which the radar starts and ends scanning the corresponding angular sector. Ranges and bearings are randomly generated between the minimum and maximum values of range and the minimum and maximum values of bearing of the corresponding scanned volume.

Figure 3 shouws the resulting tracks on the top of which have been superimposed the simulated clutter.

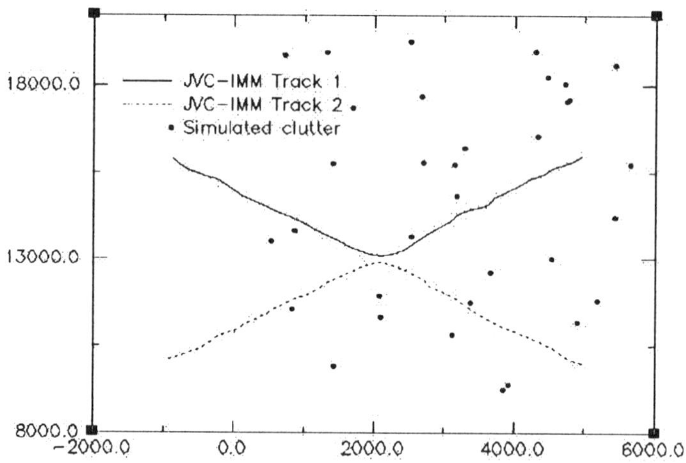

Figure 3. JVC Association, IMM_CVCA tracker with simulated clutter.

Figure 4 illustrates the life duration and status of the tracks that have been created in the track database. The line length (in x) represents the track duration while the track number in the database is given by the y value. When only one dot appears, the track has been created but has a NO-STATUS. When two, three, four and more than four dots appear, the status of the track is respectively INITIATED, TENTATIVE and FIRM. Figure 4 clearly shows that among the 22 tracks that have been created, the status of 13 of them did not go higher than NO-STATUS, 3 of them did not go higher than INITIATED at the end of the scenario (Tracks 9, 13, 14), 1 of them did not go higher than the status of TENTATIVE at the end of the scenario (Tracks 3), and only 5 of them got the ultimate status of FIRM (Tracks 1, 2, 4, 10, 11). Note that the *KillOldTrack* mechanism which removes from the track database all the tracks that have not been recently updated was deactivated for this simulation. At a scan rate of 30 rpm, tracks could be removed from the database after 5s (more than 2.5 full scan without update). Figure 4 shows that if the

KillOldTrack mechanism would have been active, no tracks except those corresponding to the two aircraft would have reached a status of FIRM (most of the Tracks would have got the status of NO-STATUS when removed, at the exception of the Track 9 being removed after having reach the status INITIATED and the Track 11 being removed after having reached the status of TENTATIVE).

Figure 5 shows a two-dimensional mapping of the number of contacts in track gates. Time is in abscissa, the 22 tracks are shown on y-axis from bottom (Track 1) to top (Track 22) as in Figure 4. The grey-level coding in this two dimensional mapping must be interpreted as follow. All pixels in black can be ignored from the analysis, the corresponding track being not selected to build the assignment matrix. Then, 3 levels of grey are used from the darker to the brighter when the corresponding track has respectively an infeasible assignment (track is selected but the probability of non-association resulting from the calculation of the statistical distance is higher than a given threshold) or 1 or 2 feasible assignments (one or two contacts in the track gate).

Finally, pixels in white represent the time at which a track has 3 contacts in its gate. This two-dimensional mapping permits to identify easily when the JVC optimization is actually challenged in the assignment.

To be more informative, this mapping must also be correlated to Figure 6 which displays the actual number of incoming contacts in each buffers. This correlation gives an evaluation of the time at which gates overlap. For example at time 43.31s, Figure 6 shows that there is three contacts in the buffer (two are true contacts from aicraft, one is due to clutter) and the two-dimensional mapping shows that two of them fall in the gate of the Track 1, two of them fall also in the gate of the Track 2 and all of them fall in the gate of the Track 3. At this time we cannot clearly establish how the true or clutter contacts are distributed in the gates of Track 1 or Track 2. However, ground-truthing of the JVC optimization output displayed in Figure 4 shows that the true contacts are correctly assigned to their aircraft track for an update without status promotion, the clutter contact being assigned to the Track 3 for an update with a status promotion from INITIATED to TENTATIVE. Notc that if the *KillOldTrack* mechanism would have been active the Track 3 would not exist at this time and the clutter point would have triggered the creation of a new track with a status NO-STATUS.

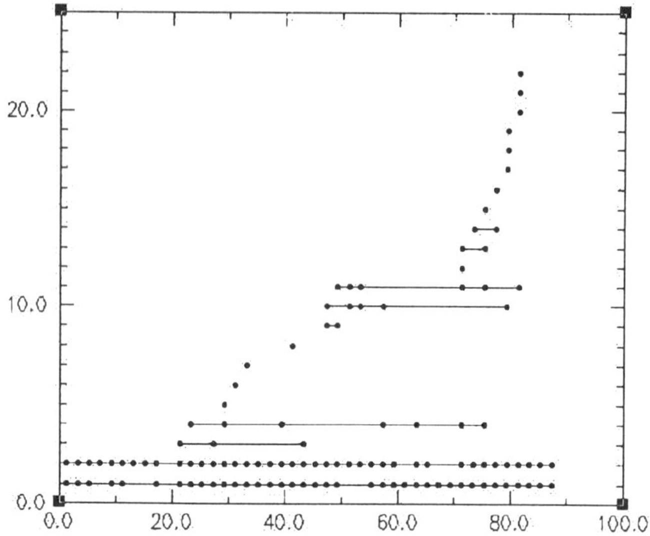

Figure 4. MSDF track creation and life duration.

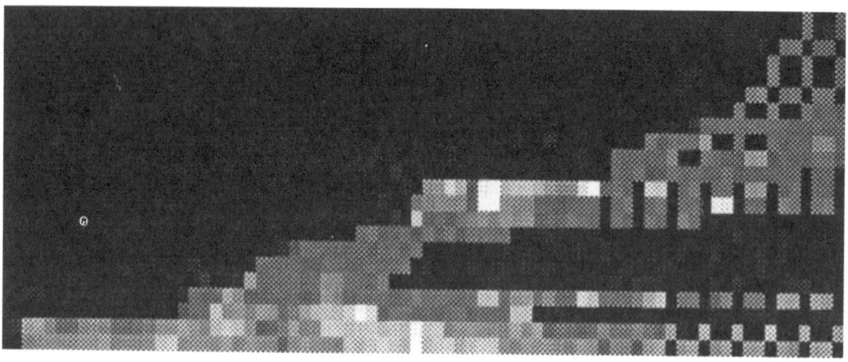

Figure 5. 2D mapping of the number of contacts in track gates.

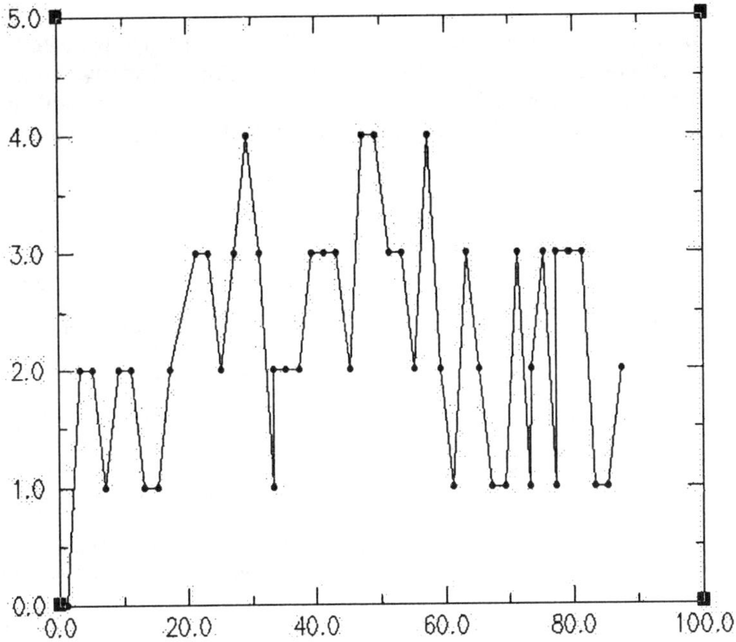

Figure 6. Number of contacts in buffers.

4. Concluding Remarks

This paper highlights the critical role played by the association algorithm for tracking closely maneuvering targets. The global assignment problem that JVC optimization is made to solve yields an excellent tracking accuracy within a reasonable computer execution time. The integration of the JVC optimization within the IMM formalism has been derived and simulated results show the excellent performance of the IMM-JVC tracker for closely maneuvering targets with clutter. Track management functionalities such as the *KillOldTrack* mechanism has also a role to play to reduce the execution time since it will lead to assignment matrices smaller in size. Some convenient analysis tools have been proposed such as the two-dimensional mapping of contacts in track gates which helps in the identification of the time at which the JVC optimization is challenged. A second version of this paper is in preparation to confirm the expected behaviour of the JVC-IMM when the *KillOldTrack* mechanism is active. In the future, the performance of the JVC-IMM tracker will be evaluated on a scenario with a high target density in such a conditions that the data preparation (bufferization, prioritization) and track management utilities (*KillOldTrack*) will have a major role to play.

5. Acknowledgements

This study has been performed within a research collaboration agreement between the Department of Electrical and Computer Engineering at McGill University and Lockheed Martin Canada and was supported by the Canadian Department of National under the Defense Industrial Research (DIR) Program W2207-8-EC01 and Lockheed Martin Canada.

6. References

1. Blom, H.A.P. and Bar-Shalom, Y.. (1988) The interacting multiple model algorithm for systems with Markovian switching coefficients. *IEEE Transactions on Automatic Control*, Vol. 33, No. 8, pp. 780-783.
2. Bar-Shalom, Y. and Li, X.R., (1993) *Estimation and Tracking: Principles, Techniques, and Software.* Boston, MA, Artech House.
3. Kastella, K. and Biscuso, M. (1966) Tracking algorithms for air traffic control applications. *Air Traffic Control Quarterly*, Vol. 3, No. 1, pp. 19-43.
4. Yeddanapudi, M. Bar-Shalom, Y. and Pattipati, K.R., (1997) IMM estimation for multitarget multisensor air traffic surveillance. *Proceedings of the IEEE*, Vol. 85, No. 1, pp. 80-94.
5. Wang, H., Kirubarajan, T. and Bar-Shalom, Y., (1999) Precision large scale air traffic surveillance using IMM/assignment estimators. *IEEE Transactions on Aerospace and Electronic Syst.*, Vol. 35, No. 1, pp. 255-265.
6. Jouan, A. Bosse, E. and al., (1998) Comparison of various schema of filter adaptivity for the tracking of maneuvering targets. *Proceedings of Aerosense'98 : Signal and Data Processing of Small Targets*, Vol. 3373, pp. 247-258.
7. Jarry, B. Jouan, A. Michalska, H., (2000) An IMM-JVC Algorithm for Multi-Target Tracking with Asynchronous Sensors. *Proceedings of the 19 IASTED Conference on Modelling, Information and Control*, in press.
8. Jonker, R. Volgenant. A., (1987) A shortest augmenting path algorithm for dense and sparse linear assignment problems. *Computing*, Vol. 38, pp.325-340.
9. de Feo, M., Graziano, A., Miglioli, R., Farina, A., (1997) IMMJPDA versus MHT and Kalman filter with NN correlation: performance comparison. *Proceedings of IEE on Radar, Sonar Navig.*, Vol. 144, No. 2, pp. 49-56.
10. Bar-Shalom, Y. and Li, X.R., (1995) *Multitarget-Multisensor Tracking: Principles and Techniques.* Storrs, CT, YBS Publishing.
11. Malkoff, D.B., (1997) Evaluation of the Jonker-Volgenant-Castanon (JVC) Assignment Algorithm for Track Association *Proceedings of the SPIE Aerosense '97 Conf. on Signal Proc., Sensor Fusion and Target Recognition*, Vol. 3068, pp. 228-239.
12. Blackman, S.S., (1986) *Multiple-Target Tracking with Radar Application.* Artech House.
13. Drummond, O.E., Castanon, D.A., Bellovin, M.S., (1990) Comparison of 2-D Assignment for Sparse, Rectangular, Floating Point, Cost Matrix. *Journal of the SDI Panels on Tracking*, Institute for Defense Analyses, Issue No.4, pp.4-81 to 4-97.
14. Bertsekas, D.P., (1998) The Auction Algorithm: A Distributed Relaxation Method for the Assignment Problem. *Annals of Operations Research*, Vol. 14, pp. 103-123.
15. Munkres, J., (1957) Algorithms for the Assignment and Transportation Problems. *J. Soc. Indust. Applied Math.*, Vol. 5, No 1, pp. 32-38.

ROBUST MULTISENSOR IMAGE REGISTRATION WITH PARTIAL DISTANCE MERITS

Y. SHENG[*a], X. YANG[□a], L. SÉVIGNY[c] AND P. VALIN[b]
[a] *Image Science Group*
Center for Optics, Photonocs and Laser
Université Laval, Dept. de Physique
Ste-Foy, Qc., Canada G1K 7P4
[b]*Lockheed Martin Canada*
6111 Royalmount Ave.
Montréal, Qc, Canada, H4P 1K6
[c] *Defense Research Establishment Valcartier*
2459 Boul. Pie XI nord
C.P. 8800, Courcelette, Qc, Canada G0A 1R0

1. Introduction

Most advanced vision systems utilize multiple imaging sensors to capture more information. The multi-modality sensors and hyperspectral imaging are widely used for remote sensing and medical imaging. Image fusion is to combine information from multiple sensors in order to refine estimates and predictions on the image information. Image registration is the first and critical step before image fusion. It has direct applications to image sequence stabilization [1] and superresolution image reconstruction [2].

In this paper we present a contour-based multisensor image registration for military passive surveillance ground imaging system. In this system two broadband cameras, one in visible and another in far-infrared band, are mounted on the top of a post of the ground vehicle Coyote. The spectral bands of the visible and far-infrared cameras are separated by a large gap. The technical challenge in this application is the feature detection in the cluttered environment of battle field image background and the feature inconsistency in the two image modalities, which is much more important than that encountered in convention remote sensing applications.

After a brief analysis on the feature inconsistency problem we present a review on our selection of contour-based approach and the multiresolutional feature extraction. Then we present two free-form curve matching processes: the adaptive hill climbing and the iterative closest point algorithm. Both approaches are designed for the robustness against outlier feature inconsistency using the partial Hausdorff distance. We introduce the partial

[1] Correspondence: Center for Optics, Photonics and Laser, Department of Physics, Laval University, Ste-Foy, Quebec, Canada G1K 7P4; Email sheng@phy.ulaval.ca; WWW: http://fourier.phy.ulaval.ca/sheng/Yunlong-Sheng.htm; Telephone: 418-656-3908, Fax. 418-656-2623

[2] Correspondence: Email: xyang@phy.ulaval.ca

A. K. Hyder et al. (eds.), Multisensor Fusion, 593–609.

mean distance as the objective function in the iterative closest point algorithm that permits to exclude a percentage of outlier feature points. A fast algorithm using the segment representation of Voronoi diagram for the nearest point transform and the distance transform is used in our algorithm.

2. Contour-Based Approaches

2.1 FEATURE INCONSISTENCY

Multiple sensors have different electromagnetic spectral responses to capture distinguished features in different spectral bands. In the design of a multiple sensor imaging system, one selects wavelengths in an electromagnetic spectral band as wide as possible. Multiple sensors act in a synergistic manner, providing complementary information: The IR camera provides data to locate the objects of warmer temperatures within the scene, while the optical camera provides background in the scene. The fusion of two images allows locating the warm objects with respect to the background. However, because the images from multiple sensors should be uncorrelated and independent from each other, the problems of radiometric data disparity and feature inconsistency arise for image registration.

The radiometric data captured by the IR passive sensor consist of 1) energy emitted by thermal radiation from the object bodies; 2) atmospheric emission reflected from object surfaces. Therefore, the gray-scale level of the IR images depends on the differences in temperature, emissivity and reflectivity of the objects in the scene. All of the scene temperature, emissivity and reflectivity contributions taken together can be represented at any point in the scene by an effective temperature at that point. The effective temperature is that at which a blackbody radiator would produce the measured irradiance near the point. There are significant gray-scale level disparities between the IR and visible image. A good thermal emitter is not necessarily a good visual reflector. A surface of high visual reflectivity (white surface) in the visible band usually has low emissivity, so that the visually bright objects may be dark in the thermal scene and vice versa. This is the reversal of contrast polarity between the visible and IR images. However, a simple reversal of contrast polarity cannot remove all the gray-scale level disparities in outdoor images. For instance, the sky is usually the brightest region in the visible image. It is a dark region in the IR image because of the low temperature and the lack of reflectance. But clouds in the sky can have the same polarity of contrast in both IR and visible images: they are brighter than the sky in the visible image because of their higher reflectivity, and they are also brighter than the sky in the IR image because of its higher reflectivity and emissivity. In addition, shadows in the visible images are absent in the IR images, which cannot be corrected by a reversal of contrast polarity.

2.2 CONTOUR-BASED REGISTRATION

The radiometric data disparities and the associated feature inconsistency in multi-sensor images give rise to difficulties for both area-based and feature-based image registration. Advantages of the area-based registration approaches are that they use full image information, can be applied to images with rich or poor structure, and they are robust against random noise. However, the area-based approaches and the optical flow approaches require the radiometric data of two images to be similar. If the two radiometric data distributions were different, then the area-based approaches would fail.

Several approaches have been proposed to address the registrability problem. One is to transform the two dissimilar radiometric intensity distributions into similar ones. The nonlinear transform of the Laplacian pyramid coefficients is such a transformation. The intensity of the Laplacian pyramid images is insensitive to gray-scale level disparities and polarity reversals of contrast in the visible and IR images, so that the Laplacian pyramid representation can be used for the area-based image registration. Another approach is the phase matching in the Fourier transform-based registration. In the multi-sensor images the differences in the radiometric intensity distributions appear mostly as slow variations over wide regions in the image, such as sky, land and forest in the case of out-door imagers, which are usually represented with low spatial frequencies and are concentrated in a narrow low frequency band. One can then whiten the Fourier spectrum and use only the spectrum phase for the registration using the cross-correlations. In this approach the contribution of the high frequencies is greatly highlighted, because this contribution is usually weighted by the very low Fourier transform modulus in the high frequency bands. The Fourier phase correlation registration method is then relatively independent of the sensors.

Both the Laplacian pyramid representation and the phase matching technique benefit from the use of high spatial frequencies of the image. The Laplacian pyramid represents detailed information, namely contours in the image. In some sense both the Laplacian pyramid representation and the phase matching area-based approaches share the same features with the edge matching approaches for image registration. We adopt the feature-based edge matching registration scheme, because in this scheme specific edge detectors can be used, which are more powerful than the Laplacian pyramid or the Fourier spectrum whitening for the feature detection in background of outdoor IR images, which are typically of low contrast and noisy.

The feature-based image registration is based on the common features extracted from the two images. When both types of images represent the same real world scene, although the multi-sensor images may appear differently, each object in the scene would appears always as an individual object, no matter what is the spectral response of the imaging sensors. As a result, the boundaries of the objects would be preserved, which are not sensitive to spectral responses of the sensors and can be used as comment features for image registration. The feature-based image registration can account for any image transformation, can use the powerful and versatile multi-resolution edge detection and edge saliency techniques, and the processing is fast. Its speed is independent of the image transformation. The limitation of the feature-based approach is the need for the presence of structure in the images, and that its reliability depends on the feature extraction process.

2.3 MULTIRESOLUTION EDGE EXTRACTION

The IR images are usually of high contrast for detecting targets in the scene, such as persons, vehicles and helicopters, which are of clearly higher temperatures distinguishable from that of the background. However, the targets are not useful for image registration because they are usually moving in the scene. Image registration requires extracting features from the immobile scene background. In most cases, the background objects in the IR images have the same thermal equilibrium temperature, so that the gray-scale level contrast in the IR image background is related only to the differences in the emissivity and reflectivity of the object surfaces and is therefore very low. In addition, the IR images are typically noisy.

In the 3-D real-world scene, objects are separated from the background by depth discontinuities which usually manifest as gray-scale level discontinuities in the 2-D images. Edges are defined as points where the modulus of gradient is a maximum in the gradient direction. Along an edge the gray-scale level can be singular in one direction while varying smoothly in the perpendicular direction. Edges can be created by occlusions, shadows, sharp changes of surface orientation, changes in reflectance properties, or illumination. In this sense edges can potentially represent structural information in the image, and are the common features for multiple sensors, useful for multiple sensor image registration. One possible structural feature in the outdoor ground images is the horizon lines which are widely used for unmanned ground vehicle navigation. Indeed, when cameras are mounted on a grounded vehicle, it is reasonable to assume that the camera axis is pointing approximately horizontally. Unless the terrain is very steep (or the vehicle is driving alongside a wall) the horizon is usually visible in the image and is usually the most distant part of the scene (except for the sky). The horizon lines are common in the IR and visible images, and are independent of the grayscale level disparities and contrast polarity reversals. The fact that the horizon is the most distant part in the image helps accuracy of image registration.

In the visible image the horizon is the boundary of the sky, which is usually the brightest region in the top of image. In the IR image the horizon is the longest curve in the image. However, noise occurs as local fluctuations of the image brightness function, which have strong derivative magnitudes, but represent unnecessary image details, which are unrelated to image structure. In the IR image background the low contrast varies cross the image, so that the structural edges may be disrupted and even completely disappear in the edge maps, if an inappropriate thresholding on the gradient magnitude is applied. In addition, the Canny edge detector is excessively reliant on the estimation of the gradient angle and so often fails to mark edge pixels at junctions, corners and even on some smooth curve portions where the contrast changes are too poorly defined [3]. This is the reason for broken edges. To extract horizon lines we use Canny edge detector with a large Gaussian smoothing filter to remove noise. A curve-length threshold is then applied to extract horizon lines in the coarse edge map. In the case of strong noise, which breaks horizon lines, we calculate saliency [4] measure in the edge map, and connect the edge segments of largest saliency measure to recover the horizon line. Finally, we apply multi-resolution hierarchical edge detection and edge focusing [5] to recover localization of edge at the original resolution level, using a coarse-to-fine edge tracking in continuous manner with the filters of decreasing scales.

However, the horizon lines along are not sufficient for image registration, which requires image features evenly distributed over the images. As a result we need to use additional features from other regions in the image, such as the ground. The passive surveillance imaging system should operate in an uncontrolled real-world environment. A typical battlefield scene contains mountains, hills, forest, a small number of buildings and water ponds, snow and bush on the ground. Most those ground clutter objects appear differently in the visible and IR images because of the effect of combination of temperature, IR emissivity and reflectivity. Their boundaries are inconsistent in the two types of images, resulting in a high percentage of outlier edges.

To ease the edge curve matching for registration, we need to design the feature detection process, that favorites structural boundaries and avoids noisy and inconsistent edges. This is difficult because in uncontrolled imaging environment and without *a priori* knowledge on the input scene we have no raison to reject any edge feature in the image. However, extraction and then the presence of the horizon lines increase the percentage of common features in the edge maps.

3. Free-Form Curve Matching

From the multiresolution edge detection we obtain two edge maps of the battlefield images for registration. The two edge maps from two sensors contain a large percentage of outliers. Furthermore, the boundaries of some significant and common structures in the scene could be still different in detail because of the radiometric data disparity in two image types, the different spectral responses of the sensors and the different parameter setting in the edge detector, such as the threshold values, the size of edge detection filters etc. In addition, the shadows in the visible images do not appear in IR images.

Approaches for registration of two feature maps can be divided into two categories. In the primitive-based approaches a set of primitives, such as curvature extrema of the curves, curve junctions and end points of straight lines, are extracted and used for matching. The primitives need to be described by the transformation-invariant descriptor in order to determine the correspondences in the two images. The primitive-based approaches are widely used and are in general preferable. But in a real-world environment it is difficult to detect and localize the primitives robustly and precisely. The battlefield scenes usually lack structural objects. Most extracted edges from the scene images are smooth and horizontal curves. In this case the location of the curvature extrema is highly sensitive to noise. The end points of the curves also depend highly on the noise. In addition, the description and matching of primitives using the information from the neighborhood of the primitives [1,6] can fail because of the radiometric data disparity in the multiple sensor images.

We then adopt the curve matching approaches. The idea is to find the transformation parameters by minimizing a distance-related criterion between two sets of points or curves. The curve is considered as free-form and as a set of chained points, which is not described by any equations or parameters. We do not regard particular primitives and do not search for one-to-one primitive correspondence between two images. This is the appropriate for images in a non-structured environment. The free-form curve matching is a search in the transformation parameter space. The type of image transformation should be known *a priori*, which is, in our case, the global affine transform.

3.1 DISTANCE MERITS

Discrepancy measures for the matching of two binary point sets should be based on some distance merits. For instance, cross-correlation is widely used for image matching. However, when the images are sets of points or curves, the cross-correlation values will be no longer a good merit. The cross-correlation peak occurs only when the two sets of points are perfectly superimposed. When the two sets of points are very close to each other but not perfectly superimposed, the cross-correlation value can decrease dramatically, because the points are only of one-pixel size and curves are only of one-pixel think. For matching the sets of points or curves the cross-correlation is replaced by the *chamfer matching*, or other merit. In these techniques, given two sets of points, A and B, we first compute the distance transform of the set A that is a new array of the same size as A, in which every pixel holds the value of the distance from this location x to a closest point in A:

$$d(x, A) = \min_{a \in A} \|x - a\|$$

where the distance metrics could be city-block, chessboard, chamfer and Euclidean distance depending on particular applications. Among these metrics only the Euclidean distance function is isotropic and rotation-invariant. Then, the chamfer matching is a correlation between the distance transform of A and the set of binary points B:

$$\sum_{x} d(x, a) B(x + t_x)$$

where t_x is the translation of the position vector x.

3.2 DISTANCE TRANSFORM

In both the following hill climbing Hausdorff distance matching and the iterative closest point algorithm we shall use a fast segment-list based algorithm for computing the Voronoi diagrams and distance transform. The purpose of the distance transform is accelerating computation for the distance merits between two sets of binary features. The nearest-neighbor transform is closely related to the distance transform. The nearest-neighbor transform of an image A assigns each pixel in a matrix of the same size as A the identity of a feature pixel in A which is closest to it. In the continuous space the Voronoi transform generates a Voronoi diagram, which is a partition of a space into Voronoi cells. All the points in a Voronoi cell shear the same nearest point. The nearest-neighbor transform is a construction of discrete Voronoi diagram.

An exhaustive search for the closest points and the Euclidean distances in an image of m by n is a procedure of the computation complexity $O(m^2n^2)$. Guan and Ma [13] proposed a linear-time algorithm with complexity of O(mn) for both Euclidean distance transform and Voronoi transform. The algorithm computes the Euclidean distance transform and gives in the same time the Voronoi diagram without extra cost. The algorithm consists of a segment creation, two vertical scans for the Veronoi transform followed by a distance map calculation. In an initialization phase, we compute the one-dimensional Voronoi transform in each row individually. A row is then divided into several segments. An individual nearest feature point owns each segment, which is itself in this row. Then, we build a list the segments in a row in the left-to-right order in which each segment is recorded by its start position (a real variable), the owner (currently-found nearest neighbor), and the pointer to its next segment. After forming the initial segment list for each row, two bottom-up and top-down scans are applied to those segment lists sequentially. Rows are scanned one by one in a scanning-propagation way, each row being scanned receives a propagated segment list from its previous row, and merges it with its own segment list. The segment list of that row is then replaced by the merged list. In this way nearest neighbor information is propagated upward and downward throughout the whole image to ensure that the owners of segments will be their nearest neighbor after these two vertical scans. In the scan process the new bisector lines of owners are computed with real parameter. In this sense the transform divide space in a continuous manner, and is therefore Voronoi transform instead of nearest neighbor transform. The Voronoi transform here can be thought of to be continuous horizontally and discrete vertically. The segment-list representation is not only compact in encoding a Voronoi diagram, but easily accessed as well.

The detailed rules of merging process are given in [7]. From the obtained Voronoi diagram the distance map is computed using the coordinates of the owners in the segment lists.

The basic idea of this algorithm is representing nearest neighbor information in segments rather than in pixels. We notice that neighboring pixels tend to have the same nearest point. The list-processing algorithm represents and processes the nearest neighbor information in the form of segments rather than of pixels. The algorithm with complexity of O(mn) realizes both Voronoi transform and distance transform without extra cost, where both nearest feature and the distance to it can be retrieved.

3.3 HAUSDORFF DISTANCE MATCHING

The curve matching using the Hausdorff distance does not require explicit point correspondences and is robust against outliers. The directed Hausdorff distance[8, 9] merit is defined as the maximum distance among all the distances from all the points $a \in A$ to their respective closest points in B

$$h(A,B) = \max_{a \in A} d(a,B)$$

where

$$d(a,B) = \min_{b \in B} d(a,b)$$

The general Hausdorff distance is the maximum of the two directed forward and reversed[9] Hausdorff distances:

$$H(A,B) = \max(h(A,B), h(B,A))$$

Originally the Hausdorff distance is extremely sensitive to outliers. If two sets of points A and B are similar, all the points are perfectly superimposed except only one single point in A which is far from any point in B, then the Hausdorff distance is determined by that point and is large. This sensitivity to outliers is not acceptable. The Hausdorff distance merit is so unstable as to be unusable.

To take care of the outliers problem we use the partial Hausdorff distance [9]. We compute all the distances of each point $a \in A$ to its nearest point of B, $d(a,B)$. Then we rank those distances. If n is the number of points in the set A then let $K = qn$ and K is q-th quantile of n. The partial Hausdorff distance is the K-th shortest distance, and is q-th quantile of all those distance values $d(a,B)$. There is a subset of K points in A that has distances less than the partial Hausdorff distance to their respective closest points in B. By ranking the distances we automatically select the K best matching points in A. The remaining $(n\text{-}K)$ points are considered as outliers.

Exact computation of the minimum Hausdorff distance is very expensive. This is an exhaustive search in the transformation space. Given a value ε, we need to find a transformation $g \in G$ in some group of transformations G for the model point set A, such that $H(G(A),B) \leq \varepsilon$. The best running time for determining a rigid body transformation which minimizes the Hausdorff distance between the sets of size n is $O(n^5 \log^2 n)$ [10]. The time complexity would be mush higher when scaling is included in the transformation. This complexity is not acceptable for real time implementation.

Many techniques, such as the transformation space cell decomposition, early rejection and skipping forward have been introduced to accelerate the search [9,11,12], while still guarantee that the best transformation can never be missed. However, the computing complexity is still much too high for real-time processing of our images, from which the feature detection returns typically several hundreds of points, when searching in the affine transformation space.

The search for minimizing the Hausdorff distance for the translation-only is reasonably fast. Using this property, a multisensor image registration using the local Hausdorff distance search has been demonstrated [13]. The basic idea of this approach is that an affine transform of an image can be approximated by translations of its block sub images. We first partitioned the images into blocks and then matched the sub images using the contour lines in each block and the partial Hausdorff distances merit, assuming that the sub images are only translated. Finally, from the displacements vectors of ensemble of blocks, we estimated the affine transformation of the image.

3.4 HAUSDORFF DYNAMIC HILL CLIMBING

The dynamic hill-climbing algorithm has been proposed for image registration using the Hausdorff distance [14]. In this approach the registration is formulated as an optimization problem for minimizing the Hausdorff distance. The objective function in the optimization process is the Hausdorff distance, which measures the similarity between the two sets of feature points. The search for the minimum Hausdorff distance with the optimization is mush faster than the exhaustive search in the entire transformation space. However, the search can fall to a local minimum.

The dynamic hill climbing [15,16] is combination of hill climbing and genetic algorithm. It consists of inner loops and outer loops. The inner loop is a search for the local optima in the vicinity of the starting point. The direction of move is selected randomly with a pre-selected step length. If the move decreases the Hausdorff distance then the step length is doubled and the move is pursued in the same direction. If the move increases the Hausdorff distance then the step length of the move in this direction is halved. Then, the next longest step is tried in its direction. This reduction of step size increases the accuracy of the minimum when the search is close to a local minimum. The process iterates until no further decrease of the Hausdorff distance is possible. The outer loop is designed to force the inner loop to explore the search space uniformly by choosing the starting points of the inner loops independently and as far as possible to the points, which are already evaluated. This outer loop maintains the diversity of the search to avoid local minimum.

The hill climbing optimization has four input parameters, which are the objective function, the starting points, range and the step of the search. We use the adaptive hill climbing optimization for our application for the multisensor image registration, where the input parameters of the hill climbing are determined using the data computed from the images to be matched. The reason for using this process is that in the multi-sensor image registration the cameras are mounted in the same platform. The cameras are not aligned on the same optical axis, the resolutions and magnifications of the cameras could be different. However, the two images should not be much different. A direct superimposition of the two images can be already a rough march as a good starting point, because from that point the global affine transformation parameters for translation, rotation and scale changes should be small. In the case where the translation between two images is large, a Hausdorff distance matching for translation-only can bring quickly the images to the initial alignment. The hill climbing results in a local minimum in the vicinity of the starting point. If the final correct solution is close to the starting point, then it would be easily achieved by the hill climbing. As a consequence, we do not need the outer loop in the dynamic hill climbing to generate new starting points.

We use the partial Hausdorff distance as the objective function, because the two feature maps contain a large amount of outliers. The fractional quartile q with $0<q\leq 1$ for computing the partial Hausdorff distance should be input as an additional parameter for the hill climbing process. The search ranges and steps for the transformation parameters are also determined adaptively from the images to be matched.

The search space for the hill climbing is spanned by the transformation parameter basis. The basis of the search space is in general an orthogonal set or is, at least, nondegenerated [15,16]. The rigid body rotation is orthogonal. The rotation and translation are correlated, in the sense that a rotation around an arbitrary point can be decomposed into a rotation around the origin plus a translation. The affine transformation is not orthogonal, but is nondegenerated. Hence, we can perform the hill climbing optimization directly in the affine transformation parameter space by moving the point, i.e. transform the model, independently and sequentially in each transformation parameter axes for translation in x and y, rotation and scale changes in x and y.

3.5 ITERATIVE CLOSEST POINT ALGORITHM

The iterative closest point algorithm is another approach for matching sets of points or free-form curves and surfaces. The approach requires no feature extraction, no point correspondence and computes no curve or surface derivatives. The algorithm is a general-purpose algorithm and is independent of data representation. Like the adaptive hill climbing, the algorithm does not ensure convergence to the global minimum. However, in the application of image registration and motion estimation the transformation between two images is in general small, so that the algorithm can converge to the global minimum, which is near by the initial solution. If a large translation occurs, we can first establish a rough alignment of images by using the fast Hausdorff distance matching for only translation, or by the adaptive hill climbing Hausdorff distance matching, described in subsection 3.3 and 3.4.

The iterative closest point algorithm [8] utilizes a large redundancy that allows a precise computation of the transformation between the two images. However, the iterative closest point algorithm is susceptible to gross outliers. The registration is definitely not useful, if there are significant numbers of data points that do not correspond to any point in the model shape.

We apply the iterative closest point algorithm to registration of visible and far-infrared images with two original modifications. First, we introduce fractional quantiles to exclude a percentage of outlier points in the computation of the mean-square distances. This provides the robustness to outliers. Second, we uses the list-processing approach describe in subsection 3.2 to compute the Voronoi diagrams and the Euclidean distance transform. After extraction of salient edges, the registration proceeds as follows:

1. Establish a rough alignment of images. In our case the superimposition of images of two cameras can give this initial registration;
2. Compute the Voronoi disgram and the distance transform of one image. Another image is then the model. The total number of feature pixels in the model is usually smaller than that in the image;
3. Input a quantile value $0<q\leq 1$, according to an estimation of the percentage of outliers between two feature maps;
4. For every point in the model we read from the distance transform $d(a,B)$ their closest points b_k in the image B. Notice that several points a_k in the model can share the same closest point in the image. Those points b_k can be at the same location. Then, we rank all the distances and choose the q-th quantile shortest distances, $a_k \in A$ with $k = 1,2,\cdots,nq$, where n is the total number

of feature pixels in A, to compute the affine transform (R, t) by minimizing the mean-square distance objective function:

$$E(\mathbf{R}, \mathbf{t}) = \frac{1}{n} \sum_{k=1}^{n} \left\| \mathbf{R}\mathbf{a}_{\mathbf{k}} + \mathbf{t} - \mathbf{b}_{\mathbf{k}} \right\|^{2} \tag{1}$$

Here we consider the closest points as the corresponding point pairs. Because the number of pairs nq is much larger than 3, the solution of the well-over-determined Eq.(1) is obtained by an least-square optimization approach.

5. Apply the transformation to whole set of points in the model A;
6. Go to step 3 with an increased quantile value q;
7. Continue the iterations until the gain for reducing object function in Eq.(1) in the iteration is small than a threshold value.

The convergence of the iterations has been proved [17,18]. The basic idea to understand the convergence is that first, the least-square evaluation of transformation $\boldsymbol{R}$ and t and step 5 reduce the mean square distance between nq closest point pairs in the two images. Then, in the next iteration, step 4 looks for the new closest points in image B for every point in the transformed model A. In this step the mean square distance among the nq closest point pairs are further decreased. At each step E decreases. Because E is positive, the convergence is guaranteed.

The iterative closest point algorithm has been used until now for image matching with rigid transformation (rotation and translation) only [17,18,19]. Besl and McKay proposed that shears and separate axis scaling could be also easily accommodated by allowing a general affine transformation in the least-square process [17]. Our experiments demonstrated the feasibility of the iterative closest point algorithm for affine transformation.

4. Experimental Results

The images to be registered in our research project are two broad band visible and far-infrared video sequences, as shown Figure 1. When soldiers are hidden behind the smoke in the visible image but they appear as high contrast hot objects in the IR images. Figure 2 shows the contour curves we extract for registration. We select the features of visible image as matching model because there are fewer feature points in it. It will be efficient for performing the affine transformation on the model in each iterative step. We superimpose the model to the Euclidean distance map (EDM) of image shown as Figure 3, where we highlight the model as white pixels and the image shown as black pixels, because the values of feature points in image is zero in EDM. We also show the matching of the IR image contours and that of the model which is affine transformed with the transformation parameters computed with the adaptive hill climbing and the iterative closest point algorithm in Figure 4 and Figure 5 respectively. In the matching procedure of the iterative closest point algorithm, we input the initial quantile value as 0.75 and increase it for 0.1 until it reaches 0.95. The result of original image is shown as Figure 6.

Figure 1. IR and visible battle field images

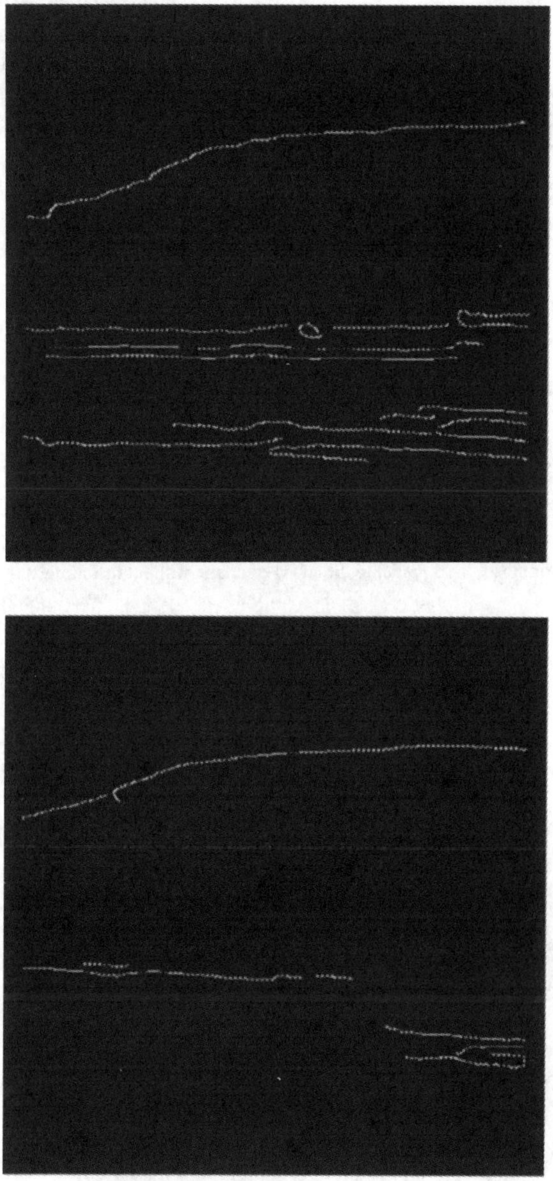

Figure 2. Edge features of IR and visible images extracted for registration

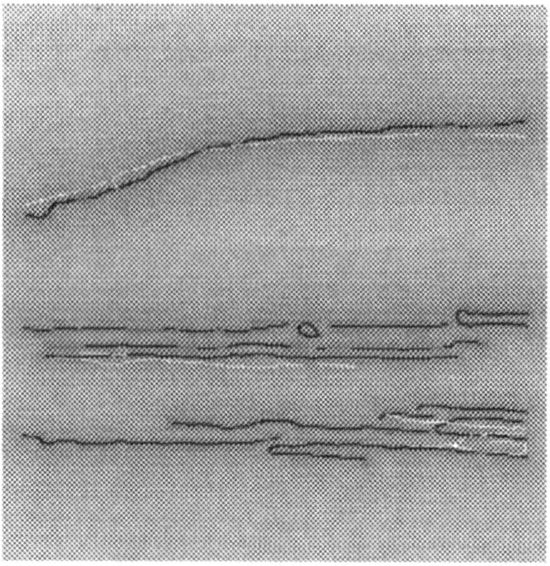

Figure 3. Superimposition of model to the EDM of image without matching

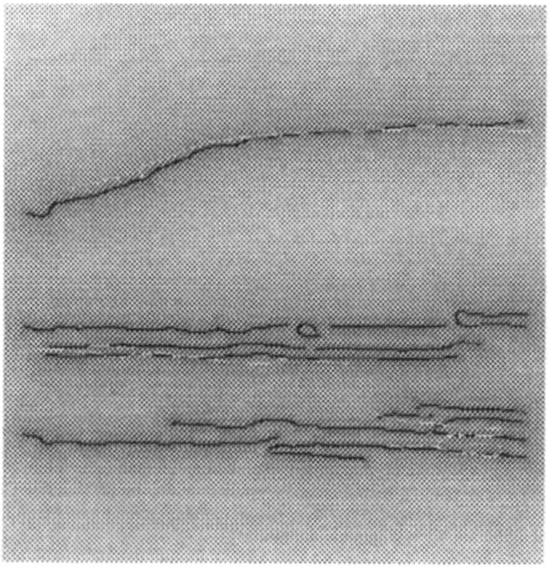

Figure 4. Superimposition of model to the EDM of image with affine transform computed with adaptive hill climbing

We use more then 500 points to estimate the six parameters of affine transformation. This large redundancy allows for a precise computation of the transformation between the two maps, but this approach requires some a priori knowledge of the transformation, such as the global affine transform in our cases.

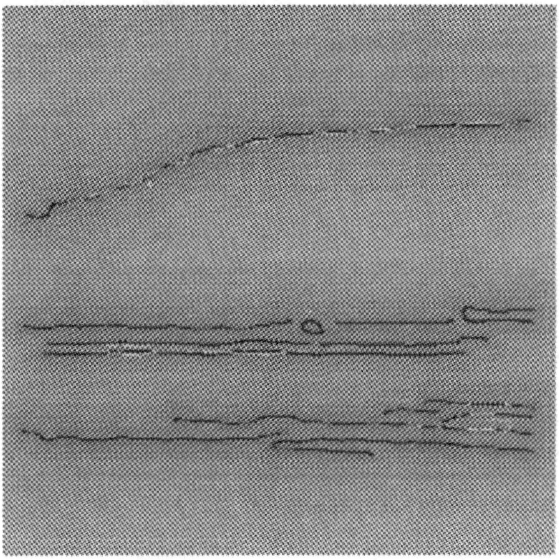

Figure 5. Superimposition of model to EDM of image with affine transform computed with ICP combination of image brightness and alignment

Figure 6. Aligned and fused IR and visible images is by fusion weighted combination of image brightness and alignment

5. Conclusion

We have analyzed the problems in the visible and far-infrared battlefield image registration. We adopted the contour-based approach, which permits to incorporate the powerful computer vision approaches for edge detection, edge saliency and curve matching. We have implemented multi-scale hierarchical edges detection and edge focusing and salience measure for extracting horizon lines.

To deal with the feature inconsistency problems we used the transformation space searching methods with the partial Hausdorff distance merit, and demonstrated free-form curve matching using 1) Image quadtree partition technique to the local Hausdorff distance matching, that dramatically reduces the size of the search space into that of the search for translations which minimize the Hausdorff distance between corresponding sub images, 2) Adaptive partial Hausdorff distance hill climbing, which is much faster than the search in the entire affine transformation space. 3) The iterative closest point algorithm using an efficient segment-list Voronoi transform and distance transform and again the mean square partial distance merit. All the three algorithms match the maps of curve features without looking for correspondence and show a large tolerance to outlier features. The condition for application of those three approaches is that the transformation should be known *a priori* and the initial match should be established as a start point for the optimization. Those conditions are fulfilled in our application.

The key issue for the successful image registration is still the extraction of salient features from the real world images using local, regional and global information and appropriate salience measures, and an efficient curve matching against outliers.

6. Acknowledgement

Supported to this project from the Natural Sciences and Engineering research Council of Canada (NSERC), the Department of National Defense of Canada (DND) and Lockheed-Martin Canada is acknowledged.

7. References

1. McReynolds, D., Marchand, P., Sheng, Y., Gagnon, L., Sévigny, L., (1999) Stabilization of infrared image sequence with differential invariants and M-estimators, In *Proceedings of SPIE Acquisition, Tracking, and Pointing XIII*, Vol. 3692, Orlando, FL, April 7-8.
2. Li, J., and Sheng, Y., Superresolution reconstruction of IR image sequences by wavelet expansion.
3. Rothwell, C., Mundy, J., Hoffman, B., and Nguyen, V., (1994) Driving vision by topology, *Report No. 2444, INRIA*, France.
4. Ullman, S., and Sha'ashua, A., (1988) Structural saliency: the detection of Globally salient structures using a locally connected network, *A.I. Memo* No. 1061 MIT.
5. Bergholm, F., (1987) Edge focusing, *IEEE Trans. PAMI-9*, No.6, 726-741.
6. Zhang, Z., Deriche, R., Faugeras, O., Quang-Tuan, L., (1995) A robust technique for matching two uncalibrated images through the recovery of the unknown epipolar geometry, *Artificial Intelligence*, Vol.78, No.1-2, 87-119.
7. Guan, W., and Ma., S., (1998) A list processing approach to compute Voronoi diagrams and the Euclidean distance transform, *IEEE Trans. Pattern Analysis and Machine Intelligence*, Vol. 21, No. 7.
8. Huttenlocher, D.P., Klanderman, G.A., and Rucklidge, W.J., (1993) Comparing images using the Hausdorff distance, In *IEEE Trans. PAMI*, Vol. 15, No. 9.
9. Rucklidge, W.J., (1996) *Efficient Visual Recognition Using the Hausdorff Distance*. Springer-Verlag.

10. Chew, L.P., Goodrich, M.T., Huttenlocher, D.P., Kedem, K., Kleinberg, J.M., and Kravets, D., (1993) Geometric Pattern matching under Euclidean motion. In *Proceeding of Fifth Canadian Conference on Computational Geometry*, pages 151-156, Waterloo, Ontario.
11. Huttenlocher, D.P., and Rucklidge, W.J., (1993) A multiresolution technique for comparing images using the Hausdorff distance. In *Proceeding of the IEEE Conference in Computer Vision and Pattern Recogntion*, pages 705-706.
12. Olson, C.F., and Huttenlocher, D.P., (1997) Automatic target recognition by matching oriented edge pixels. *IEEE Transactions on Image Processing*, 6(1):103-113.
13. Sheng, Y., Yang, X., McReynolds, D., Gagnon, L., Sévigny, L., (1999) Visible/IR Battlefield Image Registration using Local Hausdorff Distances, *International Society of information fusion, Conference Fusion99*, San Jose.
14. Escobar, A., and Laurendeau, D., (1998) 3D Tracking using the Hausdorff Distance. In *Proceeding Tenth IMDSP Workshop*, Alpbach, Austria, 12-16 July, 77-84.
15. Yuret, D., and de la Maza, M., (1993) Dynamic Hill Climbing Overcoming the Limitation of Optimization Techniques. In *Proceeding Second Turkish Symposium on Artificial Intelligence and Artificial Neural Networks*, 254-260.
16. de la Maza, M., and Yuret, D., (1994) Dynamic hill climbing. *AI Expert*, vol. 9, no. 3 26-31.
17. Besl, P.J., and McKay, N.D., (1992) A method for registration of 3D shapes, *IEEE Transactions PAMI* 14(2) pages 239-256.
18. Zhang, Z., (1994) " Iterative point Matching for registration of free-form curves and surfaces." *International Journal of Computer Vision*, Vol. 13, no. 2, 119-152,
19. Feldmar, J., (1997) Extension of the ICP algorithm to nonrigid intensity-based registration of 3D volumes. *Computer Vision and Image Understanding*. Vol. 66, no. 2, May, pages 193-206.

DETECTION, CHARACTERIZATION AND CLASSIFICATION OF RADIO TRANSMITTER TURN-ON TRANSIENTS

O. ÜRETEN[1] AND N. SERINKEN[2]
[1]*Ankara University*
Electronics Engineering Department
06100 Ankara, Turkey
e-mail: ureten@ieee.org
[2]*Communications Research Centre*
Ottawa, Ontario, Canada
e-mail : nur.serinken@crc.ca

1. Introduction

When a radio transmitter is engaged for transmission after the push-to-talk button is pressed, it is observed that its transient behavior i.e. RF power and carrier frequency changing towards their nominal values, has unique features. These transients have been attributed to variety of sources such as, phase-lock-loop systems, modulator subsystems, RF amplifiers, antenna characteristics, switch and relay characteristics. Transient state duration depends on make and model of a transmitter and can last from few microseconds to tens of milliseconds. The identification of radio transmitters in spectrum management operations is an important function for solving problems associated with deliberate or unintentional cases of frequency interference.

The turn-on transient signals are acquired from the discriminator output of a communication receiver [1]. Recording of a transient data requires triggering information to indicate the beginning of the transient. The squclch signal of the receiver can be used as a marker for the start of the transient. The problem with this approach is that there is an inherent delay between the onset of the actual transient and the trigger event. The capture system has to be continuously recording the received signal so that when a trigger is detected a portion of the pre-trigger samples are saved with the transient data. The recorded transient signal contains ambient channel noise which is followed by the start of a radio transmission similar to that shown in Figure 1a. Due to non-stationary nature of transmitter transients, though, the task of separating the transient from the channel noise is very difficult. It involves finding the exact time when the ambient channel noise stops and the transient begins.

Transients and or other non-stationary signals can be considered as outputs of chaotic dynamic systems. The dimension trajectory of such systems in phase space is often attracted to a bounded fractal object called a strange attractor. The natural approach in describing a strange attractor is to obtain its generalised dimension. A non-stationary signal, such as a transient, is not a pure fractal because its properties change with time. In the calculation of the local dimensions of a turn-on transient a sliding window technique is employed whereby the generalised dimension is estimated for successive portions of

A. K. Hyder et al. (eds.), Multisensor Fusion, 611–616.

the signal. Generalised dimensions obtained in this way are then used to construct a feature vector for classification purpose. A probabilistic neural network is then used to classify vectors after having been trained using a known set of feature vectors and tested with a test set of samples in order to evaluate the performance of the classification system.

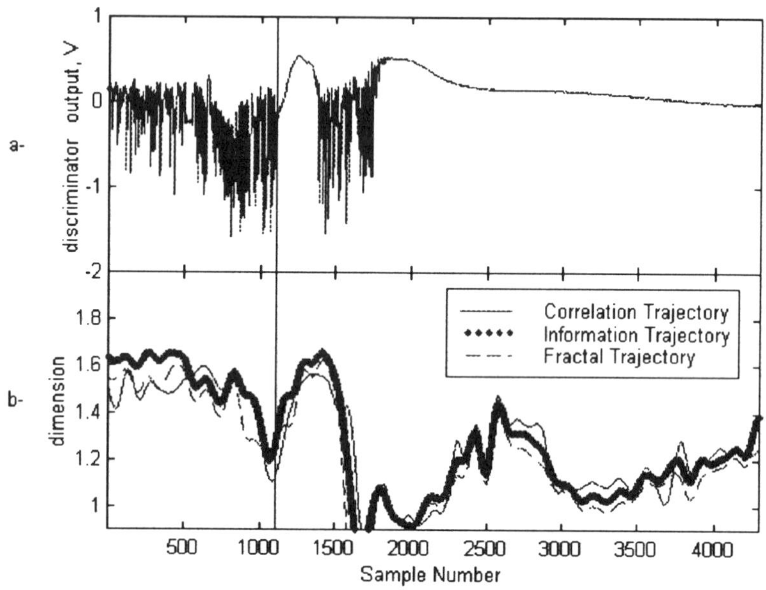

Figure 1a. Typical recorded transmitter turn-on transient data from the communications receiver.
Figure 1b. Detected turn-on transient and calculated dimension trajectories (q=0,1, and 2)

2. Transient Detection

Extracting the turn-on transient signals from the radio transmission data is difficult due to the non-stationary nature of transient signals. The problem is to find the exact time when the channel noise stops and the transient begins. However, despite being completely deterministic, many transients exhibit characteristics similar to noise due to their high degree of irregularity. Thus, to some extent, we are left with the problem of separating noise from noise having a different degree of correlation.

This problem has been approached as a change point detection problem in [2]. In this reference, transmission data are assumed to be composed of two Gaussian segments, channel noise and followed by the turn-on transient, of which mean and variance have abrupt changes at the transition point. Although the assumption that both mean and variance change, as opposed to a change in variance only, allows an analytical solution within a Bayesian framework, this method fails for some classes of transmitter transients for which the change in the mean occurs considerably later than the variance change thus delaying the detection of the start of the transient.

Reference [3] proposes an approach which transforms a change in variance into a change in the mean value based on the fractality of the sampled data. This method is based on the fact that the fractal dimension of the channel noise and the turn-on transients differ from each other. The fractal dimension, *D*, can be interpreted as the "degree of irregularity" of a fractal. A non-stationary signal, such as a transient, is not a pure fractal because its fractality changes with time. Multifractality deals with signals or objects that have varying local fractal dimensions. To calculate the local fractal dimensions of a signal, a sliding rectangular window is used whereby the fractal dimension is determined for successive portions of the signal. Fractal trajectory obtained in this way has an abrupt change at the transition point from channel noise to the turn-on transient. The calculated fractal trajectory data is modeled as:

$$d_i = \left\{ \begin{matrix} f_n + u_i, & if \ \ 1 < i < m \\ f_t + u_i, & if \ m+1 < i < N \end{matrix} \right\} \qquad (1)$$

where f_n and f_t are the fractal dimensions of the noise and transient signal respectively, *N* is the number of data points on the fractal trajectory, *m* is the change point and *u* is a zero mean Gaussian process. The actual position of the step change is then estimated using a Bayesian step change detector which is given as follows [4]:

$$p(\{m\}|d,I) \alpha \frac{1}{\sqrt{m(N-m)}} \left[\sum_{i=1}^{N} d_i^2 - \frac{1}{m} (\sum_{i=1}^{m} d_i)^2 - \frac{1}{N-m} (\sum_{i=m+1}^{N} d_i)^2 \right]^{-(\frac{N-m}{2})} \qquad (2)$$

where *I* denotes the specific choice of the signal model.

This equation shows that *a posteriori* probability density of *m* is only a function of the data. This means that there is no need to know about the probable trigger position or threshold value in order to detect the change point.

3. Generalised Dimension Characterisation

The definition of generalised dimension from [5] also known as Renyi dimension,

$$D^{(q)} = \lim_{\varepsilon \to 0} \frac{\log C^q(\varepsilon)}{\log \varepsilon} \qquad (3)$$

with

$$C_\varepsilon^{(q)} = [\frac{1}{N} \sum_i \left[\frac{1}{N} \sum_j \Theta\left(\varepsilon - \left|X_i - X_j\right|\right) \right]^{q-1}]^{\frac{1}{q-1}} \qquad (4)$$

where X_i s are co-ordinates of the vector, N is the number of points on the vector and ε is the box size. $\Theta(\varepsilon)$ is the Heaviside step function and is used to count how many pairs of points fall within the distance ε.

It can be shown that $D^{q} \leq D^{q'}$ if $q' \leq q$. For $q=0,1,2$ we obtain the fractal, information and correlation dimensions respectively. Fractal dimension is purely geometric, i.e. the number of times the dynamics visit different regions of the phase space. The information dimension represents how the entropy scales with the box size. Correlation dimension corresponds to a two-point correlation function, which calculates the probability that two randomly selected points of the attractor are within distance ε.

Generally, the dimension value is estimated by calculating the slope of the $\log C_{\varepsilon}^{(q)}$ - $\log \varepsilon$ curve. A linear regression algorithm is used in the dimension estimation. It is observed that the log-log plot is not strictly linear for the coarse and fine values of ε. The reason being that there is a noise component in the transient samples and the result of the small number of data samples that are available. An algorithm for the estimation of fractal dimensions of data with finite size is presented in [6]. This algorithm is used in the estimation of the fractal dimension (D_0), the information dimension (D_1) and the correlation dimension (D_2). Once these measures are estimated, their trajectories were used to construct feature vectors [7].

4. Experimental Results

Transients were collected from a variety 60 VHF FM transmitters which were divided into 3 different groups according to the time the tests were conducted. The first Group, Group I, contained 20 radio transmitters with 5 makes and 10 models. Group II contained 10 different transmitters with one make and model but different serial numbers. The most difficult set was Group III that contained 30 transmitters of the same make and model with different serial numbers. For each transmitter, 100 turn-on transients were recorded from the discriminator output of a general coverage communications receiver and sampled with 16-bit resolution at a rate of 44.1 kHz and each transient has 8192 samples, corresponding to less than 200 ms. Transients were separated from the noise using the described method. In order to calculate the local fractal dimensions of the transients, a sliding rectangular window is used whereby the fractal dimension is determined for successive portions of the signal. The calculated fractal trajectory has an abrupt change at the transition point from channel noise to the turn-on transient. The actual position of this step change is then estimated using a Bayesian step change detector, as indicated by the straight line in Figure 1a. In order to obtain fractal, information and correlation trajectories, a rectangular overlapping sliding window is applied to each transient signal, yielding in a feature vector of 50 components. The resulting dimension trajectories for q = 0, 1, and 2 of the transient shown in Figure 1a is displayed in Figure 1b.

A Probabilistic Neural Network was then trained to identify transmitters from each test group using 20 of the 100 transients recorded. The remaining 80 transients were used to evaluate the identification performance. For each transmitter a correct identification probability was calculated by dividing the number of correct classifications by the number of test transients. Classification results are presented as histograms in Figure 2. These histograms show the probability of correct classification for each group and the three dimension measures tested.

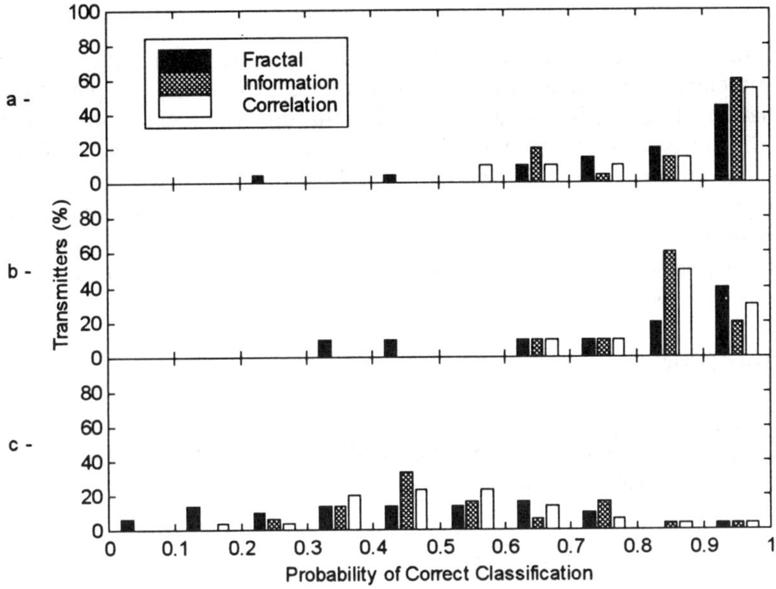

Figure 2. Probabilities of correct classifications for different methods and different transmitter sets
a Set of 20 transmitter with different make and model (Group I)
b Set of 10 transmitters of the same make and model (Group II)
c Set of 30 transmitters of the same make and model (Group III)

A summary of the experimental results is given in Table 1. According to the table the use of information and correlation dimensions gives better results than the fractal dimension. Another result is that the classification performance degrades among the transmitters of the same model. Apparently, increasing the number of transmitters in the set causes the classification rate to degrade as the complexity of the decision boundary increases.

TABLE 1. Probability of correct identification for each test group and dimension measure.

	Fractal Dimension	Information Dimension	Correlation Dimension
Group I	81.7	87.7	86.7
Group II	77.8	84.8	86.8
Group III	44.5	55.0	52.1
Average	68.0	75.8	75.2

5. Conclusions

The generalised dimension measures tested in this paper are fractal, information and correlation dimensions. Although three dimension values were tested, it is possible to use any combination of dimension values using data fusion techniques. Information and correlation dimensions gives better localisation properties than the fractal dimension because fractal dimension does not account for the frequency with which the set in question might visit the covering cells and thus properties pertaining to the neighbourhoods of individual points are not distinguishable. Experimental results show that the information and correlation dimensions give better classification performance than the fractal dimension.

6. Acknowledgement

This work is supported by NATO Research and Technology Organisation Sensors and Electronics Technology Panel in collaboration with Defense R&D Canada and Industry Canada Spectrum Engineering branch.

7. References

1. Toonstra, J. and Kinsner, W., (1996) A Radio Transmitter Fingerprinting System ODO-1, *Proceedings of Canadian Conference on Electrical and Computer Engineering, Canada, Volume 1,* pp. 60-63
2. Üreten, O. and Serinken, N. (1999) Bayesian Detection of Radio Transmitter Turn-On Transients, *Proceedings of NSIP'99, Antalya, Turkey,* pp. 830-834
3. Üreten, O. and Serinken, N. (1999) Detection of Radio Transmitter Turn-On Transients, *Electronics Letters, Volume 35, Number 23,* pp. 1996-1997
4. Ruanaidh, J.K. and Fitzgerald W.J. (1996) Numerical Bayesian Methods Applied to Signal Processing, *Springer-Verlag.*
5. Pawelzik, K. and Schuster, H.G. (1987) Generalised Dimensions and Entropies from a measured time series, *Phys. Rev. A, Volume 35, Number 1,* pp. 481-484
6. Serinken, N. and Üreten, O. (2000) Generalised Dimension Characterisation of Radio Transmitter Turn-On Transients, *Electronics Letters, Volume **36**, Number 12,* pp. 1064-1066
7. Liebovitch, L.S. and Toth, T. (1989) A Fast Algorithm to Determine Fractal Dimensions from Box Counting, *Phys. Lett. A, 141, (8,9),* pp. 386-390

OPTIMIZATION AND BENCHMARKING OF TRUNCATED DEMPSTER-SHAFER FOR AIRBORNE SURVEILLANCE

P. VALIN AND D. BOILY*
Advisory Member of R&D
Lockheed Martin Canada
6111 Royalmount ave
Montréal, QC, H4P 1K6
Canada
**Centre de Recherches Mathématiques*
Université de Montréal
C.P. 6128, Succ. Centre-Ville
Montréal, QC, H3C 3J7
Canada

1. Introduction to TDS

Evidential reasoning requirements that Decision Support Systems (DSS) must often address necessitate that one must:

(1) process incomplete information, which implies that ignorance should be defined mathematically
(2) not require a priori information, which usually rules out Bayes reasoning
(3) handle conflicts between contact/track so that formalisms such as Dempster-Shafer (DS), where conflict is precisely defined, are favored
(4) have a real-time method, which means that DS truncation is a must
(5) present the operator with the best ID, i.e. give preference to singletons
(6) present the operator with the next best thing, i.e. doublets, then triplets…
(7) be tested operationally, which translates into building complex scenarios
(8) ensure that ordinary DS explodes, which requires a large complex Platform Data Base (PDB)
(9) resist Counter Measures (CM) such that several contact reports should conflict with information contained in the PDB
(10) resist false associations, i.e. ESM reports associated to wrong track

2. Example of Large Complex PDB

The PDB used in the scenarios later described consists of about 150 platforms containing both air and naval targets. It is used in a Multi-Sensor Data Fusion (MSDF) prototype for the Canadian CP-140 maritime surveillance aircraft, which has 2 imaging sensors, a Forward Looking Infra-Red (FLIR), and a Spotlight Synthetic Aperture Radar (SAR). The size of the PDB ensures that the combination of propositions over time coming from

A. K. Hyder et al. (eds.), Multisensor Fusion, 617–623.

dissimilar sensors will generate a substantial portion of the 2^{150} possible subsets of the frame of discernment. In order to satisfy the real-time aspect of any implementation of the DS scheme, truncation at every fusion step must be applied, leading to the Truncated DS (TDS) rules listed in the next section. Each row of the ASCII file contains a different platform and each column represents an attribute that can be measured by a sensor. The ships identified in the split table below are either seen only by the Electronic Support Measures (ESM) sensor (left horizontal bars), which identifies the emitters on-board the target platform (the last column), those also imaged by the SAR (left vertical bars) and those of Russian origin (right vertical bars).

TABLE 1. Extracts from a PDB containing a priori informtion about platform attributes

ID #	NAME----------------	PLATYPE	SUBTYPE	OFFENS	CONT	V MINI	V MAXI	ACC	ALT MAXIM	LEN	HEI	WID
000001	JAHRE-VICKING-------	SURFCOM	TANKERV	HARMLE	DANM	0	35	999	0	460	33	51
000002	HALIFAX-CPF---------	SURMILI	FRIGATE	MEDIOF	CANA	0	35	999	0	130	5	16
000007	CALIFORNIA----------	SURMILI	CRUISER	STROOF	USAM	0	35	999	0	182	10	19
000008	COONTZ--------------	SURMILI	DESTROY	STROOF	USAM	0	35	999	0	156	5	16
000017	VIRGINIA------------	SURMILI	CRUISER	STROOF	USAM	0	35	999	0	178	10	19
000018	MIRKA-I-------------	SURMILI	FRIGATE	WEAKOF	RUSS	0	32	999	0	82	3	9
000019	MIRKA-II------------	SURMILI	FRIGATE	WEAKOF	RUSS	0	32	999	0	82	3	9
000025	IROQUOIS------------	SURMILI	DESTROY	WEAKOF	CANA	0	30	999	0	130	5	15
000026	ADELAIDE------------	SURMILI	FRIGATE	MEDIOF	AUST	0	30	999	0	138	5	14
000027	IMPROVED-PROVIDER---	SURMILI	SUPPORT	VEWEOF	CANA	0	21	999	0	172	9	23
000028	QUEST---------------	SURMILI	MISCELL	HARMLE	CANA	0	11	999	0	72	5	13
000029	KNOX----------------	SURMILI	FRIGATE	MEDIOF	EGYP	0	27	999	0	134	5	14
000030	IVAN-ROGOV----------	SURMILI	ASSAMPH	WEAKOF	RUSS	0	25	999	0	158	8	25
000031	KARA-KERCH----------	SURMILI	CRUISER	STROOF	RUSS	0	35	999	0	173	7	19
000032	MODIFIED-KIEV-------	SURMILI	CARRIER	STROOF	RUSS	0	32	999	0	274	10	51
000033	KIROV-ADM-USHAKOV---	SURMILI	CRUISER	VESTOF	RUSS	0	35	999	0	252	9	29
000037	TICONDEGORA---------	SURMILI	CRUISER	STROOF	USAM	0	35	999	0	173	10	17
000038	TARAWA--------------	SURMILI	ASSAMPH	WEAKOF	USAM	0	24	999	0	254	8	40
000039	SPRUANCE------------	SURMILI	DESTROY	STROOF	USAM	0	33	999	0	172	6	17
000040	NIMITZ--------------	SURMILI	CARRIER	WEAKOF	USAM	0	35	999	0	333	11	41
000041	SACRAMENTO----------	SURMILI	SUPPORT	WEAKOF	USAM	0	26	999	0	242	12	33
000042	KIROV-ADM-NAKHIMOV--	SURMILI	BATTLES	VESTOF	RUSS	0	35	999	0	252	9	29
000043	KIROV-ADM-LAZAREV---	SURMILI	BATTLES	VESTOF	RUSS	0	35	999	0	252	9	29
000044	KIROV-PYOTR-VELIKIY-	SURMILI	BATTLES	VESTOF	RUSS	0	35	999	0	252	9	29
000045	KARA-AZOV-----------	SURMILI	CRUISER	STROOF	RUSS	0	35	999	0	173	7	19
000046	KARA-PETROPAVLOVSK--	SURMILI	CRUISER	STROOF	RUSS	0	35	999	0	173	7	19
000047	KARA-VLADIVOSTOK----	SURMILI	CRUISER	STROOF	RUSS	0	35	999	0	173	7	19
000064	UDALOY-II-----------	SURMILI	DESTROY	STROOF	RUSS	0	30	999	0	164	8	19
000065	UDALOY-AND-KULAKOV--	SURMILI	DESTROY	MEDIOF	RUSS	0	30	999	0	164	8	19
000066	SOVREMENNY-II-------	SURMILI	DESTROY	STROOF	RUSS	0	32	999	0	156	7	17
000067	SOVREMENNY-OSMOTRITE	SURMILI	DESTROY	STROOF	RUSS	0	32	999	0	156	7	17
000068	UDALOY-SPIRIDONOV---	SURMILI	DESTROY	MEDIOF	RUSS	0	30	999	0	164	8	19
000074	TYPHOON-------------	SUBSURF	NUCPSTR	STROOF	RUSS	0	26	999	900000 300	165	13	25

= In scenarios = Imaged ships in scenarios = all Russians in scenarios

ID #	TD	NC	RCS SID	RCS FOR	RCS TOP	I	RP	BLADE	NE	EMITTER LIST
000001	2	99	1518000	168300	2346000	1	9		99	
000002	3	20	65000	8000	208000	1	2	4 4	8	57 58 59 60 61 6 7 8
000007	4	99	182000	190	346000	1	2	99 99	9	57 7 13 16 17 18 15 31 32
000008	4	99	72000	8000	250000	1	2	99 99	9	57 7 13 16 18 33 34 35 8
000017	4	99	178000	19000	338200	1	2	99 99	10	13 16 57 53 54 15 31 32 7 8
000018	3	99	24600	2700	73800	1	2	99 99	5	55 47 56 103 109
000019	3	99	24600	2700	73800	1	2	99 99	6	44 55 47 56 103 109
000025	1	99	65000	7500	195000	1	2	99 99	6	36 72 23 59 7 8
000026	1	99	69000	7000	193200	1	2	99 99	8	7 13 73 57 53 31 74 8
000027	4	99	154800	20700	395600	1	2	99 99	5	43 42 75 76 8
000028	2	99	36000	6500	93600	1	2	99 99	1	79
000029	4	99	67000	7000	187600	1	2	99 99	6	77 19 18 17 14 107
000030	1	99	126400	20000	395000	1	2	99 99	9	93 89 103 101 68 46 65 64 62
000031	1	99	121100	13300	328700	1	2	99 99	12	78 84 62 64 47 85 45 68 46 93 104 103
000032	4	99	274000	51000	1397400	1	4	99 99 99 99	15	88 94 95 96 97 65 98 71 46 91 99 100 101 102 106
000033	4	99	226800	26100	730800	1	2	99 99	14	77 89 90 65 67 92 84 80 45 71 46 93 101 106
000037	1	99	173000	17000	294100	1	2	99 99	9	13 110 57 53 54 32 112 7 8
000038	4	99	203200	32000	1016000	1	2	99 99	13	13 16 113 114 115 17 54 116 117 31 32 7 118
000039	1	99	103200	10200	292400	1	2	99 99	12	14 114 115 53 18 119 57 31 32 121 43 8
000040	4	99	366300	45100	1365300	1	4	99 99 99 99	14	122 16 57 115 17 124 117 125 126 127 54 121 7 8
000041	4	99	290400	39600	798600	1	2	99 99	6	13 130 33 18121 7
000042	4	99	226800	26100	730800	1	2	99 99	12	88 89 63 65 91 92 84 80 45 71 101 106
000043	4	99	226800	26100	730800	1	2	99 99	12	88 89 90 65 91 92 84 80 45 71 101 106
000044	4	99	226800	26100	730800	1	2	99 99	12	88 89 63 65 91 92 84 80 45 71 101 106
000045	1	99	121100	13300	328700	1	2	99 99	12	78 84 62 64 85 45 92 68 46 93 104 103
000046	1	99	121100	13300	328700	1	2	99 99	12	78 84 62 64 47 85 45 68 46 89 104 103
000047	1	99	121100	13300	328700	1	2	99 99	12	78 84 62 64 47 85 45 68 46 93 104 103
000064	1	99	131200	15200	311600	1	2	99 99	9	69 97 63 65 128 91 71 93 131
000065	1	99	131200	15200	311600	1	2	99 99	11	69 97 65 67 91 46 71 101 106 89 93
000066	4	99	109200	11900	265200	1	2	99 99	6	63 132 129 46 128 71
000067	4	99	109200	11900	265200	1	2	99 99	13	69 96 65 128 129 71 46 101 106 103 104 102 131
000068	1	99	131200	15200	311600	1	2	99 99	12	69 97 63 65 67 91 46 71 101 106 89 93
000074	4	99	2500	250	412500	1	2	99 99	1	138

3. Truncated DS (TDS) Rules

There are 3 parameters to the TDS algorithm as described below: MIN_BPM, MAX_BPM, and MAX_NUM. After each ID fusion that gives combined propositions with Basic Probability Assignment (BPA) calculated by the standard DS orthogonal sum, one applies the following sequential rules [1]:

(1) All combined propositions with BPA > MAX_BPM are kept
(2) All combined propositions with BPA < MIN_BPM are discarded
(3) If the number of retained propositions in step 1 is < MAX_NUM, it retains, by decreasing BPA, propositions of length 1 (singletons)
(4) If the number of retained propositions in step 3 is < MAX_NUM, it does the same thing with propositions of length 2 (doublets)
(5) Repeat a similar procedure for propositions of length 3 (triplets)
(6) If the number of retained propositions is still < MAX_NUM, it retains propositions by decreasing BPA, regardless of length.

An empirical relationship was found, such that MAX_BPM = 1 / (2.5 * MAX_NUM), and MIN_BPM = 1 / (12.5 * MAX_NUM), which was used in scenarios such as Maritime Air Area Operations (MAAO) of the CP-140 as shown below, where all 3 Russian ships have similar emitters (as can be seen in the PDB) and are imaged by the SAR [2].

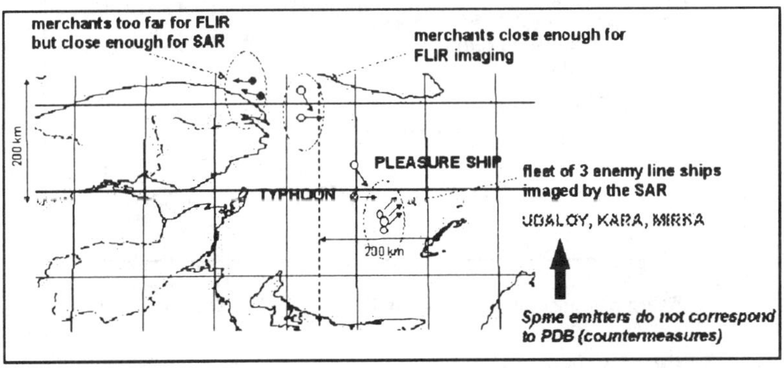

Figure 1. MAAO scenario for the CP-140 maritime surveillance aircraft (dashed line trajectory)

The following table shows that correct ID is achieved earlier and with better confidence for the Udaloy (for example) as MAXNUM is increased from 8 to the ideal maximum value of 20.

Another scenario providing Direct Fleet Support (DFS) was also treated extensively where some American ships (Coontz, Ticonderoga and Virginia) are imaged by the SAR. The Canadian contingent below is sufficiently well separated (0.5 km) and close (20 km) that tracking is relatively easy and ESM reports can associate well to ships. However for the American fleet, separated by 0.5 km also but at 100 km, tracking is at the limit of the radar bearing accuracy, and ESM reports can often associate to the wrong ship.

TABLE 2. Scenario evolution for the Udaloy in the MAAO scenario as a function of MAX-NUM

MAX NUM = 8			MAX NUM = 20	
T sec.	Max BPA	proposition	Max BPA	proposition
556	0.794	IVAN-ROGOV, KARA-KERCH, KIROV-ADM-USHAKOV, KARA-AZOV, KARA-PETROPAVLOVSK, KARA-VLADIVOSTOK, IVAN-ROGOV-ALEKSANDR, IVAN-ROGOV-MITROFAN, UDALOY-II, UDALOY-AND-KULAKOV, UDALOY-SPIRIDONOV	0.794	IVAN-ROGOV, KARA-KERCH, KIROV-ADM-USHAKOV, KARA-AZOV, KARA-PETROPAVLOVSK, KARA-VLADIVOSTOK, IVAN-ROGOV-ALEKSANDR, IVAN-ROGOV-MITROFAN, UDALOY-II, UDALOY-AND-KULAKOV, UDALOY-SPIRIDONOV
762	0.630	IVAN-ROGOV, KARA-KERCH, KARA-AZOV, KARA-VLADIVOSTOK, IVAN-ROGOV-ALEKSANDR, IVAN-ROGOV-MITROFAN	0.630	IVAN-ROGOV, KARA-KERCH, KARA-AZOV, KARA-VLADIVOSTOK, IVAN-ROGOV-ALEKSANDR, IVAN-ROGOV-MITROFAN
1169	0.5	KARA-KERCH, KARA-AZOV, KARA-VLADIVOSTOK, IVAN-ROGOV-ALEKSANDR, IVAN-ROGOV-MITROFAN	0.5	KARA-KERCH, KARA-AZOV, KARA-VLADIVOSTOK, IVAN-ROGOV-ALEKSANDR IVAN-ROGOV-MITROFAN
1369	0.467	"	0.467	"
1575	0.293	"	0.292	"
1781	0.307	IVAN-ROGOV-ALEKSANDR, IVAN-ROGOV-MITROFAN	0.294	IVAN-ROGOV-ALEKSANDR, IVAN-ROGOV-MITROFAN
1994	0.282	SOVREMENNY-BOYEVOY	0.380	SOVREMENNY-BOYEVOY
2406	0.385	SOVREMENNY-OSMOTRITE, SOVREMENNY-BOYEVOY	0.598	"
2612	0.237	UDALOY-II, UDALOY-AND-KULAKOV, UDALOY-SPIRIDONOV	0.355	"
3019	0.206	SOVREMENNY-OSMOTRITE, SOVREMENNY-BOYEVOY	0.356	"
3225	0.376	UDALOY-AND-KULAKOV, UDALOY-SPIRIDONOV	0.359	UDALOY-SPIRIDONOV
3425	0.438	UDALOY-SPIRIDONOV	0.539	"

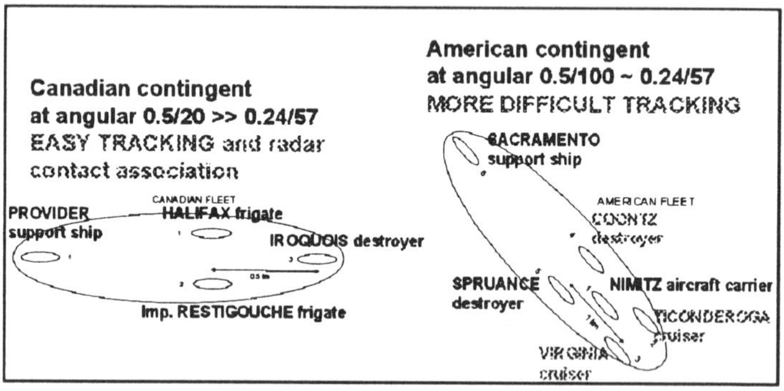

Figure 2. Canadian and American fleets involved in the DFS scenario

4. Study of 3 other Decision rules

4.1 STUDY OF THE EUI DECISION RULE

The utility function of a singleton occurring in n_j props of j elements (a parametric function) is given by the first equation below while that of a complex proposition A is given by the maximum over all subsets

$$u(\theta_i) = \sum_{j=1}^{5} c_j n_j \qquad\qquad u(A) = \max_{\theta_i \in A} u(\theta_i)$$

The lower and upper expectations of proposition A are then just

$$E_*(B) = \sum_{B \subseteq A} u(A)Bel(A) \qquad E^*(B) = \sum_{B \cap A \neq \Phi} u(A)Pl(A)$$

and an Expected Utility Interval (EUI) [E_* , E^*] similar to interval [*Bel(A)* , *Pl(A)*] can be defined by [3]

$$E(B) = E_*(B) + \rho_i \left[E^*(B) - E_*(B) \right]$$

where ρ_i is the probability that the ambiguity will turn favourably.

4.2 POSSIBILITY

The possibility mass vector P_k for a PDB of size N is calculated as follows:

(1) initialize at 1/N
(2) then perform for each sensor report E_j of BPM m containing l platforms and for each platform

$$P_k \to mP_k / l \quad if \quad k \in E_j \qquad else \qquad P_k \to (N-m)P_k /(N-l)$$

Finally one normalizes so that the sum be 1, this is achieved by dividing the P_k by the following factor:

$$\sum_{k \in E} P_k m / l + \sum_{k \notin E} P_k (N-m)/(N-l)$$

4.3 NORMALIZED PLAUSIBILITY

This rule is defined for a proposition A of length l_A by:

$$Pl_N(A) = \left(Bel(A) + Pl(A) \right) / l_A$$

Note that ALL these decision rules measures, with the exception of possibility, are used on the fusion results from TDS. However, they do NOT provide a truncation method. One should mention that there exists an EU truncation method, but it was ineffective when used on the given scenarios.

4.4 COMPARISON OF DECISION RULES

The table below shows that the EUI identifies the correct ID (68) a bit earlier for the MAAO scenario.

TABLE 3. Performance of various decision rules for the MAAO scenario

truncated DS	EUI	Normalized Plausibility	Possibility
0.794 30 31 33 45 47 48 49 64 65 68	ignorance	0.090 30 31 33 45 47 48 49 64 65 68	0.091 30 31 33 45 47 48 49 64 65 68
0.630 30 31 45 47 48 49	ignorance	0.136 30 31 45 47 48 49	0.165 30 31 45 47 48 49
0.500 31 45 47 48 49	0.157 31 45 47 48 49	0.150 31 45 47 48 49	0.198 31 45 47 48 49
0.467 31 45 47 48 49	0.143 31 45 47 48 49	0.140 31 45 47 48 49	0.198 31 45 47 48 49
0.292 31 45 47 48 49	0.143 ignorance	0.107 66 69	0.187 31 45 47 48 49
0.294 48 49	0.551 48 49	0.220 48 49	0.441 48 49
0.380 69	0.518 69	0.516 69	0.871 69
0.598 69	0.772 69	0.73 69	0.984 69
0.355 69	0.5695 69 ←	0.434 69	0.776 69
0.356 69 ←	0.693 68	0.4375 69 ←	0.827 69 ←
0.359 68	0.985 68	0.534 68	0.850 68
0.539 68	1.466 68	0.65 68	0.869 68

The following figure shows the time evolution of the various decision rules for the DFS scenario. A worst case scenario was selected to maximize the difference between the decision rules and to test their sensitivity to either counter-measures or faulty algorithms. This toughest case corresponds to when BOTH the ESM and the Image Support Module (ISM) length classifier make false ID declarations because the Virginia is small cruiser like the Spruance destroyer (solid line). It is clear that the Virginia cruiser (the correct ID, dashed line) is not selected after the wrong ISM declaration (near t=7500) but that the Virginia is preferred (or nearly so for TDS). This is due to the fact that tracking is very difficult for the American contingent as mentioned earlier. In virtually all other cases, all decision rules work equally well.

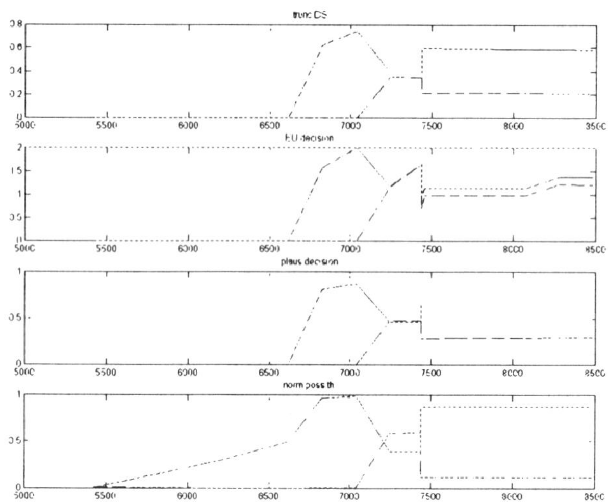

Figure 3. Time evolution of various decision rules for the DFS scenario

5. Conclusions and Outlook

This paper has shown the following:

(1) TDS is only known real-time method curing DS exponential growth
(2) MAX_NUM scales with PDB size
(3) Mixture of decision rules is ideal but CPU expensive

Remaining work will concentrate on:

(1) Search for truncation methods in other decision rules, for which measures of information uncertainty could provide a clue
(2) Analysis of April 2000 trial with real sensor data and imagery
(3) Extension of the PDB to 500 entries, and eventually 2000
(4) Tackling also complex naval scenarios
(5) Selection of a classification taxonomy tree (STANAG 4420 vs MIL-STD 2525) to provide ID at various levels (not only platform ID)

6. References

1. Valin, P. and Boily, D., (2000) "Truncated Dempster-Shafer Optimization and Benchmarking", SPIE Aerosense 2000, in *Sensor Fusion: Architecture, Algorithms and Applications IV,* **Vol. 4051**, pp. 237-246 , Orlando.
2. Jouan, A. Valin, P. and Bossé, E., (1999) "Testbed for Fusion of Imaging and Non-Imaging Sensor Attributes in Airborne Surveillance Missions", FUSION' 99 conference, **Vol. 2**, pp. 823-830, Sunnyvale, CA.
3. Cheaito, A. Lecours, M. and Bossé, E., (1999) "Study of a modified Dempster-Shafer approach using an Expected Utility Interval decision rule", SPIE Aerosense 1999, in *Sensor Fusion: Architecture, Algorithms and Applications III,* **Vol. 3719**, pp. 34-42, Orlando.

MULTITARGET TRACKING IN ELECTRONICALLY SCANNED ANTENNA RADAR

A. WITCZAK AND M. BRZOZOWSKI
Institute of Radar Technology
Military University of Technology
2 Kaliskiego
00-908 Warsaw
Poland

1. Introduction

New-generation radars are equipped with electronically scanned antennas (ESA). Such solution results with new performance of radars like low probability of interception (LPI), multifunctionality, multitarget tracking etc. Such radars have also the possibilities to match adaptively his search algorithms respectively to changes in local radar targets density, as well as to object manoeuvres. Tracking algorithms send commands to an antenna beam control unit (CU) to search a particular direction, defined by azimuth and elevation, in which is necessary to update information about tracking targets. In this way, tracking algorithm is not only the user of information but influences the total operational resources of the radar. When we add to this the pseudorandom procedure of search particular part of air space, we arrive at a complicated condition for multitarget tracking algorithms. Well known, optimal algorithms, used in old radars with mechanically scanned antennas (MSA), with near constant sampling interval, cannot be implemented directly in new ones [1].

Newly designed quasioptimal algorithms need much testing before their application in real radar, testing which should provide the best solutions for particular processes.

During such tests it is necessary to observe as much as possible the parameters of the tracking process. For such an application, we design modular test software. It is used as a base for tracking algorithms. This chapter presents the general construction of this bench test. The main ways of solving this particular function and simple example of tracking test results for a 2D case are presented. Those algorithms are matched to the performance of a new Polish industry ESA radar design.

2. General structure of simulation algorithm

Radar technical parameters directly influence the simulation algorithms' general structure as well as for particular solution. Radar uses the frequency-phase scanning antenna. In elevation, the antenna beam is moved by signal frequency change. The search area is limited to 200 km in range and up to 90^0 in azimuth. Within this element of space, the antenna beam takes pseudorandomly 90 discrete positions and goes to the next search in elevation. The average time for sequential detection in a particular azimuth position is 0.06 s. This parameter is called a sampling time. The maximum time in which

A. K. Hyder et al. (eds.), Multisensor Fusion, 625–631.

information about all elements in search volume have to be updated is limited to 7 s – so called frame time. Measurement accuracy for distance and angles are 60 m and 0.5^0 respectively. It is possible to change other parameters during the test. The main structure of test bench is presented on Figure 1.

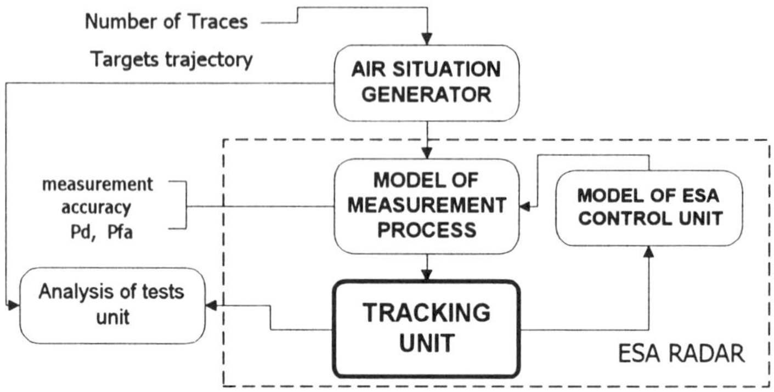

Figure 1. The general structure of simulation test software

The air situation generator calculates: position, velocity and acceleration of N targets in 3D space within a time of 0.06s. For each of them we can design Nm manoeuvres. The test scenario time is usually limited to a few minutes.

The final data file can be rewritten in the next unit (model of measurements) in which are added measurement errors in a particular axis, generated false alarms, as well as lost detection with controlled probability Pfa and Pd. Files with raw or updated data may be used several times in the simulation process. It helps to compare the results achieved with different structures or parameters of tracking algorithms.

Each record of data can be easy identified by the number of targets and time for which their position is calculated. From that data base Electronically Scanned Antenna Control Unit (ESA CU) creates, according to its own decision, the stream of data transferred to tracking algorithm.

ESA CU is an element of radar resources management and is responsible for steering the antenna for two interleaving processes – search and tracking [2]. One of the most important elements of good radar operation is to keep the correct balance between them. In our simulation, the radar resource management has been solved separately, and for tracking analysis it is not taken into consideration. The pseudorandom search scenario is interrupted only by a request for search direction (limited only to azimuth position) sent from the tracking unit with the highest priority.

The tracking unit is the main part of test bench. This unit processes all data received from the measurement unit. It will be described more precisely in the next paragraph.

Analysis of test unit is an additional part, used only in simulation compares the possibilities of all of the most important data collected in radar model: tracking errors, histograms of measurements, time spend for a particular function, etc. This unit is also used as the radar situation display.

3. Tracking Unit

This unit has a modular structure, and thanks to it, it is possible to change particular algorithms and solutions in for the most optimal one.

The main tasks of tracking unit (TU) are: data acquisition, data association, gating, track initiation, multitarget tracking, correlation problems solving and generation of parameters for ESA CU additional looks at tracking objects.

Data acquisition is a process of measurement information storage received from the measurement unit.

In presented tests, the tracking was performed in Cartesian coordinates, but it is possible to use another system. In the 2D case the state vector is 6 dimensional. The transition matrix is based on kinematic Newton or Singer model [3]. Estimation process esuses the Kalman filter concept working on-line. Innovation and covariance matrices calculated in this filter play the important role in the general logic of the algorithm.

For track initiation, ESA CU illuminates the azimuth directions in which the object is detected three times with time interval of from 1 to 2 seconds. Next, the observed time from last update and present prediction error for all tracks are determined. Those parameters are fundamental for the tracking concept. Prediction of all tracks is made in steps defined by the sampling interval. Such continuous management of tracks is useful in data association.

Continuous control of all received information also plays a key role in this process. Filtration is done when measurement of the tracking object parameters meets correlation criteria, without any difference if it is done for tracking or a search request.

In correlation process is used the Nearest Neighbour (NN) concept. A decision to illuminate a track is taken alternatively when a target's manoeuvre is detected or when coverage of the prediction error volume needs more than two or three neighbour bars. Adaptively changed time between updates helps to keep the Kalman gain at a level adequate for a medium target dynamic performance represented by a manoeuvre variation (500 m^2/s^4). In addition, the diameters of the correlation gates depend only on a target's position. Such a solution shows that by a simple mix of two different approaches it is possible to receive a satisfactory tracking performance.

4. Test Results

Many different scenarios for single or six parallel (radial or azimuthal) trajectories with separation between them from 1 to 20 kilometres have been used for tests. Examples of such trajectories are presented on Figures 2A (without manoeuvres) and 2B with 9g turns. In this figure are presented both real trajectories measurements and the results of tracking. Points between tracks show false detections. The tracking error in X-axis for trace No. 2 (see Figure 2B) is presented on Figure 3. For this trace we recorded a diagram of the time between measurements (Figure 4). It shows a dependence between: track initiation, no manoeuvre parts of trajectory, and during the turn.

Also, false correlations between close targets for which positions are in the same gates, suggest that it is necessary to find another solution, perhaps a different coordinate system.

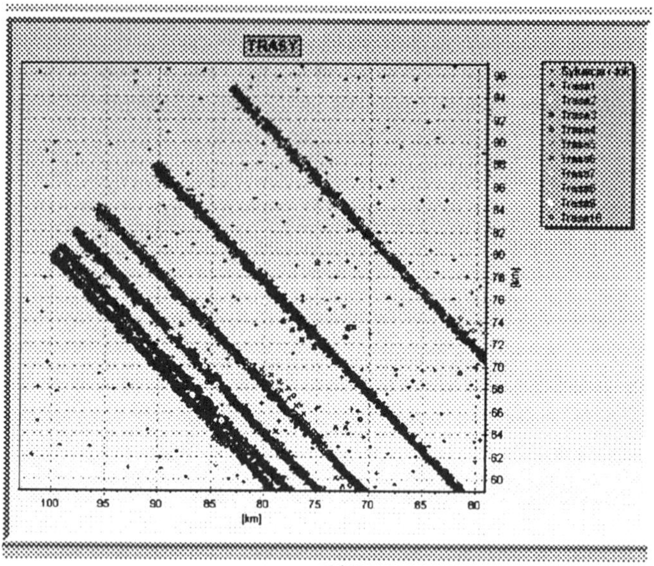

(A)

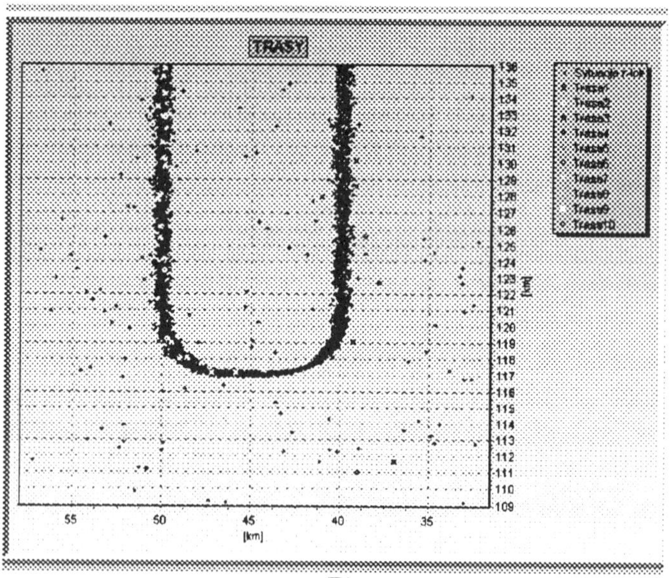

(B)

Figure 2. Examples of Test Scenarios and Results of Tracking (Zoom of Search Volume)

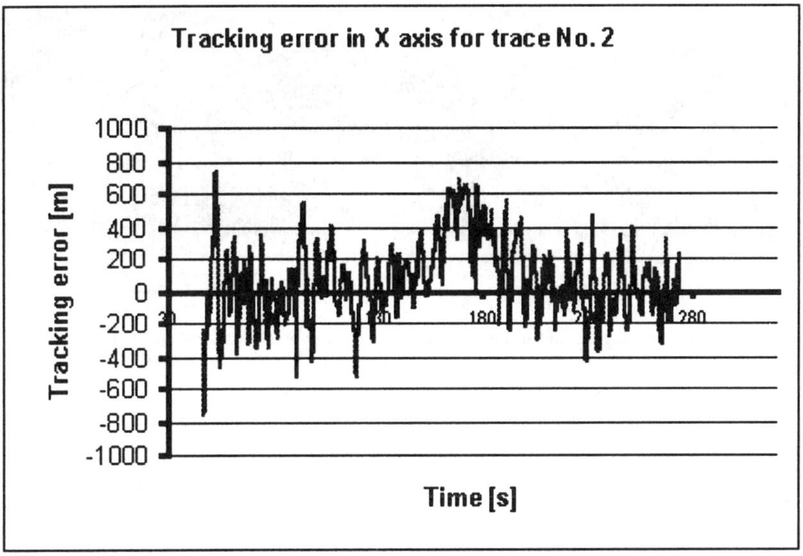

Figure 3. Filtration error in X axis for trace presented on Figure 2B

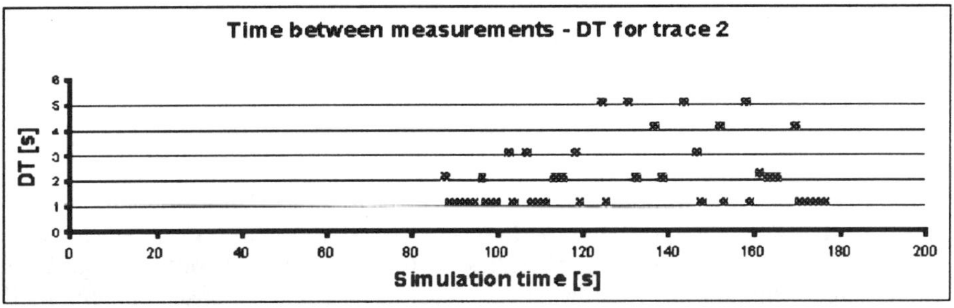

Figure 4. Diagram of time between measurements for trace presented on Figure 2B

The percentage of lost tracks at the present moment is too high. The main reason is false detection. Although a track is very quickly rebuilt, it uses additional time. It needs additional logic in association of tracks. Figure 5 presents the cumulative search patterns for radar with one tracking target.

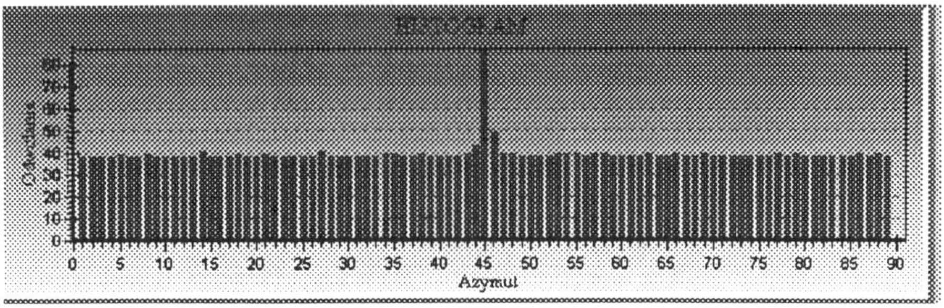

Figure 5. Histogram of number discrete azimuthally position of antenna beam (single detected target).

5. Conclusions

All elements of the model cooperate properly but the particular function needs more refinement and experimentation.

Achievement of good results in ESA Multitarget Tracking is not a simple process. In particular, the problem of crossing trajectory correlation in 2D needs more attention. Additional information about target elevation is very useful even then accuracy of this measurement is not the best. In the 3D case, results are better but of course, it demands more computational burden.

The observed too high percent of lost tracks needs a new solution adequate to work in a high-density environment including not only the real targets but also false alarms as a result of countermeasures.

The next problem is dedicated to group targets. They haven't been taken into consideration in present level of research.

When we compare the frame time with minimal time necessary for search we can easily note that the time for additional looks (as a answer for tracking unit request) is limited to 1.5 second. It offers possibilities of repeating observations of a particular azimuth only 25 times. When only a small number of targets are in a search volume, the algorithms have to become more complicated.

We expect the requirements for radar to reach up to 100 tracks. It is not possible to make all additional observations according to tracking unit expectations. So as a result the algorithm has to take into consideration the present density of targets. In the extreme case the performance of tracking will be on similar level to the MSA MTT ones.

The analysed problem is similar to a multi criterion optimisation. Solving one problem generates problems in another process.

To achieve final results, which will be good enough in different situations, we need more experiments with a different algorithms. Prepared software is a tool for such experiments.

6. References

1. Blackman S.S. (1986) Multiple–Target Tracking with Radar Applications, Artech House.

2. Van Keuck G., and Blackman S.S. (1993) On Phased-Array Radar Tracking and Parameter Control, IEEE Trans. on AES, Vol. 29, No. 1, pp. 186-194
3. Singer R.A. (1973) Estimating optimal tracking filter performance for manned manoeuvring

DATA FUSION IN REMOTE SENSING AND IMPROVEMENT OF THE SPATIAL RESOLUTION OF SATELLITE IMAGES

T. RANCHIN
Ecole des Mines de Paris
Centre d'Energétique, Groupe Télédétection & Modélisation
BP 207, 06904 Sophia Antipolis cedex
France

1. Introduction

Since the beginning of the 1990's, the availability of Earth Observation satellite data has increased dramatically. Spatial resolutions available range from a meter or less to few kilometers and with finer and finer spectral resolutions. This new situation opens new applications and new fields of research in Earth observation.

But this wealth of information is difficult to manage for the user. He can select the most representative data for his application, or he can process all the available data separately. But the most interesting approach will be to combine the available set of information in the most efficient way.

For example, a panchromatic image from any satellite such as SPOT, IRS or IKONOS is of interest for visualizing details and structures over an area. The structure and characteristic scales of a city, the buildings, the hierarchy of streets, the vehicles, even more details are detected, recognized or identified according to the high spatial resolution.

The exploitation of the multispectral set of images provided by the same satellite, but with coarser spatial resolution achieves spectral recognition and further vegetation monitoring, studies on marine or air pollution, thematic mapping, precision farming, etc., will be achieved. The use of color composition for visualizing multispectral images also facilitates the interpretation of the area of interest.

Combining both sets of information will obviously bring much more added value to the user, opening much more applications with the best spatial resolution available.

Hence, data fusion is becoming of paramount importance, in Earth Observation. Data fusion is an approach oriented to information extraction adopted in several domains. It is based on the synergetic exploitation of data originating from different sources. It aims to produce a better result than the one obtained by a separate exploitation of the same sources. According to Wald [1]:

> "The exploitation of satellite images and more generally of observations of the Earth and our environment is presently one of the most productive in data fusion. Observation of the Earth is performed by means of satellites, planes, ships, and ground-based instruments. It results into a great variety of measurements, partly redundant, partly complementary. These measurements may be punctual and time-integrated, bi-dimensional and instantaneous

A. K. Hyder et al. (eds.), Multisensor Fusion, 633–656.

(images), vertical profiles with time-integration or not, three-dimensional information (oceanic / atmospheric profiler / sounder at ground level, or satellite-borne, or ship-borne). Adding the large amount of archives and numerical models representing the geophysical / biological processes, one should conclude that the quantity of information available to describe and model the Earth and our environment increases rapidly. Data fusion is a subject becoming increasingly relevant because it efficiently helps scientists to extract increasingly precise and relevant knowledge from the available information.

The operation of data fusion by itself is not new in environment. For example, meteorologists predict weather for several tens of years. In remote sensing (i.e. Earth observation from spacecraft or aircraft), classification procedures are performed since long and are obviously relevant to data fusion. Data fusion allows formalizing the combination of these measurements, as well as to monitor the quality of information in the course of the fusion process."

In Europe and thanks to the impulsion of the European Association of Remote Sensing Laboratories (EARSeL), a Special Interest Group "Data Fusion" was created in 1996. This group contributes to a better understanding and use of data fusion in the field of Earth Observation by organizing regular meetings of its members and tackling fundamentals of Data Fusion in remote sensing. A series of bi-annual international conference called "Fusion of Earth Data – merging point measurements, raster maps and remotely sensed images" was launched in 1996 with the aim of browsing this field of research and to help the scientific community to fully understand the benefits of data fusion in the Earth Observation domain [2, 3, 4].

From the work of this Special Interest Group a set of reference terms emerged [5]. The definition of data fusion in the field of remote sensing was adopted as:

"Data fusion is a formal framework in which are expressed means and tools for the alliance of data originating from different sources."

This definition aims to conduct scientists to a formal approach of data fusion and to the benefits of a global reflection on data fusion. Data fusion aims at obtaining information of greater quality; the exact definition of 'greater quality' will depend upon the application. In this case, quality is a generic word denoting that the user is better satisfied by the results obtained through a fusion process.

In this paper, data fusion is illustrated through a specific application, the synthesis of images at the best spatial resolution available in the set of images processed. This application can be a step to further processing or applications. Some solutions proposed in this aim are described and a discussion from the quality point of view is addressed. An example is presented and exploited. Conclusions on this approach are drawn.

2. Improvement of the spatial resolution: methods

Several studies and publications have shown that merging broadband high spatial resolution images with low spatial resolution and high spectral resolution images proves to be of great benefit in many applications. Many methods have been developed in that purpose and produce multispectral images having the highest spatial resolution available within the data set. They apply on a data set comprising multispectral images B_{il} at a low spatial resolution l and images A_h at a higher spatial resolution h but with a different spectral content. Examples of such a data set are the SPOT-XS (3 bands, 20 m) and SPOT-P (panchromatic, 10 m) images, or the SPOT-4 case, with 3 bands at 20 m (XS1, XS3, and MIR) and the band XS2 at 10 m, or IRS-1C with the LISS (3 bands) at 23.2 m and a PAN at 5.8 m, or IKONOS with 4 multispectral bands at 4 m and a PAN at 1 m.

The number of methods is fairly large. Of interest in this paper are only concerned those methods which claim to provide a synthetic image close to reality when enhancing the spatial resolution, and not those which only provide a better visual representation of the image [e.g., 6].

The latter are very useful for photo-interpretation. This is particularly true when the number of spectral bands is much larger than the usual three bands for describing colors: red, green, blue. However, such methods have their limitations, especially with the new space-borne sensors and the most recent techniques, which allow the reconstruction of high spatial resolution landscapes with objects having their natural colors. Here, in this context, natural colors mean the colors that are perceived by the human eye.

These methods under discussion in this paper aim at constructing synthetic multispectral images B^*_{ih} having the highest spatial resolution available within the data set (e.g. the 3 XS bands at 10 m in the case of SPOT 1-3) which are close to reality by performing a high-quality transformation of the multispectral content when increasing the spatial resolution.

All the methods can be associated in three groups of techniques currently used:

- Projection of original data sets into another space, substitution of one vector by the high resolution image and inverse projection into the original space, such as the IHS (Intensity, Hue, and Saturation) method [6].
- Relative spectral contribution such as the Brovey transform [7] which can be applied to any set of image and the CNES P+XS method [8] dedicated to the SPOT case. It should be noted that the Brovey transform does not well represent this group because of its poor principles in construction. Nevertheless, it is often used.
- Scale by scale description of the information content of both images and synthesis of the high-frequency information missing to transform the low spatial resolution images into high spatial resolution high spectral content images. This group of methods can be presented through the ARSIS concept [9]. Two algorithms will be described, the HPF [10] and the XS-HR algorithms [9].

2.1. PROJECTION METHODS: THE IHS METHOD

The IHS method [6] is based on the projection of the original set of images into another space. It can be applied when three multispectral images and one high spatial resolution image are available. Figure 1 presents this algorithm. Each of the three bands B_i are interpolated to the spatial resolution of A_h and labeled as red, green and blue respectively. Then, these color components are converted into intensity (I), hue (H) and saturation (S) components using for example, the model for colors of "Commission Internationale pour l'Eclairage". The next step is the substitution of the intensity by the high spatial resolution image A_h. Refinements can be made which include the substitution of a linear combination of the A_h values and the original intensity. The last step performs the inverse model converting IHS components into blue, green and red components, which are the searched synthetic images B^*_{ih}. The method can apply to either digital counts or to radiances. In any case, the dynamics of the signal in each bands, including A_h, should be adjusted in order to make them similar. This may cause a distortion of the spectral content.

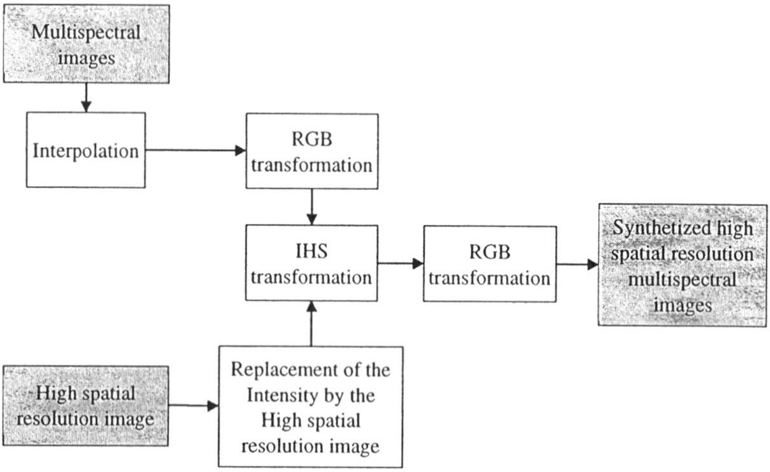

Figure 1. The IHS method

2.2. RELATIVE SPECTRAL CONTRIBUTION METHODS

2.2.1 *The Brovey Transform*

Let A_h be the high spatial resolution image, B_{il} the multispectral image, h the original spatial resolution of A and l the original spatial resolution of B_i ($l < h$). The Brovey transform applies to the digital counts of three spectral bands B'_{ih} $(i=1, 2, 3)$, where B'_{ih} is the image B_{il} resampled at resolution h, and of an image A_h of a better resolution - though it should deal with radiances and the software should request the calibration coefficients. An example is the SPOT case, where B'_{ih} are the XS1, XS2 and XS3 bands at original resolution of 20 m, resampled to 10 m, and A_h is the P band with a spatial resolution of 10 m. The synthetic bands B^*_{ih} are given by:

$$B^*_{ih} = B'_{ih}\, A_h / (B'_{1h} + B'_{2h} + B'_{3h})$$

This method can be generalized to any number of spectral bands.

2.2.2 *The P+XS Method*

The P+XS algorithm of CNES takes into account the modulation transfer function and the spectral filter of each band P and XS. It should be applied to images acquired at the same time. However, its mathematical expression is so simple that many people are using it even for non-coincident dates. In this algorithm, it is considered that the radiance of P is equal to the half-sum of the radiance of XS1 and XS2. Additionally, the ratio of the radiance of XS1 over XS2 should be equal to the ratio of the radiance of the synthesized images at 10 m called XP1 and XP2. This leads to the following equations:

$$L_{XP1} = 2L_P \frac{L_{XS1}}{L_{XS1} + L_{XS2}}$$

$$L_{XP2} = 2L_P \frac{L_{XS2}}{L_{XS1} + L_{XS2}}$$

where L_P, L_{XS1}, L_{XS2}, L_{XP1}, L_{XP2} are respectively the radiance of the P, XS1, XS2 original images and of the XP1 and XP2 synthesized images at 10 m.

The XP3 synthesized image is obtained by a simple duplication of the original pixels of XS3 image.

2.3. THE ARSIS CONCEPT

This method uses the wavelet transform and the multiresolution analysis to decompose the two images to be merge as in Figure 2.

A multiresolution analysis using the wavelet transform [11] is applied to image A_h and image B_{il}, describing image A_h and image B_{il} at different spatial resolutions and the differences of information between the successive approximations of image A_h and image B_{il}. The wavelet coefficients provided by the multiresolution analysis of the high spatial resolution image A_h, between the scale of image A_h and the scale of image B_{il}, describe the missing information for the synthesis of the image B_{il} at the same spatial resolution than the one of image A_h.

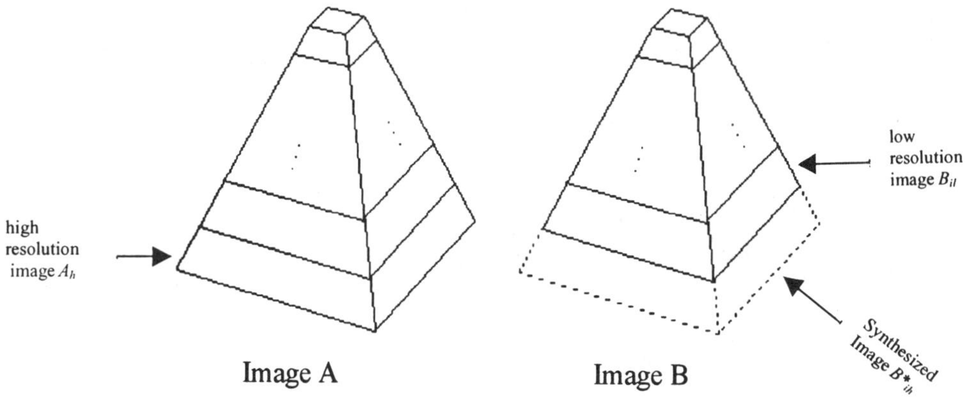

Figure 2. The use of the multiresolution analysis in the ARSIS method.

The simplest solution is to shift the wavelet coefficients from pyramid A_h to pyramid B_{il} and to use them to synthesize image B^*_{ih} at the spatial resolution of image A_h. If the wavelet coefficients provided by image A_h are used without modifications, the synthesized image B^*_{ih} will not be equivalent to "what would be seen by sensor B_{il} if it had the spatial resolution of sensor A_h". Hence, to improve the quality of the synthesized image, the model, to transform the wavelet coefficients provided by the multiresolution analysis of image A_h in the wavelet coefficients needed for the synthesis should take into account the physics of the environment.

Whatever this model is, by construction, the ARSIS method preserves the spectral content of original image. A multiresolution analysis applied to the synthesized image B^*_{ih} will allow the computation of an approximation similar to original image B_{il}.

Figure 3 presents the general scheme of the ARSIS method. First a multiresolution analysis using the wavelet transform is used to compute the wavelet coefficients and the approximations of image A_h (1). The same operation is applied to the image B_{il} (2). The wavelet coefficients provided by each decomposition are used to compute a model of transformation of the known wavelet coefficients of image A_h to the known wavelet coefficients of image B_{il}. This model takes into account the physics of both images and the correlation or anti-correlation existing between both wavelet coefficients images (3). The model can have various forms and take into account more than one scale. This model is then used to compute the missing wavelet coefficients (4). The inversion of the multiresolution analysis (WT^{-1}) allows the synthesis of the image B^*_{ih} with the spatial resolution of image A_h (5).

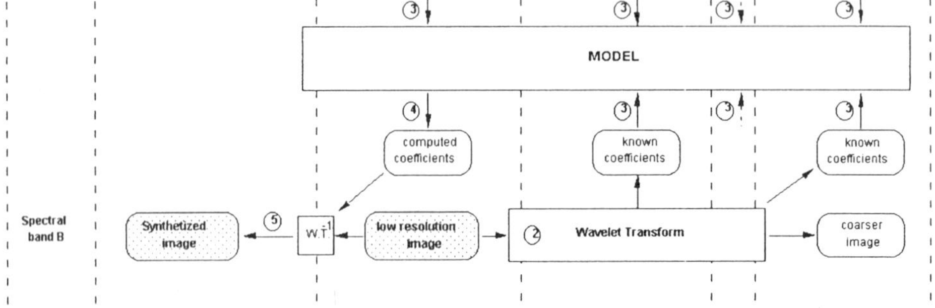

Figure 3. General scheme for the application of the ARSIS concept using wavelet transform (WT) and inverse wavelet transform (WT^{-1})

2.3.1 *The HPF method*

The High Pass Filtering (HPF) algorithm [10] was first applied to the fusion of Landsat TM and SPOT P images. It is based on the extraction of the high frequencies of the image with the highest spatial resolution through a Laplacian filtering, and on the addition of these high frequencies to the interpolated lower spatial resolution image. Chavez *et al.* [10] discussed the size of the filter and its effects on the resulting images. The filter can be derived from the computation of the second derivate of an apodisation function.

In the case of the fusion of SPOT P and XS images, the filter was limited to a 3x3 dimension. The value of the filter coefficient are:

$$h = \begin{pmatrix} 0 & -1 & 0 \\ -1 & 4 & -1 \\ 0 & -1 & 0 \end{pmatrix}$$

This filter is convoluted with the P image for the extraction of the high frequencies (corresponding mainly to the structures between 10 and 20 m). These high frequencies are added to the three XSi images interpolated at the resolution of 10 m.

2.3.2 *The XS-HR method*

In the case of the application to the P and the XS images, the ARSIS concept is implemented through the XS-HR method. Figure 4 presents the application of the ARSIS method to the case of the SPOT imagery.

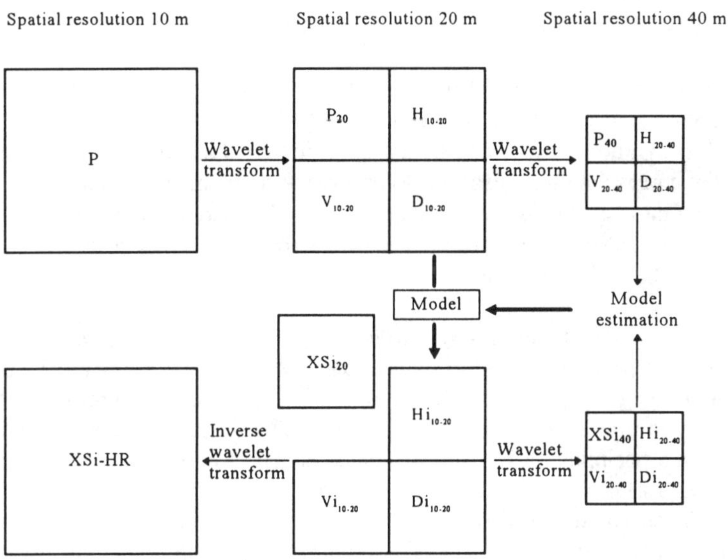

Figure 4. Application of the ARSIS method to the SPOT imagery

The set of images is composed in this case of a panchromatic image at the spatial resolution of 10 m and three multispectral images XS1, XS2, XS3 at the spatial resolution of 20 m. The aim of the XS-HR method is to compute the three XSi images at the spatial resolution of 10 m and to perform a high quality transformation of the original spectral content. Two iterations of the multiresolution analysis using the wavelet transform [11] are applied to the original panchromatic (P) image and one iteration to the original XSi image. A model of transformation, for each direction, from the panchromatic wavelet coefficient images to the XSi wavelet coefficient images is estimated at the spatial resolution of 40 m. This model can be of various types. The simplest one is the identity model. A convenient one can be an adjustment of the mean and the variance of the histogram of the wavelet coefficient images. The model must take into account the physics of both images and the correlation and anti-correlation between both images. Several models have been tested. The best results were achieved with a model taking into account the local variation between the P and the XSi wavelet coefficient images. The estimated model is inferred at the spatial resolution of 20 m. Then, it is applied to the transformation of the wavelet coefficient images representing the information between 10 and 20 m of the P image, into those corresponding to the XSi image. Finally, the multiresolution analysis is inverted and the XSi image at the spatial resolution of 10 m, called XSi-HR, is synthesized from the original XSi image and from the wavelet coefficient images computed through the model. The complete description of the implementation of the XS-HR method is provided in [9].

3. A Formal Approach for Evaluation of the Quality of resulting images

The merging methods under concern aim at constructing synthetic images B^*_h close to the reality. Wald *et al.* [12] established the properties of such synthetic images:

- Any synthetic image B^*_h once degraded to its original resolution l, should be as identical as possible to the original image B_l.
- Any synthetic image B^*_h should be as identical as possible to the image B_h that the corresponding sensor would observe with the highest spatial resolution h.
- The multispectral set of synthetic images B^*_h should be as identical as possible to the multispectral set of images B_h that the corresponding sensor would observe with the highest spatial resolution h.

Testing the first property: *Any synthetic image B^*_h, once degraded to its original resolution l, should be as identical as possible to the original image B_l.* To achieve this, the synthetic image B^*_h is spatially degraded to an approximate solution B'_l of B_l. If the first property is true, then B'_l is very close to B_l. The difference between both images is computed on a per-pixel basis. This difference image should be visually compared to the original image in order to detect trends of error, if any, possibly related to the type of landscape. Then some statistical quantities are to be used to quantitatively express the discrepancies between both images. These quantities are similar to the first and second sets of criteria described under the second property below.

There is an influence of the filtering operator upon the results, but it can be kept very small provided the operator is appropriate enough.

Testing the second property: *Any synthetic image B^*_h should be as identical as possible to the image B_h that the corresponding sensor would observe with the highest resolution h*. The second and third properties are difficult to test, because they refer to B_h, an image that would be sensed if the sensor had a better resolution. This image, of course, is not available; otherwise, all the above-cited methods would not have been developed. The difficulty is partly overcome by the following approach:

- The available images A_h and B_l are degraded to respectively A_l and B_s, respectively. For the SPOT case, the P and XS images are degraded to P_{20} (20 m resolution) and XS_{40} (40 m resolution). The images A_l and B_s are very close to what the corresponding sensor would have measured with a degraded resolution, as discussed previously.
- Then the synthesizing method under assessment is applied to A_l and B_s. It provides a synthetic image B^*_l (XS^*_{20} in the SPOT case).
- This synthetic image B^*_l is compared to the image-truth B_l (*XS* in the SPOT case) by means of some criteria described below. The numerical comparison should be made preferably in physical units and also in relative values. Thus, different tests made over different scenes may be compared.
- This comparison provides an assessment of the quality of B^*_l. It is assumed that this quality is fairly similar to that of the synthesized high-resolution image B^*_h.

Two sets of criteria are proposed to quantitatively summarize the performance of a method in synthesizing an image in one spectral band. The first set of criteria provides a global view of the discrepancies between the original image B_l and the synthetic one B^*_l. It contains:

- The bias, as well as its value relative to the mean value of the original image. Recall that the bias is the difference between the means of the original image and of the synthetic image. Ideally, the bias should be null.
- The difference in variances (variance of the original image minus variance of the synthetic image), as well as its value relative to the variance of the original image. This difference expresses the quantity of information added or lost during the enhancement of the spatial resolution. For a method providing too many innovations (in the sense of information theory), *i.e.*, "inventing" too much information, the difference will be negative because the variance of the synthetic image will be larger than the original variance. In the opposite case, the difference will be positive. In information theory, the entropy describes the quantity of information. However, we selected the variance difference because most researchers and engineers are much more familiar with variance, and entropy and variance act quite similarly for our purpose. Ideally, the variance difference should be null.

- The correlation coefficient between the original and synthetic images. It shows the similarity in small size structures between the original and synthetic images. It should be as close as possible to 1.
- The standard deviation of the difference image, as well as its value relative to the mean of the original image. It globally indicates the level of error at any pixel. Ideally, it should be null.

The error at pixel level may be more detailed. Let us compute at each pixel the absolute relative error (the absolute value of the difference between the original and synthetic values, divided by the original value). Then the histogram of these relative errors is computed. It can be seen as the probability density function. Therefore, we can compute the probability of having at a pixel a relative error (in absolute value) lower than a given threshold. This probability denotes the error made at pixel level, and hence indicates the capability of a method to synthesize the small size structures. The closer to 100 percent the probability for a given error threshold, the better the synthesis. The ideal value is a probability of 100 percent for a null relative error. Here, for reasons of computer precision, the lowest threshold "no relative error or null error" is set to 0.001 percent

Testing the third property: *The multispectral set of synthetic images B^*_h should be as identical as possible to the multispectral set of images B_h that the corresponding sensor would observe with the highest resolution h.* Visual inspection may be made through color composites of, for example, the first three principal components of the set of images. Both color composites should agree visually. Most methods for color composites are using dynamical adjustment for color coding (e.g., [13]). If the sets of images are different, even slightly, then the color coding will be different for both composites and no comparison will be possible. Practically, we recommend the following approach. For each spectral band, the B_l and B^*_l images are juxtaposed into a single computer file. The principal components analysis as well as the color coding are performed on this set of files. The projected B_l and B^*_l images are then extracted from these projected files and the color composites are displayed, simultaneously or alternatively, onto the screen. This approach guarantees that the color composites are comparable. Of course, if only three spectral bands are available as in the SPOT case, there is no need to perform a principal components analysis. The advantage of this visual assessment is that it does show trend in errors, if any, possibly related to landscape features. The drawback of it is that it is a subjective assessment and also that this assessment may be limited either by physiological factors (e.g., color contrast perception by humans), or by technical factors (e.g., when a large number of spectral bands are present). In the latter case, and if the landscape offers a large variety of objects, the color re-coding of the first three principal components reduces dramatically the differences between the B_l and B^*_l images, particularly if these differences are random, *i.e.*, not related to a peculiar landscape feature or to a spectral band.

A quantitative assessment can be made using the following three additional sets of criteria which quantify the performance of a method to synthesize the spectral signatures during the change in spatial resolution. The third set (numbered after the two sets described above for the second property) deals with the information correlation between the different spectral images taken two at a time. This dependence can be expressed by the correlation coefficients, with the ideal values being given by the set of original images B_l. As an example, for the case of SPOT, the correlation coefficient between P_{20} and $XS1^*_{20}$ is computed and compared to the correlation coefficient for P_{20} and $XS1_{20}$. This is done for every pair. The fourth set of criteria partly quantifies the synthesis of the actual multispectral n-tuplets by a method, where n-tuplet means the vector composed by each of the n spectral bands at a pixel. It comprises the number of different n-tuplets (*i.e.*, the number of spectra) observed in the original M_l and in the synthesized B^*_l sets of images, as well as the difference between these numbers. A positive difference means that the synthesized images do not present enough n-tuplets; a negative difference means too many spectral innovations.

The previous criteria do not guarantee that the synthesized n-tuplets are the same as in the original image B_l. The fifth and final set of criteria assesses the performance in synthesizing the actual n-tuplets. It deals with the most frequent n-tuplets, because they are predominant in multispectral classification. For a given threshold in frequency, only the n-tuplets having a frequency (relative number of pixels) greater than this threshold are used. The threshold is set to 0.01 percent, 0.05 percent, 0.1 percent, and 0.5 percent, successively. The greater the threshold, the lower the number of n-tuplets, but the greater the number of pixels exhibiting one of these n-tuplets. For each of the n-tuplets, the difference is computed between the original frequency and the one observed in the synthesized images. These differences are summarized by the following quantities:

- the number of actual n-tuplets, the number of coincident n-tuplets in the synthesized images, and the difference between these numbers, expressed in absolute and relative terms,
- the number of pixels in these n-tuplets, in absolute and relative terms, and
- the difference between the above number of pixels for the original and synthesized images, in absolute and relative terms.

Munechika *et al.* [14] partly quantify the performances in synthesizing the multispectral information by first computing the root mean square (RMS) of the differences, pixel per pixel, between the synthesized B^*_l and original B_l images, for each spectral band, and then by summing up these spectral RMS values to obtain a global error, which should be as low as possible. This global error can easily be computed from our first set of parameters, i.e., the bias and the standard deviation. Other criteria may be further defined.

4. Case studies: the fusion of SPOT Panchromatic and XS

In this section two case studies are presented. The first one enhances the influence of the time acquisition of the images to merge. The second one enhances the benefits of the evaluation for users, allowing a better knowledge of the different algorithms for improving the spatial resolution of images.

4.1. THE THREE GORGES DAM, CHINA: INFLUENCE OF THE TIME-GAP ON THE SYNTHESIZED IMAGES

The Three Gorges Dam is the largest water conservancy project ever built in China, and so in the world. With normal pool level (NPL) at 175 m, the total storage capacity of the reservoir is 39.3 billion m^3. The preparation of the Three Gorges project started in 1993. On Dec. 14, 1994, China Premier Li Peng declared the formal beginning of construction. Over the past three years, significant changes have taken place on both sides of the Yangtze river at Sandouping area owing to the effort of Three Gorges builders.

Due to the very large area to cover, remote sensing was selected as a mean for studying the upstream geological impact of this project, and detecting and surveying geological hazards. In order to detect geological problems such as faults, landslides, landflows, the highest spatial resolution and the most recent information are simultaneously required.

The set of data available is composed of a SPOT panchromatic (P) images from 1997 and a multispectral SPOT XS image from 1998.

In this section, the problem of the fusion of panchromatic and XS images acquired at different dates, and the influence of landscape changes on the results of the algorithms are addressed. Because of the very large changes due to the construction of the dam, this area is well-suited to this study.

Five algorithms are considered: the IHS algorithm, the Brovey and the P+XS algorithm, the High-Pass filtering (HPF) algorithm, and the XS-HR algorithm model RWM within the ARSIS concept.

Figure 5 presents the original P image from 1997. Figure 6a presents the original XS1 image from 1998. Tremendous modifications occur on the Dam site. The most visible one is the evolution of the island in the middle of the Yangtze river. The shiplock is close to completion in Figure 6a.

Figure 5. Extract of a SPOT panchromatic image from the Three Gorges site acquired in 1997. Copyright CNES SPOT Image.

Figure 6 presents the original XS1 image and the results of the different algorithms of fusion of the P and XS1 image. The $XS1^*_{10}$ images were synthesized by the IHS (Figure 6b), the Brovey (Figure 6c), the P+XS (Figure 6d), the HPF (Figure 6e) and the XS-HR (Figure 6f) methods. Recalls that the aim is to perform a synthesis of images with a high quality transformation of the missing information.

(a)

(b)

(c)

(d)

(e)

(f)

Figure 6. Original XS1 image of the Three Gorges Dam, China, acquired in November 10, 1998. Copyright CNES SPOT Image (a). XS1$^{*}_{10}$ image synthesized with the IHS (b), Brovey (c), P+XS (d), HPF (e) and XS-HR (f) algorithms. A more complete evaluation of the influence of landscape changes on some of these algorithms was achieved from a geological point of view [15].

The visual inspection over the site of the Dam of the resulting images demonstrated the failure of the IHS, Brovey, P+XS and HPF methods when the time acquisition of the images are different. On the contrary, the XS-HR method is able to fulfil the objectives of the fusion.

In the case of the IHS, Brovey and P+XS algorithms, the predominance of the structures coming from the P image is due to the inherent construction of the synthesized images. All the structures at all scales that are in the P image are introduced in the synthesized images. In the case of the P+XS algorithm, the fusion of non contemporary images is not recommended by CNES and SPOT Image. The later even refused to apply this algorithm when the two dates of acquisition are different. Figure 6c is a very good illustration of the problem occurring in this case. As the Brovey algorithm is very close to the P+XS algorithm, the resulting image presents the same characteristics due to the same behavior.

In the case of the image obtain from the HPF algorithm, it is obvious that Figures 6e presents an artificial aspect. The information added to the XS1 image is extracted from the P images by application of a Laplacian filter. Only the high frequencies comprised between both resolutions are extracted from the P images and added without any transformation to the XS1 image. The modification of the landscape from 1997 to 1998 was important in this area and the introduction of a non-contemporary information gives the impression of a superimposition of two images.

Figures 6f looks very close to Figure 6a. The XS-HR method derived from the ARSIS concept is more stable when applied to non-contemporary images. The synthesized image is still exploitable for further processing such as classification processes or interpretation of geological features and faults.

4.2. BARCELONA, SPAIN: EVALUATION OF THE QUALITY

The protocol of Wald *et al.* [12], presented section 3 is followed to assess the quality of the results of the different methods. The methods are applied to the same SPOT image of the city of Barcelona, Spain, as in their paper (Figure 7). Such an urban area has been selected for illustration because it is certainly the most difficult type of landscape to deal with according to our knowledge. Urban areas often point out the qualities and drawbacks of algorithms because of the high variability of information in space and spectral band, induced by the diversity of features both in size and nature.

In the SPOT case, the multispectral images B_{il} are the XS1, XS2 and XS3 bands at original resolution of 20 m, and the high spatial resolution image A_h is the panchromatic band P with a spatial resolution of 10 m. The synthetic bands B^*_{ih} are the XS1, XS2 and XS3 bands synthesized at 10 m. Table 1 gives the means, standard-deviations and calibration coefficient of the original images.

To assess the first property, the synthetic image B^*_{ih} made at 10 m are filtered before resampling to degrade the resolution down to 20 m: $(B^*_{ih})_l$. They are then compared to the original images B_{ih}. The filtering function is a sine cardinal (sinc) kernel truncated by a Hanning apodisation function of size 13x13.

Figure 7. SPOT P image of Barcelona used for test. The dark area in the upper left corner is the slope of a hill with Mediterranean vegetation. The size is 512x512 pixels. Copyright CNES-SPOT Image (1990).

TABLE 1. Means, standard deviations, and calibration coefficients of original images (in W m^{-2} st^{-1} μm^{-1})

	XS1	XS2	XS3	P
Mean	58	48	55	53
Standard deviation	12	15	9	15
Calibration coefficient	1.2181	1.22545	1.29753	1.39198

To test the second and third properties, the P and XS images are degraded to a resolution of 20 (A_{2h}) and 40 m ($(B_i)_{2l}$), respectively. Then, images B^*_{il} are synthesized at a 20-m resolution and compared to the original XS images B_{il} by a visual inspection on the one hand, and by performing a difference pixel per pixel. The discrepancies are analyzed and synthesized in five sets of criteria, which deal respectively with:

- each spectral band in a global way,
- the statistical distribution of errors at pixel level for each spectral band,
- information correlation between the different spectral images,
- the multispectral aspect, that is the errors in reconstructing spectral signatures,
- the reconstruction of the most frequent spectral signatures.

Wald *et al.* discussed the extrapolation of the quality assessments made at 20 m to 10 m. They underlined the unpredictability of such assessments when changing the resolution. That is, it cannot be said whether the error at 10 m is larger or lower than that at 20 m. By testing several methods on SPOT images degraded to 40 and 80 m, they found in several cases that the quality was best at 20 m than at 40 m. They suggested that one can assume that the quality of the synthetic images at 10 m may be considered as similar to that of the synthetic images at 20 m.

A visual inspection of the synthesized images B^*_{ih} is performed first, once the contrast table adjusted for each. They are visually fairly close one to the others for all methods and of satisfactory quality, except for the IHS method which produces in that case an image of bad quality in the XS3 band. The HPF images contain too much high frequencies: the contours are enforced in an excessive manner.

First property

Like the B^*_{ih} images, the contrast-adjusted different $(B^*_{ih})_l$ images are visually fairly close and of satisfactory quality, except for the IHS method in the XS3 band, the HPF images. Adjusting the contrast table for each $(B^*_{ih})_l$ accommodates for linear changes in statistical distribution, and especially mean and variance. For the Brovey and IHS methods, these parameters are strongly modified relative to the original B_{il} images.

The details of the quantitative comparison (not shown here) further demonstrate that the first property is clearly not satisfied by the Brovey and IHS methods, and also the P+XS method. In these four methods, the synthesis of the band B^*_{ih} at 10 m is influenced by the high resolution image A_h and the other spectral bands B_{jl}. This influence is irrespective of the size of the structures, that is that the large structures (*i.e.* larger than 20 m) observed in these images A_h and B_{jl} are partly included in the synthesized image B_{ih}. A mathematical analysis of these methods clearly shows that the influence of A_h and the other spectral bands B_{jl} in the synthesized image B^*_{ih} does not disappear when reducing the resolution to 20 m.

The method built within the ARSIS concept using the wavelet transform is inherently built to satisfy this first property, with reservations regarding the degradation process as discussed by Wald *et al.* [12]. On the contrary, the HPF technique does not satisfy this property, mostly due to a strong change in variance. This is confirmed by a visual inspection.

Second property

The conclusions of the visual analysis of the different B^*_{il} images are in accordance with that obtained for the first property.

Compared to the first property, we found that the drawbacks of a method are enhanced by the testing oft he second property. This is why we put an emphasis on this comparison. Tables 2 to 4 provide some statistics on the differences between the original B_{il} images XS1, XS2 and XS3 at 20 m and the synthesized B^*_{il} images. They provide a global view of the quality of a method to synthesize each individual spectral band B_i.

These tables show a very strong bias (difference between mean values) for the Brovey transform for the three bands. This bias amounts to approximately 0.65 times the original mean value. This is due to the very construction of this transform, which, briefly, written, is equal to the spectral band under concern, multiplied by the ratio of three times the panchromatic band P and the sum of the three bands. Since the method does not request the computation to be made in radiances, a difference in mean between spectral bands - as here between XS2 and the others (Table 1) - may induce a strong bias for all

TABLE 2. Some statistics on the differences between the original and synthesized images, in radiance (W m^{-2} st^{-1} μm^{-1}) or relative value, for XS1 band.

	IHS	Brovey	P+XS	HPF	XS-HR RWM
Bias (ideal value: 0)	-4.58	30.60	0.26	0.26	0.00
relative to the mean XS value	-10 %	64 %	1 %	0 %	0 %
Actual variance-estimate	31	172	-42	-597	7
relative to the actual variance	14 %	77 %	-19 %	-267 %	3 %
Correlation coefficient between XS and estimate (ideal value: 1)	0.96	0.98	0.98	0.71	0.99
Standard-deviation of the differences (ideal value: 0)	4.0	8.1	3.1	20.7	1.9
relative to the mean of XS value	8 %	17 %	6 %	43 %	4 %

TABLE 3. As Table 2, but for XS2 band.

	IHS	Brovey	P+XS	HPF	XS-HR RWM
Bias (ideal value: 0)	-5.49	35.50	0.00	0.22	0.00
relative to the mean XS value	- 10 %	65 %	0 %	0 %	0 %
Actual variance-estimate (ideal value: 0)	46	67	9	-47	8
relative to the actual variance	55 %	81 %	11 %	-57 %	9 %
Correlation coefficient between XS and estimate (ideal value: 1)	0.78	0.69	0.91	0.48	0.95
Standard-deviation of the differences (ideal value: 0)	5.8	7.0	3.8	20.8	2.7
relative to the mean of XS value	11 %	13 %	7 %	38 %	5 %

TABLE 4. As Table 2, but for XS3 band.
For XS 3, the method "P+XS" reduces to duplication

	IH S	Br ov ey	P+ XS	HP F	XSR - W H M
Bias (ideal value: 0)	-5.85	36.90	0.35	0.35	0.00
relative to the mean XS value	-10 %	64 %	1 %	1 %	0 %
Actual variance-estimate(ideal value: 0)	32	100	-50	-587	7
relative to the actual variance	22 %	70 %	-35 %	-420%	5 %
Correlation coefficient between XS and estimate (ideal value: 1)	0.92	0.97	0.97	0.66	0.99
Standard-deviation of the differences (ideal value: 0)	4.8	5.9	3.8	21.1	1.9
relative to the mean of XS value	8 %	10 %	7 %	36 %	3 %

synthesized bands. This construction also implies that the variance of a synthesized band B^*_{il} is a combination of the variances of all bands, including the panchromatic. It follows that the variance of the B^*_{il} image strongly differs from that of the original image B_{il}. This method adds too much variance by a relative amount exceeding 70 % of the original variance. The correlation between the B_{il} and B^*_{il} images is high as far as the correlation between the B_{il} and A_h images is high. The correlation between XS3 and P is only 0.35 instead of 0.97 for the two other bands, and the correlation between XS3 and $XS3^*_{20}$ is only 0.7, which is rather poor. Finally the relative error at pixel level in reconstructing the original image ranges from 10 to 17 % (standard-deviation).

As a whole, the other methods perform better, though only a few provide satisfactory results. The IHS method exhibits a relative negative bias of 10 %, which is still too large and means an overestimation of the values as a whole. This bias may be partly overcome by an *a priori* equalization of the dynamics of the images B_{il} and A_l. This would also reduce the differences in variance, and more generally would provide better results if the correlation between the images B_{il} and A_l is large. This equalization step is made at the expenses of the physical significance of the images. This remark also holds for the HPF methods.

The IHS method does not introduce enough high frequency signal in the synthesized image (the variance is too low), for which the variance is too large, except for the XS3 band. It should be noted for this substitution methods that the results are strongly dependent upon the original images. According to the mutual correlation between bands and the variance in each band, the introduction of high frequencies will be either too large or too low, and sometimes satisfactory. This is true for the other parameters under examination for this second property.

The P+XS method is unbiased but introduces too much signal from the P band into the XS*1 and 2. This method reduces to duplication for the XS3 band. Accordingly the variance of the XS*3 image is too low: there is no fusion and no addition of signal from another source.

The HPF method is rather disappointing. The amount of excessive variance is huge. All the contours are enforced but excessively. The correlation between the synthesized and original images is low for all bands. Finally the error at pixel level is extremely high.

The best results are attained by the XS-HR method using the wavelet transform. The method offers the same level of quality for the bands XS1 and XS2.

Finally it should be noted that the results are better for the bands XS1 and XS2 than for the band XS3. This is due to the fact that the band P encompasses the bands XS1 and XS2 and not the XS3.

Third property

Color composites have been created for each synthesized set of multispectral B^*_{il} images. The color coding for each set follows that used for the original set, in order to make them comparable and following the recommendations of Wald *et al.* [12]. This color composite are presented in Figure 8. Unsurprisingly, the color composite obtained by the Brovey transform does not show similarity with the original one. This also holds for the IHS method but to a lesser extent. The other methods provide color composites closer to the original with various degrees in quality, which are better described through some quantitative parameters.

Table 5 shows the performances of each method in synthesizing the multispectral information. It represents the difference between the actual number of triplets and the number found in the synthesized images for each method. These triplets may be different from the original ones; only their number is taken into account in this table. The number of original triplets is approximately 45,600 and is large compared to the number of pixels. This demonstrates the spectral diversity of urban areas.

The Brovey transform only found approximately 8,000 triplets! It means that this transform flattens out the spectral diversity of a scene. The HPF method does not perform correctly for this parameter; it provides about twice more triplets. This is due to the enforcement of structures already noted. As expected, the duplication exhibits less triplets than the original. The high discrepancy (49 %) demonstrates the changes in the statistical distribution of spectral signatures when changing the spatial resolution. The other methods perform from fairly correctly (P+XS) to very satisfactory (XS-HR).

Actually, this table 5 partly summarizes the multispectral performances of each method. Most of the triplets have a low frequency, i.e. most of them are carried by a very few number of pixels. The average number of pixels per triplet is 5.7. Many of the triplets are superfluous; they are carried by 1 or 2 pixels and are not taken into account in further classification processes, or visual analysis of the synthesized images as colored composition.

Table 6 shows the performances of each method in synthesizing the most frequent actual triplets. Each triplet under consideration has a frequency of at least 0.01 percent, which corresponds to 26 pixels in this case. The total of pixels they represent amounts to 23 percent of the total number of pixels in the image. Hence synthesizing them accurately is of primary importance in classification purposes. In this Table, for each of these

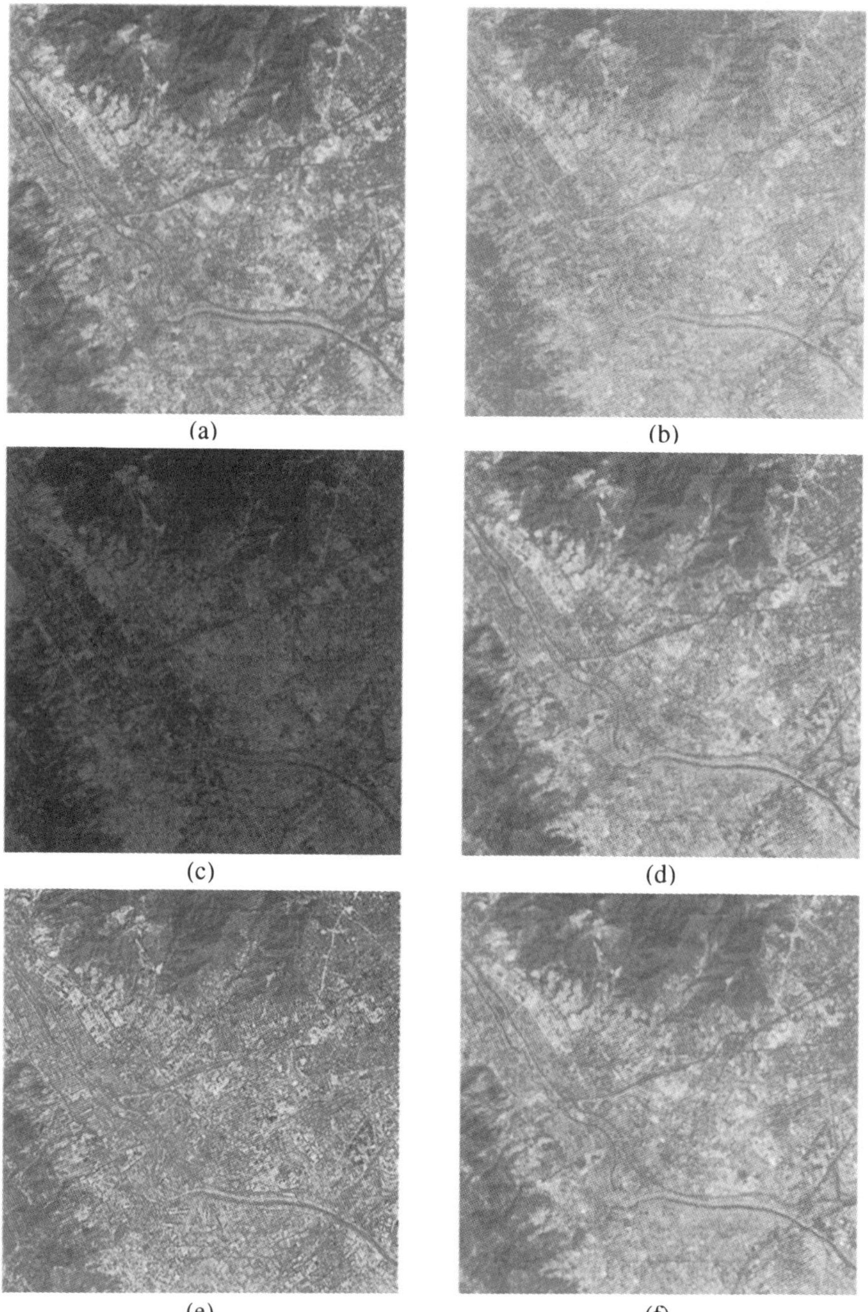

Figure 8. Color composite of Barcelona, Spain, from the original data set (a); from images synthesized by IHS (b), Brovey (c), P+XS (d), HPF (e) and XS-HR (f) methods.

triplets, the number of pixels carrying this triplet in the synthesized images is compared to the corresponding number in the original images. The differences are summed up for all the triplets, giving the "difference with original" in Table 6. A difference equal to 0 means that all the predominant triplets are exactly the same than in original images. Because of its bias in all bands, the Brovey transform is unable to retrieve any of these 1,675 triplets. Very bad results are also obtained by the IHS method: it retrieves only 721 triplets (43 %) and only 12 % of the corresponding pixels. It means that it does not synthesize correctly the triplets and even for those it retrieves, they are not correctly allotted to the pixels: this would induce errors in cartography after classification. This bad result is not contrary to Table 5, for which synthesized triplets and actual triplets may differ: only the numbers of triplets are compared.

TABLE 5. Performances in synthesizing the multispectral information. Difference between the actual frequency of a triplet (XS1, XS2, and XS3) and the estimates. Average number of pixels per triplet is 5.7.

	Original	IHS	Brovey	P+XS	HPF	XS-HR RWM
Number of triplets	**45 618**	41 052	7 965	53 162	86 700	43 918
Difference with original	-	4 566	37 653	-7 544	-41 082	1 700
(in %)		10 %	83 %	-17 %	-90 %	4 %

TABLE 6. As Table 5 but only the most frequent triplets are taken into account. Each triplet has a frequency of at least 26 pixels (0.01 percent of the total number). The total of pixels they represent amounts to 23 percent of the total number of pixels in the image.

	Original	IHS	Brovey	P+XS	HPF	XS-HR RWM
Number of predominant triplets	**1 675**	721	0	1 671	1 675	1 675
Difference with original (ideal: 0)	-	954	1675	4	0	0
(in %)		57 %	100 %	0 %	0 %	0 %
Number of pixels	**60 372**	6 961	0	35 864	28 849	60 195
Difference with original (ideal: 0)	-	53 411	60 372	24 508	31 523	177
(in %)		88 %	100 %	41 %	52 %	0 %

The other methods provide more satisfactory results. This means that each of these methods are capable of synthesizing the predominant spectral signatures. If a classification is made by using only the spectrum, these methods will provide the same number of classes. However, for most of the methods, the size of the classes will differ from original. The HPF method only retrieves 52 % (i.e. the half) of the total number of pixels belonging to these predominant classes. It follows that the resulting map may be fairly inaccurate, except if, by chance, class aggregation processes overcome this drawback. The P+XS method is also inaccurate: 41 % of ixels are missed. The XS-HR method achieves excellent results: all the triplets are exactly retrieved and the number of retrieved pixels carrying one of these triplets in both the original and synthesized images is almost exactly the same. This ensures on the one hand a good classification, and on the other hand a good accuracy in mapping from this classification.

We recall that we have dealt with the multispectral aspects of the synthesis in this chapter. In accordance with the protocol of Wald *et al.* [12], other aspects may have been considered, such as spatial gradients, forms and structures, in each spectral band and in the multispectral set. Such aspects and the corresponding criteria are of high importance in several applications such as the automatic recognition of the network of the streets [16]. They have not been considered here.

From the previous discussion, we have ranked the methods, which are now briefly discussed from the worst to the best. The conclusions drawn from this example have been validated for several other cases by the authors and are supported by other authors [10] [17,14,18,9,19,12].

Brovey transform. The Brovey transform is not relevant at all, mostly because there is a strong bias error due to its very construction. Though it can be partly corrected, it will never reproduce the spectral content in an accurate way, except in rare cases.

HPF method. The authors are actually disappointed by the HPF method. As a possible implementation of the ARSIS concept [9], they were expecting much better results. As expected from this concept, the bias is close to 0. However too much variance is introduced in the synthesis and this leads to an excessive enforcement of contours as well as to a low correlation coefficient with original. The quality of the synthesis of the predominant triplets is bad: though all these triplets are retrieved, about half of the pixels carrying these triplets are missing.

IHS method. The IHS method often produces nice-looking results but not always as in this case. The results are of poor quality: the bias is high, not enough details are injected (likely due to not taking into account calibration coefficients), and correlation coefficient is low, especially in band XS 3. Furthermore it strongly distorts the spectral content of the synthesized images.

P+XS method. The P+XS method, from CNES, is limited to the SPOT case. Belonging to the relative spectral contribution group of methods, it performs better than the projection-substitution-projection techniques. Results are not satisfactory. Of course, it performs like duplication for the XS 3 band, i.e. rather poorly. For the other bands (XS 1, XS 2) encompassed by the panchromatic band P, the results are not good. It introduces too much high frequency signal in the synthesized images. Finally the frequencies of occurrence of the predominant triplets are badly synthesized (42 % missing in total).

However the effective visual enhancement performed by the P+XS method may be recognized.

XS-HR method. XS-HR methods performs the best. It achieves very good quality products. The quality of the synthesis of the predominant triplets is impressive. Another striking feature compared to the other methods is that they are capable of achieving good results for the XS 3 band. All published comparisons with existing methods show that the ARSIS concept, combined with the wavelet transform and the multiresolution analysis leads to the best presently achievable results.

Our final conclusion is that only XS-HR method achieves satisfactory results and that further investigations are needed to improve this method or design new ones that perform better on the one hand, and to verify that that method can enter a production system delivering fused products with a controlled quality on the other hand.

5. Conclusion

This chapter has explored the application of data fusion to the Earth Observation domain. A recall of the definition widely approved by researchers was proposed, enhancing the need for a formal approach of data fusion. The benefits of such an approach is to focus on a framework and on the quality that should be reached according to the application. This has been confirmed by the example chosen: the improvement of the spatial resolution of multispectral images with the aim of being as close as possible to an image acquired at this spatial resolution.

The three groups of methods proposed in this aim were described and analyzed. The representative methods of each group have been applied to two sets of SPOT P and XS images and were quantitatively evaluated. The best results were achieved by the XS-HR method based on a conceptual approach (the ARSIS concept). The analysis of the influence of the time gap between the acquisition of the P and XS images reinforces the results of the quantitative analysis performed on the synthesized images and based on a formal approach.

Of course the IHS, Brovey, HPF methods and those derived from the ARSIS concept were applied to several other cases then the SPOT one (see *e.g.* [9]). But in this set of methods, the most robust ones are the algorithms based on ARSIS. New cases studies should confirmed this conclusion.

Finally, the users should know the benefits and drawbacks of each method to obtain the best results according to the aim of their application.

6. References

1. Wald, L. (2000) The presents achievements of the EARSeL – SIG 'Data Fusion'. To appear *In* Proceedings of 20th Symposium of EARSeL, A decade of trans-european in remote sensing cooperation. Dresden, Germany, June 13-16 2000.
2. Ranchin, T. and Wald, L. editors (1996) Proceedings of the International Conference " Fusion of Earth Data: merging point measurements, raster maps and remotely sensed images ", organised by the Société des Electriciens et Electroniciens de France, European Association of Remote Sensing Laboratories, Ecole des Mines de Paris, held in Cannes, France, 6-8 February 1996. Published by SEE/URISCA, Nice, France, February 1996.

3. Ranchin, T. and Wald, L. editors (1998) Proceedings of the 2nd International Conference " Fusion of Earth Data: merging point measurements, raster maps and remotely sensed images ", held in Sophia Antipolis, France, 28-30 January 1998, organised by the European Association of Remote Sensing Laboratories (EARSeL), Société des Electriciens et Electroniciens de France (SEE), and Ecole des Mines de Paris. Published by SEE/URISCA, Nice, France, 222 p.
4. Ranchin, T. and Wald, L. editors (2000) Proceedings of the 3rd conference "Fusion of Earth data: merging point measurements, raster maps and remotely sensed images", Sophia Antipolis, France, January 26-28, 2000, organised by the European Association of Remote Sensing Laboratories (EARSeL), Société des Electriciens et Electroniciens de France (SEE), and Ecole des Mines de Paris. Published by SEE/URISCA, Nice, France, 168 p.
5. Wald L. (1999) Some terms of reference in data fusion. *IEEE Transactions on Geosciences and Remote Sensing*, 37, 3, 1190-1193.
6. Carper, W. J., T. M. Lillesand, and R. W. Kiefer (1990) The use of Intensity Hue Saturation transformations for merging SPOT panchromatic and multispectral image data. *Photogrammetric Engineering & Remote Sensing*, 56(4):459-467.
7. Pohl, C. and van Genderen, J. L. (1998) Multisensor image fusion in remote sensing: concepts, methods and applications. *International Journal of Remote Sensing*, 19(5): 823-854.
8. Anonymous (1986) *Guide des utilisateurs de données SPOT*, 3 tomes, Editeurs CNES et SPOT-Image, Toulouse, France.
9. Ranchin, T. and Wald, L. (2000) Fusion of high spatial and spectral resolution images: the ARSIS concept and its implementation. *Photogrammetric Engineering & Remote Sensing*, 66, 1, pp. 49-61.
10. Chavez, P. S. Jr., S. C. Sides, and J. A. Anderson (1991) Comparison of three different methods to merge multiresolution and multispectral data: Landsat TM and SPOT panchromatic. *Photogrammetric Engineering & Remote Sensing*, 57(3):265-303.
11. Mallat, S. G. (1989) A theory for multiresolution signal decomposition: the wavelet representation. *IEEE Transactions on Pattern Analysis and Machine Intelligence*, 11(7):674-693.
12. Wald, L., T. Ranchin, and M. Mangolini (1997) Fusion of satellite images of different spatial resolutions: assessing the quality of resulting images. *Photogrammetric Engineering & Remote Sensing*, 63(6): 691-699.
13. Albuisson M. (1993) Codage trichrome et classification. In *Outils micro-informatiques et télédétection de l'évolution des milieux : troisièmes journées scientifiques du réseau de télédétection de l'UREF*, pp. 167-173. Presses de l'Université du Québec, Sainte-Foy, Québec, Canada, 444 p.
14. Munechika C. K., J. S. Warnick, C. Salvaggio, J. R. Schott (1993) Resolution enhancement of multispectral image data to improve classification accuracy. *Photogrammetric Engineering & Remote Sensing*, **59**(1), 67-72.
15. Yang, W., F. Cauneau, J.-P. Paris, and T. Ranchin (2000) Influence of landscape changes on the results of the fusion of P and XS images by different methods. *In* Proceedings of the third conference "Fusion of Earth data: merging point measurements, raster maps and remotely sensed images", Sophia Antipolis, France, January 26-28, 2000, Thierry Ranchin and Lucien Wald Editors, published by SEE/URISCA, Nice, France, pp. 47-56.
16. Couloigner, I., T. Ranchin, and L. Wald (1998) Benefit of data fusion to urban roads mapping. *In* Proceedings of the second conference "Fusion of Earth data: merging point measurements, raster maps and remotely sensed images", Sophia Antipolis, France, January 28-30, 1998, T. Ranchin and L. Wald Editors, published by SEE/URISCA, Nice, France, pp. 183-190.
17. Mangolini, M., T. Ranchin, and L. Wald (1993) Fusion d'images SPOT multispectrales (XS) et panchromatique (P), et d'images radar. In De l'optique au radar, les applications de SPOT et ERS, pp. 199-209. Cépaduès-Editions, 111 rue Vauquelin, Toulouse, France, 574 p.
18. Terretaz, P. (1997) Comparison of different methods to merge SPOT P and XS data: Evaluation in an urban area. *In* Proceedings of 17th Symposium of EARSeL, Future trends in remote sensing, Lyngby, Denmark, 17-20 June 1997, P. Gudmansen editor, A. A. Balkema, Rotterdam, Brookfield, pp. 435-445.
19. Raptis, V. S., R. A. Vaughan, T. Ranchin, and L Wald (1998) Assessment of different data fusion methods for the classification of an urban environment. *In* Proceedings of the second conference "Fusion of Earth data: merging point measurements, raster maps and remotely sensed images", Sophia Antipolis, France, T. Ranchin and L. Wald Editors, published by

INTRODUCTION TO DATA FUSION AND APPLICATIONS IN ASTRODYNAMICS*

D. FINKLEMAN, DR.
J. SANTIAGO, DR.
S. ALFANO, DR.
D. VALLADO
United States Space Command
Colorado Springs, Colorado, USA

1. Background

This chapter reviews fundamental precepts of multisensor data fusion and applies them to predicting the trajectories of Earth orbiting satellites.

We will first review the multisensor data fusion formalism, illuminate mathematical and physical issues, and demonstrate how the fused product of imprecise measurements can be more precise than its components. We then describe what we consider the four fundamental elements of our application of data fusion.

The internet Data Fusion Server [1] defines data fusion as a formal framework for alliance of data from different sources. The "sources" are objects which can be described uniquely by a sufficiently diverse set of characteristics, called "attributes." The complete set of attributes for a specific objects defines a "feature vector" in a space of appropriate dimensionality. These attributes are perceived by sensors. Often not all sensors can perceive all of the attributes. We must combine the sensor information in a rational and, hopefully, optimal manner to develop a perceived feature vector. Finally we can identify the object by finding the closest match (in a rigorous mathematical sense) with known objects.

Data fusion applications are distinguished by the nature of their objects and the dimensionality of their attributes. Applications range from identifying entities in an imaged scene, to determining the morphology of terrain, to estimating the future positions of heavenly bodies and satellites.

Multisensor data fusion was perhaps first invoked in astronomy. Hipparchus (c. 161-126 B.C.) developed and tested hypotheses of orbital motion based on a multitude of amazingly precise observations [2]. We accept the standard terminology: "Astrodynamics is the study of the motion of man-made objects in space, subject to both natural and artificially induced forces."

A single sensor is less desirable than many sensors for several reasons [3]. Several sensors can cover a larger volume of space or a greater period of time. Several independent, well calibrated sensors also diminish measurement ambiguity, as we will demonstrate shortly. However, when several sensors are available we face the dilemma of how best to collaborate. Opportunistic systems require that each contributor continuously monitor the evolving situation and contribute information unilaterally when

A. K. Hyder et al. (eds.), Multisensor Fusion, 657–671.

they independently determine that they can add value. For example, if a horse race is too close for the human judge to discern a winner, temporally precise photographic data is invoked. The measurements take place at different times (asynchronous); however they contribute to a collaborative, single solution. Communicative techniques are the synthesis of partial solutions, each attributed to a single sensor. For example, tracking radars might each observe a portion of a trajectory, which is described completely only by the sum of all of the pieces. Cooperative techniques develop a solution by consensus among several sensors observing the circumstance simultaneously. Often we do not have a choice. Astrodynamics employs opportunistic techniques most of the time. Unfortunately, sensors are sufficiently sparse that cooperative techniques seldom apply.

There are four fundamental considerations in any data fusion problem.

(a) Measurement. We must carefully determine measurement accuracy and precision. Accuracy is how well a measured value agrees with a reference value [4]. The accuracy of a clock is determined by how well it keeps time compared with an atomic clock maintained at a national laboratory. Precision denotes a measurement quality that describes how well repeated measurements agree with each other, rather than with a reference. It is the scatter or dispersion of the measurements. We quantify precision with standard deviations, variances, or confidence intervals. The difference between the mean of repeated measurements and a reference value is called "bias" or systematic error. Accuracy usually is a measure of bias. If bias is negligible or removed from measurements, accuracy and precision are synonymous.
(b) Hypothesis Development: Scientific principle focuses on determining the validity of hypotheses. We must have a quantitative expectation of the outcome in order to explain differences between measured and expected values. We need to know what we're looking for in order to find it. Hypotheses in astrodynamics are the mathematical statements of the laws of physics.
(c) Processing: Processing converts measurements into fused perceptions of the phenomena of interest. Data must be relevant to the hypothesis. Data from diverse sensors must be combined in a consistent and appropriate manner. For example, measurements so precise that they distinguish among external artifacts of a satellite are not relevant to determining the satellite's orbit.
(d) Topology: The density and frequency of observations must be consistent with the granularity of hypotheses and the numerical discretization of the problem. Sensors must be able to exchange data with each other or a central processing element sufficiently accurately and quickly.

2. Fundamentals of Astrodynamics

Having established the fundamentals of data fusion, we will now apply those concepts to astrodynamics. We will review the governing equations and the extent to which physical phenomena are understood and described.

We distinguish between prediction and estimation. In our discipline, prediction is deterministic. We can predict the future kinematic state of a satellite only if we have a complete and precise mathematical description of the orbit. Since many influences are not precisely known and cannot be separated individually from aggregated measurements, we can only estimate the future kinematic state of a satellite. That estimate grows increasingly imprecise as these influences act upon the satellite in the fullness of time since the last observation.

It is extremely important that we not fall prey to hueristic correction factors. For example, if a satellite is continually observed outside of the probable growth in the estimated orbit propagated into the future, it is inappropriate to increase the estimated uncertainty by a constant multiplicative factor that captures the observed positions. As scientists, we must infer physically consistent effects that might not have been included.

As in most Bayesian tests, we can only determine whether our hypotheses lead to estimates consistent with observations. We cannot determine whether these are the correct hypotheses. The greater the number of elements of the feature vector (attributes), the less likely that a physically incorrect hypothesis will match observations within measurement error. For example, the ancient Chinese believed that solar eclipses were caused by a dragon passing in front of the Sun. They hypothesized that creating a great din would scare the dragon away, revealing the Sun. It worked! The eclipse was transient. When we added the Moon's motion to the attributes of an eclipse, the dragon hypothesis evaporated. Most scientific disciplines were plagued with heuristics in the past. We must avoid them now.

Newton's Laws

$$\ddot{\vec{r}} = -\frac{\mu}{r^3}\vec{r} + \vec{a}_{perturbed}$$

$$\vec{a}_{perturbed} = \vec{a}_{nonspherical} + \vec{a}_{drag} + \vec{a}_{3\text{-}body} + \vec{a}_{SR} + \vec{a}_{tides} + \vec{a}_{other}$$

Geopotential (Gravitational) Field

$$U = \frac{\mu}{r}\left[1 + \sum_{\ell=2}^{\infty}\sum_{m=0}^{\ell}\left(\frac{R_{\oplus}}{r}\right)^{\ell} P_{\ell m}\left[\mathrm{SIN}\left(\phi_{gc_{sat}}\right)\right]\left\{C_{\ell m}\cos\left(m\lambda_{sat}\right) + S_{\ell m}\,\mathrm{SIN}\left(m\lambda_{sat}\right)\right\}\right]$$

$$U_{2\text{-}body} = \frac{\mu}{r}$$

$$\vec{a}_{nonspherical} = \nabla\left(U - U_{2\text{-}body}\right) = \nabla R$$

These equations describe two body orbital motion. Newton's laws with an idealized inverse square gravitational force are the first two terms in the first equation. This equation has a classical closed form solution that embodies conservation of energy and angular momentum. Orbits are conic sections whose characteristics depend on the constants of motion (energy and angular momentum) and the tradeoffs between

kinetic and potential energy. Unfortunately, the idealized equations do not predict satellite orbits sufficiently accurately for most modern applications.
The first non-ideal effect is that the Earth is not a sphere, its mass is not uniformly distributed, and the gravitational field is more than just the inverse squared two body attraction. The Earth's geopotential field can be described by an infinite series of Jacobi polynomials in latitude and longitude, complete and orthonormal over the sphere. If series is truncated at a sufficiently high order, we can construct a closed system of coupled ordinary differential equations. The first few terms are sufficient to predict major effects, such as orbit precession and motion of the argument of perigee. The coefficients of the series expansion of the Earth's geopotential field are determined by monitoring satellites in sufficiently stable orbit and by independent gravitational measurements. Even the most basic orbit estimates rely fundamentally upon data fusion.

$$\vec{a}_{drag} = -\frac{1}{2}\rho\frac{c_D A}{m}v_{rel}^2\frac{\vec{v}_{rel}}{|\vec{v}_{rel}|}$$

$$\vec{a}_{3\text{-}body} = -\frac{G(m_{\oplus} + m_{sat})\vec{r}_{\oplus sat}}{r_{\oplus sat}^3} + Gm_3\left(\frac{\vec{r}_{sat3}}{r_{sat3}^3} - \frac{\vec{r}_{\oplus 3}}{r_{\oplus 3}^3}\right)$$

$$\vec{a}_{SR} = -\frac{p_{SR}c_R A_{\odot}}{m}\frac{\vec{r}_{\odot sat}}{|\vec{r}_{\odot sat}|}$$

The equations above are the simplest expressions of major perturbations. Drag forces depend upon the objects "ballistic coefficient," the ratio of weight to drag coefficient multiplied by area. Measurements of satellite retardation reveal at most this inextricable convolution of these three unique parameters. Drag is proportional to the square of velocity. Three body forces vary inversely with the separation of the bodies cubed. Solar radiation pressure does not depend explicitly on distance from the Earth, although Solar intensity decreases as the square of distance from the sun. Externally imposed thrusts are often used to adjust satellite attitude and orientation or to maneuver in orbit.

The work done on or by the satellite in response to these forces transforms orbital energy into heat or other unrecoverable forms of energy. This dissipates orbital energy, so that all orbits decay. Because drag is the major perturbation, satellites in low orbits decay more rapidly than those in high orbits. Dissipated energy must be restored, usually from stored chemical reactants, in order to maintain a prescribed orbit. Even "geostationary" orbits are not geostationary in practice. Left to its own devices a satellite at what would be "geostationary" altitude ideally will drift east at a rate of about two degrees of longitude per day.

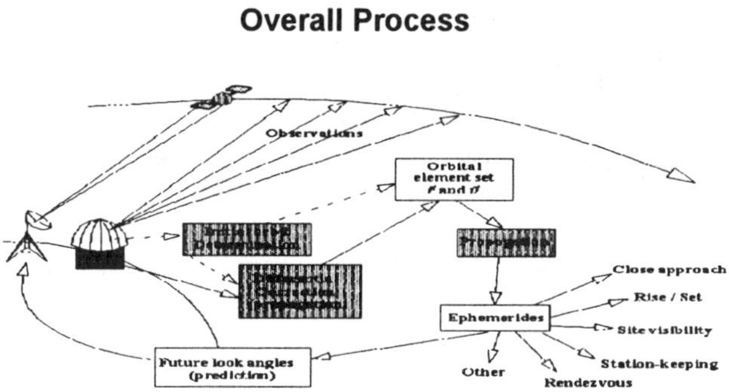

Figure 1. The Orbit Determination Process

The observation and data fusion architecture begins with sensor contact with the orbiting object. An initial estimate can be made from early observations. The orbit estimate is refined with subsequent observations through a process called differential correction. Classical orbital elements and perturbation measures result. These are employed by satellite operators and those who consume their products to predict satellite visibility from prescribed locations, to plan maneuvers or rendezvous, to determine whether satellites might pass near each other, and to predict look angles for subsequent measurements.

Depending on the type of observation measurements, there are several ways to determine orbital parameters. An active sensor, such as a radar, can determine the kinematic state of the satellite. It can determine position and velocity. These quantities alone are sufficient to develop a trajectory. Passive sensors, such as optical telescopes, do not measure range. Unless we have complete radiometric information and know source emissions, this is not sufficient to determine an orbit on just one observation. Classical texts [5,6] demonstrate that at least three observations are necessary (either from three independent sensors for the same orbit or on three successive orbits from one sensor). Occasionally both passive and active sensors (with range but not rate information) might be available. There are at least three mathematically sound approaches to orbit determination using vector position measurements over time. Finally, some sensors may lack range but perceive range rate. The different methods are distinguished by the sets of measurements they rely upon (angles, range, velocity, etc.). The approaches within each method employ different numerical algorithms for determining derivatives with differencing schemes and different sequences of numerical operations designed to preserve numerical significance. It was recognized long ago that orbit determination requires exquisite numerical precision. Many classical contributions were innovative coordinate transformations designed to amplify dimensions in which state variables vary slowly. For example, an observation period several minutes long, the range to a geosynchronous satellite may change by only a small fraction of one percent. There are many alternatives spanning opportunistic, communicative, and cooperative data fusion.

Astrodynamics data fusion suffers several physical and procedural constraints. Sensors are sparse and not often ideally situated to observe satellites of interest. Although it is possible to place satellites in orbits with repeating ground tracks, these orbits are generally not very useful. The orbits of most satellites move relative to the surface of the Earth as the Earth rotates. Depending on the orbit, a pass just out of reach to the west may be followed by a pass just out of reach to the east. Intervals between observation opportunities can be many days or weeks for some orbits and some sensors. Since active sensors provide more complete information, radars are preferred. However, few radars are dedicated to satellite observation. Radar characteristics are often optimized for different tasks. Therefore, radars on the Earth have diverse power levels, aperture sizes, and practical resolution. In fact, even the most powerful and well focused terrestrial radars are not sufficient for observing objects in geosynchronous orbits. Estimating these "deep space" orbits requires passive optical sensors. Until the modern epoch, optical sensors could not be managed as precisely as active radar sensors. They were often inaccurate and imprecise. This is changing, and in the future we may see more ground and space based optical sensors contributing to orbit estimation.

Astrodynamicists have long employed multi-sensor data fusion techniques independent of the generic data fusion community. Gauss is credited with the least squares approach to optimizing the fit of a function to a data set. Minimizing the squares of the differences between predicted and measured points determines parameters for an optimal solution within the definition of the process. Sir Ronald Fisher extended Gauss' principles to non-Gaussian statistics in the maximum likelihood method. More recently Rudolf Kalman developed (arguably) a recursive algorithm for statistical orbit determination. In the scheme of statistical orbit determination, "predicting" uses current observations to find future states. "Filtering" is determining the current state using current and past observations. "Smoothing" improves previous states based on current data. Smoothing is an approach to discarding irrelevant data or identifying anomalous measurements. "Sequential" processes use current observations to improve predictions based on past observations. "Batch" processes use the entire current and past data set simultaneously. One sees analogy with the opportunistic, communicative, and cooperative approaches to data fusion.

Astrodynamics has evolved through painful experience to the most desirable practices for different types of satellite in diverse orbit categories. As computational capability explodes, many of the most widely respected techniques become obsolete. Most were developed as numerical or analytical "tricks" to avoid computation so intensive that it was impractical. Linearization, small perturbation approximations, and asymptotic behavior analyses are most common. The problem is, however, inherently nonlinear.

3. Applications of Data Fusion in Astrodynamics

Astrodynamicists try to overcome deviations from linearized approximations with "differential correction." There techniques are most used: Non-Linear Least Squares, Kalman Filtering, and Maximum Likelihood. Non-Linear Least Squares, we linearize the problem, obtain an approximate solution, and iterate to refine the answer. We estimate *corrections* to our initial estimate of the state [7]. Practitioners have examined both fitting to all observations at once (Batch Processing) or combining observations from several short intervals (Sequential Batch Processing). Least squares updates the satellite's kinematic state but not the statistics. Kalman filters are used as well. Kalman filters apply strictly only to processes that evolve to Gaussian statistics. Finally, maximum likelihood methods are used when random processes are clearly not Gaussian. Thus far, these methods have been too computationally intensive to see wide use in time critical applications.

Many quantitative data fusion applications require the user to weight contributions from diverse sensors based on the quality of sensor observations. We use the following simple example to demonstrate an approach to consistent weighting [8].

Suppose that we have two observations of a single attribute and that the measurements are unbiased and have zero mean.

$$z1 = x + v1 \qquad z2 = x + v2$$

We seek an optimal estimate of the state, x, as a linear function of the two observations.

$$X = k1 * z1 + k2 * z2$$

In the sense in which Gauss conceived least squares, the optimal estimate is that which minimizes the mean squared differences between the observed and estimated state (residuals) Z=k*X . Since the errors are unbiased, if X is normalized to unity:

$$k2 = 1 - k1$$

The expected value of the mean squared error is:

$$E(X^2) = k1^2 * \sigma1^2 + (1 - k1)^2 * \sigma2^2$$

If we minimize this with respect to k1, the result is as follows.

$$\sigma^2 = \frac{1}{\left(\dfrac{1}{\sigma1^2} + \dfrac{1}{\sigma2^2}\right)}$$

Measurements should be weighted in inverse proportion to their uncertainty. The fused answer is no worse than the best observation, and, if additional observations convey any valid information at all the result should be better than the best observation.

We can show rigorously that multiple independent measurements with uncertainty should be weighted inversely by their uncertainty. The more precise measurements are counted more than the more imprecise measurements. The expected value of the mean squared error is a "cost function" to be minimized. If we treat the situation with mathematical rigor and we have diligently characterized our measurement instruments, the consistently fused results should be no worse than the least uncertain contribution. It should be better than that if additional, though imprecise, measurements contain any valid information at all.

Unfortunately, this rigorous scheme occasionally fails. Fusion of multiple observations may, once an independent reference is available, be worse than individual observations or smaller sets of fused observations. When this occurs we have obviously violated fundamental assumptions. Perhaps the statistics are not Gaussian. More likely, the uncertainties in contributing sensors have not been characterized sufficiently well.

The mathematical statement of batch processing is as follows.

$$\delta\hat{\boldsymbol{x}} = (\boldsymbol{A}^T\boldsymbol{W}\boldsymbol{A})^{-1}\boldsymbol{A}^T\boldsymbol{W}\tilde{\boldsymbol{b}} = \boldsymbol{P}\boldsymbol{A}^T\boldsymbol{W}\tilde{\boldsymbol{b}}$$

We develop a small (differential) correction to an initial state by minimizing the sum of the weight residuals squared. The residuals are the differences between discrete measurements and the initial prediction for the time of interest. The matrix A is composed of the values of the initial state variables. The elements of the matrix W (the weighting matrix) may be the inverse of the uncertainties in measured state variables. The matrix P depends on the initial state and the weights. Since it embodies variances and covariances, it is often called the "covariance matrix." It is square and symmetrical, and it is the uncertainty in the estimated future state. The off-diagonal, covariance, terms are correlation coefficients indicative of the dependence or independence of uncertainties in state variables.

For sequential batch processing:

$$|\delta\hat{\boldsymbol{x}}(0|k+n) = (\boldsymbol{A}_{new}^T\boldsymbol{W}_{new}\boldsymbol{A}_{new} + \hat{\boldsymbol{P}}_k^{-1})^{-1}(\boldsymbol{A}_{new}^T\boldsymbol{W}_{new}\tilde{\boldsymbol{b}}_{new} + \hat{\boldsymbol{P}}_k^{-1}\delta\hat{\boldsymbol{x}}(0|k))$$

$$\hat{\boldsymbol{P}}_{k+n} = \hat{\boldsymbol{P}}(0|k+n) = (\boldsymbol{A}_{new}^T\boldsymbol{W}_{new}\boldsymbol{A}_{new} + \hat{\boldsymbol{P}}_k^{-1})^{-1}$$

Here we update the correction sequentially with each new observation.

The Kalman filter formalism first propagates the state and covariances through time. Φ is the state transition matrix composed of partial derivatives of the state at a future time to the initial state. The prediction is updated or improved employing the Kalman Gain that depends on H, whose elements are the partial derivatives of observations with respect to state variables. Most, if not all, of these quantities are calculated numerically.

Prediction

$$\boldsymbol{PROPAGATE}\left(\hat{X}_o, t_{k+1} - t_o \Rightarrow \bar{X}_{k+1}\right) \cong \Phi\hat{X}_o \quad \textbf{Predicted State}$$

$$\delta\bar{x}_{k+1} = \Phi\,\delta\hat{x}_k \qquad \textbf{Predicted State Update}$$

$$\bar{P}_{k+1} = \Phi\hat{P}_k\Phi^T + Q \qquad \textbf{Predicted Error Covariance}$$

Update

$$K_{k+1} = \bar{P}_{k+1}H_{k+1}^T\left[H_{k+1}\bar{P}_{k+1}H_{k+1}^T + R_k\right]^{-1} \qquad \text{Kalman Gain}$$

$$\delta\hat{x}_{k+1} = \delta\bar{x}_{k+1} + K_{k+1}\left[\tilde{b}_{k+1} - H_{k+1}\delta\bar{x}_{k+1}\right] \qquad \textbf{State Update Estimate}$$

$$\hat{P}_{k+1} = \bar{P}_{k+1} - K_{k+1}H_{k+1}\bar{P}_{k+1} \qquad \text{Error Covariance Estimate}$$

$$\hat{X}_{k+1} = \bar{X}_{k+1} + \delta\hat{x}_{k+1} \qquad \textbf{State Estimate}$$

Although we have preserved a semblance of mathematical and physical rigor, astrodynamics data fusion to estimate satellite orbits can be highly subjective. For example, the weights at the core of estimation schemes are often heuristics. Certainly we expect more confident results when there is a sufficient set of high quality observations. What is sufficient depends on the physics of the phenomenon rather than the capability of the sensor. Satellites in medium altitude, circular orbits can be extremely stable. An occasional observation may be sufficient. Satellites in low orbit during Solar maximum can be extremely unstable, and very dense and frequent observations may be required. We do not know what quality or quantity of data might be best unless we have independent, accurate, and precise measurements of a satellites state of motion. Even today there are important orbit classes for which ground truth is questionable. The result is that orbit estimation and propagation is an art whose cultural characteristics match its environment: the locations of sensors, the intervals of time over which data is aggregated, the computational capability, and many other factors. Just as human cultures are affected by environments, so it is with science.

At the outset we stated that the efficacy of the products of data fusion rests upon the underlying hypotheses. We saw that numerous "perturbations" could influence satellite kinematics. However, not all possible perturbations are important for all orbits. If we incorporate an incorrect hypothesis, for example, that solar pressure is a significant element consideration for low Earth orbit satellites but that drag is not, we will always suffer large prediction uncertainties. Figure 2 shows one way of discriminating among classes or orbits based on the predominant phenomena. Perigee height and apogee height characterize the eccentricity of the orbit and its physical size. Of course, apogee and perigee are the same for circular orbits. Different combinations of perturbing phenomena have been found significant in different regions of this diagram. For example, three body effects matter for highly eccentric, large orbits. Atmospheric drag dominates for circular "small" orbits. Experience-based guidance such as this is important to modern astrodynamics.

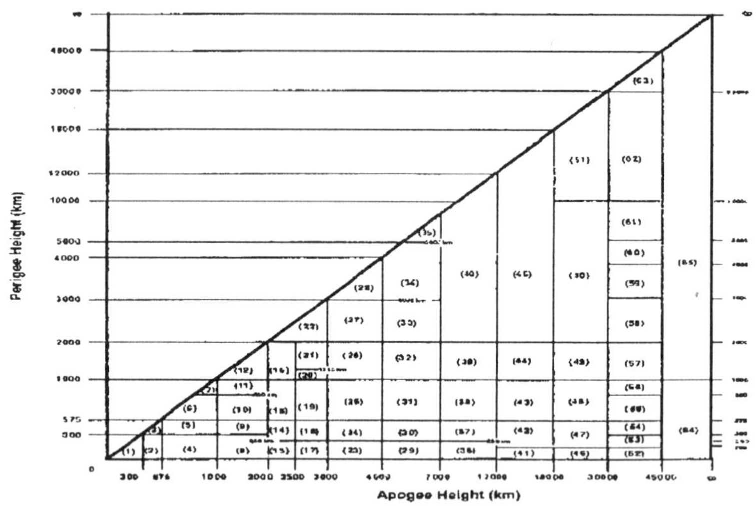

Figure 2. Definition of Orbit Classes

4. Astrodynamics Products

We conclude with examples of the efficacy of multi-sensor data fusion principles in practical astrodynamics.

What is the result of astrodynamicists meanderings over this vast space of data fusion processing schemes and topologies? Figure 3, taken from Knowles [9], (1995) categorizes the basic approaches. The ordinate is a logarithmic scale of the uncertainty in either measurement or orbit estimation as a function of orbit altitude. Horizontal lines are the precision of sensors of various qualities. For example, about 10 meters for high precision radar at low altitudes and less than a meter for high precision laser ranging. Analytical approximations, such as linearization and other schemes to avoid intensive computation, are trustworthy to within 1-10 km. They are best at mid-altitudes, where neither atmospheric drag nor three body effects are significant. Numerical techniques are trustworthy to within less than 10 meters. We must be cautious nonetheless.

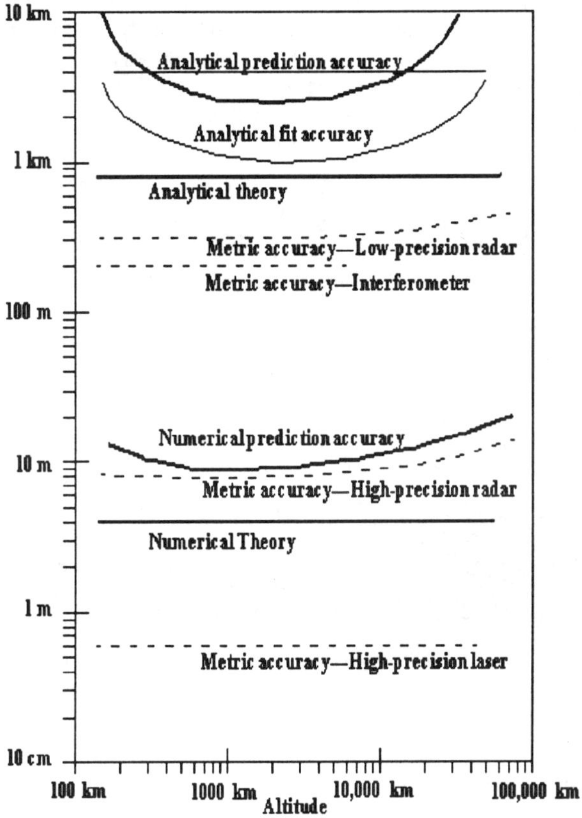

Figure 3. Quality of Different Estimation Approaches

One final illustration of the extreme significance of basing orbit predictions and statistical estimates on appropriate hypotheses.

Figure 4 is a way of depicting the energy tradeoffs associated with minimum time continuous thrust orbital maneuvers. The ordinate is the ratio of the semi-major axes of the larger and smaller orbits. The abscissa is change of inclination. The parameterization on the border is a control variable that schedules maneuvers. The parameter inside the chart is proportional to the cost of the maneuver, whether it be time to execute the maneuver or energy expended to complete the maneuver. The maneuver

occurs along a line of constant control variable. For orbit raising the maneuver begins at the origin. For this example, the control variable is -0.6. The entire transfer is completed over an interval of $\zeta=1.1$. The development of low thrust, high impulse thruster technology now makes this transfer economically feasible. The more traditional Hohmann transfer applies impulsive thrust at perigee and apogee of the transfer orbit, the transfer ellipse being tangent to both orbits at those points. For orbit raising, the Hohmann transfer is an impulsive semi-major axis change followed by a coating period. The transfer is completed with a second burn to circularize the orbit while simultaneously changing the inclination. The transfer remains entirely in the same plane, whereas the low thrust transfer does not. The low thrust transfer uses only one sixth to one tenth the reactant mass required for a Hohmann transfer.

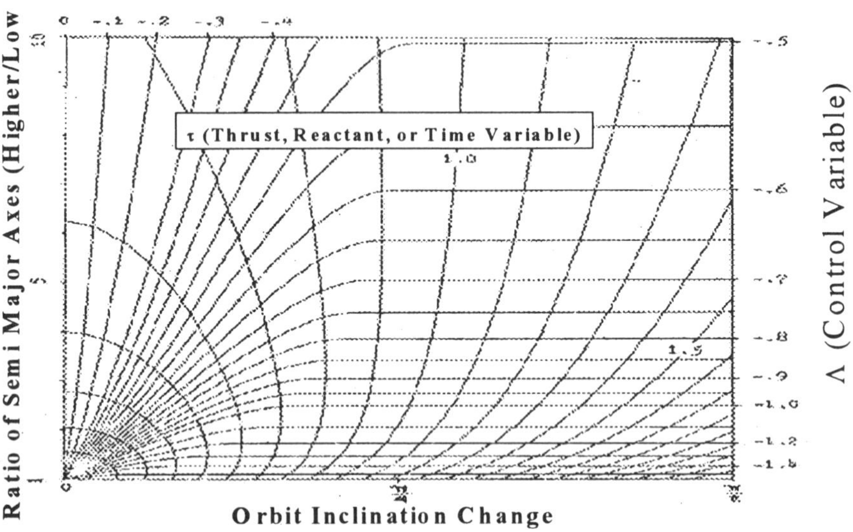

Figure 4. Alfano Low Thrust Orbit Transfer

Several satellite systems, for example Iridium, employ efficient low thrust propulsion. The same thrusters that perform orbit insertion and station keeping are used to slowly deboost the satellite near the end of its life. Because it requires no additional propulsion system for orbit transfer, reliability is increased, mass is reduced, and orbital debris is mitigated. This is more energy efficient, which saves the principal cost function for satellites -- mass. It is also less complex, since the final state is much less sensitive to small uncertainties in thrust initiation and termination.

The perturbing forces assumed for impulsive maneuvers do not describe low thrust continuous maneuvers adequately. Using the common physical hypotheses for orbital maneuvers, measurements will diverge significantly from the modeled maneuver. Covariances will be large, and many more measurements will be required to estimate the satellite's orbit in the future.

Figure 5 shows the low thrust, trajectory, while Figure 6 provides the deviation of a representative low thrust maneuver from the initial circular orbit that would be predicted with the usual force model hypotheses. If a Hohmann transfer were assumed, there would be almost no deviation from the original orbit over this period. Depending on the magnitude of energy added (delta V), the satellite could be 10's or hundreds of kilometers ahead of the predicted position and several kilometers lower. Sensors that relied upon predictions derived from inappropriate hypotheses might not even find the satellite. In densely populated regions of near Earth space, we might perceive potential collisions among satellites that were in fact not very likely.

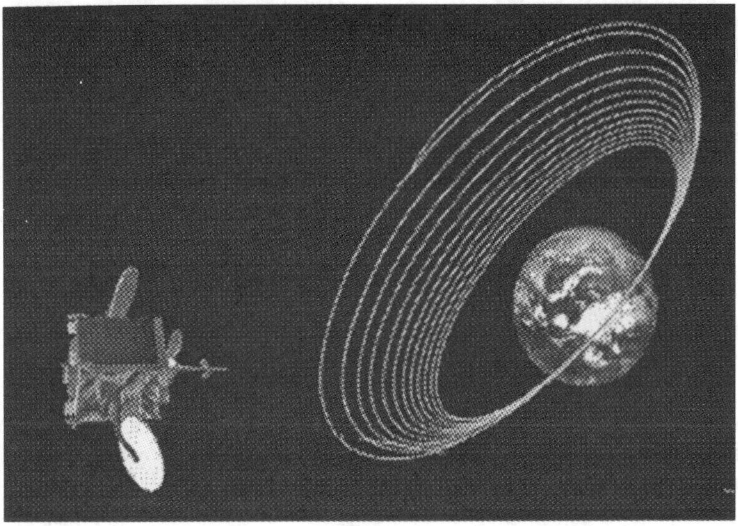

Figure 5. Low Thrust Orbit Transfer Trajectory

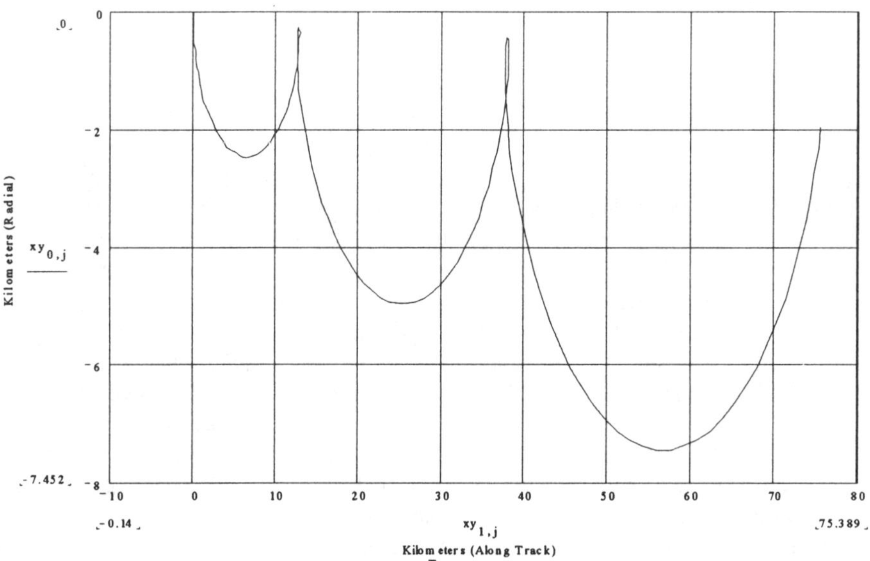

Figure 6. Relative Distance from Original Orbit for an Alfano Transfer

5. Conclusion

Astrodynamics embodies the principles and techniques of multi-sensor data fusion. As in all multi-sensor data fusion applications, astrodynamics demands consistency among:

- Experimental Practices – How we acquire observations.
- Physical Models – Hypotheses that are tested against observations
- Computational and Numerical Implementation – How physical models are represented in discrete mathematics and how quantitative observations are compared with numerical estimates.
- Observations and Data – The frequency, density, and duration of observations and observation intervals.

The manner in which we acquire data must be sufficiently precise to exploit the capabilities of exquisite models. Numerical implementation of the models must be consistent with the quality of the data. Data must be acquired frequently enough and at times chosen appropriately to capture the essential dynamics of the events. If any of these are inconsistent with the others, we corrupt results and violate fundamental assumptions. We seldom achieve the appropriate balance.

The state of the art in astrodynamics is that physical models are continuously improving, while computational capability advances rapidly. Unfortunately we seldom have sufficient sensors appropriately located to acquire data sufficient to test our physical hypotheses, and observations are often not documented precisely enough to match our physical and computational capabilities.

6. Summary

This paper reviews the precepts of multi-sensor data fusion and shows how they are applied to modern astrodynamics. Of necessity terrestrial sensor systems are sparse and diverse. Measurements include angles only optical (telescope) observations, state of motion measurements with radars, occasional capability to range on the communications links to the satellites, and (rarely) satellite position information from onboard Global Positioning System (GPS) transceivers. The hypotheses that are tested against observations are the equations of motion augmented with assumptions about the nonuniformity of the Earth's geopotential, the influence of atmospheric drag, the existence of light pressure, the influence of other heavenly bodies, and even the slow dynamics of the oceans and the Earth's mantle and core. If we model these phenomena well, we can estimate satellite orbits relatively accurately. More important, we can sustain accurate estimates with few observations. If we model these phenomena poorly, variances from measurements will be large, orbit uncertainty will be large, and many more observations might be necessary. Astrodynamics employs a rich spectrum of data fusion techniques. Above all we strive to maintain mathematical and physical rigor so that anomalies are traceable to physical or mathematical processes amenable to modification.

7. References

1. http://www-datafusion.cma.fr/
2. Vallado, David A. (1997) *Fundamentals of Astrodynamics and Applications*, McGraw Hill Astrodynamics Series, page 4.
3. Waltz, E. and Llinas, J. (1999) *Multisensor Data Fusion* (Second Edition), Artech House, page 4.
4. Langley, Richard B. (2000) *The Integrity of GPS*, GPS World, http://www.gpsworld.com/
5. Bate, Roger R., Mueller, Donald D. and White, Jerry E. (1971) *Fundamentals of Astrodynamics*, Dover Publications, Chapter 5.
6. Vallado, Ibid, Chapter 6.
7. Vallado, ibid, Page 675.
8. Gelb, Arthur, ed., (1986) *Applied Optimal Estimation*, The MIT Press, Cambridge, MA, pp 5-6.
9. Knowles, Stephen H. (1995) *A comparison of Geocentric Propagators for Operational Use*, Paper AAS-95-429, AAS/AIAA Astrodynamics Conference, Halifax, Nova Scotia, Canada.

LANDMINE PROBLEM AND MULTISENSOR DETECTION

V.M. BYSTRITSKII
Joint Institute for Nuclear Research
Laboratory of Nuclear Problems
141980 Dubna
Moscow Region, Russia

1. Introduction

The mine warfare in the 20th century has a history of near hundred years, related with countless military conflicts all over the world. As it is documented, the first systematic and large-scale application of mines took place during the siege of the Russian Army by Japanese at Port Arthur and Mukden in 1904. There the protection of Russian trenched positions by minefields with electric ignition was responsible for the major part of the near 120, 000 casualties from the Japanese side [48].

In the following wars the scale of mine usage increased even more, and in 1914-1918 during the WWI, the minefields accounted for several hundreds of thousands of casualties average monthly during the major frontal attacks [48]. The military role of the mines (sea-and land-based) during the WWII was also important, especially at the very start of the war, when in a couple of months of 1939 the British Navy lost near 60,000 tons of merchant ships from German magnetic sensitive mines. To combat the naval mines presenting threat for the convoys and merchant shipment that was comparable with those of the German submarines, up to 60 thousand of the Navy personnel were employed in the mine sweeping operations throughout the war [18]. In difference from the previous time the mines of the WWII became more specialized, being divided on the anti-personal (APM) and anti-tank (ATM) mines. The invention of the latter was triggered by the appearance of the tanks in the battlefield operations. The antitank land mine is able to deliver significant damage to the tank tracks and thus to put it out of the combat. The antipersonnel mines also became more widely used, more sophisticated, and more lethal.

The overall dynamics of the mine warfare impact on the US military personnel could be seen from the following figures and numbers of losses: 2.5% killed in Europe in WWII, 1.65% killed and 3.3% wounded in Korea, 16% wounded in Vietnam, 20% of all casualties in the Persian Gulf War, 26% of casualties in Somalia. In the past 10 years the US lost three ships with a total cost of $ 117 millions, inflicted by a simple "poor man's weapon" [33].

In present days the mine technology is characterized by further sophistication, such as: added new functions (guard systems, self destruction or self-sterilization features, "turn off and -on" devices, sensors of position, motion, etc.); the use of novel high power explosives, plastic, i.e. non-metallic shells, and their miniaturization.

The main task of all variety of mines remains the same as earlier- to "shape" terrain and the battlefield and to reduce the mobility of opposing forces [7].

A. K. Hyder et al. (eds.), Multisensor Fusion, 673–714.

In difference of all other weapon used in the battle field situation, mines represent "stay- behind" armed weapon, that remain active and dangerous after the end of the military conflicts, that results in the numerous casualties among civilian population and requires very costly remediation efforts to make mine infested lands accessible for regular social service, productive land farming, resettlement of refugees and displaced persons, etc., i.e. - to normal life. A special danger is presented by the AM. Here are some very disturbing facts about the scale of the mine problem.

Land Mines Facts [7].

- It is presently estimated that near 60 million active mines are planted in 70 countries with a bigger number stockpiled around the world waiting to be planted. Of those near 15 millions are buried in six countries- Afghanistan, Angola, Mozambique, Cambodia and Croatia. (By previous inflated estimations of the UN, these numbers were at the level of 100 and 40 millions respectively. In fact, the numbers do not matter, because amount of efforts needed to clear a terrain with 100 mines or one mine, or no mines at all is practically the same, and disposing of mines once they are found, requires significantly minor efforts compared with efforts already spent on the sweeping search of the suspected area.
- Illegal land mines usage, especially of APM has dramatically increased over the past 20 years with an emphasis on its potential as a weapon to terrorize civilians. Mines are used to deny access to or usage of farmlands, irrigation channels, roads, waterways and public utilities.
- Every year between 25-26 thousand civilians, with high proportion of women and children are maimed and killed by these mines.
- Most of the APM casualties are civilians, who are killed or maimed after hostilities have ended. Buried land mines can remain active for 50 years if not provided by a special mechanism of expiration, " trigger-off", which is the case for all deployed mines.
- According to the International medical database, land mine victims need blood transfusions twice as often as people injured by bullets or fragments, and in average 2 to 6 times more blood during surgery compared with other war casualties. Surgical and orthopaedic care cost about $3,000 per amputee in developing countries, which makes near one billion for registered by UN amputee registered worldwide by the UN at to present time.
- For every mine cleared, 20 are laid. In 1994 approximately 100,000 were removed, while more than a million additional were laid.
- APM are priced at $3 to $30 each. The cost to the international community of neutralizing them ranges from $300 to $1000 each.

Nevertheless instead of the overwhelming active negative reaction of the world community, mines remain a legitimate part of the warfare, if they are used in accordance with international conventions and agreements on warfare. Respectively, mines status becomes illegal when used in the undeclared local wars or by the terrorist groups. The boundaries between these two domains become very vague, especially in the case of inter-region civil and tribal conflicts, (such as in former Yugoslavia, Angola, Mozambique, Afghanistan, etc). This list could be prolonged. The cited above civilian casualties and respective problems with reclamation of the mine-infested lands are in their major part the legacy of these wars, where the AP and AT mines used were not provided with any type of " switch off" or timer self destructive mechanisms.

On March 1 1999 the "Convention on the Prohibition of the Use, Stockpiling, Production, and Transfer of Anti-Personnel Mines, and their Destruction " entered in the force, as signed by 133 countries, but up to now none of the major mine production countries (Russia, US, North Korea, and China) did sign the Convention on the ban for anti-personnel mines. [24].

The whole mine problem has many aspects: military, civilian, political, economical, technical, humanitarian, ethical, etc. This presentation will focus only on the technical aspects related with humanitarian demining, specifically on the conventional and emerging technical approaches for mine detection, including novel technique for multiple sensors data fusion, which seems to be the most promising.

The second section of the report is allocated for brief technical description of the mines, their specific signatures and respective environmental conditions that impose certain requirements and limitations on different mine detection sensors. We also introduce arbitrary classification of the emerging technologies, along the lines of their ability to discriminate the man made underground objects from natural anomalies, and explosives- from innocuous objects.

The third section describes and analyses strong and weak points of the conventional field mine detection techniques, that are presently being used on the large scale all over the world: mechanical prodding, simple metal detectors and sniffing dogs.

In the fourth section we describe new uni-sensor techniques at various stages of R&D that are focused on one of the many specific signatures of the land mines, and provide analysis of their promise and problems.

In the fifth section we present a brief overview on the processing algorithms of the acquired data, which are used for multi sensor systems, including data fusion that results in the enhanced discrimination of the mine signatures against clutters, and decrease of the false alarm (FA) rate.

In the sixth section we describe combined operation of two and more uni-sensors operating in a multi sensor system in series (uni-sensor first stage + uni-sensor confirming stage), and in parallel with data flow fusion.

In conclusion, we will summarize the main trends in the mine detection technologies evolution and try to formulate some prognosis and recommendations.

2. General Characteristics of the Landmines

2.1 STRUCTURE, EXPLOSIVE, APPLICATION

The common landmine consists of a bulk explosive (TNT, RDX, etc) with a weight of several tens of grams (usual for APM) to few kilograms (usual for ATM), placed in a box (metal, wood, plastic, fabrics) of a regular shape (rectangular, cylindrical, etc., pertinent to the type of delivery to the minefield or/and production line type), with a fuse (few grams of a high power explosive, placed in a small thin wall metal tube, a metal striker with a spring, or other more sophisticate type of the fuse mechanism - sounds, vibration, remote control, etc) . Some of the APM types that were used during WWII and still in the production line, are provided with an iron fragmentation body or filled with metal shrapnel (for example APM M1-6A1, etc.) or designed to be placed over the surface on a special stand, such as MON-200 [7]

In general, the shell of contemporary APM (if it is not of fragmentary type) is made of plastics, which decreases the probability of its registration even with very sensitive metal detector. For example - APM POMZ-2, which was used massively by the Soviet army in Afghanistan, had a round shape plastic case, very low metallic content and explosive weight of near 70 g. Modern APM could be disseminated directly from the air (plane, or special launcher) over the large surface area, and it becomes armed from the moment of impact with the ground. APM and ATM could be also laid along the prescribed pattern from the moving vehicle, provided with a special mechanical mean. In recent warfare the ATM are usually surrounded by several APM, to prevent their detection and removal. Table 1 gives comprehensive list of the APM and ATM with their main characteristics. Typical APM and ATM are illustrated at Figures 1 and 2.

Figure 1. APM M16 A1- antipersonnel mine. Provided with a cast iron fragmentation sleeve inside a sheet steel outer casting. Activated by a pressure or a trigger wire. The main charge is propelled in the air by black Powder fuse charge, and explodes with delay at the height of 1 m [7].

Figure 2. ATM VS22, plastic case antitank mine. Destroys the caterpillar track of tank with a propelled heavy pressure plate.

2.2 MINE SIGNATURES

Detection of the landmine that is buried in the soil, and its discrimination from the background noise and innocuous anomalies rests on registration, processing and analysis of the " answer" –signals containing specific signatures of the mine presence. These signatures are different from that of the soil, and could be provided by the following features as listed here (but not limited only by them):

- High local concentration of specific chemicals different from soil components and characteristic only for explosives and respective presence of specific vapors from the explosive, leaking through the soil.
- Anomalous physical properties of the volume occupied by the mine (such as density hardness, humidity, geometric shape and size, thermal, and electrical capacity, resistance, conductivity specific radiation, reflectivity, absorption of radiation, and scattering of various probing radiation (such as ultra sound waves, IR, RF, magnetic field pattern, neutrons, gamma, charged particles, water jet, etc).
- A regular geometric shape (cylinder, rectangle, sphere), and size (a few inches to one foot).
- Specific range of mine location depth: from a few to ten inches for APM and ATM respectively.
- Specific character of rise/decay of the "echo" signal from the mine (or natural pattern distortion) with increase/decrease of the distance from the latter, which is different from background fluctuation.

It is self-explainable that existence of the mine signatures per se does not mean their ready discrimination.

TABLE 1. Characteristics of Some Commonly Used APM and ATM [7]

Mine name and type Lethaleffect Manufacturing country	Shape Height (mm) Size/ length Weight	Shell Color Number of fuse wells	Explosive Component Explosive Weight	Action, Visibility
M18A1, APM, Directed fragmentation USA, Chile, S. Korea S. Africa, FSU	Rectangular 216x 35x83 1.58 kg	Fiberglass Olive drab 2	C-4, 0.682 kg	50 m, placed over ground, Visually detectable
PMN, APM Blast FSU, Iraq	Cylindrical 56 x 112 0.55 kg	Bakelite with rubber cover, black integral fuse	TNT, 0.200 kg	Limited to contact zone,, detectable with MD to due to large amount of metal in fuse
M 16, APM Fragmentation, blast USA, India	Cylindrical 103 x 203 0.158 kg	Metal Olive drab, green 1	TNT, 513 kg	27 m readily detectable with MD
M14, APM Blast, USA, India	Cylindrical 56 x 112, 0. 158 kg	Plastic Olive drab, green 1	TNT, 0.029	Readily Detectable
VS-50, APM, Blast, Italy, Egypt, Singapore	Cylindrical 45 x 90, 0.185 kg	Olive drab, sand integral fuse	RDX 0.043 kg	Readily Detectable
T72, APM Blast, China, S. Africa	Cylindrical, 38 x 78 0.125-0.150 kg	Green, light green Plastic body, \rubber cover	TNT/RDX 0.075-0.1 kg	Limited (contact), very difficult with MD
MON-200, APM Directed fragmentation FSU	Cylindrical 130 x 434 25 kg	Olive drab Metal 1	TNT 12 kg	200 m Visually detectable
POMZ-2, APM Fragmentation FSU, N. Korea	Cylindrical, 107 x 60 1.77 kg (POMZ-2M)	Olive drab Metal 1	TNT 0.075 kg	4 m Visually detectable
VS 1.6, ATM Pressure plate Italy	Cylindrical 90 x 215 3.2 kg	Olive drab Plastic 1	HE Comp B RDX 1.9 kg	Difficult to detect with MD (metal - 2.5 g)
VS 2.2, ATM Pressure plate Italy	Cylindrical 118 x 237 2.0 kg	Olive drab Plastic 1	HE Comp B: Booster: RDX 2.34 kg	Difficult to detect with MD (metal- 5.0 g)
M 15, ATM Pressure plate USA	Cylindrical 124 x 425 14 kg	Olive drab Steel 1	HE Comp B: Booster: RDX 11.3 kg	Easily detectable with MD
M 19, ATM Pressure plate US	Rectangular, 326 x 326 x 90 12.5 kg	Olive drab Plastic 1	HE: Comp B: Booster: RDX 9.5 kg	Difficult to detect with MD

There are several objective causes that make mine signatures identification difficult by decreasing the Signal to Noise Ratio (SNR). The most ubiquitous are:

- One side access to the buried mine and respective requirement to work only with the reflected secondary "signal-answers", that emerge through the upper layer of the soil.
- Permanent presence of the noise, generated by the semi-infinite air and soil space, and direct radiation from the probing radiation source.
- Variability of the "explosive" signatures, depending on variability of the surrounding parameters (soil composition, seasonal surface variation due to rain, snow, grass, bushes, relief, sensor positioning, human factor involved in the data analysis and processing, etc).

Of all above listed mine signatures only, the bulk presence of the chemical elements that are specific for explosives could become a real "fingerprint" of the mine. What are these elements? Almost all of the known explosives consist of four elements (with may be few exceptions), in various relative concentrations: nitrogen, oxygen, hydrogen and carbon.

Table 2 illustrates composition of explosives routinely used in mines, and for comparison - of several innocuous substances, and the soil [40,53].

As it follows from this table, with an exception of oxygen, the trio of "explosive" elements - N, H and C are absent in the common dry soil, and thus could be used as a direct signature of the man-made explosive based anomaly. In the same time it worth to be noted that there are many common man-made non-explosive compounds, containing the same tetrad of, though with relative concentration moderately to significantly different from, the explosive. Another important fact is that concentration of carbon and nitrogen in the soil could be also high in the forest zones, presenting a challenge for technologies based on discrimination of those elements.

The listed signatures could be "sensed" by various detectors of different complexity, involving analysis and processing of the acquired data. It is processing in the on-line mode that is responsible for efficient background " noise" and clutters cut-off, i.e., FA signals due to various innocuous underground anomalies that could be of man-made or/and natural origin.

Among the frequent clutters that could mimic the signatures of the buried mine are: stones, underground water puddles, burrows, buried pieces of metal, tree trunks, plastic bottles, roots of trees, cavities, etc. Among the conditions that increase the average level of the background "noise" and respectively - decrease the discrimination ability of the sensors, and mask the mine presence, are: snow, rain, vegetation, the presence of surrounding objects of natural and man made origin, uneven soil structure and shape of the soil surface, changing atmospheric conditions, variation of sensor height over the surface.

TABLE 2. Chemical composition of explosives, innocuous substances, and the soil

Substance	Chemical Formula	Density g/cm^3	Atomic abundance	Explosive /Non-explosive
TNT (Trotyl)	$C_7H_5N_3O_6$	1.65	C- 33%, H- 24%, N- 14 %, O- 29%	Explosive
Nitro-glycerine	$C_3H_5N_4O_{12}$	1.6	C- 15 %, H- 25%, N- 15 %, O- 45%	-”-
PETN	$C_5H_8N_4O_{12}$	1.75	C~ 17 %, H~ 27%, N- 14 %, O- 42 %	-”-
RDX Hexogen	$C_3H_6N_6O_6$	1.82	C ~13%, H-29 %, N ~29%, O- 29 %	-”-
Comp. B	RDX + TNT		C- 28%, H~16 %, N- 20 %, O~ 32 %	-”-
C-3 Picric Acid	77% RDX+23% $(C_6H_3N_3O_7)$	1.76	C~31%, H- 20 %, N~16 %, O~ 37 %	-”-
HMX	$C_4H_6N_8O_3$	1.8	C- 19 %, H- 29 %, N- 38 %, O- 14 %	-”-
EGDN (Liquid)	$C_2H_4N_2O_6$	1.48	C~ 14 %, H~ 29 %, N~ 14 %, O~ 43 %	-”-
Cellulose Nitrate	$C_6H_2N_2O_7$	1.4	C~ 35%, H~ 12 %, N~ 12 %, O~ 43 %	-”-
Nylon	$C_6H_{11}NO$	1.15	C-32 %, H- 58 %, N- 5 %, O- 5 %	Non explosive
Para-nitro-Phenol	$C_6H_4N_2O_5$	1.18	C- 35 %, H - 24 % N - 12 %, O - 29 %	-”-
Polyethylene	$(C_2H_2)_n$	1.1-1.2	C-50%, H- 50%, N-0 %, O- 0 %	-”-
Ethanol	C_2H_5OH	1.05	C-22 %, H- 66 %, N-0 %, O- 12 %	-“-
Soil (5 % by weight of moisture)		2.0	H~ 2 %, O ~ 61%, Si ~ 22%, Al ~ 6.4 %, Fe - 2%, Ca ~ 2%, Na ~ 2 %, Ti + K ~ 2 %	-”-

In relation with the ability to discriminate the man-made from the natural one we could arbitrary classify all mine detection technologies in four groups or levels, following.

First level: technologies that are able only to detect the presence of the buried anomaly without its classification of a man- made or natural origin, and as a mine or not a mine. Respective specific signatures could include density and hardness of the object, its anomalous electrical and EM- properties, etc. Representative technology with certain assumptions could be a simple manual prodding of the soil with a stick. In field this technology could qualify for the next level due to a combination of just a fact of detection “something” with operator guess work, based on his expertise and analysis, resulting in creation of a mental image of the hidden object with approximate shape, location and size.

In fact, many first-level technologies, especially manual, involve considerable interaction, i.e.- " fusion" with human mental work.

Simple metal detector (MD) technology, existing for at least for 50 years could also be a good example of such a "fusion".

Second level includes technologies that are able not only to detect buried anomaly, but also to discriminate its man-made nature from natural, though they are not able to discriminate the presence of explosive in the detected man made object. Second level could be provided by adding to a first level uni-sensor an additional sensor looking for another specific signature of the man-made object. It could be: a regular shape, a presence of a bulk metal, or local bulk concentration of chemical elements drastically different from that of the soil. Representative technologies include all technical approaches, providing image of the hidden object in specific EM- wave band (infra red or radar frequencies, ultrasound, etc). For example adding to a simple MD of some imaging mean would provide such a 2-nd level multi sensor.

Another example- adding to MD a sensor able to detect bulk nitrogen, which is almost absent in the natural soil. In this case nitrogen could become a signature of a man-made explosive or innocuous object, such as bulk of fertilizer, nylon or other. In rare case it could represent a buried trunk of tree containing high quantity of cellulose. Usually, the combination of second level (imaging + nitrogen detection) technique would help in declassifying the latter FA.

In general, compared with the first level, second level technologies provide significant protection against FA, especially in fields, polluted with metal shrapnel, soils with pronounced rough morphology, forests, etc.

Third level technologies must be able to discriminate an innocuous man made objects from explosives. As it is seen from the Table 2, there are a variety of materials, which have elementary composition near to that of the explosives. So, the third level technologies must include so called a " chemical nose". Beside, the ability to discriminate and reject almost all clutters, the third level must be compatible with the large scale demining operation.

Fourth level: the ultimate technology of the future that will provide full data on the detected mine (type and quantity of explosive) with required high reliability (in humanitarian operations 100 %), respective sweeping rate, (tens hundreds squared meters per minute depending on the type of the search), that is compatible with a high scale demining operation. It will be able to identify the chemical formula of the hidden explosive, its size, type and location with required accuracy (few centimeters in plane and depth).

In the following we will try to apply suggested classification to various existing mine detection approaches.

3. Conventional Technologies

Presented in the Table 3 is the trio of technologies that exists for more then half a century and up to now they remain the only few available at the market place, being used on a large scale all over the world.

TABLE 3. Conventional field technologies for mines detection

Sensor name	Maturity	Cost	Country/ application
Mechanical prodding	H	M	All, over the world
Dogs	H	HH	-"-
Metal Detectors	H	L	-"-

Legend: for maturity: L (low) - R&D laboratory tests; M (medium) - R&D prototype laboratory/field tests; H (high) - prototype field operation

For costs: L- cost < 5 k$; M - 2-5 times more expensive compared with L; H- 10 times more expensive, compared with L; HH > 10 times.

3.1 MANUAL PRODDING

Manual prodding is based on slow and cautious piercing of the upper layer of the soil - down to 1-2 feet by a rigid stick (metal or wooden) in search for a mine. This laborious procedure, when done systematically and thoroughly- provides near 100% of the mine detection probability. The average sweeping rate is in the range one-few m^2 per hour depending on environmental conditions- grass, bushes, etc. In some cases the operator is forced to work lying or sitting on the ground. The usual clutter- underground stones, tree roots, shrapnel- result in laborious repetitive checks. The mine search by prodding requires beside the physical work, also substantial guesswork and experience of the operator creating a mental image of the underground anomaly. In fact this fusion of the prodding data with the brain, guesswork plays a decisive role in the final conclusion on the nature of the discovered underground anomaly. Up to now manual prodding is frequently used as the confirmatory last stage in the mine search with MD, when each FA must be thoroughly checked manually by prodding. It is incompatible with the large scale demining, but still it is widely used in the third world countries.

3.2 SEARCH DOGS

Search dogs, as natural biosensors, have a record of being used for luggage drug and arms control. They are also efficient for land mine detection being able to detect only one- but ultimate specific signature - chemical vapors of the explosives, leaking through the soil. Recently dogs proved their perfect ability to cope with even large scale demining operation in Afghanistan. As it was mentioned earlier, in a team with experienced operator dogs provide almost 100% discrimination probability of hidden very small quantities of explosives in variety of the environment conditions. In fact the dog's nose sensitivity is high enough to detect even molecular quantities of explosive in the air. Trained dogs are able to detect the explosive vapor long after the mine was laid in the ground, and from many meters distance from the mine location. As it was reported elsewhere there are several training centers for mine search dog, and the program usually takes three year training. By various estimations, nearly 500 dogs are presently regularly used in humanitarian demining. The average sweeping rate of the dog could make up to several thousand square meters per working hour. This number varies significantly depending on the experience of the dog handler, variability of the dog's congenital sniffing abilities, and how the handler and the dog work together.

The main drawbacks of the dog application are: very high price, limited operational time, and their dependence on the climatic conditions (wind, rain, snow). The average search dog can work in the field not more than a few hours per shift.

It worth mentioning in here that a modified approach of using dogs not directly in active search in the mine fields, but as stationary chemical sniffers identifying the scent of explosives in the air samples collected from potential mine fields. Such an approach could provide information of mine presence along the big sections of the road, which supposed to be used in large-scale military operation [48]. And though it lacks may be in precision for defining precision in defining position of the buried explosive its high reliability compensate for this. A team of highly experienced operator and dog could be easily qualified as the only large scale field-operational multi sensor system of the second/third level.

3.3 METAL DETECTORS (MD)

Simple standard MD is a hand held electromagnetic induction devices based on generation of low-middle frequency EM-fields and registration of the scattered secondary electromagnetic (EM)-field. Metal detectors consist of a search head with one current carrying magnetic coil. The time varying EM-field generated by the coil interacts with soil and mine resulting in secondary EM-field, the character and pattern of which depends on different EM properties of soil and metal. As a result, the presence of the metal case under the surface provides distortion of the usual secondary EM-field, which is registered and transformed electronically in the tone in the earphones so as not to distract the visual attention of the operator to the searched surface.

MD, even in their simple version, is very selectively sensitive to metal, and able to set alarm by sensing metal shrapnel of one to a few grams, or the striker pin in the mine initiator, placed at the depth of few inches under the surface. The respective Receiver Operating Characteristics (ROC) curves are low with an average ratio of detection probability (P_d) to false alarm probability (P_{FA}) in the wide range of 100-1000 depending on the environment and expertise of the operator or the presence of, bottles, roots of trees, cavities, etc. This unselective sensitivity to metal represent the main flaw of MD. Another flaw of all MD is its very low sensitivity for plastic case mines. Due to a strong dependence of the generated EM-field amplitude on the distance from the coil and similar dependence of the secondary field distance from the soil surface, the amplitude of resulting field drops dramatically with the distance of the search head from the surface, so that operation with MD and interpretation of the audio signals require high experienced operator, and turns in some type of the "art", akin to the prodding technique

The experienced operator could demonstrate significant increase of the ROC data (one order at least) with the same MD compared with a non experienced one.

And though a simple MD could be classified as a first-level device (it detects only the metal content in the underground object, which could be an innocuous shrapnel), an experienced operator is able to get from MD signals an additional important information on the nature of the hidden object, such as depth, size and even an approximate shape, based on the amplitude and the rate of the signal decay when manually translating the head over the surface. [10,11]

With this ability plus simplicity of use, customer friendliness and low cost make the MD a primary detection device all over the world with a reliable and successful record for more than a half of a century.

4. Uni-sensor technologies (First Level)

Table 4 presents a list of uni-sensor technologies, which are currently being developed, including field tests and laboratory simulations, and comprehensive summary of their qualitative maturity and cost estimation. The data in the table are based on the published materials [10,11]. In general, uni-sensors belong to the first level technologies that are focused on one of the many specific mine signatures. With additional sensors looking for other specific signatures, we could increase the level of technology to the next stage.

TABLE 4. Cumulative List of Uni-Sensor Technologies

Sensor type	**Maturity**	**Cost**	**Specific signature**
Passive mm Wave Radiometer	H	M	Enhanced reflectivity of the mine surface (man-made object)
Infrared Thermograph	H	M	Temporal anomaly in the IR field in the mine location
Magnetometer	H	M	Change of the natural magnetic field in the mine location
Ground Penetrating Radar	H	M-H	Enhanced reflectivity of the mine (man-made object)
Sonar	M	H	Complex structure of the mine body (shell + cavity + inner parts)
Ion Mobility Spectrometer	L-M	H-HH	Specific ion mobility of various chemical compounds
Biological sensors	L	H	Specific molecular structure of the explosives
X-ray Back-scattering	L	H	Characteristic spectra of the scattered X-rays
Nuclear quadrupole Resonance	M	L	Selective absorption of RF waves, specific for explosive
STN (Scattered Thermal Neutrons)	L-M	M-H	Enhanced scattering of thermal neutrons by a mine body
TNA (capture gamma quanta)	L-M	M-H	Characteristic Gamma Lines Gamma Links
Associate particles	L	H- HH	Characteristic + Gamma Lines + α particle

Legend: for maturity: L (low) - R&D laboratory tests; M (medium) - R&D prototype laboratory/field tests; H (high) - prototype field operation. For costs: L- cost < 5 k$; M - 2-5 times more expensive compared with L; H- 10 times more expensive, compared with L; HH > 10 times.

As was mentioned earlier, such a classification is more or less arbitrary, because in reality we do not deal with a pure first level sensor, but rather in combination with human factor, especially in case of simplest uni-sensor technology, such as mechanical prodding, or a simple metal detector. The following brief analysis underlines main strong and weak points of these uni-sensor techniques.

4.1 PASSIVE MILLIMETER ELECTROMAGNETIC WAVE RADIOMETERS (PEWR)

The physics of these sensors pertains to difference between reflective and absorption properties of the soil and mine with respect to external millimetre wave band radiation that is normally provided by the sun light: the metal case has significantly higher reflectivity compared with the soil, especially when it is humid or covered with grass. Soil radiates as a "gray" body, depending on its temperature. While a metal object does reflect independently of its temperature, thus mine could become visible by subtracting normal background. [66] The difference is pronounced near the mine, especially when mines are laid on the surface or at shallow depth, and absorption of the radiation is not significant. The range of the used EM-waves could span up to 90. There is a trade-off between the wavelength (longer waves have better penetration ability) and resolution (visa versa) [67]. Being focused primarily on the metal content and secondary, on the shape of the searched anomaly, PEWR could be classified as an intermediate technology between the first and the second level.

4.2 INFRARED THERMOGRAPH

Though Infrared Radiation (IR) does not penetrate the ground, the presence of the mine body near the surface provides significant change of the thermal flow pattern in the surrounding soil volume, resulting in temporal change of the thermal field distribution around thc minc, including surfacc layer and its radiation in the IR waveband. Respectively, it could make mine marginally or reasonably visible in IR range. [17] Thus, an operator analyzing the picture could make a decision on the object nature based on its regular shape. The dynamical changes are induced by the regular day/night periodical changes of the air temperature, or they could be induced by a powerful heat pulse, applied near the soil surface. Experimental study shows that the smallest thermal contrast in the mine location when the latter is almost invisible appears during the morning and the evening, i.e. just after the equilibrium of the temperature in the soil is reached after the night and the day respectively.

IR sensor technology is very mature, and of special interest are sensors operating in the medium and long wave band range (3-5 and 8-12 microns), corresponding to "atmospheric windows" with low losses. It makes IR sensors potential for stand off applications for mines detection from the air borne devices, especially for military use [58]. Periodic flights over the same territory in different time of day or night provide the sequence of large area images, where the minefields could be expected. All this makes IR technology very promising, which explains wide efforts in this direction all over the world.

The main flaw of the IR detection stays with variability of seasonal and day-to-day atmospheric and environmental conditions. Improved versions of IR sensors use additional parameter of radiation- its polarization when reflected or emitted by the man-made objects, which could useful in case of mines laid on the surface. To improve the stability of the IR approach and its a independence on the climate variation, an intentional irradiation of the checked area with a high power microwave pulse was applied. In this case the IR contrast between buried mine and surrounding soil become dramatically enhanced. It is illustrated in Figure 3, where IR pictures taken by a standard Agema Thermovision 880 IR camera are given in a sequence of time [36].

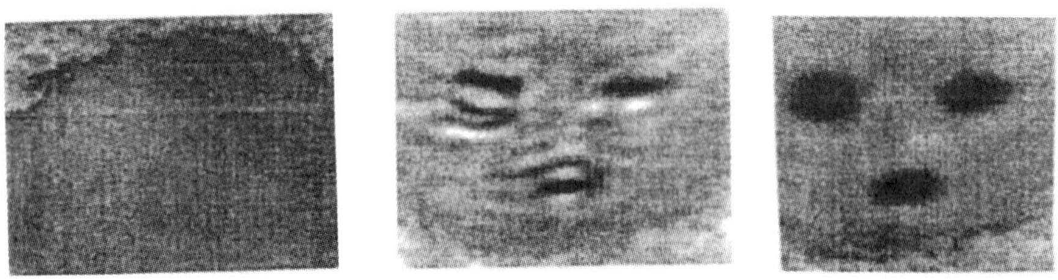

Figures 3a. b,c . IR images of 3 surrogate mines placed at 2.5 cm depth under the surface of moist sand: a) Before application of HPM (HPM- High Power Microwave pulse); b) 2 minutes after application of HPM c) 5.5. minutes after application of HPM. HPM power- 5 kW, t= 1.5 minutes [36]

4.3 MAGNETOMETRY

Magnetometer measures the deviation of the local natural magnetic field from the background level, or the spatial gradient (derivative) of the magnetic field (gradiometer). The natural magnetic field of Earth (approximately 0.4 G) experiences very slow spatial variation defined by the large scale relief, and fast very small spatial variations caused by the underground structural non-uniformity. Magnetometers have a long record of successful application in various domains, such as geology, archaeology, and medicine. Application of the highly sensitivity "squids" in magnetometers make their application promising for mine detection. Common off-the-shelf magnetometers are able to detect large deeply buried underground objects: UXO, shells, bombs, and large shrapnel. When used as a uni-sensor, the highly sensitive magnetometers suffer from a flow similar to MD flaw- high false alarm rate. The " manual" selection of the potential UXO based solely on the magnetometer data that provide only a pattern on the distorted magnetic field requires highly qualified operator to select potential UXO target, based on experience and intuition, which is time consuming and costly. [57,58].

4.4 GROUND PENETRATING RADAR

Ground Penetrating Radar (GPR) is akin to a mature technology of radio location, developed and widely used in science and technology, industry, communications, warfare, etc. for several decades. In difference from usual radar, GPR probing radiation is directed toward the soil. The promise of the radar technology for mine detection is based on its ability to provide a stand-off detection of mines more or less independently of their case material, (including plastic mines) from the distance of several meters, when using ground vehicles, and to hundreds of meters when using airborne platforms [20]. Its application for mine problem started near two decades ago, when GPR demonstrated potential to get information on size and shape of the anomaly. The typical GPR pulse is illustrated at Figure 4. Steady progress on application of GPR made it a leader among variety of emerging novel mine detection techniques. As it is now, GPR has the biggest number of partisans and has attracted the strongest R&D support under big programs such as GSTAIDS (Ground based Stand off Mine Detection System), HSTAMIDS (Hand-held Stand-off Mine Detection System), and a Short Pulse Hand held GPR [65,55].

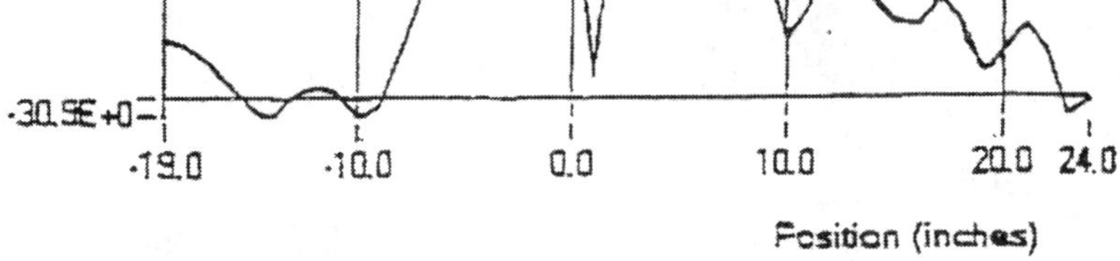

Figure 4. Typical GPR signal. Peaks correspond to edges of the mine [49]

There are three main types of GPR: the first provides mono-pulses with constant carrier frequency (time domain impulse GPR) and a square pulse envelope [22] the second (time domain GPR) operates with a pulse-trains waveform of the same square envelope amplitude, but with carrier frequency varying during each of the pulse; the third- type generates semi-continuous train of waves with frequency changing in a prescribed range (continuously or by finite steps- frequency domain GPR). In case of frequency change more than 25%, the radar is named UWB- Ultra Wide Band [21,10,55].

GPR based uni-sensors have entered a prototype field-testing stage. For example, DeTec-2, developed by US and France companies was tested recently at Cambodia [Brooks98]. Based on the wide range of the R&D and field tests by different companies and government facilities in many countries this technology is considered as highly potential for wide scale application in a multi sensor system.

4.5 SONAR

Sonar technologies, based on generating of seismic waves in the soil and registering their pattern after propagation to the sensor are also widely used in oil exploration, geothermal and environmental work. The technology of mine detection based on sonar application proved its potential for registering all types of mines, including those with non-metal case, which represent a challenging problem for all EM type detectors, such as MD detectors and partly- GPR. Sonar technology for land mines detection uses surface Rayleigh waves propagating in the near surface layer of < 0.5 m, which is characteristic for mines depth location. The conventional sonar technology provides the ratio of clutter/ATM by various estimations in the range of 10-100. By contrast, APM could not be discriminated from clutters, except for ideal conditions.[37].

To increase the discrimination ability of the sonar waves, diffraction around *the* hollow or non-uniform shell by acoustic surface waves was used. This phenomenon could provide additional specific signature for discrimination a mine from a solid body such as a stone, trunk, etc. Typical image of the reflected from the surrogate mine sonar signal in the time-frequency coordinates is illustrated at Figure 5 [39]. So far, the tests of feasibility of sonar technology for mine/clutter discrimination were done in simulated conditions of a "free" field, and needs to be extensively tested in a real field environment.

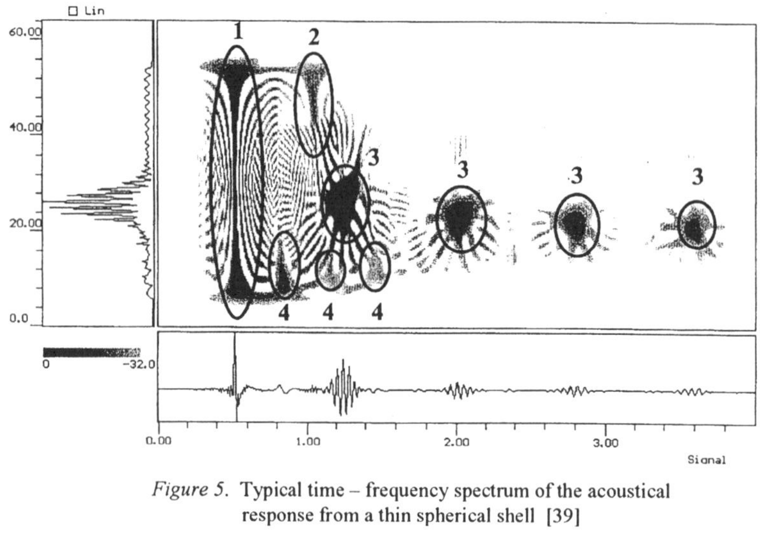

Figure 5. Typical time – frequency spectrum of the acoustical response from a thin spherical shell [39]

4.6 ION MOBILITY SPECTROMETER (IMS)

This device represents approach, which (similar to dogs) "looks" for one-step identification of specific vapors of explosive [11]. It analyses the presence of the latter in the minute sample of the air taken over the surface The technical principle of IMS is based on the different mobility of ions of respective chemical compounds. IMS registers different times of the their arrival to the collector, (that depends on the molecular weight of the ions), thus forming a specific spectrum- "fingerprint" of the compound molecule. By this sequence IMS is akin to GCE /MS analysis (gas-chromatograph/mass-spectrometry), through being several orders more sensitive than the latter, with detection limits at least on the level of 1 ppb, which corresponds to such minute quantity as 10^{-12} g of explosive vapors.

The main drawbacks of IMS are an inability to detect directly the location of explosive, increased complexity in identifying chemical "fingerprints" in case of mixed explosives, and low sweeping rate. The portable field versions of IMS with a weight of 6-10 kg, and operational time of 6-8 hours from a battery backpack, that were developed by several facilities in Germany and US, have a cost of 15-20 K$. It can work in a quasi-continuous mode with data acquisition time at each place in the range of few-to 30 s per spectrum sample. Research at Sandia National Laboratories designed a prototype for soil/vapor collection. The soil sampler employs a lightweight probe for extracting and concentrating explosive vapor from the soil, and then molecules of explosives are thermally de-sorbed from the concentrator and analyzed in the IMS. The system is capable to detect sub-part-per billion of TNT and related explosive munitions [29].

4.7 BIOLOGICAL AND CHEMICAL SENSORS

As it was discussed earlier, chemical vapors emanating from the mine represent the ultimate direct signature of the explosive presence, which could be detected by a specially trained dog. The idea of creating artificial "nose" attracted massive research efforts and respective financial support that could be comparable with the GPR. It pertains to a very high sensitivity and selectivity demonstrated by such a "nose" (parts per trillion and lower of explosive vapor in the air), customer friendliness, compared with alternative technologies using EM and nuclear radiation, bulk hard ware and sophisticated software. In contrast to them, a man-made "nose" could deliver search result in as- YES or NO in the quest for a hidden explosive. One of the studied techniques is based on using novel fluorescent polymers. When exposed to specific nitro-aromatic vapors (which are found in many landmine explosives), the intensity of light emitted by those polymers dramatically decreases (on many orders) due to the fact that a single molecule of TNT when binding to polymer chain, could completely "switch off" whole chain.[23].

4.7.1 Fluorescent bacteria

Another approach focused on the direct registration of explosive vapors uses the bacteria adapted to scavenge low concentration of nitro-aromatic compounds, and display this process by a fluorescence. Respective genetically recombinant bacteria, playing the role of a bio-reporter on the nitro-aromatics present, were created, that produce a bio-luminescent signal. Thus by finding the fluorescent bacteria one could find a mine [9].

The promising feature of this innovative approach is the possibility to substitute the fluorescent gene in bacteria for increased sensitivity. The weak point of this technique is the time scale of the bacteria growth in the field conditions. The respective full signal takes several hours after the inoculation. This technology could be perfectly applicable in humanitarian de-mining programs when there is no imposed strict time limitation.

4.7.2 Black-Carbon-Polymer-Resistors

Another method of direct sensing of the explosive compound was suggested by using a change of the electrical conductivity of carbon-black-polymer composite resistors, when being exposed to the action of the vapor emanated by the mine. This specific signature pertains to the phenomenon of reversible swelling of resistor under these conditions, the magnitude of which depends on different gas-solid partitions for various polymers. Thus in principle it is possible to get a chemical spectrum of the explosive compound using an array of different polymer resistors [38] The tests demonstrated ability of such an array to resolve common organic solvents. The approach seems to be of high potential, but still are limited by laboratory tests with simulated UXO.

4.8 X-RAY BACK SCATTERING (XRBS)

XRBS uses photons, initially directed into the ground and then scattered into the aperture of the plastic scintillator. The specific feature, used by this technique, is the difference between scattering and absorbing properties of the man-made mine material and of surrounding soil. It results in more intense back-scattering of X-rays on the light elements of the explosive compared with soil elements and make feasible generation of 2D image of the underground object even with a modest collimating degree. Figure 6 illustrates image of a plastic mine irradiated by X-rays source in simulated laboratory conditions when X-ray source was translated over the surface with buried mine. [35]. The system looks promising for discriminating light materials (which are usually man made), and imaging fuzzy shape of the underground anomaly. Its weak part is usage of X-rays generators that pertains to more strict health and safety regulations, bulky size, and large cost.

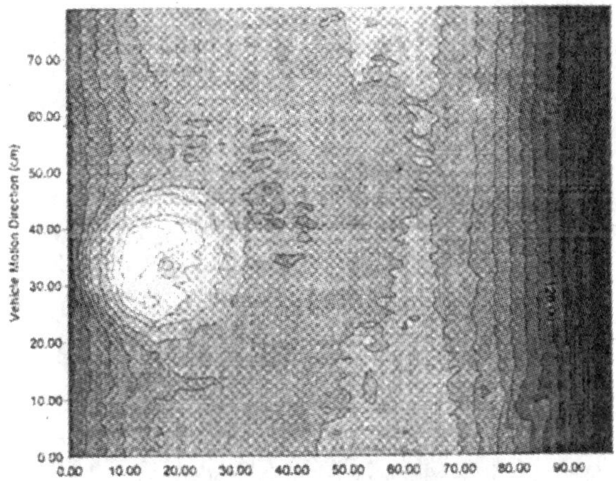

Figure 6. Image of a plastic surrogate of ATM produced by XRBS. X-ray source located at 38 cm over the sand box with a surrogate ATM Detectors: 5.1. cm thick plastic scintillator, X-ray machine – 140 kV, was transported over the sand box [35].

4.9 NUCLEAR QUADRUPOLE RESONANCE (NCR)

Phenomenon of NCR is based on a non-spherical shape of nitrogen nucleus, resulting in appearance with a specific quadrupole E-moment. The existence of the latter provides its precession in the internal E-field of the explosives crystallites with a frequency defined by the amplitude of the quadrupole moment, and atomic structure of the respective crystalline. This frequency plays a role of a specific signature: when the frequency of externally applied RF-EM is tuned to this precession frequency, it results in a peak resonant absorption of the supplied EM -energy, that could be registered and discriminated from an average background [32]. The approach is customer friendly and could be realized in the hand -held modification, as it is shown at Figure 7.

Laboratory tests in simulated conditions demonstrated a high probability of RDX detection, reaching 90% in various bulk quantities of 60g- 2.5 kg of RDX (corresponding to APM and ATM) at the depth between 5 and 20 cm. The respective ROC curve (Receiver Operational Characteristics, Figure 8) demonstrates very steep rise with respectively very low false alarm rate, (10%), which is unique, compared with other EM-based diagnostics. The main problem of application of NQR in field conditions are magneto-acoustic and piezo-electric ringing, radio frequency pickup and ground artifact signals. NFQ seems very promising for a one-stage detection of bulk nitrogen, one component of the “explosive tetrad”.

Figure 7. Nuclear Quadrupole Resonance (NQR) based gradiometer head in a water proof housing [32]

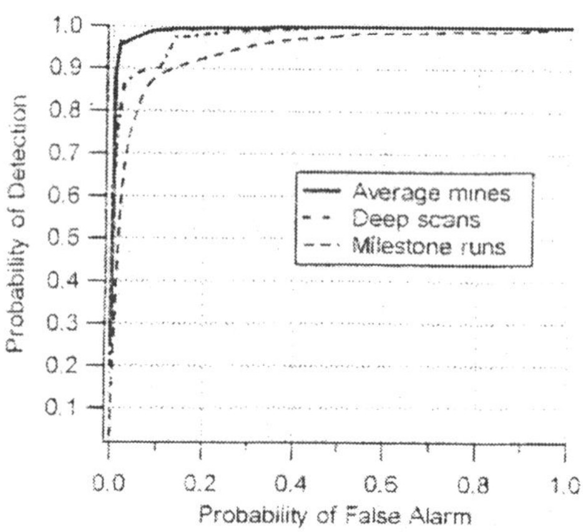

Figure 8. ROC curve for NQR blind test series data on detecting mine surrogate (1 kg mass) , placed at 15 cm depth in sand box [32]

4.10 SCATTERED THERMAL NEUTRONS (STN)

Detecting a Scattered Thermal Neutron (STN) flux could also be classified as a first level sensor, focused on only one specific mine signature- bulk concentration of light elements (primarily - of hydrogen). The latter is characteristic not only for explosives, or the plastic case of the mine, but also for natural non-uniformities present in the soil- burrows, water puddles, big trunks of trees, some innocuous man made substances, such as fertilizers, plastic waste, etc.[50]. The promise of STN is related with its sensitivity to plastic mines in the environment heavily polluted by metal, where operation of MD is very inefficient because of extremely high rate of false alarms. STN is completely insensitive to these clutters. The physics of the method is based on fast moderation and scattering of the primary fast neutrons on the bulk of light elements of present in the mine (explosive itself and plastic housing of the mine), compared with the case of a common soil. It results in fast moderation of neutrons, and their trapping inside and near the mine with formation of a "cloud" of thermal neutrons over the location of the mine, which could registered by respective neutron-sensor. It is illustrated at the Figures 9 and 10. The picture of TSN detector in operation during field trial at Nevada Test Site is given in Figure 11. Availability, reasonably low cost (~ 5-6 K$), compactness (weight – 2-3 kG) and relative radiation safety (at 1.5 meter distance from the neutron source of 4 x 10^6 n/s exposition dose satisfies to permitted limits for radiation worker) are the strong sides of this approach.

Figure 10 sketch of spatial distribution of the n-count rate over the buried hydrogen bulk anomaly. The circular curves correspond to "crow flight" distances clutter-underground area with enhanced humidity c) antitank mine [13].

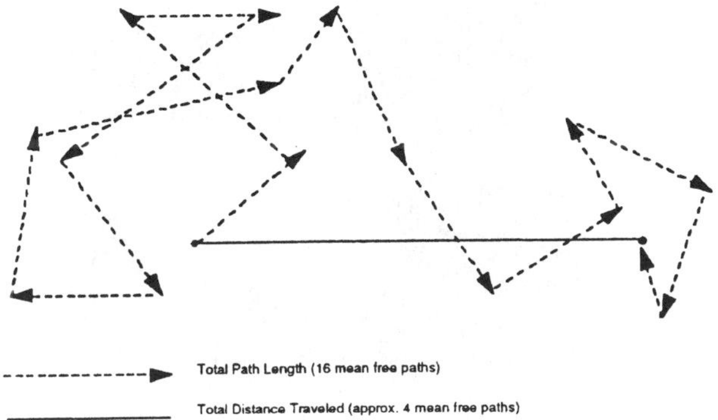

Figure 9. Neutron scattering in the matter: Path length (broken line), total distance travel (Solid line) [53]

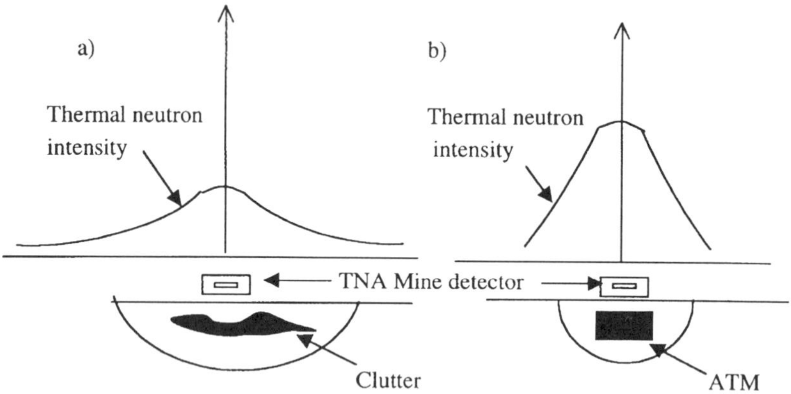

Figure 10. Neutron count rate over the buried hydrogen bulk anomaly and an antitank mine. a) clutter, b) mine [13]

Figure 11. STN detector in operation during field trial at Nevada Test Site [50]

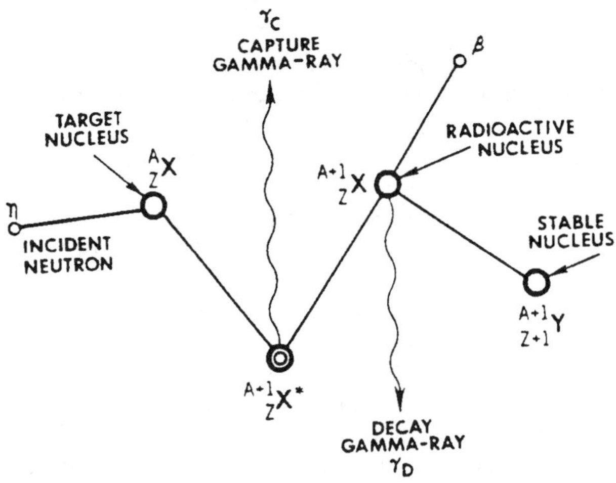

Figure 12. **Sequence of neutron/nucleus scattering and capture, accompanied with prompt and delayed gamma radiation**

Respective R&D were carried in Russia and England and several operational prototypes were tested in the field conditions [50]. The main weak points of STN are its inability to work with the wet soils of humidity higher than 5-6%, (i.e. its operation is limited by a dry climate geographical location of the place and mainl by summer season), limited depth of the screening (not more then 5-6 inches) and limited applicability in case of small mass of APM. In fact, in the latter case the STN would detect plastic case of the APM rather than explosive itself, being non-sensitive to APM with metal cases.

4.11 THERMAL NEUTRON/ACTIVATION (TNA)

This approach presents an extension of the previous method. It looks for bulk nitrogen, which is absent in the common soil) via specific gamma rays of 10.8 MeV, generated in the reaction of thermal neutron capture by the nitrogen nucleus [51,44,53]. Of all neutron reaction channels with nitrogen nucleus only 5% results in generation of this high-energy quanta. Due to its high energy, the gamma rays penetrate the upper layer of the ground almost without attenuation to be registered with off-the-shelf technology of scintillator spectrometers. The TNA is non-sensitive to metal shrapnel. Beside this, very high energy of the respective gamma rays provides TNA operation in the energy range almost free from noise. For example only 9.5×10^{-5} gamma quanta of 10.8 MeV per each thermal neutron will be generated as noise in the common soil. In the same time TNA approach as the sole first level sensor suffers from several serious flaws, such as:

* use of the neutron source, related with additional health safety rules complications

* slow accumulation of acquired counts due to a small cross section of respective reaction. The rate of data accumulation is limited by the safety consideration and possible "pile up"
* need for high quality detectors with better resolution compared with inexpensive NaI to cut off significant Si input with the energy of 10.8 MeV.

4.12 FNA/ FNA + ASSOCIATED-PARTICLE

Beside TNA, there are various approaches based on registration of prompt gamma generated during inelastic scattering of fast neutron on the nuclei of explosive [28,30]. In difference from STN the method is sensitive to all 3 elements of triad (O, C, N). Of special interest is FNA in combination with registration of associated alpha particle [56]. Using the coincidence approach with a small time window (few nanoseconds) and respective time delay, this technique provides significant improvement of SBR (signal to background ratio), and information on location of origin of registered gamma quanta. The latter could be used for imaging of the underground anomaly. FNA techniques are usually based on fast mono energetic neutrons probing fluxes generated in the DT reaction using compact neutron tubes, which operate in the DC or pulsed mode. [64,12,41]. In the latter case SBR factor could be significantly improved.

5. Data Processing And Fusion

Raw data flow pertaining to specific signatures acquired by any sensor is routinely accompanied by a noise of various origins. Such a noise masks the mine presence, thus significantly decreasing probability of mine identification. In case of active probing this noise could be generated by a primary radiation of the source itself, that leaks to the sensor; by multiple reflections of source radiation from the nearby objects surrounding mine, and from non-uniformities of the soil. There is also intrinsic noise of the registering electronics, etc. As a result field operation of the sensor could be dramatically worse compared with its performance in the simulated laboratory conditions, or literally not possible at all without processing of the raw data flow.

The procedure of processing is focused on the enhancement of the "signal/noise" ratio, related with specific signature of the mine. In general, there are four possible outcomes of this procedure: (1) mine is detected and classified as mine; (2) mine is missed or classified as a non-mine; (3) non-mine is classified as a mine; (4) non- mine is classified as a non-mine [7].

Of these four scenarios the first and fourth outcomes are correct, still the 4-th outcome could be related with some additional labor and time expense of the operator. The outcome #2 is a clear failure related with the worst consequences: possible human victims and equipment/property damage. Depending on the type and scale of the mine detection operation (humanitarian - in the post war condition or military war time de-mining in the battle field conditions) the respective permissible rates of the outcome #2 could vary from 0.2 to 0.01, which corresponds to one to several orders difference in specific expense per one detected mine.

The outcome 3 is not the worst, but still it is related with wasted laborious efforts to clear, remove and finally dispose the object, though it is not a mine.

Thus the FA problem remains the most serious obstacle for the large-scale de-mining operations. As practice shows, in general near 70% of all de-mining efforts is spent on searching the places where there is no mine at all. Beside this, in the suspicious places with rare abandoned mines the ratio of average time spent on the "true" detection of mine (outcome 1) is between 10^{-2} and 10^{-3}. The latter number is when using conventional metal detectors in heavily polluted by metallic debris former battlefields and war zones. Is it possible to improve this situation?

It is well known fact that vision provides us with near 80% of all information flow from the surrounding world. Analyzing respective images, we make conclusions, which are based on our expertise and additional available data. Presentation of the processed data flow from sensors in the form of some picture (it could be in a moment, phase, or frequency domain, i.e.- not mandatory a picture in a conventional space coordinates) significantly improves the on-line ability of the operator to make a final decision. For example, when using any type of hand-held detector over the suspicious place, generation of an image at the screen, (instead of creating a brain image of the underground anomaly based on the mental guesswork), could provide operator with unambiguous information about shape and size of the buried anomaly. This is why the general tendency of the ongoing R&D for mine detection, is to get the data from several disparate sensors "looking" for specific signatures "fused" in a form of some image via processing these data.

The spectrum of pertinent processing algorithms is very wide, and is getting wider. Here we will briefly characterize only the general principles of processing, and list types of algorithms reported in the recent publications on mine detection.

The example of the primary processing procedure in case of using MD is a subtraction of an average "noise" from signal's rate by setting an appropriate input threshold, and cutting off the average background related to the normal terrain without buried mine.

In general, processing of the digitized images generated by uni-sensors, focused on the creation of the fused image, consists of three main parts: (1) filtering the noise from the raw data; (2) emphasizing the most salient features in each of the uni-sensor's images by a special arbitrary procedure, named convolution. It is based on prescribing to each of the pixel certain "interest" and relative weight compared with the surrounding pixels. It results in a new set of pixels value; (3) data fusion.

5.1 FILTERING THE DATA

There are many available off-the-shelf filtering methods soft to suppress unwanted noise, and spurious responses, such as using smoothing Gaussian operator that blurs the image.

5.2 CONVOLUTION OF THE IMAGE

The convolution procedure in general, is a neighborhood operation, when the value (i.e. measure of interest for any pixel- ("I") is obtained using some appropriate function, selected by the user (so called a sliding neighborhood operation) [42]. For example, I of any pixel "k" could be defined as a simple sum of eight differences between the value of this pixel and those of each of the adjacent pixels. It is illustrated at Figure 13. The respective formula for I is: $Ik = \sigma (V_k - V_m)$, where V_k and V_m are the respective values of central and 8 surrounding pixels respectively . In a more flexible approach the "fuzzy" interest of the pixel I_{fuz}"-could be calculated based on Gaussian type expression [61]:

$$I_{fuz} = \exp -([V_i - <V_{av}>]. /\sigma)^P,$$

here: V_k , and $<V_m>$ are values of the central pixel "k" and the average of all eight adjacent pixels "m" respectively, p- power to which the value is raised, σ - parameter controlling the width of the function spread. The arbitrary choice for power P and spread σ provides flexibility for the user to create new families of functions for calculating I_{fuz}.

The following is an illustration of the processing steps sequence for raw digital images obtained in experiments with IR uni-sensor, [15]:

* Transform the 8-bit grayscale acquired CCD image into a false color spectrum image to make it more significant for human vision (using the jet " color" map in the MATLAB)
* Apply a Gaussian smoothing operator. It suppresses the noise and the spurious responses by blurring the image.
* Fulfill pixel convolution with a Gaussian filter having a 3x3 mask and $\sigma = 2$. It removes the detail and noise from the original image
* Fulfil Sobel type convolution to the smoothed image. It produces the edge fragments, lines and curves, forming organized geometry (via grouping the edge points in the groups).
* Discriminate the coherent regions (lines, curves) by using segmentation techniques
* Set the maximum and minimum thresholds for pixel value gradients
* Assign the limiting value1 to all pixels with values higher than maximum threshold, and zero to the all pixels with value lower than minimum threshold.

Illustration of the processing sequence for IR image is given at Figure 14.

5.3 DATA FUSION

Putting aside details, all of existing approaches for data fusion could be classified along three basic categories:

* first approach: the initial data from all sensors are combined, and only after this the specific signatures (features of interest) are extracted (so called - data level fusion) ;
* second approach: the features of interest are first extracted from the data flow of each sensor and then fused to produce a full set of the specific signatures, collected by each sensor (so called-feature level fusion) ;

* third approach: decision "yes/no" is made based on the outputs of the individual sensors, and the decisions are then fused. [45]

The data level fusion of several digitized images is illustrated here by the example from [45]. After creating convoluted images for various sensors following described above processing scenario, we could use an arbitrary rule of combination of respective pixels to generate the proportional version of fused image. Such rule of combination for example could be:

$$V_{prop}(x,y) = \frac{\Sigma\,[(I_m fuz\ V_m(x,y)\ +A_m V_m(x,y)\,]}{\Sigma\,[(I_m fuz\ (x,y)\ +A_m(x,y)\,]}$$

here : $V_{prop}(x,y)$ = resulting fused value of the pixel at the position (x,y); $I_m fuz(x,y)$, and $V_m(x,y)$ - fuzzy interest and value of pixel (x,y) for different uni-sensors to be fused; A - intensity measure for pixel, which is simply 0 and 1 for white and black pixel respectively, and using fraction -for shades of gray. The resulting image could be compared with existing database of images, and final decision on the mine presence –will be made by operator, or respective software.

In the second approach - "feature level fusion" - instead of fusing in some proportional manner all pixels from uni-sensor processed images, only salient features from the latter are fused, such as contours lines, edges, etc. The resulting fused image could be analyzed compared with an existing database. For example in case of fusing the processed images from neutron and IR sensors only the line contours of the bulk nitrogen (and/or hydrogen) and that of the buried object respectively could be used for the final image. The reasonable similarity and overlapping of those contours at the screen mean "YES"- signal for the UXO presence (it could be also a box with nitrogen-rich fertilizer). The scheme of fusing three images from IR, GPR and MD sensors is given at Figure 15 [15].

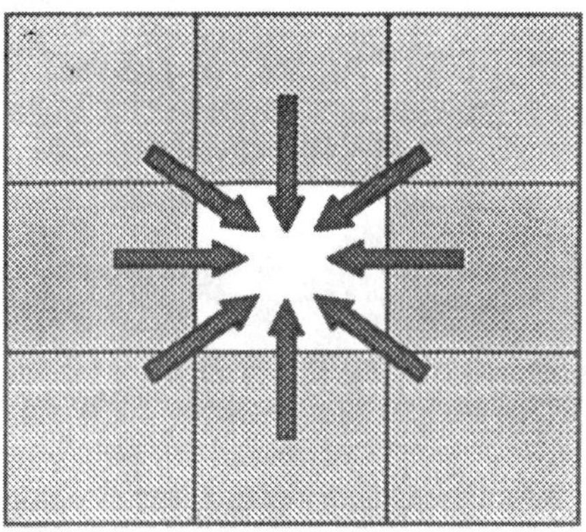

Figure 13. The pixel of concern is in the center of the square with convoluted absolute intensity calculated as sum of differences to each of the surrounding eight pixels [45]

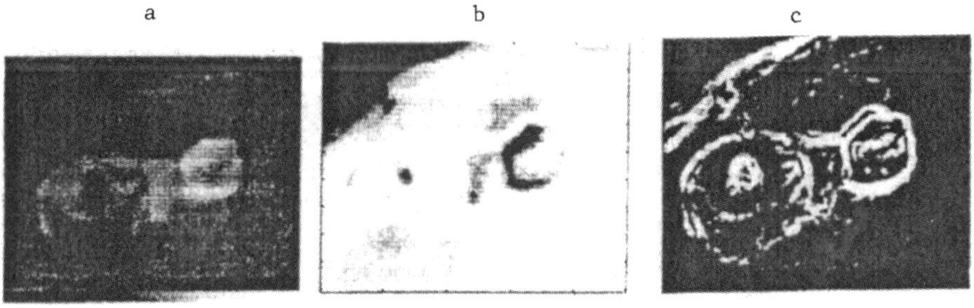

Figure 14. Sequence of IR image processing: a. Unprocessed image of 2 plastic mines, placed at the sand surface; b. Gaussian smoothing of spurious details; c. selection of Sobel egde

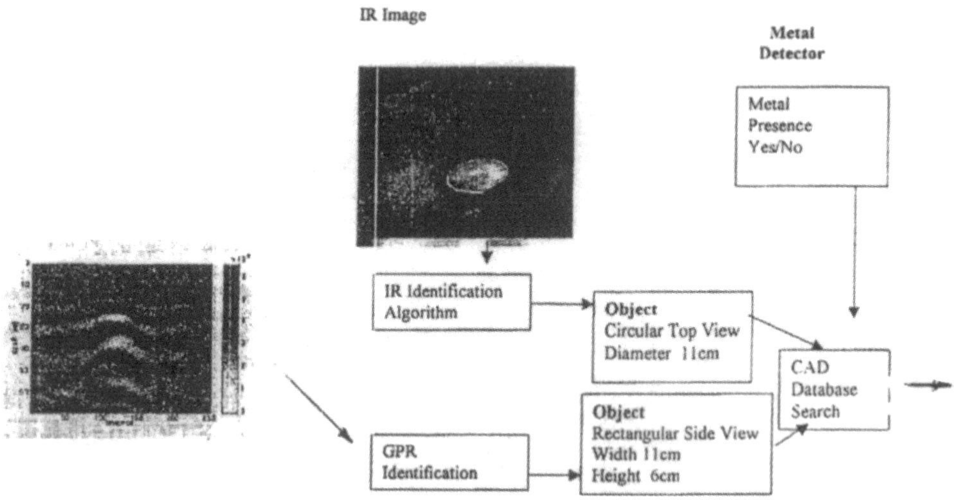

Figure 15. Data fusion scheme from IR, GPR and MD sensors [15]

6. Multisensor Technologies

Any mine detecting technology based on combination of various sensors deals with the problem of final decision, taken automatically or by a mental effort of the operator, based on the acquired data (combined/ fused). Thus, in the latter case the final decision could depend on the operator expertise, ability, will and obligation. In general, interaction of the detecting system with the operator could be analyzed in terms of OODA loop. (Observe-Orient-Decide-Act). Referring to [52], the respective function hierarchy could be considered as a sequence of levels with the first two (technical) presented by a process of data acquisition, processing and fusion, and the last one- intellectual interaction of the operator with the interface resulting with decision choice. The following discussion on multisensor systems characteristics and performance will be limited only by the technical portion of this sequence

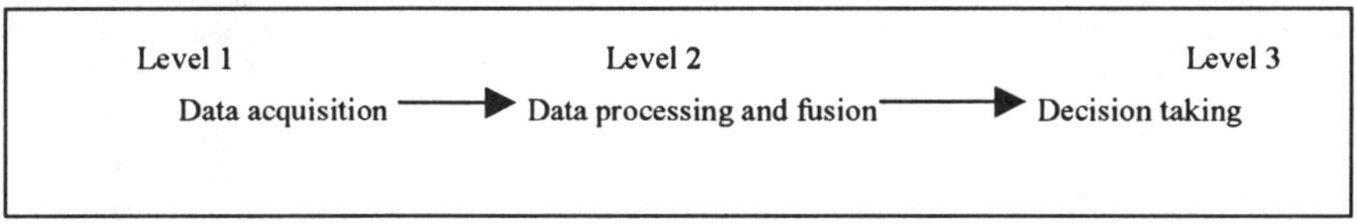

Figure 16. Sensor task and function hierarchy

The Table 5 presents a comprehensive list of reported laboratory and field trials on multisensor systems (MSS), and briefly summarizes their main characteristics and other important data. Discussion of the listed MSS follows.

6.1 PROJECT (MTADS-MULTI SENSOR TOWED ARRAY DETECTION SYSTEM)

The project was developed at the US Naval Research Laboratory and presently is in the stage of full-scale field prototype tests. The photo of the towed sensors system is presented at Figure 16. The system can detect military ordnance to its maximum self - burial depth (down to several feet) with position accuracy of 15 cm. It features 8 magnetometers, that could work in the line array of 1.75 m wide, or as four pairs with a total span of 1.5 m. Sensors are placed in the towed trailer at 4.9 m behind the tow vehicle. The magnetometer data are collected at frequency of 50 Hz with a typical survey speed of 3 m/s and sampling interval of 6 cm respectively. Pulsed inductive array includes 3 sensors positioned 3.1. m behind the tow vehicle, operating with a sampling interval of 10 Hz. Positioning of the system is controlled by GPS, providing accuracy up to 5 cm with sampling 0.2 s. The autopicker algorithm (program responsible for final decision "yes or no" without intervention of the operator) is focused on minimization of mine misses, (i.e. with minimal imposed without intervention of the operator) classifications. MTAD performance was demonstrated at two sites heavily polluted by a metal scrap. With set conditions the average FA rate was not more than 3: 1, and average accuracy of positioning was 12 cm with 95% of targets detected within 29 cm. These tests provided information on efficiency of the chosen filtering method in combination with PCA (principal characteristics analysis) for automatic mine discrimination without intervention of the operator and possible alley for its further improvement.

TABLE 5. Multisensor Systems Characteristics

Project/Start Year/ Country / Type of mines / Type of transportation/ Level of maturity / Combination of Sensors / Performance	**Sequence of Operations/ Data fusion**
Project MTADS- Multi sensor Towed Array Detection System)/94/US/unexploded UXO / Vehicle mounted, towed/Full scale tests at Badland and Blossom Point bombing sites/ Magnetometers Array (822ROV) & Gradimeters Array imaging + three Pulsed EM Sensors (Geonix EM-61) imaging + GPS controlled	Simultaneous terrain screening with Magneto and EM sensors→ selection and enhancement of specific signature (circular shape with varying diameters) for magnetometer and both EM images →PCA (Principal Component Analysis) used for fusion Mag- and two EM images with equal interest weight for each channel
Project ILDP/ 94/Canada/ ATM/ Vehicle Mounted, Remote Control/ Full scale field/road tests, IR [(Agema 1000, 640pix x 400 pix; FOV-61°/41°, trapezoid area 3m/8m/6m) + VIS (CCD)] + EMI (24 pairs of active/ passive coils) + GPR (Elta) →TNA Speed: 2.4 km/h	Operator indicates IR mine like images→ EMI produces the intensity map→ detection algorithm scan the EMI map, passes information to VCS→ data from IR and EMI compared→VCS declares detection→spatially registered detection displayed→ GPR provides map similar to EMI → data fusion fulfilled→TNA confirmation →marking the place
Project: Adjunct to MTAD ESTCP 199811/98/ US/UXO/Man-portable system/ demonstration stage	GPS + Acoustic sensor + Dual magnetometer + EMI sensor →PCA
Project: Target Identification with Weak IR activation Methods fusion of GPR, IR and MD/ POP experiment	IR image + GPR image + MD signal →CAD database search
MURI Project/96/US/Surrogate Mines/Test Stand/ Laboratory Tests/ GPR1-6 GHz, 0.1 GHz (step) + EMI (Schiebel AN-19)+ IR IRIS 160 ST, 2.2- 2.6 μm)	Separate acquisitions by each of the sensor + post-processing for features + features fusion
Project HSTAMIDS/97 US /APM/ Hand Held/Laboratory tests GPR + MD + IR	Two stages (both - first stage (MD +GPR); 2nd stage (VS + IR); 2 stages fused, one decision is provided to operator
Project Atometer X-1/00/US/ ATM, APM/Lab-tory Test/ Prompt Gamma + Alpha Particles	Measuring relative abundance of elements O, C and N, and their location: via detection of pertinent prompt gamma lines, and accompanying alpha particles; final decision on explosive presence is made via Dalitz type 2D diagram.
Project: Buried Landmine Specific Sensor98/US/ ATM/ Robotic vehicle/Field tests/ IR +GPR+ MD + TNA+ manual prodding	Primary sensors (IR, GPR, MD) → TNA→data fusion→final decision, based on database compared.
Project UCI-LLNL/98/US/ATM/Hand Held Field Tests at BODF/STN + TNA(nitrogen)Two stage system	Two sequential stages: TSN→TNA (for nitrogen) type of fusion - mental guess work
Project JINR/00/Russia/Explosives /Lab-ry Stand/Prompt Gamma + Alpha (associated particle)	Coincidence in time window of two particles (α+ γ); discrimination of chemical elements, location of hidden explosive
Project Sandia/ 97/US/ UXO/Concept level/ swarm of palm size robotic uni-sensors	A swarm of palm size robotic sensors; sensors could be independent, or communicating; data are transmitted for processing to the control station, that modify the swarm behavior
Project CRDF. USA-Russia/ATM, APM/ thermal neutron 2D imaging + STN + TNA (nitrogen)	Two stage hand held /transportable system: STN + imaging+ TNA (nitrogen)
Project PELAN/USA/99/explosives/Lab-ry prototype, remote control/ elemental analysis	Multi-elemental analysis of hidden object content, based on carbon/oxygen and carbon/nitrogen ratio.

6.2 PROJECT ILDP/CANADA (TELEOPERATED MULTISENSOR MINE DETECTOR)

The Improved Land Mine Detector Project (Canada) is an advanced model of tele-operated multi sensor, vehicle transported system for mine detection. The system includes metal detector EMI sensors array, and IR, GPR and TNA detector. Its general outlook is presented at Figure 17. The wheel system provides minimum rolling and pitching of the vehicle and constant pressure on the surface. The front EMI sensors array consists of 24 overlapping transmit/receive coil pairs mounted on a flexible tray. IR imager is located behind and above the EMI tray, and its image span near 6 m in front of the vehicle and 8 m in width. GPR is located behind the IR at the constant height of 70 cm over the ground. TNA detector is placed on a trailer behind the vehicle. During the routine scan it is positioned coarsely, and only in case of need for confirmation check, it could be positioned precisely. The summary of system trials fulfilled at the end of 1997 is presented in the Table 6.

TABLE 6. Summary for ILDP performance

Target Class	Probability of Detection %	Standard Deviation	False Alarm Rate per 10 m	Standard Deviation
All mines	78.31	14.77	2.01	0.73
All ATM	85.08	13.25	2.01	-"-
Metal ATM	100	0	2.22	-"-
Low metal ATM	77.84	20.72	2.01	-"-
APM	37.30	32.22	2.72	-"-

These average results are very encouraging compared with all other existing systems, but they are still below humanitarian deming requirements. In the same time they are quite satisfactory for military applications.

Figure 17. General view of MTAD [45]

Figure 18. General view of multisensor towed array detection system for the improved land mine detection project [44]

6.3 PROJECT HSTAMIDS

Figure 19 demonstrates an artistic view of the HISTAMIDS being operated by a soldier. HSTAMIDS is "blind" for the presence of explosives, and "looks" for metal content, structure, and shape of the man made- buried anomaly. These specific signatures are ubiquitous for the battlefield situation. . What matters- it must reliably discriminate the shrapnel from unexploded mine. HSTAMIDS involves integrated use of GPR, MD and IR sensors. GPR operates in a wide bandwidth (0.5- 3 GHz) antenna, built on the coplanar circuit boards, and receiver is sensitive to distortion of the reflected waves generated by local differences in the dielectric constant of the soil, due to any anomaly. This signal is supported by the signal from MD, sensitive to any eddy currents generated in metal object. Forward looking IR sensor registers thermal gradient of the soil and generate respective IR image and displays it on the helmet mounted miniature screen.

The data fusion concept presently considered for HSTAMIDS technology includes: sensors data correlation in space; knowledge-based prioritization of data; "fusion of the decision" algorithm.

Figure 19. Atrtistic rendention of HSTAMIDS [49]

6.4 PROJECT: BURIED LANDMINE SPECIFIC SENSOR

MSS developed by Ancore (former SAIC) uses combination of two stages: fast primary stage based on metal sensitive sensors (IR, GPR, MD) with respective processing and fusion; and slow secondary confirmation stage (TNA) focused on detection of bulk nitrogen. So, the system does not fuse the data from both stages, but relying on the high sensitivity of the first one, uses the second stage only in already marked suspicious location to confirm or to waive the first stage selection. The general view of the device is given at Figure 19. TNA sensor for specific gamma registration is of conventional type, based on PM and NaJ scintillator. The n- source is ^{252}Cf, placed in moderator and additional shielding to protect the n-sensor from direct gamma and neutron radiation of the source. The system is characterized with high reliability when working with metal containing mines. In the same time its fast first stage displays low sensitive to plastic mines. Similar two-stage approach with different first stage sensor was suggested and later tested by [50].

Figure 20. General view of the ANCORE System (courtesy of ANCORE Corp.)

6.5 PROJECT UCI-LLN: FIELD TESTS OF HAND HELD TWO STAGE SYSTEM

In this project the first stage was focused on registration of enhanced bulk concentration of the light elements, mainly hydrogen. With a hand-held detector such a check (in case of a dry soil with humidity <2-3%) could be done with a reasonably high sweeping rate (one-two meters squared per minute). The suspicious place with enhanced concentration of the secondary thermal neutrons over the surface (2 σ over the average) was marked, and in the second stage - checked by TNA on the bulk presence of nitrogen. The main advantage of this approach is that both sensors look for different elements characteristic for explosive- H and N. The ability to detect both in one location could significantly decreases level of FA, and respectively increase discrimination power of the technology. The compact size of the device, its autonomic operation, and ability to work in complicated conditions (bushy terrain, hilly landscape) adds to its potential for humanitarian demining.

The proof-of principle-experiment on this approach with two separate hand-held detectors (n-n and n-gamma) was carried on at BORF - special testing facility at Nevada Test Site in 1998. The same n-source ^{252}Cf with intensity of $4x10^6$ n/s was used for both detectors working in series. During fast scanning of the selected lane with nn- detector, detected suspicious places with anomaly high level of scattered neutrons background were marked. In the second stage the marked places were checked by the n-gamma hand transportable device. Each of the marked place check took 5 minutes. The tests were conducted in winter during the sporadic rains, with average soil humidity on the level of 5-7%. No specially trained operator participated in the test. The results are presented in Table 7.

The tested technique clearly showed such deficiencies of the first stage as:

* its inability to work with humid soil , when the mine located at depth more then 2-3 inches becomes practically invisible due to very high background level of thermal neutrons, reflected from the upper layer;
* the absence of imaging soft ware, that would substitute the mental guess work of the operator by providing him with an image of the underground anomaly shape.

Based on these results, a modification of this approach by adding neutron imaging means to the first stage was analyzed [44]. It could provide the operator with a shape of the underground H- anomaly and thus help to discriminate its man-made origin from the natural one (such as burrows filled with water, irregular patches of the soil with enhanced humidity, etc.) The 2D image of the moderated neutron cloud over the surface could be generated using set of 1D position sensitive n-counters, or alternatively - with 2D n-counter. It is illustrated at Figure 20.

TABLE 7. Mine detection results with 2 hand held (n,n) and (n,σ) devices

Total area	**120 m2**
Total number of mines in the area: (VS-1.6 + VS 2.2); M19; M15	20 6; 7; 7
Depth in inches	2
Total number of detected mines, including (VS-1.6 + VS 2.2); M19; M15	12 6; 3; 3
Number of FA	22
Probability of mine detection (all) Plastic mines (VS 1.6 +VS 2.2) M 19 and M 15	60% 100% 43%
Probability of FA	0.18 /m2

6.6 PROJECT ATOMETER X-1

Project is based on neutron probing radiation [40,41]. In difference from TNA technique, (capture of neutron followed with gamma radiation of specific energy), it uses high-energy neutron inelastic scattering on the light nuclei. The resulting prompt gamma spectra carry information on the Z of the scattering nuclei. The high energy of probing n-flux increases down to 25-30 cm penetration length compared with the use of thermal neutrons.

The usual source of the fast neutrons (14.6 MeV mono line) is DT reaction with a maximum cross section at the energy of 150 keV:

$$D + T \rightarrow n + \alpha + m \rightarrow n' + \alpha + \gamma + m$$

Here M is a scattering nucleus, radiating specific gamma quanta after collision with fast neutron DT portable/transportable accelerating tubes, with output of 10^7-10^9 n/s have a size of one-few feet. The advantage of DT accelerators compared with radioactive sources pertains to possibility of switching between turn-on, turn-off DC or pulsed mode, control on the timing and level of n-generated flux, especially in the pulse mode regime. In the same time accelerators are much more bulky and costly, require respective high voltage supply and control panel, and cannot be used in a hand held version.

The project Atometer X-1 uses the time coincidence of the prompt gamma with accompanying DT reaction α-particle. Very short time window of the coincidence electronics (7 ns) and energy window for gamma (Ge detectors with energy resolution of ~ 0.2%) provides significant cut-off of the background, making possible to get the information on the atomic ratio of three main elements of the analyzed object- C, N, O. Beside discrimination of the elemental abundance, authors consider possibility of spatial 3D imaging by restoring direction of a neutron flight before the first collision, that roughly coincide with the line of the alpha particle flight. For easy interpretation of the results on the chemical composition, the analogy with Dalitz plot is suggested, which is illustrated at Figure 21. Each side of the equilateral triangle represents one of the elemental trio, and square of the distance from any point in the triangle to any of 3 triangle sides - relative concentration of the respective element. In such a plot the positions of innocuous objects and explosives are remarkably different. This project passed proof-of-principle in laboratory conditions, and needs significant funding to carry on testing of the first prototype version in the field. The use of very expensive Ge detectors, and position-sensitive PM, in combination with the problems of pile up due to high loading of Ge and α-detector by neutron and gamma flux adds to this.

Similar technique in a more simplistic approach is presently being tested in the project JINR. [44]. In difference from Ge in Atometer X1 project the authors use commercial NaJ scintillators and different processing software for discrimination and calculation of atomic concentration of various elements in the target. The research is also in the laboratory stage and in its present configuration focused primarily on custom application for explosive detection in luggage.

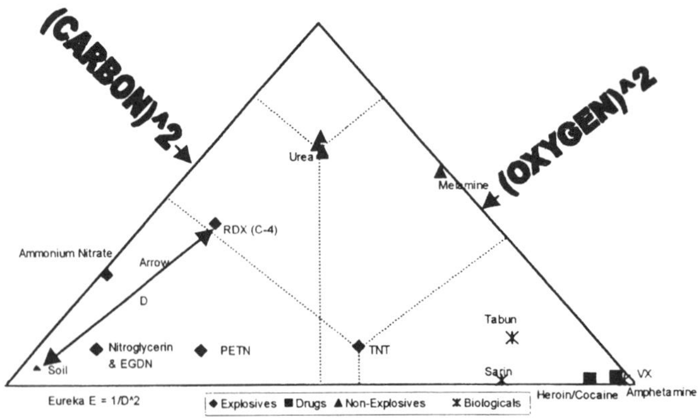

Figure 21. Dalitz plot for different compounds. The squared distance from any side is proportional to the concentraction of D, C and N [41].

6.7 PROJECTS ON SWARM OF MINI-ROBOTS

Variability of the field conditions, (swamps, forests, bushes, hilly regions, narrow paths, etc.,) limits access ability for a big robotic vehicle carrying MSS, making it impractical and/or ineffective. Beside this, the only one missed mine could deliver a fatal damage to expensive transportable system and take it out of operation. Compared with this the scenario of operational intercommunicating Swarm of inexpensive Robots (SPSR), that are able to carry only one of the sensors, and reach the places previously inaccessible (under the bushes, narrow passes, etc), looks very promising. At UCI POP was proposed with operational group consisting of 4 palm size mini robots (based on the toy-type chassis) - carriers of: neutron source, n-n sensor (looking for hydrogen), TNA (looking for nitrogen), and MD (looking for metal) [6]. The intercommunication between mini-robots is via the main station, coordinating their parallel work. Each of the robots provides the raw data flow, pertinent to respective specific signature, and its location to the main station. Low cost of each separate mini-robot and its flexible connection with the main station supports high reliability and survivability of the whole system independent of losses of few mini-robots during the operation.

A key feature of proposed distributed software technology is an integrated communication protocol that maintains information about the current set of working robots and synchronizes the robots at the same time using periodic messages of only a few bytes. This feature is essential to the success of distributed micro-miniature robot technology since robots of that size are not robust (some may fail to complete their part of a task) and they are extremely power-limited (hence efficient communication protocols are essential). The ongoing project at Sandia National Laboratory is focused on designing the program and operational protocol for such a system.

7. Conclusion

A comprehensive review on the "status quo" of the mine detection problem was presented, including brief description of the existing and emerging technologies, their strong and weak sides, and potential for supporting large scale demining operations. Based on these data we would like to outline the main facts and trends in this field and try to give some pertinent recommendations and prognosis.

The first and very encouraging important fact is a steady and considerable increase of the number of facilities involved in the anti mine activity. It includes government and non-government research programs, profit and non-profit companies, periodical international and national topical meetings, journals on related issues. In general the coalition for banning land mines by some estimations lists more than 1000 organizations. Table 8. illustrates the breakdown along the type of organizations anti mines activity, based on Lawrence Livermore National Laboratory database.

TABLE 8. Summary on types and number of organizations, involved in various activities related with mine problem all over the world

Type of Organization	Financ e	Humani Tarrian	Political	Research	Demining	Total
International	4	4	2		4	14
Non-Government Non- Profit	2	68	44	22	24	160
Government	8	3	2	27	4	44
Commerical					45	45
Academy		2		42		44

The second important, but troubling fact is that despite of these big numbers and huge variety of suggested and emerging means for mine detection, still there is a big gap between laboratory R&D studies and the technology available in the field. As has been pointed out, of all possible four levels of the mine discrimination technologies, when the large scale demining operation is in question, we are still at the first level, i.e. we could say with sufficiently high probability that: "there is something there". Among the likely reasons for this fact (beside a tremendous difficulty of the problem itself and very high standards set for the required mine detection reliability- especially for humanitarian demining - 99.6%) the following could be named:

* mines as a weapon still remain legitimate in all warfare, and respectively main R&D activity in this domain is concentrated in the defense facilities with mandatory policy of secrecy and very limited exchange on information;
* The leading developed countries do not consider this problem as of their first priority. And respectively, the government funds provided via open BAA for the mine problem, are in general far from being sufficient. It results in the contradicting perception and opinions on how these limited funds are and must be spent, and what is and what must be the first priority at the present stage: actual demining operations or related R&D and advertising? The community involved in the on-going demining in the fields, is convinced that large advertising campaigns and huge international humanitarian initiatives (the last resulted in Nobel Prize for International Campaign to Ban Landmines in 1997) "represents a double-edged sword" by distorting a real situation, and deviating limited support from a real work in the field. [24]
* private companies still do not see an adequate market for their efforts and future products, (the census of real abandoned mine fields situation in the world, and respectively- any comprehensive plans or even general strategy - what and how to deal with this problem -are absent), and as a result- there is no real market driven competition, and existing private R&D programs on emerging mine detection remain disparate and under-funded.

The third important and positive fact is that all "mine" community finally came to a consensus about the need for multisensor systems and fusion data approach. A special impetus is given to use of means sensitive to chemicals, and enabling imaging.

The reliable detection of N, H, C abundance in the buried anomaly in combination with discrimination of its shape became a prior goal for emerging multisensor technologies.

Several big programs funded by defense are focused on the development of multi million dollars, big integrated multi sensor systems, that will be able to support large scale demining operation along the roads, vast planes, as the part of military activity.

There are two options remarkably differing by the final stage of taking decision: - with or without participation of the operator. The mentioned big programs are focused on fully automatic procedure on mine detection and discrimination against background. Less costly but with not less impressive results systems involving at the final stage interaction with operator could be afforded by private small companies.

In relation with latter of special interest could be R&D directed on creating a Swarm of interacting Palm Sized mini-Robots (SPSR) remotely controlled from one remote Center, Station, where the processing and decision taking stage is being done separately on-line with data flow. Each of the mini-robots carries only one sensor for one specific mine signature. SPSR could be of high flexibility, and applicable in variety of climatic and ground conditions, (forest, bushes, stones, grass, hilly relief, etc.). It also would be far less expensive compared with integrated big transportable systems, presently under development. SPSR could also survive as a system without loss of its main functions in case of loss of several mini-robots. Various combinations of sensors for such a swarm could be suggested, focused on presence of "explosive " elements in the anomaly - specifically of H and /or N, and discrimination of its shape against background:

* GPR (shape) + STN (H content and shape) + MD (metal, shape).
* several specific chemical "noses" sensitive to different explosives + MD (metal presence, shape)
* MD + STN + TNA

This list could be extended. Combination of mini-robots will provide a third level of technology, which is sufficient for supporting the needed speed and reliability for large scale demining operations.

8. Acknowledgements

The author would like to acknowledge with many thanks the cooperation and support of his colleagues Drs. G. Pekarsky, G. Miller, Dr. A. Toor and T. Walk during collaborative research on UCI-LLNL project, included in this report.

9. References

1. Dubey, A.C., Harvey, J.F., and Broach, J.T. (1998), Proc. SPIE Conf. on Detection and Remediation Technologies for Mines and Minelike Targets III, Orlando, Florida, Edit, v. 3392.
2. Book of Abstr., 3-th Intern. Symp. on Technology and Mine Problem, Naval Post-Graduate school, Monterey, CA, June 1998.
3. Book of Abstr., 4-th Intern. Symp. on Technology and Mine Problem, Naval Post-Graduate School, Monterey, CA, June 2000.

4. Proc. 2-nd Intern. Conf. On Detection of Abandoned Land Mines, Edinburg, UK, 12-14 October 1998, London, UK, ANCORE Corp., Buried Landmine Specific Sensor based on Thermal Neutron Analysis, http://www.ancore.com
5. Baertlein B., Gunatilaka, A., *Improving Detection of Buried Mines Through Sensor Fusion*, in *Ref.1*, p. 1123-1133.
6. Blough, D., Scalable and Reliable Software for Cooperating Robots, DOE Prop., UCI, 1999
7. Bottoms, A., (1998) Mine Problem Issue, In: *Ref. III, # 4* BODF, Land Mine Facts, Briefing Materials for LLNL Press Conference at BODF, Nevada Test Site.
8. Brooks, J., Applications of GPR Technology to Humanitarian Demining Operations in Cambodia: Some Lessons Learned, in *Ref. II* p. 55
9. Burlage, R. T. Youngblood, Bioreporter Bacteria for Landmine Detection, in *Ref.II*, p. 69
10. Bruschini C., Gross, B., (1997) A Survey of Current Sensor Technology Research for the Detection of Landmines, Presented, at Intern. Workshop on Sustainable Humanitarian Demining, SusDem'97, 29 Sept., 97, Zagreb, Croatia, <http://didecs1-f.epfl.ch/w3lami/detec/susdensurvey.html
11. Bruschini C., De Bryun, K., Sahli, H., et al, (1999) *Study of the State of the Art in the EU related to Humanitarian Demining Technology*, Products and Practice, EUDEM: The EU in Humanitarian Demining, Final Report, Ecole Polyutechnique Fed., Losanne, Brussels, 30-07.
12. Bystritsky, V., Kobzev, A., Mialkovsky, V., Nikitin, V., et al, Experiments on Hidden Substances Identification with neutron probing via Associated Particle Registration, in *Ref 3*, paper # 90.
13. Bystritskii, V., Toor, A., Pekarskii, G., (1998) *Testing of Russian technology for Field Demining Operation*, UCI- LLNL Agreement # B339039, CA.
14. Bystritskii, V., Svetsov, V., (1999) Neutron Imaging for Land Mine Detection, CRDF Prop. FSU-USA UCI, CA.
15. Coutsomitros, C., Kokonozi, A., Andsritsos, F., et al., (1999) Target Identification in Civilian De-Mining, Using Weak IR Activation Methods, Technical Note, Pub. of European Commission, Joint Research Center, Ispra, Italy.
16. Castigliochello Conf. on Nuclear and Conventional Disarmament, 1997.
17. Chatterjee, C., Infrared- Based Land Mine Detection on a Vehicle, in: *Ref. I*, pp. 104-114
18. Churchill, W.C., (1989) The Second World War, Penguin Books, pp 174-6
19. Cressie, N., A. Lawson, Bayesian Hierarchial Analysis of the Mine Field Data, in: *Ref. I*, p. 941.
20. Daniels D. J., (1996) Surface penetrating Radar, IEE Radar, Sonar, Navigation and Avionics, Series 6, 300 p. ISBN 0852968620
21. DeLuca, C., Marinelli, V., Ressler, M., and Ton, T., Unexploded Ordnance Detection Experiments Using Ultra wide band Synthetic Aperture Radar, in: *Ref.I*, pp. 668-677
22. Ericsson, A., Gustafsson, A., Detection and Classification Results for an Impulse radar Mine Detecting System, in SPIE97, p. 636-642. http://www.ae.chalmers.se/~brunzell//project.html
23. Fisher, M., Cumming, C., Fox, M., et al., (2000) A Man-Made Chemical Sniffer Utilizing Novel Fluorescent Polymers, for Detection of Ultra-Trace Concentrations of Explosives Emanating from Landmines", Book of Abst., Monterey.
24. Flynn, M., (1999) Political Minefield, Bulletin of Atomic Scientist, p. 49-53.
25. Freeman, J., Skapura, D., (1997) Neural Networks, Applications, and Programming Techn., Addison Wesley.
26. Fritzsce, M., Lohlein, O., (1998) Multisensor Fusion for the Detection of Buried Landmines, In: Proc. of EuroFusio 98. International Data Fusion Conference, Great Malvern, UK, 6-7 Oct. 1998, Edited by: Bedworth, M., O'Brien, J. Malvern, UK, DERA, p. 93-100
27. Gelenbe, E., T. Kosak, Area based Results for Mine Detection, in: *Ref. 1*, p. 894-905.
28. Glascock, Practical Applications of Neutron Capture and Prompt Gamma- Rays, Instr. Phys. Conf., Ser. 62, Ch. 4, p. 641-654, 1981.
29. Gomez, B., Jones, E., Rodacy, P., et al., Trace Chemical Detection of Mines and Unexploded Ordnance, in : *Ref. I*, p. 42
30. Gozai, T., Industrial Applications of Neutrons, Physcs Methods, Proc. Intern. Conf. On Reactor Physics, Tel-Aviv, 1994, p. 199-214.
31. Harvey, J. T. Broach, Orlando, in: *Ref.* I, pp. 1111-1121
32. Hibbs, A. D., Barrall, G.A., Czipott, P.V., et al., (1998) Man Portable Detector Using Nuclear Quadrupole Resonance- First Year Progress and Tests Results, in Detection of Abandoned Land Mines , 12-14 Oct. 1998, Conference Publications, No 458, IEE.

33. Honeger, B., Symposium Overview, Demine Website, *Ref. I*
34. Huang, Q., R. Azimi-Sadjadi, S. Sheedvash, Structurally Adaptive neural network for Underwater Target Classification, in: *Ref. I*, p. 941-951
35. Jacobs, J., Dugan, E., Jacobs, A., et al., Examination of Cross-Talk Between Adjacent X-ray Generator-Detector Systems, in :Ref. I, p. 868.
36. Khanna S., M., Paquet, F., Apps, R., Seregelyi, J.S., New Hybrid Remote Sensing Method.
37. Krumhansl, P., (1998) "Seismic Sonar for landmine and UXO Detection and Classification, In: Book of Abst., Symp. on Technology and the Mine Problem, Naval Postgraduate School, Monterey, CA, p. 56
38. Lorengan,M., Severin, E., Doleman, B., Lewis, N., Array-Based Vapor Sensing Using Chemically Sensitive, Carbon Black Polymer Resistors, ibid., p. 102.
39. Magand, F., Chevret, P., (1996) Time Frequency Analysis of Energy Distribution for Circumferential Waves on Cylindrical Elastic Shells, Acustica United with Acta Acustica, p. 707-716.
40. Maglich, B., Powell, C., Kani, K., etc., (1998) Atometry, Hienergy Report, HIEN98-111.
41. Maglich, B., Development of Gamma Nose for Humanitarian De-Mining: Evidence for On-Line
42. MATLAB Image Processing Toolbox User's Guide", Mathworks, Inc., Natick, MA, 1997
43. McDonald, J.R., Nelson, H.H., Robertson, R.R., "Results of the MTADS Technology Demonstration at the Magnetic Test Range", marine Corps Air Ground Combat center, Twenty- nine Palms, CA, NRL/PU/6110-97-XXX.
44. McFee, J., Cousins, T., Jones, T., et al., Thermal Neutron Activation System for Confirmatory Nonmetallic Land Mine Detection", *In Ref. I* p. 553-564
45. McCullough C., Ulug, M., (1999) Feature and data Level Fusion of Infrared and Visual Images, presented at the SPIE AeroSense Conference, Orlando, FL.
46. McFee, J., Das, Y. " Advances in the Location and Identification of Hidden Explosive.
47. McFee, J., Aitrken, V., Chesney, R., Das, Y., Russel, K., A Multisensor, Vehicle Mounted, Tele-operated Mine Detector with Data Fusion, in: *Ref. I*, p. 1082-1093
48. Nardulli, G., Marangi, C., Trends in Landmine Warfare and Landmine Detection, http://www.ba.infn.it/~nardulli/castiglionchello.html #3.2. Presented at the VII Int.
49. Ngan, P., Data Fusion technique for Hand-Held Stand-off Mine Detection System (HSTAMIDS), *Ref. I*. P. 1150-1162
50. Pekarsky, G., (1994) Neutron technologies and Devices for Non-Destructive Testing and identification of Explosives and Chemical Warfare Agents, Proc. Intern. Conference on Nondestructive Testing in Science and Industry, Moscow, p. 198.
51. Pekarskii, G. Multisensor Technology for Buried Mines Detection, *in: Ref IV*, p.147-51.
52. Petterson, G., Stromberg, D., Roldan-Prado, R., (1998) Temporal Decision Support and data Fusion in BVR Combat, Eurofusion98, Great Malvern, UK.
53. Porter, L., Sparrow, D., (1997) Assessment of Thermal Neutron Activation Applied to Surface and near Surface Unexploded ordnance, IDA Paper P-3339.
54. Priebe, C., Cowen, L., Mine Detection via generalized Wilcoxon -Mann-Whitney Classification, ibid., p. 906-917.
55. Rhebergen, J., Zwamborn, P., Design of an Ultra Wide band ground penetrating radar System Using Impulse radiating Antennas, in *Ref IV*, p. 45
56. Rhodes, E., Dickerman, C., Peters, C., (1993) Associated-Particle Sealed Tube Neutron Probe for Characterization of Materials, Pres. at Active Probe Technologies Conf. of Intern. Symp. On Substance Identification Technologies, Insbruck, Austria.
57. Rose-Pehrsson, R., Shaffer, R., McDonald, J., et al, UXO target Detection Using Magnetometry and EM Survey data, http://chem.nrl.navy.mil/MTAD/1998/SERFP98.pdf
58. Simard, J.R., Mathieu, P., Larochelle, V., Bonnier, D., Air-borne Far-IR Imaging System: Description and Preliminary Results, in *Ref. I*, pp. 84-95
59. Trang, A., Czipott P., Waldron, D., (1997) Characterization of Small Metallic Objects and Non-Metallic Antipersonnel Mines, SPIEProc. v. 3097, pp. 372-383
60. Ulisses, Braga- Neto, On Detecting Mines and Mine like Objects in Highly Cluttered Multispectral Aerial Images by means of mathematical Morphology, in *Ref. I*, 3392, p. 987-998.
61. Ulug, M., McCullough, C., (1998) Fusion of Thermal and Vision Images, in Book: Intelligent Engineering Systems Through Artificial Neural Networks, V. 8, ASME Press, New York.
62. Van Orden, G., Van Der Pyl, T., Sims, G., and Sieber, A., (1997) Development in Demining, in Book: The Elimination of Land Mines, n. SI 3/97, UN, Ed.: Kerstin Hoffman, Geneva.

63. Visualizing Metal Detectors for Humanitarian Demining LAMI, EPFL, 1997, http://lamiwww.epfl.ch/lami/detec/visualmd.html
64. Vouvropoulos, G., Womble, P., Paschal, J., Pelan: A Pulsed Neutron Portable Probe for UXO and Land Mine identification, in: *Ref. 1*, paper # 49
65. Witten, T., Present State of the Art in GPR for Mine Detection, in: *Ref.1* p. 565-575,
66. Yujiri, L, Hauss, B., Shoucri, M., Detection of Metal and Plastic Mines using Passive Millimeter Waves, SPIE96 Proc., v. 3067, 1996, pp. 330-336
67. Yujiri L, Hauss, B., Shoucri M., Microwave/millimiter wave detection of metal and plastic mines, SPIE97 Proc., v. 3079, 1997, pp. 652-658
68. Zakharia, M., F. Magand, J. Sageloli and J. Sessarego, Time-Frequency Approaches for Sonar Target Description: Application to Fisheries, In Acoustic Signal Processing in Ocean Exploration, NATO ASI series, Jose M.F. Moura and Isabel M.G. Lourtie Editors, 541-546, Kluwer Publishers, 1993

MULTISENSING IN ROAD TRAFFIC MEASUREMENTS

R. SROKA
Faculty of Electrical Engineering, Automatics,
Computer Science and Electronics;
Department of Measurement and Instrumentation
Al. Mickiewicza 30, 30-059 Kraków
Poland

1. Introduction

The first system oriented towards road traffic control was put into use in London, in 1868. It was a simple light signalling system, using gas lamps. Subsequent, attempts in this area were undertaken in Ohio, USA, in 1914.

Much more advanced control system, called "Green Wave", whose assignment was to increase traffic continuity, was employed for the first time in Mϋnich, Germany, in 1953. Since then, vehicle traffic has increased rapidly from year to year.

Thus problems of road capacity, traffic smoothness and safety are becoming visible as well as the problems of highway infrastructure preservation and traffic influence upon the environment (noise, pollution etc.).

Simple systems of rigid traffic control appear to be insufficient. For optimal control the executive systems should be supplied with information - best of all in real time - about the situation taking place on the road. Therefore systems of traffic monitoring are coming into existence. The monitoring includes among others: presence and the number of vehicles in the detection area, velocity of individual vehicle (or mean value of velocity for specified number of vehicles), vehicle assortment, their load exerted upon road surface, occurrence of traffic jams, weather conditions (which have substantial influence upon traffic safety), etc.

The monitoring assignments are not merely aspects of direct traffic control, but its results may be of service for statistical studies, economic investigations, automatic toll collecting, highway infrastructure planning and reconditioning, as well as for protection against overloaded vehicles.

So monitoring prompts necessity to create measurement systems, capable of data acquisition and traffic parameters estimation. The basis of the latter are measurements of parameters of individual vehicle, participating in the traffic.

Nowadays multidetector systems are in use for this purpose as well as algorithms enabling multisensor data fusion. This produces an increase in efficiency and accuracy of these systems, and makes it possible to measure more and new parameters. Present-day measurement systems used for traffic purposes determine dozens of different parameters, with the use of many technologies of detection, both old and new.

A. K. Hyder et al. (eds.), Multisensor Fusion, 715–747.

Information on individual vehicle parameters, traffic parameters and environmental parameters may be used in following systems [16]:

- **MOCS (Mobile Operation Control Systems)** - These systems support administrator to operate logistics and commercial vehicle. They collect traffic information, for example location, time, of taxi and track, and provide operation administrator and driver various information got by analysing data.
- **PTPS (Public Transportation Priority Systems)** - These systems control the time required to pass the crossroad by public mass transportation vehicle such as bus by signal control. They pre-empt the lane for public transportation vehicles by setting up and cancelling priority lanes.
- **DRGS (Dynamic Route Guidance Systems)** - These systems improve the convenience of drivers, by providing the optimum route to the destination. They are able to decrease traffic jam by means of increasing the number of effective routes.
- **AMIS (Advanced Mobile Information Systems)** - These systems provide for drivers traffic informations such as: traffic jam, accidents, roadwork, travel time to the destination a.s.o. In such systems there are often used road signs with changeable information, e.g. concerning speed limit or displayed only for drivers which break a ban of road traffic.
- **DTMS (Dynamic Traffic Management Systems)** – These systems distribute traffic flows in time and space and avoid or attenuate the disturbance caused by recurrent or random perturbations (as a result of accidents and road works). DTMS depend on the real-time on-line measurements of traffic and environmental parameters. Traffic management in real time consists mainly of equipment performing three activities: data collection, data processing, and information dissemination. The data collection function of the system is carried out by the following equipment: vehicle detectors, Automatic Vehicle Classification, weather surveying equipment, traffic monitoring television cameras, emergency telephones and patrol cars. A computer in the traffic control room collects and fused detector data at short intervals, for example 1-minute, and computes such traffic data as volume, speed and occupancy. Using these traffic data, the computer judges the current state of traffic such as congestion. Based on the congestion information and the incident/accident information such as traffic accident, wrecked or disabled vehicle, and construction work loaded by the control officer through the control desk, the computer controls traffic lights and prepares information to be displayed on Variable Message Signs. From the speed data or data collected from AVC units, the computer also determines the time required to traverse each route.
- **WIM (Weigh-In-Motion)**: - These are systems which estimate the static weight of a moving vehicle, or the load of its axle(s) or wheel(s) on base of dynamic measurements. WIM is often applied to detect vehicles that exceed axle load limits.
- **ASPCI (Accommodation Systems of Pedestrian Crosswalks Inspection)** – These systems are furnished with sensors assigned for monitoring pedestrians.

They can, for example, detect the presence of pedestrians at crosswalks and, as appropriate, can prolong the duration of a green light.

2. Road Traffic Detection

The Road Traffic Detection problem may be considered within three areas [6]:

- **Vehicle detection** is defined as the sensing of the presence, passage and characteristics of individual vehicles on the road. A vehicle may be defined as a device for transporting passengers or goods, travelling on a road. Examples include cars, trucks, buses, motor-cycles a.s.o. When a trailer is attached, the entire combination is regarded as one vehicle.
- **Zone Traffic detection** is concerned with the overall condition of traffic, in a zone of detection. Such a zone may typically be 200 metres long and the full width of the road. Measured parameters might include either stationary vehicles presence, or an "incident" occurrence.
- **Environmental detection** is concerned with the airspace through which the vehicle passes. The roadside environment is particularly important to on-line capacity calculation. The detection of pollution levels may affect control strategies. Detectors may also sense visibility, precipitation and surface conditions.

The road traffic characteristics can be divided into three areas:

- **Individual Vehicle Characteristics:** measurements such as vehicle length, speed and type, all on a vehicle-by-vehicle basis. They may be further sub-divided into: **transient** or **fixed** characteristics; e.g. speed and length respectively.
- **Zone Traffic Characteristics:** data about vehicle(s) in the zone of detection: e.g. the sensing of vehicle presence, motion, and occupancy in the zone.
- **Zone Environmental Characteristics:** measurements in two sub-areas; the airspace directly above the zone of detection: e.g. visibility and pollution levels; and at the road surface: e.g. coefficient of friction and road surface temperature.

There are many parameters connected with road traffic which can be measured. The main measurements that have traditionally been used to monitor traffic operations on freeways include the following:

- Volume,
- Speed,
- Occupancy.

Volume is used to measure the quantity of traffic and is defined as the number of vehicles travelling in a given section of freeway over a period of time. The capacity of a freeway represents the maximum volume of traffic that can pass a given roadway section under prevailing roadway and traffic conditions. Volume is typically used to track bygone trends and to predict the future occurrence of congestion on given freeway sections.

Speed is an important data to determine the quality of traffic operations. Speed is frequently used to describe traffic operations because it is easy to measure in the field, and it is easy to explain and understand. Speed measurements are typically taken for individual vehicles and averaged to characterise the traffic stream as a whole. Measured speeds can be compared with optimum values to estimate the level of operations for a freeway or to detect incidents. For example, an alarm for an incident detection system might be set to go off if average speeds fall below a target value. Taking speed measurements at different points along a freeway can help in determining where congestion might exist.

Occupancy is defined as the percent of time a given section of roadway is occupied and can be used as a substitute for density. Occupancy is measured using presence detectors, and is much easier to obtain than density. When occupancy is being measured, a single lane is usually considered, with occupancy ranging from 0 percent (no vehicles passing over a section of roadway) to 100 percent (vehicles stopped over a section of roadway).

Although volume, speed, and occupancy have been the traditional types of data collected by a surveillance system, today's traffic management centers also rely on other types of data for traffic management purposes.

2.1 INDIVIDUAL VEHICLE PARAMETERS

Individual vehicle parameters may by divided as follows:

Transient Parameters:	**Fixed Parameters:**
- Time of Arrival - Lane Number - Speed - Headway Time - Headway Distance - Gap Time - Gap Distance - Direction of Travel - Acceleration - Gross Weight - Axle Weight (Static Component) - Axle Weight (Dynamic Component) - Axle Frequency - Number of Occupants	- Class - Number of Axles - Total Wheelbase - Wheel Bases - Length - Width - Area - Height - Volume - Front Overhang - Rear Overhang - Chassis Height

2.2. ZONE TRAFFIC PARAMETERS

The zone traffic parameters are as follows:

- Vehicle presence
- Zone occupancy

- Motion
- Queue
- Incident
- Vehicle flow/hour
- Axle flow/hour
- Lane occupancy

2.3. ZONE ENVIRONMENTAL PARAMETERS

These parameters may have influence upon traffic conditions (necessity of giving danger signals - information boards, radio announcements, informations for road surface maintenance services etc.), and may be used to correct results of measurements of traffic and vehicles parameters (since atmospherical phenomena have influence on characteristics of sensors used in traffic conditions). The environmental parameters are as follows:

Weather Parameters	**Surface Parameters**	**Another Environmental Parameters**
- Rate of Precipitation - Type of Precipitation - Wind Speed - Wind Direction - Air Temperature - Relative Humidity - Atmospheric Pressure - Visibility - Freezing Point - Dew Point	- Pavement temperature. - Sub-Surface temperature - Salinity - Coefficient of friction - Roadway wet/dry condition (presence of water, ice, snow) or chemicals on pavement.	- Carbon Monoxide con-centration, - Nitrogen Oxides con-centration - Level of noise, - Air dust, - Water pollution.

The specific areas where environmental detectors are used are tunnels. In these areas, special sensors are needed to monitor the level of noxious fumes. Most systems monitor the level of noxious gases and visibility by using infrared absorption techniques. Output from the monitors is used to perform the following functions:

- Control tunnel ventilation.
- Warn drivers of poor visibility.
- Set safe speed limits within the tunnel.

Until recently, most of the data listed above was difficult to measure in the field; however, due to improvements in detector technologies, these measurements can now be realised.

3. Vehicle Detection Technologies

There are many technologies available for collecting traffic data. Although inductive loop detectors (ILD) have been used more than any other method up to now, the other technologies are beginning to replace or complement loop detectors in many applications last years. Many of these new technologies are roadside (or vertical sensors) which do not require pavement cuts or the disruption of traffic for installation. Many of the new sensors are cost competitive with ILD systems. Future freeway management systems will be able to integrate a wide variety of detector technologies, which either will double or will be complementary one to another so that information redundance will ensure correct and efficient operation of the whole system.

This section provides descriptions of currently available detectors. Road traffic detectors can be divided into three groups [17]:

- Embedded Detectors,
- Non-intrusive Detectors,
- Environmental condition Detectors.

3.1 EMBEDDED DETECTORS

An embedded detector system consists of sensors in or below the surface of the roadway. These detectors are currently the most widely used form of vehicle detection. The main detectors being used include the following, described in subsequent subsections:

- Rubber tubes,
- Inductive loops,
- Magnetometers,
- Piezoelectric sensors,
- Capacitive sensors,
- Fiber optic sensors,
- Strain gauge sensors,
- Resistive sensors.

3.1.1 *Rubber tube*

This type of detector is rarely used nowadays. It is made as a rubber tube filled with the air or a liquid. Usually it mates with pressure gauges. The wheels of passing vehicle press the tube and produce a pressure pulse. The tube is positioned perpendicularly to the direction of vehicles motion, above the road way or partly plunged in it. It make possible a vehicle presence detection and velocity measurement as well as the number of axles counting and the measurement of the distance between them, thus enabling the vehicle classification. (*Low cost*)

3.1.2. *Inductive Loop Sensors*

A vehicle detector consisting of a coil of insulated wire about 2 metres square buried in the traffic lane. It is energised by an alternating current. The magnetic field is created around the loop. When a vehicle passes over the loop, the ferrous material of the vehicle absorbs some of the energy and inductance of the circuit decrease. Currently a correctly installed loop detector provides excellent performance and reliability for vehicle detection, counting, speed measurement and classification. Although installed in the road, it is a dominant technology for vehicle detection. In case of short term measurement, surface mounted loops can be used. *(Low cost)*

3.1.3. *Magnetic sensors*

Magnetic detectors indicate the presence of a metallic object by the disruption it causes in an inducted or natural magnetic field. Then these detectors may be active (magnetometers) or passive (magnetic detector) devices.

Magnetometers are active devices, excited with an electrical current in windings around a magnetic core material and measure the variation in the magnetic field when the vehicle passes.

Passive magnetic detectors (they do not radiate a field) sense perturbations in the Earth's magnetic flux produced when a vehicle passes over the detection zone. It's normally used where the only information required is vehicle passage, and require some minimum vehicle speed for detection (5 to 8 km/h).

They are used where point or small-area location of a vehicle is required, such as on bridge decks and viaducts where inductive loops are disrupted by the steel support structure or can weaken the existing structure and in heavily reinforced pavement. They exhibit excellent detection rates, and are effective for counting vehicles. *(Medium cost)*

3.1.4. *Piezoelectric Sensor*

The piezoelectric sensor is a strip sensor containing a piezoelectric material and generates an electrical charge in response to the load of a vehicle tire. It may be of two types: piezoceramic sensors (used since 1980) and piezopolymer sensors (used since 1995), very often built in the form of a coaxial cable. The third kind is piezoquartz sensor (used since 1995).

Piezo cable has been usually encapsulated in a rubber-like carrier and installed in a narrow slot across the traffic lane. The accurate time-of event data given by this sensor leads to good speed and wheelbase measurements. The sensors made in newer technology are also used in weigh-in-motion (WIM) systems. These sensors are unable to 'hold' an axle presence for a long period of time. *(Low cost)*

3.1.5. *Capacitive Sensor*

This sensor is an axle detector, made of a thin aluminium extrusion, with an inner air gap which is closed slightly when an axle passes over it. It is primarily a weigh-in-motion sensor, but also suitable for axle detection, since very accurate time-of event data is available. An in-road sensor, it can 'hold' an axle presence. While both capacitive and piezo technology have the disadvantage of being in-road sensors, they are critical for accurate axle detection and wheelbase measurement. These sensors have been in use since 1980. *(Medium cost)*

3.1.6. *Fiber optic sensor*

Fibre optic sensor (used in road traffic since 1997) is a strip sensor incorporating an optic fibre. The fibre bending that results from an applied force (by a wheel or an axle) modifies the light propagation conditions. The applied force may be derived from this modification. With regard to the operation principle, the sensors of this type may be classified as:

- amplitude sensors, reacting mainly to luminous flux change due to the measured factor,
- interferometric sensors,
- polarisation sensors,
- sensors using wave change (e.g. Bragg effect - based),
- microbend sensors. *(Medium cost)*

3.1.7. *Strain gauge sensors*
They are resistors of particular structure, appropriated to strain measurement, with proper arrangements they measure stress and forces. They are pasted on the surfaces of elements subjected to forces action. They are used to build force, mass, pressure transducers etc. In the domain of vehicles parameters measurements they are used for construction of bending plates which are placed under wheels or axles to measure their loads, either staticaly or in motion (in Low Speed WIM systems). They are used in road traffic measurement since 1980. *(Low to medium cost)*

3.1.8. *Resistive sensors*
There are sensors which measure an applied force (e.g. coming from vehicle wheels) through a variation in their electrical resistance. Made in strip form, placed on the roadway, they may serve to measure a vehicle velocity, the number of axles and the distance between them. *(Low cost)*

3.2 NON-INTRUSIVE DETECTORS

Non-intrusive detectors are also known as above ground detectors. They are mounted on a structure above the surface of the pavement.

Advantages that non-intrusive detectors have over embedded detectors include the following: modifications to pavement are not required for installation, detectors can be moved or replaced more easily, lane closure may not be required during installation and maintenance, detectors are not subject to stresses of traffic, detectors can be used during and after any reconstruction or maintenance activities.

A disadvantage of non-intrusive detectors is that they may produce unreliable results during adverse weather conditions.

Non-intrusive detectors include the following [15]:

- Microwave Radar
- Laser detector
- Passive Infrared Detector
- Ultrasonic Detector
- Passive Acoustic Detector
- Light Curtain
- Mobile Phone Technology
- Video Image Processing

3.2.1 *Microwave Radar*

Microwave radar vehicle detectors transmit electromagnetic energy at the speed of light in frequency bands between 2.5 to 24.0 GHz and listens for a return “echo”. They are able to count vehicles, measure speed, and, in some configurations detect vehicle presence. Experimental models have been used for vehicle classification by measuring the vertical profile of a vehicle. They are generally insensitive to weather (rain, fog, snow, ice, etc.), and provide day and night operation. Advanced units can determine the distance of the target from the detector, thus one unit can have multiple detection zones. Can be used with across-the-road, or along-the-road, orientation. Most current radar units do not require setup, configuration or calibration.

The waveform used to transmit the microwave energy determines the type of traffic. Continuous-wave energy, transmitted at one frequency allows vehicle speed to be measured from the Doppler shift in the received signal. The frequency shift is proportional to vehicle velocity. Stationary vehicles can not be detected with this wave form. The frequency modulated, continuous wave signal allows the measurement of both speed and presence, as well as the detection stationary vehicles. Further development is possible with digital signal processing techniques using one or two radar 'heads'. *(Medium cost)*

3.2.2 *Laser Detector*

Laser detectors operate on the same principles as microwave radar, but transmit energy at higher frequencies (shorter wavelengths) in the visible and infra-red spectrum. Active infra-red technology is included in this category. Narrow beam width is suitable for precise targeted application. The detector senses a portion of the reflected energy in its field of view. May be used along-the-road, or in an across-the-road orientation with a twin detector unit. The distance of an object from the detector is found by measuring the two-way travel time of the infrared pulse, from the detector to the target and back. Lasers provide presence, speed, volume, occupancy, and classification information in day and night conditions. IR detectors are vulnerable to weather conditions such fog, mist, dust, rain, and snow which scatter and attenuate wave energy. Problems may also arise when radar locks onto and measures the speed of the strongest backscattered signal, excluding smaller vehicles in the same area. Pulse waveforms are used with laser radar sensor technology. *(High cost)*

3.2.3 *Passive Infrared Detector*

Passive IR detectors do not transmit energy themselves, but rather measure energy emitted by objects in their field of view. A wide detection zone is possible. By measuring the difference in emitted energy (temperature) from the road and from vehicles, vehicle presence is detected. Vehicles may not be detected if they do not have a temperature difference to the background level. Passive IRs provide volume, occupancy, and presence information. These sensors have difficulty measuring speed because the extended nature of the vehicle distorts the IR signature, making velocity less clear. Weather such as fog, snow, and precipitation which scatter energy can have adverse effects on operation. *(Low to medium cost)*

3.2.4 *Ultrasonic Detector*

The development of ultrasonic devices for vehicle detection began in the mid-1950's. Similar to radar detectors, they can operate using both continuous and pulse waveforms. These sensors with continuous wave operate on the same principles as Doppler radar, except that they emit sound waves with frequencies between 20 and 200 kHz, which are above the human audible range. They are pressure waves which travel through the air at about 1200 km/h at sea level. They can measure speed, occupancy, presence, and in some configurations, queue length. Because of the complexity of the returns from moving vehicles, the effective range of the sensor (around 15 m) is smaller than the microwave detectors.

Because ultrasonic waves propagate through the air, they are subject to attenuation and distortion from a number of environmental factors including changes in ambient temperature, air turbulence, and humidity. Nearly all targets reflect ultrasonic sound waves, but textured or porous surfaces (e.g. snow on a vehicle) produce a weaker echo.

Vehicle profiling can be achieved by installing a pulsed ultrasonic detector above the roadway. The vertically aligned (downward looking) transducer measures wave travel time to the pavement or to the top of passing vehicle. A fast pulse repetition rate allows a minimum of 2 to 3 measurements with up to 1" resolution to be made. Excellent classification performance can be achieved for most vehicle types, though the sensor can have difficulty differentiating between cars and vans.

Recently developed with multiple microphones, to generate a 3D traffic image from alongside the road. *(Medium cost)*

3.2.5 *Passive Acoustic Detector*

This detector is usually configured as vertical dipole array of microphones, which listen to noise produced by approaching vehicles. The time delay between the arrival of sound at the upper and lower microphones changes with time as the vehicle emitting the sound passes under it. When the vehicle is distant from microphones the sound arrives almost instantaneously at both phones. When the vehicle is under the microphones, sound reception at the upper microphone is delayed by the intersensor distance. Vehicles are tracked using cross-correlation between phones, and best results are achieved when the data is filtered to a bandwidth of 50-2000Hz. In this band frequency content includes both engine and tire noise (though most of the acoustic noise is produced by the tires).

Interference between the noise of multiple vehicles is a key limitation to acoustic technology. Detection and speed performance are poor and fair, respectively. *(Low cost)*

3.2.6. *Light Curtain Detector*
These devices emit multiple horizontal light (visible or infrared) beams to measure vehicle presence and profile. A transmitting tower sends light beams across the lane to a receiving tower. As a vehicle breaks the light beams, a two-dimensional profile of the vehicle can be produced. Using two curtains speed and length of vehicle can be measured. More often used in toll systems than in road traffic. *(Medium to high cost)*

3.2.7. *Mobile Phone Technology*
Several local transportation authorities have implemented programs which allow drivers to call and report incidents from their mobile phones. Evaluations have found such detection to be effective, and that a the large proportion of major incidents are first identified from cellular reports. About two-thirds of the calls are originating (provide new information), while about a quarter are duplicate calls, providing previously received information. These programs have received favourable media reaction and public acceptance. Other unexpected benefits also occurred, such as the reporting of malfunctioning traffic signals, debris in the roadway, and other incident causing conditions. Cellular detection is most effective during peak periods when coverage is highest. As cellular phone ownership rates continue to increase, cellular detection promises to become a useful component in freeway surveillance systems. *(Low cost)*

3.2.8. *Video Image Processing*
A new vehicle detecting technology with wide potential applications in general-purpose sensing, scene assessment, (incident detection) and number plate reading. At least five devices are in the market now. Very important part of this technology are image processing algorithms. Algorithms in use include: target finding, character reading, scene assessment, a.s.o. The introduction of low-cost integrated camera and processing systems is increasing the development potential. The disadvantages include low presence accuracy, line of sight limitations and inherent problems in poor visibility. This may limit application to complementary roles with other technologies, except where 100% on-line time is not required. *(High cost now but expected to fall)*

Summarized information on sensors are provided in Table 1.

TABLE 1. Advantages and disadvantages of some traffic detectors

Detector Type	Advantages	Disadvantages
Inductive Loop	• Flexible design. Wide range of applications. Provides basic traffic parameters (e. g., volume, speed, presence, occupancy).	• Installation requires pavement cuts. • Installation and maintenance requires lane closure • Detectors subject to stresses of traffic.
Magnetometer	• Can be used in situations where loops are not feasible (e. g., bridge decks). • Less sensitive than loops to stresses of traffic.	• Installation requires pavement cuts. • Installation and maintenance require lane closure. • Small detection zone. • Typically used only to provide count and occupancy
Microwave Radar	• Generally insensitive to weather conditions. • Provides day and night operation.	• Requires license foroperation and maintenance. • May lock on to the strongest signal (e. g., large truck).
Infrared	• Active detector emits narrow beam allowing for accurate determination of vehicle position. • Provides day and night operation. • Provides most basic traffic parameters. • Passive detectors can be used for strategic loop replacement.	• Operation affected by precipitation (e. g., rain, fog, etc.). • Difficulty in maintaining alignment on vibrating structures
Ultrasonic	• Provides most basic traffic parameters.	• Environmental conditions (e. g., temperature, humidity, air turbulence, etc.) can affect performance. • Snow covered vehicles are difficult to detect. • High level of special maintenance capability is required.
Acoustic	• Completely passive. • Generally insensitive to weather conditions. • Provides day and night operation.	• Relatively new technology for traffic surveillance.
Video Image Processing	• Location or addition of detector zones can easily be done. • Provides basic traffic parameters. • Provides wide-area detection.	• Inclement weather, shadows, and poor lighting can affect performance. • May require significant processing power and a wide communication bandwidth.

3.3 ENVIRONMENTAL CONDITION SENSORS

There are several environmental sensors available today. Categories of environmental sensors include the following:

- Road condition sensors - measure surface temperature, wetness or dryness, presence of snow, and surface moisture conductivity,
- Visibility sensors - detect presence of fog, smog, heavy rain, snow, or sandstorms,
- Thermal mapping - detects presence of ice,
- Weather sensors - measure air temperature, pressure, humidity aso.
- Pollution sensors.

4. Automatic Vehicle Classification

Automatic Vehicle Classification Systems (AVC) are often used as a part of another systems e.g.:

- the toll collection system, with which the toll equipment is able to determine the configuration of the vehicle for the purpose of charging the appropriate toll to the customer,
- WIM systems,
- traffic data collection systems for statistical purposes.

Vehicles are divided into classes, such as cars, trucks, buses, etc. (in toll systems to charge different tolls for the different classes of vehicles).

A vehicle's class can be variously determined by the physical attributes of the vehicle, the number of occupants in the vehicle, and the purpose for which the vehicle is being used at the time of classification, or some combination of these three determinants may be employed. When a vehicle's class is dependent on the number of occupants of a vehicle, there is no practical way to determine the class by automated means. Similarly, there is no way to automatically determine the purpose for which a vehicle is being used.

As an example, a vehicle with two axles and single set of tires on the second axle may be classified as an automobile, while a vehicle with two axles and dual tires on the second axle may be classified as a truck. It is a simple matter for a human to make the distinction between the two. Determining a vehicle's class automatically, however, may be rather complex. In the above example, the two vehicles differ only in terms of the number of tires on the second axle. In other cases, two vehicles may be distinguished by the height of the vehicle or the total weight of the vehicle.

Thus, an automatic vehicle classification system may need to determine vehicle dimensions (e.g. height, length, wheel-base, height over first axle), number of axles and/or tires, presence of dual tires, and vehicle weight to distinguish between vehicles and assign the correct class.

An AVC system consists usually of lane sensor devices that record the physical characteristics of vehicles, and a processing unit, which fuses data from those devices and interprets this data to assign a class to each vehicle passing through the lane. For AVC systems used in Europe *European Classification of Road Vehicle* is recommended, which distinguishes 13 classes. This classification is presented in Table 2.

TABLE 2. European classification of road vehicle

Class No.	Vehicle type
1	
2	
3	
4	
5	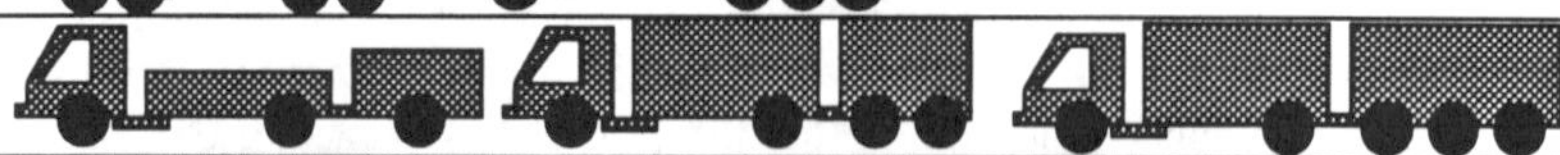
6	
7	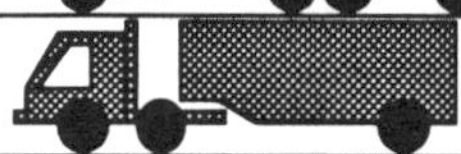
8	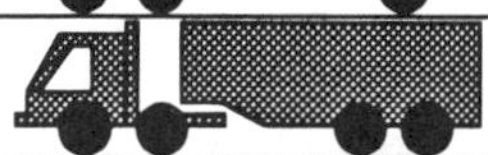
9	
10	
11	
12	
13	

There are possible and known cases, when it is necessary to distinguish greater number of classes (some USA recommendations provide even 60 classes of vehicle) or even smaller (e.g. three, when classification is realized considering the vehicle length).

AVC systems that are used to determine a vehicle's classification before the vehicle is at the toll or WIM system, are named as Pre-Classification systems, whereas those AVC systems that determine a vehicle's classification after the vehicle has passed the toll or WIM system are known as Post-Classification systems.

5. Weigh-in-Motion (WIM) Systems

The road network in every country is one of the largest investments generally supported by the public authorities or by the road users. Every road network has a limited service lifetime and needs periodic maintenance. The maintenance schedule, as well as the duration cycle or the performance, and therefore the need for financial means depend to a high degree on the mechanical wear of the road structure caused by heavily loaded vehicles.

At the beginning, static weighing of heavy vehicles was initiated by the authorities in charge of road construction and maintenance, in order to have a knowledge of traffic composition and aggressivity with respect to pavements.

Recent decades have seen a rapid increase in road and motorway traffic and a major expansion in the number of heavy goods vehicles on European roads. It is of particular concern that heavy vehicles are aggressive on bridges and pavements and that a significant number of lorries are illegally overloaded. Today more than fifty percent of the calculated aggressiveness is caused by overloaded axles. It can be stated that the road wear (or damage) depends on the number and magnitude of the applied axle loads and on the aggressiveness of the traffic. The relationship between axle load and road damage is not linear, and typically a power law with a high exponent have to be used.

To manage road networks efficiently, road managers need accurate information about vehicle loads and heavy vehicles. Traditional methods, such as weighbridges, may be time consuming and inconvenient for drivers. WIM technology delivers reliable and accurate data on the weight and types of vehicles using roads. The first WIM systems appeared in USA in the 1950’s. They used strain gauge sensors. The development of WIM in Europe has been ongoing for the last 20 years, initially in France and the UK.

Weigh-in-Motion can be defined as a process in which dynamic tire forces are sampled and then these are used to estimate the corresponding static weight or mass of the axle of the vehicle. **Weigh-In-Motion Station** is a set of mounted sensors and electronics with software which measures moving vehicle wheel and/or axle loads and gross weights, as well as other parameters such as speed, axle spacing, silhouettes, etc.

When WIM systems are installed on the road network, they collect a large number of data, and the static weighing became less useful for the traffic statistical knowledge. Random surveys of vehicle weights using static or low speed (LS)-WIM systems are expensive and the process of stopping vehicles may bias the statistics (the drivers of overloading vehicles may deliberately avoid these sites). Statistics from enforcement weight checks tend to be biased because enforcement officers are trained to select overloaded vehicles and may not weigh lightly loaded vehicles. The use of high speed WIM systems seems to be an ideal solution. These systems can weigh all vehicles travelling along a section of road and do not affect the traffic flow. Therefore, the number of static weighing operations was reduced from 193,000 in 1979 to 74,000 in 1993. Static weighing of obviously nonoverloaded vehicles was stopped [10].

Data provided by Weigh-in-Motion systems are important for:

- infrastructure design and management,
- freight/trade planning and regulation (e.g. saving time during freight transportation),
- enforcement and detection,
- traffic management,
- policy development and freight studies (e.g. the presence of a high percentage of two-axle rigid lorries (tractor with trailer) show that short distance haulage and delivery transport are predominant at that site. On the contrary, the presence of high proportions of articulated lorries or multiple-trailor road trains suggests that the long-distance transport is more important on that road section. An appreciable proportion of three-axle or four-axle rigid lorries may identify the existence nearby of construction or public works),
- toll (provide government authorities necessary information for a harmonised tax system),
- preparing basis for legislation relating to road traffic safety (an overloaded vehicle does not have the same behaviour on the road as a legally loaded vehicle, and the weight is a factor which is critical in accidents),
- competition: to ensure a fair competition between different transport modes (road, rail, air, water), but above all between transport companies, it is necessary that everybody respects the same rules and legislation,
- optimising the traffic flow,
- economic and environmental purposes (in each country economists are concerned about the movements of goods and trends in freight transport, thus forming indicators of the growth and development (or recession) of industrial activity. These statistical surveys are used also for environmental problems).

Such a system provides similar data as other equipment currently used in these fields but provides also the weight and a more elaborate classification than the actual one based on vehicle length. In systems without weighing a semi-trailer with 5 axles takes the same place and the same importance as a small car, in the safety field; a heavy lorry appears to get up hills or to brake in the downhill as easily as a 5 tonne truck, and from an environmental perspective, a heavy lorry appears to produce the same quantity of pollutant emission as a 5 tonne lorry or a petrol car. At present, Weigh-in-Motion systems are the only systems which can distinguish between such cases. Then WIM devices are often integrated into the automatic traffic data collection systems.

Generally WIM systems can be divided as follows:

- On-Board WIM
- Road WIM
  - ♦ Non-intrusive (e.g. seismic WIM)

- ♦ Embedded
  - * Bridge WIM
  - * Pavement WIM
    - ⇒ Low Speed (LS-WIM)
    - ⇒ High Speed (HS-WIM)
      - – One Sensor WIM
      - – Multi Sensor WIM

5.1 ON-BOARD WIM SYSTEM

Since 1994, on-board Weigh In Motion (WIM) technology has been developing [7]. The technique to find the weigh depends on the fundamental relationship between force, mass, and acceleration, as stated in Newton's Second Law. By accurately measuring force and acceleration it is possible to determine mass, which is equivalent to weight.

The task has been to keep all equipment on the tractor, with no equipment nor modification required on the trailer. This allows complete freedom of tractor-trailer interchangeability.

This unique technology has advantages in accuracy, calibration traceability, cost and ease of interchanging trailers. Applications can include:

- The Intelligent Transportation System, where trucks will be interrogated prior to reaching a weight inspection station. If they are maintaining a legal weight, they will be signalled to maintain their highway speed and bypass the inspection site. This will increase productivity for the trucking industry, will reduce emissions and air pollution, and will allow the enforcement officers to devote their time to the few trucks that are non-compliant.
- On-Board Cargo Theft Detection; As the vehicle rolls along, it is constantly re-weighing itself. When coupled to a datalink system, the truck can report any changes in cargo weight directly to the dispatcher. If the truck is not datalink equipped, the WIM System can keep an electronic log of the weight, date, and time for later examination in case of cargo loss.
- Brake Proportioning; By determining the weight on each individual axle group, the system can aid in brake proportioning. Knowing where the weight is, this improved braking can result in safer operation and reduced tire wear.

Weigh In Motion as we know can be divided into two broad categories depending on whether the equipment is on the vehicle, or in the roadway. Each has it's advantages. A major difference that may not be readily apparent involves the number of data samples that are available. As each axle passes over a sensor in the pavement, one weight reading is made. For multiple readings, multiple sensors are required (MS-WIM). On the other hand, equipment located on the vehicle has the opportunity to make thousands of individual weight determinations and can more easily integrate out errors and noise. Even in a perfect system, any individual measurement is subject to errors due to vehicle dynamics, so the ability to make a great number of such measurements is advantageous. On the other hand, in pavement WIM is able to make axle-by-axle measurements directly, while systems located solely on the tractor require a second (and perhaps a third) sensor to resolve axle group weights. Accuracy of such WIM system is comparable to WIM pavement systems.

5.2 ROAD WIM SYSTEM

5.2.1 *Seismic WIM System*

The seismic sensor arrays can be exploring to perform various transportation-related measurements [14]. Primarily, such system allows to realize noninvasive, inexpensive measurement of a vehicle's weight without having to stop it on a mechanical scale. An array of seismometers records ground displacement data from a moving vehicle, and that data are used to determine the signal characteristics generated by the vehicle. Signal processing algorithms are then employed to estimate vehicle weight.

A truck is moving by a 40-foot-long array of seismometers located on the shoulder of a roadway. The digitized data are recorded for subsequent computer processing. The recorded signal can be represented as the sum of high- and low-frequency terms. The low-frequency component is associated with pavement distortion and the high-frequency with vehicle suspension assembly characteristics. It has been found that there is a positive correlation between the measured signal amplitude from the sensors and the vehicle weight.

Advantages of such system are its nonintrusiveness and use of hardware that is both inexpensive and portable. An important feature of seismic system is the measurement of ground disturbances surrounding the path of the vehicle, the area generally perturbed by the moving source. This can help to overcome technical difficulties of attempting to estimate the static weight of an object in the presence of complex dynamic forces of the wheels and axles of a moving vehicle interacting with the pavement.

Because the amplitude of the sensor signals is dependent on speed, this array can also be deployed for traffic control to determine vehicle speed. A similar system could be used to study the highway degradation problem. The signals recorded from the array have propagated from the source to the receiver through the ground, and can therefore be used to noninvasively detect damage via signal distortions.

Subsoil vibrations are disturbing factors and have significant influence upon the quality of measurements for this class of WIM system. Therefore the signals acquired from seismic sensors may be employed for correction purposes.

5.2.2. *Bridge WIM System*

There are two basic concepts for weighing vehicles in motion using instrumented bridges as a large scale. The first bridge WIM system was developed in the United States (1979), as an alternative to the pavement WIM. Australia (1984) developed another bridge WIM system, but it was soon replaced with one that uses culverts. Traditionally, a bridge WIM system was only recommended for bridges with spans between 8 and 25 m. The strains measured in some of the bridge elements are used to determine the gross weights and sometimes the axle loads of vehicles crossing the bridge.

All bridge WIM systems involve an existing bridge or culvert from the road network being instrumented with reusable strain transducers and used to weigh the vehicles. The strain gauges are glued or screwed to the bottom flange of each girder, precast element or across the slab along a line normal to the direction of the bridge.

There are currently more than one hundred such systems in operation. Despite many advantages, bridge WIM systems have not been extensively explored and are at present used only in a few countries around the world.

5.2.3. *Low Speed WIM*

These systems are for weighing a vehicle in motion, but on a specific area outside of the traffic flow, on a horizontal and even pavement surface under controlled conditions, such as constant and low speed (e.g. 5 to 15 km/h) in order to avoid any dynamic effect (but the speed upper limit is not fully specified and may depends on the system and the pavement conditions).

Low speed systems provide the high level of accuracy needed for enforcement (vehicles are caught in the traffic flow and stopped on specific weighing area) or fee collection uses. In these systems load cell plates are usually used for weighing (the same type as for static weighing). However, the weighing platform should be longer (>30 m) than for static weighing (≅15 m), which increases the installation cost and requires more space at the weighing area.

However, systems of this type are encumbered with a series of shortcomings, among others: small flow capacity (long time of an individual vehicle weighing), high costs of laying on and exploitation, the traffic smoothness reduction, the coefficient of overloaded vehicles highly dependent upon the experience of the serving personnel. In connection with the later, the systems of this type often collaborate with High Speed WIM systems, serving for vehicle preselection.

5.2.4. *High Speed WIM*

These are systems weighing vehicles in motion in the traffic flow, at speed. But there is no lower speed limit. High speed (HS) WIM systems are the way to improve the efficiency of weight enforcement. Such systems may use any type of strip sensors (piezoelectric, capacitive). That allows the weighing of all vehicles and axles without stopping them or slowing the traffic flow. However, HS-WIM systems are always sensitive to dynamic effects induced by the pavement profile. They can be used to:

- select vehicles for enforcement weighing (pre-selection), thus improving the efficiency of enforcement
- monitor the level of overloading, allowing the authorities to make better decisions about the resources required for enforcement, the timing of enforcement checks, etc.;
- automate weight enforcement. Automated systems which record the weights and other details of vehicles could greatly improve the efficiency of enforcement and the probability of an individual overloaded vehicle being detected
- identify vehicles for follow-up checks;
- protect weak bridges by activating warning signs
- enforce environmental restrictions on goods vehicle movements.

The usefulness of HS-WIM systems for enforcement depends on the:

- accuracy of the weights measured using the WIM system (compared with the equivalent static weights);
- information available on the maximum permitted axle and gross weights for individual vehicles. Some types of vehicle have relatively predictable maximum permitted weights whilst others, such as 2-axle rigid vehicles, are variable;
- identification of the vehicle. This is important when the system is used to identify individual overloaded vehicles (for example, for pre-selection).

5.2.5 *Multi Sensor WIM*

A fundamental limitation on the performance of Weigh-in-Motion (WIM) systems is imposed by the dynamic tire forces due to vehicle-pavement interaction. These dynamic forces have RMS amplitudes ranging from 10% to 30% of the static axle load of heavy goods vehicles. The dynamic forces depend on a variety of factors such as the speed of the vehicle, the roughness of the road and the suspension characteristics. In order to increase the accuracy of the static weight estimation, a series of several strip sensors (Multi Sensor) may be used, and the information provided by each of them is then combined (fused). Using multiple WIM sensors the effects of the vehicle dynamics on WIM accuracy can by "smoothed out". Multiple sensor systems have been introduced in period 1993-1998.

Although there are many of relatively inexpensive WIM sensors in the market, existing multiple-sensor WIM systems suffer from errors, partly due to the dynamic tire forces and partly due to inherent sensor noise. Selection of the sensors, distance between them (uniform, non uniform) and the data processing method plays a key role in minimising the effects of these errors.

Multiple Sensor WIM is the process whereby the results from an array of strip sensors are fused to determine a more accurate estimate of the static axle weights. Most of work on multiple-sensor WIM considered a simple arithmetic average of the sensor outputs or similar averaging methods for the estimation of the static axle load. An alternative approach, which performs a least-squares fit for the dominant Fourier component, was also proposed.

5.3 ACCURACY OF WIM SYSTEM

The WIM systems are classified in six accuracy classes, each of them corresponding to a range of applications or requirements. Additional classes are given for systems which do not meet the main classes. Any level of accuracy not only refers to the performance of the WIM system used (i.e., the sensor(s) and electronic station with its software), but also to the calibration procedure and frequency, to pavement/road quality and evenness and vehicle behaviour. The following requirements are given unless otherwise stated by the customer:

- **Class A (5)** (5 to 10%): Legal purposes such as enforcement of legal weight limits and commercial weighing applications (if allowed by the legislation) and other particular needs; to provide reference weight values for in-service checks, if the classes B, C or D are required, for all the traffic flow vehicles (assuming that it is not possible to get static weights for such a large population);

- **Class B+(7)** (7 to 10%): Enforcement of legal weight limits in particular cases, if the class A requirements may not be satisfied, and with a special agreement of the legal authorities; to provide reference values for in-service checks, if the classes C or D are required, for all the traffic flow vehicles (assuming that it is not possible to get static weights for such a large population);
- **Class B (10)** (10 to 15%): Accurate knowledge of weights by axles or axle groups, and gross weights, for: - infrastructure (pavement and bridge) design, maintenance or evaluation, such as aggressiveness evaluation, fatigue damage and lifetime calculations,
  - efficient preselection of overloaded axles or vehicles,
  - vehicle identification based on the loads.
- **Classes C(15)** (10 to 20%) **or D+(20)** (20 to 30%): Detailed statistical studies, determination of load histograms with class width of one or two tonnes, and accurate classification of vehicles based on the loads.
- **Class D(25)** (25 to 30%): Rough weight indications required for statistical purposes, economical and technical studies, standard classification of vehicles according to wide weight classes.

The WIM site characteristics have some influence on the in-motion vehicle behaviour and may lead to large discrepancies between the axle impact forces and the corresponding static loads. Therefore the specified criteria about the road geometry and the pavement characteristics are given in order to reduce these discrepancies and to keep them within some limits in accordance with the required accuracy levels.

It is strongly recommended that road section between 50 m upstream and 25 m downstream of the system meets the following geometrical characteristics:

- longitudinal slope < 2% and as far as possible constant,
- transverse slope < 3%,
- radius of curvature > 1000 m (but a straight road would be preferred).

The WIM systems should be installed away from any area of acceleration or deceleration, (i.e. close to a traffic light, toll station, etc.), in order to weigh vehicles travelling at uniform speed. It is also desirable to avoid the area where drivers make gear changes, line changes etc.

The pavement characteristics directly influence the signal recorded by any WIM sensor, because of :

- the pavement/vehicle interaction leading to dynamic impact forces,
- in most cases, the road is the support for the sensor and therefore forms part of the measurement device.

5.4 CALIBRATION OF WIM SYSTEMS

After installation and general checking, an initial calibration must be performed before an operational use of any WIM system. The accuracy of WIM data depends greatly on the calibration procedure of the WIM system. Different calibration methods are commonly used, which depend on the sensor type, the application and requirements of the user and the time and means available. There are:

- **Static calibration.** Such a calibration may only be used for those WIM systems which also allow static measurements. This is the case for strain gauge and load cell scales, piezo-quartz bars, capacitive strips or fibre optic sensors, but not for piezo-electric bars. Even for the strip sensors (piezo-quartz, capacitive strips and fibre optic), the static calibration condition differs from the loading condition under traffic flow, because the integration of the signal may not be performed during a static test.
- **Use of shock or pressure variation devices.** The principle of such calibration methods is to apply to the sensor some repeatable calibrated shocks or pressure variations.
- **Use of pre-weighed calibration lorries**. The main principle consists of passing some test (calibration) pre-weighed vehicles over the WIM system, repeatedly.
- **Use of instrumented calibration lorries**. This method is of special interest if the WIM system is intended to measure instantaneous axle impact forces, instead of static loads, or to calibrate multiple sensor array.
- **Automatic self-calibration procedures and software**. The principle consists in fitting some statistics recorded and computed by the WIM system to some target values depending on the site-specific traffic. In most countries and road networks, there are some „characteristic vehicles" which have some axle(s) and/or gross weight with a low coefficient of variation and a quite constant mean (the target value). In such a case, the moving average of a certain number of these axle loads or gross weights becomes almost constant for a large enough sample size, and may be fitted to the target value. This provides a new coefficient of calibration after the passage of the required number of characteristic vehicles.

5.5. DATA STORAGE

Data storage is also very important (structure and format of the data files which contain the information recorded or computed) in WIM system. Two main types of data are used:

- detailed data vehicle by vehicle (mode 1),
- aggregated data or statistics (mode 2).

In the first mode, a complete picture of the traffic flow is given with all the information measured, but the data file size increases more or less proportionally with the number of vehicles, and therefore with the time of recording.

In the second mode, only pre-defined statistics computed in real time are stored, which makes possible a data file size more or less independent of the number of vehicles, and therefore the WIM system may be in use continuously without maintenance, and will not lose information. Aggregated data are generally summarised over fixed time periods.

Both modes may be combined simultaneously in the same WIM system.

5.6 SHORT GLOSSARY TO WIM

- **Axle**: an axle consists of two or more wheel assemblies lying approximately on a common axis oriented transversely to the nominal direction of motion of the vehicle.
- **Axle load**: sum of all the wheel loads of an axle of a vehicle (may be static or dynamic).
- **Axle scale (Axle weigher)**: device which measures an axle load at once. It generally consists of two plates or mats, weighing simultaneously a half axle (one wheel or twin wheels); but it may also be one large plate or mat, with a width larger than the wheelbase; the plate or mat longitudinal extent must be longer than the tire imprint length.
- **Bending plate**: plate instrumented with strain gauges placed under wheels or axles to measure their loads, either statically or in motion.
- **Bridge WIM**: WIM using an instrumented bridge as a large scale; the strains measured in some of the bridge elements are used to determine through software the gross weights and sometimes the axle loads of vehicles crossing the bridge.
- **Gross weight (GW)**: static gross weight = mass of a whole vehicle multiplied by the acceleration due to gravity; dynamic (in-motion) gross weight = sum of forces applied by all axles of a vehicle (may be measured at the same time or at the same place).
- **Group of axles**: set of axles in the same lorry, tractor, trailer or semi-trailer, spaced less than 2 m each from the next one.
- **Impact force**: force applied to the pavement by a wheel/axle while the vehicle is travelling at speed; by extension, sum of the forces applied by all the axles of a vehicle (may be measured at the same time or at the same place).
- **Impact factor**: ratio of the force applied to the pavement by a wheel/axle while the vehicle is travelling at speed to the static wheel/axle load.
- **Gross impact factor**: sum of the forces applied by all the axles of a vehicle (may be measured at the same time or at the same place) divided by the static gross weight.
- **Sensor**: part of a measuring instrument that is directly affected by the parameter to be measured.
- **Static load**: load applied by a stationary vehicle, axle or wheel.
- **Wheel load**: force applied by tire(s) of a wheel on the pavement, either statically (static load) or in motion (dynamic load).
- **Weighbridge**: plate instrumented with strain gauges placed under wheels or axles to measure their static loads.

6. Exemplary Measurement Methods

6.1 SPEED MEASUREMENTS

The vehicle velocity is one of more important and more often measured traffic parameters. It results from the fact that knowledge of velocity is necessary to precisely derive many other parameters, e.g. vehicle length, classification, traffic jams detection, etc. The classical method of velocity measurement in highway systems is based upon its definition. It is connected with the use of two detectors located in known distance one from another and measurement of time of passing between them. The same method of measurement is used in videodetection-based systems.

The velocity may be also measured with the use of further generation detectors (e.g. microwave radar, laser detector) based on Doppler effect. Yet these solutions are connected with considerable costs. It is often possible to limit the number of detectors and the complexity of the mating systems, to obtain reliable results at affordable costs. This may be the case of inductive loop detector, which up to the present has been treated pointwise, but in fact it is surface-distributed. This makes possible to measure the velocity of individual vehicle with the use of single sensor. Of course the algorithm of the sensor signal processing is more complex than in the case of definition-based measurement.

The single-loop method does not make possible to distinguish the vehicle motion direction, unlike the two-detector system. Exemplary signals obtained from inductive loop detector for various vehicle classes, are presented in the Figure 1.

Two algorithms, based upon the signal from single inductive loop, appropriated for a vehicle velocity estimation are presented.

The first algorithm (Algorithm 1) is based on the determination of the time of vehicle coming in and going out of the inductive loop zone. These times (i.e. the times of rising and falling sensor signal) are determined as the times of signal passing between specific thresholds, which makes it possible to determine the averaged values of derivatives over both slopes. The velocity is defined as [8]:

$$V = a \cdot \left(\frac{\Delta U_n}{T_1} + \frac{\Delta U_n}{T_2} \right) \tag{1}$$

where: V – measured velocity, a – coefficient scaling to the given velocity unit, ΔU_n – difference between the upper and lower threshold values, T_1, T_2 – signal rise time and fall time, respectively, between thresholds.

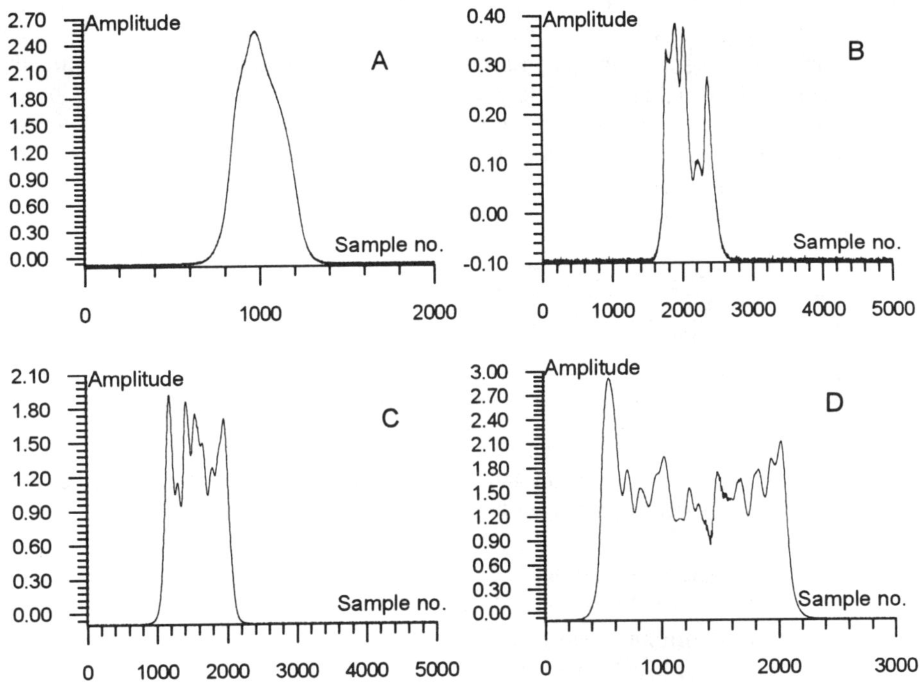

Figure 1. Signal from inductive loop detector in time domain for different classes of vehicle: A - cars, B - trailers, C - city bus, D - long bus.

A second algorithm (Algorithm 2) determines in real time the extreme values of the derivatives of the sensor time signal. Differentiation is made through the filtering of the sensor signal with a FIR filter [2]. The velocity is calculated according to the following formula:

$$V = b \cdot \left(|\max(y'(t))| + |\min(y'(t))| \right) \tag{2}$$

where: b = coefficient scaling to the given velocity unit, $y'(t)$ = derivative of the signal generated by a vehicle.

The program of study included velocity measurement according to the described algorithms and comparing them to a reference velocity. The latter was obtained by the classic method, i.e. with the use of two-loop sensors located at a known distance one from another. It should be stressed that the velocity measured in such a way is an averaged velocity of the vehicle over the segment in between the sensors.

In order to standardize the validation criteria for algorithms and compare their properties, a quality criterion known as the correlation coefficient ρ between the results obtained from Algorithms 1 and 2 and the reference results was computed. Its value ranges from 0 for the results uncorrelated with the reference results to 1 for the measurement results matching completely with the reference results.

Velocities were investigated for three types of vehicles, i.e. Fiat 126, Fiat Uno and Lada. The obtained values of the ρ index are presented in Table 3.

TABLE 3. Values of the correlation coefficient ρ for particular vehicle types

Type of the vehicle	algorithm 1	algorithm 2
all vehicles	0.8408	0.9758
Fiat Uno	0.8973	0.9979
Lada	0.9551	0.9931
Fiat 126	0.8958	0.9708

The values of the relative error defined as the difference between the initial velocity determined according to an algorithm and the value measured in the two-sensor system related to a value accepted as the range (90 km/h), are as follows:

- all vehicle classes, Algorithm 1: less than 15%,
- all vehicle classes, Algorithm 2: less than 8%,
- Fiat Uno, Algorithm 1: less than 6%,
- Fiat Uno, Algorithm 2: less than 2.5%.

The results show that algorithm 2 gives much better results and that a further advantage is the possibility of real-time realization. It should be also stressed that additional information about vehicle class provides a considerable reduction in error (in this example from 8% to 2.5%).

Taking into consideration the possibility of vehicle class recognition the Algorithms 2-based results of velocity measurement for a selected vehicle type (Fiat Uno) are presented in Figure 2. Measurement values are presented in Figure 3 as isolated points, with the solid line as the best fit in the mean square sense.

6.2 VEHICLE CLASSIFICATION

The shape of a time signal generated by the interaction of a vehicle passing over a inductive loop sensor with the magnetic field produced by the sensor can be the basis for the recognition of vehicle class [9]. Signals generated by different vehicles differ in their parameters such as e.g. amplitude, frequency spectrum, statistical parameters. The essence of the problem of the vehicle class recognition and, possibly, vehicle type, lies in the analysis of the recorded measurement signals concerning the above mentioned parameters and the comparison of the results of this analysis to the results of the analysis of the reference signals representing selected vehicle classes. This vehicle class is adopted as an identification result which is represented by a reference signal closest in the sense of

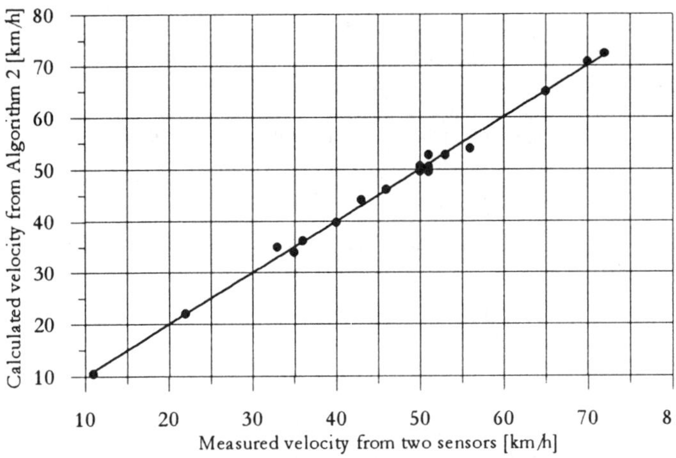

Figure 2. Velocity measured according to Algorithm 2 for a Fiat Uno.

the analyzed properties, to the recorded measured signal generated by the vehicle being classified. Reference signals are generated as results of averaging many measurement signals corresponding to the selected vehicle class and, possibly, vehicle type.

An essential problem here is to make the measurement signal parameters (and, therefore, the results of identification) independent of the vehicle velocity, vehicle suspension height and other disturbing factors.

The classification was made based on various characteristics of the registered measurement signals. They were:

- temporal shape of the signal;
- signal amplitude probability distribution;
- measurement signal amplitude versus vehicle length.

The mean effective error criterion was employed to assess the distance between the recorded measurement signal and the successive reference signals. Three classes were defined: cars, trucks and buses.

Exemplary reference signals for the **buses** class, corresponding to the three mentioned characteristics of the measurement signals generated by a vehicle being classified, are shown in figure 3.

The efficiency of vehicle type classification was assessed as follows. The mark 'very good' was assigned in the case of the full conformity of a measured vehicle with the pattern (type and class of vehicle). If the conformity concerned the class of vehicle only, the result of the classification was assessed as 'good'. All other cases were assigned a 'bad' mark. The drivers were not informed about the measurements to be made. No additional limitations, excluding those resulting from the traffic regulations, were imposed on the vehicle velocity and course.

The percentage efficiency of classification was calculated, as the ratio of the total number of very good and good identification results to the total number of vehicles. The exemplary classification results for patterns in time domain are listed in Table 4.

TABLE 4. Vehicle class identification results obtained with the use of the patterns representing the shape of measurement signals in the time domain.

Identification quality			Identification efficiency	Class of vehicle
very good	good	bad		
29	32	4	0.94	cars
29	18	12	0.80	trucks
35	14	3	0.94	buses

Unfortunately, this vehicle classification method is complicated. It was very time-consuming and needed to store vast reference patterns. An alternative method can also be used. It depends on conversion of the measured signal into a vector of numerical parameters (only a few). Vehicle classification is done by comparison of the vector of measured data to reference vectors containing values of these parameters typical for predetermined classes of vehicles. In this way the necessity of storage of long vectors of reference signals is eliminated.

Six signal parameters were defined as the classification criteria which can be used to determine vehicle classes. They are:

(a) standard deviation

$$\sigma_x = \sqrt{\frac{1}{N-1}\sum_{i=1}^{N}(x_i - \bar{x})^2} \tag{3}$$

where: x_i = i-th signal sample, $\bar{x} = \frac{1}{N}\sum_{i=1}^{N} x_i$ - average, N = total number of samples.

(b) mean square value

$$\Psi_x^2 = \frac{1}{N}\sum_{i=1}^{N} x_i^2 \tag{4}$$

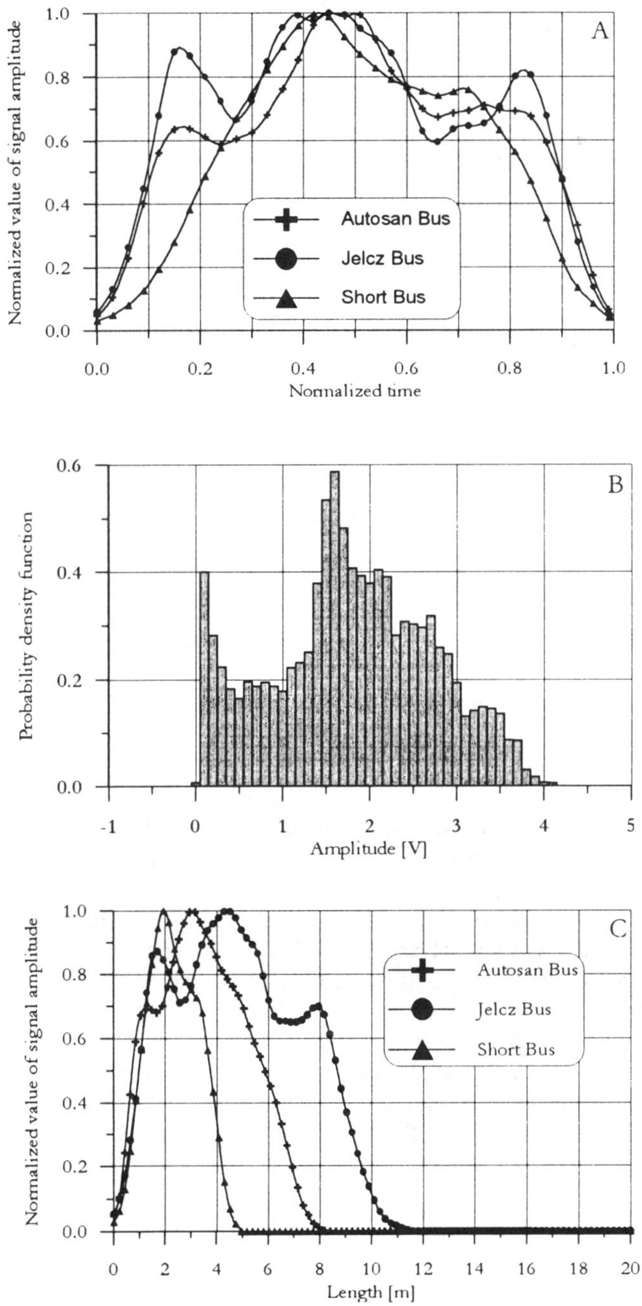

Figure 3. References signals for buses: A – in time domain; B – in the domain of probability density function, C – in the domain of vehicle length.

(c) third-order moment

$$m_{3x} = \frac{1}{N}\sum_{i=1}^{N} x_i^3 \tag{5}$$

(d) third-order central moment

$$m_{3x}^c = \frac{1}{N}\sum_{i=1}^{N} (x_i - \bar{x})^3 \tag{6}$$

(e) signal fluctuation

$$x_h = \sum_{\ell=1}^{K-1} \left| x_{\ell+1} - x_\ell \right| \tag{7}$$

where: x_ℓ for $\ell = 1,2,...K$ - values of the signal at the ends of its monotony intervals, K = number of monotony intervals.

(f) length

$$x_L = (t_d - t_u)\cdot V \tag{8}$$

where: t_u, t_d = time instants at which the measuring signal crosses the adopted threshold level when rising (t_u) and falling (t_d), V = vehicle velocity.

The efficiency of the mentioned criteria used in the vehicle classification varies with the vehicle class (recognised visually) and can be assessed using the enclosed histograms (Figure 4).

Next the hierarchical classification method is described. In the hierarchical method primary and secondary criteria are defined. The primary criterion is used to make an initial classification of vehicles into groups containing usually two or more classes (when the total number of classes is large). The secondary criteria are used to make classification within particular groups, dividing the vehicles into classes. Different criteria can be used in individual groups. Some typical schemes of the classification process with a hierarchical method are shown in Figure 5.

The efficiency of the hierarchical classification method is substantially higher than that with application of single criterion (Figure 6).

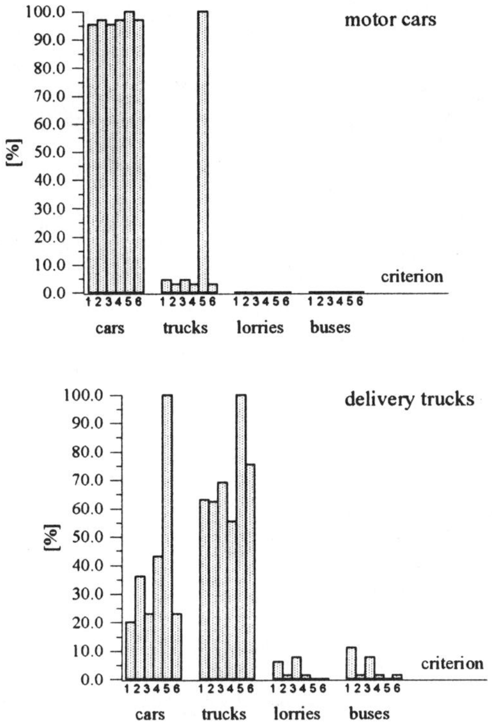

Figure 4. Assessment of the efficiency of criteria (1) - (6) of road vehicle classification where four vehicle classes are defined (cars, delivery trucks, lorries, and buses). 1,2,3,4,5,6 - numbers of classification criteria according to description in the text.

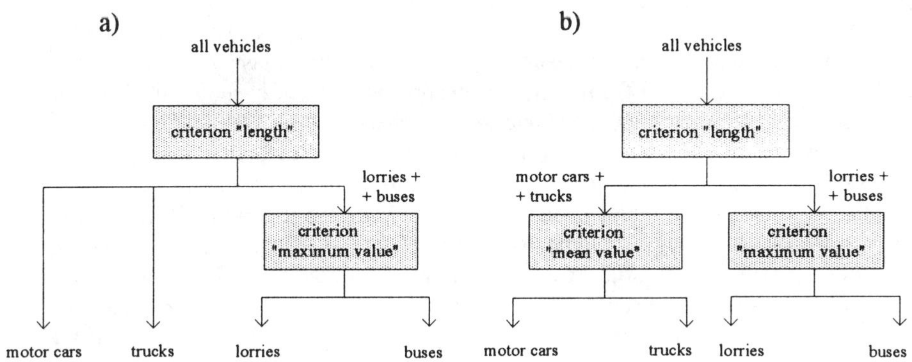

Figure 5. Typical schemes of a hierarchical classification process.

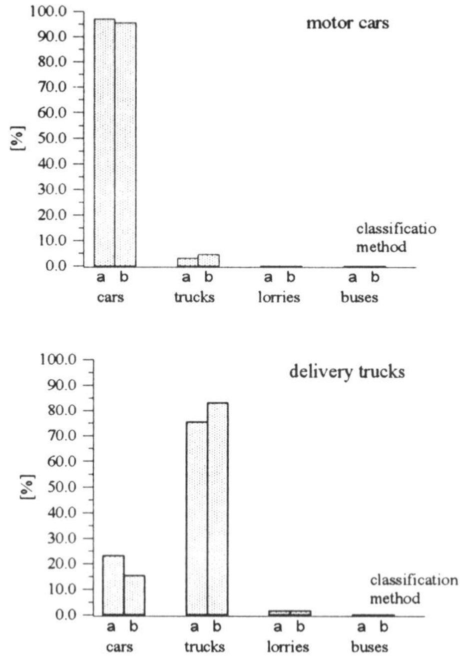

Figure 6. Efficiency of the hierarchical classification according to schemes **a** and **b**.

7. Summary

Information on individual vehicle parameters, traffic and environmental condition parameters may be used in many road traffic control systems. Today, over 60 different parameters are measured to monitor and control road traffic. A key part of such systems are sensors; there exist over 30 sensing technologies in use today.

Systems that fuse data from multiple sensors are frequently required. The basic goal of data fusion is to increase the accuracy and efficiency of the system.

Weigh-in Motion systems provide weighing information (axle loads, gross weight, inter-axle distance), classical traffic information (traffic flow, occupancy rate, gap, speed, length, etc.), and more detailed and precise vehicle classification. WIM systems avoid some identification and implementation problems associated with radar and video recording (during night time or for bad weather conditions). Similarly, classification systems are susceptible to changes in future years, with new types of vehicles or current ones. This is due to the diminution of the quantity of metallic pieces in short vehicles (such as personal cars and vans) and due to the appearance of monospace cars. Such car models have a length and a ground clearance which come closer to the values of a van. A classification based only on these characteristics could be insufficient and the weight factor could become the primary information for urban traffic management.

Time passage and speed provide traffic flow and occupation rate. This information is required for traffic management. WIM systems based on velocity measurement provide a sharper classification than the current ones in two or three classes. For the freight and economic management and tolling purposes, the gross weight and axle loads (by categories) are required. For safety and environment problems, speed linked to the gross weight by vehicle categories is a very useful tool. For economic statistical purposes, gross weight, loads of group of axles classified in several categories are a requirement to build frequency distributions.

Actually, the principal uses of WIM, in the short term, are linked to classification and weighing statistics. Such applications do not require a high accuracy on the load factor. WIM applications in the longer term will be developed only on two conditions which are slightly contradictory: a low price and a good accuracy and reliability. New applications in each country require an implementation of a WIM station network on motorways and on roads which are less important in terms of traffic but are strategically more important. In many cases, exploitation software must be developed and adapted to the WIM data. Also, the WIM system technique is new, and for the past 5 years, many countries have begun to make observations and to create traffic databases. Thus it is possible to expect more applications and lower prices in a few years when software will be developed and the needs created.

8. Acknowledgements

This work was financed by KBN (Scientific Research Committee) grant No. 8T10C01217.

9. References

1. Cebon D. (1999): Handbook of Vehicle-Road Interaction, Swets&Zeitlinger.
2. Oppenheim A. V., Schafer R. W. (1989): Discete-Time Signal Processing. Prentice-Hall.
3. Weigh-in-Motion Handbook (1997), U.S. Department of Transportation, Federal Highway Administration.
4. Clark M.A.G., Kidson P., Hodge A.(1990): Infrared detector development. Conference Proceedings, Third International Conference on Road Traffic Control, pp. 158-162.
5. Cole D.J., Cebon D.(1992): Performance and application of a capacitive strip tire force sensor. 6th International Conference on Road Traffic Monitoring and Control, IEE London, pp. 123-127.
6. Dalgleish M.J.: Vehicle Detection for Advanced Transport Telematics, Golden River Traffic Ltd., UK (Presented to the IEE in 1994).
7. Darryl H. Phillips: On-Board Weigh-in-Motion of Class 8 Trucks Using Newton's Second Law, AirSport Corporation, Copyright © 1997 Society of Automotive Engineers, Inc.
8. Gajda J., Sroka R., Stencel M., Zeglen T.(1997): Measurement of road traffic parameters using an inductive single-loop detector, IMEKO TC-4, Glasgow UK, pp. 221-224.
9. Gajda J., Stencel M.(1997): Determination of road vehicle types using an inductive loop detector, XIV IMEKO Congress, Tampere, Finland, pp. 231-236.
10. Jacob B.(1998): WIM Standards and Specification. Proceedings of 2nd European Conference on Weigh in Motion of Road Vehicles, Lisbon, pp. 79-86.
11. Koniditsiotis C., Henau A.(1998): Developments in sensors, systems and products. Proceedings of 2nd European Conference on Weigh in Motion of Road Vehicles, Lisbon, pp. 73-75.
12. Newton W.H., Frith B.A., Barbour I.A.: Use of WIM system to aid the selection of goods vehicles for enforcement weighing. Materials of Transport and Road Research Laboratory, UK.
13. European Specification on Weigh-in-Motion of Road Vehicles, COST 323, Draft 2.2, June 1997.
14. Data of Lawrence Livermore National Laboratory (http://www.llnl.gov).
15. Non Intrusive Traffic Detectors, ASIM Technologies Ltd., (http://www.asim.ch).
16. Intelligent Transportation System, (http://www.mmm.com).
17. Other Vehicle Detection Technologies, Path-California, (*http://pelican.its. berkeley*).

INFORMATION FUSION APPLIED TO SELECTED FINANCIAL PROBLEM DOMAINS

D. LOWE
Aston University
Aston Triangle
Birmingham B4 7ET, U.K.
Email: d.lowe@aston.ac.uk

1. Introduction

Technical financial analysis is an ideal domain for the development of data fusion techniques. There exists a wide base of diverse application requirements such as: the analysis of consumer credit, financial economics, options pricing, portfolio analysis, yield curves, risk management, strategic investment and trading models. Across this broad spectrum is a technical drive to incorporate methods of extracting information from data. Financial analysis invariably has to contend with multiple sources of data (news reports, company accounts, market data, mulitvariate time series), on different time scales (quarterly government indicators vs foreign exchange tick data) and with differing levels of accuracy from the qualitative to the quantitative. Hence it is an ideal medium for the development of pattern recognition and information fusion methods.

There are various frameworks for data fusion hierarchies. This book tends to concentrate upon the JDL model derived from military requirements. The UK Technology Foresight Working Party on Data Fusion [2] provided a 6-level taxonomy of generic data fusion, provided from a non-military-centric starting point. The levels in this taxonomy proceeding from the low level data processing to the high level symbolic processing were defined as Sensing, Signal Processing, Data Feature Extraction, Pattern Processing, Situation Assessment, and finally Decision Making. In the context of financial analysis we can also recognise a mapping of each of these levels to a specific requirement in a financial application. `Sensing' implies the collation and trawl for relevant raw data sources. `Signal Processing' includes the obvious basic preprocessing, labelling, enumerating and construction of the data sources into usable data streams. 'Data Feature Extraction' in finance would correspond to tasks such as selection of relevant features and specific data selection such as removal of obvious outliers. `Pattern Processing' would include the manipulation of data valid streams for the purpose of visualisation, rescaling, renormalisation, SnR enhancement, forecasting modelling, and transformations into more appropriate structures such as wavelet spaces. `Situation Assessment' would be the level at which various models would be combined with extra knowledge such as levels of risk an institution is prepared to manage, or obtaining measures of uncertainty in markets or other economic arenas, or optimisation of portfolios of investments according to different what-if criteria. The final layer of decision making would employ different types of utility function, combined with the various models' estimates of the financial instrument under study. Hence,

A. K. Hyder et al. (eds.), Multisensor Fusion, 749–764.

it is easy to see how a generic financial application domain can be mapped on to the Data-To-Decision cascade.

Despite this link to an obvious developed activity, the exploitation of fundamental data fusion techniques remains almost non existent in real financial applications. In the ASI meeting itself, I discussed 6 different financial applications spread across the different levels of the fusion hierarchy, and I demonstrated how simple applications of fusion techniques could return significant performance improvements over the more traditional approaches to these problems. In the talk I considered fusion of sources, features and models, regression, optimisation, prediction and classification, and used examples from option pricing, portfolio construction, managed pension fund selection, short and long term prediction of US bonds, FX ($/DM) data and equities (banks and electricals sector) from the UK markets. Within this diversity we observed a generic performance improvement, which makes the lack of the routine application of these methods all the more surprising.

In this chapter I will expand on just some of the approaches and results discussed at the ASI meeting itself. All examples used are based on real data obtained from collaborations with companies within various financial sectors, to whom I am most grateful for the additional comments and provision of background knowledge in addition to access to data. I also discuss a couple of interesting mathematical results appropriate to fusion domains which do not seem to be widely appreciated across the community, but give support for the use of multiple simple models, rather than sophisticated models. Technical financial analysis however has its own specific problems which create difficulties associated with traditional approaches to data fusion. We recognise that financial data is charaterised by the three `N's': Noise (and specificially non-Gaussian invariably), Nonlinearity (and hence modelling capabilities are complicated - so the Kalman Filter approach is known to be inappropriate for example), and Nonstationarity (so again we should be aware of the tradeoff between stability and plasticity in constructing short-time stationary models for example). These difficulties both constrain and motivate novel developments.

2. Fusion of Sources

We begin by considering a medium-level fusion problem: that of how to deal with multiple sources when those sources are subject to ignorance of specific form. At this level we will have already performed sensing, signal processing and data feature extraction. This is a very common situation in finance. Finance intrinsically deals with multivariate time series, multiple time scale values, and different information sources. Since we are not likely to have good knowledge of the generator which produces the sources of financial time series, we inevitably have to deal with uncertainty. This implies that we should employ a probabilistic approach to looking at the issues of source fusion under uncertainty. Hence, we first consider the mathematics of simple probabilistic approaches to combining inputs. We will see an interesting consequence on decision-making by incorporating an uncertainty of the source models into the analysis.

2.1 SIMPLE COMBINATION STRATEGIES FOR CLASSIFICATION

Ideally in finance we would want to estimate the probability distribution of a given quantity conditioned jointly upon a set of different sources of information,

$$P(\theta = \omega | x_1, x_2, \ldots, x_S)$$

Given this distribution, we would then make decisions based on a Bayesian decision rule or incorporate measures of utility to bias our decisions. Unfortunately we cannot generally estimate such a distribution in practice without severe approximations on the model form. Instead it is common to make assumptions on the statistical relationships between the information sources, and use the simpler models derived from the individual information sources. One usual method is to assume conditional independence between inputs (which could be multiple data sources or multiple models). We follow through this common assumption and also discuss an even simpler model. In practice, we often find that simpler data fusion models which are not as well `principled' as the more rigorous approximtions usually are more reliable in real world problems. The reason for this will become apparent in the analysis. We follow an elegant approach which has recently been discussed in [5] in the context of multisensor person identification, and we expand on that approach in what follows.

Let us assume that we want to obtain an estimate of a posterior probability of an event ω_k (buy/sell signal, or stock k is likely to gain most) occurrin given a pattern vector $\mathbf{x}$. The Bayesian approach would be to choose the event which maximises the posterior distribution. Under the independence assumption we have:--

$$P(\omega_k | x_1, \ldots, x_S) = \frac{P(\omega_k) \prod_{i=1}^{S} p(x_i | \omega_k)}{\sum_{k=1}^{S} P(\omega_k) \prod_{i=1}^{S} p(x_i | \omega_k)}$$

So the decision rule would be to choose the event which maximises

$$max_{k=1}^{K} P(\omega_k) \prod_{i=1}^{S} p(x_i | \omega_k)$$

In terms of posterior probabilities produced by separate source models, using $p(\omega_k | x_i) = p(\omega_k | x_i)\, P(\omega_k) \,/\, P(x_i)$ gives the decision as

$$max_{k=1}^{K} P(\omega_k) \prod_{i=1}^{S} \left[\frac{P(x_i)}{p(\omega_k | x_i)} P(\omega_k) \right]$$

$$max_{k=1}^{S} P^{1-S}(\omega_k) \prod_{i=1}^{S} P(x_i)\, p(\omega_k | x_i)$$

This is the **Product Rule**.

We will assume that $P(x_i)=$ *const* = $1/S$, i.e., we have no prior belief to select one of the sources as more important than any other.

Note an important point about this `product rule' for combining information sources: it is very severe since an event may be completely inhibited if just one of the independent source models returns an almost zero value. It also assumes we know, accurately, what the individual probability distributions of each source model is. This is most unlikely. Hence we have to assume the existence of uncertainty on the values of these source models. We can perform an error sensitivity on this product rule by assuming that the estimated aposterior probability $\hat{p}(\omega_k|x_i)$ is close to the desired value $p(\omega_k|x_i)$ by writing $\hat{p}(\omega_k|x_i) = p(\omega_k \mid x_i) + \varepsilon_{ki}$. Substituting and linearising around P assuming that the error is small, we find

$$max_{k=1}^{S} P^{1-S}(\omega_k) \prod_{i=1}^{S} P(x_i)\hat{p}(\omega_k|x_i)$$

$$\approx max_{k=1}^{S} P^{1-S}(\omega_k) \prod_{i=1}^{S} P(x_i)p(\omega_k|x_i) \left[1 + \frac{\varepsilon_{ki}}{p(\omega_k|x_i)}\right]$$

For small ε the product compoment may be separated as

$$\approx max_{k=1}^{S} P^{1-S}(\omega_k) \left[\prod_{i=1}^{S} P(x_i)p(\omega_k|x_i)\right]\left[1 + \sum_i \frac{\varepsilon_{ki}}{p(\omega_k|x_i)}\right]$$

We see from this that the effect of allowing for uncertainty in the source distributions is to allow for a modulation of the decision rule by the term

$$1 + \sum_{i=1}^{S} \frac{\varepsilon_{ki}}{P(\omega_k|x_i)}$$

It is now obvious that whenever a source model returns a very small value for the likelihood of an event, then that particular error in estimation is magnified.

Interestingly we can introduce a robustness into the decision making by incorporating a more severe simplification than the independence assumption. If we now assume that the posterior does not really deviate much from the prior of the event, so that $P(\omega_k \mid x_i) \approx P(\omega_k)\ [\ 1 + \delta_{ki}]$ where δ_{ki} is assumed a small perturbation, then we find that the decision rule can be expressed as :

$$ax_{k=1}^{S} P^{1-S}(\omega_k) \prod_{i=1}^{S} P(x_i)p(\omega_k)[1+\delta_{ki}]$$

$$\approx max_{k=1}^{S} P(\omega_k)[1 + \sum_i \delta_{ki}]$$

(assuming that P(xi) = const). Substituting in doe δ gives:-

$$\approx max_{k=1}^{S} P(\omega_k)\,[\,1+\sum_i \left\{\frac{p(\omega_k|x_i)}{P(\omega_k)} - 1\right\}]$$

or

$$max_{k=1}^{S}\left\{(1-S)P(\omega_k) + \sum_{i=1}^{S} p(\omega_k|x_i)\right\}$$

This is now a *sum rule* for decision making. Again it assumes that all source probabilities are known accurately. As before, we can examine the effect of this assumption on the decision making process by performing an error analysis. Assuming a small error of ε_{ki} in $p(\omega_k|x_i)$, the sum rule becomes:

$$max_{k=1}^{S}\left\{(1-S)P(\omega_k) + \sum_{i=1}^{S} p(\omega_k|x_i)\,[\,1+\frac{\sum_i \varepsilon_{ki}}{\sum_i p(\omega_k|x_i)}\,]\right\}$$

This expression shows that the decision rule is modulated by the error term of the form:

$$1+\frac{\sum_{i=1}^{S} \varepsilon_{ki}}{\sum_{i=1}^{S} p(\omega_k|x_i)}$$

Now note that all source probability models have to return an almost zero value to distort the error. Generally the sum over all sources of the probabilities will be greater than unity and so this will have the effect of damping down the errors. Hence, even though this is a much simpler model for source fusion, it is more robust and reliable in practice since we can never approximate the true distribution function with any degree of accuracy, and hence the decision making is usually dominated by the sensitivity to errors. We will observe this effect when we give an example of combining models later.

2.2 PRICING OF FINANCIAL OPTIONS CONTRACTS

In the previous section we discussed the consequences of the statistical distribution of separate source models. An alternative strategy occurs when there exists a deterministic (though inevitably nonlinear) relationship linking together the various information sources. The financial application domain we use is a very simple example which may be posed as a straightforward nonlinear multivariate regression problem. It is complicated by the fact that one of the required input variables cannot be measured, but must be inferred from other sources. The example considered is a problem based on the pricing of financial instruments known as *Options*. An option on a commodity is a contract which gives the owner of the contract the right (but not the obligation) to either buy or sell the underlying commodity at a predetermined price at some time in the future. The price in the contract is known as the *exercise price* or the *strike price*. The date in the contract is known as the *expiration date*, *exercise date* or *maturity*. A contract giving the owner the right to buy by a certain date is known as a `Call', and the right to sell is known as a `Put' options contract. If the contract can only be exchanged at the expiry date of the contract, then it is known as a *European* option. If the contract can be exercised at any time up to the expiry date it is known as an *American* option. Options on stocks were first traded on an organised exchange in 1973, since when there has been a dramatic growth in the options markets. The underlying assets of the contracts now include stocks, stock indices, currencies, debt, commodities and futures contracts. Although often used for balancing risk by constructing portfolios, options can be used for speculation due to their inbuilt gearing of capital. When used to speculate it is possible to gain significantly but also lose all the initial investment. Hence mispricing or incorrectly estimating the value of an options contract can have significant consequences, as the 1995 saga of the Baring Bank has emphatically illustrated.

This particular area of financial mathematics is of specific interest, as there exists a widely accepted model which estimates the price of European call or put options by an analytic formula which involves five variables. The model is known as the Black--Scholes pricing formula (these proponents were recently awarded the Nobel Prize for Economics for this work) and was derived in 1973 [1]. The model makes various assumptions, such as the existence of equilibrium markets, stock prices assumed to follow a random walk in continuous time and are lognormally distributed, the existence of constant short term risk free interest rates, and the assumption of no dividends or transaction costs. The model also assumes a rational investor.

Under these assumptions, Black and Scholes produced an analytic formula for the value of a European call or put option in terms of five quantities: the underlying stock price, S, the strike price X, the risk free interest rate r the time to expiry T and the 'volatility' σ (usually taken as an estimate of the standard deviation of the logarithm of price fluctuations). The value of a European call option under the model based approach is predicted to be

$$C = S \times N(d_1) - X \times \exp^{-r(T-t)} N(d_2)$$

where $d_1 = [\,ln(S/X) + (R + \sigma^2/2)(T-t)\,]/\sigma\sqrt{T-t}$ and $d_2 = d_1 - \sigma\sqrt{T-t}$

and $N(x)$ is the cumulative probability distribution function for a standardised normal variable. However this formula, though widely in use today, is known to consistently underestimate or overestimate the market depending upon whether the option contract is a call or a put.

Figure 1 shows the actual and predicted values for 755 Put options on \$/DM exchange rates which demonstrates the consistent bias. Despite many efforts to improve on the basic model (for example, see articles in [3]), there is still no reliable model to estimate the market value of options contracts.

As an alternative to the model based approach, we can view the problem of estimated the price of a Put contract P (say) as one of nonlinear regression by the fusion of the five driving variables, $\{S,X,r,T,\sigma\}$. Since we require a flexible semiparametric model capable of approximating the underlying (but unknown) data generator, we choose to employ a radial basis function spline network. Figure 2 depicts the simple fusion process.

As a demonstration we choose to estimate the market value of European put options on \$/DM exchange rates. Training data was taken from 42 consecutive days between February 8th 1989 and April 18th 1989. This amounted to 1592 separate contracts. The risk free interest rates were taken from the contemporary values of 3 month Treasury Bills as a good proxy for true risk free interest rates, and volatility was calculated using an optimally chosen sliding window on historic data.

Note that volatility is the only parameter not directly accessible and significant activity occurs around the financial world to estimate accurate volatilities. Volatility is generally (though not by this author) considered to be a key parameter for option pricing, to the extent that options are also often quoted in volatility values. We examined a variety of methods to estimate volatility which we could also use in out-of-sample experiments (and so implied volatility calculations are not available since for implied volatility we also need the cost of the contract in the first place. We eventually settled on a chosen sliding window calculated on historic values on the basis of overall minimum mean square error between estimated and actual option values on an out-of-sample set. All other information was easily available.

The network performance was tested against the consecutive trading days, between May 24th 1989 and June 30th 1989, amounting to 752 different contracts. The input data was prewhitened to be zero mean, unit variance on a channel by channel basis (with parameters evaluated in the training data only).

The 'fusion' in this example is simply achieved by the single RBF network acting as a universal approximator to produce an estimate of the (unknown) nonlinear regressor, or data generator, which connects the 5 input variables with the single output (the price of the option contract). Combining multiple models to improve output estimates will be demonstrated later in this chapter. Note that in this example, since we are only using a single model to achieve the fusion of the inputs, it is quite important to match the complexity of the model (as determined essentially by the number of basis functions) to the complexity of the problem. There are several ways we could do this. For the purposes of this experiment we used cross validation on a separate contiguous data set of values and then fixed the model complexity when applied to the separate test set which we used to present the results in the figures.

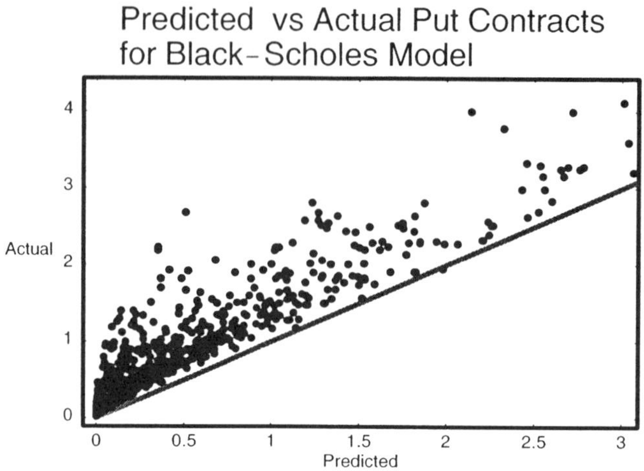

Figure 1. Scatter plot of actual and predicted Put options contracts obtained by the standard Black-Scholes model

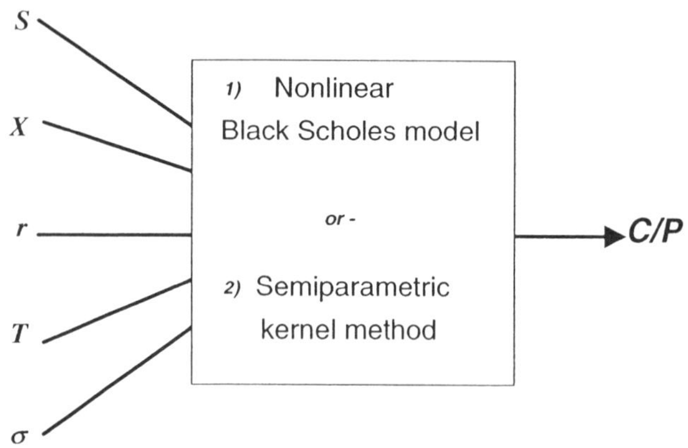

Figure 2. Option pricing by model-based or semiparametric-based regression

Figure 3 shows out of sample performance of the semiparametric approach. The scatterplots show the actual market value of the contracts plotted against the Radial Basis Function predictions. Clearly the model has produced very good estimates of the underlying values of the contracts with a Spearman rank order correlation of 0.9753, hence indicating very good correlations between the network models and the data.

What is particularly noteworthy is the robustness of estimates even for high priced contracts where the data density is clearly more sparse. These results provide a major improvement over the standard Black-Scholes model which has a consistent bias away from the true values. Results on Call contracts are similar though not as good as these obtained on Put contracts. Further details can be obtained in [7].

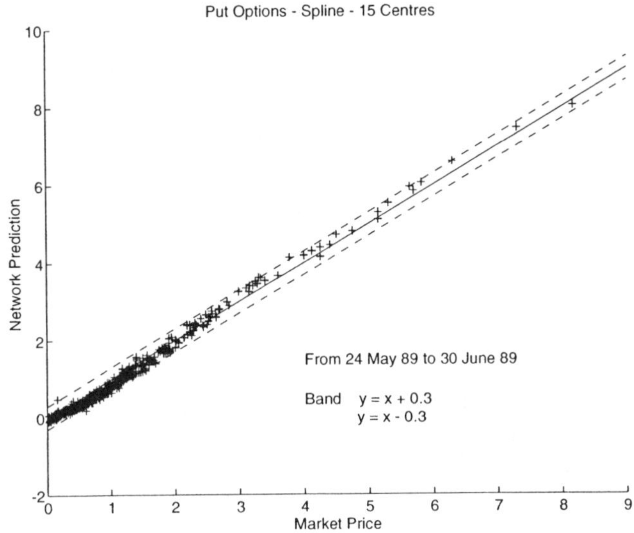

Figure 3. . Scatter plot of actual and predicted Put options contracts obtained by the simple fusion model

An interesting aside developed as part of these nonparametric models is a demonstration of the benefits of unconstrained semiparametric data fusion. Specifically, the main problem with option pricing models lies in the determination of the volatility. The remaining parameters may be directly obtained or inferred from reliable sources. Much effort is involved in attempting to produce better interpretations and estimates of volatility, as it is widely perceived as an indication of risk. It also dominates the uncertainty in the Black-Scholes formula. However, given that volatility is an inherent property of the time series, and wc employ the time series in the semiparametric model, then is it feasible that volatility is already implicitly contained within the underlying stock price?

Since the semiparametric approach is not constrained by the model based assumptions of the Black-Scholes model we may investigate this question. The results of Figure 4 indicate that a 4 parameter model excluding volatility still performs significantly better than the Black-Scholes model. These ramifications have yet to be properly investigated.

3. Model Combination

In finance, it is common practice to employ a variety of models when trying to produce forecasts or inferences, and then select those models which the user has most confidence in. Interestingly, model selection is not a particularly good method of arriving at a decision, even when some of the models give quite poor results. Instead, a simple `model fusion' result of simple averaging of the predictions of a variety of models -- especially when those models have a diverse basis -- usually returns superior performance. We now examine this higher level data fusion issue of model averaging, and demonstrate the main result on a single-equity iterated forecasting problem.

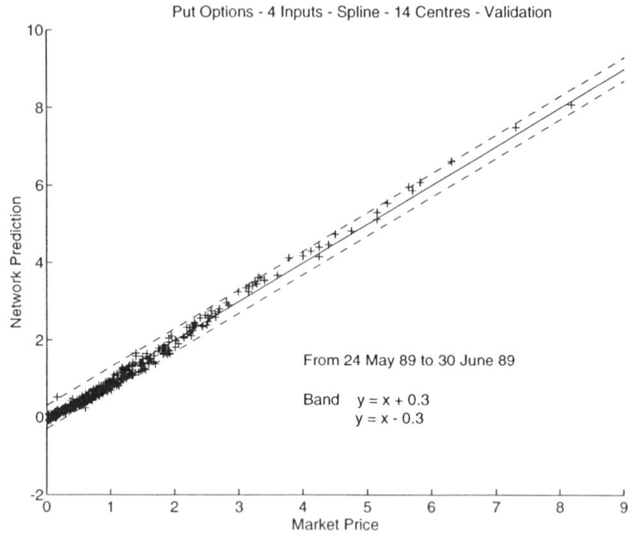

Figure 4.. Scatter plot of actual and predicted Put options contracts obtained by the simple fusion model, but only using a 4 parameter regression

3.1 ERROR AVERAGING

Let us first provide some background mathematics justifying the combination of models - even when it is known that some of the models are inferior (e.g. linear models). We first consider the case of a simple, uniform averaging of multiple models. Consider the case when we have M distinct models (we are not concerned about the specific forms here), each one of which produces an output value $y_i(\boldsymbol{x}) = r(\boldsymbol{x}) + \varepsilon_i(\boldsymbol{x})$. $r(\boldsymbol{x})$ is the true (unknown) function of the input data $\boldsymbol{x}$.

The average (squared) error produced by each model i is $E_i = \langle [y_i(x) - r(x)]^2 \rangle = \langle \varepsilon_i^2 \rangle$ where the expectation $\langle ... \rangle$ is over the distribution of the data $\boldsymbol{x}$. The average (squaared) error over these models would be $\widehat{E} = \sum_{i=1}^{M} E_i/M$.

Now consider the expected error produced by a simple average fusion over all model predictions. By averaging over all models, a modified prediction value would be just $\hat{y} = \sum_{i=1}^{M} y_i(x)/M$. This gives an error due to this 'committee' of equally voting models as $\widehat{E}_M = \langle (\sum_{i=1}^{M} \varepsilon_i/M)^2 \rangle$. If we make the assumption that each model's errors are independent at second order, then this expected committee error is simply related to the error of each committee member as:

$$\widehat{E}_M = \frac{1}{M} E_i$$

This result seems to indicate that a significant reduction (a factor of M for M models) in expected mean error can be achieved by a simple averaging over multiple models. Of course the real error reduction is much smaller than this as the assumption that the committee model errors are independent is in practice false. Nevertheless, in practice it is found that performance improvements can be made by this simple averaging approach. Since the major gains are to be made when the errors are uncorrelated then this leads to the desire of using multiple diverse models, since this generally increases the chance of the error distributions being different. It also justifies the use of even keeping weak models (e.g. linear models) in the committee scheme. This reduction in error is derived from a reduced variance. So this tells us that it is preferable to choose models with small bias and reduced the overall error by variance reduction. In terms of neural network models, this implies using a collection of models which are too complex for the problem at hand.

A word of warning, however. Committee averaging is good to improve the performance of a collection of weak models. It is an expectation on average across models. Committee averaging will reduce the effectiveness of using a set of very good models -- or equivalently, if one model is known to be the expert, it is best to use that one model rather than heed the committee!

3.2 WEIGHTED ERROR AVERAGING

Despite the previous result, we might imagine that a better fusion model exists if we allow for differentially weighting those committee members who we believe give better predictions more heavily in the averaging process. So more generally we still consider a simple linear averaging, but in the form

$$\hat{y}_W(\boldsymbol{x}) = \sum_{i=1}^{M} \alpha_i y_i(\boldsymbol{x}).$$

Under the constraint that $\Sigma_i \ \alpha_i = 1$, minimising the weighted average committee error with respect to these parameters we find a solution for the weighting coefficients in the form:

$$\alpha_i = \frac{\sum_{j=1}^{M} (C^{-1})_{ij}}{\sum_{i=1}^{M} \sum_{j=1}^{M} (C^{-1})_{ij}}$$

which gives a value for the expected minimum error as $\hat{E}_W = (\sum_{i=1}^{M} \sum_{j=1}^{M} (^{-1})_{ij})^{-}$. In these expressions C^{-1} is the inverse covariance matrix between the different model errors:

$$C_{ij} = \langle \varepsilon_i(\boldsymbol{x})\, \varepsilon_j(\boldsymbol{x}) \rangle$$

Hence, either knowing something about the correlation between the errors of the different models (or estimating the covariance matrix from sampling of the trained models), allows us to construct a better weighted average committee.

From an analysis of the error of the weighted committee network [6] we can show that the expected error by weighting exploiting the covariance knowledge will in general be better than the simple averaging of committee members. The weighted committee error may be decomposed into two terms:

$$\widehat{E}_W = \sum_i \alpha_i < (y_i(x) - r(x))^2 > - \sum_i \alpha_i < (y_i(x) - \widehat{y}_W(x))^2 >$$

This is reminiscent of the bias-variance decomposition, though the negative sign assists in reducing the overall error by allowing for a more diverse variance, provided the bias is not increased. The first term is the error due to the individual models around the true regression. The second term is the effect of the spread of predictions of the committee members around the weighted average committee prediction -- the 'variance'. Hence if we can increase the spread of predictions around the committee prediction without increasing the individual member prediction errors, then we should be able to reduce the overall committee error.

3.3 EXAMPLE OF MODEL WEIGHTING

As a simple example of a prediction-fusion model, we considered the forward iterative prediction of the quoted daily closing price of HSBC plc from the FTSE-100 Price Index based only on historic information. In this example we used the previous two years of price information. Individual predictive models were obtained based on 10 multilayer perceptron networks separately optimised by standard methods. A model with 10 hidden units was employed. Note that model order selection techniques were not emphasised as the model averaging compensates for small bias models. Each model was then initialised using data up to the end of the training period, and then the outputs of each model were recursively fed back into the inputs of the models, thus allowing each model to produce an iterated daily forecast of HSBC prices for each of 21 days in the future. These predictions were then averaged using the simple weighting model and the results displayed in Figure 5.

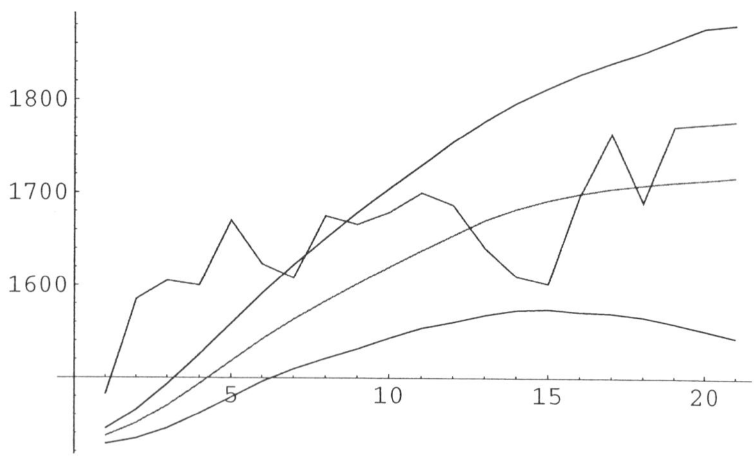

Figure 5. The mean forecast using the fusion of model outputs (the middle smooth line) iterated 21 days ahead and the true price trajectory (the top jagged line). Also shown are the confidence intervals

This figure also shows an estimate of confidence intervals. There are many techniques for obtaining estimates for the confidence of predictions using nonlinear models [8] -- some including fusion techniques as indicated in the previous variance analysis. Each has respective advantages and disadvantages depending upon the assumptions which are used in their derivations. Figure 5 employed a bootstrap technique [9] to produce the confidence bands. Results obtained using this method of committee fusion on out-of-sample performance have been demonstrated to return consistently better performance than single model estimates and is now part of a live trading system.

As a comparison, Figure 6 provides the prediction estimates for three of the individual models.

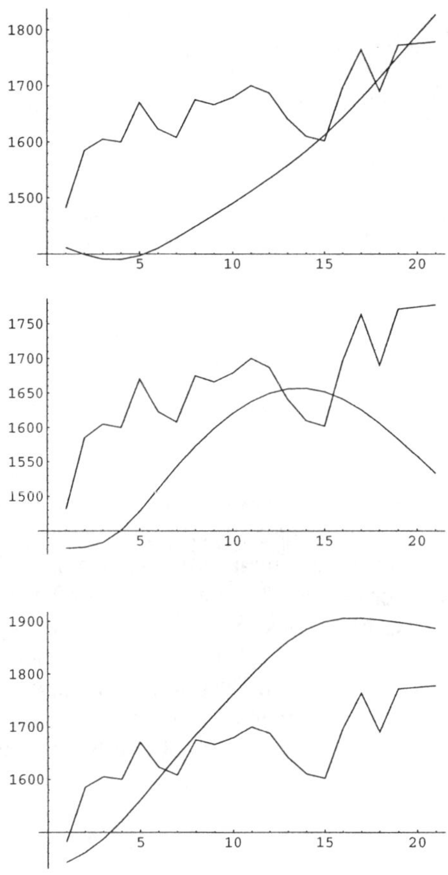

Figure 6. Exemplar prediction estimates produced by three of the individual models, demonstrating the diversity of the outputs of the models

In this example the model averaging approach produced estimates as good as the a posteriori selected best individual model, as evidenced in Figure 7 which shows the spread of sum squared errors made by the various models on out of sample performance.

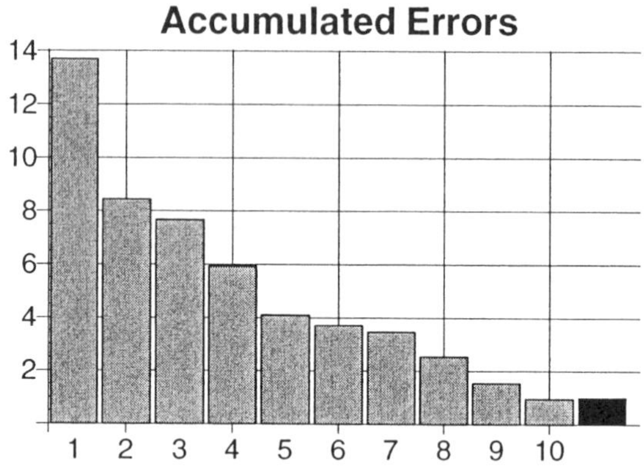

Figure 7. Bar chart of the error produced by the ten individual models and the committee model (dark-shaded bar) accumulated over all 21 predictions and normalised to the error of the committee result

3.4 BAYESIAN COMMITTEES

We can pose part of the model fusion problem, especially for semiparametric models such as neural networks, in an approximate Bayesian framework. A basic principle of a Bayesian approach to inference is that integration over models is superior to model selection. It is of no surprise therefore to note that the committee approach to fusion can be shown to emerge from a Bayesian perspective. It is instructive to provide a brief outline sketch. Each of the models in our committee or portfolio is usually characterised by a set of parameters **w** (e.g. the weights in a neural network) sampled from a corresponding probability density p(**w**/data).For sufficiently diverse models after training on historic data the chosen weights are likely to be sampled from distinct regions of parameter space. (For example, for neural networks, symmetry and redundancy indicate that there are distinct weight vectors which yield neural network models with simlar or equivalent perfromance in terms of a cost function such as mean square error or mutual information.) We assume that the complete density may be approximated by a set of effectively distinct densities - one for each model m:

$$p(w|data) \approx \Sigma_m\, p(m,\mathbf{w}|data) = \Sigma_m\, p(\mathbf{w}|m, data)\, P(m|data).$$

All observable quantities should be obtainable by averaging over the density function. In general we cannot perform this averaging. However for the simple assumption of separated model regions in parameter space we can decompose the average. In particular, the expected value of the prediction of the committee of models would be just :

$$\hat{y}_B(x) = \int y(w,x)p(w|data)\,dw$$

$$\approx \sum_m P(m|data) \int_{\Omega_m} y(w,x)p(w|m,\, data)\,dw = \sum_M P(m|data)\hat{y}_m$$

where Ω_m is the region of 'activity' of parameter space surrounding model m and $\hat{y}m$ is the expected prediction due to model m.

Note that this simple outline provides the best solution as a fusion of individual committee member predictions where the weighting depends on the likelihood of each particular committee member occurring.

It is interesting that committee averaging can be obtained from a Bayesian perspective, where a different weighting scheme is suggested for the source models. However we note that the Bayesian perspective has assumed that the individual source models are accurately known. We saw earlier in this chapter that decision-level fusion was significantly distorted if we allowed for errors in the source models. This result may account for the observation that pragmatically we have found that weightings determined by the previous methods (and usually just the simple uniform weighting) typically produce better predictions than this Bayesian separable approximation. However we recognise this as an interesting open issue which needs further study - theoretically and numerically.

4. Conclusions

In this short chapter we have merely selected a few examples of where data fusion methodologies are likely to be useful in the technical analysis of financial problems. We have completely excluded fascinating discussions of other topics. Amongst these we mention a few obvious candidates for the application of data fusion techniques. These include: high level decision making by exploiting alternative utilities under uncertainty, the exploitation of subjective information combined with data analysis for visualisation, the tradeoffs in portfolio construction, selective source selection by confidence ranking, model output moderation by the incorporation of uncertainty modelling, and indeed the whole area of uncertainty modelling in financial fusion. The potential for exploitation of fundamental data fusion techniques in this arena is enormous. Given the obvious benefit and potential gain to be derived, it is extremely surprising that there is hardly any real activity in this domain. It is common folklore that advances in financial engineering are not made public if they are a success, because they are used for profit `behind closed doors'. In which case, perhaps data fusion techniques really are being exploited within the industry successfully. The alternative hypothesis is that there are very few useful principled techniques being applied and it is the knowledge of this paucity that is being concealed. I favour the latter.

We should also stress that we have only demonstrated very simple information fusion strategies, and concentrated upon those approaches which can be readily applied to the real world problems. Yet despite the simplicity (which in itself has some theoretical support as indicated in this chapter), the experimental evidence has demonstrated the effectiveness of applying such techniques. It is our hope that this brief overview might stimulate some small interest in the development of information fusion towards technical financial analysis.

5. Acknowledgments

This work was supported in part by the EPSRC contract GR/K51815, *"Nonstationary Feature Extraction and Tracking for the Classification of Turning Points in Multivariate Time Series"*. I would also like to extend my thanks to my student Krzysztof Zapart for helping me with some of this work, and to several collaborators from various financial domains who have provided me with some small amount of their expert domain knowledge to allow me to progress slightly with some of their problems. Of specific note has been the guidance provided by Dr C Satchwell of Technical Forecasts Ltd.

6. References

1. Black, F. and Scholes, M. (1973) "The Pricing of Options and Corporate Liabilities", *Journal of Political Economy*, **81**, 637--659.
2. "Defence and Aerospace Foresight Technology Working Party: Data Fusion and Data Processing", Royal Aeronautical Society, ISBN 1857680685, (1997).
3. Hand, D.J, and Jacka, S.D. (1998) (eds), Statistics in Finance, Arnold.
4. Haussler, D. and Opper, M. (1997) "Mutual Information, Metric Entropy and Cumulative Relative Entropy Risk", The Annals of Statistics, **25**(6), 2451—2494.
5. Kittler J, (1998) "Combining classifiers: A theoretical framework", Pattern Analysis and Applications, **1**, 18--27.
6. Krogh, A. and Vedelsby, J. (1995) "Neural network ensembles, cross validation and active learning", In D S Touretzky, G Tesauro, T K Leen (eds) Advances in Neural Information Processing Systems, **7**, Cambridge MA: MIT Press.
7. Lowe, D. (1995) "On the use of nonlocal and nonpositive definite Basis functions in RBF networks",206--211, Proceedings 4th IEE International Conference on Artificial Neural Networks, Conference Publication number **409**, The Institute of Electrical Engineers.
8. Lowe, D. and Zapart, K. (1997) "Validation of Neural Networks in Automotive Engine Calibration", 5th IEE International Conference on Artificial Neural Networks, **440**, 221--226, The Institute of Electrical Engineers.
9. Zoubir, A.M. and Boashash, B. (1998) "The Bootstrap and its Application in Signal Processing", 56--76, IEEE Signal Processing Magazine, **15**(1).

COMPUTER NETWORK SECURITY

A Data Driven Approach

E. LEFEBVRE
Lockheed Martin Canada
6111 Royalmount Avenue
Montréal QC, H4P 1K6
Canada

1. Introduction

When considering computer security the most important thing should be, as for any other security system, the cost effectiveness of the security. This means that the cost of security controls should not exceed expected benefits. For example, if the backup procedure for a system is more expensive than the cost of producing the data, the backup security is clearly not cost effective. On the other hand, data could be very cheap to produce but very sensitive to an organization. In this case the cost of security may exceed the raw value of the data but should never exceed the loss encountered should the data become accessible to potentially hostile persons. Therefore, one of the difficulties when evaluating security needs is correctly assessing the system to be protected.

Another issue is that computer security requires a comprehensive and integrated approach. Consider, for example, a home with two doors. If a good lock is bought for the front door but the rear door open is left open, then the home's security is not well integrated. Likewise, if the same lock has been installed on the rear door but the keys have been forgotten inside, a comprehensive and integrated approach is still lacking. Without a firm understanding of the interdependencies of security controls, they can actually undermine one another.

Perhaps the most important aspect of computer security, from an engineering point of view, is that it should be periodically reassessed. Since systems evolve and new threats emerge constantly, security requirements are ever changing.

This paper presents a software approach that eases the integration of a computer network security system. Considering the three previous points, the tools used must fulfil the following needs:

- Since one versatile tool may be more cost effective than several specific tools for specific applications, there is a need for a minimum number of tools. This is also in accordance with the second need.
- There is a need for tools that can be integrated or that can at least communicate with each other. Tools that offer modularity and portability would improve the integrated approach.

A. K. Hyder et al. (eds.), Multisensor Fusion, 765–778.

- Finally, there is a need for the ability to improve tools in an easy and incremental fashion. Tools must be expandable.

In the next section the multi-sensing aspect of computer network security is presented, followed by security management in Section 3. Sections 4 and 5 present the data driven approach with some features specific to the blackboard application. Finally, in Section 6, a high level architecture is proposed for computer network security integration.

2. A Multi-Sensing Application

The importance of a multi-sensing aspect of security becomes readily apparent (Table 1) when the threats to computer network systems are considered. Although deliberate acts are probably the first threats that come to mind, organizational shortcomings, human or technical failures and force majeure cases must also be considered. To deal with these threats, various sensors, e.g., software, electronic, human, etc., may be used, with each performing a different task. It is then easy to see that computer network security implies a collection of several dissimilar data.

Table 1. Possible threats.

Organizational Shortcomings – Lack of rules – Insufficient knowledge – Lack of maintenance – Uncontrolled use of resource – etc... **Human Failure** – Negligent destroying of equipment or data – Improper use of the IT system – etc...	**Force Majeure** – Loss of personnel – Failure of the IT system – Water, humidity – Fire, temperature – etc... **Technical Failure** – Disruption of power supply – Loss of data – etc... **Deliberate Acts** – Theft, vandalism, attack – Trojan horse, viruses – etc...

In addition, functionality associated with incoming signals may change depending on the set of data collected. For example, if a fire is detected in a room containing electronic devices and employees, perhaps the system should not open the halon valve immediately in order to allow them time to reach the exit or gas masks. In the same manner, if a power failure is detected on one system, its functionality could be performed temporarily by another system. But if the power failure affects all systems, it is wiser to properly shut down non-essential systems and keep the spare energy for vital ones.

The last point regarding security control concerns the flow of data. Threats do not occur at a constant rate and even if sensors acquire data at fixed rates, the security controls may be in a "wait for something" state for long periods. This favours an approach that can easily deal with a stand-by state and triggering events.

3. Security Management

Without good management, the best security system is worthless. Security management consists mainly of the establishment of policies and their application. Many Computer Security Programs (CSPs) that are distributed throughout an organization have different elements performing various functions, which makes security easier to manage because it is performed at several levels. Managing computer security at multiple levels also has many benefits in that each level contributes to the overall CSP with different types of expertise, authority, and resources. Organizations have at least two levels of computer security management: the central level and system level.

The central CSP can be used to address the overall management of computer security within an organization or a major component of an organization. It should provide two quite distinct types of benefits: increased efficiency and economy of security throughout the organization, and the ability to provide centralized enforcement and oversight. But there is also a downside to the central security program; it presents the greater risk that errors in judgement will be more widely propagated throughout the organization.

Internal security-related information, such as which procedures were effective, virus infections, security problems, and solutions, needs to be shared within an organization. A CSP administered at the central (organization) level can provide a way to collect the internal security-related information and distribute it, as needed, throughout the organization.

Figure 1 shows some principal security program interactions. The ellipses represent external sources that provide information about security. Communications are bi-directional; the central security program retrieves information from external sources and also provides information or in-house solution to enrich the knowledge of the community. The central security program also distributes information among the systems belonging to the organization, and is responsible for the standardization of policies and procedures.

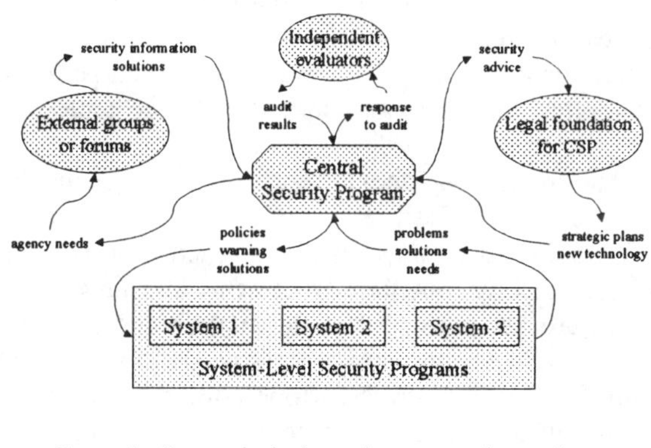

Figure 1. Some principal security program interactions.

The system-level CSP addresses the management of computer security for a particular system. It ensures appropriate and cost-effective security for each system and is responsible for the implementation of the policies, guidance, and regulations of the central security program. The process of securing a system over its life cycle is the role of the system-level security program.

4. From Goals Driven To Data Driven

In the previous section we presented two aspects of computer network security: control and management. The objective of this section is to move towards an integrated software approach for security control and management. In other words, we want a practical tool that can easily integrate the control and management required in computer network security. In this section, we present what we think is the more appropriate approach for computer network security.

In the goal driven approach, the order of functions is known in advance. The data is processed by one function which modifies the data or produces new data, which in turn is processed by the second function and so on. In this case the process is perceived as a flow of actions, with each action performed by a function. When designing systems with such an approach, the emphasis is put on the succession of functions to reach a predefined goal. Thus the function is the fundamental element for representing a system based on this approach.

In the data driven approach, however, data (events) are the fundamental elements of the design. In such an approach the relations between data are used instead of the sequence of actions. Incoming data triggers a specific action to produce new data. Data can then be created, transformed or deleted according to their interdependencies and occurrences.

To understand the differences between the goal driven and data driven approach, a discussion of each is provided in the following paragraphs. But first some function aspects need to be clarified.

Simply defined, a function transforms data or creates new data. If data is taken in the broadest sense, i.e., all kinds of information, then a function must have input and output data. The input data can be regarded as information sent to the function to trigger its activation and the output data as being the result of the action performed by a function. With this basic input/output rule, we can continue with the representation of the goal driven approach.

The easiest way (but not the only one) to represent a process involving a sequence of actions is through a Functional Flow Block Diagram (FFBD). Such a diagram is presented in Figure 2, which depicts a succession of functions performing actions. In this diagram the arrows connecting functions represent the flow of the functionality. Concurrent or exclusive processes, as well as function iteration, are represented by split arrows or reversed ones. When a decision process needs to be explicitly represented, a function box with multiple exits (i.e., several arrows pointing out of the box) is used.

This diagram allows functional threads to be identified. A functional thread is a list of functions executed for a specific process. The threads in Figure 2 include F1-F2-F3-F4-F4-F6, F1-F2-F5-F6, F1-F2-F3-F5-F4-F6, etc. Generally, a system is designed starting with a collection of threads. These threads are then allocated to the systems to obtain coherent and integrated functionalities. This is the traditional procedure when building a system from a set of established requirements. Since requirements are often

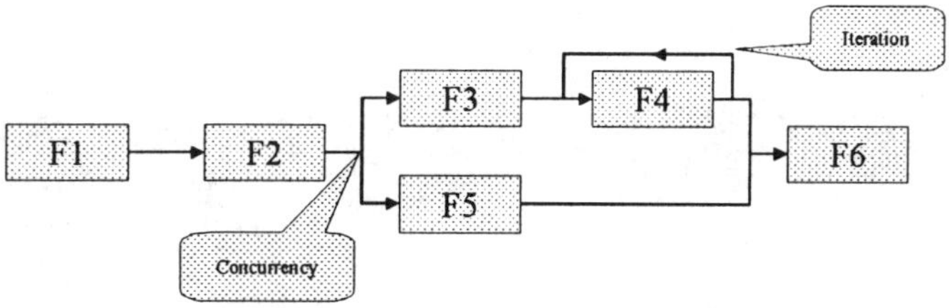

Figure 2. Functional Flow Block Diagram (FFBD).

expressed as conditional actions (if *condition* do *sequence of actions*) or a set of rules (when *event* apply *rules*), it is relatively easy to translate requirements into functional threads. The difficulties arise when these threads are allocated to the system. If new requirements involving new functions are captured after the first system design, it may be impossible to allocate the corresponding threads to the existing system. In this case, threads allocation would involve major design changes to the system. As a result, the functional analysis, which is the verification that each requirement is covered, could become a very time consuming activity, especially if systems evolve and need to be refined frequently. Thus it would be easier and faster to translate requirements into a set of independent threads without worrying about their allocation. This is what is sought with the data driven approach.

Figure (3a) shows the same system as Figure 2 but one enhanced by the addition of input and output data for all functions. In Figure (3b) output data is used to trigger the next function to be executed. The first occurrence of D1 triggers F1, which modifies D1. The modified D1 then triggers F2, which produces either D2 or D3, which in turn triggers F3 and F5, respectively, and so on. At this stage one could validly argue that the concept of the data triggering function is just a matter of representation and consequently would not affect system implementation. The next step therefore is to separate the data space from the function space and transform the flow of function into a set of rules governing relations between data and function.

Such a design following a blackboard paradigm is illustrated in Figure 4. The main components of this blackboard are the data, the agents, the blackboard itself, and the blackboard controller.

In this design, data are not only triggers; they can symbolize simple numbers, arrays or structures, or more complex objects such as messages, a computer screen, a situation, a human being, etc. These data can be created, deleted, or transformed by agents, whose role is to fulfil the computational or procedural task of the application. Each agent is designed to act on a specific data and to produce (or modify) another specific data (which can be different or similar to the input data). An agent can be a simple function, a thread or an entire system. In fact, blackboard systems can incorporate data triggering as well as functional flow. The direct mapping between the functional flow of the system in Figure 2 and the blackboard system of Figure 4 was done mainly to demonstrate the difference between the two approaches and not to signify that this direct transformation is always desirable.

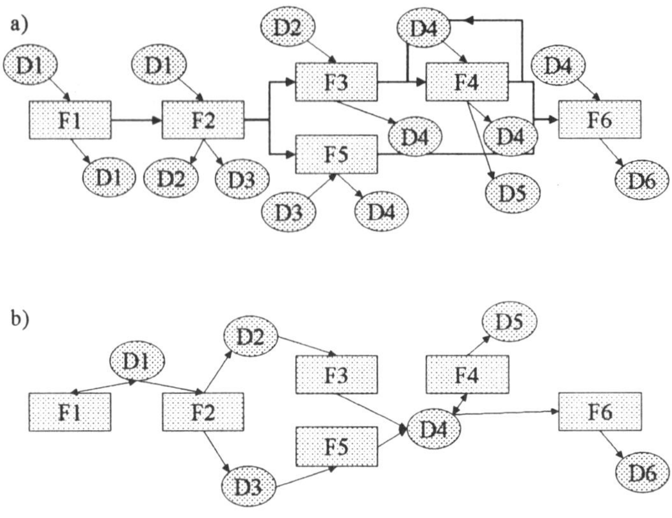

Figure 3a. Enhanced Functional Flow Block Diagram (EFFBD).
Figure 3b. Functional Flow Expressed with Data Triggers.

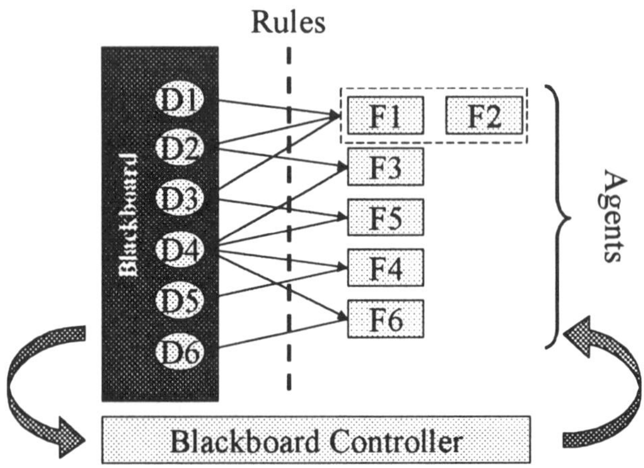

Figure 4. Blackboard representation.

The various data types representing the application are instantiated and reside on the so-called blackboard. The blackboard can be understood as a global database containing the instantiated data, which all agents can access in the system. The remaining component of this blackboard shell is the blackboard controller. Put simply, the main task of the blackboard controller is to schedule and execute the processing tasks within the application by continuously monitoring the state of the blackboard and associating the data with their corresponding agents.

Even though this particular architecture has been developed for expert systems, it offers considerable advantages for building more standard applications, both for the design and implementation phases of the capabilities and as a powerful run-time engine.

5. Blackboard Architecture Features

Lockheed Martin Canada has developed a new Knowledge-Based System (KBS), which provides fast execution speed and easy representation of abstract data types. The KBS is programmed using C++ in order to take advantage of the speed of a compiled language as well as the benefits of an object-oriented programming language such as data abstraction, class inheritance, information hiding, etc. The KBS, which is implemented on a blackboard architecture, represents a powerful environment to develop complex applications. All functions of the application are mapped into agents, and the application data manipulated by these agents reside on the blackboard. Hence, from a software development point of view, the KBS application is just a collection of (C or C++) functions and data structures. However, the developer does not need to worry about how the functions and data interact with each other; the blackboard controller takes care of this task.

Any application developed with the KBS has the advantage of being modular, versatile, and easily expandable. Modularity means that tasks to be programmed can be embedded in agents and the functionality and data are self-contained and independent from each other. It provides the software developer with ways of modifying parts of the application without jeopardizing the entire system. It also facilitates the process of rapid prototyping, which is closely related to KBS development. Finally, it simplifies parallel implementation as modules can be distributed on different processors.

Versatility means the possibility to easily and quickly reuse parts of a program during the software development. This also occurs frequently in the rapid prototyping process. An agent (or function) designed to perform a certain task can thus be reused directly in some other application.

Expandability is the ability to improve software in an easy and incremental fashion. Within the KBS environment, agents can be added or removed at will, without necessarily affecting the execution flow of the application. Hence, the KBS is particularly well suited for a spiral approach to software development, where the functionality is implemented and tested bit by bit.

All of these advantages significantly help minimize the risks associated with software development of complex applications.

Certainly one of the biggest advantages of building any application using the KBS as a run-time engine is the ability to have several programmers develop the system concurrently. Indeed, once the data types and the functionality of the application have been properly defined, each individual processing task can be mapped into an agent with its input and output data types uniquely defined. After this phase, it becomes very easy for several people to implement the individual agents without interfering with the other software developers. Also, since all agents are independent, they can be thoroughly tested *independently*.

5.1 EXECUTION SPEED OF THE RUN-TIME ENGINE

The design of the KBS shell is such that the overhead of the blackboard controller represents only a minor perturbation of the real-time execution of any given application. Data present on the blackboard are passed very efficiently to their respective agents and searches are avoided altogether. This usually represents a serious drawback of most expert systems. The speed of the KBS shell relies on the implementation of the Rete algorithm. Simply stated, the algorithm checks only those rules that are related to a given data. Although this approach seems obvious and very simple, it is not easy to implement in an expert system, and therefore most expert shells do not comply with this algorithm. Only the latest expert system shells are starting to use this approach. The Rete algorithm significantly reduces the workload of the program as the number of rules increases, and becomes a necessary requirement if real-time performance needs to be achieved.

5.2 DATA OBJECTS

5.2.1 *Base and Derived Data Types*

The data structures of the blackboard system need to be considered if a reliable and efficient blackboard system is to be constructed. In the KBS, the data structures comprise two main data types:

- Base data types
- Derived data types.

The base data type constitutes a member of the data structure set from which any other data type can be constructed. The integer data type can serve as a particular example of the base data type. The derived data types are constructed by grouping any number of existing data types, based or derived, into one new (derived) data type.

One of the main differences between base and derived data types lies in the process of data creation. For base data types, the data structures need to be declared before the main program is executed. In other words, the base data types are described and coded before the whole blackboard system starts its execution. The derived data types, on the other hand, can be created *dynamically* during the system run-time execution.

5.2.2 *Clone Data*

Let us consider two clearly distinct data types: LENGTH and WIDTH. Each represents the length and the width, respectively, of a typical two-dimensional object. Because these two data types can be characterized using the same unit [real number], the classical data type definition would not distinguish these as two different data types. Consequently, the classical model is not flexible enough to carry out most of the transformational tasks within the blackboard architecture ("task" being defined as any action performed by an agent). For instance, an agent that acts on a real number by multiplying it by a factor of 2 should be able to act on the data types LENGTH and WIDTH without having to define specific agents for each. Therefore, a new concept needs to be introduced: the clone data.

The clone data is obtained by cloning a parent data type, either base or derived. In the example above, a base data type REAL can be defined, from which the data LENGTH and WIDTH can be cloned. The clone data have the exact same properties as their parent (for instance both WIDTH and LENGTH are base data types as well). Most importantly, an agent who has been designed to act on the parent data will automatically act on the child data as well.

5.3 RELATIONS AND KNOWLEDGE REPRESENTATION

5.3.1 *Knowledge Representation*

The KBS user needs to be able to represent different types of knowledge within the system (e.g., semantic, declarative, procedural or heuristic knowledge, strategic or behavioral knowledge, etc.). Although these types of knowledge are very different in nature, they are all represented in the same fashion in the KBS shell and are referred to by the generic term "knowledge", with the understanding that this knowledge can have very different meanings depending on the user.

There are many models of knowledge representation (frames, scripts, schemata, etc.) compatible with the blackboard architecture. Most of these models use a semantic graph representation of the knowledge with the links between the nodes (data objects) representing relations between the nodes. This convenient representation has been implemented in many working expert systems. There are however a few drawbacks to this approach. For instance, there are many problems related to the representation of the context dependent information.

In the KBS, information is represented as a *directed tree-graph structure*. The nodes of the graph are composed of various data types and the links of the graph represent the relations between the graph nodes. A typical example of such a multilevel (3-level in this case) semantic graph is displayed in Figure 5, where eight data types are connected by semantic links representing the membership relation (e.g., D_5 is composed of D_0 and D_1). The leaves of this directed (as opposed to bi-directional) relation correspond to the base data types, i.e., D_0, D_1, D_2, and D_3.

In the KBS, these semantic graphs are partitioned into data types in such a way that the resulting structure is composed of trees with a maximal depth of one, that is, the resulting trees, treated as separate semantic graphs, can have at most one parent in their structures. Figure 5b illustrates how the multilevel semantic graph of Figure 5a can be partitioned into four one-level tree graphs with each node corresponding to the derived data types D_5, D_6, D_7, and D_8. This one-level depth is useful in order to increase data type reusability, since no data type is intrinsically nested inside another. It is easy to see that the low-level leaves can be retrieved from the top one through successive iterations.

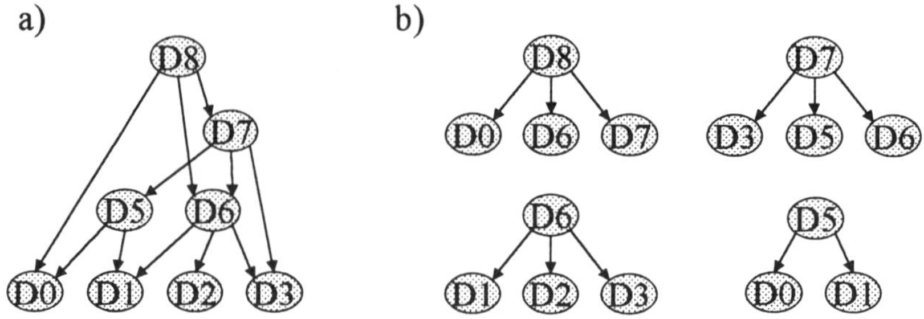

Figure 5a. Multilevel semantic graph.
Figure 5b. Partition into one-level semantic graphs.

5.3.2 *Definition of a Relation*

One relation that has already been introduced in Figure 5b, the membership relation, describes the data dependency of the derived data types. This particular relation, however, is not sufficient to describe the complex relations required between the objects in the blackboard system. A realistic blackboard system should allow the user to define and manipulate data types connected by complex semantic links. This can be achieved by redefining the relations between the data types and redefining the semantic links.

The ability to integrate rules is necessary for a good decision process.

Figure 6 shows how rules can be linked to relations between data types. These sets of rules can be based on either a simple gating method or more complex schemes like bayesian, fuzzy logic or evidential reasoning. These rules will be coded in separate C++ files, which will be called in the declaration of data type relations.

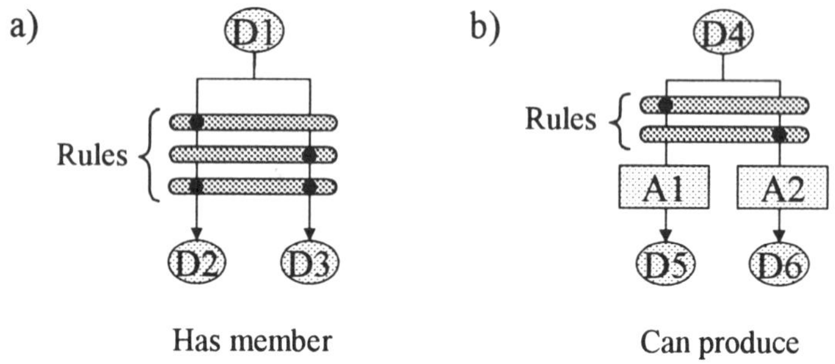

Figure 6. Basic Data relations.

As mentioned above, agents are components that contain the actual algorithms to fulfil the computational or procedural tasks of the blackboard system. These agents, known in the literature as knowledge sources, have the ability to transform the data types inside the blackboard system. The "can produce" relation presented in Figure 6b is exceptional as this is the only relation that allows the system to perform the computational tasks, i.e., to actually execute, or trigger, the agents. The "has member" relation associates a data member with a certain data type according to certain rules. In addition to the information contained in this relation, the "can produce" relation needs to specify which agents actually *produce* the data.

5.4 DISTRIBUTION AND COMMUNICATION

Modern applications often need to be distributed on multi-processor machines. Even when an application can run on a single-processor machine, it is sometimes desirable to take advantage of the speed provided by a multi-processor environment. Unfortunately, distributing a given application on several processors is not a trivial task. With the KBS shell, however, the blackboard controller assumes the burden of launching individual processes on separate threads. Since agents are independent from each other, any agent can be executed on a separate processor, allowing for a fully distributed KBS architecture. The input data of an agent is copied from one blackboard to another located on a separate processor where the agent is executed. Once the agent's task is completed, the output data is returned to the original blackboard (Figure 7). The blackboard controller is responsible for preserving the data integrity between the two blackboards. From a software point of view, the programmer need only specify that the agent be executed in *parallel* (as opposed to *sequential*).

The KBS is particularly well suited for real-time applications. Indeed, since KBS can be distributed on several processors, the time-critical task performed by some agents can be executed on their own processor. In this case, it is the blackboard controller that is responsible for redistributing the processing load on various processors.

The KBS possesses its own built-in communication layer which allows some data to be transmitted automatically to another blackboard through the Internet (TCP or UDP). Built-in agents are responsible for accepting communication requests from remote hosts, connecting with remote hosts, and sending and receiving data. A special data type with an attached Internet address automatically triggers the communication agents, and the data from one blackboard is automatically transported to the second blackboard (see Figure 8).

The KBS is intrinsically an open architecture and, as such, agents can be written to interface with any outside system (Application Program Interface (API)). Data between the KBS application and any non-KBS application is done through these interface agents and some specialized data types (Figure 9). API agents can use blocking, polling, or callback function interfaces.

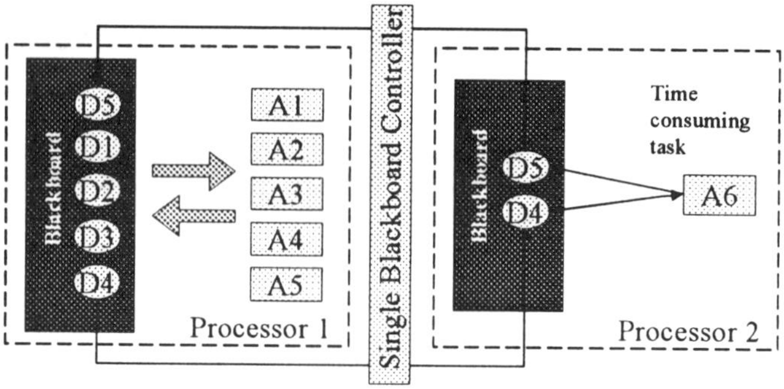

Figure 7. Multiple Blackboard architecture.

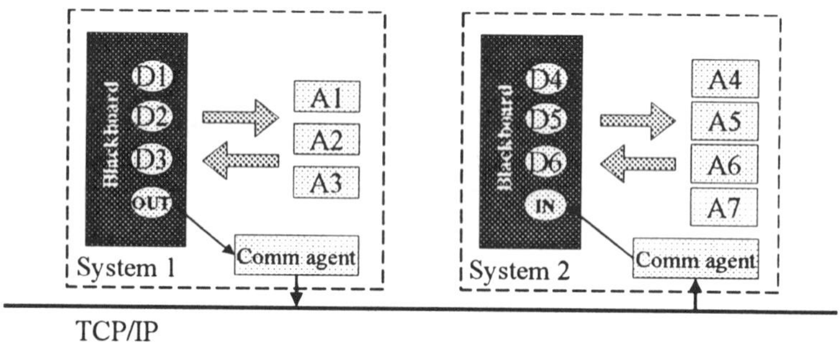

Figure 8. Communication agents.

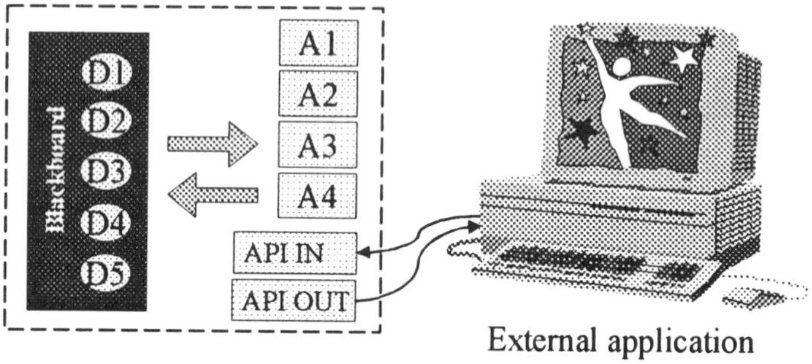

Figure 9. Application Program Interface (API).

6. Computer Network Security Integration

This section describes how a blackboard application could be used for the integration of a computer network security system. Figure 10 presents a high level architecture of this integration. Each system runs a multi-blackboard application. One blackboard is dedicated to the sensor, another is dedicated to data acquisition, data analysis and procedure, and yet another is used for system-level management, which includes user interface, communication with the central management system, database of procedures, logs, etc. Each blackboard can be distributed on a different processor, while remaining part of the same application. The communication between blackboards could simplify procedure changes and the implementation of new directives from the central security management.

The central security management system could also run a multi-blackboard application, whose individual blackboards communicate with the external world and each system. Different blackboards could also be used for the user interface, database, logs, etc.

The architecture proposed in Figure 10 maps the security program interactions shown in Figure 1. The versatility of a blackboard application facilitates the mapping of integration to the desired functional architecture.

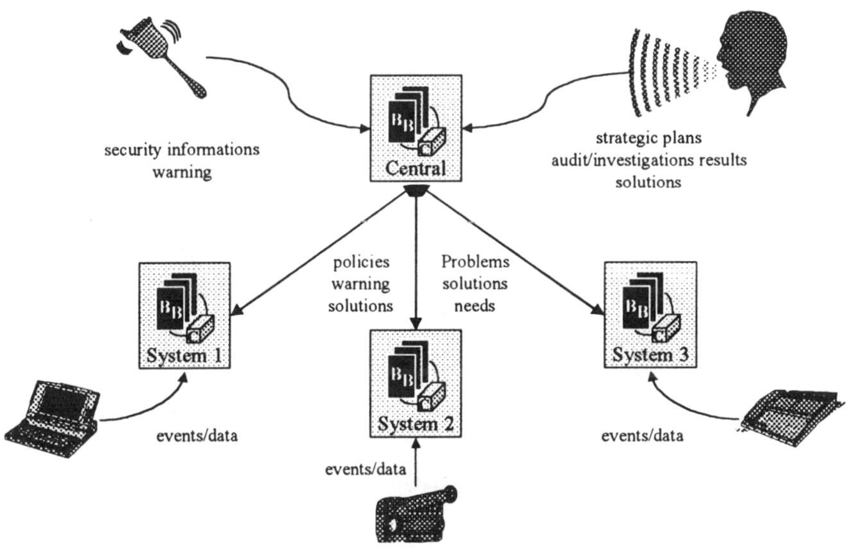

Figure 10. Computer Network Security integrated on blackboard architecture.

7. Summary and Conclusions

In this paper we presented an approach to integrate a computer network security system. The procedural goal driven approach can lead to difficulties when integrating systems which have to respond to specific events, while the data driven approach on blackboard architecture eases the difficult processes of system development and maintenance. The data driven approach is well suited for applications with functionalities selected by incoming events.

8. References

1. NIST, *An Introduction to Computer Security*: The NIST Handbook, Special Publication 800-12.
2. INCOSE, (1998) *Systems Engineer Handbook*, release 1.0.

MULTISENSING IN CHERNOBYL: THE STATE AND MONITORING OF OBJECT "SHELTER"

I.N. ONISHCHENKO
Institute of Plasma Electronics
NSC Kharkov Institute of Physics & Technology
Academic 1, Kharkov 61108, Ukraine

1. Introduction

The accident on Chernobyl NPP that happened October 26, 1986, a significant technological catastrophe. It resulted in many deaths. It is difficult to estimate material and psychological damage and the extent to which it gave rise to a wide spectrum of social, ecological, and biomedical consequences that without any doubt will affect many generations.

Unprecedented by its nature, the Chernobyl catastrophe has raised a number of new serious practical, organizational, and scientific problems to the economy, ecology, and public health institutions.

Medical, radiobiological, and dosemetric problems of post-chernobyl period have no analogs and differ from others known in the world cases of mass radiation of the peoples (Japan, USA, Brasilia, Russia) by the number of victims, the structure, the complexity of radiation sources, and the presence of the complex of negative nonconventional side-factors.

The large scale of Chernobyl accident can be evaluated from the data published in reviews [1-9] and discussed many times at the forums of world levels that were held by WOH, MAGATE, UN.

At present the world community has become anxious by the state and safety of object "Shelter" and ChNPP in a whole. The main goal of this chapter is to offer a discussion of the problems related to object "Shelter" and the necessity of using many monitoring sensors and their data fusion for continuous monitoring the situation around ChNPP.

1.1. GENERAL INFORMATION ABOUT THE CHERNOBYL ACCIDENT

In 1986, on ChNPP, placed approximately 130 km to the north from Kiev, four energy units were operating. Each of them consisted of nuclear reactor RBMK-1000 (channel high power reactor) and 2 turbines with electric generators of power 500 MW each.

The part of Ukraine where ChNPP is placed in flat forest country with a relatively small density of people. Approximately 3 km from reactor the town Pripyat is located where at the moment of accident 49 thousand inhabitants lived, and 15 km to south-east from NPP the town Chernobyl was situated with 12.5 thousand people. All together in 30 km zone around ChNPP about 135 thousand people lived.

A. K. Hyder et al. (eds.), Multisensor Fusion, 779–797.

Before the accident, the radiation levels around ChNPP didn't differ from other regions of the Ukraine.

In night from 25 to 26 April 1986, as a result of two consequent explosions, part of graphite and nuclear fuel of 4th reactor unit was thrown out. Firing fragments fell on the roof of the machine hall and on the NPP territory, causing about 30 incidents of fire. As the result of explosions in the 4th energy unit, a large amount of nuclear fission products were expelled into the atmosphere (by some estimations about 10^{18} Bq).

Because the reactor unit did not have a containment envelope the radioactive waste was transmitted directly into the atmosphere in the resulting graphite firing. In the whole 50-80 MCi of radionuclides, including about 30% radioisotopes of I and 20% of Cs were thrown into the atmosphere. [10].

Intense releases continued during 10 days. In the vicinity around reactor a large amount of shortlife radionuclides and longlife radioisotopes of Pu, Sr-90, and some others elements were released. Everywhere was contaminated by radioisotopes of cesium and iodine. Additionally, the environment was filled by so called "hot particles" consisting of fuel mixture of reactor origin and high activity products of fission, condensed on clay, sand, and others nonactive carriers, thrown into reactor from helicopters in the amount of about 5 thousands tons to minimize radioactive emissions. Among these thermoderivating and filtering materials there were in particular carbide boron, dolomite, lead, sand, and clay.

A large amount of radioactive inert gases, halogens, alkaline metals, tellurium, alkalineland, rareland, and refractory elements was also released.

There were two big emissions. The first one occurred mainly as a result of mechanical breaking of fuel during explosion. It consisted of volatile isotopes, including inert gases, iodine compounds, and cesium. The second one took place between 7th and 10th days of the accident and was caused by melting of fuel in reactor. The smaller events were observed after 6th May, but their scale equalled the previous two.

Radioactive materials entered into atmosphere as gases and aerosols. Gases - krypton and xenon – were wholly thrown out from the reactor. Iodine entered into atmosphere as vapors and compounds. It was 50-60% of total contaminant of this element in the fuel of the reactor. Cesium and tellurium were found out as particles that were not joined with fuel dust. Unexpected and an extremely unfavorable particularity of Chernbyl was the appearance of so called "hot particles", formed in particular in the process of graphite burning. To them such nuclides as cesium, zirconium, strontium, barium, lanthanum, and actinides were related.

As a result of Chernobyl accident three main sites of radioactive pollution at the CIS territory were formed – Central, Bryansko-Belorussian, and Kaluzhsko-Tulsko-Orlovski ones (see map in Figure 1) [1]. 30-km zone around reactor occurred the most polluted. There the fall down of ^{137}Cs on the earth exceeded 1500 kBq/m^2 (see map in Figure 2) [1].

In the whole of the USSR territory the area with pollution exceeding 1500 kBq/m^2 was 3100 km^2, from 600 to 1500 kBq/m^2 it was 7200 km^2, and from 40 to 200 kBq/m^2 it was 103000 km^2.

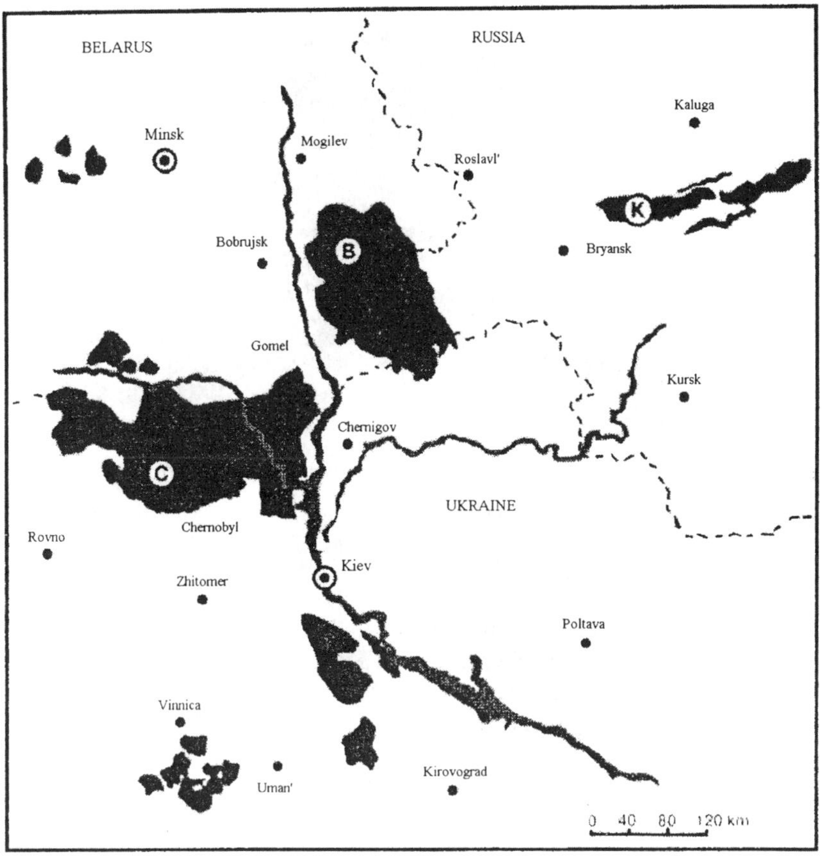

Figure 1. ^{137}Cs emissions around the Ukraine and Belorussia from Chernbyl exceeding 555 kBq/m^2

The volatile elements in particular iodine and cesium were observed even in the stratosphere. As the speed of the wind at the height 1500 m can be 8-10 m/sec 38, hours after accident the radioactive cloud reached 1250 km distant, Sweden. Indeed after the first emission, the radioactivity fell in Sweden and Poland. The second emission polluted not only in Sweden but also in the countries of Central Europe, especially in Austria, German, Northern Italy, and later in Romania and Bulgaria. On 30 April, radioactive emissions of Chernobyl accident reached Italy and Switzerland. On 1 May their traces were revealed at the east of France, in Belgium, Netherlands, and Great Britain and 2 May in Greece. At the beginning of May radionuclide pollution of atmosphere was found in Israel, Turkey, Kuwait. The Chernobyl trace had being traveled all over the Northern Hemisphere.

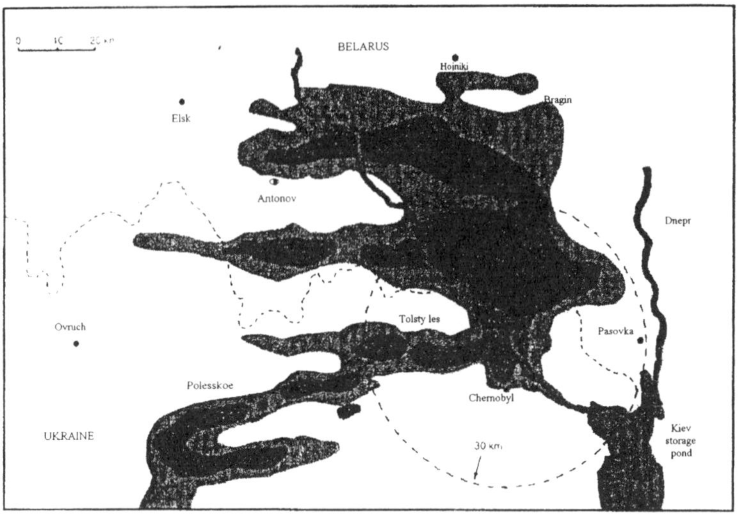

Figure 2. ^{137}Cs fall out patterns around Chernobyl. Gray area >555 kBq/m^2; black >1500 kBq/m^2

The most considerable radioactive pollution out of the USSR were marked in Austria, eastern and southern Switzerland, Scandinavia, and Southern Germany, i.e. where it was raining at the time – in contrast to Spain, France, and Portugal. So the averaged level of ^{137} Cs in some places of Austria, Salzburg, and Karintia was 59, 48, and 38 kBq/m^2 ,respectively, and in Portugal was 0.02 kBq/m^2 only.

More than 70 million peoples of Europe were affected by radioactive iodine whose dose load on thyroid gland was variable and for 150 thousands it was 0.3 Gy [1].

At the same time in the Southern Hemisphere "Chernobyl trace" was not revealed [11].

2. Structure of the radionucleides contamination spot resulting in Chernobyl accident

It is known now that as a result of the accident at Chernobyl Nuclear Power Plant (ChNPP) about 2x10^7 Ci were released. As long ago as July, 1986, the contamination maps of the earth surface by ^{137}Cs, ^{90}Sr, and ^{239}Pu were obtained. In 1991, they were published in the open press. The distinctive feature of contamination is its "spottiness". This fact was taken into consideration when people were resettled from Chernobyl in 1986. This "spottiness" can be well seen at the contamination map in Figure 3 [12].

There remain some facts corresponding to the first days of the Chernobyl accident. The release of the radionuclides lasted 10 days from 26 April, 1986, to 06 May, 1986. These phenomena were irregular: they were especially intensive during the first and the last twenty-four hours. At that time it was not raining in a 30 km diameter zone around ChNPP. The wind direction during this time interval varied considerably. As a result of the wind direction changing, three main tracks of contamination were formed: north,

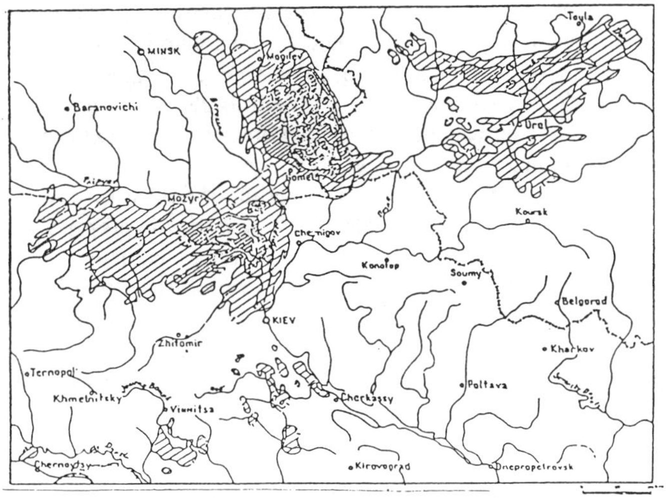

Figure 3. Spottiness of the Chernobyl fallout

west, and south. Each is characterized by "spottiness". Until now the nature of such "spottiness" has no final explanation. A priori there are a lot of factors influencing on the origin of such a structure. It is the land topography, inhomogeneity of releases and their complex component composition, and atmospheric turbulence. The latter is important in explaining the structure arising from admixture transport in a turbulent atmosphere. When considering attentively the contamination maps it is observed the analogy with the fractal objects appearing in the chaotic or turbulent systems.

Based on experimental data for ^{137}Cs, the fractal dimension of the spot was evaluated using the method of coverings. It is $d_s=1.68$ for the spot and $d_b=1.37$ for the boundary.

The following model of the "spotty" radionuclides contamination origin at the earth surface has been proposed in [12]. The essence of this model concludes to the following: From the point source (reactor) the small particles (TCM particles of sizes 3-10 μ) are being thrown up to the height 2 km. They appear in the earth atmosphere, which is characterized by the chaotic field of velocities of the turbulent air motions. The particle experiences the action of velocity field, gravity, and Stocks friction force. Naturally the landing place of the particle is accidental. The dynamics of the admixture particle is described by the system of maps of the following view:

$$x_{i+1} = x_i + a z_i^2 \xi_x ,$$

$$y_{i+1} = y_i + (d/\kappa) \ln(z_i/z_{\infty}) + b z_i^2 \xi_y ,$$

$$z_{i+1} = z_i - e + c\xi_z .$$

Here a, b are the parameters determining the amplitudes of horizontal (x, y) components of turbulent field of velocities, c is one for vertical (z) component; d, e are the amplitudes of averaged horizontal and vertical velocities; ξ_x , ξ_y , ξ_z are the random values given by the generator of random numbers. This map for the given position of the particle and realizations of accidental values ξ determines particle motion up to reaching the earth surface at the coordinate $z=0$. Thus, putting the initial position of the particle (e.g. $x_0=y_0=0$, $z_0=1$) we can evaluate the distribution of released admixture on the earth surface at the given parameters a, b, c, d, e.

In [12] it was shown that contamination covering of the earth surface has a "spotty" nature, if the field of turbulent motion velocities possesses the property of persistence. It is defined by the fractal dimension [13] or so called Hurst parameter [14] of random field of velocities that determines the characteristic sizes of the "spottiness". The fractal dimension reflects the rough features of the spot structure, for example, the "portion" of surface filling with radioactive deposits. The property of persistence for example, if for some time a positive increase of the velocity is being observed, then further increases in velocity is the most probable, and contra verse. For Hurst parameter $H>1/2$ the system is persistent and for $H<1/2$ it is antipersistent.

3. Medical and social consequences of Chernobyl catastrophe

The medical consequences [1] of the industrial nuclear failure at Chernobyl Nuclear Power Plant for some areas of Ukraine, Belorussia and Russia have common features with consequences of nuclear catastrophes in another regions of the world. In the FSU such regions are: area, Altai Territory which to a large extent was a primary test area for nuclear explosions at Semipalatinsk Test Range, some areas in Southern Urals that were affected by failures at Mayak Plutonium Plant, and circumpolar area where the Novaya Zemlya nuclear test range is found, the largest in the world. In each case, common issues were studied:

- Can reliable medical consequences of nuclear emergencies be tracked at those areas?
- If so, what are their stages, sequences and duration?
- Which is the effect of ionizing radiation as such into those processes as well as of circumstances concomitant to the environment pollution by radionuclides (precipitation of heavy and toxic elements, psychogenous stresses)? Which effects may be considered specific just for such situations?
- Which conclusions may be come to on the basis of comparison among radiation-affected areas of different origin?
- Which is a supposed mechanism (or mechanisms) for development of those effects?
- Can we predict to some degree of certainty the future development of Chernobyl disaster medical consequences?

The comparison of medical consequences between the Chernobyl disaster and Semipalatinsk Range nuclear tests («Semipalatinsk-Altai problem») allows us to separate factors because in the former case the effect of ionizing radiation was combined to a powerful psychogenous stress whereas in the latter case nuclear tests were highly classified and, therefore, psychogenous stress was absent. Thus, in that sense the affected areas of Altai Territory may be considered as reference relative to Chernobyl area.

The presented conclusions were inferred from statistical materials submitted by Ukrainian Ministry of Health, published results of medical survey of inhabitants in Altai Territory, Southern Urals and Extreme North, as well as data from literature characterising medical consequences of Chernobyl disaster.

It has been shown that Chernobyl disaster, emergencies in Southern Urals, and nuclear tests at Semipalatinsk Range' (especially the earliest, most «contaminating» explosion of 1949) exerted a versatile deep influence upon the health state of population. The question whether such phenomena take place in the Extreme North is less distinctly expressed due to local influence of some other, rather powerful negative factors-natural and social.

The consequences of those unnatural events are versatile, long-term and embrace a period of many years and decades. They include both specific direct effects, immediately connected with the action of ionizing radiation and nonspecific ones, including indirect effects. Thus, the analysis of Altai situation shows that the effect of ionizing radiation upon local population that took place at the end of 1940s and during 50s, i.e. for a period of almost 50 years have caused medical consequences that can be traced up to now. First of all, cancer is one of the most indicative pathologies. On a par with them, however, at the effected Altai areas we see an increase in cases of some other pathologies whose orgin is unclear.

The most wide spread indicative disease of those who suffered from the Chernobyl disaster is thyroid gland cancer, the number of its cases having increased dramatically. At the same time we witness an undoubted increase of cases of some nonindicative pathologies such as diseases of endocrine, cardiovascular, nervous, circulatory systems, gastroentheropathologies and chronic nonspecific lung diseases. The most vulnerable group are the so called liquidates-people who eliminated consequences of the disaster at the power station itself, especially in 1986; the second important group are evacuated civilians, the last being those who permanently live in the polluted areas.

As in Altai the effect of small doses of ionizing radiation cannot be denied, whereas any neurogenous stresses were absent it is supposed that small dosage of ionizing irradiation is a factor that contributes to emergence or manifestation of not only stochastic effects underlying indicative pathologies but also of non-stochastic ones. Hypothesis is discussed assuming that the pathogenesis of such conditions lies in disfunction of a complex of organs including primarily thyroid gland which is well known as a «target organ», and secondly thymus as a central organ of systemic immunogenesis directly connected to development of all basic types of age-caused chronic pathology. An assumption has been substantiated that a «challenge» in the form of some morphofunctional disturbances in thyroid gland connected to a local effect of incorporated iodine radioactive isotopes creates a «response» by thymus. Consequent disturbances in thymus link start cascade reactions in the sphere of neuroimmunoendocrine regulation connected to pathogenesis of actually all forms of pathologies developing against a background of metabolic immunodepression as described by V.I. Dilman (cancer, atherosclerosis, diabetes). Activation of once viruses taking place against the background of disfunctioning immune system, is an important concomitant circumstance favoring an increase of malignant tumors in people who suffered nuclear emergencies. A final consequence of all this will be a premature body aging and outcome from vital cycle.

It should be especially stressed that medical and social hygienic consequences of such events as Chernobyl disaster cannot be limited to a single effect of ionizing radiation only. The spectrum of factors induced by them is much wider and includes primarily a psychogenous stress aggravated by concomitant medicopsychological and social-hygienic consequences of evacuation arrangements. It should be pointed out that the synergic result of combined effect of pathogenic factors is not equal to their arithmetic and even algebraic sum, it is a principally new quality the essence of which still remains to be understood and evaluated. Just in evaluation of the combined effect of pathogens upon human body we see the prospect of future studies in this field.

An opinion is expressed that a comparison of medical and social hygienic situation in areas with different radioecological «history» is quite a prospective way. Though the causes of negative phenomena in the Altai and Chernobyl areas are quite dissimilar we are still able to assert that the present medical situation in Altai is a distorted, but still sufficiently adequate, prototype of what we shall see in Chernobyl area in 2020-2040.

Beginning approximately with 2000 a deterioration of the oncological situation can be predicted here due to basic, the most widely spread forms of solid tumors, more frequent autoimmune diseases and accelerated aging process. Primarily liquidators and evacuated persons have been affected; the effects will tell upon the persons who permanently live in radionuclide-polluted areas as well but it would be more difficult to detect those effects because this category is too heterogeneous. A statement is substantiated that medical and social hygienic consequences of emergency radioecological situations cannot be limited only to processes developing at a individual level. The processes tracked at population levels and envisaging starting up of natural biosocial selection mechanisms must be equally, and in some cases even more, important. One of them is a change in immunogenetic structure of population which is already observed in some radioecologically unfavourable.

In connection with the enormity of Chernobyl disaster which involved a vast number of people not only in Council of Independent States (CIS) countries but far abroad as well, it is proposed to unite resources and assets at an international level in order to establish a joint register of people who suffered radiation effects of this disaster. Such an international register under sponsorship of World Health Organization (WHO) would play the positive role.

4. The system of sensors and the problems to be solved [15]

4.1. OBJECT "SHELTER" (OS)

After the accodemt. ;ava-like fragments were observed in Unit 4 [16]. At this object, initial measures have been taken to reduce the consequences of the accident and works have been continuing to control its conditions and ensure nuclear and radiation safety.

Object "Shelter" of ChNPP Unit 4 is the totality of the construction sealing the sources of radiation released by the reactor core of block 4 as a result of the accident, including constructions in reactor hall, deaerator stack, turbine hall (which are separated from operating Unit-3 with isolating walls), and from the environment with newly constructed structures and fences, (which provided control systems, dust suppression, feeding of neutron-absorption solutions, power and water supply, and other systems for personnel protection. Figure 4 shows the general view of one variant of shelter project and the present object "Shelter" [17]. Unit-4 cross sections are represented in Figure 5. It

Figure 4. Scheme of unit showing the buttress wall as is looks today

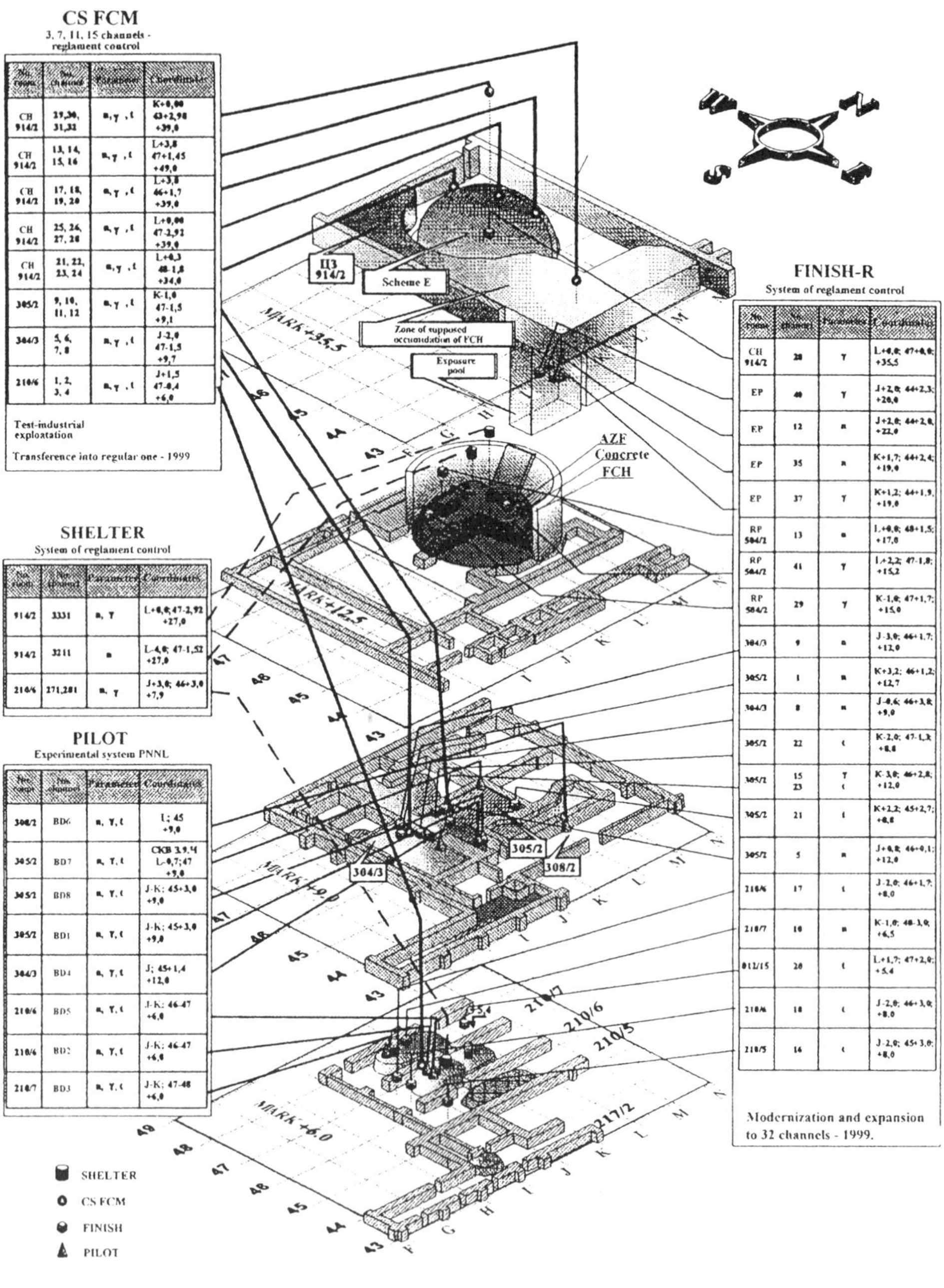

CS FCM — 3, 7, 11, 15 channels - reglament control

No. room	No. channel	Parameter	Coordinates
CH 914/2	29,30, 31,32	n, γ, t	K+0,00 43+2,98 +39,0
CH 914/2	13, 14, 15, 16	n, γ, t	L+3,8 47+1,45 +49,0
CH 914/2	17, 18, 19, 20	n, γ, t	L+3,8 46+1,7 +39,0
CH 914/2	25, 26, 27, 28	n, γ, t	L+0,00 47-2,92 +39,0
CH 914/2	21, 22, 23, 24	n, γ, t	L+0,3 48-1,8 +34,0
305/2	9, 10, 11, 12	n, γ, t	K-1,0 47-1,5 +9,1
304/3	5, 6, 7, 8	n, γ, t	J-2,0 47-1,5 +9,7
210/6	1, 2, 3, 4	n, γ, t	J+1,5 47-0,4 +6,0

SHELTER — System of reglament control

No. room	No. channel	Parameter	Coordinates
914/2	3331	n, γ	L+0,0; 47-2,92 +27,0
914/2	3211	n	L-4,0; 47-1,52 +27,0
210/6	271,281	n, γ	J+3,0; 46+3,0 +7,9

PILOT — Experimental system PNNL

No. room	No. channel	Parameter	Coordinates
308/2	BD6	n, γ, t	L; 45 +9,0
305/2	BD7	n, γ, t	CKB 3.9.4 L-0,7;47 +9,0
305/2	BD8	n, γ, t	J-K; 45+3,0 +9,0
305/2	BD1	n, γ, t	J-K; 45+3,0 +9,0
304/3	BD4	n, γ, t	J; 45+1,4 +12,0
210/6	BD5	n, γ, t	J-K; 46-47 +6,0
210/6	BD2	n, γ, t	J-K; 46-47 +6,0
210/7	BD3	n, γ, t	J-K; 47-48 +6,0

FINISH-R — System of reglament control

No. room	No. channel	Parameter	Coordinates
CH 914/2	28	γ	L+0,0; 47+0,0; +35,5
EP	40	γ	J+2,0; 44+2,3; +20,0
EP	12	n	J+2,0; 44+2,0; +22,0
EP	35	n	K+1,7; 44+2,4; +19,0
EP	37	γ	K+1,2; 44+1,9; +19,0
RP 504/2	13	n	L+0,0; 48+1,5; +17,0
RP 504/2	41	γ	L+2,2; 47-1,8; +15,2
RP 504/2	29	γ	K-1,0; 47+1,7; +15,0
304/3	9	n	J-3,0; 46+1,7; +12,0
305/2	1	n	K+3,2; 46+1,2; +12,7
304/3	8	n	J-0,6; 46+3,8; +9,0
305/2	22	t	K-2,0; 47-1,2; +8,8
305/2	15 23	γ t	K-3,0; 46+2,8; +12,0
305/2	21	t	K+2,2; 45+2,7; +8,8
305/2	5	n	J+0,8; 46+0,1; +12,0
210/6	17	t	J-2,0; 46+1,7; +8,0
210/7	10	n	K-1,0; 48-3,0; +6,5
012/15	20	t	L+1,7; 47+2,0; +5,4
210/6	18	t	J-2,0; 46+3,0; +8,0
210/5	16	t	J-2,0; 45+3,0; +8,0

Figure 5. Sensor locations in the "Shelter"

is seen four views at various levels of heights 6, 9,12.5, and 35.5 meters. The sensors location at each level is shown. There are four control systems for monitoring the state of fuel-containing masses in object "Shelter":

- "CS TSM" (control systems of fuel-containing masses), in the state of test-industrial exploitation, transference into regular one in 1999.
- "TENT" (or "SHATER") the oldest system of reglament control, is.being gone out of operation.
- "FINISH-R", the power system of reglament control, modernization and expansion to 32 channels in 1999.
- "PILOT", experimental system of PNNL (USA).

The sensors are designed for the detection of neutron flow, gamma radiation, and temperature measurement. Neutron detectors register both slow neutrons (when fission chamber of detector is installed with ^{235} U films) and fast neutrons (with ^{232} Th films in chamber). The sign of nuclear danger is the simultaneous appearance both slow and fast neutrons together with correlated gamma radiation. Besides, the intensity of gamma radiation allows evaluating the total amount of uranium.

The estimations with error of 40% indicate that from the total 200 tons of UO_2 fuel in Unit-4, about 20 tons were ejected. The estimated value of fuel-containing masses in the premises of object "Shelter" is 3,000 tons. The amount of radioactive waste is 800 thousand tons including 100 thousand tons of high activity mass. Neutron flow ranges in $10\text{-}10^6$ neutrons/cm^2 sec and is equal 200 neutrons/cm^2 sec in active zone. The α, β, γ - activities are measured by α - dosemeters, ionization chambers, scintillators etc.

4.2 FUEL CONTAINING MASSES (FCM)

Fuel containing masses characterization will concern only core fragments (CF – parts of destroyed fuel elements) located in the central hall of the former reactor building of unit 4 and in adjacent premises (including spent fuel assemblies, which are stored in SSFP – Southern Spent Fuel Pool) and lava-like fuel containing material (LFCM) located in the sub-reactor premises, which are subject to characterization SIP (Shelter Implementation Plan in the frame of program TACIS) [15] and certification in compliance with SIP. Dust and aqueous solutions with high uranium content are not considered as part of FCM characterization. Organization scheme and zoning for FCM characterization are shown in Figure 6.

Zone 1: This is the zone of the object "Shelter" (OS) with high levels of radiation which is not accessible to personnel during FCM characterization. This zone has dose rates of gamma radiation in excess of 5×10^{-5} Sv.

Zone 2: This is the zone of the OS accessible to personnel under restrictions during FCM characterization with time limitations and safety precautions. This zone has dose rates of gamma radiation between of 5×10^{-5} Sv and 2.3×10^{-5} Sv

Zone 3: This is the zone of the OS with unlimited access to personnel during FCM characterization. This zone is assumed to be clean. This zone has dose rates of gamma radiation less than 2.3×10^{-5} Sv.

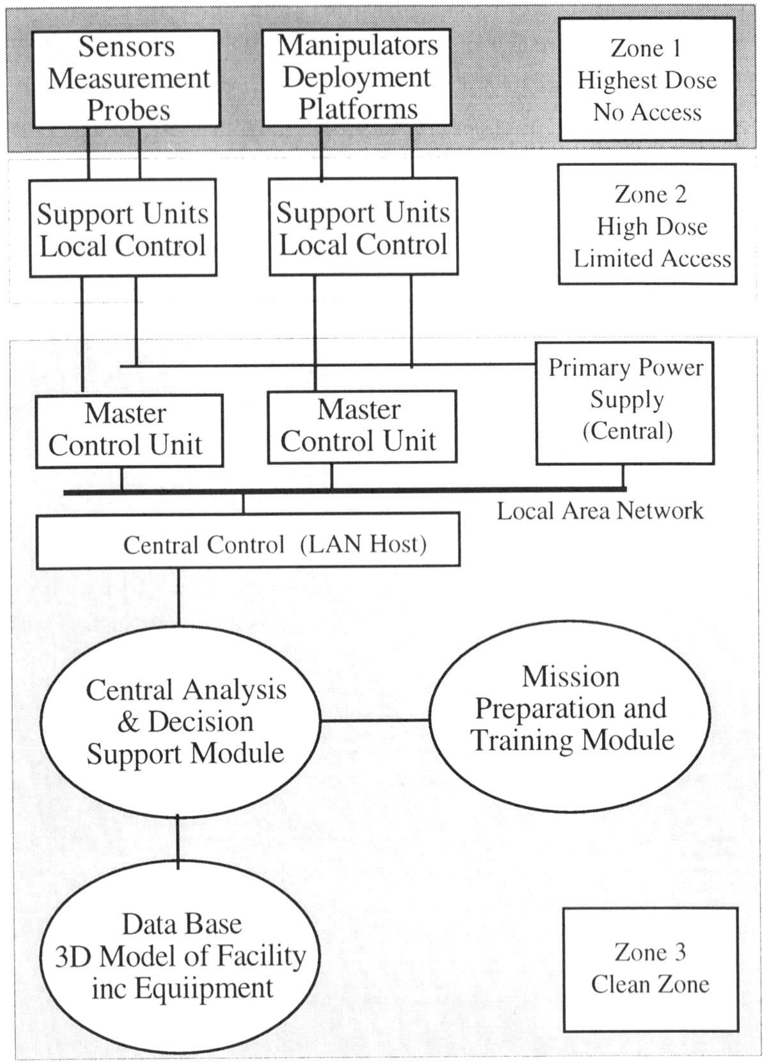

Figure 6. Organization and zoning for FCM characterization

5. Scope of FCM characterization

According to SIP, the main objective is to "characterize the FCM in the Shelter to increase nuclear safety by developing an understanding of criticality risk, and provide guidance for potential fuel removal activities, to increase worker and environmental safety. FCM characterization is necessary to more precisely define fuel/FCM locations, inventory, configuration, and characteristics."

The scope of FCM characterization is to "evaluate, select, develop and procure and test a system for location and assessment of the FCM in each major premise, including the upper elevations and possibly under the Shelter walls, at all reasonably accessible areas and deposits of FCM at the site. The major deposits of FCM in the Shelter have been previously characterized by several different techniques, but the characterization task is so difficult that the resulting knowledge still includes large uncertainties. Ongoing degradation of FCM deposits and resulting translocation of FCM means that any estimates must be continually updated. At present inaccessible FCM (such as in the Cascade Wall) should be investigated, but will be very difficult to characterize,..." [15].

The FCM characterization will survey reasonably accessible premises to determine locations, quantities of FCM and the physical form of the masses (including chemical composition, overall density, porosity, and density of fuel and moderators). The strategy for the surveys involve the identification and use of remotely operated equipment, including specific equipment for the survey purposes ..." [15].

6. FCM Characterization Objectives

FCM characterization will be focused on the:

- Determination of major FCM locations (spatial distribution in the main premises defined below),
- Determination of FCM inventory (based on radioactivity/radiation measurements mainly),
- Determination of the element composition as well as the chemical and physical state of FCM and major premises.

Based on the objectives of FCM characterization, the following measurement parameters have been identified as being of major interest:

- Determination and visualisation of the geometrical (spatial) distribution of FCM:
    - Topographic mapping of premises containing FCM,
    - Identification of FCM containing objects in major premises,
    - Determination of location and dimension of these objects with high precision;
- Determination of parameters of radiation fields in the premises, in particular:
    - Direct characterization of activities associated with objects and materials identified by visualisation process (identification and classification of radiation sources),
    - Dose rate fields for the use in activity distribution models,
    - Neutron fields for use in activity distribution models;
- Determination of the element composition of FCM;
- Location and inventory estimate of FCM that has been heavily covered by concrete and rubble.

The location of the highest priority for the FCM characterization are summarized in Table 1.

TABLE 1: Grouping of premises with highest priority (assigned to zone 1)

Group	Description	Premise Identification Number
I	Central Hall	914/2
II	Spent Fuel Pool (Southern)	505/3
III	Sub-reactor premises	305/2, 307/2
IV	FCM accumulation below and inside the Cascade Wall	

6.1. CENTRAL HALL

The area above the former reactor location has been destroyed by the accident but is still called the central hall. The upper parts can be viewed from the roof. Within FCM characterization, the only feasible access to these parts and also to the lower parts of the central hall is from above. Alternative direct access by the use of remotely operated vehicles is under consideration.

The dose rates inside mainly originate from core fragments distributed throughout the premise. They range from about 0.05 - 0.1 Gy/h under the roof increasing to 6 Gy/h further down into the central hall premise, near the surface of FCM accumulations.

The deployment of equipment can be done through tubes in the roof, of which the majority have a diameter of 100 mm and 5 tubes in the centre of the roof with 140 mm diameters. Although the dose rate on the roof is up to 0.05 Gy/h, it is classified as Zone 2. The main problem with the access option over the roof is a limitation of permissible weight load of 100 kg/m^2 (including extra shielding and personnel). The use of a remotely controlled platform moving along the roof is seen as an alternative deployment option. This could be either a magnetic platform climbing along the metal roof or one that is suspended by ropes from the building structure.

Very little systematic information is available on the location/distribution of FCM. Some dose rate measurements have been done.

6.2. PREMISES OF THE ACCIDENT LOCALISATION SYSTEM (PALS)

Positions 305/2 and 307/2 are referred to the accident localization system and located at 9.0 level. Access to these premises is by boreholes (diameter of the cross section of boreholes is from 59 to 151 mm). Dose rates up to 25 Gy/h are possible. Very little (not yet evaluated) or no systematic information is available for these premises. The following systematic investigations is planned: photo and video recording, radiation fields investigations, estimation of FCM composition (for separate areas) A full characterization (visual, radiometric and element breakdown of selected areas) is required.

During FCM characterization, the investigation of the 305/2 and 307/2 position were of highest priority.

6.3. SPENT FUEL POOL

The former spent fuel storage pool consists of two pools (North and South). The Southern pool contained some spent fuel assemblies (129) at the time of accident. All the water in the pool was lost, resulting in damage to the fuel elements. A number of boreholes (accommodating probes from 112 to 151mm in diameter) provide access to the pool and have been used in the past to take a brief look into the area and to take dose rate measurements. Dose rates of 36 Gy/h have been measured. It is planned to carry out a number of visual, radiation investigations to obtain more detailed information about the actual location and state of the fuel.

6.4 CASCADE WALL

The Cascade Wall is a structure that has been erected after the accident. It has been built on top of, and includes in its structure, waste dumps containing various of metallic and non-metallic waste. The quantities of FCM that have been buried there during and immediately after the accident are not known. The wall represents a very effective shield against radiation from the base (up to several meters of concrete). There are 4 horizontal boreholes in existence to date, but there is a possibility that new boreholes will be available in the future to support measurements. The existing boreholes are not traversing fuel containing materials and end in cavities. Information about the quantities and location of FCM buried underneath these structures is strongly required.

7. Description of the needed equipment for the FCM characterization

Ground-penetrating radar system for the inspection of FCM locations under large material layers (GPR).

The radar system is required for the inspection of concrete massive of the cascade wall for localizing FCM and to investigate the structure of the Cascade Wall itself.

Equipment to Provide Access to Premises of the Shelter with High Radiation Level - Remotely Controlled Equipment for Drilling.

The RCED (remote controlled equipment for drilling) is intended for rotary drilling of horizontal, inclined, and vertical boreholes for investigation and technological purposes. It is using hardened and/or diamond drills suitable for ferro-concrete constructions, heterogeneous constructions and materials, and passing through cavities. It has to be capable of using lining tubes, perform with and without coring and can be used for dry drilling as well as for drilling with flushing.

Equipment to Provide Access to Premises of the Shelter with High Radiation Level - Modular Manipulators.

The equipment includes manipulators that allow remote control of measurement probes and lights through boreholes and roof tubes. Specialized equipment for geodesic and endoscopic investigations of boreholes is also part of the lot.

In situ measurement equipment for the visual, radiometric and elemental characterization of Fuel Containing Material (FCM) in the Shelter.

This contains equipment for the determination and visualisation of the geometrical distribution of FCM. Other equipment in the lot serves the purpose of determining parameters of radiation fields in the premises, including information about nuclide specific activities (e.g. by collimated gamma - ray spectrometry), and dose and neutron fields for the use in activity distribution models. Finally the lot contains equipment for the in situ determination of chemical composition of FCM. A11 this characterization equipment has to be suitable for work under high radiation and provide the ability to be remotely controlled.

Equipment for the determination of geometric (spatial) distribution of FCM - 3D topographic measurements (3DMVS).

A system is to be provided that produces a complete 3D computer representation of the premise based on a capability to systematically collect information on the geometric layout of the premises as well as specific objects in the premises including accumulations of FCM and other materials. This equipment has to be suitable for work under high radiation and provide the ability to be remotely controlled.

System for the visualization and measurement of the geometric distribution of gamma activities (SVGA).

A system is to be provided that produces a 3D computer representation of the premise in terms of radioactive materials present. Nuclide specific activity data in terms of their location in the premise is required.

Remote Magnetic Vehicle (RMV).

The remote magnetic vehicle (RMV) is to enable access to the Central Hall at the inner side (ceiling) of the metallic roof construction. The RMV will carry measurement probes for FCM characterization, position them at the places of measurement, and determine the spatial coordinates of probe position and transmit them in real time to a master control unit. The RMV will be kept in suspense by electromagnets and will move on the inner, heterogeneous (metal) ceiling.

Pulsed neutron generator tube with neutron detection system (PNG).

The application of a pulsed neutron generator tube in boreholes together with a neutron detection system is a direct method to characterize the U content in FCM. The method is based on the sub-critical neutron multiplication within the U bearing material.

Equipment for characterization of existing FCM samples.

In order to enable the systematic analysis of existing FCM samples, a selection of laboratory equipment is required. The lot includes equipment for micro-mechanical test of brittle materials, for the measurement of acoustic FCM parameters, for the determination of electro-dispersive FCM parameters, a scanning electron microscope, an X-ray diffractometer, a mass spectrometer, an optical pyrometer, a synchronic detector, a column test facility as well as standard laboratory equipment and protective equipment

8. General requirements

With respect to reduced contamination and easy decontamination, all equipment shall comply with the requirements specified in the state normative document "Basic Sanitary Rules of Managing Radioactive Substances and Other Sources of Ionizing Radiation".

It has been assumed that only parts of measurement and control systems can be deployed in Zone 1, because of lifetime restrictions due to high radiation fields. Parts of the measurement and control system will have to be situated in Zones two or three, depending on their rating in terms of environmental and radiation protection. The specific access conditions in Zone 2 introduce time constraints for installation and setting up of equipment in this zone. In addition, all equipment to be used in Zone 2 is to be man-portable (modules up to 20 kg; a maximum of 50 kg has to be specially approved).

The specific working conditions are radiation-hazardous. The basic requirements include temperature range, IP rating (resistance against ingress of dust and water), and availability of power supply.

9. Noise Radar Technology and Interferometric SAR Technique for precise all weather monitoring of "Shelter"[18].

It is well known that the "Shelter" (by other words "Sarcophagus") of the 4th unit of Chernobyl's Power Station undergoes different structural changes with time that requires a continuous precise monitoring of its surface in order to have possibility to identify probable critical states of Sarcophagus construction.

For that purpose various approaches have been suggested. One of them is based upon the laser holography. This method allows a remote acquiring of the information about an illuminated surface and achieving of a very high accuracy and resolution of those measurements. However, its rather high sensitivity to bad weather conditions, such as raining, dusty air, snowing etc. does not allow its usage under realistic conditions. On the other hand, it has been shown that rather precise information about a surface being irradiated with microwave electromagnetic field can be obtained with help of Synthetic Aperture Radar (SAR) instrumentation and obtaining of an interference pattern between two SAR-images (which are essentially microwave holograms) taken from the same position, but at different times. Today, this method is referred to as Differential Interferometric SAR Technique. It provides the detection of structural changing in man-made objects such as buildings, dams, towers etc. The great importance of this method concludes in "Sarcophagous" continuos monitoring. Unlike optical fields, the microwave one, at the frequency being chosen properly, are insensitive to the above bad weather conditions and enables us to go around that drawback of laser holography. This approach has been recently suggested by scientists of Space Application Institute, JRC-Ispra to be used for monitoring of structural changes in manmade and natural objects. They have shown experimentally a very good potential ability of the method in detecting of very small changes in an irradiated surface from long distances. However, the radar method used in that SAR, as well as other basic constrains in today's conventional radar technologies do not enable designers in development of an operational and easily portable system for precise monitoring of object structural changes, having relatively short time of data acquisition.

In [18] it is proposed to use Noise Radar Technology and Interferometric SAR Technique for the design and development of method and instrumentation for precise all weather monitoring of Chernobyl's Sarcophagus.

The Noise Radar Technology (NRT) is a radar technology that uses the Noise Continuous Waveform (NCW) as a probe signal and correlation processing (or double spectral one) of radar returns for their optimal reception (matched filtration). The NRT has been shown to be both promising for short-range applications and adequate for today's microwave and electronic technologies. The main advantages of the NRT mentioned above consist in its ability of simultaneous providing of (1) optimal reception of radar returns (gives high sensitivity), (2) waveform compression (gives high range resolution) and (3) simultaneous measurement of both time-of-flight (distance) and phase shifts of the NCW. To the date, both Ka- and X-band sources of the NCW having frequency bandwidth up to 200 MHz and output power of up to 500 mW have been developed. Besides, the methods and instrumentation for the on-line correlation processing of such NCW, having a very high accuracy of range and phase measurements have been also developed. All this enables to design Interferometric SAR on the basis of the NRT with performances allowing its usage for the precise all weather monitoring of Chernobyl's Sarcophagus. A simplified prototype of the system suggested has been recently tested in X-band.

The advantages of the suggested approach compared to the known ones are as follows: (1) Weather independence of whole system and remote acquisition of the data providing a protection of the SAR hardware from damages due to radioactivity; (2) Detection of the structural changes not only in the Sarcophagus' surface, but also inside it; (3) Small sizes, light weight and simplicity of the mechanical part of the SAR hardware; (4) High electromagnetic compatibility parameters and resistance against interference; (5) On-line processing of the data acquired that simplifies the mechanical part of the SAR.

10. Summary

Elimination of ecological hazards connected with the object "Shelter" and realization of the ultimate goal of burying of nuclear fuels contained in it together with radioactive materials and constructions is an extremely complicated scientific and technological problem whose solution will involve advanced production methods and scientific and technological potential of the best national and foreign organizations and enterprises.
The diversity of the problem and the difficulties in achieving of the final goal bring to its dividing into two main tasks:

(a) Conversion of the "Shelter" encasement into a long-term ecologically safe system.
(b) Invention of technologies for dismantling, radwaste managment, transportation and burial of fuel-containing and radioactive materials which are inside of the "Shelter".

The accident on ChNPP followed by social, medical, and ecological consequences revealed the important role of sensing and monitoring nuclear power plants in general and object "Shelter" in particular by means of a system of many sensors of various types with consequent data fusion and analysis.
There are two issues of activity:

- Many parameters state detection and control of such complex and potentially dangerous energy sources as working nuclear power plants are.
- Object "Shelter" and Chernobyl Nuclear Power Plant in a whole are needed in many sensors observation and data fusion conclusions on buildings state, radiation situation, ecological safety, and even chain nuclear reactions. Neutron, gamma, temperature detectors, combined with laser diagnostics, high frequency interferometry can provide these goals. The planned term 15 December, 2000 of ChNPP closing doesn't withdraw the responsibility and enormous technical and financial efforts of the world community to continue monitoring and controlling to take needed measures in the case of necessity for preventing the dangerous consequences. Separately the problem of energy deficit compensation stands.

This work was partly supported by grant STCU # 376

11. References

1. Vasilev N.V., Maltsev V.I., Kovalenko V.N., Shubik V.M., Moskalenko V.F., Volyanski Yu.L., Kolyada T.I., Ponomarenko V.M., Paramonov Z.M. (1999) *Medico-socialnye posledstvia yadernych katasrof*, Kyiv, Zdorov'ya, (Russ.)
2. Shubik V.M. (1996) Immunity and health of peoples suffered after accident, *Monitoring*, . No. 2, 53
3. Izrael Yu.A., Petrov V.N., Avdyushin S.I. et al. (1987) Radioactive pollution of environment in accident zone at ChNPP, *Meteorology and Hydrology* . No. **2**, 9-18.
4. Il'in L.A., Pavlovski O.A. (1987) *Bulletin of MAGATE.* V.29. No. 4. 20-28.
5. Radiation Fallout in Food and Agriculture: *Review*.- Vienna, (1988).
6. Report to the General Assembly: Document of the 38th session of UNSCEAR.-P. 1-5.
7. Health and Environment Consequences of the Chernobyl Nuclear Power Plant Accident (Rep. DOE/EP-0332). Washington, (1987).
8. Environment Contamination Following a Major Nuclear Accident. Vienna, (1990). V. 1-2
9. Romanenko A.E. (1996) The state of health of Ukrainian peoples suffered by radioactive influence in result of accident at ChNPP, *Medical radiology* No. **2**, 37-39.
10. Gavrilin Yu.I., Gordeev K.I., Ivanov V.K. et al. (1992) *Vestnik AMN USSR.* No. **2**, 35-43
11. Chernobyl. Ten years after. Radioactive influence and consequences for public health. November (1995). Agency on Nuclear Energy, Economy Cooperation Organization and Development (OECO). – Paris, 1996.
12. V.G. Baryakhtar, V.Yu. Gonchar, V.V. Yanovsky (1993) Origin of complex structure of radionuclide contamination spot resulted from the accident at Chernobyl nuclear power plant, *Ukr. Phys. Journ.* .**38** 967-975.
13. Mandelbrot B.B. (1982) *The fractal Geometry of Nature*, W.H.Freeman, New York.
14. Hurst H.E., Black R.P., Simaika Y.P. (1965) *Long-Term Storage: An Experimental Study*, London: Constable.
15. Chernobyl Unit-4 Shelter Implementation Plan (SIP), Tacis services DG IA, European Commission and US Department of Energy (1997).
16. Description of the Ukrytiye Encasement and Requirements for Its Conversion. (1992) Kiev, Naukova Dumka,, 50 p.
17. Belovodski L.F. (1996) Minatom of Russia, Chernobylskaya AES. *Atom,* No.**3**,.32.
18. Lukin K.A. (1999) Project Proposal No 1954 to STCU: Microwave Method and Instrumentation for Precise Monitoring of Chornobyl's Sarcophagous (Application of Noise Radar Technology and Interferometric SAR method).

BLACKBOARD ARCHITECTURES FOR SONAR DATA ASSOCIATION*

B.A. MCARTHUR
Defence Research Establishment Atlantic
P.O. Box 1012, Dartmouth, NS, Canada B2Y 3Z7

1. Introduction

Semi-automated systems are being developed to aid naval command teams in building undersea warfare (USW) tactical pictures representing the surface and subsurface environment and containing information regarding the number of detected targets, target identities, and target positions, courses, and speeds. The fusion of multiple acoustic data sources, as well as the incorporation of supporting information from non-acoustic data sources, is recognised as a key requirement for achieving accurate and efficient USW tactical picture compilation. This paper provides an overview of work conducted at the Defence Research Establishment Atlantic (DREA) on the application of the blackboard architecture, a software architecture for distributed problem solving, to sonar data association, a specific aspect of sonar data fusion involving the association of multiple acoustic signals corresponding to a given platform.

2. The Sonar Data Association Problem

The complexity of the sonar data association problem is illustrated by considering that a combination of any of the following basic association operations may be applicable at any given point in time: temporal association of intermittent signals; intrasensor association between multiple signals from the same acoustic source and detected on a single sensor; intersensor association between signals from the same acoustic source detected on different sensors, such as multiple passive or active sonars; and, association between signals from the same acoustic source produced using different signal processing techniques, such as narrowband and broadband processing on a single passive sonar.

One approach to sonar data association is to use a data-driven architecture in which a hierarchy of composite association hypotheses is generated by executing a fixed sequence of data association operations on all detected signals. While the simplicity of the data-driven architecture is well-suited for hardware implementation, there are drawbacks to its use in applications, such as sonar data association, in which there is both a large number of signals and a high level of uncertainty in the data. First, a combinatorial explosion of association hypotheses may occur when attempts are made to apply data association operations to all hypotheses at each level in the hierarchy. In addition, the use of a fixed sequence of data association operations assumes that the information necessary for making an association decision is available at the current level in the hierarchy or is implicitly contained in lower-level associations. Given the complex, uncertain nature of

A. K. Hyder et al. (eds.), Multisensor Fusion, 799–805.

the acoustic data, efforts to control the generation of association hypotheses by deleting all but high-probability associations at each level are often unreliable.

Our solution is to adopt a more flexible association framework that permits the ordering of association classes to be determined in a situation-specific context-based fashion. In cases where association decisions are unambiguous a fixed ordering of association classes is used for efficiency. However, when conflicting or ambiguous associations are detected, additional constraints provided by other information sources are applied. This concept is based on the observation that sonar operators performing data association manually make use of diverse types of knowledge in addition to the measurement data, including knowledge of pattern similarities in narrowband or broadband data, ownship motion and of ownship position relative to the sensors, acoustic propagation and noise modelling, and target behaviour [1,2].

3. Blackboard Architecture for Sonar Data Association

The blackboard architecture [3,4] is composed of three main components: a hierarchically-organized global database, known as the blackboard, which contains all partial or complete problem solutions; a set of specialized sources of domain knowledge, or knowledge sources, which make incremental changes to the solutions stored in the blackboard database; and a control module that determines the order in which knowledge sources respond to changes on the blackboard database.

Blackboard architectures have been used in data fusion applications including the classification of passive sonar data [5] and the fusion of shipborne and airborne electromagnetic sensors [6]. Several factors make the blackboard architecture well-suited for implementation of a context-based approach to sonar data association. First, the requirement for the situation-dependent ordering of association operations is satisfied by the blackboard architecture's flexible control structure. Each knowledge source monitors the blackboard database for conditions under which it can contribute to the problem-solving process. This makes it possible for the control module to dynamically determine the order in which knowledge sources are executed. Secondly, the blackboard database and knowledge sources are capable of supporting the diverse types of knowledge which may potentially be used in sonar data association. Individual levels in the blackboard database may support different numeric or symbolic data representations and knowledge sources may be implemented using sets of heuristic rules, procedures, or less conventional techniques such as neural networks. Finally, the blackboard architecture provides a modular design that is well-suited for exploratory development. Individual knowledge sources and control modules are functionally independent and may be modified without affecting the rest of the system.

The following sub-sections describe in greater detail the design of the blackboard database, knowledge sources, and control structure for a specific example, a blackboard-based data association system for a passive towed array sonar [7]. Versions of this system have been implemented using the CLIPS expert systems shell and GBB, a Common Lisp-based blackboard systems toolkit.

3.1. BLACKBOARD DATABASE

The blackboard database for the sonar data association system has four levels, as illustrated in Figure 1. The lowest level contains the input data which are signal tracks obtained as outputs from narrowband and broadband signal tracking algorithms. In this instance narrowband and broadband data, which are produced using different signal processing techniques on time-series data from the same towed array sonar, are treated as being produced by different sensors. Higher levels in the blackboard database correspond to different classes of association hypothesis: the temporal association level contains temporal association hypotheses between signal tracks from the same sensor; the intrasensor association level contains intrasensor association hypotheses between signal tracks and/or temporal association hypotheses from the same sensor; and, the intersensor association level contains association hypotheses between signal tracks, temporal association hypotheses, or intrasensor association hypotheses, but from different sensors. Except for the intersensor association level, the blackboard database is partitioned into separate sub-blackboards for each sensor in order to increase the efficiency of database access.

Signal tracks and association hypotheses are represented by an object-oriented data representation. At each level in the blackboard database, data are represented by instances of an object class which is defined by a set of features represented as slots, and a set of attached procedures associated with each slot, termed methods. Separate object classes are defined for signal tracks and for each class of association hypothesis. Each object class contains slots which describe time extent, overall visual appearance of the signal track, temporal behaviour, and links to associated objects on other levels of the blackboard.

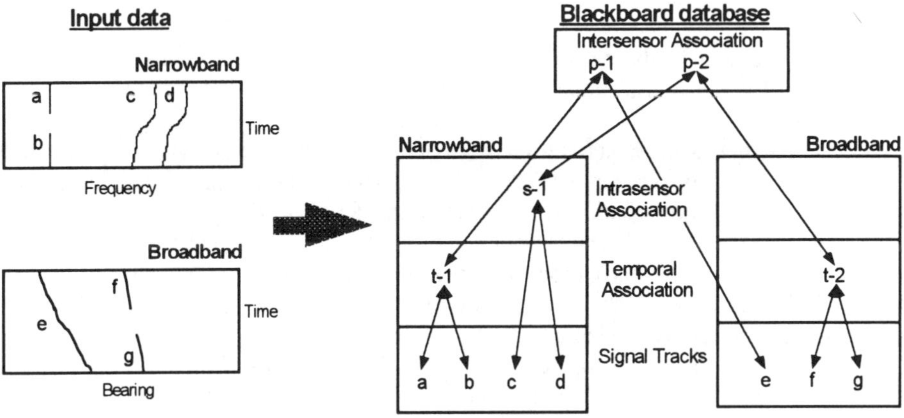

Figure 1. Blackboard database representation of associations between narrowband and broadband signal tracks

3.2. KNOWLEDGE SOURCES

Knowledge sources in the sonar data association system initiate or confirm association hypotheses and are implemented using a combination of procedures and rules. Separate knowledge sources are maintained for the temporal, intrasensor, and intersensor modes of association. The association rules used in the temporal and intrasensor association knowledge sources utilize the following general criteria for assessing the degree of association between signal tracks: frequency continuity; similarity of signal track attributes; temporal separation between tracks; temporal correlation of signal dynamics; and, harmonic relations.

Each association hypothesis on the blackboard is assigned one of three symbolic confidence levels — "unlikely", "possible" or "likely". Two approaches have been investigated for assigning confidence levels. The first approach uses logical criteria containing numeric thresholds on differences between object slot values, such as *Confidence is possible if frequency difference < 5.3 Hz.* In the second approach [8], implemented only for temporal association, confidence levels are interpreted to be fuzzy relations over the set of signal tracks. Fuzzy production rules specify criteria for assigning a given degree of confidence based on linguistically represented frequency diferences, temporal separations, and attribute value differences. An example fuzzy production rule might be *If frequency-difference is small and temporal-separation is small and attribute-value-similarity is high then confidence is likely.*

3.3. CONTROL STRUCTURE

The input data are processed on a scan-by-scan basis. At each scan the following operations are performed.

(a) Measurement Updating. Each signal tracker sends a message to its associated signal track object to update its measurement, attribute, and temporal dynamics slots.

(b) Knowledge Source Activation. Knowledge sources are activated by methods which monitor object slots for knowledge source pre-conditions. A knowledge source activation results in the generation of a Knowledge Source Activation Record (KSAR) — a data structure with fields for the type and current status of a knowledge source activation. Measurement updating will trigger the generation of a KSAR for each object which satisfies knowledge source pre-conditions.

(c) Knowledge Source Selection. Control rules are used to determine which KSAR to execute. In the absence of any other information a default ordering is used, based solely on the association mode as follows: temporal association; intrasensor association; and intersensor association. If the KSAR indicates a conflict between association hypotheses, heuristic rules are applied that search for additional constraints provided by other association classes.

(d) Knowledge Source Execution. The execution of knowledge sources is performed in pre-condition and action phases. The pre-condition phase is used to identify a set of potential matches and the existence of conflicts between potential matches which would require further information for resolution. The action phase of a knowledge source performs the actual association and will modify information on the blackboard.

Following knowledge source execution, control passes back to knowledge source selection. The cycle of knowledge source selection and execution is repeated until there are no KSARs which can be executed.

Figure 2 contains a specific scenario demonstrating how multiple modes of association may be used to resolve an association conflict. In this case, a conflict is detected when both tracks *b* and *c* are marked as being possible temporal associations with track *a*. This conflict can be resolved using the following heuristic rule, *Given signal tracks s_1 and s_2 having conflicting temporal associations with a common signal track s_3, attempt conflict resolution using a signal track s_4 which is intrasensor associated with s_3.* Application of the rule results in the selection of track *d*, which is intrasensor associated with track *a*, as a possible source of information for resolving the conflict between the two possible temporal associations. Track *d* has a higher degree of correlation with track *b* than with track *c*. Accordingly, track *b* is temporally associated with track *a*.

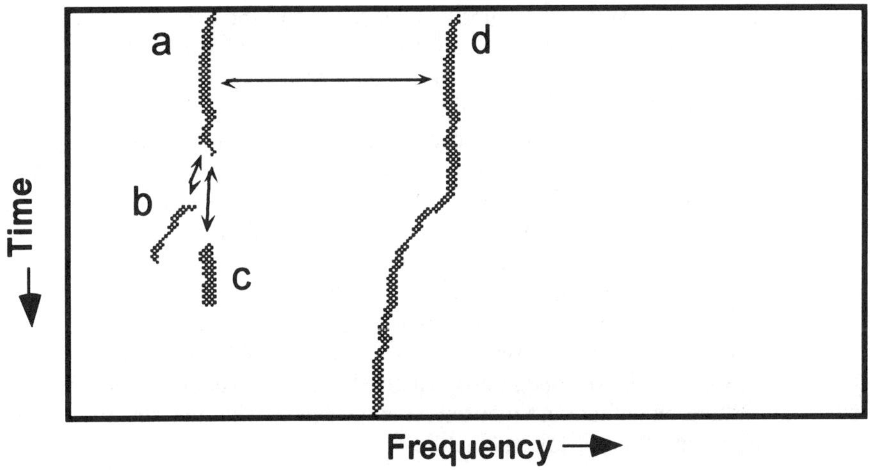

Figure 2. The application of an intrasensor association to resolve conflicting temporal associations.

4. Blackboard Architecture Extensions

One limitation of the sonar data association system described in [7] is its reliance on accurate input data provided by signal trackers. In reality, automated tracking of acoustic signals is difficult to perform reliably, particularly in high clutter environments. Consequently, the blackboard architecture has been extended to feed back constraints derived from sonar data fusion in order to modify the behaviour of lower-level signal trackers. The concept is based on the addition of knowledge sources that plan responses to specific classes of events, such as fading signals. When an event is detected for a given signal track, a specific response is determined based on additional contextual information that is available at the time of the event, such as target kinematic or classification information, environmental information, or data association information. Contextual information is then translated into numeric constraints on signal tracker parameters.

This technique has been demonstrated successfully for regaining contact on narrowband signals following ownship manoeuvres, using constraints derived from intrasensor associations [9]. Given two narrowband signals for which an intrasensor association is established before the ownship manoeuvre, signal tracks are re-detected using a multi-hypothesis beam search over pairs of signal detections falling within gate regions centered on the last known frequency and bearing of the two signal tracks. A cost function is derived from the following constraints on pairs of signal detections: equal bearings; a frequency ratio equal to the ratio measured prior to the ownship manoeuvre; and, smoothness in the bearing of the resulting signal track.

A second extension of the original blackboard system concept is the introduction of an explicit representation for the tactical picture compilation task to which data association operations are being applied. The result, termed goal-directed data integration [10], manages the selection of signals and data association operations so as to solve specific problems that may arise during tactical picture compilation. The tactical picture compilation task is decomposed into a hierarchy of goals, with each goal representing a task, an unresolved source of uncertainty in the evolving tactical picture, or a mission objective. A planning process is used to prioritise unsatisfied goals and to determine sequences of data association operations which should be capable of achieving each selected goal. Attention is therefore focused on those data association operations with the greatest potential impact on the tactical picture.

5. Conclusions

Sonar data association is a difficult task to automate because of large volumes of uncertain, ambiguous data and a complex acoustic environment. The blackboard architecture, through its flexible, adaptive control structure, its ability to incorporate diverse sources of information, and its modular design, is well-suited to the implementation of solutions to the sonar data association problem which emulate the capabilities of the human sonar operator. Work in this area being conducted at DREA provides a foundation for the development of operator aids for USW tactical picture compilation.

6. References

1. Iotek, Inc. (1992) *Expert Systems Study*, Document DN0929.
2. McIntyre, C.M., and Roger, W.A. (1993) Data association in passive acoustic tracking, in O.E. Drummond (ed.), *Signal and Data Processing of Small Targets*, Proc. SPIE vol. 1954, pp. 376-385.
3. Nii, H.P. (1986) Blackboard Systems: *The Blackboard Model of Problem Solving and the Evolution of Blackboard Architectures*, AI Magazine, pp. 38-53.
4. Nii, H.P. (1986) Blackboard Systems: *Blackboard Applications Systems, Blackboard Systems from a Knowledge Engineering Perspective*, AI Magazine, pp. 82-106.
5. Nii, H.P., Feigenbaum, E.A, Anton, J.J. and Rockmore, A.J. (1982) *Signal-to-Symbol Transformation*: HASP/SIAP Case Study, AI Magazine, pp. 23-35.
6. Shahbazian, E., Duquet, J., Valin, P. (1998) *A Blackboard Architecture for Incremental Implementation of Data Fusion Application*, Proc. Fusion '98, pp. 455-461.
7. McArthur, B.A. (1993) *An Expert Systems Approach to Track Segment Association*, Proc. TTCP GTP-1 Applications Working Group Specialists' Meeting, Portland, UK.
8. McArthur, B.A. (1994) *Sonar Track Segment Association using Fuzzy Confidence Measures,* Proc. 6th Symposium on Applications of Expert Systems in DND, Kingston, Canada.
9. McArthur, B.A. (1995) *A Knowledge-Based Approach to Processing Unmodelled Events in Contact Data Management*, Proc. TTCP GTP-1 Applications Working Group Spring Workshop, Portland, UK.
10. McArthur, B.A. (1997) *Goal-Directed Data Integration for Sonar Picture Compilation*, Proc. IEEE Intl. Conf. on Systems, Man and Cybernetics, pp. 488-493.

ON THE USE OF STATISTICAL CONSIDERATIONS IN MULTI-MODAL IDENTITY VERIFICATION SYSTEM DESIGN

P. VERLINDE AND M. ACHEROY
Royal Military Academy
Signal and Image Centre
Renaissancelaan 30
B-1000 Brussels, Belgium
{verlinde,acheroy}@elec.rma.ac.be

1. Introduction

In this chapter, we would like to show two of the possibilities of statistical analysis in the design process of a multi-modal identity verification system. The multi-modal identity verification system under consideration is built of 3 modalities in parallel, each one delivering as output a scalar number, called score, stating how well the claimed identity is verified. A decision fusion module receiving as input the 3 scores has to take a binary decision: *accept* or *reject* the claimed identity. We have solved this fusion problem using parametric and non-parametric classifiers. There are two trivial steps in the design process of such a multi-modal identity verification system in which statistical analysis can be useful.

A first (*a priori*) statistical analysis of the individual biometric identity verification experts allows to get an idea on one hand of their individual discriminatory power, and of their complementarity on the other hand. We can indeed get an idea of the independence of the different experts (and thus of the amount of *extra* information that each expert brings in), by analyzing their correlation. Furthermore, the analysis of the extreme values gives us insight into the possible use of personalized approaches.

A second (*a posteriori*) statistical analysis is performed on the results obtained by the different paradigms, in order to find out if there are or not statistically significant differences between them.

2. Characterization of An Identity Verification System

In this chapter, we will consider the verification of the identity of a person as a typical two-class problem: either the person is the one (in this case he is called a *client*), or is not the one (in that case he is called an *impostor*) he claims to be. This means that we are going to work with a binary {accept, reject} decision scheme.

When dealing with binary hypothesis testing, it is trivial to understand that the decision module can make two kinds of errors. Applied to this problem of the verification of the identity of a person, these two errors are called: False Rejection (FR): *i.e.* when an actual *client* is rejected as being an *impostor*, and False Acceptance (FA): *i.e.* when an actual *impostor* is accepted as being a *client*. The performance of a speaker

A. K. Hyder et al. (eds.), Multisensor Fusion, 807–814.

verification system is usually given in terms of the global error rates computed during tests: the False Rejection Rate (FRR) and the False Acceptance Rate (FAR). A unique measure can be obtained by combining these two errors into the Total Error Rate (TER). All verification results in this work will be given in terms of FRR, FAR, and TER. For each error the 95 % level confidence interval will be given between square brackets.

3. Identity Verification Experts

3.1 PERFORMANCES

All the experiments have been performed on the M2VTS database consisting of 37 persons, using the following three single modalities: (1) a profile image expert based on a template matching method [7], (2) a frontal image expert based on a robust correlation method [5], and (3) a vocal expert based on a sphericity measure calculated on second order [1]. The performances achieved by these three mono-modal identity verification systems are given in Table 1.

TABLE 1. Verification results for single modalities

Expert	FRR (%)	FAR (%)	TER (%)
	(37 tests)	(1332 tests)	(1369 tests)
Profile	21.6 [11.4,37.2]	8.5 [7.1,10.1]	8.9 [7.5,10.5]
Face	21.6 [11.4,37.2]	8.3 [6.9,9.9]	8.7 [7.3,10.3]
Vocal	5.4 [1.5,17.7]	3.6 [2.7,4.7]	3.7 [2.8,4.8]

3.2 STATISTICAL ANALYSIS OF THE DIFFERENT EXPERTS

3.2.1 Analysis of Correlation

An important statistical analysis is the calculation of the correlation that exists between the different experts. A popular way of seeing the importance of this is to say that the more the errors the different experts make are de-correlated, the better our fusion could get since the amount of *new* information introduced by each expert will tend to be larger. The correlation can be visualized by taking the experts two-by-two and plotting the observed results for each population. These matrix scatter plots are shown in Figure 1, and the fact that all the distributions have shapes in the form of a cloud indicates that the correlation is very low.

3.2.2 Analysis of Extreme Values

Another important point in descriptive statistics is the analysis and the handling of extreme values. Normally speaking, extreme values should be discarded from the calculation of characteristic statistical measures such as means or variances. In our work however we will not do that, since these extreme values can contain interesting information as will be shown hereafter. Indeed, in Figure 1 it can be seen that the profile expert presents a large number of extreme values, which form very specific alignments in the four sub-plots where the scores of the profile expert are plotted against one of the axis.

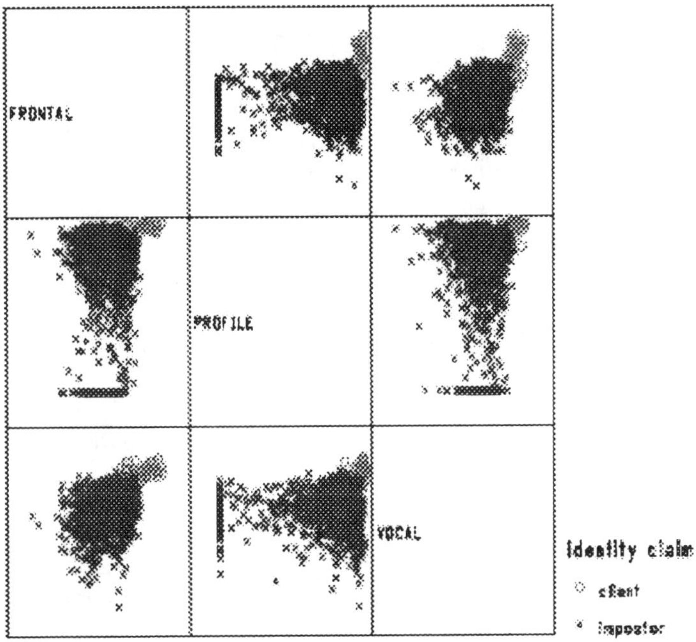

Figure 1. Visual representation of the correlation of the different experts taken two-by-two

One of the possible explanations for this phenomenon, taking into account the experimental protocol, is that the profile of one the clients is very different from the profiles of all other clients. After analyzing the scores of the profile expert it turned out that, when claiming the identity of client number eight, 22 out of the 36 other clients obtained a score equal to zero! This phenomenon is represented as matrix scatter plots in Figure 2.

In order to understand this phenomenon, it is interesting to have a look at a typical profile image of client number eight. Such an image is presented in Figure 3.

From this profile image, it is easy to see that the chin of this specific client is very pronounced, which makes it a very personal characteristic. Therefore few profiles of other clients of the database will present good results when being matched against this typical profile, which obviously leads to very good impostor rejection performances. This observation suggests that, in some specific cases, a *personalized approach* based for instance on specific characteristics of certain persons, might improve system performance substantially. This is an interesting observation, especially when seen in the light of the actual efforts to come to *robust* methods, in which extreme values, such as the ones we have been considering here, are very likely to be excluded!

In this work we did however not use such a personalized approach, since the database we used did not provide enough training data.

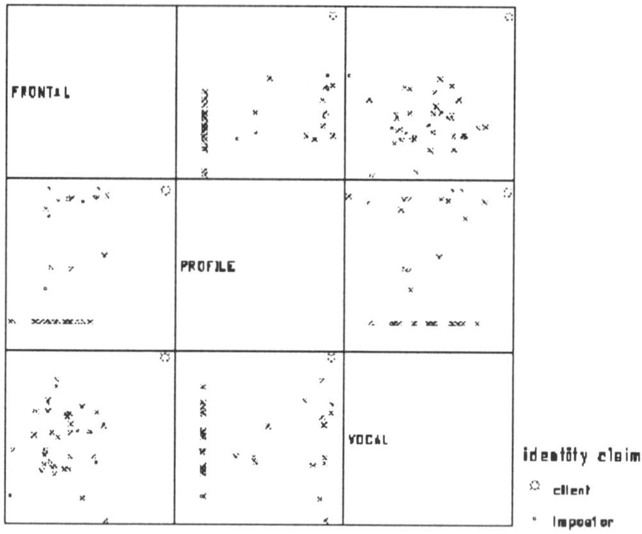

Figure 2. Matrix plots representing the scores for all 37 persons claiming the identity of client number eight (i.e. one client access and 36 impostor accesses).

Figure 3. Typical profile image of client number eight.

4. Paradigms

To solve our particular classification problem, classical parametric and non-parametric statistical pattern recognition techniques have been adapted.

In the parametric class, piece-wise linear classifiers, and classifiers based on the general Bayesian decision theory (Maximum A-posteriori Probability and Maximum Likelihood), and on a simplified version of it (the Naive Bayesian classifier, which has been applied in the case of simple Gaussians and in the case of a logistic regression model), were experimented [10]. Furthermore experiments have been done using Linear, Quadratic, Multi-Layer Perceptron [10], and Support Vector Machine classifiers [3].

Also, several non-parametric classifiers [10] have been used in this work. A first representative has been the family of the (very simple: AND, OR) voting methods. Thereafter the *k*-NN classifier and some of its possible variants have been studied. And to finish with this class, another popular paradigm, the binary decision tree, has also been used.

5. Experimental Comparison of Classifiers

5.1 TEST RESULTS

Table 2 gives an overall view of the best verification results obtained with the classifiers used in this study.

5.2 STATISTICAL SIGNIFICANCE

As can be observed in Table 2, the FAR, FRR and TER results from most of the methods used are lying very close to one another and they have confidence intervals which are overlapping each other in almost all cases. This means that, based on such FAR, FRR or TER results, it is not easy to decide without hesitation which method to use. That is why it would be very useful to have a systematic method which detects statistical significant differences between the FAR, FRR and/or TER results obtained by all the methods used.

In the case the scores obtained by the different fusion modules are independently drawn from normally distributed populations with the same variance, this problem can be solved by performing a basic analysis of variance (ANOVA), supplemented with so-called ad-hoc Tests [6]. The ANOVA tells us *if* there are statistical significant differences between the different methods, and the ad-hoc tests (Least Significant difference (LSD), Duncan's Multiple Range Test and many others) tells us *where* the statistical significant differences exactly are. It is reminded here that the statistical comparison needs to be done between all the methods at the same time (as opposed to a series of pairs-wise comparisons), to avoid the in the statistical community well-known effect of dramatically increasing the type one error [7].

Applying the ANOVA method in our case leads to a firm rejection of the hypothesis H_0, which means that there are significant differences between the presented fusion modules. To find out where these statistical significant differences are, we did use

TABLE 2. Summary table of verification results for the following fusion modules: Maximum Likelihood (ML), Maximum A posteriori Probability (MAP), Logistic Regression (LR), Quadratic Classifier (QC), Linear Classifier (LC), Multi-Layer Perceptron (MLP), Support Vector Machine (SVM), AND-voting rule (AND), OR-voting rule (OR), *k*-Nearest Neighbor (*k*-NN), *k*-Nearest Neighbor using Vector Quantization (*k*-NN + VQ), Binary Decision Tree (BDT).

Method	FRR (%)	FAR (%)	TER (%)
	(37 tests)	(1332 tests)	(1369 tests)
ML	2.7 [0,5,13.8]	0.7 [0.4,1.3]	0.7 [0.4,1.3]
MAP	5.4 [1.5,17.7]	0.0 [0.0,0.3]	0.1 [0.0,0.5]
LR	2.7 [0.5,13.8]	0.0 [0.0,0.3]	0.1 [0.0,0.5]
QC	0.0 [0.0,9.4]	2.4 [1.7,3.4]	2.3 [1.6,3.2]
LC	0.0 [0.0,9.4]	3.1 [2.3,4.2]	3.0 [2.2,4.0]
MLP	0.0 [0.0,9.4]	0.4 [0.2,0.9]	0.4 [0.2,0.9]
SVM (linear)	2.7 [0.5,13.8]	0.0 [0.0,0.3]	0.1 [0.0,0.5]
OR	0.0 [0.0,9.4]	7.4 [6.1,8.9]	7.2 [5.9,8.7]
AND	8.1 [2.8,21.3]	0.0 [0.0,0.3]	0.2 [0.1,0.6]
MAJ	0.0 [0.0,9.4]	3.2 [2.4,4.3]	3.1 [2.3,4.2]
k-NN	8.1 [2.8,21.3]	0.0 [0.0,0.3]	0.2 [0.1,0.6]
k-NN + VQ	0.0 [0.0,9.4]	0.5 [0.2,1.0]	0.5 [0.2,1.0]
BDT	8.1 [2.8,21.3]	0.3 [0.1,0.8]	0.5 [0.2,1.0]

Duncan's Multiple Range Test. The result of this ad-hoc test with the highest power (*i.e.* the method with the lowest error of type II) is that three different groups of methods are significantly distinct. These groups are the following ones:

- the first group (the one with the best performances) is formed by the following methods: LR, SVM, MAP, AND, *k*-NN, MLP, *k*-NN + VQ, BDT, and ML;
- the second group consists of: QC, LC, and MAJ; and
- finally the worst method in our case is the "OR"-vote.

However, since some of the fusion modules we use generate hard binary decisions as their output, we are, strictly speaking, not allowed to perform an ANOVA. Therefore we do have to use non-parametric methods. For the same reason as in the parametric case, it is here again absolutely necessary to compare all methods *at the same time*.

In [8], five approximate statistical tests are presented for determining whether one learning algorithm out-performs another one on a particular learning task, but unfortunately these tests only allow a two-by-two comparison.

We believe that, depending on the discrete or continuous aspect of the output of the fusion module, two different non-parametric statistical tests can be applied to test the hypothesis H_0: all used fusion methods are of equal performance, against the alternative H_1: there are variations in performance. If one or more of the outputs of the fusion modules are *binary* or *hard* decisions, then *Cochran's Q-test* for binary responses is suitable to solve this problem. If however the outputs of *all* fusion module are continuous or *soft* decisions, then *Page's test* for ordered alternatives could be used. The latter test has the advantage that it has more *power* than the former [9].

Since in this specific case there are several fusion modules with binary (in casu "0" for a reject and "1" for an acceptance) outputs, the only possibility is to use Cochran's Q-test. In conventional terms of this test, the different fusion methods are called the *treatments* (t) and the different access tests are called the *blocks* (b). The exact distribution of Q is difficult to obtain, but for large samples Q has approximately a chi-squared distribution with (t-1) degrees of Freedom [10].

The application of Cochran's test for binary responses gives in our case a Q value which is much larger than the corresponding critical value of the corresponding chi-squared distribution, which leads us to reject the hypothesis H_0. This means that there are significant differences between the presented fusion methods, which is the same conclusion as the one obtained by performing the ANOVA. And although Cochran's test does not say where *exactly* these differences are, we did however establish one thing for sure: the best method (which in our case is the logistic regression, together with the linear SVM: two "linear" and thus *robust* parametric methods) *is statistically significant better* than the worst method (which in our case is the "OR" voting scheme). This result is rather trivial, since for these two "extreme" methods the 95 % confidence intervals do not overlap at all.

To conclude this section it can be seen that the results of the ANOVA and ad-hoc tests (although strictly speaking not allowed) do reinforce our first, intuitive approach. The allowed Cochran's Q-test does confirm the results of the ANOVA, but unfortunately we did not find a non-parametric equivalent for the ad-hoc tests. Combining all this information leads us to the conclusion that there is no statistically justified evidence to prefer one specific method from the first (best performing) group above another one from the same group.

6. Conclusions

One of the most important conclusions after observing the results in Table 2 is that in our particular application fusion always improves the system performances beyond those of even the best single expert. These results have to be seen in their correct perspective, taking into account the very limited database. In any case, these performances are much better than those of verification systems using only one of the presented modalities.

With respect to the use of statistical considerations in the design of multi-modal identity verification systems, the following two comments can be made:

1. Statistical analysis *a priori* is important to increase the supplementary information input coming from additional verification experts. This can be done by analyzing the correlation of the different experts. An interesting offspring of this correlation analysis is the fact that sometimes - in a personalized approach - it might be interesting to analyze the extreme values quite carefully, before discarding them.
2. Statistical analysis *a posteriori* is important to choose a fusion method which is statistically significant the best (or which belongs to the best group). The question of which fusion method should be chosen, is indeed a difficult one to answer. A lot depends on the application. To be able to choose a number of potentially powerful fusion paradigms, it is important to have a (large) representative database of your application, which can be used for training purposes. It also helps if one is able to visualize the different populations (clients versus impostors), because in that specific case the choice of the fusion methods

could be guided by the shape of the separation frontier between the two populations one wants to obtain.

7. References

1. Bimbot, F., and Mathan, L., (1994) Second-order statistical measures for text-independent speaker identification. In *ESCA workshop on automatic speaker recognition, identification and verification*, Martigny, Switzerland.
2. Dietterich, T.G., (1998) Approximate Statistical tests for Comparing Supervised Classification Learning Algorithms. *Neural Computation*, 10(7).
3. Gutschoven, B., and Verlinde, P., (2000) Decision Fusion using Support Vector Machines (SVM). In *Proceedings of the 3rd International Conference on Information Fusion*. Paris, France.
4. Manly, B.F.J., (1994) *Multivariate Statistical Methods*. Chapman & Hall, second edition.
5. Matas, J., Johnson, K., and Kittler, J., (1997) Fast face localization and verification. In A. Clark, editor, *British Machine Vision Conference*, pages 152-161. BMVA Press.
6. Montgomery, D.C., (1997) *Design and Analysis of Experiments*. John Wiley & Sons, fourth edition.
7. Pigeon, S., and Vandendorpe, L., (1997) Profile authentication using a Chamfer matching algorithm. In *Proceedings of the First International Conference on Audio- and Video Based Person Authentication*, Lecture Notes in Computer Science, pages 185-192. Springer Verlag.
8. Siegel, S., and Castellan, N.J., (1988) *Nonparametric Statistics*. McGraw-Hill.
9. Sprent, P., (1989) *Applied nonparametric statistical methods*. Chapman & Hall.
10. Verlinde, P., (1999) *A contribution to multi-modal identity verification using decision fusion*. PhD thesis, Ecole Nationale Supérieure de Télécommunications, Paris, France.

MULTISENSORY EXPERIMENTS ON THE MESON FACILITIES

V.M. BYSTRITSKY
Joint Institute for Nuclear Research
Laboratory of Nuclear Problems
141980 Dubna
Moscow Region, Russia

1. Introduction

To know parameters of all accompanying or competing processes is a necessary condition for increasing the measurement accuracy for characteristics of physical processes. Depending on the experimental goals, this sort of information can be provided by a variety of additional detectors, such as elementary particle and nuclear radiation detectors, electromagnetic transducers, pressure gauges, temperature sensors, humidity detectors. The present paper describes some typical many-detector systems and systems for data exchange with a computer used in the experiments at the meson factories[1] PSI (Paul Scherrer Institute, Switzerland), TRIUMF (Vancouver, Canada), LAMPF (USA), RAL (England), KEK (Japan).

2. Muonic Catalysis of Nuclear Reactions

Fifty years ago it was first proposed that a negative muon could catalyze the nuclear fusion reaction of hydrogen isotopes [1, 2]. This phenomenon was experimentally found by Alvarez [3] in 1957 in a mixture of hydrogen and deuterium. A hope to create an alternative source of power to be produced at room temperature arose on the basis of the hypothesis that one muon can initiate many nuclear fusion reactions of hydrogen isotopes. Essentially muonic catalysis of nuclear reactions consists in the following. When a negative muon stops in a target filled with hydrogen, deuterium, tritium, or their mixture, an excited hydrogen mu-atom with initial kinetic energy of the order of some eV is formed and then undergoes de-excitation and thermalization. Systems like this are electrically neutral, very compact (the orbit of the pμ-atom has an about 200 times smaller radius than the orbit of normal hydrogen atom), and thus easily penetrate through the electron shell of other atoms coming within a distance of the order of a mesoatomic length unit ($a_\mu = 2.56\times10^{-11}$ cm) to their nuclei. This close approach allows formation of muonic molecules of hydrogen isotopes (depending on the composition of the mixture of hydrogen isotopes, ppμ, pdμ, ptμ, dtμ, ddμ, ttμ molecules can be formed), where nuclear fusion reactions (dd, pd, dt, pt, tt fusion) with emission of γ quanta, neutrons, and charged particles (^{3}He, ^{4}He, p, t) may occur. As is seen, mu-atomic (MA) and mu-molecular (MM)

[1] A meson factory is an accelerator to produce high-intensity π-meson and muon beams in a wide energy range. The maximum intensity of π-meson and muon beams, e.g. at PSI, is 10^{10} π/s and 5×10^8 μ/s respectively.

A. K. Hyder et al. (eds.), Multisensor Fusion, 815–837.

processes in mixtures of hydrogen isotopes form a multitude of competing processes (CP) and accompanying processes (AC) with respect to the process under study. Therefore, information on characteristics of all CP and AP is also necessary to describe correctly the kinetics of the process under study. One can gain this information by using a great variety of detectors for nuclear fusion products, mu-X-ray radiation, muons and electrons from the muon decay. In the past three decades considerable study has been given to MA and MM processes arising from stops of negative muons in pure hydrogen isotopes and their mixtures [4–7] at various temperatures and densities. In 1978 a Dubna team experimentally found the resonance mechanism [8] for formation of muonic deuterium molecules. In 1979 the same team was the first to find muonic catalysis of the fusion reaction of deuterium and tritium nuclei [9]. The muonic catalysis efficiency for the dt fusion reaction was found to be high: one muon stopped in the deuterium-tritium mixture could initiate up to 200 dt fusion reactions, which corresponded a total energy release of about 3.4 GeV/μ. So much promising data allow muonic catalysis to be seriously treated as an alternative power production method [10].

This challenge has retained its topical nature up to now, but it requires solution of some technical problems, such as generation of a magnetic field with strength ~20 T in a volume of a few cubic meters and handling of radioactive tritium of a few MCi, which themselves are sophisticated and nontrivial. Nevertheless, the investigations of MA and MM processes yielded important results of their own, characterizing a system of three bodies interacting by the Coulomb law [11–14].[2]

Before listing the most important of the results, we briefly describe some typical detecting systems used in the muonic catalysis experiments.

By way of example we consider experiments with a setup used at the meson factory at PSI (Switzerland). The setup (see Figure 1) is designed to study fusion reactions in charge-nonsymmetrical muonic molecules dHeμ [15–19] and reactions of muon capture by the He nucleus resulting in production of two and more charged products in the final state [20]:

$$d\mu{}^3He \rightarrow {}^4He\mu + p\ (14.7\ MeV), \qquad (1a)$$
$$d\mu{}^4He \rightarrow {}^5Li\mu + \gamma, \qquad (1b)$$
$$\rightarrow p + n + n + \mu, \qquad (2a)$$
$${}^3He\mu$$
$$\rightarrow d + n + \mu. \qquad (2b)$$

Figure 2 presents a scheme of muonic processes occurring in the D2+He3 mixture after the negative muon stopped in it. As seen, themain competing process with respect to the investigated processes is the muon decay $\mu^- \rightarrow e^- + \nu_\mu + \tilde{\nu}_e$ (it rate is $0.455\times10^6\ s^{-1}$).

The setup consists of a gas target filled with a mixture of deuterium and He (He concetration is about 5%) and a system of detectors comprising:

(a) three pairs of E-ΔE Si telescopes (diameter 45 mm, E detector thickness 4 mm, ΔE detector thickness 360 μm);

[2] Comparison of the experimental and calculated data will allow a test of the solution algorithm for the problem of three bodies interacting by the Coulomb law.

(b) eight plastic scintillator detectors of muon decay electrons arranged around the target;
(c) a detector of mu-X-ray radiation in the range 4–10 keV (V = 0.17 cm^3);
(d) a Ge detector (V = 122 cm^3) to record γ quanta in the energy range 10–20 MeV;
(e) a Ge detector of volume 75 cm^3 to record mu-X-ray radiation from Oμ, Nμ, and Cμ atoms in the energy range 100–500 keV (in order to reveal C, N, and O impurities in the deuterium-helium mixture);
(f) a detector based on liquid scintillator NE 213 to record neutrons from the fusion reaction in the ddμ molecule (with energy 2.5 MeV);
(g) muon beam detectors on the basis of plastic scintillators.

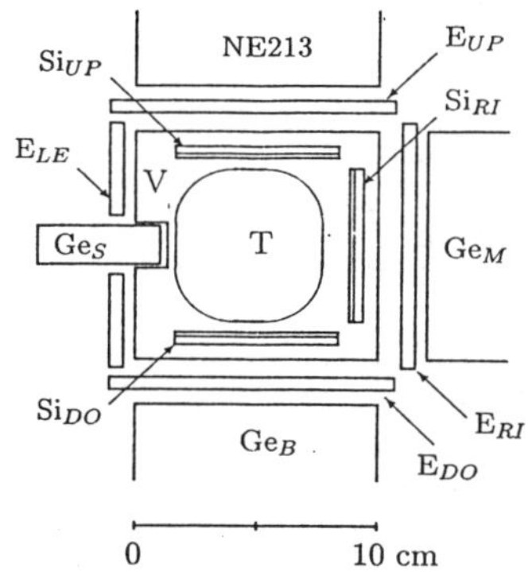

Figure 1. The Experimental setup for investigation μ-atomic and μ-molecular processes in hydrogen-helium mixtures (dHe-experiment): Si_{up}, Si_{RI}, Si_{DO}...Si (dE/dx - E) detector pairs located above, below and to the right of the muon beam; E_{UP}, E_{LE}, E_{DO}, E_{RI} - muon decay electron counters; Ge_S, Ge_B, Ge_M - three different Ge detectors to measure different γ - ray energies: 0 - 20 keV, 0 - 1 MeV; 0 - 25 MeV; NE 213 - dd fusion neutron counter with the liquid scintillator NE 213; T - gaseous target; V - vacuum chamber. Muon beam detectors 1 and 2 are not indicated.

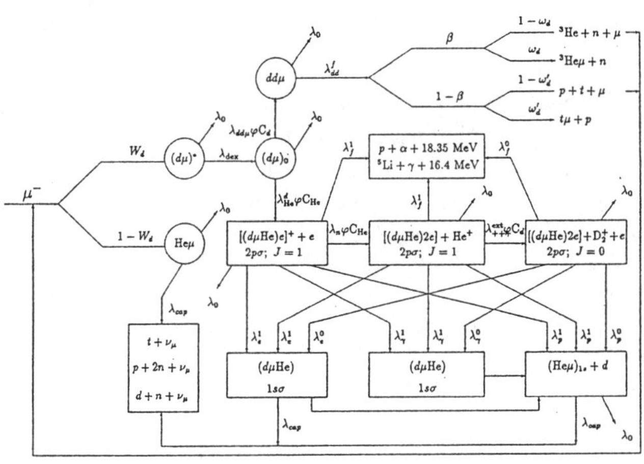

Figure 2. Coincidence timing

The setup was placed in muon channel μE4 of PSI. A beam of muons with momentum P_{μ} = 35 MeV/c passed through monitor counter 1 (plastic scintillator 4.2×4.2×0.1 cm^3), connected in anticoincidence with another scintillation detector 2 with a 3.5-cm diameter aperture centered around the beam axis, through the entrance window of the cryostat target and stops within the deuterium-helium mixture initiating reactions (1)–(2). Detectors 1 and 2 are not shown in Figure 1.

The diagram of the recording electronics trigger and the time diagram are displayed in Figures 3a, b. The trigger signal, which was composed from a pulse from the beam entrance scintillator in delayed logical coincidence with a gate enabled by any of the active detectors, was sent to the data acquisition system when a good event was detected.

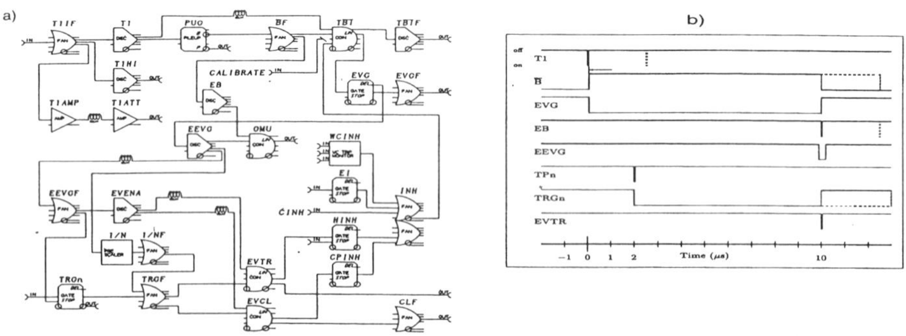

Figure 3. Top - the electronics diagram for the trigger circuit of dHe - experiment. The input to T1IF came from the beam entrance scintillator T1 and the inputs to the (generic) TRGn were supplied by the individual detectors. The critical outputs of EVGE, EVTR and EVCL were sent to the detectors electronics, data acquisition system, and CAMAC module clear inputs respectively; bottom: The timing setup for the electronics in the top. Time is indicated horizontally, and logic level, on/off, indicated by the lower/higher lines in keeping with NIM logic level standard.

The pile-up-gate(PUG) busy(B) output was set on for a 10 mks lock-out time. The barred output from the PUG(symbol B^),always set to the logical complement of the PUG out value, was used to indicate either that the system had accepted a muon, or was able to accept a muon. A general inhibit signal,I^,due either to an event trigger(HINH),the computer readout time(CINH and EI),or a failure in the high voltage supply of the photomultipliers (WCINH),was combined with T1 and B^ signals to designate the condition when the event gate, EVG, could open and good events could be accepted. The TBI^ signal was sent to the individual detector coincidences to permit acceptance of events, and to the CAMAC TDC units as a common start. If the second muon activated the entrance scintillator T1 (dashed event in T1 time line, Figure 2b),the PUG began timing the lock-out again, extending it. When any one of the n detectors recorded an event at a time when the EVG gate was true, such as pictured TPn in Figure 2b, a gate, TRGn, was opened. An event trigger would immediately initiate a hardware inhibit(HINH) to prevent loss of the data already measured by disabling the TBI^ coincidence. A computer-driven signal output system would maintain the inhibit while the data were read(CINH) and while the CAMAC modules were cleared(EI).An event clear would inhibit the trigger(CPINH) and the remove any values accidently recorded by CAMAC system.

One of the more important scalers was the GMU count, which was made by the coincidence of the end of busy, EB, and the end of event gate, EEVG, signals. These two signals in coincidence implied that there was no pileup and hence the muon was "good" in terms of the physics we wish to measure.

Information on the recorded event includes the time and amplitude of the signals from all above-mentioned charged and neutral particle detectors within 10 μs since the instant of operation of muon beam detector 1. As 14.7-MeV protons were recorded in the experiment, the following selection criteria for events recorded by Si detectors were used to appreciably suppress the background in the data analysis: arrival of a signal from at least one of six Si detectors within 10 μs since the instant of operation of the entrance muon counter; absence of another signal from the entrance muon detector within 10 μs; arrival of a signal from detector 1 in a certain amplitude range corresponding to the beam muons (because of electron and pion impurity in the muon beam);certain ranges corresponding to recording of protons and deuterons from reactions (1)–(2); arrival of an electron detector signal with an amplitude falling within a certain range at a time satisfying the criterion $t_e > t_{Si}$ (t_e, t_{Si} are the times of arrival of signals from the electron detectors and the Si detectors).

The time and amplitude distributions of the selected events (recorded by Si detectors) were compared by the χ^2 method with the theoretical Monte Carlo distributions to find the rates of processes (1)–(2). Note that selection of events and their further analysis are usually carried out with standard codes like PAW [21]. The data acquisition systems used in meson factory experiments will be discussed in more detail below.

By way of example, Figures 4 and 5 display how amplitude distributions of events recorded by the Ge_S detector and Si telescopes are transformed after application of the above criteria. As is evident, the effect-to-background ratio is appreciably improved, which in turn increases reliability of the data analysis results.

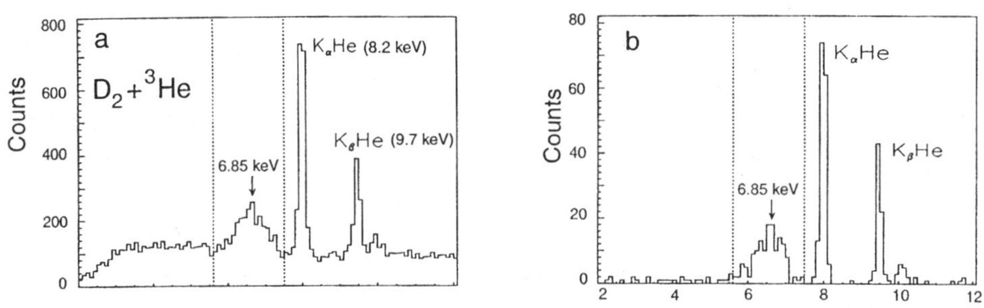

Figure 4. Energy spectrum of the registered events by Ge_S detector when the target is fill with $D_2 + {}^3He$ mixture: top, bottom - without and with detection of the delayed muon decay electrons, respectively.

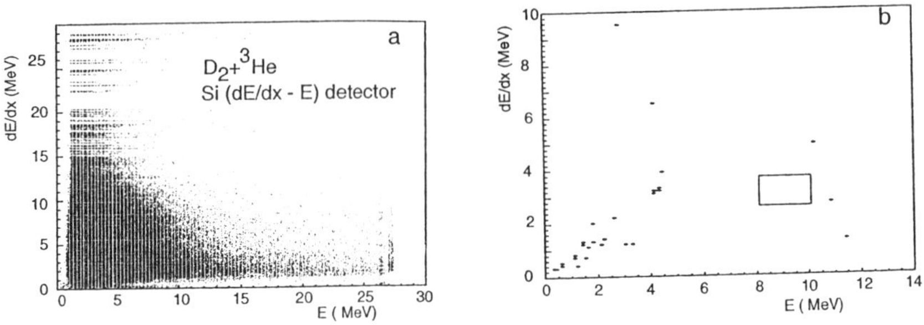

Figure 5. Energy spectrum of all three Si detectors for D_2 + ^3He mixture, without and with detection of the delayed muon decay electron, respectivelly. dE/dx is plotted versus E.

Events recorded by other detectors of the setup were processes in a similar way. The setup in question was used at the above-mentioned meson factories in a lot of experiments (250–300) that yielded important results.[3] Among the findings are the energy-resonant mechanism for formation of ddμ and dtμ molecules [8, 9], muonic catalysis of the dt fusion reaction [9], the hyperfine level structure effect in mu-molecules of hydrogen isotopes [22–23], the mechanism for molecular charge exchange of mu-atoms of hydrogen isotopes on He nuclei [24–25], The Ramsauer–Townsend effect[4] [26–27], the limit of the Bragg scattering of muonic hydrogen from the H_2 (D_2, T_2) lattice [26–27]. In addition, a mesocatalytic method of energy production is proposed [10]; the upper limit for the nuclear fusion rate in the d^3He molecule is found [16, 29]; the value of the astrophysical S-factor in the pd interaction is found with a high accuracy, S = 0.128 + 0.008 [30], which is very important for solution of some astrophysical problems [31]; nuclear fusion rates in the muonic molecules pdμ, ptμ, ttμ, ddμ, and dtμ are measured [4, 6], i.e., information on characteristics of strong interaction between light nuclei in the collision energy range beyond the reach of classical accelerators is gained[5] (this information is necessary for correct macroscopic description of the mechanism for interaction of few-nucleon systems within the framework of modern concepts [32–33]); rates for muon capture by the proton and the deuteron from the hyperfine structure states of pμ and dμ atoms are measured (these results are important for checking four-fermion interaction universality and determining weak interaction constants [34–36]); the Lamb Shift of the n = 2 level in the Heμ atom is measured [37]. Now measurements of the Lamb shift in the pμ atom are under way, which will allow the contribution to the level energy due to vacuum polarization to be found [38].

[3] Selection of recorded events for analysis followed practically the same principles in various experiments. Only the type and number of secondary radiation detectors varied in accordance with tasks set.

[4] The scattering cross section shows a deep minimum at a certain energy of collision of dμ(tμ)-atoms with the H_2 molecule. This phenomenon was first found in scattering of electrons from atoms of inert gases (Ar, Kr, Xe) [28].

[5] The astrophysical range of constituent nucleus collision energies is realized in muonic molecules.

This great variety of investigations of MA and MM processes allowed various off-line software packages to be developed on the basis of combined analysis of the energy and time distributions of muon decay electrons, products of nuclear fusion reactions in muonic molecules, mu-X-ray radiation, which in turn allows reliable precise information on characteristics of processes under study to be obtained.

3. Search for rare decays

3.1 INTRODUCTION

Study of rare processes is a priority trend in the research at meson factories. Among these processes are muonium-to-antimuonium conversion ($\mu^+e^- \to \mu^-e^+$); $\mu \to e$ conversion in the process $\mu^-A \to e^-A$; decays $\mu \to e + \gamma$, $\pi^+ \to e^+ + \nu_e$, $\mu \to 3e$, $\pi^+ \to \pi^0 + e + +\nu_e$, $\mu^- A \to e^+ A$. According to estimations, probabilities for these processes are as low as 10^{-13}–10^{-16} against the basic muon or pion decay modes. Interest in the study of these processes arises from the necessity to test modern particle physics models.

It is an aim of physics to understand the fundamental laws of matter and the forces which hold matter together. Experiments are done to investigate the existing models and to show their validity. Deviations from predictions of existing theories lead to change extensions of these or to completely new ideas about the laws of nature. The last three decades of elementary particle physics were marked by great successes of gauge theories in several domains: the Glashow-Weinberg-Salam theory [39] succeeded in unifying the electromagnetic and week interactions, and quantum chromodynamics gave a description of the strong force. This, so-called standard model, was subject to a large number of experiments of different kind, and the predictions of the standard model agreed with the measured quantities.

Below we describe some of the performed (planned) experiments on testing the Standard Model of particle physics and searching for new physics.

3.2 MUON-ELECTRON CONVERSION(SINDRUM II EXPERIMENT)

3.2.1 Introduction

Observation of muon to electron conversion would be the first evidence for a process violates muon and electron type lepton number and that cannot be explained by the Standard Model of particle physics, extended to include massive neutrinos. Tests of lepton-flavor conversation (LFC) are very sensitive probes of new physics beyond the Standard Model of particle physics [40, 41] . One of the most stringent tests is the search for the neutrinoless coherent muon electron conversion in muonic atoms (A=mass number, Z=proton number):

$$\mu^- + (A,Z) \to e^- + (A,Z). \qquad (3)$$

Only the coherent fraction of μ-e conversion is of interest because then the signature is a single electron with characteristic energy

$$E_e = m_\mu c^2 - B_\mu - R_{nucl}, \quad (4)$$

where m_μ is the muon mass, B_μ is the binding energy of the muonic atom in 1s state and R_{nucl} is the recoil energy of the nucleus. Therefore the energy of the emitted electrons from the target has to be measured in a search for $\mu^- \rightarrow e^-$ conversion.[6]

Below we describe the μ-e experiment with SINDRUM II performed at PSI [42–45].

3.2.2 *Experimental setup*

The setup of the spectrometer SINDRUM II is shown in Figure 6. The muon beam with an average intensity of $9 \cdot 10^6$ μ/s is analyzed in scintillation beam counter, degraded and stopped in a titanium target. The signals from the beam counter are used for the suppression of background from pions and electrons in the beam. A superconducting solenoid produces a magnetic field which is parallel to the beam axis in a cylindrical volume 180 cm length and 67.5 cm radius. This geometry allows to use the cylindrical detectors which facilitates the trigger electronics and the offline analyses. For a magnetic field 1.2 T a μ^- -e^- conversion electron, which has a momentum of 104.3 MeV/c, is contained within the radius of 67.5 cm. The time is measured with a cylindrical 64 element plastic scintillator hodoscope (1500mmx33.75mmx3mm) which has an inner radius of 34.6 cm and a thickness of 3mm.The scintillator hodoscope is also used for triggering.Two drift chambers measure the trajectories of the emitted electrons outside the hodoscope.The first drift chamber(DC1) makes precise measurements of two track segments per revolution of the trajectories.DC1 contains384 sense wires which measure r and ϕ coordinates. The outer wall of the chamber is covered with 192 cathode strips split in the middle (z=0) into 192 upstream and 192 downstream strips. With the induced signals of the strips three-dimensional space points of the trajectories are reconstructed. The second drift chamber (DC2) is used to connect the two track segments of a revolution of the helical trajectories of emitted particles measured by DC1 and to recognize cosmic ray background. The drift region of DC2 is separated from amplification region. The amplification region consists of 96 aluminium modules with 6 sense wires and two guard wires. Since a granularity of 96 channels is sufficient in DC2 all six sense wires of one module are electrically connected together and define one read-out cell. At both ends of spectrometer there are rings of 16 plexiglass cherenkov detectors (345 mm length with trapezoidal cross section which has baselines of 107.0 mm and 118.7 mm and a height of 30 mm) inside the scintillator hodoscope which are used for triggering. The time resolution of the hodoscope is 0.8 ns (FWHM) per track. The momentum resolution of spectrometer is 2.3 %(FWHM) at p= 100 MeV/c which is dominated by the energy losses in the target. The solid angle covered by spectrometer is $0.82 \cdot 4\pi$ Sr.

With a total acceptance of 14.5% icludings all inefficiencies, for detecting μ^- - e^- conversion a total of $7 \cdot 10^{12}$ captured muons is needed to reach a sensitivity of 10^{-12} per event.

[6] For really used Ti and Al targets the conversion electron energy is close to the muon rest mass m_μ = 106.5 MeV.

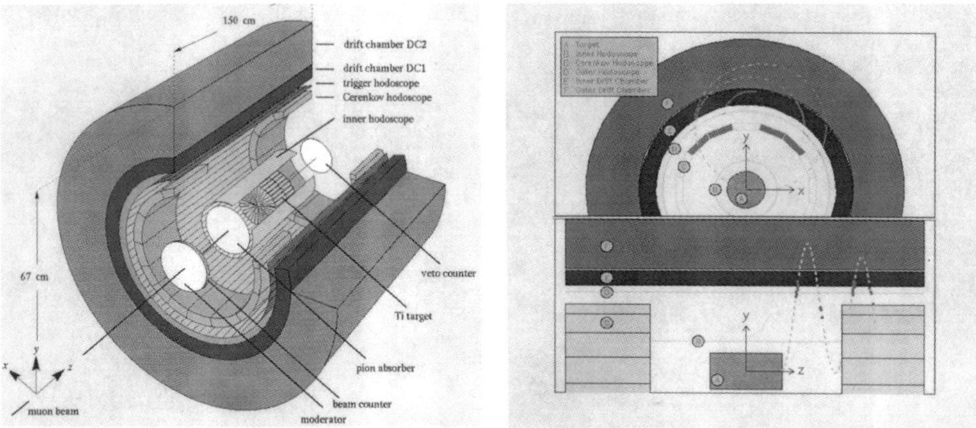

Figure 6. Setup of the spectrometer SINDRUM II for μ^--e^- conversion.

3.2.3 Electronics

Figure 7 shows a block-diagram of the read-out electronics and data acquisition computers of the SINDRUM II spectrometer. Signals coming from the scintillator hodoscope and cherenkov hodoscope are processed using CAMAC and read out by the STARBURST [46], whereas the signals from the drift chambers and the beam counter are read out by GPM [47] ("General Purpose Master") using mainly FASTBUS. The FASTBUS data are transfered to CAMAC via the CFI (CAMAC_FASTBUS) interface [48]. The CFI is also used as an online filter. The CFI consists on the FASTBUS side of a dedicated microprocessor that accepts commands from and sends data to a CAMAC I/O unit (CFICC) over a private bus. All data are collected by the STARBURST processor. The anode signals of the drift chambers DC1 and DC2 and the cathode signals of DC1 are read out with chain which consists of preamlifiers, postamlifiers, discriminators and TDCs. The 128 analogue signals of the scintillator hodoscope and 32 signals of the cherenkov counters are discriminated in the discriminator - meantimers by a "high-low" threshold method. The times of the 128 scintillator and 32 cherenkov counter discriminator output signals are measured by TDCs (LeCroy 2228A) with a resolution of 50 ns/channel. The start times for the TDCs and the ADC gate are provided by the first level hardware trigger. In the second level, programmable FASTBUS modules (the so-called "CPS") performed a check on the event topology. The second level hardware trigger makes uses of a programmable delay generator (PDG 4222). The delay generator is started by the first level hardware trigger and produces delays which are used as stop signals for the drift chamber TDSs and the beam counter waveform digitizer. The PDG produces the start signal for third level trigger which is the interrupt signal for the GPM. The third level trigger is a software filter. After a GPM start signal has arrived, the GPM reads out the FASTBUS components the frond end electronics. For accepted events, the "GPM YES" signal produces the LAM signal, which creates an interrupt in the STARBURST online computer. The analogue signals of the scintillator and cherenkov counters are integrated during a time gate of 50 ns with ADCs (LeCroy 2249A) which have a resolution of 0.25 pC/channel.

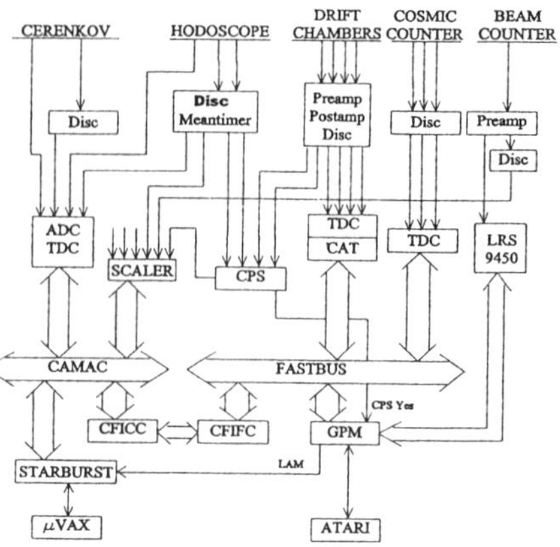

Figure 7. Block diagram of the read-out electronics and the data acquisition system in μ^--e^- experiment.

The trigger was sensitive to electrons with a momentum of about 100 MeV/c which reach DC2 and produce at least two track elements in DC1. The beam counter signal is combined by time gates with the first level hardware trigger signal, and is sent to the digital oscilloscope (LeCroy 9450) which serves as a waveform digitizer (ADC 8 Bit/channel, time base 2.5 ns/channel.

SINDRUM II is the first experiment that used the data acquisition system TANDEM which has been developed at PS1 [49]. The front end side of the SINDRUM II data acquisition splits into two branches which may be called "CAMAC" and "FASTBUS" branches according to the data bus involved. The back end computer MicroVAXII and writes it on disk or tape. An ATARI 1040 ST computer controls the high voltage (HV) power supply for the drift chambers. The data are read out via the GPIB port (IEEE-488) of the LeCroy 9450 and a specially developed interface to the parallel port of the GPM.

4. Data analysis Results

The result of a particle physics experiment is a sample of events, each of them characterized by a set of variables $\vec{x}$ (e.g. kinematical variables). The analysis has to determine the true distribution $f(\vec{x})$ of these variables and to compare these to the theory. A characteristic of such experiments is that mostly the distributions $f(\vec{x})$ cannot be measured directly. The distribution are modified by the detector acceptance and resolution. Generally one measures track parameters $\vec{y}$ and their distributions $g(\vec{y})$ instead of the original distributions $f(\vec{x})$. The experiment can still be compared with the theory by applying one of the two following procedures:

(a) One tries to unfold the measured distributions $g_{exp}(\vec{y})$ to derive the original distributions $f_{exp}(\vec{x})$. These can be directly compared with theory and with other experiments.

(b) In the second method, the theoretical distributions $f_{theo}(\vec{x})$ are folded with the detector response function $f_{theo}(\vec{x}) \rightarrow g_{theo}(\vec{y})$ and then the theoretical prediction for the distributions $g_{theo}(\vec{y})$ can be compared with measured distributions $g_{exp}(\vec{y})$.

The detector response function can only be derived with Monte Carlo methods. The complete experiment is simulated on a computer. The simulation is based on the CERN programme GEANT [50].

As a result of data analysis the upper limit for the branching ratio $B_{\mu e}$ (for μ^--e^- conversion with respect to the rate for the ordinary muon capture) at 90% confidence level is obtained:

$B_{\mu e} < 6.1 * 10^{-13}$ (90% C.L.).

This result is an independent confirmation of the previous best result ($4.6 \cdot 10^{-12}$) which was measured by experiment at TRIUMF [51].

There is a project of a μ^--e^- experiment with a specially formed muon beam to be carried out at BNL [52]. According to this project, the upper limit for $B_{\mu e}$ can be decreased by three orders of magnitude.

5. Pion Beta Experiment

5.1 INTRODUCTION

The PIBETA experiment will pursue a precise measurement of the rare pion beta decay ($\pi^+ \rightarrow \pi^0 e^+ \nu_e$) rate with an accuracy of about 0.5 %. Thanks to low theoretical uncertainties the latter can be used to constrain Cabbibo-Kobayashi-Maskawa (CKM) unitary and certain aspects of physics beyond the Standars Model [53-55]. The best experimental value at present has error limits of 4% (TRIUMF[56]) far exceeding the theoretical uncertainty of <0.15 %.

5.2 SET UP

The experiment (see Figure 8) will be performed at the πE1 beamline at PSI. This beamline turned to 116 MeV/c π^+, will give the necessary pion rate of up to $2 \cdot 10^6$ π^+/s with a low contamination of muons, positrons and protons. Due to the Michel positron background, the detector must be very efficient in the π^0 detection and have excellent background suppression. The heart of the detector is a 3π spherical calorimeter consisting of 240 pure CsI crystals (see Figure 7). The central part of the apparatus contains active beam degrader and pion stopping target counters, surrounded by a charged particle (CP) tracking system consisting of a 20-bar scintillator hodoscope and a pair of concentric cylindrical wire chambers (MWPC1 and MWPC2).

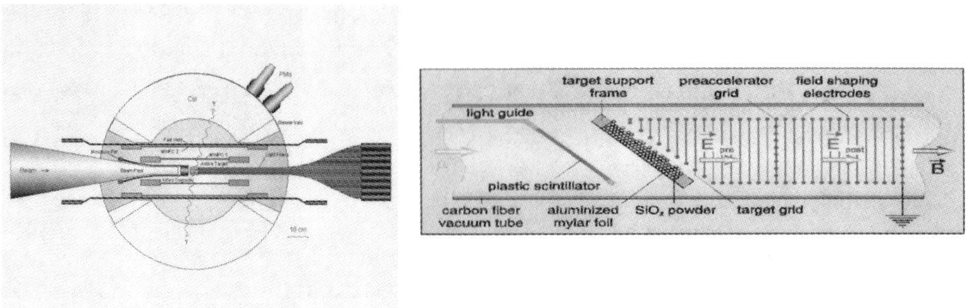

Figure 8. Layout of the Pion Beta Detector for investigation of the $\pi^+ \rightarrow \pi^0 e^+ \nu_e$ process.

5.2.1 Data acquisition System MIDAS

This section deals with the Maximum Integrated Data Acquisition System (MIDAS) [57] which will be used in the projected experiment and is being successfully used in experiments at the meson factories PSI and TRIUMF.

The MIDAS library contains routines for buffer management, a message system, a history system and an online database. MIDAS buffers are FIFOs which support multiple producers and multiple consumers. Consumers may request a certain subset of event types from a buffer. The transfer speed between a producer and a consumer is on the order of 10MB/sec for local access and 980kB/sec for remote access over standard ethernet. The history system is used to store data on a hard disk and produce value-over-time plots. The online database is a central data storage which contains all variables of an experiment like logging channel information, event definitions, slow control variables, front-end parameters and histogram definitions. It is kept entirelyin shared memory for fastest access of up to 50,000 read/writes per second locally and 500 read/writes remotely. Several applications have been written using the MIDAS library. A general purpose logger supports multiple logging channels to disks and tapes. Different logging channels can receive different types of events in different formats. Currently three data formats are supported: ASCII, an optimized MIDAS format and the YBOS format. A command-line based editor has been created which can be used to change values in the online database interactively and to perform run control. A graphical implementation of that program using the Java language is also available. This makes it possible to control an experiment from any computer running a Java capable browser.

The PiBeta Analyzer is a computer program which can analyze data produced by the PiBeta detector. The analyzer consists of two parts: a system part which is responsible for reading/writing events in various formats and a user part which actually does the experiment specific data analysis. In order to make data analysis more flexible, a multi-stage concept has been chosen for the analyzer. A raw event is passed through several stages in the analyzer, where each stage has a specific task. The stages read part of the event, analyze it and can add the results of the analysis back to the event. Therefore, each stage in the chain can read all results from previous stages. To reflect the multi-stage concept in the data structures, a bank system is used for event storage. The online system produces banks for different parts of the detector, like a ADC bank, a TDC bank, a scalar bank and so on. The first stages in the analyzer will use these banks to produce calibrated data like energy deposition in MeV in an "calibrated ADC" bank, track position in mm in a "MWPC" bank and so on. User's can at the end add private banks which contains variables they are interested in. Since the contents of banks is defined in a database, the system part of the analyzer knows how to interpret the contents of an event. This way, N-tuples can be booked automatically by the system. When running off-line, column-wise N-tuples are used. Each bank is booked as a "block" in an CWN-tuple. The id of the CWNT is the same as the event id.

Figure 9 shows the relationship between programs on different computers and the online Database/Buffer Manager. In the Pion-Beta Experiment two PCs are used as front-end computers connected to CAMAC, Fastbus and a LeCroy LRS1440 high voltage system. Using polling under MS-DOS an average trigger response time of 8us has been achieved. Several PC plug-in cards are used to read temperatures and gas flows. Two additional PCs are used as back-end computers running different analyzers under Windows NT. Data are written to two DLT tape units in parallel. The analyzer can read data from files, tapes or the online DAQ system. It can write events or specific banks inside events to output files. The analyzer supports currently three different file formats: MIDAS binary (.mid), pure ASCII files (.asc) and HBOOK rz files (.rz) with column-wise N-tupels and row-wise N-tuples. While the rz file format can only be used for the analyzer output, .mid and .asc files can also be read by the analyzer. Additionally, these files can be written and read directly in GNU-zipped format (.asc.gz and .mid.gz). The data is compressed and decompressed on-the-fly. While this method saves about 50% of disk space, it takes about 20% more CPU time to decompress .gz files.

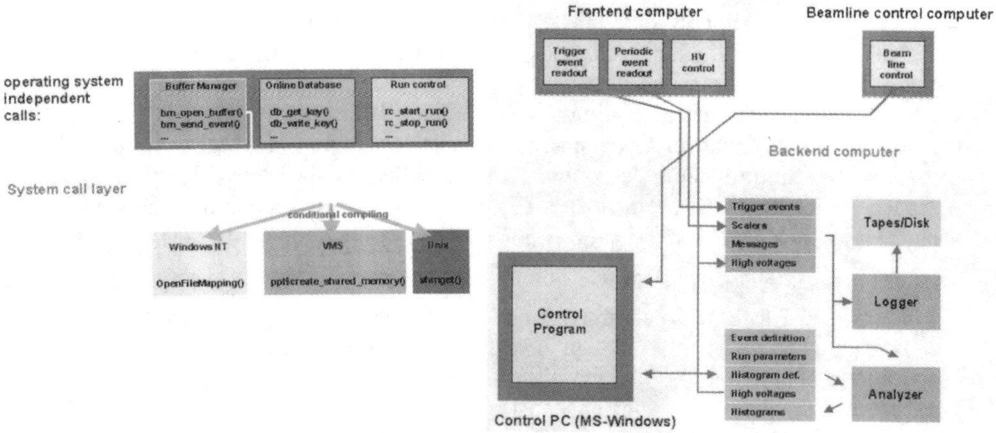

Figure 9. The relationship between programs on different computers and the Online DataBase/Buffer Manager (Maximum Integrated Data Acquisition System).

5.2.2 Results

First calibration and commissioning data with the whole PIBETA detector were taken during a six-week long summer 1998. The analysis of the commissioning run data as well as instrumental improvements and fixes are under way.

5.2.3 Muonium –Antimuonium Conversion Experiment

Introduction. The search for the spontaneous conversion of Muonium ($M = \mu^+e^-$) to Antimuonium ($\overline{M} = \mu^-e^+$) is a sensitive method for testing new physics. This reaction violate separate additive muon and electron number conservation. Consequently, it is not provided in the standard model of electroweak interactions as are all other lepton flavour violating processes. However, in many models developed to derive some of the parameters of the standard model from fundamental principles, lepton flavour violation appears to be natural [58]. In particular, some left-right symmetric models [59] and supersymmetric models with broken R-parity [60] allow muonium-antimuonium conversion with a strength G_{MM} of the order of 10^{-2} of the weak interaction strength G_F. At present, the best upper limits established in a series of experimental approaches using different techniques published for the coupling G_{MM} are 16 % and 13 % of G_F, respectively. The former experiment has been performed at LAMPF [61] the latter at Dubna [62].

The experiment at the Paul Scherrer Institute (PSI) has aimed for a sensitivity to yet undiscovered interactions with an effective coupling as small as a few permille of the Fermi coupling constant G_F, which corresponds to a sensitivity improvement by more than a factor of 1000 in comparison with the LAMPF and the Dubna experiments.

The Muonium-Antimuonium-Conversation-Spectrometer MACS. The schematic view of the MACS is presented on Figure 10. Muonium atoms in the ground state are formed by electron capture after stopping a subsurface beam of positive muons (momentum less than 26 MeV/c) in a silicon dioxide powder target with $\cong$ 60% efficiency. Some of these atoms (5%) emerge from the target with thermal kinetic energies into the surrounding vacuum where an eventual transition to antimuonium is not suppressed as it is in matter (mainly due to the removal of degeneracy between muonium and antimuonium). Both constituents of antimuonium (muonium) are identified as the atoms decay. MACS is designed to identify both constituents of the antimuonium (muonium) atom after its decay. The disintegration of the atom is due to muon decay with the dominant channel $\mu^- \rightarrow e^- + \nu_\mu + \tilde{\nu}_e$. The particle in the atomic shell has negligible ($<10^{-6}$) influence on the decay process. Electrons (positrons) from muon decay have a characteristic energy spectrum ranging up to 53 MeV. In this experiment they are detected in a magnetic spectrometer operated at 1kG which allows to reconstruct the particle's momentum and sign of charge. The spectrometer consists of 5 cylindrical concentric Multi Wire Proportional Chambers (MWPC) surrounded by a plastic scintillator hodoscope located inside a solenoid magnet, which produces a homogeneous (1%) magnetic field of 0.1 T. Five chambers are equipped with helical cathode strips to allow a measurement of axial coordinates. From the measured hit coordinates the path of the particle can be reconstructed yielding its momentum and charge sign. The hodoscope consists of 64 plastic scintillator strips, each read out by photomultipliers on both ends. It is used in the

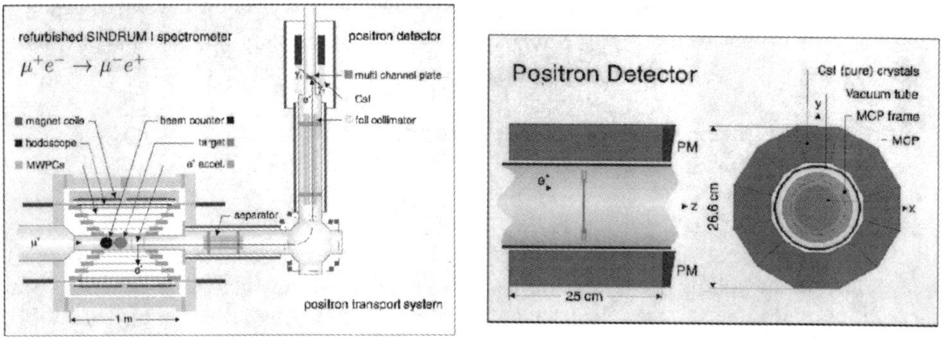

Figure 10. The Muonium - Antimuonium Conversation Spectrometer (schematic view).

trigger and as time marker for the decay process as the start of the Time of Flight measurement for the atomic shell particle. The particle from the atomic shell has a mean energy corresponding to a Rydberg energy of 13.5 eV (spectrum). It is electrostatically accelerated (≈ 8 keV) and guided by a magnetic transport system to a position sensitive "positron" detector which is able to detect electrons as well. Impact time and position are recorded with a MicroChannel Plate (MCP). For the identification of positrons a segmented crystal detector surrounds the MCP to detect their annihilation radiation. The MCP assembly consists of two plates of 89 mm diameter (sensitive area 75 mm diameter) each and with resistive anode readout. The efficiency of commercial plates of 16% for 8 keV electrons and positrons could be enhanced to 64% using a MgO-coated carbon foil emitting secondary electrons in front of it. The barrel shaped crystal spectrometer surrounding the MCP is made of pure CsI and consists of 12 segments read out with magnetic field insensitive photomultipliers. Positron annihilation photons of 511 keV are detected by two segments in opposed hemispheres of the detector. The acceptance for at least one of the annihilation photons was measured to be 79(4) % using a Na-22 radioactive source.

This experiment was carried out with the same data acquisition system MIDAS as the experiment PIBETA [57].

5.2.4 Results

An upper limit on the conversion probability in an external magnetic field of 0.1 T was obtained [63]

$$P_{MM}(0.1\ \mathrm{T}) < 8.3\ 10^{-11} \qquad (90\%\ \mathrm{C.L.})$$

representing an improvement of the LAMPF result by a factor of 2740. The new bound has been rurned into an upper limit for the coupling constant G_{MM} of a hypothetical M-$\overline{\mathrm{M}}$ interaction:

$$G_{M\overline{M}} < 3{\cdot}10^{-3}\, G_F \qquad (90\%\ \mathrm{C.L.}).$$

The new upper limit increases the domains of validity of the standard model by postponing the scale of new physics to higher energies. In the future, significant improvements in sensitivity are conceivable, of enhanced muonium production (in vacuo) becomes feasible and/or high intensity pulsed muon beams are available.

6. Precision Measurement of the Michel Parameters of Muon Decay

6.1 INTRODUCTION

The standard model of subatomic particles and interactions purports to provide an exact description of the muon decay spectrum. This is the same model which describes the interactions and indeed the existence of the quarks, and which has been incredibly successful in describing a wide variety of subatomic phenomena. In spite of this success, it is widely believed that the theory is only approximately correct. Muon decay is a particularly interesting phenomena, In particular, the quarks are not involved in this decay. Because of this, the "strong" interaction does not play a role, which makes possible an exact quantum treatment of the decay. Furthermore, the decay is thought to respect certain symmetries. In particular, the neutrinos which are emitted are expected to have an angular momentum which is always directed opposite to the linear momentum. This "feature" of the standard model has been incorporated by fiat. There is no a priori reason to expect that neutrinos violating this symmetry principle cannot exist. If they do exist, it would be taken as evidence of an additional component of the Weak Interaction, mediated by a new particle which would be referred to as a right-handed W gauge boson. It is likely that a careful examination of the muon decay spectrum would be the best test for the existence of such "right-handed" interactions. It is remarkable that to date, a precise measurement of the entire decay distribution has not been undertaken.

In TRIUMF proposal E614, "A Precision Measurement of Muon Decay Parameters " [64] it is proposed to make a precise measurement of the muon decay distribution for the purpose of determining the electroweak coupling constants which determine the underlying symmetries. Deviations from the expected decay distribution at the level of 1 part in 10,000 could reveal flaws in the standard model - such as yet undetected "right-handed" interactions.

The scientific impact of such a discovery would be enormous, competing in impact with the biggest of the collider experiments.

6.2 EXPERIMENT

The drawing of the experimental setup is shown in Figure11. The measurements will employ a large bore superconducting solenoid magnet. The ~100% polarized 4.1 MeV surface positively charged muons produced at TRIUMF will be directed into the 2T magnetic field where they will be stopped at the centre of a precision detector package. After an individual muon is stopped in the target, (aluminium foil) the tracking system is watched for several micro-seconds. If a muon decay event is observed, the positrons from the decay spiral outward in the magnetic field. By observing the angle through which the positron is.emitted and measuring the radius of curvature of the track in the magnetic field using precision wire chambers, (WC Detector Package) the angular and momentum distributions of decay events can be reconstructed.The complete array of WC Detector Package consists of a 26 planar drift chambers and 6 proportional chambers. There are two thin (~85 mg/cm) silicon counters(SiLi):the upstream one to detect the incoming muons and the downstream one to preserve symmetry. By carefully controlling systematic uncertainties in the experiment, and by collecting data from a sample of 10^9 decay events, the decay distributions can be studied at an accuracy of less than one part in 10,000. This accuracy heightens our sensitivity to various effects, including the possible existence of right-handed interactions.

MIDAS [57] is used for data collection in this experiment. At present time the experiment in progress.

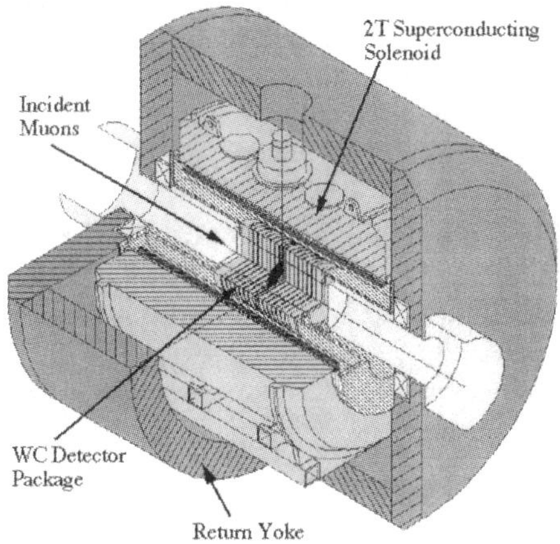

Figure 11. The Experimental Setup for measurement of the Michel Parameter at TRIUMF.

7. Proposal

A Precision Measurement of the Positive Muon Lifetime Using a Pulsed Muon Beam and μLan Detector.

7.1 INTRODUCTION

It is proposed [65] to measure the positive muon lifetime to a precision of 1 part in 10^6 giving an absolute uncertainly on the level of 2 ps. This represents more than an order of magnitude increase in precision beyond the current world average. The muon lifetime is used to determine the Fermi coupling constant, G_F, which is the fundamental quantity governing the strength of any electroweak process. Recent theoretical work [66] enables a clean extraction *of* G_F, from τ_μ and paves the way for a new high-precision measurement. Such an undertaking is challenging and in keeping with a recent trend to improve the precision on the knowledge of the fundamental parameters of the Standard Model to the extent that modem technology permits. A 1ppm determination of τ_μ requires 10^{12} muon decays. An intense muon source and a detector system capable of handling multiple events at once are required in order to acquire the data in a relatively short running period (e.g., one month). This experiment, having been designed from the start to utilize a pulsed beam, is ideally suited for a chopped surface muon beam at PSI. The experiment is designed to address specifically the known critical sources of systematic error. A depolarizing sulfur target is used to stop the μ^+ and to reduce the residual polarization of the ensemble to a few percent. Individual muon spins are dephased during the accumulation period by the inclusion of a 75 G transverse magnetic field. Decay positrons are registered in the μLan (Muon Lifetime Analysis) Detector (see Figure12) which consists of 180 independent triangular timing tiles distributed uniformly within the 20 Super Triangles of an icosahedral geometry centered on the target. Each tile consists of an inner and an outer scintillator coupled to independent photomultiplier tubes (PMT). The inner and outer scintillators have a thickness 6.35 mm and 3.175 mm, respectively. A gap of approximately 7 mm is maintained between each tile so that no positron from any edge of the target can cross from the inner scintillator of one tile into the outer scintillator of an adjacent tile. The light from the outer scintillator is transported to a PMT through a solid, UV-Absorbing (UVA) tapered lightguide whose length is sufficient to stop the most energetic decay positrons. A system of 500 MHz waveform digitizers are used to read out each PMT. This enables both timing and energy deposition to be recorded for each event. The high segmentation and the double-pulse rejection capabilities reduce the effect of pileup on the measured lifetime to a level below 1ppm. The geometry features 90 point-like symmetric tile pairs; the sum of any pair is used in the lifetime analysis. The difference in rate versus time in any tile pair reveals sources of asymmetry.

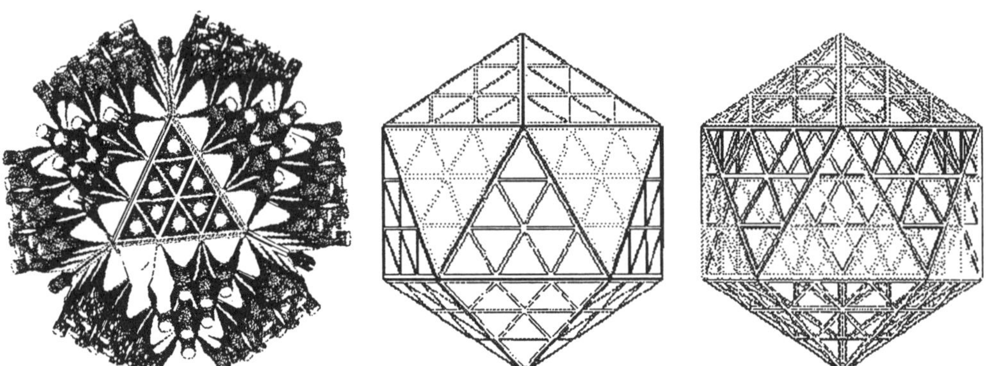

Figure 12. The Muon Lifetime Analysis Detector.

7.2 ELECTRONICS

The task of the front end electronics is to provide timing and pulse height information for signals produced by decay positrons in the two layers of scintillating tiles. All data will take using waveform digitizers connected to every channel. In addition to the basic experimental measurement, both pulse height and timing information will be used to suppress and correct for pulse pileup. The data acquisition system must be able to read out 360 channels of phototube signals, one signal from each of the two layers of 180 tiles. The signals from front and back sections of each tile will sit on the same analog line, with the back signal delayed by 40 ns. To keep the beamtime required to a minimum, front end acquisition and readout to tape will proceed in parallel. Block diagram of the waveform digitizer (WFD) with First in First Out (FIFO) memory for continuous real-time data acquisition is shown in Figure13. An analog waveform from the PMT is sent to an 8-bit 500 MHz flash ADC. Every six bytes of flash ADC information (12 ns) are accompanied by a 16 bit time word. The six ADC samples and two bytes of time information are assembled into one 64-bit word by an on-board ASIC. The time word will be latched from a counter on the ASIC, which will be driven, in turn, by the same clock which strobes the flash ADC. The 64 bits of time and ADC sample information will be written to the FIFO (First in First out) memory at 500/6 MHz. Various control signals will be required to operate the μLan WFD (waveform digitizers).The ASIC will write data to memory only when enabled by a front panel start signal and will cease to write data when it receives a front panel STOP. A VETO input will disable the writing of data without stopping the clock. A RESET input will zero the clock and memory counter. A Control/Status register will be used to control and monitor the board. The control register

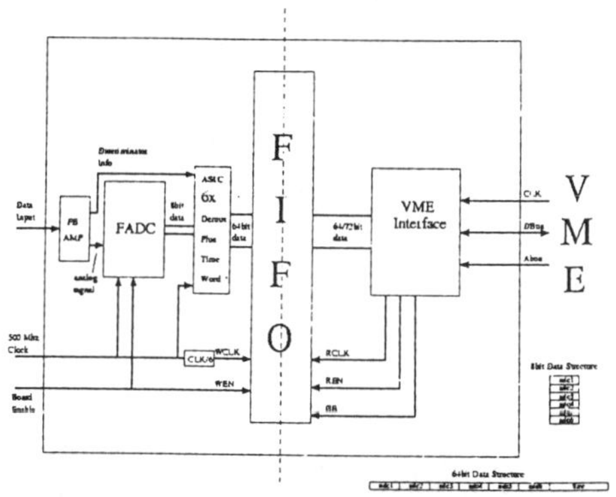

Figure 13. Block diagram of the waveform digitizer (WFD) with First in First Out (FIFO) memory.

will include software emulation of front panel hardware such as START and STOP, so that the board can be operated by a computer alone, as well as bits to set the threshold for the front-end comparator, or to activate or disable zero suppression. The status register will describe the state of the board, including error conditions, the number of data acquisition cycles since RESET, and most importantly, the location of the data pointer in memory. It is planned that. the 50 reguired WEDs will group into five VME crates Each crate will also contain a fast single-board computer running the vxWorks operating system. Approximately 10 times per second, the embedded computer will cycle through all the WFDs in the crate, draining the data from their FIFO buffers. At that point, it will process the data to determine the parameters that describe each pulse. In particular, it will compute the time, area, height, and width of each pulse. The samples for any events which seem to be anomalous in any way will be retained. The remaining data will be forwarded to an event-builder computer. The total data rate at this stage is less than 10 MBytes per second. It will be feasible to transfer this amount of data over an ordinary 100 MHz Fast Ethernet network to a industry-standard computer running the Linux operating system. This machine would collect the data from the various crates and write it to tape. In fact, it should be possible for it to fill a set of histograms for the pulses which could be fit in a straightforward manner. These histograms would be flushed to tape every minute instead of the entries from which they were made.

It is planned to build and stage the μLan Experiment over a three year period as outlined in this proposal.

8. Measurement of the transverse polarization of positrons from the decay of polarized muons

8.1 INTRODUCTION

Measurements of muon decay are low energy tests of standard model. Only a few years ago it has been shown that V-A, as one of the basic assumptions of the standard model, fellows from the results of a selected set of muon decay experiments (including unverse muon decay). The experimental limits obtained up to now, however, still allow for substantial contributions from non-standard couplings. To get better constraints especially for scalar couplings it will improved present limits [67] on the transverse positron polarization by one order of magnitude. These measurements will yield a more precise of the low energy parameter Mishel η which is needed for a better determination of the Fermi coupling constant. It will also give limits for the time reversal noninvariant component P_T.

8.2 EXPERIMENT

The experimental setup [68] is shown in Figure14. The experiment will be performed at the PSI accelerator with a high intensity muon beam. These muons are highly polarized and arrive individual banches every 20 ns at graphite stopping target in the center of a pair of Helmholtz coils. The magnetic field strength chosen so the precession frequency of the stopped muon spins equals the accelerator frequency which is responsible for the time structure mentioned above. This stroboscopic method allows to stop many muons simultaneously. In our experiment 200 muons will be in the target at any moment corresponding to a stopping rate of 100/s. Positrons from the decay of these muons and emerging parallel to the field are then detected by a plastic scintillator and a drift chamber array. They then try interaction with polarized electrons in a magnetized iron foil transverse to their line of flight. Annihilation in flight of the decay positrons with the polarized electrons is sensitive to any transverse positron polarization. In that case the resulting two-gamma distribution is not azimuthally isotropic any more. The anisotropy of the gamma quanta will then be measured by an array of 127 BGO counters. The hexagonal drift chamber are able to track the positrons at rates of $3 \cdot 10^6$ s^{-1}.It is proposed to use in this experiment the Date Acquisition System based on the TANDEM [49]. At present the experiment is going on.

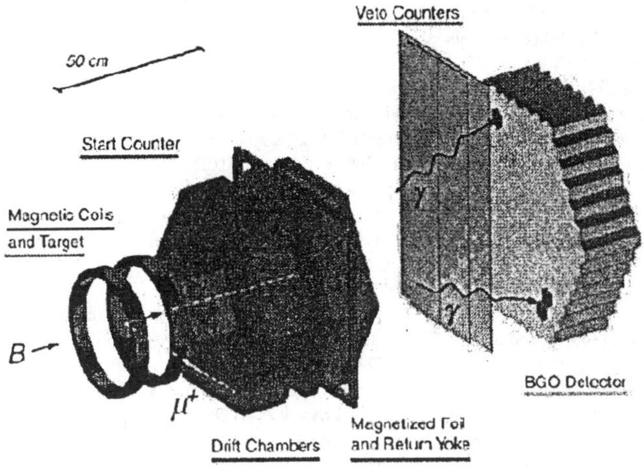

Figure 14. The experimental setup for measurement of the decoy of polarized muons.

9. Conclusion

The most information-rich experiments that are under way or have been performed at the meson factories PSI, TRIUMF, LAMPF, KEK are described above. Structure investigations of various materials with the use of muon spin spectroscopy, investigations of hadron-hadron and hadron-nucleus interactions, and research in physics of exotic systems arc also widely carried out at meson factories. A lot of detectors were used (are used) in all these experiments. The information from the detectors is processed in accordance with certain selection criteria dictated by the structural ideology of the recording electronic system and by physics of the process under investigation. Improvement of the main parameters of the many-detector systems, such as energy, space, and time resolution of the detectors, multifunctionality and operation speed of data acquisition systems, will allow (due to development of new methods for recording elementary particles and nuclear radiation and upgrading of the data reception and analysis system) an appreciable increase in reliability of the final result that is a goal of particular investigations.

10. References

1. Frank, F.C. (1947) Nature 160, 525.
2. A.D.Sakharov, A.D. (1948) Report FIAN, M.
3. Alvarez, L.W., et al., (1957) Phys. Rev.105, 1127.
4. Ponomarev, L.I. (1990) Contemp. Phys. 31 (1990) 219, and references cited therein.
5. Bystritsky, V.M. and Yadernaya Fizika (1995) 58, 808 (Physics of Atomic Nuclei 58 (1995) 746), and references cited therein.
6. Petitjean, C. in: Proc. 13-th Intern. Conf. On Few Body Problems in Physics, Nucl. Phys. A543 (1992), and references cited therein; G.M.Marshall et al. Hyperfine Interaction 82 (1993) 529; Hyperfine Interaction 118 (1999) 89 and references therein.
7. Bystritsky, V.M. and Yadernaya Fizika (1995) 58 688 (Physics of Atomic Nuclei 58 (1995) 631) and references cited therein.
8. Bystritsky, V.M. et al., (1979) Zh. Eksp. Teor. Fiz. 76 460 (Sov. Phys. JETP 49 (1979) 232).

9. Bystritsky, V.M. (1980) et al., Phys. Lett. 94B 476; Pisma Zh. Eksp. Teor. Fiz. 31 (1980) 249; Zh. Eksp. Teor. Fiz.80 (1981) 1700 (Sov. Phys. JETP 53 (1981) 877).
10. V. Petrov, Y.V., (1979) Proc. XIV LNPI Winter School, Leningrad, p.139 (in Russian); Muon Catalyzed Fusion 1 (1987) 351; Muon Catalyzed Fusion 3 (1988) 525 , and references therein.
11. Abramov, d.i., et al., (1999) Hyperfine Interactions 119 127, and references therein.
12. Kino, Y., et al., (1996) Hyperfine Ineractions 101/102 325.
13. Abrashkevich, A.G., (1996) et al., Hyperfine Interactions 101/102 381, and references cited therein.
14. Leon, M., (1993) Hyperfine Interactions 82 151 and references therein; J.S. Cohen, Hyp. Inter. 82 (1993) 15 and references therein.
15. Mulhauser, F., (1998) et al., PSI Proposal R-98-02; PSI Annual Scientific Report, 1998, p.21.
16. Del Rosso, A., (1999) et al., Hyperfine Interactions 118 177.
17. Boreiko, V.E., (1998) et al., Nucl. Instr. and Methods A416 221.
18. Bystritsky, V.M., (1999) and Pen`kov, Yadernaya Fizika 62 316(Phys. of Atomic Nuclei 62 (1999) 281).
19. Bystritsky, V.M., (1999) et al., Nucl. Instr. and Methods A432 188.
20. Kuhn, S.E., (1994) et al., Phys. Rev. C 50 1771.
21. Programme PAW, CERN Program Library Long WriteUp 121, Geneva, 1995.
22. Kammel, P., et al., (1982) Phys. Lett. 112B 319; Phys. Rev. A28 (1983) 2611.
23. Breunlich, W.H., (1987) et al., Muon Catalyzed Fusion 1 67, and references therein.
24. Aristov, Y.A., et al., (1981) Yad. Fiz. 33 1066 (Sov. J. Nucl. Phys. 33 (1981) 564).
25. Bystritsky, V.M., (1983) et al., Zh. Eksp. Teor. Fiz 84 1257.
26. Mulhauser, F., et al., (1999) Hyperfine Interactions 119 35.
27. Wozniak, J., et al., (1999) Hyperfine Interactions 119 63.
28. Adamczak, A., et al., (1996) At. Data Nucl. Tables 62 255.
29. Maev, E.M., et al., (1999) Hyperfine Interactions 118 171.
30. Olin, A., et al., (1999) Hyperfine Interaction 118 163.
31. Belyaev, V.B., et al., (1995) Nucleonika 40 85, and references therein.
32. Kharchenko, V.E., et al., Yad.Fiz. 55 (1992) 86.
33. Friar, J.L., et al., (1990) Phys.Lett. B251 11.
34. Bardin, G., et al., Phys.Lett. 104B (1981) 320; Nucl.Phys. A352 (1981) 365.
35. Bystritsky, V.M., et al. Zh.Eks.Teor.Fiz, 66 (1974) 43 (Sov.Phys. JETP 40 (1974) 811).
36. Bertin, A., et al., (1973) Phys.Rev. 8D 3774.
37. A.Bertin, A., et al., (1974) Nuovo Cim. 23B 489.
38. Taggu, D., et al., (1999) Hyperfine Inter. 119 311; PSI Proposal R-98-03. 1 (1998).
39. Glashow, S.L., (1961) Nucl.Phys. 22 579; A.Salam in Elementary Particle Theory: Relativistic Groups and Analiticity Proc. Nobel Symposium, ed. By N.Svartholm (Almquist and Wiksell, Stockholm), nr.8 (1968) 367; S.Weinberg, Phys.Rev.Lett. 19 (1967) 1264; 27 (1971) 1688.
40. Depommier, P., and Leroy, C., (1995) Rep.Prog.Phys. 58 61.
41. Barbieri, R., et al., (1995) Nucl. Phys. B445 219.
42. SINDRUM II Collaboration, PSI Proposal R-87-03, 1987.
43. Muheim, F., (1991) Ph.D.Thesis, Zurich University.
44. Kaulard, J., (1997) Ph.D.Thesis, RWTH Aachen.
45. SINDRUM II Collaboration, C.Dohnmen et al., Phys.Lett. B317 (1993) 631.
46. CES ACC2180 “STARBURST” processor, Creative Electronic Systems SA, 70 Rte. du Pont-Butin, CH-1213 Petit – Lancy (1987).
47. The GPM general purpose Fastbus Master/ Slave, User Manual, CERN doc.nr.EP85-02 (1985).
48. CFI service manuals, Dr.B.Struck Company, Hauptstr. 95, D-2000 Tangstadt (1984).
49. PSI computing division, TANDEM (set of documents).
50. Brun, R., et al., (1987) GEANT 3 User`s Guide, CERN doc.nr.DD/EE/84-1.
51. Ahmad, S., et al., (1988) Phys.Rev. D38 2102.
52. Molzon, W., et al., (1999) BNL Proposal Experiment MECO.
53. Pocanic, D., et al., (1991) Proposal PSI R89-01.1.
54. Frlez, E., et al., (1997) PSI Annual Report, p.12.
55. Frlez, F., et al ., (1998) PSI. Scientific Report v.I, p.6.
56. McFarlane, W.K., et al., (1985) Phys.Rev. D32 547.
57. Ritt, S., and Amaudruz, P.A., (1955) MIDAS, Program Library of PSI and TRIUMF.
58. Chang, D., and Keung, W.K., (1989) Phys.Rev.Lett. 62 2583.
59. Herczeg, P., and Mohapatra, R.N., (1992) Phys.Rev.Lett. 69 (1992) 2475.

60. Halprin, A., and Masiero, A., (1993) Phys.Rev. D 48 R2987.
61. Mattias, B.E., (1991) Phys.Rev.Lett. 66 2716.
62. Gordeev, V.A., et al., (1954) JETP Lett. 59 589.
63. Willmann, L., et al., (1999) Phys.Rev.Lett. 82 49; PSI Scientific Report v.1, 1998, p.7.
64. Gill, D.R., Selivanov, V., et al., (2000) TRIUMF Proposal E 614.
65. Carey, R.M., et al., (2000) PSI Proposal.
66. Ritbergen, T.V. and Stuart, R.G., (1999) Phys.Rev.Lett. 82 488; Phys.Rev. D 437 (1998) 201.
67. Burkand, H., et al., (1985) Phys.Lett., B160 343.
68. Bodek, K., et al., (1997) PSI Proposal R94-10.1, Collaboration ETH Zurich-PSI-CRACOW-KATOWICE; PSI, Annual Report 1997, Annex I, p14; PSI, Scientific Report 1998, p.8.

DEVELOPMENT OF REMOTE SENSING TECHNOLOGY FOR DETECTION OF NUCLEAR EXPLOSIVES

V. CHORNY
BAE Systems
Kharkiv National University
4 Svoboda Sq., 61077 Kharkiv, Ukraine

1. Introduction

To support the signed International Treaties for the nuclear weapon reduction demands development of reliable universal and trustworthy means of control over their location and quantity. The same means are required to solve the problem of control over non-proliferation of nuclear weapons, including illicit transportation of Uranium and Plutonium isotopes, used in nuclear warheads. The existing detecting systems provide the reliable information mainly on the static-based nuclear facilities.

It is traditionally recognized that control over nuclear weapons that have been loaded on space objects or on board of aircraft performing their duty operations or training flights, is very difficult to be realized. These difficulties may be accounted for the fact that the existing methods of control cannot ensure the *on-line* and *long-range* operation mode. The review and analytical treatment of these methods which can be found in [1] is likely to be one of the first concerning this problem. Thus, in [2, 3], for example, mainly passive methods of control have been reported, which can be accounted for some difficulties to construct compact sources of probe particles (neutrons or photons) with the required energy and/or intensity, and difficulty to eliminate the source effect on the recording equipment.

2. Remote Detection Method

The active method of remote non-destructive detection of weapon fissile materials (WFM) presence on the chosen object, being under development, is based on registration of the induced radiation. The method combines use of a pulsed source of probe neutrons at a sub-MeV energy level and a system of detectors for registration of characteristic radiation of the object response. Schematically such a "detecting complex" is shown in Figure 1.

This method belongs to those that are widely used in laboratory and industrial conditions and are given a common name "method of neutron probing" [4]. In this method the source of probing and the controllable object are emitting neutrons and γ-quantums, and the task can be solved only on condition of independent registration of primary radiation (monitoring) and secondary radiation (detecting). The discriminating registration of secondary radiation, induced by the primary radiation is achieved due to their energy discrimination and serial kinetics.

A. K. Hyder et al. (eds.), Multisensor Fusion, 839–849.

Sensitivity of the method is determined directly by the probing source power and the efficiency of the induced radiation detectors, whose properties should be used fully to create effective, safe and compact detecting complexes. Usage of moderate-energy neutrons was of paramount importance what was made for the reasons described below.

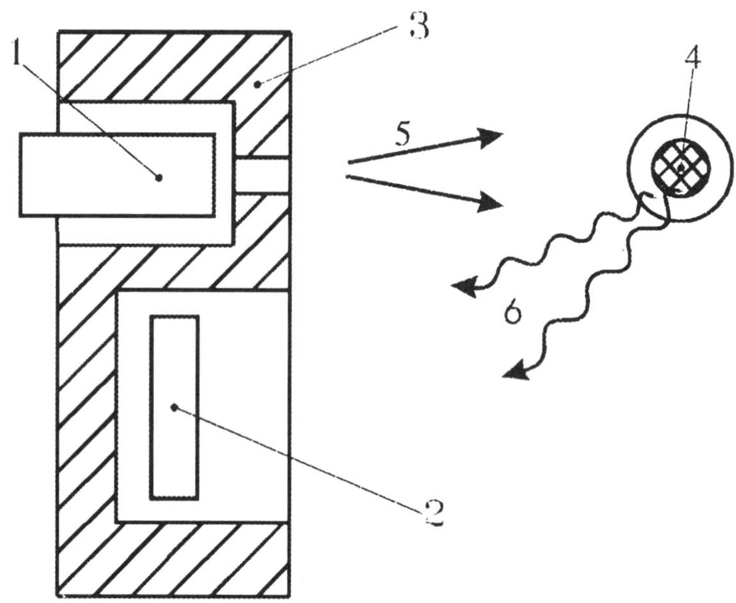

Figure 1. Scheme of active method of remote detection of nuclear warheads
1 – neutron generator; *2* – detector; *3* – collimating and detecting system;
4 – the model; *5* – probe neutrons; *6* – induced radiation.

3. Physical Concepts

As it is generally known, the process of neutron interaction with a matter occurring within a compound nucleus, results in excitation of the final nucleus, which is de-excited by emission of photon train. Such reactions include non-elastic scattering (n,n'), γ-capture (n, γ) and neutron multiplication (n,N_n). The spectrum of such radiation is consistent with the structure of the exited nuclear levels of a chemical element participating in the interaction. Besides, the fission reaction (n,f) is accompanied with intensive production of photons; the fission photon spectrum is also well known. All the specified reactions are applicable for fast neutrons, what means, that irradiation of the real object by neutrons with such energy will result in generation of γ-response of a complex spectral composition. Besides, the weight of "ballast" of the nuclear weapon carrier is supposed to exceed sufficiently the weight of the fissile material (from the standpoint of the task of detection). As cross-sections of the reactions (n,n') and (n, f) for fast neutrons differ insignificantly, the γ-response amplitude during probing by fast neutrons is expected to depend insignificantly on the amount of the fissile material. Thus, with using fast neutrons the task of detection may be solved just by analyzing the γ-response spectrum. The complexities of such technique are described in detail in [1].

Quite different situation develops with decreasing of the probe neutrons energy to the moderate energy band. Irradiation of a real object will be accompanied by generation of γ -response, the value of which to a considerable extent depends on the presence of fissile materials on the object. That happens on the following reasons: in the given power range the reaction of non-elastic scattering on ballast is impossible, as the first nuclear levels of excitation for the most of chemical elements that are included in high technology materials exceed 0.5 MeV [5]. Besides, at such energy level the reaction of neutron multiplication, which is also accompanied by photons emission, is impossible too. That means, that the γ-response of an object without fissile materials is determined by (n,γ) reaction whose cross-section for constructional materials in moderate energy range of the probe neutrons does not exceed tens - hundreds mbarn. The presence of ^{235}U-like matter on the diagnosed object will result in production of γ-response composed of photons of (n,γ) reactions on ^{235}U and on the ballast, (n,f) reactions on ^{235}U, N_γ – 8 - 9 per act of interaction, and (n,n') on ^{235}U and on the ballast, N_γ – up to 5 per act of interaction, as the energy of ^{235}U fission neutrons is mainly more than E′. Thus, the use of moderate-energy neutrons allows to effectively discriminate between the two considered cases as to the γ-response amplitude.

In a number of early works usage of a technique of probing by neutrons at a sub-MeV energy level for detection of fissile materials has been reported. The block diagram of a device for detection of ^{235}U composition using sub-MeV-energy neutrons has been offered in [6]. To produce neutrons with energy below that of the ^{238}U fission threshold, a low-energy electron linear accelerator, $(\bar{e},\gamma)$ converter and $(D(\gamma,n))$ converter were used. Boron counters wrapped in polyethylene moderator served as detectors of fission neutrons. Using prompt neutrons the authors of the work have obtained a discriminative relation *50* between ^{235}U and ^{238}U.

The experiment, similar to those described in [6] and planned to be performed, as to its parameters and concepts, is described in [4]. To irradiate the fuel elements composed of PuO_2 and UO_2 Van de Graaf accelerator was used. The probe neutrons with maximum energy - 350 keV and average energy - 150 keV were generated in a thick Lithium target, to which a proton beam accelerated up to 2.1 MeV was directed. The neutrons with proper penetrating ability yielded 3×10^8 c^{-1} without inducing fission of ^{238}U. The prompt fast fission neutrons were registered by proportional counters with ^4He radiator designed to register recoil nuclei, which were shielded by Cadmium. Control samples of the nuclear fuel wastes were installed directly in the leaden containers designed to protect detectors from the γ-background of the fission products. The fast neutrons were registered during the accelerator operation, whereas those delayed – only after breaking the proton beam. In the book the following results of checking of one of the samples have been reported:

maximum content detection range for $^{235}U = 109 \pm 3$mg., for $^{239}Pu = 13 \pm 3$mg.

The above mentioned works describe methods and technology of fissile materials detection in the laboratory condition, when the neutron source, the object to be detected and the detector are arranged close to each other. The amount of the fissile material is determined by the fission neutron yield. The proposed technique realizes distant detection of fissile materials, which is achieved by using a neutron source with high pulse intensity and by registration of both neutron and γ-response of the object.

4. Pulsed Neutron Generator

The intensive development of pulse high-current accelerators, stimulated by researches on controlled fusion, has resulted in creation of powerful compact sources of high-current ion beams. Utilization of such sources has good prospects for development of pulse power neutron sources.

Experiments with the magnetic-insulation diodes (MID) and electron beam pinching diodes whose anode and cathode were covered with deuterium or tritium [8] have shown a potentiality to produce neutron flows with integral amount of neutrons per pulse of about 10^{12} - 10^{14}, depending on the power of the generator to be used and the nuclear reactions, in which the neutrons are produced.

The generator of high-power probing neutron pulses to be used in the diagnostic complex according to the described principle can be constructed on the basis of a pulsed high-current proton accelerator with using the reactions: $^{7}Li(p,n)^{8}Be$ and $T(p,n)^{3}He$. To match these reactions efficiency the yield of neutrons at energy E_n = (0.1 – 0.5) MeV from generators having the same (10 kJ) pulse energy of the accelerated protons was calculated.

For $T(p,n)$ reaction the following conditions were realized:

I_p = 100 kA – maximum pulsed current of protons;

Ep= 1.4 MeV – maximum beam energy;

1 μm thick target with tritium concentration of 8.4×10^{18}cm^{-2} (n_0=8.4×10^{22}cm^{-3}).

And for $^{7}Li(p,n)$ reaction the following conditions were realized:

Ip = 58.3 kA;

Ep= 2.4 MeV;

13.2 μm thick target with LiF concentration of 8.0×10^{19}cm^{-2} (n_0 = 6.1×10^{22} cm^{-3}).

The temporal mode for the both generators was supposed to be the same: trapezoidal pulse with radix width – 100 ns, and at the top – 50 ns.

In Table 1 the following output characteristics of generators are summarized:

the neutron flux density F_{25} at the distance R = 25 m from the (p,n)-target;

L – maximum detection range according to the criterion sited in [1], m;

τ – duration of the model exposure;

t_d – duration of the response of prompt fission neutrons with E = (0.5 - 2) MeV in the detector, ns.

TABLE 1.

Reaction	F_{25}, n/cm^2 × pulse	L, m	t, ns	$t_{□}$, ns
$T(p,n)^3He$	3464	25	3159	4437
$^7Li(p,n)^7Be$	2.308×10^4	42	5307	7454

The calculations show, that the probe neutron source on $^7Li(p,n)^7Be$ reaction is more favorable to be used for remote detection of the fissile materials.

5. Equipment for the Experiment

5.1 To determine the potential realization of the technique in practice the hardware for the modeling experiments on detection of fissile materials has been prepared. The experiments will be carried on the basis of the accelerator-diagnostic complex at the High-Current Electronics Laboratory of KhNU. High-current electron accelerator "Nadia", which is utilized as a part of the complex, is the base installation for improving the technology of neutron pulses generation.

This accelerator comprises a high-voltage pulse generator used to charge the double forming line (DFL) up to 1 MV within 6×10^{-7} sec. The DFL forms accelerating voltage impulse of $\cong 10^{11}$ W which is supplied to the vacuum diode for the ion beam generation and acceleration.

The accelerator provides the following beam parameters:

- ion energy – up to 1.3 MeV;
- maximum current of the ion beam – up to 50 kA;
- pulse duration – 80 ns;
- operation mode – single pulses at the intervals of 5 minutes.

Proton beam of the required power is generated by the magnetic-insulation diode, see Figure 2.

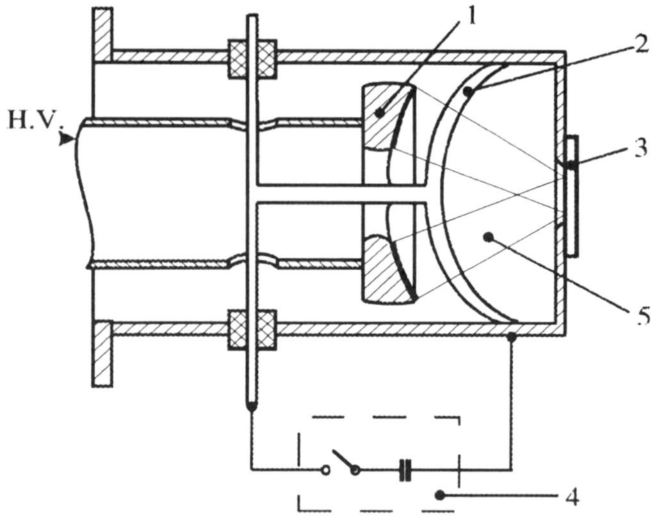

Figure 2. Basic elements of the MID
1 – anode; *2* – cathode; *3* – target; *4* – current generator; *5* – ion beam.

The anode *1* was made in the form of a thick copper ring whose outer diameter is 120 mm and inner - 50 mm. To produce anode plasma by breaking down over the anode surface its emitting area was covered with polyethylene. The anode was joint to the DFL output with a hollow copper tube. To direct the emerging ions into the drift chamber and onto the target *3*, the cathode is made in the form of radially arranged blades *2*. In the diode ballistic focusing of the ions on the target is realized owing to the spherical area of the anode and cathode. Magnetic insulation is realized by passing the pulsed current from generator *4* through the cathode blades *2*.

Proceeding from the potentialities of the device, on which the modeling experiments will be carried out, $T(p,n)^3He$ reaction whose threshold is lower, than that of the reaction on the Lithium target, will be used as a source of neutrons. Besides, to set the operation modes of the ion diode and to calibrate the neutron detectors the $^{12}C(d,n)^{13}N$ reaction is supposed to be used as a source of neutrons. The Deuteron energy threshold for this reaction is 328 keV (E_n=3.4keV). At 1MeV-energy of deuterons the neutron energy makes 693 keV, the reaction cross-section being 120 mbarn.

The process of remote detection of fissile materials is developed in time as sequence of events consisting of:

- generation of probing neutron pulse;
- travel of probing neutrons to the model;
- arising of the model response;
- registration of the response signals.

If to arrange an universal and fast (*n,g*) detector as compared to the pulse duration, at a given distance from the neutron generator, then, all the process of detection can be registered by using only one measuring channel. Scintillation detector may be fast enough and universal for this purpose. The choice of the detector scintillator depends on the distances at which the detection is carried out. At distances of several tens meters, comparatively slow, but highly effective *NaJ (Tl)*-scintillators, may be used. At distances of some meters organic scintillators and fast photo-multipliers should be used.

The scheme of equipment disposition during the experiments is shown in Figure 3. The organic scintillator detector ***3*** registers step-by-step: the MID operation γ-signal; travel of the probe neutron flow (with the help of the B^{10} – target); the model γ-response; the model neutron response. The technique of time-of-flight measurement realized on the detector ***3*** allows to get complete information both about the probe pulse and about the Model response signal.

To calculate the expected signal a model was designed in which the real shape of the current pulse and the accelerator voltage were considered formation of the response signal with the model real geometry taken into account, which is shown in Figure 4.

The model is a hollow thick-wall cylinder made of Aluminum whose outer diameter is 200 mm and inner diameter - 100 mm. The height of the cylinder is 500 mm. The insert composed of a weapon fissile material (WFM) with dimensions 70×80×10 mm is placed close to the geometrical center of the case.

Figure 5. shows expected signals from the detector system.

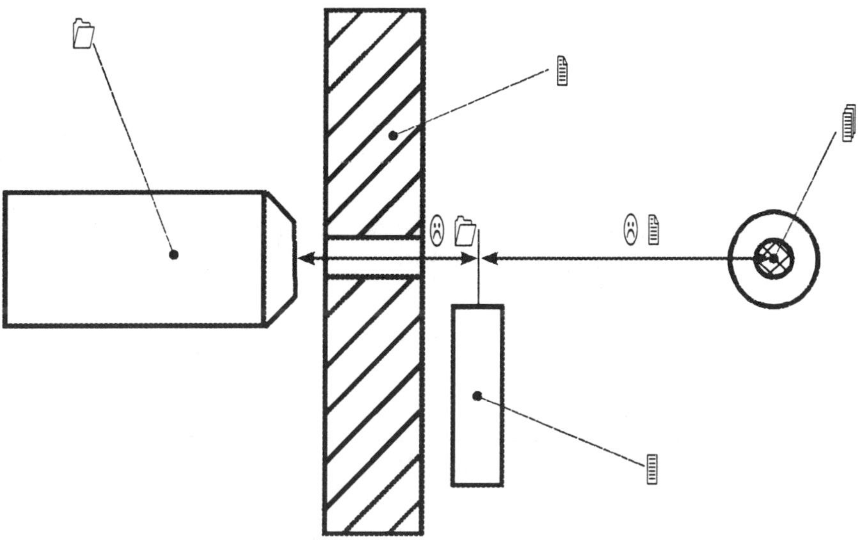

Figure 3. Scheme of equipment disposition for the modeling experiments on fissile material detection.
1 – neutron generator; ***2*** – collimating and protecting system; ***3*** – detector; ***4*** – model.
L_1 – diagnostics base, L_2 – detection range.

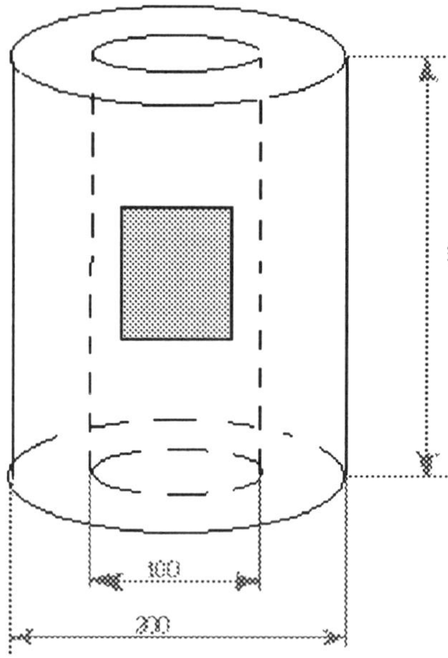

Figure 4. Sketch of the Model

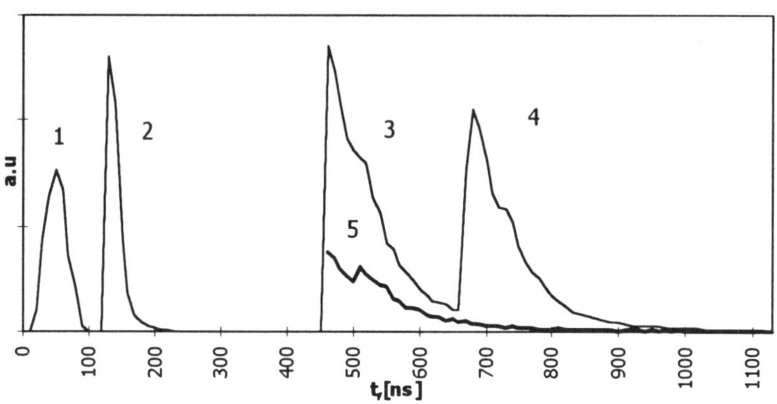

Figure 5. Neutron and gamma-oscillograms of the probing pulse and the object response signals.
L1 = 0.7 meters, *L2* = 3.3. meters (see Figure 3), E_{fn} = 1 MeV

The first signal (1) is induced by Bremsstrahlung radiation which accompanies the operation of the neutron generators, and also by γ-quantum from the $p(T,\gamma)He^3$ reaction. The signal may be very strong, and can bring_the detector into saturation mode, therefore special measures to the signal limitation are applied: the detector is passively shielded by *Pb* and neutrostop of *C3*-type, and an electronic circuit to monitor the photomultiplier sensitivity by lowering the photomultiplier amplification factor for the time of the MID operation is also used.

To register the second signal (*2*) indicating the passage of the probe neutron pulse beside the detector, the ^{10}B -target is placed close to the detector. The detector registers γ-radiation from the $^{10}B(n,\alpha\gamma)^7Li$ reaction.

Signals ***3, 5*** are induced by the γ-response of the model to the probe neutron pulse. The γ-response of the model without fissile materials is caused by (n,γ) reaction on the nuclei of the container material – curve ***5***. The presence of WFM in the model will result in generation of γ-response consisting of (n,γ) reaction photons on the WFM and on the case, and (n, f) on ^{235}U, the curve ***3***. The signal ***4*** arises due to prompt fission neutrons from the reaction on the WFM.

The analysis of time-amplitude response of the model with using the mathematical model of process allows to make a conclusion about either presence or absence of the fissile materials in the object.

To conduct the experiments several versions of the detectors have been prepared.

A silver-foil-activation-based detector comprising of silver-foil-wrapped Geiger tubes placed in polyethylene moderator was manufactured to register the neutron flow at the MID output, see Figure 6.

To provide step-by-step registration of the entire detection process a detector based on an organic scintillator and a fast photomultiplier was developed. UPS-92 organic scintillator (polystyrene with 0.02 % of POPOP and 1.5.% of paraterphenyle) manufactured by the Kharkov Monocrystal Institute was used in the detector. It represents a parallelepiped with dimensions of 500×500×50 mm. The scintillator characteristics were as follows:

- Radiation wave length – 420 nm;
- Half-height flash duration – 3 ns.

Figure 6. Activation detector of neutrons

To provide maximal yield of light all the surface of the scintillator was covered with metallized mylar. The only not covered side of the scintillator was that for fixing to the photo-multiplier with the help of silicone glue. The detector used FEU-85 with diameter of the photocathode working surface of 25 mm. Time of pulse rise was 4 ns.

To keep the detector of photomultiplier, being under the influence of the light pulse produced by the MID Bremsstrahlung, from reacting the state of saturation an electronic device designed to change quickly the photomultiplier multiplying factor was manufactured. By a monitoring signal the device resets the photomultiplier to the mode of maximum multiplication in 2 ns and keeps it in this mode during 10 μs. The contrast range of the device is $\approx 10^4$.

The detector was assembled in a rectangular steel case, which also served as a screen.

According to mathematical modeling of the experiment it follows, that the most easy for detection and informative Model response to a pulse neutron irradiation will be γ-yield of the radiation capture and that of the fission of the WFM. And consequently the WFM γ-radiation is accompanied by yield of the case radiation capture, as well as capture radiation of the neighboring objects

The detector of the Model γ-response to the neutron probing pulse represents an assembly of a number of spectrometric blocks of 6932-20 type, see Figure 7.

Figure 7. Detector of γ-response

The detector spectrometric block 6932-20 is designed to convert γ-quantum energy into pulses of proportional amplitude. Constructionally the detecting blocks are made as a unit comprising a scintillation block, a matching board, a unit for connecting FEU.

The scintillation block is a vacuum-tight unit, in which *NaJ (Tl)* scintillator with 150 mm diameter and 100 mm height, and FEU-49B are bound to each other with silicon-organic glue.

Registration efficiency of γ-radiation in the peak of its full absorption is:

$E\gamma$ = 662 keV;
ε = 25 %; $E\gamma$ = 1800 keV;
ε = 14%;

its original background level - 33.3 s^{-1}.

To provide the extension of the photomultiplier output current range of linearity a conventional circuit of the voltage divider was modified. Since the major limitation of the output is caused by the space charges at several outlet photomultiplier cascades, the voltage divider is modified to make the difference of potentials at the outlet cascade of about 3.6 times as much as at the foregoing cascades. Besides, several decoupling capacitors with large capacity have been installed in the divider to maintain the cascade voltage while working with high-current pulses.

6. Conclusion

The approach to the method of active remote detection of WFM presence on the chosen object offered in the paper point out wide range for potential researches in this field. Numerical simulation of the processes showed a real feasibility of the method for detection at distances of a few of meters. Hardware for the experimental verification of the method is being prepared at present. The experiments are planned to be performed in the forth quarter of the current year.

7. References

1. Ogawa, *Detection of Nuclear Warheads by Radiation*: a Brief Review, Rep. KEK, p 149-157 (1990).
2. Ewing R. and Marlow, K. (1990) *A Fast Neutron Detector Usen in the INF Treaty*, Nucl. Instrum. And Methods. Phys. Res., Vol A299, p 559-561.
3. Trayball, S. and Mackpheit, P. (1990) *Ternistui Put' k Dogovoru po OSSV*, V Mire Nauki: Rus. Transl from: Scientific World, **2**, p 6-16.
4. Frolov, V.V, (1989) Yaderno-fizicheskie metodu Kontrolya delyaschihsya veschestv, Energoisdat, Moskva, p. 93.
5. Tablitsy Fizicheskih Velichin: Spravochnik, Red.: I. Kikoin, Atomizdat, Moskva (1976)
6. Kovalev, V.P. (1979) *Vtorichnye izlucheniya uskoritelei elekronov*, Atomizdat, Moskva, p 166.
7. Bystritsky, V.M. and Didenko, A.N. *Sil'notochnye Ionnye Puchki*, Ukrainskii Fizicheskii Zhurnal (Sov.), Vol 132

MULTISENSOR FUSION DATA AND EARTHQUAKE PREDICTION

P. POKHOLKOV, B.S. RYBAKOV, N.N. KHORSOV
Tomsk Polytechnic University (TPU)
Tomsk, Russia

1. Introduction

When and where will an earthquake occur? What will be the power of this earthquake? How should you refer to predictions, which have no 100 per cent guarantee to take all necessary measures to decrease a negative influence of earthquakes? For a long time all these questions are under deep consideration in many countries all over the world, and as a result earthquake predictions can be classified using the following parameters.

Time of a prediction:
(a) Long-range;
(b) Medium-range;
(c) Short-range.

Place of a prediction:
(a) In the epicenter;
(b) In seismically active areas;
(c) In geographical region.

Energy of a prediction:
(a) Magnitude in epicenter;
(b) Shock's intensity in the definite place.

At present, a large number of earthquake precursors of various types have been discovered. These are precursors based on changes of physical parameters of the medium, which cause earthquakes: compression, strain and shift of some parts of rocks. And as a result of such processes there may be changes of underground water level, micro seismic fluctuations, changes of gas exudation from rocks (radon, for example) that in some cases display themselves as earthquake precursors.

The next type of earthquake precursors belong to those associated with the influence of the lithosphere-mechanical tension on variations of physical fields of different nature (magnetic, electromagnetic, electrotelluric) and on animal behavior.

Time of earthquake prediction using short-range precursors is ranged from one to tens of hours.

A. K. Hyder et al. (eds.), Multisensor Fusion, 851–863.

Some geophysical parameters, where precursors of short-range prediction appear, are given below:

(a) The level of underground waters;
(b) Strain and deformation of rock;
(c) Concentration of radon;
(d) Micro seismic activity;
(e) Geomagnetism;
(f) Resistance of rock;
(g) Velocity ratio of cross and longitudinal seismic waves;
(h) Foreshocks;
(i) Telluric current;
(j) Radio waves propagation along the path of preparing earthquake;
(k) Natural Pulse Electromagnetic Field of the Earth (NPEMFE).

Earthquake precursors using of NPEMFE parameters was started no long ago. First, measuring of NPEMFE parameters for precursor's detection and development of earthquake prediction methods were begun at Tomsk Polytechnic University 30 years ago.

The possible reasons of electromagnetic earthquake precursor's exhibition may be the following.

(a) Change of electromagnetic signal generation conditions
(b) Accumulation of volume charges in the earth crust and it's action on electromagnetic field of natural and artificial origin.
(c) Change conditions of electromagnetic waves propagation in region of impending earthquake.

The precursor itself is not enough to make a prediction of an impending earthquake. Methods and algorithms of prediction should be formulated for this. All present methods of prediction are made using precursor parameters retrospective analysis.

It can be said that at present there are no any reliable methods of prediction, and the literature reasons for the failure in earthquakes predictions.

Firstly, many efforts to search for earthquake precursors were based upon intuition and engineering. All of them are connected with revealing during the anomalies in a measured parameter before the quake and further detection of parametric anomalies during the earthquake in real time. In the structure of any measurement parameter there are some components, which are not connected with the generation of an earthquake (noise). Frequently it is very difficult to separate useful signal from noise. That is why the method and algorithm of earthquake prediction must be described using characteristics of possibility, as for the precursors' patterns; they must be done using stochastic models.

Another reason is the ambiguity between the relation precursors and the earthquake parameters. At present the idea of "self-organizing criticality" (SOC) to explain the mechanism of earthquake preparation is very popular. It means that the distribution of strains, deformations and strengths in crisp rocks are heterogeneous in all character scales. With increase of mechanical strains in the region of preparing earthquake they are redistributed through forming back connections so that systems everywhere approached a condition close to local crisp failure. The word "criticality" means that system is between two conditions of behavior. The word "self-organizing" means that the system makes itself in the definite condition. Inside of such a system the reaction of any chosen volume to changes of mechanical strains on its boards can be predicted only statistically. Because of the fact that the system is always on the edge of stability, a small change of stress can cause loss of stability and snowballing transition of the system into another quasi-stable condition that causes an earthquake. Then the strain again increases and the process can continue for years, decades and more. That is why the short-range prediction of earthquakes, as some authors consider, is impossible.

The third reason for failure in prediction is in not using the proper spatial precursor characteristic, which could lead to an increase signal-to-interference ratio. For such a purpose it is necessary to make a measurement of one parameter in different points, it is to say, to use multisensor system for measurement and treatment.

There are some multisensor systems of earthquake prediction in the world. The system of taking a sight on earthquake preparation in Japan is known, where different types of inclinometers and strainmeters are installed over large areas. This system was created many years ago. The practical data have been collected, but a more or less reliable method of prediction has not been worked out yet.

Another system of prediction was created in Greece (so called VAN system). It includes units to measure telluric currents. The method, as its authors believe (Varotsos, Alexopoulos, Nomikos), predicts the magnitude and the time of an earthquake with the parameters of telluric current, which is associated with and fore-runs earthquakes events.

The authors of the method have published optimistic prediction results of some years. But the analysis made by many seismologists shows that the method mentioned above is not very effective during its statistical testing, because to make predictions the authors had to enlarge the time window from some days or hours to weeks. Besides, during the testing of hypotheses in probability it turned out that the role of chance predictions is rather large.

The system of prediction made in China where the reliability of prediction was not large can be mentioned here too.

What is the reason of failure in forecast using these systems of earthquake prediction?

From our point of view the first reason is that the distances between sensors of homogeneous information sources of the multisensor systems of observation were installed without appreciating of spatial correlation characteristics. Many data were received about the fact that the intervals of spatial correlation for different types of signals are different [2]. That is why sensors for measurement of different geophysical fields parameters should be installed in a distance as firstly to estimate spatial precursor characteristics of an incipient earthquake and secondly to filter signals from local noises, which spatial correlation interval is less than the distance between sensors.

The earthquake prediction is not only a scientific-technical challenge but also a social problem. The successful prediction of an earthquake can help to avoid catastrophes at the expense of evicting people from their houses, turning off gas and electricity to prevent from fire, etc. In the case of a false alarm economical and social negative consequences can be very important. Some specialists consider that a false alarm triggering administrative measures can have serious negative societal consequences.

If the time window is rather large (weeks, months) no doubt that the safety will increase, but there will be big financial loses. Besides, social problems will happen. But if the time window is small, the reliability of prediction will decrease.

2. TPU Scientific-Research Work on Investigation of NPEMFE Parameters for Geodynamic Phenomenon Prediction

Starting from the 1970s, on TPU ex-rector A.A. Vorobiev's initiative, for the first time research was done on studying the natural pulse electromagnetic field of the Earth (NPEMFE) to search for earthquake [5,8,7,9]. At those times seismologists regarded with some doubts the fact that parameters of existing NPEMFE can be earthquake precursors. Subsequently, NPEMFE was studied as a candidate to precursors in our country and overseas.

In Tomsk Polytechnic University a series of NPEMFE parameters observation stations were established. The field station of "Katyusha" type was the first. The net of three such stations was installed near with the Issyk-Kul Lake in the Central Asia. The magnetic components of pulse current intensity (PCI), i.e. the number of pulses higher than the given threshold in the time unit, was measured with two magnetic antennas orientated with the magnetic current north-south and west-east. The measurement was held at 15 kHz frequencies. Later, the station "Katyusha-5" was configured for registration of many parameters at the same time: intensity of pulse current on electric (at wide frequency band) and magnetic (at 15 kHz frequency) components of electromagnetic field, pulse amplitude, telluric currents, intensity of atmosphere electric field, micro seismic waves. (Gordeev V.F., Malyshkov Yu.P. Eliseev V.P., 1995). Round-the-clock measurements on the north of Bajkal Lake have been held with the two stations of this type since 1997.

In 1989 the stations of the other type were elaborated: pulse current analyser (PCA). They permit to define the amplitude of each registered pulse, higher than the given threshold, its duration at the level of this threshold and the time of its coming. (Mastov Sh.R., Rybakov B.S., Khorsov N.N.). The last modification PCA-4 permits not only registration of the above-mentioned parameters of the current, but also formulation of the database in month cycles of the measurement.

The PCA-4 main technical characteristics are given below:

The amplitude of the input signals	(50-300) mkV;
The duration of the input signals	$(10^{-5}$-$10^{-2})$ sec;
The maximal proceeding frequency	100 pulses/sec.

The block-diagram of the device is shown on Figure 1.

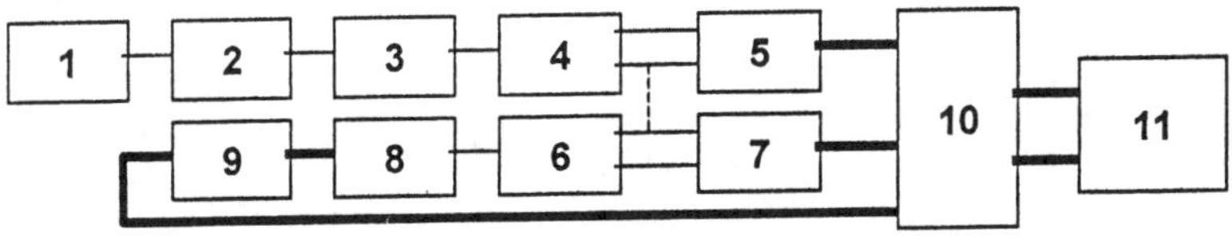

1 – antenna; 2 – amplifier; 3 – filter; 4 – peak detector; 5 – analog-digital converter;
6 – threshold register; 7 – digital-analog converter; 8 – amplitude selector;
9 – interval measurer; 10 – interface block; 11 – computer.

Figure 1. Block-diagram of PCA-4

The PCA-4 operation principle is the following. The NPEMFE signals are perceived with the magnetic or electric antenna (1), transformed into the electric signal and amplified (2). The frequency band is formed with the filter (3). To limit the intensity of the pulse current the pulse amplitude selection is provided. The threshold due to pulses, is established with those amplitudes lower than the threshold not registered. The level of the threshold is adjusted from the computer control panel, through the interface block (10) it goes to the threshold register (6), with using digital-to-analogue conversion (7) it is transformed into the analogue signal and goes to the amplitude selector (8). With the excess of the selection threshold by the pulse, peak detector (4), which records the pulse peak meaning, is activated; analog-digit converter (5) transforms it into the digital mode. The amplitude selector activates the interval measurer (9), in which the pulse duration on the level of the threshold is measured digitally. The pulse amplitude and its' duration through interface block go to computer (11).

The database is formed in the computer in which the amplitude, the duration of each pulse, and the time of its arrival are save. Besides, the information about PCA-4 number is given, if there are several observation stations. Five PCA-4 stations were allocated near Alma-Ata city, Kazakh republic for round-the-clock observation.

According to the Ministry of Emergency Situations assignment for the employment on Kamchatka 4-channel movable regional observation system "MMS PEMF" (the movable measurer of signals of pulse electromagnetic field) was established in 1997. The station has two channels to receive signal on magnetic component of the field, the channel to register electric component and seismic channel to register micro seismic noises [8]. The attached microprocessor provides preliminary processing and saving of the information.

According to the long years of observation, the following conclusions were obtained.

It turned out that the most information in the pulse electromagnetic current is its intensity of the magnetic component at limited frequency band. The obtained precursor is a reduction of the natural diurnal variation of PCI in several days before the earthquake occurrences. Immediately before the earthquake occurrence the PCI diurnal variation was restored or was even higher than the average meaning. Schematically The pattern of an earthquake precursor is shown on Figure 2.

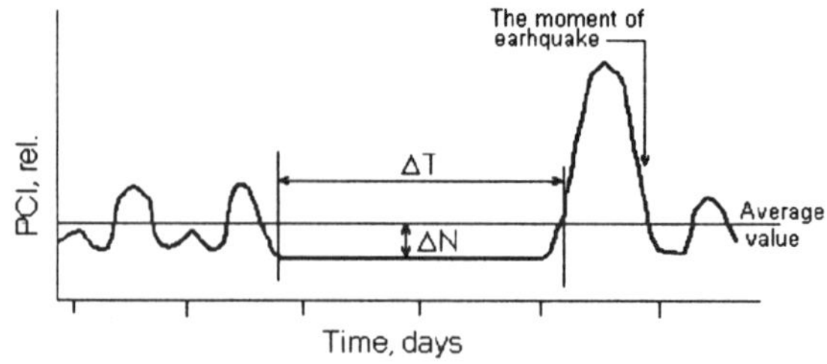

Figure 2. The pattern of earthquake precursor in NPEMFE structure

Information obtained from the resulting precursors parameters analysis shows that the PCI reduction ΔN with regard to average meaning is correlated to the distance to impending earthquake epicenter, and the interval of the reduction ΔT with its magnitude. Quantitative statistical evaluations of the correlation of PCI reduction duration with the magnitude were executed. The correlation equation is: K=1 ::: 5+6:8 log (ΔT). Where K is energy type of the earthquake, and ΔT is the interval of the calm. Figure 3 shows regression line dependence of this correlation.

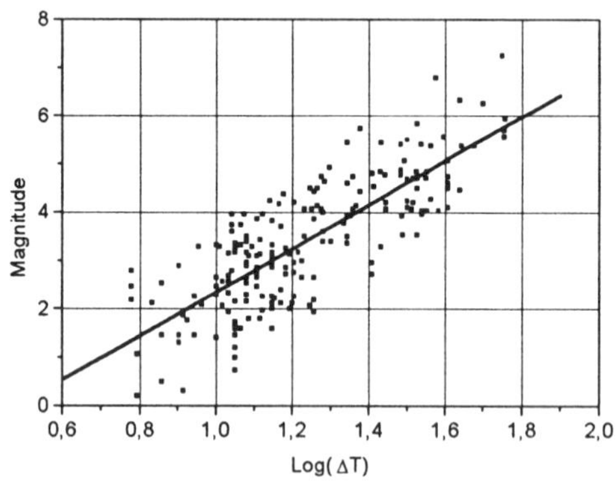

Figure 3. Regression precursor's duration (ΔT) on earthquake magnitude.

The reliability of the prediction using the mentioned precursor with its strict appraisal is not sufficient to find practical implementation because we do not yet fully appreciate the statistical nature of the noise.

In TPU, laboratory research continues on electromagnetic emission (EME) from the samples with mechanical loading, both for the no destructive control of the materials, and for the possibility of using the obtained data for the search of the electromagnetic earthquake precursors were held.

A schematic diagram and results of the experiments on the research of the EME parameters of the granite sample in the conditions of quasi-statistic loading are shown on Figure 4.

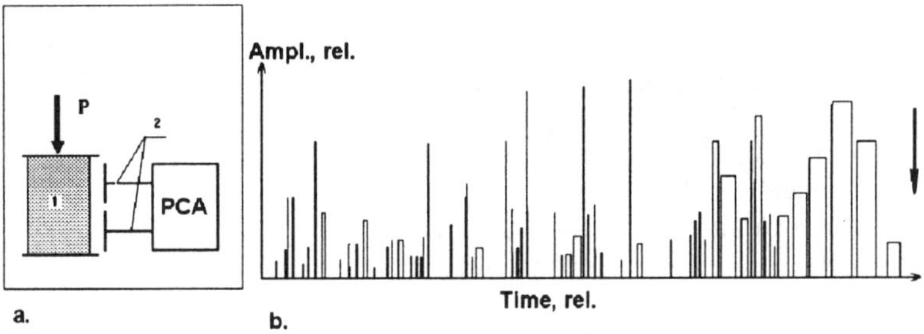

Figure 4. Parameters of electromagnet pulse currency from of granite sample (4b) before it's destruction under quasi-static loading and diagram (4a) of the experiment. 1 - sample; 2 - differential capacitance sensor.

The experiment was carried out in the following manner: a graduated load of monoaxial pressure **P** was applied to the sample (1) and held invariable for sometime on each grade (fig. 4a); the pulse current parameters on electric component using differential capacitance sensor (2) were measured with the PCA. At some level of loading (the loading of destruction) the micro zones of destruction, determined by parameters of electric pulses (amplitude, duration, PCI) appeared in the sample, soon after the sample was destroyed. On Fig. 4b temporal correlation of pulse current parameters before the destruction of the sample is shown. Pulses are shown as rectangles, the width of which qualitatively reflects duration, and height - amplitude. As is seen from the Fig. 4b the character of the pulses changes with the approach of destruction time. The arrow down marks point of time of sample's destruction. It is stated, that for the different grades of partial destruction of the sample, the pulse current has different characteristics.
Certainly, the described experiment can not be considered to be the analogue of earthquake preparation, but some common features from the point of view of the prediction of events can be found in both processes: the destruction of the sample and earthquake.

3. Self-Learning, Regional Adaptive Information and Advising System for the Short-Term Earthquake Predictions

In our opinion, low efficiency of earthquake prediction with the help of short-grade precursors lies not in their total informational "weakness", but in the lack of a system approach both in the organisation of the monitoring and in the application of optimal methods of spatial-temporal processing of measurements, based on statistical decisions theory. That is why it is necessary to change the paradigm in the approach to earthquake prediction problems.

The main task of new approach may be formulated as follows.

"To create self-learning, regional adaptive information and advising system for the short-term earthquake predictions and reduction of negative aftermath of the earthquake through the fusion of various informational sources databases.

(a) Principles of the information and advising system creation are following.
(b) Foundation of spatial-temporal networks for real time monitoring of information flows variation parameters of various sources.
(c) Allocation of spatial-temporal supervision networks in the region, interested in earthquakes predictions.
(d) Forming of parameters' database for current earthquakes with their classification.
(e) Forming of various informational sources' databases.
(f) Development and detection of earthquakes precursor based on seismic events database and various informational sources database fusion.
(g) Provision of system's adaptability and self-teaching.
(h) Impending earthquake's statistical parameters estimation and recommendations on administrative decision-making.

The decision-making algorithm, which defines class and time of the seismic event, constitutes the highest hierarchical level of earthquake prediction system. It is illustrated with block-diagram on Figure 5.

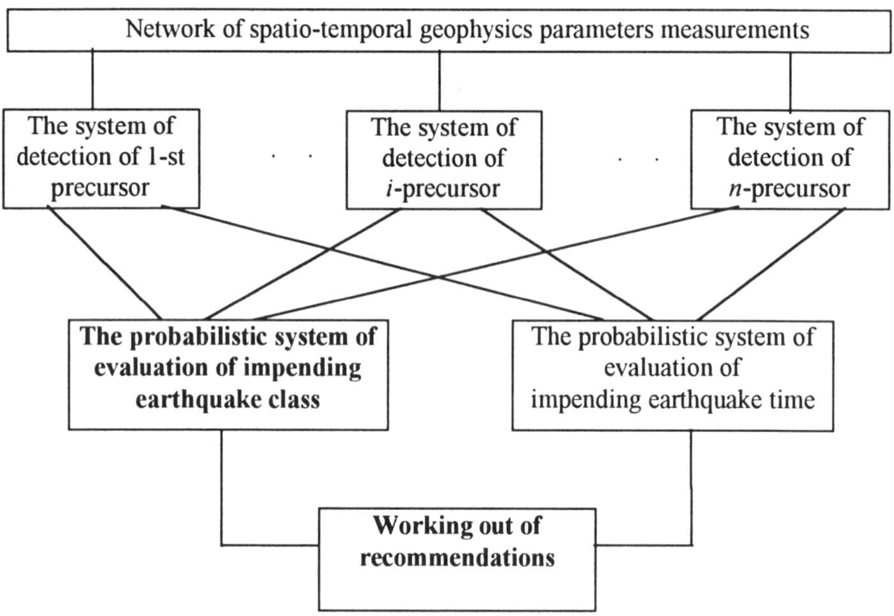

Figure 5. Block-diagram of recommendations formation

According to geophysics parameters' spatial-temporal observations network data the multichoice discovery system makes decision with each precursor about the type of impending seismic event and about the moment of its' origination. These results are used for determination of general decision of event's class and its time occurrence.

Make the general decision that seismic event of *k*–class will occur, through the event A_k, hypothesis, that along with it the prediction was made with the help of i-precursor through H_i. All earthquake intensity intervals we quantize into 3 to 4 classes. The probability of general decision-making A_k is evaluated by using equation

$$P(A_k) = \sum_{i=1}^{n} P(H_i)P(A_k/H_i),$$

where n is the number of precursors; $P(A_k/H_i)$ is probability of k-class earthquake correct prediction by using *i*-precursor; $P(H_i)$ is hypothesis probability. That decision is taken as A_k, the probability of which is maximal.

In an analogous way it is established the single time of the earthquake prediction on time t_i for every of precursors.

The central unit of the above-described algorithm is *i*-precursor's multichoice detection system.

The long, laborious (not less than one year long) survey work of the region with the purpose of creating prediction models as a basis of its' synthesis, should be prior to the elaboration of the algorithm.

In essence, this process is continuous: as the precursor measurement data accompanying impending seismic events accumulates, more reliable models' parameters assessment occurs and therefore constant training of the system takes place.

The prediction will begin with the help of the simple determined model, which demands lesser volume of measurements to evaluate its' parameters. Further, the parameters of more precise (adequate) statistical model are evaluated, more powerful algorithm of statistical hypotheses are used and results in a constant growth of prediction efficiency [12]. Necessarily developing statistical model is due to time-varying process of background noise reduction with expressed days and season variations in structure of measured parameters.

For the synthesis of prediction algorithms it is necessary to take into account both temporal performances of precursors, and singularity of their spatial development. Let's consider singularities both temporal, and spatial information processing for development of prediction models.

Temporal handling. Block diagram of earthquake prediction using either of several precursors is shown on the Figure 6.

Preliminary it is determined duration of precursor's display of k-class earthquake, which is quantized on temporal segments (usually on hour duration).

The measurement data proceed on two blocks: the block polynomial smoothing and the block of estimation.

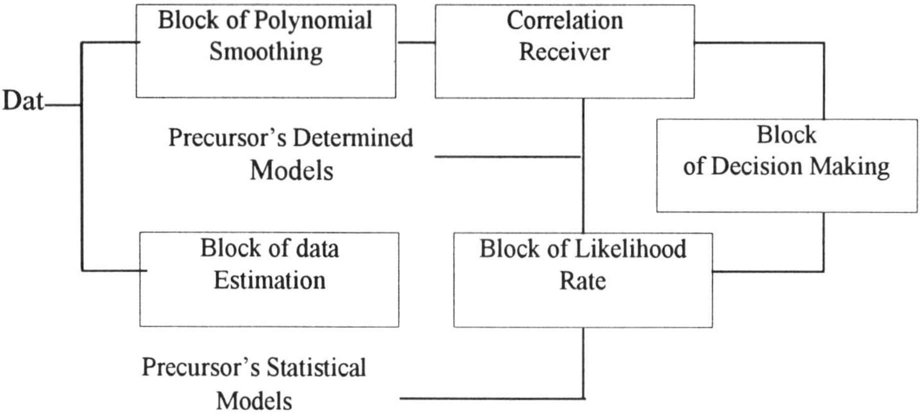

Figure 6. Block-diagram of earthquake prediction using one of precursors

In the block of polynomial smoothing average values of measurable quantities m_j of a parameter s in selected temporal segments on a sliding window of precursor's display duration are calculated.

$$m_j = \frac{1}{d}\sum_{i=1}^{d} s_{i,j} \, ,$$

where j is index of a time segment; d is amount of measured quantities inbounds of a temporal segment; i=1...d.

The sequence of average values is smoothed with polynomial function and moves on one input of the correlation receiver. Sequences of the polynomials-precursors of different class earthquakes by turn move on other input. In the block using cross-correlation function compares patterns of sequences. By results of a comparison a class and time of the future earthquake is determined.

In the block of data estimation parameters of distributions of measurable quantities in bounds of temporal segments on the sliding window are evaluated.

Using obtained distributions it is evaluated regression parameters of an available known statistical distribution. For so measuring parameter as PCI of NPEMFE (intensity of pulse current of natural electromagnetic field of earth) it was used Waybull distribution, having two parameters α and λ.

$$W(N / \alpha,\lambda) = \alpha\lambda N^{\alpha-1} e^{-\lambda N^{\alpha}}$$

(Hereafter we shall apply this distribution as an example).

Evaluated on the anomalous display period before earthquake occurrence regression distribution is consists of signal (s) mixed with noise (n)

$$W_{s+n}(N_{i+j}/\alpha_{j,s+n}, \lambda_{j,s+n}).$$

Here i is an index of day's hour (i=1, 24); j - current index of the precursor's display period (j=1, ..., l); 1 is amount of temporal segments on duration of the slide window; In case of absence precursor's display, parameters of distributions are determined by only noise and can be expressed by

$$W_n(N_{i+j,n}/\alpha_{i+j,n}, \lambda_{i+j,n}).$$

Co-operative probability density of distributions assemblage within interval of precursor's display is expressed by follow equation

$$W_{s+n}(N_i) = \prod_{j=1}^{l} W_{s+n}(N_{i+j,s+n} / \alpha_{j,s+n}, \beta_{j,s+n}).$$

The equation for noise is:

$$W_n(N_i) = \prod_{j=1}^{l} W_n(N_{i+j,n} / \alpha_{i+j,n}, \beta_{i+j,n})$$

The last stage of theory of hypothesis test is to evaluate likelihood ratio λ (coefficient of likelihood).

$$\lambda_i = \ln \frac{W_{s+n}(N_i)}{W_s(N_i)}.$$

The last expression is precursor's statistical model of k-class earthquake for one measured parameter in the given space point. Index i appoints on time of day.

In process of prediction through a defined interval of time (the time segment) it is calculated λ, which is compared with corresponding precursor's coefficients of likelihood for either of earthquake classes. Evaluation of earthquake probability is determined by choose of either of decision strategies (usually we use the Neyman-Pearson criteria in which is fixed probability of false alarm). If evaluated $\lambda > \lambda_0$ it is concluded, that earthquake will occur.

In the block of a decision-making the final evaluation of probability of making dispositions earthquake, its class and time is produced.

On Figure 7 is presented block for forming precursors parameters. In process of system operation continuous database of every parameter is formed. When earthquake is occurred, the classifier determines class of event. In the analyser patterns of precursors of each parameters is determined by using its meanings before earthquake occurrence period.

Spatial handling. The network of observations on space should be redundant that the spatial characters of geophysical fields, switching on NPEMFE, practically are not investigated. Data about "patchiness" and migration of precursors' display, i.e. about a modification it of spatial characters in time first of all are important.

Let is present l of observations points with coordinates x_i, y_i (i=1, 2..., l). The outcomes of measurements for fixed time represent a scalar field $n\ (x_i, y_i)$.

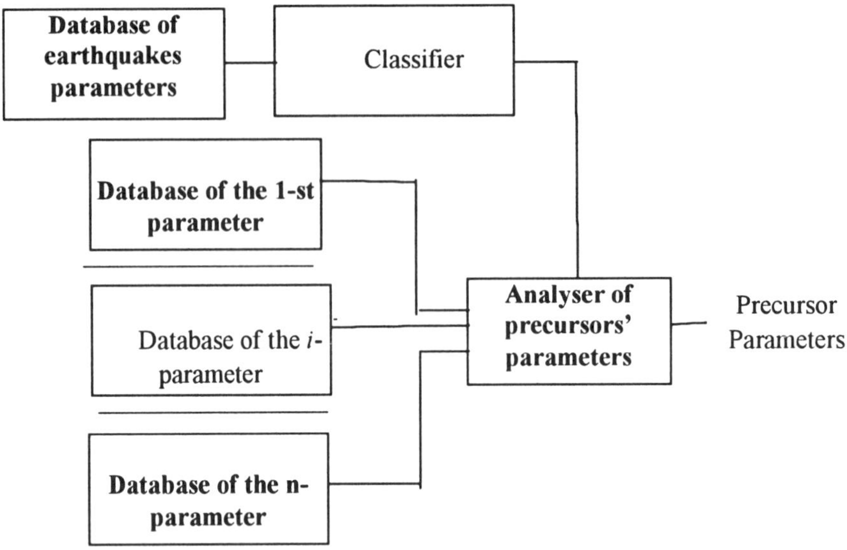

Figure 7. Block- diagram for forming precursors' parameters

The spatial handling consists of the following:

(a) The search of minimum and maximum is carried out using the simple ordering of set $n\ (x_i, y_i)$ on basis of i=1, 2..., l

(b) Fixation of seemed minimums and maximums.

(c) The smoothing out seemed extremums occur within a region is carried out with the help of regressive surface or lines of regression in cross-sections.

(d) The field of vector quantity (the field of gradients) is built within a region of extremums using smoothing curves (surfaces).

(e) The coordinates of extrimums are fixed and improved using the criteria of steepest descent.

(f) The global extremums and their priorities are distinguished.

(g) The data of time treatment in points, surrounding points of global and local extremums are requested.

(h) The system of seismic danger point priorities is constructed using mosaic structure pictures taking into account-discovered classes of seismic events on the basis of time treatment data.

(i) In accordance with accumulation of data the spatial statistical models of precursors are developed.

To synthesis prediction algorithm for sensors space disposition it is necessary to take into account a space correlation functions. We hypothesise that space and temporal distribution functions are independent of each other. Then spatial-temporal multidimensional precursor's probability density may be expressed as the product of one-dimensional distributive laws

$$W(X/\alpha)=\prod_{i=1}^{n}W_{s+n}(\overset{*}{X}_{i,j}/\alpha_{i,s+n})\cdot\prod_{j=1}^{m}W_{s+n}(\overset{**}{X}_{i,j}/\alpha_{j,s+n}),$$

where $\overset{*}{X}_{i,j}$ is sampling precursor's meanings in time (i=1,2,...,n) in the j-point of space; $\overset{**}{X}_{i,j}$ is sampling precursor's meanings in space (j=1,2,...,n) in the i-point in time.

4. Conclusion.

The realization of the approaches, included in offered information and advising system, will help, on our sight to approximate a solution of a problem of the short-term earthquakes prediction as from a point of view of the definition precursors' parameters, and development of a system of measures on a diminution of their negative consequences.

The offered system of the prognosis of earthquakes can be used for development of prediction algorithms of other geodynamic appearances, such as rock bumps, landslips.

5. References

1. Acada T. (1982) *Earthquake Prediction Techniques*. Their Application in Japan (Edited by Toshi Asada). University of Tokyo press.
2. Leary, P.C. (University of Edinburgh): *Rock as a critical-point system and the inherent implausibility of reliable earthquake prediction.*
3. Alexandrov, M.S. et al., (1992) *Fluctuations of electromagnetic Field of Earth in ULF.* M.: Nauka, h. 195.
4. Alexandrov, M.S. et al., (1976) Anomalous change of intensity of EPEMFE in Tashkent's region before earthquake. Uzb. Geol. Journ. No. 2. p. 9 - 11.
5. Vorobjov A.A., Mastov S.R., Rybakov B.S., et al. Temporal characteristics of natural electromagnetic field of Earth. VINITI, pp 1020 -81 Dep., p.12.
6. Gorjainov, V.T., Guravlov, A.G., Tichonov, V.I. (1980) *Statistical radioelectrotechnical engineering//* Editor Tichonov V. I., Sov. Radio.
7. Malyshkov, J.P., Dzumabaev K. (1987) *Prediction of earthquakes using parameters of natural elecrtomagnetic field of Earth,* Vulk. and Seysm. (Russia). No. 1. p. 97 - 03
8. Malyshkov, J.P., Gordeev, V.F. et al. (1984) *Regularities of generate of electromagnetic signal by solids under mechanical forces.* Journ. of Th. Phys, v.54, issue. 2, p. 336 - 341
9. Mastov, S.R., Rybakov, B.S. (1990) *Determination precursors and earthquake prediction by statistical test of hypothesis method.* Izv. AN USSR. Phys. of Earth. No. 7. p. 70 -75.
10. Mastov, Sh., Rybakov, B.S., et al., (1986) *Statistical method of short-time prediction of crust earthquake time using pulse electromagnetic activity.* Izv. AN USSR. Phys. of Earth, No. 5. p. 72 -75.
11. Movlanov, G.A. et al. (1979) *Research of variations of natural electromagnetic field parameters for earthquake.* //Uzb. Geol. J. No. 5. p. 11 -15.
12. Pokholkov, Ju. P., Rybakov B.S., Khorsov N.N. (2000) *Statistical method of short-time prediction of crust earthquake time using parameters of pulse electromagnetic activity.* Izv TPU. Vol. 303. issue 3. p. 171 - 179.
13. Pugachev, V.S. (1976) *Probability's theory and mathematical statistic,* Nauka, p. 196.
14. Remezov, L.T. (1985) *Natural stray,* Nauka. p. 200.
15. Helstrom, K. (1963) *Statistic Theory of signals' detection,* Izdat. inostr. lit. p. 340.

THE NEW THEORY ABOUT ION-SELECTIVE ELECTRODES

E. PUNGOR
Institute for General and Analytical Chemistry,
Technical University
Budapest, Hungary
Bay Zoltán Foundation for Applied Research
Budapest, Hungary

1. Introduction

This Intitute should address many of the questions of application and theoretical aspects needed for the development of reliable sensors.

The field of application of these sensors has been extended far beyond the early application in industry and environment protection during the last 30 years and is now being applied to biochemical uses.

This growth of knowledge, having been started from the sixties, has resulted in the effort to try to fit in this knowledge with the former theories of the glass electrode based on its data and theory, that is why, in the field of sensors, anomaly explanations have been established and the rejection of these is very important.

In this chapter we do not intend to show the fields of application of chemical sensors but to highlight misinterpretations that are embraced by many researchers in fields where the validity of statements can not be proved.

It is first necessary to elucidate the details of the basic reactions, otherwise, the facts may not be enough for the understanding of the behaviour of the system. This kind of interpretation of phenomena and its consequences have resulted in much non-productive research.

The subject at hand is the interpretation of the phenomena observed for glass electrodes, and later for all the ion-selective electrodes. At the beginning of the 1900s, the phenomenon observed by Haber and Clemensiewith was a great surprise among electrochemical phenomena, namely that the glass electrode produced an electrochemical signal in response to the acidity of solutions (the definition of the pH was given later). It was also a surprise that this potential signal could only be measured by electrometers of high input resistance (impedance), in contrast to the method developed earlier by the Nernst school.

For electrodes of the first kind, the transfer of electrons was the potential-determining reaction, but this interpretation did not seem to be applicable for glass electrodes. A few years after the discovery of this phenomenon, a new experiment was made by Donnan. He separated a solution of an alkali halide from a solution of a protein by a membrane, and found that there was an equilibrium, if the protein, which did not diffuse through the membrane, and the halide, which diffused through the membrane, established an electrochemical equilibrium. As a consequence, an electrical potential

A. K. Hyder et al. (eds.), Multisensor Fusion, 865–878.

could be measured across the membrane. Such an explanation of this experiment was plausible and clear. But it was not clear why this explanation could be applied to the phenomena of the glass electrode and why it was assumed that the glass electrode worked on the principle that ions diffused into and through the glass electrode membrane, i.e. why the transport mechanism was introduced for the interpretation of the operation of the glass electrode.

When other ion-selective electrodes were also developed besides glass electrodes, this transport mechanism was applied for them, too. In Figures 1 and 2 we illustrate the application of this theory for the interpretation of the working mechanism of electrodes. In Figure 2 the trajectories are given, but they have never been measured through the membrane profile.

Potential profile is shown for a membrane containing exchangeable cations, with lower cation concentration in the membrane than in either bathing solution. The membrane is bathed on one side at constant activity a_{in} and at a variety of activities on the sample or 'outer' side (theoretical investigation). The a_{out} and a_{in} curve labels correspond to +1 ions for negative ions (-1 charge); the lower three labels correspond to 0.1 a_{in}, 0.01 a_{in}, and 0.001 a_{in}

A further question of interpretation was raised by Guggenheim, who deduced the electrochemical potential from the chemical potential by applying the new term of the Galvani potential, which changes proportionally to the electrochemical potential. Hereby the only real measurable term in the equation was the chemical potential. He assumed that on both sides of the membrane (inside and on the solution side) the electrochemical potentials have to be defined, and in the case of equilibrium we can obtain a measurable electrical potential.

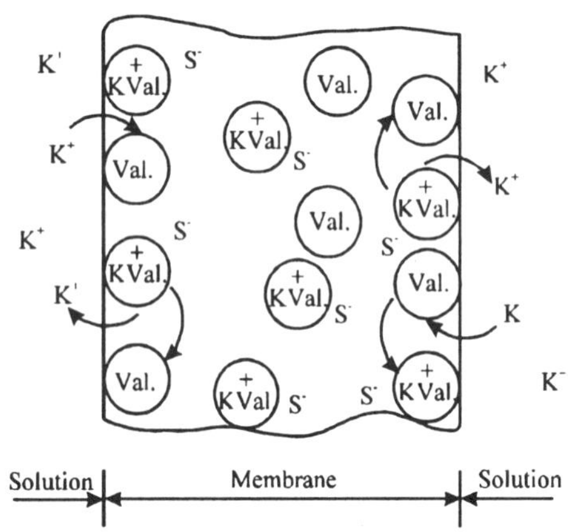

Figure 1. Negatively-sited membrane (mobile or fixed sites S^-) containing dissolved neutral carrier, valinomycin, exposed to bathing solutions of counter ion K^+

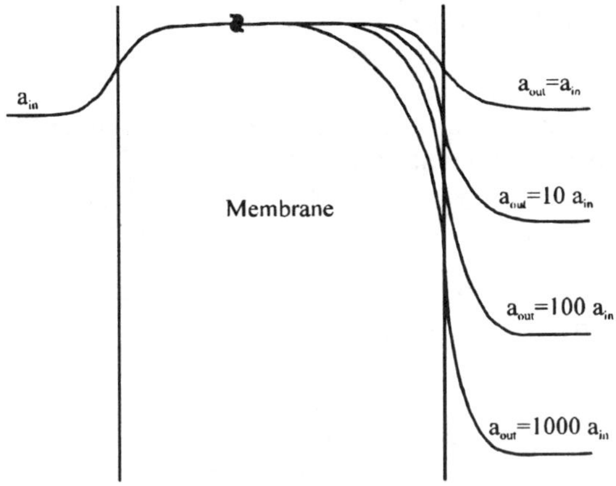

Figure 2. Trajectories through the membrane profile

2. Experimental Investigations

Using the Donnan potential concept, various types of investigations were conducted on glass electrodes, e.g. studies on the transfer of Na+, Li+, etc., but even so, it was difficult to understand the behaviour of glass electrodes. Just therefore, to clarify the phenomena, we investigated the physical and chemical characteristics of membrane electrodes containing different compounds. The aim of one of these fundamental experiments was to examine whether the transport of ions can really be measured. For this purpose we prepared "sandwich electrodes" as follows: we put silicon-rubber based silver iodide electrode membranes on both sides of a metal plate. The sandwich electrode thus prepared was immersed in an electrolyte solution, and the potential of the electrode was measured *versus* a reference. We changed the metal, and applied platinum, silver or aluminium (the surface of the latter was covered by an oxide layer), and in each case, the measured potential values were equal, i.e. the quality and thickness of the metal plate did not influence the potential values. We applied the three different metals because, as a first approximation, it was claimed by some researchers that the silver plate covered by the silver iodide membrane could take part in the silver transport. It became clear that this could not be the case with the other metals.

In the 1960s we developed an instrument for measuring the response times of ion-selective electrodes. With this instrument real electrode response times could be measured (in contrast to systems where the response time was measured in an electrochemical cell assembly in which internal eddy currents influenced the measured response times). The solution was jetted onto the surface of the electrode at a high velocity. For the iodide electrode, the concentrations of the solution used were 10^{-2} and 10^{-3} mol/l. The change of solution at the electrode surface took a very short time, less than 1 ms. We measured the response time of a silver plate electrode, of silver halide electrodes, and later on the response times of potassium ion-selective electrodes which contained a potassium-measuring component built into a polymer membrane. To our surprise, we obtained in each case practically the same response time of 20 ms. This response time indicated that after changing the concentration of the electrolyte, the components of the new solution need a certain time to get onto the surface of the electrode through the stagnant layer of solution (Figures 3a, 3b, 3c)

The Components in Figure 3a are: *1* Nitrogen tank; *2* high precision reduction valve; *3, 4* sample solutions of different concentrations; *5* salt bridge electrolyte; *6* pneumatic pistons; *7* optical detectors; *8* jets; *9* mobile clamping arrangement for the jets; *10* ISE; *11* reference electrode; *12* electrochemical cell; *13* drainage holes for sample solutions; *14* pH-meter; *15* high impedance amplifier (Keithley 604); *16* experimental control interface; *17* HP 82941A BCD interface; *18* HP-85 desk-top computer; *19* HP 7470A digital plotter; *20* HP-226 microcomputer .

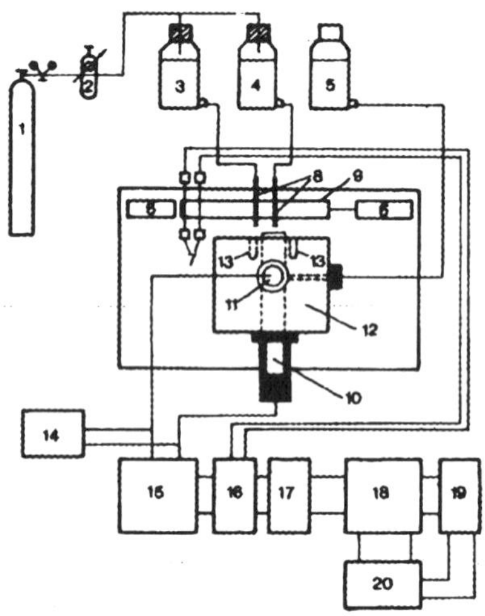

Figure 3a. Measuring set-up for response time studies.

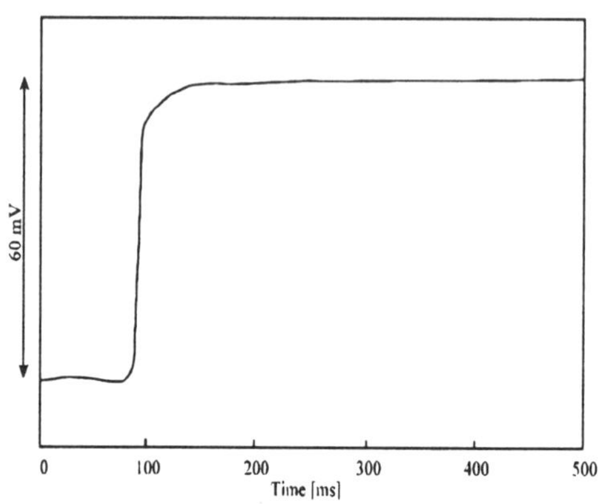

Figure 3b. Response-time curve of an AgI electrode: change of activity from 10^{-3} to 10^{-2} Mol/l

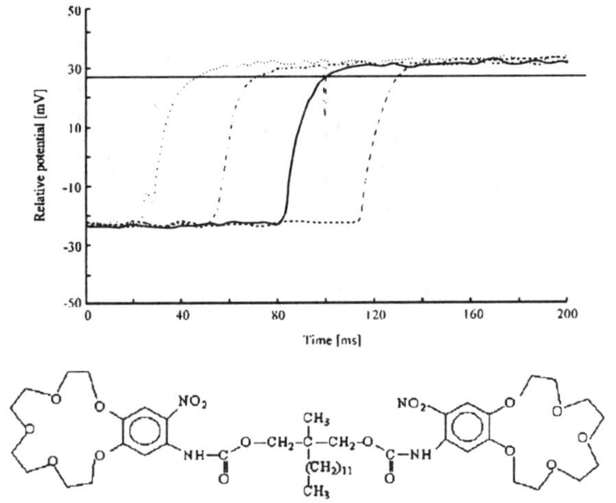

Figure 3c. Effect of conditioning 10^{-3} mol/l KCl solutions on the response time curves of BME-44/DOS based potassium selective electrodes. Activity step: 10^{-2} mol/l KCl⇒10^{-3} mol/l KCl. Ionic strength: 0.1 mol/l (NaCl); flow rate: 115 ml/min. (1) no conditioning; (2) 67-h cond.; (3) 10-d cond.; (4) 13-d cond.

This finding shows that the diffusion of ions into the membrane should be excluded, since diffusion of ions in the membrane is much slower than in the solution, and the establishment of the potential would take a much longer time than the value measured.

The next question was whether the electrode potential could be influenced by changing the electrode surface. The copper sulfide precipitate based copper ion-selective electrode was an excellent example in this respect. When we immersed the electrode with a Nernstian response into the solution of an oxidant, the electrode gradually failed to produce a Nernstian response as a function of the time of oxidation (Figures 4a, 4b, 4c).

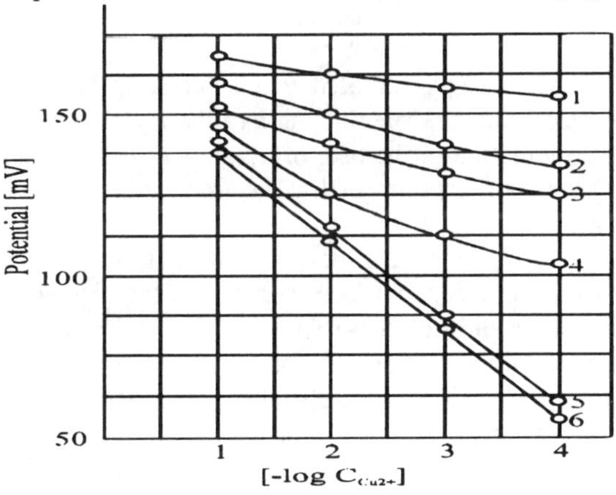

Figure 4a. Copper(II) response of a pressed pellet electrode after soaking in cerium(IV) solutions and regenerating in ascorbic acid solutions for different times. (1) soaking for 30 min in Ce(IV); (2) soaking for 10 min in Ce(IV); (3) regenerating in ascorbic acid for 1min; (4) regenerating in ascorbic acid for 30 min; (5) regenerating in ascorbic acid for 80min; (6) untreated electrode.

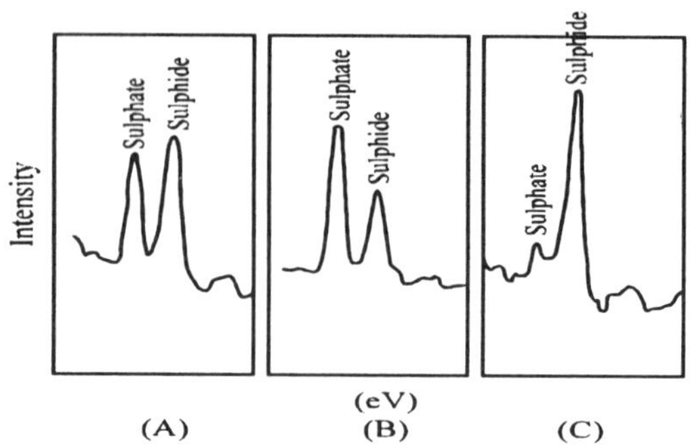

Figure 4b. Photoelectron spectra. Mg K_α x-ray, 6 keV, 40 mA, 240 W, $1{,}3*10^{-4}$ Pa. (a) Untreated CuS electrode; (b) electrodes treated in $KMnO_4$; (c) after treatment with $KMnO_4$ and then ascorbic acid

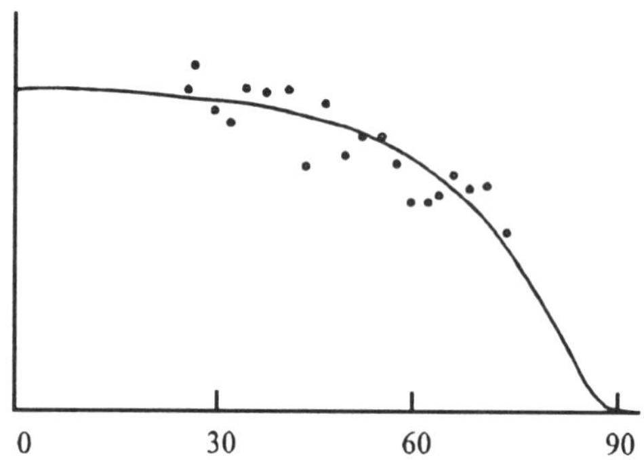

Figure 4c. Determination of the thickness of the sulphate layer by application of the variable take-off angle technique.

Photoelectron spectroscopy revealed that a blocking layer containing sulfate was formed on the electrode surface. The thickness of the layer was measured as 0.8 nm. For this purpose the determination of the take-off angle was used, which indicated molecular dimensions. The Nernstian response of the electrode was restored by treatment with a reducing agent, e.g. ascorbic acid.

The investigation of a lead sulfide-based lead-selective electrode seemed to be another interesting example, for which, in contrast to the copper-selective electrode, the response of the electrode did not cease entirely after the oxidation treatment: only the slope of the electrode calibration curve decreased below a certain solution concentration limit. In order to interpret this finding, we studied the surface by electron microscopy and determined the composition of the layer formed on the surface of the electrode. We stated that part of the electrode surface was covered by lead sulfate. At low concentrations of lead ions in the solution, dissolution of the lead sulfate assured the lead ion concentration in the solution at the electrode surface, which was higher than that originally present in the solution. We obtained a Nernstian response only at higher lead ion concentrations in the solution (Figures 5a, 5b).

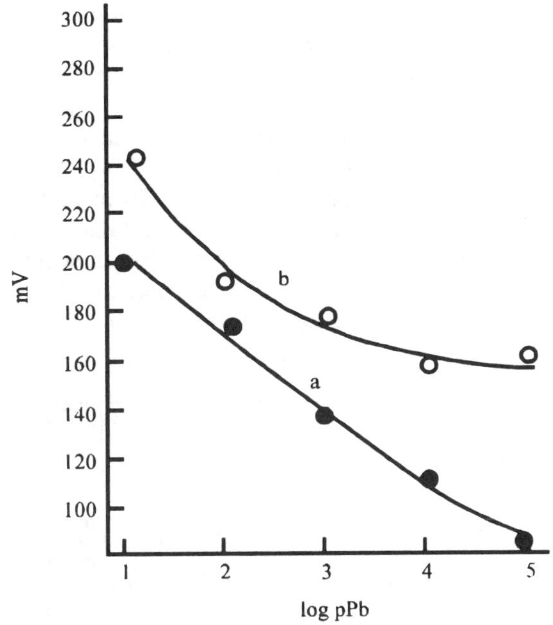

Figure 5a. Potential of a lead ion-selective electrode vs. a saturated calomel electrode plotted against the logarithm of lead ion concentration: (a) new electrode, without any pretreatment; (b) electrode treated with 2% hydrogen peroxide for 20 min. The electrode was prepared from Ag_2S/PbS (2:1 Molar ratio) pressed pellet membrane (for all Pb electrodes)

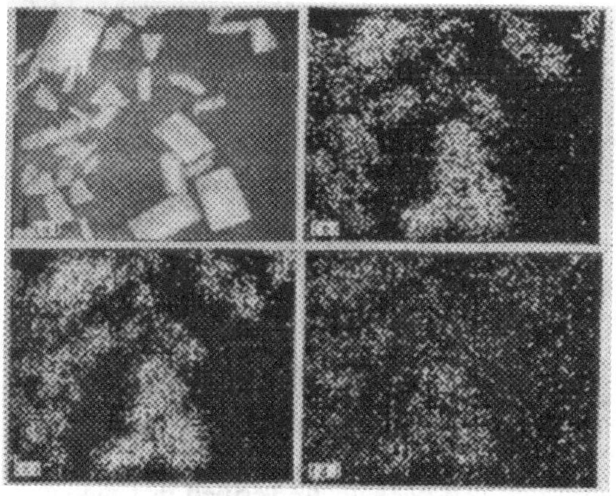

Figure 5b. Micrographs of a lead-selective electrode taken, from graphite replicas, after oxidation in 3% H_2O_2 for 3,5 h. (a) Secondary electron image; (b) sulphur S K_α element-specific image; (c) lead Pb M_α element-specific image; (d) oxygen O K_α element-specific image. The same surface was used in each case. (Magnification 3550×)

A further interesting experiment seemed to be to measure the depth of penetration of the primary ion into the electrode membrane. We conducted experiments with two types of electrodes: one was a solvent-polymetric membrane responding to potassium ions, and the other was a glass electrode.

The experiment was carried out by an ATR-IR technique on polymer membrane electrodes containing a selective comlexs of potassium. The electrode was soaked for a long period of time in a solution containing either KSCN or KCl. In the presence of the lyophobic SCN^- ions, K+ ions entered the membrane and during the relatively long time of soaking reached a depth exceeding that of the penetration depth of the IR rays. In the presence of the lyophilic Cl^-, K+ ions only enter the outermost surface layers, down to a depth of only 5-10 nm. Thermal diffusion may also have influenced the determination of the penetration depth (Figure 6a).

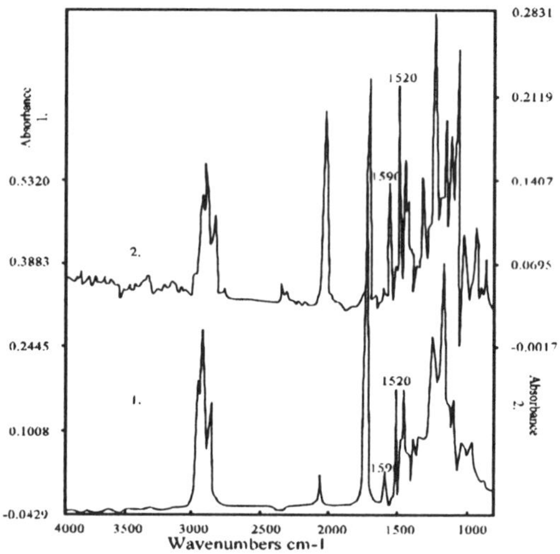

Figure 6a. Spectrum of the K-salt of the bis-crownether, in the case of 1. KCl; 2. KSCN. Peaks of the complex are at 1590 and 1520 cm^{-1}, the peak of SCN^- at 2300 cm^{-1}.

Composition: 60 % DOS (dioctylsebacate) in PVC with 3% bis crown etherFor a better indication of the phenomena, in the experiments with glass electrodes, instead of protons, Ag+ ions were used, to which glass electrodes also respond. An electrode was immersed in a 10^{-3} mol/l solution of Ag+ for 20-30s, until the potential was stabilized. The electrode was then dried, and covered by a gold layer. The removal of the surface layers was studied by means of a SIMS technique. It was found that the depth distribution of silver and gold was the same. (Figure 6b).

The comparison of the two phenomena verified that the surface of the electrodes is not uniformly plain. All the above findings speak against the phenomenon of ion transfer through the electrode, and we have to reject the assumption that an ion transfer mechanism can be applied for ion-selective electrodes on the basis of the Donnan phenomenon.

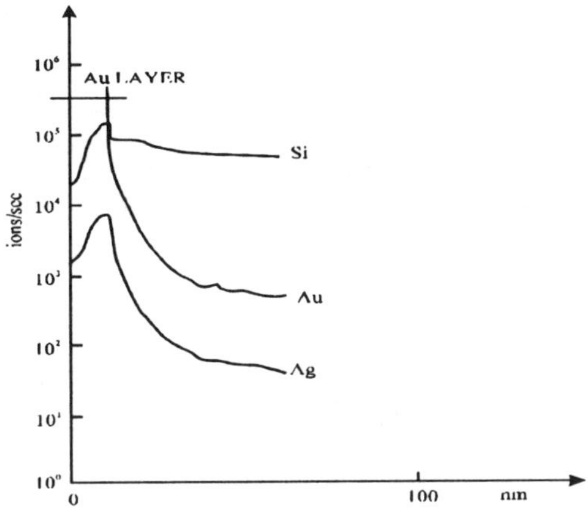

Figure 6b. Results of experiments with a glass electrode

A further question that can be raised is whether there may occur any transport of ions through electrode membranes. Yes, there may, but only if we apply an external voltage greater that the potential of the membrane electrode. The electrode then operates as a resistance, through which the current flows. In case of transfer, there can also be a selective partner (ionophor) conducting the current through the electrode, as the experiment of Simon excellently proved. If we do not produce a coulometric system, i.e., we do not apply an external voltage, the transfer cannot be established. In an ideal case we cannot measure any current across the electrodes, if the resistance of the electrode can be considered negligible compared to the input impedance of the measuring system. When we apply an electrometer for measuring the potential of electrodes, we achieve a situation where negligible polarization current flows in the system. Practically, the impendance of the measuring system should be about 10000-fold that of the impedance of the electrode.

3. Super Nernstian Potential Response

There are a number of examples of misinterpretations defining super-Nernstian potentials for ion-selective electrodes. Many of these papers have been published with claims which contradict the rules of thermodynamics. The phenomenon misinterpreted here can be traced back to the phenomena appearing at the surface of electrodes. We approached the interpretation of this phenomenon as follows:

We applied a microcell (Figure 7a), in which the electrode surface to volume of solution ratio was great, and the equilibrium was set up at the electrode in a few minutes.

We used various sorts of electrodes in these experiments. The (super-Nernstian) phenomenon was published in the literature for silver halides and silver sulfide. Our experiments focused on silver iodide and silver sulfide-based electrodes (Figures 7b, 7c).

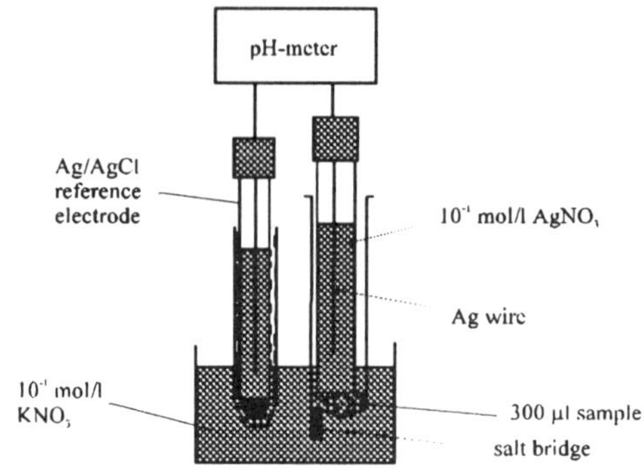

Figure 7a. Arrangement for the measurement of the electrode potential

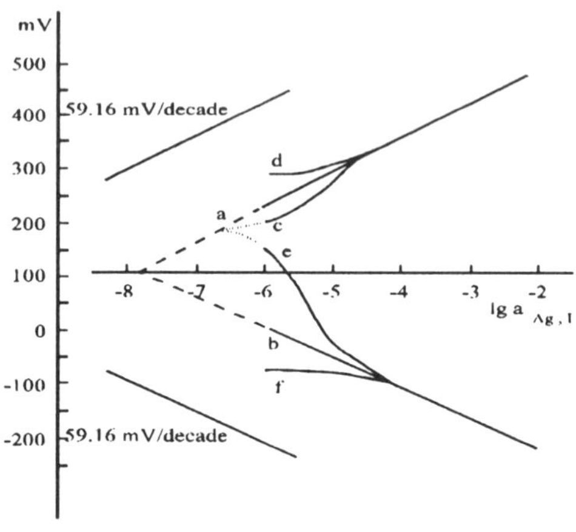

Figure 7b. Calibration curves (see text)

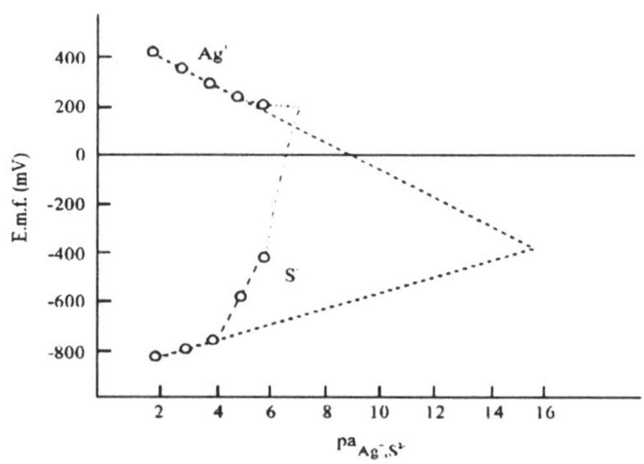

Figure 7c. Calibration curves for silver and sulphide ions measured

Calibration curves were taken with the silver iodide ion selective electrode:

(a) silver calibration curve in 50 mL solution volume; (b) iodide calibration curve in 50mL solution volume; (c) silver calibration curve taken in 300 μL solution volume, the elctrode conditioned in distilled water (d) silver calibration curve taken in 300 μL solution volume, the elctrode conditioned in 10^{-1} mol/L silver nitrate solution;(e) iodide calibration curve taken in 300 μL solution volume, the electrode conditioned in distilled water; (f) iodide calibration curve taken in 300 μL solution volume, the elctrode conditioned in 10^{-1} mol/L potassium iodide solution. Composition: 5% AgI in polysilicone rubber.

The results represented by the figures verified that the pretreatment of the electrodes, i.e. whether the electrode had been soaked in the solution of one or the other ion of the components of the membrane material, was reflected by the potential of the electrode, i.e. a sort of memory effect was measured. After the equilibrium was set up, i.e., a stable electrode potential was measured, we analyzed the solution in the microcell by means of an independent method and found that the potential data corresponded to the actual concentrations measured, assuming a Nernstian response.

If this is true, i.e. the measurable potential depends only on the ionic composition of solution at the surface of the electrode after equilibrium, then we can influence the concentration at the surface by means of a coulometric method, by applying an external current. It can be seen in the figure that the limit of electrode response is shifted to lower concentrations (Figure 8).

Accordingly, the super Nernstian relation is a false interpretation of the measured data. Of course, a sub-Nernstian relation can also be obtained. In the latter case, further parallel reactions can produce this effect.

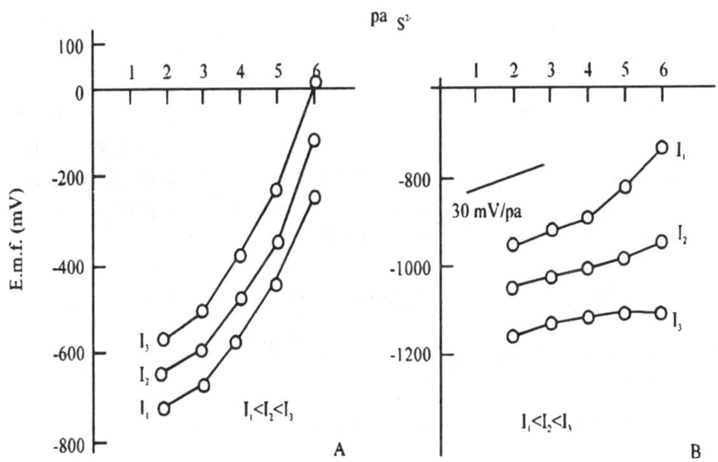

Figure 8. Calibration curves for sulphide: (A) with anodic current; (B) with cathodic current; Applied current: I_1=5,9 μA; I_2=11,8 μA; I_3= 17,7 μA. Composition: Ag_2S dispersed in polysilicone rubber

4. How is the Electrode Potential Produced ?

The establishment of an electrode potential is caused by charge separation at the surface of the electrode. The interpretation is connected with the chemisorption of the primary ion from the solution phase onto the surface of the electrode. In this case counter ions accumulate in the solution phase, and this is the charge separation. A further informative finding was that by using potassium electrodes containing different complexing agents (valinomycin or a bis-crown-ether compound) we measured the same potential for the same potassium ion concentration of the solution. This can be understood only if the chemical potential of the solute is responsible for the electrochemical signal. To express the thermodynamics, the Gibbs-Duham relation can be given as:

$$-\Delta G = nFE \quad (1)$$

where G is the change of chemical potential of the solute, n the number of charges, F the Faraday constant and E the electrical potential.

In the field of ion-selective electrodes, the experience and interpretation of Nicholsky brought about an enormous change, i.e. the ion exchange conception was introduced for glass electrodes and later, of course, for all electrodes. The selectivity coefficients of the electrodes were practically calculated for the first time using precipitate-based electrodes, where the solubility equilibrium was known. For other electrodes the selectivity coefficient could not be deduced from such thermodynamic data. In this case, the selectivity coefficients did not contain any other factors; they contain only the thermodynamic data and no other factors which include transfer coefficients.

This really brilliant approach of Nicholsky had a lot of negative effects, too. One of them was that a great number of studies set about investigating ion-exchange systems, for which they thought to prepare electrodes. From the end of the 1930s, the literature records such investigations by the hundreds. But "selective" electrodes could not be invented in this way. The other negative effect was caused when, besides the Nicholsky equation, textbooks and other books presented a formula standing opposite to the principles of physics. Knowing that if the quantity of charges in the system does not change, one must conclude that no potential change can appear, i.e. the following equation cannot be given as the basic equation describing the operation of an electrode:

$$A^+ + \text{polymer } B \leftrightarrow \text{polymer } A + B^+ \quad (2)$$

5. Classification of the Electrodes

The electrodes used currently can be divided into two gorups:

(a) basic electrodes suitable for measurement of ions and
(b) electrodes suitable for measurement of molecules.

The signal is provided by a so-called basic electrode and the measured componenet is produced, by some chemical (e.g. enzyme reaction) or physical (diffusion etc.) reactions, from the molecules.

According to our present knowledge, the basic electrodes can also be divided into two groups:

(a) Electrodes giving chemosorption, the three types of which are as follows:
    (1) electrodes based on the principle of acid-base reactions
    (2) electrodes based on the principle of precipitation reactions
    (3) electrodes based on the principle of complexation reactions
(b) Electrodes whose operation is based on the lyotropic series, for which the ideal Nernstian relation is not always valid, the potential is shifted in time etc.

The Donnan exclusion effect can be understood if we realise that, if electrode is in contact with a solution concentrated enough for the cation to saturate the active sites of the surface, then the anion may enter the membrane, and the electrode will respond to the anion.

According to the above, the electroanalytical definition of ion-selective electrodes by IUPAC must be reworded, and in the same way the determination technique of selectivity coefficients could also be touched upon.

This paper has criticised the old theory of transfer reactions as producing the potential signal of ion-selective electrodes. It presented experimental results providing definitely the occurrence of chemisorption reactions. In case of certain electrodes, the operation can be interpreted in terms of the lyotropic series. Thus this paper places not only the interpretation, but also the possibility of invention of new electrodes, on a new basis.

6. References

1. Habber, F. and Clemensiewicz, Z. (1909) *Phys. Chem.*, 67, 385 .
2. Donnan F.G. (1911) *Electrochem,* 17, 572.
3. Guggenheim, resp. Kortum, G. and Bockris, I.O'.M. (1951) *"Textbook of Electochemistry"*, Elsevier, Amsterdam.
4. Data (Unpublished paper).
5. Pungor, E., and Hollos-Rokosinyi, E. (1961) *Acta. Chim. Hung.*, 27, 63
6. Tóth, K., Gavallér, I., and Pungor, E. (1971) *Anal. Chim. Acta*, 57, 131
7. Stefanac, Z., and Simon, W. (1966) *Chimica*, 20, 436
8. Pungor, E. Tóth, K., Pápay, M.K., Pólos, L., Malissa, H., Grasserbauer, M., Hoke, W., Ebel, M.F., and Perry, K. (1979) *Anal. Chim. Acta*, 109, 279
9. Pungor, E., Tóth, K., Nagy, G., Pólos, L., Ebel, M.F., and Wernisch, I. (1983) *Anal. Chim. Acta,* 147, 23
10. Tóth, K., Lindner, E., Pungor, E., Zippel, E., and Kellner, R. (1988) *Fresenius' Z. Anal. Chem.* 331, 448
11. Pungor, E. (1996) *Electroanalysis*, 8, 348
12. Simon, W., and Morf, W.E (1972) *Ion-Selective Electrodes,* Symp. held in Mátrafüred, Hungary, 23-25 October, Akadémiai Kiadó, Budapest
13. Harsányi, E.G., Tóth, K., Polos, L. and Pungor, E. (1982) *Anal. Chem.*, 54, 1094
14. Harsányi, E.G., Tóth K., and Pungor, E. (1984) *Anal. Chim. Acta,* 161, 333
15. Nicholsky, B.P. (1937) Acta Physicochem. USSR, 7, 797
16. Eisenman, E. (1968) *Anal. Chem.* 40, 310

17. Buck, R.P., and Lindner, E. (1994) Recommendations for nomenclature of ion-selective electrodes, *IUPAC recommendations* 1994. *Pure Appl. Chem.* 66, 2527
18. Umezawa, Y., Umezawa, K. and Sato, H. (1995) Selectivity coefficients for ion-selective electrodes: Recommended methods for reporting Kpot values, *Pure Appl. Chem.* 67, 507

FINANCES AND PHYSICS

V.M. BARYSHEVSKY
Professor, Director
Research Institute of Nuclear Problems
Bobruiskaya str. 11
220050 Minsk
Belarus

A physicist opening a modern book on economic analysis (for example, [1, 2]) will discover with surprise that it is not only very interesting reading. It will turn out that the description and study of financial market behaviour are similar to the natural phenomena analysis in physics at their initial perception when the intrinsic fundamental laws are not revealed and correspondent basic equations are not formulated. (The situation looks like the one that existed when Faraday observed electromagnetic phenomena, but Maxwell's equations had not yet been discovered) So, it is no wonder that in the recent decade the idea to apply the experience gained by physicists in natural investigations to financial market analysis has been stimulating intense interest [3, 5].

Now, the methods of economic analysis are proceeding gradually from concept definition and consideration of simple mathematical correlations to the discovery of laws underlying the market behaviour in order to predict it in future. Until recently the probabilistic characteristics (mean value, dispersion) were especially studied and formalistic mathematical methods have been used (i.e. some people do believe it is possible to study past market data and use chaos theory, fractals and so on to predict future prices behaviour).

The possibility to profit much more from the financial market (and, particularly, in stock markets) than from bank deposits becomes a major stimulus for this study. But the above methods can not render the necessary assistance to the traders to gain this extra-profit (i.e. profit exceeded over that obtained from bank deposits). Market participants, even being equipped with the most advanced tools (from trade-stations to neural networks), always encounter absolutely unexpected situations and only cool-headed and experienced traders succeed in market operations. "Housewives" (i.e. unprofessional market participants) can profit in the mean only when the whole market rises. What is the matter? Why is it that advanced models such as chaos theory, fractals and so on cannot forecast anything in practice, but can only finely explain what has already happened? This becomes understandable if we consider that the market, as the Universe, is developing according to laws that are not quite clear to us. Only knowledge of these laws can enable us to make accurate forecasts.

In contrast with physics describing the behaviour of the material world, the more detailed analysis of market behaviour can even prevent gaining extra-profit. This is because a person (market participant) should be taken into consideration as market constituent. Suppose one states equations that describe price behaviour in the market. It is

A. K. Hyder et al. (eds.), Multisensor Fusion, 879–882.

easy to find out that these equations should include not only fundamental economic indicators (such as profit, dividends ...) but also psychology of market participants (sellers and buyers). Due to the feedback, which appears at the analysis of price behaviours made by market participants, their actions become synchronised just as at a spiritualistic seance. (It is important to note that these actions also determine the price of a company.) One should also consider the thought-out actions of infuential players and even misinformation frequently proposed as the recommendations of market experts. It can be easily illustrated by the following examples (see Figure 1): though market analysts recommended to buy and hold shares of At Home Corporation and E Trade Group Inc., the price of these shares fell three times and twice, respectively. From the above, we see that we are not dealing with the physics of the material world but with the psycho-physics of a complicated materially-mental object. For the analysis of the latter, it seems, the methods similar to those being developed for lie detectors are more suited.

And now the most significant concept can be formulated as a principle of complementarity for the forecast of market prices: the more thorough and accurate our theory of stock market prices is developed for the trader, the more errors it will give. This concept can be easily illustrated by the following example. Suppose that somebody develops such a theory and all market participants begin to apply it. Then, everyone increase their profits, but this is absolutely impossible – someone must lose.

Therefore, the more perfected the theory, the greater the influence on the behaviour of market participants, disturbing the theory of operation.

The knowledge of the laws governing the Universe in physics is based on the principles of beauty and harmony. The fundamental laws of nature are stable: the physical parameters vary over such long time periods that they can be considered as time independent.

At the same time, for economics as a whole and, especially, for the stock market, rapid changes are typical. Now, in contrast with past decades, the stock market displays the properties of detached individuals. And as it was mentioned, the collective psychological factors are of special significance. Therefore, the major premise of physics, that fundamental laws do not depend on the individual behaviour, can not be applied to the stock market description because the "world constants" of the latter continuously and unpredictably change. To describe and perform quantitative analysis of the events occurring in the stock market, the combined use of physical and psychological methods is necessary. Actually, the market tries to deceive the trader. Nevertheless, in spite of this pessimistic consideration, the laws of price change can be discovered and used as a tool by the trader.

At Home Corp (NasdaqNM:ATHM) **1-Jun-2000**

Number of brokers recommending as:	Month ago 0	1	2	3	Average Recommendation	
					Strong Buy) 1.00 - 5.00 (Strong Sell)	
Strong Buy	7	7	7	7	This Week	**1.85**
Moderate Buy	8	6	8	10	Last Week	1.83
Hold	4	4	3	3	Change	-0.0
Moderate Sell	0	0	0	0	Industry: INTERNET SERVCS	
Strong Sell	0	0	0	0	Ranked 79 of 95	
Mean*	1.85	1.83	1.76	1.7	Earnings Date (Approx.): N/A	

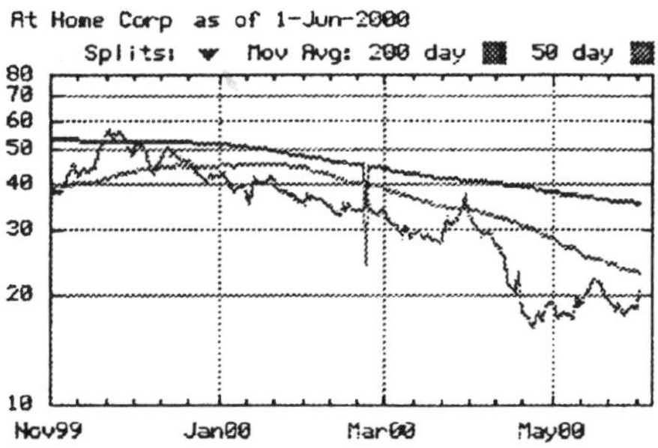

Trade Group Inc (NasdaqNM:EGRP) **1-Jun-2000**

Number of brokers recommending as:	Month ago 0	1	2	3	Average Recommendation	
					Strong Buy) 1.00 - 5.00 (Strong Sell)	
Strong Buy	5	6	4	4	This Week	**2.05**
Moderate Buy	7	7	6	6	Last Week	1.97
Hold	7	6	6	6	Change	-0.1
Moderate Sell	0	0	0	0	Industry: FIN-INVEST BKRS	
Strong Sell	0	0	0	0	Ranked 20 of 29	
Mean*	2.05	1.97	2.15	2.09	Earnings Date (Approx.): N/A	

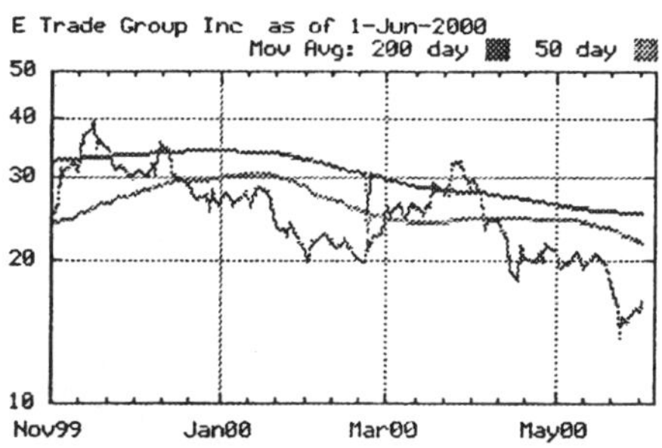

E Trade Group Inc (NasdaqNM:EGRP) **1-Jun-2000**

Number of brokers recommending as:	Month ago 0	1	2	3	Average Recommendation	
					(Strong Buy) 1.00 - 5.00 (Strong Sell)	
Strong Buy	5	6	4	4	This Week	**2.05**
Moderate Buy	7	7	6	6	Last Week	1.97
Hold	7	6	6	6	Change	-0.1
Moderate Sell	0	0	0	0	Industry: FIN-INVEST BKRS	
Strong Sell	0	0	0	0	Ranked 20 of 29	
Mean*	2.05	1.97	2.15	2.09	Earnings Date (Approx.): N/A	

Figure 1. During long time market analysts recommended to buy and hold shares of At Home Corporation and E Trade Group Inc., the price of these shares fell three times and twice, respectively.

References

1. Campbell, C.D., Campbell, R.G., Dolan, E.G., (1995) "Money, Banking and Monetary Policy".
2. Sharpe, W.F., Alexander, G.J. Bailey, J.V. (1995) "Investments" Prentice Hall International, Inc.
3. Chashyhair, S., Breymann, W., Peinke, J. Talkner, P., Dodge, Y., (1996) "Turbulent cascades in foreign exchange markets" Nature v.381, 767-770.
4. Adamchuk, A.N., Esinov, S.E., (1997) "Collective fluctuating assets in the presence of arbitrage opportunities and option pricing", Sov. Uspehi Phys. Nauk v.167, N12.
5. Pimbley, J.M., (1997) "Physicists in Finance" Physics Today, January, 42-46.

REMOTE MULTISENSING OF BIOLOGICAL SYSTEMS

B.A.K. CHIBBER*[1], P. FAY[2], G.H. BERNSTEIN[2],
F.J. CASTELLINO[3] AND J.G. DUMAN[3]
[1]*Department of Biological Sciences,*
[2]*Department of Electrical Engineering, and,*
[3]*Department of Chemistry and Biochemistry*
The University of Notre Dame
Notre Dame, IN 46556 USA
chibber@nd.edu

1. Introduction

The promise and potential of biotechnology rests, ultimately, on our understanding of biological processes at a molecular level. Such understanding allows manipulation and modulation of complex biochemical, metabolic and genetic processes in microorganisms, plants, animals and humans. The vast beneficial potential in manipulating biological processes and systems derives from the diversity of biological organisms and their inherent versatility and adaptability, as reflected in their validation of Darwinian paradigms concerning evolution and natural selection. Within living systems, the obviously chaotic nature of diversity based biological environments is transformed into complex and intricate physiological structures and biochemical pathways based on spatial and temporal organization of an organism's biochemical components. Utilizing structures and pathways, the myriad possible chemical consequences inherent in a collection of molecules in living systems reduces to organized metabolic processes. This ordered biochemical architecture in living systems is based almost entirely on molecular recognition and biospecificity. It is, therefore, through intervention of selective effectors, inhibitors, substrates, hormones and other pairs of transmitter and receptor molecules that living organisms maintain control over the course of chemical change within their systems. Examples of such control range from control of gene expression to self/non self discrimination via the immune system. Biotechnology, in essence, attempts to exploit the known factors affecting the flow of chemical information in biological systems for human use.

2. Rationale

Traditional techniques in the study of metabolic change in living systems have utilized a "snap shot" approach to evaluating their metabolic status. These are laborious methods and techniques of separating & purifying metabolic component molecules, and subsequently reconstituting such components to discover and understand their function(s). Continual, real time monitoring of biological processes representing metabolism and its control, its biochemical means of adaptation, and its versatile responses to biological

A. K. Hyder et al. (eds.), Multisensor Fusion, 883–887.

environmental and stress factors would represent a quantum leap in our ability to both understand and exploit the strengths and weaknesses of biological systems.

Specific shared properties of biological systems and their chemical characteristics provide a means of developing such real time monitors. In the first instance, regulation of inter and intra-cellular biochemical interactions in biological systems generally takes place in an aqueous fluid matrix such as the cytoplasm in all cells, or the hemolymph in insects, or blood in higher vertebrates including Man. Further, biochemical interactions and reactions are very specific, being based on molecular recognition among metabolites; and finally, metal surfaces can be suitably modified by chemically attaching a variety of chemicals, including biochemicals such as proteins, metabolites and other biologically active molecules.

3. Design

A general utility design for a biological monitor is proposed upon the preceding principles. In essence, this monitor is an electrochemical detector coupled to a silicon chip, the latter capable of electronically relaying the detector's response to remote sensor(s). The monitor consists of two microelectrodes mounted on a silicon chip designed appropriately to respond to voltage fluctuations across those electrodes, and capable of electronically relaying its responses to remote sensors. Further, one or both electrodes are chemically modified with molecules that specifically bind to, or otherwise interact with, constituents of interest (insulin, coagulation factors, hormones, etc.) in the biological fluid, such that those specific interactions modulate the electrical potential across those electrodes. Thus, such biological probes act as monitors for bio-molecules of interest and continuously relay the relative levels of those molecules to remote sensors.

Clearly, such a monitoring and sensing system is applicable to a vast number of biological organisms as well as to whole ecosystems. It is also equally clear that such monitors/remote sensors can be integrated into information feed back loops for the control and maintenance of the target biological organisms and ecosystems.

4. Implementation

A critical piece of technology that must be developed in order to realize a fully-wireless biological testing architecture is the ability to remotely sense, via low-power wireless techniques, the outcome of real time biochemical sensing. The operation of this scheme is described scematically in Figure 1, and consists of an antenna and L-C resonant circuit is constructed as a single silicon integrated circuit. The resonant frequency of the resonator is set by appropriate choice of L and C. In our sensing scheme, an electrochemical potential is sensed and converted to a shift in resonant frequency through the well-known bias-dependent change in diode capacitance. For this work, we anticipate that a varactor-based resonator, coupled to metal probes, will be used. Several resonators attached to separate locations and antennae allow the results at each location to be determined wirelessly by exciting the resonators with an excitation transient (supplied externally through the monitoring antenna), and monitoring the frequencies radiated by

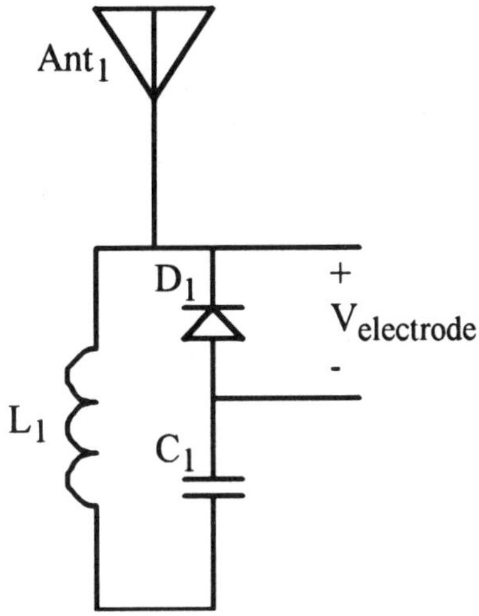

Figure 1. Bioelectochemical Resonant Sensor

the resonators. Since the resonant frequency of the circuit associated with each individual test to be performed is unique, the outcome of all tests can be determined simultaneously by examining the frequencies generated and re-radiated by the chips. An external transmitter will excite the resonators. The excitation transient from the transmitter is a short burst from a swept-frequency microwave source.

This approach to sensing the test outcomes has several features worthy of note. In particular, it requires no power supply except for that provided by the stimulation transient, and so eliminates the need for batteries or connections to the test chip, making it feasible to attach to small biological systems. The very simple, purely passive, circuitry required to implement this scheme also makes it very simple to fabricate, providing a convenient means of monitoring biological fluids in real time.

The resonant circuitry will be monolithically integrated with the micromachined needle probes as a system on a silicon chip. The complete device will be one millimeter square, with two probes extending out in the plane of the wafer. The varactor diode and inductor structure will be lithographically defined on the area of the square. Metal electrodes will extend off of the chip along the surface of the needle probes. The thickness of the square will be at most the original thickness of the silicon wafer, which is approximately 0.4 mm. This can be thinned if necessary. The length of the needles can be as much as 200 microns, with a width of as little as 10 microns. The thickness of the needles can be modified through bulk micromachining techniques using backside wafer etching with membrane etch stops on the surface to a minimal thickness. Further study will reveal the optimum dimensions of the needles for strength.

5. Applications

In situ, continuous biochemical monitors can be applied to the study of a very wide variety of biological systems. Two such disparate applications of such remote sensors are provided as examples of the diverse nature of information that they can provide.

1. Applications in Insect Biology

The Pyrochroid beetle Dendroides canadensis over winters in various larval stages under the bark of partially decayed logs. The life cycle requires 2-3 years to complete, and therefore each larva over winters at least twice. The size of the last larval instar is ~20 by 4 mm. We have studied this species for nearly 25 years, and therefore there is a considerable volume of information on their subzero temperature adaptations, especially concerning the role of antifreeze proteins.

Antifreeze proteins (AFPs) lower the freezing point of water by a non-colligative mechanism, but do not affect the melting point. Thus they produce a characteristic difference between the freezing and melting points which has been termed thermal hysteresis, which can be easily measured, and which reflects the antifreeze activity of the AFPs present in a sample.

D. canadensis AFPs are the most active AFPs that have been identified. They function to prevent the larvae from freezing by (a) inhibiting inoculative freezing initiated by external ice across the cuticle and (b) promoting supercooling by inhibiting ice nucleating sites in the hemolymph, gut, etc. AFP activity in the hemolymph varies seasonally. Low levels of activity are present in the summer, but activity generally rises throughout the autumn, peaks in mid-winter, and slowly decreases through the spring. Temperature, photoperiod and thermoperiod act to cue these changes. It is difficult to attribute all of the seasonal changes in AFP activity to variations in AFP concentration, although this certainly plays a major role. The ability to instantaneously and continuously monitor AFP concentrations in several individuals in a population over an extended period, and to compare these to temperatures within the hibernaculum (which often vary considerably from air temperatures) would be most useful.

Data will be obtained continuously from four separate remote sensors, viz. ambient air temperatures, hibernaculum temperatures, hemolymph conductivity, and AFP levels. Processing such data will provide a clear picture of this organism's response to its environment. Further input may be obtained by tracking photoperiods and diurnal temperature variations to augment these observations.

2. Applications in Blood Coagulation and Fibrinolysis

A complex group of biochemical mechanisms regulate hemodynamics and hemoistasis in circulating blood. In particular, components of the blood coagulation and fibrinolytic systems function in tandem to maintain structural and functional integrity of the circulatory system. Application of real time monitoring of the multiple components which orchestrate these processes are of immense importance to the management of disparate disease states, trauma and as adjuncts to pharmacological and surgical treatments which impact or otherwise impose physiological stress on the circulation system.

A schematic outline of the coagulation and fibrinolytic systems is described in Figure 2. Coagulation or blood clotting is initiated intrinsically or extrinsically following external or internal injury through activation of factors XI or VII. These activated factors (XIa, VIIa) in turn activate other factors, leading ultimately to formation of a fibrin clot. The clot, in turn, is contained by its breakdown to Fibrin Degradation Products by Plasmin. The various feedback mechanisms known to be in place for a dynamic control of these enzyme systems is partially shown on this figure. As an initial application of the multisensors proposed herein, real time monitoring is proposed for the generation of Coagulation Factors XIa, VIIa, Xa, VIIIa, Va, Thrombin, as well as of the fibrin clot, plasmin and activated Protein C. Such investigations on the time dependent changes and stress responses in circulating blood will be invaluable in providing information for the management of hemostatsis in Man.

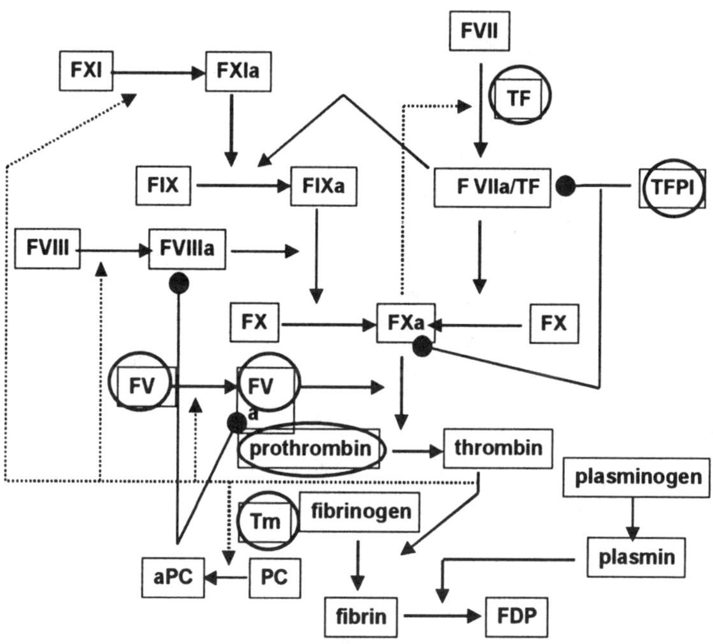

Figure 2. Coagulation Cascade, Showing Positive and Negative Controls

EXTRACTING INFORMATION FROM MULTICHANNEL VERSUS SINGLE CHANNEL EEG DATA IN EPILEPSY ANALYSIS

C. JAMES AND D. LOWE
Neural Computing Research Group
Aston University
Birmingham
United Kingdom

1. Introduction

The EEG is one of the major tools used in the diagnosis of epilepsy. EEG recorded during an epileptic seizure reflects the disturbed ongoing background EEG that exhibits a mixture of evolving rhythmic activity and heavy artifactual contamination. The electrographic recordings yield information about the seizure origin and the extent of spread of the activity over time. However, analysis of epileptiform EEG with traditional methods is severely hampered by contamination of the ongoing EEG with artifacts from various sources – muscle artifacts (EMG) being one of the major contributors.

In previous work we have introduced a methodology for analysing epileptiform activity recorded in the EEG at long term epilepsy monitoring units by first decomposing the recorded EEG into its fundamental components; Kobayashi et al. [1], James and Lowe [2]. In this paper we show the results obtained in extending our investigation into contrasting the use of multichannel data versus single channel data in such methods. Dynamical systems theory is applied to separate single channels of data where a dynamical embedding matrix is constructed for each single channel. ICA is then applied to each single channel in order to extract a number of independent components (ICs). ICA is also applied directly to the multichannel data. In order to assess the results given by ICA for each case, the IC's are clustered into a 2-D mapping space using a topography preserving clustering technique known as Sammon Mapping. This mapping, coupled with visual examination of the IC's and their derived spectrograms, allows for a comparison between the information extracted using single and multichannel analysis techniques.

2. Methods

2.1 EEG DATA

Multichannel EEG data was recorded according to the international 10-20 electrode placement system over 25 electrodes as shown in Figure 1; Lagerlund et al [3]. The EEG is sampled at a rate of 200 samples/sec, digitised to 12 bits and stored for off-line analysis. The scalp EEG data was obtained from a patient of the Montreal Neurological Institute and Hospital. The data was recorded with a reference at electrode position FCz

A. K. Hyder et al. (eds.), Multisensor Fusion, 889–895.

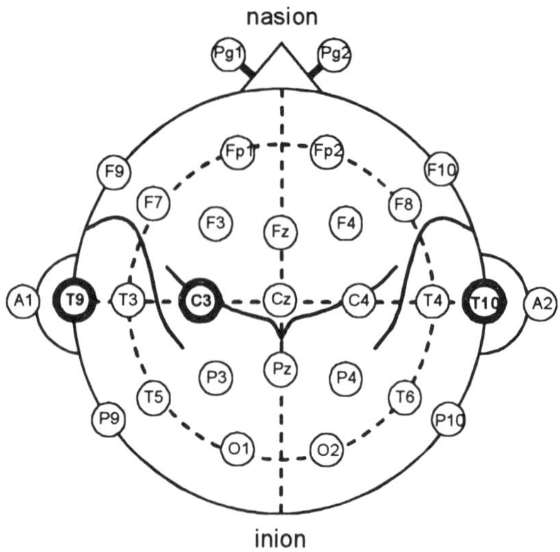

Figure 1. The modified 10-20 electrode placement system.
The highlighted electrodes correspond to those chosen for single channel analysis.

and the data matrices in each case were mean corrected in the columns (i.e., an average referential montage was assumed) and the rows. The recordings consisted of 20 s of seizure EEG with the seizure onset (as determined visually from the scalp EEG) occurring approximately 5-10 s into the segment (see Figure 2).

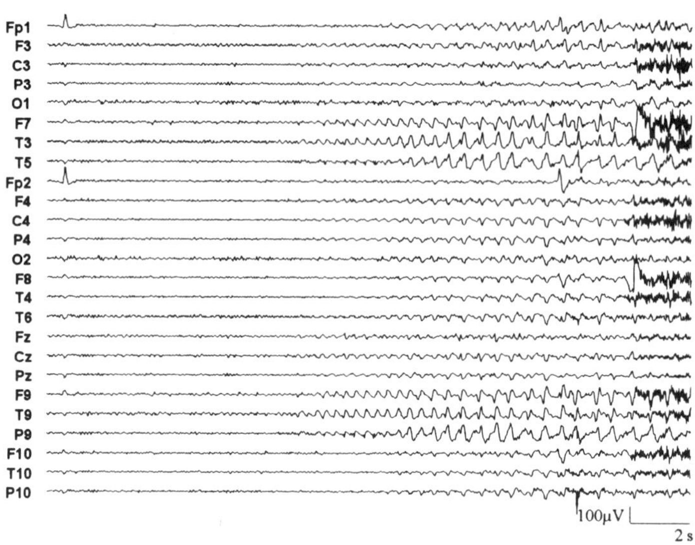

Figure 2. The 20 s segment of EEG (average reference) for patient #1 with the seizure onset occurring 8-10 s into the segment. The seizure has a focal left temporal onset and spreads to wider regions over the next few seconds.

Dynamical embedding is a method for extracting information from a sampled time series. The information regards the underlying generators of that series; Broomhead and King [4]. This is based on the principle that the measured signal is due to the nonlinear interaction of just a few degrees of freedom with additive noise suggesting the existence of an unobservable deterministic generator of the observed data. For an underlying system with the number of degrees of freedom given by D, then the measure D can be used as one coarse measure of system complexity. Takens' theorem [5] allows for the reconstruction of the unknown dynamical system with D degrees of freedom that generated the measured time series, simply be reconstructing a new state space out of successive observations of the time series. Applied to EEG signals, we thus assume that we can capture the dynamics of the generator simply from the observed (recorded) EEG.

A dynamical embedding is constructed from a series of delay vectors taken from the observed data, for a given lag τ and a given number of lags or embedding dimension (m). This delay vector describes (scalar) observations of the underlying system states, assuming that the data are generated by a finite dimensional nonlinear system. Takens showed that the embedding dimension $\hat{m}$ must be at least at large as D, but in practice $\hat{m}$ must be such that: $\hat{m} > 2D + 1$. When applied to real world data the vector delay size m actually used needs to be a lot larger than the Euclidean embedding dimension ($\hat{m}$) because of dependencies in the time series data. The vector delay size m needs to be 'big enough' to capture the information content necessary and if the time series data is heavily correlated, then more time series samples are needed to make up the required information content of the delay vector. An embedding matrix is finally constructed out of a number (N) of consecutive delay vectors. Provided the choice of lag term τ, delay vector size m and number of lag vectors N are adequate, the embedding matrix contains enough information to extract a value for the observed degrees of freedom D which is linked to the complexity of the sequence generating the measured data.

Based on the above dynamical embedding is performed on a number of single channels separately (delay vector size m=90, number of delay vectors N=4000-m, where 4000 samples represent 20 s of data sampled at 200 Hz).

2.2 INDEPENDENT COMPONENT ANALYSIS

ICA performs a blind separation of statistically independent sources using higher-order statistics. Several implementations of ICA can be found in the literature; Comon [6], Bell and Sejnowski [7], Makeig et al [8] and Hyvärinen and Oja [9]. We use the Fast ICA algorithm, [9] and [10], because of its ease of implementation and speed of operation. However ICA has some fundamental restrictions, which are mainly: (a) neither energies nor signs of the ICs can be calculated, and (b) there is no ordering between the ICs. In our implementation of ICA for scalp EEG analysis we make assumptions that are in keeping with the general assumptions governing the application of ICA. In particular we assume that:

(i) The scalp EEG is a linear summation of the electrical activity from various brain regions; Nunez and Katznelson [11].

(ii) The potential field distribution is spatially fixed and only the electrical 'strength' is changing within these regions; Scherg and Von Cramen [12].
(iii) Any epileptic activity is independent of the ongoing background EEG - this holds true at least early on in the evolution of a seizure.

In order to compare the fundamental components of the multichannel EEG versus those extracted from single channels of EEG, Fast ICA was performed on (a) the multichannel EEG directly (for n=25 channels) and (b) 3 separate channels, independently, following the construction of an embedding matrix from each. The 3 channels were chosen thus: one channel over the focus of the seizure (as determined visually from the scalp recordings), the 2nd on the contralateral side and the 3rd chosen to be near the focus but be relatively "electrically-quiet".

2.3 SAMMON MAPPING

Fast ICA yields as many independent components as there are measurement channels. Although, by definition, all components are statistically independent, for mixtures with relatively large numbers of channels the components tend to cluster into groups which relate to similar origins. In order to compare (n+3m) ICs we perform a cluster analysis of the ICs using a technique known as Sammon Mapping. Sammon Mapping allows for the visualisation in lower dimensional space (2-D in this instance) of data residing in a higher dimensional data space [13]. Sammon Mapping works by minimising an error measure of the inter-point distances also called the STRESS measure. The stress measure requires a measure of distance between the ICs in the higher dimensional space. We tentatively derive a distance matrix giving the "similarity" of the components as the maximum (normalised) correlation coefficient derived from the cross-correlation between IC's allowing for a lag of 500 ms, either way, in the components. This measure is intended to capture those IC's which are statistically independent yet are representing the same underlying neural activity.

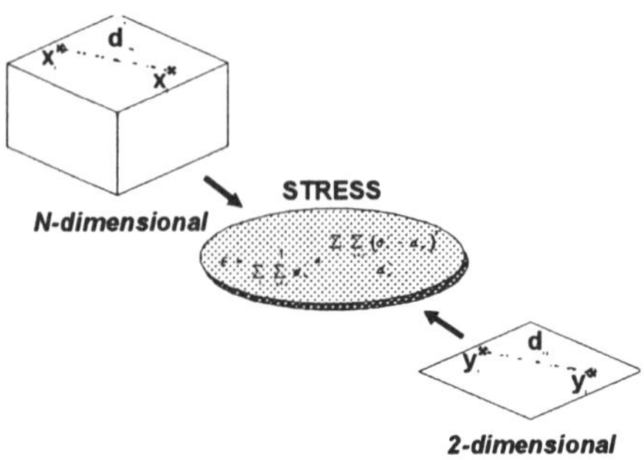

Figure 3. Sammon Mapping Clusters Data From A High Dimensional Space (N) Into A Lower Dimensional Space (2-D In This Case) Whilst Preserving The Topography. The Stress Term Is Optimised To This Effect.

3. Discussion

Preliminary results indicate that both single channel analysis as well as multiple channel analysis extract more than a single seizure component from the EEG data. In each case artifacts are also readily identifiable (where they are present in the original signal). For the single channel case the method of dynamical embedding results in a data matrix describing the underlying variability of the signal in question – providing the size of the delay vector is large enough and a sufficient number of delay vectors are extracted. The biggest problem in applying ICA to both types of data matrices is the identification of the ICs. There are a number of techniques to visually (i.e., subjectively) identify seizure or artifactual components as outlined in this paper, however an objective method is still required to separate the "relevant" components from the noise. Furthermore, for cases where there are a large number channels to be analysed, ICA results in a large number of components which are statistically similar but belong to the same 'category'. Sammon Mapping tackles this drawback and provides an intuitive visual representation of the components extracted in each case and gives the relation of each single channel with the multichannel data. The results of the clustering show definitive broad clusters given by the components of each channel/multichannel as the case may be. Within clusters ICs of an artifactual nature form definitive clusters in each case. The multichannel ICs span a broader space than the individual channel ICs. The individual channel ICs show similarities with different components of the multichannel data, with the channel closest to the focus best representing the multiple seizure components extracted by the multichannel case.

Overall the Sammon Map results in an intuitive clustering given the data, and there is still room or error when one considers the distance measure used to compare the similarity of the ICs, a better similarity measure could probably result in a better clustering of the components. This notwithstanding, the Sammon Map provides an objective method to compare and contrast the many components yielded but multichannel analysis and single channel analysis followed by dynamical embedding.

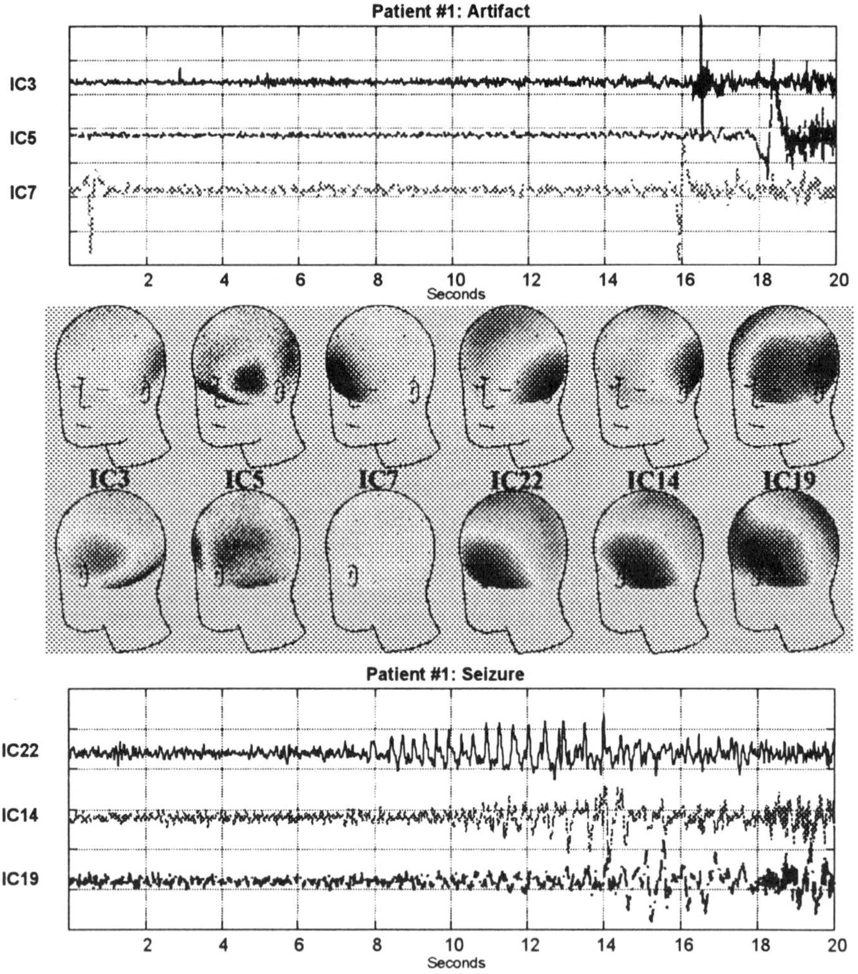

Figure 4. Some of the independent components (ICs) for the multichannel case and their corresponding topographies. The ICs and their corresponding topographies indicate that IC 3 represents electrode artifact and ICs 5 & 7 ocular artifact. ICs 14, 19 & 22 depict seizure elements. The topographies indicate dissimilar fields in the region of interest.

4. Acknowledgements

This work is funded by EPSRC Grant #GR/L94673. The authors would like to thank Dr Katsuhiro Kobayashi of Okayama University Medical School, Japan and the EEG Department, Montreal Neurological Institute, Canada for the epileptiform data.

5. References

1. Kobayashi, K., James, C.J., Nakahori, T., Akiyama, T. and Gotman, J., (1999) "Isolation of epileptiform discharges from unaveraged EEG by independent component analysis", *Electroenceph. clin. Neurophysiol.*, 110, pp.1755-1763
2. James, C.J., Lowe D., (2000) "Using Dynamical Embedding to Isolate Seizure Components in the Ictal EEG", (Submitted), *Invited Paper - Special Issue of IEE Proceedings – Science, Measurement and Technology: Advances in Medical Signal Processing.*
3. Lagerlund, T.D., Sharbrough, F.W., Jack, C.R., Erickson, B.J., Strelow, D.C., Cicora, K.M., and Busacker, N.E., (1993) "Determination of 10-20 system electrode locations using magnetic resonance image scanning with markers", *Electroenceph. clin. Neurophysiol.*, 86, pp.7-14
4. Broomhead, D.S. and King, G.P., (1986) "Extracting qualitative dynamics from experimental data", *Physica*, 20(D), pp.217-236
5. Takens, F., (1981) "Dynamical Systems and turbulence, Lecture Notes in Mathematics", (Eds: Rand, D.A. and Young L.S, Springer), 898, pp.366-381
6. Comon, P., (1994) "Independent component analysis, a new concept?", *Signal Processing*, 36, pp.287-314
7. Bell, A.J. and Sejnowski, T.J., (1995) "An information-maximization approach to blind separation and blind deconvolution", *Neural Computation*, 7, pp.1129-1159
8. Makeig, S., Jung, T.P., Bell, A.J., Ghahremani, D. and Sejnowski, T.J., (1997) "Blind separation of auditory event-related brain responses into independent components", Proc. Natl. Acad. Sci., USA, 94, pp.10979-10984
9. Hyvärinen, A. and Oja, E., (1997) "A fast fixed-point algorithm for independent component analysis", *Neural Computation*, 9, pp.1483-1492
10. The FastICA MATLAB Package, 1998, Available at: http://www.cis.hut.fi/projects/ica/fastica/
11. Nunez, P.L., and Katznelson, R.D., (1981) "Electric Fields of the Brain: The Neurophysics of EEG", New York: Oxford University Press.
12. Scherg, M. and Von Cramon, D., (1985) "Two bilateral sources of the late AEP as identified by a spatio-temporal dipole model", *Electroenceph. clin. Neurophysiol*, 62, pp.32-44
13. Sammon, J.W, (1989) "A nonlinear mapping for data structure analysis", *IEEE Transactions on Computers*, C-18(5), pp.401-409

RECONSTRUCTION OF BOTTOM AND WATER PARAMETERS IN SHELF ZONE ON THE BASIS OF FUSION OF GEOPHYSICAL AND ACOUSTICAL DATA

B.G. KATSNELSON
Associate professor of Voronezh State University
Universitetskaya sq.1,
Voronezh 394693
Russia
V.G. PETNIKOV
Head of Laboratory of General Physics Institute
38, Vavilova str.
Moscow, 117333
Russia

1. Introduction

Setup of experiment concerning determination of bottom and water properties in shallow water can be presented in the following way. Radiated acoustical signal from the fixed (the so called stationary track) or from the moving source, is being received by the fixed vertical array of hydrophones placed on rather long distance from the source (from tens up to hundreds km). Besides acoustical field, produced by our sources we can register noise field excited by wind waves. Set of characteristics of the sound field, obtained as a result of experiment will be denoted by vector $\mathbf{A}_{exp}$ with dimension determined for example by the total number of hydrophones measuring the sound filed of the vertical interference structure, or by the number of frequency points in the spectrum of the broadband signal. This dimension can be of the order from several tens up to several hundreds in dependence on situation. During this experiment we independently conduct measurements of different parameters of environment. We mean measurements by CTD (Conductivity, Temperature, Depth) – probe, which allows us to know sound speed profile, ADCP (Acoustical Doppler Current Profiler) for registration of water motion, different sorts of echo-sounders, wind speed meter for registration of surface waves, analysis of samples of sediment obtained by the drilling, etc. Some of these parameters can be measured rather accurately (for example depth of water or sound speed profile), another ones have more high degree of uncertainty (sound speed and absorption coefficient in sediment) so we could suppose these parameters as unknown. We denote set of unknown parameters by the vector **p** with comparatively small dimension (about ten) On the basis of mentioned independent measurements of environment characteristics we construct model of the waveguide within the framework of which we can calculate the same sound field on the vertical array. These calculations give vector of the sound field depending on vector of parameters $\mathbf{A}_{th}(\mathbf{p})$. The main idea of the approach to determination

A. K. Hyder et al. (eds.), Multisensor Fusion, 897–903.

of parameters of model is minimization of the mismatch between experimental and theoretical vectors σ $\|\mathbf{A}_{exp} - \mathbf{A}_{th}(\mathbf{p})\|$ or maximization of coefficient of the matching K on the basis of selection of the corresponding vector **p**. The choice of expression for the norm (mismatch) σ or coefficient K can be different [1-4], in dependence on algorithm of the searching optimal parameters. This method is known as Matched Field Tomography (Processing) or MFT(P), it was applied for different tasks [1-10]. Below we consider examples of the actualization of this scheme.

2. Theoretical model

Model of the shallow water waveguide constitutes two (or many-) layered media consisting of water layer of the depth $H(x, y)$ (here (x, y) are horizontal coordinates, z is vertical coordinate directed down) and absorbing layered bottom. This waveguide is characterized by the set of parameters: sound speed profile $c(z, x, y)$ and density $\rho(z, x, y)$ in water, bottom parameters: number and thickness of possible layers, their densities, sound speeds, absorption coefficients etc. Within the framework of this model sound field on the receiving array produced by the source placed at the point $r = 0$, $z = z_s$ ($r = \sqrt{x^2 + y^2}$ is distance from the source along the acoustical track) can be presented as the sum of adiabatic waveguide modes:

$$P(r,z) = |P|\exp[i\varphi] = \sum_{l=1}^{M}(\xi_l r)^{-1/2}\psi_l(0,z_s)\psi_l(r,z)\exp\left(i\int_0^r \xi_l dr\right) \quad (1)$$

where $\psi_l(r,z)$ and $\xi_l(r)$ are eigen functions and eigen values of the problem:

$$\frac{d^2\psi_l}{dz^2} + \left[\frac{\omega^2}{c^2} - \xi_l^2(r)\right]\psi_l = 0,\ \psi_l(0) = 0,\left[\psi_l + g\frac{d\psi_l}{dz}\right]_{z=H} = 0 \quad (2)$$

Here g is coefficient conditioned by bottom properties and depending upon frequency f and eigen values ξ_l. We can get explicit expression for different specific bottom models – liquid absorbing bottom, hard bottom, many-layered bottom etc. [5], in other words g depends on vector **p**. In expression (1) the constant factor is omitted, as usually we consider sound intensity in logarithmic (decibel) scale:

$$\delta = 20\ \lg\ |P/P_0| = 10\ \lg\ (I/I_0) \quad (3)$$

where P_0, I_0 are reference amplitude (intensity) at the distance r_0^*, for example 1 km.

3. Horizontal interference structure

In this case vector of the sound intensity **A** can be obtained on the basis of measurement of range dependence of the sound intensity, or (in decibel scale) function $\delta(r,z)$ as a function of r for fixed depth of hydrophone z. This dependence is experimentally measured using fixed array of hydrophones and towed source on distances up to several hundreds km. Mentioned value $\delta(r,z)$, considered as a function of coordinates, is very changeable due to interference of space beating with different scales. For the comparison of experimental $\delta_e(r,z)$ and theoretical $\delta_{th}(r,z)$ dependencies we introduce coefficient of matching (connected with mismatch) in the following way: let's present $\delta(r,z)$ as the sum of averaged over range interval (greater than scale of interference beating) part $\overline{\delta}$ and oscillating part $\widetilde{\delta}$: $\delta(r,z) = \overline{\delta}(r,z) + \widetilde{\delta}(r,z)$. Correspondingly intensity can be divided: $\widetilde{I}(r,z) = I(r,z) - \overline{I}(r,z)$, in turn matching coefficient is also sum of fluctuating and average parts:

$$K = \overline{K} + \widetilde{K}, \qquad \delta = \overline{\delta} + \widetilde{\delta},$$

$$\overline{K} = \int_0^r \overline{\delta}_e \overline{\delta}_{th} dr \Bigg/ \left[\int_0^r \overline{\delta}_e^2 dr + \int_0^r \overline{\delta}_{th}^2 \right] \quad \widetilde{K} = \int_0^r \widetilde{\delta}_e \widetilde{\delta}_{th} dr \Bigg/ \left[\int_0^r \widetilde{\delta}_e^2 dr + \int_0^r \widetilde{\delta}_{th}^2 dr \right] \tag{4}$$

$0 \le K \le 1$, maximal K corresponds to the best fitting of experiment and theory.

As an example of this approach we show some results of low-frequency (100 and 230 Hz) acoustical sounding at the long-range track in Barents sea [6]. In Figure 1 horizontal interference structure for the frequency 100 Hz (after some averaging) is shown.

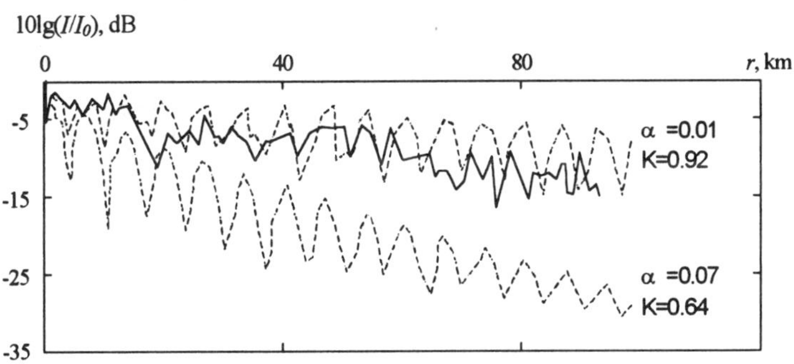

Figure 1. Range dependence of relative intensity. Solid line denotes experimental data, dotted lines denote calculated dependencies

Upper theoretical curve corresponds to the best matching, where vector of parameters is $\mathbf{p}$ = (c_1, ρ_1, σ), where c_1 =1630 m/s, ρ_1 =2 g/cm^3 are sound speed and density of sediment, α is attenuation coefficient in bottom and can be defined through imaginary part of wave vector in the bottom: $k_1 = 2\pi f / c_1 (1 + i\alpha / 2)$.

4. Vertical interference structure

For the matching process we have another form mismatch, if we deal with the so called vertical interference structure, e.i. with depth dependence of the amplitude and phase of the sound field. In this case vector $\mathbf{A}$ is determined by the sound field on separate hydrophones. The processing of the sound field is the selection of the amplitude $|P_j| = |P(z_j, t)|$ and phase $\varphi(z_j, t)$ of the sound field on hydrophone with the number j. For the matching of experimental and theoretical data we use expression

$$\tilde{\sigma}_\varphi = \frac{1}{J_h}\sqrt{\sum_j^{J_h}\left(\hat{F}_{e\,j} - \hat{F}_{th\,j}\right)^2},\quad \tilde{\sigma}_P = \frac{1}{J_h}\sqrt{\sum_j^{J_h}\left(\left|P_{e\,j}\right|^2 - \left|P_{th\,j}\right|^2\right)^2} \qquad (5)$$

where $\hat{F}_j \equiv \cos(\varphi(z_j, t))$, J_h is total number of hydrophones. Minimal value of $\tilde{\sigma}$ (ideally zero) corresponds to the best fitting between experimental and theoretical data.

Below, the example is shown of the using of this criterion for experiment at the stationery track in Barents sea of the length about 14 km. On the basis of these experimental data we have found set of bottom parameters, in other words components of vector $\mathbf{p}$ = (c_1,ρ_1, H, α). In the Figure 2 vertical distributions of the amplitude and phase are shown, corresponding to the best fitting (parameter s, $\tilde{\sigma}_\varphi \sim \tilde{\sigma}_\rho \sim 0.1$) i.e., to the parameters: H = 166.4 m, c_1 = 1800 m/s, α = 0.032, ρ_1 = 1.8 g/cm^3.

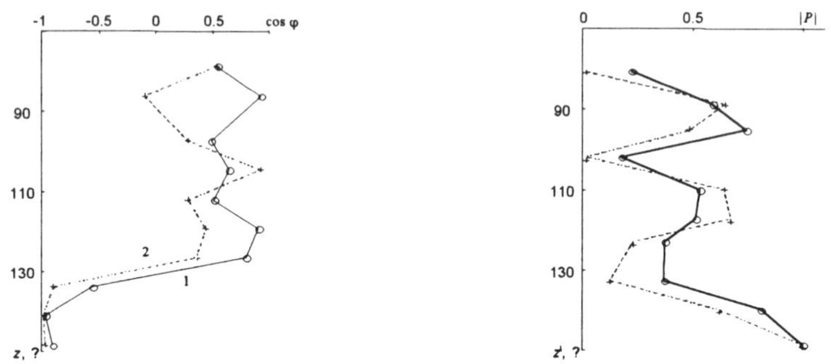

Figure 2. Depth dependence of amplitude (right) and phase (left) of the sound field at the frequency 100 Hz. Solid lines correspond to experiment , dotted lines – theory

5. Interference structure in frequency domain

The third case of the matching we have using broadband source. In this case we can use spectrum of received signal as a vector **A**, more exactly vector **A** is formed by the set of frequency points within the band of signal. Let's consider experiment with Amplitude Modulated (AM) signal at the stationary track in Barents sea. Here we have another vector of model parameters **p** due to another waveguide model. Namely, in our model we consider bottom as two-component media, consisting of water and mineral particles. It means that the main parameter of bottom is porosity κ. Density and sound speed in sediment are expressed through κ:

$$c_1 = c_H(1.631 - 1.78\kappa + 1.2\kappa^2), \quad \rho_1 = \rho_H(2.604 - 1.606\kappa)$$

The second idea of this model is the detailing mechanisms of loss. We suppose that there are two main processes responsible for the attenuation of the sound – absorption of energy (transform into heart) and volume scattering by inhomogeneities, it leads to existence of new parameters, describing attenuation coefficient

$$\alpha = \alpha_a + \alpha_v = \alpha_a + \alpha_v^0 (f / f_0)^{b-1}$$

where b = 2÷4 in dependence on relationship between scale of volume inhomogeneities and wavelength, α_a, α_v^0 are specific coefficients of absorption and volume scattering.

and experimental dependencies. Let us have experimental $|P|_i^{exp}$ and calculated $|P|_i^{theor}$ spectra represented M frequency samples. $|P|_i = |P(\mathrm{r}, \mathrm{z}; f_i)|$; $i = 1,\ldots,M$. Thus

$$\sigma = \sqrt{\sum_{i=1}^{M} \left(|P|_i^{exp} - |P|_i^{theor} \right)^2}, \tag{6}$$

A minimum of σ (zero in ideal case) signals the best selection of bottom model.

For the matching experiment with source emitting AM-signal of band 25-95 Hz was carried out. Distance of the sounding was 14 km. In the Figure 3 results of the experiment and calculations, corresponding to the best matching are shown.

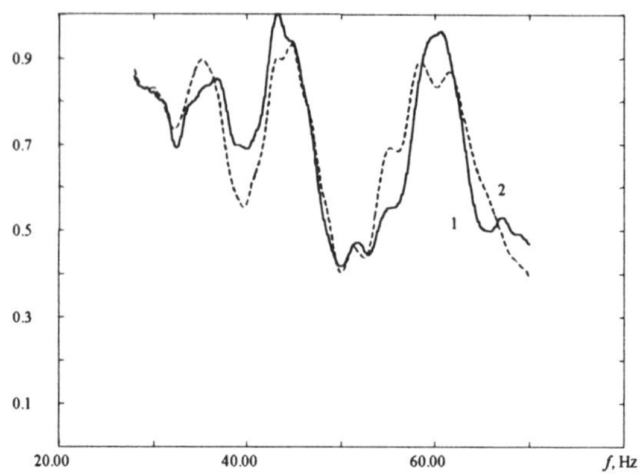

Figure 3. Frequency dependence of intensity of the received signal. Solid line denotes experiment, dotted line – theory, corresponding to the best matching

Parameters of sediment, corresponding to the best fitting are $\kappa = 0.3$, correspondingly $\rho_1/\rho_H = 2.1$, $c_1 = 1765$ m/s, $\alpha_a = 0.005$; $\alpha_v = 9.3 \cdot 10^{-8} f^3$ or $\alpha = 0.005 + 9.3 \cdot 10^{-8} f^3$ (Hz) We can see that for the frequency lower than approximately 40 Hz absorption dominates in attenuation, at the same time for the frequency upper than 40 Hz volume scattering dominates.

In conclusion we remark that the same approach can be applied for the problem of source localization [2], authors suppose also to use this method for the detecting of internal waves (solitons) on the shelf on the base of new acoustical phenomena conditioned by internal solitons (resonance absorption [11] and significant horizontal refraction [12])

This work was supported by RFBR, grants 99-02-17671, 00-05-64752.

6. References

1. Diachok, O., Caiti, A., Gerstoft, P., and Schmidt, H. (eds) (1995) *Full field inversion methods in Ocean and Seismo-Acoustics*, Kluwer Academic .Publisher., Dordrecht, 419 p.
2. Special issue on detection and estimation in matched-field processing., IEEE Oceanic Engineering, 1993, v.18, N3
3. Baggeroer, A.B., and Kuperman, W.A. (1992) Matched field processing in ocean acoustics. In J.M.F.Moura, I.M.G.Lourtie (eds) *Acoustic signal processing for ocean exploration*,. Kluwer Academic Publishers, Dordrecht, Boston, London, P.79-114
4. Daugherty, J. R., and Lynch, J.F. (1990) Surface wave, internal wave, and source motion effects on matched field processing in a shallow water waveguide.- *J.Acoust. Soc.Am.*, , v.87(6), pp.203-2526.
5. Katsnelson, B.G., and Petnikov, V.G. (1997) *Shallow water acoustics*, Nauka publ.,Moskow, , 191p.
6. Bunkin, F.V., Katsnelson, B.G., et. al (1989) Average estimates of sound absorption in shallow ocean waveguides. *Sov.phys. Acoust.*,**35**, N1, pp.1-4

7. Katsnelson, B.G., Kulapin, L.G. Migulin, A.N., and Petnikov, V.G. (1992) Effect of the hydrodynamic fluctuations on vertical interference structure of a sound field in waveguide. *Sov.phys.acoust.*,.**38**, N2, p.164-168
8. Katsnelson, B.G., and Petnikov ,V.G. (1995) Bottom parameters estimations in shallow water by means of field inversion method. In: Diachok at al (eds) *Full field methods in ocean and seismo acoustics.* Ed.. Kluwer Academic Publishers, Dordrecht, , p.359-364
9. Andreev, M.Yu., Katsnelson, B.G. at al. (1996) Influence of hydrodynamic variation in shallow sea on the sound field phase. *Acoustical physics*, **42**, N4, p.459-464
10. Grigoryev, V.A., Katsnelson, B.G. and Petnikov, V.G. (1996) Frequency dependence of sound absorption coefficient in Barents sea bottom. *Acoustical physics*,.**42**, N5, p.459-464
11. Zhou, J., Zhang, X., and Rogers, P.H. (1991) Resonant interaction of sound wave with internal solitons in coastal zone.-*J. Acoust. Soc.Am*, **90**(4), pp.2042-2054
12. Katsnelson, B.G., and Pereselkov, S.A. (1998) Sound fluctuations due to soliton like internal waves crossing shallow water acoustic track.

USE OF WAVE LENGTH SHIFTING FIBRES IN MULTI-SENSOR HIGH ENERGY PHYSICS DETECTORS

V. KRYSHKIN
Institute for High Energy Physics
Protvino, Moscow reg., Russia
Russia

1. Introduction

A requirement to collect light from large area scintillators lead to wave length shifting (WLS) technique based on the following principle. Light produced by ionising particle in a scintillator plate after some internal reflections exits the scintillator and hits a WLS bar. Inside the bar the blue light of the scintillator is re-emitted isotropically into longer wave length. Part of the light which lies within angles of internal reflection is trapped. For a polystyrene bar the trapping efficiency is 20-30% (in both directions).

The main advantage of the approach is the possibility to squeeze light into a small cross-section (ratio is about 200). The disadvantage is the reduction of light collection about 20 times. The final result is the gain an order of magnitude in comparison with traditional waveguides.

Next step – use of optical fibres that drastically eased construction, long term stability and maintenance. For optical fibres (double cladding) the trapping efficiency is 11%. The smaller trapping efficiency is more than compensated by the large attenuation length (>7 m). Besides fibres are robust and easy to handle. The first use of the technique is related with rectangular bars surrounding large scintillation counters.

2. Application

2.1 SCINTILLATION COUNTERS

Even for a small size scintillation counter a need to guide the light far from the scintillator appears due to requirements to place a photodetector outside of a strong magnetic field etc. Figure 1 shows the geometry of such counter. The thickness of the scintillator is determined by the number of fibres used for the light collection. As a thumb rule one fibre corresponds to 3 photoelectrons (exact value depends on a particular scintillator, WLS fibre and photodetector quantum efficiency). Hence for~100% detection efficiency 3-4 fibres are needed. That corresponds to ~ 5 mm thick scintillator for 1 mm diameter fibre. For a small scintillator several turns of a fibre is acceptable without substantial increase of light pulse length. Typical time of light re-emission of WLS fibre is 10 ns. Therefore the total WLS fibre length comparable with this value (about 2 m) will not change drastically the pulse length. There is a substantial requirement: high efficiency to detect a single particle. The number of the photoelectrons must be at least 10. The optimal geometry of

A. K. Hyder et al. (eds.), Multisensor Fusion, 905–909.

fibre placement is on the periphery of the scintillator. The uniformity of light collection in this case is much better (~4%) than with a traditional light guides type of "fish tail".

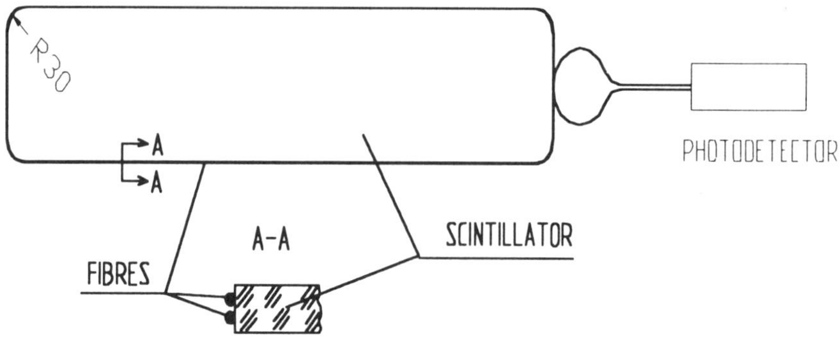

Figure 1. Layout of scintillation counter with WLS fibre read out

2.2 SAMPLING CALORIMETERS

In sampling calorimeters the main requirement is that number of the photoelectrons ($N_{p.e.}$) per scintillator per relativistic particle must not contribute substantially to an energy resolution. Therefore this number depends on the energy resolution of the calorimeter and as a consequence the efficiency can be rather low. Intrinsic resolution of hadron and electromagnetic calorimeters is differ by order of magnitude.

2.2.1 Hadron calorimeter

Typical energy resolution of a hadron calorimeter is about $100\%/\sqrt{E}$. In this case the required number of photoelectrons is $N_{p.e.}\geq 1$. The transverse shower dimension (radius) is about one absorption length (15-20 cm). It defines the dimension of scintillator (tile). Figure 2 shows a typical layout of a scintillator for CMS hadron End Cap calorimeter [1]. To minimise the gap between scintillators the WLS fibres are placed into a groove machined in the scintillator. The bending radius of the fibres must be about 30 mm. Thus such approach is acceptable only for comparatively large scintillators. One end of the fibre is diamond machined, aluminised and spliced with clear fibres outside of the scintillator to guide the light to a photodetector.

2.2. 2 Electromagnetic calorimeters

Energy resolution and shower dimensions of electromagnetic calorimeters are much smaller than the hadronic ones. To realize good space resolution that is defined by scintillator dimensions other design was propose [2]. This approach is presented in Figure 3. Layers of absorber material (lead plate with thickness less than 1 mm) are interspersed with scintillator plates (thickness about 1 mm which is equal the used fibre diameter). Between two adjacent scintillators a WLS fibre is placed. One end is aluminised and the other one is contacted with a photodetector. Two scintillator plates with a fibre between them defines a tower width (dimension of the square). Between the scintillators and lead plates there is aluminised mylar and scintillator sides not contacted with fibre are painted with reflective paint. To minimise the "channelling" effect the angle between the calorimeter axis and a particle trajectory is set about 150 mrad.

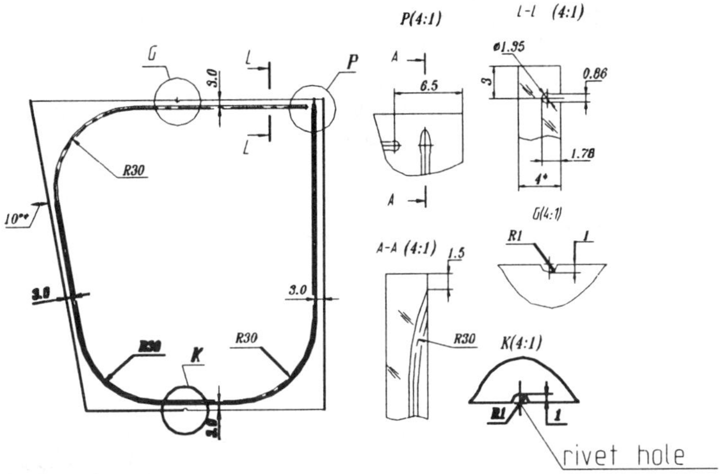

Figure 2. The structure of a scintillator for hadron end cap calorimeter

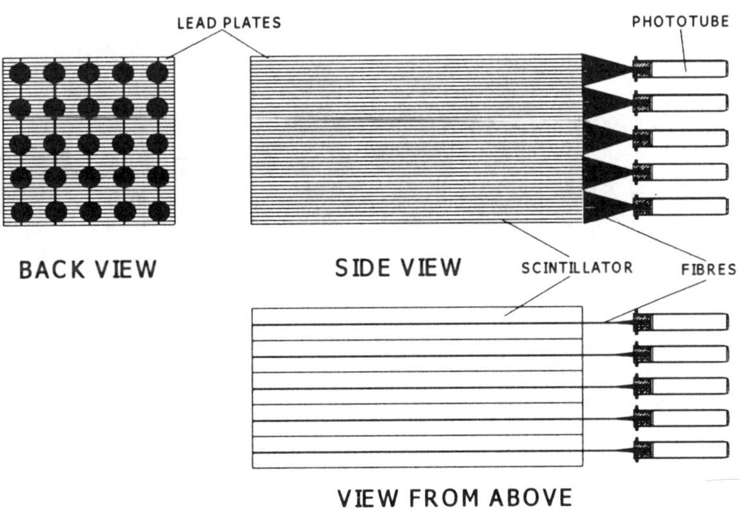

Figure 3. Sampling Electromagnetic Calorimeters with WLS Fibres Light Collection and Scintillators Placed Parallel to Particle Trajectory

3. Multi-Sensor use of the Detectors

Calorimeters are used in high energy physics to measure energy and direction of jet of particles originated from fragmentation of partons. Figure 4 shows the layout of Compact Muon Solenoid – apparatus for LHC. Blue colour denotes hadron calorimeters – Barrel calorimeter and End Cap calorimeter. Between them there is a gap for cables and services.

When particles fly in the direction of the gap from interaction point situated in the centre of the apparatus part of the particle energy is missed. Figure 5 presents the dependence of the value of the missing energy on the conditions of data analysis [4]. As one cam see the additional information from electromagnetic calorimeter and tracker allows to improve reconstruction on particle energy.

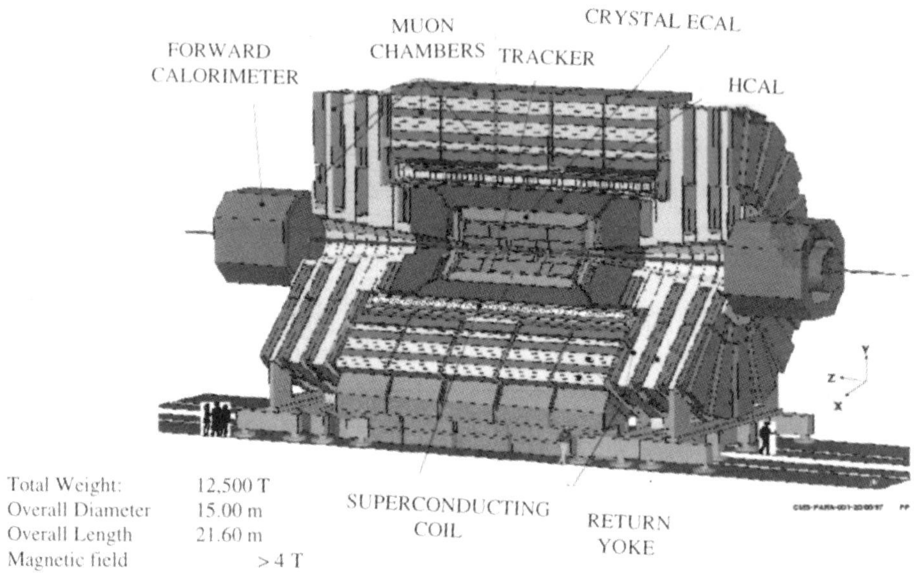

Figure 4. General view of Compact Muon Solenoid.

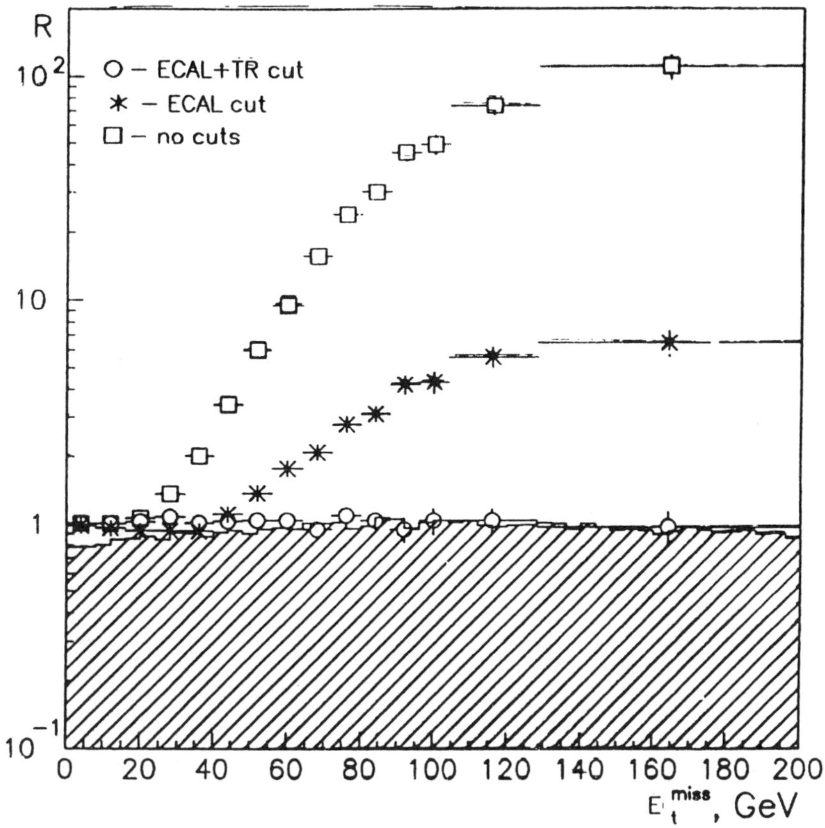

Figure 5. Dependence of the R=ratio dN/dE_t^{mis} measured by hadron calorimeter with crack to that without crack for three cases: a) shaded histogram is for ideal case (no crack between the calorimeters); b) square points are for the case when only the information from hadron calorimeters is used; c) histogram with • points is for the case when information from electromagnetic calorimeters is used also; d) histogram with ◦ points is for the case when information from hadron calorimeter, electromagnetic calorimeter and trackers is used.

4. References

1. The hadron calorimeter project. Technical design report. CERN/LHCC 97-31.
2. Kryshkin, V. and Ronzhin, A., (1986) Nuclear Instruments and Methods, A247, 583.
3. Dolgopolov, A.V., Inyakin, A.V., Lapshin, V.G., et al, (1995) Nuclear Instruments and Methods, A363, 557.
4. Roinishvili, V.N., (1994) CMS week report.

REAL-TIME NETWORK DISTRIBUTED IMAGE PROCESSING AND ANALYSIS FOR A SECURITY APPLICATION

B.W. HAN[1], I.V. LEVASHEVICH[2],
D.N. MATSUKEVICH[2], O.N. METELITSA[2]
Tongwon College, Seoul, Korea[1]
Institute of Nuclear Problems,
Belarusian State University,
Minsk, Belarus[2]

1. Introduction

Last years network based video surveillance systems are becoming to be widely used for different applications like security systems (see, for example, [1]), road traffic analysis [2] and even weapon control [3]. The possibility of the modern video & audio Internet streaming systems are about 15 frames per second with the frame size 360*240 pixels, some times 180*120 pixels. However, recent progress in area of computer systems, communications and modern lossy compression methods allows us to expect also changes in quality of transmitted through the Internet video and audio.

This work is devoted to creation the distributed system containing many video servers (computer with attached cameras) and one global server with multimedia database (see Figure 1). Both video servers and global server must be accessible through Internet/Intranet for getting video, audio and for administration. Each video server periodically sends pictures to global server for collecting and analysis. In case of alarm caused by motion, sound detection or external alarms all cameras begin to record sound and pictures. At the same time images are transmitted to global server (real-time video). The volume of transmitted information is limited by a line bandwidth. After the end of alarm, the quality of pictures in the global server database is improving by sending the skipped frames. The global server is expected to analyze and recognize moving objects to exclude part of alarms.

2. Example of Operation of Video-Server

At present time we partially realized a video server with one fish-eye lens camera for Linux operating system. An example of this server operation is shown in Figure 2. The fish-eye lens works such a way that the same polar angles are transferred to the same radius difference in a final picture. Therefore, we see a round picture captured by fish-eye camera. The points on the border of the picture correspond to the angle of view about 180 degrees.

A. K. Hyder et al. (eds.), Multisensor Fusion, 911–916.

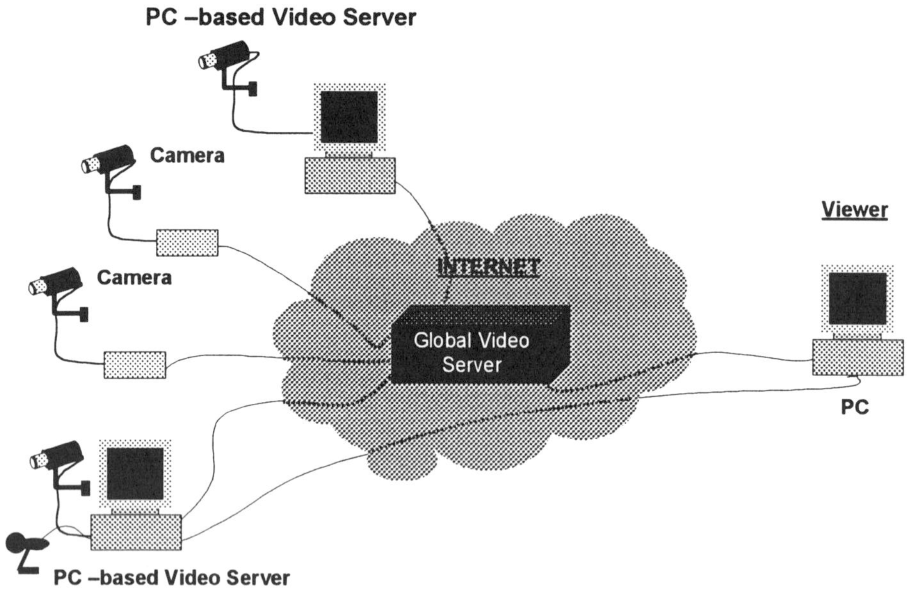

Figure 1. A network based video surveillance system

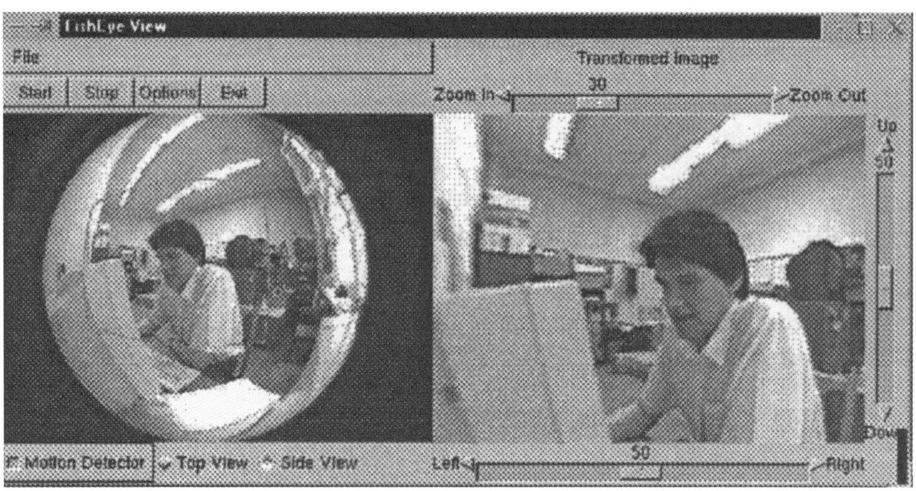

Figure 2. Example of the application operation. Left hand part of image contains a captured frame. Right part of image contains a corrected from fish-eye lens aberration moving object (one of the author of the article).

We have to made a picture correction module, because a fish-eye lens distorts a picture. It is not necessary to correct the picture as a whole, only moving object is inserted for a viewer. On the right hand part of Figure 2 we see corrected image of the moving object. The area inside the orange motion-tracking box at the left image corresponds to the right picture. Digital zoom ratio was chosen such a way to fit all moving objects into right picture box. We can increase or decrease zoom ratio using a scroll bar. Our algorithm tracks a moving object. In case of two or more moving objects are presented we expand the tracking box and change zoom ratio to admit all moving object.

In future versions of the s/w we are going to use a separate box for each moving object and additional (not fish-eye) cameras with pan-tilt-zoom mechanical parts to track moving objects. The role of fish-eye camera in this case is very important, because it allows to determine the direction and distance to moving objects [4].

3. Image Processing Algorithms in Video-Server

The first block in image processing part of the program is a fast motion detection algorithm (see Figure 3). There are different types of motion detection algorithm [5-7] but we use the not very complicated one, because we exclude a big part of false alarms later in motion tracking procedure. For motion detection we divide the picture obtained from camera on 8*8 pixel "superpixels". For each superpixel we calculate a mean intensity μ and its variance σ. Then we use maximum likehood principle to compare the same superpixels in the present frame and the frame obtained *N* frames ago:

$$\frac{(\sigma - \sigma_N)^2 + 0.5(\mu - \mu_N)^2}{2\sigma\sigma_N} < T \qquad (1)$$

where index *N* denoted the values obtained *N* steps ago (N~10-20), T is a threshold. We calculate a mean intensity and a variance to reduce noise interference. The "*N* frame ago value" is used to make algorithm more sensitive for slow motion. It is possible to use *2N* frame average instead of "*N* frame ago value" to save memory. The maximum likehood principle allow us to answer the question whether μ, σ and μ_N, σ_N describe the different distributions or the same one.

The motion-tracking algorithm is used for moving object tracking and for changing the size of the motion tracking area (zoom in or zoom out) to include all moving objects. The algorithm consists of the following steps:

- Coordinates of a geometrical center of a moving part of the image are calculated as a mean values of pixel coordinates where motion was detected;
- The horizontal and vertical sizes of the moving part of the image are calculated as variances of the pixel coordinates where motion was detected;
- The M-frame mean values of the moving area sizes and center are calculated. It is necessary to smooth sudden oscillations of the center.
- We change the coordinates of the tracking box center only if the displacement of the moving area center is greater than 30% of the area size;

- We estimate the area of moving part of the picture and compare it to the tracking box area. If the difference is greater than a threshold, the tracking box is changed on the designated value;
- We test whether all borders of the tracking box are inside the picture and correct them if necessary.

Our tracking algorithm keeps all moving objects in the tracking box. It also allows us to "take close-up picture" in case of single moving object. The algorithm guarantees a smooth change of tracking box sizes and position of the center.

Image enhancement module contains two algorithms related to correction of a fish-eye lens distortion and zooming. At first we calibrated the lens by taking a picture from squared paper using our fish-eye lens and making a table of dependence between the real coordinates and coordinates on the image. The fish-eye lens transfers coordinates in the plane that is parallel to the image plane such a way that equal polar angles between points are converted to equal radius difference in the picture. It is possible to show that dependence between distance from a small object to the optical axis of camera $\boldsymbol{R}$ and the distance from the image of this object to the center of picture $\boldsymbol{r}$ is:

$$r = A \cdot Arc\tan\left(\frac{2\beta R}{R_{max}}\right) \tag{2}$$

where $R_{max} = A\ Arctan\ (\beta)$.

We calculated parameters $\boldsymbol{A}$ and $\boldsymbol{\beta}$ by fitting Eq.(2) using the table of dependence between coordinates of object and their images and the chi squared criterion (See Figure 4). We obtained a chi-square value $\chi^2=3*10^{-5}$. That means very good agreement between Eq.(2) and results of measurements.

We developed a fast algorithm for fish-eye image correction based on Eq.(2) and use bilinear interpolation for exact calculation of transformed pixel values and for zooming. The slowest part of our algorithm is getting pixels values from initial image, because the pixels that correspond to a row in corrected picture belong to the different rows in initial picture. We used Assembly language to optimize this part of the program.

4. Real-Time Video and Audio Transmission Through Internet / Intranet

Usually glossy compression methods are used to reduce a necessary line bandwidth for transmission audio and video information through the Internet or local net. In security system it is very important that the compression methods would have the possibility to get every separate frame. Therefore we use Motion Jpeg s/w compression. MPEG-4 video compression is under realization.

Network protocols are also very important. The most often used in the Internet TCP protocol is not very suitable for multimedia transmission. The reason is its reliability. At first, this protocol needs in setting up connection before information transmission and it takes time. To avoid the disadvantage the big files containing many video frames are usually prepared.

TCP protocol requires repetition of lost packets. For real-time video it is much better to skip lost information than waiting a repetition of a lost frame. Therefore, the better

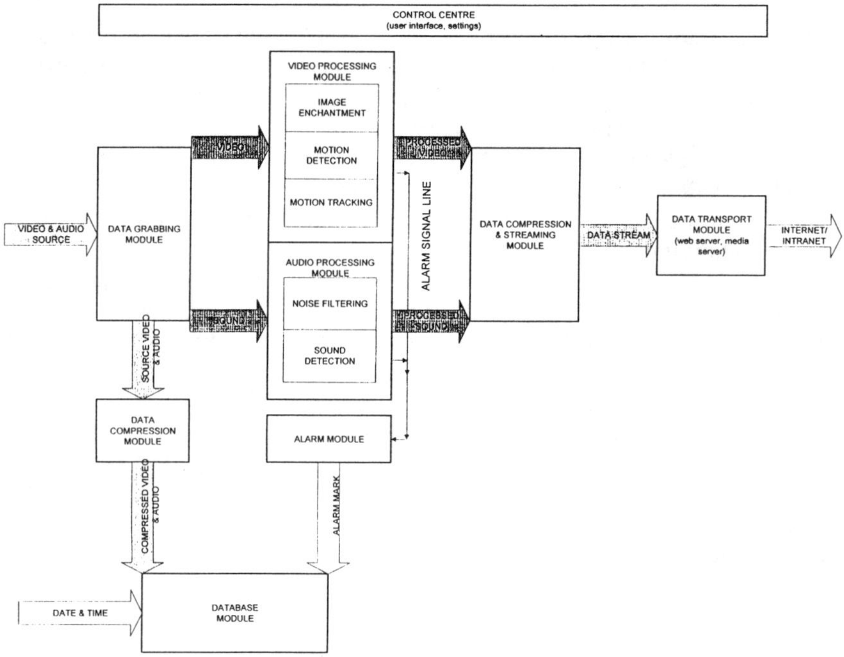

Figure 3. The block-scheme of data streams in the application

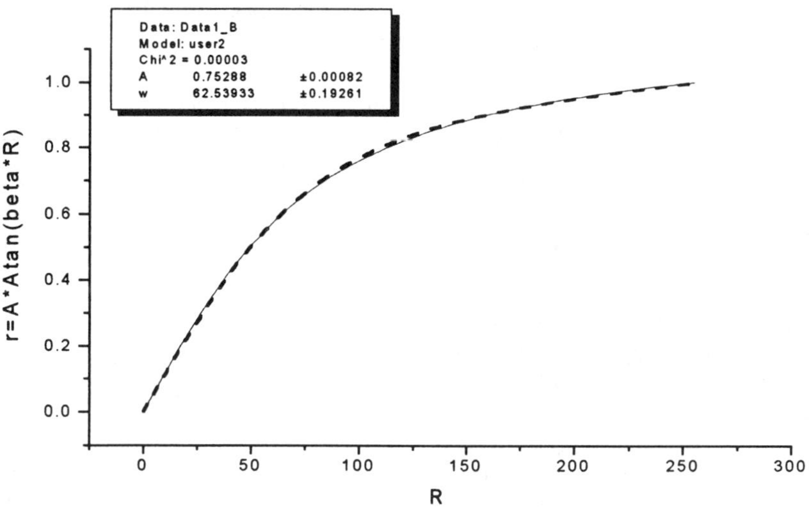

Figure 4. Calibration of the fish-eye lens. Dotted line contains measured data and solid line was obtained after fitting parameters A and β in Eq.2.
part of the near 120, 000 casualties from the Japanese side

solution is using UDP network protocol. As well as TCP protocol it also bases on IP protocol. UDP protocol does not require to set-up connection before data transmission. In case of UDP protocol we need to process all possible situations related to packet losses and change the packets order by our application.

Also special protocols (RTP, RTSP) for real-time data transmission are exists. Unfortunately, not all network equipment supports them.

In our application we realized http based multimedia data transmission. UDP and RTP based transmission is under realization now.

Audio and video information is processed as different streams in our program (See Figure 3). There are two possible methods of video-audio synchronization during network transmission. The first is based on synchronizing features of real-time network protocols. We use the second way and synchronize audio and video streams inside the file for network transmission. The format of this file is made according to open Microsoft ASF format specifications. Unfortunately, Microsoft does not publish detailed specifications that it uses for its products. Therefore, our solutions are not compatible with Microsoft Media Player. We cannot use Microsoft media format SDK to obtain the compatibility because this SDK is available only for Windows platforms but not for Linux. Now we consider the possibility of using other open formats, particularly real video format.

5. Conclusion

The working model of video server is realized and tested on Pentium II-450 computer and 10T based network. We have up to 15 frames per second with frame size 320*240 pixels and motion jpeg compressions. After compression, the frame size is about 15K. Now we are working on the realization of the image storage database. Possibility of including RTSP protocol for multimedia data transmission and MPEG-4 compression is under consideration.

The system can be used for surveillance papers both as stand along system and in combination with a server supporting alarm image storage and analysis for big number of cameras.

6. References

1. http://www.dii-megachips.com
2. http://www.highwaynet.com/cameras
3. Ondrik, M., Kadner, S., Kraus, S., Thompson, V., Horak, W., Reisman, A., Nadezhdinskii, A., Rudov, S., Medvedev, L., and Spiridonov, M., *Motion Detection Technologies*, http://159.226.136.75/can/html/reference/safeguards/motion.htm
4. Gutchess, D., and Jain, A.K. (2000) *Automatic Surveillance Using Omnidirectional and Active Cameras*, ACCV http://www.cse.msu.edu/~gutchess/papers/accv00/index.html
5. Jain, R.C., and Nagel, H., (1979) *On the analysis of accumulative difference pictures from image sequences of real word scenes.* – IEEE Transactions on Pattern Analysis and Machine Intelligence, 1(2), 206.
6. Daniilidis, K., Krauss, C., Hansen, M., and Sommer, G., (1998) *Real-time tracking a moving objects with an active camera, - Real-time Imaging*, **4** 3.
7. Rowley, H., Baluja, S., Kanade, T., (1998) - *Neural network-based face detection.* IEEE Transactions on Pattern Analysis and Machine Intelligence, 20(1), 23.

NAVIGATIONAL-RADAR TRACKING OF A MARITIME TARGET IN CLUTTER

A Comparison of IMM-NN and IMM-PDA Filtering Algorithms

Z. KOWALCZUK
Technical University of Gdansk
Ul. Narutowicza 11/12
80-952 Gdansk, Poland
M. SANKOWSKI, P. URUSKI
Telecommunications Research Institute
Ul.Hallera 12
80-401 Gdansk, Poland

1. Introduction

The observed development of the sea traffic makes it necessary to apply precise navigating and manoeuvring techniques in order to ensure the safety of sailing in difficult conditions. A modern maritime navigation system, which is an important part of equipment of any vessel, usually employs an on-board navigational radar [4].

Tracking algorithms (TAs) are a principal part of radar data processing systems[5]. Such an algorithm processes plots (validated returns with their estimated parameters) in order to maintain a track representing target's trajectory. Two basic parts that constitute every TA algorithm are: a plot-to-track association (PTA) procedure and an algorithm of estimation of a state vector, which includes the parameters of the target's trajectory [5].

The quality of the maritime-target's tracking depends strongly on the presence of environmental disturbances, often called sea clutter, that are a source of plots not originating from the objects of interest (false alarms, FA). In the case of intensive sea clutter the importance of the PTA rule increases, because incorrect plot-to-track assignments lead to overall degradation of the TA's performance.

In this chapter the problem of tracking of a manoeuvring maritime target in clutter is considered. Two methods of PTA are examined: nearest-neighbour (NN) method [3, 5] and probabilistic data-association (PDA) one [2, 3]. Each of them cooperates with an interacting multiple-model (IMM) adaptive state-estimation algorithm [2]. The tracking performance of the two TAs is evaluated by simulation. The results of tests with different false alarm densities (FAD) are referred to the performance standards defined by the International Maritime Organisation (IMO) [6], and the capacity of tracking of manoeuvring targets in clutter is analyzed.

2. Tracking Algorithm

The tracking filter as a part of TA is a state estimator of the object being tracked. Its task is to process radar measurements in order to to estimate the current and future parameters of

A. K. Hyder et al. (eds.), Multisensor Fusion, 917–922.

motion by means of time averaging. The basic problem of tracking of manned moving objects lies in the unpredictability of object manoeuvres which concerns the type of the manoeuvre trajectory and the time of its occurrence. Many approaches to tackle this problem can be found in the vast literature on the subject [5].

2.1 MODELS OF OBJECT MOTION

For the following discussion two types of targets' trajectories, associated with two hypotheses $\{H_j^n\}$, j in $\{0,1\}$, are considered: the uniform (H_0^n) and manoeuvring (H_1^n) motion trajectories, with n denoting the instant when H_j^n is valid. The two trajectories are assumed to be described by the following discrete-time state-space models:

$$\boldsymbol{x}[n+1] = \boldsymbol{A}_j \boldsymbol{x}[n] + \boldsymbol{w}_j[n] \qquad (1)$$

$$\boldsymbol{y}[n] = \boldsymbol{C}\boldsymbol{x}[n] + \boldsymbol{e}[n] \qquad (2)$$

where

In the above $x[n]$ and $y[n]$ denote position, $v_x[n]$ and $v_y[n]$ - velocity, while $a_x[n]$ and

$$\boldsymbol{C} = \begin{bmatrix} 1 & 0 & 0 & 0 & 0 & 0 \\ 0 & 0 & 0 & 1 & 0 & 0 \end{bmatrix}, \quad \boldsymbol{x}[n] = \begin{bmatrix} x[n] & v_x[n] & a_x[n] & y[n] & v_y[n] & a_y[n] \end{bmatrix}^{\mathrm{T}} \qquad (3)$$

$$\boldsymbol{A}_j = \begin{bmatrix} \boldsymbol{A}_j^d & 0 \\ 0 & \boldsymbol{A}_j^d \end{bmatrix}, \quad \boldsymbol{A}_0^d = \begin{bmatrix} 1 & T & 0 \\ 0 & 1 & 0 \\ 0 & 0 & 0 \end{bmatrix}, \quad \boldsymbol{A}_1^d = \begin{bmatrix} 1 & T & \dfrac{\alpha T - 1 + \exp(-\alpha T)}{\alpha^2} \\ 0 & 1 & \dfrac{1 - \exp(-\alpha T)}{\alpha} \\ 0 & 0 & \exp(-\alpha T) \end{bmatrix}. \qquad (4)$$

$a_y[n]$ - acceleration coordinates of the moving object, $\boldsymbol{x}[n]$ is a state vector, $\boldsymbol{y}[n]$ is a measurement vector, $\{\boldsymbol{w}_j[n]\}$ is a stationary Gaussian white noise vector sequence with a covariance matrix $\boldsymbol{W}_j$, $\{\boldsymbol{e}[n]\}$ is a stationary Gaussian white noise sequence with a covariance matrix $\boldsymbol{E}$ that imitates additive measurement noise. It is also assumed that sequences $\{\boldsymbol{w}_j[n]\}$, $\{\boldsymbol{e}[n]\}$ and an initial Gaussian-distributed state $\boldsymbol{x}[0]$ are mutually uncorrelated. Parameter T denotes a constant sampling period while α is the reciprocal of the manoeuvre time constant [5].

2.2 VALIDATION OF MEASUREMENTS

There are multiple measurements incoming to the TA. Based on a suitable validation test an m_n -member-set $\boldsymbol{Y}_n$ of measurements $\boldsymbol{y}_i[n]$ is obtained at the time instant n [3, 5]:

$$Y_n = \{\boldsymbol{y}_i[n]\}_{i=1}^{m_n} \qquad (5)$$

2.3 NEAREST-NEIGHBOR KALMAN FILTER

The systems (2.1) plus (2.1) for j={0,1} constitute a design basis for tracking of targets by the Kalman filter [1] with a nearest-neighbor rule (NN-KF):

$$\hat{\boldsymbol{x}}_j[n|n-1]=\boldsymbol{A}_j\hat{\boldsymbol{x}}_j[n-1|n-1] \tag{6}$$

$$\hat{\boldsymbol{x}}_j[n|n]=\hat{\boldsymbol{x}}_j[n|n-1]+\boldsymbol{L}_j[n]\boldsymbol{\nu}_j[n] \tag{7}$$

$$\boldsymbol{\nu}_j[n]=\boldsymbol{y}_{NN}[n]-\boldsymbol{C}\hat{\boldsymbol{x}}_j[n|n-1] \tag{8}$$

$$\boldsymbol{P}_j[n|n-1]=\boldsymbol{A}_j\boldsymbol{P}_j[n-1|n-1]\boldsymbol{A}_j^{\mathrm{T}}+\boldsymbol{W}_j \tag{9}$$

$$\boldsymbol{P}_j[n|n]=\boldsymbol{P}_j[n|n-1]-\boldsymbol{L}_j[n]\boldsymbol{C}\boldsymbol{P}_j[n|n-1] \tag{10}$$

$$\omega_j[n]=\boldsymbol{C}\boldsymbol{P}_j[n|n-1]\boldsymbol{C}^{\mathrm{T}}+\boldsymbol{E} \tag{11}$$

$$\boldsymbol{L}_j[n]=\boldsymbol{P}_j[n|n-1]\boldsymbol{C}^{\mathrm{T}}\omega_j^{-1}[n] \tag{12}$$

where $\boldsymbol{y}_{NN}[n]$ in $\boldsymbol{Y}_n$ denotes the measurement that is the 'nearest' to the predicted position of the target at time n. For other purposes, let $\boldsymbol{Y}_{\mathrm{NN}}{}^n$ = $\{\boldsymbol{y}_{NN}[l]\}_{l=1}{}^n$ denote set of all measurements associated with the NN-KF up to the instant n [5].

2.4 PROBABILISTIC DATA-ASSOCIATION KALMAN FILTER

The PDA-based filter incorporates all measurements that might have been originated from the object in track rather than selecting one of them as it is in the NN rule. Let $\boldsymbol{Y}_{\mathrm{PDA}}{}^n$ = $\{\boldsymbol{Y}_l\}_{l=1}{}^n$ denote the set of all validated measurements up to the instant n. Each validated measurement $y_i[n]$, i=1,...m_n, at the time instant n can be characterised by the conditional probability

$$\beta_{j,i}^n = \mathrm{P}\{\chi_{j,i}^n | Y_{\mathrm{PDA}}^n\} \tag{13}$$

where $\chi_{j,i}{}^n$ denotes the hypothesis meaning that $y_i[n]$ is the correct measurement for the H_j^n -filter. The above along with some additional assumptions lead to a filtering scheme (PDA-KF) that is the KF of (Enqs. 6-12) modified by replacing Eqns.(8) and (9) with Eqns. (14, 15 and 16) respectively. The weighted innovation and covariance matrix of filtered estimate can be approximated as

$$\mathrm{v}_j[n] = \sum_{i=1}^{m_n} \beta_{j,i}^n \mathrm{v}_{j,i}[n] \tag{14}$$

$$\mathrm{v}_{j,i}[n] = y_i[n] - C\hat{x}_j[n|n-1] \tag{15}$$

$$P_j[n|n] = P_j^0[n|n] + L_j[n]\left(\sum_{i=1}^{m_n} \beta_{j,i}^n \mathrm{v}_{j,i}[n]\mathrm{v}_{j,i}^{\mathrm{T}}[n] - \mathrm{v}_j[n]\mathrm{v}_j^{\mathrm{T}}[n]\right)L_j^{\mathrm{T}}[n] \tag{16}$$

where $\boldsymbol{P}_j^0[n|n]$ is the covariance of the update when only one validated measurement is available[2, 3].

2.5 INTERACTING MULTIPLE-MODEL ALGORITHM

The TA algorithm based on an IMM filter scheme is presented in Figure 1.

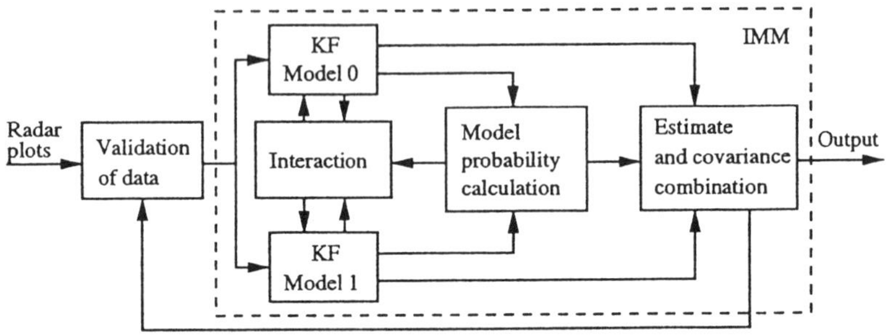

Figure 1. The TA Scheme Consisting of IMM and PTA Algorithms

The basic idea of IMM estimation is to yield a current average value of conditioned state estimate $\hat{x}[n|n]$ by linear combination of partial estimates $\hat{x}_j[n|n]$ obtained from interacting KFs (either NN-KF or PDA-KF):

$$\hat{x}[n|n] = \sum_{j=0}^{1} \mu_j^n \hat{x}_j[n|n] \tag{17}$$

$$P[n|n] = \sum_{j=0}^{1} \mu_j^n \{P_j[n|n] + (\hat{x}_j[n|n] - \hat{x}[n|n])(\hat{x}_j[n|n] - \hat{x}[n|n])^{\mathrm{T}}\} \tag{18}$$

where $\mu_j^n = \mathrm{P}\{H_j^n | Y_{\mathrm{NN/PDA}}^n\}$ denote an a-posteriori probability of the model (hypothesis) H_j^n while $\boldsymbol{P}[n|n]$ is the covariance matrix of the estimation error.

The interaction of the KFs is accomplished as a re-initialisation procedure prior to the predictive stage of the KF scheme:

$$\hat{x}_j[n|n-1] = A_j \hat{x}_j^0[n-1|n-1] \tag{19}$$

where re-initialisation estimates and covariance matrices are as follows:

$$\hat{x}_j^0[n-1|n-1] = \sum_{k=0}^{1} \mu_{j,k}^{n-1} \hat{x}_k[n-1|n-1] \tag{20}$$

$$P_j^0[n-1|n-1] = \sum_{k=0}^{1} \mu_{j,k}^{n-1} \{ P_k[n-1|n-1] +$$
$$(\hat{x}_j[n-1|n-1] - \hat{x}_k[n-1|n-1])(\hat{x}_j[n-1|n-1] - \hat{x}_k[n-1|n-1])^{\mathrm{T}}\} \tag{21}$$

with $\mu_{j,k}^{n-1} = \mathrm{P}\{H_j^n | H_k^{n-1}, Y_{\mathrm{NN/PDA}}^{n-1}\}$ denoting transition probabilities of the Markov chain for $j,k = \{0,1\}$ [2].

3. Verification of Algorithms

The above described TAs cooperating with a model of the CRM-100 FMCW naval radar [7] have been parametrised in order to fulfil the IMO's performance standards required for automatic radar plotting aid systems [6] and to expose capabilities of tracking of manoeuvres. The TA algorithms have been examined for different clutter densities (FAD) in order to estimate their performance in various environmental conditions.

The CRM-100 radar provides 2D measurements (range, bearing) using mechanically rotating antenna with the scan period T=2 [s]. Nonlinearities, which occur as a result of using measurements from the polar coordinate system in a KF based on the Cartesian coordinate system, have been managed by means of the extended KF technique[1].

3.1 PROCEDURE OF TESTING

Let v_x denote the relative velocity, ψ_x is the relative course of the target being tracked, while $\hat{v}_{x,k}[n]$ and $\hat{\psi}_{x,k}[n]$ stands for the estimates of these parameters, where x=1,…,4 is an index of a tested IMO scenario and k=1,…,K denotes the index of statistically independent realisation of the scenario. The IMO defines the accuracies $2\sigma_{v,x}(nT)$ and $2\sigma_{\psi,x}(nT)$ for certain time periods of steady-state tracking that cannot be exceeded with the probability p=0.95.

In order to determine a test for the relative velocity, RMS values of estimation errors can be calculated for the whole duration of testing trajectory as

$$\mathbf{RMS}_{v,x}[n] = \sqrt{\frac{\sum_{k=1}^{K}(\hat{v}_{x,k}[n] - v_x)^2}{K-1}} \tag{22}$$

and compared to standard deviations $\sigma_{v,x}(nT)$. A similar test can be made for the object course. The algorithm is claimed to pass the test if the RMS values do not exceed the respective standard deviations for the entire duration of the trajectory.

3.2 SIMULATION RESULTS

In order to evaluate the accuracy of estimates yielded by the TA, testing routes compliant with the IMO-1 to 4 navigational scenarios have been generated. In this paper, however, solely the results for the most severe scenario IMO-2 are presented. The radar carrying (own) ship moves North at velocity 10 [kn], while the target moves South with velocity 14 [kn] and turns 135 [deg] left at angular velocity 1 [deg/s]. The relative-motion scenario after the manoeuvre is compliant with the IMO-2 scenario (see Figure 2a).

The accuracy and manoeuvre-tracking capabilities of the TAs have been tested for the assumed probability of target detection P_D=0.9 and for FAD=0-40 [km^{-2}]. 50 statistically independent realisations of the route have been processed, and then RMS values of position, velocity and course estimation errors have been calculated. The results of the tests for FAD=0 and 30 [km^{-2}] are shown in Figures 2b and 3 respecti vely.

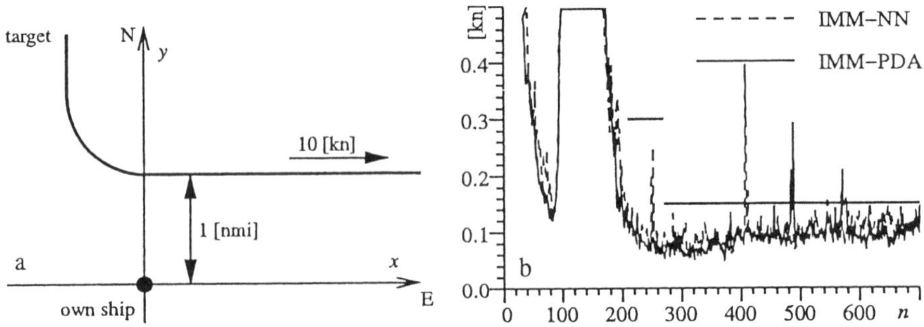

Figure 2. Relative scenario (a). Velocity estimation error $RMS_{v,2}$ for FAD=0 [km^{-2}] (b).

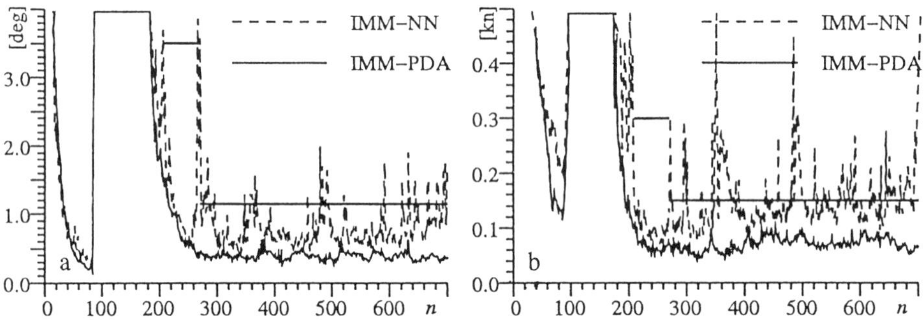

Figure 3. Estimation errors of course $RMS_{\psi,2}$ (a) and velocity $RMS_{v,2}$ for (b) FAD=30 [km^{-2}].

4. Conclusion

In this work the IMM-NN and IMM-PDA filtering algorithms have been applied to tracking of manoeuvring maritime targets. The algorithms have been examined for different clutter densities. It has been observed that the IMM-PDA filter provides estimates being in line with the IMO standards for FAD$\leq$ 30 [km^{-2}] (over 90 [%] maintained tracks) while the performance of the IMM-NN filter degrades for FAD> 10 [km^{-2}]. As the accuracy satisfying IMO requirements has been the essential demand for the TA, the capability of tracking of manoeuvres limited to 0.13 [m/s^2] results.

5. References

1. Anderson, B.D.O. and Moore, J.B. (1979) *Optimal Filtering*. Prentice-Hall, Englewood Cliffs, USA.
2. Bar-Shalom, Y. (1990) *Multitarget-Multisensor Tracking*. Artech House, Norwood, USA.
3. Blackman, S.S. (1986) *Multiple-Target Tracking with Radar Applications*. Artech House, Dedham, USA.
4. Bole, A. and Jones, K. (1981) *Automatic Radar Plotting Aids Manual*. Heinemann, London, UK.
5. Farina, A. and Studer, F.A. (1985) *Radar Data Processing*. Research Studies Press, Letchworth, UK.
6. International Maritime Organisation (1995) *Performance Standards for Automatic Radar Plotting Aids (ARPAs)*. Resolution A.823 (19).
7. Telecommunications Research Institute (1999) Quiet naval radar CRM-100, http://www.

APPENDIX A: ORGANIZING COMMITTEE

Anthony K. Hyder, Director of ASI
Associate Vice President,
Graduate School
Professor of Physics
University of Notre Dame
501 Main Building
(574) 631-8591
(574) 631-6630 facsimile
e-mail: ahyder@nd.edu

Vitaly Bystritskii, Co Director of ASI
Research Physicist
Dept. of Physics and
Astronomy University of California at Irvine
Irvine, CA 92697
(714) 824-2611
(714) 824-2914 facsimile
e-mail : vbystrit@uci.edu

Alain Appriou
Deputy Director, Modelling and
Information Processing Department
ONERA
BP 72
92322 Chatillon Cedex
France
33 1 46 73 41 67 facsimile
e-mail: appriou@onera.fr

Phil Greenway
Senior Principal Scientist
British Aerospace (Operations) Ltd.
Sowerby Research Centre
FPC 267, P.O. Box 5
Filton, Bristol BS12 7QW
United Kingdom
(44) 119 936 6892
(44) 119 936 3733 facsimile
e-mail: Phil.Greenway@src.bae.co.u

Elisa Shahbazian
Director of Research and Development
Lockheed Martin Canada, Inc.
6111 Royalmount Avenue
Montreal, Canada H4P1K6
(514) 340-8343
(514) 340-8318 facsimile
e-mail: elisa.shahbazian@lmco.com

Giorgio Tacconi
Universita'di Genova
Dipartimento
Ingegneria Biofisica
Via all'Opera Pia 11/A
16145 Genova, Italy
(39) 0187 779054 Cell (39) 0339 5731653
(39) 010 3532777 facsimile
e-mail: jg.tacconi@tin.it

Guenter van Keuk
FGAN, FFM Dept. SDF
Newenahrer Str. 20
D-53343 Wachtberg-Werthhoven
Germany
(49) – (0) 228/9435-373
(49) - (0) 228/348953 facsimile
e-mail: vkeuk@fgan.de

Ed Waltz
Veridian Systems
Manager, Information Technologies
P.O. Box 134008
Ann Arbor, MI 48113
1975 Green Rd.
Ann Arbor, MI 48105
(734) 994-1200 Ext. 2618
(734) 662-3384 facsimile
e-mail: waltz@erim-int.com

Peter J. Lombardo, Administrative Director
Graduate Alumni Director
Alumni Association
University of Notre Dame
Notre Dame, Indiana 46556
(574) 631-7005
(574) 631-8648 facsimile
e-mail: Lombardo.1@nd.edu

APPENDIX B: LECTURERS

Alain Appriou
3 rue de la Truie qui file
91400 SACLAY
ONERA-DTIM, BP 72
92322 Chatillon Cedex
Phone: 33 1 46 73 4919
Email: Alain.Appriou@onera.fr
FRANCE

Mark Dwight Bedworth
21 Windrush Cr.
Malvern
WR14 2XG
Worcestershire
Phone: 44 0 1684 567605
Email: mark@bedworth.clara.co.uk
UK

Isabelle Bloch
Ecole Nationale Superieure des Telecommunications
Dept. TSI CNRS URA 820
46 rue Barrault 75013 Paris
Phone: 33 458 7585
Email: bloch@tsi.enst.fr
FRANCE

Eloi Bosse
Defence Research Establishment Valcartier
2459 Pie-XI North
Val-Belair, QC G39 1X5
Phone: (418) 844-4400 Ext. 4478
Email: eloi.bosse@drev.dnd.ca
CANADA

Vitaly Bystritskii
Dept. Of Physics and Astronomy
University of California
Irvine, CA 92697
Phone: 949-824-2611
Email: vystrit@uci.edu
USA

Vincent Jean Henri Chalmeton
Thomson-CFS/DETEXIS
1 Avenue Jean Moulin
78852-Elancourt
Phone: (33) 1 34 59 74 00
Email: vincent,chalmeton@detexis.thomson-csf.fr
FRANCE

Christodoulos Chamzas
Cultural and Educational Technology Institute
Tsimiski 58
Xanthi, 67100
Phone: 30 541 79571
Email: chamzas@ee.duth.gr
GREECE

David Finkleman
Senior Executive Service
Orth American Aerospace Defense Command and US Space Command
Peterson Air Force Base
Colorado
Phone: 719-554-5071
Email: David.Finkleman@Peterson.af.mil
USA

Gian Luca Foresti
Assistant Professor
Department of Mathematics and Computer Science
University of Udine
Via Delle Scienze 206
33100 Udine
Phone: 39 0432 558406
Fax: 39 0432 558499
Email: foresti@dimi.uniud.it
ITALY

Qiang Gan
Department of Computer Science
University of Essex
Colchester C04 3SQ, UK
Phone: 44 1206 873487
Fax – 44 1206872788
Email: gg@ecs.sotan.ac.uk
UK

David L. Hall
Associate Director
Applied Research Laboratory
The Pennsylvania State University
P.O. Box 30
Room 228. ARL Bldg.
State College, PA 16804-0030
Phone: 814-863-4155
Email: dlh28@psu.edu
USA

Roger Von Hanwehr
CEO Solus, Inc.
700 New Hampshire Ave, N.W. #1008
Washington, DC 20037

Martin Holena
Institute of Computer Science
Pod Vodarenskou vezi 2
18207 Prague
Phone: 420 2 6605 4921
Email: martin@uivt.cas.cz
CZECH REPUBLIC

Anthony K. Hyder,
Associate Vice President,
Graduate School
Professor of Physics
University of Notre Dame
501 Main Building
(574) 631-8591
(574) 631-6630 facsimile
e-mail: ahyder@nd.edu
USA

Jean-Pierre Le Cadre
Directeur de recherche CNRS
Campus de Beaulieu
35042 RENNES CEDEX FRANCE
Phone: 33 2 99 84 72 24
Email: lecadre@irisa.fr
FRANCE

Eric LeFebvre
Lockheed Martin Canada
6111 Royalmount Ave,
Montreal H 4P 1K6
Phone: 514-340-8310
Email: eric.lefebvre@lmco.com
CANADA

James Llinas
Research Professor
Director, Center for Multisource
Information Fusion
State University of New York
at Buffalo
Buffalo, NY 14214
Phone: 716-645-2357 Ext. 2105
Email: llinas@eng.buffalo.edu
USA

David Lowe
Neural Computing Research Group
Aston University
Aston Triangle
Birmingham B4 7ET
Phone: 44 0 121 333 4631
Email: d.lowe@aston.ac.uk
UK

Larry Luessen
Senior Combat Systems Engineer,
Code T202
NAVSEA Dahlgren
Surface Warfare Center Division
17320 Dahlgren Rd.
Dahlgren, VA 22448-5100
Phone: 540-653-6034
Email: lluesse@nswc.navy.mil
USA

Rabinder N. Madan
Program Officer
Office of Naval Research
800 N. Quincy St.
Arlington, VA 22217
Phone: 703-696-4217
Email: madanr@onr.navy.mil
USA

William Miceli
Associate Director
Office of Naval Research International Field Office, DOD
London
Phone: 44 (0) 171 514-4954
Email: wmiceli@onreur.navy.mil
USA

Vincent Nimier
ONERA
29 Avenue de la Division
Leclerc 92322
Chatillon Cedex
Phone: 33 1 46 73 49 98
Email: nimier@onera.fr
FRANCE

Ivan Onishchenko
Gatseva St. 7A Apt. 33
Kharkov 310108
Phone: 38 0572 356140
Email: onish@kipt.kharkov.ua
UKRAINE

Yury Petrovitch Pokholkov
3 Kuleva Str. Apt. 91
Tomsk, 634034
Phone: 7 3822 415 620
Email: Rector@tpu.ru
RUSSIA

Erno Pungor
H 1116 Budapest
Fehervari ut 130
Phone: 36 1 463 0500
Email: Pungor@Bzaka.hu
HUNGARY

Thierry Ranchin
Ecole des Mines de Paris
BP 207
06904 Sophia Antipolis cedex
Phone: 33 4 93 957453
Email: thierry.ranchin@cenerg.cma.fr
FRANCE

Nageswara S. V. Rao
Mailstop 6355, Bldg 6010
Oak Ridge National Laboratory
Oak Ridge, TN 37831-6355
Phone: 865-574-7517
Email: raons@ornl.gov
USA

Carlo S. Regazzoni
Dept. of Biophysical and Electronic Engineering
University of Genoa
Via All'Opera Pia
11A I-16145 Genoa
Phone: 39 010 3532792
Email: carlo@dibe.unige.it
ITALY

Roger Reynaud
IEF Bat 220
University of Paris
11 F91405
Orsay Cedex
Phone: 33 1 69 15 41 23
Email: Roger.Reynaud@ief.u-psud.fr
FRANCE

Michaela Sanielevici
Montral Neurological Institute
McGill University
4042 Dorchester Blvd.
Westmont, Quebec
Phone: 514-932-8766
Email: mickey@sanielevici
CANADA

Emil Atanassov Semerdjiev
Central Laboratory for Parallel
Processing
25A Acad. G. Bonchev Str.
1113 Sofia
Phone: 359 2 979 67 47
Email: Signal@bas.bg
BULGARIA

Elisa Shahbazian
Director of Research and Development
Lockheed Martin Canada
6111 Royalmount Ave,
Montreal H 4P 1K6
Phone: 514-340-8343
Email: elisa.shahbazian@lmco.com
CANADA

Ryszard Sroka
University of Mining And Metallurgy
Al. Mickiewicza 30, 30-059
Krakow
Phone: 4812 617 39 72
Email: rysieks@galaxy.uci.agh.edu.pl
POLAND

Giorgio Tacconi
Italian National Research Center
Institute for Naval Automation
Via dei Marini 6
16149
Genoa
Phone: 39 10 6475 610
Email: jg.tacconi@tin.it
ITALY

Pierre Valin
Lockheed Martin Canada
6111 Royal Mount Ave.
Montreal H4P 1K6
Phone: 514 340 8310 Ext. 8703
Email: pierre.valin@lmco.com
CANADA

Paul Vangasse,
Senior Group Leader, Advanced
Embedded Systems,
BAE SYSTEMS Advanced Technology
Centres - Sowerby,
FPC 267, PO Box 5, Filton,
Bristol, BS 34 7QW,
Phone: 44 117 936 6822
Email: Paul.Vangasse@baesystems.com
UK

Pramod K. Varshney
121 Link Hall
Syracuse University
EECS Dept.
Syracuse, NY 13244
Phone: 315-443-4013
Email: varshney@cat.syr.edu
USA

Ed Waltz
Veridian-Systems
P.O. Box 134008
Ann Arbor, MI 48113-4008
Phone: 734-994-1200 Ext. 2618
Email: waltz@erim-int.com
USA

APPENDIX C: PARTICIPANTS

Vladimir Baryshevsky
Institute of Nuclear Problems
Bobruiskaya Str. 11
Minsk 220050
Email: bar@inp.minsk.by
BELARUS

Tulkum Bekmuratov
70 Acade.
Yahyo Gulomov Street
Tashkent, 700047
Email: gulamov@physic.uzsci.net
UZBEKISTAN

Kaouthar Benameur
Defence Research Establishment
Ottawa
3701 Carling Ave.
Ottawa K1A 0Z4
Ontario
Phone: 613-991-6707
Email: kaouthar.benameur@dreo.dnd.ca
CANADA

Christian M. Birkemark
Danish Defence Research Establishment
Ryvangs Alle 1
P.O. Box 2715
DK-2100
Copenhagen
Phone: 45391517-46
Email: Cmb.ddre.dk
DENMARK

Bruce Bishop
Dean, College of Engineering
Utah State University
Logan, UT 84322-4100
Phone: 435-797-2776
Email: abbiship@engineering.ueu.edu
USA

Lars Brandt
Danish Defence Research Establishment
Ryvangs Alle 1
DK-2100 Copenhagen
Email: lbt@ddre.dk
DENMARK

Vitaly Bystritskii, Co Director of ASI
Research Physicist
Dept. of Physics and
Astronomy University of California at Irvine
Irvine, CA 92697
(714) 824-2611
(714) 824-2914 facsimile
e-mail : vbystrit@uci.edu

Vyacheslav Bystritsky
Laboratory of Nuclear Problems
Joint Institute of Nuclear Research
Dubna, Moscow Region
Phone: 0 8453002000
Email: Bystvm@nusun.jinr.ru
RUSSIA 141980

Bakshy Chibber
Research Assistant Professor
Biosciences Core Facility
446 Stepan Chemistry
University of Notre Dame
Notre Dame, IN 46556
Phone: 219-631-6754
Email: chibber.1@nd.edu
USA

Valentyn Chorny
Kharkov National University, Kharkov, Ukraine
P.O. Box 2096
Kharkov-108
Phone: 380-573-352722
Email: chorny@pht.univer.kharkov.ua
UKRAINE 61108

G. Gallagher-Daggitt
9 Tullis Close
Sutton Courtnay
Oxfordshire
UNITED KINGDOM

Robert DeWitt
Physics Department
Penn State University
2428 2nd Ave.
Altoona, Pennsylvania 16602
Email: rnd3@psu.edu
USA

Reuben Dobkin
RAFAEL, IDF
P. O. Box 2250
Haifa 31021
Phone: 054 248199
Email:
rostikd@techunix.technion.ac.il
ISRAEL

Susan E. Durham
USAF
835 Potomac Ridge Ct.
Sterling, VA 20164
Phone: 703-444-6775
Email: pitbulldurham@aol.com
USA

Dwight Duston
Vice President of Technologies
The Egg Factory, LLC
59 FairlaneRoad
Laguna Niguel, CA 92677
Email: dduston@eggfactory.com
USA

Murat Efe
Ankara University,
Faculty of Science,
Electronics Engineering Department
Ankara
Phone: 90 312 2126720
Email: efe@science.ankara.edu.tr
TURKEY

George Emert
President, Utah State University
Room 116 Old Main
Logan, UT 84322
USA

Muret Eren
Aselsan Inc.
Microelectronics, Guidance and
Elektro-Optics
P.O. Box 30, Etlik
06011 Ankara
TURKEY

Zacharaki Evangelia
Skiathou 141
11255 Patisia
Athens
Email: czachar@central.ntua.gr
GREECE

Brent Gerity
Dept. of Special Education
Utah State University
Logan, UT 84322
Email: bgerity@cc.usu.edu
USA

Peter Gerity
Vice President for Academic Affairs
New Mexico Tech
801 LeRoy Place
Socorro, NM
Phone: 505-835-5646
Email: pgerity@admin.nmt.edu
USA

Lee N. Gilbert
Naval Science Advisor to the
Commander, U.S. Naval Forces, Europe
PSC 802 Box 9
FPO AE 09499
Phone: 44 20 7514 4770
Email: nsapcne@nosc.mil
USA

Charles Glover
Oak Ridge National Laboratory
POET/ORNL Suite 1100
1745 Jefferson Davis Hwy.
Arlington, VA 22202
Phone: 703-412-5217
Email: cqg@ornl.gov
USA

Lars Heen
Norwegian Defence Research
Establishment
P.O. Box 25 N-2027 Kjeller
Phone: 47 63807242
Email: Lars-Trygve.Heen@ffi.no
NORWAY

Chris Helleur
Surface Radar
Defence Research Establishment
Ottawa
3701 Carling Avenue
Ottawa, Ontario K1A 0Z4
Phone: 613 998 2395
Email: Chris.Helleur@dreo.dnd.ca
CANADA

Christopher James
Neural Computing Research Group
Aston University
Aston Triangle
Birmingham B4 7ET
Phone: 44 0 121 359 3611
Ext. 4652
Email: jamescj@aston.ac.uk
UK

Mirzaahmedovich Kamilov
70, Acad. Yahyo Gulomov Street
Tashkent
UZBEKISTAN

Irini Karanasiou
National Technical University of
Athens
9, Iroon Polytechniou Str.
157 73 Zografou
Athens
Email: ikaran@esd.ece.ntua.gr
GREECE

Boris Katsnelson
Voronezh University
Universitetskaya Sq. 1
394693 Voronezh
Email: katz@mph.vsu.ru
RUSSIA

Paul J. Koskey
Ballistic Missile Development
Organization
US Department of Defense
Pentagon
Washington D.C. 20301-7100
Email: paul.koskey@bmdo.osd.mil
USA

Victor Kryshkin
Institute for High Energy Physics
142284, Protvino, Moscow Region
Russian Federation
Phone: 0 845 300200
Email: kryshkin@mx.ihep.su
RUSSIA

Amelie Litman
Etudes et Productions Schlumberger
26, Rue de la Cavee – BP 202
92142 Clamart Cedex
Phone: 33 0 1 45372940
Email: litman@clamart.srpc.slb.com
FRANCE

Pierre Loonis
Universite de La Rochelle
Laboratoire d'Informatique et d'Imagerie Industrielle
Avenue Marillac
17042 La Rochelle Cedex 01
Email: pierre.loonis@univ-lr.fr
FRANCE

Rodoula Makri
9, Iroon Polytechniou Srt.
15 773 Zografou
Athens
Email: rodia@esd.ece.ntua.gr
GREECE

Dominique Maltese
SAGEM-SA – Defense and Security Division
72-74 rue de la Tour Billy
B.P. 72-95101
Argenteuil Cedeoc
Phone: 33 01 34262807
Email: dominique.maltese@sagem.com
FRANCE

Lucio Marcenaro
Department of Physical and Electronic Engineering
University of Genoa
Genoa I-16145
Phone: 39 060 3532968
Email: mlucio@dibe.unige.it
ITALY

Bruce McArthur
Defence Research Establishment Atlantic
P.O. Box 1012
Dartmouth, Nova Scotia
B2Y 3Z7
Phone: 902 426 3100 Ex. 359
Email: Bruce.mcarthur@drea.dnd.ca
CANADA

Oleg Metelitsa
Research Institute of Nuclear Problems
11 Bobruiskaya Str.
Minsk 220050
Phone: 375 17 2066238
Email: metelitsa@inp.minsk.by
BELARUS

Ole Martin Mevassvik
Norwegian Defence Research/FFI
P.O. Box 25, N-2027
Kjeller
Phone: 47 63807423
Email: ole-martin.mevassvik@ffi.no
NORWAY

Hannah Michalska
Dept. of Electrical and Computer Engineering
McGill University
3480 University St.
Montreal, Quebec
H3A 2A7
Email: hanna.michalska@mcgill.ca
Phone: 514 398 3053
CANADA

H.B Mitchell
ELTA Electronics Industries Ltd.
Section 3424
Ashdod
Phone: 00972 8 8577337
Email: Mitchell@is.elta.co.il
ISRAEL

Prasanna Mulgaonkar
Director, Advanced Automation Technology Center
SRI International
333 Ravenswood Ave.
Menlo Park, CA 94025
Phone: 650 859 6530
Email: Prasanna@erg.sri.com
USA

David Nicholson
Advanced Technology Centres
BAE Systems - Sowerby
FPC 267, PO Box 5
Filton Bristol BS34 7QW
Email:
David.Nicholson@nadcu7.net.bae.co.uk
UK

Maria de Fatima Alves Nunes
Senior Engineer, Portugese Air Force Academy
Phone: 351 962327497
Email: mariaalves@yahoo.com
PORTUGAL

Gerald O'Connor
Colonel, United States Air Force Commander
EOARD, PSC 802, Box 14
FPO, AE 09499-0200
Phone: 44 0 207 514 4376
Email: gerald.oconnor@london.af.mil

Levent Ozbek
Ankara University Faculty of Science
Department of Statistics
Ankara 06100 Tandogan
Phone: 90 312 2126720 11920
Email: ozbek@science.ankara.edu.tr
TURKEY

Jiri Patera
Centre de Recherches Mathematiques
Universite de Montreal
C.P. 6128 Succ. Centre-ville
Montreal, Quebec H3C 3J7
Phone: 514 - 343 - 6419
Email: patera@crm.umontreal.ca
CANADA

David Reinitz
RAFAEL
P.O. Box 2250
Haifa 31021
Phone: 972 4 8795166
Email: rntz@rafael.co.il
ISRAEL

Miroslaw Sankowski
Telecommunications Research Institute
Hallera 13 80-401
Gdansk
Phone: 4858 341 8007
Email: msankowski@pit.gda.pl
POLAND

Emile Schweicher
Royal Military Academy
Dept. OMRA
8, Rue Hobbema
B-1000 Brussles
Phone 32-2-7376560
Fax: 32-2-7376212
Email: emile.schweicher@rma.ac.be
BELGIUM

John Sekula
Lockheed Martin Corporation
19681 Junipero Way
Saratoga, CA 95070
Phone: 650-424-2765
Email: john.a.sekula@lmco.com
USA

Yunlong Sheng
Center of Optics, Photonics and Laser,
Department of Physics
Laval University
Ste-Foy, Quebec G1K 7P4
Phone: 418 656 2131 Ext. 3908
Email: Sheng@phy.ulaval.ca
CANADA

Gonul Turhan-Sayan
Associate Professor of Electrical Engineering
Middle East Technical University
Ankara 06531
Phone: 90 312 2102371
Email: gtsayan@hotmail.com
TURKEY

David Tedford
University of Strathclyde
Glasgow G11XW
Phone: 0141 548 2071
Fax: 041 552 2487
Email: k.ogorman@eee.strath.ac.uk
UK

Oktay Ureten
Ankara University
Department of Electronics Engineering
Ankara Tandogan 06100
90 312 212 6720 Ext. 1160
Email: ureten@ieee.org
TURKEY

Patrick Verlinde
Royal Military Academy
ELTE Department
Renaissancelaan 30
1000 Brussels
Phone: 32 2 737 6621
Email: verlinde@elec.rma.ac.be
BELGIUM

Andrzej Witczak
Institute of Radar Technology
Military University of Technology
Kaliskiego 2
Warsaw 00-908
Phone: 4822 6859646
Email: awitczak@wel.wat.waw.pl
POLAND

INDEX